STRESS AND STRAIN

Other titles in the series

*

Force Systems and Equilibrium
J. G. A. CROLL

Virtual Work and Energy Concepts
J. RHODES

Simple Bending
J. R. TYNE

Torsion
A. C. WALKER

The Buckling of Struts
A. C. WALKER

Modular Textbooks in Engineering

Editor: A. C. WALKER, Ph.D.
Department of Civil and Municipal Engineering,
University College, London

Stress and Strain

T. R. GRAVES SMITH
M.A., Ph.D., C.Eng., M.I.C.E.
Department of Civil Engineering
University of Southampton

1974

CHATTO & WINDUS
LONDON

Published by
Chatto & Windus Ltd
42 William IV Street
London WC2N 4DF

*

Clarke, Irwin & Co. Ltd
Toronto

ISBN 0 7011 1950 0

Printed in Great Britain by
T. & A. Constable Ltd
Hopetoun Street
Edinburgh EH7 4NF

CONTENTS

1

INTRODUCTION

The concepts of stress and strain are fundamental to the theory of structures. In order to see why, we must consider one of the basic functions of structures, which is to distribute loading from one point to another. Thus a bridge, for example, must distribute wheel loads from span to abutments, while an engine connecting rod (which can be regarded as a small structure) must transmit thrust from piston to crankshaft. The behaviour of structures under load is then seen to be of critical importance, for if the bridge were to sag alarmingly or the crankshaft snap, they would of course be useless. The theory of structures is thus primarily concerned with the deformation and failure of bodies subject to forces.

When determining structural behaviour we have to ascertain the internal forces set up at each point within a body and the corresponding local deformation of the material. This is broadly speaking, because (1) failure occurs when the internal forces reach critical values, and (2) the overall deformation of the body is the summation of the local deformation. The internal forces are then generally described as *stress*, while the local deformation is described as *strain*.

This book is concerned with how stress and strain are specified and how they are related.

We shall begin in this first chapter by considering two useful introductory concepts: (1) the continuum concept and (2) distributed loading.

1.1 The continuum concept

Let us consider the internal structure of a body made from an elastic material such as steel. Such a body is composed of millions of microscopic crystals* each of which takes the form of a symmetrical three-dimensional array of atoms. These crystals are generally of random size and orientation, unless the steel has been heavily worked during manufacture. Now the atoms are held in position by so-called metallic bonds*, which can be thought of as springs connecting neighbouring atoms in the arrays. We can therefore imagine that forces applied to the surface of the body are distributed internally in some particular way by the compression and extension of millions of tiny springs. The 'internal forces' of the previous

* L. H. Van Vlack, *Materials Science for Engineers* (Addison-Wesley, 1970).

section can be regarded as the forces set up in the springs, while the 'local deformation' can be regarded as their corresponding compression or extension. It then becomes clear that internal forces and local deformation, or stress and strain, are always closely associated in an elastic body.

It is apparent from the above that at the atomic level a body is made up of discrete particles. Its properties also vary widely between neighbouring crystals. However, when considering a body of any practical size we ignore this discontinuity and variation at the microscopic level and treat the material as a *continuum*. Specifically this means that in calculations where we wish to consider small elements of the material, we assume that however small we make them, even of differential size relative to the body as a whole, these elements still contain sufficient crystals for the microscopic variation in properties to cancel out. An example of such a calculation arises in the next section. Therefore, although it is convenient to introduce the concepts of stress and strain in terms of atomic behaviour, we shall in fact go on to describe them mathematically in terms of the behaviour of larger elements of material within the body.

1.2 Distributed loading

The classical theory of statics describes loading on bodies in terms of concentrated vectors* called forces. The basic unit of force is the *newton* (giving 1 kilogramme mass an acceleration of 1 m/s^2), and each force is associated with a point on the structure called its point of action. We shall now consider two types of loading that occur on practical structures.

(a) *Surface stresses*

The characteristic of the forces acting on the surfaces of structures caused by, say, the wheels of a lorry on a bridge deck, water on a dam or wind on a high building, is that they are always distributed over the finite areas of contact between the structures and the agencies causing them. We describe these distributed forces by means of surface stress vectors as shown in the following simple example.

Consider the pile of books resting on a horizontal table top shown in Fig. 1.1 (a). As far as the table is concerned the books are equivalent to vertical forces distributed over the area ABCD, strictly taking the form of a practically infinite number of minute forces acting on the surface molecules. (Fig. 1.1 (b) depicts the resultant forces acting on small elements of the area ABCD.) Let us next consider an element of the surface surrounding a point P (Fig. 1.1 (c)) whose area dA is of differential size compared with the area ABCD. We recall that according to the con-

* The definition and representation of vectors is described in Appendix 1.

tinuum concept this element contains a very large number of surface molecules. The resultant of the forces acting on these molecules is a vector of differential magnitude which we call d$\boldsymbol{F}$. Now if we divide d$\boldsymbol{F}$ by dA, we produce a *vector* whose direction is the same as that of d$\boldsymbol{F}$ but whose magnitude being the quotient of two differential quantities $|\,\mathrm{d}\boldsymbol{F}\,|$ and dA, is *finite*. We denote this vector by $\boldsymbol{S}$ and call it the *surface stress vector* at P. $\boldsymbol{S}$ then has the dimensions of force per unit area and gives a measure of the average intensity of the distributed forces in the

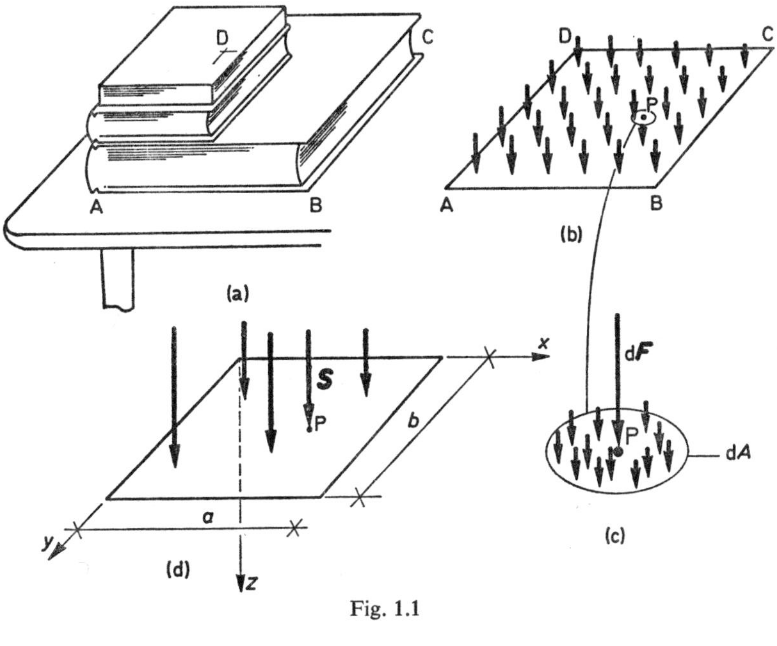

Fig. 1.1

immediate neighbourhood of P. If, as in this example, it is directed towards the surface of the body it it called *compressive* and, if away from the surface, it is called *tensile*.

Stress vectors can be defined at every point in the region ABCD, and a few of them are shown in Fig. 1.1 (d). Relative to the x, y, z coordinate system their scalar components are $S_x = S_y = 0$, $S_z = |\,\boldsymbol{S}\,|$. Thus a possible non-uniform distribution of stress over the surface might, for example, be specified by three equations such as

$$S_x = S_y = 0, \quad S_z = \left(1.5 - 0.3\frac{x}{a} + 0.4\frac{y^2}{b^2}\right) \mathrm{N/m^2} \qquad (1.1, 1.2, 1.3)$$

which express the components of $\boldsymbol{S}$ as functions of the coordinates of its point of action.

The interaction between solid bodies is expressed as above by stress vectors acting at points within their areas of contact. However, we should perhaps note that if a structure is subjected to forces caused by fluids, such as water on a dam or air on an aircraft wing, then $(\mathrm{d}F/\mathrm{d}A)$ is more commonly called the *pressure* vector at each surface point.

$\boldsymbol{S}$ need not necessarily be normal to the surface of the structure. Thus if we were to apply a horizontal force to the books in Fig. 1.1 and friction were sufficient to keep them stationary, the distributed forces exerted on the table would become non-vertical as in Fig. 1.2. The corresponding stress vectors, defined in exactly the same way as before, would then be inclined to the surface, and S_x and S_y in (1.1, 1.2)* would no longer be zero. S_z in this case is called the *normal component* of $\boldsymbol{S}$, whilst S_x and S_y are called the *tangential components*.

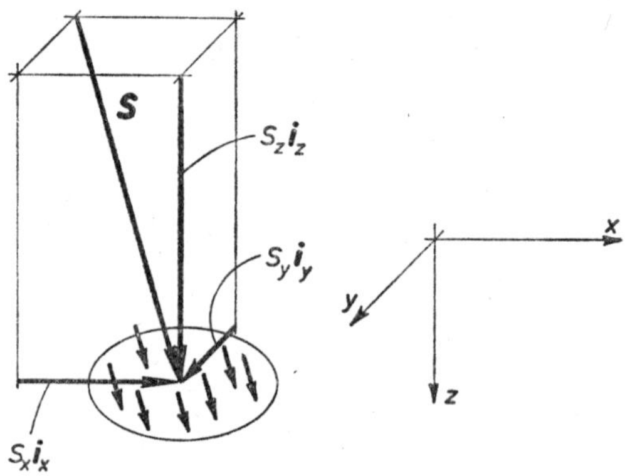

Fig. 1.2

Concerning the magnitudes of surface stress vectors acting on practical structures, the loading on a warehouse floor is typically of the order of 10 $\mathrm{kN/m^2}$, and the wind loading on a high building is of the order of 1 $\mathrm{kN/m^2}$.

(b) *Body forces*

In addition to surface loading, a structure can also be subjected to forces acting at points *within* its surface. Thus forces are exerted at all points within a structure situated in the Earth's gravitational field. In order to describe such loading, we define body force vectors acting at internal points in much the same way as the surface stress vectors in the previous section.

* Bracketed numbers in the text refer to the corresponding equations.

Thus suppose the body in Fig. 1.3 (a) is subjected to distributed forces due to gravity. We consider an internal element of volume surrounding P (Fig. 1.3 (b)) whose volume dV is of differential size. The resultant of the forces acting on the element is again a vector $d\boldsymbol{F}$ of differential magnitude. The *body force vector* $\boldsymbol{B}$ at P is then defined as the vector $(d\boldsymbol{F}/dV)$ and has the dimensions of force per unit volume. Like the surface stress vector, $\boldsymbol{B}$ can be specified by three equations similar to (1.1, 1.2, 1.3) which express its components in terms of the coordinates of its point of action.

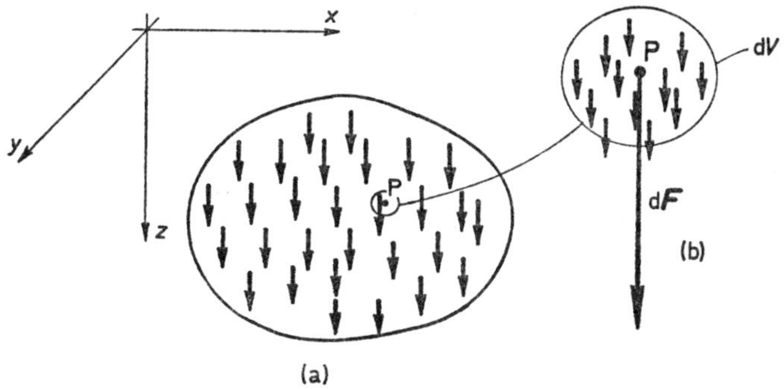

Fig. 1.3

The body force vectors caused by the Earth's gravity take a particularly simple form. Thus the force acting on the element in Fig. 1.3 (b) is directed vertically downwards and is equal to $(\rho g\, dV)$ N, where ρ kg/m³ is the density of the material, g is the gravitational acceleration (9.81 m/s²). Therefore

$$d\boldsymbol{F} = (\rho g\, dV)\boldsymbol{i}_z$$

and

$$\boldsymbol{B} = \frac{d\boldsymbol{F}}{dV} = (\rho g)\boldsymbol{i}_z. \qquad (1.4)$$

The corresponding distribution of the body forces is then seen to be uniform in a body of uniform density, and the scalar components of $\boldsymbol{B}$ are given by

$$B_x = B_y = 0, \quad B_z = (\rho g)\ \text{N/m}^3. \qquad (1.5, 1.6, 1.7)$$

Values of ρ for practical materials are given in Table 1.1, whence it is apparent that the magnitudes of the body force vectors due to gravity range between about 5 and 80 kN/m³.

The surface stresses and body forces discussed in this section are called *external forces* acting on a structure. The means whereby body forces are produced is still obscure, but in this book we shall be content to recognise external forces by the fact that agencies *outside* the structure experience equal and opposite reactions to them. Thus books on a table experience upward vertical forces from the table, and a magnet producing body forces in a nearby piece of iron is attracted to the iron. Even the Earth experiences an attraction to a body within its gravitational field, although this obviously has negligible effect on the Earth.

TABLE 1.1

Material	water	steel	aluminium alloy	concrete	glass	timber
ρ kg/m^3	1 000	7 840	2 800	2 410	2 580	576

PROBLEMS

Problem 1.1 The wind pressure on the front face of the building in Fig. 1.4 varies with height as follows

$$\boldsymbol{P} = -(528+6.39y^2)\boldsymbol{i}_z \ \mathrm{N/m^2}.$$

Calculate the total force on the face, and the overturning moment about the base line AB. (Ans. $-138\boldsymbol{i}_z$ kN, $-1\,806\boldsymbol{i}_x$ kNm)

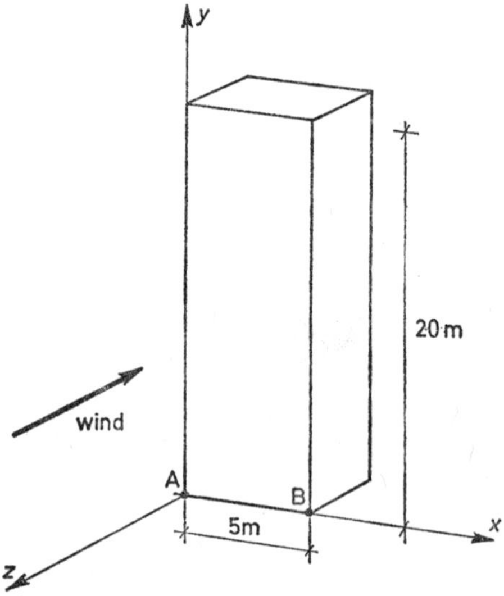

Fig. 1.4

Problem 1.2 The concrete anchor block shown in Fig. 1.5 is loaded by two equal horizontal forces of 1 500 kN. Assuming that the normal component of stress on the foundation ABCD is uniform in the z direction and varies linearly in the x direction, and that the tangential component is uniform, calculate the stress distribution on the area ABCD.

$$(\text{Ans. } \boldsymbol{S} = 40.0\boldsymbol{i}_x - (37.9 + 10.7x)\boldsymbol{i}_y \text{ kN/m}^2)$$

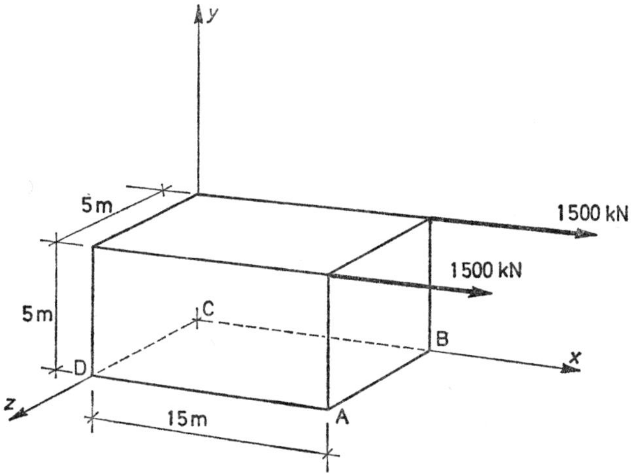

Fig. 1.5

2

INTERNAL STRESS

2.1 Introduction

In Chapter 1 we described the action of distributed forces on structural bodies. We shall now consider the effects produced by these forces in the bodies themselves, beginning in this chapter with the internal stress.

2.2 The internal stress vector

Suppose a body is subjected to the loading shown in Fig. 2.1 (a) and we wish to describe the corresponding internal forces at a particular point P. Imagine first a plane surface within the body passing through P.

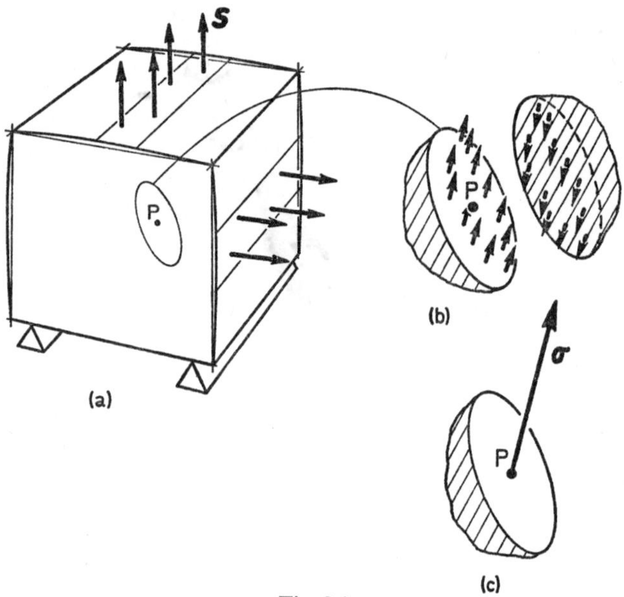

Fig. 2.1

Now the forces exerted between atoms across this surface can be represented by the two equal and opposite sets of forces shown in the expanded view of Fig. 2.1 (b) (where, as in Fig. 1.1 (b), each vector represents the resultant force on a small element of surface). If we then consider the forces acting on the material to the left in the figure, we note that they can be described by stress vectors in exactly the same way as in § 1.2. The stress vector at P then gives the required measure of the average intensity of the internal forces acting across the particular surface at P. We denote it by $\boldsymbol{\sigma}$ and call it the *internal stress vector.* If $\boldsymbol{\sigma}$ is directed away from the material as in Fig. 2.1 (c) it is called *tensile,* and if towards the material it is called *compressive.*

Now the magnitude and direction of the internal stress vector at P clearly depends on the orientation of the chosen surface passing through P. Indeed it is quite possible for $\boldsymbol{\sigma}$ to change from tensile to compressive at any particular point in the body as indicated in Fig. 2.2. (In addition $\boldsymbol{\sigma}$ also depends on which side of the surface the material is situated, for if we were to consider the forces on the material to the right in Fig. 2.1 (b), then $\boldsymbol{\sigma}$ would have the same magnitude as before but *reversed in direction.*) Thus, at first sight, it appears that the problem of defining all the internal forces at P entails specifying the stress vectors on an infinite number of surfaces of different orientation passing through P. However, in the next sections we shall show that this difficulty can, in fact, be avoided.

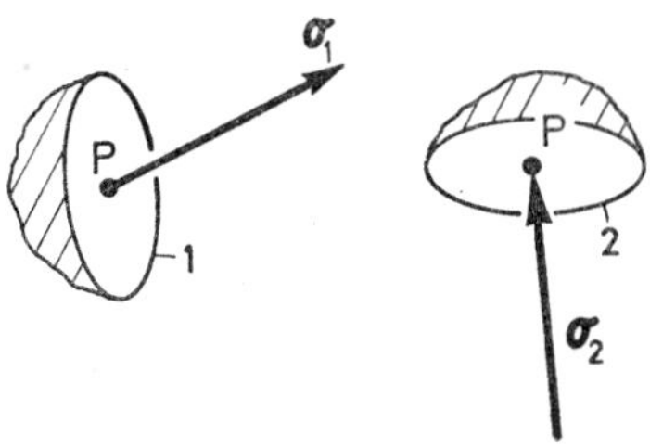

Fig. 2.2

Concerning magnitudes, the maximum value for $\boldsymbol{\sigma}$ in the most highly stressed regions of a steel structure under working loads is typically of the order of 100 MN/m^2.

2.3 Components of stress

Let us consider the internal stress vectors at P acting on three particular surfaces called the *positive coordinate surfaces.* These are defined as follows. The positive x coordinate surface is the surface parallel to the

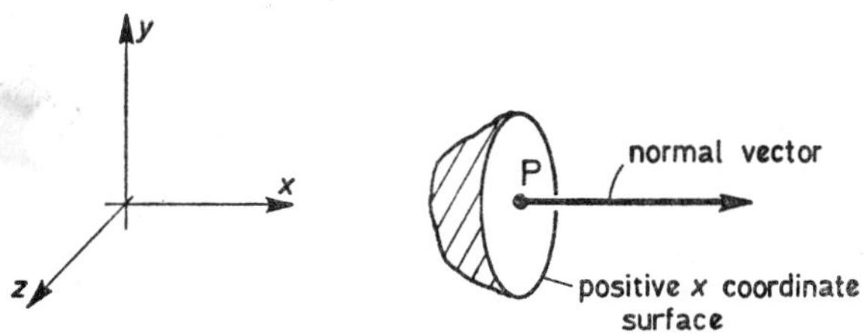

Fig. 2.3

$y-z$ plane of an x, y, z coordinate system, with the material situated so that a vector directed outwards from the material and normal to the surface is in the *positive* direction of the x coordinate line (Fig. 2.3).

The positive y and z coordinate surfaces are defined similarly. The internal stress vectors acting on these surfaces are then distinguished by appropriate subscripts. Thus $\boldsymbol{\sigma}_x$ is considered to act on the positive x coordinate surface, whilst $\boldsymbol{\sigma}_y$ and $\boldsymbol{\sigma}_z$ respectively act on the y and z surfaces. Further, the scalar components of these vectors are quite consistently distinguished by *two* subscripts. Thus the components of $\boldsymbol{\sigma}_x$ are σ_{xx}, σ_{xy}, σ_{xz} and are shown in Fig. 2.4 (a). Similarly the components of $\boldsymbol{\sigma}_y$ are σ_{yx}, σ_{yy}, σ_{yz} and $\boldsymbol{\sigma}_z$ are σ_{zx}, σ_{zy}, σ_{zz} and are shown in Fig. 2.4 (b) and (c).

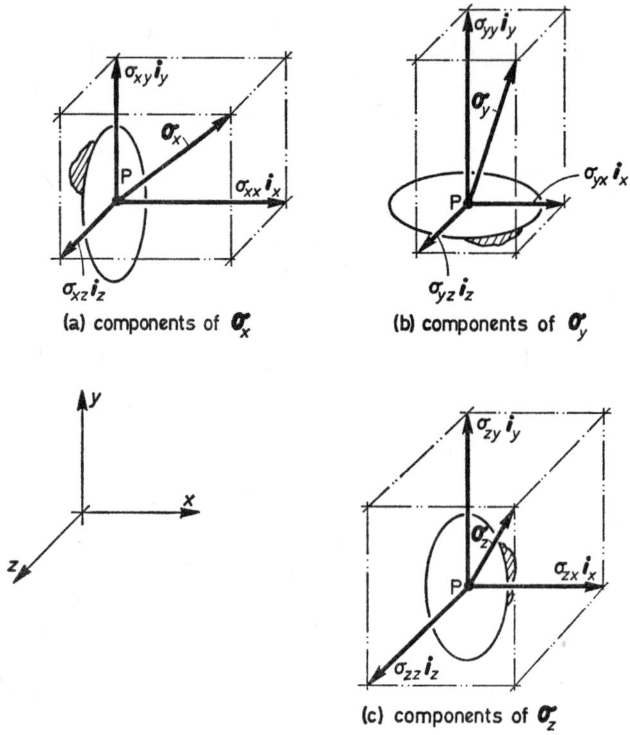

Fig. 2.4

σ_{xx}, σ_{yy}, σ_{zz} are called the *direct stress components* at P in the x, y and z directions respectively, while σ_{xy}, σ_{xz}, σ_{yx}, σ_{yz}, σ_{zx}, σ_{zy} are called the *shear stress components*.

In engineering texts, it is conventional to emphasise the difference between direct and shear stress components by denoting the latter by the Greek letter τ plus appropriate subscripts, and we shall follow this convention here. Thus σ_{xy} will be written τ_{xy}, etc. But in order to avoid confusion, the reader should take great care to remember the basic definitions of the stress components, and that τ_{xy} and τ_{xz} for example, are in fact components of the vector $\boldsymbol{\sigma}_x$.

The nine stress components are important because once their values are known, the state of stress at P can be regarded as being completely defined. This is because it is then possible to use equilibrium conditions to calculate the components of the internal stress vector acting on any other surface passing through P. We shall discuss this calculation for a special state of stress in a later section (§ 2.4 (a)).

(a) *Rotational equilibrium*

Equilibrium conditions also allow us to make a further simplification in the definition of stress. Thus suppose we consider the rectangular parallelepiped of material containing P shown in Fig. 2.5 (a). Since the

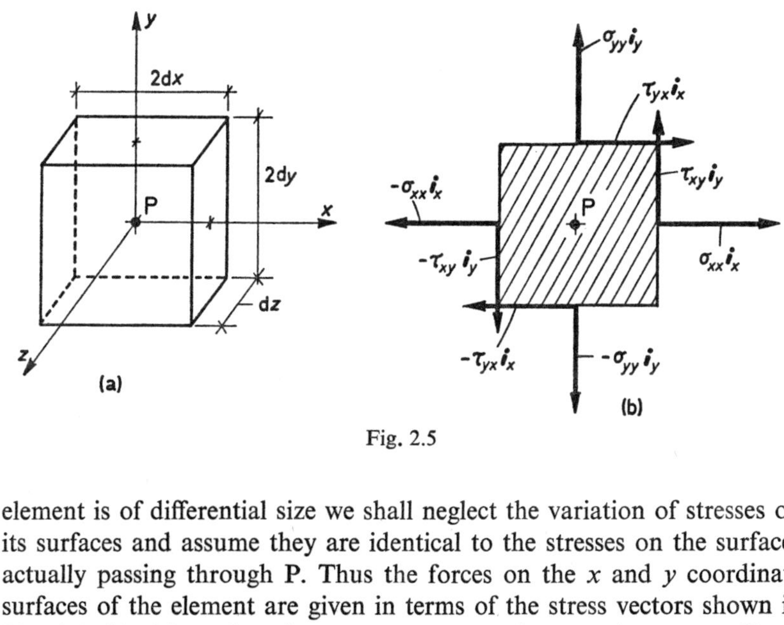

Fig. 2.5

element is of differential size we shall neglect the variation of stresses on its surfaces and assume they are identical to the stresses on the surfaces actually passing through P. Thus the forces on the x and y coordinate surfaces of the element are given in terms of the stress vectors shown in Fig. 2.5 (b). (Note that the stress vector on the negative x coordinate surface is $-\boldsymbol{\sigma}_x$.) Now considering rotational equilibrium of the element about the z axis we equate to zero the sum of the moments of the forces about this axis. Each moment is obtained by multiplying a stress component by: (1) the area over which it acts and (2) the appropriate lever arm. Thus it is apparent that only the shear stress components τ_{xy} and τ_{yx} do in fact exert moments on the element, and that for equilibrium

$$\tau_{xy}2dy\,dz\,dx-\tau_{yx}2dx\,dz\,dy+\tau_{xy}2dy\,dz\,dx-\tau_{yx}2dx\,dz\,dy=0.$$

Since the size of the element is arbitrary we can conclude that

$$\tau_{xy} = \tau_{yx}. \tag{2.1}$$

Similarly taking moments about the x and y axes leads to

$$\tau_{yz} = \tau_{zy}, \quad \tau_{zx} = \tau_{xz}. \tag{2.2, 2.3}$$

Thus for bodies in equilibrium, three equalities between the shear stress components must always be satisfied. The two components τ_{xy} and τ_{yx} for example, are then called *complementary shear* stresses. In the remaining text we shall assume that (2.1, 2.2, 2.3) are indeed always satisfied and define stress exclusively in terms of the *six* components σ_{xx}, σ_{yy}, σ_{zz}, τ_{xy}, τ_{yz}, τ_{zx}.

2.4 Plane stress

Unfortunately further detailed discussion of the general state of stress represented by the six stress components is beyond the scope of the present text and in this section we shall consider the properties of a restricted state in which only σ_{xx}, σ_{yy} and τ_{xy} are non-zero. This is called a state of *plane stress.*

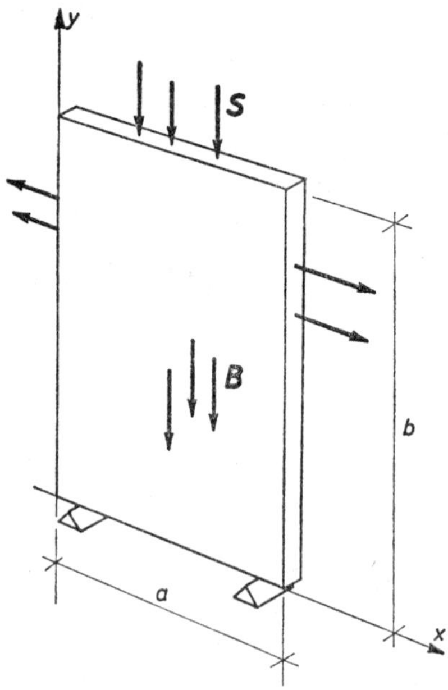

Fig. 2.6

Plane stress is of considerable practical importance since it occurs widely in structures made of thin plates, such as steel box girders. Thus we find that in the plate loaded by in-plane edge stresses and body forces shown in Fig. 2.6, we can assume that all internal stress vectors are parallel to the plane of the plate, so that specifically $\sigma_{zz} = \tau_{yz} = \tau_{xz}(= \tau_{zx}) = 0$. We can also assume that the stresses are uniform across the thickness so that σ_{xx}, σ_{yy} and τ_{xy} are independent of z. Thus the variation of σ_{xx} from point to point in the plate might, for example, be defined by an equation such as

$$\sigma_{xx} = 50.0+20.0\frac{x^2}{a^2}+30.0\frac{x}{a}\cdot\frac{y^2}{b^2}\ \text{MN/m}^2. \tag{2.4}$$

(a) *Relationship between stresses*

We first consider the problem of deriving the components of the stress vector at an internal point P, acting on an arbitrarily oriented surface normal to the plane of the plate; this being a special case of the general problem which we briefly mentioned in the previous section. For clarity we shall denote this vector by $\boldsymbol{\sigma}_n$ and its components by σ_{nx}, σ_{ny}. (Note that for plane stress $\sigma_{nz} = 0$.)

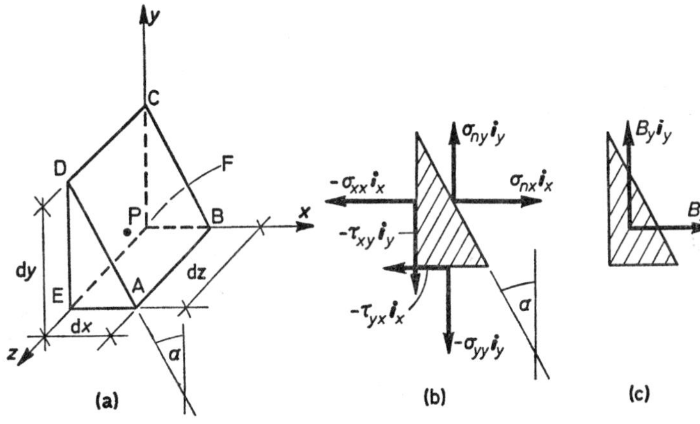

Fig. 2.7

Let us consider the equilibrium of the wedgeshaped element of material containing P shown in Fig. 2.7 (a) of which the side ABCD is parallel to the surface of interest. We suppose this surface subtends an angle α anticlockwise to the positive x coordinate surface. Since the element is of differential size we can, as in § 2.3 (a), assume that the forces on the surfaces of the element are given in terms of the stresses on the surfaces actually passing through P, Fig. 2.7 (b). In this calculation we shall also consider the possible effect of the body forces shown in Fig. 2.7 (c). Thus the resultant force $\boldsymbol{R}$ on the element is found by multiplying the stress vectors by the areas over which they act, the body force vector by the volume, and summing as follows

$$\begin{aligned} \boldsymbol{R} = \{-\sigma_{xx}\,\mathrm{d}y\,\mathrm{d}z - \tau_{xy}\,\mathrm{d}x\,\mathrm{d}z + \sigma_{nx}\,(\mathrm{d}x^2+\mathrm{d}y^2)^{\frac{1}{2}}\,\mathrm{d}z + \tfrac{1}{2}B_x\,\mathrm{d}x\,\mathrm{d}y\,\mathrm{d}z\}\boldsymbol{i}_x \\ +\{-\tau_{xy}\,\mathrm{d}y\,\mathrm{d}z - \sigma_{yy}\,\mathrm{d}x\,\mathrm{d}z + \sigma_{ny}\,(\mathrm{d}x^2+\mathrm{d}y^2)^{\frac{1}{2}}\,\mathrm{d}z + \tfrac{1}{2}B_y\,\mathrm{d}x\,\mathrm{d}y\,\mathrm{d}z\}\boldsymbol{i}_y. \quad (2.5) \end{aligned}$$

The quantities in curly brackets are the two scalar components R_x, R_y. Now, since the element is stationary within the plate, linear static equilibrium requires that $\boldsymbol{R} = 0$. Equating the two scalar components of $\boldsymbol{R}$ to zero in turn then leads to equilibrium equations in the x and y directions. Thus $R_x = 0$ yields, after dividing through by $(\mathrm{d}x^2+\mathrm{d}y^2)^{\frac{1}{2}}\,\mathrm{d}z$,

$$\begin{aligned} \sigma_{nx} &= \frac{\sigma_{xx}\,\mathrm{d}y}{(\mathrm{d}x^2+\mathrm{d}y^2)^{\frac{1}{2}}} + \frac{\tau_{xy}\,\mathrm{d}x}{(\mathrm{d}x^2+\mathrm{d}y^2)^{\frac{1}{2}}} - \frac{B_x\,\mathrm{d}x\,\mathrm{d}y}{2(\mathrm{d}x^2+\mathrm{d}y^2)^{\frac{1}{2}}} \\ &= \sigma_{xx}\cos\alpha + \tau_{xy}\sin\alpha - \tfrac{1}{2}B_x\,\mathrm{d}x\cos\alpha. \end{aligned}$$

The term containing the body force vector is of differential magnitude and can be ignored so that we have finally

$$\sigma_{nx} = \sigma_{xx}\cos\alpha + \tau_{xy}\sin\alpha. \quad (2.6)$$

Similarly $R_y = 0$ yields

$$\sigma_{ny} = \tau_{xy}\cos\alpha + \sigma_{yy}\sin\alpha. \quad (2.7)$$

(2.6) and (2.7) then determine the components of $\boldsymbol{\sigma}_n$ on any surface through P normal to the plane of the plate.

Example 2.1

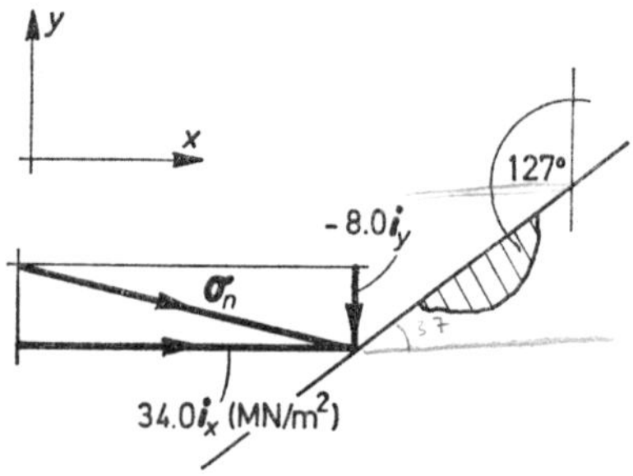

Fig. 2.8

The components of stress at a particular point in a plate are given by

$$\sigma_{xx} = -30.0, \quad \sigma_{yy} = 5.0, \quad \tau_{xy} = 20.0\ \mathrm{MN/m^2}.$$

Derive the components of the stress vector on the surface for which $\alpha = 127°$.
From (2.6) and (2.7)

$$\begin{aligned} \sigma_{nx} &= (-30.0)(-0.6) + (20.0)(0.8) = 34.0\ \mathrm{MN/m^2} \\ \sigma_{ny} &= (20.0)(-0.6) + (5.0)(0.8) = -8.0\ \mathrm{MN/m^2}. \end{aligned}$$

$\boldsymbol{\sigma}_n$ is then seen to be the compressive stress vector shown in Fig. 2.8.

(b) *Transformation of stresses*

Hitherto we have considered the state of plane stress specified by its components (σ_{xx}, σ_{yy} and τ_{xy}) relative to a particular coordinate system x, y, z, but it is important that we can relate these components to the ones corresponding to some other set of axes passing through the same point. Suppose now we define a new coordinate system x', y', z' with z' coinciding with z, and the x' and y' coordinate lines rotated α anticlockwise relative to the x and y coordinate lines as shown in Fig. 2.9 (a).

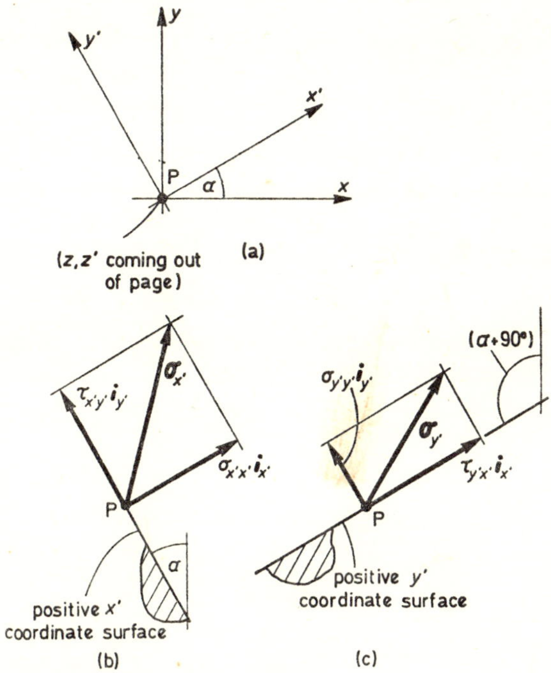

Fig. 2.9

Then the state of stress in this new system is specified by three components $\sigma_{x'x'}$, $\sigma_{y'y'}$ and $\tau_{x'y'}$ at each point P in the plate, and these components are defined in exactly the same way as before. Thus $\tau_{x'y'}$ for example, is the component in the y' direction of the stress vector $\boldsymbol{\sigma}_{x'}$ which itself acts on the positive x' coordinate surface through P, Fig. 2.9 (b) and (c). It is the purpose of this section to establish an important relationship between $\sigma_{x'x'}$, $\sigma_{y'y'}$, $\tau_{x'y'}$ and σ_{xx}, σ_{yy}, τ_{xy}.

Let us consider the stress vector $\boldsymbol{\sigma}_{x'}$; since the positive x' coordinate surface subtends an angle α anticlockwise to the positive x coordinate surface, the components of $\boldsymbol{\sigma}_{x'}$ relative to the x, y coordinate system ($\sigma_{x'x}$, $\sigma_{x'y}$) are given directly by (2.6) and (2.7) as

$$\sigma_{x'x} = \sigma_{xx} \cos \alpha + \tau_{xy} \sin \alpha, \tag{2.8}$$

$$\sigma_{x'y} = \tau_{xy} \cos \alpha + \sigma_{yy} \sin \alpha. \tag{2.9}$$

The relationship between $\sigma_{x'x'}$, $\tau_{x'y'}$ and these components is then apparent from the geometry of Fig. 2.10 (a). Thus

$$\sigma_{x'x'} = \sigma_{x'x} \cos \alpha + \sigma_{x'y} \sin \alpha,$$

$$\tau_{x'y'} = \sigma_{x'y} \cos \alpha - \sigma_{x'x} \sin \alpha.$$

Therefore substituting (2.8) and (2.9) we have

$$\sigma_{x'x'} = \sigma_{xx} \cos^2 \alpha + \sigma_{yy} \sin^2 \alpha + 2\tau_{xy} \sin \alpha \cos \alpha, \tag{2.10}$$

$$\tau_{x'y'} = -(\sigma_{xx} - \sigma_{yy}) \sin \alpha \cos \alpha + \tau_{xy}(\cos^2 \alpha - \sin^2 \alpha). \tag{2.11}$$

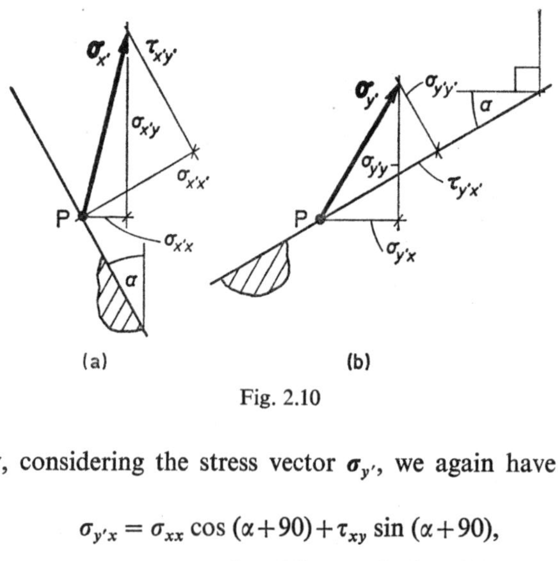

Fig. 2.10

Similarly, considering the stress vector $\boldsymbol{\sigma}_{y'}$, we again have from (2.6) and (2.7)

$$\sigma_{y'x} = \sigma_{xx} \cos (\alpha + 90) + \tau_{xy} \sin (\alpha + 90),$$

$$\sigma_{y'y} = \tau_{xy} \cos (\alpha + 90) + \sigma_{yy} \sin (\alpha + 90),$$

and from Fig. 2.10 (b)

$$\sigma_{y'y'} = \sigma_{y'y} \cos \alpha - \sigma_{y'x} \sin \alpha,$$

so that

$$\sigma_{y'y'} = \sigma_{xx} \sin^2 \alpha + \sigma_{yy} \cos^2 \alpha - 2\tau_{xy} \sin \alpha \cos \alpha. \tag{2.12}$$

Collecting (2.10), (2.11) and (2.12) and simplifying, we obtain the following equations giving the required relationship between the stress components in the two coordinate systems

$$\sigma_{x'x'} = \tfrac{1}{2}(\sigma_{xx}+\sigma_{yy})+\tfrac{1}{2}(\sigma_{xx}-\sigma_{yy})\cos 2\alpha+\tau_{xy}\sin 2\alpha, \quad (2.13)$$

$$\tau_{x'y'} = \qquad -\tfrac{1}{2}(\sigma_{xx}-\sigma_{yy})\sin 2\alpha+\tau_{xy}\cos 2\alpha, \quad (2.14)$$

$$\sigma_{y'y'} = \tfrac{1}{2}(\sigma_{xx}+\sigma_{yy})-\tfrac{1}{2}(\sigma_{xx}-\sigma_{yy})\cos 2\alpha-\tau_{xy}\sin 2\alpha. \quad (2.15)$$

These are called the *transformation equations of stress.*

(c) *Principal stresses*

For a certain value of α the shear stress component $\tau_{x'y'}$ becomes zero. We call this value α_p and the corresponding directions of the x' and y' coordinate lines the *principal directions of stress* at the point P. From (2.14) we see that α_p is given by

$$0 = -\tfrac{1}{2}(\sigma_{xx}-\sigma_{yy})\sin 2\alpha_p+\tau_{xy}\cos 2\alpha_p,$$

or

$$\tan 2\alpha_p = \frac{2\tau_{xy}}{(\sigma_{xx}-\sigma_{yy})}. \quad (2.16)$$

Thus

$$\alpha_p = \tfrac{1}{2}\tan^{-1}\left(\frac{2\tau_{xy}}{\sigma_{xx}-\sigma_{yy}}\right). \quad (2.17)$$

Suppose we let the particular coordinate system subtending α_p to x and y be denoted by X and Y. The corresponding direct stress components σ_{XX} and σ_{YY} are then called the *principal stress components* at P. Now from (2.16) and Fig. 2.11 it appears that

$$\cos 2\alpha_p = \frac{(\sigma_{xx}-\sigma_{yy})}{[(\sigma_{xx}-\sigma_{yy})^2+4\tau_{xy}^2]^{\frac{1}{2}}},$$

$$\sin 2\alpha_p = \frac{2\tau_{xy}}{[(\sigma_{xx}-\sigma_{yy})^2+4\tau_{xy}^2]^{\frac{1}{2}}},$$

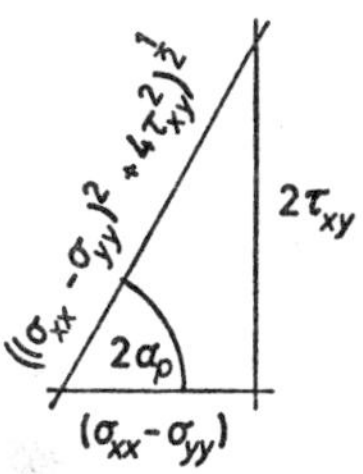

Fig. 2.11

so that substituting into (2.13) and (2.15) gives

$$\sigma_{XX} = \tfrac{1}{2}(\sigma_{xx}+\sigma_{yy})+\tfrac{1}{2}[(\sigma_{xx}-\sigma_{yy})^2+4\tau_{xy}^2]^{\frac{1}{2}}, \quad (2.18)$$

$$\sigma_{YY} = \tfrac{1}{2}(\sigma_{xx}+\sigma_{yy})-\tfrac{1}{2}[(\sigma_{xx}-\sigma_{yy})^2+4\tau_{xy}^2]^{\frac{1}{2}}. \quad (2.19)$$

(d) *Mohr's circle of stress*

The transformation equations (2.13, 2.14, 2.15) relate the components of stress in two systems separated by an angle α. We shall now adapt them to relate the components $\sigma_{x'x'}$, $\sigma_{y'y'}$ and $\tau_{x'y'}$ to the *principal* stress components, supposing that x' and y' are rotated θ anticlockwise relative to the principal directions (Fig. 2.12). Thus noting that τ_{XY} is by definition zero we have

$$\sigma_{x'x'} = \tfrac{1}{2}(\sigma_{XX}+\sigma_{YY})+\tfrac{1}{2}(\sigma_{XX}-\sigma_{YY}) \cos 2\theta, \tag{2.20}$$

$$\tau_{x'y'} = -\tfrac{1}{2}(\sigma_{XX}-\sigma_{YY}) \sin 2\theta, \tag{2.21}$$

$$\sigma_{y'y'} = \tfrac{1}{2}(\sigma_{XX}+\sigma_{YY})-\tfrac{1}{2}(\sigma_{XX}-\sigma_{YY}) \cos 2\theta. \tag{2.22}$$

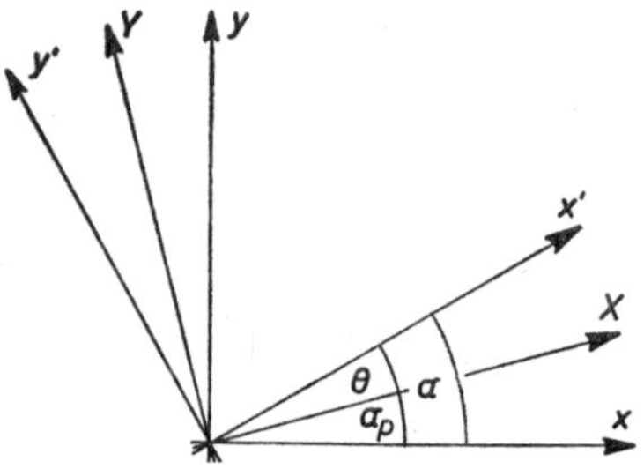

Fig. 2.12

The above transformation equations can be represented by a simple graphical construction; suppose we draw two axes, the vertical representing shear stress components $\tau_{x'y'}$, and the horizontal, direct stress components $\sigma_{x'x'}$, $\sigma_{y'y'}$, and construct a circle whose centre is on the horizontal axis at a distance $\frac{1}{2}(\sigma_{XX}+\sigma_{YY})$ from the origin and whose radius is $\frac{1}{2}(\sigma_{XX}-\sigma_{YY})$. Then the line through the centre of the circle at an angle 2θ *clockwise* to the positive direction of the direct stress axis, intersects the circle at $\sigma_{x'x'}$, $\tau_{x'y'}$ as shown in Fig. 2.13. Produced backwards it intersects the circle at a point whose abscissa is $\sigma_{y'y'}$. This construction was devised by Otto Mohr in 1882 and the circle is called *Mohr's circle of stress*. Perhaps we should emphasise that if $\sigma_{XX} > \sigma_{YY}$ and θ is positive, then the construction indicates that $\tau_{x'y'}$ will be *negative*, as required by equation (2.21).

We have described the construction of Mohr's circle in terms of the principal stresses, but in fact, if we know the stress components σ_{xx}, σ_{yy} and τ_{xy} relative to the x, y coordinate system, the circle can still be constructed immediately. Thus since the x, y system is located at the angle α_p *clockwise* to the principal directions as in Fig. 2.12, the corresponding stresses on Mohr's circle are as shown in Fig. 2.14. It is then apparent that the circle can also be defined as that which has its centre at the point $(\frac{1}{2}(\sigma_{xx}+\sigma_{yy}), 0)$ and which passes through the point (σ_{xx}, τ_{xy}). (We note that the radius of the circle is $\frac{1}{2}[(\sigma_{xx}-\sigma_{yy})^2+4\tau_{xy}^2]^{\frac{1}{2}}$, whence equations (2.18) and (2.19).)

Finally, let us consider again the original relationship between the stress components in the $x'y'$ and the x, y coordinate systems. Recalling that x' is rotated α anticlockwise relative to x and that $\alpha = (\alpha_p + \theta)$ (Fig.

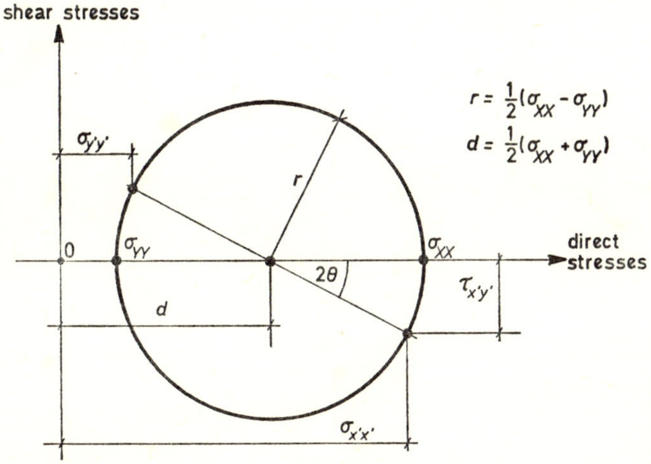

Fig. 2.13

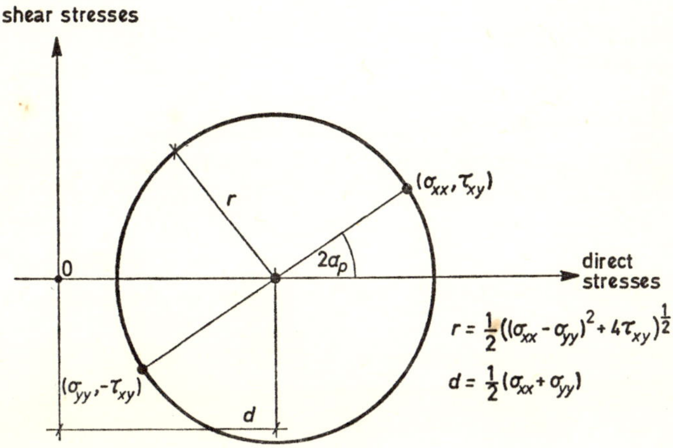

Fig. 2.14

2.12), it is apparent from Figs. 2.13 and 2.14 that the stress components $\sigma_{x'x'}$, $\sigma_{y'y'}$ and $\tau_{x'y'}$ lie on the line drawn through the centre of Mohr's circle subtending a clockwise angle 2α to the line joining the centre and (σ_{xx}, τ_{xy}), as shown in Fig. 2.15.

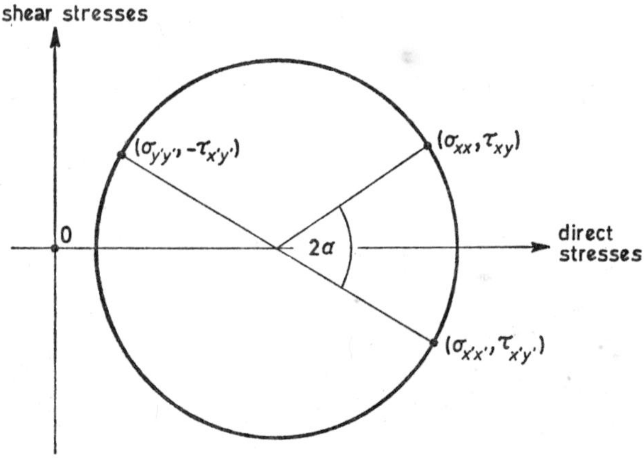

Fig. 2.15

Example 2.2

A state of stress at a point is given by

$$\sigma_{xx} = 10.0,\ \sigma_{yy} = 2.0,\ \tau_{xy} = 3.0\ \text{MN/m}^2.$$

1. Draw Mohr's circle and assuming that the principal directions are such that $\sigma_{XX} > \sigma_{YY}$ calculate the corresponding values of α_p and σ_{XX}, σ_{YY}.
2. What are the values of the components $\sigma_{x'x'}$, $\sigma_{y'y'}$ and $\tau_{x'y'}$ in the x', y' system rotated through 30° anticlockwise to the x, y system?

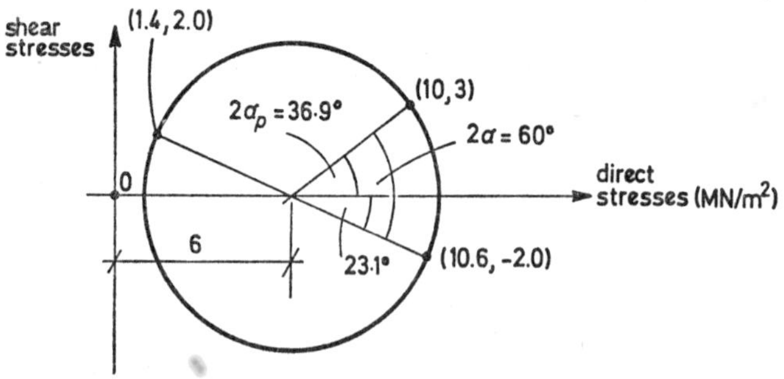

Fig. 2.16

The corresponding Mohr's circle is shown in Fig. 2.16. Thus from the geometry $2\alpha_p = 36.9°$ and $r = 5.0$ MN/m². Whence

$$\alpha_p = 18.45°, \quad \sigma_{XX} = 11.0, \quad \sigma_{YY} = 1.0\ \text{MN/m}^2.$$

Now since $2\alpha = 60°$, the line joining $(\sigma_{x'y'}, \tau_{x'y'})$ to the centre of the circle subtends 23.1° to the direct stress axis as shown. Thus

$$\sigma_{x'x'} = 6.0 + 5.0 \cos 23.1° = 10.6\ \text{MN/m}^2$$

$$\tau_{x'y'} = -5.0 \sin 23.1° = -2.0\ \text{MN/m}^2$$

$$\sigma_{y'y'} = 6.0 - 5.0 \cos 23.1° = 1.4\ \text{MN/m}^2.$$

Considering Mohr's circle in general, it is apparent that there is a particular coordinate system in which the shear stress component $\tau_{x'y'}$ has a maximum value. Thus we see from Fig. 2.13 that this is the case when $2\theta = \pm 90°$, that is when the angle between the x' and X coordinate lines is $\pm 45°$. The corresponding directions of the x' and y' coordinate lines are called the *directions of maximum shear stress.* The maximum shear stress is itself equal to the radius Mohr's circle, so that

$$(\tau_{x'y'})_{\max} = \tfrac{1}{2}[(\sigma_{xx}-\sigma_{yy})^2+4\tau_{xy}^2]^{\frac{1}{2}} = \tfrac{1}{2}(\sigma_{XX}-\sigma_{YY}). \qquad (2.23)$$

(e) *Surface equilibrium*

If a point P is very close to the edge of the plate, as shown in Fig. 2.17, it is possible to choose a wedge-shaped element of material containing P whose surface ABCD corresponds to the actual surface of the plate.

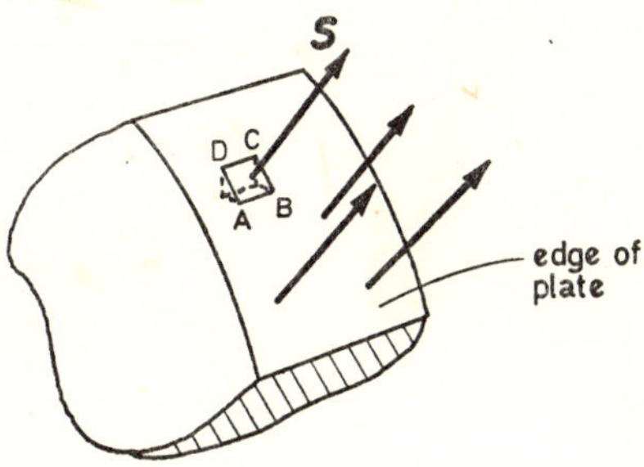

Fig. 2.17

σ_n in (2.6) and (2.7) can then be replaced by the external stress vector $\boldsymbol{S}$ acting at this point and the equations of equilibrium for the element take the form

$$S_x = \sigma_{xx} \cos \alpha + \tau_{xy} \sin \alpha \qquad (2.24)$$

$$S_y = \tau_{xy} \cos \alpha + \sigma_{yy} \sin \alpha, \qquad (2.25)$$

where α is now the angle between the edge surface at P and the positive x coordinate surface. (2.24) and (2.25) are called the *surface equilibrium equations* and must be satisfied at every point on the edge surfaces of the plate.

(f) *Internal equilibrium*

We have assumed that the three components of plane stress generally vary with the x and y coordinates of points in the plate (see equation (2.4)). We now ask whether the requirements of internal equilibrium restrict this variation in any way.

Let us consider again the equilibrium of a rectangular parallelepiped of material containing P shown in Fig. 2.18 (a), but now taking into account the differences in the stresses on its faces. Now the stresses at the mid points of the faces are related to the components of stress at P by means of partial derivatives.* Thus at the point A for example, which is dx from P on a positive x coordinate surface, the components of the stress vector are

$$\left(\sigma_{xx}+\frac{\partial\sigma_{xx}}{\partial x}\,\mathrm{d}x\right),\quad \left(\tau_{xy}+\frac{\partial\tau_{xy}}{\partial x}\,\mathrm{d}x\right),$$

while at the point B, which is $-\mathrm{d}y$ from P on the negative y coordinate surface, the components of the stress vector are

$$-\left(\sigma_{yy}-\frac{\partial\sigma_{yy}}{\partial y}\,\mathrm{d}y\right),\quad -\left(\tau_{xy}-\frac{\partial\tau_{xy}}{\partial y}\,\mathrm{d}y\right),$$

as shown in Fig. 2.18 (b). A detailed analysis of the equilibrium problem does in fact demonstrate that the stresses on each individual face can be considered to be uniform without introducing error. Thus for conciseness

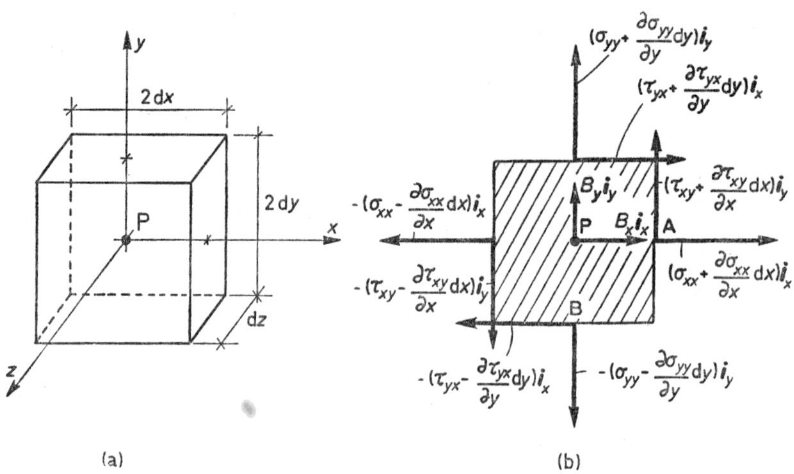

Fig. 2.18

we shall assume that the equations of linear equilibrium of the rectangular element can be found by multiplying the stresses shown in Fig. 2.18 (b) by the areas of the faces on which they act and equating the resultant force $\boldsymbol{R}$ to zero, as in § 2.4.(a). Therefore in the x-direction (noting that $\tau_{xy}=\tau_{yx}$) we have

* A brief outline of partial differentiation is given in Appendix 2.

$$\left(\sigma_{xx}+\frac{\partial\sigma_{xx}}{\partial x}\,\mathrm{d}x\right)2\mathrm{d}y\,\mathrm{d}z+\left(\tau_{xy}+\frac{\partial\tau_{xy}}{\partial y}\,\mathrm{d}y\right)2\,\mathrm{d}x\,\mathrm{d}z$$

$$-\left(\sigma_{xx}-\frac{\partial\sigma_{xx}}{\partial x}\,\mathrm{d}x\right)2\,\mathrm{d}y\,\mathrm{d}z-\left(\tau_{xy}-\frac{\partial\tau_{xy}}{\partial y}\,\mathrm{d}y\right)2\,\mathrm{d}x\,\mathrm{d}z+B_x(4\,\mathrm{d}x\,\mathrm{d}y\,\mathrm{d}z)=0,$$

and it is apparent that in this case the first-order terms cancel out, leaving us with the equilibrium equation

$$\frac{\partial\sigma_{xx}}{\partial x}+\frac{\partial\tau_{xy}}{\partial y}+B_x=0. \qquad (2.26)$$

Similarly in the y direction we have

$$\frac{\partial\sigma_{yy}}{\partial y}+\frac{\partial\tau_{xy}}{\partial x}+B_y=0. \qquad (2.27)$$

Equations (2.26) and (2.27) are called the *internal equations of linear equilibrium* and must be satisfied if individual elements within the plate are to be in static equilibrium.

(g) *Plane stress analysis*

The main problem in structural analysis is that of obtaining values of the internal stress components corresponding to particular loading on the structure. The equilibrium equations (2.26) and (2.27) are an essential

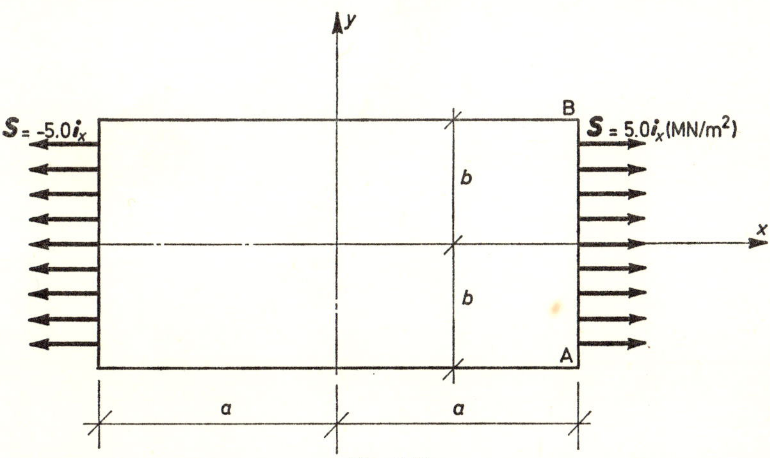

Fig. 2.19

part of the analysis. We can illustrate this by considering the simple example of the rectangular plate shown in Fig. 2.19 which is loaded by equal and opposite stresses on its end edges so that

$$S_x = \pm 5.0,\ S_y = S_z = 0\ \mathrm{MN/m^2}.$$

If all parts of the plate are to be in static equilibrium then the solution for σ_{xx}, σ_{yy} and τ_{xy} must satisfy the surface equilibrium equations at all points on the edges, and the internal equilibrium equations at all points within the plate. Suppose we consider from this point of view the following uniform stress distribution

$$\sigma_{xx} = 5.0, \quad \sigma_{yy} = \tau_{xy} = 0 \text{ MN/m}^2. \tag{2.28}$$

Now the partial derivatives of the stress components in (2.28) are all zero, and in the absence of body forces it is clear that the internal equilibrium equations (2.26) and (2.27) are indeed satisfied. Let us next consider surface equilibrium on the edge AB, where $S_x = +5.0$, $S_y = 0$ MN/m^2. Since $\alpha = 0$, (2.24) and (2.25) require that $\sigma_{xx} = 5.0$ and $\tau_{xy} = 0$ MN/m^2 and this condition is also obviously satisfied by (2.28). It is equally easy to show that the surface equilibrium equations are satisfied on the other three edges of the plate. Therefore the stress distribution (2.28) fulfils all the requirements of a satisfactory equilibrium solution to the problem.

It is tempting to assume without further discussion that the stress distribution (2.28) actually arises in the plate. It is, however, quite possible to find other equally satisfactory equilibrium solutions for the given loading. One such is given by

$$\sigma_{xx} = \left(-\frac{x^6}{6a^5} + \frac{x^4}{2a^3} - \frac{x^2}{2a} + \frac{a}{6}\right)\left(-\frac{5y^4}{b^5} + \frac{6y^2}{b^3} - \frac{1}{b}\right) + 5.0,$$

$$\tau_{xy} = \left(+\frac{x^5}{a^5} - \frac{2x^3}{a^3} + \frac{x}{a}\right)\left(-\frac{y^5}{b^5} + \frac{2y^3}{b^3} - \frac{y}{b}\right),$$

$$\sigma_{yy} = \left(-\frac{5x^4}{a^5} + \frac{6x^2}{a^3} - \frac{1}{a}\right)\left(-\frac{y^6}{6b^5} + \frac{y^4}{2b^3} - \frac{y^2}{2b} + \frac{b}{6}\right) \text{ MN/m}^2. \tag{2.29}$$

Thus since

$$\frac{\partial \sigma_{xx}}{\partial x} = \left(-\frac{x^5}{a^5} + \frac{2x^3}{a^3} - \frac{x}{a}\right)\left(-\frac{5y^4}{b^5} + \frac{6y^2}{b^3} - \frac{1}{b}\right),$$

$$\frac{\partial \tau_{xy}}{\partial y} = \left(\frac{x^5}{a^5} - \frac{2x^3}{a^3} + \frac{x}{a}\right)\left(-\frac{5y^4}{b^5} + \frac{6y^2}{b^3} - \frac{1}{b}\right),$$

it is clear that (2.29) satisfies the internal equilibrium equation (2.26). Similarly it can be shown to satisfy (2.27). On the edge AB we again require $\sigma_{xx} = 5.0$ and $\tau_{xy} = 0$ MN/m^2, and noting on this face that $x = a$, it is apparent that (2.29) also satisfies the surface equilibrium equations (2.24) and (2.25) as indeed it does on the remaining three edges of the plate.

inciple first enunciated by Barré de Saint-Venant in 1855 and called Venant's Principle, which states that if two sets of loading which are atically equivalent are applied in turn to a body, then the stresses at oints remote from the loaded surfaces are the same in the two cases.

ROBLEMS

Problem 2.1 Draw Mohr's circles for the following states of stress, and ssuming that $\sigma_{XX} > \sigma_{YY}$ calculate the values of α_p and σ_{XX}, σ_{YY} for each case.

(a) $\sigma_{xx} = 10.0$, $\sigma_{yy} = 2.0$, $\tau_{xy} = 3.0$ MN/m^2

(b) $\sigma_{xx} = 10.0$, $\sigma_{yy} = 18.0$, $\tau_{xy} = 3.0$ MN/m^2

(c) $\sigma_{xx} = 10.0$, $\sigma_{yy} = 2.0$, $\tau_{xy} = -3.0$ MN/m^2

(d) $\sigma_{xx} = 10.0$, $\sigma_{yy} = 2.0$, $\tau_{xy} = 0$ MN/m^2

(e) $\sigma_{xx} = 10.0$, $\sigma_{yy} = 10.0$, $\tau_{xy} = 0$ MN/m^2

(f) $\sigma_{xx} = -10.0$, $\sigma_{yy} = -2.0$, $\tau_{xy} = 3.0$ MN/m^2.

(Ans. (a) $\alpha_p = 18.5°$, $\sigma_{XX} = 11.0$, $\sigma_{YY} = 1.0$ MN/m^2

(b) $\alpha_p = 71.5°$, $\sigma_{XX} = 19.0$, $\sigma_{YY} = 9.0$ MN/m^2

(c) $\alpha_p = -18.5°$, $\sigma_{XX} = 11.0$, $\sigma_{YY} = 1.0$ MN/m^2

(d) $\alpha_p = 0$, $\sigma_{XX} = 10.0$, $\sigma_{YY} = 2.0$ MN/m^2

(e) α_p indeterminate, $\sigma_{XX} = 10.0$, $\sigma_{YY} = 10.0$ MN/m^2

(f) $\alpha_p = 71.5°$, $\sigma_{XX} = -1.0$, $\sigma_{YY} = -11.0$ MN/m^2.)

Problem 2.2 The maximum permissible direct stresses in concrete in any direction must not exceed 15.0 MN/m^2 compression and 1.5 MN/m^2 tension. At a particular point in the web of a prestressed beam, which is in a state of plane stress, the vertical direct stress is always zero and the horizontal direct compressive stress can be chosen in advance to be any desired value. What is the maximum vertical shear stress that the beam can be designed to resist at the point, and what is the corresponding horizontal compressive stress?

(Ans. 4.74 MN/m^2, -13.50 MN/m^2.)

Problem 2.3 A square plate with edges parallel to the x and y coordinate lines is subjected to the edge stresses shown in Fig. 2.23. Show that equilibrium is satisfied by the following uniform shear stress distribution

$$\tau_{xy} = 100 \text{ MN/m}^2, \quad \sigma_{xx} = \sigma_{yy} = 0,$$

It is apparent from the above that the equilibrium equations alone are insufficient to yield a unique solution to the problem of the plate under edge loads, even though it is well known from the observation of loaded structures that a unique solution does in fact exist. The insufficiency arises because there are only *two* equilibrium equations at each point in the plate to give a solution for the *three* unknown components σ_{xx}, σ_{yy} and τ_{xy}. An additional equation is necessary, and this is found by considering the deformation of the material within the plate. It is called the *compatibility equation* and its derivation will be discussed in the next chapter. However, in this particularly simple problem, we can reasonably deduce from symmetry, that if the material of the plate is homogeneous, the stress distribution set up by the loading will in fact be uniform as given by (2.28).

2.5 Three-dimensional stress

Transformation equations and equilibrium equations for the general state of stress in which the six stress components in § 2.3 take values, can be obtained in a similar way as for the state of plane stress.*

However, the only aspect of the general theory of stress that we need to consider for later reference in this present text is concerned with principal stresses in three dimensions. Thus it can be shown that it is always possible to find an orthogonal coordinate system x', y', z' oriented relative to the x, y, z system for which the shear stress components $\tau_{x'y'}$, $\tau_{y'z'}$, and $\tau_{z'x'}$ are zero. Denoting this system by X, Y and Z the directions of the X, Y and Z coordinate lines are again called the *principal directions of stress* and the corresponding direct stress components (σ_{XX}, σ_{YY}, σ_{ZZ}), the *principal stress components.* If we then consider the x', y', z' coordinate system formed by a rotation of θ anticlockwise about the Z coordinate line shown in Fig. 2.20 (a), the stress components $\sigma_{x'x'}$, $\sigma_{y'y'}$ and $\tau_{x'y'}$ are located on a Mohr's circle drawn between the two principal stresses σ_{XX} and σ_{YY} as shown in Fig. 2.21. Similarly the stresses $\sigma_{y'y'}$, $\sigma_{z'z'}$, $\tau_{y'z'}$, and $\sigma_{z'z'}$, $\sigma_{x'x'}$, $\tau_{z'x'}$, relative to the coordinate systems formed by a rotation about the X and Y coordinate lines (Figs. 2.20 (b) and (c)), lie on the other two Mohr's circles shown in Fig. 2.21. The circles in Fig. 2.21 identify the limiting stress states at a point in the body, for the stresses in any other arbitrarily oriented coordinate system must lie within the shaded area in the figure. We note that the maximum shear stress component at a point is equal to the radius of the greatest of the three Mohr's circles, and is therefore equal to one half the greatest difference between the principal stresses.

* S. P. Timoshenko and J. N. Goodier, *Theory of Elasticity*, 2nd Edn. (McGraw-Hill, New York 1951).

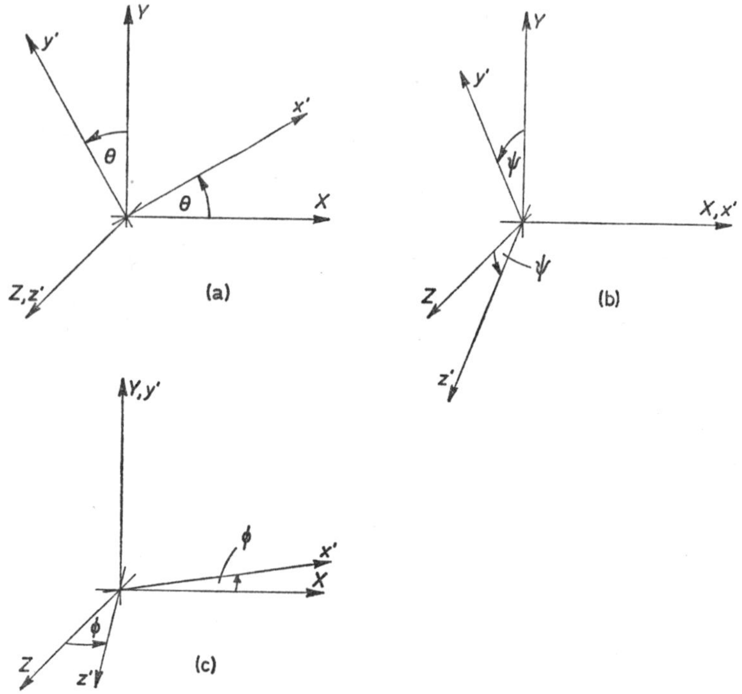

Fig. 2.20

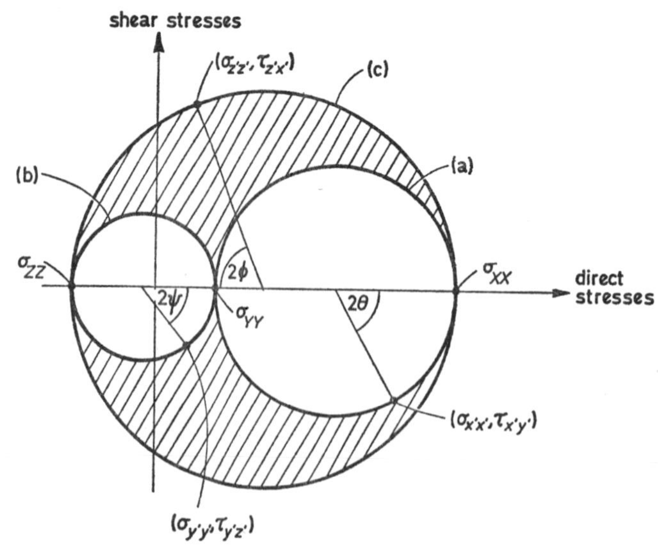

Fig. 2.21

2.6 Stresses in a cylindrical bar

We conclude this chapter by considering the simple b
very important stress system which occurs in a cylindrica
to end loads.

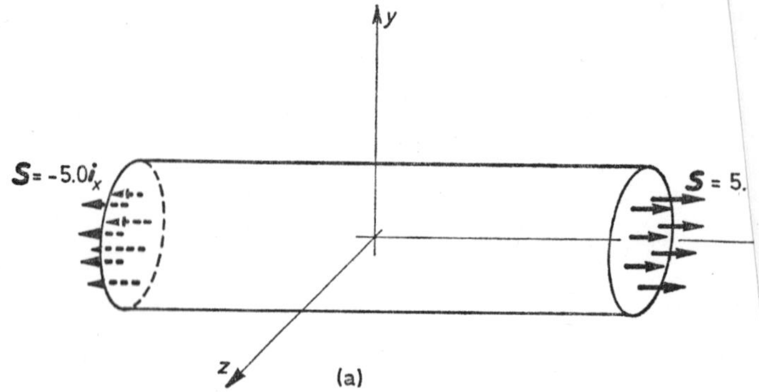

Fig. 2.22 (a)

Consider such a bar, shown in Fig. 2.22 (a), subject to unifor
tributed surface stresses on its ends, for which $S_y = S_z = 0$ and

$$S_x = \pm 5.0 \text{ MN/m}^2$$

say. Then as in the case of the rectangular plate discussed in §
can be shown that the uniform stress distribution

$$\sigma_{xx} = 5.0, \quad \sigma_{yy} = \sigma_{zz} = \tau_{xy} = \tau_{yz} = \tau_{zx} = 0 \text{ MN/m}^2$$

completely satisfies the internal and surface equilibrium conditi
the body forces are negligible. Again from symmetry we conclude th
is the actual system that occurs if the material of the bar is homogen

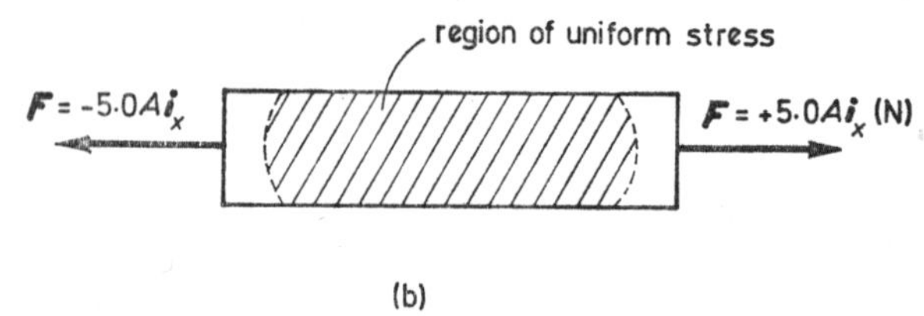

Fig. 2.22 (b)

Now, when subjecting a bar to axial forces it is very difficult to ens
that these forces are uniformly distributed over the ends as required
the solution (2.30). However, it is found that even if the forces are co
centrated and of magnitude $\pm 5.0A$ MN, where A is the sectional ar
of the bar, the uniform stress distribution (2.30) is still set up over mo
of its length as shown in Fig. 2.22 (b). This is an example of a well-know

and that this is equivalent to a direct tensile stress of 100 MN/m² in the direction of one diagonal and a direct compressive stress of -100 MN/m² in the direction of the other. How would you expect the plate to deform eventually, if the loads causing τ_{xy} were gradually increased?

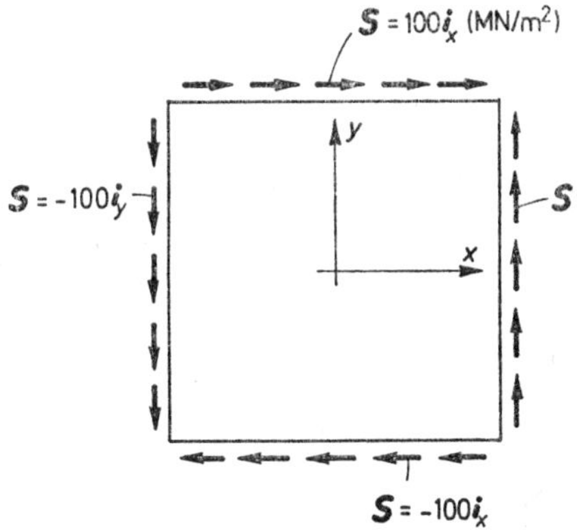

Fig. 2.23

Problem 2.4 A cylindrical compressive specimen is loaded by an axial stress σ. A radial pressure p is then applied by a fluid so that $p = n\sigma$, where n is a factor less than unity. Show that the maximum shear stress in the specimen is reduced by the factor n.

Problem 2.5 Show that for a system of plane stress in which the body forces are zero, any distribution of stress which satisfies

$$\sigma_{xx} = \frac{\partial \phi^2}{\partial y^2}, \quad \sigma_{yy} = \frac{\partial^2 \phi}{\partial x^2}, \quad \tau_{xy} = -\frac{\partial^2 \phi}{\partial x \partial y},$$

where ϕ is some function of x and y, automatically satisfies the internal equilibrium equations.

Problem 2.6 A cylindrical vessel shown in Fig. 2.24 (a), of radius r and thickness t (where $r \gg t$), is closed by rigid end panels and contains fluid under pressure p. The stresses at any point are described by the local coordinate system x, y, z shown in the figure. (1) By considering the linear and rotational equilibrium of the section shown in Fig. 2.24 (b) show that $\sigma_{xx} = \frac{pr}{2t}$, $\tau_{xy} = 0$. (2) By considering linear equilibrium in the radial direction of the section shown in Fig. 2.24 (c) show that $\sigma_{yy} = \frac{pr}{t}$. Hence, assuming that the stresses at right angles to the cylindrical surface are of the same order of magnitude as p, show that the stress system is approximately one of plane stress. What is the maximum shear stress in the cylinder? $\left(\text{Ans. } \dfrac{pr}{2t}\right)$.

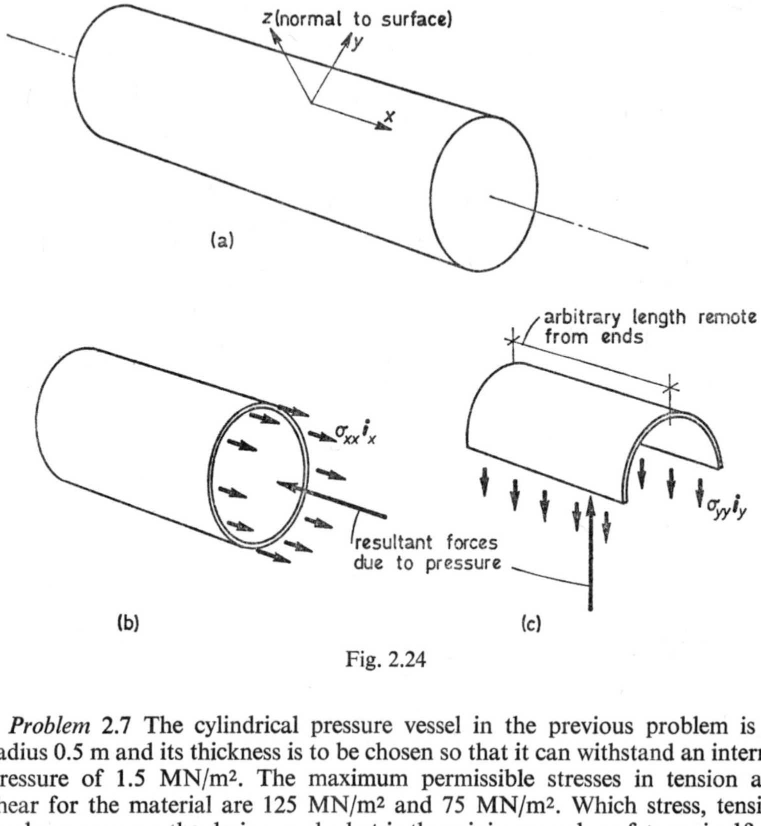

Fig. 2.24

Problem 2.7 The cylindrical pressure vessel in the previous problem is of radius 0.5 m and its thickness is to be chosen so that it can withstand an internal pressure of 1.5 MN/m². The maximum permissible stresses in tension and shear for the material are 125 MN/m² and 75 MN/m². Which stress, tension or shear, governs the design, and what is the minimum value of t required?

(Ans. tension, 0.006 m.)

Problem 2.8 A spherical vessel of radius r and thickness t contains fluid under pressure p. By considering the equilibrium of a hemispherical section show that $\sigma_{xx} = \sigma_{yy} = \dfrac{pr}{2t}$, $\tau_{xy} = 0$, where x, y, z is a local coordinate system chosen so that z is normal to the surface of the vessel at any point.

3

STRAIN

3.1 Displacements

In the previous chapter we discussed the internal stresses set up by forces acting on a body. In this chapter we shall consider the deformation of the material associated with these stresses and introduce the concept of strain.

Let us first consider the displacement of a particular point P in the body. Before loading P occupies its initial position P_i say, and after loading its final position P_f, and the line joining P_i to P_f, having a particular magnitude and direction, is a vector. We denote this vector by $\boldsymbol{u}$ and call

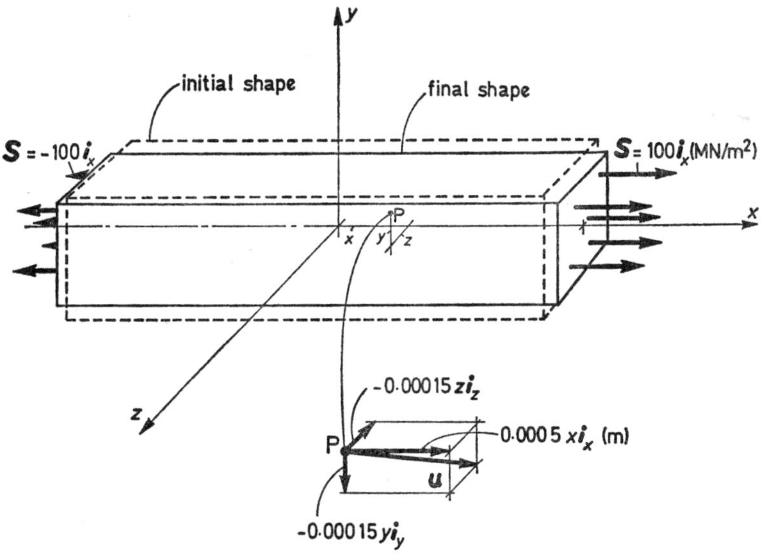

Fig. 3.1

it the *displacement vector* at P; $\boldsymbol{u}$ is then defined at every point in the body, and like, for example, the body force vector, its components can be expressed in terms of the coordinates of P.* For example, it is found that

* In most cases $\boldsymbol{u}$ is so small that the coordinates of P do not change appreciably during the loading.

if a steel bar is subjected to end stresses equal to say 100 MN/m², as shown in Fig. 3.1, then

$$u_x = 0.000\,5x, \quad u_y = -0.000\,15y, \quad u_z = -0.000\,15z \text{ metres.} \qquad (3.1, 3.2, 3.3)$$

The bar therefore expands longitudinally and contracts laterally. Expressions such as (3.1, 3.2, 3.3) completely specify the displacements of a body under loading. We next consider how to describe the corresponding deformation of the material.

3.2 Plane Strain

As in the previous chapter, we shall simplify the discussion by considering only two-dimensional theory, which in this case applies to a system of displacements called *plane strain.* This is said to exist when all the displacement vectors within a body lie in parallel planes. If, therefore,

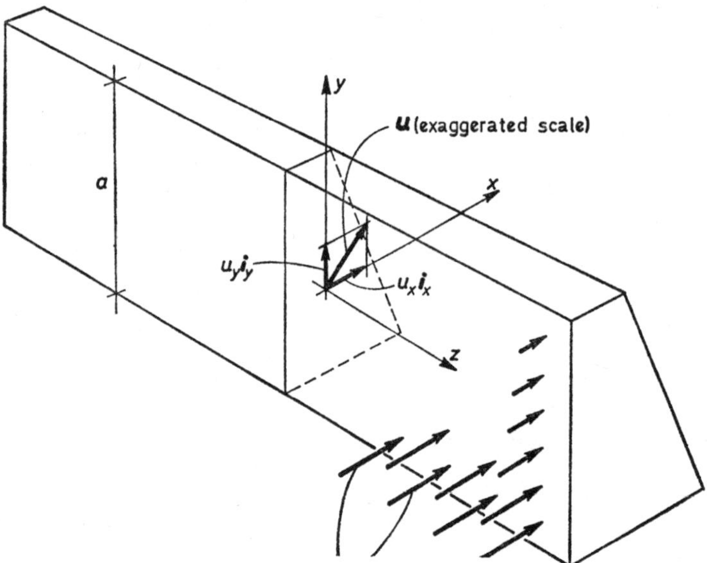

Fig. 3.2

these planes are the z coordinate surfaces of an x, y, z coordinate system then $u_z = 0$ at every point in the body.

From symmetry, we deduce that plane strain occurs in the central region of a body which (1) is very long in the z direction, (2) is of uniform cross section and (3) is subjected to loading in the z plane that is uniformly distributed along its length (Fig. 3.2). Thus it typically occurs in structures such as gravity dams, tunnel linings or retaining walls. We also deduce from symmetry that the two non-zero components of displacement u_x and u_y are uniform along the length of the body, so that they vary with x and y only. For example, we might have

$$u_x = 0.000\,5\frac{x^2}{a} + 0.000\,2y \text{ m}, \tag{3.4}$$

$$u_y = -0.000\,3\frac{xy}{a} \text{ m}. \tag{3.5}$$

(a) *Relative displacements*

In order to define the deformation of the material caused by displacements such as (3.4) and (3.5) we need to be able to ascertain the corresponding *relative* displacements of points within the body. A simple expression for these relative displacements in the immediate neighbourhood of a point P can be found in terms of the partial derivatives of the displacement components as follows.

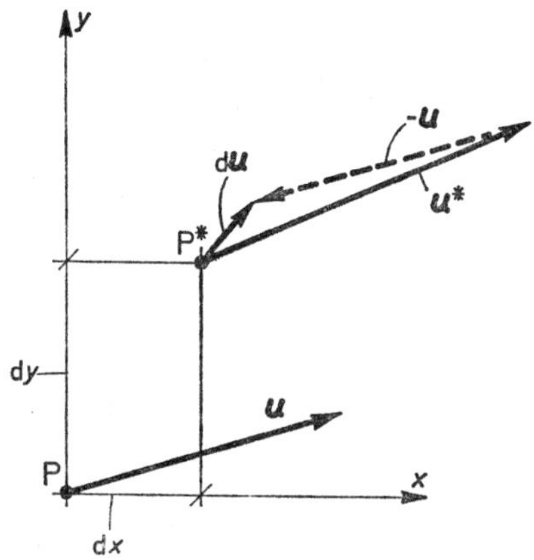

Fig. 3.3

Consider the point P and a neighbouring point P* which is dx, dy from P as shown in Fig. 3.3; if P moves through $\boldsymbol{u}$ and P* moves through $\boldsymbol{u}^*$ say, where $\boldsymbol{u}^*$ generally differs slightly from $\boldsymbol{u}$, the difference between these two displacement vectors is a vector giving the displacement of P* *relative* to P. Calling this vector d$\boldsymbol{u}$ we have (Fig. 3.3)

$$\mathrm{d}\boldsymbol{u} = \boldsymbol{u}^* - \boldsymbol{u}. \tag{3.6}$$

Further, the scalar components du_x and du_y of $d\boldsymbol{u}$ are equal to the differences between the corresponding components of $\boldsymbol{u}^*$ and $\boldsymbol{u}$. Thus

$$du_x = u_x^* - u_x, \quad du_y = u_y^* - u_y. \tag{3.7, 3.8}$$

Now u_x is a function of x and y, and as shown in Appendix 2 (A2.4)

$$u_x^* - u_x = \frac{\partial u_x}{\partial x}\,dx + \frac{\partial u_x}{\partial y}\,dy.$$

Similarly

$$u_y^* - u_y = \frac{\partial u_y}{\partial x}\,dx + \frac{\partial u_y}{\partial y}\,dy.$$

Thus substituting into (3.7) and (3.8) we see that $d\boldsymbol{u}$ is given by

$$\begin{aligned} d\boldsymbol{u} &= du_x\boldsymbol{i}_x + du_y\boldsymbol{i}_y \\ &= \left(\frac{\partial u_x}{\partial x}\,dx + \frac{\partial u_x}{\partial y}\,dy\right)\boldsymbol{i}_x + \left(\frac{\partial u_y}{\partial x}\,dx + \frac{\partial u_y}{\partial y}\,dy\right)\boldsymbol{i}_y. \end{aligned} \tag{3.9}$$

The relative displacement of any point in the neighbourhood of P can thus be determined by (3.9), if the partial derivatives of the displacement components at P are known. An idea of the numerical form of these derivatives can be obtained by differentiating (3.4) and (3.5). Thus

$$\frac{\partial u_x}{\partial x} = 0.001\,\frac{x}{a}, \quad \frac{\partial u_x}{\partial y} = 0.000\,2, \quad \frac{\partial u_y}{\partial x} = -0.000\,3\,\frac{y}{a}, \quad \frac{\partial u_y}{\partial y} = -0.000\,3\,\frac{x}{a};$$

and it is apparent that they are dimensionless and of the order of 10^{-4} in magnitude. We shall now use the expression for $d\boldsymbol{u}$ in a more detailed discussion of the deformation of the material at P.

(b) *Components of strain*

In order to obtain a concise description of the deformation at P it has historically been found convenient to define four components ε_{xx}, ε_{yy}, ε_{xy} and ω_{xy} by the following equations

$$\varepsilon_{xx} \equiv \frac{\partial u_x}{\partial x}, \quad \varepsilon_{yy} \equiv \frac{\partial u_y}{\partial y}, \quad \varepsilon_{xy} \equiv \tfrac{1}{2}\left(\frac{\partial u_x}{\partial y} + \frac{\partial u_y}{\partial x}\right), \tag{3.10, 3.11, 3.12}$$

$$\omega_{xy} \equiv \tfrac{1}{2}\left(\frac{\partial u_x}{\partial y} - \frac{\partial u_y}{\partial x}\right). \tag{3.13}$$

We then note from these definitions that, since

$$\frac{\partial u_x}{\partial y} = (\varepsilon_{xy} + \omega_{xy}) \quad \text{and} \quad \frac{\partial u_y}{\partial x} = (\varepsilon_{xy} - \omega_{xy}),$$

by specifying values of ε_{xx}, ε_{yy}, ε_{xy} and ω_{xy} at P we completely define the values of the four displacement derivatives. The physical meaning of these components becomes clear when we consider the deformation of the rectangular element of material containing P shown in Fig. 3.4 (a) for each component in turn.

Firstly, if $\varepsilon_{xx} \neq 0$ and $\varepsilon_{yy} = \varepsilon_{xy} = \omega_{xy} = 0$ then

$$\frac{\partial u_x}{\partial x} = \varepsilon_{xx}, \quad \frac{\partial u_x}{\partial y} = \frac{\partial u_y}{\partial x} = \frac{\partial u_y}{\partial y} = 0.$$

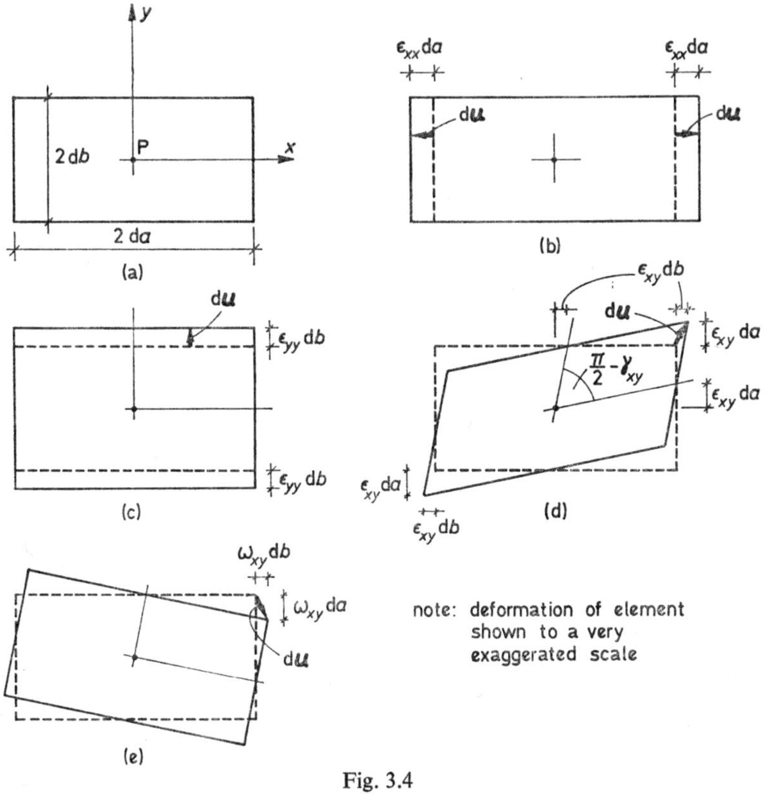

Fig. 3.4

Thus from (3.9) the displacement vector of a point in the element relative to P is given by

$$d\boldsymbol{u} = (\varepsilon_{xx}\, dx)\boldsymbol{i}_x,$$

where dx is the distance in the x direction of the point from P. It is then clear that the element deforms as in Fig. 3.4 (b); ε_{xx}, corresponding to this type of longitudinal deformation is called the *direct strain component* in the x direction at P. If it is positive it is called *tensile* and the element lengthens, and if negative, it is called *compressive* and the element shortens.

Similarly if $\varepsilon_{yy} \neq 0$ and $\varepsilon_{xx} = \varepsilon_{xy} = \omega_{xy} = 0$ the element deforms as in Fig. 3.4 (c) and ε_{yy} is called the *direct strain component* in the y direction at P.

If, next, $\varepsilon_{xy} \neq 0$ and $\varepsilon_{xx} = \varepsilon_{yy} = \omega_{xy} = 0$, then

$$\frac{\partial u_x}{\partial y} = \varepsilon_{xy}, \quad \frac{\partial u_y}{\partial x} = \varepsilon_{xy}, \quad \frac{\partial u_x}{\partial x} = \frac{\partial u_y}{\partial y} = 0,$$

and

$$\mathrm{d}\boldsymbol{u} = (\varepsilon_{xy}\,\mathrm{d}y)\boldsymbol{i}_x + (\varepsilon_{xy}\,\mathrm{d}x)\boldsymbol{i}_y.$$

Thus the element deforms as in Fig. 3.4 (d). Deformation of this type is called shear strain, and ε_{xy} is called the *mathematical shear strain component* at P. We use the adjective 'mathematical' to distinguish between this and the frequently defined *engineering shear strain* at P which is denoted by γ_{xy} and is equal to the closure of the angle between the x and y coordinate lines. Noting that ε_{xy} will be of the same order of magnitude as the displacement derivatives, that is of the order of 10^{-4}, it is clear from the geometry of Fig. 3.4 (d) that the two shear strain components are in fact related simply by

$$\gamma_{xy} \equiv 2\varepsilon_{xy}. \tag{3.14}$$

Finally, if $\omega_{xy} \neq 0$ and $\varepsilon_{xx} = \varepsilon_{yy} = \varepsilon_{xy} = 0$, then

$$\frac{\partial u_x}{\partial y} = \omega_{xy}, \quad \frac{\partial u_y}{\partial x} = -\omega_{xy}, \quad \frac{\partial u_x}{\partial x} = \frac{\partial u_y}{\partial y} = 0$$

and

$$\mathrm{d}\boldsymbol{u} = (\omega_{xy}\,\mathrm{d}y)\boldsymbol{i}_x - (\omega_{xy}\,\mathrm{d}x)\boldsymbol{i}_y.$$

Thus the element rotates *without deformation* about the z axis as in Fig. 3.4 (e).

The deformation of the material in the neighbourhood of P can thus be completely defined by specifying the values of the three strain components ε_{xx}, ε_{yy} and ε_{xy} at P. Equations (3.10), (3.11) and (3.12) are called the *strain-displacement* relations, and we see, for example, that the strain components corresponding to (3.4) and (3.5) are

$$\varepsilon_{xx} = 0.001\frac{x}{a}, \quad \varepsilon_{yy} = -0.000\,3\frac{x}{a}, \quad \varepsilon_{xy} = 0.000\,1 - 0.000\,15\frac{y}{a}.$$

(c) *Uniform strain*

Let us consider a situation in which the displacement components are linear functions of the coordinates of P, for example

$$u_x = 0.001x + 0.000\,3y, \quad u_y = 0.000\,4x + 0.001\,2y. \tag{3.15, 3.16}$$

The corresponding strain components are then given by (3.10, 3.11, 3.12) as

$$\varepsilon_{xx} = 0.001, \quad \varepsilon_{yy} = 0.001\,2, \quad \varepsilon_{xy} = 0.000\,35. \tag{3.17, 3.18, 3.19}$$

A distribution such as (3.17, 3.18, 3.19) where the components are independent of x and y is called a state of *uniform strain*.

The deformation of the section of a rectangular body undergoing the displacements (3.15, 3.16) is shown in Fig. 3.5, and it is apparent that originally straight lines and plane surfaces in the body remain straight

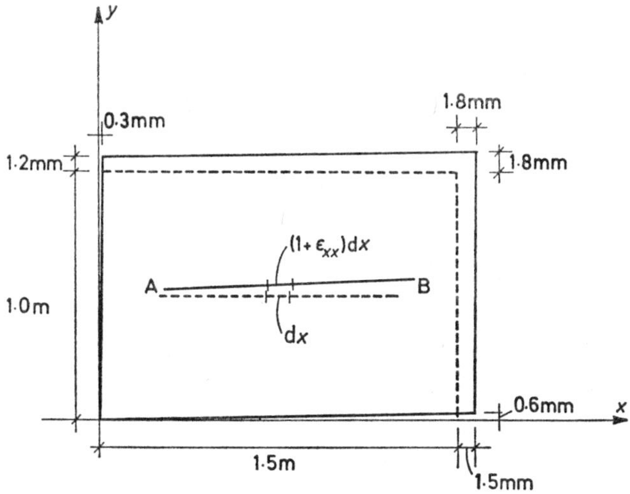

Fig. 3.5

and plane after the deformation. The overall changes in geometry are in this case simply related to the uniform strain components. Thus suppose we consider a line AB in the body of length l, which originally coincides with an x coordinate line. The increase in the length of a small element of the line dx is given by $\varepsilon_{xx}\,\mathrm{d}x$ (see Fig. 3.4 (b)), and the total increase Δl in the length of AB is given by

$$\Delta l = \int_0^l \varepsilon_{xx}\,\mathrm{d}x = \varepsilon_{xx} l,$$

(ε_{xx} is the same at all points on AB). Therefore

$$\varepsilon_{xx} = \frac{\Delta l}{l}. \tag{3.20}$$

ε_{yy} is similarly equal to the ratio of the extension to length of any line coinciding with a y coordinate line in the body. Further, we see from (3.14) that the shear strain ε_{xy} is equal to one half the closure of the angle between two lines coinciding respectively with the x and y coordinate lines in the unstrained body.

Example 3.1

The deformation of a body of rectangular cross section is shown in Fig. 3.6. Determine the uniform strain components.

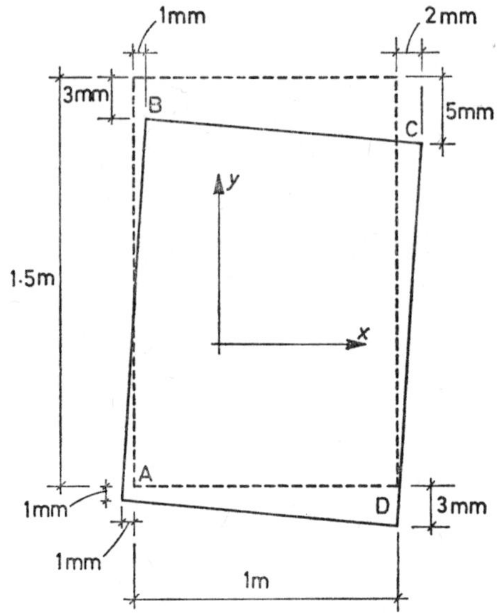

Fig. 3.6

The increase in the length of the line BC (which originally coincided with an x coordinate line) is 0.001 m. Therefore

$$\varepsilon_{xx} = \frac{0.001}{1.0} = 0.001.$$

The decrease in the length of the line AB is 0.002 m. Therefore

$$\varepsilon_{yy} = \frac{-0.002}{1.5} = -0.001\,33.$$

Finally the closure of the angle BAD is $\left(\frac{0.002}{1.5} - \frac{0.002}{1.0}\right)$ rad. Therefore

$$\varepsilon_{xy} = -0.000\,67/2 = -0.000\,33.$$

(d) *Transformation of strains*

As in the discussion of internal stresses, we next derive an important relationship between the strain components at a point relative to various coordinate systems. Suppose, as before, we consider the x', y' and z' coordinate system with z' coinciding with z, and the x' and y' coordinate lines rotated α anticlockwise relative to the x and y coordinate lines as in Fig. 3.7 (a). The displacement vector $\boldsymbol{u}$ then has components $u_{x'}$, $u_{y'}$

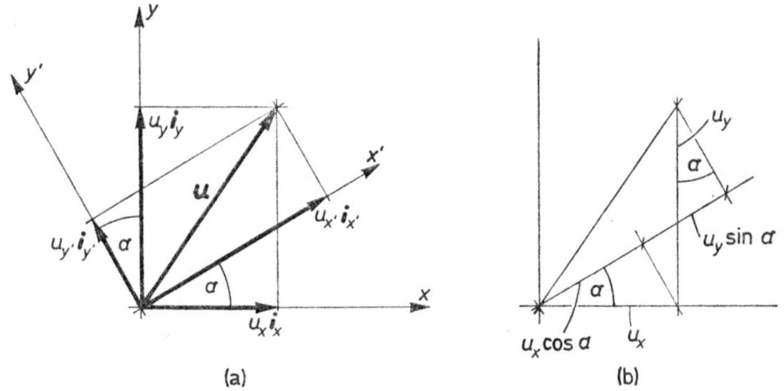

Fig. 3.7

relative to the new coordinate system as shown and the components of strain $\varepsilon_{x'x'}$, $\varepsilon_{y'y'}$ and $\varepsilon_{x'y'}$ are given by the following strain-displacement relations

$$\varepsilon_{x'x'} = \frac{\partial u_{x'}}{\partial x'}, \quad \varepsilon_{y'y'} = \frac{\partial u_{y'}}{\partial y'}, \quad \varepsilon_{x'y'} = \tfrac{1}{2}\left(\frac{\partial u_{x'}}{\partial y'} + \frac{\partial u_{y'}}{\partial x'}\right) \qquad (3.21, 3.22, 3.23)$$

The relationship between $\varepsilon_{x'x'}$, $\varepsilon_{y'y'}$, $\varepsilon_{x'y'}$ and ε_{xx}, ε_{yy}, ε_{xy} can then be found by linking the partial displacement derivatives in the two systems as follows.

Let us first consider the partial derivative $\dfrac{\partial u_{x'}}{\partial x'}$. Suppose a point P* lies at a distance dx' from P on the x' coordinate line through P as in Fig. 3.8. Calling the displacement vector at the two points $\boldsymbol{u}^*$ and $\boldsymbol{u}$ respectively their corresponding components in the x' direction are $u^*_{x'}$ and $u_{x'}$. Then from the definition of partial derivatives (Appendix 2)

$$u^*_{x'} - u_{x'} = \frac{\partial u_{x'}}{\partial x'}\,\mathrm{d}x'. \qquad (3.24)$$

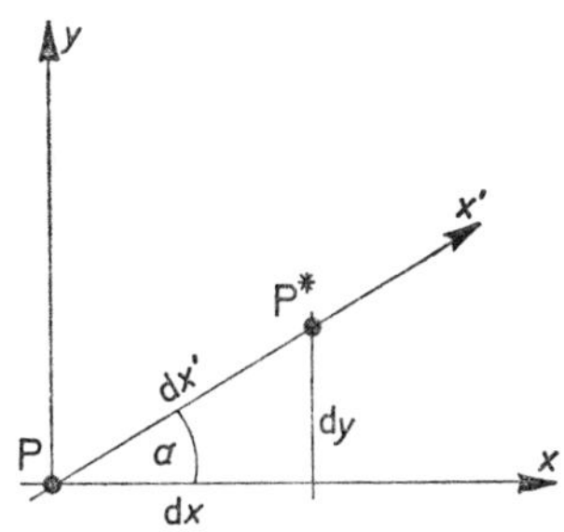

Fig. 3.8

Now $u_{x'}$ can be treated as a function of x and y. Thus from (A2.4) we have also

$$u^*_{x'} - u_{x'} = \frac{\partial u_{x'}}{\partial x}\,\mathrm{d}x + \frac{\partial u_{x'}}{\partial y}\,\mathrm{d}y, \qquad (3.25)$$

where dx = dx' cos α and dy = dx' sin α (Fig. 3.8). Comparison of (3.24) and (3.25) leads to

$$\frac{\partial u_{x'}}{\partial x'} = \frac{\partial u_{x'}}{\partial x}\cdot\frac{\mathrm{d}x}{\mathrm{d}x'} + \frac{\partial u_{x'}}{\partial y}\cdot\frac{\mathrm{d}y}{\mathrm{d}x'} = \frac{\partial u_{x'}}{\partial x}\cos\alpha + \frac{\partial u_{x'}}{\partial y}\sin\alpha. \qquad (3.26)$$

Now from the geometry of Fig. 3.7(b) it is clear that

$$u_{x'} = u_x \cos\alpha + u_y \sin\alpha, \tag{3.27}$$

so that

$$\frac{\partial u_{x'}}{\partial x} = \frac{\partial u_x}{\partial x}\cos\alpha + \frac{\partial u_y}{\partial x}\sin\alpha; \quad \frac{\partial u_{x'}}{\partial y} = \frac{\partial u_x}{\partial y}\cos\alpha + \frac{\partial u_y}{\partial y}\sin\alpha.$$

Whence substituting into (3.26) gives

$$\frac{\partial u_{x'}}{\partial x'} = \frac{\partial u_x}{\partial x}\cos^2\alpha + \frac{\partial u_y}{\partial y}\sin^2\alpha + \left(\frac{\partial u_x}{\partial y} + \frac{\partial u_y}{\partial x}\right)\cos\alpha\sin\alpha.$$

Finally we can use the strain-displacement relations (3.10, 3.11, 3.12) and (3.21) to produce an expression for $\varepsilon_{x'x'}$ in terms of ε_{xx}, ε_{xy} and ε_{yy}. Thus

$$\varepsilon_{x'x'} = \varepsilon_{xx}\cos^2\alpha + \varepsilon_{yy}\sin^2\alpha + 2\varepsilon_{xy}\cos\alpha\sin\alpha. \tag{3.28}$$

A parallel argument leads to similar expressions of $\varepsilon_{y'y'}$ and $\varepsilon_{x'y'}$; these are

$$\varepsilon_{y'y'} = \varepsilon_{xx}\sin^2\alpha + \varepsilon_{yy}\cos^2\alpha - 2\varepsilon_{xy}\sin\alpha\cos\alpha, \tag{3.29}$$

$$\varepsilon_{x'y'} = (\varepsilon_{yy} - \varepsilon_{xx})\sin\alpha\cos\alpha + \varepsilon_{xy}(\cos^2\alpha - \sin^2\alpha). \tag{3.30}$$

The reader is asked to deduce these expressions noting the following results, corresponding to (3.26) and (3.27),

$$\frac{\partial u_{x'}}{\partial y'} = -\frac{\partial u_{x'}}{\partial x}\sin\alpha + \frac{\partial u_{x'}}{\partial y}\cos\alpha,$$

$$\frac{\partial u_{y'}}{\partial x'} = \frac{\partial u_{y'}}{\partial x}\cos\alpha + \frac{\partial u_{y'}}{\partial y}\sin\alpha,$$

$$\frac{\partial u_{y'}}{\partial y'} = -\frac{\partial u_{y'}}{\partial x}\sin\alpha + \frac{\partial u_{y'}}{\partial y}\cos\alpha,$$

$$u_{y'} = -u_x\sin\alpha + u_y\cos\alpha.$$

Equations (3.28), (3.29) and (3.30) can be simplified to

$$\varepsilon_{x'x'} = \tfrac{1}{2}(\varepsilon_{xx} + \varepsilon_{yy}) + \tfrac{1}{2}(\varepsilon_{xx} - \varepsilon_{yy})\cos 2\alpha + \varepsilon_{xy}\sin 2\alpha, \tag{3.31}$$

$$\varepsilon_{x'y'} \qquad -\tfrac{1}{2}(\varepsilon_{xx} - \varepsilon_{yy})\sin 2\alpha + \varepsilon_{xy}\cos 2\alpha, \tag{3.32}$$

$$\varepsilon_{y'y'} = \tfrac{1}{2}(\varepsilon_{xx} + \varepsilon_{yy}) - \tfrac{1}{2}(\varepsilon_{xx} - \varepsilon_{yy})\cos 2\alpha - \varepsilon_{xy}\sin 2\alpha, \tag{3.33}$$

and these are called the *transformation equations of strain*. We note that they are similar in form to the transformation equations of stress (2.13, 2.14, 2.15).

(e) *Principal strains*

For a certain value of α the shear strain component $\varepsilon_{x'y'}$ becomes zero, we call this value α_p and the corresponding directions of the x'

and y' coordinate lines the *principal directions of strain* at the point P. For most materials α_p in this context *has the same value* as α_p in § 2.4 (c) so that at a given point the principal directions of stress and strain coincide. From (3.32) we see that

$$\alpha_p = \tfrac{1}{2} \tan^{-1}\left(\frac{2\varepsilon_{xy}}{\varepsilon_{xx}-\varepsilon_{yy}}\right). \tag{3.34}$$

Letting the particular coordinate system subtending α_p to x and y be denoted by X and Y as in § 2.4 (c), the corresponding direct strain components ε_{XX} and ε_{YY} are called the *principal strain components* at P. Then by comparing with (2.18) and (2.19) we have

$$\varepsilon_{XX} = \tfrac{1}{2}(\varepsilon_{xx}+\varepsilon_{yy})+\tfrac{1}{2}[(\varepsilon_{xx}-\varepsilon_{yy})^2+4\varepsilon_{xy}^2]^{\frac{1}{2}} \tag{3.35}$$

$$\varepsilon_{YY} = \tfrac{1}{2}(\varepsilon_{xx}+\varepsilon_{yy})-\tfrac{1}{2}[(\varepsilon_{xx}-\varepsilon_{yy})^2+4\varepsilon_{xy}^2]^{\frac{1}{2}}. \tag{3.36}$$

(f) *Mohr's circle of strain*

The transformation equations between $\varepsilon_{x'x'}$, $\varepsilon_{x'y'}$, $\varepsilon_{y'y'}$ and ε_{XX}, ε_{YY} take the form

$$\varepsilon_{x'x'} = \tfrac{1}{2}(\varepsilon_{XX}+\varepsilon_{YY})+\tfrac{1}{2}(\varepsilon_{XX}-\varepsilon_{YY}) \cos 2\theta \tag{3.37}$$

$$\varepsilon_{x'y'} = -\tfrac{1}{2}(\varepsilon_{XX}-\varepsilon_{YY}) \sin 2\theta \tag{3.38}$$

$$\varepsilon_{y'y'} = \tfrac{1}{2}(\varepsilon_{XX}+\varepsilon_{YY})-\tfrac{1}{2}(\varepsilon_{XX}-\varepsilon_{YY}) \cos 2\theta, \tag{3.39}$$

where θ is the angle between the x and X coordinate lines. They can be represented by a circle constrcuted relative to axes for shear strains and direct strains as shown in Fig. 3.9. This is called *Mohr's circle of strain*, and its construction and interpretation are analogous to Mohr's circle of stress discussed in § 2.4 (d).

Example 3.2

A state of strain at a point is given by

$$\varepsilon_{xx} = -5.0\times 10^{-4}, \quad \varepsilon_{xy} = 4.0\times 10^{-4}, \quad \varepsilon_{yy} = 7.5\times 10^{-4}.$$

1. Draw Mohr's circle of strain and assuming that the principal directions are such that $\varepsilon_{XX} > \varepsilon_{YY}$, calculate the corresponding values of α_p and ε_{XX}, ε_{YY}.
2. Calculate the values of the strain components in a coordinate system which is rotated through 50° clockwise to the x, y system.

Mohr's circle is shown in Fig. 3.9. Its radius is 7.42×10^{-4} and the angle β is 32.6°. Thus

$$2\alpha_p = 147.4°, \quad \alpha_p = 73.7°$$

$$\varepsilon_{XX} = (1.25+7.42)\times 10^{-4} = 8.67\times 10^{-4}$$

$$\varepsilon_{YY} = (1.25-7.42)\times 10^{-4} = -6.17\times 10^{-4}.$$

The coordinate system at 50° clockwise to x and y is located at 100° anti-clockwise on Mohr's circle along the line PCQ in Fig. 3.9. Thus γ is given by

$$\gamma = 100.0 - 32.6 = 67.4°.$$

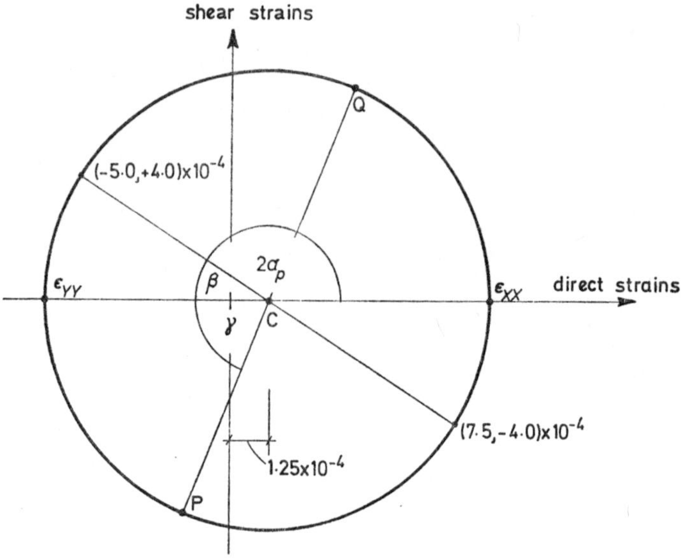

Fig. 3.9

Therefore

$$\varepsilon_{x'x'}(= \text{abscissa of P}) = (1.25 - 7.42 \cos 67.4) \times 10^{-4} = -1.60 \times 10^{-4}$$

$$\varepsilon_{x'y'}(= \text{ordinate of P}) = -7.42 \sin 67.4 \times 10^{-4} = -6.85 \times 10^{-4}$$

$$\varepsilon_{y'y'}(= \text{abscissa of Q}) = (1.25 + 7.42 \cos 67.4) \times 10^{-4} = 4.10 \times 10^{-4}.$$

Finally, the particular coordinate system for which the shear strain component is a maximum occurs when the angle between the x and X coordinate lines is $\pm 45°$. The corresponding direction of the x' and y' coordinate lines are called the *directions of maximum shear strain* and in fact coincide with the directions of maximum shear stress. The maximum shear strain is itself equal to the radius of Mohr's circle so that

$$(\varepsilon_{x'y'})_{\max} = \tfrac{1}{2}[(\varepsilon_{xx} - \varepsilon_{yy})^2 + 4\varepsilon_{xy}^2]^{\frac{1}{2}} = \tfrac{1}{2}(\varepsilon_{XX} - \varepsilon_{YY}). \qquad (3.40)$$

(g) *The compatibility equation*

The complete state of deformation of a body can be defined by specifying the components of strain as functions of the x, y coordinates of every point within it. However, in the same way that the stress components cannot be specified independently but have to satisfy equilibrium conditions, so the strain components cannot be specified independently but have to satisfy so-called compatibility conditions. In effect we must restrict the components so that they are related to a single system of continuous displacements.

To clarify the problem, let us reconsider the strain-displacement relations, but from the point of view of differential equations giving the displacements in terms of known values of strain, that is

$$\frac{\partial u_x}{\partial x} = \varepsilon_{xx}, \quad \frac{\partial u_y}{\partial y} = \varepsilon_{yy}, \quad \frac{\partial u_x}{\partial y} + \frac{\partial u_y}{\partial x} = 2\varepsilon_{xy}. \qquad (3.41, 3.42, 3.43)$$

These can be regarded as *three* simultaneous equations for *two* unknowns u_x and u_y and as with ordinary algebraic equations, a relationship can be found between the coefficients which must be satisfied if the equations are to be mutually consistent. This relationship is found by eliminating u_x and u_y from (3.41, 3.42, 3.43) as follows. Partially differentiating (3.41) with respect to y twice,* (3.42) with respect to x twice and (3.43) with respect to x and y leads to

$$\frac{\partial^3 u_x}{\partial x \partial y^2} = \frac{\partial^2 \varepsilon_{xx}}{\partial y^2}, \quad \frac{\partial^3 u_y}{\partial y \partial x^2} = \frac{\partial^2 \varepsilon_{yy}}{\partial x^2}, \quad \frac{\partial^3 u_x}{\partial x \partial y^2} + \frac{\partial^3 u_y}{\partial y \partial x^2} = 2\frac{\partial^2 \varepsilon_{xy}}{\partial x \partial y}.$$

Whence eliminating the displacement derivatives we obtain

$$\frac{\partial^2 \varepsilon_{xx}}{\partial y^2} + \frac{\partial^2 \varepsilon_{yy}}{\partial x^2} = 2\frac{\partial^2 \varepsilon_{xy}}{\partial x \partial y}, \qquad (3.44)$$

which must be satisfied by the strain distribution if (3.41, 3.42, 3.43) are to be mutually consistent. If it *is* satisfied, then a single solution for the displacements u_x and u_y can be found; (3.44) is called the *compatibility equation for strain.*

The compatibility equation is of great importance in plane stress analysis because it supplies an additional equation to the two equations of static equilibrium (2.26) and (2.27) thus enabling the three unknown stress components σ_{xx}, τ_{xy} and σ_{yy} to be obtained. In practice it is commonly converted into an equation between the stress components by means of the stress-strain relations to be discussed in the next chapter.

3.3 Three-dimensional strain

For the general three-dimensional system of displacements in which $u_z \neq 0$, we define six strain components by the following strain-displacement relations

$$\varepsilon_{xx} = \frac{\partial u_x}{\partial x}, \quad \varepsilon_{yy} = \frac{\partial u_y}{\partial y}, \quad \varepsilon_{zz} = \frac{\partial u_z}{\partial z} \qquad (3.45, 3.46, 3.47)$$

* See Appendix 2.

$$\varepsilon_{xy} = \tfrac{1}{2}\left(\frac{\partial u_x}{\partial y} + \frac{\partial u_y}{\partial x}\right), \quad \varepsilon_{yz} = \tfrac{1}{2}\left(\frac{\partial u_y}{\partial z} + \frac{\partial u_z}{\partial y}\right), \quad \varepsilon_{zx} = \tfrac{1}{2}\left(\frac{\partial u_z}{\partial x} + \frac{\partial u_x}{\partial z}\right). \qquad (3.48, 3.49, 3.50)$$

The physical meaning of these components in terms of the deformation of the material is the same as for plane strain.

The directions of principal strain in three dimensions are the same as the directions of principal stress discussed in § 2.5. The direct strain components ε_{XX}, ε_{YY} and ε_{ZZ} in the X, Y, Z coordinate system are therefore the principal strain components. The strain components $\varepsilon_{x'x'}$, $\varepsilon_{x'y'}$ and $\varepsilon_{y'y'}$ relative to an x', y', z' system formed by a rotation about the Z coordinate line as in Fig. 2.20 (a) are located on a Mohr's circle drawn between the two principal strains ε_{XX} and ε_{YY}. Similarly, the strains in coordinate systems formed by rotation about the X and Y coordinate lines are located on Mohr's circles drawn between ε_{YY} and ε_{ZZ} and between ε_{XX} and ε_{ZZ} respectively. The maximum shear strain is therefore equal to one half the greatest difference between the principal strain components.

3.4 Measurement of strain

It is obviously impossible to devise an instrument for directly measuring the strain components at internal points in a body, because we can only measure the displacements of points on the surface. However, surface measurements of strain are of practical interest in the large number of structures which are composed of relatively thin members.

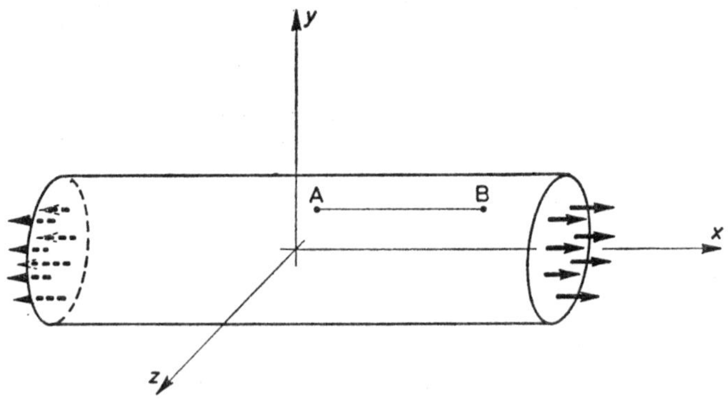

Fig. 3.10

Suppose first that we can assume that a state of uniform strain exists within the body. If, then, we select two points on the surface of the body which both lie on an x coordinate line, as for example A and B on the cylindrical bar in Fig. 3.10, the direct strain component ε_{xx} is equal to the increase in the distance between the points divided by the original distance between them (3.20). Instruments have been devised to measure this ratio directly. For example, a typical *mechanical strain gauge* has two probes located at points on the surface of the body which are a certain distance apart called the *gauge length*. Any relative movement of the probes due to the deformation of the body, is then magnified by gearing

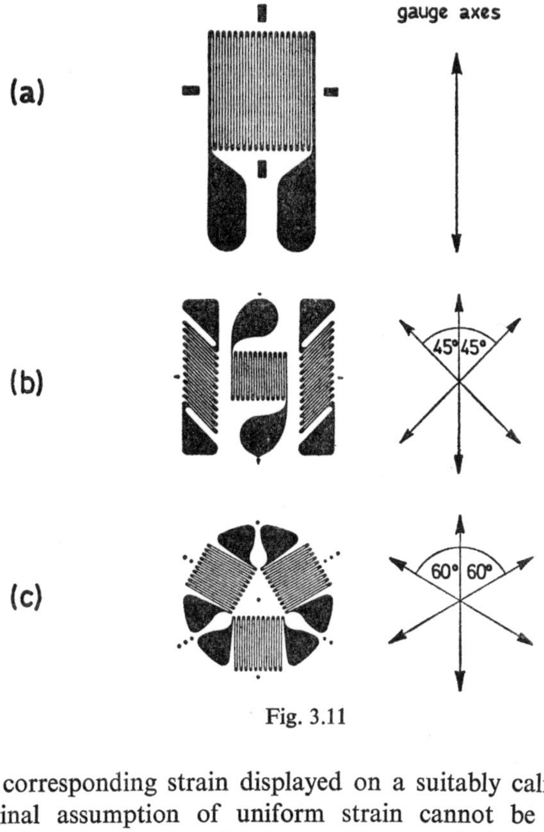

Fig. 3.11

and the corresponding strain displayed on a suitably calibrated dial. If the original assumption of uniform strain cannot be made, then a mechanical gauge measures the *average* direct strain between its probes.

Localised measurements of direct strain on the surfaces of bodies can be obtained by *electric resistance strain gauges*. These use the fact that the resistance of a wire alters when it undergoes strain. Thus a typical layout for a commercial gauge is shown in Fig. 3.11 (a), and by sticking the gauge onto the surface of the body, the direct strain in the direction of the axis of the gauge can be obtained using a suitable transducer meter.

In order to interpret surface strain measurements, we first note that the unloaded surface of a body can be regarded as being in a state of plane *stress*. This is because there are no forces that could give rise to stress components normal to the surface. If then we consider both the stress and strain components relative to the x', y', z' coordinate system shown in Fig. 3.12, it is apparent that the z' coordinate line is a principal direction of stress (and strain), while the other two principal directions, X and Y,

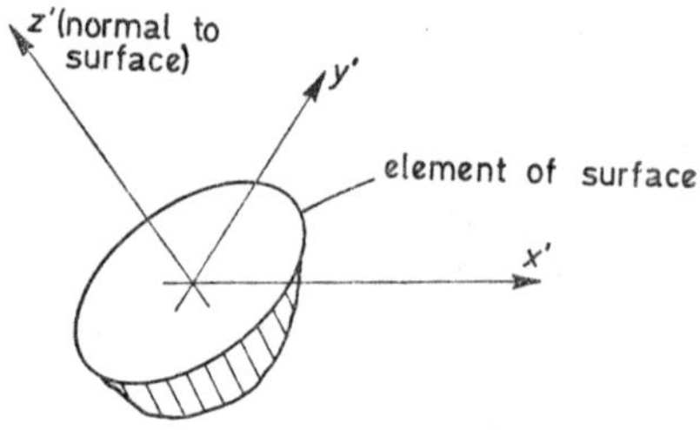

Fig. 3.12

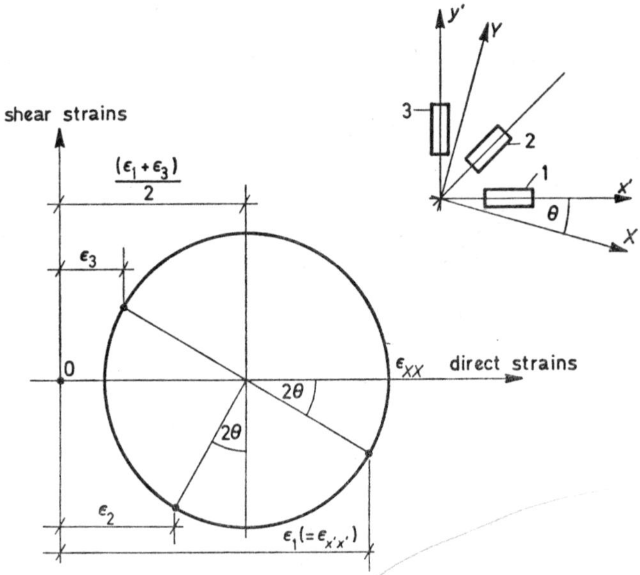

Fig. 3.13

are tangential to the surface. Therefore as discussed in § 3.3, the strain components $\varepsilon_{x'x'}$, $\varepsilon_{x'y'}$ and $\varepsilon_{y'y'}$ are related to ε_{XX} and ε_{YY} by Mohr's circle of strain. Now this circle of strain can be constructed by measuring the direct strains in three distinct directions at P. Thus suppose for example we have three strain gauges arranged as a *45° rosette* shown in Fig. 3.11 (b) and the direct strains are ε_1, ε_2 and ε_3. Then the three strains subtend 90° clockwise to each other on Mohr's circle as shown in Fig. 3.13. Let the principal direction X be at $\theta°$ clockwise to the x' coordinate line and the radius of the circle be r; from the geometry of the figure we see that

$$r\cos 2\theta = \varepsilon_1 - \frac{(\varepsilon_1+\varepsilon_3)}{2} = \frac{(\varepsilon_1-\varepsilon_3)}{2},$$

$$r\sin 2\theta = \frac{(\varepsilon_1+\varepsilon_3)}{2} - \varepsilon_2.$$

Therefore

$$\tan 2\theta = \frac{\varepsilon_1+\varepsilon_3-2\varepsilon_2}{\varepsilon_1-\varepsilon_3} \tag{3.51}$$

and

$$r = \tfrac{1}{2}\{(\varepsilon_1-\varepsilon_3)^2+(\varepsilon_1+\varepsilon_3-2\varepsilon_2)^2\}^{\frac{1}{2}}. \tag{3.52}$$

The two principal strains in the surface are then given by

$$\varepsilon_{XX} = \frac{(\varepsilon_1+\varepsilon_3)}{2}+r, \quad \varepsilon_{YY} = \frac{(\varepsilon_1+\varepsilon_3)}{2}-r. \tag{3.53, 3.54}$$

Example 3.3

The strains measured by the three gauges of the 45° rosette shown in Fig. 3.11 (b) are respectively

$$\varepsilon_1 = -5.0\times10^{-4}, \quad \varepsilon_2 = +3.0\times10^{-4}, \quad \varepsilon_3 = +1.0\times10^{-4}.$$

What are the principal strains at the point and the orientation of the principal direction X, to the x' coordinate line?

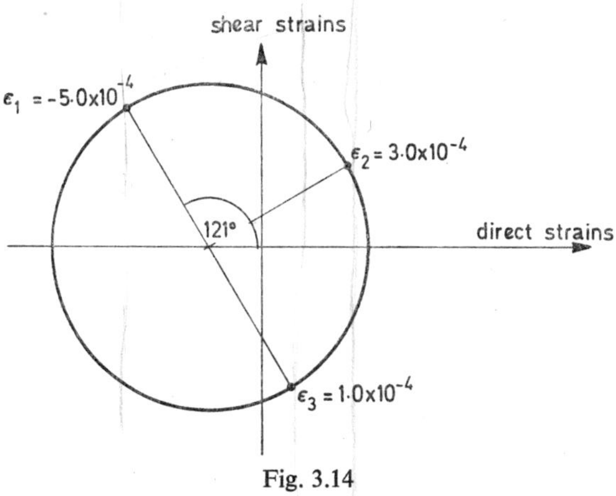

Fig. 3.14

From (3.52)

$$r = \tfrac{1}{2}\{(-5.0-1.0)^2+(-5.0+1.0-2\times3.0)^2\}^{\frac{1}{2}}\times10^{-4}$$

$$= 5.8\times10^{-4}.$$

Thus $$\varepsilon_{XX} = 3.8\times10^{-4},\quad \varepsilon_{YY} = -7.8\times10^{-4}.$$

From (3.51)

$$\tan 2\theta = \frac{(-5.0+1.0-6.0)}{(-5.0-1.0)} = 1.667.$$

Thus $$2\theta = 59.0° \text{ or } -121.0°.$$

The ambiguity as to whether we should take the solution for $2\theta >$ or <0 can be resolved by examining the position of the strains on the Mohr's circle as shown in Fig. 3.14. It is clear that in this case 2θ must be less than 0. The X coordinate line is therefore directed at 60.5° anticlockwise to the x' coordinate line.

Another common layout for strain gauges is the *60° rosette* shown in Fig. 3.11 (c). The method of calculating the principal directions and strains from the three measured direct strains is similar to that for the

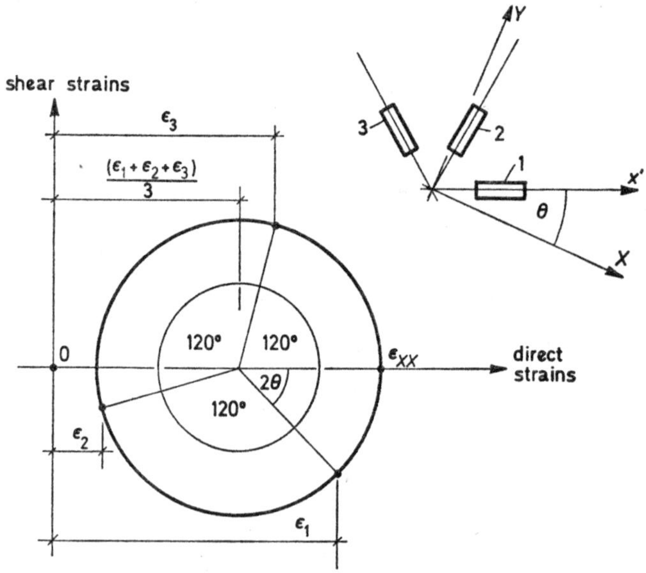

Fig. 3.15

45° rosette except that in this case the three strains subtend 120° to each other on Mohr's circle as shown in Fig. 3.15. Thus from the geometry of the figure it can be shown that

$$\tan 2\theta = \frac{\sqrt{3}(\varepsilon_3-\varepsilon_2)}{(2\varepsilon_1-\varepsilon_2-\varepsilon_3)}, \tag{3.55}$$

and $$r = \tfrac{2}{3}(\varepsilon_1^2+\varepsilon_2^2+\varepsilon_3^2-\varepsilon_1\varepsilon_2-\varepsilon_2\varepsilon_3-\varepsilon_3\varepsilon_1)^{\frac{1}{2}}. \tag{3.56}$$

Thus $$\varepsilon_{XX} = \frac{\varepsilon_1+\varepsilon_2+\varepsilon_3}{3}+r, \quad \varepsilon_{YY} = \frac{\varepsilon_1+\varepsilon_2+\varepsilon_3}{3}-r. \qquad (3.57, 3.58)$$

PROBLEMS

Problem 3.1 The displacement vector in the central region of a long body is given by

$$\boldsymbol{u} = 0.005y \sin\left(\frac{\pi x}{a}\right)\boldsymbol{i}_x + 0.001\frac{y^2}{a}\cos\left(\frac{\pi x}{a}\right)\boldsymbol{i}_y.$$

Determine the components of strain at the points $x = a/4$, $y = a$.

(Ans. $\varepsilon_{xx} = 0.011\,1$, $\varepsilon_{yy} = 0.001\,41$, $\varepsilon_{xy} = 0.000\,657$.)

Problem 3.2 The beam of length l subject to end couples C shown in Fig. 3.16 undergoes the following displacements

$$u_x = \frac{C}{EI}xz, \quad u_y = \frac{-\nu C}{EI}yz, \quad u_z = \frac{C}{2EI}\left(\frac{l^2}{4}-x^2-\nu z^2+\nu y^2\right)$$

where E, I and ν are constants. Show that the components of strain are given by

$$\varepsilon_{xx} = \frac{Cz}{EI}, \quad \varepsilon_{yy} = \varepsilon_{zz} = \frac{-\nu Cz}{EI}, \quad \varepsilon_{xy} = \varepsilon_{yz} = \varepsilon_{zx} = 0.$$

Sketch the deformed shape of the beam in cross section and side elevation.

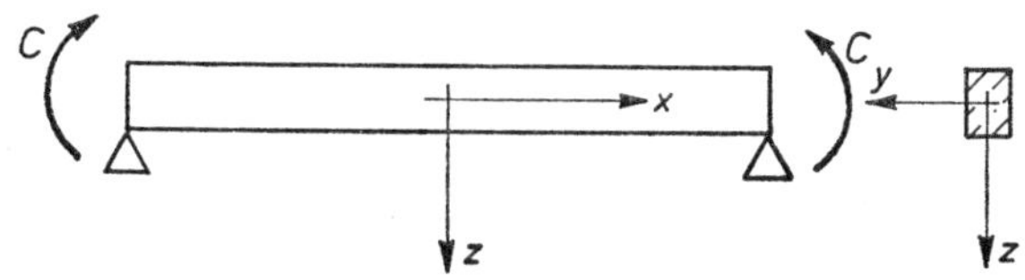

Fig. 3.16

Problem 3.3 The gravity dam sketched in Fig. 3.17 is in a state of uniform strain. The base line CD is fixed and point A moves 15 mm vertically downwards and 25 mm horizontally to the right. What are the vertical and horizontal movements of B and the components of strain?

(Ans. 9.6 mm, 16 mm, $\varepsilon_{xx} = 0$, $\varepsilon_{yy} = -0.001\,2$, $\varepsilon_{xy} = 0.001$.)

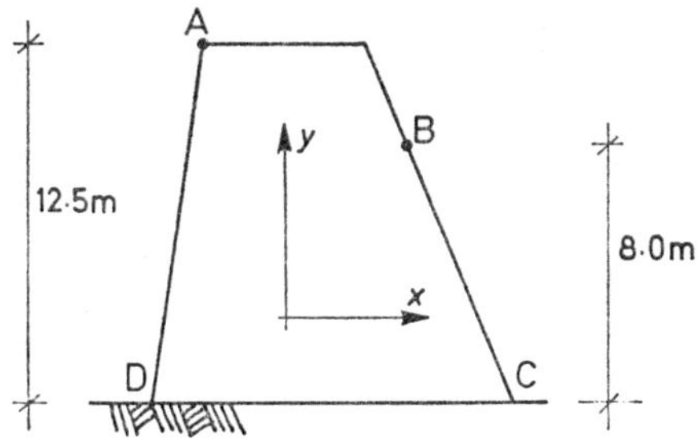

Fig. 3.17

Problem 3.4 A region in a body is in a state of uniform plane strain given by the components ε_{xx}, ε_{yy} and ε_{xy}. A line AB in the region is inclined at $\alpha°$ anti-clockwise to the x coordinate line. By considering the movement of B relative to A, show from first principles that the direct strain ε in the direction of the line is given by

$$\varepsilon = \varepsilon_{xx}\cos^2\alpha+\varepsilon_{yy}\sin^2\alpha+2\varepsilon_{xy}\cos\alpha\sin\alpha.$$

Problem 3.5 The direct strains measured at a surface point on a body by three strain gauges arranged in a 60° rosette are

$$\varepsilon_1 = -1.0\times10^{-4},\quad \varepsilon_2 = +5.0\times10^{-4},\quad \varepsilon_3 = +1.0\times10^{-3}.$$

What are the strain components measured relative to the x, y coordinate system shown in Fig. 3.18?

(Ans. $\varepsilon_{xx} = -1.0\times10^{-4}$, $\varepsilon_{xy} = -2.89\times10^{-4}$, $\varepsilon_{yy} = 1.03\times10^{-3}$.)

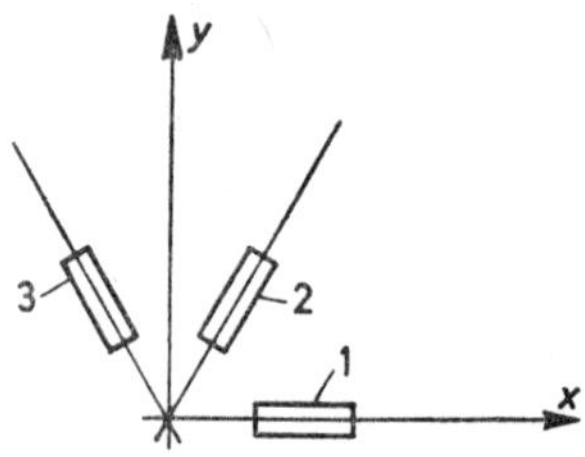

Fig. 3.18

Problem 3.6 The strains measured at a point on a structure by a 45° rosette are ε_1, ε_2, ε_3. If, due to an increase in loads, ε_1 increases by $d\varepsilon_1$, show that the corresponding increase in the greater principal strain ε_{XX} at the point is

$$d(\varepsilon_{XX}) = \{\tfrac{1}{2}+(\varepsilon_1-\varepsilon_2)[(\varepsilon_1-\varepsilon_3)^2+(\varepsilon_1+\varepsilon_3-2\varepsilon_2)^2]^{-\frac{1}{2}}\}\,d\varepsilon_1.$$

g the question of yielding, a typical plot of σ_{xx} versus ε_{xx} steel tensile specimen takes the form shown in Fig. 4.2 (a). straight portion of the curve of slope equal to E corresponds it at a certain stress of the order of 250 MN/m^2, the strain ramatically with little or no increase of load. This stress is

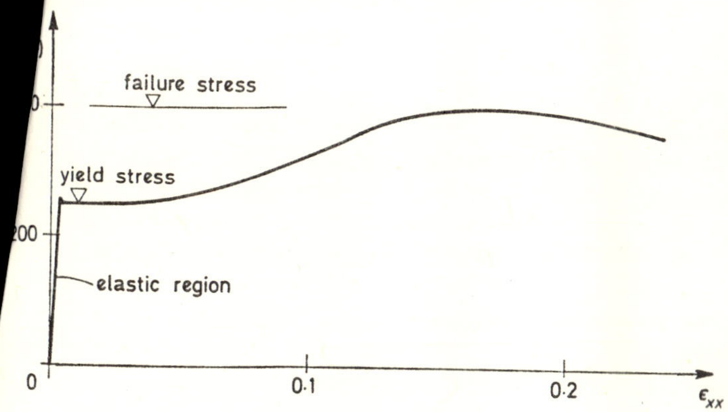

(a)

Fig. 4.2 (a)

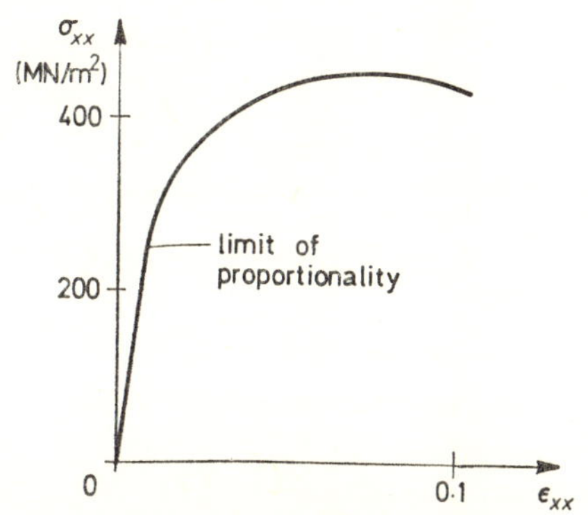

(b)

Fig. 4.2 (b)

called the *uni-axial yield stress* of the material. Subsequently the stress–strain curve indicates that the specimen will in fact support larger stresses up to a maximum value of the order of 400 MN/m^2, which can be regarded as the *ultimate stress* of mild steel. After yielding, the steel loses it property of elasticity and the specimen displays a permanent set when the loading is removed. The stress–strain curve for an aluminium alloy specimen

4

STRESS–STRAIN RELATIONS

4.1 Introduction

In the two previous chapters we have shown that the internal forces at a point P in a body can be completely specified by the six components of stress σ_{xx}, σ_{yy}, σ_{zz}, τ_{xy}, τ_{yz}, τ_{zx}, while the deformation of the material in the neighbourhood of P can be specified by the six components of strain ε_{xx}, ε_{yy}, ε_{zz}, ε_{xy}, ε_{yz}, ε_{zx}. In this chapter we shall describe the way in which the internal forces and local deformation are related. We shall in fact, derive the so-called *stress–strain relations* which express the strain components in terms of the stress components at every point in the body.

The stress–strain relations depend on the nature of the interatomic forces in the body, and these in turn depend on the material of which the body is made. Though they could, in principle, be derived by considering the atomic structure of the material, the calculation has proved impossibly difficult in practice because the structures of real materials such as steel or concrete are far too complicated for analysis. Thus the stress–strain relations have to be deduced from experimental evidence obtained by testing specimens made from particular materials of interest.

We can, however, make one obvious assumption regarding the stress–strain relations before resorting to tests. This is that if a body is made from a *homogeneous material*, then the stress–strain relations are the same at every point within it. (It is then interesting to note that this assumption justifies the conclusion in § 2.4 (g) that the uniform stress distribution arises in a plate under end loading, for we now see that a uniform stress distribution must be associated with a uniform distribution of strain if the plate is homogeneous. It is clear that such a distribution automatically satisfies the compatibility equation (3.44) so that all the requirements of the unique solution to the problem discussed in § 2.4 (g), are fulfilled.)

The most frequently used test for finding the stress–strain relations is based on the problem of the bar subject to end forces F (§ 2.6). If the bar is homogeneous, we know that the internal stress distribution is uniform and given by

$$\sigma_{xx} = \frac{|F|}{A}, \quad \sigma_{yy} = \sigma_{zz} = \tau_{xy} = \tau_{yz} = \tau_{zx} = 0 \qquad (4.1, \text{ cf. } 2.30)$$

and that the corresponding strain distribution is also uniform. Therefore experimentally we can produce internal stresses in a bar by the application of external forces, and determine the corresponding strains by measuring the relative displacements of points on its surface (§ 3.4). The details of the test are discussed in the next section.

4.2 The tensile test

A typical test specimen is shown in Fig. 4.1 (a); it is carefully machined cylindrical over a central length and the ends A and B can be screwed or clamped into the jaws of a testing machine. We assume, following

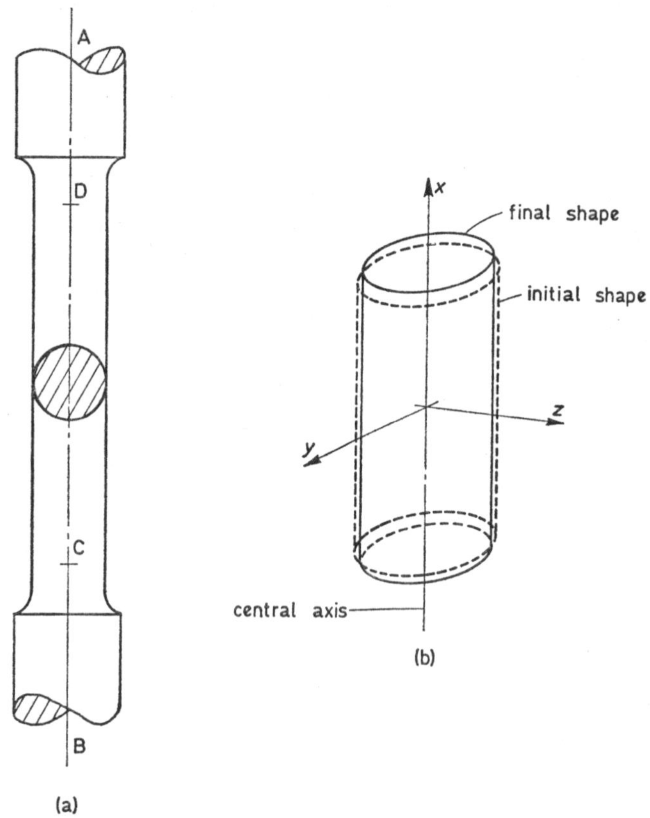

Fig. 4.1

St Venant's principle, that under end loading the stress and strain distributions over the length CD are uniform. Therefore the direct strain in the axial direction can be obtained by a mechanical gauge placed on the surface parallel to the axis of the specimen, while the direct strains

normal to the axis can be obtained b
in the diameter of the specimen.

On loading the specimen with tensile fo
tion of the central section takes the form
corresponding to the single positive di
certain positive direct strain ε_{xx}, and neg
In addition the shear strains ε_{xy}, ε_{yz} and ε_z
then carry out a series of tests we find that
forces the strains are directly proportional to
the following equations

$$\varepsilon_{xx} = \frac{\sigma_{xx}}{E}, \quad \varepsilon_{yy} = -\frac{\nu\sigma_{xx}}{E}, \quad \varepsilon_{zz}$$

(This relationship also applies for compressive
constant having the dimensions of stress and is
after its discoverer Thomas Young (1773-1829), a
constant called *Poisson's ratio* after the French
Poisson (1781-1840). Typical properties for practic
in Table 4.1.

TABLE 4.1

Material	E GN/m²	ν	α
Steel	200	0.31	1
Aluminium alloy	70	0.30	2.
Concrete	25-36	0.20	1.
Glass	60	0.26	0.7
Timber (with grain)	7	—	0.6

4.3 Hooke's Law

Materials having stress–strain relations such as (4.2, 4.3,
to be *elastic* and *linear*. The essential feature of elastic mate
if a set of stresses are induced at a point together with co
strains, then when the stresses are removed the strains always
disappear. The fact that most elastic materials have linear str
relations, so that doubling the forces on an elastic body dou
deformation, etc., was discovered by Robert Hooke and enunc
1678 in the form '*ut tensio sic vis*' (as the tension so the force).

Hooke's Law holds for a wide range of loading on bodies made
most practical materials. It becomes invalid when either the intera
bonds in the materials break down—called *yielding*—or when
deflections occur due to the *geometrical instability* of the bodies.
latter situation arises for example when a strut buckles in compressio

Concernin
for a mild
The initial
to (4.2), b
increases

shown in Fig. 4.2 (b) does not display a marked yield stress and the material is linear elastic up to a stress called the *limit of proportionality*, which again is of the order of 250 MN/m^2. Further increments of stress produce progressively greater increments of strain until the maximum value of the order of 450 MN/m^2 is reached.

4.4 Three-dimensional stress–strain relations

(a) *Direct stresses and strains*

The tensile test gives information about the state of strain corresponding to the single stress component σ_{xx}. Now most practical materials are found experimentally to be *isotropic*, which means that there is no directional variation in the way they respond to stresses. Specifically this means that the single stress component σ_{yy} at a point would produce the corresponding strains

$$\varepsilon_{yy} = \frac{\sigma_{yy}}{E}, \quad \varepsilon_{zz} = \varepsilon_{xx} = -\frac{\nu\sigma_{yy}}{E},$$

where E and ν have the same values as for σ_{xx}, while the single stress component σ_{zz} would produce

$$\varepsilon_{zz} = \frac{\sigma_{zz}}{E}, \quad \varepsilon_{xx} = \varepsilon_{yy} = -\frac{\nu\sigma_{zz}}{E}.$$

Further, if the three direct stresses act simultaneously at a point the strains given above are superimposed. Therefore the corresponding total direct strains are given by

$$\varepsilon_{xx} = \frac{1}{E}\{\sigma_{xx} - \nu(\sigma_{yy} + \sigma_{zz})\} \tag{4.5}$$

$$\varepsilon_{yy} = \frac{1}{E}\{\sigma_{yy} - \nu(\sigma_{zz} + \sigma_{xx})\} \tag{4.6}$$

$$\varepsilon_{zz} = \frac{1}{E}\{\sigma_{zz} - \nu(\sigma_{xx} + \sigma_{yy})\}. \tag{4.7}$$

Example 4.1

The thin aluminium alloy plate in Fig. 4.3 is subject to the edge stresses with normal components shown. The thickness of the plate is 10 mm. Calculate the corresponding strains in the plate, and its change in volume.

The reader should verify, as in § 2.4(g), that static equilibrium is satisfied by the plane stress distribution

$$\sigma_{xx} = 50.0, \ \sigma_{yy} = -30.0 \text{ MN/m}^2, \ \sigma_{zz} = \tau_{xy} = \tau_{yz} = \tau_{zx} = 0.$$

Using (4.5, 4.6, 4.7) we have

$$\varepsilon_{xx} = \frac{1}{70\times 10^9}\{50.0\times 10^6 - 0.30(-30.0\times 10^6)\} = +0.000\,84$$

$$\varepsilon_{yy} = \frac{1}{70\times 10^9}\{-30.0\times 10^6 - 0.30(50.0\times 10^6)\} = -0.000\,64$$

$$\varepsilon_{zz} = \frac{1}{70\times 10^9}\{-0.30(+50.0-30.0)\times 10^6\} = -0.000\,086.$$

(It is of interest to note that though the plate is subjected to a state of plane stress, the fact that the normal strain component ε_{zz} is non-zero indicates that the corresponding state of strain is *not* one of plane strain.)

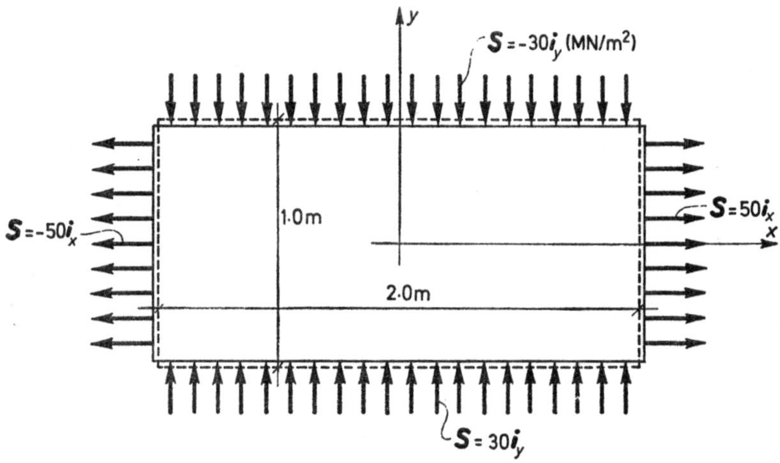

Fig. 4.3

Now the strain system is uniform and the edges of the plate are parallel to the coordinate lines, thus from (3.20) the final dimensions of the plate are as follows

length $= 2.00 + 0.000\,84\times 2.00$ m

width $= 1.0 - 0.000\,64\times 1.0$ m

thickness $= 0.01 - 0.000\,086\times 0.01$ m,

and neglecting second order strain terms, the final volume is approximately $2.00\times 1.00\times 0.01(1+0.000\,84-0.000\,64-0.000\,086)$ m³. Thus the change ΔV in the volume of the plate is given by

$$\Delta V = 2.00\times 1.00\times 0.01(0.000\,84-0.000\,64-0.000\,086) = 2.3\times 10^{-6}\ \text{m}^3.$$

The final result in the above example can be put in a more general form. Thus the change ΔV in the volume V of a body undergoing uniform direct strains ε_{xx}, ε_{yy} and ε_{zz} is given by

$$\Delta V = V(\varepsilon_{xx}+\varepsilon_{yy}+\varepsilon_{zz}). \tag{4.8}$$

The ratio $(\Delta V/V)$, which is then seen to be equal to the sum of the direct strains, is called the *volumetric strain* of the body.

(b) *Shear stresses and strains*

In the tensile specimen the shear components of stress and strain relative to the x, y and z coordinate system in Fig. 4.1 are zero. But, perhaps surprisingly, it is possible to deduce the relationship between shear stresses and shear strains from the relations (4.5, 4.6, 4.7) without having to resort to another test procedure.

Suppose we have a system of plane stress at a point in a body such that

$$\sigma_{xx} = +p, \quad \sigma_{yy} = -p, \quad \sigma_{zz} = \tau_{xy} = \tau_{yz} = \tau_{zx} = 0,$$

where p is a particular stress value. Then from (4.5, 4.6, 4.7) the corresponding strains are

$$\varepsilon_{xx} = \frac{(1+\nu)p}{E}, \quad \varepsilon_{yy} = -\frac{(1+\nu)p}{E}, \quad \varepsilon_{zz} = \varepsilon_{xy} = \varepsilon_{yz} = \varepsilon_{zx} = 0,$$

and we see that in this case, the strain system is also one of plane strain. Next let us consider the components of stress and strain relative to a new coordinate system x', y', z', where z' coincides with z and x' and y' are rotated 45° relative to x and y, as shown in Fig. 4.4 (a). $\sigma_{x'x'}$, $\tau_{x'y'}$, $\sigma_{y'y'}$ and $\varepsilon_{x'x'}$, $\varepsilon_{x'y'}$, $\varepsilon_{y'y'}$ are then found from Mohr's circles of stress and strain (Figs. 4.4 (*b*) and (c)) to be

$$\tau_{x'y'} = -p, \qquad \sigma_{x'x'} = \sigma_{y'y'} = 0, \quad (\sigma_{z'z'} = \tau_{y'z'} = \tau_{z'x'} = 0)$$

$$\varepsilon_{x'y'} = -\frac{(1+\nu)p}{E}, \quad \varepsilon_{x'x'} = \varepsilon_{y'y'} = 0, \quad (\varepsilon_{z'z'} = \varepsilon_{y'z'} = \varepsilon_{z'x'} = 0).$$

Thus in the x', y', z' system the only non-zero stress component is the shear component $\tau_{x'y'}$, and the only non-zero strain component $\varepsilon_{x'y'}$ is related to it by the equation

$$\varepsilon_{x'y'} = \frac{(1+\nu)}{E}\tau_{x'y'}. \tag{4.9}$$

Equation (4.9) then provides the necessary relationship between shear stresses and shear strains, because if the material is isotropic we can conclude that the shear stress component τ_{xy} would in turn produce the corresponding shear strain

$$\varepsilon_{xy} = \frac{(1+\nu)}{E}\tau_{xy}, \tag{4.10}$$

while the components τ_{yz} and τ_{zx} would produce

$$\varepsilon_{yz} = \frac{(1+\nu)}{E}\tau_{yz}, \quad \varepsilon_{zx} = \frac{(1+\nu)}{E}\tau_{zx}. \tag{4.11, 4.12}$$

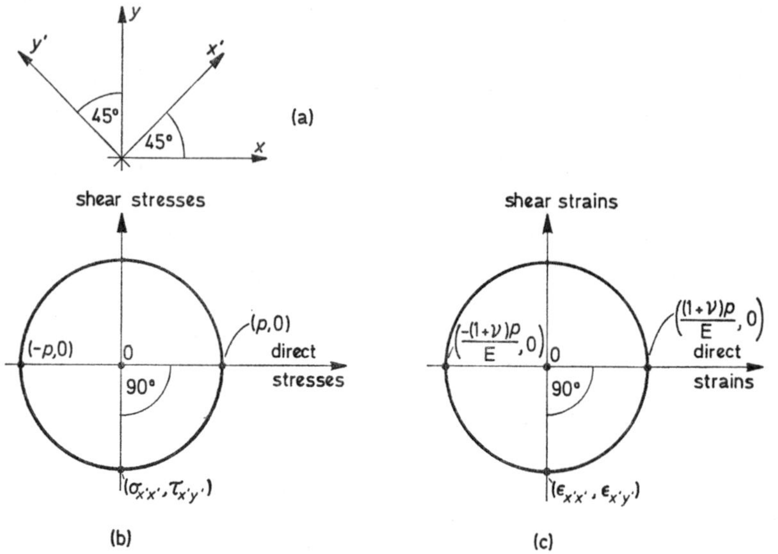

Fig. 4.4

Equation (4.10) is often cast in the form of a relationship between the engineering shear strain γ_{xy} and τ_{xy} as follows

$$\gamma_{xy} = \frac{\tau_{xy}}{G}, \tag{4.13}$$

where G is a constant called the *shear modulus* of the material. Noting that $\gamma_{xy} = 2\varepsilon_{xy}$, and comparing (4.10) and (4.13), we see that

$$G = \frac{E}{2(1+\nu)}. \tag{4.14}$$

A direct experiment for obtaining the shear modulus is briefly described in the following example.

Example 4.2

A thin walled tube shown in Fig. 4.5 (a) has a mean diameter D of 50 mm, and a thickness t of 1 mm. When the tube is subjected to a torque of 280 Nm, the angle of twist in a gauge length l of 200 mm is 0.45°. Estimate the magnitude of the shear modulus of the tube material.

Let us describe the problem in terms of the local coordinate system shown in Fig. 4.5 (a) for which x is an axial coordinate, y a tangential coordinate and z a normal coordinate, at all points on the surface of the tube.

Now the torque of magnitude T, say, is statically equivalent to the uniformly distributed edge stress S, where

$$S_x = S_z = 0, \quad S_y\left(\pi Dt \,.\, \frac{D}{2}\right) = T.$$

Static equilibrium for the problem is then satisfied by the plane stress distribution

$$\tau_{xy} = S_y, \quad \sigma_{xx} = \sigma_{yy} = \sigma_{zz} = \tau_{yz} = \tau_{zx} = 0.$$

The corresponding engineering shear strain is given by (4.13) as

$$\gamma_{xy} = \frac{\tau_{xy}}{G} = \frac{2T}{G\pi D^2 t}.$$

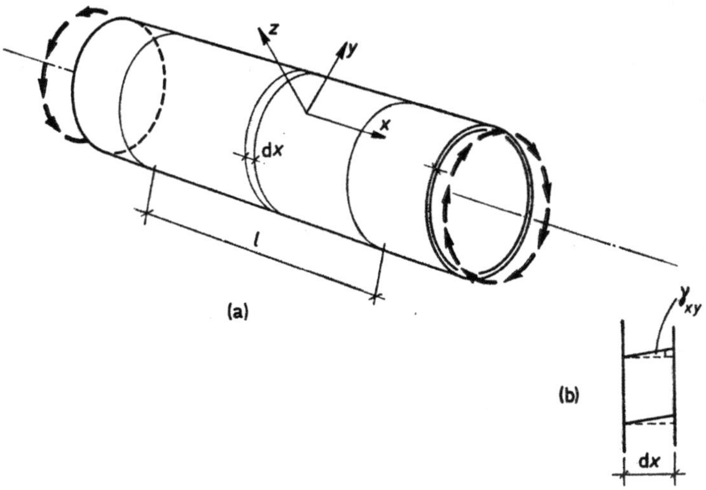

Fig. 4.5

The presence of this shear strain results in relative axial rotation of normal sections of the tube. Thus consider an element of tube of length dx. It is clear from Fig. 4.5 (b) that one edge moves relative to the other through γ_{xy}dx and that the corresponding angle of twist dθ is given by

$$\mathrm{d}\theta = \frac{2\gamma_{xy}}{D}\,\mathrm{d}x.$$

Thus the total twist θ of a gauge length l is given by

$$\theta = \frac{2\gamma_{xy} l}{D}\ \text{(rad)},$$

and substituting for γ_{xy} we have

$$\theta = \frac{2l}{D} \cdot \frac{2T}{G\pi D^2 t}.$$

Therefore

$$G = \frac{4lT}{\pi D^3 t\theta},$$

and using the given experimental values

$$G = \frac{(4 \times 200 \times 10^{-3} \times 280)}{(\pi \times (50 \times 10^{-3})^3 \times 10^{-3} \times 0.007\,85)} = 72.7 \text{ GN/m}^2.$$

(c) *Bulk modulus*

Let us consider the particular state of stress at a point P in a body for which $\sigma_{xx} = \sigma_{yy} = \sigma_{zz} = p$ and $\tau_{xy} = \tau_{yz} = \tau_{zx} = 0$. By definition the x, y and z coordinate lines are principal directions, but in this case the Mohr's circles giving the stress components in any other coordinate system x', y', z' contract to a single point which is located at p on the direct stress axis. Thus the direct stresses $\sigma_{x'x'}$, $\sigma_{y'y'}$ and $\sigma_{z'z'}$ are also equal to p while the shear stresses are zero. If p is negative then this state of stress is called *hydrostatic pressure*, since it always arises in fluids at rest.

The corresponding direct strain components are given by (4.5, 4.6, 4.7) as

$$\varepsilon_{xx} = \varepsilon_{yy} = \varepsilon_{zz} = \frac{(1-2\nu)}{E}\, p. \tag{4.15}$$

If we then consider a differentially small rectangular element of material surrounding P over which these strains can be considered to be uniform, the volumetric strain of the element is given by

$$(\varepsilon_{xx} + \varepsilon_{yy} + \varepsilon_{zz}) = \frac{3(1-2\nu)}{E}\, p.$$

The ratio of p to the volumetric strain is called the *bulk modulus* of the material and is denoted by K. Thus

$$K = \frac{E}{3(1-2\nu)}. \tag{4.16}$$

If, as we should expect in real materials, the volumetric strain has the same sign as p, so that the element decreases in volume if p is compressive and vice versa, then K must be positive. Thus for all real materials

$$\nu < \tfrac{1}{2}. \tag{4.17}$$

(d) *Temperature effects*

It is well known that if a body is subjected to temperature changes, it undergoes displacements which are quite independent of the loading on the body. The corresponding strains are again found by experiments.

When the specimen shown in Fig. 4.1 is heated and its temperature raised by ΔT, it is found that the three direct strains ε_{xx}, ε_{yy}, and ε_{zz} are equal and are proportional to the temperature rise, while the three shear strains are zero, that is

$$\varepsilon_{xx} = \varepsilon_{yy} = \varepsilon_{zz} = \alpha\Delta T, \tag{4.18}$$

where α is a constant called the *temperature coefficient of expansion*; α depends on the material from which the specimen is made and its values for particular materials are given in Table 4.1.

The temperature strains are superimposed on the direct strains given by (4.5, 4.6, 4.7), and combining these results with the shear strain relations (4.10, 4.11, 4.12) we produce the following complete stress–strain relations for a linear elastic isotropic material

$$\varepsilon_{xx} = \frac{1}{E}\{\sigma_{xx} - \nu(\sigma_{yy} + \sigma_{zz})\} + \alpha\Delta T, \tag{4.19}$$

$$\varepsilon_{yy} = \frac{1}{E}\{\sigma_{yy} - \nu(\sigma_{zz} + \sigma_{xx})\} + \alpha\Delta T, \tag{4.20}$$

$$\varepsilon_{zz} = \frac{1}{E}\{\sigma_{zz} - \nu(\sigma_{xx} + \sigma_{yy})\} + \alpha\Delta T. \tag{4.21}$$

$$\varepsilon_{xy} = \frac{(1+\nu)}{E}\tau_{xy}, \quad \varepsilon_{yz} = \frac{(1+\nu)}{E}\tau_{yz}, \quad \varepsilon_{zx} = \frac{(1+\nu)}{E}\tau_{zx}. \qquad (4.22,\ 4.23,\ 4.24)$$

The corresponding inverse stress–strain relations can be found by solving (4.19 to 4.24) for the stresses and are as follows

$$\sigma_{xx} = 2\mu\varepsilon_{xx} + \lambda(\varepsilon_{xx} + \varepsilon_{yy} + \varepsilon_{zz}) - (3\lambda + 2\mu)\alpha\Delta T, \tag{4.25}$$

$$\sigma_{yy} = 2\mu\varepsilon_{yy} + \lambda(\varepsilon_{xx} + \varepsilon_{yy} + \varepsilon_{zz}) - (3\lambda + 2\mu)\alpha\Delta T, \tag{4.26}$$

$$\sigma_{zz} = 2\mu\varepsilon_{zz} + \lambda(\varepsilon_{xx} + \varepsilon_{yy} + \varepsilon_{zz}) - (3\lambda + 2\mu)\alpha\Delta T, \tag{4.27}$$

$$\tau_{xy} = 2\mu\varepsilon_{xy}, \quad \tau_{yz} = 2\mu\varepsilon_{yz}, \quad \tau_{zx} = 2\mu\varepsilon_{zx}, \qquad (4.28,\ 4.29,\ 4.30)$$

where

$$\mu \equiv \frac{E}{2(1+\nu)}, \quad \lambda \equiv \frac{\nu E}{(1+\nu)(1-2\nu)}.$$

Example 4.3

The steel plate in Fig. 4.6 is held between two rigid but frictionless surfaces (platens). It is subjected both to edge stresses on its longitudinal edges, such that

$$S_x = S_z = 0 \text{ and } S_y = \pm 50.0 \text{ MN/m}^2$$

as shown, and to a general temperature rise of 25°C. Calculate the edge stresses exerted by the surfaces, and the components of strain.

Since they are frictionless, the surfaces exert stresses for which $S_y = 0$. Denoting by $\pm S_x$ MN/m² the unknown components of the stresses in the x direction, the following uniform stress distribution satisfies static equilibrium for the problem

$$\sigma_{xx} = S_x, \quad \sigma_{yy} = -50.0 \text{ MN/m}^2, \quad \sigma_{zz} = \tau_{xy} = \tau_{yz} = \tau_{zx} = 0.$$

The corresponding direct strain ε_{xx} is then given by (4.19) as

$$\varepsilon_{xx} = \frac{10^6}{200 \times 10^9}\{S_x + 0.31 \times 50.0\} + 1.25 \times 10^{-5} \times 25$$

$$= \frac{S_x}{200 \times 10^3} + 39.0 \times 10^{-5}.$$

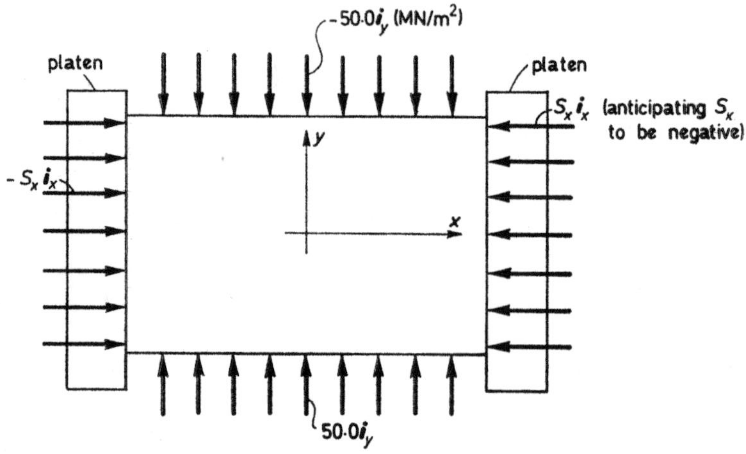

Fig. 4.6

Now the surfaces, being rigid, resist elongation in the x direction with stresses that are sufficient to reduce ε_{xx} to zero. Thus

$$S_x = -200 \times 10^3 \times 39.0 \times 10^{-5} = -78.0 \text{ MN/m}^2.$$

The corresponding strains are then obtained from (4.19 to 4.24) as follows

$$\varepsilon_{xx} = 0,$$

$$\varepsilon_{yy} = \frac{10^6}{200 \times 10^9}\{-50.0 - 0.31(-78.0)\} + 31.3 \times 10^{-5} = +18.4 \times 10^{-5},$$

$$\varepsilon_{zz} = \frac{10^6}{200 \times 10^9}\{-0.31(-78.0 - 50.0)\} + 31.3 \times 10^{-5} = +51.1 \times 10^{-5},$$

$$\varepsilon_{xy} = \varepsilon_{yz} = \varepsilon_{zx} = 0.$$

PROBLEMS

Problem 4.1 Show that for a state of plane stress for which $\sigma_{zz} = 0$,

$$\varepsilon_{xx} = \frac{1}{E}(\sigma_{xx} - \nu\sigma_{yy}), \quad \varepsilon_{yy} = \frac{1}{E}(\sigma_{yy} - \nu\sigma_{xx})$$

$$\varepsilon_{zz} = \frac{-\nu}{E}(\sigma_{xx} + \sigma_{yy}) = \frac{-\nu}{(1-\nu)}(\varepsilon_{xx} + \varepsilon_{yy})$$

(neglecting temperature effects).

Problem 4.2 Show that for a state of plane strain for which $\varepsilon_{zz} = 0$,

$$\sigma_{xx} = (2\mu + \lambda)\varepsilon_{xx} + \lambda\varepsilon_{yy}, \quad \sigma_{yy} = \lambda\varepsilon_{xx} + (2\mu + \lambda)\varepsilon_{yy}$$

$$\sigma_{zz} = \lambda(\varepsilon_{xx} + \varepsilon_{yy}) = \nu(\sigma_{xx} + \sigma_{yy}).$$

Problem 4.3 A sheet of material of Young's modulus E and Poisson's ratio ν, is in a state of plane stress. At a certain point the principal stress in one direction in the plane of the sheet is $-p$, while the direct strain at right angles to this direction in the plane is zero. What are the principal stresses at the point and the corresponding principal strains?

$$\left(\text{Ans.} \ -p, \ -\nu p, \ 0; \ -(1-\nu^2)\frac{p}{E}, \ 0, \ (1+\nu)\nu\frac{p}{E}.\right)$$

Problem 4.4 A 45° strain gauge rosette is placed on the web of a concrete beam with gauge 1 horizontal, gauge 3 vertical and gauge 2 bisecting the angle between 1 and 3. Under load the gauges are found to give readings of

$$\varepsilon_1 = -6.2 \times 10^{-4}, \quad \varepsilon_2 = -5.3 \times 10^{-4}, \quad \varepsilon_3 = 0.6 \times 10^{-4}.$$

Determine the principal stresses at this location, the maximum shear stress, and the direct strain at right angles to the plane of the web. ($E = 25 \text{GN/m}^2$)

(Ans. 0.04 MN/m^2, -17.5 MN/m^2, 8.8 MN/m^2, 1.40×10^{-4})

Problem 4.5 A long thin pressure vessel of radius r is subjected to an internal pressure p and a torque kpr^3. Show that if k is greater than $(\sqrt{2\pi})$ a compressive stress exists in the walls of the cylinder.

Problem 4.6 Show that the strains in the cylindrical pressure vessel of Problem 2.6 are given by

$$\varepsilon_{xx} = \frac{pr}{Et}(\tfrac{1}{2} - \nu), \quad \varepsilon_{yy} = \frac{pr}{Et}\left(1 - \frac{\nu}{2}\right).$$

Whence, noting the worked Example 4.1 deduce that the change ΔV in the internal volume V of the cylinder is given by $\dfrac{\Delta V}{V} = \dfrac{pr}{Et}(\tfrac{5}{2} - 2\nu)$.

Problem 4.7 A spherical vessel made of aluminium is 1 m in diameter and 5 mm thick. It is full of water at atmospheric pressure. If the temperature of the whole system increases by 10°C, what is the increase in the internal pressure? (The bulk modulus of water is 2 GN/m^2, and its volume increases by 0.01%

for each degree temperature rise. Hint: equate the change in the volume of the water to the change in volume of the vessel, taking into account *both* Δp and ΔT.)

(Ans. $\Delta P = 15.5$ kN/m^2.)

Problem **4.8** Show that for the state of plane stress of Problem 4.1, the compatibility equation (3.44) can be expressed in the form of the following equation in stress components

$$\frac{\partial^2}{\partial y^2}(\sigma_{xx}-\nu\sigma_{yy})+\frac{\partial^2}{\partial x^2}(\sigma_{yy}-\nu\sigma_{xx}) = 2(1+\nu)\frac{\partial^2\tau_{xy}}{\partial x\partial y}.$$

Given that

$$\sigma_{xx} \equiv \frac{\partial^2\phi}{\partial y^2}, \quad \sigma_{yy} \equiv \frac{\partial^2\phi}{\partial x^2}, \quad \tau_{xy} \equiv -\frac{\partial^2\phi}{\partial x\partial y},$$

where ϕ is a function of x and y, show that the compatibility equation can also be expressed as

$$\frac{\partial^4\phi}{\partial x^4}+\frac{2\partial^4\phi}{\partial x^2\partial y^2}+\frac{\partial^4\phi}{\partial y^4} = 0.$$

SUGGESTIONS FOR FURTHER READING

The following references contain valuable discussions of stress and strain, as well as accounts of the applications of the theory of elasticity. They are listed in order of increasing level of difficulty and scope. (The particular notation used for the components σ_{xx}, τ_{xy}, ε_{xx}, ε_{xy} is given in brackets.)

1. W. A. Nash, *Strength of Materials* (Schaum's Outline Series) (McGraw-Hill, New York, 1957). (s_x, s_{xy}, ε_x, $\frac{1}{2}\gamma$)
2. J. Case and A. H. Chilver, *Strength of Materials and Structures* 2nd Edn. (Edward Arnold, London, 1971). (σ_x, τ_{xy}, ε_x, $\frac{1}{2}\gamma_{xy}$)
3. S. P. Timoshenko, *Strength of Materials, Part 1* 3rd Edn. (Van Nostrand, Princeton, 1955). (σ_x, τ_{xy}, ε_x, $\frac{1}{2}\gamma_{xy}$)
4. D. S. Dugdale, *Elements of Elasticity* (Pergamon Press, Oxford, 1968). (σ_{xx}, σ_{xy}, e_{xx}, e_{xy})
5. S. P. Timoshenko and J. N. Goodier, *Theory of Elasticity* 2nd Edn. (McGraw-Hill, New York, 1951). (σ_x, τ_{xy}, ε_x, $\frac{1}{2}\gamma_{xy}$)
6. R. L. Bisplinghoff, J. W. Mar and T. H. H. Pian, *Statics of Deformable Solids* (Addison-Wesley, Reading, 1965). (σ_{11}, σ_{12}, γ_{11}, γ_{12})
7. A. E. H. Love, *The Mathematical Theory of Elasticity* 4th edn. (Dover Publications, New York, 1944). (X_x, X_y, e_{xx}, e_{xy})
8. I. S. Sokolnikoff, *Mathematical Theory of Elasticity* (McGraw-Hill, New York, 1956). (τ_{11}, τ_{12}, e_{11}, e_{12})
9. A. E. Green and W. Zerna, *Theoretical Elasticity* 2nd Edn. (Oxford, 1968). (τ_{11}, τ_{12}, γ_{11}, γ_{12})

APPENDIX 1

VECTOR REPRESENTATION

A vector, such as for example a force acting at a point P on the surface of a body, is a quantity possessing both magnitude and direction. It is denoted in text by a bold letter, $\boldsymbol{F}$ say, and in figures by a bold line passing through P which contains an arrow to indicate direction. The magnitude of $\boldsymbol{F}$ is denoted by $|\boldsymbol{F}|$.

$\boldsymbol{F}$ can also be represented by its components relative to a particular coordinate system. Thus suppose a region of space is covered by a right-angled coordinate system x, y, z. We then imagine that at every point P in the region, there exists a set of base vectors $\boldsymbol{i}_x$, $\boldsymbol{i}_y$ and $\boldsymbol{i}_z$ (Fig. A1.1) which are parallel respectively to

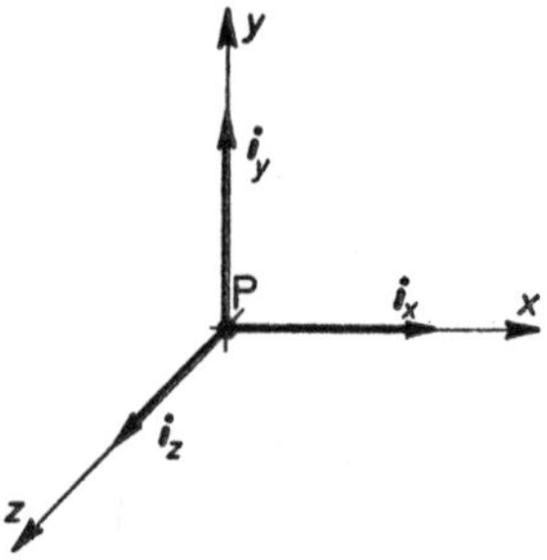

Fig. A1.1

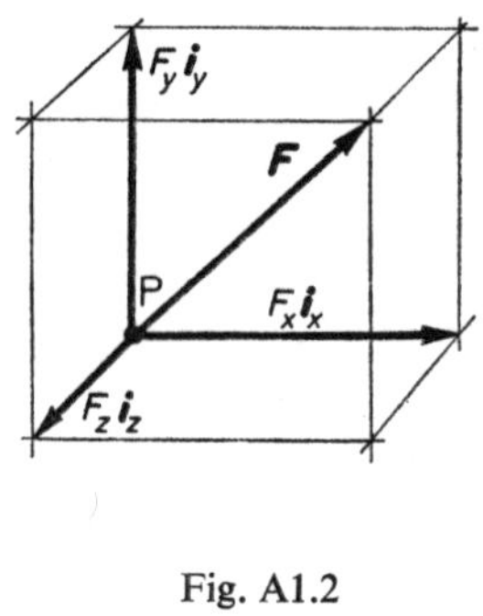

Fig. A1.2

the x, y and z coordinate lines and are of unit magnitude. Now by the law of vector addition, $\boldsymbol{F}$ is equal to the sum of three vectors which are parallel to $\boldsymbol{i}_x$, $\boldsymbol{i}_y$ and $\boldsymbol{i}_z$, these vectors being the sides of the parallelepiped whose diagonal is $\boldsymbol{F}$ as shown in Fig. A1.2. They are scalar multiples of $\boldsymbol{i}_x$, $\boldsymbol{i}_y$ and $\boldsymbol{i}_z$, where by 'scalar' we mean a quantity possessing magnitude but not direction. Let these scalar multiples be F_x, F_y and F_z. Then

$$\boldsymbol{F} = F_x\boldsymbol{i}_x + F_y\boldsymbol{i}_y + F_z\boldsymbol{i}_z \tag{A1.1}$$

F_x, F_y and F_z are then called the scalar components of $\boldsymbol{F}$ in the x, y, z coordinate system.

Occasionally for clarity in the text we shall use an additional coordinate system x', y', z' and its corresponding base vectors $\boldsymbol{i}_{x'}$, $\boldsymbol{i}_{y'}$, $\boldsymbol{i}_{z'}$. The components of $\boldsymbol{F}$ in this system are then denoted by $F_{x'}$, $F_{y'}$ and $F_{z'}$.

Example

A force vector F of 23.5 N acting at a point P in a region covered by an x, y, z coordinate system is directed at 45° to the positive direction of the x coordinate line, at 60° to the positive direction of the y coordinate line and at 60° to the negative direction of the z coordinate line. What are its components?

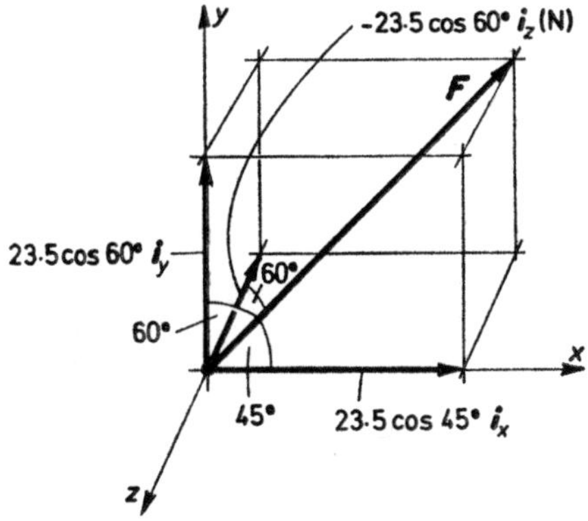

Fig. A1.3

The vector is shown in Fig. A1.3, from the geometry of the figure we have

$$F_x \boldsymbol{i}_x = 23.5 \cos 45^\circ \, \boldsymbol{i}_x,$$

$$F_y \boldsymbol{i}_y = 23.5 \cos 60^\circ \, \boldsymbol{i}_y,$$

$$F_z \boldsymbol{i}_z = -23.5 \cos 60^\circ \, \boldsymbol{i}_z.$$

Hence,

$$F_x = 16.63 \text{ N}, \quad F_y = 11.75 \text{ N}, \quad F_z = -11.75 \text{ N}.$$

APPENDIX 2

PARTIAL DIFFERENTIATION

A2.1 Definitions

Suppose f is a function of two variables x and y, for example

$$f = 2x^2y + y^3. \tag{A2.1}$$

If the variables are the coordinates of points in a plane, then f changes as we move from point to point according to (A2.1). Consider f at a particular point P. The *partial derivative of f with respect to x* at P is then defined as the rate of change of f with respect to x as we move along a line through P parallel to the x coordinate line (so that y remains constant). The partial derivative is denoted by the term $\dfrac{\partial f}{\partial x}$, which can be found merely by differentiating the expression (A2.1) with respect to x, treating y as a constant. Thus

$$\frac{\partial f}{\partial x} = 4xy. \tag{A2.2}$$

A similar definition for $\dfrac{\partial f}{\partial y}$ leads to

$$\frac{\partial f}{\partial y} = 2x^2 + 3y^2. \tag{A2.3}$$

It is apparent that the partial derivatives are themselves functions of x and y and can be differentiated again to produce second order partial derivatives. Thus

$$\frac{\partial^2 f}{\partial x^2} = \left(\frac{\partial\left(\dfrac{\partial f}{\partial x}\right)}{\partial x}\right) = 4y, \quad \frac{\partial^2 f}{\partial y^2} = 6y,$$

$$\frac{\partial^2 f}{\partial x \partial y} = \left(\frac{\partial\left(\dfrac{\partial f}{\partial y}\right)}{\partial x}\right) = 4x.$$

In the last case, it can be shown that the order of partial differentiation is always immaterial, so that $\dfrac{\partial^2 f}{\partial x \partial y} = \dfrac{\partial^2 f}{\partial x \partial y}$.

Higher order derivatives are produced by continued partial differentiation, for example

$$\frac{\partial^2 f}{\partial x^2 \partial y} = 4, \quad \frac{\partial^3 f}{\partial y^3} = 6, \text{ etc.}$$

A2.2 Values of *f* in the neighbourhood of P

Consider a point P* which lies on the x coordinate line through P and is a differential distance dx from P (Fig. A2.1 (b)). The value of the function at P*

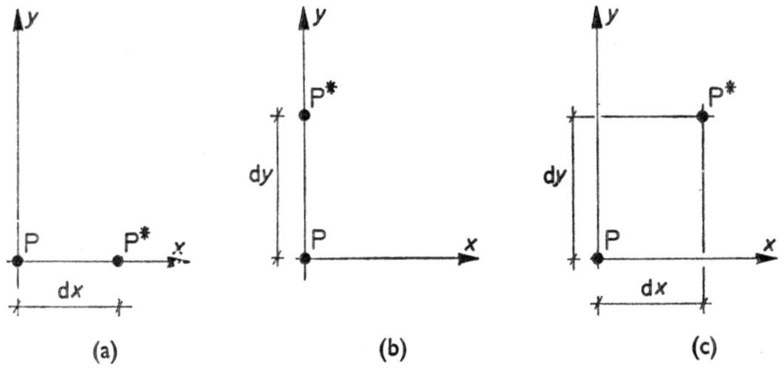

Fig. A2.1

generally differs from its value at P. Suppose we call the two values f* and f respectively, then by definition

$$\frac{f^* - f}{\mathrm{d}x} \equiv \frac{\partial f}{\partial x},$$

so that

$$f^* - f = \frac{\partial f}{\partial x} \,.\, \mathrm{d}x. \tag{A2.4}$$

Similarly if P* lies on the y coordinate line through P and is dy from P (Fig. A2.1 (a)), then

$$f^* - f = \frac{\partial f}{\partial y} \,.\, \mathrm{d}y. \tag{A2.5}$$

Combining these results, if P* is in an arbitrary position relative to P and is dx, dy from P (Fig. A2.1 (c))

$$f^* - f = \frac{\partial f}{\partial x} \mathrm{d}x + \frac{\partial f}{\partial y} \mathrm{d}y. \tag{A2.6}$$

INDEX

MICHAEL
MARSHALL

LA TRILOGIE DES HOMMES DE PAILLE

LES HOMMES DE PAILLE
LE SANG DES ANGES • LES MORTS SOLITAIRES

LES HOMMES DE PAILLE

Traduit de l'anglais
par Jean-Pascal Bernard

À Jane Johnson.

Nous venons trop tard pour les dieux
et trop tôt pour l'Être.
L'homme est un poème
que l'Être a commencé.

Martin HEIDEGGER
L'Expérience de la pensée

Palmerston, Pennsylvanie

Palmerston n'est pas une grande ville, ni de celles dont on puisse sérieusement vanter le dynamisme. Elle est simplement là, telle une trace sur le trottoir. Comme toutes les villes, elle a un passé, et elle avait jadis un avenir, même si cet avenir ne lui a réservé qu'un peu plus de poussière et de torpeur, achevant de la placer en marge de l'histoire : elle évoque un vieux robinet au bout d'un conduit rouillé qui, à force d'accumuler les fuites, finira par ne plus produire une seule goutte d'eau.

La ville borde la rivière Allegheny, à l'ombre de collines rebondies, et possède plus d'arbres qu'on n'en peut compter, à moins d'avoir beaucoup de temps à perdre et d'être gravement atteint. La voie ferrée passait tout près, de l'autre côté du fleuve, mais au milieu des années soixante-dix la gare fut condamnée et la plupart des rails démontés. Il n'en reste plus grand-chose, hormis des souvenirs et un musée fatigué que même les groupes scolaires ont déserté. Quelques touristes s'y arrêtent de loin en loin, contemplent avec une indifférence consommée les clichés grisâtres, puis regagnent leur voiture en estimant avoir assez traîné comme ça. Trente ans plus tard, les résidents de longue date (à Palmerston ils le sont tous, de longue date, et non sans une certaine fierté) souffrent encore de l'absence de la voie ferrée, comme d'un membre amputé qui les démangerait de temps en temps. Certains avaient lutté contre sa disparition à coups de pétitions et de meetings, d'autocollants à pare-chocs et de collectes ; pour d'autres, le changement s'était fait en douceur, d'un œil résigné, comme participant de quelque virage entropique de l'histoire. Dans une ville un peu plus grande et animée, cette ancienne voie envahie de mauvaises herbes serait l'endroit parfait pour acheter de la drogue ou se faire agresser. Mais à Palmerston, elle permet surtout aux parents de promener leurs enfants le week-end, de leur montrer les oiseaux et les arbres ; puis quelques années plus tard, à ces mêmes enfants de mettre une grossesse en route afin de goûter à un nouveau style d'aliénation.

La ville est bâtie autour d'un carrefour en T, motif brouillé par des rues adjacentes à l'utilité douteuse. Le tronc est jalonné de stations-service, d'un lavage de voitures, d'un vidéo-shop, de deux petits motels et d'une supérette qui brade tout un stock de CD des plus grands succès du Marshall Tucker Band. Dans l'angle se dresse une vieille église en bois à la peinture écaillée qui, sur un fond de ciel bleu, demeure pittoresque. La branche droite s'enfonce dans les collines vers l'État de New York et des Grands Lacs.

Si vous prenez à gauche, comme pour suivre la route 6 vers l'ouest jusqu'au réservoir de l'Allegheny – et c'est quasiment la seule raison qui puisse vous mener par là –, vous déboucherez sur Main Street et trouverez quelques banques et boutiques. Les vitres des premières sont réfléchissantes et anonymes ; celles des secondes, qui mériteraient un bon nettoyage, proposent des antiquités de valeur limitée, dont l'arrangement brouillon suggère qu'elles auront tout le temps de se bonifier sur place. Suivent deux petits cinémas, dont l'un s'essaya aux films d'art et d'essai voilà une dizaine d'années, et fut récompensé par une audience si faible qu'il décida, dans un accès de dépit, de mettre la clé sous la porte ; l'autre, resté fidèle aux canons de la culture populaire, fourgue des rêves inaccessibles aux accros du pop-corn. Du côté sud de la rue, au milieu d'un vaste terrain isolé, s'élève une splendide maison victorienne. Elle est vide depuis des années, et si la plupart des carreaux sont intacts, la peinture est encore plus écaillée que celle de l'église, et quelques planches commencent à se détacher.

Si vous avez faim, vous avez toutes les chances d'échouer au McDonald's, juste à côté du musée du Chemin de fer. C'est ce que font la plupart des gens. Palmerston n'est pas un vilain endroit. La vie y est paisible, les habitants aimables. C'est un coin du monde agréable, qui jouit d'une faible criminalité et de sa proximité avec la forêt d'État de Susquehannock. Vous pourriez y naître, y élever vos gosses et y mourir sans vous sentir particulièrement floué par le destin.

L'ennui, c'est qu'il n'y a pas grand-chose à y faire entre-temps.

Le mercredi 30 octobre 1991, à l'heure du déjeuner, le McDo était bondé. Les tables étaient presque toutes occupées, et quatre files d'attente s'étiraient depuis le comptoir. Deux fillettes de quatre et six ans, en sortie avec leur mère, réclamaient à cor et à cri des Chicken McNuggets. Les autres clients étudiaient les menus affichés avec tout le respect qui leur était dû.

Trois non-résidents étaient présents, une aubaine pour l'industrie touristique de Palmerston. Le premier était un homme en costume entre deux âges, attablé seul dans un coin. Il s'appelait Pete Harris et rentrait à Chicago au terme d'une tournée commerciale fort décevante. De son siège, il considérait, tout en mastiquant, la tour italienne de la maison victorienne, consterné que personne n'ait entrepris de la retaper.

Les deux autres étaient un couple de touristes anglais, que le hasard avait installés à la table voisine de Pete. Mark et Suzy Campbell avaient sauté le petit déjeuner pour parcourir trois cents kilomètres dans la matinée ; autrement dit, ils étaient très en appétit. Ils avaient ratissé la ville en quête de cuisine du terroir avant de se rabattre sur le fast-food. Entamant leurs burgers d'un air farouche, ils furent d'abord alarmés puis somme toute satisfaits de se découvrir assis à côté d'un autochtone sachant parler. Le dénommé Trent était grand, âgé d'une quarantaine d'années, et bien pourvu en cheveux roux. En apprenant que le couple entamait tout juste la traversée du pays, il acquiesça du bout des lèvres devant une pratique qu'il pouvait comprendre mais qui n'éveillait en lui aucun intérêt, comme de collectionner les boîtes d'allumettes, pratiquer l'escalade ou avoir un boulot. Il connaissait l'Angleterre de nom, lui savait une sacrée longue histoire et une grosse industrie rock, et dans les deux cas il était à fond pour.

La conversation finit par se tarir, s'abîmant dans les limbes de leurs expériences réciproques. Suzy fut un brin déçue : elle avait apprécié leur échange. Mark était préoccupé : il souhaitait faire des emplettes. La veille, le réceptionniste de leur hôtel avait passé un bon moment à sonder les ondes en quête d'un truc à diffuser à plein volume. Il était accidentellement tombé sur une fréquence de musique classique, et l'espace d'un instant bref et merveilleux, quelques mesures des *Variations Goldberg* avaient traversé le bar. Mark s'était représenté cette station de radio comme l'œuvre d'un individu perché sur sa montagne, la porte barricadée contre des hordes armées jusqu'aux dents de disques de Garth Brooks. Les notes de Bach avaient résonné dans le crâne de Mark tout au long des heures suivantes – baignées, elles, de ballades sirupeuses opposant la fragilité du mariage à la fidélité des chiens –, et il brûlait d'acheter le CD pour l'écouter dans la voiture. Palmerston ne possédait pas de magasin de disques.

Trent fut vite rejoint par une troupe d'adolescents dégingandés et disgracieux, et en tendant l'oreille Suzy comprit que le rouquin tâchait de convaincre ces jeunes de l'aider à déplacer un gros tas de terre qui gênait sa caravane, près de l'ancienne voie ferrée. En revanche, la raison pour laquelle il fallait s'en débarrasser séance tenante lui échappait. Les garçons se montrèrent naturellement sensibles à l'idée d'une rétribution, laquelle prit la forme d'une caisse de bière. Puisqu'il leur fallait patienter encore trois ans pour être en âge d'acheter de l'alcool, le marché fut vite conclu. En attendant que Trent ait terminé de barboter dans ses sandwiches, ils restèrent tapis comme une bande de mouettes véreuses, à échanger les insultes affectueuses et les injonctions ineptes qui fondent le discours des garçons. Il émanait de leurs propos qu'en dépit des marques de surf agrémentant leurs tee-shirts, aucun n'avait jamais surfé, peu avaient déjà quitté l'État, et un seul avait vu la mer.

En entendant cela, les Campbell mesurèrent ce qu'il y avait d'exceptionnel à explorer un pays différent du leur, à la faveur d'une lubie surgie par une nuit arrosée dans un pub à dix mille kilomètres de là. Soudain ivres de fierté et édifiés par la dimension de leur entreprise, ils sirotèrent leur café d'un air méditatif, s'accordant cinq ou six minutes de plus. Sans cette pause, ils seraient ressortis vers 12 h 50. Avec, ils auraient repris la route vers 12 h 56 au plus tard. Alors Suzy aurait eu envie d'une cigarette, ce que les affiches murales prohibaient au moyen de phrases courtes et faciles à lire, et d'une iconographie à portée universelle. Pete Harris, quant à lui, n'était pas vraiment pressé, et se serait trouvé là de toute façon : toujours absorbé par la maison isolée, à se demander vaguement combien elle pouvait coûter, mais conscient que, même s'il parvenait à rassembler la somme nécessaire, son épouse destinerait déjà ce pécule à d'autres projets.

À 12 h 53 une femme hurla au milieu du restaurant.

Ce fut une interjection brève et puissante, qui n'exprimait rien d'autre que l'urgence. Les gens s'écartèrent d'instinct, créant un vide dans l'allée centrale. Il apparut alors que deux hommes – l'un approchant de la vingtaine, l'autre en plein dedans, tous deux vêtus de longs manteaux – étaient la cause de cette détresse. Les cheveux du second étaient courts et blonds, ceux du premier moins clairs et plus longs. Bien vite il apparut aussi qu'ils portaient des fusils semi-automatiques.

L'éclairage de la pièce sembla soudain très vif, et les sons anormalement clairs et secs, comme si l'on venait d'évacuer un gaz cotonneux. Quand vous êtes assis chez McDonald's à midi en semaine, devant un café qui atteint juste la bonne température, et que vous voyez d'un coup le ciel vous tomber sur la tête, le temps bascule dans un lent moment de lucidité. Telle la longue seconde précédant l'impact entre deux voitures, ce sursis ne vous sera d'aucun secours. Ce n'est pas une planche de salut, ni la main tendue de Dieu, et tout effort sera vain sinon celui d'essayer d'accueillir la mort et de se demander ce qui l'a retenue si longtemps.

L'une des larves de la bande à Trent eut à peine le temps de lancer : « Billy ? » d'un ton idiot, et les deux hommes ouvrirent le feu.

Plantés dans l'allée centrale, ils tiraient avec calme et rapidité, la crosse de leur arme fermement calée contre l'épaule. Alors que la première victime tombait à la renverse, le visage empreint d'une stupeur muette, les tueurs poursuivaient leur besogne : ils semblaient soucieux de montrer à une autorité supérieure qu'ils étaient dignes de cette mission et y mettaient tout leur savoir-faire.

Après environ une seconde, et deux morts supplémentaires, la foule quitta son état d'hébétude. Le temps repartit tambour battant, et les cris jaillirent. Les gens tentèrent de fuir, de se cacher ou de s'abriter les uns derrière les autres. Certains s'élancèrent vers les portes, mais les fusils virevoltèrent à l'unisson et raflèrent les déserteurs. La ligne de tir balaya les étrangers de passage, et Mark Campbell reçut un coup direct dans la nuque à l'instant où le visage de sa femme se répandait sur la vitre blindée qui arrêta les deux balles. Trent mourut furieusement peu après, à moitié debout, comme il s'élançait de son siège dans le fol espoir de se jeter sur l'ennemi. Peu furent suffisamment maîtres d'eux-mêmes pour envisager une telle action, et ceux qui le firent périrent sur-le-champ. Les deux canons pivotaient en symbiose, mus par un même fil, et nos héros de comprendre que si le tabagisme passif était mauvais pour la santé, l'absorption passive de balles vous emportait bien plus vite.

La majorité chercha simplement à courir. À s'enfuir. Le vice-président de Bedloe Insurance s'y essaya, tout comme sa bonne à rien de secrétaire. Douze écoliers s'y essayèrent. Tous à la fois, de sorte qu'ils se gênèrent mutuellement. Beaucoup se prirent les pieds dans les corps des blessés et moururent sans grâce, se déboîtant le genou ou la hanche dans leur chute. Ceux qui avaient la voie libre furent happés dans leur course, s'écrasant contre les tables, les murs ou le comptoir sous lequel la dernière employée en vie était recroquevillée, consciente de croupir dans une flaque de sa propre urine. De son poste elle apercevait les pieds convulsés de Duane Hillman, le jeune homme qu'elle avait suivi dernièrement le long de la voie ferrée. Il s'était montré tendre, et avait proposé d'utiliser un préservatif. Sachant qu'il n'avait pas seulement reçu une balle, mais s'était effondré alors qu'il portait un bac d'huile bouillante, elle était peu encline à poser les yeux sur lui. Elle espérait plutôt qu'en ne regardant nulle part, et en se faisant toute petite, elle serait épargnée. Un peu plus tard une balle perdue traversait l'épaisseur du comptoir pour se loger dans sa colonne vertébrale.

Il y a ceux qui ne tentèrent même pas de s'échapper, mais se figèrent, les yeux grands ouverts, quittés par leur âme avant même que les projectiles ne perforent leurs poumons, bassin ou ventre. Au moins l'une des victimes, chez qui on venait de diagnostiquer le cancer qui avait tué son père à petit feu, ne vit pas la tournure des événements sous un jour totalement négatif. C'était idiot, car le jeune médecin de l'hôpital, dont elle se méfiait en grande partie parce qu'il ressemblait vaguement au méchant dans son feuilleton préféré, aurait su la sauver si elle avait vécu et suivi ses conseils.

Rien ne motivait chez les autres un tel détachement. Ils étaient juste incapables de bouger, du moins tant que ce choix leur appartenait encore.

Dans une salle remplie de cadavres, les meurtriers ressemblent à des dieux. Les hommes continuaient de tirer, leurs fusils girouettant de concert pour mitrailler un coin inattendu de la salle. Ils rechargèrent à plusieurs reprises, mais jamais au même moment. Ils étaient très efficaces. Aucun n'ouvrit la bouche pendant toute la durée de l'opération.

Sur les cinquante-neuf personnes présentes au McDonald's ce midi-là, seules trente et une entendirent le bruit étouffé de la dernière détonation. Douze de celles-là moururent avant la nuit, portant le bilan à quarante morts. Parmi les survivants se trouvait la fille du comptoir, qui ne recouvrerait jamais l'usage de ses jambes et deviendrait alcoolique avant de trouver la foi puis de la perdre. L'une des fillettes eut elle aussi la vie sauve. Elle fut confiée à une tante dans l'Iowa, et mena une existence relativement paisible. L'un des copains de Trent s'en sortit, et devint quatre ans plus tard garde-côte à Laguna Beach.

Pete Harris survécut lui aussi. En toute justice il aurait dû mourir très tôt, dès la première salve frappant le côté gauche du restaurant, mais le corps de Suzy Campbell s'était écrasé sur lui alors même qu'il tentait de se nicher sous la table. Le poids de la femme l'avait fait riper de son siège, pour le plaquer la tête la première sur le sol. Ils furent rejoints quelques instants plus tard par le mari de Suzy, qui était déjà mort. Aucun des deux Campbell n'eût été identifiable d'après les photos de leurs passeports (précautionneusement rangés dans leurs poches respectives, au cas où l'on eût forcé la voiture pendant qu'ils déjeunaient). En revanche, les vêtements que portait le couple – certains provenant d'Angleterre, d'autres d'un déstockage Gap dans le vieux Back Bay de Boston – étaient quasiment immaculés. Un simple époussetage, et ils auraient pu franchir la porte, remonter dans leur voiture de location et reprendre la route. Peut-être, dans un monde meilleur, cette issue-là eût-elle été permise, et Mark aurait trouvé par un coup de chance les *Variations Goldberg* lors de la prochaine halte, et ils auraient suivi toute la journée une longue voie rectiligne plantée d'arbres dont les feuilles semblaient éclairées de l'intérieur, avalant les crêtes et les creux de l'autoroute qui les entraînait dans l'après-midi puis dans la soirée, sans jamais remarquer qu'ils roulaient tout seuls.

Dans ce monde-ci ils s'en tinrent à sauver une vie, celle de Pete Harris qui gisait sous eux, sonné par la rencontre de sa tête avec le sol carrelé. Tout autour de lui s'étalaient des membres, et ce qu'il voyait n'était que mort et chaos ; tout ce qu'il sentait n'était que l'élancement de ses entrailles et un mal de crâne glacial qui évoluerait en commotion si aiguë qu'il aurait parfois l'impression qu'elle ne cesserait jamais. Une jeune infirmière, qui semblait lui vouer une superstitieuse vénération car il était l'un des rares survivants, passa la nuit à le maintenir éveillé à l'hôpital de Pipersville, quand il aurait nettement préféré dormir.

Mais cela viendrait un peu plus tard, comme la crise cardiaque qui accomplirait ce que les balles avaient manqué. Il ne chercha jamais à savoir si la maison victorienne était à vendre. Il se contenta de travailler jusqu'à épuisement.

Derrière les *pan !* rythmés de la pétarade et les râles et cris des mourants, s'éleva au loin le son caractéristique de sirènes à l'approche. Les tueurs prolongèrent le feu une vingtaine de secondes, liquidant une petite poche près du comptoir où la maman et ses filles avaient provisoirement trouvé refuge. Puis ils s'arrêtèrent.

Ils balayèrent la salle du regard, leurs visages ne trahissant aucune réaction devant le résultat obtenu. Le plus jeune des deux – le dénommé Billy – recula d'un pas et ferma les yeux. Le deuxième homme lui tira dans le visage à bout portant. Pendant que le corps étendu de Billy rendait ses derniers spasmes alanguis, l'homme s'accroupit et trempa ses mains dans le sang. Il se releva pour écrire quelque chose sur la porte vitrée, avec des gestes posés, en grandes lettres dégoulinantes, puis scruta la pièce à nouveau, calme et décontracté. Il n'eut pas un regard pour les voitures de flics qui déboulaient sur Main Street, bien trop tard pour empêcher un fait divers qui propulserait enfin Palmerston sur le devant de la scène.

Puis, lorsqu'il fut prêt, l'homme plongea à travers la vitre éclatée derrière les corps des Campbell et disparut dans la nature – le long de l'ancienne voie ferrée, supposa-t-on. Il ne fut jamais appréhendé. Personne n'en fournit une description précise, et avec le temps ce fut comme s'il s'était dégagé de l'événement. Au final on mit tout sur le dos de Billy, ce jeune garçon qui n'avait fait qu'obéir aux ordres d'un homme qu'il prenait pour un nouvel ami.

En entendant les véhicules de police s'arrêter sur le parking, Pete Harris tenta de se redresser, de déployer assez de force pour repousser les corps des Campbell. Il échoua, mais parvint à relever la tête, juste assez pour déchiffrer ce qu'on avait inscrit en lettres de sang sur la porte. L'écriture avait coulé, et sa vision était embrumée par une lueur blanche dans son crâne, mais les mots étaient tout à fait lisibles. « Les Hommes de Paille ».

Onze années passèrent.

PREMIÈRE PARTIE

De la colline, et non sur la colline...

Frank Lloyd Wright,
à propos de l'architecture
de Taliesin

CHAPITRE 1

L'enterrement fut des plus satisfaisants, dans la mesure où l'assistance était nombreuse et vêtue de manière appropriée, et où personne ne se leva pour lancer : « Vous comprenez que ça signifie qu'ils sont morts ? » La cérémonie eut lieu dans une église située au bout de la ville. J'ignorais de quelle obédience elle relevait, et plus encore pourquoi cette précision eût figuré dans les instructions laissées à Harold Davids. À ma connaissance, mes parents n'avaient aucune foi religieuse, hormis quelque athéisme de bon aloi et la conviction tacite que si Dieu existait il devait conduire une belle voiture, très certainement de fabrication américaine.

Le cabinet de Davids avait organisé les funérailles avec maestria, de sorte que je n'avais pas grand-chose à faire, à part attendre. Je passai l'essentiel de ces deux jours au Best Western. J'aurais dû me rendre à la maison, je le savais bien, mais c'était au-dessus de mes forces. Je lus la majeure partie d'un mauvais roman et feuilletai une flopée de magazines de style hôtelier, sans rien apprendre, sinon que l'on peut dépenser une somme astronomique pour une montre. Chaque matin je quittais l'hôtel de bonne heure, décidé à remonter la grand-rue, mais je ne dépassais jamais le parking. Je savais ce que recelaient les vitrines de Dyersburg, Montana, or je ne recherchais ni matériel de ski, ni objets d'« art ». Je prenais mes dîners au restaurant de l'hôtel et déjeunais d'un sandwich au comptoir. Chaque repas était assorti de frites, dont la texture suggérait que de nombreux processus industriels étaient intervenus entre la terre et mon assiette. Il était impossible de ne pas avoir de frites. Je soumis la question par deux fois aux serveuses, mais capitulai devant leurs regards paniqués.

Après que le pasteur nous eut expliqué en quoi la mort n'était pas le désastre complet qu'elle paraissait de prime abord, nous sortîmes de l'église en file indienne. J'étais navré de partir.

Je m'étais senti en lieu sûr là-dedans. Dehors il faisait un froid de canard, dans un air piquant et sourd. Derrière le cimetière se dressaient les contreforts de la chaîne de Gallatin, aux lointains sommets estompés comme s'ils étaient peints sur du verre. On avait creusé deux caveaux voisins. Une quinzaine de personnes allaient assister à l'inhumation. Davids était là, flanqué d'une femme qui devait être sa secrétaire. Mary se tenait à mes côtés, sa chevelure blanche ramenée en un chignon austère, les rides de son visage lissées par le froid. Je crus reconnaître vaguement deux autres individus.

Le pasteur proféra encore quelques mensonges de réconfort permettant d'emballer le tout. Possible que certains y eussent été sensibles. Pour ma part je les entendais à peine, trop occupé à empêcher mon crâne d'exploser. Puis deux hommes – dont c'était le métier, qui répétaient ce genre de choses toutes les semaines – descendirent avec des gestes experts les deux cercueils sous terre. Les cordes coulèrent entre leurs mains, et les cercueils s'arrêtèrent exactement six pieds sous la plaine où se tenaient encore les vivants. Suivirent quelques dernières phrases de baume, débitées à la va-vite – comme si l'Église admettait que son temps de parole était révolu. On ne peut enterrer des gens dans de grandes boîtes sans que le public trouve ça suspect.

Une ultime déclaration à mi-voix, et voilà. C'était fini. Rien n'arriverait jamais plus à Donald et Beth Hopkins. Rien qu'on pût décemment envisager, en tout cas.

Certains participants s'attardèrent un moment, désormais sans but. Puis je me retrouvai seul. Deux êtres en un. Le premier à la gorge scellée comme de la pierre, qui se voyait déjà paralysé à vie ; le second ayant conscience de sa posture très cliché face à ces deux tombes, comme du fait que, un peu plus loin, les gens passaient en voiture, écoutaient les Dixie Chicks et avaient des soucis d'argent. Chacune de mes deux moitiés trouvait l'autre ridicule.

Je savais que je ne pouvais demeurer planté là éternellement. Ils ne l'auraient pas souhaité. Cela n'aurait aucun sens, ne changerait rien, et il faisait vraiment très froid. Relevant enfin les yeux, je vis que Mary était également restée, postée un ou deux mètres derrière moi. Ses yeux étaient secs, endurcis par la certitude de connaître le même sort d'ici peu, et qu'il n'y avait pas lieu d'en rire ni d'en pleurer. Je pinçai mes lèvres, elle posa la main sur mon avant-bras. Nous restâmes silencieux un certain temps.

Lorsqu'elle m'avait appelée, trois jours plus tôt, je me trouvais sur la terrasse d'un joli petit hôtel sur De la Vina Street à Santa Barbara. Temporairement sans emploi, ou *de nouveau* sans emploi, je consacrais mes maigres économies à des vacances non méritées. J'étais assis face à un bon merlot du cru, que je m'employais à assécher. Ce n'était pas la première bouteille de la soirée, aussi quand mon portable sonna je fus tenté de laisser tourner la messagerie. Mais, en jetant un coup d'œil sur l'appareil, je vis qui m'appelait. J'enfonçai la touche verte.

— Salut, dis-je.

— Ward, répondit-elle.

Puis rien.

Je perçus un faible bruit au bout de la ligne. Un son doux et visqueux.

— Mary ? Tu vas bien ?

— Oh, Ward... dit-elle d'une voix brisée, marquée par l'âge.

Je me redressai sur mon siège, dans le vain espoir qu'une pseudo-préparation, qu'un ressaisissement de dernière minute puissent d'une façon ou d'une autre limiter la force avec laquelle allait s'abattre la massue.

— Qu'y a-t-il ?

— Il vaudrait mieux que tu viennes, Ward.

Au bout du compte je parvins à la faire parler. Un accident de voiture en plein centre de Dyersburg. Tués l'un et l'autre.

Je crois que je m'attendais à quelque chose de cet ordre. Il fallait que mes parents soient tous deux concernés pour que Mary m'appelle. Mais même là, alors que nous nous tenions ensemble dans ce cimetière, les yeux baissés sur leurs cercueils, j'étais incapable d'envisager la réalité de leur disparition. En outre, il m'était désormais impossible de répondre au message laissé par ma mère une semaine plus tôt. Je n'avais pas eu le temps. Je n'avais pas prévu qu'ils seraient subitement rayés de la surface de la terre et placés sous celle-ci, là où ils ne pouvaient m'entendre.

Soudain je n'avais plus envie de rester près de leurs dépouilles. Je reculai d'un pas. Mary plongea la main dans sa poche et produisit une chose attachée à une languette en carton. Un jeu de clés.

— J'ai sorti la poubelle ce matin, dit-elle, et débarrassé quelques produits du réfrigérateur. Le lait, par exemple. Pour qu'il ne répande pas une sale odeur. J'ai laissé le reste tel quel.

J'acquiesçai, les yeux rivés sur les clés. Je n'en possédais aucun double. Pas besoin. Ils avaient toujours été là lors de mes rares visites. C'était la première fois, pensai-je, que je croisais Mary ailleurs que dans la cuisine ou le séjour de mes parents. Ça se passait comme ça, avec mes vieux : c'est toujours eux qui recevaient. Ils se plaçaient toujours au centre des choses.

— Ils parlaient de toi, tu sais. Souvent.

J'opinai de nouveau, sans trop savoir si je la croyais. Le plus clair de ces dix dernières années, ils n'avaient même pas su où je me trouvais, et tout ce qu'ils avaient à raconter concernait un homme plus jeune, un enfant unique qui avait grandi et vécu chez eux dans un autre État. Ce n'était pas une question de manque d'amour. On s'était aimés, à notre façon. Disons que je ne leur avais pas souvent donné l'occasion de parler de moi, les privant de ces joies qui permettent aux parents de fanfaronner devant amis et voisins. Pas de femme, pas d'enfants, pas de job sérieux. Je vis que Mary avait toujours la main en l'air. J'empochai les clés.

— Tu restes combien de temps ? demanda-t-elle.

— Le temps que ça prendra. Peut-être une semaine. Peut-être moins.

— Tu sais où me trouver. Ne te sens pas obligé de jouer les étrangers, d'accord ?

— Promis, soufflai-je dans un sourire gêné.

Je regrettais de ne pas avoir un frère ou une sœur qui eût assuré cette conversation à ma place. Un être responsable et doué pour les civilités.

Elle me rendit mon sourire, mais avec réserve, comme si elle savait que le choses ne sont jamais aussi simples.

— Au revoir, Ward, dit-elle avant de s'éloigner vers la pente.

À soixante-dix ans elle était un peu plus âgée que mes parents, et sa démarche s'en ressentait. Elle avait toujours vécu à Dyersburg, mené une carrière d'infirmière, et c'est tout ce que je savais d'elle.

J'aperçus Davids posté de l'autre côté du cimetière, tuant le temps avec sa secrétaire, sûrement en train de m'attendre. Il avait l'air d'un type décidé à agir vite et bien, à régler tous les détails.

Un dernier coup d'œil sur les tombes, et je descendis le chemin d'un pas lourd, pour affronter les formalités administratives occasionnées par la disparition de ma famille tout entière.

Davids avait chargé l'essentiel de la paperasse dans sa voiture, et il proposa de tout m'expliquer devant un déjeuner. J'ignore si le résultat s'avéra moins déplaisant que dans le cadre d'un bureau, mais je fus sensible à de tels égards venant d'un homme qui me connaissait à peine. Nous nous rendîmes dans le centre historique de Dyersburg, dans un établissement baptisé Auntie's Pantry. On s'était donné du mal pour décorer ce « garde-manger de Tantine » façon cabane de bois, avec un mobilier taillé par des lutins. Le menu offrait un choix terrifiant de soupes bio et de pains maison, accompagnés de salades à base de germes de soja. Au risque de paraître vieux jeu, je ne considère pas les germes de soja comme de la nourriture. Ils n'ont même pas l'air comestible. On dirait des larves mutantes. Il n'existe rien de pire, hormis peut-être la semoule de couscous, elle aussi présente en force sur la carte. Je ne connais pas de tante sur cette planète qui mange ce genre de merde, mais le personnel comme les clients semblaient ravis. Limite illuminés.

Après une brève attente un peu embarrassée, on nous attribua une table près de la devanture. Cela contraria la jeune famille proprette postée derrière nous, qui lorgnait la table depuis un moment et ne comprenait pas qu'être premier dans une file vous confère certains avantages. La femme fit part de son mécontentement à la serveuse, en remarquant à voix haute que la table pouvait accueillir quatre personnes quand nous n'étions que deux. En temps normal ce genre de situation éveille ce qu'il y a de meilleur en moi, en particulier si mes adversaires portent tous le même caban bleu marine, mais là mes ressources étaient à sec. Si le mari était hors compétition, les deux enfants étaient blonds et sérieux, comme une paire de juges séraphiques. Je ne voulais pas me froisser avec le ciel. La serveuse, l'une de ces femmes bronzées et plutôt jolies quoiqu'un peu dodues qui se massent dans les endroits comme Dyersburg en période de sports d'hiver, choisit de rester neutre, préférant fixer intensément un point du sol à peu près équidistant des deux groupes de belligérants.

Davids dévisagea brièvement la matriarche. Du même âge que mes parents, grand, et affublé d'un pif conséquent, il avait tout l'air du type que Dieu dépêche pour nous faire vivre l'enfer. Il ouvrit sa sacoche et tira de nombreux documents, sans chercher à dissimuler le type d'événement auquel ils renvoyaient. Il les disposa devant lui comme lors d'un rendez-vous d'affaires, puis saisit la carte et l'étudia. Lorsque j'eus terminé d'observer ses gestes, je m'aperçus que la petite famille détournait scrupuleusement les yeux. J'attrapai mon propre menu, et tentai de me figurer en quoi son contenu pouvait m'intéresser.

Davids était l'avocat de mes parents, et ce depuis qu'ils avaient quitté la Californie. Nous avions échangé quelques mots une ou deux fois par le passé, lors d'un pot de Noël ou de Thanksgiving à la maison, mais dans mon esprit il ne représentait qu'une de ces multiples relations qui allaient connaître un point final. D'où, en mon for intérieur, un curieux mélange de distance et de désir de prolonger le contact, que je peinais toutefois à concrétiser sur le plan conversationnel.

Par chance, Davids prit les devants dès qu'arrivèrent nos bols de soupe à la courge et aux lichens. Il récapitula les circonstances du décès de mes parents, qui en l'absence de témoins se réduisait à un fait unique : vers 23 h 05 le vendredi précédent, au retour d'une partie de bridge chez des amis, leur voiture avait embouti de front un véhicule stationné le long du trottoir, au croisement des rues Benton et Ryle. L'autopsie avait révélé un taux d'alcoolémie équivalent à environ une demi-bouteille de vin chez mon père, qui était le passager, et à beaucoup de jus de myrtille chez ma mère. La chaussée était glissante, le carrefour n'était pas très bien éclairé, et un accident s'était produit au même endroit pas plus tard que l'an dernier. Voilà tout. Une de ces choses qui arrivent et contre lesquelles on ne peut rien, à moins de se lancer dans une procédure civile perdue d'avance, ce qui n'était pas mon intention. Il n'y avait rien à ajouter.

Puis Davids passa aux choses sérieuses, qui consistaient à me faire signer toute une série de documents, par lesquels j'acceptais d'acquérir la maison et son contenu, ainsi que quelques terrains en friche et le portefeuille d'actions de mon père. La transmission soulevait une légion de questions fiscales qui furent dûment expliquées puis expédiées au moyen de quelques paraphes supplémentaires. Ce petit cours de finances m'entra par une oreille pour ressortir par l'autre, et je n'accordai à ces papiers qu'un bref coup d'œil. Mon père avait tenu Davids en haute estime, or Hopkins senior n'était pas homme à faire confiance au tout-venant. Ce qui convenait à papa me convenait.

Sur la fin j'écoutais à peine un mot sur deux, car à vrai dire j'appréciais ma soupe – maintenant que j'avais amélioré la recette en ajoutant une bonne dose de sel et de poivre. Je regardais les cuillérées monter à ma bouche, savourant leur goût d'une manière studieuse, réfléchie, et les encourageant à occuper le maximum de mon esprit. Je ne refis surface que lorsque Davids mentionna UnRealty.

Il expliqua que l'agence de mon père, spécialisée dans l'immobilier haut de gamme, était liquidée. La valeur des derniers actifs serait transférée sur le compte de mon choix, sitôt la procédure achevée.

— Il a fermé UnRealty ? demandai-je en relevant les yeux. Quand ça ?

— Non, répondit Davids en secouant la tête, tout en essuyant son bol avec un morceau de pain. Il a souhaité qu'il en soit ainsi à sa mort.

— Sans que j'aie mon mot à dire ?

Il regarda par la fenêtre, et se frotta les mains dans un mouvement économique qui délogea quelques miettes de ses phalanges.

— Il s'est montré très clair à ce sujet.

Ma soupe s'était subitement refroidie, et avait un goût d'épis d'eau bouillis. Je repoussai le bol. Je comprenais à présent pourquoi Davids avait tenu à traiter la paperasse après les obsèques, et non avant. Je ramassai les exemplaires qui me revenaient et les rangeai dans l'enveloppe fournie par ses soins.

— C'est tout ? fis-je d'une petite voix sèche.

— Je crois bien. Je suis navré de vous avoir infligé tout ceci, Ward, mais il fallait en finir une fois pour toutes.

Il sortit un portefeuille de sa veste et considéra l'addition comme si, en plus de douter du total, il prenait en grippe l'écriture de la serveuse. Son pouce s'arrêta sur une carte de paiement, avant d'opter pour du liquide. J'y vis la décision de ne pas faire passer ce repas en note de frais.

— Très aimable de votre part, dis-je.

Davids balaya le compliment d'un revers de main, avant de laisser un pourboire de dix pour cent tout ronds.

Nous nous levâmes et zigzaguâmes entre les tables de touristes diserts pour quitter le restaurant. Je voulus détourner le regard au moment de croiser la table du bataillon atomique en cabans bleus, mais trop tard : je l'avais déjà sous les yeux. Père et mère discutaillaient pour savoir où dormir à Yellowstone, tandis que le petit garçon tâchait de reproduire au moyen de sa cuiller et de sa soupe l'effet d'un astéroïde s'écrasant dans le Pacifique. Les deux mains sur une timbale en plastique, sa sœur rêvait béatement, le regard vague. Puis elle sourit à mon approche, comme devant un gros chien. C'était sûrement un sourire charmant, mais sur le moment j'eus envie de l'effacer.

Dehors nous restâmes plantés côte à côte quelques instants, à regarder des bandes affamées de bourgeoises nanties errer dans College Street, la carte de crédit frémissante.

Finalement Davids rangea ses mains dans son manteau.

— Vous repartez bientôt, je présume. Si je puis faire quoi que ce soit d'ici là, n'hésitez pas à me joindre. Je n'ai pas le pouvoir de ressusciter les morts, bien sûr, mais dans d'autres domaines je pourrai peut-être me rendre utile.

Nous échangeâmes une poignée de main, et il s'éloigna dans la rue d'un pas plutôt leste, le visage délibérément fermé. Et là je compris, avec un retard impardonnable, que Davids n'avait pas seulement été l'avocat de mon père, mais aussi son ami, et que je n'étais peut-être pas le seul à avoir connu une matinée éprouvante.

Je serrai les poings jusqu'à mon hôtel, et vers 21 heures j'étais complètement soûl. Les portes de l'établissement ne s'étaient pas refermées derrière moi que j'avais les deux mains sur mon premier *boilermaker*. Je sus dès la première gorgée que je commettais une erreur. Je l'avais su durant tout le trajet du retour, l'avais su au cimetière et depuis l'instant où j'avais ouvert les yeux ce matin-là. Ce n'était pas comme si je réduisais à néant des années de sevrage, tournais le dos à mon libre arbitre et faisais tout pour me réveiller entre deux ogresses dans un pays inconnu, certes. Mais se cuiter, c'est un peu comme de passer la nuit dans un autre lit pour se venger d'une infidélité : un geste qui n'apporte rien d'autre que des souffrances, en même temps qu'il tronque le piédestal moral auquel, pour une fois dans votre vie, vous auriez pu prétendre. Le problème, c'est que je ne voyais aucune autre réponse sensée à la situation.

Je commençai perché au bar, puis au bout d'un moment je m'exilai vers l'une des tables longeant la baie vitrée. Un gros pourboire préventif me garantissait de ne pas devoir attendre, ni même bouger, pour refaire les niveaux. Une bière, puis un scotch. Une bière, puis un scotch. Un moyen rapide et efficace de se torcher, et le dévoué barman maintenait la cadence comme demandé.

Je sortis les documents de l'enveloppe kraft de Davids et les étalai devant moi, obnubilé par un point précis.

Durant toute mon enfance, j'avais compris une chose au sujet de mon père : c'était un homme d'affaires. Cela résumait sa fonction et son identité. *Homo sapiens affairis*. Le matin il partait faire des affaires, et rentrait le soir après en avoir conclu de fameuses. Mes parents ne parlaient jamais de leurs premières années de mariage, et rarement de choses importantes, mais j'en connaissais un rayon sur UnRealty. Mon père avait travaillé de nombreuses années dans une petite agence, avant d'inviter ma mère dans un restaurant gastronomique pour lui annoncer qu'il se lançait en solo. C'est en ces termes exacts qu'il aurait annoncé la nouvelle, comme s'il tournait une pub pour des prêts bancaires. Il avait rencontré des gens et noué des contacts, imbibé de la mythologie entreprenariale qui reconnaît à chacun le droit de prendre un jour la parole au country club pour dire : « Je me suis fait tout seul. » Il en avait sûrement bavé, mais mon père avait une certaine force de caractère. Garagistes et plombiers, contractuelles et guichetières, tous savaient au premier regard qu'il ne fallait pas le chercher. Quand il poussait la porte d'un restaurant, le personnel recevait pour consigne de se tenir droit et de cesser de cracher dans la marmite. Son entreprise, et l'histoire de celle-ci, étaient l'aspect du personnage que je connaissais le mieux.

Et pourtant, en rédigeant son testament, il avait décidé qu'Un Realty s'éteindrait avec lui. Au lieu de confier cette décision à son fils, il avait tranquillement tiré un trait sur vingt années de labeur.

Quand Davids m'avait appris cela, j'en avais tout de suite compris la raison : mes parents n'avaient pas voulu que je reprenne l'agence. Ce qui pouvait s'expliquer. J'avais vendu beaucoup, beaucoup de choses, et brassé toutes sortes de biens, mais jamais de maisons de luxe, c'est vrai. Je m'y connaissais, pourtant. Et comment. Je connaissais le magazine *Unique Homes*, le Registre DuPont et *Christie's Great Estates*. Je pouvais disserter sur les sites protégés et les ranches, savais apprécier l'artisanat d'antan, une vue plongeante sur le fairway du trou numéro 15, un havre d'intimité où la sérénité abonde. Comment y aurais-je échappé ? Mon sang en était imprégné. J'avais même fait deux années de fac d'archi, avant de changer de voie suite à un fâcheux incident. Et malgré tout, mon père n'avait pas voulu, ou jugé déraisonnable, que je reprenne le flambeau. Plus j'y pensais, et plus j'avais mal.

Je continuai de boire, pour voir si cela arrangeait quoi que ce soit. Cela n'arrangeait rien. Je continuai quand même. Le bar resta calme jusqu'en milieu de soirée. Puis, sur le coup de 22 heures, il y eut un soudain afflux d'hommes et de femmes en costumes et tailleurs, réchappés de quelque séminaire casse-couilles avec polycopiés, transparents et micros-cravates. Ils se répandirent au milieu de la salle, fourmillant frénétiquement, excités comme des puces à l'idée de se lâcher un bon coup et de s'envoyer une ou deux bières allégées. À ce stade ma cervelle était lourde et froide. Le vacarme ne fit qu'empirer, comme si j'étais cerné de forçats charriant des galets.

Je rongeai mon frein dans mon box, foudroyant les envahisseurs du regard. Deux types tombèrent la veste. Un larron desserra même sa cravate. Les sous-fifres se rapprochaient de leurs chefs, les collaient tels des bécasseaux, comme s'ils pensaient ainsi s'infiltrer dans leurs petits papiers. Je tiendrais bon. Je laisserais passer l'orage. Ces gens-là savaient peut-être se servir d'un tableur et redresser des entreprises, mais soumis à un test d'endurance au bar, ils avaient encore besoin de flotteurs. J'étais confiant. J'étais dans mon élément. J'étais aussi, rétrospectivement, encore plus rond que je ne pensais.

Trois types apparurent dans l'entrée. Ils s'immobilisèrent, regardèrent autour d'eux.

L'instant d'après les gens criaient, et les endimanchés plongeaient pour s'abriter. Tout d'abord je pris peur, avant de comprendre que c'était moi qu'ils fuyaient.

Je titubais au milieu de la salle, les vêtements imprégnés d'éclaboussures de bière. J'avais un flingue dans la main, braqué sur les visiteurs de l'entrée, et leur aboyais une longue suite d'ordres contradictoires. Ils avaient l'air mort de trouille. Sûrement parce que face à un type qui pointe un flingue sur vous, vous avez envie d'obéir. Mais ce n'est pas facile quand lui-même ne sait pas ce qu'il veut.

Je finis par me taire. Je comptai soudain six hommes à la porte, puis de nouveau trois. La salle était redevenue silencieuse, mais j'avais l'impression que mon cœur allait couler une bielle. Tout le monde attendait que la situation s'apaise ou s'aggrave.

— Désolé, murmurai-je. Simple malentendu.

Je rangeai le flingue dans ma veste, récupérai mes papelards et filai vers la sortie. J'atteignis le milieu du vestibule avant de trébucher, entraînant dans ma chute une table, un grand vase et une centaine de dollars de fleurs.

À 3 heures du matin, refroidi à coups d'eau glacée, j'étais allongé, de dos, sur le lit de ma chambre.

J'avais rencontré la direction de l'hôtel et les autorités locales, qui s'étaient montrées compréhensives, malgré leur désir de confisquer mon arme jusqu'à la fin de mon séjour. J'invoquai les funérailles. Je possède un permis de port d'arme, ce qui parut les surprendre. Mais ils firent remarquer, avec un certain bon sens, que ce document ne m'autorisait pas à l'agiter dans les bars.

Les papiers du cabinet de Davids, ceux qui me déclaraient détenteur de 1,8 million de dollars de liquidités, étaient soigneusement étendus sur le radiateur. J'avais ravalé ma rancœur. Le fait que les dernières volontés de mon père sentent désormais la bière semblait lui donner raison.

Au bout d'un moment je me retournai, décrochai le téléphone et composai un numéro. Six sonneries, puis le déclenchement d'un répondeur. Une voix que je connaissais mieux que la mienne indiquait que M. et Mme Hopkins étaient navrés de ne pouvoir répondre, mais qu'ils m'invitaient à laisser un message. Ils me rappelleraient.

CHAPITRE 2

À 10 heures le lendemain matin je me trouvais, pâle et piteux, devant l'allée de mes parents. Je portais une chemise propre. J'avais pris un petit déjeuner. Je m'étais excusé auprès de tous ceux que j'avais croisés à l'hôtel, jusqu'au gars qui nettoyait la piscine. J'étais stupéfait de ne pas avoir passé la nuit en cellule. Je me sentais tout merdeux.

La maison était située vers le bout d'une route étroite et pentue, dans la partie à flanc de montagne du principal secteur résidentiel de Dyersburg. Elle m'avait un peu déconcerté lors de l'emménagement de mes parents. Le terrain était de belle taille, environ deux mille mètres carrés, avec deux vieux arbres qui ombrageaient le mur latéral. Elle était flanquée de demeures victoriennes fin d'époque, de volume comparable, que personne ne semblait pressé de repeindre. Une haie impeccable délimitait le terrain de part et en part. Mary occupait la maison suivante, et cette femme ne roulait pas sur l'or. De l'autre côté résidaient depuis peu un professeur d'université et sa doctorante d'épouse. Je crois d'ailleurs que c'est mon père qui leur avait vendu leur habitat. Des gens charmants, eux aussi, mais pas vraiment le genre à prendre des bains au champagne. La maison elle-même possédait un étage, un superbe porche circulaire, un atelier en sous-sol et un garage qui mordait sur le jardin. C'était sans conteste une belle baraque, jolie et bien pourvue, dans un quartier agréable. On voudrait vous installer ici, vous n'iriez pas râler. Mais *Baraques de stars* n'y consacrerait pas un numéro spécial pour autant.

J'agitai la main par-dessus la clôture au cas où Mary serait à sa fenêtre, et remontai lentement le chemin. J'avais l'impression d'approcher une usurpatrice : la vraie maison de mes parents, celle où j'avais grandi, se trouvait loin dans le temps et cinq cents kilomètres plus à l'ouest. Je n'avais jamais remis les pieds à Hunter's Rock depuis le déménagement, mais je me souvenais de cette maison-là comme si j'y vivais encore. La disposition des chambres avait scellé ma conception de l'espace domestique. La maison que je contemplais à présent était comme une seconde épouse, arrivée trop tard pour obtenir de ses beaux-enfants davantage qu'une tiède cordialité.

Une poubelle en fer galvanisé reposait à côté de l'entrée, le couvercle soulevé par un sac plein. Pas de journal sur le perron. Davids avait dû résilier l'abonnement. À juste titre, sauf qu'ainsi la maison paraissait déjà couverte d'un drap. J'extirpai de ma poche le trousseau de clés et ouvris la porte.

La maison baignait dans un silence d'agonie. Je ramassai le courrier, principalement des prospectus, et le posai sur la console. Puis je flânai un moment, de pièce en pièce. On aurait dit des show-rooms annonçant je ne sais quel vide-grenier, où chaque objet provenait d'un logis différent et affichait un prix bien en deçà de sa valeur. Même les choses qui allaient par lot – les livres dans le bureau de mon père, les poteries anglaises des années trente de ma mère, soigneusement alignées sur le vieux buffet en pin du salon – paraissaient irrémédiablement soustraites à mon toucher comme au temps. Je n'avais pas la moindre idée de ce que j'allais en faire. Les mettre dans des boîtes et les entreposer quelque part pour qu'elles prennent la poussière ? Les vendre et conserver la recette, ou la reverser à une bonne cause ? Ou bien vivre parmi tous ces objets, en sachant que je n'aurais jamais à leurs yeux qu'un regard de seconde main ?

La seule solution qui me parût un tant soit peu avisée était de tout laisser tel quel, de sortir de là et de ne plus jamais revenir. Ce n'était pas ma vie. Ce n'était la vie de personne, plus maintenant. En dehors de l'unique photo de mariage du hall, il n'y avait aucun cliché nulle part. Il n'y en avait jamais eu dans notre famille.

Je finis par me retrancher au salon. Il donnait sur le jardin, face à la route, à travers de grandes fenêtres qui transformaient en chaleur la lumière froide de l'extérieur. Il y avait un divan et un fauteuil assortis, aux motifs raffinés. Un petit téléviseur seize neuvièmes, posé sur un meuble à vitres fumées. Sans oublier le fauteuil de mon père, une vieille antiquité en bois foncé et tissu vert, l'unique rescapé de l'ancienne maison. Une nouvelle biographie de Frank Lloyd Wright était posée sur la table basse, la page marquée par un reçu de Denford's Market. Huit jours plus tôt, l'un de mes parents avait acheté de la charcuterie, un gâteau à la carotte (ça alors), cinq grandes bouteilles d'eau minérale, du lait écrémé et un tube de vitamines. La plupart de ces produits devaient figurer parmi ceux que Mary avait éliminés du frigo. Il restait peut-être de l'eau minérale, ainsi que les vitamines. J'irais peut-être me servir après.

Dans l'immédiat je me carrai dans le fauteuil de mon père. Je frottai mes mains sur les accoudoirs élimés, avant de les reposer sur mes genoux tout en contemplant le jardin.

Et là, par vagues violentes, je pleurai longuement.

Bien plus tard, je me souvins d'une soirée qui remontait loin en arrière. Je devais avoir dix-sept ans, du temps où nous habitions encore la Californie. C'était un vendredi soir, et j'avais rendez-vous avec les potes dans un bouge situé sur une petite route à l'orée de la ville. Lazy Ed's était l'un de ces bars à bière de type boîte-à-chaussures-plus-parking que l'on croirait conçus par les Mormons pour donner de la boisson une image non seulement impie, mais morne, triste et désespérée. Ed savait qu'il n'était pas en position de jouer les difficiles, et puisque nous ne faisions jamais d'histoires et alimentions en pièces de vingt-cinq son billard et son juke-box – Blondie, Bowie et ce bon vieux Bruce Stringbean, à la grande époque de Molly Ringwald et des couleurs à la Mondrian –, notre clientèle de jeunots était bienvenue chez lui.

Ma mère était de sortie, partie chez une copine pour faire ce que peuvent bien faire les femmes quand leurs hommes ne sont pas là à semer le désordre, à prendre un air blasé ou à ne pas écouter avec suffisamment de gravité les mésaventures de gens qu'ils ne connaissent ni d'Ève ni d'Adam, et qui de toute façon ont l'air assez nazes, à en juger par les problèmes en question. À 18 heures, papa et moi étions assis à la grande table de la cuisine, devant des lasagnes que maman avait laissées dans le frigo, ignorant la salade verte. J'étais perdu dans mes pensées. Lesquelles, je ne puis dire. Je ne sais pas davantage me glisser dans ma tête à dix-sept ans que dans celle d'un indigène de Bornéo.

Je finis par constater que papa avait terminé de manger, et qu'il m'observait.

— Quoi ? demandai-je, relativement affable.

Il repoussa son assiette.

— Tu sors, ce soir ?

J'opinai lentement, pétri d'embarras adolescent, et me remis à précipiter la bouffe dans ma gorge.

J'aurais tout de suite dû saisir le sens de sa question. Mais celui-ci m'échappa, de même que m'échappait la raison de cette petite portion de salade dans son assiette par ailleurs vide. Je ne voulais pas de cette merde verte, alors je n'en prenais pas. Lui non plus n'en voulait pas, mais il en prenait quand même – bien que maman fût absente. Je comprends aujourd'hui que le but du jeu était de réduire le volume du saladier, pour éviter les sermons sur nos mauvaises habitudes alimentaires. En jeter une partie directement à la poubelle eût paru déloyal, mais si elle transitait d'abord par son assiette, la morale était sauve. À l'époque, je n'y voyais qu'une absurdité sans nom.

Je vidai mon auge, et constatai que papa était resté assis. Ça ne lui ressemblait pas. D'ordinaire, sitôt le repas terminé, il fallait qu'il s'active. Charger le lave-vaisselle. Sortir la poubelle. Mettre le café en route. Puis passer à la suite. En abattre, toujours en abattre.

— Et toi, qu'est-ce que tu vas faire ? Regarder la télé ? demandai-je au prix d'un effort que j'estimais très adulte.

Il se leva et débarrassa son assiette. Il y eut un silence, puis il répondit :

— Je me demandais.

Cela ne s'annonçait guère passionnant.

— Tu te demandais quoi ?

— Si ça te dirait de tirer quelques bandes avec un vieux bonhomme ?

Mon regard se planta dans son dos. Le ton de sa requête était aux antipodes de son assurance habituelle, en particulier cette tentative affectée d'autodénigrement. Comment pouvait-il croire que je mordrais à l'hameçon ? Il n'était pas vieux. Il faisait du jogging. Au tennis et au golf il écrasait des plus jeunes que lui. Surtout, c'était la dernière personne que j'imaginais penchée sur une table de billard. Il n'avait tout simplement pas le profil. Si vous aviez dû dessiner un diagramme de Venn comprenant les cercles « individus ayant l'air de joueurs de billard », « individus ayant peut-être l'air de joueurs de billard » et « individus n'ayant pas l'air de joueurs de billard, mais l'étant peut-être », il n'aurait même pas figuré sur la feuille. Il portait ce soir-là, comme si souvent, un pantalon de coton sable repassé avec soin et une chemise toute propre en lin blanc, qui ne provenaient ni l'un ni l'autre d'une marque aussi grand public que Gap. De belle taille, il avait la peau mate, les cheveux foncés et grisonnants, et le genre d'ossature qui donne envie aux gens de voter pour vous.

Il paraissait destiné à se prélasser sur le pont d'un grand yacht, dans les eaux de Palm Beach ou de Jupiter Island, à discuter d'art. Et tout particulièrement de l'art qu'il essaierait de vous vendre. J'étais pour ma part blond et maigre, vêtu des réglementaires Levi's et tee-shirt noirs. L'un et l'autre semblaient avoir servi à régler des moteurs de voiture. Ils en avaient aussi l'odeur, sûrement. Celle de papa devait être la même que d'habitude, dont je n'avais pas conscience à l'époque mais que je sais retrouver aujourd'hui comme s'il se tenait juste derrière moi : une odeur sèche, propre, convenable, comme du bois de chauffage soigneusement empilé.

— Tu veux venir jouer au billard ? demandai-je, vérifiant que je n'avais pas perdu la raison.

Il haussa les épaules.

— Ta mère est sortie. Il n'y a rien dans le poste.

— Tu n'as pas une cassette en réserve ?

C'eût été inconcevable. Papa vouait à son magnétoscope les mêmes sentiments qu'un maître à son vieux chien, et des étagères entières de vidéos soigneusement étiquetées s'alignaient aux murs de son bureau. J'en ferais tout autant, bien entendu, si j'avais une adresse fixe. Je leur attribuerais même des codes-barres si j'en avais le temps. Mais, à l'époque, ce hobby était l'aspect de son personnage qui m'évoquait le plus facilement des images d'État policier.

Il ne répondit rien. J'éliminai les reliefs de ma propre assiette, avec une application à la fois franche et inconsciente, du fait que j'étais à l'âge où montrer de l'affection envers sa mère est difficile, et qu'éviter que la merde ne bouche son précieux lave-vaisselle était un geste qui passait inaperçu, y compris à mes propres yeux. Je ne voulais pas que papa vienne au bar. C'était aussi simple que ça. Mes sorties obéissaient à une certaine routine. J'appréciais le trajet en voiture. C'était du temps *pour moi*. Et puis les potes allaient trouver ça bizarre. C'*était* bizarre, sans déconner. Mon copain Dave, qui avait toutes les chances d'arriver complètement murgé, allait faire une attaque en me voyant accompagné de ce suppôt de l'autorité, du conservatisme et de la ringardise.

Je le dévisageai, cherchant une façon d'exposer mon point de vue. Les assiettes étaient rangées dans le lave-vaisselle. Le reste de salade dans le réfrigérateur. Papa avait essuyé le comptoir. Si d'aventure une brigade de la police scientifique s'infiltrait en milieu de soirée pour chercher des traces d'activité alimentaire, elle repartirait bredouille. Et cela avait le don de m'agacer.

Mais quand papa replia le torchon et le suspendit à la poignée du four, j'eus un avant-goût de ce que j'éprouverais pour de bon, presque vingt ans plus tard, le jour où je me retrouverais assis en larmes dans son fauteuil, dans une maison vide de Dyersburg. La découverte que sa présence n'était ni incontournable ni acquise ; qu'un jour le saladier demeurerait plein et les torchons en boule.

— Si ça te dit, répondis-je.

Redoutant aussitôt la réaction des copains, j'avançai de quarante minutes notre départ de la maison. Ainsi nous disposerions peut-être d'une heure entière avant d'être confrontés au regard des autres, car les potes étaient toujours en retard.

Nous roulâmes jusque chez Ed, moi au volant, papa sur le siège passager, peu bavard. À l'approche du bar il rapprocha son visage du pare-brise.

— C'est ici que tu joues ?

Je confirmai, un peu sur la défensive. Il grogna. Comme nous traversions le parking, je songeai que débarquer avec mon père risquait de nourrir les doutes d'Ed au sujet de mon âge, mais je ne pouvais plus faire marche arrière. Et puis on ne se ressemblait pas tant que ça. Peut-être me croirait-on simplement copain avec un type plus âgé. Un sénateur, ou un truc dans ce goût-là.

La salle était quasi déserte. Deux viocs inconnus étaient affalés sur une table du coin. Cet endroit ne s'animait que très tard, et encore, dans une ambiance des plus précaires, que deux mauvais choix consécutifs sur le juke-box pouvaient suffire à plomber. En attendant qu'Ed daignât quitter son arrière-boutique, papa s'adossa au comptoir et scruta les lieux. Il n'y avait pas grand-chose à voir. Des tabourets délabrés, une couche de poussière antédiluvienne, un billard, une pénombre et des néons. Je ne souhaitais pas que ça lui plaise. Ed apparut enfin, souriant à ma vue. D'ordinaire je buvais ma première bière tout en lui faisant la conversation, et il s'attendait sûrement qu'on remette ça ce soir.

Puis il remarqua mon père et s'arrêta. Pas comme s'il avait heurté un mur, mais il eut un mouvement d'hésitation, et son sourire fit place à un masque impénétrable. Mon père n'était pas le genre de type à fréquenter un tel endroit, et je suppose qu'Ed se demandait quelle impensable erreur d'orientation l'avait amené jusqu'ici. Papa se retourna vers lui et hocha la tête. Ed lui rendit la politesse. J'avais hâte d'en finir.

— Mon père, fis-je.

Ed opina de nouveau, et sur cette note s'acheva ce grand moment de socialisation virile.

Je commandai deux bières. En attendant qu'elles arrivent je regardai papa s'éloigner vers le billard. Gamin, je m'étais habitué à ce que les gens l'interpellent dans les magasins, pensant qu'il en était le directeur, et à ce titre la seule personne capable de démêler je ne sais quelle bagatelle montée en psychodrame. Paraître tout aussi à l'aise dans ce bar miteux était une sorte de gageure, qui déclencha en moi un accès d'admiration. Il s'agissait certes d'une forme de respect bien spécifique et limitée, celle que l'on éprouve devant une qualité dont il n'est pas exclu qu'on la convoite un jour, mais il n'empêche qu'elle était bien là.

Je le rejoignis au pied du billard, après quoi notre session de resserrement affectif partit en eau de boudin. Je gagnai les trois parties. Toutes trois longues et lentes. Il n'était pas si nul que ça, mais il lui manquait toujours cinq pour cent de précision, et j'avais la maîtrise de la table. Nous parlâmes très peu, occupés à nous pencher sur la table, à jouer nos coups, à encaisser nos ratés. Au terme du deuxième jeu, il partit chercher une seconde bière pendant que je piégeais les billes dans le triangle. J'avais secrètement espéré qu'il s'en tiendrait à une seule, aussi avais-je à peine touché à la mienne. Nous disputâmes la dernière partie, qui fut un brin meilleure, mais non moins atroce. Puis papa raccrocha sa queue au râtelier.

— Tu t'arrêtes là ? demandai-je d'un ton qui se voulait nonchalant.

J'étais tellement soulagé que je pris le risque de brandir une nouvelle pièce de 25 cents.

Il secoua la tête.

— Je ne te complique pas beaucoup la tâche.

— Alors, tu ne vas pas dire : « T'es drôlement doué, fiston », ou un truc de ce genre ?

— Non, répondit-il calmement. Parce que ce serait faux.

J'écarquillai les yeux, mortifié comme un môme de cinq ans.

— Ah, d'accord, parvins-je finalement à articuler. Merci pour ces encouragements.

— Ce n'est qu'un jeu, dit-il en haussant les épaules. Ce qui me dérange, ce n'est pas que tu sois mauvais. C'est que tu t'en satisfasses.

— Quoi ? m'étranglai-je. Tu sors ça d'où, d'un bouquin de développement personnel ? Balancez une vacherie au bon moment et votre gosse finira P-DG ?

Toujours aussi calme :
— Arrête tes conneries, Ward.
— C'est toi qui déconnes, répliquai-je. Tu *pensais* que je serais mauvais, et que tu pourrais venir me donner une leçon, alors que tu ne touches pas une bille !
Il resta figé un moment, les mains dans son pantalon de coton, à me considérer. Il avait un drôle de regard, froid et songeur, mais non dénué d'amour. Puis il sourit.
— Peu importe, fit-il.
Et sur ces mots il s'en alla – pour rentrer à pied, je suppose.
Je me retournai, attrapai ma bière et la vidai d'un trait. J'essayai d'expédier l'une de ses dernières boules au fond d'une poche d'angle, et mis un kilomètre à côté. Jamais je n'avais haï mon père comme à cet instant.
Je fondis sur le bar, pour découvrir qu'Ed m'avait déjà servi un nouveau demi. Je plongeai la main dans ma poche, cherchant ma monnaie, mais il secoua la tête. Une grande première. Je me perchai sur un tabouret, et restai muet pendant quelques minutes.
Puis la discussion s'engagea malgré tout, sur divers sujets : les opinions d'Ed concernant la politique et le féminisme – en gros, les deux lui déplaisaient –, et la cabane qu'il envisageait de bâtir dans les bois. Je voyais mal Ed peser de quelque façon que ce fût sur les deux premiers, ou retrousser ses manches pour édifier sa planque, mais je l'écoutai quand même. Lorsque Dave arriva je parvins à feindre la plus parfaite routine.
Ce fut une soirée convenable. Bavardage, boisson, affabulations. Nous jouâmes quelques parties, sans grand talent. Puis je regagnai la voiture, et m'arrêtai en découvrant un mot pincé sous l'un des essuie-glaces. C'était l'écriture de mon père, mais bien plus petite que d'habitude.
« Si tu n'arrives pas à lire ceci du premier coup, fais-toi raccompagner. Je te ramènerai demain pour que tu récupères la voiture. »
Je chiffonnai et jetai le papelard, avant de prendre le volant – en redoublant de prudence. À mon retour maman était couchée. Il y avait de la lumière dans le bureau de papa, mais la porte était fermée, alors je montai droit dans ma chambre.

Je me levai une fois, en fin de matinée, pour me préparer une tasse de café soluble. À part ça, je stagnai dans le fauteuil jusqu'au milieu de l'après-midi, quand le soleil vint m'éblouir à travers la vitre. Cela me sortit de ma léthargie, et en me relevant je sus que je ne m'assoirais plus jamais là. D'une, c'était inconfortable. Le coussin était râpé et bosselé, et après plusieurs heures passées dessus, j'avais drôlement mal au cul. Je retournai à la cuisine, rinçai ma tasse, la reposai à l'envers sur le bord de l'évier. Puis je me ravisai : l'essuyai et la rangeai dans le placard. Je repliai le torchon et l'accrochai à la poignée du four.

Je restai paralysé dans le couloir, ne sachant que faire ensuite. Une partie de mon être estimait que l'attitude la plus filiale eût consisté à rendre ma chambre d'hôtel pour passer la nuit ici. Mais tout le reste de mon âme s'y opposait. Catégoriquement. Je voulais des lumières vives, un hamburger, une bière, et quelqu'un qui me parle d'autre chose que de mort.

Redevenu d'un coup triste et grincheux, je me ruai dans le salon pour récupérer mon téléphone sur la table basse. J'avais mal dans le bas du dos, sûrement à cause de ce fauteuil pourri.

Le fauteuil. Peut-être était-ce dû au changement de lumière : le soleil avait traversé toute la largeur du jardin, créant de nouvelles ombres. Ou bien, plus probable, ces quelques heures de larmes m'avaient un peu éclairci les idées. Quoi qu'il en soit, maintenant que j'avais les yeux dessus, le siège présentait un drôle d'aspect. Tout en rempochant mon Nokia, j'ajustai mon regard. Le coussin, qui était solidaire de la structure, se renflait bel et bien au milieu. J'y posai la main, palpai la bosse. Elle était un peu dure.

Peut-être papa l'avait-il fait retapisser, ou rembourrer de je ne sais quoi. De cailloux, peut-être. Je me redressai, prêt à oublier ce meuble et à quitter la maison. La gueule de bois commençait à me reprendre. Mais un autre détail retint mon attention.

Il existe tout un art d'agencer les objets entre eux, surtout lorsqu'ils sont volumineux. Certaines personnes ne voient pas ça. Elles se contentent de disposer leur mobilier au petit bonheur la chance, collé contre les murs, en angles droits, ou de telle sorte que toute la famille puisse voir la télé. Mais mon père avait toujours veillé à l'équilibre esthétique entre les meubles et se mettait en rogne si on en déplaçait un seul. Or le fauteuil de papa n'était pas à sa place. Il ne manquait pas grand-chose, et je doute que personne d'autre l'eût remarqué. Il était trop d'équerre avec le reste, et paraissait trop en avant. Ça ne rendait pas bien, tout simplement.

Je m'accroupis pour examiner la jointure entre le coussin et l'ossature. Elle était longée d'une ganse effilochée. Je l'attrapai par un bout et tirai. Elle vint facilement, révélant une ouverture qui semblait autrefois cousue.

J'y glissai ma main. Mes phalanges naviguèrent dans une matière sèche et molle, sûrement des copeaux de mousse. Parvenues au centre, elles se heurtèrent à un corps dur. Je l'exhumai.

C'était un livre. Un polar à succès, en édition poche, d'aspect récent, le genre d'ouvrage que ma mère avait pu choisir sur un coup de tête et parcourir en un après-midi. Sauf qu'il n'avait pas l'air entamé. La tranche n'était pas cassée, or ma mère n'avait jamais accordé un soin particulier à l'état des couvertures. Je ne comprenais pas. Il n'avait pu arriver là-dedans par accident.

Je feuilletai les pages. Vers le milieu du bouquin se trouvait un bout de papier. Je le retirai. Un petit mot, une seule phrase, écrite de la main de mon père :

« Ward, nous ne sommes pas morts. »

CHAPITRE 3

Un ruisseau dans le sud du Vermont, à l'eau claire et froide, courant sur un lit de galets pâles entre les flancs d'une vallée nichée dans les Green Mountains. Le ciel semble commencer quelques mètres à peine au-dessus de la cime des arbres, un drap de barbe à papa figé dans le gris du jour déclinant. Les feuilles au sol, bris d'ampoules aux couleurs de vitrail, sont inégalement saupoudrées de neige. De part et d'autre du cours d'eau, réuni par deux vieux ponts de pierre espacés d'une cinquantaine de mètres, s'étend le petit village de Pimonta. Il doit comprendre une vingtaine d'habitations en tout, quoiqu'une bonne douzaine d'entre elles semblent réservées aux vacances d'été, voire simplement abandonnées. Contre l'une d'elles repose la carcasse d'une antique Buick, colorée par l'oxydation dans des tons orageux. Quelques véhicules sont parqués dans les allées, des modèles robustes laissant deviner que leurs propriétaires possèdent plusieurs enfants et au moins un chien. Il règne un silence parfait, hormis le bruit du ruisseau, qui court depuis si longtemps que sa clameur tient plus de la couleur que du son. De la fumée s'échappe indolemment de quelques cheminées, dont celles du Pimonta Inn, un confortable bed and breakfast adossé au ruisseau et quasi complet en cette dernière semaine de défoliation.

Sur l'un des ponts se tient un homme, appuyé contre le muret, le regard plongé dans les remous de l'eau. Il s'appelle John Zandt. Il mesure un mètre quatre-vingts et porte un épais manteau pour se protéger du froid. Ce vêtement accentue sa carrure, large et compacte. Il a l'air capable de transporter deux valises sur une longue distance ou de cogner quelqu'un extrêmement fort. Les deux sont vrais. Ses cheveux sont courts et sombres, ses traits rugueux mais harmonieux. Une barbe de deux jours lui recouvre les joues et le menton. Il a passé la semaine au Pimonta Inn, dans une suite composée d'une chambre, d'une salle de bains et d'un petit salon avec cheminée, le tout onéreusement douillet, dans un style rustique négligé. Il a consacré ses journées à arpenter les monts et les vallées de la région, en évitant les sentiers balisés et leurs chapelets de randonneurs bariolés qui redoutent les ours. Il est parfois tombé sur des vestiges de fermettes, réduites à des tas de fagots disséminés dans les sous-bois. On a beau tendre l'oreille, aucun écho ne parvient jusqu'ici, et des sites jadis en bordure de sentier ont à présent disparu des cartes. Les routes ont trouvé de nouveaux chemins, transformant certains coins en destinations et laissant les autres à la nature, peut-être pour toujours.

Zandt aime s'asseoir un moment dans ces lieux, à imaginer ce qu'ils avaient pu être. Puis il se remet en marche, jusqu'à ce que survienne la fatigue et qu'il soit temps de regagner l'auberge. En début de soirée il s'installe dans le grand séjour chaleureux, en évitant poliment la conversation des autres pensionnaires et des propriétaires. Les ouvrages de la petite bibliothèque traitent de fossiles et d'épanouissement personnel. Une quarantaine de personnes a dû le saluer au cours des deux dernières semaines, sans connaître son nom ni rien apprendre sur son compte.

Après le dîner, d'ordinaire excellent malgré la lenteur du service, il réintègre sa suite, fait du feu et reste éveillé aussi longtemps qu'il le peut. Il a beaucoup rêvé ces derniers jours. Parfois de Los Angeles, d'une vie révolue pour de bon et qui dès lors ne peut être fuie. Par le passé il s'est essayé à l'alcool comme à l'héroïne, sans y trouver grand secours, même à doses excessives. À présent il se contente de veiller, allongé sur le dos, attendant le matin, songeant au vide. Il n'a jamais tenté de se tuer. Ce n'est pas dans sa nature. Sinon il serait déjà mort.

Là, appuyé sur le muret du pont dans le jour déclinant, il s'interroge sur la suite des événements. Il a de l'argent, dont une partie lui reste d'un été de travail manuel intensif. Il se demande si le temps n'est pas venu de reprendre son baluchon, pour repartir vers une grande ville. Peut-être quelque part dans le sud, bien qu'il se soit découvert un goût pour le froid et les forêts sombres. Sa volonté est bridée par le fait qu'il n'a besoin d'aucune rentrée d'argent supplémentaire ni, pour celui qu'il possède déjà, de projets particuliers. Et que, après toute une vie passée au milieu des buildings, ceux-ci ont soudain perdu tout intérêt à ses yeux. Les routes désertes et les espaces vierges lui parlent davantage.

Il relève la tête en distinguant le bruit d'une voiture s'approchant par le nord. Au bout d'un moment une lumière de phares, allumés plus tôt que ne le veut la coutume locale, pointe au sommet de la colline, puis précède la voiture jusque dans le village, croisant la petite épicerie et la vidéothèque. Il s'agit d'une Lexus, très noire et récente. Elle s'arrête en douceur devant l'auberge.

L'automobile cliquette tandis que le moteur refroidit. Dans un premier temps, personne n'en sort. Zandt finit par comprendre que les deux passagers ont les yeux rivés sur lui. Son propre véhicule, une chose étrangère et bon marché qu'il a dégottée sur un parking désolé du Nebraska, est garé devant la dépendance qui abrite sa suite. Les clés de la voiture sont dans sa poche, mais impossible de l'atteindre sans se rapprocher de la Lexus. Il pourrait se retourner, traverser le pont, s'enfoncer dans l'autre partie du village et gagner les hauteurs, mais il n'en a pas le courage. Il sait qu'il aurait dû payer son séjour en liquide. Comme il le fait d'habitude. Mais il est arrivé les poches vides, et il était tard. Retirer de l'argent au distributeur de la ville la plus proche aurait laissé une trace tout aussi visible. Cela fait donc deux semaines que la confrontation est inévitable. Alors il contemple de plus belle le ruisseau sous ses pieds, et attend.

La portière passager s'ouvre et une femme descend. Cheveux bruns mi-longs, tailleur vert bouteille, taille moyenne. Son visage est saisissant, ce qui signifie qu'on la jugera soit laide, soit splendide. La plupart des gens se rallient à la première opinion, ce qui lui convient tout à fait. Son silence durant le voyage a agacé l'agent Fielding, qui la connaît depuis seulement trois heures – et, n'eût-il été chargé de la conduire à Pimonta, il aurait retrouvé ses pénates depuis belle lurette. Fielding ignore encore le motif de ce périple, ce qui n'est pas plus mal, car c'est à peine si cette mission peut être considérée comme officielle. Il fait juste ce qu'on lui dit, talent assez sous-estimé.

La femme referme la portière avec un clac qu'elle sait perceptible par l'homme sur le pont. Ce dernier ne bouge pas, ne relève même pas la tête avant qu'elle ait atteint l'auberge puis l'échoppe murée d'un défunt potier, pour s'engager sur le pont.

Elle s'arrête à quelques mètres de lui, se sentant un peu bête et assez frigorifiée.

— Bonjour, Nina, dit-il toujours sans la regarder.

— Quel flegme ! Je suis épatée.

Il se tourne vers elle.

— Joli tailleur. Très Dana Scully.

— On veut toutes lui ressembler ces derniers temps. Même certains mecs.

— C'est qui, dans la voiture ?
— Un local. De Burlington. Le gentil monsieur a bien voulu m'emmener.
— Comment tu m'as trouvé ?
— Carte de crédit.
— Évidemment. Ça fait un sacré voyage.
— Tu le vaux bien.
Il considère d'un air sceptique la femme qu'il avait autrefois trouvée splendide et juge à présent quelconque.
— Alors, qu'est-ce que tu veux ? On gèle. J'ai faim. Ça m'étonnerait qu'on ait quelque chose à se dire.
L'espace d'un instant elle paraît de nouveau splendide, et blessée. Puis, comme si rien de tout cela n'avait d'importance à ses yeux, ou n'en avait jamais eu :
— Ça vient de se reproduire, dit-elle. J'ai pensé que tu préférerais le savoir.
Elle tourne les talons et se dirige vers la voiture. Le moteur tourne déjà quand elle rouvre sa portière, et en deux minutes la vallée redevient vide et silencieuse, autour d'un homme sur un pont, la bouche entrouverte, le visage blême.

Il la rattrapa trente kilomètres au sud, roulant à tombeau ouvert sur d'étroites routes de montagne, la voiture ballottant dans les lacets. Le sud du Vermont n'est pas conçu pour la vitesse, et par deux fois le véhicule chassa sur le verglas. Zandt n'y prêta aucune attention, pas plus qu'à la poignée d'automobilistes du cru qui eurent à peine le temps de ciller avant qu'il ne fonde sur eux et les laisse pantelants dans son sillage. À Wilmington, la route partait en fourche. La Lexus n'était visible dans aucune des deux directions. Présumant que Nina filait vers le premier endroit d'où elle pourrait regagner la civilisation par les airs, il prit à gauche sur la route 9 en direction de Keene, juste derrière la frontière du New Hampshire.
La chaussée plus large lui permit d'augmenter sa moyenne, et il ne tarda pas à distinguer au loin les feux arrière caractéristiques de la Lexus, qui vacillaient derrière les arbres quand la route sinuait, ou s'éteignaient au passage d'un col. Il finit par la rejoindre dans une ligne droite à l'orée de Hardsboro, où la route longeait un lac lisse et froid qui réfléchissait un ciel couvert de nuages.
Il fit un appel de phares. Aucune réponse. Il se rapprocha, recommença. Cette fois-ci la Lexus reprit un peu de vitesse. Zandt

enfonça l'accélérateur. Il vit Nina se retourner et s'aviser de sa présence à travers la lunette arrière. Elle s'adressa au chauffeur, qui ne ralentit pas.

Pied au plancher, Zandt déboîta, doubla et se rabattit avant de freiner à fond. Le moteur n'était pas encore arrêté qu'il bondissait hors de l'habitacle, imité par Fielding, qui sortait la main de son blouson.

— Range ça, conseilla Zandt.

— Va te faire foutre.

L'agent tenait son arme à deux mains. Nina s'extirpa de la voiture à son tour, en tâchant d'éviter la boue.

— Je vous préviens, lui dit Fielding d'une voix calme, vous devriez rester en retrait.

— Tout va bien, dit Nina. Merde, mes pompes...

— Tout va bien, mon cul ! Il a essayé de nous mettre dans le fossé.

— Peut-être qu'il avait juste envie de parler. On se sent facilement seul là-haut.

— Qu'il parle à mon cul, rétorqua Fielding. Vous, vos mains sur la voiture !

Zandt resta là où il était jusqu'à ce que Nina ait contourné la Lexus et foulé la chaussée.

— Tu es sûre que c'est lui ? demanda-t-il.

— Tu crois que j'aurais fait tout ce chemin si ce n'était pas le cas ?

— Je n'ai jamais rien compris à ta manière d'agir. Jamais. Alors réponds seulement à ma question.

— Tu vas foutre tes mains sur ce toit, oui ou merde ? hurla Fielding.

Tinta alors le son mécanique d'une sécurité qu'on débloque.

Zandt et Nina se tournèrent vers l'agent. Il était hors de lui. Nina scruta le route, où approchait une grosse Ford blanche fleurant bon « location », qui roulait au pas de sorte que ses occupants puissent apprécier la vue du lac dans ce qu'il restait de lumière.

— Du calme, fit Nina. Vous avez vraiment envie d'expliquer à votre chef que vous avez descendu un collègue ?

Fielding jeta un œil par-dessus son épaule. Vit la voiture s'immobiliser sur un refuge du bas-côté, à une centaine de mètres. Il baissa son arme.

— Vous allez enfin me dire ce qui se passe, bordel ?

Nina secoua dédaigneusement la tête puis revint à Zandt :

— Oui, j'en suis sûre, John.

— Alors pourquoi tu n'es pas là-bas ?

Elle haussa les épaules, comme à son habitude.

— Pour tout te dire, je ne sais pas. Je ne devrais pas être ici, et encore moins te parler. Tu comptes m'envoyer paître, ou tu préfères qu'on aille discuter ailleurs ?

Zandt détourna le regard vers la surface lisse du lac. Elle était noire par endroits, gris gelé par d'autres. De l'autre côté se trouvaient une petite clairière et une maison de vacances en bois, avec un tas de bûches empilées sur le côté. La structure ne paraissait ni préfabriquée ni choisie sur catalogue – c'était plutôt comme si quelqu'un ou « quelques deux » avaient passé de longues soirées dans un endroit invivable à dessiner les plans sur des blocs rapportés du bureau, brûlant de se transporter dans une tout autre dimension. Une fois de plus, Zandt regretta de ne pas être un autre. Le type qui vivait là, par exemple. Ou l'un des touristes arrêtés plus haut sur la route, qui étaient descendus dans les fourrés au bord de l'eau pour admirer les arbres, et ressemblaient dans leurs anoraks chamarrés à une harde de feux tricolores.

Enfin il hocha la tête. Nina retourna auprès de Fielding et lui parla quelques instants. Au bout d'une minute, le pistolet de l'agent avait regagné son étui. Le temps que Zandt détache ses yeux du lac, Fielding était remonté dans le véhicule, l'air décrispé.

Nina attendit Zandt à sa voiture, un épais dossier sous le bras.

— Je lui ai dit que je partais avec toi, expliqua-t-elle.

Comme Nina s'installait sur le siège, Zandt s'avança vers la Lexus. Fielding l'observa par le carreau avec une expression impénétrable, et démarra. Puis il pressa le bouton pour baisser sa vitre.

— Allez, je passe l'éponge pour cette fois, dit-il. À la prochaine.

Zandt se fendit d'un sourire. Un sourire mince, qui n'avait rien d'une manifestation de joie.

— Il n'y aura pas de prochaine fois, répondit-il.

Fielding se raidit.

— Et qu'est-ce que je suis censé comprendre ?

— Que si jamais on se recroise et que tu braques un flingue sur moi, la surface d'un joli lac sera jonchée de petits morceaux de fédéral. Et je me branle de savoir si ça nique l'écosystème.

Là-dessus, Zandt tourna les talons, laissant l'agent bouche bée.

Fielding fit une marche arrière nerveuse, soulevant une giclée de gravier. Il fit vrombir son moteur et détala, ne s'arrêtant que pour se pencher du côté passager et tendre le majeur de sa main droite.

En regagnant sa voiture, Zandt découvrit Nina assise, les bras croisés, qui le dévisageait en levant un sourcil.

— Tes facultés relationnelles ne cessent de progresser, dit-elle. Tu devrais, je sais pas, moi, donner des conférences. Ou écrire un livre. Sérieusement. C'est un don. Ne le réprime pas, partage-le. Va au bout de ton véritable ego.

— Ferme-la, Nina.

Il conduisit en silence jusqu'au retour à Pimonta. Nina tenait son dossier sur les genoux. La nuit était tombée lorsqu'ils retrouvèrent le village, et de nouvelles autos étaient apparues sur le parking des pensionnaires. Plusieurs fenêtres étaient éclairées. Zandt se gara devant l'entrée de l'auberge, coupa le contact. Il ne fit aucun geste pour ouvrir sa portière, alors Nina resta immobile.

— Tu as toujours faim ? finit-elle par demander.

L'habitacle se refroidissait. Deux couples avaient déjà croisé la voiture en se dirigeant vers le bâtiment principal, le visage rond à l'heureuse idée de manger.

Zandt s'étira, comme au terme d'un long voyage.

— C'est comme tu veux.

Elle tenta l'enthousiasme :

— Ça m'est égal.

— Détrompe-toi. Ici, on dîne entre 18 h 30 et 21 heures. Alors c'est maintenant ou demain matin. Le petit déj' est servi entre 7 et 8. Et plutôt chiche.

— Tu veux dire qu'il n'y a pas moyen de bouffer un hamburger dans l'intervalle ? Ou qu'on ne te servira nulle part un petit sandwich ?

Zandt tourna la tête, et cette fois-ci son sourire parut presque sincère.

— Tu n'es pas du coin, n'est-ce pas ?

— Dieu merci, non. Et toi non plus. Chez nous on peut manger quand on veut. On donne de l'argent, et on reçoit de la nourriture en échange. C'est moderne et pratique. Ou bien es-tu resté trop longtemps chez les ploucs pour t'en souvenir ?

Il s'abstint de répondre. Sans prévenir, elle laissa tomber le dossier sur le tapis et ouvrit sa portière.

— Attends-moi, dit-elle.

Zandt attendit, la regardant s'éloigner d'un pas décidé vers le bâtiment. La faim qu'il avait ressentie après sa promenade s'était dissipée. Il se sentait congelé, dedans comme dehors. Il avait perdu l'habitude de côtoyer des gens qui le connaissaient bien et il se sentait déboussolé, ses pensées et sentiments déphasés. Il avait passé un sacré temps sur la route, à se fondre dans le décor : l'homme au bar qui commande un deuxième verre ; le gars qui fait des extras pendant quelques jours ; le type dans une station-service fouettée par le vent, qui contemple le vide par-dessus le toit de sa voiture pendant qu'il fait le plein, puis repart aussitôt et disparaît au loin. Pendant de longues périodes il avait pensé à très peu de choses, aidé en ce sens par l'absence de liens avec son passé. La présence de Nina changeait sensiblement la donne. Il regrettait de ne pas avoir bougé un jour plus tôt, pour qu'elle se casse les dents sur une chambre vide. Mais Zandt connaissait son obstination mieux que personne : une fois décidée à le débusquer, elle n'aurait jamais renoncé.

Il regarda le dossier plutôt épais abandonné sur le plancher. Il n'avait aucune envie de le toucher, encore moins de découvrir son contenu. Il en connaissait déjà l'essentiel. Et le reste ne serait pas bien différent. Il lui inspirerait un mélange d'hébétude et d'horreur, des lames de rasoir dissimulées dans des boules de coton.

Au bruit d'une porte qu'on referme, il leva les yeux pour voir Nina qui revenait de l'auberge. Elle tenait quelque chose dans une main. Il sortit de la voiture. La température avait sensiblement chuté. Un ciel de plomb. De la neige.

— Bon sang, fit-elle, le visage embrumé par son souffle. Tu n'avais pas tort. Nourriture réservée aux heures de repas. J'ai quand même trouvé ceci. (Elle brandit une bouteille de whisky irlandais.) J'ai dit qu'on en avait besoin comme pièce à conviction.

— Je ne bois plus vraiment, tu sais.

— Moi si. Tu n'auras qu'à me regarder.

Elle ouvrit sa portière afin de ramasser le dossier. Zandt vit qu'elle vérifiait son emplacement sur le tapis, pour voir s'il y avait jeté un coup d'œil en son absence.

— Qu'est-ce que tu fais ici, Nina ?

— Je suis venue te sauver. Fêter ton retour dans le monde.

— Et si je ne veux pas revenir ?

— Tu es déjà revenu. Tu l'ignores encore, c'est tout.

— Ça veut dire quoi, ça ?

— Écoute, John, il fait plus froid que dans le slip d'une bonne sœur, ici. Allons à l'intérieur. Je suis sûre que ton nouveau regard-qui-tue-à-mille-mètres fonctionne aussi bien entre quatre murs.

Surpris, il émit un rire rauque :

— Où sont passées tes bonnes manières ?

Elle haussa les épaules.

— Tu connais les règles du jeu. Couche avec une femme, et elle gagne le droit de te malmener jusqu'à la fin de tes jours.

— Même si c'est elle qui a pris l'initiative ? Puis tout arrêté ?

— Autant que je me souvienne, tu ne t'es pas défendu bec et ongles, ni au début ni à la fin. Laquelle de ces granges rustiques est ton domicile actuel ?

De la tête, il indiqua son bâtiment, et elle se mit en marche. Après avoir envisagé un instant de remonter en voiture pour filer, il la suivit.

CHAPITRE 4

Il prépara le feu pendant qu'elle l'observait, assise dans l'un des fauteuils râpés, les pieds sur la table basse. Il la sentait qui jaugeait la pièce dans la lumière électrique : les tapis délicieusement usés, le mobilier chic minable, les tableaux que seul un hôtelier pouvait trouver à son goût. Le plancher était peint en blanc cassé, et un bouquet de fleurs locales se dressait gaiement dans un vase à quelques centimètres des pieds de Nina.

— Alors, à quelle heure Martha Stewart doit-elle passer ?

— Sitôt que tu seras partie, répondit-il en se rendant à la salle de bains pour trouver des verres. Elle et moi, c'est comme un truc animal.

Nina sourit, et contempla le petit bois dans l'âtre. Le feu crissait et craquait, ravi d'être éveillé, prêt à se consumer. Il lui semblait qu'elle n'avait pas vu de vrai feu depuis une éternité. Cela lui rappela ses vacances d'enfant, et lui donna des frissons.

Zandt revenu, elle ouvrit la bouteille et versa deux mesures. Il resta debout quelques instants, comme s'il ne pouvait encore se résoudre à la suivre, puis finit par prendre le second fauteuil. La pièce commença lentement à se réchauffer.

Des deux mains, elle porta le verre à sa bouche, sans quitter Zandt des yeux.

— Alors, John, quoi de neuf ?

Il garda le regard rivé sur le feu.

— Viens-en au fait, répondit-il.

Trois jours plus tôt, la jeune Sarah Becker s'était trouvée assise sur un banc de Third Street Promenade à Santa Monica, en Californie. Elle écoutait un minidisc sur le lecteur qu'elle avait reçu pour ses quatorze ans. Elle s'était fabriqué une belle petite étiquette avec l'ordinateur familial, de sorte que ses nom et adresse figuraient au dos de l'appareil, recouverts de Scotch transparent pour empêcher l'encre de s'effacer. Bien qu'elle s'en fût voulu de gâter ainsi le chrome brossé du lecteur, elle aurait eu plus de remords encore de le perdre. Quand celui-ci fut retrouvé, il apparut que l'album qu'elle écoutait s'intitulait *Generation Terrorists*, d'un groupe anglais baptisé Manic Street Preachers. Sauf que, comme le savait Sarah, on disait simplement les Manics. Ils n'avaient pas trop la cote dans son école – raison de plus pour les écouter. Tous ses camarades se pâmaient devant des princesses pop aboyeuses et des boys-bands insipides, ou balançaient la tête pendant qu'un rappeur de supermarché braillait l'argot de l'année dernière sur une mélodie archiconnue depuis sa villa fortifiée de Malibu. Sarah préférait les artistes qui donnaient l'impression que, à un moment donné, ils avaient voulu exprimer quelque chose. Elle y voyait un gage de maturité. À quatorze ans, on n'était plus une gamine, loin de là. Plus aujourd'hui. Pas à L.A. Pas en 2002. Ses parents peinaient à se mettre à la page, mais même eux savaient qu'il en était ainsi. À leur manière ils se faisaient à cette idée, tels des hommes de Neandertal regardant avec circonspection les premiers Cro-Magnon descendre de la colline en sifflotant.

Là où elle était assise, près de la fontaine face au Barnes and Noble, le bout de la Promenade était assez vide en ce milieu de soirée. Quelques personnes entraient dans la librairie, et l'on distinguait les clients à travers la baie vitrée sur deux étages : feuilletant intensément livres et journaux, salivant devant les fiches techniques des ordinateurs dernier cri ou cherchant des formules magiques dans des manuels d'écriture scénaristique. L'an passé, elle avait séjourné quinze jours à Londres en famille, et elle avait été frappée par l'indigence des librairies de la capitale anglaise. Elles étaient carrément bizarres. Genre il n'y avait que des livres. Pas de café, zéro magazine, même pas de toilettes. Rien que des rangées et des rangées de bouquins. Les gens se servaient, payaient, et repartaient aussi sec. Si sa mère semblait trouver ça plutôt cool, Sarah y voyait l'un des aspects les plus craignos de l'Angleterre. Puis ils avaient déniché un grand Borders flambant neuf, et elle s'y était précipitée, pour découvrir les Manics sur l'une des bornes d'écoute. Les groupes anglais étaient cool. Les Manics étaient particulièrement cool. Londres était cool dans l'ensemble. C'est clair.

Opinant de la tête tandis que le chanteur se proclamait bruyamment « chien maudit », elle embrassait la Promenade du regard. À l'autre bout de cette zone piétonne longue de trois pâtés de maisons, s'enfilaient les restaurants. Son père l'avait déposée vingt minutes plus tôt, et passerait la reprendre à 21 heures précises, comme tous les mois. Elle devait retrouver sa copine Sian au Broadway Deli. Ces demoiselles dînaient en ville. Ce rendez-vous gastronomique était né de l'imagination de la mère de Sian, qui accompagnait l'adolescence de sa fille en ouvrant toutes les portes imaginables, de peur que le fait d'en laisser une seule fermée – et la mauvaise – ne ruine leur relation privilégiée. La mère de Sarah s'était facilement laissé convaincre par ce projet : d'une part tout le monde se laissait facilement convaincre par Monica Williams, d'autre part Zoe Becker demeurait suffisamment à l'écoute de la jeune fille qu'elle avait été pour savoir combien elle aurait aimé faire la même chose au même âge.

Le père de Sarah avait toutefois un droit de veto occasionnel, et pendant un long et pénible moment elle avait craint qu'il ne l'exerce. On avait en effet assisté quelques mois plus tôt à une recrudescence des meurtres entre gangs, liée aux mouvements de restructuration saisonniers dans l'industrie du crack. Mais au bout du compte, après avoir proposé et conclu avec sa fille un contrat déclinant un arsenal de mesures sécuritaires – la déposer et la récupérer à des lieux et heures fixes, exiger qu'elle présente une batterie de portable pleine, et lui faire réciter les principaux gestes de bon sens pour éviter l'impondérable intrusion du mauvais sort –, papa avait donné son feu vert. Ce rituel faisait désormais partie de leur calendrier mondain.

Le problème, c'était que Sian n'était pas là quand la voiture s'arrêta au carrefour habituel. Michael Becker se dévissa le cou, scrutant la rue de haut en bas.

— Où est donc passée la fameuse miss Williams ? marmonna-t-il tout en pianotant sur le volant.

Il y avait un os dans la série qu'il concevait en ce moment pour la Warner, et c'était le stress des grands jours : un calme lourd ponctué de sautes d'humeur. Sarah ne savait quel était le problème au juste, mais elle connaissait le credo paternel selon lequel, dans le Business, les choses avaient mille façons d'aller de travers, et une seule d'aller droit. Il lui avait soumis quelques idées en vue de l'épisode pilote, il avait même fait appel à ses lumières sur certains points, considérant sa réaction comme représentative d'une certaine frange de l'audience visée. À vrai dire, et non sans surprise, Sarah avait estimé que la série s'annonçait plutôt cool. Mieux que *Buffy* ou *Angel*, en tout cas. En son for intérieur, elle voyait en Buffy une petite emmerdeuse, et jugeait que le vieil Anglais qui la cornaquait tenait beaucoup moins de Hugh Grant qu'il ne semblait le croire, que ce soit dans la voix ou dans l'allure. L'héroïne de *Dark Shift* était en revanche plus indépendante, moins frimeuse, et moins prompte aux pleurnicheries. Elle était aussi, sans que Sarah en eût conscience, librement inspirée de la fille de Michael Becker.

— La voilà, avait lancé Sarah en pointant du doigt.

Son père fronça les sourcils.

— Je ne la vois pas.

— Mais si, regarde : sous le réverbère là-bas, devant Hennessy and Ingels.

À cet instant un connard derrière eux klaxonna, et son père se retourna vivement pour le foudroyer du regard. Il se mettait très rarement en rogne au sein du cercle familial, mais il lui arrivait de passer ses nerfs sur le monde extérieur. Sarah savait, pour l'avoir récemment étudié en classe, que ce comportement relevait de la hiérarchisation des espèces, avec la jungle des rues comme lieu de confrontation – mais elle redoutait qu'un de ces jours son père ne choisisse le mauvais gorille pour affirmer son ascendant. Il n'avait pas l'air de comprendre que les pères peuvent eux aussi attirer le mauvais sort et que l'âge influe peu sur la violence du châtiment.

Elle ouvrit sa portière et sauta sur le trottoir.

— Je cours la rejoindre. Ne t'inquiète pas.

Les mâchoires serrées, Michael Becker regarda la LeBaron du type pressé contourner leur voiture.

Puis il se retourna, et son visage se métamorphosa. L'espace d'un instant, on aurait oublié que derrière ses yeux défilaient en boucle des trames narratives et des taux d'audience, comme s'il voyait le monde à travers une grille de distributions et de cessions de droits étrangers. Il ressemblait juste à un type fatigué, qui aurait eu besoin de caféine chaude, et à son papa.

— À plus, dit-elle avec un clin d'œil. Fais une crise cardiaque sur le chemin du retour.

Il consulta sa montre.

— Pas le temps, désolé. Peut-être un petit ennui de prostate, alors. À 9 heures ?

— Tapantes. Je suis toujours en avance. C'est toi qui es en retard.

— Voyez-vous ça ! Touche Dubois, petite dame.

— Touche Dub', papa.

Elle claqua la portière et le regarda s'engager dans le trafic. Il la salua d'un petit signe de main, puis s'éloigna, rappelé par son monde intérieur, à la merci de gens qui lui achetaient du texte au mètre et attendaient le jour de la diffusion pour savoir ce qu'ils voulaient vraiment. Comme la voiture s'éloigna, Sarah se répéta le serment qu'elle s'était fait : jamais elle ne céderait aux sirènes du Business.

Bien entendu, elle n'avait pas aperçu Sian sous le réverbère. Elle avait juste fait semblant, pour ménager son père, lui permettre de regagner au plus vite la maison et sa table de travail. Sian continua de ne pas être là pendant dix minutes, puis le téléphone de Sarah sonna.

C'était Sian. Elle se trouvait sur Sunset, plantée à côté de la voiture maternelle, écumant de rage. Sarah distinguait en fond sonore la voix de Mme Williams qui déchargeait vertement sa bile sur un pauvre mécano, lequel avait dû, devant cette image d'une mère et de sa fille en détresse, se projeter mentalement le synopsis d'un porno en direct live. Sarah espérait qu'il avait compris que non seulement il pouvait ravaler ses fantasmes, mais qu'il était un homme mort s'il ne réparait pas cette bagnole fissa.

Quoi qu'il en soit, Sian n'allait pouvoir honorer le rendez-vous. Ce qui plongeait Sarah dans l'embarras. Son père ne serait pas encore rentré, et quand il atteindrait l'allée de la maison, son cerveau ne serait plus qu'un maelström d'impacts de balles et de pannes d'intrigue – peut-être même serait-il déjà en ligne avec son coauteur Charles Wang, à se demander par quels tours de passe-passe sortir le projet de la zone de turbulences. Un important petit déjeuner d'affaires avec le studio était prévu le lendemain matin, une réunion fatidique devant décas et omelettes sans cholestérol. Son père redoutait comme la mort ce type de rendez-vous, pour la simple raison qu'il ne mangeait jamais le matin et détestait picorer ses toasts pour éviter de tripoter ses couverts. Bref, elle ne voulait pas ajouter à son stress, sachant que sa cadette Melanie ferait assez de boucan pour lui taper sur le système.

Puis elle se dit qu'au fond rien ne l'obligeait à appeler. Elle avait un peu moins de deux heures devant elle. L'étendue de la Promenade était le paradis du lèche-vitrine, avec ses innombrables boutiques dont la plupart étaient encore ouvertes. Elle pourrait simplement traîner, munie d'un Frappuccino. Musarder chez Anthropologie, à la recherche d'idées de cadeaux. Vérifier les bornes d'écoute chez Barnes and Noble, au cas où ils se seraient décidés à proposer du neuf. Ou même déguster sa salade Cobb en solo au Deli. N'importe quoi, en fait, du moment qu'elle pourrait revenir au carrefour à l'heure dite. Puis, selon l'humeur de son père, elle révélerait que Sian ne s'était pas pointée ou prétendrait que tout s'était déroulé au poil.

Elle composa le numéro de Sian pour s'assurer que son plan ne serait pas compromis pas un coup de fil inopiné de Mme Williams à sa mère. Mais elle ne put obtenir la communication. La voiture devait être réparée et se trouver momentanément isolée au milieu d'un canyon. Cela dit, Sarah serait déjà au courant si sa mère avait été alertée. Le ciel serait investi d'hélicos, et Bruce Willis descendrait à sa rescousse au bout d'une corde.

Elle laissa un message à Sian, se mit en marche et poussa la porte du Starbucks. Il lui était venu à l'esprit qu'en se rendant au Deli toute seule, elle pourrait savourer le menu de ses envies, et déroger à cette maudite salade Cobb qu'elles commandaient invariablement, s'imposant un régime vingt ans plus tôt que de besoin. Elle pourrait carrément s'offrir un hamburger. Un bon gros burger, saignant, avec du fromage. Et des frites.

Elle se disait que c'était peut-être ça, la vie de grande, et que ça pouvait s'avérer somme toute intéressant.

Elle était parvenue au bout de son Frappuccino, et les Manics de leurs braillements, quand elle vit un grand type sortir de la librairie. Il fit quelques pas, puis s'arrêta pour examiner le ciel. La nuit n'était pas complètement tombée, mais le crépuscule déclinait. Le type mit la main dans sa poche, sortit un paquet de cigarettes et tenta d'en extraire une sans reposer ce qui était à l'évidence un gros sac de livres. La scène dura un bon moment, sans que l'homme remarque le regard amusé de Sarah. À sa place, songeait-elle, elle aurait sans doute posé le sac par terre, mais cette solution ne l'avait visiblement pas effleuré.

Au bout du compte, exaspéré, il se rendit à la fontaine pour caler son sac sur le rebord. Une fois sa cigarette allumée, il porta les mains à sa taille tout en considérant l'allée, avant de la considérer, elle.

— Salut, dit-il d'une voix douce et gaie.

Maintenant qu'il s'était approché, elle lui donnait une quarantaine d'années, voire un peu moins. Elle-même ignorait le pourquoi de cette déduction, car, dans le contre-jour du réverbère, son visage demeurait obscur. Il avait juste ce je-ne-sais-quoi du mec plus vieux.

— Répétez, pour voir ?

— Euh... Salut ?

Elle opina d'un air grave.

— Vous êtes anglais.

— Mon Dieu, ça crève les yeux, alors ?

— Vous avez un accent anglais, quoi.

— Ah, oui. Bien sûr.

Il tira de nouveau sur sa cigarette, puis avisa le banc.

— Je peux m'asseoir à côté de vous ?

Sarah haussa les épaules. Le haussement d'épaules était un bon plan. Ça ne voulait pas dire oui, ça ne voulait pas dire non. Peu importe. Le banc était assez grand pour deux, de toute façon. D'ici quelques secondes elle serait en partance pour une salade. Ou un burger. Elle hésitait encore.

L'homme s'assit. Il portait un pantalon en velours côtelé, pas particulièrement neuf, mais sa veste légère paraissait de première qualité. Il avait de grandes mains soignées. Sa blondeur naturelle avait été rehaussée, un travail de pro, et son visage était plutôt pas mal. Style prof de physique dans le coup – ou de sciences sociales. Le genre qui par principe ne couchait pas avec ses étudiantes, mais pourrait facilement s'il en avait envie.

— Alors vous êtes acteur, ou un truc de ce style ?

— Oh, non. Rien de si original. Je suis un simple touriste.

— Et vous restez combien de temps ici ?

— Deux semaines.

Il plongea la main dans sa poche et produisit un petit objet en métal brillant. Il en ouvrit la partie supérieure, révélant un petit cendrier de poche.

Sarah l'observa avec grand intérêt.

— Les Anglais fument beaucoup, n'est-ce pas ?

— Je l'admets, répondit l'homme, qui n'était pas anglais. (Il écrasa son mégot et rangea le cendrier dans sa veste.) C'est-à-dire que nous n'avons pas peur.

Ils bavardèrent quelque temps. Sarah évoqua ses souvenirs de Londres. L'homme parvint à lui donner la réplique, car deux jours plus tôt il se trouvait encore là-bas. Il se garda d'expliquer que son sac Barnes and Noble contenait des livres qu'il possédait depuis des années, et qu'il était resté assis une heure entière au rayon Économie et Politique, tournant le visage aux autres clients, à guetter par la baie vitrée l'arrivée de Sarah. Au lieu de ça, il s'enquit des curiosités de la ville, après avoir énuméré les hauts lieux de Los Angeles qu'il avait déjà visités, une sélection des pièges à touristes habituels.

Endossant avec sérieux cette lourde responsabilité, Sarah conseilla les fosses goudronneuses de La Brea, ainsi que Rodeo Drive et la tour Watts, un panel qui donnait à ses yeux un bon éclairage tant sur les origines de L.A. que sur son avenir. En outre, songea-t-elle, Rodeo Drive permettrait à ce type de troquer son velours côtelé contre une chose un peu plus *bon marché*[1] ainsi que Sian – qui avait séjourné à Antibes l'été précédent – aimait à répéter.

Puis l'homme se tut. Sarah jugea venu le moment de lécher l'enfilade de vitrines jusqu'au lieu de son dîner. Elle s'apprêtait à lui souhaiter une bonne soirée quand il pivota pour la dévisager.

— Vous êtes très jolie, dit-il.

Que ce fût vrai ou faux – ces temps-ci, Sarah était partagée sur ce point –, ce compliment sortait tout droit de la boîte estampillée « attention, barjo ! ».

— Merci, répondit-elle avec la tiédeur qui s'imposait.

Le soir parut soudain fraîchir, puis se stabilisa à mesure qu'elle prenait la situation en main.

— Eh bien, ce fut une discussion sympa, conclut-elle.

— Je suis désolé, lâcha-t-il. C'était une déclaration pour le moins incongrue, je le sais bien. Mais vous me rappelez tellement ma propre fille ! Vous avez à peu près le même âge.

— Entendu, fit Sarah. Y a pas de souci.

— Elle est restée à Blighty, poursuivit-il bille en tête. Avec sa mère. J'ai hâte de les revoir, vous pensez bien. Une grande hâte. Par tous les lords de l'île. Lady Di, reposez en paix...

Son regard obliqua pour scruter rapidement les alentours. Sarah y vit un signe dc gêne. En fait, il estimait que d'ici vingt secondes tous les paramètres seraient réunis, c'est-à-dire que toutes les lignes de vision dévieraient de ce banc. Il était très fort à ce jeu – savoir quand il serait repéré, et entrevoir les petits pas qui le ramèneraient dans l'ombre. C'était un talent inné chez lui. Il se rapprocha de quelques centimètres de la fille, qui se leva au même moment.

— Bon, dit Sarah, je dois y aller.

L'homme se mit à rire, en même temps qu'il voyait tous les feux passer au vert. Il attrapa la main de Sarah et tira avec une force inattendue. Elle glapit et retomba sur le banc, trop hébétée pour se débattre.

---

1. En français dans le texte. (*N.d.T.*)

— Lâchez-moi, ordonna-t-elle, luttant pour garder son calme.

Le sol semblait se dérober sous ses pieds – sensation fluide et vertigineuse. Comme si on la prenait en flagrant délit de triche, ou de vol.

— Une très jolie fille, vraiment. (Il lui pressa la main plus fort.) À garder précieusement.

— Par pitié, lâchez-moi.

— Ferme ta gueule, marmonna-t-il, toute trace d'accent anglais disparue. Sale petite pute...

D'un geste ramassé, il lui flanqua son poing au visage.

La tête de Sarah partit en arrière, les yeux écarquillés, hagards. Oh non... gémit une voix intérieure. Oh non...

— Vise un peu, Sarah, chuchota l'homme d'un ton fébrile. Vise un peu tous ces veinards. Tous ces gens qui ne sont pas toi.

De la tête, il indiqua la Promenade. Un pâté de maisons plus loin, la rue était bondée. Va-et-vient continu à la porte des commerces, haltes devant la carte des restaurants. Mais autour de Sarah et du type, personne.

— Autrefois il n'y avait que des buissons, ici. Tu t'en rends compte, au moins ? Une côte déchiquetée, des rochers, des coquillages. De rares traces dans le sable. Si tu te tais, tu peux entendre comment c'était avant toute cette merde.

Clignant des yeux pour réprimer ses larmes, Sarah tâcha de deviner les intentions du type. Peut-être y avait-il quelque chose à tenter, une question subsidiaire à laquelle répondre, un moyen d'arracher un sésame.

— Mais les gens ne voient rien de rien, continua-t-il. Ils ne font même pas l'effort de regarder. Tous aveugles. Délibérément aveugles. Prisonniers de la machine.

Lui attrapant les cheveux, il lui tourna la tête, de sorte qu'elle voie le Barnes and Noble. Là aussi les gens étaient nombreux. Lisaient, flânaient, bavardaient. Pourquoi auraient-ils regardé dehors ? Et quand bien même, qu'auraient-ils vu de plus que deux silhouettes sur un banc ? Rien que de très normal.

— Je devrais te régler ton compte sur-le-champ, grogna-t-il. Juste pour montrer que c'est possible. Et que tout le monde s'en moque. Quand on passe sa vie entouré d'inconnus, comment repérer ce qui cloche ? Dans cinq kilomètres carrés de maladie, qui se soucie du sort d'un pauvre virus isolé ? Personne, sauf moi.

Sarah comprit qu'il n'y aurait pas de question de repêchage, ni maintenant ni jamais, aussi s'apprêta-t-elle à crier. Mais l'homme la sentit gonfler ses poumons, et plaqua une main sur son visage.

Il saisit à deux doigts sa lèvre supérieure, par au-dessus, et tira fort. Le cri ne quitta pas la gorge de Sarah. Elle voulut se débattre mais cette main la paralysait, couplée au poids du bras appuyant sur son crâne.

— Personne ne regarde, promit le type avec le même calme haineux. J'y ai veillé. Je sais marcher sans être vu.

Des sons confus sortirent de la bouche de la jeune fille, comme si elle voulait parler. Il parut la comprendre et répondit :

— C'est faux. Ils ne sont pas en route. Ils sont à la maison. Maman est belle comme un Jackson Pollock dans sa cuisine. Papa est au jardin, avec la petite sœur. À poil tous les deux. Ils forment un tableau intéressant. Obscène, même, diraient certains.

En fait, la maman de Sarah et sa sœur Melanie étaient au même moment plantées devant une rediffusion des *Simpson*. Il s'agissait, Zoe Becker s'en souviendrait toujours, de l'épisode où George Bush s'installe à Springfield. Michael Becker pianotait furieusement dans son bureau, ayant enfin trouvé, espérait-il de tout son cœur, une trame cohérente. S'il pouvait simplement modifier les dix premières minutes, et imposer l'idée que certains personnages avaient dépassé l'adolescence, tout fonctionnerait au poil. Et sinon, rien à foutre, il les ferait tous ados – et rétablirait tous ces panoramiques à la con sur le fronton du lycée, comme le réclamait Wang. À quelques kilomètres de là, Sian Williams venait d'écouter le message de Sarah, et jalousait un peu l'aventure solitaire de sa copine.

— Si tu continues à gesticuler, je t'arrache les dents, menaça le type. Je n'hésiterai pas, crois-moi. Ce n'est pas très facile, mais ça vaut le détour. Le bruit est très insolite.

Sarah se figea, et tous deux restèrent immobiles un instant. L'homme semblait prendre plaisir à cette position, assis derrière la fille, lui infligeant à la bouche une douleur indicible, comme s'ils partageaient un moment d'intimité au milieu d'une rue peuplée.

Puis il soupira, comme un lecteur captivé contraint de reposer son magazine. Il se leva, entraînant Sarah avec lui. Le baladeur se fracassa au sol. Le type y jeta un œil, sans y toucher.

— Bonsoir et bonne nuit, braves gens, susurra-t-il en regardant au fond de la rue. Vous allez tous pourrir en enfer, et j'adorerais vous y conduire moi-même. (Ramenant son coude dans la nuque de Sarah, il lui couvrit la bouche d'une main, et de l'autre ramassa son sac de livres.) Mais j'ai un rancard, et on doit y aller.

Là-dessus, par de longues foulées rapides, il emmena Sarah de l'autre côté de la rue, jusqu'à l'allée où il était garé. Elle n'avait d'autre choix que de le suivre. Il était grand et costaud.

Il ouvrit la portière arrière, puis lui agrippa de nouveau les cheveux pour lorgner son visage. L'horrible gros plan sur la figure du type paralysa l'esprit de Sarah.

— Viens, mon cœur, notre carrosse nous attend.

Et de lui flanquer un violent coup de tête, juste au-dessus des yeux.

Lorsque ses genoux cédèrent, l'adolescente eut une dernière pensée des plus terre à terre. Dans sa table de chevet se trouvait un cahier où elle avait consigné ses réflexions. Les plus récentes traitaient entre autres de sexe : d'exaltantes rêveries sur un domaine de la vie qu'elle n'avait pas encore expérimenté, mais qu'elle savait plus ou moins imminent. La plupart étaient le fruit des confidences de Sian, mais aussi de sa propre imagination, comme de divers éléments glanés à la télé, au ciné, et dans une revue assez soft dénichée sous la Jetée.

Le cahier était planqué, mais grossièrement. À sa mort, son père et sa mère mettraient vite la main dessus, et sauraient alors qu'elle avait tout fait pour s'attirer de tels ennuis.

Nina ne connaissait pas tous ces détails, mais elle put décrire à Zandt les faits dans leurs grandes lignes. Quand elle eut terminé, elle remplit son verre. Zandt n'avait pas touché au sien.

— Quatre témoins situent Sarah Becker sur ce banc entre 19 h 12 et 19 h 31. Leur description de l'homme à ses côtés va de « ordinaire, plutôt grand » à « putain, aucune idée », en passant par « eh bien, c'était, disons, un mec. » Nous ne disposons même pas d'un âge ou d'une couleur de peau que je puisse entrer dans la banque de données, même si « blanc » et « blond » reviennent par deux fois dans les déclarations. Deux témoins se souviennent d'un long manteau, un autre d'un blouson. Personne ne les a vus s'en aller, alors que le banc se trouvait sous le nez de millions et de millions de personnes. Si le type est passé par la librairie avant de l'aborder, personne ne l'a remarqué. Un autre témoin dit avoir aperçu une voiture de couleur et de marque indéterminées dans la première rue latérale. Possible qu'on ait déplacé une poubelle pour cacher la plaque d'immatriculation – ce qui est plutôt astucieux, mais dénote un aplomb digne de Zeus. N'importe qui aurait pu bouger la poubelle, sans compter que c'était un stationnement interdit. Vers 20 h 15, le véhicule avait disparu.

« Le père de l'adolescente est arrivé par le sud de la Promenade à 21 h 07. Il s'est garé à l'endroit habituel, a patienté. Puis, voyant au bout de plusieurs minutes que ni sa fille ni Sian Williams n'arrivaient, il est entré dans le restaurant. Les serveurs avaient bel et bien enregistré une réservation bidon au nom de Williams mais n'avaient vu aucune des deux filles. Là-dessus il a appelé la mère de l'autre gamine, pour apprendre que le dîner avait été annulé au dernier moment à cause d'une panne de voiture. Nos experts ont examiné celle-ci, et ne peuvent affirmer qu'elle n'a pas été sabotée.

« Michael Becker a tenu à interroger personnellement la copine. Elle a fini par avouer que Sarah avait laissé un message disant que, pour éviter de déranger son papa, elle allait tuer le temps jusqu'à son retour. M. Becker a fouillé la rue de long en large, sans trouver la moindre trace de sa fille. Puis il s'est rendu tout au bout de la Promenade, et après avoir inspecté le Barnes and Noble, il a repéré un lecteur de minidisc Sony à moitié caché sous le banc. L'identité de sa propriétaire était doublement attestée : par l'étiquette que Sarah y avait collée, et du fait qu'il avait lui-même acheté l'appareil. Il renfermait un CD de son groupe préféré, dont elle avait le poster dans sa chambre. Alors Becker a contacté le bureau du shérif, le département de police de Los Angeles, puis la nana qui lui sert d'agent – va comprendre. Il a dû se dire qu'elle aurait plus d'emprise sur les flics que lui. Enfin il a appelé sa femme, pour lui dire ne pas bouger au cas où leur fille arriverait en taxi.

« Tout le secteur a été passé au crible. Rien. Le lecteur ne comporte pas d'autres empreintes que celles de Sarah. Il y avait bien une centaine de mégots autour du banc, mais nous ne savons même pas si le coupable fumait. L'un des témoins a toutefois déclaré que ce n'était pas exclu, et c'est pourquoi, à l'heure où je te parle, un pauvre gus procède au fond de son labo à une recherche ADN sur tout un sac de clopes.

— Le père n'est pas suspect ?

— En aucun cas. Ils étaient très proches, mais dans le bon sens du terme. Certes, pendant un ou deux jours, les gens se sont posé la question. Mais non. On ne pense pas que ce soit lui, et puis ça ne collerait pas du tout avec son emploi du temps. On a aussi écarté son associé Charles Wang. Il se trouvait à New York.

Zandt leva lentement son verre, le vida, le rabaissa. L'histoire ne s'arrêtait pas là.

— Et ensuite ?

Nina ôta ses pieds de la table, se pencha pour saisir le dossier posé par terre. Outre une liasse de documents photocopiés, il contenait un mince paquet enveloppé de kraft. Nina n'en retira qu'une photographie.

— C'est arrivé au domicile des Becker le lendemain en fin d'après-midi. Entre 16 h 30 et 18 heures. Ils l'ont découvert posé dans l'allée.

Elle tendit le cliché à Zandt. On y voyait un pull couleur lilas, soigneusement plié en carré. Il était entouré d'une sorte de ruban strié attaché par un nœud à boucles.

— Il a été ficelé avec des cheveux tressés. Ceux de Sarah étaient longs, et la couleur correspond. Les techniciens ont prélevé des échantillons sur sa brosse et nous donneront très vite la confirmation.

Zandt vit que son verre s'était rempli tout seul. Il but. Le whisky piquait sa gorge sèche et lui donnait la nausée. Il avait l'impression d'avoir un ballon à la place de la tête, légèrement surgonflé, flottant à cinq centimètres de son cou.

— L'Homme Debout, dit-il.

— Eh bien, nous avons interrogé les familles des victimes d'il y a deux et trois ans, ainsi que chaque agent ayant participé aux enquêtes, et nous sommes quasi persuadés que la nature des colis reçus n'a jamais été divulguée. Peut-être avons-nous tout de même affaire à un imitateur. Mais j'en doute. J'ai fait lancer une veille multimédia, Internet compris, pour toute occurrence des expressions « Garçon de Courses » et « Homme Debout ».

— Internet ?

— Ouais, une sorte de gadget informatique. Ça fait fureur ces temps-ci.

— C'est bien lui, affirma Zandt.

Lui seul mesurait toute l'ironie qu'impliquait son assurance.

Elle l'observa, puis rouvrit le dossier à contrecœur. Une seconde photo montrait le pull déplié et posé à plat. Le prénom de Sarah était brodé sur le devant en grandes lettres majuscules.

— Les cheveux qui forment ce nom sont très bruns. Et bien plus secs que ne devaient l'être ceux de Sarah, ce qui laisse à penser qu'ils ont été coupés depuis un moment déjà.

Elle se tut et regarda Zandt plonger lentement la main dans sa poche. Il sortit un paquet de Marlboro et une boîte d'allumettes. Il n'avait pas encore fumé depuis qu'ils s'étaient installés dans cette pièce dépourvue de cendrier. La main qui attrapa une cigarette était assez stable. John ne regarda pas Nina, mais seulement l'allumette qu'il craqua d'un air concentré et résolu, comme si ce geste lui était nouveau mais qu'il en saisissait instinctivement la finalité. Si l'allumette ne s'embrasa qu'à la troisième tentative, c'est peut-être qu'elle était humide.

— J'ai obtenu que les cheveux foncés soient analysés en premier, poursuivit Nina avant de prendre une longue inspiration. Ça colle, John. Il s'agit bien des cheveux de Karen.

Elle le laissa seul quelques instants, se retirant dans le froid pour écouter le silence de la nuit. Des rires assourdis s'échappaient du bâtiment principal, dont les vitres révélaient des couples d'âges divers, emmitouflés dans de gros chandails, planifiant leurs randonnées du lendemain. Derrière une porte béante de l'autre côté du bâtiment, quelqu'un lavait des assiettes qui ne lui appartenaient pas. Survint un léger bruissement vers les sous-bois, derrière la route, mais il demeura sans suite.

À son retour, elle trouva Zandt dans la position où elle l'avait laissé, tétant une nouvelle cigarette. Il ne releva pas les yeux.

Elle remit du petit-bois dans la cheminée, un peu au hasard, incapable de se rappeler s'il fallait l'empiler au milieu ou le disposer sur les côtés. Elle s'assit dans le fauteuil, se versa un autre verre. Et passa la nuit assise près de lui.

CHAPITRE 5

Dans l'après-midi je m'étais entretenu avec la police et les médecins de l'hôpital, et avant cela avec les voisins des deux côtés. J'avais prêté une oreille très attentive à ce qu'on me disait.

J'avais appelé les flics depuis la maison et on m'avait mis en contact avec un certain Spurling – par chance, il n'était pas de ceux qui m'avaient interrogé après l'esclandre au bar de l'hôtel. Le soir de l'accident, Spurling et son coéquipier McGregor étaient arrivés les premiers sur place, alertés par l'appel d'un automobiliste qui passait par là. Ils avaient attendu les secours, puis aidé à désincarcérer les corps. Ils avaient suivi l'ambulance sur le chemin de l'hôpital, et c'est sous l'œil de Spurling qu'on avait prononcé le décès de Donald et Elizabeth Hopkins. On avait identifié les victimes grâce à leurs permis de conduire, et aux confirmations de Harold Davids (avocat) et de Mary Richards (voisine), recueillies dans les deux heures suivantes.

L'officier Spurling comprenait mon désir d'éclaircir les circonstances entourant le décès de mes parents. Il me communiqua le nom du médecin qui les avait examinés, et me demanda si j'avais songé à la psychothérapie. Je supposai qu'il me conseillait là un traitement, non une reconversion. Je le remerciai de m'avoir accordé de son temps, et il me souhaita les meilleures choses. Je raccrochai en priant pour ne pas le croiser en allant récupérer mon arme au commissariat, même s'il y avait fort à parier qu'il était déjà au courant. Son allusion à la psychothérapie n'était pas vide de sous-entendus.

Intercepter la toubib fut une autre paire de manches. Elle n'était pas de service quand j'atteignis l'hôpital, et vu le temps qu'il me fallut pour obtenir ce simple renseignement, après avoir subi un bataillon d'infirmières à cran et de standardistes blasées, je compris que j'allais devoir me déplacer. Les urgences sont conçues pour les vivants. Une fois mort, vous n'êtes plus qu'un souvenir incommodant, et n'êtes plus de leur ressort.

Je me rendis là-bas et patientai pendant une heure des plus silencieuses. Puis le Dr Michaels daigna enfin quitter son bunker pour me parler. Proche de la trentaine, elle cultivait un air surmené et une intense autosatisfaction. Après m'avoir sévèrement pris de haut, elle me répéta ce que je savais déjà. Traumatismes majeurs à la tête et dans le haut du corps. Morts de chez morts. Et à présent, si c'était tout, voulais-je bien l'excuser. C'était une grande fille maintenant, et elle avait des patients à examiner. Je fus plus que ravi de renoncer à sa compagnie, voire tenté de faciliter son départ d'un bon coup de pied où je pense.

Je ressortis de l'hôpital. Le jour s'était fondu dans un soir précoce d'automne. Quelques voitures étaient dispersées au hasard du parking, monochromes et anonymes sous la lumière des lampadaires. Une jeune femme fumait et pleurait en silence un peu plus loin.

Je me demandais que faire ensuite. Après avoir trouvé le message, j'étais resté assis sur la table basse un bon bout de temps. Ni le vertige, ni la sensation grouillante dans mon estomac n'avaient voulu cesser. Après un examen minutieux, le reste du livre s'était révélé vide. Il ne faisait aucun doute que ce billet portait l'écriture de mon père.

« Ward, était-il noté dans une écriture à nul égard différente de ce que j'aurais pu attendre, ni trop grande ni trop resserrée, ni forcée ni brouillonne, nous ne sommes pas morts. » Mon père avait inscrit cela sur un morceau de papier, l'avait glissé dans un livre, puis il avait planqué ce dernier dans son vieux fauteuil en prenant soin de replacer la ganse recouvrant la jointure. Mes vieux avaient laissé un message niant leur mort dans une cachette qui ne puisse être découverte qu'après leur disparition. Car en quelle autre occasion me serais-je trouvé seul dans cette maison ? Et pourquoi diable aurais-je choisi ce fauteuil ? À en juger par l'emplacement du billet, celui qui l'avait dissimulé avait prévu que j'opterais pour ce siège – quand bien même je le savais le moins confortable de la pièce. Eh bien, il ne s'était pas trompé. Je m'étais carré là-dedans, et pour un sacré bout de temps. Il leur avait paru logique que j'agisse ainsi à leur mort, tout du moins que j'y songe, ne serait-ce qu'en passant la main sur le tissu. C'était tout à fait le genre de chose qu'un fils en deuil était susceptible de faire.

Mais, et voilà qui me turlupinait, cela signifiait qu'avant de mourir l'un ou l'autre, ou les deux, avaient pris le temps d'imaginer ce qu'il adviendrait après leur décès. Ils avaient considéré la situation en détail, parié sur mes faits et gestes. Mais pourquoi ces pensées morbides, déjà ? C'était incompréhensible. C'était absurde.

À supposer qu'ils soient bien morts...

L'idée que ces derniers jours n'aient été qu'une sombre farce, que mes parents ne soient pas morts en fin de compte, était dure à avaler. Une partie de moi frémissait à cette perspective, celle-là même qui me réveillait toutes les nuits depuis le coup de fil de Mary. Même si je m'étais peu soucié d'eux, et regrettais surtout de ne pouvoir les incendier au sujet d'UnRealty, je voulais retrouver mes parents. Mais quand la chair est endommagée, l'organisme se met en branle dans les secondes qui suivent. Des globules blancs déferlent sur la zone accidentée, réparent, colmatent, empilent tous les sacs de sable à leur disposition. Le corps se protège, et il en va de même pour l'esprit. Mollement et imparfaitement, tel un travail bâclé par des ouvriers négligents, mais en quelques minutes une concrétion de mécanismes défensifs se déploie autour de la blessure, émoussant ses bords jusqu'à la sceller dans une cicatrice. Telle une écharde logée au fond d'une entaille, l'événement ne disparaîtra jamais, et au gré de certains mouvements il pressera une terminaison nerveuse et fera un mal de chien. Mais quelle que soit la douleur occasionnée, pour rien au monde on ne voudrait prendre un couteau pour rouvrir la plaie.

Je quittai la maison, la verrouillant avec soin, et allai sonner à côté, chez Mary. Avec une joie mêlée de surprise, elle me reçut dans une débauche de café et de gâteau. Abusant de sa gentillesse, je parvins par des moyens détournés à lui faire dire que mes parents étaient tout à fait eux-mêmes dans les jours et les semaines précédant l'accident, et qu'elle avait identifié les corps – comme allait ensuite me le confier l'officier Spurling. Je savais déjà tout ça. Elle me l'avait expliqué au téléphone, alors que je bullais amorphe à Santa Barbara. J'avais juste besoin de l'entendre à nouveau. Certes, rien ne m'aurait empêché d'aller me recueillir devant leur dépouille aux pompes funèbres, au lieu de croupir à l'hôtel pendant deux jours. Mais je ne l'avais pas fait, et maintenant j'avais honte. Je m'étais dit sur le moment qu'il valait mieux rester sur l'image de ce qu'ils avaient été, plutôt que sur deux bandes de chair broyée. Ce n'était pas faux. Mais j'avais surtout cédé à la peur, à la gêne, au refus.

Après avoir pris congé de Mary, je me rendis chez les autres voisins. Une jeune femme ouvrit presque instantanément. L'air sûr d'elle, pleine de vitalité, les vêtements couverts de peinture. Le couloir derrière elle était à moitié repeint dans un ton que j'estimais discutable. Je déclinai mon identité, avant d'exposer ce qui était arrivé à ses voisins. Elle était déjà au courant, comme je m'y attendais. Elle me présenta ses condoléances, et nous discutâmes un peu. Rien dans son attitude ne parut suggérer que l'accident n'était pour elle qu'une demi-surprise, ni que l'un ou l'autre des Hopkins ne tournait plus rond. Soit.

J'appelai les flics puis me rendis à l'hôpital. Comme je restais planté sur le parking après avoir parlé au médecin, je décidai que trois confirmations suffisaient. Mes parents étaient bel et bien morts. À ce stade, seul un fou aurait soutenu le contraire. Je pouvais toujours interroger Davids demain si l'envie m'en prenait – l'ayant raté à son cabinet, j'avais laissé un message –, mais je savais que tout ce qu'il dirait aboutirait à la même conclusion. Le petit mot n'était pas ce qu'il prétendait être. Ce n'était pas une carte « quittez la case chagrin ». Il ne pouvait défaire ce qui était.

Pourtant il y avait forcément une raison à son existence, fût-ce que l'un des deux n'avait plus toute sa tête. Cette missive avait une signification, et je voulais savoir laquelle.

Je fouillai le garage, puis l'atelier de mon père au sous-sol. Je pressentais qu'il fallait chercher une chose bien particulière, mais j'ignorais laquelle. Alors je me contentai de fouiner un peu partout. Perceuses, étaux, divers coffrets de bricolage à la finalité obscure. Clous et vis de toutes tailles soigneusement ordonnés. Moult chutes de bois, que la mort rendait inutiles et inexplicables. Aucun désordre apparent, mais toute la rigueur et la netteté que j'avais escomptées. Si l'ordre extérieur reflète l'état d'esprit, mon père n'avait pas varié d'un iota.

Je remontai dans la maison et m'attaquai d'abord au rez-de-chaussée. La cuisine et la buanderie, le salon, le bureau de mon père, la salle à manger, et la portion de porche convertie en véranda. Ici je m'appliquai davantage. Je regardai sous chaque coussin, sous les tapis, derrière chaque meuble. Dans le placard hi-fi et sous le téléviseur, où je ne trouvai rien de plus que de la technologie et une paire de DVD. Je vidai les placards de la cuisine, passai la tête dans le four et le cellier. Je secouai tous les livres que je trouvai, qu'ils fussent rangés dans la bibliothèque du couloir ou bien, méthode originale de ma mère, au milieu des paquets de pâtes. Il y avait beaucoup de livres. J'y consacrai un temps fou. Surtout dans le bureau, à l'entresol, que j'inspectai en tout premier. Je brassai le contenu des tiroirs, des étagères, et feuilletai les dossiers suspendus du classeur en chêne. J'allai jusqu'à allumer l'ordinateur pour survoler quelques fichiers, bien que cette intrusion me parût déplacée. Je n'aimerais pas que mes proches viennent passer en revue le contenu de mon portable. Les pauvres, ils auraient envie de me déterrer pour me faire rôtir. Je compris vite que je n'aurais jamais le temps de lire tout ce que renfermait la machine, qui a priori ne consistait qu'en une somme de factures et de courriers anodins. Je laissai l'ordinateur en marche, dans l'idée d'y revenir si je faisais chou blanc ailleurs, mais en tout état de cause mon père n'était pas un mordu de la bécane, et je doutais qu'il eût consigné un autre message sur un support si peu tangible.

Je me sentis rapidement lessivé. Non par l'effort physique, négligeable, mais par la charge émotionnelle. Décortiquer la vie de mes parents les ramenait de manière brutale, notamment à travers les babioles les plus triviales. La photocopie sous verre du tout premier acte de vente conclu par UnRealty, frappée d'un logo qui, je m'en apercevais seulement, semblait tracé à la main. Par ma mère, sans doute. Un cahier garni de recettes familiales, dont les fameuses lasagnes que je pouvais humer à la seule lecture des ingrédients.

Je fis une pause et passai les quinze minutes suivantes assis dans la cuisine, à boire leur eau minérale. J'essayai une nouvelle fois de me mettre à leur place, d'imaginer une deuxième étape évidente. En supposant qu'ils aient laissé ce mot dans le fauteuil pour attirer mon attention, il semblait logique que le message ou indice suivant fasse lui aussi résonance. Mais pas moyen de mettre le doigt dessus. J'avais remué toute la maison. Il n'y avait rien.

L'étage se révéla tout aussi décevant. Je regardai sous leur lit, fouillai tous les tiroirs. Prenant mon courage à deux mains, je parcourus leur garde-robe, en m'arrêtant sur les articles que je reconnaissais – les vieilles vestes de mon père, les sacs racornis de ma mère. Je fis quelques découvertes : des reçus, des talons de tickets, une poignée de petite monnaie – rien qui me semblât significatif. Je m'attardai devant une collection d'anciennes cravates, soigneusement rangées dans une boîte au fond de la penderie, de son côté à lui. Pour la plupart, je ne les avais jamais vues.

Je jetai même un coup d'œil sous le toit, en me hissant à travers la trappe étroite du couloir. Mon père y avait suspendu une ampoule, mais rien de plus. Les combles ne renfermaient rien d'autre que de la poussière et deux valises vides.

Je finis par regagner le rez-de-chaussée, et le fauteuil de mon père. L'après-midi touchait à sa fin. Je n'avais rien trouvé, et je me sentais idiot. Peut-être m'échinais-je simplement à arracher au chaos un ordre chimérique. Assis dans le fauteuil, je repris leur billet. Il n'avait pas plus ni moins de sens qu'auparavant, même à la centième lecture.

Mes yeux se posèrent une fois de plus sur le téléviseur. Il se trouvait dans le prolongement exact du fauteuil, ce qui me donna une idée. Si j'avais vu juste en estimant que le siège n'était pas tout à fait en place, peut-être sa position n'était-elle pas uniquement destinée à m'attirer vers le coussin, mais aussi à aiguiller mon attention vers une tout autre direction.

Je me relevai pour ouvrir les portes vitrées du meuble télé. J'y découvris la même chose que précédemment : un magnétoscope, un lecteur DVD, et deux vieux films sur DVD. Rien d'autre.

Aucune cassette. Bizarre.

Je n'avais pas vu une seule vidéo dans cette maison. Deux étagères de DVD dans le bureau, et une troisième dans la chambre d'amis. Mais pas la moindre bande.

Mon père était un téléspectateur quasi professionnel. Aussi loin que je me souvienne, la maison avait toujours grouillé de cassettes. Où étaient-elles donc passées ?

Je rebroussai chemin jusqu'au bureau. Pas davantage de cassettes ici, malgré la présence d'un second magnéto sur une étagère basse. Je renonçai à reparcourir les tiroirs et les placards. Je savais qu'ils n'en contenaient aucune. Pas plus que la maison, l'atelier ou le garage. Je tâchai de me remémorer l'avant-dernier Thanksgiving, où j'avais daigné leur tenir compagnie pendant vingt-quatre heures. Là non plus, je ne me souvenais pas d'avoir remarqué des cassettes. Ni leur absence, du reste. J'avais passé le plus clair du temps à picoler.

Mon père aurait-il donc vu dans le DVD l'aube tant attendue d'un nouvel âge du divertissement familial, déclaré la cassette vidéo morte, et érigé un bûcher dans le jardin ? Je n'y croyais guère. Dyersburg possédait sûrement une décharge quelque part, mais je n'adhérais pas davantage à ce scénario. Même si au fil des ans une large partie de sa collection avait cessé de l'intéresser, il n'aurait pas jeté tous ses vieux classiques comme ça. Je commençais à me demander si créer l'absence d'un objet banal n'était pas un moyen subtil d'attirer l'attention d'un proche, d'une personne ayant la notion des choses censées peupler votre environnement.

Soit ça, soit je perdais mon objectivité, m'emballais sur la mauvaise pente. J'avais retourné toute la maison. À quoi bon désormais savoir ce que je cherchais vraiment ? Ce n'était pas là. La faim commençait à poindre, et la colère. Si mon père avait quelque chose à me dire, pourquoi ce subterfuge ? Pourquoi ne pas m'avoir passé un coup de fil ? Laissé une lettre à Davids ? Envoyé un e-mail ? C'était à n'y rien comprendre.

Je savais cependant que lorsque je quitterais cette maison, ce serait pour de bon. Alors il valait mieux en avoir le cœur net. Autant garder une cicatrice aussi résistante que possible.

J'allumai l'éclairage extérieur pour faire un tour sur le porche. Aucune latte n'était décollée, et je ne voyais pas comment on aurait pu se faufiler là-dessous. Une grande caisse de bois reposait sur le côté, mais au bout de deux pénibles minutes il s'avéra qu'elle n'abritait que des bûches et des araignées. Je descendis les quelques marches menant au jardin, m'éloignai sur la pelouse à reculons et considérai la maison, exaspéré.

Cheminée, planches horizontales, fenêtres. Les pièces du haut. Leur chambre. La chambre d'amis.

Je regagnai l'intérieur. Comme je dépassais le bureau de mon père, un détail me frappa le coin de l'œil. Je m'arrêtai, revins sur mes pas et scrutai la pièce, sans trop savoir ce que j'avais vu. La réponse me vint au bout d'une seconde ou deux : le magnéto.

Comme un imbécile, je n'avais pas regardé à l'intérieur des deux engins. J'inspectai d'abord celui du salon. Il était vide. Puis je regagnai le bureau et me penchai sur le second, à la recherche de la touche d'éjection. En la pressant, j'obtins un ronronnement contrarié, qui demeura sans effet. Je découvris alors que la fente était obturée par une bande d'adhésif noir. Pour dissuader les gens d'y introduire une cassette, ou empêcher mon père de le faire par inadvertance ? Peu probable : si la machine était foutue, il en aurait simplement racheté une neuve.

Je tentai d'arracher l'adhésif, mais il aurait pu souder deux planètes tant il était costaud. Je sortis mon canif de ma veste. La première lame, longue et pointue, sert à couper. L'autre est un tournevis. On ne soupçonne pas comme il est fréquent d'avoir besoin de l'un juste après l'autre. Je dépliai la lame coupante et tranchai la bande noire en son milieu.

Le compartiment n'était pas vide. J'élaguai le reste de l'obstacle jusqu'à ce que l'éjection s'opère. L'appareil rugit et se vida.

Il recracha une cassette VHS standard. Je la saisis et l'examinai un long moment.

Comme je me redressais lentement, mon père appela d'en bas :

— Ward ? Ward ?

Après un instant de violente hébétude, mon corps voulut filer vers un lieu sûr dont il connaissait visiblement l'existence. Y disparaître sur-le-champ. Ailleurs. J'ignorais où. Dans l'Alabama, peut-être. Par acquit de conscience, j'essayai toutes les directions à la fois.

Je bondis en arrière, lâchant la cassette et manquant m'étaler de tout mon long sur la moquette. Je ramassai l'objet et le fourrai dans ma poche, à peine conscient de mes gestes, mais me sentant piégé, coupable et menacé. Des pas gravirent les dernières marches de l'escalier, marquèrent une pause, puis s'approchèrent de la porte du bureau. Je ne voulais pas regarder.

Il ne s'agissait pas de mon père, bien sûr. Juste une voix d'un registre similaire, surgie de nulle part dans une maison silencieuse. L'homme que j'avisai sur le palier était Harold Davids. L'air vieux, tendu et de mauvaise humeur.

— Seigneur, vous m'avez fichu une de ces peurs ! dit-il.

— Et vous, donc ! expirai-je en toussant presque.

Davids posa les yeux sur mes mains, et je vis que je n'avais pas rangé mon canif. Je le repliai et tentai de le rentrer dans ma poche, avant que mes doigts ne butent sur la cassette.

— Qu'est-ce qui vous amène ? demandai-je en m'efforçant d'être aimable.

— J'ai eu votre message, répondit-il en relevant lentement les yeux. J'ai appelé l'hôtel. Vous n'étiez pas dans votre chambre, alors j'ai pensé vous trouver ici.

— Je n'ai pas entendu sonner.

— La porte d'entrée était entrouverte, expliqua-t-il avec un brin d'irritation. J'ai craint qu'un rôdeur ne se soit introduit après avoir appris que la maison était vide.

— Ce n'était que moi.

— Je vois ça. L'incident est donc clos.

Il leva un sourcil bonhomme, et mon cœur reprit peu à peu son rythme normal.

De retour dans le couloir, il me demanda l'objet de mon appel. Une broutille, répondis-je, un vague problème de jargon juridique que j'avais fini par élucider seul. Il hocha la tête avec tiédeur avant de flâner jusqu'au salon.

— Quelle jolie pièce ! dit-il au bout d'un moment. Elle me manquera. Je passerai de temps à autre, si vous le voulez bien, pour vider la boîte aux lettres.

— Parfait.

Je n'avais rien contre lui, mais je ne souhaitais pas rester plus longtemps dans cette maison. Je remontai au bureau pour éteindre l'ordinateur. Ayant remarqué plus tôt que mon père possédait un lecteur ffiz !, je décidai sur un coup de tête de faire une copie du disque dur.

Lorsque j'eus tout éteint, je regagnai le rez-de-chaussée et retrouvai Davids dans l'entrée, l'air revigoré.

Nous descendîmes l'allée ensemble. Il ne paraissait guère pressé de vaquer à ses affaires et il s'enquit de mes intentions pour la maison. J'avouai ne pas savoir encore si j'allais la garder ou la vendre, et acceptai son offre implicite de services, valable dans un cas comme dans l'autre. Nous nous attardâmes cinq minutes devant sa grosse berline noire, à parler de choses diverses. Peut-être me recommanda-t-il quelques adresses de restaurants. Je n'avais plus tellement faim.

Pour finir, il se cala dans son siège et attacha sa ceinture avec tout le sérieux d'un homme qui n'a pas l'intention de mourir, jamais. Il eut un dernier regard pour la silhouette sombre de la maison, puis me fixa en opinant gravement. Flairant quelque chose de changé entre nous, je me demandai si Davids s'était mis en tête d'établir ce que le fils de Don Hopkins fabriquait là-haut avec un couteau si peu décoratif.

J'attendis sagement qu'il ait passé le virage, puis courus à ma voiture et partis en sens inverse.

CHAPITRE 6

Un peu d'argent et un doigt de flagornerie me valurent la livraison d'un magnétoscope dans ma chambre. Soit l'hôtel était d'un meilleur standing que je ne l'avais cru, soit mes acrobaties au bar avaient convaincu la direction que je faisais partie de ces clients dont les besoins méritent d'être satisfaits. Je regardai avec une impatience grandissante un jeune crétin se mélanger les pinceaux devant un problème de connexion enfantin, avant de le flanquer dehors.

Je sortis la cassette de ma poche et l'examinai. Elle ne portait aucune indication. À en juger par l'épaisseur de la bobine, elle devait durer entre quinze et vingt minutes, maximum une demi-heure.

J'attendis le café que j'avais commandé. Il me fallait un environnement optimal. La tasse finit par arriver, encore assez chaude. Par miracle, elle n'était pas accompagnée de frites.

J'enfonçai la cassette dans la machine.

Quatre secondes de purée cathodique, le son blanc de l'information zéro.

Puis un bruit de vent, et une vue sur un alpage. Au loin, un paysage de carte postale montrant une chaîne de sommets blancs – trop furtive pour être identifiée. Au premier plan s'étendait une douce montée enneigée, interrompue par un immeuble austère : pas le moindre salon de thé, ni boutique de vêtements de ski. Personne dans les parages, aucune voiture sur le petit parking. Hors saison. La caméra tourna pour montrer une deuxième structure d'aspect administratif, sous de menaçants nuages gris. Le plan dura quelques secondes, avec en fond sonore le bruit d'une manche claquant au vent.

Enchaîné sur un intérieur. La caméra était basse, comme si elle était cachée, et la scène ne durait à nouveau que quelques secondes. Je revins en arrière, pour figer le plan le plus net.

Ce n'était pas le meilleur magnétoscope au monde, et l'image sautillait un peu, mais je parvins à discerner un espace collectif dans ce qui ressemblait à un chalet de ski, avec un plafond mansardé. Un long comptoir courait sur le côté, sûrement la réception, mais il était désert. Derrière, un grand tableau ornait le mur, dans le style habituel : abstrait sans peine, signé de quelque fumiste aussi sous-doué que surpayé. Je distinguai la partie gauche d'une imposante cheminée en gros galets. En bas, un feu d'agrément dégageait une ambiance policée. Des fauteuils en cuir noisette étaient soigneusement disposés autour de tables basses, chacune arborant une sculpture en bois vernissée célébrant la glorieuse et regrettée vie sauvage du vieil Ouest : un aigle, un ours, un Amérindien – que des espèces disparues, en somme.

Je laissai repartir la cassette et vis à la deuxième lecture que quelqu'un allait pénétrer dans le champ au moment de la coupure : on décelait une ombre sur le mur d'un corridor desservant la partie supérieure de la pièce, et des bruits de pas sur la pierre.

Puis un extérieur final, de retour sur le parking. Un certain temps avait dû s'écouler depuis la première prise – à supposer qu'elle datât du même jour. Le vent avait molli et le ciel était d'un bleu clair et perçant. Un plan moyen sur l'immeuble austère, a priori celui que nous venions de quitter. Plusieurs personnes se tenaient devant, dans la neige. Peut-être sept ou huit, difficile à dire, car elles étaient toutes vêtues de couleurs sombres et agglutinées les unes aux autres, comme en pleine discussion. Aucun visage n'était visible, et je n'entendais que le vent – sauf à la toute fin, quand le cameraman prononça quelques mots, une courte phrase. Je l'écoutai trois fois de suite. Elle demeurait inintelligible.

Enfin, comme l'un des protagonistes semblait sur le point de se retourner vers la caméra, l'image revint au son blanc.

J'appuyai sur « pause » et fixai l'écran tremblotant. Que penser de ce que j'avais vu ? Je ne m'étais pas attendu à ça. D'après la qualité de l'image, on aurait dit que la séquence avait été tournée au moyen d'un Caméscope numérique. Or je n'avais rien déniché de tel dans la maison. La vidéo avait pu être réalisée n'importe où entre les Rocheuses moyennes et septentrionales : Idaho, Utah, Colorado... Mais la logique désignait plutôt le Montana, et dans un secteur proche. Je connaissais le genre de lieux que montrait ce film. De grands complexes pour nantis, les plus beaux endroits du pays façonnés en résidences privées afin que les riches puissent dévaler les montagnes sans craindre de percuter un quidam aux revenus moyens. Certains étaient clôturés, mais la plupart n'avaient pas besoin de ça. En y posant un pied, vous saviez tout de suite si vous étiez le bienvenu ou non. Celui qui nourrissait des projets de cambriolage rebroussait chemin illico, ébranlé dans sa chair.

Mes parents connaissaient sûrement des gens qui s'étaient offert un chalet au pied des pistes. Peut-être même était-ce mon père qui le leur avait vendu. Et alors ?

Je remis la cassette en route.

Du vrai bruit. De la musique, des cris, des conversations bruyantes. Un visage flou, en très gros plan, qui riait aux éclats, puis disparut du cadre pour révéler un bar pris dans les turbulences d'une fiesta. La salle était bordée d'un grand comptoir, de rangées de bouteilles et d'une longue glace. Hommes et femmes y étaient accoudés, par grappes, qui braillaient entre eux, à l'adresse du barman ou en direction du plafond. Si la plupart semblaient jeunes, d'autres étaient nettement plus mûrs. Ils avaient tous l'air de fumer, et l'éclairage jaune pisseux était tout embrumé. Les murs étaient tapissés d'affiches arc-en-ciel ou d'un strict noir et blanc. En arrière-plan, un juke-box faisait des heures sup, réglé si fort qu'il faisait saturer aussi bien ses haut-parleurs que le micro de la caméra, m'empêchant même de reconnaître la chanson.

À l'évidence, cette scène était beaucoup plus vieille que la précédente. Non seulement l'image semblait repiquée sur du 8 mm, mais, à moins qu'il ne se fût agi d'une soirée rétro méticuleusement reconstituée, la tenue des gens renvoyait au début des années soixante-dix. Couleurs atroces, jeans atroces, coiffures atroces. Le look « soigné » avait dû se faire recaler à l'examen. Ma réaction n'était sans doute guère éloignée de celle des parents de l'époque : mais qui sont ces Martiens ? Que veulent-ils au juste ? Sont-ils aveugles ?

La caméra glissait et ondoyait à travers le bar, avec un entrain laissant penser que le réalisateur était soit sous l'emprise de substances hallucinogènes, soit soûl comme une bourrique. À un moment donné, l'image se resserra précipitamment, comme s'il (ou elle) avait manqué de se casser la figure. S'ensuivit un rot puissant et long, qui dégénéra en une violente quinte de toux, pendant que la caméra baissée montrait un carré de sol barbouillé de bière. Puis elle se redressa d'un coup et fonça dans le tas, comme fixée à l'avant d'une auto tamponneuse. Mes sourcils migrèrent lentement vers le haut de mon front, tandis que dans un embarras stupéfait je tâchais d'admettre que ce pouvait être mon père aux commandes. Quelques individus saluaient ou conspuaient l'objectif qui les frôlait, mais sans jamais prononcer un nom.

Parvenue à un coin de la pièce, l'image vira brusquement, débouchant sur une salle annexe peuplée sur tout le pourtour de gens assis ou debout. Au milieu se trouvait une table de billard. Un type était couché dessus, préparant son coup. Épais, doté d'un gros nez, le visage presque entièrement dissimulé sous ses cheveux, moustache et favoris. On aurait dit un ours galeux. Derrière lui végétait une blonde aux cheveux longs, agrippée à une queue comme si c'était la seule chose qui la maintenait debout. Elle tâchait tant bien que mal de se concentrer sur la partie, le visage ridé par l'effort, mais elle donnait l'impression de s'éteindre doucement. Son partenaire ne paraissait pas plus frais, qui mettait un temps fou à ajuster sa position. De l'autre côté de la table, dos à la caméra, un autre couple se tenait par la taille, chacun muni d'une queue. Tous deux étaient coiffés de longs cheveux bruns. La fille portait un ample chemisier blanc et une longue jupe bordeaux striée de vert ; le type arborait un jean à pattes d'eph fatigué et un gilet afghan qui paraissait tout juste apprivoisé.

Détachant ses yeux de la table, la blonde remarqua la caméra. Elle glapit, et montra l'appareil d'un geste du doigt à la fois vigoureux et extrêmement vague, comme si elle devait choisir entre trois images différentes mais oubliait sans cesse laquelle elle avait retenue. Le joueur jeta un regard sur l'objet désigné, roula des yeux et revint à son coup. Puis le couple aux cheveux bruns se retourna, et je découvris que ma gêne précédente était infondée.

Ce n'était pas mon père qui filmait. Je pouvais l'affirmer, dans la mesure où les deux bruns étaient mes parents.

Comme je fixais l'écran bouche bée, mon père grimaça un sourire torve et montra son majeur à la caméra. Ma mère tira la langue. L'image se déporta brusquement vers le joueur du billard, qui se décida enfin à tirer. Il mit complètement à côté.

J'arrêtai l'image, rembobinai.

Mes parents se retournèrent. Mon père sourit, fit un doigt. Ma mère sortit sa langue.

Re-pause. Je contemplai.

Ma mère n'est jamais devenue grosse, mais disons bien en chair, et se mouvait dans mon souvenir avec la grâce d'un paquebot tiré par un remorqueur. La femme figée devant moi pesait quant à elle une petite cinquantaine de kilos, et ils étaient drôlement bien répartis. Sans bien mesurer la portée de mes pensées, je savais que si j'étais entré dans un bar pour découvrir cette nana au bras d'un autre type, il y aurait eu de la bagarre. On redeviendrait homme des cavernes pour côtoyer cette femme. Non que mon père parût incapable d'en remontrer : il était un peu plus lourd que ce que j'avais connu, mais bougeait avec aisance et économie. Il aurait pu faire du cinéma. Leur duo respirait l'accord, la santé, le glamour. On voyait deux êtres vrais, vivants, un couple qui faisait l'amour. Mais surtout, ils paraissaient jeunes. Bon sang, ce qu'ils avaient l'air jeunes...

La scène dura encore cinq minutes. Il ne se produisit rien de particulier, sinon que je pus voir mon père jouer au billard, à une époque où, me découvrant tel qu'aujourd'hui, il m'aurait vu plus vieux que lui. Et il savait jouer. Un peu qu'il savait jouer ! Quand l'autre ours se recula de la table après avoir loupé son coup, mon père tourna le dos à la caméra et se pencha sur le tapis. Il ne s'embêta pas à contourner la table pour chercher la facilité : il joua la bille la plus proche, et l'envoya au fond. Alors il se mit à bouger, rôdant autour du jeu, le regard empreint de cette nonchalance appliquée propre à ceux qui comptent blouser un max de boules, qui se sont présentés à la table dans cette intention. Il réussit le point suivant, la bille parcourant trente centimètres de bande avant de dégringoler, puis encore le suivant – comme si ses boules étaient rabattues dans les poches par des élastiques. Triomphante, ma mère le félicita d'une tape sur le cul. Il accomplit alors un audacieux doublé rétro dans la poche centrale, puis rentra la noire depuis le milieu de la table, se retournant sans même attendre qu'elle soit tombée. *Game over*, mec.

Il adressa un clin d'œil au grizzli, qui roula les yeux de plus belle. De même que ce dernier était rompu à la connerie du cameraman, il avait visiblement l'habitude de se faire mater par mon père au billard. Rien de nouveau sous le soleil. Voilà des gens qui se connaissaient très bien.

Il ne se produisit rien de particulier, sinon que ma mère se mit à danser avec la blonde. Puis elle se lança dans une sorte de tortillement désarticulé, fouettant bras et jambes dans des directions opposées, tout en claquant des doigts. J'avais déjà vu ce mouvement dans des films, à la télévision. Exécuté par des danseurs professionnels. Mais je n'en avais jamais perçu toute l'essence avant de voir ma mère s'y frotter, se trémoussant en rythme, la bouche entrouverte et les paupières mi-closes.

T'es chaude, poupée, me surpris-je à penser. T'étais vraiment chaude.

Il ne se produisit rien de particulier, sinon que, pendant que l'ours rassemblait laborieusement les billes dans le triangle, je vis mon père se percher sur un tabouret et s'envoyer quelques lampées de bière. Ma mère – dansant toujours – lui fit une œillade, qu'il lui retourna, et je m'aperçus qu'ils étaient beaucoup moins ivres que le reste de l'assemblée. Ils se payaient du bon temps, mais ils avaient chacun un boulot, et le lundi venu ils seraient opérationnels. Tout bien réfléchi, mon père devait déjà être dans l'immobilier, malgré l'afghan de week-end et le tee-shirt débraillé. À vrai dire, ses quelques kilos supplémentaires lui allaient plutôt bien. Grâce à sa carrure, ils passaient pour du muscle et non du gras. Rajoutez-en une bonne louche et il glisserait vers le difforme, mais pour l'heure c'était surtout le type qu'on évite de heurter s'il traverse la salle lesté d'un plateau de bières. Je devinais pourtant que cette prise de poids était une acquisition assez récente, et qu'il n'y était pas encore acclimaté. Régulièrement il roulait ses épaules en arrière, à première vue dans le but de lisser les plis accumulés à force de se coucher sur le tapis pour disperser les billes aux quatre coins de la table. Mais aussi, soupçonnai-je, pour s'assurer que ses épaules étaient bien droites. Par la suite il découvrirait le jogging, et le club de gym, et il n'aurait plus jamais ce physique-là. En attendant, sur le film de cette soirée, je le vis faire quelque chose. Une chose triviale, anodine, certes. Mais vue d'ici, dans cette chambre d'hôtel de Dyersburg, elle m'arracha l'air des poumons, comme si j'avais reçu un coup au ventre.

Tout en allumant sa cigarette – j'ignorais qu'il avait été fumeur –, il décolla machinalement le tee-shirt de son diaphragme, puis le relâcha – de sorte qu'il retombe un peu mieux sur son ventre à peine rebondi. Je rembobinai pour me repasser la scène. Puis me la repassai encore, penché en avant, en butte au grain de l'image vidéo. Ce geste était sans appel. Je l'avais pratiqué moi-même. De tout le temps que j'ai connu mon père, je doute de l'avoir jamais vu commettre un geste aussi spontané, révélateur et personnel. C'était le geste d'un homme qui a conscience de son corps, et le sait imparfait, même au plus fort d'un soir de fête. Un petit ajustement qu'il avait déjà expérimenté, sans que ce fût déjà un tic. Plus encore que le tee-shirt lui-même, que les pichets de bière et les torrides encouragements, que les déhanchements de ma mère et le fait que jadis mon père avait manifestement su tirer le meilleur d'une queue de billard, ce petit geste rendait leur mort inconcevable.

Quand la table fut enfin prête pour une nouvelle partie, mon père se releva et se prépara à casser le triangle, armant son bras comme pour infliger à la blanche une gifle dont elle se souviendrait tout le reste de sa petite vie sphérique. À cet instant la scène s'interrompit d'un coup, comme parvenue en fin de bobine.

Avant que j'aie pu faire « pause », on enchaîna sur tout autre chose.

Un nouvel intérieur. Une maison. Un salon. Sombre, éclairé par des bougies. Le rendu était médiocre, la pellicule n'étant pas adaptée à une lumière tamisée. Une musique en sourdine, que je reconnus cette fois comme provenant de la bande originale de *Hair*. Une collection de bouteilles de vin jonchait le sol, plus ou moins vides, et je dénombrai plusieurs cendriers pleins à ras bord.

Avachie sur un canapé bas, ma mère chantonnait *Good Morning Starshine*. Elle avait plus ou moins la tête de l'homme-ours sur les genoux, lequel roulait un joint sur son torse.

— Remets celle qui parle de sodomie, grognait-il. Vas-y, remets-la.

La caméra panoramiqua doucement sur le côté, révélant un deuxième homme allongé face contre terre. La blonde était assise derrière lui, veillant sur une rangée de bougies alignées sur le dos du type au moyen de coupelles. À l'évidence il gisait depuis assez longtemps pour faire partie des meubles, et je fis l'hypothèse qu'il s'agissait du cameraman du bar. La fille se pliait tout doucement, dans un angle improbable, à partir de la taille, maintenue en équilibre par sa seule volonté. Loin de la frénésie du bar, elle paraissait plus âgée. Une grosse vingtaine d'années, voire trente – un peu trop vieille pour cadrer avec cette scène. Mais, si nous étions au tout début des années soixante-dix, calculai-je, mes parents avaient peu ou prou le même âge.

Ce qui signifiait que j'étais déjà né.

— Remets-la, insistait l'ours, sur qui la caméra vint aussitôt zoomer. Remets-la.

— Non, répondit une voix hilare toute proche du micro, confirmant à mes oreilles que c'était bien mon père aux commandes.

Il se débrouillait mieux que le comateux avant lui.

— On a dû l'écouter un million de fois, ajouta-t-il.

— C'est parce qu'elle est cool, fit l'ours en hochant vivement la tête. Tu vois, ce qu'elle raconte est si... ah, merde ! (Le champ s'élargit pour montrer qu'il avait laissé tomber le joint. Il avait l'air catastrophé.) Merde ! J'ai plus qu'à recommencer. J'ai passé ma vie à rouler ce putain de pétard, mec. Je le roule depuis que je suis né. Cet enfoiré de Thomas Jefferson a commencé ce pétard, et me l'a légué dans son testament. Disait que je pouvais finir son joint ou hériter de sa baraque à Monticello. J'ai dit va chier avec ta baraque, moi j'veux le kif. J'ai passé ma vie à le rouler, comme un serviteur brave et dévoué. Et maintenant il en reste plus rien.

— Plus rien, répéta la blonde avant de pouffer.

Sans escamoter une demi-croche de *Starshine*, ma mère se redressa pour prendre le matos des pattes empotées de l'ours. Avec des gestes experts, elle tint le papier d'une main, répartit le tabac avec l'index puis s'empara de la drogue.

— Allez roule, Beth ! trompetta l'ours, ragaillardi par la tournure des événements. Roule ma poule, roule...

La caméra zooma brièvement sur le joint. Il était presque prêt.

À ce stade, mes sourcils étaient dressés si haut qu'ils flottaient au-dessus de mon crâne. Ma mère venait de rouler un pétard.

— Vas-y, mets-la ! quémanda l'ours. Mets la chanson sur la sodomie. Allez, Don, mets-la, Don, qu'on s'bidonne, Don !

En fond sonore ma mère continuait de fredonner.

La caméra pivota et s'éloigna vers le couloir. Une pile de manteaux reposait au sol, jetés en vrac. Je repérai une cuisine sur la gauche, et un escalier sur la droite. C'était notre ancienne maison, à Hunter's Rock. Le mobilier et la déco contredisaient mes souvenirs, mais le reste n'avait pas varié.

Les yeux grands ouverts, je suivis la progression de l'objectif dans le couloir puis jusqu'à l'étage. Pendant quelques secondes le film ne restitua qu'une obscurité mouvante, agrémentée des mugissements sourds du poilu :

— Sodomie... fellation... cunnilingus... pédérastie... scandait-il sans même tâcher de respecter la mélodie.

Mon père atteignit le palier et fit une brève halte, marmonnant quelques mots. Puis se remit en marche, et je compris dès le virage suivant où il se rendait. Le rez-de-chaussée s'était tu ; je n'entendais que son souffle et ses pas feutrés lorsqu'il poussa la porte de ma chambre.

Il faisait d'abord tout noir, puis peu à peu la lumière du hall dévoila mon lit contre le mur, et moi dormant dedans. Je devais avoir cinq ans. On ne distinguait que le sommet de ma tête, une pommette, et un bout d'épaule couverte d'un pyjama sombre. Le mur était d'une sorte de vert marbré, et la moquette marron, comme je les avais toujours connus.

Mon père resta là deux bonnes minutes, sans parler ni bouger, se contentant de tenir la caméra, de me regarder dormir.

Je regardai de même, osant à peine respirer.

Le fond sonore changea, comme si l'on avait changé de disque au salon. Suivit une série de bruits faibles, peut-être des pas sur la moquette. Ils cessèrent, et là je sus, sans attendre de confirmation visuelle ou auditive, que ma mère venait de rejoindre mon père.

La caméra s'attarda sur le garçonnet couché, sur moi, un petit moment encore. Puis elle s'écarta, lentement, par la gauche. Je crus d'abord qu'il repartait, mais elle ne fit que pivoter.

Elle vira de cent quatre-vingts degrés et se stabilisa.

Mes parents regardaient droit dans l'objectif. Leurs visages remplissaient le cadre : pas collés l'un à l'autre, juste rapprochés. Aucun n'avait l'air soûl ou shooté. On aurait dit qu'ils me fixaient dans le blanc des yeux.

— Salut, Ward, susurra ma mère. Je me demande bien quel âge tu as aujourd'hui.

Elle jeta un œil par-dessus la caméra, sûrement vers la forme endormie sous les draps.

— Je me demande bien quel âge tu as, répéta-t-elle avec un je-ne-sais-quoi dans la voix de triste et de décalé.

Mon père restait rivé sur l'objectif. Il avait peut-être cinq ou six ans de moins que ma pomme aujourd'hui. Lui aussi prit la parole, d'une voix douce, mais sans affection débordante.

— Et je me demande ce que tu es devenu.

Le son blanc. Quelqu'un frôla la porte de ma chambre avec un chariot.

Je n'arrêtai pas la bande. J'étais paralysé.

La dernière scène provenait elle aussi d'un 8 mm, mais les couleurs étaient plus diluées, passées, blanchies en lumière pure. Des empreintes noires de cheveux, de poussières et de taches qui pétillaient sur tout l'écran réduisaient d'autant l'impact des scènes filmées.

Un rayon de soleil jaunâtre traversant une grande vitre. Dehors, des arbres qui défilent, les frissons du feuillage noyés dans le bruit. Le battement régulier d'un train, et d'autres sons plus faibles que je ne savais identifier.

Le visage de ma mère, encore plus jeune. Les cheveux moins courts, noircis par la laque. Contemplant le paysage par la fenêtre. Elle tourna la tête, avisa la caméra. Son regard semblait lointain. Elle sourit du bout des lèvres. La caméra se baissa lentement.

Enchaînement brutal sur une grande avenue. Je ne pouvais situer la ville, et je fus absorbé par la forme et la couleur des véhicules stationnés de part et d'autre de la chaussée, comme par la tenue vestimentaire des rares passants. Les bagnoles avaient de l'allure, pas les costumes, et les robes se portaient courtes. Je ne m'y connaissais pas suffisamment pour arrêter une date précise, mais je tablai sur la fin des années soixante.

La caméra avançait au rythme d'un promeneur. De temps en temps le crâne de ma mère surgissait par la gauche du cadre, comme si mon père marchait à sa droite, un peu en retrait. L'objet du tournage ne sautait pas aux yeux. Cette rue n'avait rien d'exceptionnel. Je crus distinguer un grand magasin sur la droite, et un petit square en face. Les arbres avaient toujours leurs feuilles, mais elles paraissaient fatiguées. Mon père maintenait la caméra à hauteur de visage, droit devant lui, sans le moindre mouvement vertical ou horizontal. Maman et lui ne pointèrent rien de particulier, n'échangèrent aucun mot. Au bout d'un moment ils traversèrent une route puis tournèrent dans une rue de traverse.

Enchaîné sur une tout autre rue, un peu plus étroite, comme si l'on s'était éloigné du centre-ville. Ils semblaient gravir une colline. Ma mère précédait mon père, filmée à partir des épaules. Elle s'arrêta.

— Pourquoi pas ici ? dit-elle en se retournant.

Elle avait chaussé des lunettes noires. La caméra resta figée un instant, puis vacilla, comme si mon père décollait son œil du viseur pour étudier les alentours.

Sa voix :

— Un peu plus loin.

Ils poursuivirent leur ascension, une minute environ. Puis stoppèrent de nouveau. La caméra tourna sur elle-même, offrant un bref et séduisant panorama sur ce qui ressemblait au sommet d'une côte dans une ville vallonnée. Des deux côtés de la rue s'élevaient de hauts immeubles. Au niveau du bitume, des pancartes indiquaient la présence d'épiceries et de petits restaurants, mais les fenêtres au-dessus étaient celles de logements. Des gens étaient plantés devant les boutiques, coiffés de chapeaux, examinant les étalages ; d'autres entraient ou ressortaient. Un quartier animé à l'heure du déjeuner.

Ma mère se retourna vers l'objectif et hocha la tête. C'était son tour. Elle s'exécuta, à contrecœur.

Plus tard le même jour. Une vue légèrement différente, mais le sommet de la même colline. Les nuages s'étiraient davantage. Une fin d'après-midi, des rues quasi vides. Ma mère se tenait debout, les bras le long du corps. Un étrange gargouillis retentit hors champ, le même que dans le train.

La caméra bougea un peu, comme si mon père avait tendu le bras vers quelque chose. Puis ma mère s'avança de quelques centimètres, à moins que lui n'ait reculé. Une lourde expiration de mon père.

Et de son fils, trente-cinq ans plus tard.

Ma mère donnait la main à deux très jeunes enfants, un de chaque côté. Ils paraissaient du même âge, habillés pareil, sauf que l'un portait un haut bleu, l'autre jaune. Ils avaient un an, dix-huit mois à la rigueur, et savaient tout juste marcher.

La caméra zooma sur eux. L'un avait les cheveux courts, l'autre un peu plus longs. Leurs visages étaient identiques.

Le champ s'élargit de nouveau. Ma mère lâcha la main de l'un des bambins. Celui aux cheveux plus longs, avec le haut jaune et le petit cartable vert. Elle s'accroupit à côté du second.

— Dis au revoir, fit-elle. (L'enfant en bleu la considéra sans comprendre.) Dis au revoir, Ward.

Les deux gamins se regardèrent. Puis celui aux cheveux courts, c'est-à-dire moi, leva les yeux vers ma mère pour être rassuré. Elle saisit ma main et la souleva.

— Dis au revoir.

Elle me secoua le poignet, puis me prit dans ses bras et se releva. L'autre enfant dévisagea ma mère, sourit, tendit les bras pour me rejoindre là-haut. Je ne pouvais à coup sûr déterminer son sexe.

Ma mère se mit à descendre la rue.

Elle marchait d'un pas tranquille, sans se presser, mais sans se retourner. La caméra s'attarda sur l'autre enfant, même quand mon père emboîta le pas de ma mère. Ils le laissèrent là.

L'enfant s'éloignait de plus en plus, immobile au sommet de la pente. Il ne pleura même pas – tant qu'on pouvait l'entendre, en tout cas.

Puis la caméra prit un virage et il sortit du champ.

L'image s'échoua sur le son blanc, et y resta pour de bon. Au bout d'une minute, la cassette s'arrêta toute seule, me laissant seul face à mon reflet dans l'écran.

Je trouvai la télécommande à tâtons, rembobinai, appuyai sur « pause ». Et fixai l'image arrêtée d'un enfant abandonné en haut d'une colline, en me couvrant la bouche des deux mains.

CHAPITRE 7

La trappe s'ouvrit, projetant d'en haut une faible lueur.

— Bonjour, mon cœur, dit l'homme. Je suis de retour.

Sarah ne distinguait pas son visage. Au son de sa voix, il était assis sur le plancher juste derrière sa tête.

— Bonjour, répondit-elle d'une voix aussi posée que possible.

Elle voulait se rétracter, intercaler ne fût-ce que deux centimètres supplémentaires entre eux, mais elle ne disposait pas d'une telle marge. Elle lutta pour garder son calme, pour s'en tenir à son plan de fausse indolence.

— Comment ça va aujourd'hui ? lança-t-elle. Toujours aussi barje, j'imagine.

L'homme rit doucement.

— Tu n'arriveras pas à me mettre en colère, dit-il.

— Pourquoi voudrais-je te voir en colère ?

— Pourquoi ces paroles, alors ?

— Mon père et ma mère vont mourir d'angoisse. J'ai peur. Il se peut que j'oublie les bonnes manières.

— Je comprends.

Il se tut alors, un long moment. Sarah patienta.

Au bout d'une poignée de minutes, elle vit une main descendre vers son visage. Munie d'un verre d'eau. Sans prévenir, l'homme l'inclina doucement. Sarah ouvrit la bouche à temps, et but autant qu'elle pouvait. Puis la main disparut.

— C'est tout ? demanda-t-elle.

Son palais lui procurait une sensation bizarre, à la fois sèche et mouillée. L'eau possédait cette saveur que les adultes devaient prêter au vin, vu la façon dont ils s'en gargarisaient, à le garder en bouche comme s'ils n'avaient jamais rien goûté de meilleur. Même si, de sa propre expérience, le vin avait surtout un goût de truc pas net.

— Tu espérais quoi, poupée ?

— Si tu veux me garder en vie, il faudra plus qu'un peu d'eau.

— Qu'est-ce qui te fait croire que je veux te garder en vie ?

— Parce que sans ça tu m'aurais tuée dès le début, avant de m'étendre à poil dans un endroit où tu pourrais me mater tout en t'astiquant.

— Ce n'est pas très gentil.

— Je te renvoie à mes réponses précédentes. Je ne suis pas d'humeur très gentille, et tu es un malade, alors rien ne m'y oblige.

— Je ne suis pas malade, Sarah.

— Ah non ? Et comment te décrirais-tu ? Original ?

Il rit de plus belle, avec délectation.

— Sans aucun doute.

— Original... comme ce connard de Ted Bundy ?

— Ted Bundy était un imbécile, rectifia l'homme d'une voix vidée de tout humour. Un idiot d'opérette doublé d'un imposteur.

— OK, fit-elle dans un souci d'apaisement, tout en le jugeant non seulement fêlé mais prétentiard. Je suis désolée. Moi non plus, je ne suis pas très fan. Tu vaux bien mieux que lui. Alors, j'ai le droit de manger, oui ou non ?

— Plus tard, peut-être.

— Génial. J'ai hâte d'y être. Pense à couper de petits morceaux, que je puisse les attraper.

— Bonne nuit, Sarah.

Elle l'entendit se lever, et son aplomb de façade se délita. Le plan n'avait pas fonctionné. Du tout. Il la savait terrifiée.

— Par pitié, ne remets pas le couvercle ! Je ne peux pas bouger, de toute façon.

— Je crains de ne pas avoir le choix.

— Je t'en supplie...

La trappe tomba, et Sarah retrouva le noir.

Elle entendit ses pas s'éloigner, une porte se refermer en douceur, puis ce fut de nouveau le silence.

Elle lécha frénétiquement le contour de sa bouche, traquant les moindres restes d'humidité. Passé le choc initial, elle se rendit compte que l'eau n'avait pas le même goût qu'à la maison. Elle devait provenir d'un autre réseau, ce qui signifiait qu'elle se trouvait loin de chez elle. Comme lorsqu'on part en vacances. Bon, elle savait au moins une chose. Plus elle en saurait, mieux ça vaudrait.

Puis elle songea qu'il pouvait s'agir d'eau minérale, d'un truc en bouteille, auquel cas le goût ne prouvait rien du tout. Il s'agissait peut-être d'une marque différente, tout simplement. Mais peu importait. Cela méritait quand même réflexion. Plus elle aurait d'idées, mieux ça vaudrait. Celle-ci, par exemple : quand elle avait mentionné l'inquiétude de ses parents, le type n'avait pas répété qu'il les avait tués. Alors qu'au moment de la capture, il avait tenu à lui décrire les supplices qu'il leur avait infligés. Il y avait peut-être une raison à cela. Avec un peu de chance, cela signifiait qu'ils étaient toujours en vie, et qu'il avait proféré ces horreurs juste pour l'effrayer.

Ou peut-être pas. Sarah reposa dans le noir, les poings serrés, se retenant de crier.

DEUXIÈME PARTIE

Peu de gens parviennent au bonheur sans une personne, une nation ou un credo à haïr.

Bertrand Russell

CHAPITRE 8

Le vol atteignit Los Angeles à 22 h 05. Nina n'était chargée que d'un sac à main et du dossier, et Zandt pouvait porter sur une épaule tout ce qu'il possédait sans paraître bancal. Une voiture les attendait. Rien de luxueux ni d'officiel. Juste un taxi que Nina avait réservé depuis l'avion, pour déposer Zandt à Santa Monica avant de rentrer chez elle.

Des lumières et des panneaux dans la nuit, des visages entraperçus, le crépitement et les klaxons de la vie : une soirée ordinaire dans une ville dont le cœur ne semble jamais battre là où l'on se trouve, mais toujours au détour d'un virage, au bout de la rue ou de l'autre côté d'immeubles mastocs, dans quelque nouveau night-club dont les heures de gloire seront passées avant même qu'on en apprenne l'existence. Sur le chemin se succèdent des hôtels bas de gamme, des boutiques d'alcool empoussiérées, des parkings remplis de véhicules d'occasion de provenance douteuse. Une triste palette d'individus postés aux carrefours, sans rien de très positif en tête, au milieu de bunkers où siègent des entreprises qui engloutiront un nombre incalculable de mornes vies sans être jamais cotées au Nasdaq. Transition progressive vers les quartiers résidentiels, puis Venice. Vu de l'extérieur, en empruntant les bonnes rues, Venice paraît vouloir regagner le haut du panier. L'immobilier y est par endroits très cher, façon « International Style » merdique. Vous tomberez ici ou là sur une relique des années cinquante, une affiche ultrakitsch commise au temps des ampoules de flash et du glamour froid. Aujourd'hui la plupart ont été arrachées et remplacées par des panneaux d'affichage agressifs imprimés en helvetica, la police de caractères officielle du purgatoire. L'helvetica n'a pas été conçu pour vous inspirer quoi que ce soit d'agréable, vous promettre de l'aventure ou vous mettre du baume au cœur. L'helvetica sert à vous annoncer que les bénéfices sont en baisse, que le photocopieur est hors service et au passage que vous êtes viré.

Enfin, Santa Monica. De plus jolies maisons, de petits bureaux, des commerces qui proposent des plats japonais et le *London Times*. La mer avec sa jetée bâtie au temps regretté des photos sépia. Les Palisades juste au-dessus, la bouillonnante Ocean Avenue, puis une première rangée d'hôtels et de restaurants. La sensation diffuse que cette banlieue était jadis une ville à part entière. Peut-être est-ce la présence de la mer qui suggère que cette agglomération se trouve là pour une raison précise. Par endroits, c'est toujours vrai : on perçoit un rapport à la nature qui dépasse le simple fait de l'avoir aplanie. Des boutiques, des cafés, des lieux de sortie, des sites à visiter et à consommer. On pouvait vivre ici et y trouver un sens, comme la famille Becker jusqu'à une date récente. Ce n'est pas un endroit réel, mais Los Angeles l'est elle-même très peu, et les coins réels sont ceux où l'on ne veut pas aller. La réalité est le domaine des flingues et des lendemains de cuite. La réalité est ce que l'on souhaite éviter. L.A. se croit pleine de magie, et l'on peut même s'y laisser prendre, mais cela repose avant tout sur une arnaque mutuellement consentie. Elle vous fera miroiter un destin de rêve, puis deux rues plus loin vous condamnera à une mort imminente. C'est de la poudre aux yeux, on le sait bien, mais on veut quand même y croire. Les plans de la ville vous indiqueront où rencontrer les stars mais jamais comment en devenir une : il ne vous reste plus qu'à sillonner les rues et à hanter les bars, en espérant que la chance vous sourira. L.A. est une ville qui s'est entichée de la Destinée, lui a payé de nombreux verres et a obtenu son numéro de téléphone après lui avoir fait de l'œil de longues soirées durant – mais qualifier cette relation d'amour vache serait encore trop tendre. La Destinée tient davantage de la starlette malveillante sur une mauvaise pente de cocaïne, qui se fend d'une relative bonne action par semaine des fois qu'un producteur passerait par là. La Destinée n'agit pas toujours au mieux de vos intérêts. La Destinée n'en a simplement rien à battre.

— Alors, heureux de rentrer ? lança Nina.

Zandt grogna.

Le taxi le laissa devant le Fountain, une tour de dix étages en stuc jaune délavé, plantée sur Ocean Avenue entre les boulevards Wilshire et Santa Monica, qui déversent les gens sur la plage. L'immeuble possède une touche Art déco qui lui donne l'air plus chic qu'il n'est vraiment. Composé à l'origine d'appartements de standing, il avait fait une longue carrière d'hôtel avant de se reconvertir dans le meublé. L'ancienne piscine repose sous un vaste « espace détente » peu fréquenté : malgré la véranda, les plantes et les fauteuils à l'ombre, on sent par trop qu'il y manque quelque chose. Zandt connaissait bien le vestibule depuis un certain homicide de 1993 : un acteur européen de seconde zone et une jeune prostituée, une saynète ayant mal tourné. Comme de bien entendu, l'acteur avait pris le large. Zandt ne savait plus dans quelle chambre s'était produit le drame. Sûrement pas dans la suite qui lui était allouée, en tout cas : spacieuse, bien équipée, avec une belle vue sur la mer. Il posa son sac dans le coin séjour, examina la kitchenette. Placards vides, très peu de poussière. Il n'avait pas faim et se voyait mal se mettre aux fourneaux. Le Fountain ne proposait ni bar, ni restaurant, ni service en chambre. Ce n'était pas une vraie destination, raison pour laquelle il avait choisi de s'y établir. Ça, et son emplacement.

Il quitta la suite, reprit l'ascenseur et flâna un moment à l'extérieur du bâtiment. Nina était repartie avec le taxi, et ils avaient rendez-vous demain en fin de matinée. Elle avait joint l'antenne du Bureau de Westwood depuis l'avion, et avant cela depuis Pimonta, mais elle était tenue de faire au minimum acte de présence. Une intuition retint néanmoins Zandt dehors, à étudier les véhicules garés le long du trottoir. Nina aurait bien été du genre à faire le tour du pâté de maisons pour revenir l'épier. Non parce qu'il était expert en quoi que ce soit. Mais juste pour savoir. Nina aimait savoir.

Au bout de cinq minutes il avança jusqu'au carrefour, prit à droite sur Arizona et longea deux blocs d'immeubles jusqu'à Third Street Promenade. C'était sur Arizona Street que Michael Becker avait déposé sa fille le soir de la disparition.

Il vira à gauche et remonta la partie ouest de la Promenade, jusqu'à l'extrémité où Sarah Becker avait été vue pour la dernière fois. Les 23 heures approchaient, soit bien plus tard que l'heure où la fille avait été enlevée. Les commerces étaient presque tous fermés. Les artistes et musiciens de rue avaient plié bagage depuis un moment, même l'imitateur de Frank Sinatra qui ne lésinait pas sur les heures sup. Mais peu importait. En l'absence du kidnappeur et de sa victime, impossible de travailler en conditions réelles, de toute façon.

Il gardait un œil ouvert sur les badauds. Les tueurs reviennent souvent sur leurs pas, notamment ceux pour qui le meurtre ne se réduit pas à une expérience ponctuelle. Ils remontent le temps, rembobinent afin de rejouer leurs souvenirs. Zandt n'escomptait pas remarquer quelqu'un en particulier, mais il demeurait aux aguets. Parvenu à la rue latérale mentionnée dans l'exposé de Nina, là où l'on avait repéré une voiture en stationnement interdit, il s'immobilisa un bon moment. Sans rien chercher de précis. Juste pour être sur place.

— Z'attendez quelqu'un ?

Zandt se retourna sur un jeune homme svelte et gracile. Un adolescent, dix-huit ans à tout casser.

— Non, répondit-il.

Le garçon sourit.

— Vous êtes sûr ? Je crois bien que si. Je me demande si ce n'est pas moi que vous attendez.

— Je t'assure que non. Mais il y aura forcément quelqu'un. Pas ce soir, pas ici, mais ça ne saurait tarder.

Le sourire s'effaça.

— Vous êtes flic ?

— Non. Je t'explique juste comment ça marche. Va te trouver un mec dans un lieu éclairé.

Il poussa la porte du Starbucks et commanda un café, à emporter, puis alla s'asseoir sur le banc où Sarah avait été enlevée. Le garçon n'était plus là.

On pourrait s'attendre que le théâtre d'un drame se charge d'une énergie mortifère. Il n'en est rien. Si le cerveau humain sait reconnaître les visages, sa compréhension des lieux laisse à désirer. Avec le physique, c'est relativement simple : plus les gens connaissent votre bobine, plus vous êtes célèbre. Pas besoin de papiers d'identité. Vous n'êtes pas un inconnu, mais un membre de notre famille élargie : des frères costauds et des sœurs ravissantes, un père ou une mère attentionnés... autant de fausses parentèles censées nous faire oublier que nos réseaux de relations se sont réduits à une peau de chagrin. Avec les lieux, il s'agit de savoir de quels événements ils ont été le théâtre. Mais quand cette information est coupée du présent et que l'on s'attarde un peu, on ne perçoit plus rien. On retrouve l'endroit tel qu'il était avant qu'il n'arrive quelque chose, tel qu'il était le soir du crime. Ce sera là le point limite de votre voyage dans le temps, dans l'avant, le summum de votre capacité à tenir un couteau tel qu'il se présentait au sortir d'un tiroir de cuisine, avant d'être barbouillé de sang – quand il ne possédait rien d'autre qu'un potentiel.

Zandt resta sur le banc jusqu'à que celui-ci devienne un endroit comme les autres, et plus longtemps encore.

Au long de ses années de brigade criminelle, Zandt avait eu affaire à un nombre étonnamment élevé de tueurs en série. La règle veut que ce genre de dossier échoie au FBI, qui possède à cet effet un laboratoire de Sciences comportementales à Quantico, des procédures de profilage, des Jodie Foster et des David Duchovny impeccablement coiffés et habillés. Comme les tueurs eux-mêmes, les fédéraux passent pour des êtres à part. Mais en l'espace de huit ans, le simple mortel John Zandt avait été confronté à plusieurs séries de meurtres qui allaient s'avérer l'œuvre d'individus tout à fait dignes du label serial killer. Deux de ces hommes avaient été appréhendés, et Zandt avait joué un rôle non négligeable dans leur arrestation. Il avait ça dans le sang, et tous en convenaient. Le premier cas concernait un habitant de Venice Beach ayant liquidé quatre vieilles dames, et Zandt avait été mis sur le coup un peu par hasard. Lors de la seconde enquête, il avait collaboré avec le FBI et rencontré Nina Baynam.

Au cours de l'été 1995, on avait retrouvé les restes de quatre jeunes Noirs à moitié enterrés dans divers quartiers de la ville. La méthode de démembrement utilisée, ainsi que la présence d'une cassette vidéo auprès de chaque cadavre, suffisait à imputer ces assassinats à un unique auteur. Ces quatre victimes avaient toutes été enlevées sur des terrains vagues, et trois donnaient déjà dans la drogue et la prostitution de rue. Les deux premiers décès furent largement ignorés par le reste de la population, qui se contenta d'y voir un phénomène social cyclique : l'écrémage naturel du sous-prolétariat. C'est uniquement par leur caractère répétitif que ces meurtres quittèrent peu à peu le domaine du bruit de fond pour celui de l'événement. Les cassettes abandonnées près des corps contenaient entre une et deux heures de vidéo amateur grossièrement montées, qui illustraient combien les derniers jours des victimes avaient été déplaisants. Chaque vidéo était rangée dans un boîtier montrant une photo du garçon, accompagnée de son nom et de la mention « Démo ».

La presse surnomma le tueur l'« Agent de Casting », ce que tout le monde trouva fort drôle – sauf les parents des quatre victimes, mais, leur chagrin constituant un rappel gênant de la réalité qui sous-tendait le spectacle, il fut soigneusement occulté sauf lorsqu'il fallut doper l'intérêt du public. Et puis les proches étaient avant tout spectateurs du drame, or c'est toujours l'acteur que l'on préfère. Quelqu'un que l'on puisse apprendre à connaître, un visage en contre-jour dans les journaux ou aux infos. On veut une personnalité. Une star.

Zandt enquêta sur le meurtre du premier garçon, puis à la deuxième victime le FBI entra en scène. Nina était un jeune agent aguerri, qui avait passé l'année précédente sur une sale affaire dans le Texas et en Louisiane. L'instinct et le travail de terrain de Zandt, associés à l'analyse de Nina sur l'emplacement des corps, et à la chance qui voulut que le meurtrier ait déclaré la caméra à son nom, permirent de confondre le coupable. Il s'agissait d'un Blanc âgé de trente et un ans, qui officiait comme graphiste en périphérie de l'industrie du vidéo-clip et avait tourné lorsqu'il était enfant dans des films tombés aux oubliettes. Au bout d'une série d'entretiens avec Zandt, il avoua les crimes, précisa quelques points annexes et révéla la localisation de ses talismans, à savoir la main droite de ses victimes – chacune enfoncée dans un pot contenant à l'origine le produit d'un éminent fabricant de café soluble. Pour finir, il conduisit la police jusqu'aux cadavres de deux victimes précédentes, sur lesquelles il avait peaufiné sa technique. Pour justifier ses agissements, il expliqua qu'on l'avait violé sur les plateaux de cinéma de son enfance, allégation qui satisfaisait à merveille le besoin du public d'un début et d'un milieu à chaque fable. La véracité de ses propos fut impossible à établir, et la fin de l'histoire survint lorsqu'un codétenu armé d'une cuiller aiguisée égorgea le tueur avant le début du procès. C'est qu'on trouve des victimes aux deux bouts de la chaîne alimentaire : même les violeurs et les assassins ont besoin de souffre-douleur, et les tueurs de gamins remplissent très bien cet office. Au bout du compte l'histoire de l'Agent de Casting entra dans la légende, fit l'objet d'un bouquin d'écrivaillon au succès modéré et d'innombrables sites Internet. Un logiciel de traitement d'images capricieux baptisé CastingAgent connut un bref succès, ainsi qu'un magasin d'Atlanta qui commercialisa un canapé tacheté de rouge sombre, sous le nom de Casting Sofa.

L'enquête dura treize semaines. John et Nina vécurent les huit dernières en tant qu'amants. Leur liaison s'acheva peu après l'arrestation du suspect. Nina avait beaucoup donné en termes de premiers pas. Puis elle avait cessé de prendre l'initiative, et ça n'était pas allé loin. Zandt n'en parla jamais à son épouse, avec qui il entretenait une relation dans l'ensemble cordiale et harmonieuse, bien qu'elle eût traversé une zone d'aridité. Il ne voulait pas la perdre, ni elle ni leur fille, aussi accueillit-il la fin de cette liaison comme un soulagement.

Au cours des cinq années suivantes, Nina et lui déjeunèrent ensemble à l'occasion, alors que Zandt traitait son lot habituel de guerre des gangs, de crimes passionnels, et de tartempions retrouvés criblés de balles dans des ruelles, suffoquant comme des poissons hors de l'eau et déclarés morts à l'arrivée aux urgences dans l'indifférence générale. Il élucidait certains crimes. En ratait d'autres. La loi du genre. Nina enquêta sur un double meurtre très médiatisé à Yellowstone, une série de disparitions dans le nord de la Californie, et une autre dans l'Oregon, qui toutes demeurent à ce jour irrésolues. Pendant ce temps, du côté du monde réel, derrière le rideau de mort et de méfaits qui entoure les représentants de l'ordre, les affaires allaient bon train. La Bosnie implosait ; les cigares du Président lui valaient des ennuis ; nous découvrions les joies de l'e-mail et de *Frasier,* de la Playstation et de Sheryl Crow.

Puis, le 12 décembre 1999, une adolescente disparut à Los Angeles. Josie Ferris, seize ans, venait de fêter l'anniversaire d'une amie devant un hamburger au Hard Rock Café de Beverly Boulevard. À 21 h 45, après avoir pris congé devant l'entrée du restaurant, elle marcha seule en direction de l'hôtel Ma Maison, dans l'intention d'y héler un taxi. Beverly Boulevard n'est pas une route de campagne ni une ruelle. C'est une voie large et très empruntée, et l'avant-cour de l'hôtel tout comme le hall du centre commercial étaient fréquentés ce soir-là. Malgré tout, quelque part sur ce tronçon de trois mètres, elle disparut.

Le non-retour de Josie au foyer fut signalé au commissariat vers 0 h 50. Devant une réponse qu'ils jugèrent d'une promptitude insuffisante, ses parents décidèrent d'aller remplir les papiers sur place. M. et Mme Ferris avaient des tempéraments bien trempés, et il ne fallut pas longtemps pour que les policiers prennent l'incident avec plus de sérieux, du moins tant qu'ils étaient à portée d'oreille des parents. Hélas ! cela ne changea rien au problème. On ne revit jamais leur fille vivante.

Deux jours plus tard, quelqu'un laissa un pull devant chez eux. On avait brodé le prénom de Josie sur le devant, avec, les analyses l'attesteraient, sa propre chevelure. Sa meilleure copine le lui avait offert pour ses seize ans ; elle avait cousu sur la manche les lettres « APE » : Amies Pour l'Éternité. C'était bien vu. L'éternité se révélerait juste un peu courte. Aucune demande de rançon n'accompagnait le vêtement. La police commença à considérer la situation avec le plus grand sérieux, même quand les parents avaient le dos tourné. Un groupement spécial fut mis sur pied, coordonné par le chef de l'antenne locale du FBI, Charles Monroe. La presse eut vent de la livraison du pull, mais pas des altérations qu'il avait subies. Au bout d'un mois, on n'avait accompli aucun progrès d'aucune sorte pour retrouver l'adolescente.

Fin janvier et début mars 2000, deux autres filles se volatilisèrent. Elyse LeBlanc et Annette Mattison ne rentrèrent jamais, du cinéma pour la première, de chez une copine pour la seconde. Toutes deux présentaient d'évidentes similitudes avec Josie Ferris. Elles avaient pratiquement le même âge (quinze et seize ans) et les cheveux longs. Les LeBlanc et les Mattison vivaient dans l'aisance, leurs filles étaient séduisantes et d'une intelligence supérieure à la moyenne. Mais cela ne suffisait pas à établir un lien ferme entre les trois disparitions, qui étaient survenues dans des quartiers fort éloignés.

L'arrivée de deux autres pulls le permit, toutefois. Là encore, ils furent déposés devant le domicile familial, en plein jour, le prénom de la fille brodé avec ses cheveux. Il n'y eut pas d'autre message. La gravité de la situation amena le FBI à taire ces deux nouvelles disparitions. En général, les kidnappeurs en série cultivent la discrétion. Aussi, le choix d'adolescentes dont l'absence serait aussitôt remarquée, tout comme le regain d'attention que promettait la livraison des pulls, laissait à penser que l'on avait affaire à un individu hors norme. Qui voulait faire parler de lui, et tout de suite.

Les enquêteurs lui refusèrent cet égard.

Une semaine après la disparition d'Annette Mattison, des piqueniqueurs retrouvèrent dans Griffith Park le cadavre d'une jeune femme habillée. Bien que chauve, salement brûlé et entamé par la faune locale, il fut vite identifié grâce à des soins dentaires récents et à un bijou distinctif. Elyse LeBlanc. On estima qu'elle était morte depuis à peu près la moitié du temps écoulé depuis son absence, mais n'avait été laissée dans ce parc que tout récemment. On lui découvrit plusieurs traumatismes crâniens, mais aucun n'avait provoqué le décès. Le corps fut aussitôt expédié au laboratoire fédéral de l'État de Washington ; ni ses vêtements ni sa dépouille ne livrèrent d'indice matériel quant à l'identité du tueur. La fouille entière du parc menée par les autorités locales et le FBI ne permit pas de mettre au jour les corps de Josie Ferris et d'Annette Mattison, pas même sous forme de morceaux.

Le black-out vis-à-vis de la presse fut levé. Un appel à témoin déboucha sur les traditionnels canulars, confidences de timbrés et fausses pistes. Les parents d'élèves prirent leurs dispositions pour que leurs adolescentes se déplacent en groupe.

Le corps de Josie apparut dix jours plus tard. Il reposait dans des buissons bordant une route de Canyon Laurel, dans un état comparable à celui de la jeune LeBlanc. Sauf qu'il présentait en outre des traces de violences sexuelles très poussées.

Le tueur venait d'hériter d'un surnom. Les médias l'appelaient le « Garçon de Courses ». L'idée émanait en sous-main de l'agent spécial Monroe, qui estimait que rabaisser le coupable de cette manière, par l'emploi dépréciatif du mot « Garçon », pouvait rapporter quelque avantage stratégique ; qu'une personne ayant réussi à enlever trois filles brillantes et dégourdies dans des rues passantes, à les tuer et à abandonner leur dépouille dans des lieux publics, sans jamais être vu ni laisser le moindre indice derrière lui, se laisserait peut-être démonter par ce coup bas lexical.

Qu'il serait vexé comme un pou, et s'effondrerait aussi sec.

Nina n'était pas de cet avis. Pour diverses raisons, elle s'était entretenue du dossier avec John Zandt, en dépit du fait qu'il ne participait pas à l'enquête officielle. Mais ils avaient fait du bon boulot ensemble sur l'affaire de l'Agent de Casting. Elle voulait connaître son point de vue.

Zandt se prêta au jeu, sans grand enthousiasme. Nina déployait dans ces enquêtes un zèle et une ardeur dont il ne se sentait plus capable. Son mariage était reparti de l'avant, et sa fille avait grandi, passant du statut d'enfant à celui de jeune femme, ce qui consolidait la famille. Elle avait la chevelure de sa mère, d'un brun sombre, riche, presque auburn, et les yeux de son père, marron tacheté de vert. Elle écoutait la musique à s'en crever les tympans, sa chambre était un foutoir, elle passait trop de temps sur Internet et sentait le tabac de temps en temps. Il y avait des prises de bec. Mais elle participait aux expéditions shopping de sa mère, alors que c'était hyper-chiant, car elle savait que Jennifer adorait sa compagnie. Elle veillait à écouter son père lorsqu'il parlait, réprimant les bâillements qui la gagnaient. Ses parents ignoraient qu'elle avait tiré sur un joint à plusieurs reprises, essayé la coke et volé une paire de boucles d'oreilles de valeur. S'ils l'avaient su, ils lui auraient botté le cul jusqu'à la mettre en orbite, mais sans s'inquiéter outre mesure. Tout cela demeurait dans l'éventail des égarements acceptables pour le lieu et l'époque qui étaient les siens.

Mais surtout, Zandt avait pris un peu d'âge, et ne souhaitait pas consacrer plus de temps que nécessaire aux ténèbres dont le monde pouvait accoucher. Il vaquait à son travail, puis rentrait chez lui et vaquait à sa vie. Après s'être immergé le temps de deux enquêtes dans les basses œuvres de meurtriers en série, décrypter les rouages de leur esprit ne l'intéressait plus. Arrivé à un certain point, on ne pouvait continuer sans avoir la nausée.

Les serial killers ne ressemblent pas aux portraits flamboyants que dresse d'eux le cinéma : des génies charmeurs, pétris de cruauté, les adeptes charismatiques d'un art sanglant. Ils doivent davantage à l'alcoolisme ou à la folie légère. Impossible à approcher ou à raisonner, reclus du monde dans une vision indicible, inaccessible à ceux qui ne la partagent pas. On en trouve de toutes formes, de toutes tailles et de toutes sortes. Certains sont des monstres, d'autres des individus tout à fait corrects – n'était, bien entendu, leur propension à tuer des innocents et à détruire la vie des êtres qui les ont aimés. Jeffrey Dahmer a d'abord lutté de toutes ses forces pour refouler les pulsions dont il savait qu'elles plaçaient ses désirs à mille lieues de la normalité. Mais il échoua, et dans les grandes largeurs. Il n'implora aucune clémence quand il fut pris, ne joua pas au plus fin avec la police, ne fit rien d'autre qu'admettre sa culpabilité et se lamenter d'avoir mal agi. Dans les limites mêmes de sa pathologie meurtrière, il a eu une attitude exemplaire. Reste qu'il avait ôté la vie à seize jeunes hommes au bas mot, dans des circonstances à peine croyables tant elles étaient atroces.

D'autres tueurs se vautrent dans leur notoriété, s'arrogent publicité et privilèges en manipulant médias et autorités, se jouent de la douleur des gens qu'ils ont amputés d'une chose irremplaçable. Ils se délectent de leurs exploits, de leurs grands secrets. Ils dévorent dans les journaux le feuilleton de leur procès, profondément satisfaits d'avoir enfin capté toute l'attention qu'ils ont toujours pensé mériter. Cela ne les rend pas plus redoutables. Seulement différents. Ted Bundy. L'Agent de Casting. John Wayne Gacy. Philippe Gomez. L'Éventreur du Yorkshire. Andrei Chikatilo. D'aucuns sont très beaux, d'autres plus efficaces, certains intelligents, d'autres un peu limités, voire débiles. Les uns ont l'air de types ordinaires ; les autres sont immédiatement repérables comme cinglés, même au milieu d'une rue bondée. Aucun n'est un surhomme ni l'esclave du Mal, ou alors au sens le plus superficiel du terme. Il ne s'agit que d'individus qui brûlent d'ôter la vie, d'enrichir leur propre expérience sexuelle par la torture et la mutilation. Ce ne sont pas des démons. Rien que des hommes – rarement des femmes – qui commettent l'inacceptable, sous l'emprise d'une obsession névrotique. Ils ne s'inscrivent pas dans un système binaire opposant le bien au mal, mais sur un large spectre comprenant également ceux qui vérifient leurs verrous dix fois par nuit ou qui ne peuvent souffler après le repas tant que la cuisine n'est pas impeccable. Les tueurs en série ne sont pas effroyables en soi. L'effroi réside dans le constat que l'on peut être humain sans éprouver la même chose que les autres humains.

Zandt connaissait les facteurs présidant à l'apparition d'un serial killer. Une mère violente et dominatrice, un père abusif ou faible. Des expériences sexuelles précoces et conflictuelles, en particulier avec les parents, les frères et sœurs, ou des animaux. Être né aux États-Unis, en ex-URSS ou en Allemagne, trois contrées où la population comprend un taux record de meurtriers multiples. L'exposition à des cadavres à l'âge où l'esprit se forme. Des lésions à la tête ou une intoxication au métal – l'élément, pas la musique. Un événement déclencheur, quelque chose qui permette au potentiel de se réaliser. Aucune de ces conditions n'est nécessaire ou suffisante, mais elles constituent quelques aspects d'un syndrome qui fournit parfois un terreau suffisamment noir pour produire une gerbe de pulsions insanes : un individu névrotique et violent, incapable de mener une vie normale. L'ombre dans nos rues. Le croque-mitaine.

Il les avait bien assez vus. Il ne voulait rien savoir de plus. Dans le secret de ses songes, il considérait l'Agent de Casting comme tel : l'Agent de Casting. Il se donnait du mal pour ne pas penser à son vrai nom, pour lui assigner l'irréalité cartoonesque dans laquelle le tueur tenait à l'évidence ses victimes. Puisque ce dernier s'était montré incapable de reconnaître à ces jeunes Noirs la dignité de leur individualité, la moindre des choses était de le loger à la même enseigne.

À cette époque, Zandt traitait les meurtres procédant du triptyque habituel : drogue, amour, argent. Il trinquait avec ses collègues, écoutait Nina conter ses efforts pour relier les disparitions de Josie Ferris, Elyse LeBlanc et Annette Mattison. Il dînait avec sa femme, conduisait sa fille à droite et à gauche, fréquentait le club de gym.

Le 15 mai 2000, Karen Zandt sortit du lycée en fin d'après-midi. Elle ne rentra pas à la maison.

Dans un premier temps ses parents imaginèrent le meilleur. Puis le pire. Ils reçurent un pull au bout de trois interminables semaines.

Zandt contacta Nina. Elle arriva toutes affaires cessantes, flanquée de deux collègues. On déballa le colis. Ce coup-ci, il n'y avait aucun nom brodé sur le pull, et ce n'était pas celui de Karen. Le sien était couleur pêche ; celui-ci, noir.

Il y avait un mot glissé à l'intérieur, imprimé au laser en police courier, sur un papier qu'utilisaient les entreprises et les foyers à travers tout le pays.

M. Zandt,

Une « course ». Vous devrez attendre pour la suite.
J'ai vu ta misère et la fatigue de tes mains, et je te condamne.

L'Homme Debout

Un mois plus tard, le corps d'Annette Mattison fut découvert dans un canyon des collines d'Hollywood. Même état que celui d'Elyse LeBlanc, même absence d'indices matériels. Les enlèvements d'adolescentes cessèrent – du moins, ceux suivis de la livraison d'un vêtement.

On ne retrouva jamais d'autre corps.

Au bout de deux heures, la Promenade était quasi déserte. Barnes et le Starbucks avaient fermé. Des gens s'approchaient du banc de temps en temps, des poivrots qui se traînaient vers les Palisades pour la nuit, tirant leurs petits chariots remplis de leurs effets personnels. Ils voyaient là un homme assis, les bras le long du corps, les paumes au ciel, le regard plongé au fond de la rue. Aucun n'osa faire un écart pour lui demander une pièce. Ils passèrent leur chemin.

Zandt se leva enfin, alla jeter son gobelet vide à la poubelle. Il songea qu'il aurait pu entrer dans la librairie et chercher les coins ayant offert à l'Homme Debout un poste idéal pour guetter Sarah Becker. Car, même en l'absence de preuves matérielles, Zandt était convaincu que le type surveillait étroitement ses victimes avant de frapper. Quelques-uns ne prenaient pas cette peine, mais la plupart, si. Et si, avec Karen, l'Homme Debout avait dérogé à cette règle, histoire de faire passer un message ? Zandt en doutait. Les filles étaient trop semblables, et leurs disparitions trop rondement orchestrées.

Barnes pouvait attendre, peut-être même pour toujours. Il s'était laissé convaincre par Nina de revenir. Il avait voulu croire que cette fois ce serait différent, qu'il saurait faire mieux que chasser sa propre queue à travers toute la ville, criant dans la nuit, sans jamais trouver l'homme qui lui avait pris sa fille. Qui avait capturé la vie de Zandt dans sa main folle et enragée, pour la réduire en miettes. Mais ce soir il n'y croyait plus.

Il regagna le Fountain à pied, acheta quelques provisions en route. Le hall de l'immeuble était vide, même à la réception. Zéro muzak, et peu de raisons de croire qu'il n'était pas la seule âme dans les murs. L'ascenseur s'éleva lentement, par à-coups, comme pour souligner la pénibilité de sa tâche.

En attendant que l'eau bouille, il se planta debout devant la télé, tandis que CNN réduisait la complexité du monde en une série de phrases choc qu'un businessman puisse recracher à l'heure du déjeuner. Après quelques minutes, ce résumé fit place à une info toute fraîche. En fin de matinée, un homme d'une cinquantaine d'années avait traversé la grand-rue d'une petite ville anglaise. Il était armé d'un fusil, avec lequel il avait abattu huit adultes et blessé quatorze autres.

Personne ne savait pourquoi.

CHAPITRE 9

J'étais assis sur le siège passager avec la portière ouverte. Il était 8 heures et des poussières, le matin. J'avais un café au lait dans une main, une cigarette dans l'autre. Mes yeux étaient grands ouverts et secs, et je regrettais déjà d'avoir allumé cette cigarette. J'avais été fumeur. Un gros fumeur, pendant longtemps. Puis j'avais arrêté. Mais au cours de cette nuit, que j'avais passée à errer lentement et sans but sur des routes noires, comme si je cherchais la sortie d'un réseau infini de tunnels, j'en étais venu à l'idée que le tabac serait ma seule planche de salut. Quand on a fumé un certain temps, il est des situations où l'on aura toujours une sensation de manque, faute d'un bâtonnet d'herbes en combustion pincé entre les doigts. Sans cigarette on se sent mal-aimé, désemparé et seul.

J'étais garé dans la rue principale de Red Lodge, une petite ville située un peu moins de deux cents kilomètres au sud-est de Dyersburg. J'étais assis dans ma voiture car le snack où j'avais acheté le café – une boutique proprette où le personnel arborait tabliers et petites fossettes rieuses – montrait une résistance implacable aux arts tabagiques. De nos jours, la qualité du café est inversement proportionnelle à la probabilité qu'on vous laisse l'accompagner d'une clope. Leur *latte* était succulent : ils avaient accroché au mur des têtes de fumeurs empaillées. J'étais ressorti en bougonnant avec mon gobelet pour regarder à travers le pare-brise Red Lodge s'éveiller doucement. Les gens allaient et venaient, ouvraient de petites échoppes pleines d'objets qu'on achète pour prouver qu'on est parti en vacances. Une bande de types arrivèrent munis de pots de peinture et entreprirent de rehausser le teint d'une maison sise sur le trottoir d'en face. Quelques touristes apparurent, emmitouflés dans des tenues de ski jusqu'à devenir quasi sphériques.

Je grillai la moitié d'une seconde cigarette, grimaçai, et la jetai à terre. Ça ne m'aidait pas. Ce n'était qu'un motif de culpabilité.

Je crois savoir, en outre, que cela nuit à la santé. Conscient que ma volonté est à peu près aussi faible que la lumière de l'étoile la plus éloignée par une nuit couverte, j'attrapai le paquet sur le tableau de bord et le lançai en direction d'une poubelle clouée à un poteau et bardée de vertueux slogans civiques. Le projectile tomba dans le trou sans même toucher le cerceau. Personne n'était là pour le voir. C'est toujours pareil. Ça doit faire bizarre, d'être basketteur professionnel : les gens regardent quand on la met au fond.

Je n'avais pas rendu ma chambre d'hôtel. J'avais juste sorti la cassette du magnéto avant de descendre. J'envisageais sans doute de m'installer au bar, mais cette fois-ci même mon sens atrophié de la bienséance jugea cette idée incongrue. Je m'étais donc retrouvé sur le parking, à m'enfourner dans la voiture, et à prendre la route. J'avais roulé au pas dans Dyersburg, traversé à deux reprises le carrefour où la berline de mes parents s'était emplafonnée, la vidéo posée sur le siège de droite. Au second passage devant le lieu de l'accident, je jetai un coup d'œil sur la cassette, comme si cela pouvait m'aider en quoi que ce soit. Ce ne fut pas le cas, et je n'y gagnai qu'un frisson, petit spasme glacé invisible à l'œil nu.

Au bout d'un moment, j'avais atteint ma vitesse de libération et quittai la ville. Je progressai sans carte, me contentant de suivre la route et de tourner quand le cœur m'en disait.

Je finis pas me retrouver sur l'Interstate 80, alors que l'aube commençait à poindre. Sentant que j'avais besoin de café, ou de quelque chose, je pris l'embranchement qui me mena à Red Lodge à peu près à l'heure où les premiers commerces ouvraient.

Je me sentais vide et vaseux. Affamé, peut-être. C'était difficile à dire. Mon esprit était engourdi, comme si je l'avais poussé trop longtemps au mauvais régime.

Il ne faisait aucun doute que c'étaient bien mes parents sur les deux séquences les plus anciennes de la cassette. Et tout portait à croire que mon père tenait la caméra lors de la première, la plus récente. Individuellement ou collectivement, ces trois scènes avaient nécessairement une signification. Pourquoi les compiler sur cette bande, sinon ? C'est à peine si j'osais songer au dernier passage, celui montrant l'abandon d'un enfant dans un centre-ville. Ma première impression, comme quoi ce bambin était mon frère ou ma sœur, tenait toujours la corde. Tout dans les geste de ma mère, et dans la façon dont nous étions habillés, plaidait dans ce sens. Soit ce gosse et moi étions jumeaux, soit c'est ce que mes parents voulaient me faire croire. Cette dernière option semblait grotesque. Mais pouvais-je décemment admettre que j'avais eu un frère ou une sœur, et qu'il ou elle avait été abandonné(e) de sang-froid ? Que notre petite famille avait entrepris un voyage – détail suggéré à dessein par ces brèves images d'intérieur de train – pour revenir diminuée d'un quart ? Et que mon père avait tout filmé ? Je ne voyais qu'une seule raison à son attitude : la prévision qu'un jour il souhaiterait me révéler ce qui s'était passé, et que seul un film serait à même de vaincre mon incrédulité. J'avais rejoué la séquence dans ma tête toute la nuit, dans l'effort d'y lire autre chose. Mais c'était impossible, et ce qui me taraudait le plus, en définitive, était le caractère dépassionné de l'événement. Ils avaient cherché le bon endroit pour abandonner l'enfant, s'étaient arrêtés une première fois avant de se raviser, avaient poursuivi leur chemin sur quelques mètres. Ils avaient choisi un endroit fréquenté, où la densité de commerces et d'habitations augurait que l'enfant serait vite repéré. En un sens, c'était encore pire. Cela rendait leur acte plus réfléchi, plus délibéré, plus réel. Ils n'avaient pas tué l'enfant. Ils s'en étaient juste débarrassés. Ils avaient mûri leur projet, avant de passer à l'acte.

L'épisode du milieu était moins insolite. Passé ma stupeur devant ces bribes de jeunesse d'un couple que je n'avais jamais vraiment compris, j'avais vu une soirée somme toute assez banale. Je n'avais reconnu aucun de leurs copains, mais ce n'était pas surprenant. Les cercles d'amis évoluent avec les années. On change, on déménage. Ceux qui paraissaient indispensables le deviennent un peu moins, jusqu'à n'être plus que des noms sur une liste de cartes de vœux. Puis un beau jour on remarque en grognant qu'on n'a pas vu untel et untel depuis au moins dix ans, alors les cartes cessent, et l'amitié n'existe plus qu'en souvenir, à travers quelques phrases clés et une poignée d'expériences communes à moitié oubliées. Elle sommeille jusqu'à la toute fin, quand on regrette de ne pas avoir maintenu les liens, ne serait-ce que pour le plaisir d'entendre parler quelqu'un qui vous aura connu jeune, qui saura que votre déchéance pré-tombale est une blague d'un millésime récent, et que vous n'avez pas toujours été aussi décati.

Le hic, c'était la façon dont ils s'étaient adressés à la caméra. Et ce qu'ils avaient dit. Comme s'ils avaient su que je regarderais ce film un jour. À leur place, je me serais imposé un ton plus guilleret. « Salut, fiston, comment ça va ? Bons baisers du passé ! » Le ton de ma mère était à mille lieues de ça. Elle m'avait paru triste, résignée. La dernière phrase de mon père résonnait encore dans mon crâne : « Je me demande ce que tu es devenu. » À quoi rime une telle question quand son destinataire n'a que cinq ou six ans et se trouve endormi dans la même pièce que vous ? En un sens, cela collait assez bien avec la liquidation d'UnRealty : une profonde défiance vis-à-vis d'un être qui était leur fils. Je ne suis pas particulièrement fier de mon existence, mais indépendamment de ce que je suis ou ne suis pas devenu, je n'ai encore jamais abandonné un bébé en plein cœur d'une ville, tout en filmant la scène pour la postérité.

Autant que je me souvienne, je n'avais jamais vu de caméra à la maison ou entre les mains de mon père. Ni visionné ce type d'archives. Or pourquoi s'embêter à filmer sa petite famille si l'on n'a pas l'intention de se réunir un soir pour savourer le spectacle, rire des coiffures et des vêtements, et souligner combien tout le monde a pris des centimètres, en hauteur comme en largeur ? Et si mon père avait un jour donné dans ce cinéma-là, pourquoi avait-il arrêté ? Et où étaient passées les bobines ?

Restait la toute première scène, celle tournée en vidéo, bien plus récente. Par sa brièveté et son aspect abscons, ce segment semblait fournir la clé. En montant sa petite bombe à retardement, mon père avait dû commencer par ces images pour une raison précise. On l'entendait tout à la fin prononcer quelques mots, une courte phrase couverte par le vent. Il fallait que je la décrypte. Peut-être comprendrais-je ainsi la finalité de cette cassette. Ou peut-être pas. Mais au moins j'aurais entre les mains toutes les pièces du puzzle.

Je refermai la portière et sortis mon téléphone. J'avais besoin d'aide, alors j'appelai Bobby.

Cinq heures plus tard j'avais regagné mon hôtel. Entre-temps je m'étais arrêté à Billings, l'une des rares villes du Montana à peu près dignes de ce nom. Conformément aux conseils reçus, et contrairement à mes attentes, on y trouvait bien une boîte à copies où je pus obtenir ce que je cherchais. En vertu de quoi je rentrais avec un nouveau DVD-ROM en poche.

En pénétrant dans le hall de l'hôtel je me rappelai n'avoir réservé que pour deux jours après les funérailles. Je m'arrêtai donc à la réception afin de prolonger mon séjour. La fille hocha distraitement la tête, sans décoller les yeux d'un téléviseur réglé sur une chaîne d'information continue. Le journaliste ressassait les maigres éléments connus à cette heure concernant la tuerie survenue en Angleterre, que j'avais apprise à la radio sur la route de Billings. Il n'y avait apparemment rien de neuf. On répétait en boucle la même litanie, comme un rite, comme pour fabriquer un mythe. Le type s'était barricadé quelque part pendant deux heures, avant de se donner la mort. Les flics devaient en ce moment même retourner toute sa maison, à la recherche d'une explication, de quelqu'un ou de quelque chose à incriminer.

— Quel drame affreux ! dis-je, surtout pour vérifier si j'avais capté l'attention de la réceptionniste.

Dans le hall, des pancartes indiquaient que l'hôtel allait accueillir des stages et séminaires d'entreprise tout au long de la semaine, et je ne voulais pas me retrouver à la rue.

La jeune femme ne répondit pas sur le coup, et j'allais rouvrir la bouche quand je m'aperçus qu'elle pleurait. Ses yeux étaient trempés, et une larme isolée s'échappait en douce sur sa joue.

— Ça va ? demandai-je, un peu surpris.

Elle se tourna vers moi d'un air rêveur, hocha lentement la tête.

— Deux jours supplémentaires, chambre 304. C'est noté, monsieur.

— Parfait. Mais vous allez bien ?

Elle s'essuya la joue d'un rapide revers de main.

— Oui, oui. C'est juste... tellement triste.

Et sur ces mots elle retourna à son téléviseur.

Je l'observai depuis l'ascenseur en attendant que les portes se referment. Le hall était désert. Elle continuait de fixer l'écran, inerte, comme si elle regardait par la fenêtre. Elle n'aurait pas semblé plus chamboulée par l'événement – survenu à des milliers de kilomètres de là, dans un pays où elle n'avait sans doute jamais mis les pieds – si elle y avait elle-même perdu un proche. J'aimerais pouvoir dire qu'il m'inspira le même degré de compassion spontanée, mais ce serait mentir. Ce n'était pas de l'indifférence, plutôt l'incapacité à laisser l'émotion migrer de ma tête vers mon cœur. Ce n'était pas comme le World Trade Center, un acte ignoble et ahurissant survenu sur notre propre sol, frappant des êtres qui, gamins, avaient rempli leur tirelire avec les mêmes pièces que nous tous. Je savais, sur le plan intellectuel, que cela ne faisait aucune différence, mais c'est ainsi que je sentais les choses. Je ne connaissais pas ces gens-là.

De retour dans la chambre, je sortis mon ordinateur de la penderie, le posai sur la table et l'allumai. Pendant qu'il démarrait, je pris le DVD-ROM dans ma poche. La vidéo de mon père était cachée sous la roue de secours de la voiture de location. Le disque en contenait une version numérisée. Quand mon PowerBook eut accompli sa petite routine de réveil – une douche, une gorgée de café, un coup d'œil rapide sur le journal, et je ne sais quel autre rituel interminable –, j'insérai le disque dans la fente latérale. Une icône ronde apparut sur le Bureau. La vidéo avait été dupliquée sous forme de quatre gros fichiers MPEG, car elle était trop longue pour tenir tout entière en résolution maximale dans un seul fichier. Aussi, profitant de ce que personne dans la boutique ne regardait par-dessus mon épaule, j'avais copié les première et dernière séquences en haute définition, ainsi que l'extrait du milieu qui se déroulait chez mes parents. Quant à la longue scène du bar, je l'avais convertie dans un plus gros grain. L'opération n'en fut pas rapide pour autant. Tout le bazar tenait de justesse sur un disque de dix-huit gigas.

J'essayai d'abord CastingAgent, un vieux programme de traitement de l'image qui bogue à mort mais permet parfois d'obtenir ce que d'autres logiciels interdisent. Il se planta de façon si nette que je dus redémarrer. Je me rabattis alors sur les logiciels classiques et je vis défiler mon film.

J'avançai jusqu'à la fin de la première séquence, celle prise dans la montagne. Je sélectionnai les dix dernières secondes et les transférai sur le disque dur. Puis je recourus à MPEGSplit pour éliminer la composante vidéo du fichier, de manière à ne conserver que la piste son. Je connaissais déjà l'image : un groupe plus ou moins compact d'individus en manteaux noirs. Ce que je voulais percer, c'étaient les paroles du cameraman.

Je sauvegardai le fichier, quittai l'éditeur vidéo et lançai une batterie d'applications professionnelles de traitement du son – SoundStage, SFXLab, AudioMelt Pro. Je passai les trente minutes suivantes à triturer la piste, la soumettant à différents filtres pour voir ce qu'il en ressortait. Augmenter l'amplitude n'arrangeait rien, sinon le volume, tandis que le déséchantillonnage aléatoire et la réduction de bruit la rendaient plus trouble. Je découvris tout au plus qu'elle était formée de deux ou trois mots.

Alors je passai aux choses sérieuses, en commençant par prélever un nouvel extrait, juste avant le début des paroles. J'analysai les fréquences du vent en fond sonore, puis bricolai un filtre de bandes passantes. Je l'appliquai sur le premier extrait, et le son devint plus net. Quelques ajustements complémentaires, et il se mua en mots. *Ram ne bal* ? *Vom te pam* ? Résolument bloqué, je sortis de la sacoche une paire d'écouteurs et l'enfilai. Je programmai la piste en boucle et fermai les yeux.

Vers la quarantième écoute je trouvai la solution : « Les Hommes de Paille ».

J'arrêtai le défilement, ôtai les écouteurs. J'étais assez sûr de mon coup. Les Hommes de Paille. Le seul problème, c'est que ça ne voulait rien dire. On aurait cru le nom d'un groupe de rock indé – même si je doutais que les gens montrés à l'image aient vécu du commerce de braillements sous-enregistrés. Les membres d'un groupe de rock ne cohabitent pas dans une station de ski. Ils se font bâtir des demeures en simili-Tudor aux quatre coins du monde et ne se retrouvent que pour empocher de l'argent. En somme, je n'avais réussi qu'à ajouter un ultime voile de mystère au contenu de cette cassette. Je la visionnai en entier une dernière fois, d'après le DVD, au cas où le changement de format m'eût apporté un élément nouveau. Mais rien ne me frappa.

Je m'éternisai dans mon fauteuil, le regard vague, sentant la nuit me rattraper. De loin en loin, j'entendais quelqu'un longer ma porte dans le couloir ; dehors, le passage occasionnel d'une voiture, ou les fragments diffus de lointaines conversations entre des gens que je ne connaissais ni ne connaîtrais jamais. Cela non plus n'avait aucun sens à mes yeux.

Sur le coup de 6 heures, la sonnerie de mon portable m'arracha à la somnolence. Je décrochai, tout engourdi.

— Yo, lança une voix sur fond de conversations et de musique sourde. Ward, c'est Bobby.

— Mon sauveur, répondis-je tout en me frottant les yeux. Merci pour le tuyau. Ton adresse à Billing était parfaite.

— Cool. Mais ce n'est pas pour ça que j'appelle. Je me trouve dans cet endroit – putain c'est quoi son nom déjà ? Le Sacagawea, je crois. Un genre de bar. Genre. Sur l'avenue principale. Avec une enseigne monstrueuse.

Soudain j'étais fin réveillé.

— Tu es à Dyersburg ?

— T'as tout compris. Un saut d'avion, et hop !

— Mais qu'est-ce qui t'as pris ?

— Ben, tu vois, après ton appel, je me faisais un peu chier. Alors j'ai repensé à un truc que tu m'avais dit, et j'ai jeté un œil à gauche, à droite.

— Jeté un œil sur quoi ?

— Des trucs. Allez, ramène ton cul, Ward. Il y a une bière qui t'attend. J'ai un truc à t'annoncer, mon pote, et je le ferai pas au téléphone.

— Pourquoi ?

Je remballais déjà l'ordinateur.

— Parce que ça va te faire flipper.

CHAPITRE 10

Le Sacagawea est un vaste motel de la grand-rue. Il possède une énorme enseigne au néon multicolore, que l'on distingue à plus d'un kilomètre dans chaque direction, et qui attire les imprudents tel un aimant. J'y avais résidé une dizaine de minutes, la première fois que j'étais venu rendre visite à mes parents. La chambre qu'on m'avait attribuée déclinait un mobilier bas de gamme digne d'un musée des années soixante, avec des tapis ressemblant à un chien mal-aimé. De prime abord j'avais trouvé cela plutôt funky, avant d'y regarder de plus près et de constater qu'elle n'avait en fait jamais été redécorée depuis l'année de ma naissance. C'est en découvrant qu'elle n'offrait pas de service d'étage que j'avais pris mes cliques et mes claques. On ne me fera pas dormir dans un hôtel sans service d'étage. Je ne le supporterais pas.

Étroit et humide, le vestibule empestait le chlore, sans doute à cause de la minuscule piscine qu'abritait la pièce d'à côté. Une vieille dinde derrière son comptoir me dirigea vers l'étage sans recourir à la parole mais au moyen d'un étrange regard. Une fois au bar, je compris pourquoi. Le lieu ne débordait pas de vie. J'y découvris un comptoir central, une serveuse esseulée, et une rangée de machines à sous placidement alimentées par des gens aussi démodés qu'elles. Vraiment, notre espèce sait vivre. De grandes vitres donnaient sur le parking et sur le trafic qui perlait de part et d'autre de la grand-rue. Quelques couples étaient dispersés dans la salle, conversant à grand bruit, comme s'ils espéraient insuffler à ce lieu un semblant d'atmosphère. En vain.

Attablé contre la baie vitrée, m'attendait Bobby Nygard. Ses premiers mots furent :

— C'est quoi, cette connerie de « Sacagawea » ?

Je m'assis en face de lui.

— Sacagawea est le nom d'une jouvencelle amérindienne qui traînait avec Lewis et Clark. Elle les a aidés à négocier avec les gars du coin, à ne pas se faire tuer, ce genre de trucs. Leur expédition est passée à quelques encablures d'ici, en remontant vers la chaîne de Bitterroot.

— Merci, professeur. Mais est-on autorisé à dire « jouvencelle » de nos jours ? Ce n'est pas un peu sexiste, comme on dit ?

— Sûrement. Et tu sais quoi ? J'en ai rien à branler. C'est toujours mieux que « squaw ».

— Ah oui ? C'est pas sûr. C'est peut-être comme « nègre ». Repris comme un étendard. Appropriation de la terminologie de l'oppresseur.

— Si tu veux, Bobby. Ça fait du bien de te revoir.

Il me lança un clin d'œil, et nous choquâmes nos verres. Bobby n'avait pas bougé d'un trait, bien que notre dernier tête-à-tête remontât à plus de deux ans. Un peu plus petit que moi, un peu plus large. Le cheveu ras, le visage toujours un brin empourpré, et l'air du type que vous pourriez passer à la batte de base-ball sans l'incommoder outre mesure. Un ancien des Forces, qui donnait parfois l'impression d'en être encore – mais pas dans le genre d'armée qu'on voit aux infos.

Après avoir trinqué, il reposa son verre sur la table et parcourut la pièce du regard.

— Un vrai trou à rats, ce bouge.

— Alors pourquoi t'es là ?

— J'ai aperçu un panneau d'enfer. Il m'a aspiré dans son champ d'attraction. Mais pourquoi ? Y a un meilleur hôtel en ville ?

— Non, je veux dire : qu'est-ce qui t'amène à Dyersburg ?

— Chaque chose en son temps. Dis-moi d'abord comment tu te sens. Je suis désolé pour le décès, mec.

Soudain, peut-être parce que j'étais en présence d'un être que je considérais comme un ami, la disparition de mes parents m'atteignit de nouveau. En plein cœur, sans crier gare, comme ce serait probablement le cas jusqu'à la fin de mes jours, quoi que ces deux-là aient pu commettre. J'allais dire quelque chose, puis renonçai. Je me sentais trop fatigué, perturbé, triste.

Bobby a fait de nouveau tinter nos verres et nous avons bu. Il a laissé planer un petit silence puis changé de sujet :

— Alors, tu fais quoi à présent ? Tu ne me l'as pas dit.

— Pas grand-chose.

Il a levé un sourcil.

— C'est si tabou que ça ?

— Non. Rien qui mérite qu'on en parle, c'est tout. Il y a peut-être un ou deux jobs que je n'ai pas encore essayés, mais je doute qu'ils soient très différents. Il semble que j'aie gardé un faible pour les activités parallèles, mais les employeurs ne veulent pas voir en quoi c'est un rôle clé dans une économie moderne.

— Frilosité et absence de vision commerciale, a opiné Bobby avant de commander d'un geste deux autres bières. Toujours la même rengaine.

Après que la serveuse d'aspect jeune et déprimé nous eut servis, nous bavardâmes un moment. J'avais officié neuf ans à la CIA, et c'est là que j'avais rencontré Bobby. On avait tout de suite accroché, lui et moi. J'avais surtout fait du terrain, sauf à la fin, où je m'étais occupé de surveillance média. J'avais démissionné quand l'Agence avait instauré un test annuel au détecteur de mensonges, voici quelques années. Bon nombre d'agents avaient rendu leur tablier à la même époque, furieux de ce manque de confiance implicite après qu'ils eurent risqué leur peau pour le pays. Moi, j'étais parti parce que j'avais commis certaines choses. Rien d'affreux, devrais-je préciser. Juste le genre de choses pour lesquelles on vous fout en taule. La CIA n'est peut-être pas l'organisation la plus réglo du monde, mais dans la mesure du possible elle préfère que ses employés évitent de verser dans la criminalité. J'avais utilisé certains contacts pour lever un peu d'argent, pomper un peu de liquide à travers les mailles du filet. Il y eut quelques incidents. Un type y perdit la vie. Voilà tout.

S'il résidait maintenant en Arizona, Bobby assurait encore des missions ponctuelles pour la maison, et gardait le contact avec une poignée de vieux amis communs. Deux d'entre eux s'attelaient en ce moment à infiltrer des milices paramilitaires, et en entendant ça je me félicitai de mon choix. Qui voudrait d'un tel job ? Personne qui tienne à la vie, en tout cas. L'un de ces gars, un frappadingue dénommé Johnny Claire, vivait présentement au sein d'un de ces groupes, une brochette de fanas d'armes à feu désocialisés, terrés dans une forêt de l'Oklahoma. Je ne l'enviais vraiment pas, bien que Johnny fût assez allumé pour s'imposer dans n'importe quel milieu.

— Bon, lança Bobby derrière sa nouvelle bière. Tu vas enfin m'expliquer pourquoi tu t'éternises dans cette cambrousse, et d'où vient ce besoin pressant de numériser un bout de vidéo familiale ?

— Peut-être, répondis-je, admiratif devant sa façon de me tirer les vers du nez sans révéler ce que lui-même avait derrière la tête – comme un savoir-faire devenu seconde nature.

À l'époque de notre rencontre, il passait beaucoup de temps dans les salles d'interrogatoire avec des ressortissants du Moyen-Orient. Ils finissaient tous par parler. De là il avait bifurqué vers la surveillance.

— Mais je ne promets rien, poursuivis-je. Pas tant que tu ne m'auras pas dit pourquoi tu as sauté dans l'avion et traversé trois États afin de me payer une bière.

— OK. OK. Alors laisse-moi d'abord te poser une question. Où es-tu né ?

— Bobby...

— S'il te plaît, Ward.

— Tu sais parfaitement où je suis né. County Hospital, à Hunter's Rock, Californie.

Ce lieu sortait de ma bouche aussi facilement que mon prénom. C'est l'une des premières choses qu'on apprend.

— En effet, je me souviens que tu m'as dit ça. T'as même piqué une crise parce que plus personne ne respecte l'apostrophe de Hunter's.

— Ça me fout hors de moi.

— Tu as raison. C'est un pur scandale. Bon. Tout à l'heure au téléphone, tu m'as parlé de tes vieux et de cette fameuse cassette, comme quoi elle avait un rapport avec ton enfance, etc. Voilà donc où j'en suis quand on raccroche. Je n'ai rien à faire de particulier. Je suis entouré d'ordinateurs, j'ai surfé sur le Web jusqu'à plus soif et j'ai déjà eu ma branlette du jour.

— Charmant, Bobby. J'espère seulement que ce n'était pas pendant que je te parlais.

— Rêve pas trop, fit-il avec un sourire un coin. Alors je me suis dit : « Tiens, et si je fouillais un peu dans le vie de Ward ? »

Je le dévisageai, sans oublier qu'il était mon ami et pouvait à ce titre se permettre ce genre de choses, mais avec malgré tout une certaine sensation d'intrusion.

— Je sais, je sais, poursuivit-il en levant une paume pacifique. Mais je m'emmerdais, et que veux-tu que je te dise ? Je suis désolé. Bon, bref, j'active mes bécanes et j'ouvre quelques bases de données. Je te préviens tout de suite : je n'ai rien trouvé que je ne sache déjà. Interrogé sur quelques affaires survenues au fils des ans, et patati et patata, puis relaxé pour manque de preuves. Plus un témoin qui s'est rétracté. Et celui qui a disparu. L'opération anti-drogue dans la Grande Pomme en 1985, quand tu as accepté de dénoncer un groupe d'étudiants de Columbia...

— C'étaient des connards, plaidai-je. De sales connards racistes. En plus, l'un d'eux couchait avec ma copine.

— Par pitié, mec. Tu m'as déjà raconté tout ça et je ne porte aucun jugement sur cette histoire. Et puis, sans elle tu ne te serais jamais retrouvé à l'Agence et on ne se connaîtrait pas, ce que j'estimerais fort dommage. Je disais donc que tes dossiers ne contiennent rien que je ne connaisse déjà. Autrement dit, s'il y a autre chose, tu l'as bien caché. Drôlement bien. J'avoue que ça ne me déplairait pas de savoir quoi, par simple curiosité.

— Je te le dirai pas. Un type doit avoir ses petits secrets.

— En tout cas, tu les conserves bien. Je t'accorde au moins ça.

— Ça veut dire quoi ?

— Attends. Au bout d'une heure ou deux, je suis assez déçu de n'avoir rien découvert, alors je décide de vérifier des trucs du côté de Hunter's Rock – avec apostrophe. J'obtiens l'adresse de la maison de tes parents, et leurs dates d'arrivée et de départ. Ils ont emménagé là-bas le 9 juillet 1956, ce qui devait être un lundi. Ont payé leurs impôts, mené leur petit bonhomme de chemin, etc. Ton père gagnait sa vie chez Golson Realty, ta mère bossait à mi-temps comme vendeuse. Une bonne dizaine d'années plus tard, tu as vu le jour là-bas. C'est exact ?

— Exact, répondis-je tout en me demandant où il voulait en venir.

Il secoua la tête.

— Faux. Le County Hospital de Hunter's Rock n'a aucune trace d'un Ward Hopkins né à cette date.

Le monde parut exécuter un petit pas chassé.

— Pardon ?

— Ils n'en ont pas davantage au General de Bonville, ni au James B. Nolan, ni dans aucun hôpital situé dans un rayon de trois cents bornes.

— Normal. Je suis né au County. À Hunter's.

Il secoua la tête de plus belle.

— Je te dis que c'est faux.

— Tu es en sûr ?

— Non seulement j'en suis sûr, mais j'ai épluché les cinq années précédentes et les cinq années suivantes, des fois que tu aurais menti sur ton âge par coquetterie ou nullité en calcul mental. Pas le moindre Ward Hopkins. Aucun Hopkins, quel que soit le prénom. Je ne sais pas où tu es né, mon pote, mais je te garantis que ce n'était ni à Hunter's Rock ni dans ses environs.

J'ouvris la bouche. La refermai.

— Ce n'est peut-être pas grave, déclara-t-il avant d'ajouter habilement : y aurait-il un rapport avec tes besoins numériques ?

— Repasse-le un coup, dit-il.

— Je doute d'en avoir la force.

Il leva les yeux vers moi. Il était assis dans l'un des fauteuils de la chambre d'hôtel, penché sur mon ordinateur. Je lui avais montré les MPEG, et j'avais atteint mon point de saturation pour la journée. Pour le reste de ma vie, même.

— Crois-moi, Bobby. Tout ce qu'il y a à voir est là dès le premier visionnage.

— OK. Alors passe-moi la piste audio.

Je tendis le bras, posai le curseur sur le fichier et l'ouvris d'un double clic.

Il écouta la version filtrée plusieurs fois, puis l'arrêta de lui-même. Il hocha la tête.

— Oui, « Les Hommes de Paille », ça m'a tout l'air d'être ça. Et tu ne sais pas ce que ça peut signifier ?

— Je connais la définition du dico, « prête-nom », mais ça ne nous avance pas des masses. Et toi ?

Il attrapa son verre. Il nous restait à ce stade une demi-bouteille de Jack Daniel's.

— La seule chose qui me vienne à l'esprit est « achat de paille ».

J'acquiesçai, considérai l'idée. Il faisait référence à la combine permettant aux personnes interdites d'armes à feu – pour raison d'âge, de casier judiciaire ou d'absence de permis – de s'en procurer malgré tout. Il suffit de se rendre chez un armurier en compagnie d'un acolyte présentant les qualités requises. Vous discutez avec le vendeur, trouvez votre bonheur. Puis, quand arrive le moment de payer, c'est votre copain – l'acheteur de paille – qui tend l'argent et se porte officiellement acquéreur. Bien entendu, le marchand n'est pas censé accepter, à partir du moment où il sait que l'arme finira entre vos mains, mais la plupart du temps il fermera les yeux. Les affaires sont les affaires. Une fois que vous êtes ressorti de sa boutique, il se contrefiche de savoir ce que vous mijotez. Tant que vous n'allez pas zigouiller sa mère, il n'en a rien à cirer. Certes, on trouve beaucoup de gens honnêtes et droits parmi les armuriers. Mais on en trouve aussi beaucoup qui jugent en leur for intérieur que tous les Américains, tous autant qu'ils sont, y compris les petites dames, devraient porter une arme à feu dès la naissance. Qui ne voient rien à redire au fait que ces engins ne sont qu'un moyen commode d'ôter la vie d'autrui, qui pensent que les flingues sont moralement neutres et que seuls leurs utilisateurs ont le pouvoir de les rendre mauvais. Les utilisateurs à la peau noire, notamment, ou ces bons à rien de punks drogués que nous ne servons pas dans cet établissement, ça non.

— Tu crois que c'est ça ?

— Peu probable, convint Bobby. Remarque, il y a eu un truc à ce sujet ces deux dernières années. Les fédéraux et quelques municipalités ont tenté de remettre un peu d'ordre, de traquer les revendeurs trop complaisants. Une forte proportion des armes qui circulent dans les quartiers y sont arrivées par ce biais, via des types qui achètent en grosses quantités puis revendent à des trafiquants de rue. Il y a un ou deux procès en cours, et je crois même que l'un d'eux a fait jurisprudence voilà un an. Je ne me souviens plus du verdict. Mais, quoi qu'il en soit, je ne vois pas le rapport avec tes vieux.

— Moi non plus. Autant que je sache, mon père n'a jamais possédé d'arme. Je ne me souviens pas l'avoir entendu se prononcer sur la question, mais en général les pro-flingues ont une armoire bien remplie. Rien de tel à la maison.

— Et t'as fait une recherche ?

— Où ça ? Dans le Grand Dictionnaire des Phrases Courtes ?

Il leva les yeux au ciel.

— Sur le Net, voyons !

— Tu m'as bien regardé ?

J'aime Internet. Si, si, vraiment. Dès que je suis en quête d'un programme vérolé, de la météo de Bogota ou d'une photo de femme avec un mulet, je suis le premier à faire chauffer le modem. Mais comme source d'informations, ça craint. Parmi ces milliards de données qui luttent pour être entendues, lues ou téléchargées, ce qui m'intéresse semble chaque fois mourir piétiné sous la foule. Curieusement, il suffit que je cherche une chose particulière pour récolter direct un message d'erreur 404.

— T'es un putain de réac, Ward.

Déjà Bobby raccordait le câble du téléphone. Je le laissai à ses distractions, tout en regrettant d'avoir jeté mon paquet de cigarettes à la poubelle.

Au bout de cinq minutes, il secoua la tête.

— Les principaux moteurs de recherche ne donnent rien, les moins connus non plus. Idem avec une série de netcrawlers spécialisés auxquels j'ai accès, dont certains requièrent des autorisations maousses.

— Je te l'avais bien dit, le Web, c'est comme un oracle sourd-muet frappé d'amnésie.

— Ça ne signifie pas qu'il n'y a rien, Ward. Ça signifie seulement que, si le terme apparaît sur un site, ce dernier n'est pas répertorié par les moteurs de recherche.

— Mais enfin, Bobby, il n'y a aucune raison de croire qu'on trouvera quoi que ce soit là-dedans. Tout ce qui est survenu sur terre n'y est pas encore répertorié. Et puis ce n'est qu'une phrase, bon sang. Trois mots. Laisse évoluer une bande de singes pendant un certain temps, et l'un d'eux alignera cette expression bien avant d'être venu à bout de *Macbeth*. Mais de là à en déduire qu'il va te concocter du HTML et te le balancer sur un serveur avec des bandeaux de pub et un compteur de visites... Et même si c'était le cas, pourquoi voudrais-tu qu'il y ait un rapport avec le contenu de la cassette ?

— Tu as une meilleure idée ?

— Oui, clamai-je. La bouteille est presque vide, je suis crevé et j'ai encore très soif.

— On verra ça après.

— Après quoi ? Tu sais déjà qu'il n'y a rien là-dedans.

Bobby pianota des doigts sur la table, le regard obliquant vers les rideaux. Je pouvais presque entendre sa cervelle mouliner. J'étais rongé d'ennui et le whisky rendait ma tête lourde et froide. Ce trop-plein d'informations depuis deux jours me donnait envie de tout oublier.

— Il doit rester quelque chose dans la maison, énonça-t-il enfin. Un truc qui t'aura échappé.

— Seulement si c'est planqué dans une putain d'ampoule. J'ai fouillé les moindres recoins. Il n'y a rien de plus.

— Tout est différent quand tu sais ce que tu cherches, insista-t-il. Tu croyais d'abord chercher un second message. Alors c'est ça que tu as cherché. C'était ton fil conducteur. Tu n'as songé à la vidéo que par hasard.

— Non. J'y ai songé parce que la maison avait été arrangée dans ce but. Je pense que mon père s'est donné du mal pour...

Je laissai ma phrase en suspens. Me levai, farfouillai dans la sacoche de l'ordi.

— Quoi ?

— J'ai copié son disque dur sur une carte ffiz !. C'est la seule chose que je n'aie pas examinée.

Je repris place à côté de Bobby et insérai la minuscule cartouche dans la machine. Là-dessus, j'ouvris un bandeau de recherche et tapai « hommes de paille ». Pressai la touche retour. L'appareil crépita et ronronna quelques instants.

L'OBJET DE LA RECHERCHE N'A PU ÊTRE TROUVÉ.

Je réessayai avec « paille » uniquement. Même résultat.

— Eh bien, nous voilà fixés, dis-je. Viens, le bar nous appelle.

Je me relevai, pensant qu'il allait m'imiter. Au lieu de quoi il entreprit une nouvelle recherche.

— Qu'est-ce que tu fabriques encore ?

— Je demande à ton explorateur de lister le contenu de tous les fichiers textes, expliqua-t-il. Si cette histoire de paille est importante, il semblerait logique de ne trouver aucun fichier sous ce nom. Simple mesure de précaution. Par contre, le terme peut figurer à l'intérieur d'un fichier.

C'était plutôt sensé. Alors j'attendis. Le ffiz ! offre une grande vitesse de traitement, si bien que l'opération ne prit que quelques minutes.

Le texte demeurait introuvable.

Bobby poussa un juron.

— Mais pourquoi il ne t'a pas laissé une putain de lettre pour te dire ce qu'il avait à dire, bordel ?

— Je me suis posé cette question des milliards de fois, et la réponse est : je n'en sais rien. Allons-y.

Il refusait toujours de se lever.

— Écoute, répliquai-je, je sais que tu fais tout ça pour moi, et je t'en suis très reconnaissant. Mais au cours des dernières vingt-quatre heures j'ai appris que j'avais soit des parents chtarbés qui m'avaient séparé d'un jumeau, soit des parents vraiment chtarbés qui voulaient juste me mener en bateau. Je n'ai rien avalé depuis des jours. Comme un idiot, j'ai allumé une cigarette ce matin et maintenant j'aimerais en griller toute une cartouche, et je dois mobiliser toute mon énergie mentale pour résister à l'envie. C'est tout ce que j'avais à dire. Je file au bar.

Il se tourna vers moi, mais son regard était lointain. Je lui avais déjà vu ces yeux-là. Ils signifiaient que Bobby ne pouvait pas bien entendre ce qu'on lui racontait, et n'y parviendrait qu'après avoir poursuivi son idée.

— On se retrouve en bas, dis-je avant de sortir.

CHAPITRE 11

Je me souviens d'une chose dont j'étais très fier étant gamin : les moustiques ne me piquaient jamais. Quand nous partions en vacances dans les bons coins ou que nous effectuions un voyage scolaire à la mauvaise saison, la plupart des gens se retrouvaient couverts de petits boutons rouges qui démangeaient à mort – et ce, quels que soient leurs efforts en matière de crèmes, de sprays et de moustiquaires. Moi, non. Au pire, j'écopais d'une seule morsure, à la cheville. Curieux objet de fierté, penserez-vous peut-être, mais vous savez comment on est à ces âges-là. Une fois que l'on a compris qu'on n'est pas le centre du monde, on éprouve une telle soif de se distinguer qu'on se contente d'à peu près n'importe quoi. Moi, j'étais le garçon qui ne se faisait pas piquer par les insectes. Prenez note, messieurs dames, et un peu de respect, je vous prie : voici Anti-Mosquito Boy, le Gosse impiquable. Puis un beau jour, à l'approche de la trentaine, je me rendis compte que j'avais tout faux. Que j'avais, selon toute vraisemblance, reçu le même lot de piqûres que les autres. La seule différence étant qu'elles déclenchaient chez moi une réaction allergique moins forte, ce qui expliquait l'absence d'indurations. Je restais toutefois « à part » – même si en grandissant j'avais compris que mon titre n'avait rien de mirobolant, et me souciais surtout, désormais, de ne pas être trop différent des autres –, mais moins que je ne l'avais cru. Je me faisais piquer comme vous tous, et Anti-Mosquito Boy était régulièrement vaincu.

Assis au bar en attendant Bobby, je peinais à déloger ce souvenir. Ma famille, ma vie étaient deux choses que je ne comprenais plus. Comme si j'avais soudain remarqué que l'on apercevait sans cesse les mêmes immeubles en arrière-plan de mon existence, où que je me trouve, au point que j'en venais à me demander s'il ne s'agissait pas d'un décor de cinéma. À vrai dire, je voyais bel et bien les mêmes immeubles jour après jour. Depuis l'Agence, je n'avais jamais pris le chemin d'une existence normale, et revoir Bobby rendait cette vérité criante. Je faisais un peu de ci, un peu de ça ; une partie de ci était illégale, une partie de ça était violente. La plupart du temps, je me rappelais à peine en quoi cela avait consisté. Ça devenait flou. Je passais ma vie dans des motels, des restaurants, des aéroports régionaux, à parler à des inconnus, à lire des écriteaux adressés à la foule en général et jamais à moi en particulier. Je me sentais cerné de gens dont la vie avait une épaisseur, qui ressemblaient aux quidams qu'on voit à la télé. Contextualisés. Parties prenantes d'une histoire, avec ses phases habituelles. La mienne paraissait n'en avoir aucune. Le segment « voici d'où tu viens » venait d'être tronçonné, laissant place à un nombre non divulgué de pages vides.

Mon barman était de service, qui se montra une fois de plus un allié capable et efficace. Il expédia l'aspect « incident précédent » de nos retrouvailles en abordant le sujet d'emblée :

— Et après, vous comptez sortir votre flingue ?

— Pas si vous me filez des noix de cajou.

Il m'en fournit. C'était un bon barman, décrétai-je. La salle était vide de robots-cadres, et les seuls autres clients étaient une paire de couples très âgés, installés dans un coin. Ils avaient accueilli mon arrivée d'un air hostile. Je ne leur en voulais pas. Moi aussi, à leur âge, je maudirai les jeunes gens. Je les maudis déjà, d'ailleurs, ces petits trouducs d'échalas aux joues roses. Ça ne m'étonne pas que les ultra-vieux soient si bizarres et si ronchons. La moitié de leurs amis sont morts, ils se sentent merdeux les trois quarts du temps, et le prochain événement majeur de leur vie sera le dernier. Ils n'ont même plus l'espoir d'améliorer leur état dans un club de gym, de rencontrer une personne charmante dans les premières heures d'un samedi matin, ou de voir rebondir leur carrière et de finir au bras d'une star de cinéma. Ils sont passés de l'autre côté, vers une morne plaine de douleurs et de vue déclinante, là où le froid ronge les os et où il n'y a pas grand-chose à faire sinon regarder ses enfants et petits-enfants commettre bille en tête les erreurs contre lesquelles on les avait mis en garde. Je ne leur reproche pas d'être un peu déphasés. Je me demande simplement pourquoi les anciens ne sont pas plus nombreux à arpenter les rues en bandes pour beugler des insanités et se finir à la bière. Au train où va la démographie, ce sera peut-être ça, le prochain grand phénomène de société. Des gangs d'octogénaires, courant comme des dératés entre deux prises de crack. Quoique *marchant* comme des dératés paraisse mieux adapté – avec peut-être une heure de sieste de dératés dans l'après-midi.

Au bout d'un moment, le groupe du coin parut admettre que je n'allais pas me mettre à jouer d'un instrument ultramoderne ni braver les bonnes mœurs sexuelles. Ils retournèrent à leurs moutons, et je revins aux miens : nous coexistions, telles deux espèces animales partageant avec circonspection le même point d'eau.

Près de deux heures plus tard, Bobby fit son apparition. M'apercevant avachi dans mon box, il fit signe au barman d'apporter deux fois ce que j'avais pu commander, et me rejoignit.

— T'es murgé comment ?

Il me regardait d'un air bizarre.

— Sur une échelle de un à dix, répondis-je gaiement, je me mettrais au moins F.

— Parfait. J'ai trouvé quelque chose. Enfin presque.

Je sentis la tension revenir au galop. En me redressant je vis qu'il tenait une petite liasse de papiers.

— J'ai pu avoir accès à l'imprimante de la réception, dit-il. Mais où sont nos boissons, bordel ?

Elles arrivèrent au même moment dans les mains du barman.

— Pas d'autre noix ? demanda ce dernier.

— Non, non, répliquai-je. Juste lui et moi.

Je ris longuement. Oui, je suis sûr d'avoir ri. Le garçon s'éloigna. Bobby attendit que je reprenne mes esprits. Il me fallut du temps.

— OK, déclarai-je enfin. Crache.

— Pour commencer, j'ai refait un tour sur la toile. Toujours aucune trace des Hommes de Paille en tant qu'entité, mais j'ai trouvé sur des pages encyclopédiques d'autres explications à l'expression : une histoire de types qui au siècle dernier se plantaient à l'extérieur des tribunaux avec de la paille dans les grolles – j'ai pas trop compris ce passage – pour indiquer qu'ils produiraient de faux témoignages moyennant finance. Et une autre entrée évoquant l'absence de conscience – un truc du style paille contre chair.

— Autrement dit, des prête-noms trempant dans l'illégalité. On le savait déjà, merci.

— Puis j'ai examiné le disque, poursuivit-il en ignorant mon commentaire. Lancé un balayage de bas niveau, cherché des fichiers cachés, des partitions, des travaux. Rien du tout. Alors j'ai regardé ses logiciels, qui sont peu nombreux.

— Papa n'était pas un fêlé de la bécane, dis-je. C'est pour ça que je n'ai pas pris la peine d'explorer l'ordinateur sur place.

— Soit. Mais il utilisait le Net.

Je haussai les épaules :

— L'e-mail, de temps en temps. Et sa boîte possédait un site, même si la maintenance avait été confiée à un tiers. J'allais parfois y faire un tour.

Cela m'avait en effet semblé plus facile que de téléphoner à mes vieux. Depuis que j'avais laissé tomber la fac, ils n'avaient jamais vraiment su ce que je faisais, et encore moins les raisons qui m'avaient poussé à lâcher mes études, pas plus que le sigle de mon nouvel employeur. Mes parents n'avaient jamais paru très politisés, mais ils avaient pleinement vécu les années soixante, ainsi que la vidéo l'attestait. Ceux qui avaient connu le Summer of Pattes d'Eph en gardaient toujours quelques séquelles. À ce titre, découvrir que leur fils bossait pour la CIA ne les aurait pas exactement emballés. Je le leur avais donc caché, sans me rendre compte que cela revenait à leur cacher tout le reste. Bien sûr, cette pensée avait l'air un peu décalée aujourd'hui, sachant ce qu'eux m'avaient dissimulé...

Bobby secoua la tête.

— Il avait Explorer et Navigator dans sa bécane, et de toute évidence il en faisait grand usage. Une mémoire tampon énorme, et des trilliards de signets dans l'un comme dans l'autre.

— Pour quel type de contenus ?

— Tout ce qu'on veut. Références, commerce en ligne, sport...

— Du cul ?

Il sourit.

— Non.

— Dieu merci.

— Je les ai tous visités, sans exception. Même ceux qui ne m'inspiraient rien, juste au cas où il aurait renommé le signet pour masquer la nature du lien.

— T'es vicieux, mon pote. Je l'ai toujours dit.

— Ton père aussi, gars. Il en avait effectivement renommé un, noyé dans un dossier de cent soixante signets qui représentent à mes yeux la face la plus austère du monde de l'immobilier. Le lien s'appelait « Ventes récentes Mizner/Intercoastal ». Ça t'évoque quelque chose ?

— Addison Mizner était un architecte des années vingt et trente. Il a bâti une série de demeures de prestige à Miami et à

Palm Beach. Style villas italiennes. Très recherchées et incroyablement hors de prix.

— T'en sais des choses dingues, toi. Sauf que voilà : ce lien ne débouche sur aucun site traitant de terrains ou de maisons. Il mène à une page blanche. Alors je me suis dit : « Merde, un cul-de-sac ». Il m'a fallu plusieurs minutes pour découvrir que cette page était en fait recouverte d'un graphique transparent contenant de l'hypertexte caché. Après avoir pigé le truc, j'ai atterri sur une nouvelle série de pages, comportant des liens assez zarbis.

— Zarbis comment ?

— Zarbis, répéta-t-il en secouant la tête. Ça ressemble à des pages d'accueil ordinaires, avec tout plein de fioritures, une ponctuation approximative et des couleurs pourries, mais le contenu est insipide au possible. Ça fait assez toc, comme s'il s'agissait de faux.

— Mais quel intérêt de fabriquer de fausses pages d'accueil ?

— C'est la question que je me suis posée. La plupart des liens m'ont conduit vers des impasses et des 404. Mais la ligne continuait à se dérouler, à travers des pages entières de liens – et sur chacune d'elles un seul lien semblait conduire au moins deux pages plus loin. Puis je me suis mis à entrer des mots de passe. D'abord des trucs en java fastoches que je savais craquer tout seul, à l'aide de quelques joujoux que tu planquais sur ton disque dur. À ce propos, tu manques de RAM. Ce salaud m'a planté à peu près cinq fois. Là-dessus – ne m'en veux pas, mais j'ai passé quelques appels longue distance sur le téléphone de la chambre –, j'ai demandé de l'aide à des amis spécialistes. Il fallait que je parvienne à forcer des portes de service UNIX et autres merdes. Un petit malin qui connaissait très bien son affaire avait brouillé les pistes à mort.

— Mais à quoi ça sert ? pestai-je. J'imagine que n'importe qui peut marquer d'un signet le site final, où qu'il se trouve, pour s'y rendre directement la fois suivante. Pourquoi s'emmerder à concevoir un jeu de pistes alors que le principe même du Web est l'accès non linéaire ?

— Mon opinion, c'est que l'adresse de destination change régulièrement. Mais toujours est-il que je suis parvenu au bout.

— Et t'as trouvé quoi ?

— Rien.

Je le dévisageai.

— Tu peux répéter ?

— Rien. Il n'y avait plus rien.

— Putain, Bobby, elle est nulle, ton histoire. Ça craint, mec. Qu'est-ce que t'entends par « rien » ?

Il me fourra sa liasse de papiers sous le nez. La page de garde était vide, hormis une courte phrase en plein centre : « NOUS NOUS DRESSONS. »

— C'est tout ce qu'il y avait, dit Bobby. L'équivalent de deux ou trois heures de boulot pour camoufler une page contenant zéro lien et trois petits mots. Les autres feuilles retracent simplement l'itinéraire que j'ai suivi pour arriver là, ainsi que certaines bidouilles dont j'ai eu besoin. J'ai aussi extrait l'adresse IP de la dernière page pour remonter à sa source.

Le plupart des adresses Internet ont un libellé qui, à défaut d'aller toujours de soi, se présente au moins sous forme de mots. En fait, l'ordinateur lui-même ne connaît d'adresses que numériques – 118.152.1.54, par exemple. En utilisant cette forme plus basique, on peut plus ou moins établir la provenance géographique de la page.

— Qui se trouve où ? demandai-je.

— En Alaska.

— Et plus précisément ? Anchorage ?

Il secoua la tête.

— Juste l'Alaska. Puis Paris. Puis l'Allemagne. Puis la Californie.

— Mais de quoi tu parles ?

— Elle changeait sans cesse. Elle migrait d'un endroit à l'autre, et à vrai dire je doute qu'elle ait jamais résidé dans aucun de ces lieux. Elle était déguisée. Je ne suis pas Mister Bécane, mais je sais ce que je fais, et je te jure que je n'ai jamais rien vu de tel. J'ai mis deux copains sur l'affaire. En résumé, il se passe des choses vraiment pas claires.

— Sans blague.

— Il n'y a pas que tes mésaventures, Ward. Ce genre de truc fait partie de mon boulot. Je dois découvrir comment ils s'y prennent. Et qui ils sont. (Il but une grande gorgée, puis me fixa d'un air grave.) Et toi ? Que comptes-tu faire ? À part te soûler.

— Eh bien, la cassette comporte trois parties. Pour ce qui est de la dernière, je suis impuissant. Comment veux-tu que je retrouve... l'autre enfant ? (J'avais prévu de dire « mon jumeau », avant de me dégonfler.) J'ignore de quelle ville il s'agissait, et la scène remonte à plus de trente ans. Il ou elle peut se trouver n'importe où sur terre. Ou sous terre. Quant à la deuxième section, elle ne semble mener nulle part. Alors il ne me reste plus qu'à chercher cet endroit à la montagne.

— Sage décision, estima-t-il. Et je vais t'aider.

— Bobby...

Il secoua la tête.

— Sois pas débile, Ward. Tes parents n'ont pas péri dans un accident. Tu le sais comme moi.

Oui, je devais le savoir, et depuis un moment déjà, même si je m'étais interdit de formuler clairement l'idée, de l'habiller de mots.

Bobby le fit pour moi :

— Ils ont été assassinés.

CHAPITRE 12

Nina était assise au jardin avec Zoe Becker. Dans la nuit fraîche, elle regrettait d'avoir décliné la tasse de thé qu'on lui avait proposée d'un voix distraite. Zandt s'était déjà entretenu avec madame, puis planté dans la chambre de la fille, et il se trouvait à présent en compagnie du mari, à l'intérieur. Aucun des époux Becker n'avait paru surpris de trouver deux enquêteurs sur le pas de leur porte, même à une heure aussi tardive. La rupture était consommée entre leur existence et ce qu'ils pouvaient accepter comme réalité. Les deux femmes bavardèrent un moment, de façon discontinue, puis rapidement le silence reprit ses droits. Zoe regardait son pied frétiller au bout de sa jambe croisée. Telle était du moins la direction de son regard. Car Nina doutait qu'elle vît quoi que ce fût, la devinant prisonnière d'un néant où le mouvement d'une cheville devenait aussi prégnant que tout autre événement. Nina bénissait ce silence, car elle connaissait l'unique sujet de conversation qui intéressait cette femme. Sa fille était-elle en vie ? Nina pensait-elle qu'on la reverrait un jour ? Ou bien y aurait-il désormais, dans cette maison que Zoe avait tellement bichonnée, une pièce que le vide et le silence allaient noircir jusqu'à former un cristal sombre au centre de leur existence ? Au mur de cette chambre se trouvait le poster d'un groupe qu'aucun d'entre eux n'avait jamais entendu, sauf par accident. Alors à quoi servait-il maintenant ?

Nina ne connaissait pas la réponse à cette question, ni à aucune autre de cet acabit. Voyant que la femme s'apprêtait à parler, elle releva les yeux avec appréhension. Pour découvrir en fait que Zoe s'était mise à pleurer, des larmes d'épuisement qui ne ressemblaient ni à un début ni à une fin. Nina ne lui ouvrit pas les bras. Certains acceptaient le réconfort d'inconnus, d'autres le refusaient. Mme Becker était de ces derniers.

Elle préféra se renverser dans son fauteuil pour regarder à travers les portes-fenêtres du séjour. Michael Becker était assis tout au bord d'un fauteuil. Zandt, debout derrière le divan. Nina avait passé toute la journée en sa compagnie sans l'entendre prononcer plus de cinq phrases, lesquelles étaient sans rapport avec l'affaire. Ils avaient arpenté le lieu de l'enlèvement en début de matinée, avant les heures de shopping. Puis visité l'école de Sarah Becker, pour que Zandt voie comment le bâtiment s'inscrivait dans son environnement. Il avait repéré les angles de vue et les points d'accès, les endroits d'où l'on pouvait guetter l'objet à aimer. Il s'y attarda un temps considérable, comme s'il espérait tomber sur une nouvelle perspective qui lui permette de capter l'image d'un homme en plein jour. Il était de sale humeur en repartant.

Ils n'avaient pas rencontré les familles des précédentes victimes de l'Homme Debout. Mais ils possédaient la transcription des dépositions, alors qu'auraient-ils pu apprendre de plus ? Nina savait que Zandt gardait ces entretiens en mémoire, et qu'il aurait pu rappeler aux familles des détails qu'elles-mêmes avaient oubliées. Leur parler n'aurait fait qu'ajouter de la confusion à la douleur. En outre, elle pensait en son for intérieur que, si Zandt parvenait à les rapprocher du tueur, ce serait moins par accumulation d'infos que par intuition.

Nina avait une autre raison de laisser Zandt à l'écart des familles : il ne s'agissait pas qu'un parent s'emballe et appelle la police ou le Bureau pour savoir où en était l'enquête. Personne ne savait que Nina avait remis John Zandt sur le coup. Et si cela venait à s'ébruiter, c'était la catastrophe assurée. La sanction ne serait plus seulement disciplinaire ; ce serait tout bonnement la fin de sa carrière. Présenter Zandt aux Becker était cependant un risque nécessaire. De toute façon, avec tous les flics qui avaient défilé chez eux depuis la disparition, pourquoi se souviendraient-ils de lui en particulier, ou mentionneraient-ils sa venue à d'autres ? Pourvu que cela n'arrive pas... Et pourvu que l'échange en cours entre les deux hommes fasse mouche dans la cervelle de Zandt.

Et, le cas échéant, qu'il lui en parle.

— Je peux recommencer, si vous voulez.

Michael Becker avait récapitulé son emploi du temps deux fois de suite, en répondant aux questions de manière vive et concise. Zandt savait que ce type n'avait rien d'utile à lui apprendre. Il comprenait aussi que, au cours des semaines précédant la disparition, le boulot de Becker l'avait tellement accaparé qu'il n'avait pu remarquer grand-chose autour de lui. Il secoua la tête.

Becker baissa les yeux et se prit la tête entre les mains.

— Vous n'avez vraiment rien d'autre à me demander ? Il doit bien y avoir quelque chose. Il y a forcément quelque chose.

— Il n'existe pas de question magique. Ou, s'il en existe une, je ne la connais pas.

Becker releva les yeux. Ce discours n'était pas celui qu'avaient tenu les autres policiers.

— Vous croyez qu'elle est encore en vie ?

— Oui, affirma Zandt.

Becker fut surpris par l'assurance qu'il lut sur le visage du flic.

— Tout le monde fait comme si elle était morte, pourtant. Ils ne le disent pas. Mais ils le pensent.

— Ils se trompent. Pour le moment.

— Pourquoi ? fit-il d'une voix blanche, d'un souffle erroné, le son d'un homme pris à vouloir y croire.

— En général, quand un tueur de cette espèce se débarrasse d'une victime, il cache le corps et fait son possible pour maquiller son identité. En partie pour compliquer la tâche de la police. Mais aussi parce que nombre de ces individus voudraient se dissimuler leurs crimes à eux-mêmes. Or les trois victimes précédentes ont été retrouvées en terrain fréquenté, vêtues de leurs propres habits et en possession de leurs effets personnels. Ce type ne se cache de personne. Il voulait nous faire savoir qui elles étaient, et qu'il en avait terminé avec elles. Ce « terminer » suppose une période au cours de laquelle il préfère les garder vivantes.

— Préfère les garder...

— Une seule de ses victimes a été violée à ce jour. Hormis quelques blessures bénignes à la tête, les autres ne présentaient aucune trace de sévices, sinon leur crâne rasé.

— Et leur meurtre, bien sûr.

Zandt secoua la tête.

— Dans ce type de situation, les meurtres ne constituent pas des sévices, mais ce qui abrège les sévices. Si les autopsies ne disent jamais tout, elles laissent penser que les filles ont vécu plus d'une semaine après leur enlèvement.

— Une semaine... répéta le type d'une voix sans timbre. Ça fait déjà cinq jours.

Zandt attendit avant de répondre. Durant l'entretien, son regard avait balayé l'essentiel de la pièce, mais il remarquait à présent un détail qui lui avait échappé. Une petite pile de livres scolaires, sur une table basse. Ils étaient d'un niveau trop avancé pour appartenir à la sœur cadette. Il se rendit soudain compte que son interlocuteur le dévisageait.

— J'en suis bien conscient, monsieur.

— Vous sembliez avoir une autre raison...

— Oui. Je ne crois pas qu'il l'ait déjà tuée.

Becker lâcha un rire amer.

— Vous ne « croyez pas » ? C'est ça, l'argument ? Super. On est vachement rassuré.

— Mon boulot n'est pas de vous rassurer.

— Non, répondit Becker, le visage inexpressif, j'imagine bien que non.

Un moment de silence, puis il ajouta :

— Ces choses-là existent vraiment, n'est-ce pas ?

Zandt comprenait le sens de cette phrase : certains événements, que l'on rencontre en temps normal sur écran ou sur papier, peuvent se produire pour de bon. Tels la mort violente, le divorce et les lésions à la colonne vertébrale ; tels le suicide, la toxicomanie, et le cercle de figures grises et floues penchées sur vous qui murmurent : « Le conducteur ne s'est pas arrêté. » Ces choses-là arrivent. Elles sont aussi réelles que le bonheur, le mariage ou la sensation du soleil dans le dos, et elles s'effacent plus lentement. On ne retrouvera pas toujours la vie d'avant. On ne fera pas toujours partie des veinards. Parfois, ça durera indéfiniment.

— Oui, elles existent.

Il toucha discrètement la couverture d'un manuel scolaire. Passa son doigt sur la surface rayée.

— À combien estimez-vous nos chances de la revoir vivante ?

La question était posée avec une simplicité et une maîtrise qui forçaient l'admiration. Zandt se détourna de la table.

— Partez du principe que vous n'en avez aucune.

Becker parut choqué, ouvrit la bouche pour répondre. Il n'en sortit rien.

— Chaque année, des centaines de personnes meurent entre les mains de ces types-là, expliqua Zandt. Des centaines, pour ne pas dire plus. On ne les coince pratiquement jamais. On fait tout un foin quand on en prend un, comme si on avait rentré le tigre dans sa cage. Mais ce n'est pas le cas. Il en naît un tous les mois. Les rares que l'on attrape sont malchanceux, ou idiots, ou parvenus au stade où ils commencent à faire des erreurs. L'immense majorité passe entre les gouttes. Ces gens-là ne sont pas des aberrations. Ils incarnent une facette de ce que nous sommes.

C'est toujours la même histoire. La survie des plus forts. Des plus malins.

— Le Garçon de Courses est malin ?

— Il ne s'appelle pas comme ça.

— C'est ce nom-là qu'employaient les journaux. Et les flics.

— Il s'appelle l'Homme Debout. C'est le nom qu'il s'est choisi. Oui, il est malin. Et c'est peut-être ce qui le perdra. Il est assoiffé de reconnaissance. D'un autre côté...

— Il se peut qu'on ne l'attrape jamais, auquel cas on ne reverra jamais Sarah.

— Si vous la retrouvez, répondit Zandt tout en rangeant carnet et stylo dans sa poche intérieure, ce sera un cadeau des dieux, soyez-en conscient. Aucun de vous ne sera plus jamais le même. Ce n'est pas forcément une mauvaise chose. Mais c'est la vérité.

Becker se leva. De mémoire, Zandt n'avait jamais vu de visage si crevé et si insomniaque à la fois. Sans qu'il le sache, Michael Becker pensait la même chose sur son compte.

— Vous allez quand même essayer ?

— Je ferai tout mon possible. Si je peux le trouver, je le ferai.

— Alors pourquoi vous me conseillez de prévoir le pire ?

À cet instant sa femme rentra par la porte-fenêtre, talonnée par l'agent du FBI, ce qui évita au policier de répondre.

Nina remercia les Becker de leur accueil et promit de les tenir informés. Elle parvint aussi à sous-entendre que leur visite n'était qu'une formalité, sans lien direct avec le déroulement de l'enquête.

Michael Becker les regarda s'éloigner dans l'allée. Lorsqu'ils furent hors de vue, il ne ferma pas la porte tout de suite, mais resta un moment à contempler la nuit. Dans son dos, il entendit Zoe monter à l'étage pour veiller Melanie. Il doutait que sa seconde fille soit endormie. Les cauchemars de l'année passée avaient repris, mais comment lui en vouloir ? Le peu de sommeil qu'il grappillait le mettait au supplice, lui aussi. Il savait qu'elle entonnait encore avant de se coucher la petite incantation qu'il avait composée, et cela le glaçait d'effroi. L'ironie n'était pas une armure à toute épreuve, quoi qu'ils aient pu en penser, lui, Sarah et les nouveaux producteurs de films d'horreur. Dans un univers d'os et d'hémoglobine, l'ironie ne faisait pas le poids. Michael se rappelait avoir discuté des angoisses nocturnes avec Sarah, voilà plusieurs années. La gosse curieuse qu'elle avait toujours été lui demandait pourquoi les gens avaient peur du noir. Il lui avait répondu que c'était une réminiscence des temps primitifs, lorsqu'on dormait en plein air ou dans des grottes, et que les bêtes sauvages risquaient de nous dévorer pendant la nuit.

Sarah avait paru sceptique :

— Mais ça fait une éternité, ça ! (Elle avait réfléchi un instant, avant d'ajouter, du haut de ses dix ans :) Non. Ce doit être autre chose qui nous fait peur.

Michael s'était finalement rangé à cet avis. Ce ne sont pas les monstres qui nous effraient. Les monstres sont de sympathiques chimères. En revanche, nous savons de quoi notre propre engeance est capable : c'est de nous-mêmes que nous avons peur.

Il se résolut enfin à refermer la porte et se rendit à la cuisine. Il prépara le café, son nouveau rituel à cette heure de la soirée. Il acheminerait le pot sur un plateau jusqu'au salon, accompagné de deux tasses et d'un pichet de lait chaud. Voire d'un cookie ou deux, le seul aliment que Zoe semblait encore tolérer. Ils s'installeraient devant ce que la télévision aurait à leur offrir, et attendraient que le temps passe. Le mieux, c'étaient les vieux films. Des reliques d'un autre âge, quand Sarah n'était pas née et que tout ceci ne pouvait être vrai. Parfois ils échangeaient quelques mots. En général, très peu. Zoe garderait le téléphone à portée de main.

Comme il sortait deux tasses du nouveau vaisselier – en vieux pin, importé d'Angleterre dans la foulée de leur récent voyage –, Michael rumina les paroles du policier, les soupesant avec soin. Et de découvrir que, pour la première fois depuis la disparition, il sentait les balbutiements d'une chose qui devait être l'espoir. Oh, ils auraient disparu d'ici le matin, mais tout répit était bienvenu. Pourquoi ce soupçon d'optimisme ? Parce qu'il pensait avoir décelé ce qu'il fallait lire entre les lignes, à savoir que les paroles du policier importaient moins que ses silences.

La nana du FBI avait décliné son identité, mais celle du type n'avait jamais été précisée. Avec l'application de celui qui croit en la magie de la sémantique, convaincu que décrire et encadrer les faits par des mots permet de les dominer, Michael Becker avait lu tout ce qu'il trouvait concernant les précédents crimes du ravisseur de sa fille. Il avait surfé sur Internet, exhumé les articles de presse, et même acheté le bouquin sur les crimes irrésolus vendu au supermarché du coin. Tout cela au détriment, entre autres, de son travail. Il n'avait pas touché à *Dark Shift* depuis la nuit de l'enlèvement. Au fond de lui, il doutait fort de s'y remettre un jour, même s'il n'en avait encore rien dit à son partenaire. En attendant, il repoussait sans cesse la réunion avec le studio. Mais Wang avait de l'argent, et des relations inépuisables. Il était introduit dans la ville à un point dont Michael n'osait même pas rêver. Il survivrait.

À la faveur de ses recherches, Michael avait appris, ou réappris, qu'outre la petite LeBlanc, Josie Ferris et Annette Mattison, une quatrième adolescente avait disparu à la même période. La fille d'un policier ayant pris part à l'arrestation de deux précédents tueurs en série. D'aucuns avaient alors suggéré, à demimot, que son kidnapping visait à sanctionner les succès de son père. Contre l'avis du FBI, celui-ci s'était investi dans les recherches pour la retrouver, et au moins un journal laissait entendre qu'il avait accompli quelque progrès là où les fédéraux piétinaient. Puis il avait disparu de la circulation. Son nom était John Zandt. Le Garçon de Courses – Michael Becker était bien placé pour le savoir – n'avait pas été appréhendé. Un long article publié un an après la série d'enlèvements rapportait qu'une Mme Jennifer Zandt était retournée en Floride, auprès de sa famille. Mais le journaliste ignorait ce qu'était devenu le policier.

Michael prit une décision. Quel que soit le programme télévisé, lui et sa femme devaient parler ce soir. Il lui ferait part de ses hypothèses concernant le type sans nom, et proposerait qu'à la prochaine visite des autres policiers, ces hommes et femmes dévoués qui devenaient tragiquement familiers, nulle mention ne soit faite de la rencontre de ce soir.

Une dernière chose. Bien que sa foi dans le langage fût sérieusement ébranlée, il s'accrochait à l'idée que les mots et les noms sont à la réalité ce que les colonnes et l'architecture sont à l'espace. Ils l'humanisent. De la même façon que l'ADN fabrique à partir d'éléments aléatoires des ensembles identifiables, le langage sait former, à partir de phénomènes inexpliqués, une situation sur laquelle on peut discourir, et donc peser.

Il ne penserait plus au Garçon de Courses. Ce serait désormais l'Homme Debout. Dans le même temps, il se préparerait au pire. Le flic avait raison sur ce point. Et puis c'était sûrement ce que Sarah aurait voulu.

Au diable M. Touche Dubois. Si les Parques exigeaient une telle offrande, elles pouvaient aller se faire foutre.

Ils étaient à la terrasse du Smorgas Board, un repaire de surfeurs situé à huit mètres environ de l'endroit où la petite Becker s'était volatilisée. Cela faisait une heure qu'ils étaient là, et le café s'apprêtait à fermer. Ils étaient les derniers clients, avec deux jeunes gens affalés sur une table plus loin, qui aspiraient d'un air éteint le contenu d'immenses gobelets.

— Tu cogites, ou tu regardes seulement ?

Zandt ne répondit pas sur-le-champ. Assis à côté de Nina, il observait la rue. Il avait à peine bougé. Son café était froid. Il n'avait fumé qu'une cigarette, dont la majeure partie s'était consumée toute seule. Il était vraiment ailleurs. Devant ce spectacle, Nina pensait à un chasseur, humain ou autre. Un animal capable de se tapir et d'attendre, aussi longtemps que nécessaire, sans qu'aucun ennui, aucune rage ou aucune douleur réussisse à le distraire.

— Ils ne reviennent pas tous, dit-elle d'une voix lasse.

— Je sais, répondit-il aussitôt. Je ne regarde rien de particulier.

— Tu parles ! ricana-t-elle. Alors tu nous fais une attaque ?

Il la surprit en esquissant un sourire.

— Je cogite.

Elle croisa les bras.

— J'ai le droit d'en profiter ?

— Je pensais : « Quelle perte de temps, et pourquoi m'a-t-elle fait venir ici ? »

Nina comprit que le sourire n'en était pas tout à fait un.

— Parce que j'espérais que tu pourrais nous aider, répondit-elle avant de pivoter tant bien que mal sur son siège. Qu'est-ce qu'il y a, John ? Tu sais très bien pourquoi. Parce que tu m'as aidée par le passé. Parce que ton avis m'est précieux.

Il sourit de plus belle, à lui donner la chair de poule.

— Et quel fut mon bilan, la dernière fois ?

— Je ne sais pas, admit-elle. Je t'écoute. Que s'est-il passé ?

— Tu le sais comme moi.

— Non, je ne le sais pas ! Tu m'as juste dit que tu avais une piste. Puis tu es devenu muet, inaccessible, alors que tu t'étais toujours appuyé sur moi pour connaître les infos du Bureau. Des tuyaux que tu n'aurais jamais dégottés autrement, puisque ta propre hiérarchie t'avait écarté de l'investigation. Je t'ai rendu un service, et tu m'as envoyée paître.

— Tu ne m'as rendu aucun service, objecta Zandt. Tu as fait ce qui servait au mieux tes intérêts.

— T'es vraiment trop con.

Sur ces mots, les deux légumes de l'autre table se redressèrent d'un bond, tels deux pantins dont le manipulateur se serait réveillé en sursaut. Chaude ambiance.

Elle baissa d'un ton et débita :

— Si c'est vraiment comme ça que tu me vois, tu n'as qu'à te barrer d'ici et rentrer dans ton Vermont de merde. Ce sera bientôt les grosses neiges, là-bas. Va donc t'y enterrer.

— T'es en train de dire que tu m'as tuyauté dans le seul intérêt de ma famille ?

— Oui, bien sûr. Et pour quelle autre raison ?

— Malgré le fait que tu m'aies aidé à jouer les maris infidèles ?

— Tu es pathétique, John. Je ne suis pas responsable des errements de ta queue.

Elle le fusilla du regard. Il lui rendit la pareille. Un silence s'installa, puis les yeux de Nina piquèrent vers le sol.

Il émit un rire bref.

— C'est censé me faire croire que j'ai le dessus ?

— Quoi ? dit-elle tout en se maudissant.

— Ce regard fuyant. Un vieux truc du règne animal. On flatte l'ego du mâle par un signe de soumission. Et maintenant que je suis redevenu le roi de la butte, je ferai tout ce que tu me demanderas, c'est ça le calcul ?

— Mon pauvre John, tu es complètement parano, dit-elle pour ne pas reconnaître qu'il avait mis dans le mille. (Elle en déduisit qu'elle avait passé trop de temps en compagnie d'imbéciles.) Ça ne m'intéresse pas d'ergoter avec toi.

— À ton avis, c'est quoi, l'idée des cheveux ?

Elle fronça les sourcils, déroutée par ce revirement.

— Quels cheveux ?

— L'Homme Debout. Pourquoi coupe-t-il les cheveux ?

— Eh bien, pour les pulls. Afin de broder les prénoms.

Zandt secoua la tête, alluma une cigarette.

— Pas besoin d'une chevelure entière pour ça. Toutes les filles avaient les cheveux longs. Pourtant on les a toutes retrouvées avec la boule à zéro. Pourquoi ?

— Pour les déshumaniser. De sorte qu'il soit plus facile de les tuer.

— Possible. C'est ce que nous pensions tous. Mais aujourd'hui je m'interroge.

— Et si tu me disais enfin ce que *toi*, tu penses ?

— Je me demande s'il ne s'agissait pas d'un châtiment.

Nina envisagea l'hypothèse.

— En réponse à quoi ?

— Je l'ignore. Mais je pense que ce mec a choisi ces filles, ces filles d'un type bien particulier, pour une raison précise.

Je pense qu'il projetait certaines choses avec elles, et que d'une manière ou d'une autre elles n'ont pas été à la hauteur. Alors en guise de punition il leur a confisqué un attribut qu'il devinait primordial à leurs yeux.

Il but une gorgée de café, sans s'arrêter à sa froideur.

— Tu sais ce qu'on faisait aux collaborateurs à la fin de la Seconde Guerre mondiale, en France ? lança-t-il.

— Bien sûr. Les femmes qu'on soupçonnait d'avoir été un peu trop accueillantes avec l'envahisseur allemand étaient exhibées dans les rues avec le crâne rasé. Un grand pas pour l'humanité, n'est-ce pas ? (Elle haussa les épaules). Je peux admettre qu'il s'agisse d'une punition, comme tu dis, mais je ne vois pas le rapport avec la guerre. Ces adolescentes n'ont pas pactisé avec l'ennemi.

— Tu as peut-être raison.

Zandt semblait déjà se désintéresser du sujet. Renversé sur son dossier, il regardait vaguement à travers la terrasse. L'un des légumes croisa son regard par mégarde. Zandt ne détourna pas les yeux. Le légume, si. Fissa. Il fit un signe à son pote, signifiant qu'il était justement temps d'aller cirer leurs planches. Ils se levèrent et se fondirent dans la nuit.

Zandt parut satisfait de ce dénouement.

Nina tenta de raviver sa concentration.

— Et où nous mènerait cette hypothèse ?

— Nulle part, peut-être, reconnut-il tout en écrasant sa cigarette. Disons simplement que je n'ai pas assez réfléchi la dernière fois. J'étais focalisé sur la méthode qu'il avait employée pour les trouver. Comment leurs routes étaient venues à se croiser. À présent je juge cela très étrange. La façon dont elles ont failli à ses yeux. Ce qu'il attendait vraiment d'elles...

Nina ne répondit rien, espérant qu'il allait développer. Mais, lorsqu'il reprit la parole, cela ne concernait plus l'enquête :

— Pourquoi as-tu cessé de coucher avec moi ?

Prise de court, elle hésita.

— Nous avons cessé de coucher ensemble, tu veux dire.

— Non. Me raconte pas de salades.

— Je n'en sais rien, John. C'est arrivé comme ça. Ça ne te faisait ni chaud ni froid à l'époque, je te signale.

— Disons que je l'acceptais.

— Qu'essaies-tu de me dire ? Que maintenant tu ne l'acceptes plus ?

— Bien sûr que si. Il y a tellement longtemps ! Disons que je pose les questions que je n'avais jamais posées. C'est marrant, une fois qu'on est lancé, elles se mettent à jaillir de partout.

Nina ignorait quoi répondre à ça.

— Alors, que veux-tu faire à présent ? s'enquit-elle.

— Je veux que tu t'en ailles, rétorqua-t-il. Je veux que tu rentres chez toi et que tu me laisses tranquille.

Elle se leva.

— À ta guise. Tu as mon numéro. Appelle-moi si jamais tu décides de remuer ton cul.

Il tourna lentement la tête, pour la regarder droit dans les yeux.

— Tu veux savoir ce qui s'est passé ? La dernière fois ?

Elle se figea, l'observa. Son visage était froid et distant.

— Oui, dit-elle.

— Je l'ai trouvé.

Un frisson parcourut la nuque de Nina.

— Trouvé qui ?

— Je l'ai traqué pendant deux semaines. J'ai fini par me rendre chez lui. Je l'avais vu épier d'autres filles. Je ne pouvais plus supporter ça.

Elle hésitait entre se rasseoir et rester debout.

— Que s'est-il passé ?

— Il a nié. Mais je savais que c'était lui, et il savait que je l'avais chopé. C'était notre homme, mais je n'avais aucune preuve, et il se serait enfui. Je suis resté deux jours avec lui. Il refusait d'indiquer où elle était.

— Tais-toi, John.

— Je l'ai tué.

Elle vit à son regard que c'était vrai. Elle ouvrit la bouche, la referma.

— Deux jours plus tard, on a reçu le pull et la lettre.

Soudain très las, il détourna le visage. Quand il reprit la parole, ce fut d'une voix neutre.

— J'ai buté le mauvais type. Libre à toi de donner suite à cette info.

Elle traversa la Promenade en s'interdisant de se retourner, concentrée sur la cime des palmiers qui s'inclinaient dans la brise du soir, à deux pâtés de maisons de là.

Mais, parvenue au trottoir d'en face, elle s'arrêta, virevolta. Il avait disparu. Elle attendit un peu, en se mordant la lèvre, mais il ne refit pas surface. Alors elle se remit en marche, à pas lents.

Il y avait quelque chose de changé. Jusqu'à ce soir Zandt avait paru coopératif, mais ce tête-à-tête en terrasse l'avait déçue. En définitive, il ressemblait moins à un chasseur qu'à un boxeur. Un boxeur capturé par l'objectif une heure avant le début du match. Pendant ce moment où l'on met le show-business entre parenthèses et où le combattant paraît se mouvoir dans un monde à lui, se soustrait au regard des autres pour se fondre dans son archétype. D'autres pouvaient prendre les paris, enfiler un costard, s'éclater aux frais de la princesse. Le reste plaiderait avec un mépris affecté pour l'interdiction de ce sport, depuis leur petite vie douillette où personne à leur place ne chercherait l'issue de secours. Pour les gars du ring, c'était une autre histoire. Ils se battaient pour de l'argent, mais pas seulement. Ils le faisaient parce que c'était leur truc. Eux non plus ne cherchaient pas la sortie. Ils cherchaient une entrée, une route qui les ramène vers ce lieu qu'ils sentaient palpiter en eux.

Le coup des parents s'était révélé une erreur. Mais Zandt en savait très peu, et il s'était mis à jouer les difficiles. Les seuls éléments nouveaux ne pouvaient provenir que des Becker. D'où la nécessité de les lui présenter. Mais elle avait su, sitôt rentrée du jardin, que cette rencontre avait ouvert des portes qu'il eût mieux valu garder fermées.

Elle n'avait pas besoin de ça. Elle n'avait jamais voulu de chasseur, ni de tueur. Elle estimait que la seule chose qui sortirait l'Homme Debout de sa tanière serait un adversaire qu'il ait envie de dominer.

Elle voulait un appât.

CHAPITRE 13

L'homme trônait dans son fauteuil, au centre du salon. La pièce saillait de la façade de la maison, avec des fenêtres sur trois murs. Deux étaient protégées par des rangées d'arbres, la dernière donnait sur une pelouse pentue, en terrasse. En cet après-midi, tous les rideaux étaient tirés, d'épaisses tentures qui ne laissaient pas filtrer le moindre soupçon du dehors. L'homme les préférait parfois fermées, parfois ouvertes. Il était à cet égard tout à fait imprévisible.

Le fauteuil était positionné dos à la porte. L'homme aimait la sensation que cela lui procurait. Une légère tension, un sentiment de vulnérabilité. En théorie, l'on pouvait ainsi se glisser derrière lui et lui assener un coup sur le crâne. Il faudrait certes au préalable déjouer l'important système de protection, mais le principe demeurait valide. Il montrait à quel point il dominait son environnement. Le monde extérieur ne lui inspirait aucune peur. Depuis un très jeune âge il avait été contraint de s'y frayer un chemin, pour ne pas dépérir. C'était comme ça et pas autrement qu'il aimait ses intérieurs.

Son visage était lisse et sans rides, grâce à l'usage intensif de crèmes et autres produits pour la peau. Son regard était clair et perçant, ses mains légèrement hâlées, les ongles impeccables. Il était nu de la tête aux pieds. La chaise formait un léger angle avec les lattes du parquet ciré qui traversait la pièce en rangées régulières. Une tasse de café noir fumait sur une petite table jouxtant le fauteuil, à côté d'une soucoupe remplie de perles de verre. Une mince publication reposait là, elle aussi. La tasse était placée en sorte que la quasi-moitié de sa base dépasse du plateau. Le fauteuil était vieux, revêtu de cuir râpé. En temps normal, on aurait trouvé un exemplaire du *New York Times* replié sur un accoudoir, et un larbin s'affairant derrière, prêt à servir des sandwichs sans croûte. L'homme avait recouvert toute une bibliothèque de hachures faites au stylo vert, bleu et rouge, dont la largeur n'excédait jamais trois millimètres, jusqu'à obtenir un noir délicatement marbré. Il lui avait fallu dix-sept stylos et plusieurs semaines de travail. À l'autre bout de la pièce, un beau bureau de chez Arts and Crafts était entièrement tapissé de minuscules photos de Madonna découpées dans des magazines, dont aucune n'était postérieure à sa période *Material Girl*, après laquelle l'homme avait cessé d'être fan. Il avait appliqué sur son collage de nombreuses couches de vernis foncé, jusqu'à rendre au meuble un simple aspect plaqué noyer. Comme dans le cas de la bibliothèque, seul un examen très rapproché permettait d'expliquer l'effet rendu.

Son projet du moment concernait la petite table volante dressée à côté du fauteuil, qu'il recouvrait de perles de verre. Celles-ci faisaient à peu près un millimètre de diamètre, dans quatre coloris : rouge, bleu, jaune et vert. Des couleurs génétiques. Les assembler requérait un grand soin, notamment du fait qu'elles n'étaient pas alignées au hasard, mais selon un motif séquentiel complexe. Cela terminé, il appliquerait plusieurs couches d'épaisse laque noire, jusqu'à noyer le moindre semblant de texture. Il ne viendrait à personne l'idée de se demander ce qui se cachait sous la surface, de même que personne ne remarquerait qu'une, et une seule, latte du parquet se composait d'une myriade d'allumettes collées les unes aux autres puis poncées et cirées jusqu'à se fondre parfaitement dans le décor. La collecte des allumettes avait pris plus de six mois. Sauf improbable coïncidence, chacune avait été craquée par une personne différente. L'homme croyait profondément en l'individualité, en son apport crucial à l'humanité. De nos jours, tout le monde suivait les mêmes émissions, lisait les mêmes magazines de luxe, était sommé par les médias de former une jolie file indienne pour regarder les mêmes films idiots. Les gens arrêtaient de fumer suite aux mises en garde de bonnes âmes qui, dans le même temps, se gavaient de graisses. Pour le confort et le bien-être d'autrui. Ils menaient leurs vies selon les règles édictées par l'autrui en question, par des êtres qu'ils n'avaient jamais rencontrés. Ils vivaient à la surface des choses, dans un monde CNN et MTV où l'Histoire se réduisait aux cinq dernières minutes. Le triomphe du momentané. Ils n'avaient aucune notion d'« hier » et se vautraient dans un présent perpétuel.

La publication couchée sur la table était un article universitaire récent, qu'il avait reçu au courrier ce matin-là. Il en avait lu le résumé sur Internet, avant de commander la version intégrale. On y traitait d'un sujet pointu, mais il était fort capable d'en saisir le sens. Cela faisait des années qu'il consacrait ses lectures aux disciplines qui le passionnaient : la génétique, l'anthropologie, les cultures préhistoriques. En dépit d'une scolarité très brève, il était intelligent et avait beaucoup appris de la vie. La sienne et celle des autres. Les gens disaient de ces choses, in extremis. Il y avait beaucoup de vérité dans leurs propos, une fois passé le stade du plaidoyer, quand le corps parle sans que l'esprit interfère.

Avant d'ouvrir sa revue il se leva, fit quelques pas et enchaîna trois séries de pompes. La première, les paumes à plat, écartées dans la largeur des épaules. La deuxième, les paumes toujours à plat, mais éloignées sur les côtés. La troisième, les mains rapprochées, en s'appuyant sur les poings. Trois séries de cent, entrecoupées d'une courte pause.

Il transpira à peine. Il était content.

Sarah Becker entendait sous forme de bruits sourds les efforts rythmés du type, sans prendre la peine de deviner leur nature. Elle ne voulait pas savoir. Elle ignorait l'heure qu'il était et ne s'en souciait pas davantage. Son horloge interne suggérait qu'on était en journée, peut-être l'après-midi. En un sens, c'était pire que la nuit. Il se passait de vilaines choses la nuit. Mais on s'y attendait. Les gens avaient peur du noir parce que le noir évoquait la nuit, or c'est avec la nuit que les malheurs nous tombent dessus. Ainsi tourne le monde. Le jour se veut plus positif. Le jour, on va à l'école, on déjeune, le ciel est bleu et la vie est relativement sûre tant qu'on évite les coins où les gens sont pauvres. Mais si la journée n'était plus sûre, il valait mieux ne pas y penser. Elle ne voulait pas savoir.

En étirant son cou vers le haut, elle parvenait à toucher avec le front le plafond de sa prison. Il y faisait noir comme dans un four. Couchée sur le dos, elle pouvait remuer ses mains et ses pieds d'environ deux centimètres dans chaque direction. Elle gardait cette position depuis un long moment, au moins quatre jours d'après ses calculs, peut-être même six. Elle ne se souvenait de rien entre le moment où elle se trouvait sur Third Street Promenade et celui où elle reposait sur le dos avec une étroite fenêtre en face du visage. Elle s'était vite rendu compte qu'on y apercevait le plafond d'une pièce, et que la fenêtre en question était un trou dans le plancher, sous lequel elle occupait un espace à peine plus large que son corps. La fenêtre mesurait approximativement douze centimètres de long sur dix de large, et courait du sommet de ses sourcils jusqu'au menton.

Elle s'était mise à crier et quelqu'un avait fini par arriver. Le type lui avait susurré des choses. Encore quelques cris, et il avait bouché le trou au moyen d'un petit panneau. Elle avait entendu ses pas s'éloigner, et un seul événement s'était produit depuis. Sarah s'était réveillée d'un somme durant ce qui ressemblait à la nuit, pour découvrir qu'on avait rouvert la trappe. La pièce au-dessus baignait dans l'obscurité, mais elle discernait le contour d'une tête penchée sur elle. Elle tâcha de parler au type, de le supplier, de négocier, mais il resta muet. De guerre lasse, elle se tut, et se mit à pleurer. Apparut alors une main, tenant un gobelet. Le type fit pleuvoir quelques gouttes sur son visage. Elle tenta d'abord de détourner la tête, puis, découvrant combien elle avait soif, ouvrit la bouche pour en avaler le maximum.

Après un laps de temps indéterminé, il était revenu, et ils avaient eu leur petit échange au sujet de Ted Bundy. Cette fois elle avait tout bu.

Avec le temps, son esprit avait paru se dégourdir, à mesure que son organisme éliminait la drogue qu'on avait dû lui injecter. Le revers de la médaille, c'était la difficulté à prolonger l'insouciante léthargie des débuts. Elle avait tenté de soulever la trappe à l'aide du nez et de la langue, en tirant à fond sur son cou, mais son emplacement était finement calculé et il était impossible de la déplacer de cette façon. À l'instar de cette cache, l'ouverture était prévue pour quelqu'un de sa taille, à se demander s'il ne l'avait pas conçue spécialement pour elle. Sarah était robuste, une pro du Rollerblade, et plus forte que la plupart des filles de sa stature. Malgré cela, sa geôle lui résistait depuis le début, si bien qu'elle avait jeté l'éponge. Son père disait souvent que les gens avaient tant de problèmes parce qu'ils gaspillaient leur énergie à vouloir changer ce qui ne pouvait l'être. Elle n'était pas en âge de comprendre ce qu'il entendait précisément par là, mais elle saisissait l'idée générale. Elle n'avait rien avalé depuis une éternité. Tant qu'elle n'était pas assurée de refaire bientôt des réserves d'énergie, il ne rimait à rien de gâcher le peu qu'il lui restait. Se débattre était stupide. Alors elle resta immobile, et songea à Touche Dubois.

M. Dubois était une invention d'elle et de son père. Du moins le croyaient-ils. En fait, ce personnage devait son existence, quoique de manière indirecte, à la mère de Sarah. Zoe Becker croyait en beaucoup de choses. Elle ne les prenait pas forcément au pied de la lettre, mais dans le doute, elle n'allait pas tenter le diable. L'astrologie ? Des fadaises, bien sûr, mais ça ne coûte rien de jeter un œil à son horoscope, et le plus étonnant, c'est que ça dit souvent vrai. Le Feng shui ? Un ramassis d'évidences, bien sûr, mais les carillons sont de beaux objets et leur timbre est joli, alors pourquoi les bouder ? Et si d'aventure un type particulier d'oiseau croise votre route, ce que certains verront comme un mauvais présage, il existe quelques parades verbales ou gestuelles qui de toute façon ne peuvent pas faire de mal.

Comme souvent dans les familles, Zoe tenait ses superstitions de sa grand-mère, et non de sa mère – une ex-éditrice très terre à terre qui croyait surtout au footing. Michael Becker n'avait, lui, aucun penchant pour ces histoires de poisse et de fatum, pas plus que sa fille. M. Dubois fut à l'origine une *private joke* entre eux deux, une réponse aux superstitions qui mettaient Michael Becker hors de lui. Dès que l'on prononçait sous son toit la moindre phrase qui puisse de près ou de loin tenter le mauvais sort, Zoe Becker la complétait aussitôt d'un « je touche du bois » aussi prompt et mécanique qu'un « à tes souhaits » après un éternuement. Si quelqu'un déclarait : « Je ne finirai jamais comme ça », elle ajoutait : « Je touche du bois » tout en frappant ses phalanges sur la table. S'il affirmait : « Mon père est en bonne santé », elle le disait aussi – de moins en moins fort, consciente avec le temps que son époux trouvait cela passablement crispant, mais elle le disait. Elle le disait même, et c'est dans ces moments-là que son mari se sentait prêt à bouffer le piano, après une phrase du style : « Je ne me suis jamais cassé la jambe. » Michael Becker faisait valoir qu'il s'agissait là d'une assertion factuelle, et non d'un bras d'honneur à la fatalité. C'était le simple énoncé d'un état du monde objectif, et lui adjoindre un mantra superstitieux était ridicule. On n'aurait pas idée de dire : « Deux plus deux égale quatre – touchons du bois », soulignait-il patiemment, alors pourquoi placer cette expression après d'autres vérités premières ? Cette manie s'expliquait, et demeurait à peu près supportable, dans le cas d'une bravade contre le pouvoir destructeur du monde. Mais face à un putain de fait...

Zoe écouterait cet exposé, comme souvent par le passé. Elle rappellerait ensuite qu'il s'agissait d'une tradition bien ancrée dans de nombreuses régions du monde ; qu'elle n'était peut-être pas infondée dans la mesure où les arbres possédaient un certain pouvoir ; et qu'en tout état de cause ça ne pouvait pas faire de mal. Alors Michael opinerait du bonnet, quitterait la pièce en silence, et se jetterait sur le piano pour le bouffer.

Sarah avait pris le parti de son père, et ils avaient au fil des ans développé le personnage de « Touche Dubois », un elfe maléfique, d'extraction scandinave sans doute, dont le seul boulot consistait à repérer les gens qui tentaient le mauvais sort en se livrant sans vergogne à des observations factuelles. Puis il se faufilait, invisible, dans leurs maisons au milieu de la nuit et changeait le cours de leur existence, pour le pire. Mieux valait ne pas se sentir en veine quand Touche était dans les parages, parce qu'il l'apprendrait et vous le ferait payer !

Avec les années ce concept était devenu leur rituel de séparation, lorsqu'ils se souhaitaient mutuellement du mal afin que Touche sache en les écoutant que sa présence n'était pas requise. Ce jeu avait également fait des merveilles l'année précédente lorsque Melanie, la petite sœur, s'était mise à avoir des cauchemars la nuit. Sur une idée de Sarah, Michael lui avait raconté que ces mauvais rêves étaient l'œuvre de Touche Dubois, qui rôdait autour du lit, à l'affût de gens à attaquer. Il suffisait alors de réciter une petite incantation – sur laquelle son père avait passé un temps fou, la fignolant comme peu de ses scripts –, grâce à quoi Touche Dubois comprendrait qu'on n'avait pas besoin de lui ici et irait embêter quelqu'un d'autre. Melanie testa ce remède, d'abord à reculons, puis elle prit vite l'habitude de réciter ce court texte en allant au lit – et petit à petit les cauchemars disparurent, et elle se soucia moins de savoir si les placards étaient fermés de façon complètement hermétique. Sa mère ne raffola pas de la plaisanterie, et pour sa part n'invoqua jamais Touche, mais il lui arrivait de sourire quand celui-ci surgissait dans la discussion. À table, il expliquait certaines choses sur le monde qui nous entoure et il figurait désormais au casting de *Dark Shift* parmi les démons secondaires ligués contre l'héroïne. Le coscénariste de Michael avait tiqué, question palette de pouvoirs et antécédents du personnage, mais Michael n'en avait pas démordu.

Couchée sous le plancher d'une maison habitée par un cinglé, Sarah se demandait si sa maman n'avait pas raison, tout compte fait. Peut-être existait-il un tel monstre, un tel esprit. Et peut-être avait-il su qu'ils s'étaient moqués de lui, qu'ils avaient joué les fiers-à-bras. Et peut-être s'était-il mis en colère. Et peut-être était-ce lui qui était venu la chercher, et qui revenait à présent la voir, dans le noir, parce qu'il n'avait pas de visage sous le masque qu'il avait revêtu pour la capturer.

Sarah se tenait immobile, les yeux grands ouverts.

L'article, qui s'intitulait « Le site de Krüniger et la société de Mittel-Baxter », décrivait une fouille archéologique menée récemment dans une région d'Allemagne que l'homme connaissait mal. Il l'avait localisée dans son atlas, puis il avait calculé qu'aucun de ses contacts ne résidait assez près pour réaliser des observations sur place. Il était donc contraint de s'en remettre aux données de l'article.

On venait d'exhumer un cimetière non loin des vestiges d'un village du Néolithique. La datation au carbone 14 des squelettes, ainsi que les indices complémentaires fournis par des objets personnels retrouvés dans les tombes, permettait de situer cet endroit dans la seconde moitié du huitième millénaire avant notre ère. Il y avait donc dix mille ans. Assis dans son fauteuil, l'homme savoura cette idée, tâchant de concevoir une image de cette période. Alors qu'aucun langage recensé à ce jour n'était encore parlé, longtemps même avant l'édification des pyramides – à moins de croire aux thèses des archéologues New Age avec leurs collectes sélectives d'indices et leurs extrapolations grotesques –, ces humains avaient vécu, péri, été enterrés, fait l'amour, mangé et déféqué à même le sol. L'homme trempa les lèvres dans son café, en veillant à reposer la tasse en équilibre sur le bord de la table. Puis il poursuivit sa lecture.

On avait dénombré vingt-sept jeux d'ossements. Des femmes, dont certaines avaient atteint la cinquantaine, des enfants, quelques hommes jeunes, et un plus mûr. Des annexes détaillées décrivaient l'état de chaque squelette et présentaient les techniques utilisées tant pour déterminer leur âge que pour déduire leur régime alimentaire et leurs conditions de vie. Les auteurs de l'article soulignaient que les squelettes avaient été étendus sur une grille, système d'inhumation que l'on n'avait observé sur aucun autre site européen de la même période. Suivaient alors des diagrammes montrant comment l'orientation des grilles répondait à l'intérêt supposé de cette époque pour les solstices d'hiver et d'été. Dieu merci, on échappait aux éternelles digressions sur l'astronomie primitive. En contrepartie, les chercheurs produisaient une série d'arguments expliquant en quoi cette disposition renforçait l'hypothèse qu'ils défendaient depuis des années, à savoir que cette région d'Allemagne avait hébergé une forme hybride d'organisation sociale qu'ils appelaient société de Mittel-Baxter (d'après leurs propres noms), une culture éphémère et isolée qui représentait un intérêt scientifique mineur et une influence historique négligeable.

L'homme étudia l'article jusqu'à la fin puis se plongea dans les annexes. Après avoir parcouru le compte-rendu sur les autres squelettes, hochant la tête devant une conclusion parfaitement étayée, il se rendit à la section concernant l'homme âgé. La position de sa dépouille – au centre d'une grille de cinq sur cinq – laissait penser qu'il avait été le premier enterré à cet emplacement, et les auteurs d'en déduire logiquement que l'homme était une figure importante du village voisin. On supposait de même qu'il était né dans une autre partie du pays, car le creusement bilatéral à l'intérieur de ses orbites – un phénomène baptisé *cribra orbitalia* – semblait révéler une carence prolongée en fer. La quantité de fer dans la végétation dépend des propriétés géologiques du sol, et son absorption est conditionnée par la teneur en plomb ; ainsi deux organismes issus de régions différentes présentent deux taux différents. Or une analyse en coupe des dents de l'homme, et une évaluation des isotopes de plomb et de strontium leur avaient permis de situer son lieu d'origine à quelque quatre cents kilomètres de là. Dans un encadré, on évoquait une lésion crânienne qui ne lui avait pas été fatale – car l'os avait eu tout le temps de cicatriser avant qu'il ne rende son dernier soupir. On voulait y voir la conséquence d'une bataille ou d'une lutte de pouvoir, et le signe qu'il avait mené une vie longue et trépidante. Les auteurs allaient jusqu'à former l'hypothèse qu'il avait lui-même introduit la culture Mittel-Baxter dans un milieu sous-développé et que sa sépulture témoignait d'un statut exceptionnel.

L'homme relut ce passage, puis reposa l'article sur ses genoux. Il était aux anges. C'était la meilleure découverte à ce jour. Elle remontait encore plus loin que les sept tombes exhumées tout là-haut à Cahuachi, dans la plaine de Nazca, où les défunts avaient la bouche pleine d'excréments fossilisés. Ces Mittel et Baxter lui inspiraient de la pitié, même s'il doutait que l'imbécillité crasse de leurs conclusions soit un jour mise en lumière. Peut-être même que leur papier allait sauver leur chaire dans cette vague université du Middle West où ils trimaient. Il pouvait toujours, a priori, prendre langue avec eux pour leur expliquer les choses. Il ne se sentait pas sûr d'être compris, cependant, même si la vérité finissait par s'imposer à ceux qui avaient des yeux. Les archéologues étaient imbattables dès qu'il s'agissait d'interpréter des indices sur la base d'idées préconçues. Peu importait s'ils avançaient au flair, comme Hancock et Baigent, ou à la sueur comme Klaus Mittel et George Baxter : tous ne voyaient que ce qu'ils voulaient voir. Les traditionalistes ne voyaient que des chemins de procession ; les adeptes du New Age, des pistes d'atterrissage pour Martiens. Il arrivait qu'ils soient tous dans le vrai, mais ils ne sauraient jamais à quel moment – car dans leur esprit ils avaient toujours raison. On n'accédait à la vérité qu'en étant prêt à examiner les faits de manière dépassionnée.

La lésion crânienne dénotait à l'évidence un coup à la tête, mais Mittel et Baxter étaient loin d'en saisir toute la portée. Une blessure d'enfance suffisamment profonde pour activer une zone du cerveau qui chez la plupart d'entre nous demeurait hélas en sommeil. De même, le *cribra orbitalia* observé n'était pas forcément un syndrome importé. Certes, celui-ci était souvent synonyme d'un manque de fer, voire d'une anémie congénitale ou hémolytique, mais il pouvait avoir une origine autrement intéressante. Une surexposition au plomb produisait cet effet. Or notre homme savait que celle-ci ne constituait en aucun cas un « empoisonnement », mais un véritable don qui, combiné à d'autres facteurs, entraînait une mutation d'ordre génétique, laquelle réveillait des segments neutralisés du génome humain et leur permettait de s'exprimer.

La vraie faute de Mittel et Baxter n'était pas tant leur erreur d'interprétation que leur incapacité à établir la véritable nature du site. L'homme retrouvé au milieu du cimetière n'était pas mort le premier. Bien sûr que non. Il était mort en dernier. À son heure, et par sa propre main.

Au centre de sa création.

CHAPITRE 14

L'agent immobilier posa les coudes sur le bureau, se pencha en avant et ouvrit sa petite bouche :

— Et dans quelle tranche envisageriez-vous d'acheter ? Dites-le-moi franchement. J'ai bien conscience que nous en sommes aux tout premiers jours de notre relation, monsieur, euh... Lautner, à l'aube de notre recherche d'une résidence potentielle, mais je vais aller droit au but en vous confiant que notre entente grandira sur le mode du bénéfice mutuel si je sais exactement combien vous comptez investir en matière d'immobilier à l'heure d'aujourd'hui.

Il se renversa dans son fauteuil et me fixa d'un air entendu, fier comme tout d'avoir joué cartes sur table. Ce gars-là n'aimait pas qu'on le mène en bateau, constatai-je avec dépit. Si je n'avais que huit dollars et quelques pièces en poche, ou comptais le payer avec des cailloux, il préférait le savoir d'entrée de jeu. Maigre, roux, la cinquantaine, il répondait à l'improbable nom de Chip Farling. Je m'étais déjà coltiné moult individus de cet acabit, et mon seuil de tolérance était en chute libre.

— J'aimerais plafonner dans les six, répondis-je brusquement. Pour l'instant. Un truc vraiment exceptionnel, et je serai prêt à monter.

Il s'illumina.

— Ce serait un paiement comptant ?

— Tout à fait.

Je lui rendis son sourire.

Il opina du bonnet, ses petites mains soignées déplacèrent deux ou trois documents sur son bureau.

— Bien, dit-il, dodelinant encore. Excellent. La partie s'annonce intéressante.

Puis il pointa un doigt sur moi. Je me raidis, avant de comprendre que ce geste introduisait le suivant, qui consistait à porter sa main à son menton afin de gratter celui-ci tout en fixant un point imaginaire situé à mi-distance. Comprendre : il réfléchissait.

Au bout d'une demi-minute de ce numéro, il s'ébroua.
— Bon. Mettons-nous au travail.
Il bondit de son siège et fila à l'autre bout de l'agence tout en claquant des doigts. Je soupirai dans mon café, résigné à l'attendre.
Je m'étais d'abord rendu chez UnRealty, bien entendu. C'était fermé. Une affichette remerciait les clients de leur confiance et leur indiquait que l'entreprise cessait son activité suite au décès de son propriétaire. C'est tout juste s'il n'était pas ajouté que la nullardise de l'unique héritier avait concouru à cette décision. Je m'étais collé à la vitre pour scruter l'intérieur. Peu importe si les bureaux et classeurs sont encore là, les ordinateurs à leur place et le calendrier de l'imprimeur local toujours suspendu au mur, signalant les périodes de vacances fixées unilatéralement par quelque sous-fifre constipé : on voit au premier coup d'œil si une boîte a de l'air dans les poumons. UnRealty n'en avait plus. Je savais qu'il en serait ainsi, mais le spectacle me saisit malgré tout. Réflexion faite, je ne m'étais même pas demandé si les découvertes de ces dernières quarante-huit heures éclairaient de quelque façon que ce soit la décision de mon père concernant UnRealty. Mes réflexions ne menaient nulle part.
Alors j'avais choisi de m'activer physiquement, en faisant le tour des agences accessibles à pied. Le standing d'une petite ville peut être grossièrement estimé au nombre d'agences immobilières qui jalonnent ses rues. À Cowlick, dans le Kansas, il vous faudra chercher longtemps. Tout le monde rêve d'en partir, non d'y venir, avec pour seule exigence que ce ne soit pas les pieds devant. Tant qu'à faire. Dans un coin d'opulence modérée, vous trouverez peut-être une ou deux agences, agglutinées à d'autres entreprises en vertu d'un processus commercial de type brownien. Et dans un endroit comme Dyersburg, on croule sous les agents immobiliers. Avant même ses foulards, ses galeries et ses petits restaurants, ce type de ville vend d'abord un concept : l'idée que vous pourriez mener ce mode de vie toute l'année, être de ceux qui mettent la main sur l'un des bons morceaux et l'entourent d'une solide clôture ; que vous pourriez, vous aussi, vous prélasser dans un chalet construit sur mesure, sous un haut plafond mansardé, et vous sentir en communion avec Dieu et ses saints. Dans toute l'Amérique, les riches creusent leurs nids. Des ranchs jadis consacrés au bétail ou simplement à la nature sont rachetés puis divisés en propriétés de dix hectares où vous jouissez d'une belle vue et de voisins exactement comme vous. Ce n'est pas pour critiquer : moi aussi, je veux cette vue-là, et

cette vie-là, nichée au creux de la montagne dans l'un des plus beaux paysages au monde. Juste, je ne veux pas ce qui va avec. Le golf. L'achat collectif d'un jet. Les caves à cigares. Les androïdes inexpressifs d'une insolente sérénité qui peuplent ces country-clubs et ces loges : des types carrés au bronzage parfait et aux poignées de main fermes, des femmes aux yeux d'acier et aux joues liftées ; des conversations faites d'une dose d'avidité, de deux doses d'autosatisfaction et de trois doses de silence sinistre. Tout cela me rendrait dingue.

Au bout de quelques instants, Chip réapparut, muni d'une liasse de prospectus et de deux cassettes vidéo.

— Monsieur Lautner ? susurra-t-il. C'est l'heure de trouver son rêve...

Je regardai sagement les films, en prenant soin d'émettre de-ci de-là quelques râles ou moues d'approbation. Mais je ne vis rien qui ressemblât à ce que je cherchais. Je feuilletai les brochures, qui montraient de faux chalets de bois décorés par je ne sais quel cow-boy shooté, ou des box tout blancs, d'une stérilité si Art moderne qu'on les aurait crus importés de la Lune. Le seul aspect intéressant était la force comique des prix affichés. Comme dans chacune des agences précédentes. Je m'apprêtais, en bon client, à demander la carte de Chip avant de me retirer – puis je joindrais peut-être Bobby pour lui demander où il en était de son côté –, quand au milieu des publicités glacées je découvris une feuille toute simple.

« Les Halls, était-il écrit dans une police élégante. Pour ceux qui veulent davantage qu'une maison. »

Suivaient trois paragraphes d'une étonnante sobriété, décrivant un complexe perché dans la chaîne de Gallatin. Au pied des pistes, bien entendu. Un endroit calme et retiré, cela va sans dire. Une étendue de deux cents hectares en pleine montagne, transformée en un lotissement d'une indicible perfection, au point que Zeus lui-même y avait sûrement acheté un pavillon sur catalogue – et pourtant ce tract ne faisait aucun effort pour appâter le chaland. Il n'y avait aucune photo, ni même de prix, ce qui raviva ma curiosité.

Je sélectionnai une autre brochure plus ou moins au hasard, en veillant simplement à ce que le prix soit astronomique.

— J'aimerais voir celle-ci.

Chip vérifia mon choix, et d'acquiescer d'un air ravi.

— C'est une merveille, promit-il.

— Tenez, et puisqu'on est dans ce secteur, ajoutai-je comme une idée de dernière minute, allons faire un saut là-bas.

Je lui tendis le tract par-dessus le bureau. Il le reconnut, puis joignit les mains et me considéra.

— Voyez-vous, monsieur Lautner, avec les Halls le maître mot est sélectivité. Nous irions alors sur du très haut de gamme, financièrement parlant. Avec six millions, nous serions loin du compte. Et de beaucoup.

Je lui servis mon plus beau sourire d'homme riche.

— Comme je vous l'ai dit : montrez-moi un truc exceptionnel.

Une heure plus tard, j'écoutais Chip parler golf. L'écoutais encore. Et encore. Jusqu'à la fin des temps, parti comme ça. Dès le début du trajet, avant même que nous n'ayons quitté Dyersburg, il s'était enquis de mon intérêt pour ce sport. J'avais spontanément avoué que je n'y jouais pas, me ressaisissant juste à temps pour ne pas ajouter : « Et pourquoi diable ferais-je une chose pareille, nom d'un chien ? ». Il m'avait fixé si longuement, d'un air si consterné et si ahuri, que j'avais déclaré vouloir m'y mettre sitôt que je serais installé – à vrai dire, ce projet était même la première des raisons m'amenant à chercher ce type de propriété. Alors il avait lentement hoché la tête, et pris sur lui de m'offrir un cours accéléré sur les secrets de cet art. Je pensais tenir ainsi encore une quinzaine de minutes, après quoi je n'aurais d'autre choix que de l'achever sur place.

J'avais déjà enduré la visite de la première maison, à Big Sky, avec son électroménager de chez Sub-Zero, son parquet en érable du Honduras et sa cheminée en gros galets faite main par un maboule. À la fin du circuit je m'étais contenté de secouer la tête. Chip m'encouragea d'une tape amicale sur l'épaule – à ce stade nous étions quasiment potes – et nous retournâmes à la voiture. Nous reprîmes la route principale, pour nous enfoncer plus avant dans la montagne, cependant que Chip me tuyautait sur deux petites faiblesses qu'il croyait avoir décelées dans le jeu de Tiger Woods – et qu'il attribuait l'une et l'autre au facteur racial. Le ciel, pourtant clair au petit matin, avait pris la couleur de la route. Vive et glaciale, la rivière Gallatin courait sur notre gauche. De l'autre côté s'étirait une étroite vallée peuplée d'arbres, comme une entaille dans les Rocheuses qui s'élevaient à pic de part et d'autre. En poursuivant dans cette direction, vous atteignez un haut plateau, puis redescendez à l'est dans le parc de Yellowstone, la caldeira d'un super-volcan en sommeil dont la dernière éruption remonte à six cent mille ans. De la roche en fusion s'accumule depuis lors sous sa croûte, et d'après mon père la légende locale fait état d'un léger bruissement sur les rives du lac de Yellowstone– celui de la pression qui s'accroît lentement dans les profondeurs de la roche. On dit que le bazar pourrait à nouveau péter d'un jour à l'autre, nous replongeant aussitôt à l'Âge de pierre. La tuile, quoi. Vu mon état après une heure de Chip, j'aurais pu déclencher la catastrophe rien qu'avec les crépitements de ma cervelle.

Au bout de trente kilomètres, Chip se rangea sur la droite, sans raison apparente. Il sauta de la voiture et se hâta vers une barrière, où je décelai un modeste portail. Surprenant. À l'instar de Big Sky, les endroits de ce genre possédaient en général une entrée principale imposante, façonnée dans des arbres qui avaient déjà belle taille du temps où personne n'avait jamais entendu parler de Chip Farling. Ce portail ne semblait protéger rien de plus qu'une route de service. Chip se pressa sur la droite de la grille, et je vis remuer ses lèvres, avant de remarquer un Interphone encastré dans le poteau. Il se redressa et attendit un instant, les yeux au ciel. Quelques gouttes de pluie commençaient à tomber. Puis Chip se retourna, tendit l'oreille et revint à la voiture.

Le temps qu'il boucle sa ceinture, le portail s'était ouvert. Il se referma sur notre passage. Chip s'engagea sur une piste matérialisée par deux bandes parallèles d'herbe aplatie. Il conduisait avec prudence, ce qui ne m'empêchait pas d'être ballotté dans tous les sens.

— Plutôt du genre rustique, non ? grimaçai-je.

Il sourit.

— Vous allez voir.

La piste filait sur presque un demi-kilomètre, perpendiculairement à la route principale, en direction d'un épais bouquet d'arbres. Comme nous contournions ceux-ci, le sol changea d'un coup : les deux lignes creusées firent place à un macadam étroit et immaculé. Virevoltant sur mon siège, je constatai que la grande route n'était plus visible, masquée par les arbres.

— Astucieux, dis-je.

— Rien n'est laissé au hasard dans les Halls, récita Chip. Ceux qui y élisent domicile sont assurés d'y trouver le nec plus ultra de la tranquillité.

Le chemin s'écartait à nouveau de la rivière, serpentant derrière un bosquet pour gravir la pente raide d'une ravine, puis slalomait encore pour rester hors de vue de la route. Au bout de quelques minutes, on peinait à croire que la nationale ait seulement existé. Les Halls étaient un endroit très pensé. Et j'étais assez impressionné.

— Depuis quand ce lieu existe-t-il ?

— C'est sa septième année de développement, répondit Chip, qui louchait sur la route à travers son pare-brise inondé. Quel dommage de le visiter par un temps si peu clément ! Une bonne chute de neige ici, et vous avez l'impression d'être monté au paradis.

— Vous en avez vendu beaucoup ?

— Pas une seule. Mais ils n'ont que dix maisons à céder, et ils n'éprouvent aucune urgence à remplir les dernières. Pour être honnête, j'avoue que leur prospectus ne leur rend pas service. Je leur ai pourtant conseillé d'insérer des photos.

Nous approchions du sommet d'une colline, après que les interminables zigzags nous eurent élevé d'au moins cent cinquante mètres.

— Aucun des autres agents que j'ai rencontrés ne semblait connaître cet endroit.

Chip secoua la tête.

— Nous avons l'exclusivité. Du moins, nous l'avons aujourd'hui.

Il me fit un clin d'œil, et l'espace d'une seconde j'entrevis l'homme que pouvait être M. Farling lorsqu'il regagnait ses pénates le soir. Je détournai la tête, convaincu d'avoir bien fait de me présenter sous un faux nom. Mon petit doigt me disait que Chip aurait reconnu le nom d'Hopkins plus vite que celui de tel ou tel architecte de Los Angeles, même si ses immeubles avaient servi de décor dans des centaines de films.

Une entrée se dessina comme nous passions un dernier virage. Elle n'était pas faite de bois, mais d'énormes blocs de roche, et se dressait en haut d'une petite butte, de sorte que ce qu'elle protégeait demeurait invisible. À mesure que nous approchions, je déchiffrai les mots « Les Halls » gravés à la main, dans le même style que sur le prospectus.

— Nous y voilà, pépia Chip inutilement.

De l'autre côté de la butte, la route bifurquait à gauche toute. À un kilomètre au loin, je crus deviner une chaîne de montagnes plus hautes, mais un nouveau bouquet d'arbres nous barrait la vue. Derrière eux s'étirait de part et d'autre une palissade. Sa hauteur empêchait de voir derrière. La pluie s'intensifiait, et le ciel noir semblait prêt à éclater.

— Le golf se trouve de l'autre côté, expliqua Chip, comme passé subrepticement en pilotage automatique. Neuf trous, signé Nicklaus *père et fils, bien sûr*[1]. Bien entendu, il est recouvert en cette période de l'année, mais qui en aurait besoin, avec ceux de Thunder Fall et de Lost Creek à quelques petites minutes d'ici ? Imaginez : des équipements de classe internationale, accessibles d'un saut de voiture, pour le bonheur du plus exigeant et sophistiqué des acquéreurs.

En effet, pensai-je. Et imagine que je te fourre mon doigt dans le nez.

— Vous voyez ici le centre-accueil, dit Chip. (Dans l'obscurité apparut une grappe de bâtiments en bois.) Salle de réunion, bar non-fumeur, et un excellent restaurant.

— Vous y avez déjà mangé ?

— Non. Mais je me suis laissé dire que c'est, disons, vraiment succulent.

Il se gara dans un espace bordant l'entrée, au bout d'une rangée de voitures très chères. Nous sortîmes de la sienne, et il me dirigea vers la porte. Je tâchai de scruter les alentours, de me faire une idée du reste de la cité, mais la visibilité était nulle, et nous ne pouvions nous éterniser sous cette pluie qui martelait chaque surface plane.

— Putain de flotte, marmonna Chip. (Devant ma surprise, il s'excusa d'un haussement d'épaules :) Désolé, mais la pluie est le pire ennemi de l'agent immobilier.

— Quoi, pire que des voisins latinos ?

Il s'esclaffa, me tapa dans le dos et me guida à l'intérieur.

Il régnait un parfait silence. Sur la gauche se trouvait une sorte de salle de cocktail, avec des fauteuils en cuir entourant des tables de bois sombre. Déserte. Tout au fond, une fenêtre qui d'ordinaire devait dispenser une vue époustouflante. Elle n'était aujourd'hui qu'un pauvre rectangle gris. Sur la droite, une imposante cheminée, où crépitait un feu discipliné. En sourdine filtraient quelques notes de Beethoven, une sonate pour violon et piano. Le comptoir de la réception était un plan lisse d'un très beau bois, et le mur de derrière arborait une œuvre d'« art ». Comme nous attendions d'être reçus suite au coup de sonnette de Chip, je plongeai la main dans ma veste et pressai une touche de mon téléphone. Si le réseau

1. En français dans le texte. (*N.d.T.*)

captait jusqu'ici, Bobby allait recevoir un appel. Nous étions convenus que je procède ainsi si ma recherche était fructueuse.

Elle l'était.

L'étape introductive dura une demi-heure. Une svelte et séduisante femme d'une petite quarantaine d'années, rehaussée par de grosses dépenses d'ordre capillaire, nous installa dans le salon pour nous décrire tous les trésors que les Halls avaient à offrir. Elle était vêtue d'un impeccable tailleur gris et possédait de jolis petits yeux bleus ainsi qu'une belle peau, aussi fallait-il sûrement prendre toutes ses paroles pour argent comptant. Elle ne déclina pas son nom, ce qui m'intrigua. En langage commercial américain, on commence toujours par le blaze : d'entrée de jeu, dans la foulée de la poignée de main. Comme une preuve d'engagement. Tu sais comment je m'appelle, alors je ne peux vouloir que ton bien. Comment imaginer que je t'embobine – moi, ton ami ?

Au cœur du concept des Halls, expliqua Mme Anonyme, se trouvait le désir de reproduire l'idéal traditionnel de la « communauté » – mais en mieux. Du personnel était disponible vingt-quatre heures sur vingt-quatre pour offrir une assistance dans tous les domaines, même les plus complexes. Les résidents considéraient d'ailleurs ces employés comme de véritables amis – le genre d'amis, présumai-je, qui devaient faire tout ce qu'on leur disait, à n'importe quel moment, et quelle que soit la difficulté de la tâche. Le chef du restaurant avait officié auparavant dans une cantine chicos de L.A. dont moi-même je connaissais le nom, et les résidents pouvaient se faire livrer des repas à domicile entre 9 heures du matin et minuit. La cave à vin, jura-t-elle, défiait l'imagination. Chaque pavillon était bourré d'électronique, avec l'Internet haut débit généralisé. En plus du club de golf si longuement vanté, il existait un club de gym, un club de gastronomes, et plusieurs autres que je ne pris pas la peine de retenir. L'adhésion à chacun d'eux était automatique et coûtait environ un demi-million de dollars. Par année. Par club. Pendant tout ce temps, je voyais Chip opiner fébrilement du bonnet, comme s'il n'en revenait pas de trouver une si bonne affaire. Je sirotai mon centième café de la journée – au moins, celui des Halls était bon – tout en m'efforçant de ne pas blêmir.

La fille conclut en soulignant qu'il ne restait plus que trois maisons à vendre au sein de la communauté, à des prix allant de onze et demi à quatorze millions de dollars – des sommes obscènes, même pour de l'immobilier de luxe. Elle termina par une touchante envolée sur les joies de la copropriété, que je vis Chip recopier dans un coin de sa tête.

— Cool, fis-je quand le topo prit fin. (Je reposai ma tasse.) Allons voir ça de plus près.

La femme me dévisagea poliment.

— Vous pensez bien que c'est impossible.

— J'ai déjà été mouillé, la rassurai-je. Plein de fois. Je suis même allé à la piscine, un jour.

— Le temps n'a rien à y voir. Nous ne laissons jamais visiter les Halls tant que la validité de la candidature n'a pas été établie.

Elle lança un regard à Chip, qui se planquait derrière un masque inexpressif.

— La validité, répétai-je.

— Financière et autre.

Je levai les sourcils, souris gentiment.

— Quoi ?

Chip intervint :

— C'est-à-dire que, si je peux me permettre, comme je vous l'ai expliqué en venant ici, les Halls pratiquent une politique très...

— Je sais tout ça, le coupai-je. Dois-je donc comprendre, madame... ? (Je laissai un blanc, mais elle n'y glissa pas son nom ; cette femme n'avait aucune hâte d'être mon amie.) Dois-je donc comprendre que je n'irai pas plus loin que cette pièce tant que je n'aurai pas sauté à travers une série de cerceaux dressés par vos soins ?

— C'est exact. (Elle me gratifia d'un beau sourire, comme devant un enfant ayant enfin compris, au prix de longs et pénibles efforts, comment la position relative de la petite et de la grande aiguille indique le temps qu'il reste avant d'aller au lit.) Ainsi que M. Farling aurait dû le préciser.

— Et en quoi consisterait cet examen ?

La femme extirpa d'une chemise une feuille de papier. La plaçant sous mes yeux, elle énuméra :

— Le remise du montant intégral de l'achat désiré, ainsi qu'une provision couvrant les adhésions aux clubs pendant cinq ans. Aucune hypothèque ni option de paiement différé n'est envisageable. Une garantie d'accès à votre comptable ou à tout autre représentant agréé, dans le but de brosser un tableau général de vos finances. Une rencontre entre vous et l'ensemble du bureau de la communauté, qui comprend le personnel d'encadrement et un représentant de chaque propriété occupée, laquelle rencontre sera suivie si nécessaire d'un examen en comité restreint.

La désignation par vous-même de deux personnalités emblématiques – et par « emblématiques », comprendre qu'elles le soient au niveau de la société au sens large – auprès desquelles notre bureau puisse se renseigner concernant votre situation passée et présente. En supposant que tout ceci se déroule sans accroc, vous serez invité à l'intérieur de la propriété pour découvrir la cité plus en détail, et procéder à votre choix.

— J'espère que vous plaisantez.

— Je vous assure qu'il n'en est rien.

Je tentai le bluff :

— Avez-vous seulement idée de qui je suis ?

— Non. (Elle sourit, joignant ses lèvres telle une plaie à peine cicatrisée.) Et c'est là toute la question.

J'avais plus ou moins conscience que le réceptionniste, un jeune homme qui avait passé beaucoup de temps au club de gym, nous observait. Je soutins le regard de la femme un instant, puis lui retournai son sourire.

— Excellent, dis-je.

Après un temps d'hésitation, elle fronça les sourcils.

— Je vous demande pardon ?

— C'est exactement ce que j'espérais. Je vois que M. Farling avait parfaitement cerné mes besoins. (Mon ton devenait un peu sec, sans doute pour coller à mon nouveau personnage.) Quelqu'un dans ma position a besoin de certaines garanties, et je suis ravi de constater que vous y pourvoyez.

Mme Anonyme recouvra des traits amicaux.

— Nous sommes donc en pleine convergence de vues ?

— Absolument. Serait-il possible de voir les plans des propriétés disponibles ?

— Bien entendu.

Elle rouvrit la chemise et produisit deux épais documents. Elle les déplia sur la table et je les examinai vite fait. Ils étaient détaillés et abondamment annotés. Ce que je voyais m'intéressait encore plus que prévu.

— Fascinant, dis-je. Je suis navré de ne pouvoir les apprécier en vrai dès aujourd'hui, mais ceci suffit déjà à piquer mon intérêt.

Je commençai à replier les plans, avant de songer qu'un homme riche comme moi laisserait à d'autres une tâche aussi triviale. Alors je me levai. La soudaineté du mouvement les prit tous deux par surprise, et ils se dépêchèrent de m'imiter. Je tendis ma main à la dame, et serrai la sienne vigoureusement.

— Merci de votre accueil, dis-je en faisant mine d'avoir déjà l'esprit ailleurs. J'imagine que pour toute question complémentaire je devrai m'en remettre à M. Farling ?

— C'est la voie normale, en effet. Puis-je vous demander comment vous avez entendu parler des Halls ?

J'hésitai un instant, car avouer que j'étais tombé par hasard sur leur papelard faisait un peu léger.

— Des amis, répondis-je.

Elle opina, presque imperceptiblement. Bonne réponse.

Je la saluai du bonnet et traversai le hall d'entrée, sans attendre Chip. Dehors, je stagnai quelques instants sous la marquise, à regarder l'averse. Quand bien même l'envie m'eût pris de désobéir, je constatais que les bâtiments étaient agencés de manière à soustraire la communauté à tous les regards extérieurs. Chip n'avait rien exagéré en parlant d'extrême tranquillité.

Ce dernier émergea peu de temps après et me ramena à la voiture. Au moment de monter, je vis qu'un autre véhicule venait de franchir l'entrée et suivait le chemin à vive allure. Noir, mastoc, un de ces monstres tout-terrain. Ses pneus barbouillèrent un arc autour du petit parking, et il se rangea à cinq ou six mètres de nous.

Je pris tout mon temps pour ouvrir la portière, m'engouffrer dans l'habitacle, me caler dans mon siège, allant même jusqu'à laisser un pied dehors afin de prolonger la manœuvre. J'attachai ma ceinture quand un type sortit du bâtiment que nous venions de quitter. Blond, il faisait à peu près ma taille et marchait d'un pas décidé, la tête baissée. À aucun moment il ne nous regarda. Je crus deviner qu'il avait les traits plutôt rugueux. Comme il se dirigeait vers le 4x4, le conducteur descendit et contourna l'engin pour ouvrir le coffre. Dos à nous, le premier type y enfourna un gros sac d'un ton bleu pétrole. La poignée était baguée d'un papillon des douanes, mais je ne pus déchiffrer les lettres. Les deux types grimpèrent dans la voiture.

Entre-temps, Chip avait démarré. Il recula avec précaution, passa la marche avant, et nous laissâmes les Halls derrière nous.

Chip fut quasi muet jusqu'à notre retour en ville. Je le soupçonnais de s'être fait cuisiner par Mme Anonyme après ma sortie et de se mordre les doigts pour ne pas avoir su fournir des réponses satisfaisantes. Qui j'étais, d'où je venais... Même moi, je savais que c'était là le b.a.-ba de l'immobilier, les acides aminés du génome transactionnel. Mon père aimait à dire, dans nos rares discussions, que pour inciter un homme à mettre la main à la poche, il fallait lui servir ses propres arguments, c'est-à-dire l'avoir suffisamment cerné pour faire mouche.

Chip me demanda néanmoins ce que j'avais pensé de ces visites. Je lui dis que Big Sky ne m'intéressait pas, surtout après avoir vu ce que proposaient les Halls. Il ne parut guère surpris. Je lui demandai à mon tour combien de personnes il avait emmenées là-bas. La réponse était huit, en trois ans. Toutes avaient franchi les étapes de sélection imposées par la direction. Aucune n'avait été admise à acheter.

Je le dévisageai.

— Ces gens-là ont rassemblé quinze ou vingt millions, ouvert leurs livres de comptes, et on les a quand même déboutés ? Ils veulent les vendre, leurs baraques, ou quoi ?

— Sélectivité, monsieur Lautner. Tel est le maître mot. (Il jeta un œil sur moi, pour s'assurer d'avoir toute mon attention.) Nous vivons dans un drôle de monde, c'est un fait. Nous possédons le plus beau pays du globe, les gars les plus travailleurs, et cependant nous côtoyons au quotidien des individus dont personne ne voudrait dans son hémisphère. Il y a une dimension historique derrière tout ça. Nous avons ouvert les portes au tout-venant, pour les refermer trop tard. Nous avons dit : « Venez donc, braves gens, rejoignez-nous. Nous avons besoin de sang neuf. Nous avons plein d'espace à occuper. » Mais nous avons omis de vérifier s'il s'agissait vraiment du bon sang. Nous n'avons pas suffisamment pensé à l'avenir. C'est pour cette raison que les gens comme vous viennent dans l'Ouest. Pour échapper aux grandes villes, aux hordes, pour se retrouver entre gens de la même espèce. Pour recréer un mode de vie authentique. Je ne vous parle pas de races, même si elles jouent leur rôle dans ce que je décris. Je veux parler d'attitude. De qualité. De personnes qui sont faites pour être ensemble, et d'autres qui ne le sont pas. Voilà pourquoi les gens choisissent des coins comme Dyersburg. C'est une sorte de filtre, qui la plupart du temps fonctionne très bien – mais on retrouve toujours des êtres qui n'y ont décidément par leur place. Des étudiants. Des skieurs du dimanche. Les petits Blancs massés près de l'autoroute. Des gens qui ne comprennent rien. Mais que voulez-vous faire ? On ne va pas les empêcher de s'installer, c'est un pays libre. Alors il ne reste plus qu'à se prendre en main.

— C'est-à-dire ?

— Vous resserrez le filtre de manière radicale. Vous trouvez des gens qui vous correspondent, et vous érigez une grande muraille.

— C'est ça, le principe des Halls ?

— On peut le voir comme ça. Mais cela demeure avant tout un lieu de résidence unique et inestimable.

— Si vous aviez l'argent, vous iriez là-dedans, vous ?

Il lâcha un rire bref et amer.

— Oui, monsieur, sans aucun doute. Mais d'ici là, je vais me concentrer sur mes commissions.

Nous redescendîmes la colline jusqu'au petit plateau. Quand nous atteignîmes Dyersburg, il faisait nuit noire, et la pluie commençait à décliner. Chip se gara devant l'agence, avant de se tourner vers moi.

— Alors ? sourit-il. Que décidez-vous ? Vous voulez réfléchir à ce que vous avez vu, ou bien on retourne dans mon bureau et je vous montre quelques possibilités pour demain ?

— Je voulais vous poser une question, dis-je sans ôter les yeux du pare-brise.

Le trottoir était désert.

— Je vous écoute.

Il semblait fatigué mais vaillant. Ma mère répétait souvent que l'immobilier n'était pas un métier pour qui aimait les horaires réguliers.

— Vous disiez avoir depuis peu l'exclusivité sur les Halls. J'en déduis qu'il y avait une autre boîte sur le coup ?

— C'est exact, fit-il d'un air méfiant. Pourquoi ?

— À votre connaissance, elle n'a vendu aucune maison ?

— Non, monsieur. À vrai dire, elle n'a pas gardé le dossier bien longtemps.

— Et comment se fait-il qu'elle ne soit plus de la partie ?

— Le patron est mort et la boîte a fermé. Pas facile de vendre des maisons quand on n'est plus de ce monde.

Je hochai la tête, gardant tout mon calme.

— De combien serait votre commission pour l'un de ces logements ? Une coquette somme, j'imagine.

— Rondelette, admit-il avec prudence.

Je laissai courir un silence.

— Assez pour tuer quelqu'un ?

— Quoi ? !

— Vous avez très bien entendu.

Mon sourire s'était envolé.

— Je ne vois pas de quoi vous voulez parler. Vous croyez que... quoi ? Qu'est-ce que vous me chantez, bon sang ?

Il y avait je ne sais quoi de déplaisant dans ses dénégations, or vous seriez étonné, et attristé, de savoir combien les gens sont doués pour mentir, même dans les circonstances les plus difficiles. J'avais attendu. Je m'étais montré gentil. Maintenant j'en avais marre de jouer la comédie.

Je plongeai la main derrière son crâne et le projetai en avant, écrasant son front contre le volant. J'avais ajusté l'angle de sorte que le plastique central lui fracasse l'arête nasale. Puis je le redressai.

— Je vais te poser une question, dis-je tout en rabaissant sa tête à quelques centimètres de la colonne de direction. (Il gémit faiblement.) Cette fois, il faut que je te croie. Il faut que je sache que tu dis la vérité, et c'est ta dernière chance de me convaincre. Autrement, je te tue. Pigé ?

Je perçus un acquiescement fébrile. De nouveau, je lui tirai les cheveux en arrière. Il saignait du nez, et son front était traversé d'une marque blafarde. Il écarquillait les yeux.

— Tu as tué Don Hopkins ?

Il secoua la tête. La secoua encore et encore, avec toute la frénésie d'un gosse paniqué. Je l'observai quelques instants. J'ai eu affaire à de nombreux menteurs dans ma vie, l'ai moi-même été pendant de longues périodes. J'ai l'œil pour ça.

Chip n'avait pas tué mon père. Du moins, pas de ses propres mains.

— OK, fis-je avant qu'il ne se brise le cou à gesticuler comme un sourd. Mais je crois que tu sais des choses au sujet de sa mort. Alors voilà le marché. Je veux que tu prennes un message. Tu ferais ça pour moi ?

Il hocha la tête. Cligna des yeux.

— Tu diras aux nazis sur leur montagne que quelqu'un s'intéresse à eux. Dis-leur que je ne crois pas que mes parents soient morts dans un accident, et que je leur ferai payer ce qu'ils ont fait. C'est noté ?

Il opina de nouveau. Je lâchai sa tignasse, ouvris la portière et sortis sous la pluie.

Une fois dehors je me penchai pour l'observer. Sa bouche était tordue de stupeur et d'effroi, et le sang lui coulait le long du menton.

Je me retournai, les mains tremblantes, et partis à la rencontre d'un être humain.

CHAPITRE 15

Bobby était accoudé au comptoir chez mes parents, devant un verre d'eau minérale. À mon entrée, il releva les yeux, me regarda ruisseler sur le sol. J'avais fait tout le chemin sous la pluie.

— Qu'est-ce qui t'est arrivé ? demanda-t-il d'une voix posée.

— Rien.

— Soit, dit-il après un bref silence.

J'attrapai son verre d'eau et le vidai d'un trait. Seulement après avoir dégluti, je me souvins qu'elle provenait des dernières courses de mes parents.

— Il en reste ? demandai-je.

— Un peu.

— Ne la bois pas.

Je reposai le verre sur le comptoir et m'assis à la table. Puis décidai de retirer mon manteau, comme si une petite voix me prédisait une pneumonie. Par la vitre j'aperçus de la lumière dans le salon de Mary. Pourvu qu'elle ne sache pas que j'étais encore en ville. Elle m'aurait trouvé grossier de ne pas lui rendre visite. Puis je pris conscience que cette maison-ci avait plusieurs lampes allumées et une voiture dans l'allée : alors Mary devait bien être au courant. Décidément, je n'avais pas les idées très claires.

Bobby attendait, les bras croisés.

— Alors, lançai-je, ta journée ?

— À toi l'honneur, Ward.

Je secouai la tête. Il haussa les épaules et capitula :

— J'ai inspecté le lieu de l'accident. Vu la position de la voiture emboutie, il est plausible que ta mère ait simplement loupé son virage. Il est plutôt serré, il faisait nuit, et il y avait de la brume par-dessus le marché.

— Je vois, soupirai-je. Et ça ne faisait jamais que... quoi ? quarante-ans qu'elle conduisait ? Elle n'avait jamais dû rencontrer un virage serré, ni traverser ce carrefour une seule fois durant toutes les années qu'elle a passées ici. Le jus de myrtille et de la brume auront concouru à sa perte, c'est ça ? Je comprends enfin. Et c'est un miracle que la voiture ne soit pas partie en vrille au-dessus des premiers immeubles pour rebondir jusqu'à l'océan.

Bobby poursuivit sans relever :

— Il y a une petite station-service en face du lieu de l'accident, et une vidéothèque un peu plus loin. Il va sans dire qu'aucun des types à qui j'ai parlé n'était présent le soir de l'accident. Le vidéo-shop est une boîte indépendante gérée par deux frangins. Celui que j'ai rencontré affirme que son frère n'a rien remarqué avant qu'une voiture de police ne débarque.

— Il n'a pas entendu le fracas d'un lourd objet métallique qui s'écrase dans un autre, le genre de truc un rien inhabituel ?

— Tu connais ces boutiques-là. Un gros téléviseur suspendu au plafond, un film de John Woo à plein volume, et le type derrière son comptoir qui occupe sa soirée avec des bières et un joint de la taille d'un burrito. Tu lui aurais foutu un coup de marteau sur le carafon qu'il aurait à peine cillé. Alors je me suis rendu à la station-service, et l'employé m'a donné le numéro de son manager. Je l'ai appelé et il m'a fourni l'adresse du type qui était d'astreinte ce soir-là.

— En lui racontant quoi ?

— Que j'assistais la police dans son enquête.

— Génial, dis-je. À présent je vais avoir tout le comico sur le dos.

— Qu'est-ce que ça peut bien foutre, Ward ?

— Je ne fais plus partie de l'Agence, Bobby. Ici, dans le monde réel, les flics peuvent te faire chier.

Bobby balaya cette objection d'un revers de main.

— Je me suis donc pointé chez lui. Lui aussi déclare n'avoir rien vu. Il a entendu un bruit, mais s'est dit qu'un gus devait faire le con derrière la station. Il s'est tâté pour savoir s'il devait appeler les flics ou non, mais le temps de découvrir qu'il y avait eu un accident et que sa boutique était intacte, la police était déjà sur place.

— OK. (À vrai dire, je n'avais rien espéré de ces vérifications, mais Bobby tenait à les mener.) Quoi d'autre ?

— Là-dessus, je suis venu ici pour jeter un œil, comme convenu.

— T'as trouvé quelque chose ?

Il secoua la tête.

— Que dalle.

— Je te l'avais dit.

— C'est vrai, répliqua-t-il. Non seulement tu es beau, Ward, mais tu as toujours raison. Ah, si seulement j'étais gay... je n'irais pas chercher plus loin. T'es le meilleur. Alors à toi de parler, maintenant.

— L'endroit qui figure au tout début de la cassette s'appelle les Halls et se trouve en haut d'une ravine de la Gallatin Valley. Il faut être très riche pour en être, et ils ne te laissent même pas visiter tant que tu n'as pas prouvé que tu étais bien comme il faut.

— Les Halls ? C'est quoi, ce nom ?

— Je ne sais pas, soupirai-je. Un clin d'œil au Walhalla, peut-être. Ces gens se prennent peut-être pour des dieux. Possible qu'ils le soient, pour avoir autant de fric.

— Tu es certain qu'il s'agit du même endroit ?

— Ça ne fait aucun doute. Le vestibule était en tous points identique à celui de la vidéo, jusque dans le choix des tableaux. C'est cet endroit-là, Bobby. Et ils sont très, très sélectifs.

— Pourquoi n'as-tu pas appelé, dans ce cas ?

— J'ai bien essayé. On ne doit pas capter le réseau là-haut. Mais je n'en savais rien : le téléphone était planqué dans ma poche.

— Et c'est comment ?

— Au poil. Je n'ai vu aucun résident, sauf à la fin, un type qui passait en coup de vent, mais je ne saurais même pas le reconnaître. En gros, si t'as du pognon et que tu ne veux pas être emmerdé par les Terriens de base, c'est l'endroit qu'il te faut. On m'a montré les plans, cela dit, et ça n'a rien du pavillon de banlieue. Ils ont fait appel à quelqu'un de fortiche, qui avait une idée très précise en tête.

— Quel genre ?

Je sortis un stylo de ma poche pour esquisser un croquis.

— Une disposition éclatée. Des espaces de vie surélevés par rapport au terrain. Des cheminées disposées à la confluence des pièces. Des vitraux sur les fenêtres opposées à la cheminée, et sur les lucarnes des couloirs. De longs avant-toits, des vitres zébrées à l'horizontale, de gigantesques terrasses.

Bobby scruta mon dessin.

— Et alors ? À mon avis, mon ami, ça m'a tout l'air d'une maison ordinaire.

— Nombre de ces éléments sont entrés dans l'architecture moderne, convins-je. Mais la façon dont ils étaient assemblés sur ces plans était du pur Frank Lloyd Wright.

— Ils l'ont peut-être engagé.
— Peu probable. À moins qu'ils n'aient aussi embauché un médium.
— Alors ils connaissent quelqu'un qui dessine comme lui. Il doit en exister des centaines. La belle affaire.
— Sans doute. Mais ce genre de truc n'est plus à la mode, et ne l'a jamais été pour ce type d'habitation. On s'attendrait plutôt à trouver de grands escaliers façon roi du pétrole, des chambres organisées en suites, et regarde-un-peu-comme-je-suis-riche.
— Plutôt séduisant.
— Mais artificiel. À l'origine, nos lieux d'habitation étaient forgées dans l'environnement naturel, et non construites ex nihilo. Voilà pourquoi l'architecture moderne paraît si souvent aride : elle n'a aucun rapport organique au site. Les maisons de Wright sont différentes. Le chemin d'accès est tortueux pour symboliser la retraite vers un havre de paix, et l'âtre de la cheminée est disposé au centre de la structure pour rappeler un feu au milieu d'une grotte. Les pièces sont modulables afin de permettre une perspective intérieure, synonyme de défense ultime ; elles évoquent aussi l'apprivoisement d'un espace naturel. Les fenêtres sont zébrées pour révéler l'extérieur sans exposer l'intérieur. Les vitraux évoquent aux habitants un mur de végétation à travers lequel ils peuvent voir, mais qui les cache en même temps. L'humain se sent bien quand il possède à la fois la perspective et le refuge – quand il possède une jolie vue tout en se sentant protégé des autres et des regards. C'est précisément ce qu'offrent ces structures.
Bobby me dévisagea.
— Tu es un drôle d'oiseau.
Je haussai les épaules.
— J'ai juste écouté en classe. Tout ça pour dire que si tu me trouves un autre village qui ressemble à ça, je te lèche le cul sur-le-champ.
— C'est tentant, mais je préfère te croire sur parole.
— Ce doit être l'une des raisons pour lesquelles ils refusent de laisser visiter au début. Ce n'est pas le genre d'édifice pour lequel on raque des millions en temps normal. Ce qui signifie qu'ils ont dû choisir ce style dans un but précis.
— Le promoteur est donc un fana de Wright. Ou bien ils ont dégotté un archi qui écoutait en classe, lui aussi. Je ne vois pas du tout où ça nous mène. Mais ça me ferait vraiment plaisir si tu me disais comment ça s'est terminé.

— J'ai agressé l'agent immobilier.
— Là-bas ?
Je secouai la tête.
— Tu me prends pour qui ? À notre retour en ville. Il n'y avait personne dans les parages.
— Il est mort ? demanda Bobby d'un ton routinier.
— Bien sûr que non !
— Et pourquoi t'as fait ça ?
— Sa tête me revenait pas. Et puis il y avait auparavant deux agences au service des Halls. Il n'y en a plus qu'une.
Bobby hocha lentement la tête.
— La boîte de ton père étant la perdante.
— T'es brillant, comme mec.
— J'en déduis également, vu que tu l'as pas buté, que tu ne soupçonnes pas cet agent d'avoir tué tes vieux. Malgré l'intérêt financier qu'il y aurait trouvé.
Je secouai la tête.
— Ce n'est pas lui personnellement. Mais il est de mèche avec ceux qui ont fait le coup. Pour quelle autre raison verrait-on ce lieu sur la cassette ?
D'un coup j'étais sur mes deux jambes, quittant la cuisine d'un pas vif. Comme je traversais le couloir, un détail me perturba, mais je ne parvins à l'identifier, aussi continuai-je d'avancer. Bobby me suivit au salon, où je m'arrêtai devant la table basse.
Je ramassai le livre qui s'y trouvait et l'agitai sous ses yeux.
— Un ouvrage consacré au grand architecte susnommé, constata-t-il. Et alors ? Ton père était agent immobilier. Les maisons, c'était son domaine. Et c'était un type âgé. Les vieux raffolent de biographies. C'est ça et la chaîne Découverte qui les maintiennent en vie.
— Bobby...
— OK, concéda-t-il. C'est une coïncidence intéressante. Si on veut.
Je rebroussai chemin vers le couloir puis m'arrêtai de nouveau. Je ressentais comme une boule d'énergie qui me poussait à l'action, mais je ne savais quelle direction prendre.
— Tu as fouillé la maison de fond en comble ? demandai-je.
— J'ai décollé la moquette, soulevé le plancher, passé la tête sous le toit et glissé une lampe torche dans la citerne. J'ai démonté les téléphones. Je n'ai rien trouvé. Évidemment, je ne saurais te dire si des trucs ont disparu.

— Moi non plus. Je ne venais pas assez souvent. J'ai juste remarqué l'absence de vidéos. (Je fronçai les sourcils.) Attends un peu. La dernière fois que je suis venu, j'ai posé le courrier ici. Il a disparu.

Je fixai Bobby intensément, persuadé d'avoir levé un lièvre.

— T'emballe pas, Sherlock. Un vieux mec est passé le prendre il y a environ deux heures. Avec un nez crochu. S'est présenté comme l'avocat de tes vieux. Je l'ai laissé entrer, en expliquant que j'étais un de tes amis. Ça ne l'a pas flippé plus que ça, même s'il avait l'air de vouloir recompter l'argenterie.

— Harold Davids. Il m'avait dit qu'il passerait.

Bobby sourit.

— Écoute, Ward, il y a assez de bizarreries autour de toi pour que tu n'ailles pas en inventer d'autres. Ne sois pas si parano.

Retentit un bruit de déflagration en provenance du salon. Nos corps réagirent, mais pas assez vite.

Ce n'est pas tant le bruit qu'une sensation d'immense pression, aussi brutale que de se faire gifler, enfant, par un adulte qui ne vous avait encore jamais frappé. Si vous êtes suffisamment proche d'une explosion, vous percevez avant tout le coup mat qu'essuient la tête et la poitrine, un impact qui transforme le vacarme en onde profonde, avec l'impression que l'univers s'est fait dégommer de son axe. Le son lui-même paraît secondaire, comme si vous l'entendiez avec deux jours de retard.

Je dus aussitôt cogner le mur, violemment, et m'écraser la tête la première sur une rangée de cadres. Comme je tombais à terre, le crâne gorgé de lumière blanche, assailli de bris de verre, une nouvelle explosion résonna, moins forte, puis je soulevai Bobby et l'entraînai jusqu'aux décombres de la porte d'entrée.

Nous détalâmes ensemble dans l'allée, glissant et chutant sur les dalles mouillées. Il y eut une troisième détonation, autrement puissante que la première. Cette fois j'entendis siffler les objets projetés tout autour, et le *fuip-pah !* de l'air comprimé puis libéré. Bobby continuait de gratter la terre, s'aidant de ses mains pour nous faire avancer. Je ruinai ses efforts en me retournant pour observer la maison, ce qui nous déséquilibra et nous projeta de dos sur l'herbe trempée. Le mur extérieur du salon était tombé, et l'intérieur était déjà en flammes. Je ne pouvais en détacher mes yeux. Voir une maison en feu, c'est comme de regarder brûler une âme, la voir bouffer par des vers de vingt mètres de haut.

Le temps que je me relève, Bobby avait dégainé son téléphone et s'éloignait derrière la clôture. Je refis quelques pas en direction de la maison. J'envisageais peut-être de rentrer et d'éteindre l'incendie. Ou de sauver quelques objets. Je ne sais pas. Je me disais qu'il devait bien y avoir quelque chose à faire.

De nouveau, une petite détonation, suivie de bruits de casse au fond de la maison. La chaleur grimpait très vite. La pluie n'était plus qu'une faible bruine, et je me dis que c'était vraiment typique. Il avait flotté tout l'après-midi. Pourquoi pas maintenant ?

Bobby me rejoignit en courant et referma son portable. Il saignait un peu au front.

— Ils sont en route, annonça-t-il.

Je n'avais aucune idée de qui il voulait parler.

— Qui ça ?

— Les pompiers. Allez, on s'en va.

— Je ne peux pas partir. C'est leur maison.

— Non, rétorqua-t-il. C'est le lieu du crime.

Bobby fit rapidement le tour de ma voiture en scrutant le sol. Puis se mit à quatre pattes dans la boue et regarda sous le châssis. Il se releva, se frotta les mains et déverrouilla la portière. Il se pencha pour jeter un œil sous le siège conducteur, débloqua le capot, contourna l'aile et examina le moteur.

— OK, dit-il. Ça se tente.

Il referma le capot et regagna le coté conducteur. Il introduisit la clé dans le démarreur, me regarda en grimaçant, donna un coup de poignet. Le moteur démarra, sans que rien explose. Bobby poussa un long soupir, tapota le toit de la voiture.

— Mais on n'a rien entendu, m'étonnai-je. Aucune bagnole.

— Je ne suis pas surpris, fit-il avec dans la voix quelques trémolos soulagés. Dans un coin comme celui-ci, il est plus discret de passer par les jardins. Personnellement, j'aurais planqué ma caisse en bas de la colline et parcouru les quatre cents derniers mètres à pied. Sauf que, si ç'avait été moi, nous n'aurions pas cette discussion à l'heure qu'il est. Tu as entendu comme ça continuait de fuser après la première explosion ? Ils ont monté leur coup dans la précipitation et fait un travail de porcs.

— Qu'est-ce que ça change ? C'est bien la première détonation qui a fait péter le reste, non ?

— La charge du détonateur a niqué les connexions. Ils ont voulu monter une putain de sucette, mais elle s'est désagrégée avant d'avoir pu se déclencher normalement.

— On serait restés au salon, on y passait. (Je me frottai vigoureusement le visage.) Il semble que Chip ait transmis le message.

— Tu l'as dit.

— Auquel cas... (Je consultai ma montre.) Ils ont assemblé l'engin en à peine une heure, trajet compris !

Je remarquai une belle estafilade au dos de ma main. Je l'épongeai avec mon blouson.

— C'est ce que je dis. Ils ont bâclé.

— Ils foirent peut-être les finitions, mais ils sont très motivés, n'est-ce pas ? (Au loin, je perçus des sirènes à l'approche, et les portes s'ouvraient de l'autre côté de la rue.) Ils ont plastiqué la baraque de mes vieux, articulai-je, incrédule, les yeux rivés aux flammes. Avec une bombe...

La maison en feu paraissait bizarre, comme une aberration dans cette rue aux logis impeccables. Je me tournai vers celle de Mary, de l'autre côté de la haie. Quelques lampes étaient allumées, et la porte d'entrée ouverte.

— Tu as affaire à des enculés de première classe, confirma Bobby en plaquant à nouveau sa paume sur la carrosserie. Et maintenant, on se tire d'ici.

Mais j'étais déjà loin, courant et pataugeant en direction du portail. J'entendis Bobby jurer et s'élancer à ma suite. Au bout de l'allée, je coupai directement à travers la haie pour atteindre le jardin de Mary. À peine y avais-je posé le pied que Bobby m'agrippait l'épaule et me faisait virevolter.

Je me libérai, repris ma course. Il me rattrapa, mais lâcha prise quand il vit ce que je vis, puis il me dépassa.

Elle était allongée de moitié sur le bord du porche, la tête et les épaules renversées sur les marches, un bras étiré sur le côté. Je songeai d'abord à une crise cardiaque, avant de remarquer tout le sang autour d'elle, la mare brunâtre qui engluait le bois rongé des planches. Bobby posa un genou à terre pour lui soulever la tête.

— Mary... murmurai-je. Oh, mon Dieu !

La saisissant de chaque côté nous la fîmes pivoter en douceur pour la ramener à l'horizontale. Elle respirait péniblement. Le feu voisin projetait assez de lumière pour creuser ses rides tels des canyons. Bobby fouilla les plis de ses vêtements, découvrant trou sur trou, espérant étancher le sang dont le flux paraissait anormalement lent. Elle toussa, et une masse sombre affleura dans sa bouche.

Avant ce jour je n'avais vu en elle que la vieille dame, l'un de ces êtres qui encombrent les caisses au supermarché et attendent le bus debout, qui savent quel cadeau l'on doit offrir pour tel anniversaire, qui ont l'air frêles et refroidis comme s'ils avaient toujours été ainsi. Des gens qui n'auraient jamais pu se soûler, ni escalader des barrières interdites, ni se bousculer, hilares, pour que l'autre se coltine la zone mouillée du lit. De vieux bâtons secs dont on ignore s'ils ont jamais aimé – personne qui ne soit encore de ce monde, en tout cas, personne qui ne soit davantage qu'un souvenir, et dont la dernière demeure n'arbore que quelques fleurs fanées qu'eux seuls pensent à remplacer. Mais à présent je voyais quelqu'un d'autre. La femme qu'elle avait été et demeurait sûrement, sous la patine des cellules mortes, de la peau sèche et parcheminée, des cheveux gris, courts et bouclés. Sous le déguisement des années, derrière l'idée fausse qu'en raison de son âge elle n'avait jamais été, et n'était toujours pas, un être réel.

Puis sa gorge émit un claquement, laissant jaillir une pleine pochée à l'odeur chaude et âcre. Ses pupilles parurent s'assécher dans la seconde, comme en avance rapide. Peut-être était-ce dû à la fraîcheur extérieure, mais on aurait dit que le ciel la reprenait sous nos yeux, d'un seul coup.

Bobby releva lentement la tête. Je croisai son regard. Je n'avais rien à dire.

— Qu'est-ce qui s'est passé ? demandai-je, la seule phrase qui voulût sortir en dix minutes. Qu'est-ce qui s'est passé, bordel ?

Bobby fixait la route, se démanchant d'avant en arrière pour inspecter les rues adjacentes que nous croisions à toute allure. Toutes baignaient dans le calme ordinaire d'un début de soirée. Le corps de Mary reposait deux kilomètres derrière nous, à l'endroit où nous l'avions découverte. Les secours arriveraient plus vite que nous n'aurions pu gagner l'hôpital, et sa mort était de toute façon sans appel. Bobby et moi le savions tous deux.

Il haussa les épaules.

— Elle s'est trouvée sur leur chemin. Comme je te le disais, le plastiqueur est passé par les jardins. Mary aura entendu du bruit, sera sortie. Alors ils lui auront vidé un demi-chargeur dans la peau. Je suis désolé, mec.

— Un type vient ici pour m'exploser, muni d'un flingue et d'un silencieux juste au cas où. Une vieille dame sans défense apparaît, alors il la liquide, comme ça ?

— Ces gens ne plaisantent pas, Ward, et ils ne t'aiment pas du tout.

Il s'engagea dans un virage serré à gauche qui nous ramena vers le centre-ville. Un camion de pompiers nous doubla sur la grand-rue, sirène et gyrophares en marche, dans la direction opposée à celle de la maison.

— Mais qu'est-ce qu'il fabrique, ce con ?

On nous klaxonna par-derrière. Nous nous retournâmes comme un seul homme. Le conducteur d'un pick-up nous fit signe que le feu était passé au vert et que nous aimerions peut-être bouger. Bobby redémarra et prit les pompiers en chasse.

— Le camion file dans la mauvaise direction, Bobby !

— Je leur ai pourtant transmis l'adresse que tu m'avais donnée. J'ai trouvé sans problème, moi.

— Mais qu'est-ce que... ?

Je n'achevai pas ma phrase. Nous percevions une lueur orange un peu plus loin.

Bobby se rangea d'un coup, sans mettre son clignotant. Cela nous valut un nouveau coup de klaxon du vieux en pick-up, qui nous fixa longuement en nous dépassant. Ni moi ni Bobby ne lui prêtâmes beaucoup d'attention. Nous découvrions que le Best Western, en partie du moins, était en flammes. Je contemplai le spectacle, ahuri, me demandant comment Dyersburg s'était soudain retrouvé sur l'un des cercles de l'enfer.

— Rapprochons-nous, dis-je d'une voix étranglée.

Nous roulâmes lentement, et au carrefour suivant quittâmes l'avenue pour contourner l'hôtel par une rue latérale en pente. Nous nous arrêtâmes en haut, à une trentaine de mètres du bâtiment à vol d'oiseau. De là nous vîmes que le feu était assez limité, qu'il ne ravageait qu'une petite dizaine de mètres dans une seule aile. L'hôtel se remettrait à temps pour accueillir de nouveaux séminaires. Quatre camions de pompiers étaient à l'œuvre, bientôt rejoints par un cinquième. Le bas de la rue se remplissait de badauds, et d'autres frôlaient la voiture d'un pas pressé pour mieux profiter du spectacle. La moitié du commissariat semblait sur le pont.

— Ta chambre correspond à peu près au foyer de l'incendie, non ?

Je ne pris même pas la peine de répondre. J'avais envie de vomir. Curieusement, l'attaque de l'hôtel me meurtrissait davantage que celle de la maison. Je me demandais si mes voisins de palier étaient présents.

— Ce message que tu leur as adressé, Ward, il disait quoi, au juste ?

— C'est grotesque. Et complètement disproportionné. Et la maison, alors ? Qu'est-ce que les pompiers attendent pour... ?

— Ils ont déjà des hommes sur place, j'en suis sûr. Les gens du voisinage les auront aussi alertés. Au fait, avant que la question ne t'étreigne, ton barda est sauf.

— Quel barda ?

— Pas tes fringues, disons. Regarde un peu derrière.

Me retournant, j'avisai la sacoche de mon ordinateur sur la banquette arrière.

— Ne jamais se croire à l'abri, énonça Bobby tout en pianotant sur le volant, les yeux fixés sur l'incendie. Je suis moi-même du genre prévoyant. Toujours garder le nécessaire à portée de main. Je pense que c'est le moment parfait pour mettre les voiles.

Je voulais filer dans la montagne pour tuer quelqu'un. Mais Bobby devina ma pensée et fit non de la tête.

— Quand cet incendie-ci sera maîtrisé, les flics verront bien dans quelle chambre il s'est déclaré. Avec un peu de chance, tes ennemis auront pris le temps de le rendre un minimum crédible. Mais entre ça et la maison, tu deviens l'ennemi numéro un de Dyersburg.

— C'est quoi, c'est conneries ? J'ai rien fait, moi.

— La maison de tes vieux est assurée ?

— Oui.

— C'est une grosse assurance ?

— Sûrement, soupirai-je. Ça m'a jamais intéressé. Puis ils vont trouver Mary et un flic futé va décider de la passer au pinceau, à tout hasard. Avec tout ce sang, ils pourraient bien trouver quelque chose. Tes empreintes sont fichées, Bobby ?

— Tu sais bien que oui.

— Tout comme les miennes. T'as raison, mec. Il est temps de partir.

Vingt minutes plus tard, nous étions à l'aéroport de Dyersburg.

CHAPITRE 16

Zandt atteignit Beverly Boulevard sur le coup de 21 heures. Épuisé, les pieds endoloris. Et soûl.

À 3 heures du matin, il s'était posté devant le cinéma où Elyse LeBlanc avait été vue pour la dernière fois. À l'instar des boutiques et des restaurants, les cinémas ont un drôle d'aspect dans les petites heures de la nuit. Ils font décalés, arbitraires, pour ne pas dire décoratifs – comme si nous étions des explorateurs ayant raté d'une ou deux décennies la mystérieuse civilisation qui les a créés. Quelques heures plus tard, Zandt contemplait la maison où la jeune Annette Mattison avait passé sa dernière soirée entre copines. Il reconnut la femme en tailleur strict qui en émergea vers 7 heures pour rallier la chaîne télévisée où elle officiait. Zandt avait interrogé Gloria Neiden plus d'une fois. Elle avait vieilli en l'espace de deux ans. Avait-elle gardé le contact avec Frances Mattison ? Leurs filles avaient passé beaucoup de temps l'une chez l'autre, et elles avaient toujours parcouru les trois petits blocs du retour à pied. Après tout, elles habitaient un quartier fort agréable, à Dale Lawns, 90210 – et la possibilité de se promener sous les étoiles en toute sécurité était sans doute l'une des raisons qui vous poussaient à signer un chèque à sept chiffres. Zandt devinait que les rapports entre les deux mères s'étaient distendus, pour ne pas dire rompus. De la même façon que la voix de Zoe Becker s'était teintée de gris en évoquant Monica Williams – alors que cette dernière n'y pouvait rien si Sarah avait voulu attendre son père jusqu'à l'heure dite. Mais face à un drame de cette ampleur, vous cherchez une cause, des responsables. Ceux qui habitent dans les murs sont les plus accessibles.

Zandt se retourna quand la voiture de Mme Neiden le croisa. Outre qu'elle aurait pu le reconnaître, lui-même aurait eu l'impression, à s'éterniser ainsi sur ce trottoir, d'imiter un autre type, celui qui avait fait le guet devant cette même maison, peut-être au même endroit, deux ans plus tôt.

Il reprit son chemin. En fin de matinée, il s'était aventuré à Griffith Park, là où le corps d'Elyse était réapparu. Rien ne marquait cet emplacement, même s'il avait longtemps été recouvert de fleurs, et qu'y stagnaient encore les débris d'un vase de verre. Zandt s'y attarda un bon moment, contemplant la vue sur la ville brumeuse, là où des millions de gens travaillaient, dormaient et mentaient, devenant la mauvaise herbe du sol urbain.

C'est sur ces entrefaites qu'il poussa la porte d'un bar. Puis, un peu plus tard, d'un second. Entre-temps, et ensuite, il avait continué de marcher, quoique moins vite, à mesure que sa motivation s'effilochait. Il avait arpenté ces rues de nombreuses fois. Elles ne lui avaient offert que sang et déchirement. Il entendait encore les voix qui l'avaient arraché à son siège après le départ de Nina ; ces voix et les cris des disparues – mais trop faibles dans la lumière du jour et de la raison pour le conduire où que ce soit. Sa chemise lui sortait du pantalon, et il remarquait les regards obliques des passants. On prétend qu'il est possible de reconnaître un policier rien qu'à ses yeux : un regard qui pèse et teste, qui juge du haut de ses soupçons et de son pouvoir. Zandt se demandait si l'on pouvait aussi repérer ceux qui n'étaient plus flics, pour ce qu'ils suggéraient de castration ou de fuite. Il avait connu cette ville autrefois, de l'intérieur. Il avait foulé ce pavé dans la peau d'un de ces hommes vers qui l'on se tourne en temps de désordre. Un maillon du système immunitaire. Le voilà désormais privé de cette auréole. Fini l'uniforme, fini la gloire – ou son équivalent dans le métier. Il n'était plus qu'un badaud dans une ville où l'on marchait très peu – et où les rares piétons le regardaient d'un drôle d'œil. C'était là un domaine aussi réel que n'importe quelle steppe ou vallée ombragée ; cet endroit était à la campagne ce que la vallée de la Mort était au Vermont, ou le Kansas aux fonds marins, ni plus ni moins. La seule distinction provenait des gens, ceux couverts de smog et ceux fatigués de se battre. Les gens dans leur ensemble.

En fin d'après-midi, titubant légèrement déjà, il s'était arrêté au bord d'une petite rue de Laurel Canyon. Les buissons d'hier avaient été déracinés et remplacés par une dalle rectangulaire mesurant une cinquantaine de centimètres de plus que le corps d'Annette. Malgré son ébriété, Zandt demeurait assez lucide pour repérer qu'on l'observait depuis le joli pavillon de l'autre côté de la rue.

Au bout de quelques minutes, un homme en sortit, vêtu d'un pantalon de jogging et d'une veste gris clair. Il respirait la santé.

— Je peux vous aider ?

— Non, répondit Zandt.

Il tenta de sourire, mais l'homme resta de marbre. Si Zandt avait vu sa grimace dans un miroir, il ne lui en aurait pas voulu.

Le type renifla.

— Vous êtes soûl ?

— Je ne fais que passer. Rentrez chez vous. Je n'en ai plus pour longtemps.

— C'est quoi, d'abord ? demanda l'homme tout en pivotant légèrement, révélant un téléphone dans sa main cachée.

Zandt le dévisagea.

— Quoi ?

— Ces dalles, là. Qu'est-ce que ça fait ici ? Ça ne sert à rien.

— Une fille est morte ici. Y fut retrouvée morte, du moins.

Le visage du type s'adoucit un brin.

— Vous la connaissiez ?

— Pas de son vivant.

— Alors qu'est-ce que ça peut bien vous faire ? C'était quoi, une professionnelle ?

La gorge de Zandt se serra. Ah, la fameuse échelle des causes de la mort... Comme si les putes, les toxicos et les jeunes Noirs valaient à peine mieux que des chiens mal dressés, comme s'ils ne s'étaient jamais jetés dans les bras de leurs parents en gazouillant, n'avaient jamais prononcé leur premier mot, ni passé de longues nuits à se demander ce qu'ils trouveraient dans leur botte de Noël.

L'homme recula d'un pas vif.

— Je vais appeler les flics, menaça-t-il.

— Ils n'arriveraient pas assez vite. Vous auriez peut-être le temps de cracher sur une deuxième plaque de ciment, mais à votre place je n'y compterais pas trop.

Et sur ces mots il s'en alla, laissant l'homme à ses préjugés.

Enfin parvenu à Beverly Boulevard, il passa devant le Hard Rock Café, en même temps qu'il rentrait sa chemise dans son pantalon, lissait sa veste et redressait les épaules. Il pénétra dans l'hôtel Ma Maison sans problème et fondit sur les toilettes du bar. Après un bon coup d'eau sur son visage, seul un barman saurait qu'il n'avait pas sa place ici. Il regagna la salle et s'installa derrière une table basse d'où l'on voyait la rue. Avec les kilomètres qu'il avait dans les jambes, le moelleux du canapé lui donna l'impression d'être assis sur un nuage. Un jeune homme affable promit de lui apporter un verre.

En attendant, Zandt examina le boulevard où Josie Ferris avait disparu. Il ne s'agissait pas du dernier lieu du crime en date, mais il n'avait aucune envie de se planter devant l'ancienne école de Karen, ni devant la maison qu'il avait occupée avec sa petite famille. Et que serait-il allé faire dans l'autre lieu, la destination finale ? Un endroit qu'il avait lui-même créé et qui ne serait d'aucune aide aujourd'hui – pas plus qu'hier, du reste. Se pencher au-dessus du cadavre de l'homme qu'il avait tué ne lui avait rien apporté, sinon souligner la subtilité des nuances dont on fait des lois.

Jennifer avait appris, pour le meurtre. Il s'en était expliqué le surlendemain, à l'arrivée du pull. Ça n'avait pas sonné le glas de leur couple. Pas tout de suite, du moins. Elle avait compris son geste, lui avait tout pardonné sauf sa méprise. Ils avaient essayé de tenir le coup. Sans succès. Zandt s'était retrouvé en porte-à-faux. Soit il supportait l'horreur de la disparition de Karen et restait fort pour le bien de son épouse, tout en pressentant qu'il allait se désagréger en mille éclats coupants, soit il lui confiait la douleur qui le consumait à petit feu. En choisissant cette deuxième option, il renonça à l'image du mâle solide, sans rien gagner à l'aveu de sa souffrance, réputé l'apanage des femmes. À elle de s'épancher ; à lui de résister.

Il décida qu'il avait assez fait semblant d'être flic, en même temps qu'elle décidait de rentrer chez ses parents. On avait volé leur œuf d'or, et la poule qui l'avait pondu était morte.

À présent, avec le recul, il estimait avoir manqué de discernement. Seule sa rigidité avait creusé le sillon de ses fautes. Jennifer aurait bien fini par accepter sa faiblesse. Les femmes sont les plus raisonnables quand il s'agit d'assouplir certaines règles. La vie à deux exige de la flexibilité, a fortiori dans les périodes d'angoisse, ces phases où elle prend le goût d'un pacte désespéré contre un monde horriblement noir. Les couples solides lutteront pour préserver leur équilibre, quels que soient les tiraillements du moment. C'était là une consolation à double tranchant, mais cette prise de conscience lui avait permis de rester en vie. Parfois c'est en reconsidérant un traumatisme passé et en reconnaissant sa part d'erreur que l'on reprend pied. Avant cela on se sent floué, blessé, sans trouver le moindre répit. Mais « c'est pas juste » est le cri de l'enfant, de celui qui refuse de voir que les relations causales opèrent dans les deux sens. Du jour où l'on comprend que les torts étaient partagés, peu à peu la douleur se dissipe. Comme on fait son lit on se couche, et l'on s'y couche alors plus facilement, aussi dur et souillé soit-il.

Arriva sa Budweiser, qu'il sirota du bout des lèvres, les yeux rivés à la fenêtre. En fait il tentait, et ce depuis qu'il était levé, de considérer un faisceau d'éléments sous un nouvel angle. Face à des meurtres dépourvus d'indices véritables, le mieux à faire est de chercher de nouvelles façons d'imbriquer les infos disponibles. La quintessence d'un crime se réduit généralement à une simple phrase. Des empreintes, une liaison, un couteau dissimulé en hâte, des dettes et un alibi foireux : tout cela concerne les tribunaux, sert à boucler le réquisitoire. Mais le vrai crime, dans toute sa splendeur, tient en quatre mots : les gens s'entre-tuent. Les maris tuent leur femme. Les femmes tuent leur compagnon, et les parents leurs enfants, les enfants leurs parents, les inconnus d'autres inconnus. Les gens prennent ce qui ne leur appartient pas. Les gens incendient des maisons pour de l'argent, ou parce qu'il y a du monde à l'intérieur. Quand chaque déclinaison a été rangée dans la bonne case juridique, la vérité demeure entière. On peut prendre deux personnes au hasard et glisser le mot « tue » entre elles.

Zandt ne parvenait pas à élucider ce que l'Homme Debout attendait de ses victimes. Pourquoi les punissait-il ? L'avaient-elles déçu en refusant de l'aimer, de céder à ses avances ? Avaient-elles manifesté une peur excessive, ou bien insuffisante ? Avaient-elles failli en craquant devant lui, en se montrant dépourvues de cette force qu'il convoitait et souhaitait leur dérober ?

Son verre vide, il se tortilla sur son siège, cherchant le garçon dans la salle. Introuvable. Pourtant, les autres clients semblaient servis depuis peu. Il les examina brièvement. Des inconnus, s'abreuvant d'alcool pour se sentir mieux. Pour arrondir les angles de l'anxiété. Tout le monde s'y adonnait. Les Américains – sauf lors d'une brève expérience qui entraîna une explosion de la criminalité unique à ce jour. Les Allemands et les Français, avec cœur. Les Russes, avec une gravité mélancolique. Sans oublier les Anglais, fanas de bière. Tous passaient leur temps libre dans les bars, les pubs ou à la maison, à s'embrumer l'existence. Il leur fallait le masque d'écume, la glu.

Quelqu'un finit par arriver. Costume noir et chemise blanche, exactement comme son collègue – mais avec dix ans de plus et une mine moins épanouie. Si le précédent semblait s'accrocher au fol espoir de vendre un scénario ou de crier « coupez ! » un jour, celui-ci se rangeait à l'idée que les actrices de Hollywood allaient très bien survivre sans son amour. Il étudia Zandt d'un air suspicieux, lisant sur son radar de serveur que ce client n'avait ni chambre ni rendez-vous particulier dans cet hôtel.

— La même chose, monsieur ?

Cela prononcé en inclinant la tête, un geste de cordiale ironie : nous savons tous deux que Monsieur n'appartient pas à notre milieu favori, qu'il est plus ou moins rond et n'arbore pas la tenue adéquate.

— Où est passé l'autre gus ?

— L'autre « gus », monsieur ?

— Le type qui m'a servi.

— Je le remplace. Ne vous inquiétez pas. La bière aura le même goût.

Comme il repartait d'un pas languide, en faisant rebondir son plateau sur son genou, Zandt envisagea de l'abattre. En guise de leçon pour tous ces loufiats qui osaient prendre de haut ceux qui les faisaient vivre. Un rappel à l'ordre qui n'avait que trop attendu. Peut-être le message parviendrait-il jusqu'à la corporation des vendeurs, même à ceux de Rodeo Drive. Zandt n'oublierait jamais l'incident survenu le jour de son anniversaire de mariage, six ou sept ans plus tôt, lorsqu'il avait emmené son épouse dans une boutique chic pour lui offrir un corsage. Ils étaient vite ressortis, Jennifer cramponnée gauchement à un sac, Zandt frémissant de rage contenue. Le chemisier n'allait jamais quitter l'armoire, à jamais taché de l'humiliation infligée lors de l'achat.

Aujourd'hui ce souvenir affectait Zandt comme jamais. Pour ne plus y penser, il attrapa le bloc à en-tête de l'hôtel dans l'idée de prendre quelques notes – ce qui lui passerait par la tête –, mais il n'alla pas plus loin. Il apercevait le serveur derrière son comptoir, en train de lui verser une seconde bière.

Il s'agissait d'une Budweiser. Comme la première. Logique : le collègue précédent avait dû laisser un mot signalant ce que le client avait bu et combien il devait.

Une indication sur ses désirs, en somme.

Sur ses préférences.

Quand le serveur revint avec le demi, il trouva un siège vide et un billet de dix dollars.

CHAPITRE 17

La maison était perchée sur les hauteurs de Malibu. Petite et singulière, une enfilade de pièces comme dans un minuscule motel. Pour se rendre de l'une à l'autre, il fallait ressortir et emprunter un passage couvert. Le pavillon se dressait au bord d'une falaise desservie par une route sinueuse, pentue et mal éclairée. Ce n'était pas l'endroit où l'on échoue par accident. Le loyer était bas, malgré l'emplacement, car il reposait sur un sol instable et se trouvait à deux doigts de l'évacuation forcée. La partie séjour-cuisine, vaste et lumineuse avec sa large baie vitrée, constituait son principal atout, en dépit de la faille qui traversait la dalle de béton. On pouvait y glisser les trois quarts d'un poing, et les deux plaques étaient dénivelées d'au moins cinq centimètres. Dehors, côté terre, se trouvait une petite piscine. Elle était vide, les tuyaux ayant fondu dans un feu de broussailles l'année précédant l'arrivée de Nina. Il fallait un certain cran pour trouver le sommeil, ici.

Nina avait passé la soirée sur le patio à l'arrière de la maison, adossée au mur, bras et jambes étendus devant elle. La vue donnait sur l'océan, avec juste quelques arbres et buissons avant le précipice. On n'apercevait aucune autre habitation. Ce soir, même la mer était invisible, voilée par une brume que Nina pouvait presque toucher du bout des pieds. C'était un phénomène régulier, et Nina se demandait si ce n'était pas ce qu'elle préférait. Un lieu au bout du monde, où tout devenait possible. Elle avait prévu de se munir d'un verre de vin, avant d'oublier. Sitôt assise, elle s'était sentie lourde comme un navire encalminé, sans trouver le courage de repartir vers le frigo.

Elle avait perdu sa journée à chercher Zandt. Il n'était ni à son hôtel, ni sur la Promenade, nulle part. En début de soirée, elle avait même roulé jusqu'à son ancien domicile, mais le propriétaire avait changé et Zandt n'était pas apparu. Alors elle était rentrée. Elle n'avait rien d'autre à faire ce soir que de rester assise.

Le séjour situé dans son dos était longé d'étagères pleines de documents, de paperasse et de notes. Elle n'avait aucune envie de les consulter. Elle ne souhaitait parler à personne du Bureau. Sa position là-bas n'était plus celle d'autrefois. L'affaire de l'Homme Debout l'avait mise sur la touche. Non parce qu'il courait encore, même si cela n'arrangeait rien, mais plutôt parce qu'elle avait persisté à tuyauter un policier expressément écarté de l'enquête suite à la disparition de sa fille. D'autres avaient pris la porte pour moins que ça. Elle était parvenue à sauver sa tête, mais ce n'était plus comme avant. On l'avait connue comme la petite protégée de Monroe ; leurs rapports étaient désormais empreints de défiance et de lassitude.

Elle se sentait seule, et menacée. Mais sa peur était indépendante de sa solitude. Nina était habituée au célibat et le vivait plutôt bien, en dépit d'une nature qui aspirait à autre chose. Elle avait abrégé sa liaison avec Zandt pour une seule raison : plus elle s'était attachée à lui, moins elle avait voulu détruire sa vie de famille. Mais le fait que celle-ci ait péri malgré tout lui interdisait de fournir une telle explication à Zandt, quoi qu'il réclame. Certes, elle aurait pu distiller sa pensée sous forme de phrases habiles. Mais elle eût risqué de se trahir. De trahir le fait que, deux semaines après la disparition de Karen, Nina avait surpris dans un coin de sa tête une pensée licencieuse : quitte à ce que Zandt perde ses proches, autant que ce soit pour elle.

Il y avait eu d'autres hommes après lui, et il y en aurait sûrement d'autres. Trouver des hommes n'était pas sorcier, du moins ceux que l'on ne tenait pas à garder. C'était plutôt le désespoir qui la minait, et cet inépuisable cortège d'atrocités. Si nous étions vraiment comme ça, la partie était peut-être perdue d'avance. À voir ce que notre espèce s'infligeait à elle-même comme aux animaux, on en venait à se demander si nous ne méritions pas tout ce qui nous arrivait, quel que fût notre zèle en matière d'autopunition ; si les bêtes qui rampaient vers Bethléem [1] n'étaient autres que nos enfants prodigues rentrant à la maison.

Vers 21 h 30 elle se releva pour rentrer. Tout en ouvrant son frigo, qui ne contenait qu'une bouteille de vin à moitié vide, elle jeta un œil sur le petit téléviseur posé sur le comptoir. On y traitait encore de la tuerie en Angleterre – même si, ayant coupé le son, elle ignorait ce qu'on en disait, révélait ou alléguait.

1. Allusion à *La Seconde Venue*, célèbre poème de W.B. Yeats. (*N.d.T.)*

Des faits sordides, en tout état de cause, et un motif d'abattement supplémentaire.

Elle referma le réfrigérateur, sans avoir touché à la bouteille, et colla son front quelques instants contre la porte fraîche.

Elle redressa la tête en percevant un bruit dehors. Celui-ci se précisa peu à peu : des pneus crépitant sur les gravillons de la route. Elle traversa la pièce en trombe, enjambant la faille, et sortit un pistolet de son sac.

La voiture s'arrêta, et elle décela le murmure étouffé d'une conversation. Puis un claquement de portière, suivi de nouveaux bruits de pneus, qui rebroussaient chemin dans l'allée. Quelques pas, puis trois coups frappés à la porte. Elle alla ouvrir, l'arme cachée dans son dos.

Zandt se tenait sur le perron. L'air pantelant et un peu soûl.

— Mais où étais-tu passé, bon sang ?

— Un peu partout, répondit-il en s'avançant dans la pièce d'un air contemplatif. J'adore ce que tu as fait de cet endroit.

— Je n'y ai rien fait de particulier.

— Justement. Ça n'a pas bougé d'un poil.

— Toutes les nanas ne sont pas des maniaques de la déco, tu sais.

— Bien sûr que si. Sous tes habits de femme, doit se cacher un homme.

— Zut, je suis découvert. (Elle croisa les bras sur sa poitrine.) Qu'est-ce que tu veux, John ?

— Seulement te dire que j'ai buté le bon type, en fin de compte.

Quand elle gagna le patio avec la bouteille à la main, Zandt était déjà lancé :

— Le problème, c'est que nous ne pouvions traiter cette affaire comme les autres. Les procédures d'investigation classiques ne donnaient rien. Quand plusieurs personnes disparaissent, on part de leurs relations, puis on tire sur chaque fil. On rencontre les familles, les amis, l'entourage. On cherche une intersection. Un bar qu'ils fréquentaient tous, à différents moments de la semaine. Un club de gym dont ils étaient membres. L'ami d'un ami d'un ami. Un point de confluence prouvant que ces disparus ont davantage en commun que le fait d'être morts. Un élément antérieur, qui aura lui-même décidé de leur sort. Or, avec l'Homme Debout nous avions de multiples disparitions, mais des similitudes superficielles. Même sexe, à peu près le même âge.

Toutes jolies. Et après ? La ville regorge de garçons qui rêvent de nanas mignonnes du fond de leur piaule. De femmes, quoi. C'est un désir consensuel, non une pathologie. Sauf pour les cheveux longs, peut-être. C'est la seule caractéristique évidente, le seul critère patent – outre le fait qu'elles venaient toutes de milieux où l'argent n'était pas un réel souci. Ce n'étaient ni des fugueuses, ni des junkies. Cela nous apprend seulement que la tâche du tueur n'en était que plus ardue, car ces filles-là sont plus difficiles à capturer. Mais ce n'est pas une piste en soi.

Il se tut. Nina attendit. Il ne la regardait pas. C'est tout juste s'il remarquait sa présence. Il se tenait tout au fond du patio, et depuis l'embrasure de la porte sa silhouette paraissait floue. Il reprit la parole, dans un débit plus lent :

— Un homme poursuit une quête. Il est rongé d'angoisse, dévoré par un trouble qu'il ne peut surmonter qu'à travers une certaine ligne de conduite, qu'il a découverte soit par hasard, soit par tâtonnements. Mais il tient bon, depuis un moment déjà. Il s'est bien comporté. Il n'a pas replongé. Il s'est tenu à l'écart, sans nuire à personne. Ce n'est pas un faible. Il ne recommencera pas. Il n'en a plus besoin. Ni maintenant ni demain. Peut-être qu'il ne recommencera jamais. Peut-être qu'il pourra tirer un trait sur tout ça. Peut-être qu'il est guéri.

« Mais petit à petit... il se sent rechuter. Ça devient plus difficile. Il peine à se concentrer. Il n'est plus dans son assiette. Il n'a plus la tête à son travail, à sa famille, à sa vie. Il est tendu. Des idées lui reviennent en boucle, les vieux démons d'hier. L'anxiété le gagne, aggravée du fait qu'il en sait la cause. Il connaît l'unique moyen d'y remédier. Il commence à se remémorer ses exploits antérieurs, mais ça ne l'aide pas. Il ne se souvient pas forcément des détails. Et ceux-ci n'atténuent en rien ses sentiments actuels. C'est de l'histoire ancienne. On ne peut vaincre les angoisses du présent par les gestes du passé : le bon vieux temps n'efface pas le malheur d'aujourd'hui. Cet homme a besoin d'un objectif, d'un projet inédit. Même les talismans sont sans effet, ces reliques qu'il a conservées comme preuves de ses antécédents. Ils lui rappellent juste que c'est possible. Il a tellement envie de récidiver et sait qu'il ne peut vivre sans – de toute façon, quels que soient ses efforts, il a déjà versé le sang, ce qui exclut d'avance toute rédemption. Son existence est viciée, il ne peut revenir en arrière.

« Alors un jour, de façon quasi accidentelle, il se remet à rôder. Il se raconte peut-être que ça n'ira pas plus loin, qu'il ne fait qu'observer. Qu'il se maîtrise mieux, désormais. Qu'il évitera de toucher. Mais quand il franchit ce pas, il n'y a qu'une seule issue possible. Il oubliera les remords qui l'avaient assailli la dernière fois, de la même façon que se souvenir d'une gueule de bois n'empêche personne de re-boire le vendredi suivant. Ou alors, il l'a fait si souvent qu'il n'éprouve plus aucun complexe. C'est peut-être la seule chose qui ait du sens à ses yeux. Il retournera dans un endroit qu'il connaît, ou dans un lieu similaire. À ce stade il disposera d'un plan. C'est une activité dangereuse, alors il aura peaufiné quelques techniques pour limiter les risques. C'est ici que les intersections entrent en jeu, car ces intersections sont en l'homme lui-même, au carrefour de ses itinéraires. Elles proviennent des lieux où il se sent en sécurité, où il peut flâner tout en restant lui-même. Tel tueur considérera cet ensemble comme son terrain de chasse. Tel autre y verra un univers où il peut se fondre, où personne ne le regarde, où il devient invisible. Là où notre homme n'est pas faible, mais puissant ; où il n'appartient pas à la foule, mais la domine. Une cachette, là où la proie s'aventure à la faveur du soir, pour plonger dans la nuit qu'il lui réserve. Dans un premier temps il se contentera d'observer, puis arrivera le jour J. La fille tournera au coin de la rue et sentira soudain une présence dans son dos, et là l'homme trouvera enfin du répit, jusqu'à ce que vienne l'heure de nettoyer, de se sentir minable et de promettre à Dieu, ou à quiconque daigne l'écouter, qu'on ne l'y reprendra plus, plus jamais.

— Et c'est comme ça que tu l'as coincé, anticipa Nina.

— Non. Mais laisse-moi continuer. Nous n'avons rien trouvé qui relie toutes les filles entre elles. Nous n'avons jamais pu nous approcher du type car nous étions incapables de dire où il les avait vues pour la première fois. C'est pourquoi, après la disparition de Karen, je n'ai cessé d'arpenter les lieux des disparitions. C'étaient les seuls endroits que nous savions en rapport avec le tueur. Il ne me restait que ça. Zéro lien. Aucun moyen d'en trouver un. Jusqu'à la dernière fois, quand il est revenu sur ses pas. Je l'ai vu visiter un site, et j'ai pensé qu'il souhaitait revivre ce qu'il y avait commis. Après l'avoir repéré dans deux des coins en question, je me suis dit que c'était notre homme. Alors je l'ai suivi, et je l'ai trouvé.

— Mais ensuite, dit Nina en veillant au choix de ses mots, tu t'es rendu compte que ce n'était pas lui, en définitive.

— Erreur. Le type que j'ai buté a bel et bien enlevé certaines de ces filles.

— Essaies-tu de me dire que celui-ci n'est qu'un imitateur ?

— Non, j'essaie de te dire que j'ai tué le serveur, et non le client qui commande la bière.

— Je ne comprends pas.

— Le type qui a expédié les colis n'a pas commis les enlèvements.

Nina le dévisagea.

— L'Homme Debout décide qu'il a besoin d'une fille, et hop ! il passe commande ? Comme pour une vulgaire pizza ?

— Voilà pourquoi les disparitions ont cessé après Karen, bien qu'on ait reçu le paquet avec le pull. Le type qui les enlevait n'était plus là. Le tueur, si.

— Sauf que les serial killers ne fonctionnent pas comme ça. On en a connu qui agissaient en tandem, c'est vrai. Leonard Lake et Charles Ng. John et Richard Darrow. Même les West, dans une certaine mesure. Mais rien qui ressemble à ce que tu dis.

— C'est un fait nouveau, je te l'accorde. Mais nous vivons dans un monde en mutation, où tout devient plus grand, plus brillant, plus rapide. Pratique. À la carte...

— Alors comment expliques-tu l'absence de lien entre toutes ces filles ? Si l'on suit ton raisonnement, le kidnappeur devait avoir un mode opératoire fixe. On aurait su l'établir.

— Seulement s'il s'agissait toujours du même gars.

Nina ouvrit de grands yeux.

— Il y aurait donc *deux* ravisseurs ?

— Peut-être même plus. Pourquoi pas ?

— Mais parce que... parce que l'Homme Debout n'a fait qu'une seule victime en deux ans, John. Sarah Becker.

— Qui te dit qu'il n'y a que lui ? (Il attrapa la bouteille, constata qu'elle était vide.) Il doit te rester du pinard quelque part.

Nina le suivit à l'intérieur de la maison. Il ouvrit le réfrigérateur et fixa d'un air stupéfait le compartiment désert.

— Je n'ai plus rien à boire, John. Comment ça : « Qui te dit qu'il n'y a que lui » ?

— Dis-moi, combien la Californie compte-t-elle de tueurs en série en activité, au jour d'aujourd'hui ?

— Disons entre sept et onze. Ça dépend de ce qu'on entend par...

— Exactement. Et il ne s'agit là que de ceux dont on connaît l'existence. Dans un seul État, et un État qui arrive loin quant au classement national. Alors admettons qu'il y en ait cent cinquante à travers le pays, et que dix ou quinze d'entre eux puissent se payer une victime vingt mille dollars pièce. Voire plus. Voire beaucoup plus. Ça fait un vrai fichier clients, ça. Et un gros. On pourrait décrocher un putain de prêt bancaire avec un tel business.

— À supposer que tu aies raison, ce qui reste à prouver, en quoi cela nous aidera-t-il à retrouver Sarah Becker ?

— En rien, concéda-t-il, et son ardeur retomba d'un coup. (Il se frotta vigoureusement le front.) J'imagine que les fédéraux continuent de sonder l'entourage ?

Nina hocha la tête.

— Bon, soupira-t-il. Alors je suppose qu'il ne reste plus qu'à attendre.

Il fixa le téléviseur muet, qui diffusait un sujet sur les grandes tueries de ces dernières années, en guise d'illustration au carnage anglais.

— Tu as suivi cette histoire ? demanda Zandt.

— J'ai essayé de résister.

Ils regardèrent ensemble un bout de l'émission, debout dans la cuisine. Rien de nouveau. On ignorait toujours les motivations du forcené. La perquisition de son domicile avait mis au jour de la littérature raciste classique, une deuxième arme à feu, un ordinateur truffé d'images pornos, et un tableau médiocre montrant une nuée de silhouettes sombres sur un fond blanc, tel des spectres dans la neige.

Rien de cela n'était jugé probant.

CHAPITRE 18

— Il me faut plus qu'un peu d'eau, avait dit Sarah.

Elle parlait d'une voix faible, même à ses propres oreilles. Ce n'était pas la première fois qu'elle prononçait cette phrase. C'était devenu son leitmotiv chaque fois que la trappe s'ouvrait.

— Tu n'aimes pas cette eau ?

— Si, j'aime cette eau. Merci pour cette eau. Mais il me faut davantage. Il me faut davantage que de l'eau.

— Qu'est-ce que tu veux ?

— J'ai besoin de nourriture. D'un truc à manger.

Elle toussa. Ça la prenait de plus en plus souvent, et lui donnait la nausée.

— Nous mangeons trop, de nos jours, dit l'homme. Beaucoup trop. On tue des bêtes pour nous, on cultive à la tonne, puis on dépose le tout devant notre porte et nous nous jetons dessus comme des porcs sur leur auge. Nous ne sommes même plus des chasseurs. Rien que des charognards. Des hyènes munies de coupons de réduction, qui picorons les restes sous cellophane de parfaits inconnus.

— Si tu le dis. Mais j'ai besoin de manger.

— J'ai besoin de manger, j'ai besoin de manger, j'ai besoin de manger... railla-t-il.

Visiblement séduit par la sonorité de ces mots, il poursuivit ainsi pendant quelques minutes. Puis il se tut un instant, avant de déclarer :

— Il fut un temps où l'on se privait parfois plusieurs jours de suite. Nous étions plus maigres.

— Je sais, la crise de 29, les grandes sécheresses des années trente, etc.

L'homme se mit à rire.

— Mais c'était hier, ça ! Et ça n'a strictement aucun intérêt. Je te parle d'avant l'invasion.

— L'invasion, releva Sarah, tout en songeant : « OK, nous y voilà. Les petits hommes verts. Les Russes, les Juifs et je ne sais quoi d'autre. »

Elle toussa de plus belle, et l'espace d'un instant elle vit tout blanc. Lorsque le type répondit, on aurait dit que sa voix venait de très loin ou bien qu'il utilisait l'un de ces appareils comme Cher sur la chanson *Believe*.

— Oui, l'invasion. Comment tu appellerais ça, toi ?

Elle déglutit, en comprimant ses paupières, puis rouvrit les yeux.

— Je ne l'appellerais pas. J'ai trop faim.

— Tu ne vas pas pouvoir manger.

Quelque chose dans la voix de l'homme lui fit soudain très peur. Il n'avait pas l'air de sous-entendre qu'elle allait juste jeûner aujourd'hui, mais à jamais. Elle avait réussi, avec une remarquable rapidité, à s'adapter aux circonstances du moment, aidée en cela par une sensation de dislocation grandissante. Mais cette menace de diète absolue, définitive, suffit à la ramener en pleine réalité.

— Écoute, fit-elle d'une voix vacillante, tu dois bien attendre quelque chose de moi. Tu as forcément une raison d'agir ainsi. Je t'en supplie, fais ce que tu as à faire, mais tue-moi ou nourris-moi. *J'ai besoin de manger.*

— Ouvre la bouche.

Elle s'exécuta fébrilement, la salive affluant aussitôt sous sa langue. D'abord il ne se passa rien, puis une main apparut. Elle ne tenait rien qui ressemble à de la nourriture, mais juste un morceau de papier blanc. Le type le pressa quelques instants sur la langue de Sarah, puis le reprit. Sarah fondit en larmes.

Il se tut un moment, puis fit claquer sa langue en signe de réprobation.

— Aucun changement, dit-il. Ah, ce petit génome têtu. (Le bout de papier voleta pour atterrir à côté de Sarah.) Tu n'as pas appris grand-chose, n'est-ce pas ?

Elle renifla.

— Tu ne m'as rien enseigné.

— Je commence à m'interroger sur ton compte. Je te croyais différente. Capable de changer. Je suis venu te chercher, toi personnellement. J'avais des projets pour nous deux. Mais je commence à me demander si tu feras l'affaire, en fin de compte.

— Ah bon ? Pourquoi ça ?

— Tu es paresseuse, gâtée, et tu n'accomplis aucun progrès.
— Et toi, t'es qu'un malade mental.
— Et toi une sale petite conne.
— Je t'emmerde, lâcha-t-elle. T'es qu'un pauvre taré, et moi je vais m'échapper et te fracasser la gueule !

Elle ferma la bouche lorsqu'il lui versa son verre d'eau.

Ça faisait longtemps que son geôlier n'était pas revenu.

CHAPITRE 19

Nous atteignîmes Hunter's Rock à 3 heures du matin, au terme d'un court saut d'avion et d'une longue route. Après notre atterrissage dans l'Oregon, nous avions franchi la frontière de l'État par l'autoroute, puis emprunté des axes dont je conservais un lointain souvenir – à vrai dire, j'avais davantage l'impression de marcher dans les pas d'un aventurier dont j'aurais lu les exploits que de fouler mon ancienne terre. Puis nous traversâmes des coins de plus en plus familiers, et le trajet se compliqua. L'itinéraire que j'avais choisi n'était pas très direct, et je pense que Bobby s'en est rendu compte. Mais il ne fit aucun commentaire.

Je coupai enfin le moteur devant un vieux motel que je ne reconnaissais pas, trente kilomètres avant la ville. J'avais été partisan de dormir dans la voiture, mais Bobby estimait, avec son indéfectible pragmatisme, que nous serions plus productifs après quelques heures de sommeil dans un lit. Nous gagnâmes le bâtiment et tambourinâmes à la porte de la réception. Au bout d'une longue attente, apparut un homme en tee-shirt et pantalon de pyjama, qui exprima sans détour son hostilité. Nous reconnûmes qu'il était très tard, mais maintenant qu'il était debout, autant qu'il rentabilise l'incident en nous louant une chambre double.

Il nous examina longuement.

— Vous êtes pas un couple de pervers ?

Nous lui rendîmes son regard, et il sembla considérer qu'il valait mieux louer à deux homos en puissance que de se faire passer à tabac par ces mêmes homos au beau milieu de la nuit. Il me remit une clé.

Bobby s'allongea sur l'un des lits et s'endormit sur-le-champ. Je voulus faire de même, mais le sommeil ne venait pas. Je finis par me relever et quittai la chambre. J'achetai un paquet de cigarettes au distributeur et sortis dans la vieille cour, où une clôture rouillée entourait les vestiges d'une piscine. J'installai une chaise délabrée à un bout et m'assis dans le noir. Il n'y avait aucune lumière, hormis le rose terni du signal CHAMBRES LIBRES au fronton de la loge, un embryon de lune et quelques reflets sur les murs lépreux. Je sortis le flingue que Bobby m'avait confié et l'examinai un instant. Il n'avait pas grand-chose à me raconter, alors je le rangeai dans mon blouson.

Je scrutai ensuite les ombres de la piscine, me demandant depuis quand elle était vide. Depuis un bon bout de temps, à en juger par son aspect : ses parois étaient fissurées, et les quinze centimètres de vase qui croupissaient au fond auraient pu présider à l'apparition de la vie. Cette piscine avait un jour abondé d'eau fraîche, et les parents y auraient envoyé leurs bambins de bon cœur, ravis d'un tel répit après un long voyage. La pancarte du motel, aussi délavée que délabrée, situait sa construction à la fin des années cinquante. Je pouvais me projeter dans ce temps-là, mais uniquement sous la forme d'images arrêtées : des instantanés d'une époque glorieuse aux couleurs légèrement passées, chaque scène singeant une réclame pour le genre de vie qu'on nous a toujours vendue comme inéluctable. Un jardin de douceur et de lumière, de barbecues et de viriles poignées de main, de dur labeur, de grand amour et de franc jeu. La vie telle qu'elle devrait être. Au lieu de quoi on tourne en rond, sans charisme ni direction ni script – et quand le cirque finit par s'arrêter, on se rend compte que de toute façon personne ne regardait. Nous avons tellement l'habitude de voir les événements présentés sous un angle particulier que, lorsqu'ils nous arrivent à nous et que notre vie ne colle pas au modèle fourni, nous ne savons trop comment réagir. Notre propre existence nous devient méconnaissable. Faut-il poursuivre tête baissée la quête du bonheur, quand tout paraît si imparfait, détraqué et grisâtre ? Comment sommes-nous censés nous réjouir, quand tout est tellement mieux à la télé ?

Je me doutais bien que Bobby disait vrai en affirmant que ma naissance n'était pas enregistrée à Hunter's Rock, mais je devais le vérifier par moi-même. Durant tout le temps où Chip Farling m'avait trimballé en voiture, mon enfance m'avait étreint d'une main glacée. Mes parents se seraient donc rendus ailleurs pour ma naissance – était-ce si important ? Peut-être étaient-ils partis en week-end, le dernier en amoureux avant l'élargissement de la famille, quand je m'étais déclaré ? Mais n'était-ce pas typiquement le genre d'anecdote que l'on raconte à son gamin, le genre d'anecdote qui rend chaque vie unique ? Je ne voyais qu'une seule explication : ils n'avaient pas révélé mon lieu de naissance afin de me cacher l'existence d'un jumeau. Quant à savoir pourquoi ils allaient se séparer de lui, je n'en avais toujours pas la moindre idée. Peut-être était-ce là l'origine du vide autour duquel j'avais inconsciemment bâti mon existence. Tout le monde ressent ce type d'absence par moments. Dans mon cas, c'était souvent. Peut-être venais-je enfin d'en découvrir la cause.

J'ignore depuis combien de temps durait ce bruit. Quelques secondes, je dirais. Je décelai comme un faible clapotement. Il semblait tout proche, si proche qu'il me fit pivoter sur mon siège. Mais il n'y avait rien derrière moi. En me retournant à nouveau, je constatai mon erreur d'appréciation : le son provenait du fin fond de la piscine. Il faisait trop sombre pour le vérifier, mais le bassin paraissait se remplir d'eau. Surpris, je me penchai en avant. Le niveau montait, lentement mais sûrement. La profondeur n'était plus de quinze centimètres, mais au moins du double. Alors seulement, je m'aperçus qu'il y avait deux personnes dans la piscine. À l'autre bout. L'une était un peu plus grande que l'autre, et je ne distinguai d'abord que deux ombres épaisses. Elles se tenaient par la main tout en luttant pour avancer, poussant dans l'eau visqueuse qui continuait de monter. Le clapotement s'intensifia, le bac s'emplit de plus en plus vite, et les silhouettes s'activèrent pour gagner la partie émergée, de mon côté.

À présent la lune révélait leurs traits, et je sus qu'il s'agissait de ma mère et de mon père. Leurs efforts eussent mieux payé s'ils s'étaient lâché la main, mais ils n'en firent rien. Même après que le niveau eut atteint leur taille, ils demeuraient unis sous l'eau. Je crois qu'ils m'avaient vu. Ils regardaient dans ma direction, en tout cas. Les lèvres de mon père remuaient, mais le son ne me parvenait pas. Leurs bras libres taillaient dans les flots, sans même éclabousser, et ils continuaient de s'enfoncer. Leur progression n'y changeait rien : la pente de la piscine ne réduisait pas la profondeur ; l'eau ne cessait de monter. Même quand elle eut dépassé leur menton, qu'elle se fut mise à déborder, à se répandre à mes pieds telle une flaque de mercure sombre. Le regard de ma mère fut stoïque jusqu'au bout – c'est mon père que je vis paniquer, pour la seule fois de ma vie, et c'est sa main qui disparut en dernier, alors qu'ils avaient presque atteint le bord. Ils continuaient de couler, mais essayaient de m'agripper.

Mes paupières s'ouvrirent d'un coup. C'était l'aurore, et Bobby était penché sur moi, à secouer la tête.

Je me redressai, hagard, et vis que mon paquet de cigarettes avait quitté mes genoux pour sombrer dans les quinze centimètres de vase du bassin. Bobby me fit un clin d'œil.

— Z'ont dû valdinguer dans ton sommeil.

En fin de matinée, c'était confirmé. Aucun Ward Hopkins, aucun Hopkins d'aucune sorte n'avait vu le jour à Hunter's Rock. J'avais été reçu par une chouette fille derrière un bureau, qui s'était proposé de regarder si elle trouvait autre chose. Je ne voyais pas quelle autre information aurait pu m'aider, et il apparut qu'elle non plus. Elle essayait juste de me rendre service, à la fois par compassion et par ennui. Je lui laissai mon numéro et la remerciai.

Bobby m'attendait sur le trottoir, suspendu au téléphone. Je scrutai bêtement la rue de haut en bas jusqu'à ce qu'il ait raccroché. Même si j'avais su que ce moment arriverait, je me sentais dépossédé. C'était comme de s'entendre dire qu'on n'est pas sorti du ventre de sa mère, en fait, mais qu'on a bien été déposé sous un buisson par une cigogne. Je m'étais fait opérer des amygdales dans ce petit hôpital, et recoudre un genou écorché à deux reprises. À chacune de ces visites j'avais cru revoir l'endroit où j'avais poussé mes premiers cris.

— Alors, mon ami, dit enfin Bobby, les valeureux hommes et femmes du commissariat de Dyersburg aimeraient drôlement savoir où tu te trouves. Il semblerait, tu seras heureux de l'apprendre, qu'ils soient juste inquiets à ton sujet. Pour l'instant, en tout cas.

— Et la maison ?

— Gros dégâts dans le salon et le couloir, l'escalier du bas est partiellement détruit. Mais ce n'est pas un tas de cendres.

— Et maintenant ?

— Montre-moi ton ancienne maison.

J'ouvris de grands yeux.

— Pourquoi ?

— Parce que tu es grand, beau et blond, mon bichon, et que je veux *tout* savoir de toi.

— Va te faire foutre, lui conseillai-je, illustrant mon propos d'un geste nonchalant. C'est une idée stupide, sans intérêt.

— Tu as mieux à proposer ? Cette ville n'a pas l'air délirante, question distractions.

Je pris donc la grand-rue. Je n'aurais su dire ce qui me paraissait le moins familier : les nouveaux ou les anciens trucs. Le plus frappant était la disparition du vieux Jane's Market, remplacé par un modeste Holiday Inn paré d'une de ces petites enseignes carrées à la mode. Je suis très nostalgique des grosses pancartes dodues d'antan. Sincèrement. Et je ne vois pas en quoi les rectangles seraient meilleurs.

À l'approche de notre destination, je ralentis l'allure, puis me rangeai le long du trottoir d'en face. Cela faisait dix ans que je n'avais pas vu cette maison, peut-être même plus. Elle correspondait à mon souvenir, malgré sa façade ravalée et la végétation transformée. Un break de marque asiatique était garé dans l'allée, et trois vélos rangés sur le côté.

Au bout d'une minute, je distinguai une silhouette derrière la fenêtre. Ce n'était qu'un banal pavillon de banlieue, mais il me faisait l'effet d'une maison de pain d'épice dans un conte de fées. Sa réalité était trop forte, trop imposante, comme surchargée de glutamate. Je tâchai de me rappeler la dernière fois que j'y avais pénétré. Pourquoi n'étais-je pas venu lui faire mes adieux avant qu'elle ne change de mains ? N'avais-je pas encore compris que rien n'est immuable ?

— Tu es prêt ?

Je vis que mes doigts tremblaient un peu. Je me tournai vers Bobby.

— Prêt à quoi ?

— À visiter.

— Jamais de la vie.

— Oh que si !

— Enfin, Bobby, t'es cinglé ou quoi ? Il y a des gens qui vivent là. Je vais quand même pas m'incruster.

— Bon, écoute-moi. Il y a deux ans, mon vieux est mort. Je m'en foutais un peu – on s'était jamais entendus. Ma mère a insisté pour que j'assiste à l'enterrement. Mais j'avais des choses à faire. Je me suis défilé. Puis, six mois plus tard, je me suis rendu compte que j'étais pas dans mon assiette. Rien de visible à l'œil nu. Disons que je me sentais stressé. Tout le temps. Je m'angoissais sans raison particulière. Des crises de panique, en quelque sorte. Comme si le sol se dérobait sans cesse devant mes pieds.

Je ne savais que dire. Bobby regardait droit devant lui, à travers le pare-brise.

— Pour finir, le boulot m'a conduit un jour tout près de la maison, alors j'ai fait un saut chez ma mère, à Rochford. On n'était pas non plus les meilleurs amis du monde, tu sais.

Mais c'était quand même bon de la revoir. « Bon » n'est peut-être pas le terme exact. Utile, disons. Je l'ai trouvée changée. Tassée. En reprenant la route, j'ai fait un arrêt au cimetière, et je me suis planté devant la tombe de mon vieux. C'était un après-midi ensoleillé, il n'y avait personne dans les parages. Et là, son fantôme, oui, son fantôme est sorti de terre d'un seul coup et m'a dit : « Eh, Bobby, tu flippes ? »

Je le regardai. Il pouffa doucement.

— Non, je déconne. Je n'ai ni senti sa présence, ni éprouvé d'indulgence particulière pour l'homme qu'il avait été. Mais depuis, je suis moins anxieux. Il m'arrive de penser à la mort, je suis plus prudent, et je me fais peu à peu à l'idée de me poser d'ici quelques années. Mais le truc zarbi est parti. J'ai retrouvé la terre ferme. (Il se tourna vers moi.) Les soucis non réglés sont un poison, Ward. Tu penses te protéger, mais en vérité tu creuses tout un tas de petites fissures. Laisse-les s'ouvrir toutes à la fois, l'édifice tombera en miettes et tu te retrouveras comme un chien affamé qui erre dans les rues la nuit. Et crois-moi, mon ami, tu as un sacré paquet de fissures qui ne demandent qu'à péter.

J'ouvris la portière et m'extirpai de la voiture.

— S'ils me laissent entrer...

— Ils le feront, promit Bobby. Je t'attends ici.

Je me figeai. J'avais dû penser qu'il m'accompagnerait.

— C'est ta baraque, Ward. Si on va sonner ensemble, la maîtresse de maison se croira bonne pour une figuration dans *Profession légistes*.

Je remontai l'allée et frappai à la porte. Le porche était en ordre, pas un gramme de poussière.

Une femme apparut, tout sourire.

— Monsieur Hopkins ? dit-elle.

D'abord surpris, je pigeai. D'un même élan, je maudis et bénis le nom de Bobby. Il avait préparé le terrain avec un coup de fil, en se faisant passer pour moi, puis il avait déroulé son scénario. Je me demandais ce qu'il aurait fait si j'avais refusé.

— C'est exact, répondis-je, prenant le train en marche. Vous êtes sûre que je ne vous dérange pas ?

— Pas le moins du monde. (Elle s'écarta pour m'inviter à entrer.) Vous avez eu de la chance de me joindre tout à l'heure. Hélas, je crains de devoir repartir bientôt.

— Bien sûr, dis-je. Quelques minutes me suffiront amplement.

La femme, une quadragénaire suffisamment jolie et gentille pour jouer la maman dans une série télé, me proposa du café.

Je déclinai, mais le jus était déjà prêt et au final il était plus facile de dire oui. Pendant qu'elle l'apportait, je restai dans le hall, à scruter les lieux. Tout avait changé. Cette dame (je ne pouvais lui demander son nom, l'ayant théoriquement eue au téléphone) avait un goût prononcé pour le pochoir. C'était assez réussi, comparé à ce que j'avais connu, si on aimait le style Pottery Barn.

Nous fîmes le tour du propriétaire. La femme n'avait pas à justifier sa présence. Je trouvais assez inhabituel qu'on laisse entrer un inconnu sur la seule foi d'un coup de fil, et comprenais d'autant mieux qu'elle veuille garder un œil sur ses biens. Mais je parvins à émettre assez de commentaires sur l'ancienne disposition des lieux pour qu'elle baisse totalement sa garde et préfère vaquer à ses tâches domestiques. Je me promenai dans chaque pièce, puis à l'étage. Je jetai un rapide coup d'œil dans l'ex-chambre des parents et l'ex-chambre d'amis, qui avaient toujours été deux territoires plutôt étrangers. Puis je me préparai pour l'objectif ultime.

Quand la porte de mon ancienne chambre s'ouvrit, je déglutis malgré moi. Je fis quelques pas avant de m'immobiliser. Des murs verts, une moquette marron. Quelques cartons et de vieilles chaises, un ventilateur cassé et la majeure partie d'un vélo d'enfant.

Je m'aperçus que la femme se tenait derrière moi.

— Je n'ai pas touché à cette pièce, expliqua-t-elle. Ma fille a préféré l'autre chambre, qui offre une meilleure vue, bien que celle-ci soit un peu plus grande. On s'en sert comme d'un débarras. Bon, je vous laisse.

Je m'attardai quelques minutes dans la pièce, tournant sur moi-même pour l'apprécier sous différents angles. Elle mesurait dans les six mètres carrés, et paraissait à la fois minuscule et immense comme l'Afrique. L'espace où l'on grandit diffère de l'espace normal. On le connaît si intimement, on en a pratiqué chaque recoin assis, allongé, debout. C'est là qu'une foultitude de pensées nous viennent pour la première fois, et dès lors cet espace s'étire à l'image du temps qui nous sépare de Noël, à mesure qu'on y vit et qu'on attend de devenir grand.

— C'est ma chambre, murmurai-je à mi-voix.

La revoir en vidéo m'avait fait bizarre. Mais plus ici. L'endroit d'où je venais n'avait pas bougé. Toute ma vie n'était pas effacée. En ressortant je pris soin de fermer la porte, comme pour conserver un trésor.

En bas, mon hôtesse était appuyée à la table de la cuisine.

— Merci, lui dis-je. C'était très chic de votre part.

Elle balaya le compliment d'un chuchotis, et j'étudiai brièvement la cuisine. Les appareils avaient rajeuni, mais les placards étaient d'origine. Robustes, un bois de qualité, pourquoi les remplacer ? La belle ouvrage de mon père lui survivait.

C'est alors que me revint en mémoire ce tête-à-tête père-fils autour d'un plat de lasagnes. Un torchon suspendu à la poignée du four, une partie de billard infructueuse. J'ouvris la bouche mais ne dis rien.

Ressortir de cette maison fut une expérience insolite : quitter mon ancien monde pour celui où je vivais à présent. Je fus presque surpris, en repérant notre berline blanche de l'autre côté de la rue, de remarquer combien les bagnoles d'aujourd'hui ressemblaient à de grosses puces.

J'adressai un dernier signe à la femme et redescendis l'allée d'un pas léger. Le temps que j'atteigne la portière, la maison derrière moi s'était refermée, pour de bon.

Bobby m'attendait plongé dans le contrat de location du véhicule.

— Ces trucs sont d'un chiant ! gémit-il. Je te jure. Ils devraient embaucher des écrivains. Histoire de pimenter un peu.

— Tu es un homme vil, lui dis-je. Mais merci.

Il rangea les papiers dans la boîte à gants.

— Bon, j'imagine que nous en avons fini avec Hunter's Rock ?

— Je ne crois pas, non.

— À quoi tu penses ?

— Et si, en fait, ils avaient déjà su, au moment de notre naissance, qu'ils allaient prendre cette décision-là ? Qui sait, ils s'estimaient peut-être incapables d'élever deux gosses à la fois.

Bobby parut sceptique.

— D'accord, acquiesçai-je. Mais admettons qu'ils aient prévu à l'avance d'abandonner l'un de nous deux. Ils savaient aussi qu'ils allaient mourir un jour, et que je risquerais d'entreprendre ce que j'entreprends aujourd'hui. De revenir à la maison, de mettre mon nez dans leurs affaires. Et de découvrir auprès de l'hôpital l'existence d'un frère ou d'une sœur.

— Alors ils s'arrangent pour que tu naisses ailleurs : au pire, tu buteras sur un mystère mineur quant à ton lieu de naissance, mais tu n'auras jamais vent d'un jumeau abandonné.

— C'est bien ce que je me dis.

— Dans ce cas, comment expliques-tu que l'Agence n'ait pas tiqué lors de ton recrutement ?

— Je rendais de fiers services, à l'époque. À mon avis, ils ont expédié l'enquête de moralité pour gagner du temps, et après je faisais déjà partie de l'équipe, alors quelle importance ?

Bobby réfléchit.

— J'ai pas mieux, conclut-il. Mais ça reste étrange. Si tes parents se sont donné tout ce mal pour cacher cette histoire, pourquoi laisser traîner une cassette montrant leur geste ?

— Il sera peut-être arrivé quelque chose, suite à quoi ils auront changé d'avis et décidé que j'avais le droit de savoir.

Songeant que la femme pouvait nous épier depuis sa fenêtre, je démarrai et repris la route.

— Je me dis qu'on a peut-être cherché dans la mauvaise direction, Bobby. Il y a trois parties sur cette vidéo. La première montre un endroit que je pouvais trouver. Les Halls. La dernière m'apprend quelque chose que j'ignorais. Et celle du milieu se déroule dans deux lieux différents. D'abord la maison, que je viens de revoir, grâce à toi. Rien à signaler. Puis un bar, que je n'ai pas reconnu, où je n'ai jamais mis les pieds.

— Et donc ?

Nous étions arrivés à un croisement.

— Un peu de patience, dis-je avant de prendre à gauche.

Une direction qui nous mènerait, à supposer qu'il existe encore, vers un rade que j'avais fréquenté jadis.

CHAPITRE 20

Ça n'avait jamais été le genre d'endroit où l'on se rend à dessein, à moins que le hasard n'en eût fait votre repaire. Je lui prédisais deux destins possibles : rehaussé d'une salle à manger et d'une armée de serveuses enjouées toutes de rouge et blanc vêtues, ou bien rasé au profit de logements sociaux où les gens criaient fort la nuit. En fait, le progrès semblait avoir tout bonnement ignoré le Lazy Ed's, contrairement au pourrissement des ans, qui l'avait imprégné tel un dégât des eaux.

L'intérieur était vide, silencieux. Le bois du comptoir et les tabourets étaient toujours aussi éraflés. La table de billard était à sa place, tout comme l'essentiel de la poussière – peut-être même la mienne. On notait quelques ajouts de-ci de-là, signes d'un mieux évident. Le néon MILLER avait été évincé par Bud Lite, et le calendrier mural montrait de jeunes femmes plus proches de l'état naturel qu'à mon époque. Naturel, en tout cas, dans leur degré d'habillement, à défaut de la forme ou de la constitution de leurs seins. Quelque part, sûrement très bien cachée, se trouverait une plaque avertissant la femme enceinte des dangers de l'alcool – même si la présence de cette dernière semblait inconcevable ici, sauf cécité ou pure démence. Les femmes ont des exigences plus élevées. C'est d'ailleurs pour cette raison qu'elles ont une influence civilisatrice sur les jeunes hommes. Il faut trouver un endroit décent pour les faire boire.

Bobby s'accota au billard, tout en embrassant la salle du regard.

— Tout est comme avant ?

— Comme si je n'étais jamais parti.

Je remontai jusqu'au comptoir avec une certaine appréhension. Autrefois, il me suffisait de déclamer le nom d'Ed pour être servi. Mais vingt ans s'étaient écoulés, et recommencer aujourd'hui serait un peu comme de retourner à l'école en pensant que les profs vont nous reconnaître. La dernière chose qu'un individu ait envie d'apprendre, c'est qu'il n'a jamais été qu'« un des gamins ».

Un type émergea de l'arrière-salle, s'essuyant les mains sur un torchon qui ne pouvait que les salir davantage. Il nous salua d'un coup de menton cordial, mais avec un enthousiasme mesuré. Il avait mon âge, peut-être un peu plus. Obèse et déjà frappé de calvitie. J'adore voir mes contemporains perdre leurs cheveux. Ça me dope.

— Salut, dis-je. J'aurais voulu voir Ed.

— L'avez devant vous, répondit-il.

— Celui auquel je pense aurait à peu près trente ans de plus.

— Lazy, alors. L'est pas là.

— Me dites pas que vous êtes Ed junior.

Ed n'avait pas d'enfants. Ni même de femme.

— Jamais de la vie, se récria le type, comme choqué par cette idée. Juste une coïncidence. Je suis le nouveau proprio. Depuis qu'Ed a pris sa retraite.

Je m'efforçai de cacher ma déception.

— Sa retraite, répétai-je.

Je ne voulais pas paraître trop insistant.

— Ouais, ça fait bien deux ans. M'a évité de devoir changer la pancarte.

— Cette salle n'a pas bougé d'un poil, hasardai-je.

Le type secoua la tête d'un air las.

— Je suis au courant, merci. Quand Lazy me l'a vendue, il a posé une condition. L'a dit qu'il vendait une affaire, pas sa résidence secondaire. Et que ça devait rester tel quel jusqu'à sa mort.

— Et vous avez marché ?

— Je l'ai eue pour une misère. Et Lazy se fait très vieux.

— Comment sait-il si vous respectez l'accord ?

— Il continue de venir. Presque tous les jours. Si vous êtes pas trop pressé, z'avez toutes les chances de le voir arriver. (Il avait dû remarquer mon sourire, car il ajouta :) Je vous préviens quand même. Vous risquez de le trouver pas mal changé.

Je passai une première commande et retrouvai Bobby à sa table. Nous vidâmes nos chopes et jouâmes au billard pendant un bon moment. Ce fut Bobby qui gagna.

Nous enchaînâmes les bières et, quand j'en eus marre de perdre, Bobby s'entraîna tout seul pendant une heure. Tant de discipline aurait plu à mon père. Nous eûmes longtemps le bar pour nous seuls, puis quelques âmes arrivèrent au compte-gouttes.

En fin d'après-midi Bobby et moi représentions encore un tiers de la clientèle. J'avais gentiment sondé Ed sur les horaires de Lazy, mais ceux-ci était apparemment des plus imprévisibles. Je songeai à demander son adresse, mais quelque chose me disait que ce type ne la cracherait pas, et que la question éveillerait ses soupçons. Le début de soirée connut un rush quand quatre clients débarquèrent d'un coup. Aucun d'eux n'était Ed.

Puis à 19 heures, il se produisit quelque chose.

Bobby et moi étions penchés sur le billard. Il me battait à présent avec moins de facilité. Quelqu'un avait sélectionné un vieux Springsteen dans le juke-box, et on aurait aussi bien pu se trouver vingt ans en arrière, à l'époque du gel dans les cheveux et des bras de chemise. J'avais suffisamment bu pour frôler la nostalgie des années quatre-vingt, ce qui n'est jamais bon signe.

Du coin de l'œil je vis s'ouvrir la porte du bar. Toujours penché sur la table, je guettai le visiteur. Je n'eus qu'un aperçu furtif. Un visage, très vieux. Qui me fixa droit dans les yeux. Puis le type tourna les talons.

J'alertai Bobby, mais il avait vu. Il traversa la pièce en trombe et enfonça la porte avant même que j'eusse lâché ma queue.

Le jour déclinait et une voiture s'ébranlait en catastrophe. Une vieille Ford déglinguée, qui souleva des gravillons tout en chassant du cul vers la sortie. Bobby poussait des jurons de première classe, et j'en compris vite la cause : un connard bloquait notre voiture avec son gros camion rouge. Bobby se retourna.

— Pourquoi il s'est enfui ? demanda-t-il.

— Aucune idée. Tu sais par où il est parti ?

— Non.

Il pivota et flanqua son pied dans le pick-up le plus proche.

— Démarre la caisse, dis-je.

Je regagnai le bar et fondis sur le comptoir.

— C'est à qui, le camion ?

Un type tout en jean leva la main.

— Vous dégagez votre bahut de merde ou on le vire du parking manu militari.

Il me considéra un instant, puis se leva et sortit. Je me tournai vers Ed.

— C'était lui, hein ? Le type qui s'est taillé.

— Faut croire qu'il avait pas trop envie de vous causer, finalement.

— Eh bien, c'est fort dommage, répliquai-je. Parce qu'il va le faire, de gré ou de force. Je dois lui parler du bon vieux temps.

Ça me travaille tellement que je pourrais en chier. Alors, il habite où ?

— Ça, je vous le dirai pas.

— Joue pas au con avec moi, Ed.

Le type glissa la main sous le comptoir. Je dégainai mon flingue et le braquai sur lui.

— Joue pas à ça non plus. Ça vaut pas le coup.

Le jeune Ed releva les mains. Je sentais les regards des clients converger sur nous, et priai pour qu'aucun ne soit d'humeur bagarreuse. Les mâles peuvent devenir très protecteurs vis-à-vis de ceux qui leur servent la bière. C'est un lien fort.

— Vous seriez le genre de gars à tuer quelqu'un, vous ?

Je le dévisageai.

— Qu'est-ce que tu crois ?

Un instant de flottement, puis Ed soupira.

— J'aurais dû me douter que vous alliez faire des ennuis.

— Je veux juste discuter.

— Là-bas sur Long Acre, dit-il. La vieille caravane près du ruisseau, de l'autre côté du petit bois.

Je plaquai quelques billets pour les consos et ressortis à toutes jambes, manquant de renverser le type qui revenait d'avoir déplacé son camion.

Bobby avait manœuvré la voiture, prêt à mettre les gaz. Maintenant que je savais où nous allions, le chemin me paraissait assez familier. Long Acre est une route interminable qui serpente de l'orée de la ville jusque dans les collines. Il n'y a pas beaucoup d'habitations par là, et le ruisseau en question se situait bien au-delà, de l'autre côté d'une épaisse futaie.

Nous y fûmes au bout d'une dizaine de minutes. Il faisait très sombre, et Bobby roulait très vite. Pas de phares rouges à l'horizon.

— Peut-être qu'il ne rentrait pas chez lui, suggéra Bobby.

— Il y sera contraint tôt ou tard. Ralentis un peu. Ce n'est plus très loin. Et puis tu me fous les jetons à conduire si vite.

Peu après nous distinguâmes la surface moirée du ruisseau, grise sous le ciel bleu-noir. Bobby freina d'un coup aussi violent qu'un choc frontal et bifurqua sur une piste à peine marquée. Tout au fond se dessinait une vieille caravane jouissant d'une tranquillité parfaite. Pas de voiture en vue.

— Merde, pestai-je. Bon, range-toi dans un coin où on ne puisse pas nous voir depuis la route.

Au bout d'une demi-heure je commençais à perdre patience. Même en supposant que Lazy ait pris un autre chemin pour éviter d'être suivi, il serait déjà arrivé. Bobby était d'accord sur ce point, mais il en tirait une tout autre conclusion.

— Non, protestai-je. On se connaissait, lui et moi. Pas question de fouiller chez lui.

— Je te proposais pas d'y aller toi-même. Allez, Ward. Dès la seconde où ce type t'aperçoit, il prend ses jambes à son cou. T'avais raison : la séquence du bar était censée te rappeler quelqu'un, et ce vieux mec sait des choses.

— Il m'aura peut-être confondu avec quelqu'un d'autre.

— Tu es sans doute un peu plus épais qu'à l'époque, mais ce n'est pas comme si tu avais pris cinquante kilos ou changé de couleur de peau. Il savait que c'était toi. Et pour un type soi-disant croulant, il nous a semés drôlement vite.

J'hésitai, mais pas longtemps. J'en avais passé, des soirées en compagnie de Lazy Ed. Je n'étais alors qu'un client parmi d'autres, c'est certain, et plusieurs générations de buveurs mineurs avaient défilé après moi. J'avais seulement espéré un accueil plus amical.

Nous sortîmes de la voiture et marchâmes jusqu'à la porte de la caravane. Bobby crocheta la serrure, se glissa à l'intérieur, et au bout de quelques instants une lueur dansa à travers les vitres.

Je m'assis sur la marche et montai la garde. Mes parents avaient-ils prévu qu'on en arriverait là un jour ? Leur fils à moitié bourré forçant la caravane d'un vieillard... Mais cet acte n'avait rien de gratuit : c'était le souvenir de cette lointaine soirée avec mon père, et de la réaction d'Ed en tombant sur lui, qui m'avait conduit vers ce bar, puis ici. Malgré tout, j'avais l'impression, comme je surveillais le bout du chemin tout en écoutant Bobby s'affairer à l'intérieur, d'entendre à nouveau la voix de mon père : « Je me demande ce que tu es devenu. »

Bobby réapparut dix minutes plus tard avec quelque chose dans la main.

— Qu'est-ce que t'as trouvé ? demandai-je en me relevant, les jambes endolories.

— Je te montrerai au chaud. Tu dois te geler les miches.

De retour dans la voiture, j'allumai le plafonnier.

— Alors voilà, commença Bobby. Ton Lazy Ed traverse le crépuscule de sa vie à l'aide de boissons alcoolisées, et il en est au stade où il se cache les cadavres de bouteille à lui-même.

Soit ça, soit il porte bien son surnom[1] et n'est même pas foutu de les jeter dehors. C'est une vraie porcherie là-dedans. Je n'ai pas pu tout regarder. Mais j'ai découvert ceci.

Il me tendit une photo, que j'inclinai à la lumière.

— Je l'ai dénichée dans une boîte glissée sous ce qui lui sert de lit. Le reste n'était qu'un ramassis de babioles.

Le cliché montrait un groupe de cinq adolescents, quatre garçons et une fille. La photo était sombre, et le photographe avait oublié de dire « cheese ». Seul l'un des sujets, debout en plein milieu, semblait conscient d'être immortalisé. Les autres étaient pris de trois quarts, la majeure partie du visage dans la pénombre. On ne pouvait dire où se déroulait la scène, mais les vêtements ainsi que la qualité de l'image plaidaient pour la fin des années cinquante, début soixante.

— C'est lui, dis-je. Le mec au centre.

J'avais un peu honte de piquer les souvenirs d'un autre, surtout quand ils ne me concernaient pas.

— Tu veux sûrement parler de Lazy Ed ?

— C'est ça. Mais ce truc date d'au moins cinquante ans. Il n'était pas si fringant quand je l'ai connu. Loin de là.

— OK, fit Bobby avant de montrer la femme sur la gauche. Et elle, c'est qui ?

Je me rapprochai un peu. Je ne distinguais qu'un demi-sourcil, des cheveux et l'essentiel d'une bouche. Un visage fin, jeune, plutôt joli. Je haussai les épaules.

— Va savoir. Personne que je connaisse.

— Vraiment ?

— Où veux-tu en venir, Bobby ?

— Il est possible que je me trompe, et je ne voudrais pas t'induire en erreur.

J'examinai le cliché de plus belle. Étudiai avec soin les autres visages afin de me rafraîchir les yeux. Puis les reposai sur la femme. Elle ne m'évoquait toujours rien.

— Ce n'est pas ma mère, si c'est à ça que tu penses.

— Non, non. Cherche encore.

Je m'y employai, et sentis poindre une idée. Je la laissai venir. Elle mit quelques secondes avant de tomber comme une brique.

— Putain de merde...

— Ça y est ?

---

1. *Lazy* : fainéant. (*N.d.T.*)

Je continuai de la dévisager, pensant être vite repris par le doute. Mais il n'en fut rien. Une fois qu'on l'avait vu, c'était indiscutable. Bien que son visage fût largement caché, tout était là dans son regard, et dans la courbure de l'arête nasale.

— C'est Mary, dis-je. Mary Richards. La voisine de mes parents. À Dyersburg.

Je rouvris la bouche pour ajouter quelque chose – je ne sais trop quoi – avant de la refermer comme un clapet, cloué par un nouveau flash.

Bobby ne remarqua rien.

— Mais que pouvait bien fabriquer ce bon vieux Ed dans le Montana ? Ou que faisait-elle ici ? demanda-t-il.

— Tu tiens vraiment à attendre son retour ?

— Sauf si tu as une meilleure idée.

— J'ai peut-être autre chose à te montrer. Sans compter que ça caille et que je doute qu'Ed rentre chez lui ce soir. On devrait retourner en ville.

Mes mains tremblaient, ma gorge était sèche.

— Ça me va.

Je quittai la voiture, gagnai la caravane et forçai la porte. Je griffonnai un mot au dos de la photo, m'excusant pour cette intrusion, et la plaçai en évidence au centre d'une table de bridge. J'ajoutai mon numéro de portable, puis ressortis en prenant soin de coincer la porte avec un magazine.

Bobby retourna vers le centre-ville les phares éteints, mais nous ne vîmes personne, et lorsque nous croisâmes le bar, la vieille Ford n'était pas sur le parking. Ni, songerais-je après coup, le camion rouge.

CHAPITRE 21

Nous prîmes deux chambres au Holiday Inn. Je me douchai puis soufflai un peu en attendant Bobby. Ma chambre était propre, fraîche et rassurante. J'avais un pichet de café sous la main, tout juste livré par une soubrette arborant un chouette uniforme blanc et un sourire de façade, celui que je préfère. Je suis né sans le gène jovial. Et j'aime autant que les gens ignorent mon nom.

Je regrettais de m'être séparé de la photo. J'aurais aimé la réexaminer, car je n'étais pas loin de croire à un effet de lumière trompeur. Le visage sans vie de Mary restait gravé dans ma mémoire. Son cadavre devait reposer dans un casier de la morgue à l'heure qu'il était, et personne ne comprendrait ce qui lui était arrivé. J'estimais qu'on avait le droit de le savoir, et notre fuite de Dyersburg me pesait encore sur la conscience. Je me disais qu'un coup de fil chez les flics de Dyersburg aurait pu les mettre sur la voie. Ils auraient voulu connaître mon identité, mais j'aurais pu inventer quelque chose. Je suis plutôt fort à ce jeu-là.

Je tendais le bras vers le téléphone quand Bobby frappa à ma porte. Je laissai l'appareil où il était et m'extirpai du fauteuil.

— Tout va bien ? demanda-t-il tout en refermant derrière lui.

— Quelles journées, Bobby !

J'ouvris mon ordinateur et le posai au milieu de la table. Je fis signe à Bobby de prendre l'autre siège, puis insérai le DVD-ROM dans la fente et chargeai la scène du bar extraite de la vidéo.

De la musique forte. Le chaos. La progression titubante du type qui filmait. La quinte de toux, puis le passage dans la salle où l'on jouait au billard. Un jeune couple debout, dos à la caméra, et un gros barbu et sa copine qui s'apprêtaient à jouer leur coup.

La caméra chancela plus avant, et la fille aux cheveux longs fixa l'objectif. Je fis pause sur le logiciel de lecture. En deux, trois coups de souris, je sauvegardai l'image, puis lançai Photoshop. De là, j'opérai un zoom sur le visage de la fille. Je prélevai un peu de fond et en appliquai sur le bas de la chevelure pour la raccourcir. Je clonai sa texture de peau et la repiquai sur le contour des joues, afin de les vieillir et les épaissir. Je prélevai un paquet de cheveux pour leur restituer le style que prisaient les vieilles dames en 2002. Je délimitai grossièrement la coiffure obtenue, l'injectai de gris acier, et pour finir brouillai légèrement les parties modifiées pour masquer la différence de grain, avant de lisser les angles saillants à l'aide d'un flou gaussien. J'opérai un zoom arrière jusqu'à réduire l'image de moitié par rapport à sa taille normale, et le trucage devint ainsi plus discret.

Il fallait faire abstraction du second plan tout morcelé, bien sûr, mais ce n'était pas difficile, vu la puissance du verdict. Je m'en doutais depuis la visite de la caravane d'Ed, mais la preuve à l'écran me suffoqua.

— OK, murmura Bobby. C'est encore elle. En compagnie de tes parents.

— Mais ils n'ont rencontré Mary qu'en arrivant dans le Montana...

— Et ils t'ont dit un truc du genre : « Elle, Mary. Complètement nouvelle dans notre vie. Jamais-jamais rencontrée auparavant. »

— On aurait juré qu'ils se connaissaient depuis quelques années seulement. (Je me sentais gagné par un insidieux vertige.) J'entends encore ma mère m'expliquer au téléphone qu'elle avait fait la connaissance de Mary, la voisine, qui leur avait apporté des cookies le jour de leur emménagement.

— Ils se connaissaient depuis plus de trente ans.

Pendant que je parlais, Bobby avait avancé le clip jusqu'à l'image de la fille qui ondoyait, assise en tailleur, dans le séjour de mes parents.

Je hochai la tête. La façon dont la lumière soulignait ses pommettes et son nez rendait tout bidouillage inutile. C'était Mary.

— Et ce brave Ed, alors ? Ce pourrait-être lui, le cameraman ?

— La seule fois que je l'ai vu dans la même pièce que mon père, on aurait dit de parfaits étrangers. (J'avais raconté la scène à Bobby sur le chemin vers le bar.) Mais il faut croire qu'ils se connaissaient. Ils se connaissaient tous, en fait. Réfléchissons. Pour une raison X, Mary déménage dans le Montana. Peu de temps après le tournage de ce film, si ça se trouve. Ça faisait sûrement un bail qu'elle vivait là-bas quand mes vieux l'ont rejointe. Entre-temps eux et Ed restent ici, mais se perdent de vue, et la seule fois où je les réunis par hasard, mon père et Ed font mine de ne pas se connaître.

Je me remémorai les fois où j'avais croisé Mary chez mes parents, mais elles ne faisaient que renforcer mon impression : s'ils s'étaient tous connus avant le Montana, ils s'étaient donné un mal fou pour le dissimuler. Mais pourquoi s'évertuer à me cacher cette vérité ? me demandai-je avant de juger la question égocentrique et stérile.

Une seule conclusion : mes parents n'avaient pas migré là-bas par hasard.

— Ils ont convergé vers le Montana car ils pensaient ou savaient qu'il allait se produire quelque chose, et voilà pourquoi ils faisaient semblant de ne pas se connaître d'avant.

— Tu extrapoles un peu, là.

— Tu crois ? Peut-être que Mary n'a pas été tuée juste parce qu'elle était dans le passage. Peut-être que l'individu qui s'est rendu chez mes parents avait deux cibles en tête et que Mary était l'une d'elles.

Bobby réfléchit.

— Et quand tu réapparais à Hunter's Rock, Ed détale comme un lapin.

— On aurait dû rester devant sa caravane.

Il secoua la tête.

— Il ne doit pas être pressé d'y retourner. Il aura appelé le gars du bar, et appris qu'on avait son adresse. Et puis tu m'as pas l'air bien frais pour te lancer dans une activité qui puisse impliquer de courser des gens. Tu lui as laissé ton numéro. S'il rentre chez lui, il saura comment te contacter. Demain on retourne au bar et on cuisine le proprio. Savoir si l'autre croulant a des associés ou d'autres repaires.

— Une aiguille dans une botte de foin, en somme.

— Mais l'aiguille est toujours là. Pour peu qu'elle ait été placée au hasard, ça pourrait être la première chose qu'on trouvera.

— Très profond, Bobby. Je crois que je vais la noter, celle-là.

— C'est bien. Pendant ce temps, je vais faire un tour sur la Toile. (Il posa les yeux sur le cellulaire couché sur la table.) Et si tu espères un coup de fil de Lazy, tu ferais peut-être mieux de rallumer ton machin.

Tandis qu'il reliait l'ordi à la ligne de la chambre, j'examinai l'écran de mon téléphone. Bien vu, Bobby. Au bout de quelques secondes, l'indicateur de messages s'illumina.

— Tu as reçu quelque chose ?

Je composai le numéro du service et tendis l'oreille. Une voix de femme.

— Ce n'est pas lui. C'est la fille qui m'a renseigné à l'hôpital. Elle avait dit qu'elle consulterait ses dossiers et me préviendrait si elle découvrait quoi que ce soit.

— C'est le cas ?

— J'en sais rien, dis-je en raccrochant. Elle me demande simplement de l'appeler demain.

— Regarde, Ward, t'as un mail.

Je me penchai par-dessus son épaule. Un bref message occupait l'écran :

GROSSES RENTRÉES DE $$$ GARANTIES !!!

Nous sommes une petite société offrant un service EN PLEIN ESSOR. Grâce à notre produit, vous pourrez transformer votre univers, en travaillant à votre rythme et en toute liberté. Heureux les purs quand ils verrons notre site Web !

Revenez vous procurer les informations qui pourraient changer nos vies à tous ! Commencez sans attendre – avec une entreprise qui se développe de jour en jour. Des centaines d'entre vous gagnent déjà des sommes qu'ils n'auraient jamais osé imaginer. Alors pourquoi pas vous ?

N'hésitez pas plus longtemps – cette offre prend fin à minuit.

— Tu cherches « mail foireux » dans le dico, commentai-je, et ils auront sûrement reproduit ce texte in extenso.

— Mais regarde, dit Bobby. Il n'y a aucun bon de commande. L'adresse de l'expéditeur a l'air bidon, ils font mention d'un site Web sans donner son URL, et ils te fixent un délai qui expire dans trois heures. Comment comptent-ils t'escroquer, à ce compte-là ? Et tu vois ces deux phrases ponctuées d'un point d'exclamation ? La première est drôlement tournée – genre citation biblique – et la seconde dit « revenez ». Mais revenir où ?

Je réfléchis un instant.

— Si je reçois ça, c'est parce que j'ai visité un site qui a enregistré mon adresse IP.

— Parfois, Ward, on jurerait que tu as presque toutes tes facultés mentales.

Il double-cliqua sur un signet du Bureau, et son navigateur téléchargea. Au bout de quelques secondes, la page s'afficha, qui comportait les trois mots : NOUS NOUS DRESSONS.

Mais cette fois-ci ils étaient soulignés, et quand Bobby y plaça son curseur, ce dernier se transforma en main pointeuse.

— C'est devenu un lien, dit-il.

Il cliqua dessus et une boîte de dialogue jaillit, réclamant un mot de passe.

— Et merde...

— Les Hommes de Paille, proposai-je.

Il essaya. Apparut une page blanche coiffée des mots ACCÈS NON AUTORISÉ. Bobby jura et cliqua sur l'icône « précédent ».

— Remontre-moi le mail, dis-je.

Il réduisit la fenêtre du navigateur pour ramener le message au premier plan.

Je le parcourus rapidement.

— Essaie *verrons*, tel que c'est écrit dans la phrase avec « heureux les purs ».

— Pourquoi ?

— C'est la seule faute d'orthographe du texte, et elle se trouve dans la phrase qui parle du site.

Il cliqua et entra le mot. Accès refusé.

— On va bientôt se faire éjecter, grommela Bobby en revenant de nouveau sur ses pas.

— Essaie avec la bonne orthographe.

Il cliqua et tapa *verront*. Il y eut une pause. Puis une autre page emplit l'écran. Noire, avec en plein milieu le mot « BIENVENUE » écrit en lettres blanches.

— Très bien, dit Bobby, très concentré.

Il plaça le curseur sur BIENVENUE et obtint une nouvelle main. Je me rapprochai encore un peu, et il cliqua.

Un temps d'arrêt, et l'écran devint vert mousse, rempli de texte blanc.

LE MANIFESTE HUMAIN

[image : paillelogo.jpg]

--

VOICI LA VÉRITÉ

Certains réfutent la Théorie de l'Évolution. Ils ont tort. Si l'on nous a répété si longtemps que l'Évolution était un tissu de mensonges, c'est pour NOUS EMPÊCHER de voir la vraie vérité. Mais maintenant que nous l'avons Vue, elle ne peut plus être confisquée pas les Politiciens et autres MENTEURS.

Vous croyez connaître la Vérité mais c'est faux : vous ne connaissez que des MENSONGES.

L'HISTOIRE DE L'HUMANITÉ

Au temps jadis nous étions tous des singes. Puis un beau jour, voilà 5 Millions d'années, une nouvelle branche s'est séparée pour créer trois nouveaux types de singes : les gorilles, les chimpanzés et les « hominidés » – qui allaient devenir nous. Quiconque a vu un documentaire montrant combien les chimpanzés sont intelligents croira facilement en cette VÉRITÉ. 2,5 Millions d'années plus tard sont apparues les premières créatures formant la véritable Humanité. On se réfère parfois à elles sous le nom d'Habilis, bien que les noms concernant cette période soient controversés : nous vivons une phase sombre de notre Évolution, et les chercheurs emploient des MOTS LONGS lorsqu'ils en savent moins qu'ils voudraient nous le faire CROIRE.

Il y a environ 1 Million d'années, nous avons vu arriver un nouveau type d'hommes, baptisé Erectus, ainsi nommé car il se tenait Debout. C'est cette faculté qui nous distingue du singe, et de tout autre animal. Cette lignée déboucha pour partie sur l'Homme de Neandertal, qui régna pendant une longue période. Au cours des centaines de milliers d'années qui suivirent, cette espèce devint meilleure marcheuse, perfectionna ses outils, et apprivoisa le FEU. L'ÉVOLUTION se poursuivit alors en Afrique, aboutissant à l'Homo Sapiens. La taille de notre cerveau a grandi, ainsi donc que notre intelligence, qui est unique. L'Homo Sapiens a ainsi supplanté les Néandertaliens.

Au cours de ces différentes périodes, l'humanité et ses précurseurs étaient des CHASSEURS-CUEILLEURS. Nous vivions en petits groupes unis par des liens de Parenté et de Coopération. Nous nous nourrissions du gibier que nous CHASSIONS ainsi que des baies et racines que nous GLANIONS, puis nous nous remettions en route.

CE QUE TU SÈMES REGRETTERAS

Il y a environ quinze mille ans, tout se mit à changer. Cela peut paraître lointain, mais pas quand on raisonne en termes de Millions d'années. Nous avons alors abandonné la chasse et la cueillette, qui nous étaient pourtant naturelles. Pourquoi ?

D'aucuns ont attribué cela à l'accroissement de la population, qui aurait entraîné une réduction des ressources et de la mobilité. Ou au facteur climatique, car l'Âge de Glace était révolu, et à diverses autres choses. J'ai lu toutes les explications soi-disant scientifiques, et il s'avère que personne ne détient la réponse. Il fut un temps où des millions de bisons peuplaient les plaines d'Amérique. Ils ne dépérissaient pas pour autant. Ils devaient migrer pour trouver de nouvelles sources de nourriture, mais telle est la Voie Naturelle. Les Humains, qui ont la faculté de se tenir Debout, sont CONÇUS pour marcher sur de longues distances. Alors pourquoi avons-nous soudain cessé de nous déplacer – après avoir passé des millions d'années à évoluer dans un autre sens ?

La raison est que NOUS AVONS INVENTÉ L'AGRICULTURE. En conséquence, les gens se sont peu à peu établis dans des endroits fixes, et ont agrandi leurs groupes jusqu'à inclure des centaines puis des milliers de personnes. Une fois ce processus enclenché, il ne pouvait être arrêté. L'Agriculture procure davantage de nourriture, mais cette méthode est MOINS EFFICACE pour alimenter des groupes restreints. Elle ne fonctionne qu'au sein de grosses communautés. L'Agriculture entraîna en outre une augmentation des naissances, ce qui accéléra l'élargissement des groupes. Une fois en présence d'une forte population, vous ne pouvez changer d'avis et revenir au glanage. Vous êtes Piégés.

De ces bouleversements naquirent les villages et les villes, qui accentuèrent encore l'accroissement démographique. Cela entraîna des INÉGALITÉS, puis l'apparition de DIRIGEANTS et de la RELIGION. Ce fut aussi la naissance de la MORALE. Si vous vous installez dans un endroit pour un certain temps, vous croiserez demain les mêmes gens qu'hier. Cela signifie que vous devrez vous comporter d'une certaine façon vis-à-vis d'eux, sans quoi ils vous TUERONT. De là, les gens en ont déduit qu'ils DEVAIENT adopter certaines attitudes – même face à des inconnus. Et pour la première fois nous avons assisté à une tendance humaine peu sympathique mais très prononcée, une chose qui nous différencie de toute autre espèce vivante : ALTÉRER LA TERRE. Jusqu'alors, nous avions vécu comme des éléments de la nature : depuis l'avènement de l'agriculture nous VIOLONS la terre en la soumettant à nos propres fins.

Et pourtant les peuples agricoles étaient en MOINS BONNE SANTÉ que les glaneurs. À effort égal, la culture offrait un rendement plus faible. Les chasseurs-cueilleurs avaient davantage de LOISIRS et TRAVAILLAIENT MOINS que les agriculteurs. Ils avaient un régime alimentaire plus équilibré que les cultivateurs, qui privilégiaient les racines ou les céréales. Les agriculteurs étaient plus exposés aux infections et aux épidémies – à cause de la promiscuité. Les gens vivaient moins vieux, et tombaient plus souvent MALADES.

ALORS POURQUOI CE MODE DE VIE A-T-IL DÉFERLÉ SUR LE MONDE, EN À PEINE QUELQUES MILLIERS D'ANNÉES ? Pourquoi, à travers la quasi-totalité de la planète, notre Espèce tout entière a-t-elle changé de mode de vie après des Millions d'années – surtout si dans un premier temps ce fut pour vivre MOINS BIEN ?

LE GÉNOME INHUMAIN

Les virus sont minuscules, mais lorsqu'ils sont en vous ils s'emparent de votre organisme pour le plier aux volontés de la Maladie. De nombreux virus vous rendent malade, tels que les rhumes. D'autres vous tuent, comme le SIDA. Mais les virus les plus malins ne font ni l'un ni l'autre – car ils cherchent en vous un HABITAT.

Il y 20 000 ans NOUS AVONS ÉTÉ INFECTÉS. L'Homo Sapiens a importé le virus d'Afrique, ce qui explique l'extinction des Néan-

dertaliens. Ces derniers savaient mieux s'adapter aux conditions difficiles, comme à l'Âge de Glace – et pourtant, en quelques milliers d'années à peine, ils ont disparu.

Ce Virus nous a poussés à vivre en groupes et dans les villes pour qu'il lui soit plus facile de se RÉPANDRE PARMI NOUS. *Nous n'avons pas suivi cette voie parce qu'elle était meilleure. Nous l'avons suivie parce que nous étions pris au piège. Lorsque le Virus nous a capturés et fait de nous sa maison, notre nature s'était modifiée et nous ne pouvions pas revenir en arrière.*

Désormais le Virus fait tellement partie de nous-mêmes que les gens de l'Establishment Scientifique ne le décèleront jamais, aussi malins se croient-ils.

Voilà pourquoi la patrie est une notion primordiale pour tant de personnes, y compris les JUIFS *: si nous migrons, le Virus va croire que nous retournons à notre vrai mode de vie, et nous causer des ennuis.*

Voilà pourquoi nous n'avons que faire des habitants d'autres pays : ils ne sont rien à nos yeux.

Voilà pourquoi les Terroristes et les Meurtriers tuent d'innocents Américains : nous ne sommes rien à leurs yeux.

Voilà pourquoi nos villes sont pleines de Violence : nous sommes obligés de vivre dans la crasse des autres, pareils à des rats dans des boîtes, ce qui constitue pour nous une pure infamie.

Voilà pourquoi des événements tels que l'holocauste Nazi, la Bosnie et le Rwanda surviennent : les autres tribus sont nos ennemies, et si on nous rapproche trop nous allons nous battre.

Voilà pourquoi nos dirigeants sont des Menteurs et des Idiots. L'État signifie la privation de notre LIBERTÉ*, au nom des prétendus droits d'individus que nous ne connaissons même pas.*

Voilà pourquoi les gens Assassinent et Tuent : parce que la seule chose qui puisse nous retenir est la Morale, qui fut inventée par le Virus.

Ils veulent sans cesse nous faire croire que nous sommes les mêmes, et racontent que le même sang coule dans nos veines, mais même ça c'est faux : il existe différents groupes sanguins – à cause de la génétique.

Même à un niveau si basique nous sommes incompatibles entre nous. Jusque dans notre sang nous sommes différents.

QUE POUVONS-NOUS FAIRE ?

Nous devons passer à l'action dans les villes, au milieu de vous autres, les Noirs. Peut-être qu'on ne vous aime pas beaucoup – parce que nous ne sommes pas du même type, et que nous cohabitons par la seule faute du Virus – mais vous aussi êtes des Victimes. Vous avez été arrachés à votre Foyer naturel et parqués dans des lieux sans espoir pour votre Espèce. Vous devez être les premiers à vous dresser. Le monde suivra.

Nous ne sommes pas censés vivre dans de grands groupes. Nous ne sommes pas conçus pour nous soucier de gens que nous ne connaissons pas. Nous sommes faits pour être libres, et non agglutinés dans de grandes villes sous la coupe de personnes qui se moquent bien de nous et ne s'intéressent qu'à l'ARGENT. La seule façon d'empêcher cette chose de nous détruire est d'en TUER les porteurs. Les Politiciens ne seront d'aucune aide, car ils se nourrissent de cet environnement Maléfique. À l'image du virus, sans la « civilisation », ils n'ont plus d'habitat. Le choix nous appartient.

Ceux qui Tuent seront Libres.
Ceux qui ne Tuent pas sont Infectés.

Assainissez la planète.

Tuez le virus.

Les Armes vous rendront forts.

Je parvins au bout du texte quelques secondes avant Bobby.

— Sauvegarde-le, dis-je. Ce ne sera plus là demain.

Quand Bobby se fut exécuté, je remontai tout en haut de la page et la relus attentivement. On aurait dit la synthèse de centaines de délires, de tous ces tracts diffusés à la sauvette, que l'on parcourt pour tuer l'ennui sur le chemin de la maison ; de ces complaintes marmonnées la nuit dans la pénombre d'un bar, la voix alourdie par l'alcool, l'ignorance et la hargne. Mais il y avait là quelque chose de différent. Je me renversai sur mon siège et tâchai de mettre le doigt dessus.

— Reconnaissons-lui un mérite, dit Bobby après sa seconde lecture. Il a pris la peine d'aller en bibliothèque pour consulter quelques bouquins. Mais ça n'en reste pas moins un truc de cinglé. J'imagine que tu es d'accord ?

— Oui et non. Des mots comme « supplanté » cadrent mal. Ou « infamie ».

— Ce ne sont pas deux mots un peu savants qui en font l'œuvre d'un génie. Il les a peut-être pompés dans un livre.

— Chaque virgule est à sa place, Bobby. Il y a des tas de zozos là-dehors qui prendraient ce truc pour parole d'Évangile. Ceux des milices paramilitaires, pour commencer. M'étonnerait pas qu'ils soient derrière tout ça.

Cela fit rire Bobby.

— J'en doute fort. Tu connais ces types comme moi. Des vétérans grisonnants et de pauvres gosses qui ont maté tellement de vidéos sur le Vietnam qu'ils finissent pas croire qu'ils y étaient aussi. Ils montent un camp dans les bois, où ils astiquent leur matos et se disputent les femmes.

— Ce ne sont pas tous des hommes de Cro-Magnon. Ni des crétins.

— Bien sûr que non. Mais on parle de mecs qui dévorent *Soldat de fortune* d'un bout à l'autre et achètent des manuels où l'on trouve la recette du napalm et des conseils pour installer des pièges dans ton jardin. Des gens qui ont rempli leur baignoire de boîtes de conserve en vue du prochain millénaire – et ont été super déçus de voir qu'il ne se passait rien, que la civilisation poursuivait son petit bonhomme de chemin. Ils enfilent un treillis et te rabâchent que le monde part en couille et que c'est la faute aux Juifs et aux Latinos, sans parler de Washington et de Saddam Hussein. Tu ferais mieux de t'inquiéter pour les Noirs des bas-quartiers, comme dit l'autre. Les mecs des gangs sont *très* remontés, et tous n'attendent pas de se prendre une gifle pour dessouder un mec.

— C'est la même chose, Bobby : des gens qui ne se sont jamais sentis intégrés dans une communauté, sauf celle où tout le monde se connaît par son prénom.

— Tu vas me faire pleurer, Ward.

— C'est ça, fous-toi de ma gueule. Tu mets toute ta confiance dans ton pays et tu l'aimes de tout ton cœur, comme on te le demande, puis tu découvres que ce n'était qu'une ruse pour que tu la boucles, et que le vrai message c'est : « Tout le monde peut tout avoir, sauf vous, les gars. On vous comptait pas dans le lot. » Sans blague, comment tu réagirais, toi ?

— OK, OK. Admettons qu'ils soient animés de sentiments authentiques, que leur QI soit à peu près dans la moyenne de la Chambre des représentants et qu'ils n'aient pas toujours dit que des conneries, à la rigueur. Mais ce que tu ne me feras jamais avaler, Ward, c'est qu'ils puissent être structurés. La plupart de ces bandes ont un mal de chien à maintenir trente mecs sur la même ligne. Alors tu les imagines discuter stratégie et objectifs avec d'autres groupes plantés à des centaines ou des milliers de kilomètres de là ? Pas dans ce monde-ci, Ward.

— Avant l'Internet, nuançai-je.

— On trouve des trucs sur la Toile, c'est vrai. Assez de vomi psychotique, en tout cas, pour occuper tous les thérapeutes du pays. Des antis de tout poil, des apocalyptiques, les Illuminati qui brûlent des effigies de chouettes à Bohemia Grove – et la face visible de Mars est une base de missiles de contrôle social, et les ogives sont braquées directement *sur toi*. Mais moi qui passe mes journées à parcourir cette daube, je te garantis un truc : aucune organisation mondiale n'est dirigée par ces mecs-là. Ils détestent tous ceux qui ne sont pas comme eux. Mets-les ensemble dans une pièce et ils se feront tous exploser.

— Tu ne peux pas surveiller chaque page de chaque site, rétorquai-je. Il pourrait exister tout un Web alternatif, qui utiliserait les mêmes bécanes, les mêmes lignes téléphoniques et les mêmes disques durs, mais grouillerait de tueurs, de barbarie et de projets sordides – et tant que tu ne sauras pas où chercher, tu ne trouveras même pas le sommaire. (Bobby leva les yeux au ciel, et je pris la mouche.) Écoute-moi, putain. *On est faits comme ça*. Tu ne l'as toujours pas compris ? Les scientifiques ont créé la Toile sur leur temps libre afin d'échanger des données et de passer titulaires à *Star Trek*. Et voilà que du jour au lendemain tu ne peux plus te connecter sans qu'on te bombarde de pubs, et le moindre cordonnier possède son propre site. Même avant ça, c'était la pornographie à tous les étages et des hommes et femmes prostrés dans des pièces sombres qui s'écrivent combien ils aiment se déguiser en Shirley Temple et se faire fouetter jusqu'au sang. Voilà ce que le Net va devenir : un moyen de se terrer dans l'anonymat pour cesser de jouer à monsieur et madame Voisins Modèles, et être vraiment soi ; pour cesser de prétendre se soucier d'un village global de mes deux, alors que nos listes de vœux font la taille d'une minuscule tribu préhistorique et qu'on a envie d'en oublier la moitié.

— C'est touchant, autant d'estime pour ses contemporains. Tu m'as l'air fin prêt à rejoindre la cause. (Il se frotta le visage.) Écoute, Ward, ce n'est peut-être que le trip d'un mec isolé.

— Arrête ton char. On est arrivés ici à partir d'un signet sur l'ordi d'un homme qui a tourné une vidéo où il évoque des « Hommes de Paille ». Un homme et sa femme viennent de mourir, ainsi qu'une voisine qu'ils connaissaient depuis des lustres. Une menace lancée à ceux qui habitent sur le lieu du film a entraîné le plastiquage d'une maison et d'un hôtel. Sans déconner, même l'architecture des Halls colle avec tout ça. Ils bâtissent des grottes à plusieurs millions de dollars pour des chasseurs-cueilleurs !

— OK, dit Bobby en levant les mains. J'entends bien ce que tu me dis. On fait quoi, maintenant ?

— Bon, on vient de découvrir ça. La suite ? Il n'y a ni liens, ni mail, rien du tout. À quoi peut bien servir ce truc s'il ne mène nulle part ?

Bobby inclina le portable vers lui et pressa une combinaison de touches. Apparut alors à l'écran le HTML de la page, ce langage multi-plateforme caché qui permet d'afficher un contenu quel que soit le système d'exploitation de la bécane. Bobby examina lentement la succession de lignes.

Puis s'arrêta.

— Attends un peu...

Il re-bascula sur l'affichage normal, qu'il fit défiler jusqu'à la toute fin du document.

— OK, fit-il en hochant la tête. C'est pas grand-chose, mais c'est un début. (Il indiqua l'écran.) Tu ne vois rien ici ? En dessous du texte ?

— Non. Pourquoi ?

— Parce qu'il y a bien un truc. Quelques mots écrits exprès de la même couleur que le fond. On ne remarque leur présence qu'en regardant le code ou en imprimant le document.

— Si on est réceptif à ce baratin, en somme. Et quels sont ces mots ?

Il repassa en mode HTML, puis sélectionna un bref segment tout en bas. Au milieu du charabia on pouvait lire :

‹font colour="#339966"›L'Homme Debout‹/font›

— L'Homme Debout, articulai-je. Qu'est-ce que c'est que ce blaze ?

TROISIÈME PARTIE

Nous avons l'histoire sur les talons.
Elle nous suit comme notre ombre,
comme la mort.

Marc Augé
Non-lieux : introduction à
une anthropologie
de la surmodernité

CHAPITRE 22

Quand l'avait-elle entendu pour la première fois ? Sarah ne le savait plus. Peut-être un ou deux jours plus tôt. Il s'approchait lentement, attendait le bon moment. Elle croyait savoir qu'il était venu la nuit précédente, avant de s'éclipser en devinant qu'elle flairait son approche. Elle s'était demandé s'il ne passait pas aussi dans la journée, parfois, mais dans ces heures-là elle avait l'esprit plus alerte et parvenait à se convaincre qu'elle se faisait des idées. Puis, par une fin d'après-midi, elle l'entendit juste au-dessus de sa tête et se dit que, s'il passait désormais en journée, la situation devenait critique.

Le psychopathe lui avait rendu visite une heure ou deux avant que ça ne se produise. Il lui avait parlé sans relâche. Parlé, parlé, parlé. De cueillette. D'un fléau. Puis d'un coin d'Italie baptisé Castenedolo, un nom qui sonnait comme une destination de vacances, où l'on boirait des cocktails raffinés, où l'on pourrait peut-être même *manger*, des spaghettis, du salami, un steak, du calmar ou de la soupe, mais à l'évidence il ne s'agissait pas de ça. C'était juste un endroit où l'on avait découvert les restes d'un type, et le fait qu'il se trouve là prouvait qu'il était fait de Plasticine ou de Pliocène, et qu'il avait au moins deux millions d'années et qu'est-ce qu'elle pensait de ça, hein ?

Franchement, Sarah n'en pensait pas grand-chose. Elle faisait de son mieux pour suivre ce qu'on lui racontait, mais depuis la veille elle avait la nausée les trois quarts du temps. Elle avait renoncé à quémander de la nourriture : elle n'était même plus affamée. Elle émettait de petits bruits, des borborygmes, lorsque le type se taisait comme s'il attendait une réponse. Dans l'ensemble elle jugeait ses méthodes d'enseignement assez efficaces, et ses profs de collège auraient gagné à s'en inspirer. La moitié de ses camarades n'apprenaient rien de la journée et voyaient l'école à mi-chemin entre une amicale de quartier et un podium de défilé. Les séquestrer sous un plancher pour les inonder de paroles redresserait peut-être leurs priorités, songea Sarah.

Les listes de vocab' espagnol rentreraient toutes seules, qui sait ? Elle soumettrait peut-être l'idée à maman en vue de la prochaine A.G. de parents d'élèves. Mais attention : il fallait quand même avaler quelque chose de temps en temps, sans quoi l'élève ne pouvait rester concentré.

Patient, le type la laissa surmonter une interminable quinte de toux, avant de reprendre son verbiage. Le sujet était Stonehenge, à présent, et Sarah écouta un moment, car Stonehenge était en Angleterre, et même si elle n'avait pas visité ce coin-là elle savait qu'elle aimait ce pays. L'Angleterre était cool et possédait de bons groupes. Mais quand il s'embarqua dans un laïus comme quoi Stonehenge n'était pas tant un observatoire qu'une carte de l'ADN humain, elle décrocha.

À la fin il lui donna un peu d'eau. La période où elle disait non avait fait long feu. Quand bien même elle eût voulu camper dans son refus, son corps aurait passé outre. La troisième fois sa bouche s'était ouverte sans même l'intervention du cerveau. L'eau avait un goût propre, pur et délicieux. Elle se souvenait d'un temps où ce goût lui avait semblé nouveau. Cela paraissait une éternité.

— Voilà une gentille fille, avait dit l'homme. Tu vois, tu n'es pas maltraitée ici. J'aurais pu te pisser dessus et il aurait quand même fallu que tu boives. Écoute ton corps. Écoute ce qu'il y a à l'intérieur.

— Il n'y a rien à l'intérieur, grinça-t-elle, avant de le supplier une dernière fois : je t'en prie, n'importe quoi. Même des légumes crus. Des carottes, du chou ou des câpres.

— Tu persistes, hein ?

— Je t'en prie, répéta-t-elle, sentant que ses tempes s'atomisaient. Je me sens pas bien et tu dois me nourrir, sinon je vais mourir.

— Tu as de la suite dans les idées, toi. C'est la seule chose qui me fasse encore espérer.

Il n'avait pas explicitement refusé sa demande, mais embrayé sur le végétarisme, pour expliquer que cette pratique était une hérésie car les humains ont une denture d'omnivores, et que le refus de la viande était dû au fait que les gens écoutaient trop leur esprit, qui était infecté, et pas assez leur corps. Sarah le laissa disserter à sa guise. Elle-même n'était pas trop copine avec les végétariens, notamment parce que ceux qu'elle connaissait se donnaient toujours de grands airs, comme cette Yasmin Di Planu qui s'épanchait sur les droits des animaux à longueur de temps mais possédait la plus belle collection de chaussures du bahut, pour la plupart élaborées à partir de choses qui avaient jadis pu se mouvoir de leur propre chef avant d'être attachées à ses jolis petits pieds.

Après l'avoir laissée boire, il referma la trappe et s'en alla. Sarah était restée deux heures durant dans un état de parfaite lucidité, ce qui rendit la suite encore plus effrayante. Elle savait qu'elle était restée lucide car elle avait songé à la fuite. Non à s'échapper vraiment – elle avait cessé d'y croire, même si cela avait occupé la majeure partie de ses cogitations au début. Elle s'était d'abord imaginée trouvant la force de crever d'un coup le plancher au-dessus d'elle, comme une personne rongée de haine d'avoir été inhumée trop tôt. Puis elle avait envisagé de parler au type, de l'amadouer – elle avait du charme et le savait : à l'école, il y avait des garçons qui buvaient ses paroles, sans parler du serveur du Broadway Deli qui n'avait cessé de revenir à leur table sans raison, et pour une fois ce soir-là ce n'était pas de Sian Williams qu'un être à pénis tentait d'attirer l'attention. Enfin, elle avait caressé l'idée de débattre avec le type de façon rationnelle, ou tout bonnement de lui ordonner de la relâcher. Chacune de ces options avait lamentablement échoué. Pour finir, Sarah s'était mise à rêver que son père venait à sa rescousse. De temps en temps, elle se prêtait encore à cette pensée, mais moins souvent.

Et puis elle avait entendu du bruit dans la salle. Elle l'avait d'abord attribué au type, avant de comprendre que c'était impossible : il y avait bien trop de pieds. Ces pieds avaient arpenté la pièce de long en large, sillonné le plancher juste au-dessus de son front. Puis avaient stoppé pile au-dessus de la trappe. S'étaient alors répandus des bruits semblables à des rires, tantôt aigus, tantôt profonds ou retenus. La chose avait remué d'avant en arrière pendant un certain temps, émettant des sons désagréables, des grognements et une espèce d'aboiement bizarre ; des parties de sa carcasse s'étaient posées, tandis que d'autres glissaient en crissant. Un gémissement, enfin, qui ne semblait pas l'œuvre d'une seule gorge mais de plusieurs à la fois, comme si la créature était dotée de plusieurs bouches.

Un long silence, puis la chose était repartie.

Sarah reposait avec les yeux grands ouverts. La situation prenait à l'évidence une mauvaise tournure. Très mauvaise. Il ne s'agissait pas du type, ou alors il avait muté. La créature qu'elle avait entendue incarnait ses pires angoisses, et voilà qu'elle se pointait en plein jour, sans attendre le bon moment.

Pas de doute possible. Ce ne pouvait être que Touche Dubois lui-même.

CHAPITRE 23

Nina quitta la maison de bonne heure, en laissant un mot disant qu'elle appellerait. Zandt passa la matinée à faire les cent pas dans le patio. Chaque matin, les chances de retrouver Sarah Becker saine et sauve s'amenuisaient. Mais savoir cela n'était d'aucun secours.

Il reconsidéra la théorie qu'il avait soumise à Nina, sans parvenir à lui trouver de faille. Elle reposait avant tout sur des spéculations, il en était conscient, et il s'y accrochait pour des raisons toutes personnelles. Si le type qu'il avait tué était bien l'auteur des enlèvements, s'il avait kidnappé les filles pour les remettre à un tiers tout en sachant qu'elles allaient mourir, Zandt entrevoyait la possibilité d'accepter son propre geste. Ces deux années de solitude lui avaient appris une chose essentielle : quand tu es en paix avec toi-même, le jugement des autres ne t'atteint plus. L'Homme Debout pensait sans doute la même chose, mais cela n'en était pas moins vrai.

Une bonne dose de café et la vue imprenable du patio triomphèrent petit à petit de sa gueule de bois, qui se mua en un mal-être général qu'il était capable d'ignorer. Les douleurs cervicales et lombaires consécutives à sa nuit sur le divan s'étaient elles aussi dissipées. La mer avait ce pouvoir, même à cette distance.

À la mi-journée, il avait ratissé la maison en quête de nourriture. Rien dans le frigo. Rien dans les placards ni au congélo. De mémoire, Zandt ne connaissait aucune femme n'ayant pas même un paquet de biscuits chez elle, ou des tranches de pain au freezer, prêtes à griller. Il avait l'impression que la plupart des nanas se nourriraient exclusivement de toasts si elles en avaient la possibilité. Dans son infortune, il se retrouva à errer dans le salon, où il examina le contenu des étagères. Des ouvrages sur les crimes en série, grand public ou plus spécialisés ; des séries de documents sur la psychologie criminelle ; des notes d'enquêtes photocopiées, rangées dans des dossiers et classées par États – une infraction caractérisée. Quelques romans, dont aucun n'était récent, la plupart dus à certains Harris, Thompson, Connelly et King. Bref, rien qui ne se rapporte pas, de près ou de loin, à la face sombre du comportement humain. La maison était un cadre familier, suite aux après-midi que Zandt y avait passés en 1999, durant ces heures où la criminologie avait été le cadet de ses soucis. Il s'était réconcilié avec cette phase de son existence depuis un bout de temps. Jennifer n'en avait jamais eu vent, et cette liaison n'avait affecté ni ses sentiments à son endroit ni l'issue de leur mariage.

Il attrapa l'un des recueils de notes et le feuilleta machinalement. La première partie détaillait les activités d'un dénommé Gary Johnson, qui avait violé et tué six vieilles dames en Louisiane au milieu des années 1990. Un billet épinglé en première page indiquait que Johnson purgeait actuellement six perpétuités cumulées dans une prison que Zandt savait sortie tout droit de l'enfer : un donjon rempli d'hommes dangereux dont les rares restes de douceur étaient généralement réservés à leurs pauvres mères. À vrai dire, c'était un miracle si ce Johnson était encore en vie. Un point pour les gentils. La partie suivante concernait une investigation en Floride qui, à la date des derniers commentaires, était toujours en cours. Sept jeunes hommes disparus.

Et un point pour les tueurs. Un sur combien...

Il piocha un nouveau dossier.

Deux heures plus tard, il était assis par terre, cerné de papiers, quand il entendit frapper à la porte. Il releva la tête, désorienté. Il fallut quelques coups supplémentaires pour qu'il reconnaisse leur nature.

Il se leva et ouvrit à un petit homme mal peigné, derrière lequel stagnait une voiture qui avait jadis eu de l'allure.

— Taxi, annonça-t-il.

— Je n'ai pas réservé de taxi.

— Je sais, merci. C'est la dame qui l'a fait. Elle m'a dit de venir vous prendre ici. Le plus vite possible. Vous emmener sur-le-champ.

— Quelle dame ?

Zandt se sentait perdu, l'esprit retenu par tout ce qu'il venait de lire. Quelque chose là-dedans le titillait.

Fulminant, le type plongea la main dans sa poche, brandit un papier froissé qu'il lut à voix haute.

— La dame, c'est Nina. Elle me dit de vous dire de vous dépêcher. Vous avez peut-être trouvé un truc, ou alors c'est elle qui a trouvé un truc, un homme à bout – je n'ai pas trop compris ce passage. Mais on part tout de suite.

— Où ça ?

— À l'aéroport, m'sieur. Elle m'a dit de faire très vite et m'a payé trois fois la course, et moi j'ai besoin de cet argent, alors on peut y aller, oui ?

— Attendez-moi ici, dit Zandt.

Il retourna au salon, décrocha le téléphone et composa le numéro de portable de Nina.

Elle répondit à la deuxième sonnerie. Il y avait du bruit autour d'elle, ponctué par une voix sourde et impérieuse de type annonce publique.

— Qu'est-ce qui se passe ? demanda-t-il.

— Tu es dans le taxi ?

Elle avait l'air tout excitée, ce qui pour une raison quelconque agaça Zandt.

— Non. Qu'est-ce que tu fabriques à LAX ?

— J'ai reçu un coup de fil du type que j'ai préposé au Web. On a un résultat avec l'« Homme Debout ».

— Ce ne sont que trois mots, Nina. Ça pourrait être le titre d'une expo photo de Robert Mapplethorpe. J'imagine que les fédéraux sont déjà sur le coup, de toute façon.

— Ce n'était pas une recherche officielle, avoua-t-elle avec un brin de gêne. J'ai fait ça en parallèle.

— Je vois. Tout s'explique.

— Il a repéré l'adresse IP de l'ordinateur qui a émis la requête et réussi à identifier sa ligne d'accès. Allez, John ! C'est la première fois que ça arrive en deux ans. Je n'ai jamais diffusé la lettre que tu as reçue, alors pour la terre entière, c'est toujours le Garçon de Courses.

Une nouvelle annonce publique envahit la ligne. Zandt attendit la fin pour répondre :

— Je l'ai dit à Michael Becker.

— La piste ne se situe pas à L.A., répliqua Nina.

— Où ça, alors ? *Où ?*

— Dans le nord de l'État. Un patelin près de la frontière de l'Oregon. Dans un Holiday Inn.

— Tu as appelé le bureau local ?

— Le directeur du coin me hait. Aucune chance qu'il m'envoie quelqu'un.

Génial, songea Zandt. Et dans l'hypothèse, certes peu probable, où il ne s'agirait pas d'une fausse alerte, tu tiens à procéder toi-même à l'arrestation... Dans l'embrasure de la porte il voyait trépigner le chauffeur de taxi.

— Trop risqué, Nina.

— Y aura toujours les flics du coin pour m'escorter. Peu importe. Écoute, Zandt, il y a un avion qui décolle dans quarante minutes. Je serai dedans, et j'ai acheté deux billets. Tu viens, oui ou non ?

— Non, dit-il avant de raccrocher.

Il regagna l'entrée, expliqua au type qu'il n'irait nulle part, et lui remit assez d'argent pour qu'il décampe.

Puis il poussa un juron, rafla son manteau et une poignée de dossiers, et parvint à se jeter devant le taxi avant que celui-ci n'eût quitté l'allée. Il estimait en avoir assez gros sur la conscience pour ne pas y ajouter Nina.

Et ce n'était pas du tout pour la protéger.

CHAPITRE 24

En ouvrant les yeux à 9 heures le lendemain matin, étalé sur le lit, comme tombé du haut d'une falaise, je vis que Bobby m'avait laissé un mot sur la table de chevet. Il m'invitait à le retrouver dans le hall dès que possible. Après un passage sous la douche qui me rendit un semblant d'humanité, j'enfilai les couloirs tel un paresseux contraint de marcher sur ses pattes arrière, un paresseux dans un jour sans. Après cette nuit de sommeil, je me sentais différent, à défaut de me sentir mieux. Mon esprit demeurait brumeux, engourdi, comme rempli de glace pilée et d'un alcool inconnu.

La réception était quasi vide : juste un couple appuyé au comptoir. De la musique douce suintait en fond sonore. Bobby était prostré au milieu d'un large sofa, plongé dans le journal local.

— Yo, murmurai-je en me plantant devant lui.

Il releva les yeux.

— T'as une mine à gerber, mon ami.

— Et toi t'es frais comme un putain de gardon. Sérieux, comment tu fais ? Chaque soir tu grimpes dans un œuf pour renaître le lendemain ? Tu pratiques certains exercices ? Dis-moi tout. Je veux devenir comme toi.

Dehors le ciel était clair et dégagé, et c'est tout juste s'il ne me fit pas gémir. Je suivis péniblement Bobby à travers le parking, me recouvrant les yeux avec la main.

— Ton téléphone est allumé ? Et rechargé ?

— Oui, dis-je. Mais entre nous, je n'en vois pas trop l'utilité. Soit Lazy Ed n'est pas rentré, auquel cas on perd notre temps à retourner là-bas, soit il est rentré et il n'a pas envie de parler.

— Fous êtes très négatif, Vard, répondit Bobby en singeant l'accent allemand. File-moi les clés, je vais conduire.

— Je me sens négatif, oui. Heureusement, j'ai un joyeux robot à mes côtés. Mais si tu refais cette voix-là, je te saigne.

Je lui tendis les clés.

— Ne faites plus un geste.

Cette phrase fut prononcée d'un ton net et ferme, et ne provenait pas de la bouche de Bobby. Nous échangeâmes un regard avant de nous retourner.

Quatre individus se tenaient derrière nous. Deux flics en uniforme, des locaux – l'un maigre et soigné, proche de la soixantaine ; l'autre avait trente ans de moins, et un bon mètre de tour de taille. Sur le côté, un peu en retrait, un type vêtu d'un long manteau. Plus près de nous, à trois mètres environ, une femme gracile dans un joli tailleur. De tous, c'était de loin la plus intimidante.

— Posez les mains sur le toit de la voiture, dit-elle.

Bobby répondit d'un sourire sinistre, sans bouger ses mains d'un iota.

— S'agirait-il d'une sorte de blague ?

— Vos mains sur cette putain de voiture, tonna le jeune flic.

Il rapprocha la main de son arme, brûlant de s'en servir. De l'empoigner, du moins.

— Lequel d'entre vous est Ward Hopkins ? demanda la femme.

— Les deux, rétorquai-je. Une sorte de clonage bizarre.

Le jeune flic fondit sur nous. Je levai une main à hauteur de torse, qu'il vint percuter.

— On se calme, dit la femme.

L'adjoint ne pipa mot, mais cessa d'avancer, se contentant de soutenir mon regard.

— OK, fis-je sans baisser ma paume, mais sans pousser non plus. N'envenimons pas les choses. Commissariat de la ville, je présume ?

— Exact, répondit la femme tout en dépliant son badge. En ce qui les concerne. Moi, je suis agent fédéral. Alors on se détend, et on pose gentiment ses mains sur la voiture.

— Je ne crois pas, non, rétorqua Bobby d'un air détaché. Vous savez quoi ? J'appartiens à l'Agence.

La femme cligna des yeux.

— Vous êtes de la CIA ?

— Oui, m'dame, railla-t-il avec une fausse courbette et un accent péquenaud. Manque plus que les gars de la marine et on pourra faire une chouette parade.

Il y eut un instant de flottement. Le jeune flic se tourna vers son collègue, qui à son tour leva un sourcil en direction de la femme. Plus aucun n'affichait l'assurance de la seconde précédente. Au deuxième plan, le type au manteau secoua la tête.

Je décidai de baisser mon bras.

— Il est de la CIA, mais pas moi, dis-je en choisissant pour une fois d'être coopératif. Je suis un simple élément de la société civile. Nom : Ward Hopkins. Qu'est-ce que vous me voulez ?

— Attends un peu, intervint Bobby tout en désignant de la tête le jeune flic. Et si tu reculais de quelques pas, Rambo ?

— Je t'emmerde, riposta le flic d'une voix égale.

La femme me regardait toujours.

— Hier soir nous avons intercepté une requête Internet, expliqua-t-elle. Portant sur les mots l'« Homme Debout ». On a remonté jusqu'à votre compte, puis jusqu'à cet hôtel. Nous cherchons quelqu'un qui réponde à ce surnom.

— Je croyais que c'était moi qui vous intéressais.

— Hier encore, je ne savais même pas que vous existiez.

— Et pourquoi cherchez-vous l'Homme Debout ?

— Mêlez-vous de vos affaires, protesta le jeune flic. Dites, madame, vous comptez arrêtez ces deux trous du cul ou non ? Parce que si c'est pas le cas, je m'en tape, de leurs salades.

— Tu fais comme tu le sens, répondis-je. Tu peux essayer de nous choper, ou tu peux faire demi-tour. Si tu choisis la première solution, eh bien, fais comme chez toi, mais personnellement je te le conseille pas.

Le flic âgé sourit.

— Tu nous menaces, fils ?

— Non. Je suis bien trop gentil pour ça. Mais Bobby n'est pas très civilisé. Il y aura bientôt du sang à travers tout le parking, et ce ne sera pas le nôtre.

Super-manteau marmonna ses premiers mots :

— Génial... Mille bornes pour rencontrer deux petits merdeux.

La femme ignora la remarque.

— L'Homme Debout a tué au moins quatre jeunes filles, peut-être davantage. En ce moment même, il en détient une autre qui est peut-être en vie, mais on a très peu de temps pour la retrouver.

Bobby la dévisageait, la bouche entrouverte.

— Quoi ? réagit-elle. Ça vous évoque quelque chose ?

— Tu vas te faire pigeonner, Nina, avertit Super-manteau. Tu connais les barbouzes.

Cette remarque ramena Bobby sur terre – suffisamment pour qu'il rabatte sa mâchoire, mais pas assez pour déclencher une bagarre. La femme me regarda.

— Dites-moi ce que vous savez.

— OK, fis-je. Il se pourrait qu'on ait des choses à se raconter.

Le vieux flic en uniforme se racla la gorge.

— Je me demandais si vous aviez encore besoin de Clyde et moi, mademoiselle Baynam.

Nous prîmes une table près de la fenêtre du pseudo-bar de l'hôtel. La pièce était de bonne taille, et d'aspect neuf, mais elle avait l'ambiance d'une jarre à biscuits vide. Bobby et moi nous assîmes côte à côte, en face de la femme. Le type au manteau – qu'on nous avait finalement présenté comme un membre du LAPD, sans révéler son nom – s'installa un peu à l'écart, montrant bien que dans un monde idéal il ne serait même pas assis dans cet État. Les autorités du coin s'étaient déjà taillées dans leur véhicule pour s'envoyer des pancakes et fanfaronner sur la façon dont ils nous auraient rossés s'ils avaient pu.

Je saisis la liasse de papiers de Bobby et l'étalai sous les yeux de la femme.

— Si vous voulez savoir pourquoi nous cherchions l'Homme Debout, la réponse est ici. En fait, nous cherchions tout autre chose, mais voilà ce que nous avons trouvé.

Elle parcourut en vitesse les trois feuillets. Sa lecture terminée, elle les tendit au type.

— Et que cherchiez-vous, alors ? demanda-t-elle.

— Un groupe baptisé les Hommes de Paille. Bobby a débusqué un site Web qui répondait à cette requête. L'« Homme Debout » était l'étape logique suivante. On n'en sait pas plus.

— Et ceci concerne l'Agence ?

— Non, répondis-je. C'est personnel.

— Il y avait un bouton LIENS en bas de la dernière page, reprit-elle. Ça débouchait sur quoi ?

— Quel bouton ? m'étonnai-je.

— Oui, je l'ai découvert après que tu t'es endormi, expliqua Bobby d'un air penaud. Dissimulé dans un segment de java planté. J'aurais dû le repérer plus tôt.

— Et ça débouchait sur quoi ?

— Des tueurs en série. (L'homme au manteau releva les yeux.) Justes des sites de fans. Des pages remplies d'infos sur des assassins, laborieusement tapées par des neuneus qui n'ont pas assez d'ambition pour menacer réellement la société.

— Vous pourriez me remontrer la première page ? demanda la femme.

Bobby secoua la tête.

— Elle n'y est plus. J'ai voulu y retourner quand j'en ai eu marre de mater ces tronches de cinglés toutes floues. Dossier disparu du serveur. Retranché on ne sait où.

— Vous n'avez pas marqué les liens proposés ?

Bobby haussa les épaules.

— Je n'en voyais pas l'intérêt. Ce n'était qu'un ramassis de paranoïaques fascinés par les serial killers.

— Il y a eu une fuite, déclara le type au manteau en rendant les documents à la femme. Des sites de fans, effectivement. Et ça s'arrête là. Le vrai nom du Garçon de Courses a fini par s'ébruiter, et un psychopathe du dimanche l'aura repris pour baptiser cette daube. Un univers interactif pour ceux qui veulent prendre leur panard sur les statistiques de tueurs, avec des adresses de sites qui se déplacent comme des fantômes. Le Net regorge de cette merde. Des clubs de cannibales montés par des branquignols même pas foutus de gagner quelques biffetons chez McDo.

Je l'interrogeai du regard.

— Le Garçon de Courses ?

— C'est ainsi que la presse a surnommé le type que nous recherchons.

— Non, vous le cherchez encore ? m'étonnai-je.

— Et on le fera jusqu'à sa mort. Je vais m'en griller une, Nina. Puis je propose qu'on regagne la civilisation.

Il se leva et quitta la pièce.

— Il voulait dire « arrestation », murmura la femme après son départ. Dans sa tête, il pensait « arrestation ».

— Mais bien sûr, railla Bobby. Si vous voulez mon avis, votre copain mérite d'être tenu avec une très bonne laisse.

— C'est quoi, cette histoire d'Hommes de Paille ? demanda-t-elle.

— Raconte-lui, Bobby, dis-je en me levant.

— Vas-y mollo de ton côté, m'enjoignit-il. Pense à ce que je viens de dire.

Je les laissai pour regagner la réception. Le type au manteau était dehors, à quelques mètres de l'entrée principale.

— Vous avez une cigarette ?

Il me dévisagea longuement, puis fouilla dans sa poche. Ma tige allumée, nous restâmes un moment silencieux.

— Vous êtes flic, pas vrai ? finis-je par demander. (Il ne répondit rien.) Pas vrai ?

— J'étais flic. Je ne le suis plus.

— Admettons. Mais j'habitais San Diego à l'époque. Je lisais le journal. Je me souviens notamment d'un certain flic censé être la terreur des serial killers. Il a raté sa proie, et on ne l'a plus jamais revu. Personnellement, je dirais que c'est vous.

— Vous semblez avoir d'excellents souvenirs sur cette affaire. Vous seriez pas partie prenante, des fois ? Vous vouliez peut-être regarder combien vous aviez de fans ? Vérifier que vous étiez toujours une vedette ?

— On n'aurait pas cette discussion si vous me soupçonniez vraiment. Pas à moi, d'accord ?

Il tira un dernier coup sur son mégot, puis le jeta à travers le parking.

— Et vous, qu'est-ce que vous faites ?

— Je cherche ceux qui ont tué mes parents.

Il me fixa.

— Ces Hommes de Paille dont vous parliez ?

— Apparemment. Ce que j'ignore, c'est s'ils ont un rapport avec le gars que vous poursuivez.

— Aucun, assena-t-il tout en promenant son regard sur la parking. Cette histoire est un ramassis de conneries, une pure perte de temps – un temps dont on ne dispose même pas.

— Votre amie n'a pas l'air de cet avis. Honnêtement, je m'en fiche. Mais il semble qu'à l'intérieur de cet hôtel nous ayons deux personnes liées à des agences fédérales. Qui ont donc certains moyens d'action. De l'autre côté, il y a vous et moi, qui sommes actuellement liés à peau de balle. Alors on peut rester chacun dehors à pisser dans la tente de l'autre, ou bien on peut regarder où mène ce truc, en tâchant de pas trop se marcher sur les pieds.

Il réfléchit un instant.

— Honnête.

— Alors c'est quoi ton nom, mec ?

— John Zandt.

— Ward Hopkins, répondis-je, et nous scellâmes notre pacte d'une poignée de main avant de réintégrer l'hôtel.

À la porte du bar, mon téléphone sonna. D'un signe, je pris congé de Zandt et me rabattis vers la réception.

J'hésitai avant de décrocher, cherchant les mots justes pour accueillir un vieillard terrifié. Mais je n'en trouvais pas. Le mieux à faire était d'écouter ce qu'il avait à dire. Et d'éviter de lui gueuler dessus, sans doute.

Je décrochai et tendis l'oreille, mais ce n'était pas Ed. S'ensuivit un bref échange, puis je remerciai mon correspondant et rangeai mon téléphone.

De retour dans la salle, je les trouvai tous trois attablés, et Zandt semblait un peu plus loquace. La femme posa les yeux sur moi, mais c'est à Bobby que je m'adressai.

— Je viens de recevoir un coup de fil.

— Lazy Ed ?

— Non. La fille de l'hosto.

— Et donc... ?

— Elle a passé l'après-midi d'hier à remuer des archives.

— Tu as dû lui faire une sacrée impression, mon fripon. (Voyant que je ne relevais pas, il ajouta :) Tu veux me dire ce qu'elle a trouvé ?

— Elle a remonté la piste de mes parents jusqu'à leurs lieux de naissance respectifs. Aucun des deux n'était celui qu'on m'avait indiqué.

Ma voix se fendillait. Zandt me jeta un regard intrigué.

— Je n'en suis pas encore arrivé là, expliqua Bobby aux autres, mais il y a un frère ou une sœur dont les parents de Ward ne lui ont jamais parlé.

— Je crois qu'ils ne m'ont jamais dit grand-chose, tu sais. Rien qui soit vrai, en tout cas.

Je sentais les yeux de la femme posés sur moi ; et je songeais que Hunter's Rock et tout ce que j'avais toujours cru savoir n'étaient qu'une sorte de conte qu'on m'avait lu et relu, mais dont je ne me rappelais que le titre.

— Qu'y a-t-il ? demanda la femme.

— Ma mère ne pouvait pas avoir d'enfants.

— Après vous, vous voulez dire ?

— Non. À aucun moment de sa vie.

CHAPITRE 25

Ils nous accompagnèrent au bar. Le jeune Ed nous accueillit sans effusions, se contentant d'affirmer qu'il n'avait pas revu le vieux et ne savait pas où il pouvait se terrer. Il maintint cette version même après que Zandt l'eut pris à part. Je ne pouvais entendre ce que l'ex-flic racontait, mais le langage corporel d'Ed suffit à me convaincre que Zandt maniait un style conversationnel des plus persuasifs.

— Votre ami a vraiment envie d'attraper ce tueur, glissai-je à Nina.

Elle détourna les yeux.

— Si vous saviez...

Zandt prit enfin congé du barman, qui retourna fissa s'abriter derrière son comptoir.

— On perd notre temps ici, maugréa Zandt de retour sur le parking. Le prenez pas mal, les gars, mais je vois pas comment un vieux poivrot va nous aider, Nina et moi, à trouver notre tueur. C'est peut-être une piste pour vous, mais nous, ça ne nous avance pas d'un poil, et chaque minute qui passe rapproche Sarah de la mort.

— Alors qu'est-ce que tu comptes faire, John ? demanda la femme. Rentrer à L.A. et carrer ton cul dans un fauteuil ?

— Ouais. C'est exactement ce que j'avais en tête. Si tu crois que je me suis juste chatouillé le poireau quand j'étais chez toi. Je pense que...

Il secoua la tête.

— Quoi ? fit-elle d'un air agacé.

— Je te dirai dans l'avion.

— C'est bon, je vous laisse seuls un instant, offris-je.

Je les quittai pour retrouver Bobby au pied de notre voiture.

— M'est avis que la sainte alliance prend l'eau, dis-je.

— Alors, c'est quoi, le plan ?

— Arpenter les rues, vérifier les bars, les restaus, la bibliothèque et les lieux publics. En bons professionnels. C'est pas New York, ici. Il n'y a pas trente-six mille planques possibles.

— Tu as bien connu ce mec. Tu n'as pas la moindre idée de l'endroit où il aurait pu se cacher ?

— On n'était pas vraiment intimes, rectifiai-je tout en considérant l'édifice miteux. Je venais ici quand j'étais ado. On échangeait quelques mots, et il me servait de l'alcool. Ça s'arrêtait là.

Je revis à nouveau cette soirée en compagnie de mon père, et la façon dont Ed m'avait offert une bière après son départ, ce qui m'avait mis mal à l'aise. Je comprenais aujourd'hui qu'il y avait peut-être un deuxième niveau de lecture à cette soirée, comme un filigrane. Cette bière qu'Ed m'avait mise sous le nez d'un geste franc : peut-être n'était-ce qu'un élan spontané, mais ça ne lui ressemblait pas. N'était-ce pas plutôt une façon de me dire : « Ouais, je sais comment ton vieux peut être chiant par moments » ? Cela renforçait l'hypothèse selon laquelle c'était Ed qui maniait la caméra sur la deuxième séquence du film, puis comatait pendant qu'on l'utilisait comme bougeoir. Et cela rendait encore plus surprenant le fait que, retombant nez à nez après plus de dix ans, ces deux-là n'aient rien laissé paraître de leur accointance. Il avait dû se produire quelque chose à Hunter's Rock, un événement qui aurait brisé une bande d'amis, puis réuni trois d'entre eux à deux mille kilomètres de là, où ils auraient feint de ne pas se connaître – pas depuis longtemps, en tout cas, et pas avant d'emménager.

Même à moi, ils avaient joué cette comédie, mais comment s'en étonner désormais ? Après tout, si ma mère n'avait jamais pu enfanter, j'étais qui dans l'histoire ?

Par-delà le bâtiment, le ciel était chargé, qui rendait les arbres faméliques et froids. Peut-être était-ce cela, ou bien les senteurs de pin dans l'air frais, qui me restituait un souvenir si fidèle de cette fameuse nuit. Les odeurs ont ce pouvoir-là, plus encore que les images ou les sons, comme si les plus vieilles reliques de notre esprit, celles qui nous figent dans le temps et la mémoire, naviguaient encore à travers les effluves.

— Attends un peu... dis-je en sentant poindre une lueur au fond de mon crâne.

Je fermai les yeux pour laisser germer l'idée. Une chose que Lazy Ed m'avait confiée, le genre de projet qui passait pour une lubie dans la bouche d'un homme peu réputé pour la propreté de ses surfaces.

La lumière se fit enfin.
— Il y a un autre endroit qu'on pourrait essayer.
— Allons-y, dit Bobby.
Je me retournai vers les deux autres. Je voyais que dans sa tête Zandt avait déjà regagné le terminal d'embarquement. La femme semblait moins décidée. Je tranchai la question pour elle. Il y avait un coup à tenter, et je n'avais ni le temps ni la patience de l'expliquer à des tiers.
— Bonne chance, lançai-je avant de monter en voiture et de filer avec Bobby.

La Mare perdue n'est pas perdue, bien sûr. On y accède en s'enfonçant d'un petit kilomètre à pied dans la forêt qui s'étire au nord de Hunter's Rock : un domaine d'État, peu pratiqué sauf par les riverains et quelques randonneurs. C'était un endroit où vous iriez avec la classe vous instruire sur les bestioles et autres merveilles de la nature : un trajet en bus jusqu'à l'orée de la forêt, puis vous marcheriez au milieu des arbres et des feuilles mortes, ravis d'être loin de l'école. Les professeurs s'efforceraient de recentrer les élèves sur le but de la sortie, mais sans acharnement : vous devineriez, au relâchement de leurs épaules, qu'eux aussi apprécient d'échapper au carcan quotidien. Je me souviens avoir vu l'un d'eux ramasser un caillou, quand il pensait ne pas être vu, et le lancer loin devant lui en visant un tronc mort. Il toucha la cible et eut un sourire victorieux. Ce jour-là, je compris que, contrairement aux apparences, les profs étaient aussi des gens normaux.
En grandissant on n'avait plus droit à ces escapades. Les cours privilégiaient dorénavant des choses qu'on pouvait mémoriser, non expérimenter. Mais les gamins continuaient d'aller là-bas à l'occasion, juste pour voir, et c'est là que le nom du lieu prenait tout son sens. Car peu importait combien de fois on vous y avait emmené flanqué de trente camarades jappant en rang deux par deux : dès que vous tentiez d'atteindre la Mare, seul ou avec des copains, elle n'était jamais où vous pensiez. Vous progressiez au milieu des talus boisés, assez sûr de votre coup, puis, au bout de quelques centaines de mètres, la piste disparaissait. Un ruisseau partait en diagonale vers les petites collines, et la plupart des gens atteignaient ce cap. On suivait ensuite ce cours d'eau jusqu'à ce qu'il se jette dans un plus gros, et à partir de là on était condamné à prendre une mauvaise décision. Que l'on pense connaître l'itinéraire par cœur, ou convenir à l'unanimité que ce

devait être *par là*, deux heures plus tard on était revenu au parking, assoiffé, vanné, simplement heureux d'avoir retrouvé la sortie avant la tombée de la nuit, et sans avoir croisé d'ours.

Sauf moi. Un été où je n'avais pas grand-chose à faire, je m'étais assigné comme défi de localiser cette mare. Je devais avoir une quinzaine d'années, c'est-à-dire deux ans avant la soirée billard avec mon père. J'avais procédé de manière scientifique, en épuisant méthodiquement tous les chemins possibles jusqu'à débusquer la mare – et le moyen d'y accéder. Je me perdis en beauté plus d'une fois, mais ce n'était pas une façon désagréable de remplir quelques semaines. Quand on sait où l'on va, une forêt est un endroit sympa. On s'y sent en sécurité, et on s'y sent à part. Le hic, c'est qu'après avoir atteint mon objectif une bonne dizaine de fois, je me rendis compte que le charme était rompu. À quoi sert une mare perdue qui ne l'est pas ? Elle redevient une banale étendue de flotte. Alors je cessai d'y aller. Et puis, avec le temps, je m'intéressais davantage aux endroits pour peloter, or il n'y avait pas moyen d'entraîner une fille dans les bois la nuit – et sûrement pas en quête d'un coin humide que vous saviez plus ou moins retrouver. Ce n'est pas le genre de truc qui attire les nanas. Ou alors c'était moi. Allez savoir.

Bobby et moi avancions l'un derrière l'autre, longeant un bras du ruisseau. Plus de vingt ans s'étaient écoulés, et l'environnement s'était métamorphosé. Les branchages formaient un toit inégal, percé par endroits de rais de soleil froids.

Nous arrivâmes à une nouvelle subdivision du cours d'eau, qui s'enfonçait de part et d'autre de grandes buttes. Je m'arrêtai au sommet d'une d'elles, gagné par le doute. Le secteur n'était pas familier. Et ça commençait à grommeler dans les rangs :

— Et tout ce cirque parce que le gus a dit qu'il *envisageait* de fabriquer une hutte de chasseur, il y a vingt ans de ça...

— Tu peux rentrer à la maison si tu veux.

— Quoi, sans mon fidèle enfant du pays ?

En y regardant à deux fois, je compris les changements survenus dans la végétation. Un des arbres qui me servaient de repère était tombé. Et ça ne datait pas de la veille : ses restes étaient tout pourris et couverts de mousse. Je révisai ma position et m'engageai dans la ravine.

Ses parois étaient raides et nappées de feuilles glissantes. Nous descendîmes avec précaution. Parvenu au fond, je pris à gauche et suivis la faible pente.

— On y est presque, dis-je en pointant devant moi.

Deux cents mètres plus loin, la ravine partait d'un coup à droite.

— Ce doit être juste après ce virage.

Bobby resta muet, et j'en déduisis qu'il s'était, comme moi avant lui, pris au jeu. Les forêts font partie de ces choses que l'on perd de vue pendant un temps, avant de devenir parent et de redécouvrir certains plaisirs, de les voir renaître dans les yeux d'un gamin – comme la crème glacée, les petites voitures et les écureuils. Je me demandai si cela expliquait aussi mon goût pour les hôtels. Leurs couloirs n'étaient-ils pas des sentiers entre les arbres, leurs bars et restaurants de petites clairières pour se réunir et se restaurer ? Des nids de taille et de standing variables, tous hébergés dans la même structure, une forêt privée.

À croire que les écrits de l'Homme Debout ne me laissaient pas tout à fait de marbre...

— Quelqu'un nous observe, dit Bobby.

— Où ça ?

— Sais pas, fit-il en scrutant les buttes qui nous cernaient. Mais il est là, quelque part.

— Je ne vois personne, répondis-je sans tourner la tête. Mais je te crois sur parole. Qu'est-ce qu'on fait ?

— Ne t'arrête pas. Si c'est lui, soit il va péter un câble, soit il va rester cool et décider s'il a envie de nous parler ou non. Si je vois dépasser sa tête du parapet, je lui cours après.

Nous parcourûmes les dernières centaines de mètres en silence, résistant à l'envie de lever les yeux. Au détour du virage le sol s'élevait d'un coup, et nous grimpâmes d'un petit mètre.

Là, devant nous, s'étendait la Mare perdue. Une centaine de mètres sur soixante, assez encaissée, malgré quelques plages boueuses. Une volée de canards nageait au centre, et les arbres surplombaient les trois quarts du plan d'eau peu profond. Je m'avançai tout au bord et y plongeai les yeux. C'était comme de se regarder dans un miroir et surprendre son visage de quinze ans.

— Tu sais où se trouvait la planque ? demanda Bobby.

— Je sais seulement qu'il projetait d'en construire une. Il en a parlé deux, trois fois maximum. Pas pour chasser. Juste pour passer du temps. Ed était un ermite dans l'âme.

— Doublé d'un pervers, peut-être ?

— Non, dis-je en secouant la tête. Personne ne vient là pour niquer. C'est trop lugubre, la nuit.

Il considéra l'espace, examina le sol.

— Si je devais construire une cabane, je le ferais ici. (Il désigna un carré d'arbres et de buissons touffus dominant la pente ouest de la mare.) Point de vue et protection.

Alors j'ouvris la marche le long de la rive, dans la direction indiquée par Bobby. Mon imagination me jouait peut-être des tours, mais le centre de la zone paraissait effectivement plus dense, comme si l'on avait assemblé des matériaux.

C'est alors que retentit le premier coup. Un craquement aigu, précédé d'un sifflement.

Bobby m'agrippa et se mit à courir. Un deuxième tir fusa dans les feuillages moins d'un mètre au-dessus de nos têtes. Trouvant refuge derrière des troncs, je me dévissai le cou à chercher l'origine des tirs.

— C'est quoi, son problème, à ce mec ?

— Attends, dis-je. Regarde là-bas.

Je montrais l'épaisseur dans les fourrés. Une tête en émergeait, celle d'un vieillard. Bien loin de la provenance des coups de feu.

— Merde, grogna Bobby, qui avait déjà sorti son flingue.

Deux types en treillis couraient sur la pente en direction de la mare. Un autre tout en jean approchait par l'autre côté.

— C'est le mec d'hier soir au bar, dis-je. Celui qui nous a bloqués sur le parking.

Les types en kaki avaient atteint l'autre rive. Le plus épais des deux se laissa tomber à genoux et tira droit dans le bouquet d'arbres – des tirs précis, maîtrisés. L'autre contournait la mare à toute allure, obliquant vers le haut de la pente. L'homme en jean faisait feu lui aussi.

— Qui sont ces mecs, putain ?

— Bobby ! Y en a un qui fonce sur Ed.

— Je l'ai, dit-il. Couvre-moi.

Il piqua un sprint. Je dégainai mon flingue, me décalai d'un pas et me mis à tirer.

Le type agenouillé accomplit une belle roulade latérale et se glissa derrière un tronc couché. Je coupai en biais à travers les arbres. Je tirais dans une lumière froide et oblique qui me criblait le visage à travers les branches tordues, m'efforçant de ne pas trébucher dans les racines. Au bout de dix secondes je perçus un cri, et l'homme en jean virevolta avant de tomber sur le dos.

Bobby se frayait un chemin dans les fourrés, tout en visant le type qui redescendait la butte après avoir franchi le col. Il semblait nous ignorer, en dépit des tirs de Bobby ; il n'avait d'yeux que pour la planque de Lazy Ed, qu'il canardait sans relâche.

Je stoppai, me stabilisai, tirai.

La première balle l'atteignit à l'épaule. Bobby logea la suivante une demi-seconde plus tard, et le type fut projeté contre un arbre. Mais il continuait de faire feu, et pas sur nous.

Je pressai de nouveau la détente, et l'atteignis en pleine poitrine. Bobby s'était arrêté lui aussi, qui vida trois autres cartouches. Le type disparut de notre vue.

Je fis un pas en avant, mais Bobby leva sa paume, m'intimant de rester où j'étais. Il s'éloigna avec prudence.

— Ed ? appelai-je. Tu n'as rien ?

Soudain le type en treillis refit surface. Il avait progressé de quelques mètres vers le bas de la pente, caché par la végétation. Sous nos yeux ébahis, il remonta sur ses genoux, serrant dans ses mains un pistolet-mitrailleur.

Avant que je puisse esquisser un geste, il rouvrit le feu. Il mourait sous nos yeux, mais trouvait le temps de loger une quinzaine de balles dans les buissons. Il n'envisagea même pas de nous abattre. Comme si nous n'existions pas.

Puis il bascula la tête la première et se tut à jamais.

Bobby rebroussa chemin, le temps de recharger. Je m'élançai en avant, flanquai mon pied dans le mort par acquit de conscience, et me frayai un chemin dans les broussailles.

En plein milieu je découvris les restes d'une cabane. Un ensemble éclaté de bois pourri, d'herbes sèches et de vieilles branches noueuses. Un œil innocent y aurait vu une fantaisie de la nature, tout au plus une vieille ruine, mais jamais la création d'un homme qui aimait se poser dans les bois et contempler la mare. Au centre gisait Lazy Ed.

Je m'agenouillai devant lui, conscient qu'il ne quitterait pas la forêt vivant. Son corps était criblé de trous. Son visage était moins touché, bien qu'une oreille fût arrachée, laissant voir l'os.

— Qu'est-ce qui se passe, Ed ? Qu'est-ce qui se passe, bordel ? Pourquoi ils vous butent les uns après les autres ?

Il tourna la tête de quelques centimètres, me dévisagea. Difficile de le reconnaître derrière les rides et les veines explosées.

— Je t'emmerde, grinça-t-il assez distinctement. Toi et ta putain de famille.

— Ma famille est morte.

— Tant mieux, dit-il, et ce fut son tour.

La hutte n'abritait rien d'intéressant. Quelques canettes vides, un sachet de tabac, une demi-bouteille de tequila bon marché. Je songeai à fermer les yeux d'Ed, puis changeai d'avis, impatient de quitter les buissons.

Lorsque j'atteignis la mare, et le cadavre du type en jean, Bobby redescendait un talus.

— S'est barré, grommela-t-il.

— Et il avait l'air de connaître son boulot. Tu vas bien, toi ?

— Ouais, sauf que j'ai failli me paumer sur le chemin du retour.

— C'est une mare perdue, Bobby. (Mes mains tremblaient.) Seigneur...

— On ne s'attendait pas à ça.

— À qui le dis-tu ! soupirai-je, hébété de me retrouver dans un lieu d'enfance une arme au poing.

Bobby s'accroupit devant le corps, palpa ses poches et trouva un portefeuille. Ni permis de conduire, ni timbres, ni reçus, ni photos – rien du bazar habituel. Juste une quarantaine de dollars.

— Tu as vérifié l'autre mec ?

— Je me suis juste assuré qu'il n'allait pas réactiver son flingue, dis-je. Il portait un gilet pare-balles, mais je reste impressionné par sa résistance. Ce type a mené sa mission avec une ferveur incroyable. Laquelle mission n'avait rien à voir avec nous. Ils auraient pu nous buter sans problème. Non, ils en voulaient à Lazy Ed. On s'est juste trouvés dans le passage. (Bobby opina.) Ed non plus ne portait aucun élément distinctif. Mais vraiment aucun. J'ai retourné le col de sa veste, et regardé dans son froc. Pas d'étiquettes. Elles ont été coupées.

— Ce sont les Hommes de Paille, dit Bobby. Ils les liquident un par un.

— Mais pourquoi ? Et comment nous ont-ils dénichés ?

Il haussa les épaules.

— La minette du FBI y est bien parvenue. Ils s'y sont peut-être pris comme elle. C'est leur site, après tout : chaque connexion leur est notifiée en temps réel, sans qu'ils aient à pirater quoi que ce soit. Ou bien ils étaient sur le coup avant nous, Ward. Dans le cadre d'une vaste opération de nettoyage.

Il me regarda d'un air las, furieux de notre échec.

— En tout cas, ils sont parvenus à leurs fins. Il n'y a plus rien pour nous ici, à part des ennuis, et on a eu notre compte pour aujourd'hui.

Sans un mot nous repartîmes.

CHAPITRE 26

Nina attendait encore que Zandt lui confie le fond de sa pensée : dès l'instant où les deux autres étaient partis, il s'était muré dans son silence. Lorsqu'il avait émergé du taxi à LAX, à défaut d'être aimable il avait au moins paru présent. Mais sitôt établi que les deux pensionnaires du Holiday Inn de Hunter's Rock – quelles que fussent leurs intentions, et des doutes demeuraient à ce sujet – n'avaient aucun rapport avec l'Homme Debout, Zandt n'avait plus pipé mot. Nina se sentait stupide de l'avoir ainsi traîné aux confins de l'État, mais l'erreur valait toujours mieux que l'inaction. Le temps qui passait l'obnubilait, la torturait telle une main lui dépeçant le visage. Alors elle éprouvait le furieux besoin de parler, de faire ou dire quelque chose, n'importe quoi, comme si l'on pouvait hâter le dénouement rien qu'en l'invoquant. Chez Zandt, l'effet semblait inverse : il ne fallait plus grand-chose, devinait-elle, pour que son mutisme franchisse le point de non-retour.

L'avion était quasi vide, et cependant Zandt ne s'était pas assis à côté d'elle. Il était installé de l'autre côté de l'allée, plongé dans de vieux dossiers qu'il avait pris à la maison. Elle contacta le bureau de Brentwood, pour apprendre qu'il n'y avait rien de nouveau là-bas, sans dévoiler à ses collègues qu'elle ne se trouvait pas tout à fait au coin de la rue.

Puis elle détourna la tête vers le hublot, observa la terre qu'ils survolaient pour regagner L.A., se demandant s'ils passaient au-dessus de la résidence, de la tanière ou de la cabane, quel que soit le terme qu'il préférait, de l'Homme Debout. L'idée que Sarah Becker se trouvait peut-être là, dans ce paysage, était insoutenable. Alors elle attrapa devant elle le magazine de la compagnie et s'efforça de lire.

Zandt avait à peine conscience de se trouver à bord d'un avion, et ne pensait même pas à Sarah Becker. Il méditait sur quatre disparitions survenues à divers endroits du pays en l'espace de trois ans. Elles avaient peu de chose en commun, sinon qu'elles étaient présentement répertoriées sur les genoux de Zandt. Mais, dans l'hypothèse d'une sorte de sous-traitance – sa dernière théorie en date –, les méthodes applicables aux crimes en série perdaient de leur validité. Face à des enlèvements multiples ou à des cadavres regroupés dans une zone réduite, il semblait avisé de limiter à ladite zone la recherche d'indices ou de faits corollaires. La plupart des tueurs possédaient leur territoire de chasse, une poignée de kilomètres carrés où ils se sentaient en confiance. Certains restreignaient leur rayon d'action à quelques pâtés de maisons, voire deux ou trois rues – notamment s'ils sévissaient dans des milieux sociaux n'inspirant aux autorités qu'un intérêt mesuré. Zandt se rappelait les images de la démolition de l'immeuble de Jeffrey Dahmer, là où des garçons noirs et asiatiques avaient été, dans un ordre ou dans un autre, démembrés, adorés et mangés. Les familles des victimes assistaient à l'événement, la plupart en silence, quelques-unes en pleurs – mais bien peu demandaient des comptes à qui pouvait les écouter, ou cherchaient une raison d'accepter que leurs enfants soient assassinés sans que personne s'en émeuve.

Deux disparitions constatées dans des régions diamétralement opposées étaient rarement mises en regard, même après l'entrée en scène du FBI, a fortiori si elles s'étaient produites peu ou prou au même moment. On n'allait pas kidnapper une personne à San Francisco le mardi soir, puis une autre à Miami dans les premières heures du jeudi.

Du moins, pas tout seul. Zandt s'était mis en quête d'enlèvements présentant des similitudes avec ceux imputés à l'Homme Debout, et survenus à la même époque. Mais il ne s'attendait pas à trouver d'autres étoffes brodées d'un prénom de fille : l'Homme Debout était assez malin pour faire croire que ses crimes dans la Cité des Anges formaient un ensemble indépendant.

Voilà le constat qui l'avait saisi au moment où le taxi était venu le prendre pour LAX : ces pulls étaient trop voyants. Et s'ils procédaient, en définitive, moins d'un fantasme morbide que du désir d'isoler quelques cas parmi une multitude ? L'Homme Debout aurait-il fait le pari que la police serait tout autant impressionnée par cette signature que le public face aux films montrant des chrysalides dans la gorge des macchabées, et aux séries télé où les méchants crient sur tous les toits leurs plus intimes psychoses ? Vous avez un pull avec un nom dessus, alors c'est l'un des nôtres. Vous n'en avez pas, ça ne nous intéresse pas.

Notre gars présente une pa-tho-lo-gie. C'est l'un des rares indices à notre disposition, alors nous l'explorons à fond, et vous seriez gentils de nous laisser travailler, maintenant.

Zandt jugeait fort probable que l'Homme Debout ne souffre d'aucune pathologie particulière, et dès lors échappe à toute tentative de profilage. Peut-être continuait-il en ce moment même à cueillir des victimes n'importe où dans le pays. Voire n'importe où dans le monde. Juste parce qu'il en avait envie.

Ses victimes ne relevaient d'aucune catégorie précise. On surestime la beauté parce que la beauté rend les gens reconnaissables, leur donne l'air célèbre. Et d'après Zandt les cheveux longs n'étaient pas un meilleur critère. S'il voyait juste en considérant les pulls comme un écran de fumée, la longueur des cheveux n'était plus une fin, mais un moyen. Ne restaient alors que deux éléments objectifs. L'âge, d'abord. Les jeunes enfants disparaissent par légions, et l'on ne compte plus les vieillards qui se font agresser chez eux. Ces deux groupes sociaux croisent malgré eux la route des statistiques en raison de leur faiblesse physique. Parmi le reste de la population, la plupart des femmes kidnappées ont autour de vingt ans : suffisamment jeunes pour mener une vie indépendante ; des femmes souvent célibataires, que l'on verra rentrer seules le soir, qui auront la fraîche insouciance de venir en aide à un brave homme handicapé par un bras en écharpe, à demi caché dans le coin sombre d'un parking. Des femmes de tous âges se volatilisent, mais c'est dans cette tranche-là que se situe le pic de la courbe. Pourtant, les victimes avérées de l'Homme Debout, ainsi que celles dont Zandt avait le nom sous les yeux, n'étaient âgées que d'une quinzaine d'années. Des filles assez grandes pour opposer une résistance physique à leur agresseur, et trop jeunes pour hanter les lieux à hauts risques. Mais de là à considérer toute une génération comme potentiellement victime, il y avait un pas. Un peu partout dans le pays, des filles de cet âge passaient la nuit dehors, sur le trottoir. En d'autres termes, si l'Homme Debout ou son entremetteur ne regardait que l'âge, il aurait affrété un camion vers ces quartiers pour le remplir jusqu'à ras bord. Au lieu de ça, il avait choisi des filles non seulement moins vulnérables que la moyenne, mais issues de milieux assez protégés. La famille d'Elyse LeBlanc était un peu moins fortunée, mais tout de même ancrée dans la classe moyenne. Cela pour dire que l'Homme Debout ne cherchait pas une simple viande. Il cherchait ce qu'il considérait comme de la qualité.

Zandt fixait le portrait photocopié des filles mortes. Son esprit semblait tourner de plus en plus vite, amalgamant les données listées sous ses yeux à celles qu'il avait assimilées deux ans plus tôt. Les lieux, les noms, les visages... Il tâchait d'y voir un tout, en excluant toutefois sa propre fille qui, il en était désormais persuadé, avait été choisie à seule fin de le punir. Zandt avait tenté par le passé d'ôter Karen de l'équation, mais sans succès. Le poids de sa disparition avait conditionné ses moindres gestes depuis que Jennifer et lui avaient ramassé la lettre sur leur perron. Mais il se penchait aujourd'hui sur un tout nouveau groupe de filles, se demandant si elles étaient liées par davantage que des spéculations, s'efforçant d'oublier le lieu où cet avion le ramenait, où il avait vécu la majeure partie de sa vie, cette étrange ville des marchands de rêves, de la pauvreté, des bouts d'essai, du meurtre et de l'argent... Il s'efforçait de l'oublier pour d'autres endroits, d'autres nuits, d'autres terrains de chasse. D'autres villes, d'autres machines, forêts, immeubles et rivières de béton où d'autres hommes et femmes se languissent de ciels étoilés, cultivent de petites plantes au rebord des fenêtres, et s'entourent de petits chiens qu'ils promènent dans des couloirs où se succèdent à l'infini boîtes, carrefours et lumières artificielles ; où ils louent des mètres carrés pour avoir un endroit où dormir, avant de se lever pour accomplir des tâches à finalité capitalistique dont ils ne comprennent ni ne questionnent le sens, dans le seul but de gagner les jetons nécessaires pour louer l'espace où ils dorment, râlent et regardent la télévision, jusqu'au jour où certains s'esquivent par les fenêtres pour courir les rues sombres en hurlant, rejetant la torpeur imposée par une société elle-même minée de fractures, de trahison et de désespoir ; des fous solitaires dans une culture qui vire babiole de Noël, oripeau criard enveloppant un vide qui gonfle à vue d'œil dans les parkings, les centres commerciaux, les salles d'attente et autres forums virtuels – des non-lieux où personne ne sait plus rien sur personne.

Puis d'un coup le tourbillon cessa.

CHAPITRE 27

La nuit tombait lorsque nous rejoignîmes l'hôtel. Deux messages attendaient Bobby. Pendant qu'il rappelait ses correspondants, j'allumai la télé, sans le son, sur la chaîne d'infos locales, me demandant avec une curiosité malsaine combien de temps il faudrait à nos mésaventures pour devenir un titre. Selon toute probabilité, des marcheurs auraient entendu les coups de feu, puis découvert les cadavres. Même si aucun élément ne nous accusait dans cette affaire, il me tardait de quitter Hunter's Rock.

Je m'activai dans la chambre, rassemblant mon maigre barda.

— Putain de merde... articula Bobby d'une voix bizarre, toujours au téléphone. Allume la télé, Ward.

— C'est fait.

— Pas ces merdes régionales. CNN ou un truc de ce genre.

Je zappai jusqu'à lui donner satisfaction.

La prise de vue était manuelle et tremblante. Un grand bâtiment gris dans un décor urbain. Une école. Visiblement filmée plus tôt dans la journée, car il faisait encore jour.

— On l'a, dit Bobby dans l'appareil. Je te rappelle.

Je remis le son. Le commentaire portait à trente-deux le bilan des victimes, plus de nombreux disparus, car la moitié de l'immeuble restait à fouiller. On ignorait encore si les deux élèves abattus par la police étaient les uniques instigateurs du carnage, ou s'ils avaient bénéficié de la complicité d'un tiers. Ils s'étaient servis de fusils-mitrailleurs et d'un gros engin incendiaire artisanal.

La caméra tournait autour du sinistre, capturant au vol des grappes d'enfants et d'enseignants, le visage hagard dans la lumière blanche des projecteurs. Le fond sonore avait été réduit au montage, mais on percevait encore des sirènes et des sanglots. On vit passer une femme chancelante, soutenue par deux secouristes, le visage entièrement recouvert de sang.

— Où est-ce que c'est ?

— Evanston, dans le Maine.

Bobby ferma les yeux, prit une longue inspiration.

Retour au direct. Sur place, le calme semblait revenu, malgré quelques badauds bloqués devant l'école par un ruban de plastique. Un reporter en manteau marron tenait un micro, brossé par le halo intermittent des gyrophares bleus. On avait exhumé deux corps supplémentaires. Jane Mathews et Frances Lack, toutes deux âgées de onze ans.

De nouvelles images enregistrées. Camions de pompiers, ambulances. Des blessés, enfants comme adultes, étendus par terre, recevant les premiers soins. D'autres dans la même position, mais sans personne qui leur tienne la main. Ceux pour qui l'on ne pouvait plus rien.

— Nom de Dieu ! m'écriai-je en pointant mon doigt sur l'écran.

La caméra balayait le côté opposé de la rue, d'où les passants découvraient ce trou béant sur l'enfer. Parmi eux se trouvait un grand type blond avec un gros sac à l'épaule, vu de dos. Contrairement aux autres, il ne se contorsionnait pas pour mieux voir, mais se tenait impassible. Le cameraman ne le remarqua pas, qui poursuivit son lent panoramique d'épouvante.

— J'ai déjà croisé ce type, dis-je.

Un homme blond, aux Halls, avec un sac de voyage bleu.

Bobby passa une bonne partie du vol suspendu au téléphone. Je l'entendis organiser auprès de trois correspondants successifs un acheminement de cassettes vers l'aéroport de Dyersburg. Puis il raccrocha et contempla son café en silence.

Je me tournai vers lui.

— On est sûr que ces gamins ont agi seuls ?

— Les perquisitions continuent à leur domicile. On n'a encore rien dégotté. Il ne s'agirait pas d'un attentat pour des idées, ce coup-ci, mais de l'œuvre de deux jeunes Américains parfaitement intégrés. Bref, c'est pas la joie.

Je le croyais volontiers : l'atmosphère était morose dans les travées, et le commandant de bord nous avait servi un discours de bienvenue des plus tièdes.

— Je ne t'ai pas entendu raconter nos exploits du jour, glissai-je.

Il émit un rire sardonique.

— Il ne manquerait plus que ça ! « Allô, les gars, on vient de buter deux types dans les bois, et en rentrant à l'hôtel le pote avec qui je me trouve a cru reconnaître un troisième larron à la télé. » Il ne s'agit pas d'un jeu virtuel, Ward, et tu ne leur as pas exactement laissé un bon souvenir. L'Agence a fait un peu de ménage dans ses rangs. Et ils me foutraient à la porte encore plus volontiers que toi.

— Ils ne m'ont pas foutu à la porte. Je me suis barré.

— Poussé au cul par un détecteur de mensonges.

— Peu importe, rétorquai-je. C'était lui, Bobby.

— Tu dis toi-même que tu l'as à peine aperçu là-haut. Sans distinguer son visage.

— Je sais. Mais c'était lui.

— OK, je te crois. (Il prit soudain l'air grave.) Ce qui est bizarre, c'est que moi aussi, j'ai l'impression de le connaître.

— Ah bon ? Mais d'où ?

— Que veux-tu que j'en sache ? Le temps que je comprenne ce que tu me montrais, il avait disparu. Mais il avait quelque chose de familier.

Nous atterrîmes de nuit. La voiture que j'avais laissée au parking n'était plus là, sûrement récupérée par le loueur. Bobby se rendit à un autre guichet pour en retirer une nouvelle. Il ne restait qu'une énorme Ford. Je la sortis du parking avant de faire un crochet pour attendre Bobby devant la sortie principale.

Il émergea enfin du terminal avec une petite boîte sous le bras.

— Cool, fit-il d'un ton morne en grimpant à bord. On pourra caser les gosses et toute une semaine de shopping. Allons trouver un drugstore.

— Au moins, on pourra y dormir en cas de besoin.

— Alors là, n'y compte pas.

— Tu mollis, soldat.

— Absolument, et ça veut dire que je ne suis plus obligé de bouffer des brocolis, pour paraphraser un ex-président regretté.

— Regretté par qui ?

— Par sa mère.

Bobby avait conservé les clés de sa chambre au Sacagawea. S'étant assuré qu'elle ne grouillait pas de bagages inconnus, il partit négocier avec la direction.

Je dénichai deux canettes de thé glacé et les rapportai dans la chambre. Davantage encore que la piscine du motel de Hunter's Rock, celle-ci évoquait les vacances d'antan : plus de cinquante années de voyageurs s'étaient succédé entre ces murs, posant leurs bagages à mi-chemin d'un long périple. Le fauteuil. Je me carrai là où jadis un touriste avait pu découvrir *L'Île aux naufragés* à la télévision, sans que le générique sonne à ses oreilles comme un morceau de folklore rance. Un jour ce sera peut-être au tour d'un être en combinaison spatiale siliconée de s'installer ici, muni d'un soda lunaire sans sucre, sans caféine et sans goût, et de penser la même chose devant un épisode de *Friends* : « C'est quoi, cette bande de rachitiques ? Et qu'est-ce qu'ils fabriquaient avec leurs cheveux ? »

Bobby réapparut avec un magnétoscope mastoc sous le bras.

— Ce vieux chnoque n'avait même pas remarqué mon départ, dit-il. Mais il est reste assez lucide pour exiger une caution pour cette antiquité. J'ai peur qu'il faille la remonter avec une clé.

Une fois l'appareil relié au poste quasi collector de la chambre, Bobby s'assit au bord du lit et creva le colis réceptionné à l'aéroport. Deux cassettes VHS. Il vérifia rapidement les étiquettes, et glissa l'une d'elles dans la machine.

— C'est du matos inédit, prévint-il tout en pressant la touche PLAY. Discrétion de rigueur, donc.

Le cameraman avait atteint l'école dévastée très tôt après l'explosion initiale. Dans la plupart des villes américaines il existe un marché pour les équipes de tournage free-lance, ces tandems qui errent dans les rues comme deux chiens sans collier. Ils se branchent sur la fréquence des flics et arrivent souvent les premiers sur les lieux de suicides, de carambolages et de bars criblés de balles, en quête de rushes sensationnels qui aideront les networks et les chaînes câblées à satisfaire un quota d'images par minute en constante augmentation. La médiocrité des prises de vue suggérait une telle provenance, mais je pouvais me tromper. Face à de telles scènes, je ne jurerais pas que mes mains soient beaucoup plus stables. Quand on découvre des horreurs à la télé, on oublie vite – malgré l'impression de vérité qu'elles dégagent – que l'info a déjà été filtrée pour notre bien. On regarde des gens plantés devant des charniers en Bosnie, et l'aspect brut de l'image nous aide à oublier qu'on ne montre pas ce qu'il y a en dessous, ni ce que ces bribes empoussiérées signifient pour ceux qui sont réellement là-bas, et non protégés par une épaisse paroi de verre dans un living à l'autre bout de la planète. Même la couverture non-stop du cauchemar du World Trade Center nous épargna ce que virent les secouristes. Nous sommes tellement rompus aux documents montés, tellement conditionnés par les tours de passe-passe des médias, que nous repérons mieux les ajouts que les coupes. Peu importe combien de « making of » nous aurons visionnés : dans le feu de l'action, le monstre en latex nous effraiera toujours, et l'on ne se demandera jamais pourquoi le plan de tel reportage prend fin à tel moment, ni ce qui jonchait la surface de tel paysage caché. Il s'agit d'infos allégées, édulcorées avec les moyens du bord. Nous avons le droit d'entendre les cris, mais à un volume raisonnable – tout en écoutant une voix off dont la gravité sera déjà un début de réconfort. « C'est mal, dit-elle en substance. C'est affreux. Mais ça n'arrive pas souvent, et ça va s'arranger. Ça va passer, et tout finira bien. »

La vidéo était livrée sans commentaires ni coupes. Elle ne disait rien. Se contentait de montrer.

La déflagration avait arraché la façade du vaste bâtiment. Ce faisant, elle avait expulsé des tonnes de béton, de verre et de métal à des vitesses très élevées. Ces corps s'étaient heurtés à d'autres du même type, ainsi qu'à des matières autrement molles. Une bonne partie de ces dernières furent pulvérisées dans l'atmosphère. Quand le cameraman était arrivé – flanqué d'un preneur de son qu'on entendait gémir à intervalles réguliers –, il s'était simplement frayé un chemin à travers le parking dévasté bordant l'école. L'objectif virait parfois à droite, vers la loge du bâtiment, ou de l'autre côté du parking pour saisir les premières ambulances et véhicules de police. Mais pour l'essentiel la caméra enregistrait ce qui se trouvait devant elle.

Une fille, visiblement ignorante du fait qu'elle avait perdu un bras, courait en hurlant un prénom. Des restes de corps, des têtes. Un garçon, dont le visage était à ce point ensanglanté qu'on aurait dit un nouveau-né, errait dans la fumée en poussant une sorte de miaulement. Une longue traînée de chair, comme une énorme flaque de vomi sanguinolent, parsemée de quelques membres ou organes identifiables. La majeure partie d'un homme entre deux âges, convulsant à même le sol, la face carbonisée, réduite en une masse rose percée d'un orifice aussi muet qu'inutile. La moitié d'une jolie jeune femme, les yeux ouverts, sans rien en dessous de la cage thoracique hormis un moignon de colonne vertébrale et le capot de la voiture où elle avait atterri.

Peu à peu le fond sonore se modifia : les cris de détresse s'atténuaient au profit des sanglots et des hurlements qui montaient crescendo. Un semblant d'ordre se répandit parmi les gens montrés à l'image. Les mouvements désordonnés faisaient place à des gestes précis, à mesure que les lymphocytes de la société investissaient le terrain et tentaient de le structurer. Certains d'entre eux se montrèrent efficaces : pointant, dirigeant, pansant. D'autres auraient aussi bien pu faire partie des victimes.

Puis nous le vîmes.

À ce stade, l'équipe de tournage avait eu son lot de gore et s'était retranchée vers l'entrée du parking, côté rue. Le preneur de son avait vomi deux fois, le réalisateur une. La foule n'avait pas eu le temps de s'attrouper sur le trottoir d'en face, mais le ruban de signalisation se déroulait déjà, isolant l'événement de notre réalité, le consignant dans le domaine des circonstances exceptionnelles.

Notre homme était déjà présent, toutefois, plus ou moins là où je l'avais aperçu plus tôt. Grand, les cheveux blonds et courts, les pieds plantés dans le sol, il observait les dégâts, considérait le panache de fumée qui s'élevait d'un incendie loin d'être maîtrisé. Bobby fit PAUSE.

L'homme n'avait pas le sourire aux lèvres. Qu'on ne se méprenne pas. L'image sautillait dans tous les sens, et il était impossible de détailler son visage. Disons simplement qu'il regardait.

Nous restâmes muets. Bobby attrapa sa canette, essaya d'en avaler une gorgée, s'aperçut qu'elle n'était pas ouverte. Y remédia, en engloutit la moitié.

— OK, murmura-t-il. La suite, c'est juste au cas où.

Il éjecta la cassette, la maniant du bout des doigts comme un produit contaminé. Il enfourna la suivante et appuya sur PLAY.

— Je me suis procuré ça auprès d'un de nos techniciens en analyse médias. C'est à usage interne, un pense-bête pour les gens de Washington. Un outil marketing, en quelque sorte. Un florilège de certains événements survenus au cours des dix, quinze dernières années, constamment réactualisé.

Je reconnus aussitôt le sujet de la première séquence, pour y avoir eu droit plusieurs fois la semaine précédente. La tuerie en Angleterre. La luminosité laissait à désirer, une lueur de petit matin, mais la caméra était remarquablement fixe, sans doute l'œuvre d'un pro de la BBC. Des groupes qui s'étreignaient. Des secouristes massés autour d'une porte par où l'on sortait des corps, certains recouverts d'un drap, d'autres baignant dans leur sang. Deux autres équipes de reporters endimanchés. Un cordon de police à un carrefour bondé. Peu de cris et de pleurs. On entendait surtout le bruit de la circulation : des gens attendus en réunion, rentrant du club de gym, ou partant livrer des litres et des litres de Coca Light.

Nous n'eûmes pas longtemps à attendre, mais le plan était flou et peu concluant. Un travelling sur le cordon de sécurité, depuis l'intérieur de celui-ci, montrant les badauds attroupés. Parmi eux, un grand blond. Bobby arrêta l'image, la fit avancer, reculer. Le visage était trop petit. Et l'objectif passait trop vite.

— C'est lui, affirmai-je néanmoins.

Pendant les deux heures suivantes nous regardâmes la suite, une tapisserie de mort tout en points lumineux. Je finis par perdre le compte, mais au moins trente carnages défilèrent sous nos yeux, jusqu'à ce que leurs particularités respectives – lieu, sons, mode vestimentaire – s'effacent devant leurs similitudes. La plupart ne nous avançaient pas, mais quelques-unes comportaient un élément suffisamment proche pour qu'on l'ajoute à la liste dressée par Bobby sur le bloc à en-tête de l'hôtel.

Une cafétéria de Panama City, Floride, 1996. Une grand-rue du nord de la France, 1989. Un centre commercial de Düsseldorf, 1994. Une école du Nouveau-Mexique, l'année précédente. Une ruelle dans un lotissement de La Nouvelle-Orléans, en 1987, où un présumé trafic de drogue avait dégénéré jusqu'à faire seize morts et trente et un blessés.

— C'est lui, répétai-je encore et encore. C'est lui.

Enfin la cassette s'arrêta, sans cérémonie. A priori très peu la regardaient jusqu'au bout.

— Il nous en faut d'autres, dit Bobby.

— Oh non. Je t'assure que non.

— Si. Pour toutes les dates où il n'apparaît pas ici.

— C'est qu'il ne devait pas y être. Ils sont forcément plusieurs. Ils doivent se partager le boulot.

Je me rendis à la salle de bains et bus environ deux litres d'eau tiède au moyen d'un verre minuscule.

— Les crashes d'avions, reprit Bobby à mon retour. Les attentats en Irlande du Nord, en Afrique du Sud. Les guerres civiles des dix dernières années. Les épidémies de grippe. Il faut bien que quelqu'un les déclenche. On a peut-être cherché du mauvais côté. Il ne s'agit peut-être pas de fondamentalistes d'un camp ou d'un autre, mais de gens qui détestent tout le monde.

Je fis une moue dubitative, mais il avait peut-être raison.

Bobby sortit la cassette de l'appareil et la fit tourner entre ses doigts.

— Mais pourquoi reste-t-il planté là ? s'interrogea-t-il. Et quelle chance a-t-il de se faire piéger par la caméra si souvent ?

— Ce n'est pas une question de chance, répondis-je. Il s'agit d'une signature, adressée à ceux qui savent. Une façon de dire : « Ce sont les Hommes de Paille qui ont fait le coup. »

— Mais on le tient, maintenant.

— Ah bon ? Un type blond, des plans trop courts et trop lointains pour qu'on distingue quoi que ce soit, et une série d'événements sans liens entre eux, répartis sur une dizaine d'années et à travers la moitié du monde occidental ? Tu veux appeler Langley, pour voir si ça intéresse quelqu'un ? Ou bien tu préfères essayer CNN ? On n'est pas Woodward et Bernstein, et ce truc aura l'air d'un délire de conspirationnistes tant qu'il se fondera sur de simples coups d'œil. Même après une journée entière sur ta bécane, tu ne pourras pas poser le moindre nom sur ces images.

— Et la page Web, alors ? Le Manifeste ?

— Il a disparu, Bobby. On aurait aussi bien pu le taper nous-mêmes.

— Alors on laisse tomber, c'est ça ?

— Non. (Je m'assis au bord du lit et décrochai le téléphone de la chambre.) Il y a peut-être une personne qui pourra nous aider. Deux, en fait. La paire qui nous a débusqués à Hunter's Rock.

— Pourquoi ? Ils courent après un serial killer.

— Et quelle définition donnes-tu à ce terme ?

— C'est pas pareil de tuer plein de gens d'un coup.

— Peut-être. Mais personne n'a dit qu'on ne pouvait faire les deux à la fois. Ce mec est un organisateur. Logisticien, incitateur, prêcheur – le type qui choisit les pigeons, mène le projet à bien. Du terrorisme sans revendication. Tuer pour tuer. Puis il se pose et regarde les pompiers extraire les restes humains. Et tu dis que ce genre de type ne donnerait pas dans le meurtre en série ? Je crois que ce type est bien leur tueur. Que c'est lui, l'Homme Debout.

— Enfin, Ward, tu ne pourrais même pas lui coller une prune de stationnement avec de tels arguments !

— Peut-être pas. Mais on a besoin d'aide, et Nina est la seule personne qui me vienne à l'esprit. Ces enfoirés ont tué mes parents. Peu importe ce que je devrai lui raconter pour gagner ses faveurs.

Bobby me dévisagea un instant, puis hocha la tête.

— Vas-y, appelle.

CHAPITRE 28

Par moments, c'était comme d'être morte. Par moments, c'était comme d'être quelque chose d'autre, un poisson ou un arbre ou un nuage ou un chien, un sale clebs. Les chiens sont des maniaques qui prêchent dans les rues mais il vaut mieux être un sale clebs que morte. La plupart du temps, c'était comme de n'être rien du tout, un petit paquet de néant emporté par une rivière sous un ciel où aucun oiseau ne chantait.

Sarah était très malade à présent. Elle se rappelait où et qui elle était de manière épisodique. Son estomac avait renoncé aux crampes. Elle ne le sentait même plus. Elle supposait qu'il comptait toujours parmi ses organes, ainsi que ses bras et ses jambes. Parfois elle en obtenait une preuve accablante, une douleur indicible qui jaillissait de ses orteils pour remonter tout son corps. On aurait dit des aiguilles ou des épingles, sauf qu'elles étaient en fer rougi et longues de trente centimètres, et que quelqu'un les lui enfonçait sous la peau en poussant de toutes ses forces, puis les abandonnait là. La douleur finissait par se dissiper, mais Sarah n'était jamais présente pour cette phase-là : entre-temps elle avait repris le chemin de la rivière, dérivant au gré du courant.

Il arrivait qu'on lui parle tandis qu'elle flottait. Elle percevait des voix, en tout cas. Elle entendait ses amis, parfois sa grand-mère et sa sœur, le plus souvent son papa et sa maman. Ils parlaient généralement de choses futiles, comme s'ils bavardaient pendant qu'elle faisait ses devoirs sur la table du salon. On distinguait rarement tout ce qu'ils disaient. C'étaient des bribes de phrases de-ci, de-là. « Charles pense que Jeff va exploser avec cette version... OK pour un brunch, mais celle-ci pourrait valoir le coup... Ce n'est qu'un truc de troisième acte. » Sa mère évoquait sa journée, disait où elle était allée et qui elle avait vu : « On peut se faire tirer le visage, mais les mains ça trompe jamais. » Puis son père lâchait une pensée qui venait d'éclore dans sa tête, et là Sarah en captait l'intégralité : « Tu sais ce que je ferais si j'étais célèbre ? Je harcèlerais les gens. Je choisirais une cible et je passerais mon temps à m'immiscer dans sa vie. Qui va la croire ? "Eh, monsieur l'agent, Cameron Diaz n'arrête pas de m'embêter." Ou bien... "Écoutez, je croule sous les lettres de Tom Cruise. Sérieux. Il me poursuit. Regardez, c'est son écriture. Je vous jure !" Je pourrais pousser quelqu'un complètement à bout. En un rien de temps. »

Avait-elle vraiment entendu ces mots dans la bouche de son père avant de partir à vau-l'eau ? Elle en doutait. Elle y voyait plutôt une attention d'aujourd'hui, un moyen de lui tenir compagnie pendant qu'elle voguait au fil de l'eau. Papa lui parlait beaucoup, lui confiait les trucs qui lui passaient par la tête. Maman ne comprenait pas toujours qu'il s'agissait de plaisanteries, et ne les trouvait pas souvent drôles. Sarah, si.

Puis les voix finissaient par s'éteindre.

À d'autres moments elle entendait des pas et savait que ses parents venaient la sauver. Elle les écoutait s'approcher, s'approcher encore, prête à les abreuver de paroles dès que la trappe s'ouvrirait sur le visage de son père. Ils se tenaient juste au-dessus d'elle, frottant leurs pieds sur le plancher. Mais ils ne la trouvaient jamais. Les pas s'éloignaient, puis elle repartait à la dérive.

De temps en temps la douleur physique s'emparait d'elle, en général après les visites de Touche. Des haut-le-cœur qui lui tranchaient l'estomac comme un couteau glacé, menaçant de sectionner tout son corps. On ne pouvait espérer aucun remède, pas même l'eau, car son organisme l'absorbait à toute vitesse. Celui-ci s'était mis au diapason. Parfois il décidait de lui parler, de l'asticoter. Il faisait son possible pour se tenir tranquille, mais cette situation le contrariait fort. On ne pouvait décemment lui imposer ça. Son corps avait une voix comme celle de Gillian Anderson. Il était très réfléchi et s'exprimait au moyen de longues phrases qu'il avait dû peaufiner à l'avance. Mais il n'était pas content, et avait cessé de croire en des jours meilleurs. Sarah écoutait ce qu'il avait à dire et tâchait de s'y intéresser, mais elle ne voyait pas du tout comment l'aider.

Touche était son seul véritable ami, et même lui ne venait plus très souvent. Sarah craignait de l'avoir déçu. Il continuait de parler et de lui donner à boire, mais elle sentait qu'il le faisait surtout par intérêt personnel. Il prétendait être une vraie personne, et avoir rencontré, plus jeune, des gens constitués de foin. C'est eux qui l'avaient trouvé, ou bien l'inverse. Il avait beaucoup appris d'eux, et à présent ils apprenaient de lui. Touche recevait des gens, désormais. C'est ce qu'il prétendait, en tout cas. Elle savait ce qu'ils étaient : des lutins. Ils suivaient ses ordres et ratissaient un vaste territoire, guettant les sots qui avaient la faiblesse de se croire chanceux, comme Sarah autrefois. Ils surveillaient les gens à l'aide de micros et de chauves-souris espionnes qui survolaient toutes les maisons du monde. Certains de ces lutins étaient gigantesques, et capables de déclencher des séismes et des éruptions volcaniques rien qu'en tapant du pied. D'autres étaient carrément minuscules : ils volaient dans les airs et s'infiltraient par les pores de la peau pour agiter les cellules et faire pousser des masses noires dans les poumons, le cœur et le foie. Les grands lutins avaient des voix tonitruantes. Les petits parlaient comme des Gallois. Quand Sarah toussait, elle gardait la bouche fermée pour éviter qu'ils ne s'introduisent en elle. Quelques-uns de ces lutins avaient taille humaine. Mais ils étaient rares. Elle n'en voyait jamais aucun, mais devinait leur présence. Elle cognait son front contre la trappe pour les chasser.

Puis tout le monde s'en allait, le noir revenait et elle flottait de plus belle. La première fois, elle avait eu l'impression de dériver sur le dos, maintenue à la surface de l'eau. Une sensation fort agréable, à vrai dire. Mais il lui semblait désormais s'enfoncer de plus en plus, comme si elle coulait. Ses oreilles étaient déjà immergés, et ce seraient bientôt les yeux.

Quand le bout de son nez aurait sombré, elle saurait qu'elle ne flottait plus.

CHAPITRE 29

Zandt était campé devant une porte d'entrée de Dale Lawns. Son premier coup de sonnette demeurant sans effet, il récidiva, s'appuyant de tout son poids sur le bouton jusqu'à distinguer, à travers la vitre granulée de l'imposte, une silhouette s'avançant à contre-jour d'une lumière blanche.

Gloria Neiden portait un ensemble de grand couturier, pour une banale soirée à la maison. Il apparut à ses premiers mots qu'elle était ivre. Non pas gaie, ou pompette, ni même ivre morte. Ivre comme une âme esseulée.

— Z'êtes qui, vous ?

— Je m'appelle John Zandt. On s'est rencontrés il y a deux ans.

— Je crains de ne pas m'en souvenir. En tout cas, je n'ai pris aucune disposition pour renouer le contact.

Cela débité sans heurt, à peine une ou deux syllabes escamotées. Elle commençait à refermer la porte. Zandt intercala sa main.

— Je faisais partie des policiers qui ont enquêté sur la disparition d'Annette Mattison.

Mme Neiden cligna des yeux, et ce mouvement parut répandre un pigment gris sur son visage, comme un baume inégal.

— Mais oui, fit-elle en croisant les bras. Ça me revient maintenant. Un beau succès. Une affaire rondement menée, n'est-ce pas ?

— Non. Et c'est pourquoi que je viens vous voir.

— Ma fille est sortie avec des amies. Et même si elle était là, je lui interdirais de vous parler. Il nous a tous fallu du temps pour surmonter cette tragédie.

— Je n'en doute pas. Et ça a marché ?

Elle le considéra, momentanément dégrisée.

— Qu'entendez-vous par là ?

— Ce que j'entends par là, c'est que ma fille a disparu elle aussi, et que le verbe surmonter ne fera jamais partie de mon vocabulaire. J'aimerais prendre un tout petit peu de votre temps, pour que vous m'aidiez à trouver celui qui a détruit nos vies.

— Pourquoi ne pas vous adresser aux Mattison plutôt qu'à moi ?

— J'ai une question à vous poser, une seule.

Elle se déroba, rabattant la porte plus fort qu'auparavant.

Zandt la bloqua de plus belle et n'écouta que ses tripes :

— Une question qui pourrait empêcher votre mari d'entamer ou de poursuivre une liaison. Qui pourrait éviter à votre fille d'avoir honte quand elle ramène ses copines à la maison. Ce qui signifie, si vous préférez, que vous risqueriez moins de lancer votre voiture contre un mur un de ces quatre parce que vous auriez mal évalué le virage ou que l'idée vous aurait séduite.

Gloria Neiden le dévisagea. Il lui fallut quelques secondes pour recouvrer la parole.

— Allez vous faire foutre, grogna-t-elle. Vous n'avez aucun droit de me parler comme ça. Vous auriez dû le trouver. Ce n'est pas ma faute. Rien n'est ma faute dans cette histoire.

— Je sais, admit Zandt tout en observant les transformations de ce visage : successivement animal, fille apeurée, puis femme à nouveau, tel un masque de glaise trituré par un gosse vicieux. Rien de ce qui est arrivé n'est votre faute, madame. Je le sais bien. Votre famille le sait. Tout le monde le sait à part vous. Car vous avez beau l'affirmer, vous n'y croyez pas vraiment. Et c'est ce qui vous tuera.

Ils tinrent la pose pendant un moment, de part et d'autre du seuil, à pousser chacun sur la porte. Puis ils ne poussèrent plus.

Il appela Nina en chemin vers Santa Monica. Il lui trouva une voix distraite, mais elle accepta de le retrouver à Bel Air. L'adresse figurait au dossier.

Michael Becker ouvrit à Zandt, et accepta de le suivre sans connaître la raison. Ils abandonnèrent Zoe sur le perron, qui tenait leur fille cadette par la main. L'épouse ne fit pas d'histoires, n'exigea aucune explication. Zandt comprit qu'il en eût été de même si c'était Zoe qu'on venait débaucher, et Michael qu'on laissait disparaître dans le rétroviseur de la voiture familiale. Les Becker se faisaient mutuellement confiance pour garder la maison, et se transféraient les responsabilités selon les impératifs du moment. Quand plus rien n'a de sens, seule votre relation avec un être, et un seul, peut encore vous protéger du monde.

Zandt regrettait de ne pas avoir eu cette révélation avant le départ de Jennifer.

Vu l'approche post-euclidienne de Becker quant à la géométrie de L.A., il leur fallut près de quarante minutes pour gagner l'autre bout de la ville, après quoi ils grimpèrent dans les hauteurs et longèrent des maisons qui s'élargissaient à chaque virage, jusqu'à devenir si grandes qu'on ne les distinguait même plus depuis la route.

Ils arrivèrent enfin dans un cul-de-sac. Des deux côtés se dressaient de hautes grilles. Les phares révélèrent un autre véhicule discrètement garé un peu plus haut. Nina y était accotée, les bras croisés et le sourcil levé. Nina dans toute sa splendeur.

— Nous y sommes, dit Michael Becker. C'est ici qu'il habite.

Becker n'avait rien d'un idiot. Il avait parcouru une partie du chemin dans sa tête, même si celui-ci demeurait enfoui dans son inconscient.

— Qu'est-ce que je lui dis ?

Zandt sortit de la voiture. Nina avait une question à poser, mais il l'en empêcha d'un geste.

— Faites-nous entrer, c'est tout, suggéra-t-il à Michael.

Becker s'approcha de la grille et appuya sur un bouton. Il prononça quelques mots, et le portail s'ouvrit au bout d'une ou deux secondes.

Zandt remonta l'allée d'un pas vif, talonné tant bien que mal par sa suite.

La porte de la maison était ouverte, et un homme longiligne les attendait dans le halo des lumières intérieures. Le terrain alentour s'étendait à perte de vue. Zandt attrapa le bras de Michael et le ramena devant lui pour parcourir les derniers mètres.

— Salut, Michael, lança le type. Qui sont tes amis ?

Surgissant de derrière Becker, Zandt jeta sa main à la gorge de Charles Wang. De l'autre, il lui colla deux directs au visage.

Nina écarquilla les yeux.

— Mais qu'est-ce qui te prend, John ?

— Ferme la porte.

Zandt repoussa Wang dans l'immense entrée. Il le boxa de nouveau, l'envoyant s'écraser dans le mur de marbre blanc. Puis le ramassa et le projeta contre une grande psyché, dont la moitié supérieure se brisa.

Un jeune garçon en veste blanche accourut d'une porte sous l'escalier menant à l'étage. Il constata que Zandt avait un flingue et qu'il le lui pointait en pleine figure.

— Retourne à l'intérieur, Julio, dit Wang d'une voix égale.
— C'est ça, Julio, fit Zandt. Tu te casses et tu la boucles, pigé ? Si tu oses décrocher le téléphone, dès que j'ai fini avec ce connard je te retrouve et je fais sauter ta putain de cervelle.
Le garçon disparut sans demander son reste.
Zandt ramena l'arme sur Wang, à demi couché au pied du miroir, tordu comme s'il s'était rompu le dos.
— Tu ne vas pas détaler ? demanda Zandt avant de lui envoyer son pied dans les côtes. Essayer de te sauver ?
— Arrête ça ! tonna Nina. Dis-moi ce qui se passe.
Tout à coup, Wang s'animait, se relevait d'une poussée fluide. Zandt lui abattit son canon en pleine face, stoppant net son ascension. Wang émit un claquement guttural avant de retomber.
Zandt lui releva la tête de force. Les yeux de Wang le fixèrent à travers le filet de sang qui lui coulait du front. Zandt n'y lut que peur et duplicité.
— On a merdé, expliqua-t-il aux deux autres. On a cherché au niveau un. On a loupé le niveau deux. Et on n'a même pas osé imaginer un niveau trois.
Wang lui sourit, comme s'il se demandait combien acheter cet adversaire. Zandt lui lâcha la gorge pour le gifler.
— C'est lui qu'il faut regarder ! lui cria-t-il. Pas moi. Regarde Michael.
Un instant Wang parut envisager la fuite, mais la pression du flingue sur sa trachée le convainquit de rester. Il tourna lentement les yeux vers Michael Becker.
— Nous n'avons jamais coincé l'Homme Debout, expliqua Zandt, parce que nous cherchions le ravisseur des filles. Et la raison pour laquelle nous n'avons pas trouvé le ravisseur, c'est qu'il n'y a aucun lien entre les victimes, pour la bonne et simple raison qu'elles ont été kidnappées par des types différents. Aujourd'hui, je me suis penché sur des cas analogues, des filles du même âge disparues à la même période. Deux d'entre elles ont particulièrement retenu mon attention. Deux petites New-Yorkaises, qu'on ne pouvait rattacher à l'Homme Debout puisqu'elles ont disparu à l'autre bout du pays alors même qu'il sévissait ici.
Wang cligna des yeux, pour fuir ceux de Becker. Zandt renfonça le canon dans sa pomme d'Adam et le regard se rétablit.
— Le père d'une de ces filles est directeur du développement pour la côte est chez Miramax. La mère de l'autre fille est cadre dans une société de courtage qui traite principalement avec

des banques privées suisses, mais elle arrondit ses fins de mois – je l'ai vérifié cet après-midi même – en cherchant dans ses fichiers clients des bailleurs de fonds pour produire de petits films en Europe. Mais il s'agit là de New-Yorkaises, n'est-ce pas ? Or ce sont des filles de la côte ouest qui nous intéressent. Alors je suis passé voir Gloria Neiden avant de t'appeler. Je lui ai demandé de lister toutes les personnes avec qui elle a bossé au cours des douze mois précédant l'assassinat de la fille de sa meilleure amie. Le moindre partenaire, semi-partenaire, agent, directeur, financier, fifre ou sous-fifre. Ça a pris du temps, parce que Mme Neiden est un peu à cran ces temps-ci, et c'est difficile de demander à quelqu'un de remuer ses souvenirs. Mais un nom a fini par sortir.

Michael se tenait quelques mètres derrière Zandt, les yeux rivés sur un homme qu'il avait côtoyé dans des bureaux ensoleillés, abreuvé d'e-mails rigolos, serré dans ses bras après des essais quasi transformés dans l'arène télévisuelle. L'homme qui était venu à la maison des centaines de fois, qui avait dîné à la table familiale, qui s'était assis dans la chambre de sa fille pour bavarder de son fantastique séjour britannique. Qui avait compris que l'Angleterre serait un bon moyen de retenir son attention, en attendant l'instant propice pour la kidnapper.

Wang ne dit rien.

— Charles ne tue pas les filles, poursuivit Zandt. Pas plus qu'il ne les enlève. Ce serait trop risqué. Charles n'aime pas le vrai danger. Il veut du pouvoir, du bon temps, une vie empreinte de mystère. Tout ce que fait Charles, c'est transmettre des renseignements. Charles peut trouver des filles particulières, des filles de qualité. Charles travaille à la commission, sans doute, mais Charles travaille surtout pour le plaisir.

— Charles... balbutia Michael. Dis quelque chose. Dis-moi que c'est faux.

— Dis-nous plutôt combien tu touches par fille, reprit Zandt. Explique-nous pourquoi, alors qu'il serait tellement plus facile de ramasser des filles à la rue, ils préfèrent piocher dans les bonnes familles. Anéantir des gens censés être tes amis. Explique-nous pourquoi c'est si bandant, parce qu'on a une putain d'envie de savoir, Charles.

Sans prévenir, il recula d'un pas et aplatit sa semelle sur le torse de Wang. Puis il se rapprocha de son visage en hurlant :

— Qui les prend ? Qui les enlève ? Où est-ce qu'on les emmène ?

Sans quitter Michael Becker des yeux, Wang se lécha les lèvres.

— Vous croyez que je connais leurs noms ?

Zandt :

— Décris-les.

— Et si je refuse ?

Zandt écarta le flingue de trois centimètres et pressa la détente. La balle s'écrasa dans le marbre, juste derrière la tête de Wang, puis ricocha à travers la pièce. Des bris de pierre et de verre saupoudrèrent son crâne. Le pistolet retrouva aussitôt sa gorge.

— J'en connais trois, débita Wang. Il y en avait un quatrième, mais il a disparu voilà deux ans. Ils ont tous l'air différent... Que voulez-vous que je vous dise ? Vous croyez peut-être qu'on buvait des bières ensemble ?

— Décris celui qui a pris la fille de Michael. Tu dois être en contact avec lui.

— Tout s'est passé par mail et par téléphone.

— Mon cul. Les mails peuvent être lus et les téléphones écoutés. Mais deux types qui se rencontrent au bar d'un hôtel quelconque, personne n'y prêtera attention, pas vrai ?

Wang se lécha de nouveau les lèvres. Zandt rehaussa l'extrémité du flingue, lui dessinant un rectangle au milieu du front. Wang regarda le doigt serrer la détente. Quand il rouvrit la bouche, le flic leva l'index de sa main libre.

— Ne me dis pas ce que tu penses que j'aimerais entendre, prévint-il. Si j'ai l'impression que tu mens, t'es mort.

— C'est un grand gaillard. Blond. Costaud. Son nom est Paul.

Zandt se redressa, essuya de sa main la sueur du type. Puis recula d'un pas vers Nina, laissant Michael face à Wang.

— Alors c'est vrai ? demanda le père de Sarah d'une voix à peine audible. Comment... Comment peux... Mais pourquoi ? Pourquoi, Charles ? Je veux dire... (Abasourdi, perdu dans une maison qu'il n'aurait jamais les moyens de s'offrir quel que soit le nombre de culs qu'il lécherait, il se raccrocha à une chose triviale mais concrète :) Me dis pas que c'est pour ce putain de pognon ?

— Tu es un homme médiocre, avec des ambitions médiocres, répondit Wang d'un ton amer, tout en frottant sa lèvre ensanglantée du dos de la main. Des petites idiotes qui n'ont jamais été baisées. Une imagination de vieilles filles. Tu n'as jamais rien touché de grand, et tu ne le feras jamais. Tu ne la toucheras pas, *elle*, en tout cas, plus maintenant. (Il lui fit un clin d'œil.) Tu ne sauras jamais ce que tu rates.

Zandt fut le plus rapide. Il intercepta Becker, l'attrapant par les épaules pour le faire virevolter d'un coup. Zandt était bien plus lourd que Becker, mais il parvint de justesse à le contenir.

— Il vous raconte des conneries, Michael. Des conneries.

Au bout d'un moment, l'énergie de Becker sembla retomber. Par-dessus l'épaule de Zandt, qui le maintenait fermement, il toisait le sourire de l'homme à terre.

— On ne va pas le tuer, compris ? (Zandt tourna la tête de Becker pour lui parler droit dans les yeux. Ceux-ci étaient écarquillés, aveugles.) Je ne peux pas promettre de ramener votre fille vivante. Elle est peut-être morte, auquel cas cet homme porte une responsabilité écrasante. Mais nous allons quitter cette maison sans faire d'histoires. C'est la seule chose que je sois sûr de pouvoir vous offrir : vous empêcher de repartir dans la peau d'un meurtrier.

Le regard de Becker se rajusta lentement. Son corps mollit quelques instants, puis se raidit de plus belle. Il finit par reculer d'un pas, relâchant ses bras le long du corps.

Zandt rangea son flingue. Le trio considéra le type allongé par terre.

— Tu auras de la compagnie très bientôt, promit Zandt. Des flics, des fédéraux. Munis de mandats de perquisition. À ta place, je ferais un peu de ménage.

Là-dessus ils tournèrent les talons, laissant dans leur sillage un visage blême, interdit.

Aucun mot ne fut échangé jusqu'à la voiture. Michael regarda la maison.

— Que suis-je censé faire ?

Nina ouvrit la bouche, mais Zandt la prit de vitesse :

— Rien du tout. N'en parlez pas à la police. Ni à votre femme. Je sais que vous en mourez d'envie. Mais pas pour l'instant. Et surtout, ne revenez pas ici. Ce qui doit être fait le sera.

— Par qui ?

— Montez dans la voiture, Michael.

— Je ne peux pas vous laisser faire ça pour moi.

— Remontez dans la voiture, je vous dis.

Becker finit par obéir et reprit la route, s'éloignant d'un train lent et sinueux.

Nina sortit son téléphone et composa un numéro. Zandt lui dégomma l'objet des mains, l'envoyant ricocher sur l'asphalte.

— Laisse ça, dit-il.

Elle le fusilla du regard, sans ramasser l'appareil.

— Tu as vraiment appelé les flics ? demanda-t-elle.
— Tu sais bien que non.

Zandt alluma une cigarette et ils patientèrent. Au bout de dix minutes ils entendirent le bruit que Zandt avait escompté, la détonation assourdie sans laquelle il serait retourné à l'intérieur pour commettre ce qui s'imposait, quels que soient les efforts de Nina pour l'en empêcher.

Et pourtant, dès qu'il perçut ce bruit, Zandt fut pris d'une immense lassitude, loin de tout triomphalisme. Comme si, en se rapprochant de la source des événements, il était juste parvenu à se compromettre davantage ; comme si la vase tapie sous la surface de l'humanité dégageait une odeur si forte qu'il ne pourrait jamais s'en défaire.

Nina se retourna vers lui.

— Alors il est mort.

— Il se contentait de sélectionner les filles pour les échelons supérieurs. On aurait pu passer des jours entiers à l'interroger, et il nous aurait menés en bateau.

— Je ne te reproche rien. Je te demande simplement ce que tu comptes faire ensuite.

Zandt haussa les épaules.

— Parfait, dit-elle en se baissant pour récupérer son téléphone. (Des porches s'illuminaient de l'autre côté de la rue.) Parce que les flics ne vont pas tarder à rappliquer, et je n'ai pas très envie d'être là pour les accueillir.

Elle s'éloigna à grands pas vers sa voiture, ajoutant par-dessus son épaule :

— Et j'ai deux personnes qui pensent avoir localisé un grand blond, un peu du genre de celui qu'on vient de te décrire.

Zandt ouvrit de grands yeux.

— Quoi ?

— Hopkins et son acolyte. Ils m'ont appelé juste avant toi. Ils possèdent une vidéo montrant ce type sur les lieux de toute une série de boucheries top niveau, la moitié de celles qui ont marqué ces dix dernières années, y compris celle de ce matin dans une école du Maine. Ward croit justement l'avoir croisé dans ce fameux endroit à la montagne.

— Si tu savais tout ça, pourquoi m'as-tu laissé Wang ?

Elle le considéra par-dessus le toit de la voiture.

— Je ne tenais pas plus que toi à l'épargner.

CHAPITRE 30

Ni Zandt ni Nina ne savaient que, juste avant de se supprimer, Wang avait passé un dernier coup de fil.

D'abord il était laborieusement remonté sur ses jambes, les mains patinant dans son sang. Il était incapable de se tenir vraiment droit. Il s'était déjà fait rouer de coups, en avait même réclamé plus d'une fois, mais là c'était différent. Le flic n'avait pas agi en songeant au plaisir de Wang, et il y avait de la casse.

Il resta debout un moment, se regarda dans les restes du miroir sous lequel il avait livré son plus grand secret. Son visage était boursouflé, entaillé. Pire, il paraissait vieux. Le dispendieux vernis des régimes et de l'exercice, des onguents et de l'obsession de soi s'était soudain dissous. Wang faisait son âge, d'une façon dont seul un homme ayant commis ce qu'il avait commis, et gardé ses secrets aussi longtemps, était capable.

Il n'avait jamais tué. Ni même blessé. Pas de ses propres mains. Mais il avait assisté à certaines séances où de jeunes hommes étaient laissés gisants dans des mares d'urine et d'autres sécrétions. Où d'autres types de son espèce s'esquivaient dans des voitures de luxe, heureux de ne pas être poursuivis pour complicité de meurtre. Il possédait une collection pléthorique de vidéos relatant de tels événements. Si pléthorique, à vrai dire, qu'il doutait de pouvoir toutes les rassembler, sans parler de les détruire, avant l'arrivée de la police.

Son père ne comprendrait jamais.

Pas plus, présumait-il, que les hommes et femmes avec lesquels il commerçait de manière plus honnête – même s'il savait que certains possédaient leurs propres secrets, que la flamme intérieure qui les menait à la gloire et au succès les poussait aussi à des actes plus sombres, comme moyen de se prouver qu'ils étaient au-dessus de la mêlée. Car la gloire ne suffit jamais. Tôt ou tard tout le monde a besoin de s'idolâtrer soi-même, sans quoi les regards extérieurs versent dans l'absurde. Marchandises et matières premières avaient été obtenues, des femmes en pleurs avaient été rétribuées, parfois par Wang lui-même, qui se montrait toujours enclin à devenir l'ami des gens. Un confident pour ceux dont les désirs outrepassaient les normes fixées par la société. Pour ceux qui voulaient vivre plus fort, plus vite, plus intensément. Pour qui le sexe n'était jamais aussi bon que dans la peur.

C'était l'un d'eux, un homme qui savait combien Wang pouvait se montrer dévoué, qui l'avait mis en relation avec plusieurs de ses collègues. Le représentant de ce groupe était un grand blond. Le dénommé Paul. Leur rencontre n'avait eu lieu qu'après plusieurs années de collaboration, et il fallut plus de temps encore à Wang pour comprendre que cet homme n'était pas tout à fait ce qu'il paraissait, et que lui et ses amis visaient davantage que le simple plaisir. Il n'avait jamais été convié parmi eux, ce dont il concevait une certaine amertume. Mais il avait accepté de leur pourvoir des objets de distraction, d'aider ces entremetteurs à dégotter des joyaux particuliers, et cette brute de flic ne s'y était pas trompé : l'argent n'avait rien à voir là-dedans.

Chacun a son propre cheminement, et vient au monde deux fois. La seconde naissance de Wang s'était produite voilà trente-cinq ans. Il en avait alors dix, et avait surpris une servante nue derrière une vitre. Un matin de printemps dans un autre pays, une image qui lui avait coupé les jambes, ébloui par la conscience subite que le monde regorgeait de trésors cachés. Son père se trouvait alors dans le bureau de la maison, d'où s'échappait un filet de musique baroque, juste et cadencée, claire et joyeuse. Wang s'était figé un instant, perdu dans ces quelques secondes de délice. Chez la plupart des gens, une telle expérience serait demeurée sans suite, mais Charles ne fut plus jamais le même. Les plus petites graines donnent parfois les arbres les plus sombres.

Il avait alors goûté au voyeurisme, puis aux revues, aux cassettes, et aux virées en solo dans des recoins peu connus de Hong-Kong et plus tard de Los Angeles. Une panoplie qui, là encore, aurait contenté – pour ne pas dire écœuré – la majorité des gens. Le péché ne résidait pas dans l'objet, ni même dans sa convoitise, mais dans son besoin. En avoir besoin à tout prix, avant même d'en connaître l'existence, au point qu'on l'aurait inventé s'il n'existait pas. Accuser la pornographie, c'est comme accuser un pistolet. Aucun des deux ne s'est créé lui-même. Aucun n'est capable de presser sa propre détente. Il faut une main pour cela. C'est l'esprit de l'homme qui fournira cette main chercheuse, aux doigts suffisamment fins pour déceler les petites brèches, et suffisamment forts pour extraire ce qui s'y cache. La comparaison vaut jusque dans les durillons qui se forment parfois, à force de manipulations, et qui réduisent la sensibilité du toucher. Un cal qui impliquera de recourir, pour le même résultat, à des expédients plus chauds ou plus perçants – et vient ainsi le temps où, pataugeant si profond dans le sang, la portée de vos prochains pas devient indifférente.

Cette semaine-là, Wang n'avait songé qu'une seule fois au sort de la fille de Michael Becker, en formant le vœu que Michael se remette vite au boulot, car le studio semblait de plus en plus réceptif au projet *Dark Shift*. Aussi risible que fût Michael à bien des égards, il travaillait dur et il avait des idées. Mieux, des idées acceptables pour un esprit lambda. Wang possédait sa propre version de *Dark Shift*, qu'il avait rédigée pour s'amuser. Mais elle eût paru moins acceptable.

Rien de tout cela n'eût paru acceptable. Rien de ce qu'il avait fait par goût ou par plaisir. Or, privé de ces choses, il lui restait bien peu à découvrir, et rien qui puisse motiver son existence. Privé du souvenir et de l'héritage d'un certain matin de printemps, d'une vision bercée de musique et des chuchotis d'une fontaine toute proche, il ne lui restait rien.

À l'heure où Zandt allumait sa cigarette dehors, Wang s'était traîné jusqu'à son bureau. Le choc initial commençait de se dissiper, et ses côtes le mettaient au supplice. Il composa un numéro et prévint un ami que quelqu'un était à deux doigts de lire dans leur jeu, si ce n'était déjà fait.

Puis il reprit son fauteuil. Aucun signe de Julio, qui devait pourtant savoir que les visiteurs étaient partis. L'espace d'un instant, Wang regretta de ne pas avoir, pour une fois, une compagnie ne relevant pas de la seule domesticité. Le gosse avait sans doute quitté la propriété par la grille du fond, pour dévaler la route vers une nouvelle vie. Comme un sourire d'hier, il avait disparu.

Wang déverrouilla le tiroir central de son bureau et sortit son revolver. La crosse était en cerisier sculpté.

Un objet magnifique. C'était toujours ça de pris.

CHAPITRE 31

À 8 h 45 le lendemain matin, nous poireautions dans la voiture au bout de la rue d'Auntie's Pantry. Il faisait froid et les nuages noirs crachaient de la neige fondue depuis deux heures. J'avais acheté un paquet de cigarettes, que je fumais à la chaîne. Bobby n'y trouvait rien à redire. Son flingue sur les genoux, il regardait droit à travers le pare-brise.

— Alors, ils arrivent à quelle heure ?

— Je ne te garantis pas qu'ils viendront, répondis-je.

Il secoua la tête.

— Un flic sans insigne et une meuf. Putain, on est invincibles. Envahissons l'Irak.

— On n'a personne d'autre, Bobby.

Une auto quelconque déboucha au coin de la rue. La conductrice, une femme d'âge mûr, nous croisa sans même remarquer notre présence. Nous attendions qu'une certaine personne se pointe à son bureau, et ce depuis 8 heures précises. Nous étions sur les nerfs. Ni lui ni moi n'avions très bien dormi.

— OK, fit Bobby en indiquant le trottoir d'en face. Une grande asperge aux cheveux roux. C'est notre homme ?

Nous attendîmes que Chip ait pénétré dans l'agence, puis nous quittâmes la voiture sans la fermer à clé. La rue était relativement déserte. Ce n'était pas une artère principale, et le temps ne se prêtait pas au lèche-vitrine.

J'enfonçai la porte et foulai la moquette de Farling Realty, talonné par Bobby. Chip avait disparu dans le bureau du fond. La grande pièce centrale était bordée de quatre bureaux, dont deux étaient occupés par des quadragénaires très bien coiffées vêtues de petits tailleurs carrés, l'un vert et l'autre rouge. Elles nous accueillirent avec des yeux brillants, fin prêtes à nous vendre notre rêve.

— On cherche Chip, annonçai-je.

L'une d'elles se leva.

— M. Farling sera à vous dans un instant, gazouilla-t-elle. Puis-je vous offrir un café pour vous faire patienter ?

— M. Hopkins ne compte pas s'attarder.

Chip se tenait au seuil de l'autre bureau. Une vilaine ecchymose lui courait de la joue au front.

— Je crois même qu'il va repartir très vite.

— C'est bien ce que nous avions en tête, Chip. Mais tu viens avec nous. On va voir les Halls, et on a besoin de quelqu'un pour nous faire entrer. En vertu de ton exclusivité si chèrement acquise, tu tiens la pole position. Alors soit tu nous suis de ton plein gré, soit on te traîne dans la rue en t'étranglant.

— Je ne pense pas, dit-il d'un air agacé.

À cet instant, la sonnette signala une nouvelle visite. Je me retournai pour découvrir deux flics. L'un était grand et brun. L'autre plus petit et blond. Ce dernier prit la parole :

— Bonjour, monsieur Hopkins.

— On se connaît ?

— Je vous ai eu au téléphone.

— À quelle occasion ?

— Vous avez contacté le commissariat. On a discuté de la mort de vos parents.

Derrière moi, j'entendis Bobby fureter dans son blouson.

— Officier Spurling, devinai-je.

— Il est ici à ma demande, expliqua Chip. Je vous ai vu posté dehors avec votre ami. Je lui ai rapporté que vous m'aviez agressé.

— Je parlerais plutôt d'une légère divergence de vues. Puis votre corps a été pris d'étranges convulsions.

— Je n'ai pas vu ça comme ça, moi. Ni la police, du reste.

— Il bluffe, Ward, murmura Bobby.

Chip se tourna vers les deux femmes, qui suivaient l'échange telle une paire de chattes curieuses.

— Doreen ? Julia ? Si vous alliez dans le bureau du fond quelques minutes ?

— C'est vous qu'on est venus chercher, Chip. Personne d'autre n'a à bouger.

— Maintenant, ordonna Chip à ses deux employées, qui se levèrent et se hâtèrent vers l'autre pièce.

Il referma derrière elles.

— Il serait préférable que vous nous suiviez au poste, dit Spurling sur un ton calme et conciliant. Vous l'ignorez peut-être, mais la maison de vos parents a subi des dégâts, et un hôtel a brûlé. Il semblerait que les deux incidents soient liés. L'officier McGregor et moi aimerions vous aider.

— Le problème, voyez-vous, c'est que je ne suis pas convaincu sur ce dernier point, répondis-je.

— Et c'est quoi, le problème avec votre partenaire ? ajouta Bobby. Il est pas très causant, dites-moi.

Le second flic considéra Bobby, mais sans dire un mot. C'est là que j'ai commencé à tiquer. Un type qui soutient le regard de Bobby d'un air tout sauf respectueux est soit inconscient, soit extrêmement dangereux, soit les deux.

— Simple répartition des tâches, lançai-je, espérant que la situation était encore récupérable. McGregor est peut-être un as de la paperasse.

— Vous êtes un trou du cul, Hopkins, dit Chip. Il faut croire que c'est de famille.

— Alors, monsieur Hopkins, fit Spurling en ignorant l'interruption, vous voulez bien me suivre ?

— Non, répondit Bobby.

Chip sourit. McGregor dégaina son flingue.

— On se calme, plaidai-je, dans mes petits souliers.

L'officier Spurling semblait encore plus surpris que moi, les yeux rivés sur l'arme de son équipier.

— Euh, George... commença-t-il.

Mais McGregor ouvrit le feu.

Nous étions déjà en mouvement quand Chip réitéra son petit rictus hautain, mais c'était encore trop lent. Il n'y avait nulle part où se réfugier. Se planquer n'était pas à l'ordre du jour.

Pistolet en main, Bobby toucha McGregor dans la cuisse et la poitrine. Mais les impacts ne sonnaient pas comme d'habitude, et je compris que le flic portait du Kevlar. Sous la force des coups, il heurta un fauteuil et bascula sur le dos, mais il se remit vite debout. Pendant tout ce temps, Spurling restait figé, bouche bée.

J'eus trente centimètres d'avance sur la balle de McGregor, m'étant jeté à terre en roulant. Je me rétablis derrière le bureau de Doreen et retournai le coup, l'atteignant à l'épaule. Quelque chose siffla tout près de mon oreille, et je vis que Chip était, lui aussi, en possession d'un petit flingue. Après ça, mes souvenirs sont confus. J'ai simplement vidé mon arme sur tout ce qui bougeait. Dans une bataille au beau milieu d'une plaine, vous avez peut-être le temps de cogiter, d'analyser la situation entre deux tirs. Mais prenez le temps de réfléchir dans un espace confiné face à deux gars qui vous canardent, et vous n'irez jamais au bout de votre raisonnement.

Au bout de dix secondes, la fusillade cessa. J'étais alors coincé derrière le bureau de Julia, la joue et le front en feu, comme lacérés. Non par une balle, a priori, mais plutôt par les fragments d'une explosion. J'étais surpris de m'en tirer à si bon compte. La tête de Chip était répandue sur le mur du fond. McGregor n'était plus en vue, et la porte de l'agence étaient grande ouverte.

Spurling avait reçu une balle dans la jambe avant de passer par-dessus un bureau. Il remuait, mais pas très vite. Sa tête était toujours en place. Je la laissai comme ça.

Bobby était adossé au mur près de l'entrée, la main pressée sur son bras, ses doigts dégoulinant de sang. J'accourus et l'agrippai.

Nous déboulâmes sur le trottoir, trottant jusqu'à celui d'en face. J'ouvris la portière de droite et poussai Bobby à l'intérieur. Deux passants en combinaison de ski orange promenèrent leurs regards éberlués entre nous et les vitres étoilées de l'agence.

— On tourne un film, dit l'homme. Je ne vois que ça.

— Ça va aller, murmura Bobby tandis que je sautais derrière le volant avant de mettre les gaz. Ça va aller.

— Tu t'es fait tirer dessus, imbécile.

— Ralentis...

Il y avait un stop juste en face, et un trafic avec lequel il fallait composer. Je relâchai la pédale et parvins par chance à m'insérer dans un trou sur la file de gauche.

— Où tu vas ?

— À l'hosto, Bobby.

— T'es fou. Pas après ça.

— Spurling nous défendra.

— Tout ce qu'il sait, c'est que ça a tiré dans tous les sens. Les deux keufs ont été touchés, et un civil a trouvé la mort.

— Il sait que McGregor a tout déclenché. Si tu préfères, je vais prendre l'autoroute pour rallier le premier hôpital hors de la ville.

— Où les toubibs devront quand même déclarer l'incident, et où on aura toujours tué quelqu'un.

— Mais tu as reçu une balle, Bobby ! C'est la dernière fois que je te le répète.

Comme je poursuivais ma percée vers l'ouest, slalomant entre les voitures, Bobby ôta délicatement la main de son bras. Un filet de sang frais s'en échappa, mais moins nourri que je ne l'avais craint. Avec une grimace de dégoût, il écarta le tissu entourant la plaie pour mieux l'examiner.

— Il manque un bout, avoua-t-il. Ce qui n'est pas idéal. Mais je survivrai. Et il y a plus urgent que les soins médicaux.

— Quoi donc ?

— De l'artillerie, dit-il en se carrant dans son siège. Une putain de grosse artillerie.

Laissant Bobby dans la voiture, je me ruai vers la boutique. Il pleuvait à verse et les nuages viraient au noir. Avant de pousser la porte, je pris le temps de me recomposer. La plupart des armuriers cultivent une illusion selon laquelle ils ne vendent que des pistolets à eau. Inutile de leur montrer qu'on a la gâchette qui démange.

À l'intérieur, un espace tout en longueur. Un comptoir en verre où des armes de poing s'alignaient tels des bijoux ; derrière, des râteliers de fusils à perte de vue. Ni clients, ni guichet blindé. Juste un gros gars aux cheveux blancs en chemise bleu foncé, attendant la prochaine transaction.

— J'peux vous aider ?

Il plaqua deux grosses mains sur le comptoir. Au mur derrière lui s'affichaient les visages de deux célèbres terroristes du Moyen-Orient. « Mort », requérait chaque légende, « ou vif » ayant été barré d'une croix.

— J'aimerais acheter des armes, dis-je.

— On ne vend que du yaourt glacé, ici. Depuis le temps que je dois démonter cette foutue pancarte.

Je ris de bon cœur. Il rit aussi. C'était vraiment très cool. On se payait un chouette moment.

— Bon. Vous recherchez quel genre de truc ?

— Deux fusils avec huit cents cartouches, quarante chargeurs de 45, n'importe quelle marque, ce que vous avez de moins cher. Et deux gilets pare-balles, de bonne qualité, un L et un M.

— Ben mon vieux, fit-il, toujours goguenard. On veut déclencher une guerre ?

— Non. Mais pour sûr qu'on a de sérieux ennuis.

Son sourire s'évanouit, et je vis qu'il observait ma joue. J'y passai ma main, et recueillis une fine trace de sang.

— Comme vous le voyez, ça commence vraiment à dégénérer, ajoutai-je.

Il ne rit pas ce coup-ci.

— Je ne sais pas si je peux vous vendre tout ça.

Alors je brandis une American Express Gold, et son sourire réapparut. Il calcula l'addition à la main, en m'accordant une ristourne sur les munitions. Quand on achète en gros, le prix unitaire de huit cents morts potentielles est à vrai dire très raisonnable.

Il m'annonça le total et j'agitai la main, pressé d'en finir. Je jetai un coup d'œil sur Bobby par la fenêtre. Il avait retiré son blouson et déroulait une bande autour de son biceps. J'avais acheté celle-ci dans une pharmacie vétérinaire en arrivant dans cette ville, ainsi que des épingles à nourrice et de la gaze. Il grimaçait beaucoup. Je me retournai juste à temps.

— Ne faites pas ça, dis-je en braquant mon flingue sur la poitrine du type.

Il s'immobilisa, les yeux rivés sur moi, la main à quelques centimètres du téléphone.

— Laissez-moi deviner. Il y a deux jours, un flic s'est pointé et vous a dit de ne rien vendre à un certain Ward Hopkins ?

— C'est exact.

— Mais vous allez quand même le faire, pas vrai ?

— Non, monsieur. Je refuse.

Je m'avançai d'un pas, visai son front. J'étais vanné et j'avais peur. Il secoua la tête, rapprocha sa main du téléphone.

— Je vous vendrai rien, je vous dis.

Le combiné était un très vieux modèle, qui émit un son remarquable quand la balle le perfora. Effaré, le type recula d'un bond.

— Oh que si ! rétorquai-je. Autrement je vous descends et je me sers. Et puis vous n'êtes pas en position de pleurnicher parce que le flingue que j'ai entre les mains provient précisément de cet établissement. Vous savez quoi ? C'est à ça que ça sert.

Le type resta figé un instant, se demandant de quel côté plonger. J'espérais de tout mon cœur qu'il allait m'obéir, parce que je n'avais aucune intention de l'abattre et il devait s'en douter.

Puis son regard cilla. Je repérai du coin de l'œil un jeune type qui s'approchait de la boutique. Il portait un sac de sandwiches et le même genre de chemise que son copain ventru.

Je poussai un juron, me penchai en avant et raflai un maximum de boîtes.

— Vous ne m'avez vraiment pas aidé, pestai-je avant de franchir la porte en trombe, précipitant de plein fouet le jeunot dans une flaque.

Je sautai dans la voiture, jetai les boîtes de munitions sur les genoux de Bobby.

— Ça a merdé, mec.

— Je vois ça, répondit-il en avisant le gros lard qui sortait de la boutique en agitant un tromblon.

Je lançai la voiture en marche arrière, tandis qu'un premier tir frôlait le toit. Le jeune collègue remonta sur ses jambes et fonça dans la boutique en bousculant l'autre. J'écrasai le frein, réussis mon tête-à-queue et mordis la route alors qu'une balle explosait l'une des vitres arrière.

— Le vendeur avait mon nom sur une liste. (Je virai brusquement à droite, sans but précis, sinon m'éloigner du centre-ville.) Voilà au moins une première question de réglée : comment les Hommes de Paille ont rappliqué si vite chez mes vieux après que j'ai brutalisé Chip. Pas besoin de quitter leur montagne. Ils avaient McGregor dans la place.

— Ça se tient.

— Une autre chose qui se tient : McGregor et Spurling étaient les premiers sur les lieux quand mes vieux se sont emplafonnés. Sauf que McGregor est peut-être arrivé un peu plus tôt...

— Et se trouve en ce moment même au commissariat de Dyersburg, ruisselant de sang, à balancer nos noms. On s'est fait mettre profond, Ward – très profond. Qu'est-ce qu'on va faire, maintenant ?

Je ne voyais qu'une personne en ville qui puisse éventuellement nous aider. Je prononçai son nom.

— Bonne pioche, opina Bobby, qui grimaçait en se rajustant sur son siège. Vu la tournure des événements, un avocat ne sera pas du luxe.

Selon la carte qu'il m'avait remise après l'enterrement, Harold Davids habitait de l'autre côté de la ville. À la différence du quartier de mes parents, tout en collines et rues serpentines, celui-ci formait un quadrillage régulier – aux cases néanmoins vastes et plantées de pavillons coquets.

Le perron était éclairé, ainsi que l'intérieur. Une voiture semblable à celle que conduisait Davids était garée un peu plus bas dans la rue. Nous restâmes quelques instants dans la nôtre, le temps de vérifier que nous n'étions pas suivis, puis nous sortîmes.

J'actionnai la sonnette. Pas de réponse. Évidemment.

— Merde, pestai-je. Qu'est-ce qu'on fait ?

— Téléphone-lui, suggéra Bobby tout en surveillant la rue.

Je sortis le cellulaire, essayai le numéro de son cabinet ; puis celui d'ici, au cas où Davids ignorerait les coups de sonnette nocturnes ou serait captivé par une émission géniale à la télé.

Au moins deux appareils retentirent sur divers paliers de la maison, et à la huitième sonnerie un répondeur s'enclencha. L'annonce indiquait ses coordonnées professionnelles, mais aucun numéro de portable.

— On ne peut pas rester plantés là, dis-je. Vu le lotissement, tu peux être sûr qu'un voisin va appeler les flics.

Bobby tourna la poignée de la porte. Elle était fermée à clé. Il fouilla dans sa poche et sortit un petit outil. Je faillis protester, mais on n'avait nulle part où aller. Il venait de glisser l'outil dans la serrure quand nous entendîmes tourner le verrou. Nous sursautâmes.

La porte s'entrouvrit d'une dizaine de centimètres, sur le visage de Harold Davids.

— Harold, dis-je.

— Ward ? C'est vous ? (Il ouvrit la porte un peu plus grand, l'air agité comme une puce.) Seigneur, mais que lui est-il arrivé ?

— On lui a tiré dessus, expliquai-je.

— Tiré dessus, répéta-t-il à mi-voix. Qui donc ?

— Des sales types. Écoutez, je sais que vous n'aviez pas ça en tête quand vous me proposiez vos services. Mais on est dans le pétrin. Et il ne me reste plus personne.

— Ward, je...

— S'il vous plaît. Si vous ne le faites pas pour moi, faites-le pour mon père.

Il me dévisagea longuement, puis s'écarta et nous laissa entrer.

Sa maison était sensiblement plus petite que celle de mes vieux, mais à lui seul le vestibule semblait contenir trois fois plus de choses. Des estampes, des objets d'art régional, des livres rangés dans une petite bibliothèque en chêne qui semblait construite sur mesure. En arrière-fond sonnait du piano classique.

— C'est tout droit, dit-il. Et faites attention au tapis. Vous pissez le sang. L'un comme l'autre.

Des reproductions de tableaux ornaient les murs du salon. Je n'en reconnaissais aucun. L'éclairage était faible, fourni par deux lampes à abat-jour. Pas de téléviseur, mais un poste CD d'aspect coûteux d'où s'échappait la musique. Il y avait un vieux piano coiffé d'une série de photos, certaines encadrées, d'autres simplement posées à nu. Au pied du canapé s'étendait un tapis ornementé dont les bords s'effrangeaient un peu.

— Je vous apporte une serviette, dit Davids.

Il hésita un instant dans l'embrasure de la porte, puis disparut dans le couloir.

Pendant son absence, Bobby resta au milieu de la pièce, la main sur son bras, s'assurant que tout ce qu'il perdait atterrissait sur le parquet. J'embrassai la pièce du regard. Les affaires des autres sont toujours inexplicables. Surtout celles des vieux. Je me souviens, un Noël, d'avoir offert sur un coup de tête une vieille calculatrice à mon père, repérée dans la vitrine d'un antiquaire ; je lui avais trouvé de la gueule, et m'étais dit qu'elle pourrait lui plaire. Quand il l'eut déballée, il me remercia en me considérant d'un air bizarre. Je lui avouai ne pas avoir l'impression que c'était le plus chouette cadeau qu'il ait jamais reçu. Alors, sans ajouter un mot, il m'emmena dans son bureau, ouvrit un tiroir. Là, sous un amas de stylos et de trombones accumulés au fil des ans, se trouvait une vieille calculette. C'était carrément le même modèle. L'intérieur de Davids était une brocante à mes yeux ; ce que je voyais rétro, mon père l'avait vu moderne. On est isolé de ceux qu'on aime par d'incompressibles décennies, un mur de verre immaculé mais incassable, épais de trente centimètres. Vous croyez vous tenir tout près, mais essayez de les toucher et votre main sera immédiatement bloquée.

Davids réapparut avec un linge que Bobby s'enroula autour du bras, puis s'assit dans un fauteuil et contempla le sol. Il paraissait fatigué, pâle, et bien plus âgé qu'à notre dernière rencontre. La lampe jouxtant son siège creusait les lignes de son front et marquait les plats de son visage.

— Il va falloir que vous me racontiez ça, Ward. Et je ne promets pas d'être d'un grand secours. Mon domaine, c'est les contrats, non les... fusillades.

Il se passa les mains dans les cheveux puis releva les yeux vers moi, et c'est là qu'une étincelle jaillit au fond de mon crâne.

Je me retournai, examinai le dessus du piano, puis Davids de nouveau.

— Vous me dévisagez, Ward.

J'ouvris la bouche pour dire quelque chose, mais rien ne vint. Je la refermai.

— Qu'y a-t-il ? Si vous me disiez enfin pourquoi vous êtes venu ?

Ce choix de mots, bien que sûrement fortuit, acheva de me convaincre. La façon dont ils rimaient avec « ce que tu es devenu ».

Je recouvrai l'usage de la parole.

— Quand avez-vous connu mes parents, au juste ?

— En 1995. L'année de leur emménagement.
— Pas avant ?
— Non. Comment aurait-ce été possible ?
— Vous auriez pu les croiser à un autre moment de votre vie. Allez savoir. Les routes des individus se recoupent parfois de façon mystérieuse, répondent à certaines mécaniques dont ils n'ont pas toujours conscience.
Son regard retomba.
— Vous m'intriguez, Ward.
— Depuis combien de temps habitez-vous à Dyersburg ?
— Depuis que je suis né, comme vous devez le savoir.
— Alors le nom de Lazy Ed ne vous dit rien ?
— Non, répondit-il sans quitter le sol des yeux, mais sans hésitation ni fausse note. C'est un drôle de nom, si vous voulez mon avis.
Bobby me fixait d'un air intrigué.
— C'est terrible, continuai-je. Je n'ai même pas connu son nom de famille. Juste Lazy. On fait mieux comme épitaphe, mais j'imagine que ça n'a plus grande importance maintenant qu'il est mort.
— Je suis désolé d'apprendre que vous avez perdu un ami, Ward, mais je ne vois vraiment pas où vous voulez en venir.
Je pris une photo sur le piano. Ce n'était pas une vue de groupe. Il y en avait deux de ce type-là, en noir et blanc, des traces déclinantes d'anciens aïeux, roides devant une technologie dont ils se méfiaient un peu. Le cliché que j'avais entre les mains était en couleur et figurait un seul individu, dans une pose informelle, pris par quelque ami voilà bien longtemps, avec ces teintes pastel et délavées, où les rouges conservent leur éclat et les bleus leur richesse, mais où le reste semble retenu dans une autre époque, comme si la lumière reflétée par ces surfaces-là s'effaçait, n'ayant plus la force de se maintenir dans le présent ; comme si cette ère se disloquait à mesure que s'éteignaient ses survivants, ceux qui se souvenaient de son soleil sur le visage. Un jeune homme, dans une forêt.
— « Mets celle qui parle de sodomie, récitai-je tout en examinant le Harold d'antan. Allez, Don, mets-la, Don, qu'on s'bidonne, Don. »
— Taisez-vous, Ward.
Sa voix chevrotait un brin.
Bobby m'emprunta la photo.

— Elle doit dater de quelques années plus tôt, dis-je. Harold est plus jeune et plus fin que sur la vidéo. Il n'avait pas encore laissé pousser ses cheveux.

Je me tournai vers Davids.

— Vous deviez avoir, quoi ? cinq ou six ans de plus qu'Ed et mes parents, comme Mary. Et maintenant il ne reste plus que vous. Voilà pourquoi vous n'avez pas répondu à notre coup de sonnette et ne décrochez pas votre téléphone.

Davids soutenait mon regard. On aurait dit un centenaire. Un centenaire terrorisé.

— Et merde... expira-t-il.

Je brûlais de l'empoigner, de le secouer jusqu'à ce qu'il parle, jusqu'à ce qu'il m'explique tout ce cirque, qu'il me donne les clés de ma vie. Mais de la même façon qu'il avait fondu de quarante kilos en l'espace de trente ans, en vingt secondes son visage perdit tout ce que j'y avais lu précédemment, cet air forgé par une vie passée à dire aux gens où ils se situaient vis-à-vis de la loi. Je le trouvais maigre, frêle, et plus effrayé que moi encore.

— Racontez-moi, furent mes seuls mots.

Ce fut somme toute assez rapide.

Il me dit que, voilà bien longtemps, il y avait un groupe de cinq amis.

CHAPITRE 32

Harold, Mary et Ed étaient nés à Hunter's Rock, où ils avaient grandi ensemble. Une enfance paisible dans l'Amérique profonde – on faisait bien pire. Parvenus à l'âge adulte, ils firent la connaissance de deux nouveaux venus dans un bar, et de ce jour tous les cinq devinrent inséparables.

Mes parents, qui étaient déjà mariés, apprirent très tôt qu'ils ne pouvaient avoir d'enfants. Faisant contre mauvaise fortune bon cœur, ils décidèrent que ce n'était pas la fin du monde. Ils étaient là l'un pour l'autre, et savouraient la vie en tant qu'amis et amants. Il y avait des tas de choses à entreprendre et à découvrir : les années n'allaient pas s'écouler lentement, et ils n'allaient pas déprimer au seul motif que lorsqu'ils refermeraient la porte le soir ce ne serait jamais que sur eux deux dans leur grotte. Ils allèrent de l'avant, tâchèrent d'accepter les cartes qu'on leur avait distribuées. Et traversèrent ainsi deux années de travail, de sommeil et de vendredis soir, de longues parties de billard où personne ne perdait.

Puis le monde changea, et ils s'aperçurent que transmettre un capital génétique n'est pas la seule façon d'imprimer sa marque à l'univers. On assistait alors à l'avènement d'une ère que je n'ai jamais dû bien comprendre. En plein désert culturel, des montagnes et des ravines apparurent, éventrant le sol où dormaient les masses. Des manifestations de rue. Des sit-in sur les campus, l'alliance inédite entre les étudiants et le corps professoral. Des bagarres dans les restaurants qui refusaient d'installer Noirs et Blancs au même comptoir. La police qui tirait sur les citoyens, les enfants qui se révoltaient contre leurs parents. Les marches. Les invectives baiseur-de-nègres, fasciste, pédé, coco. Des idées érigées en missiles. De longues soirées dans des maisons à se défoncer, à discuter de ce qu'il fallait faire, à discuter de modes de vie alternatifs, à discuter de discuter de discuter.

Ils étaient plus vieux que la moyenne des activistes. Ils avaient du temps et de l'énergie à dépenser – et davantage de perspectives que les adolescents ou les opprimés en colère. Beth Hopkins s'investit dans la syndicalisation des employés de maison noirs. Harold dispensa des conseils juridiques gratuits aux démunis, ou à ceux dont la couleur de peau les plaçait toujours sur le mauvais plateau de la balance. Don Hopkins lança une campagne pour empêcher la démolition de quartiers entiers au profit de ceintures autoroutières qui constituaient le premier pas vers la ville américaine post-moderne, où les indésirables sont refoulés du centre par des rivières à six voies charriant des bolides d'acier, et où les inégalités sont inscrites dans le paysage. Mary et Ed étaient surtout des sympathisants, mais ils donnaient un coup de main dès qu'ils le pouvaient, chaque fois qu'Ed était sobre. Mary était amoureuse d'Harold, et Ed cherchait simplement des gens avec qui traîner. Ils allaient gagner leur vie puis travaillaient sur leur temps libre, ces vieux combattants, ces gens qui, ayant franchi le cap fatidique de la trentaine, étaient capables d'arrêter leurs priorités : se concentrer sur les activités les plus socialement utiles, plutôt que de céder à l'ivresse de la foule et aux occasions d'emballer une jeune chose sexy empourprée par l'adrénaline de la protestation.

Deux années durant ils levèrent poings et banderoles, firent don de leur temps, de leur argent et de leur cœur. Certaines choses changèrent. La plupart, non. Le statu quo a de la ressource. La guitare bruyante et l'amour libre ne peuvent pas tout. Petit à petit le goût de l'époque tourna, alors que nos cinq infatigables rempilaient pour une année. Ce fut Harold qui le premier prit la mesure de la situation. Il remarquait que les gens qui venaient solliciter ses conseils, ces vétérans des après-midi ensoleillés passés à insulter les flics, lui arrivaient dans un état de plus en plus déplorable. Que leur résistance pacifique commençait à faire du grabuge, et que les hématomes et cicatrices qu'il constatait n'étaient pas tous le fait de la police. Qu'il existait plusieurs factions au sein des *beautiful people*, et que leur confrontation devenait plus âpre que celle avec les autorités. Qu'il y avait certains groupes dont les objectifs semblaient plus simplistes et rétrogrades, dont le programme ne comprenait aucune revendication, seulement de la noirceur.

Au début les autres ne furent pas du même avis. Il ne fallait y voir, d'après eux, que l'essoufflement d'un rêve, tel que Don l'avait toujours prédit. Les clivages naturels réapparaissaient, voilà tout : l'ardeur retombait devant le triste constat que la République populaire d'Amérique n'avait jamais été aussi lointaine. Mais c'est alors que la mort s'en mêla. Des manifestations où l'on retrouvait aussi bien des flics que des étudiants gisant à terre, un tesson de bouteille planté dans le visage. Les rixes en pleine rue qui semblaient éclater spontanément. Les concerts de rock qui se transformaient en pugilats, et où l'on retrouvait des corps et un revolver après dispersion de la foule. Les explosions qui ôtaient la vie d'innocents sans servir la moindre cause. Certains incidents étaient dus à des militants sincères, persuadés que la lutte armée était le seul moyen d'avancer. Mais les pires atrocités étaient l'œuvre d'individus servant un tout autre projet. Les gens avec les flingues et la dynamite étaient mieux organisés que les guérilleros, et combattaient tout autant ces derniers que leurs causes. Il y avait un coucou dans le nid à fleurs, qui s'échauffait les ailes et préparait son envol.

Beaucoup jetèrent l'éponge à cette époque. L'Été de l'Amour se dissolvait dans l'Automne de l'Apathie, et la drogue en avait mis plus d'un sous terre. Ed disait stop. Mary aussi. Ils ne s'étaient impliqués que pour le plaisir, après tout, pour partager quelque chose avec leurs amis. La politique comme vie sociale, le slogan comme accessoire de mode. Même Harold vacillait. Il était avocat. Son âme aspirait à l'ordre.

— Mais Beth et Don, dit Davids d'une voix sèche et faible, ils ne voulaient pas lâcher prise.

Ils posèrent des questions, cherchèrent les lignes de démarcation. De certains tracts haineux ils retrouvèrent les imprimeurs, ainsi que les auteurs, et découvrirent que les fautes de grammaire et le ton démentiel étaient souvent délibérés. Ils pistèrent l'ami d'un ami d'un ami, celui que d'aucuns soupçonnaient d'avoir apporté le flingue à la manif, ou qui avait brisé la première bouteille, ou qui pouvait vous introduire auprès de ceux qui ne se contentaient pas de palabrer mais agissaient vraiment. Ils cherchèrent, et commencèrent à trouver.

Ils finirent par recevoir des menaces. Deux de leurs amis furent tabassés, laissés pour morts à l'arrière d'une voiture. Un autre disparut un après-midi pour ne jamais revenir. Harold se retrouva au chômage, preuve que ces gens-là avaient de bien meilleures relations que les étudiants et hippies dont ils sabotaient les actions.

Un soir, ma mère fut suivie, enlevée, conduite loin de la ville et retenue sur une banquette sous la menace d'un couteau, pendant qu'un individu au visage masqué lui expliquait que s'ils persistaient à fouiner, leur prochaine demeure serait exiguë, définitive, et logée au cœur d'une forêt où personne ne s'aventurait jamais. Elle fut ensuite violée, par quatre hommes, avant d'être jetée de la voiture à l'orée de la ville, nue et les cheveux taillés.

Après cela mon père changea. Il se lança à leurs trousses. Pendant quatre mois, lui et ma mère prirent congé du monde et de ses habitants, s'enfonçant dans les ténèbres jusqu'à trouver la bougie qui en éclairait le centre. Les autres ne surent jamais les détails de cette phase, constatèrent seulement que mes parents n'étaient plus les mêmes après. On voyait toujours les Hopkins, mais maintenant qu'ils ne menaient plus le bon combat, les liens du groupe s'effilochaient. Don se mit à parler de choses étranges, d'une conspiration vaste et floue ourdie par des gens qui voulaient détruire la société de l'intérieur. Les trois camarades n'y prêtèrent pas attention, au début tout du moins. Ils y voyaient juste le délire d'un couple qui n'était plus en phase avec la réalité.

Un soir, Beth et Don se rendirent au bar où se retrouvait la petite bande. Ivre suite à une dispute avec Davids, Mary ne leur adressa même pas la parole. Mon père prit Harold à part pour lui parler de toute urgence. D'abord réticent, celui-ci accepta de repartir avec eux, abandonnant Mary et Lazy Ed au comptoir. Ce qui devait arriver arriva : ces deux-là se finirent à la bière avant de gagner les bois où ils couchèrent ensemble. Au bord de la mare perdue, d'ailleurs. Harold et Mary mettraient fin à leur concubinage peu de temps après.

Les trois autres avaient roulé pendant quatre heures jusque dans les collines du sud de l'Oregon. Armés, ils approchèrent leur cible en silence. À ce stade, mon père et ma mère avaient tourné le dos à leurs convictions, même s'ils pensaient au contraire les avoir trouvées – ayant compris, au prix fort, qu'à l'heure du combat entre les partisans de la vie et les partisans de la mort, la bataille se menait sur le terrain de cette dernière.

Le camp se trouvait dans une clairière distante d'un petit kilomètre de la route, au beau milieu de la forêt. Une grappe de cabanes, bâties à la main et disposées en cercle, comme à une lointaine époque. Après que ma mère eut observé chaque homme et confirmé sa participation au viol, tous trois lancèrent l'offensive et tirèrent dans le tas.

Le silence emplit le salon de Harold.

— Vous avez abattu tout le monde ? Mes parents ont tué des gens ?

— Pas les femmes ni les enfants, rectifia Davids. Et le but n'était pas de les tuer. Mais on a tiré sur les hommes. Tous. Dans la jambe. Ou dans l'épaule. Ou dans les couilles. Ça dépend.

— Je ne leur jette pas la pierre, dis-je. (Le pensais-je vraiment ? J'imagine que oui.) Si ce que vous me dites est vrai, je ne peux rien leur reprocher, ni à l'un ni à l'autre.

— Bien sûr que c'est vrai. J'y étais. Le dernier homme que nous ayons trouvé était celui qui avait maintenu le couteau sous la gorge de votre mère. On n'en avait pas encore conscience, mais il ne s'agissait pas d'une vulgaire bande de culs-terreux livrés à eux-mêmes. Ils défendaient une cause. Ils ont toujours existé. Donc, vos parents ont débusqué cet homme assis tout seul dans sa cabane. Et là, votre père, le grand Don Hopkins, jeune agent immobilier, a posé un flingue sur son visage et l'a abattu.

J'essayai de me figurer cette nuit-là, de voir mon père dans cette position. En fait, je ne l'avais jamais vraiment connu.

Cet afflux d'informations semblait me déborder des yeux.

— Puis ils ont entendu du bruit dans l'autre pièce de la cabane, et Beth est allée voir. La femme du type l'avait quitté, ou alors il l'avait tuée. Quoi qu'il en soit, elle avait laissé ses enfants derrière elle. Des jumeaux, âgés de six mois tout au plus, enveloppés l'un contre l'autre dans un petit lit, et désormais orphelins. Deux jolis bébés, c'est-à-dire ce que Beth avait tant désiré mais ne pouvait avoir. (Davids secoua la tête.) Du moins, c'est ainsi qu'ils me l'ont raconté. Je n'ai pas assisté à cet épisode-là. Peut-être ont-ils d'abord aperçu les enfants. Beth aura trouvé les bébés, et votre père aura peut-être entrevu un moyen de la venger. Peut-être se sont-ils dit qu'ils avaient droit à une balle mortelle.

— Mes parents n'étaient pas des menteurs, protestai-je.

— Alors vous saviez tout ça ?

— Ce n'étaient pas des menteurs, répétai-je inutilement. Toute cette histoire est un tissu de conneries.

— Que sont devenus les enfants ? demanda Bobby.

— On les a ramenés à Hunter's Rock. Don et Beth les ont élevés pendant quelque temps. Mais finalement il fut décidé qu'il valait mieux les séparer. Beth n'était pas trop d'accord, pas plus que votre père, mais le reste du groupe jugeait trop risqué de les garder ensemble. Outre ces bébés, nous avons mis la main sur des tas de papiers et de bouquins. Certains étaient très, très vieux. Ils fournissaient la preuve que vos parents avaient vu juste. Il y avait bel et bien conspiration. Ces hommes dans les bois en faisaient partie. Beth et Don se sont dit qu'ils parviendraient à changer votre nature, que l'environnement était le facteur le plus déterminant. C'était alors très en vogue, ce genre d'hypothèses. Ça l'est beaucoup moins aujourd'hui, bien sûr, avec tout ce foin sur l'ADN et la génétique. De nos jours, on prétend tout expliquer par la chimie.

— Donc, les bébés ont été séparés, relança Bobby.

— Ils en ont gardé un, et l'autre a été emmené loin. L'idée était de mettre toutes les chances de leur côté en évitant qu'ils se renforcent mutuellement tels qu'ils étaient. Ou peut-être s'agissait-il d'une jolie petite expérience, Ward, mitonnée par votre père. Nature contre culture. Je n'ai jamais très bien compris.

— Mais quelle nature, Harold ? À quoi rime cette angoisse autour de la nature des bébés ?

— Eh bien, à cause de vos gènes, voyons. Parce que vous étiez tellement non-viraux. Si purs.

— Bon sang de merde, me dites pas que vous croyez à toutes ces conneries ! Vous ne pensez pas sérieusement que...

Je m'interrompis, frappé d'un éclair.

— Attendez un peu. Y aurait-il un rapport avec ce fameux concept de « virus social » ?

— Tout à fait. Mais comment êtes-vous au courant ?

— Nous avons trouvé le site Internet des Hommes de Paille.

— Et comment avez-vous entendu parler d'eux ?

— Papa a laissé une vidéo. Je venais de la trouver quand vous êtes arrivé à la maison. Vous figurez tous dessus, même si je ne l'ai pas vu tout de suite. Il m'a aussi laissé un mot. Disant qu'ils n'étaient pas morts.

Davids secoua la tête dans un léger sourire.

— Ah, Don... Toujours ce besoin d'anticiper.

Son visage dénotait de l'affection, mais pas seulement.

— Attendez, intervint Bobby, si tout cela s'est déroulé à Hunter's Rock, comment se fait-il que vous ayez tous atterri ici ?

— Nous avons continué à nous voir pendant quelques années. Il y eut de bonnes soirées, mais ce n'était pas comme avant. Au bout de quelque temps je suis parti. Pour Dyersburg. Prendre un nouveau départ. Mary m'a rejoint l'année suivante. Ça n'a pas marché. Mais elle est restée en ville. Nous allions demeurer longtemps sans nouvelles des autres. En partie parce que nous pensions que c'était mieux ainsi. Et aussi, eh bien... nous avions commis des choses fort répréhensibles. Sur le coup, cette nuit-là, nous étions pourtant convaincus d'avoir raison. Nous nous sommes laissé prendre dans un engrenage, je suppose. La frustration de voir que rien ne changeait en ce bas monde, en dépit de tous nos efforts, et puis d'être à la merci de ce genre de types.

Mais, après coup, aucun de nous ne tenait vraiment à s'en souvenir. Pour Mary et Ed, ce n'était pas si terrible. Ils n'avaient pas participé. Mais c'étaient nos amis, si bien qu'ils durent eux aussi endosser une part de culpabilité. Ils étaient au courant, et ils ont gardé le secret.

— Mon père et Ed sont tombés nez à nez un jour. Il y a longtemps de ça. J'étais là. Ils ont fait mine de ne pas se connaître.

— Ça ne m'étonne pas, dit Davids. Je ne pense pas que votre père avait totalement confiance en Ed pour tenir sa langue. Il l'a fait, pourtant.

— Vous saviez qu'il était mort ?

— Pas avant que vous n'en parliez. J'étais au courant pour Mary. Je ne pensais pas qu'ils remonteraient jusqu'à lui. Il n'avait même pas assisté à la scène.

Une voiture passa dans la rue, et la tête de Davids se tourna, comme tirée par une ficelle. Il attendit jusqu'au bout que le son se dissipe. Jamais un homme ne m'avait paru si sûr de déguster.

— Si vous aviez décidé de rester dispersés, pourquoi mes parents se sont-ils installés ici ?

— Après plus de vingt ans sans souci particulier, sans que personne vienne nous chercher, Don a dû se dire que c'était terminé. Il passait parfois dans le secteur pour son travail, et il m'a rendu visite une ou deux fois. On a fait un petit billard, on s'est souvenus du bon vieux temps. D'avant cette sale nuit. Combien on s'était amusés. Cette période où l'on se voyait déjà transformer le monde. Cette évocation a été d'abord étrange, mais très vite on a eu l'impression que ces dernières décennies n'avaient jamais existé. Don est revenu ici passer un week-end avec votre mère, et ils ont finalement décidé de déménager. De reconstituer la vieille bande. De retrouver la jeunesse.

— Alors pourquoi ne m'ont-ils jamais dit que vous vous connaissiez d'avant ?

— Parce que... (Il soupira.) Parce que la construction des Halls a commencé juste avant leur arrivée ici, et Don en a entendu parler. Il a pris contact, les a démarchés. Il voulait ce contrat. Il l'a obtenu. Et au bout d'un moment il a commencé à flairer des choses louches. Alors il a décidé que nous devions à nouveau faire semblant, comme avant. Don n'a jamais vraiment vieilli. Pas comme nous autres. Ni comme votre mère, je suppose. En général il vient un âge où l'on devient plus coulant, où l'on accepte de lâcher prise. Mais pas Don. On lui mettait un secret sous le nez, et il fallait qu'il le perce. Il fallait qu'il comprenne.

Je hochai la tête. C'était tellement vrai.

— Et qu'est-il arrivé ?

— Il s'est mis à fureter. À chercher qui se cachait derrière ce projet, ce qu'ils manigançaient. Il a acquis la conviction qu'il s'agissait bien des types qu'il avait traqués jadis dans l'Oregon. Enfin, pas les mêmes types, mais leurs frères d'armes, cette fois-ci mieux organisés. Et qu'ils appartenaient à une sorte de mouvement planétaire. Un groupe clandestin, agissant en sous-main.

Il secoua la tête.

— Vous n'étiez pas de cet avis ?

— Je ne sais pas quel était mon avis. Je voulais juste qu'il laisse tomber. Certaines personnes accordent trop d'importance à la vérité, Ward. Il arrive que la vérité ne soit pas bonne à connaître. Qu'il vaille mieux la laisser tranquille.

— Et ils l'ont démasqué.

— Ils ont découvert l'existence d'un fouineur. Rien ne l'accusait directement, mais la taupe se cachait dans un cercle très restreint. Les choses ont commencé à se compliquer pour Don. De petites choses. Je pense qu'ils ont quelqu'un ici, en ville.

— Je vous le confirme. L'homme qui a tiré sur Bobby. C'est un policier.

— Seigneur ! glapit Davids. Dites-moi qu'il est mort.

— Mais qu'est-il arrivé à mes parents, Harold ? Que s'est-il passé cette nuit-là ?

— Don a décidé qu'ils devaient partir, disparaître. Il ne pouvait soumettre cette histoire à personne. Quand bien même on l'aurait cru, c'eût été signer son arrêt de mort. Je le soupçonne d'avoir également voulu négocier avec eux, en face. Je n'ai aucune idée de la façon dont il comptait s'y prendre. Mais à nous quatre, nous avions un âge cumulé d'environ deux cent cinquante ans. Alors nous avons décidé, en fin de compte, de simuler leur mort, de faire croire qu'ils n'étaient plus de ce monde. On avait tout organisé.

Mon cœur fit un bond ; je songeai au billet caché dans le fauteuil, et comprenais que papa aurait pu fermer UnRealty pour faire croire aux Hommes de Paille que c'était terminé, avant de revenir les liquider d'une manière ou d'une autre. Il avait agi ainsi pour me protéger. Ce n'était pas par défiance vis-à-vis de moi, et ça ne voulait pas dire qu'ils...

Voyant mon expression, Davids secoua la tête.

— Ils l'ont eu avant, dit-il. Deux jours avant la date qu'on avait retenue. Ils auraient dû se rendre au lac Ely le dimanche, et faire du bateau l'après-midi. Avoir un accident. Corps jamais retrouvés. Mais voilà que le vendredi... bref, vous connaissez la suite.

Ils sont morts, Ward. Je suis navré. Ils n'auraient pas dû. Mais ils sont vraiment morts. Et bientôt, ce soir sans doute, ce sera mon tour. Et alors ce sera fini pour de bon.

— C'est ce qu'on va voir, lança Bobby. C'est ce qu'on va voir tout de suite. (Il détacha la serviette de son bras. Elle était souillée de sang, mais plus rien ne suintait.) Je suis prêt à repartir. Allons là-bas et niquons tous ces braves gens.

Davids se contenta de secouer la tête. Il paraissait à cran.

— On est aussi bien ici.

— Sauf votre respect, monsieur, je ne suis pas de cet avis. Ces deux derniers jours ont vu l'élimination en règle de votre vieille bande de potes. Or, si ces mecs étaient rencardés sur Lazy Ed, ils le sont forcément sur vous.

C'est à peine si j'entendais leur échange. J'essayais d'assimiler ce qu'on venait de m'apprendre, de réajuster tout ce que j'avais cru savoir sur ma famille. Sur moi-même. Davids me regarda.

— Tout cela est vrai, Ward. Et je peux le prouver. Accordez-moi une minute, et je vous apporte les preuves.

Il se leva et quitta la pièce.

— Quelle histoire de merde ! murmura Bobby. Tu y crois, toi ?

— Pourquoi pas ? répondis-je, ne sachant trop sur quel pied danser. Ça colle, plus ou moins. Pourquoi mentirait-il ? On sait que c'est lui sur la vidéo, ce qui prouve qu'il les connaissait à l'époque. On sait que je ne suis pas né à Hunter's Rock. Et je l'imagine mal improviser un tel récit.

Une nouvelle voiture passa dehors, sans suite. Je fixai le mur jusqu'à ce qu'il se mette à pétiller sur mes pupilles.

— Ma mère m'a contacté, à peu près une semaine avant l'accident.

— Elle n'a fait aucune allusion à tout ceci ?

— Je ne lui ai pas parlé. Elle a laissé un message. Je n'ai pas pris la peine de rappeler. Mais elle ne téléphonait pas, d'habitude. C'était papa qui s'en chargeait, et le plus souvent ils attendaient que ce soit moi qui les joigne.

— Alors tu crois que...

— Je ne sais pas ce que je dois croire, Bobby, et c'est trop tard pour le découvrir.

— Alors qu'est-ce qu'on fait, maintenant ?

— Aucune idée.

Bobby se leva à son tour.

— Je vais voir si je peux dégotter du café, annonça-t-il. Ce putain de bras devient une torture.

J'écoutai ses pas s'enfoncer dans le couloir. Une partie de moi-même, contre toute raison et toute évidence, avait visiblement espéré que cette sombre farce, depuis que Mary m'avait arraché à la torpeur de Santa Barbara, n'était qu'une méprise. Des mensonges. Cette partie de moi-même avait créé le rêve de la piscine, tenté de me convaincre que j'avais raison de foncer, que des vies pouvaient encore être sauvées. Je savais désormais que c'était faux, qu'il n'y avait pas de place pour un ultime effort. Mon père avait échafaudé un plan, bien entendu. Comme toujours. Mais le mot que j'avais trouvé en était l'unique reste.

La sonnerie du portable me fit bondir. Je ne reconnus pas le numéro affiché.

— Qui est à l'appareil ?

— Nina Baynam. Tout va bien ? Vous avez une drôle de voix.

— Plus ou moins. Qu'est-ce que vous voulez ?

Je me sentais las, et pas d'humeur à écouter de pauvres histoires de serial killers.

— Nous sommes à Dyersburg. Où êtes-vous ?

— 34, North Batten Drive, répondis-je.

Elle marqua un temps d'arrêt.

— Pourriez-vous répéter ? dit-elle sur un ton bizarre. J'ai cru entendre 34, North Batten Drive.

— C'est bien ça.

— Mais ceci est l'adresse d'un certain Harold Davids, déclara-t-elle.

Mon cœur dédoubla ses coups.

— Comment vous savez ça, bordel ?

— Restez où vous êtes, dit-elle. Faites attention à vous. On arrive.

Elle raccrocha. Je me tournai vers la porte quand Bobby réapparut. Son visage me cloua le bec.

— Davids n'est pas là, déclara-t-il. Il est parti.

— Où ça ?

— Parti. Il y a une porte à l'arrière.

Je courus à la fenêtre côté rue, écartai le rideau. Là où j'avais repéré la berline noire, ce n'était plus qu'une place vide.

Nous retournâmes toute la maison. Il n'y avait rien à trouver – rien qui nous intéresse. Ce n'était qu'une vieille baraque en ordre, pleine de vieilleries.

Au bout de dix minutes, on entendit tambouriner en bas.

CHAPITRE 33

Nina martelait encore la porte quand je l'ouvris d'un coup sec. Zandt s'engouffra aussitôt dans la maison, sautillant de pièce en pièce. Je pivotai d'un mouvement gourd et lent. J'avais l'impression de dormir, comme si un rêve s'était imbriqué dans un autre.

— Qu'est-ce qu'il fait ?

Nina ignora ma question.

— Où est Davids ?

— Parti.

Ses yeux s'écarquillèrent, bordés de larges cernes. À croire qu'elle avait enchaîné les nuits blanches.

— Parti ? s'étrangla-t-elle. Mais pourquoi l'avez-vous laissé filer ?

De colère, elle frappa du pied. Bobby émergea de la cuisine.

— Ça ne s'est pas passé comme ça, expliqua-t-il. Il s'est juste volatilisé. Et puis, qu'est-ce que ça peut bien vous foutre ? Comment se fait-il que vous le connaissiez, d'abord ?

Elle sortit un calepin de son sac à main, l'ouvrit et le lui fourra sous le nez.

— Les promoteurs des Halls se cachent derrière une myriade de sociétés écrans. Mais dans l'avion j'ai remonté la piste, et on est arrivés tout près. Il semble que la société gestionnaire soit Antiviral Global Inc., domiciliée aux Caïmans. Le sieur Harold Davids qui habite ici en est le représentant légal pour le Montana.

— Merde, pesta Bobby.

Il repartit rageusement vers la cuisine.

Je fixai Nina.

— Vous vous trompez. Je viens de lui parler. À Davids. Il m'a dit... eh bien, il m'a dit pas mal de choses. Il a entendu parler des Halls, c'est certain. Mais de l'extérieur. Il n'est pas avec eux. Il a justement voulu aider mes parents à leur échapper.

— J'ignore ce qu'il vous a raconté, dit Nina avant de se tourner vers Zandt, qui émergeait de la pièce du fond. (Il secoua la tête puis s'élança dans l'escalier.) Mais je doute que M. Davids soit ce qu'il prétend.

— Que cherche Zandt ?

— Un corps, répondit-elle sans ambages. Vivant, tant qu'à faire.

Sa voix était un rien trop plate, et je compris qu'elle tremblait presque, tant elle était tendue. Son détachement de façade ne trompait personne.

— Vous ne la trouverez pas ici. Harold n'est pas votre tueur. Ce n'est qu'un vieux bonhomme. Il...

— Vous avez le numéro des Halls, Nina ? lança Bobby depuis l'embrasure de la porte, le téléphone de maison à la main.

Elle feuilleta son calepin, tourna une page.

— Nous avons un 406-555-1689. Mais on n'obtient qu'un message enregistré et un système de menus interminable. Pourquoi ?

Bobby sourit, si l'on peut dire. Il eut une expression faciale, en tout cas.

— Harold a joint ce numéro, dit-il. C'est inscrit sur le journal d'appels, il y a vingt minutes de ça. Pendant que nous étions ici.

— Mais... (Un court moment, ma bouche remua à vide, muette, comme j'essayais de formuler mes objections.) Davids avait l'air pétrifié. Tu l'as vu comme moi, Bobby. Il restait assis, à attendre, persuadé qu'ils allaient venir lui régler son compte. Comme ils l'ont fait à Mary et Ed. Tu l'as vu comme moi, nom d'un chien ! Tu as bien vu la tronche qu'il tirait !

— C'est sûr qu'il avait peur, Ward. Mais de nous. De *nous*. Il pensait qu'on l'avait démasqué. Que *nous* allions le dérouiller.

Zandt redescendit l'escalier.

— Elle n'est pas ici.

Davids m'avait surpris un canif à la main. Il savait que nous avions des flingues. Mais j'y voyais toujours aussi flou.

— Pourquoi m'aurait-il révélé quoi que ce soit, s'il était avec eux ?

— Tu as découvert qu'il faisait partie de la clique de Hunter's Rock, dit Bobby. Tu as mentionné une vidéo, et le message du fauteuil. Tu l'as reconnu. Il ignorait ce que tu savais, et tu aurais pu le bluffer. Alors le plus simple était de te confier les trois quarts de la vérité, puis de modifier la fin.

Il conclut d'une bordée de jurons bien sentie : il prenait ce revers comme un échec personnel.

Le visage de Nina était une suite de points d'interrogation.

— C'est quoi, cette clique de Hunter's Rock ?

— Plus tard, dis-je. Il faut d'abord retrouver Davids.

Un portable sonna. Nous dégainâmes à l'unisson, comme dans un western. C'était celui de Zandt.

— Ouais ? fit-il.

— Bonjour, officier, dit une voix suffisamment forte pour qu'on l'entende tous.

Zandt croisa le regard de Nina.

— Qui êtes-vous ?

— Un ami, répondit la voix. Même si nous n'avons pas eu le plaisir de nous rencontrer. Mais ce n'est pas ma faute. C'est toi qui n'étais pas assez doué.

Zandt se tenait parfaitement immobile.

— Qui êtes-vous ?

On entendit pouffer à l'autre bout de la ligne.

— Je pensais que tu l'aurais deviné. Je suis l'Homme Debout, John.

La mâchoire de Nina tomba.

— Foutaises.

— Non, pas foutaises. Bien joué pour Wang. Et pour l'avoir encouragé à faire le bon geste. On t'est redevables, sur ce coup-là. Il aurait pu devenir gênant.

La bouche de Zandt était sèche, et claqua quand il parla.

— Si vous êtes l'Homme Debout, prouvez-le.

Bobby et moi le dévisageâmes.

— Je n'ai rien à prouver, répliqua la voix. Mais je vais te confier quelque chose pour ton bien. Si vous n'avez pas quitté cette maison d'ici deux minutes environ, vous serez morts. Tous autant que vous êtes.

Sur ces mots il raccrocha.

— On se tire d'ici, dit Zandt. Maintenant.

De retour dans la rue, nous entendîmes approcher des sirènes. Un paquet de sirènes. Je déverrouillai la voiture et sautai derrière le volant.

Nina resta plantée dehors.

— Je suis du FBI. Rien ne nous oblige à partir.

— Ben voyons, railla Bobby. Je vous signale qu'on a tiré sur deux flics tout à l'heure. Ils ne sont pas morts, mais on leur a quand même tiré dessus. Si tu veux rester au milieu de la route à brandir ton badge, fais comme chez toi. Mais on n'est pas dans une série de HBO, princesse. Ils vont t'éclater ta putain de cervelle.

La police n'ayant pas prévu d'approche en souricière, nous atteignîmes la grand-rue sans encombre. Puis je virai à droite et posai le pied au plancher.

Vingt minutes plus tard, nous avions quitté la ville et suivions la route qui sinuait lentement vers les contreforts. Personne ne demanda où j'allais. Tout le monde savait.

Nina nous expliqua ce qui s'était passé à L.A. Je lui rapportai les propos de Davids. Et Zandt retraça, avec une économie de détails, son histoire avec l'Homme Debout.

— Merde alors, expirai-je.

Bobby fronça les sourcils.

— Mais comment s'est-il procuré ton numéro ? demanda-t-il à Zandt.

— S'il est de mèche avec les Hommes de Paille, il n'aura pas eu grand mal. Ces gens-là ont monté une véritable filière d'approvisionnement en chair humaine. Ils font péter des choses à gauche, à droite et au milieu. Alors dénicher un numéro de portable est pour eux un jeu d'enfant.

— OK, mais pourquoi t'appeler ? Pourquoi te sauver la mise avant l'arrivée des flics ?

— On ne peut jamais deviner ses intentions. Mais je n'étais pas le seul en cause. Vous étiez avez moi.

— Davids leur a dit qui se trouvait chez lui, suggérai-je. Il nous a balancés. (J'étais si furax que je pouvais à peine articuler.) Et puis, vous ne trouvez pas cocasse que les Hommes de Paille mettent la main sur mes parents deux jours avant leur projet de disparition ? Ils ont tout planifié, tout préparé, mais voilà qu'au moment de tourner la page McGregor fomente un accident fatal.

— Davids les aurait tuyautés ? Pourquoi ?

— Il connaissait depuis le début la vraie nature des Halls. Puis papa en entend parler, y voit une opportunité juteuse, avant de se rendre compte que ce n'est pas ce qu'il croyait. Ça met Davids dans une position délicate. Supposons qu'il s'agisse des gars qu'ils ont combattus trente ans plus tôt. Davids a prétendu que seul leur chef avait été abattu. Si les autres ont survécu, ils ont pu raconter leurs malheurs à quelqu'un. La bande qui a créé les Halls aura ainsi découvert que Davids a participé à cette descente – c'est peut-être même pour cette raison qu'ils l'ont pris comme avocat.

— Mais s'ils ont leurs entrées partout, pourquoi faire appel à Davids ? Ils auraient pu embaucher n'importe qui.

— Certes. Mais les stars du barreau sont elles aussi bien introduites. Certaines se piquent même d'intégrité. Les Hommes de Paille peuvent pousser Davids du haut d'un ravin à la première occasion, et il le sait. « Travaille pour nous ou on ébruite ce que tu as fait une certaine nuit en forêt. Ou bien, plus simple, on te crève direct. » Que veux-tu qu'il dise ? Il est vieux, il flippe, et il a tout à perdre. Il a également du talent. La recrue parfaite, en somme.

— Puis ton père s'approche un peu trop, et Davids sait qu'il risque gros s'il n'avertit pas les Hommes de Paille. Alors il leur révèle que les Hopkins s'apprêtent à mettre les bouts.

Un long silence emplit la voiture.

— Il les a fait tuer, conclut Nina. Le seul être qui leur inspirait une confiance totale.

— C'est un homme mort, déclarai-je. Vous pouvez me croire.

Le temps d'aborder la montagne, il s'était mis à pleuvoir, de froides traînées argentées qui rayaient l'obscurité derrière nos vitres. La rivière bordant la route devenait un torrent. Nous étions les seuls sur la chaussée.

— On n'est que quatre, observa Nina.

Je me tournai vers elle.

— T'as qu'à appeler des renforts.

— Si tu crois qu'ils vont envoyer les hélicos pour mes beaux yeux ! Je n'obtiendrai pas mieux qu'une paire d'agents blasés qui mettront deux heures à arriver et dont le seul souci sera de me faire passer pour une gourdasse. (Elle regarda par sa vitre quelques instants.) Personne n'aurait une cigarette ? Je me disais que c'était le bon moment pour commencer.

Je plongeai la main dans ma poche, sortis le paquet cabossé et le plaquai sur le tableau de bord.

— Entre nous, je te le conseille pas, dis-je.

Elle me rendit faiblement mon sourire, mais laissa les cigarettes où elles étaient.

Cinquante minutes après avoir fui la maison de Davids, nous prîmes un long virage. J'avais ralenti l'allure et Bobby s'était redressé pour scruter la paroi des collines bordant la route.

— On y est presque, promis-je.

Nina regarda Bobby et Zandt charger leurs flingues, puis à contrecœur elle vérifia le sien. Ses doigts tremblaient. Les deux autres paraissaient moins troublés, mais j'aurais pu lui dire combien les pensées d'un garçon sont impénétrables parfois. Il n'y a pas un type de notre génération qui ne puisse réciter le fameux « Si tu veux savoir, dans tout ce bordel j'ai pas très bien compté non plus » de *L'Inspecteur Harry*. On s'imagine tous capables de demander à des paumés s'ils veulent tenter leur chance ou pas, tous capables d'être son propre Clint de poche. Et on croit tous que quelqu'un, quelque part, va nous regarder de haut si on n'assure pas.

Puis Zandt croisa son regard. Il lui fit un clin d'œil, et le visage de Nina montra qu'elle comprenait sa méprise. Le cinéma nous dicte peut-être notre comportement, mais ce sentiment-là se logeait bien plus profond, remontait aux jours où personne ne portait de vêtements, où chacun avait sa fonction, où les uns gardaient le feu tandis que d'autres traquaient le gibier. La seule différence réside dans la taille du groupe auquel on se sent appartenir, et dans l'intensité de l'amour qu'on porte à ceux qu'on défendrait jusqu'à la mort. Zandt était aussi tendu qu'elle. Et moi donc.

Je rangeai la voiture sur le bas-côté.

— Nous y sommes.

Cinquante mètres plus loin se trouvait le portail.

— Personne dans les parages, releva Bobby. Redis un peu comment c'est foutu.

— Tu franchis la grille, puis tu roules sur de l'herbe. Tu vires à gauche et tu tombes sur une route cachée, dissimulée derrière les arbres. Elle tortille jusqu'au plateau.

— Autrement dit, il pourrait y avoir des gens dans les arbres, ou n'importe où sur le trajet.

— En effet.

— Faisons vite, alors.

J'acquiesçai.

— Tout le monde est prêt ?

— Comme jamais, répondit Zandt.

Je pressai l'accélérateur.

La voiture bondit en avant, dérapant sur la route mouillée. Je parcourus la distance restante à tombeau ouvert, puis virai droit sur le portail.

— On baisse la tête, commanda Bobby.

Nina et Zandt s'exécutèrent. Bobby se cala dans l'angle du dossier et de la portière, son arme au poing. La seconde d'après nous défoncions le portail, dont quelques lattes voletantes étoilèrent la vitre de Nina. Le véhicule s'enfonça dans les hautes herbes, menaça de s'embourber. Je me débattis et parvins à nous dégager.

Je relâchai la pédale pour reprendre le plein contrôle, puis mis le cap sur le bouquet d'arbres en regagnant un peu de vitesse.

Au passage d'une bosse je vis Nina décoller de son siège. Elle était à peine retombée qu'elle s'élevait à nouveau. Un grognement à l'arrière : Zandt connaissait le même sort. Bobby semblait quant à lui vissé à son siège.

Une dernière bosse, plus modeste et ferme, puis le sol devint soudain tout lisse.

Je fonçai au milieu des arbres, les paupières plissées.

— Tu vois quelqu'un ?

— Non, répondit Bobby. Mais ne ralentis pas.

Au bout de cent mètres la route obliquait serré sur la droite, puis nous gravîmes la pente. Bobby balayait le paysage du regard tandis que je ballottais la voiture de virage en virage, mais les coups de feu se faisaient toujours attendre. Quand la tête de Zandt émergea doucement, au mépris de la consigne, Bobby la lui rabattit vers le plancher. Le geste lui arracha une grimace, mais son épaule ne semblait pas trop l'handicaper. Pour l'instant.

— Mais où sont-ils ? demandai-je.

— Tous regroupés au sommet, sûrement, alignés en rang d'oignons.

— T'es rassurant, comme mec. Mais je suis bien content de t'avoir, tu sais.

— Ce doit être une sorte de truc amical, répondit Bobby. Mais je te préviens : si ça finit mal, je reviendrai te hanter.

— C'est déjà le cas. Ça fait des années que j'essaie de me débarrasser de toi.

Les pneus raclèrent le dernier virage, et l'imposante entrée des Halls nous toisait au fond de la montée.

— Toujours personne, dis-je en ralentissant.

— Il y a quoi, maintenant ?

— Après l'entrée, la route part à gauche. Deux grands bâtiments. L'accueil, et ce qui ressemble à une remise. Il y a une grande palissade tout autour du pré. Les maisons se trouvent derrière.

Nos deux compagnons relevèrent la tête avec précaution.

— Alors ? fit Zandt.

— La grille de l'entrée, trancha Bobby. Impossible de franchir cette palissade.

— Mais c'est à l'entrée qu'ils nous attendront, objectai-je.

— On n'a pas le choix.

La voiture dépassa l'arche en pierre et fila droit vers les bâtiments en bois. Un projecteur teintait le parking d'un blanc lunaire et morbide. Je levai le pied. La voiture arrêta sa course au centre de parking. Il était complètement vide. Je coupai le moteur, laissant les clés sur le contact.

— Quoi ? demanda Nina.

— Pas la moindre bagnole. C'était pourtant plein quand je suis venu l'autre jour.

Zandt ouvrit sa portière et sortit sans attendre les instructions. Bobby jura avant d'émerger de l'autre côté, flingue en main. La lumière blafarde faisait d'eux des cibles faciles, mais révélait aussi qu'il n'y avait personne sur les toits. Pas de comité d'accueil. Juste deux gros chalets, reliés par un mur de planches.

Nina et moi sortîmes à notre tour. Son flingue semblait énorme et lourd dans sa main.

— On entre par ici, dis-je désignant de la tête le bâtiment de droite.

Ils me suivirent et se répartirent des deux côtés des portes vitrées. Bobby avança la tête, scruta l'intérieur.

— Personne à la réception, dit-il.

— On y va ?

— Ma foi, oui. Après toi.

— Comme c'est gentil !

Je tendis le bras, poussai doucement l'un des battants. Aucune alarme ne retentit. Personne ne me tira dessus. J'ouvris la porte plus avant et m'engageai à l'intérieur sur la pointe des pieds, talonné par les autres.

La salle d'accueil baignait dans le silence. La musique de fond s'était tue, et le feu était mort dans l'âtre de la cheminée en galets. La toile abstraite avait quitté le mur derrière le comptoir. La pièce entière semblait mise au garde-meuble.

— Merde, pestai-je. Ils sont partis.

— Tu parles, dit Bobby. Ça ne fait qu'une heure. Ils n'auront pas eu le temps de débarrasser le plancher.

— Peut-être que si, objecta Zandt. Quand on a laissé Wang, cinq ou dix minutes se sont écoulées avant qu'il se flingue. Il aura eu le temps de les appeler.

— Ça fait quand même court. Trop court pour tout empaqueter.

— Ou alors ils étaient déjà sur le départ, suggéra Nina. Tu as collé une raclée à leur agent immobilier. Le message leur a peut-être suffi, ce qui leur a laissé deux jours pour se retourner. Mais peu importe. On va quand même voir ce qu'il y a là-dedans.

Elle s'éloigna vers la porte du fond, qui débouchait logiquement au cœur des Halls. Elle semblait animée d'une sorte de rage affligée, épouvantée à l'idée qu'elle puisse arriver trop tard, que le fantôme qu'elle avait poursuivi jusqu'à en faire l'unique lueur au bout de son tunnel se dérobe à nouveau.

Nous restions immobiles. À l'évidence, elle se fichait qu'on l'accompagne ou non. Il fallait qu'elle y aille. Il fallait qu'elle voie.

Elle n'entendit pas le coup de feu.

Le temps que la détonation parvienne à ses oreilles, elle tombait déjà, projetée de guingois sur le plateau d'une table basse. Elle ouvrit la bouche pour crier, mais il n'en sortit rien. Zandt se rua auprès d'elle.

Je virevoltai pour découvrir un type dans l'embrasure des portes battantes. McGregor. Bobby vit pour sa part une femme derrière le comptoir, et dans son dos un jeune malabar surgissant d'une porte dérobée, camouflée dans le lambris.

Tous trois portaient des flingues. Tous trois en firent usage.

Le garçon mourut le premier. Sa technique n'était que pure télévision : l'arme tenue sur le côté, façon gang-bang. Bobby le descendit d'une seule balle.

Je me glissai derrière un pylône pour resurgir de l'autre côté, touchant McGregor d'abord à la cuisse, puis à la poitrine. J'évitai de justesse d'en prendre une en pleine face, la sentant siffler tout près. Je posai un genou à terre et me carapatai derrière un coin du comptoir, priant pour que la femme n'ait rien vu. Je rechargeai fébrilement, en laissant tomber la moitié des cartouches.

Zandt s'agenouilla à côté de Nina, étalée par terre, qui levait une main tremblante vers le trou dans sa poitrine. Assez haut, juste sous la clavicule droite.

— Nina... murmura-t-il, indifférent aux claquements qui criblaient l'air au-dessus de lui.

Elle toussa, le visage pris entre stupeur et déni.

— Ça fait mal, dit-elle.

McGregor continuait de tirer. La femme du comptoir faillit avoir Bobby avant que, respirant un grand coup, je me relève et lui vide un demi-flingue dans le corps. Ce n'est qu'après sa culbute par-dessus le malabar que je reconnus celle qui m'avait exposé les conditions d'admission bidons. J'ignorais toujours son nom.

Bobby maintenait McGregor au sol, écrasant le poignet du flic avec sa botte. Un pistolet gisait un ou deux mètres plus loin.

— Où est-ce qu'ils sont partis ? demanda Bobby. Et il y a combien de temps ? Dis-moi tout ce que tu sais, ou il va faire tout noir.

— Je t'emmerde, dit le flic.

— Comme tu voudras, fit Bobby en haussant les épaules.

Et il l'acheva.

Pendant qu'il vérifiait les autres corps, s'assurant qu'aucun d'eux n'allait se réveiller pour rouvrir les hostilités, j'accourus au chevet de Nina. Zandt comprimait la plaie.

— On se tire d'ici, décrétai-je.

— Non, protesta Nina d'une voix étonnamment forte.

Elle tenta de se redresser.

— T'es dans un sale état, Nina. Il faut te conduire à l'hôpital.

D'une main elle agrippa un pied de table, et de l'autre mon poignet.

— Faites vite. Mais allez voir.

J'hésitai. Je voulus interroger le regard de Zandt, mais je restais bloqué sur celui de Nina.

Bobby arriva.

— Merde, Nina !

— Je reste ici, vous entrez là-dedans, dit-elle à Zandt.

Elle en bavait, mais n'était pas au bord du coma.

— S'il te plaît, John. Emmène-les. Allez-y tous. Regarde si elle est là. Il le faut, je t'en supplie. Puis on ira à l'hosto. Promis.

Zandt tergiversa une seconde ou deux, puis se pencha pour l'embrasser sur le front et se releva.

— On fait ce qu'elle dit.

Je rechargeai mon flingue.

— Bobby, tu restes avec Nina. (Il se mit à protester, mais je continuai de parler.) Essaie de stopper l'hémorragie, et zigouille tous ceux que tu croises en dehors de nous. Tu seras plus utile que nous ici.

Il s'accroupit à côté d'elle.

— Fais gaffe à toi, mec.

Zandt et moi gagnâmes en hâte les portes du fond.

— Quoi qu'il arrive, on reste ensemble. Pigé ?

Il hocha la tête, ouvrit la porte. Elle donnait sur un chemin. La lumière blanche du parking en révélait une cinquantaine de mètres, suggérant en arrière-plan les contours de grands pavillons. Aucun ne semblait éclairé.

Nous nous élançâmes.

CHAPITRE 34

— On aurait dû prendre une lampe torche.

— On aurait dû prendre plein de choses, répliquai-je. De plus gros flingues, des renforts, une idée de ce qu'on fabrique...

Nous étions arrêtés à la première fourche du chemin. On aurait dit la grand-rue d'une minuscule ville où personne ne possédait de voiture. De part et d'autre le gazon était impeccable. Cette pâture encaissée dans les montagnes, une zone de quatre ou cinq hectares seulement, était conçue pour garantir à chaque habitation de l'intimité et un paysage au déroulé harmonieux. Je ne voyais pas comment caser un golf, ce qui signifiait que même leur agent préféré – le regretté Chip – n'avait jamais été admis dans cette enceinte. En retrait de chaque côté se dressait une maison. Le chemin s'enfonçait dans l'obscurité, où de nouvelles ramifications desservaient de nouveaux foyers, que l'on ne distinguait pas encore.

— Toi, tu vas voir à gauche.

— T'as écouté ce que je t'ai dit ? On ne se sépare pas.

— Mais t'as vu le nombre de maisons, Ward ? Nina en mène pas large, là-bas.

— C'est pas en se faisant tuer qu'on pourra l'aider. Tu veux inspecter ces machins ? OK, mais on le fait ensemble. On commence par où ?

Zandt opta pour la droite. À l'approche du pavillon, je révisai mentalement les plans qu'on m'avait montrés. Cet édifice semblait importé d'Oak Park, la banlieue de Chicago où Wright avait signé de nombreuses réalisations dans la première moitié de sa carrière. C'était une demeure splendide, et j'en voulais à ses auteurs d'avoir ainsi dévoyé la grammaire du maître. Celle-ci célébrait la vie et la communauté, pas l'individu et la mort.

Zandt était moins sensible à l'architecture :

— Où est cette putain de porte ?

Je le guidai à travers la terrasse du bas, puis suivis une allée de jardin qui longeait le flanc gauche, sous un balcon. Quelques pas supplémentaires nous menèrent au coin suivant, où se trouvait une large porte en bois. Elle était entrouverte.

— Entrée principale ?

J'opinai du bonnet. Pris ma respiration, et poussai doucement du bout du pied. Il ne se passa rien.

Je hochai de nouveau la tête. Zandt me précéda à l'intérieur.

Un petit couloir, une lueur filtrant du plafond à travers un vitrail à l'effet vert et froid. Au fond, une autre vitre bigarrée, faisant lucarne sur la pièce suivante.

Nous la contournâmes prudemment, pour découvrir une salle profonde et dénivelée. De nouveaux vitraux, et des fenêtres à claire-voie perchées en hauteur. Une cheminée sur la gauche. Des étagères à livres et un coin détente. Les étagères étaient vides. Le mobilier n'avait pas bougé, mais il n'y avait aucun tapis en vue.

Nous arpentâmes la pièce sur la pointe des pieds. Il régnait un silence parfait. Je levai la main, pointai du doigt ; Zandt avisa l'entrée d'une autre pièce, à moitié cachée derrière un panneau de bois. Il me rejoignit, et nous l'approchâmes ensemble, Zandt surveillant les arrières.

La porte débouchait sur une cuisine. Il y faisait plus sombre sans la lumière des hautes fenêtres. Elle était répartie sur deux niveaux, avec un coin repas. Sur la table trônait une unique tasse, pile au milieu. L'intérieur était sec et l'anse cassée. J'ouvris un placard, puis un tiroir. Rien.

— La maison a été vidée.

Zandt hocha la tête.

— Possible. Mais on va quand même vérifier.

Nous fouillâmes le reste de la maison.

— Il y a quelqu'un dehors, dit Nina pendant ce temps-là.

Elle était avachie dans un gros fauteuil en cuir, Bobby accroupi derrière elle. L'accueil baignait dans l'obscurité. Bobby s'était posé un dilemme à ce sujet. D'un côté, la lumière du parking étant allumée à leur arrivée, l'éteindre signalerait leur présence à quiconque rôdait dans le secteur. D'un autre côté, la bonne minute de fusillade aurait déjà rempli cet office. Pour finir, il avait farfouillé derrière le comptoir et coupé les lampes une à une. Cela paraissait ainsi plus sûr, à défaut d'être parfait. Le mur du fond n'était qu'à demi vitré, et là où ils étaient Nina et lui devaient échapper aux regards, même s'ils demeuraient des proies faciles. La loge était vaste, sombre et peuplée de trois cadavres.

— J'ai bien entendu quelque chose il y a une minute, avoua-t-il. J'espérais que c'étaient eux qui revenaient.

Nina secoua la tête.

— John va vérifier toutes les maisons. Ça va leur prendre un peu de temps, même s'il n'y a rien à trouver. *Surtout* s'il n'y a rien à trouver. Et le son provenait plutôt de devant.

Il opina.

— Ward va me tuer s'il apprend que je t'ai laissée toute seule, mais il va falloir que j'aille voir.

— Si tu dis rien, je dirai rien. Mais ne traîne pas trop.

Bobby s'assura qu'elle avait un flingue chargé, puis s'éloigna vers le mur, qu'il longea en se courbant. Parvenu aux portes vitrées, il avança lentement la tête. La voiture demeurait isolée sur le parking. Aucun signe de vie. Très bien.

Puis il entendit un nouveau bruit. Faible, mais il ne pouvait être dû aux seuls éléments. Ce n'était pas la pluie, plutôt un bref *pop* mécanique. Qui semblait provenir de l'autre côté du parking, près du second bâtiment.

— C'était quoi ?

Maintenant qu'il ne la veillait plus, Nina laissait la douleur envahir son esprit. En conséquence de quoi la tête lui tournait, et sa voix se brisait.

— J'en sais rien, répondit-il.

En pivotant il vit que Nina était bien cachée dans la profondeur du gros fauteuil. On ne pouvait mieux faire.

— Maintiens la pression sur la plaie.

Toujours accroupi, il poussa la porte vers l'extérieur. Il sentit s'engouffrer un souffle d'air glacé, sous le roulement de la pluie.

Le reste de la maison était vide. Quatre chambres, un bureau, une bibliothèque, une salle de musique. Vides et lessivées. Débarrassées de tout signe distinctif, alors qu'à l'évidence elles avaient été occupées tout récemment encore. Aucune poussière. Je redescendis avec Zandt par l'escalier central, moins soucieux du bruit désormais, et nous gagnâmes l'arrière du rez-de-chaussée. Un deuxième salon s'étendait là, vaste et un peu moins sophistiqué que celui du devant. Une large baie vitrée révélait un jardin paysagé d'environ deux mille mètres carrés. Je rétablis la sécurité de mon arme.

— La suivante ?

Il était clair que cette maison ne contenait rien d'intéressant. J'en avais fait mon deuil. J'étais disposé à aider Zandt à retrouver la fille, si tel était son souhait, mais pour ma part je brûlais de dégotter un ou deux Hommes de Paille en vie. De les asseoir par terre, et de leur soutirer quelques explications. Je ne pensais qu'à ça. Mais il semblait déjà trop tard.

— Je jette un œil derrière, répondit Zandt. Puis on fera la suivante, d'accord. Même si ça s'annonce mal.

Il ouvrit la porte qui coupait la baie vitrée en deux, et disparut sous l'averse. Je lui emboîtai le pas, mais restai contre le mur. J'adhérais de plus en plus à la thèse de Nina : ce fameux Wang avait peut-être hâté le mouvement, mais l'évacuation avait débuté dès que j'avais corrigé Chip. J'avais merdé, en somme. Je leur avais donné l'alerte, et le temps de fuir. Je ne m'étais pas attendu à une telle réaction. Ils vivaient dans un bunker ; ils étaient riches et puissants ; c'était leur territoire. Pourquoi fuir ? N'empêche que j'avais bien merdé. Il n'avait pas abordé la question, mais je soupçonnais Zandt d'être du même avis. Le regard de cet homme se remplissait d'une fureur animale.

Comme il s'affairait dans le noir, je remarquai un long câble courant au pied du mur. Il émergeait du coin de la baraque, et l'autre bout semblait s'enfoncer dans les massifs. La télé, peut-être. Ou bien l'accès ADSL tant vanté. Je m'apprêtais à y regarder de plus près quand Zandt émit une sorte de toussotement.

Je me ruai dans le jardin. Zandt se trouvait en plein milieu, droit comme un piquet.

— Quoi ?

Il ne dit rien. Pointa du doigt.

Je ne compris pas tout de suite, avant de découvrir que le sol était légèrement bombé sur notre droite.

Je me rapprochai. Déglutis.

— Dis-moi que c'est juste un animal là-dessous.

Mais Zandt secoua la tête, et je vis qu'il avait toujours le bras en l'air. Il m'indiquait un autre coin. Un autre monticule.

— Nom de Dieu, fis-je d'une voix étranglée. Vise un peu ça...

Je distinguais les autres, à présent. Trois courtes rangées de monticules. Douze au total.

Zandt posa un genou à terre, s'attaquant à la première bosse. L'herbe lui glissait des doigts, mais il parvint à extraire une motte. Le sol en dessous était lourd et humide.

Je me laissai choir pour l'aider à creuser. La tâche était âpre et il fallut deux bonnes minutes pour arriver au point où nous touchions soudain autre chose que de la terre, et où l'odeur devenait atroce. J'eus un geste de recul, mais Zandt arracha deux nouvelles poignées. Avant de renoncer.

— Il nous faut une pelle, dis-je.

Il secoua la tête.

— Tout ce qu'il y aura dans ces trous est mort. Sarah est peut-être en vie ailleurs.

— Arrête, mec. Il y a toutes les chances qu'elle soit enterrée ici.

Mais déjà Zandt rebroussait chemin vers la maison. Je le suivis, en tâchant d'éviter les monticules, même si je dus en piétiner au moins un.

De retour à l'intérieur, il fila droit au premier salon.

— Il faut recommencer la fouille, dit-il. On a loupé quelque chose.

— Je ne vois pas où.

— Alors commençons par ici.

Nous prîmes chacun un côté de la pièce, retournant les étagères, déplaçant les meubles. Je fus vite persuadé qu'il n'y avait rien pour nous, mais Zandt tenait à scruter chaque centimètre carré.

— Il y en a pour des heures. Franchement...

Je m'interrompis. Zandt releva les yeux.

— Quoi ?

Ce n'était pas dans la maison, mais derrière les baies vitrées de devant. Zandt me rejoignit.

— Tu vois, là-bas ?

Je désignai la fourche du chemin, à une vingtaine de mètres. Là, à l'intersection des allées menant aux différentes maisons, reposait une forme, pas très large, impossible à identifier à cette distance. Un petit fagot, peut-être.

— Oui, je vois, dit Zandt.

Ce n'était pas là précédemment.

J'ôtai de nouveau la sécurité du flingue et ressortis par la porte de devant. Je remontai lentement le chemin, couvert par Zandt qui restait à l'entrée pour surveiller les autres bâtiments.

Cela ressemblait en effet à un tas de bâtons. Courts, recourbés, très blancs. Très propres. Mais à deux mètres j'avais déjà une petite idée de leur nature. Je m'accroupis devant, en ramassai un. Me retournai pour appeler Zandt.

Pendant qu'il accourait, c'est moi qui le couvris, prêt à tirer sur la première silhouette venue. Car il y avait bien quelqu'un, ça ne faisait pas un pli. Qui savait où nous étions, de surcroît.

Après un bref examen Zandt déclara :
— Ce sont des côtes.
— C'est ce que je me suis dit. Humaines ?
— Ouais.
— Qui les a mises là ?
— Ward, regarde.
Environ cinq mètres en amont du sentier se trouvait un autre bâton. Je m'y rendis, me baissai pour le saisir.
— Fille ou garçon ?
Zandt me prit le fémur des mains. À l'instar des côtes, cet os était immaculé, comme préparé pour être exposé dans un muséum.
— J'en sais rien. Mais quelqu'un de pas bien vieux. Un ado.
Nous restâmes immobiles, à scruter les deux côtés du chemin.
— On veut nous conduire quelque part, dis-je.
— La question est de savoir si on dit oui.
— Je doute qu'on ait le choix.
— Mais on a déjà trouvé la maison aux cadavres.
— Une maison, rectifiai-je. La première qu'on ait visitée. Soit c'est une chouette coïncidence, soit il y en a davantage.
Au carrefour suivant nous attendait un nouvel os, sur le côté gauche du sentier, comme s'il nous indiquait la maison la plus proche. Nous la vérifiâmes en vitesse. Ici les tombes étaient dispersées sur le côté, et mieux – ou plus dignement – camouflées. Il fallut que Zandt s'étonne que les petites dalles ne forment pas un chemin sensé pour que nous comprenions qu'il s'agissait de stèles.
De l'autre côté du pavillon nous trouvâmes un nouvel os, qui invitait à s'enfoncer plus avant dans les Halls. Il s'agissait d'un demi-pelvis. Aucun de nous deux ne sut déterminer le sexe de son propriétaire, bien que l'état et la largeur de l'échancrure sciatique eussent probablement suffi à Nina pour reconnaître une fille de l'âge de Sarah Becker.

Cela faisait au moins dix minutes que Bobby était tapi dans l'ombre de la voiture. Pas de nouveau bruit depuis sa sortie, ni le moindre signe de mouvement. Mais le problème demeurait entier : les sons précédents avaient bien une cause, et il doutait qu'elle se soit envolée comme par magie. Il gardait sa planque pour voir si cette cause allait se découvrir ; la laissait venir à lui avant qu'il n'aille à elle. Possible qu'il s'agisse simplement d'un animal. Un cerf, par exemple. Peu probable, mais possible.

Encore deux minutes et il s'ébroua. Nina allait s'inquiéter s'il s'éternisait dehors, et il grelottait sous ses vêtements trempés. Son épaule l'élançait méchamment. Mais pas question de faire demi-tour. Il fallait d'abord vérifier le second bâtiment. Il remonta la suite de petits plots délimitant les places de parking. Il s'exposait dangereusement, mais il n'y avait pas d'autre accès possible. Le bâtiment ressemblait à un vaste hangar, sans les finitions du précédent, et sans fenêtres. Il longea la façade jusqu'au coin gauche, où il trouva enfin une porte.

Un gros cadenas, détaché. Bobby voulut appeler le nom de Ward, au cas où il eût atterri dans ce bâtiment, mais c'était impossible : il serait d'abord repassé par l'accueil. Non, il y avait forcément quelqu'un d'autre là-dedans. Il poussa la porte avec le coude et posa le pied à l'intérieur.

Il se retrouva dans un petit couloir, dont les murs ne dépassaient le sommet de sa tête que de cinquante centimètres avant de laisser place au vide. Comme dans une sorte d'écurie. Mais l'odeur puissante ne rappelait en rien celle des chevaux. Une faible lumière suintait d'une autre partie de l'édifice. Trois mètres plus loin, le couloir en croisait un autre, en angle droit.

Avant ce carrefour il trouva deux portes, qu'il ouvrit l'une après l'autre. L'une abritait le genre de fournitures nécessaires aux structures collectives, ainsi qu'un long mur couvert de dossiers. L'autre, plus petite, devait être une cave à vin. Les casiers étaient vides, et cela ne présageait rien de bon. S'ils avaient eu le temps de remballer le château-lafite, c'est qu'ils étaient partis depuis longtemps. L'abandon des dossiers n'en était que plus surprenant. Il regagna la première pièce, ouvrit une boîte d'archives au hasard. Elle ne contenait pas le moindre document, seulement deux cartouches Zip étiquetées « Scottsdale ». Il les mit dans sa poche et rangea la boîte.

Il reprit le couloir à pas feutrés jusqu'au carrefour. Avant de s'y risquer il s'immobilisa un instant, laissant sa mâchoire tomber. On entend mieux de cette façon, jusqu'aux sons les plus faibles – une histoire de trompes d'Eustache. Il ne capta aucun bruit, mais remarqua un câble tendu au sol devant lui. S'il alimentait l'éclairage, il lui faudrait le couper. Il semblait toutefois plus récent que le reste de la structure, comme un ajout ultérieur. En avançant la tête il vit qu'il longeait le couloir de gauche. Il entreprit de le suivre. Mais après deux ou trois pas son esprit s'arrêta sur tout autre chose.

Cette partie du bâtiment était bel et bien organisée en écurie : une succession de petits box de part et d'autre du couloir, divisés en cages d'un demi-mètre carré environ. Dans la première reposait une forme. On aurait dit une personne. De petite taille.

Bobby s'agenouilla à ras des barreaux. La forme était un garçon. Cinq, six ans à la rigueur. Nu. Pieds et mains ligotés avec de l'adhésif. Sa bouche semblait recouverte de la même matière, mais ce n'était pas évident car il ne restait plus grand-chose de sa tête. Le sang qui imbibait la paille était encore frais. On avait scotché aux barreaux la photo d'un mignon petit garçon dans un pays chaud. Il ne fixait pas l'objectif, ne semblait même pas conscient d'être photographié. Bobby comprit qu'il s'agissait du même enfant, dans une vie antérieure. Il s'appelait Keanu.

Bobby détourna les yeux. S'aida de ses mains pour quitter la stalle, et passer à la suivante. Un autre garçon, un peu plus âgé, mais tout aussi mort. Une autre étiquette sur la cage. Celui-ci souriait au photographe, mais d'un regard voilé par le doute. Comme si un inconnu l'avait abordé dans la rue au retour de l'école, en lui demandant s'il voulait bien être pris en photo, et qu'il avait accepté tout en jugeant cela un peu louche.

Bobby crut défaillir en percevant un léger bruissement. Il se figea, avant de comprendre que celui-ci provenait de l'autre côté du couloir, quelques mètres plus loin.

Dans cette troisième cage reposait une fille d'à peu près huit ans. Elle aussi était dûment référencée. Ginny Wilkins. Elle n'était pas tout à fait morte, malgré la balle qu'elle avait reçue dans l'œil. Son autre pupille était sèche et vitreuse, mais ses membres inférieurs remuaient faiblement. Une fraction de son système nerveux fonctionnait encore, pour un petit moment encore.

Bobby savait qu'il y avait d'autres box. Au moins deux de plus. Il savait aussi qu'on n'aurait pas laissé ce lieu ouvert par inadvertance. Que même du temps où les Halls étaient en activité, on n'en aurait autorisé l'accès qu'à une poignée d'individus triés sur le volet. Mais il ne pouvait détacher son regard de cette fillette dans son clapier, dans cette prison où elle avait été livrée et conservée, à la disposition du pensionnaire des Halls qui l'avait commandée.

Il se sentait idiot, petit, et malade. Il se sentait ignorant et naïf. Il avait cru connaître les horreurs du monde, s'être colleté les pires en arpentant la face sombre de la planète. L'amitié de Ward l'avait préservé. Il savait que Ward le respectait, lui et son curriculum sulfureux, et cela l'aidait à accepter l'idée de grandir, de se ranger des voitures un jour. Mais là, comme il fixait l'intérieur d'une cage, et un morceau de viande à peine animé qui rendait ses dernières contractions, Bobby se rendit compte qu'il avait à peine gratté la surface de l'imaginable – que les guerres et les meurtres montrés aux infos tenaient du résultat sportif, de la mort spectacle ; que la plupart des terroristes qu'il avait interrogés n'étaient que des amateurs au royaume des ténèbres. Eux voulaient au moins faire connaître leurs crimes. Ils les commettaient au nom d'un dieu, d'une idéologie, de camarades tombés ou de griefs ancestraux. Ils n'agissaient pas par pur égoïsme. Bobby comprit que cela faisait une différence, et que si nous appartenons tous à la même espèce, comme on le dit, nos chances de salut sont bien minces : ce que nous accomplissons le jour ne rachètera jamais ce que nous commettons la nuit. Certains aspects du comportement humain étaient inéluctables, mais ceci ne l'était assurément pas. Ou c'eût été admettre que nous ne possédions aucune limite. Le fait que nous soyons doués pour l'art n'autorisait pas à qualifier ce spectacle-ci de simple aberration ; à ranger sous la bannière « humain » ce que nous admirions, et le reste dans la catégorie « monstrueux ». Les deux étaient l'œuvre des mêmes mains. L'intelligence n'endiguait pas la barbarie, elle la perfectionnait. En tant qu'espèce nous en étions totalement responsables, et chacun portait en lui son frère des ténèbres.

Un nouveau bruissement, dans son dos, mais il ne releva pas les yeux tout de suite. Il se demandait encore s'il devait achever Ginny pour abréger son supplice – ou son supplice à lui – ou si elle survivrait à son retour dans la civilisation (et, le cas échéant, ce qu'il y aurait à sauver).

Il tergiversa trop. Ce n'était pas un autre enfant derrière lui. C'était Harold Davids, qui abattit Bobby d'une balle dans la nuque.

Les jambes de Ginny Wilkins remuaient encore, se tortillant lentement à des milliers de kilomètres de sa maison et de ceux qui la pleuraient, plusieurs minutes après que Robert Nygard eut trouvé la mort.

Zandt avait traversé la dernière maison sans même regarder les chambres. Indifférent à la pluie, il suivait seulement la piste d'ossements, muet depuis cinq minutes.

Je cavalais sur ses talons. La piste avait cessé de jouer avec nous, de nous faire crânement tourner en rond ; elle se contentait de nous guider vers le bout du sentier. Un petit carré, formé par des métacarpes, avec une rotule au centre. Une suite serpentine de vertèbres, une tous les soixante centimètres, a priori dans le bon ordre. L'Homme Debout avait dû tracer l'essentiel du parcours longtemps à l'avance, n'ajoutant le fagot du début qu'à la dernière minute, quand il nous avait sus prêts pour le voyage. Le reste avait exigé temps et minutie. Le tueur ne nous avait pas extirpés de chez Davids pour notre bien, mais pour ceci : il avait déjà préparé la rencontre, et ne voulait pas voir son travail gâché. Il avait, d'une façon ou d'une autre, dicté notre conduite depuis un bon bout de temps.

Enfin, une paire de clavicules, disposées en V inversé, transformées en flèche par le second fémur. Une flèche indiquant la dernière bifurcation du sentier, vers une maison distante de trente mètres.

Je rattrapai Zandt et l'agrippai par l'épaule.

— On ne va pas là-dedans, dis-je.

Il m'ignora, se libéra de ma main et fonça vers les marches de la terrasse.

Je lui empoignai le bras.

— Il nous attend, John, tu le sais comme moi. Il a déjà tué la fille, et maintenant c'est nous qu'il va tuer, puis il ira chercher les autres et les tuera aussi. Je sais que tu tenais à cette fille mais je ne vais pas te laisser...

Il fit volte-face et me frappa au visage.

Je m'étalai sur le sentier mouillé, plus hébété que blessé, me demandant si je n'avais pas loupé un épisode. On aurait dit que Zandt ne me voyait même plus, n'avait pas la moindre idée de qui j'étais.

Je remontai sur mes jambes et me lançai à sa poursuite.

Zandt gravit les marches d'un pas lourd. Au contraire des autres, cette maison possédait une porte en plein milieu de sa façade. L'escalier de la terrasse y menait tout droit, comme un long entonnoir jeté spécialement à notre intention. La mâchoire de Zandt semblait lui rentrer dans le cou, retenue par les seuls spasmes de son visage et de sa peau.

On avait posé quelque chose en haut des marches.

Cette fois je ne tentai pas de ceinturer Zandt, je restai seulement derrière lui. Je n'avais qu'un pas de retard lorsqu'il atteignit la pièce finale de la piste.

Un pull de fille, gorgé de pluie. Mais plié avec soin, un nom brodé sur le devant. Le pull était couleur pêche. Le nom était Karen Zandt.

CHAPITRE 35

Le petit doigt de Nina lui dit de se taire. De ne plus faire un bruit quand elle entendit s'ouvrir la porte de l'accueil. Elle avait mal au thorax, et la douleur irradiait dans son corps, lui rongeait le ventre et se propageait dans son bras droit qui tenait l'arme. Elle n'osait imaginer le résultat si la balle l'avait perforée un peu plus bas.

La porte se referma dans un souffle bref. Quelques pas sur le sol, mais personne ne prononça son nom. Elle sut alors que Bobby était mort. Elle ne pouvait distinguer cette partie de la pièce sans se redresser et tourner la tête, mouvement qui l'aurait à la fois suppliciée et trahie. Alors elle essaya de s'enfoncer dans le coussin du fauteuil. Les pas continuèrent, accompagnés d'un autre bruit. Comme un roulement. Puis on posa quelque chose au sol. S'ensuivit un silence.

— Je sais que vous êtes là, dit une voix.

Son cœur fit un bond, et elle faillit répondre. Faillit avouer. Faillit admettre que oui, elle était bien là, comme on l'avait obligée à faire durant son enfance. Mais ce temps-là était bien loin, et aujourd'hui elle serrait ses lèvres et le flingue de toutes ses forces. Sa main semblait moins opérationnelle que d'habitude.

— Ils n'auraient pas laissé Bobby ici, déclara la voix, s'il n'y avait pas quelqu'un à veiller.

De nouveaux pas. D'exploration. Le type ne l'avait pas localisée. Mais elle serait là, et il obéirait aux ordres. Comme à son habitude. S'il renvoyait une image d'homme fort, capable et meneur, en vérité il avait toujours suivi. Cela faisait si longtemps qu'il vivait dans la culpabilité et sous la menace ; le reste n'avait plus grande importance. Voilà ce que lui avait infligé le père d'Hopkins : la mise à sac d'une vie honnête et rangée. On ne savait rien refuser au Don d'autrefois. On était aspiré dans sa roue. On se réunissait chez lui, on buvait sa bière, on se laissait gagner par ses idées. Jusqu'au jour l'on ne se reconnaissait plus soi-même, mais alors il était trop tard. On en venait à haïr le suiveur qui sommeillait en soi, et on savait qui blâmer.

— Peut-être avez-vous déjà succombé, lança-t-il. Mais j'en doute. Quoi qu'il en soit, je dois m'en assurer.

Nina tenta de sombrer dans le coussin, mais c'était douloureux. Et le moindre geste efficace aurait fait chuinter le cuir.

— Bobby est mort, répéta l'homme d'une voix âgée mais ferme, qui ne laissait aucune place au doute. Et ce sera bientôt le tour des autres. On pourrait vous épargner. Mais il ne faut rien laisser au hasard, et c'est à cela que je sers.

Ses pas se firent plus glissants : l'homme progressait par foulées minuscules pour maquiller la direction de son approche. Nina avait si peur qu'elle fondit en larmes, réaction surgie des profondeurs de son corps et de son passé, sans crier gare.

Elle recula lentement son bras gauche, le calant contre le bord du fauteuil. Puis ramena ses pieds vers elle, millimètre par millimètre. Sa main tremblait, et les nerfs de son bras semblaient en feu.

— C'est une bonne nuit pour mourir, susurra l'homme d'un peu plus près. Il ne s'agit pas d'une fin. Loin de là. C'est un nouveau départ. Un nouveau monde tout propre qui commence par un *bang*. (Il eut un petit rire.) Elle n'est pas mauvaise, celle-là.

Le patinage cessa.

Nina déploya tout ce qu'il lui restait d'énergie. Son corps partit vers l'avant. Entravé, verrouillé, basculant du fauteuil pour s'écraser sur le verre de la table basse. Elle savait qu'elle avait foiré, mais elle pouvait au moins distinguer l'ombre sur sa droite.

Davids hocha la tête.

— Ah, vous voilà.

Elle souleva sa main et pressa la détente. Une, deux, trois fois.

Elle ne suscita qu'un tir en retour, qui ne l'atteignit pas.

Un moment long comme l'éternité, dans l'attente du deuxième coup. Il ne vint jamais. Elle s'agenouilla, se hissa sur ses pieds, pivota.

À deux mètres d'elle, gisait un corps. Maintenant qu'elle bougeait, la perspective d'un mouvement prolongé devenait presque crédible, même assommée par un voile blanc de douleur.

Elle planta ses pieds dans le sol et s'ébranla.

Un vieil homme aux cheveux gris. Il n'était pas tout à fait mort. Elle le toisa, le dos voûté. Harold Davids croisa son regard.

— Vous n'y changerez rien, dit-il avant de céder au silence.

Mais Nina ne l'écoutait pas. Elle observait une chose posée près du comptoir de la réception. Incapable d'en déterminer la nature, elle se rapprocha de quelques pas.

Un petit tambour. Relié à un câble. Lui-même relié à une série de fiches encastrées dans le comptoir, qui ressortaient ensuite par la porte.

Le début d'un nouveau monde.

Elle se pencha par-dessus le comptoir, mais ne vit rien qui puisse servir de détonateur. Le déclenchement devait se commander ailleurs.

Elle parvint jusqu'au parking avant qu'une de ses jambes ne lâche, la rabattant violemment sur l'asphalte.

Le regain de douleur, mêlé au rebond de la pluie glaciale sur son visage, coupa court à sa confusion. Elle se mit à ramper vers la voiture.

Je retins Zandt de franchir la porte d'entrée de la maison. À ce stade, il était impossible de le raisonner, mais je savais qu'il ne fallait pas passer par là. Je l'avais déjà empêché de rebrousser chemin pour ramasser certains os, contraint de coller mon front contre le sien en hurlant le nom de Sarah Becker, pour lui rappeler qu'on pouvait encore trouver quelqu'un de vivant. Il ne m'importait plus vraiment qu'elle soit morte, à vrai dire. Il fallait juste la trouver. J'avais changé d'avis. Nous allions pénétrer à l'intérieur. Quoi qu'il arrive. Si le type était là, tant mieux, mais il fallait à tout prix terminer le parcours.

Je repoussai Zandt sur le côté de la maison, où nous trouvâmes une autre porte. Verrouillée. Dommage que Bobby ne soit pas là. Il aurait su l'ouvrir sans bruit. J'en étais incapable, aussi avertis-je Zandt d'un signe de la main avant d'enfoncer la lourde d'un coup de pied.

Personne pour nous accueillir. Nous nous engloutîmes sur la gauche dans un demi-escalier menant vers l'avant de la maison, là où quelqu'un nous aurait guettés derrière la porte principale. Mais la pièce était vide : juste un vieux fauteuil adossé à la porte et un luxueux bureau en bois étrangement marbré. Nous cavalâmes, nous couvrant l'un l'autre, à travers un agencement de pièces devenu familier. Puis, parvenus au vestibule du fond, nous fûmes pris d'une hésitation. Il faisait sombre et froid. Mais le silence était imparfait.

Nous perçûmes un bruit au-dessus de nos têtes. Comme des battements, sourds et lointains.

Nous revînmes sur nos pas, prîmes par la cuisine jusqu'à l'escalier central. Là-haut se trouvaient quatre chambres, avec des tapis. Rien à signaler. Salles de bains. Rien non plus. Bureau. Rien. Mais toujours ce bruit sorti de nulle part.

Retour dans la première chambre, où le son était plus fort. Il semblait à présent provenir d'en bas. Retour dans la deuxième chambre : le bruit faiblissait, mais s'élevait toujours du bas. Je tournai sur place, le pistolet dressé, conscient que d'un instant à l'autre quelqu'un allait surgir de l'ombre, que personne n'aurait tendu un tel piège sans tenir à le voir opérer.

Zandt fonça dans la première chambre, se laissa tomber à genoux.

— Ça vient de là-dessous.

— Du rez-de-chaussée, objectai-je, mais les battements suivants suffirent à me convaincre.

Nous dégageâmes le tapis. Découpée dans le parquet, une petite trappe. Zandt s'écorcha les doigts jusqu'à la soulever.

En dessous, le visage d'une fille. Livide, émacié. Son front était violet d'avoir cogné contre le plancher pendant Dieu sait combien de temps. Elle était en vie.

Sarah cligna des paupières. Dans le tréfonds de sa conscience, elle avait l'impression qu'on lui relevait la tête, juste assez pour que l'eau cesse de lui rentrer dans le nez. Elle remua les lèvres.

Zandt plongea la main et lui toucha le visage. Il prononça son nom et elle acquiesça, à peine capable de bouger la tête. Elle avait les yeux rouges et gonflés. Zandt approcha l'oreille. Elle articula quelque chose, mais je n'entendis qu'un chuintement éraillé.

— Qu'est-ce qu'elle dit ?

— « Faites gaffe à tache de bois. »

Zandt pressa son front contre le sien, comme pour le réchauffer. La fille se mit à pleurer.

J'agrippai l'extrémité des lattes bordant son cou et tirai. Elles résistèrent.

— Il les a clouées. Putain de Dieu. Aide-moi, Zandt.

Elles cédèrent, mais lentement, une à une. La fille voulut pousser, pour aider, mais elle était bien trop faible, et si elle avait pu faire quoi que ce soit dans cette position elle l'aurait fait depuis longtemps.

Les deux dernières lattes brisées, Zandt tendit les bras, glissa ses mains sous le dos de Sarah et la souleva. Il la cala sur son épaule, et c'est alors qu'elle hurla en voyant mon visage.

Il fallait que Nina se lève. Il le fallait, elle le savait. D'ici, elle ne pouvait atteindre la poignée, encore moins ouvrir la portière, sans parler de monter à bord. Elle avait déjà remarqué, de son point d'observation rasant, que le câble déroulé par Davids courait à travers le parking jusque dans l'autre bâtiment. Celui où devait reposer le cadavre de Bobby. Et elle devinait qu'il se prolongeait dans toute la résidence, en guise d'ultime défense, voire davantage.

Elle laissa retomber sa tête sur le bitume. Son bras droit, ce bras qui lui avait rendu de si fiers services durant toutes ces années, qui avait toujours accompli ce qu'elle lui avait demandé, s'était soudain mis en grève. Il appartenait désormais à une autre, à une personne qui n'était pas de son côté et refusait de l'écouter. La sensation évoquait tantôt un gant de latex rempli de gelée, tantôt une pince calcinée. Ce qui, a priori, n'était pas bon signe.

Nina déglutit deux fois, leva la tête. Le sol paraissait sec sous la voiture – du moins, plus sec que tout autour. Peut-être pouvait-elle s'y glisser le temps de reprendre quelques forces. Oui, bonne idée, lui dit son corps. Très, très bonne idée. Même son bras droit semblait reprendre vie à cette perspective.

Alors elle roula sur son coude droit et se hissa sur sa main gauche. L'espace d'une seconde, l'éclair de douleur lui lava l'esprit, et l'instant d'après elle se tenait sur ses pieds. De la main gauche elle se débattit avec la poignée, en vain, alors elle essaya la droite, et fut sidérée de voir qu'elle obéissait à nouveau. La portière s'ouvrit.

Elle tomba en avant, tâcha de se hisser sur le siège conducteur. Impossible. Elle se remit sur ses pieds, attrapa le volant et tira. Cette fois-ci, quand elle chuta, ce fut au moins sur le siège.

Elle parvint plus ou moins à se redresser, claqua la portière. Chercha les clés.

Elles n'étaient pas là.

— Écoute-moi, John, plaidai-je. Elle est malade. Elle ne sait pas ce qu'elle dit.

Zandt s'éloignait dans les escaliers à reculons, me tenant en joue avec son arme. Sarah s'abritait derrière lui, lui enserrait la taille, tant pour se protéger que pour se maintenir debout. Elle trébucha, manqua de tomber. Zandt dut se tourner pour la rattraper, enveloppant ses épaules d'un bras en la pressant contre lui. Elle avait fini de crier, mais uniquement parce que sa voix s'était tarie, réduite en un simple grincement. Le bruit résonnait encore dans son crâne.

À mon tour, je descendis lentement les marches. J'avais les mains levées, je parlais d'une voix basse et calme :

— Je ne l'ai pas enlevée. Je n'étais pas à Santa Monica à cette période. J'étais à Santa Barbara, et je peux le prouver. J'ai la facture de l'hôtel.

— C'est à une demi-heure de voiture.

— Je sais, John. Je sais bien. Justement, pourquoi je te raconterais ça si je mentais ? Je te dirais que j'étais en Floride, putain ! Qu'est-ce qui te prend, John ? Tu crois que je t'amènerais ici, tu crois que je courrais après ces types si j'étais l'un d'eux ?

Zandt atteignit le palier. Sans lâcher Sarah, qui se cachait toujours dans son dos, il traversa le large couloir à rebours, jusqu'au vestibule de l'entrée. C'est par là qu'ils allaient sortir.

— On ne sait jamais de quoi les gens sont capables, répliqua Zandt. Y compris moi. Tu fais un geste et je te fais sauter la cervelle.

— Ce n'est pas moi.

— Elle dit que si. Elle dit que c'était toi à Santa Monica.

Je m'immobilisai.

— OK, alors voilà ce qu'on va faire. Je reste ici et toi tu te casses. Tu la sors, puis tu reviens et on discute.

— Ça, je reviendrai, répondit-il. Mais ce sera pas pour discuter.

Sarah se sentit flancher, mais le type gentil la redressa une fois de plus. Elle s'éloignait enfin de Touche Dubois, resté au pied de l'escalier. C'était une ruse, elle le savait. Il leur laissait croire qu'ils étaient libres, mais il les rattraperait. Il n'avait pas besoin de marcher. Il pouvait transpercer le toit et planer dans les airs. Il pouvait survoler les maisons, puis piquer du nez et les tuer par au-dessus. Ce n'était pas un être comme les autres. Il ne ressemblait à personne.

Elle voulait dire ça au type gentil, mais c'était trop dur. Alors elle essaya de lui dire d'abattre Touche, mais les mots ne sortaient pas et il n'en fit rien. Il se contentait de la porter vers la pièce située à l'avant de la maison. Sarah ne pouvait choisir sa destination. Ses jambes ne fonctionnaient plus. Elle était obligée d'aller là où on l'emmenait.

Nina pensait qu'il ne serait plus là. Tout le temps qu'elle s'était traînée sur le parking, avait rouvert la porte de la loge, et navigué entre les carcasses abandonnées des fauteuils et canapés géants, Nina s'était attendue que Davids se soit volatilisé, à ne retrouver qu'un emplacement vide. Mais ça ne changeait rien au problème. Elle ne pouvait faire démarrer la voiture sans les clés. Soit Bobby les avait, soit c'était Davids. Elle ignorait où était passé le premier. Restait à mettre la main sur Davids, en commençant par l'endroit où il s'était effondré.

Et c'est bien là qu'il reposait. Ébahie, Nina se courba pour fouiller ses poches. Il eût été plus simple de s'agenouiller, mais elle craignait de ne jamais pouvoir se relever. Elle avait réussi à regagner le bâtiment, mais ne connaissait pas l'état de ses réserves. Elle glissa les doigts dans la veste.

La main de Davids jaillit, lui captura le poignet. Il ouvrit la bouche.

— Mary, dit-il.

Terrifiée, elle fixa son visage. Il l'attira et la fit chuter.

Son genou écrasa le visage de Davids, dont le cou se tordit en craquant, mais elle s'en aperçut à peine, sa propre tête heurtant simultanément le sol.

Elle s'agita, sans trouver d'appui, avant de constater que rien ne la retenait. Elle tourna la tête. Replongea la main dans la veste. Il ne bougeait plus.

Elle devait impérativement trouver les clés. Fût-ce la dernière action de sa vie.

Elle les dénicha dans la poche droite du pantalon. Trois trousseaux. Elle les empocha et s'éloigna du cadavre, à quatre pattes, jusqu'au bord d'un fauteuil. Celui qu'elle occupait précédemment, peut-être, elle ne savait plus trop. Cela paraissait déjà lointain.

Grisée par son succès, elle mit à peine trente secondes à se relever. Puis elle traversa la réception, enjamba le corps d'un policier mort, poussa la porte et retrouva le parking. Son second souffle refluait, et elle le savait – non parce que la douleur augmentait, mais parce qu'elle ne la percevait plus. L'effet du choc et de l'hémorragie. Son corps remontait le pont-levis. Il avait besoin d'énergie, et elle la gaspillait.

Elle gagna la voiture, heureuse d'avoir laissé la portière ouverte. Grimpa sur un siège gorgé d'eau.

Le deuxième trousseau convenait au démarreur. Alors seulement elle referma la portière, sachant qu'elle n'aurait pas à chercher Bobby.

Le moteur démarra du premier coup, et elle bénit Ford et ses petits concepteurs futés. Ce n'était pas comme lorsqu'elle était môme, et qu'il fallait convaincre ces bestioles de se réveiller, à force de quoi on finissait par les aimer et leur donner des noms. Aujourd'hui, qu'il pleuve ou qu'il vente, elles démarraient au quart de tour. Pas besoin de les baptiser pour en tirer quelque chose. Suffisait de savoir où l'on voulait aller.

Elle appuya son front contre le volant, rien qu'un instant, et se sentit aussitôt sombrer. Releva la tête en sursaut, passa la marche arrière et recula d'un bond.

Puis enclencha la marche avant, enfonça la pédale, et fonça droit sur la palissade.

Je tins parole, malgré la peur et la perplexité, et ma répugnance à rester seul dans cette maison. Je restai en bas de l'escalier, à contempler l'épais câble qui s'y engouffrait, jusqu'à ce que me parvienne la voix de Zandt depuis la pièce de devant.

— Oh, mon Dieu ! disait-il, et la fille poussait un nouveau cri.

Suivit un bruit mat. J'accourus.

Dans la salle brillait à présent une lampe, qui projetait une lueur jaunâtre près de la fenêtre. Recroquevillée dans le coin, la fille poussait des miaulements plaintifs. Zandt était allongé sur le dos, à plusieurs mètres de son flingue. Le visage du flic arborait une expression insolite.

Planté au-dessus de lui, un homme armé, qui visait directement la tête de Zandt.

— Lâche-le, criai-je en tenant mon flingue à bout de bras, prêt à tirer. Lâche-le, fils de pute !

— Sinon quoi ? rétorqua le type sans même se retourner. Sinon quoi ?

— Sinon je t'explose le crâne.

— Tu crois ça ? (Il finit par me regarder.) Salut, Ward. Ça fait une paye, dis-moi.

Je vis mon propre visage. Le monde bascula, se disloqua.

Ses cheveux étaient plus longs, et leur blondeur rehaussée. Ses traits semblaient également altérés – pour avoir été animés par un esprit différent. Si vous m'aviez vu certains jours, ou dans certaines situations, vous auriez vu ce visage-là. C'étaient nos seules dissemblances. Même la carrure était identique. Je clignai des yeux.

— Eh oui, opina l'Homme Debout d'un ton indulgent. Alors, tu t'en crois toujours capable ? Assassiner le seul être qui ait ton sang ? (Ses doigts se raffermirent sur son arme.) Ta réponse m'intéresse au plus haut point, et le fait que cela revienne aussi à éliminer John ne doit pas t'influencer, d'accord ?

Il s'adressa de nouveau à Zandt :

— Je t'avais dit que je te la rendrais, un jour ou l'autre.

Et de lui flanquer son pied dans la mâchoire.

Je crus que la nuque de Zandt allait rompre, tant le coup était violent. Je voulus presser la détente. Mais j'en étais incapable.

— Tu as tué un de mes collègues, espèce d'enfoiré, poursuivit l'Homme Debout. Tu as liquidé des types qui valaient bien mieux que toi. Pour ta gouverne, j'ai essayé de changer Karen. Pendant longtemps. Ça n'a pas marché. Alors je l'ai bouillie. Mais maintenant je te la rends. Tu as aimé la « course » que j'ai laissée dehors ?

Zandt redressa son cou, écumant de douleur.

— Tu te donnes le nom que tu veux, répliqua-t-il avec sang-froid. J'en ai jamais rien eu à foutre. Tue cette merde, Ward.

Ma bouche était béante, les muqueuses asséchées. Mes bras ne tremblaient pas – ils étaient raides comme de la pierre. Pas moyen de bouger mes doigts. J'avais l'impression d'être lobotomisé, de ne vivre qu'avec les yeux.

Mon regard fit sourire l'Homme Debout.

— Ça fait drôle, hein ? On en a, des choses à se dire. Mais je sais que tu es un peu à cran, sans compter qu'il faut quitter les lieux. Alors pour te prouver ma bonne foi, je vais épargner une de ces deux salopes. T'as le droit de choisir laquelle et de buter l'autre. Le nombre de morts à ton actif est dramatiquement faible, mon vieux. Tu as besoin de quelques séances de rattrapage.

— Les fédéraux sont en route, dis-je d'un ton vague, faible et vide, même à mes oreilles.

— M'étonnerait, répondit l'Homme Debout. Ils seraient déjà là.

— Pourquoi t'as fait ça ? Pourquoi t'as tué mes parents ?

— C'étaient pas tes parents, abruti. Tu le sais très bien. Ils ont tué notre père et gâché nos vies. On était faits pour être ensemble, dès le début. Imagine tout ce qu'on aurait réalisé à ce jour. Les Hommes de Paille ont l'argent, frangin, mais nous, on a le sang. On est purs. On est le cœur de toute chose. On est le vrai.

Pelotonnée dans son coin, Sarah se bouchait les oreilles et serrait les paupières. La voix du type lui parvenait malgré tout. Cette voix tellement haineuse, cette voix qu'elle avait subie à n'en plus finir, qui l'avait inondée de paroles et de paroles et de paroles jusqu'à la convaincre que c'était ça qui la tuerait, et non la faim ; qu'il allait prononcer le mot de trop, ce mot qui, au lieu d'exciter ses tympans, lui fendrait la cervelle en deux.

— Si j'étais toi, je tuerais plutôt John, disait Touche. Il n'a plus aucune raison de vivre, de toute manière. Et de cette façon, tu peux garder la fille. Elle est pas au top de sa forme, mais on pourrait bien s'amuser, non ?

Sarah rouvrit les yeux.

— Flingue-le, Ward, disait l'homme à terre. Vas-y, flingue-le.

— Tu commences à me taper sur le système, John, répondit Touche en le frappant à nouveau. Toi aussi, Ward. Il est temps de passer à autre chose. J'ai bouclé ma mission dans cette montagne. Il est temps de décoller.

Sarah était perdue. Le type qui aurait pu être son père ne l'était pas, et il était couché par terre. Et l'autre type... elle ne savait pas qui c'était. Un homme miroir.

Touche s'adressait à l'homme miroir, qui ne bougeait pas.

— Allez, mec, finissons-en. Qu'est-ce que t'as ? Tu sais que tu en crèves d'envie. Tu as déjà tué, merde. Et ça n'a rien d'un hasard.

Touche braquait son pistolet sur la tête de l'homme au sol. Il allait le tuer puis décoller. C'est ce qu'il venait de dire. Or, si le type couché n'était pas son père, son père se trouvait peut-être à la maison avec sa mère et sa sœur. Le seul truc, c'est que leur maison avait un toit. Or, si la maison avait un toit, Touche pouvait le traverser, et vu ce qu'il lui avait fait à elle, comment imaginer ce qu'il leur ferait à eux ?

Sarah ôta les mains de ses oreilles. Elles ne bloquaient rien, de toute façon.

— C'est inscrit dans ton sang, disait Touche. Je sais que tu as lu le Manifeste. Tu l'as lu, et tu verras qu'il dit vrai.

— C'est de la merde en barre, lança le dénommé John.

Le pied de Touche riposta aussitôt, lui écrabouillant la main.

— Tant pis, Ward, je retire mon offre, lança le diable avec, pour la toute première fois, des fluctuations dans la voix. Si tu veux tuer quelqu'un, il ne reste que la fille. Ce type est à moi depuis trop longtemps.

Il rapprocha son flingue du visage de Zandt.

Puis releva brusquement la tête. Comme s'il percevait du bruit dehors.

Sarah ne réfléchit même pas. Elle s'arracha de son coin.

Je vis la fille jaillir de nulle part. Mais son corps n'était pas en état, et son élan périclita avant même qu'elle ne soit sur pied. L'impulsion la projeta néanmoins au-dessus des jambes de Zandt, et droit sur l'Homme Debout. Il tomba à la renverse, talochant la tête anguleuse de la fille, et ses dents qui tentaient de lui ronger le visage.

Un bon direct dans les yeux et elle repartit en arrière, et soudain mon sortilège se brisa.

Je fis feu, le ratai, puis Zandt était sur lui et je ne pouvais plus tirer.

Les deux hommes roulèrent, échangeant coups de poing et de pied. Je me postai sur le côté, préparant la balle que j'allais sûrement tirer, que j'allais devoir tirer à tout prix. Puis j'entendis le bruit dehors, celui d'un moteur à plein régime, sous une bordée de klaxons. Dieu merci, Bobby.

J'avisai la fille, gisant à terre, le nez dégoulinant de sang. Je me précipitai sur elle, convaincu de respecter les priorités de Zandt. Je la relevai en lui ceinturant la taille, et la traînai jusqu'à l'entrée.

J'ouvris la porte dans un bain de lumière. Ne compris rien à ce bordel avant de reconnaître les phares de la voiture louée la veille à l'aéroport.

J'entraînai la fille en bas des marches, me demandant à quoi jouait Bobby mais bénissant son nom. Puis je vis qu'il n'y avait qu'une personne dans la voiture et que ce n'était pas lui mais la nana du FBI et qu'elle avait une gueule de zombie.

Je courus à sa vitre.

— Où est Bobby ?

— Montez, répondit-elle seulement. C'est elle ?

— Ouais. Où est Bobby ?

— Où est Zandt ?

— À l'intérieur. Tu vas me dire où est Bobby, oui ou merde ?

— Bobby est mort, cria-t-elle. Davids l'a tué. Je suis désolée, Ward, mais je t'en supplie, va chercher John. Il faut partir. Tout le village est piégé et faut se barrer d'ici.

Des câbles. Dans tous les sens.

J'ouvris la portière de derrière et enfournai la fille délicatement. Sans refermer, je regagnai l'entrée en sprintant, scandant le nom de Zandt.

Tout en pensant : Bobby est mort.

Le vestibule était vide. Le pistolet de Zandt avait également disparu. Je courus dans la maison, sans cesser de crier, le flingue en avant, dévoré de honte par la part de mon âme retenue deux minutes plus tôt. Je ne l'avais pas fait, je n'avais pas abattu mon frère. Mais maintenant j'en étais capable. Je m'en savais capable. Maintenant j'en étais capable.

J'entendis soudain courir derrière moi, puis sur le côté. Je virai à fond vers l'arrière de la maison, où je tombai nez à nez avec Zandt. Je me souvins in extremis :

— C'est moi, John ! Ce n'est pas lui, c'est moi !

Son visage était injecté de sang. Il stoppa, le flingue à deux centimètres de ma tête, la gâchette à moitié relevée.

— Regarde mes fringues, John, mes putains de fringues !

Un temps d'arrêt, puis il me repoussa pour continuer son chemin. Je l'agrippai d'une clé au cou.

— Nina est dehors. Bobby est mort. Il faut partir.

Zandt se libéra en m'envoya son coude dans le ventre. Je le rattrapai aussitôt, pressai sa tête contre la mienne en hurlant :

— Cet endroit entier est sur le point de sauter ! Si on part pas, il nous aura tous ! Il aura Sarah !

Un bref instant sa rigidité faiblit, et je pus le tirer jusqu'à la porte d'entrée, où filtrait la lumière des phares.

Nina faisait vrombir le moteur, mais Zandt continuait de résister, se débattait tel un ours contre mon bras qui l'étranglait. L'espace d'une seconde, je crus voir filer une ombre dans l'embrasure d'une autre pièce, puis plus rien.

De retour dehors, Zandt semblait découvrir un nouveau monde, comme devant une lucarne révélant d'autres têtes que celle de l'homme qu'il devait tuer. Je le pressai vers la voiture, tout en me penchant pour ramasser quelque chose.

Zandt grimpa derrière à contrecœur, pestant et jurant, boxant le dossier du fauteuil d'en face. Nina s'était avachie sur le flanc, à demi juchée sur le siège passager. Je grimpai derrière le volant, la repoussai et l'attachai.

Je trouvai l'accélérateur, l'enfonçai de toutes mes forces, comme si je voulais me dresser. La voiture bondit en arrière, creusant un sillon circulaire dans l'herbe mouillée. Nina se cogna contre la portière et se mit à gémir, dans une plainte régulière mais faible.

— Tiens bien Sarah ! lançai-je à Zandt tout en bouclant ma ceinture.

Puis nous rebroussâmes chemin à travers les Halls, filant entre les maisons désertes et leurs secrets.

Il me semblait entendre craquer des os sous les pneus, mais ce devait être mon imagination, et j'espérais que Zandt n'entendait pas la même chose. J'espérais aussi ne pas avoir vu cette image furtive : les silhouettes d'un petit groupe posté sur la colline surplombant le pré, en train de nous observer.

Non, je roulais trop vite. Je n'aurais pu distinguer une telle chose. Quand je relevai les yeux, ils avaient disparu.

Je visai le trou que Nina avait enfoncé à l'aller, et mis presque dans le mille. Le pare-brise éclata quelques planches supplémentaires. Pire, j'entendis un vilain bruit de tôle là où le châssis percuta l'un des plots du parking, mais la voiture roulait toujours. Je manquai de partir en tonneau en virant à gauche vers la sortie, craignant déjà d'avoir vécu tout ça pour rien, mais les quatre roues retrouvèrent le plancher des vaches et je m'élançai dans l'allée, franchis l'arche, puis attrapai à droite la route menant en bas de la montagne. Je faillis de nouveau bousiller la bagnole au détour du virage, en tombant nez à nez avec la voiture stationnée de Harold Davids, mais je l'évitai d'un dérapage.

Une série d'épingles à cheveux, enfilées de plus en plus vite, puis la longue pente rectiligne menant à la futaie qui cachait la route. Je ne tentai même pas de négocier le dernier virage, conscient que c'était illusoire, mais coupai à travers les arbres, trouvant des espaces assez larges pour y engager la voiture, jusqu'à atteindre, après moult secousses et sueurs froides, la zone herbeuse précédant la sortie. Quelque part entre là et la chaussée, une pierre fit éclater un pneu arrière, et je parvins de justesse à ne pas gîter tandis que la voiture bondissait et chassait le long de la descente finale, jusqu'à défoncer la basse clôture dans un cri de métal rompu.

La voiture coupa la route verglacée et piqua droit dans les eaux vives, froides et peu profondes de la rivière Gallatin. Il y eut un instant figé, le temps de comprendre que nous étions tous vivants, puis la terre entière parut exploser.

Contorsionné comme j'étais, le volant dans les reins, je ne vis qu'un nouveau soleil, jaillissant de la montagne comme l'aube.

Houma, Louisiane

C'est un petit motel, sans service d'étage. Je dispose d'une petite chambre, au bout d'une aile de chambres tout aussi petites, qui s'étirent d'un bureau poussiéreux et minuscule. Le téléviseur est vieux et merdique. Il y a de l'eau dans la piscine, mais personne ne s'y baigne. Surtout pas moi.

Demain matin, de bonne heure, je repartirai. Je me rappelle le nom de la ville où vit la mère de Bobby, et me souviens vaguement l'avoir entendu décrire la rue de son enfance. Je saurai peut-être la trouver. J'aimerais pouvoir parler de son fils à cette dame. Lui dire comment il était, que c'était un homme bon, et comment il est mort. Avec un peu de chance, je retrouverai même le cimetière où repose son père, et je le dirai à lui aussi. C'est le seul hommage auquel mon ami aura jamais droit.

Il y a dix jours, assis dans une voiture à Santa Monica, j'ai regardé John et Nina ramener une fille à la porte d'un pavillon. Chacun lui donnait la main : pour Nina, la gauche, puisqu'elle avait le bras droit en écharpe. Sarah demeurait blafarde et très affaiblie, mais avait bien meilleure mine que lorsqu'on les avait conduites, Nina et elle, dans un hôpital de l'Utah. L'interne de garde avait voulu prévenir les flics. D'après ce qu'il voyait, Sarah n'avait rien avalé d'autre que de l'eau infestée de plomb et de diverses substances chimiques, dont certains agents biologiques utilisés en thérapie génique. Quant à la finalité de ceci, outre l'empoisonnement pur et simple, il n'osait même pas l'imaginer. John la devinait, cependant : il comprenait désormais – si l'on avait correctement interprété les indices – que les corps des autres victimes de l'Homme Debout dénotaient la même volonté de créer un être à son image, via les traumatismes crâniens et les violences sexuelles.

Nina se servit de son badge pour empêcher l'histoire de faire les unes du pays. Les médecins admirent Sarah et Nina pour une semaine, mais dès le lendemain John et moi sommes venus les enlever. Certes, il leur fallait des soins. Mais les laisser toutes deux dans le même lieu était bien trop risqué. Zandt appela Michael Becker pour le prévenir de son arrivée, puis nous montâmes en voiture.

Nous traversâmes d'une traite l'Utah, le Nevada et la Californie, puis L.A. jusqu'à Santa Monica, en nous relayant au volant, Zandt et moi. Bien qu'elle dormît l'essentiel du trajet, je fis un peu connaissance avec Sarah. Elle se montra adorable, et me dit que j'étais très différent, ce qui faisait du bien à entendre. Je pense qu'elle se remettra avec le temps, et je prends le pari que la prochaine fois que mademoiselle dînera en ville (vers l'an 2045, probablement, si son père a son mot à dire), elle ne commandera pas une salade Cobb mais le gros hamburger de ses envies.

Une fois sur le perron de la maison Becker, Nina lâcha la main de Sarah et appuya sur la sonnette. Tous trois restèrent figés comme des images, puis la porte s'ouvrit et ce fut une explosion d'amour qui m'obligea à détourner les yeux. Je fixai un point à travers le pare-brise, en pensant aux derniers mots que m'avait glissés la fille.

Quand j'ai regardé de nouveau, Nina revenait vers la voiture, tête baissée. Zandt restait auprès des Becker. Sarah finit par le lâcher pour retrouver ses parents. Michael Becker a serré la main de Zandt et quelque chose est passé entre eux, mais j'ignore quoi.

John a reculé d'un pas pour laisser la famille regagner ses quartiers. Il est resté planté là un petit moment, même après que la porte se fut refermée. Puis il a redescendu l'allée, est remonté en voiture et nous avons repris la route. Il se trouve actuellement en Floride, en visite chez son ex-femme.

Quand j'ai vu sa réaction devant le pull, j'ai regretté de ne pas avoir ramassé un os à la place. Mais je n'avais pas les idées bien claires – ç'avait été une réaction spontanée, l'idée qu'il souhaiterait peut-être rapporter quelque chose de cette montagne. J'imagine qu'un os aurait mieux fait l'affaire, quelque chose qui ait vraiment appartenu à Karen. Mais je pense que ce pull suffira à leur travail de deuil. On a prévu de se revoir plus tard. On a échangé nos numéros de portable. Apparemment, il ne me tient pas rigueur de ne pas avoir su tirer, là-haut dans les Halls.

Mais nos retrouvailles attendront un peu, je pense. J'espère qu'il a pour priorité de retrouver Nina, quand elle aura viré son barda de L.A. En les observant tous deux sur ce perron, j'ai vu quelque chose qu'ils comprendront un jour, je l'espère. Ils sont déjà ensemble.

Quand je conduis, mon regard reste de longs moments bloqué droit devant, sans voir ce qu'il y a derrière la vitre ; un flot d'images me traverse la tête comme un film. Parfois je songe aux Hommes de Paille, et je tâche de démêler le vrai du faux à leur sujet. Je veux croire que ce qui sous-tend leur entreprise est plus profond que le Manifeste ; que les idées contenues là-dedans ne sont que les théories d'un psychotique au sujet des clivages qui semblent nous gouverner. Puis je me rappelle que cet ouvrage que beaucoup considèrent comme le tout premier roman, le *Journal de l'année de la peste* de Daniel Defoe, fut écrit au lendemain d'un fléau qui ravagea l'Europe entière et aurait pu être imputé à notre manière de vivre, les uns sur les autres ; et que nos principales industries du divertissement, le cinéma et la télévision, ont chacune pris leur essor dans la foulée d'une guerre mondiale. J'en viens à me demander si les paysages de fiction et les utopies n'ont pas fleuri sitôt que nous avons fondé les villages et les villes, et si cela n'explique pas l'émergence des religions à peu près à la même période. Plus nos espaces sont bondés, plus nous sommes interdépendants, et plus nos rêves deviennent importants – à croire que tout ceci n'est là que pour nous souder les uns aux autres, nous aider à éprouver un manque, et nous acheminer ainsi vers une humanité qui va au-delà du simple fait humain. À l'heure où l'Internet quadrille le globe, resserrant davantage encore les mailles du filet, je me demande si c'est par pure coïncidence que cela survient au moment où nous déchiffrons notre code génétique, et commençons à le bricoler. Plus nous nous rapprochons les uns des autres, plus nous cherchons à comprendre notre nature. J'espère sincèrement que nous savons ce que nous faisons avec nos gènes, et qu'en entreprenant de retirer ce qui passe pour des erreurs, des imperfections, nous n'éliminons pas ce qui nous rend viables. J'espère que c'est notre avenir, et non le passé, qui motive nos décisions. Et j'espère que dorénavant, quand je prendrai conscience qu'il manque une chose à ma vie, je continuerai de la chercher – même si je sais qu'il s'agira parfois d'une simple promesse, non d'une pièce à capturer. Sans ça nous devenons des hommes de paille, des femmes d'ombre, abandonnés dans des champs désolés que même les oiseaux désertent, à attendre un éternel été quand c'est déjà l'hiver. Vu notre façon de vivre, si éloignée de celle d'hier, il semble miraculeux qu'on s'en tire aussi bien. On fait des rêves pour préserver sa santé mentale, et aussi pour demeurer en vie. Comme l'avait dit un jour mon père, l'important n'est pas de gagner, mais de croire qu'il y a quelque chose à gagner.

Souvent je pense à lui, et à ma mère, deux êtres qui ne sont plus là. Leur mort, comme toutes les morts, n'est pas de ces choses que l'on peut arranger. On ne peut attraper la mort pour lui flanquer une correction, pas plus que la tristesse ou le dépit, pas plus que l'Homme Debout ni l'organisation dont il a pris la tête. Peut-être y parviendrons-nous un jour, peut-être pas. Peut-être y aura-t-il toujours des gens comme lui. C'est trop tôt pour le dire ; j'ignore encore si la destruction programmée des Halls était une tentative archi-réussie pour effacer toutes les traces, ou si l'explosion était censée libérer l'immense poche de roche en fusion qui s'accumule dans les sols de Yellowstone – anéantissant ainsi notre culture et les fermes du monde occidental, pour nous ramener au mode de vie que prisent tant les Hommes de Paille. Nous renvoyant, ou nous précipitant, dans les ruines.

Nina penche pour la deuxième solution, à cause d'une phrase qu'aurait prononcée Davids : d'après elle, les Hommes de Paille se voient comme une nouvelle race de chasseurs-glaneurs, dont la fortune serait moins due à la destinée qu'à une « pureté » intérieure, et qui seraient capables de survivre dans n'importe quel milieu. Je ne sais pas. Il ne me reste personne avec qui en débattre.

Le message parvint à mon hôtel de Los Angeles le lendemain du retour de Sarah. Il provenait de mon frère. J'ignore comment il m'avait trouvé. Je rendis la clé de ma chambre dans l'heure, et depuis je ne cesse de bouger.

Ce message prenait la forme d'une cassette vidéo. La première séquence était postérieure à notre rencontre. Sa colère était patente, mais à l'évidence il ne désespérait pas d'un futur *rapprochement* [1]. Il me mettait au parfum sur certaines périodes que nous n'avions pas partagées. Sa découverte dans les rues de San Francisco, un petit enfant sans identité, hormis un prénom cousu dans son pull. Des familles d'accueil, un premier meurtre. Toute une phase sur laquelle il demeurait vague. Son dernier boulot d'entremetteur pour les riches et les sociopathes, la révélation d'un lien entre ses employeurs et son passé, son admission au sein d'un groupe clandestin et son premier triomphe en 1991 dans le McDonald's d'une petite ville de Pennsylvanie. Ses propres

1. En français dans le texte. (*N.d.T.*)

expérimentations d'évolution accélérée, à travers la violence et le viol ; son projet de se créer une compagne pure avec qui fonder une lignée non virale. Un projet dont il parlait avec une émotion dérangeante, pour ce qu'elle ressemblait à de l'amour.

Le reste de la vidéo est moins facile à décrire. Il y a quelque chose de très perturbant dans ces images, indépendamment de leur sujet ou de ce qu'implique leur existence. Voir mon sosie commettre toutes ces horreurs, c'est comme d'accéder à des ténèbres virtuelles, à un monde où ce que je crois être m'est refusé, où je deviens celui que j'espère ne pas être. Bobby et moi n'avions vu que de longs plans flous. Mais l'Homme Debout, peut-être devrais-je l'appeler Paul, avait veillé à laisser des images parfaitement nettes de ses exploits. Des bandes réalisées pas ses propres soins, le montrant tout sourire sur des parkings en feu, en train de dérouler des câbles et de poser des bombes, ou dans des pièces obscures en Amérique, en Angleterre et en Europe, distribuant armes et plans de bataille à des pigeons. Le montrant nu, accroupi au-dessus des viscères de jeunes disparus. Ou en train de manger certaines choses.

Le montrant lui, et donc moi. Moi, trônant sur une vertigineuse pyramide d'abominations, une demi-heure de preuves.

Même cette cassette ne m'est d'aucune aide. Je ne peux la montrer à personne, et pas seulement parce que les autorités de Dyersburg – voire toutes celles du Montana – ont grossi ma liste de gens à éviter. Cette cassette n'a qu'un seul pouvoir : m'incriminer. L'existence de mon jumeau n'est consignée nulle part. Ma propre existence n'est consignée nulle part, sauf sur cette bande et dans ma tête.

Avant de quitter la voiture, à Santa Monica, Sarah Becker s'est penchée pour me susurrer quelques mots :

— Tu dois le faire. Il n'y a que toi qui puisses tuer Touche Dubois.

Elle a raison. Je ne puis rien faire d'autre que ce qu'il attend de moi. Je ne puis rien faire sinon partir à sa recherche.

Dans le même temps, tandis que j'engrange les kilomètres, toujours sur le départ, j'écoute les voix du passé et je pense à ce que l'on m'a offert, à l'amour que j'ai reçu. Je ne saurai dire au juste ce que je suis devenu, et n'y parviendrai peut-être jamais, mais je sais au moins que ç'aurait pu être pire. Le billet laissé par mon père à mon intention, disant qu'ils n'étaient pas morts, reste vrai dans un sens qu'il n'avait pas prévu. Ces deux-là ne mourront pas tant que je serai de ce monde. J'aurais aimé mieux les connaître, mais les regrets de ce genre se heurtent à une double impasse : non seulement ils arrivent trop tard, mais ils ne seraient jamais apparus assez tôt.

L'image dont je me souviens le mieux est une image que je n'ai pas vue, sauf à travers un écran de télévision. Un couple de jeunes gens, tenant chacun une queue de billard, enlacés dos à la caméra. Quand ils se retournent, le sourire de mon père, le doigt qu'il brandit, ma mère qui tire la langue.

Et, plus tard, sa façon de danser.

REMERCIEMENTS

Merci à Doug Winter d'avoir prononcé une phrase qui a fait turbiner ma cervelle ; à Susan Allison, Chris Smith et Jim Rickards d'avoir alimenté la turbine, ainsi qu'à Linda Shaughnessy, Bob Bookman, Jonny Geller, et David et Margaret Smith pour leurs commentaires ; à Nicholas Royle, Howard Ely, Conrad Williams, Stephen Jones et Adam Simon pour leur soutien durant ce travail et d'autres.

Comme toujours, mille pensées affectueuses à ma famille, avec un grand merci à Paula pour m'avoir aidé à voir les failles dans le mur où je me cognais la tête, et pour sa présence quand j'ai cessé.

LES MORTS SOLITAIRES

Traduit de l'anglais
par Jean-Paul Mourlon et Évelyne Chatelain

Yakima, Washington

Nous avions rendez-vous sur le parking du centre commercial de Yakima, une petite ville de l'État de Washington, sur la côte Pacifique. J'emploie le terme de « ville » car c'est ce que Yakima prétend être, et de « centre commercial », pour la seule raison qu'on y fait ses courses à l'écart de l'extérieur. En trois heures, seules deux personnes s'y étaient risquées, des ados en maillot de footballeur. De toute évidence, ni l'un ni l'autre n'avaient les moyens de remettre le lieu à flot. Ils étaient d'ailleurs repartis les mains vides. Au deuxième étage, d'énormes calicots de toile annonçaient des réductions. Le rezde-chaussée était désert : mauvais signe.

J'étais dans la voiture, à siffler des cocktails que j'achetais au café d'en face – le seul commerce de l'avenue qui semblait croire à l'avenir. Les boutiques voisines paraissaient avoir déjà confectionné un panneau « À louer », histoire de mettre de côté quelques dollars quand surviendrait l'inévitable. Dans mon ennui, je croyais entendre le maire tambouriner sur son grand bureau luisant, perdant pied dans la sclérose de sa ville. N'empêche, cette localité survivrait – même un bled aussi mort avait besoin d'un détaillant de pneus et d'un quota de fastfoods –, mais elle avait peu de chances d'enrichir qui que ce soit. Monter à Seattle ou descendre à Portland, voilà ce qui venait d'emblée à l'esprit. Qu'est-ce qu'on pouvait bien fabriquer à Yakima ? Je n'en avais pas la moindre idée.

John Zandt arriva au volant d'un gros GMC rouge, crasseux et pas franchement récent. À voir l'état du flanc droit, on aurait dit qu'un troupeau de vaches avait voulu le défoncer. Il roula au pas sur le parking, jusqu'au niveau de ma Ford Generic flambant neuve. Nous baissâmes nos vitres. L'air était froid.

— Salut, Ward ! C'est tout ce que tu as trouvé, à l'agence de location ? Dans le genre « je veux me faire remarquer », tu ne pouvais pas faire mieux.

— Va te faire foutre ! Ce n'est pas possible d'être aussi en retard ! répliquai-je. Et où t'es-tu dégoté cette bagnole de péquenot ?

— Je l'ai volée sur le parking de l'aéroport, avoua-t-il. Allons-y.

Je sortis de la voiture en laissant les clés sur le tableau de bord. Hertz se remettrait de sa perte. Personne ne pourrait me retrouver à partir de l'identité que j'avais donnée à Spokane. En grimpant dans le camion de Zandt, j'aperçus deux armes sur le plancher. J'en pris une, l'examinai, puis la fourrai dans ma poche.

— C'est loin ?

— Près d'une heure de route. Ensuite, il faudra marcher.

Zandt sortit du parking, suivit l'avenue et dépassa le nouveau centre commercial, qui avait contribué à la ruine du premier sans paraître en avoir profité lui-même. Il tourna à droite pour emprunter la 82 en traversant Union Gap, longea des bâtiments, puis il n'y eut plus que la route. À Toppenish, il fit une brutale embardée et enquilla sur la 97 en direction du sud-ouest. Désormais il n'y aurait plus rien avant un patelin nommé Goldendale, quatre-vingts kilomètres plus loin. Ensuite, on roulerait une trentaine de bornes jusqu'à l'un des tronçons les moins attrayants du fleuve Columbia, en amont du barrage Dalles.

La veille, j'avais passé la soirée en compagnie d'un barman assez loquace du Rodney's Lounge, le prétendu bar du plus grand hôtel de Yakima. Je savais que nous nous trouvions dans la réserve des Yakama, et que sur cent trente kilomètres il n'y avait rien de part et d'autre de la route, les autochtones s'entassant dans deux ou trois villages délabrés. Je savais aussi qu'Union Gap s'appelait autrefois Yakima, avant que la compagnie de chemin de fer ne contraigne les Indiens à déplacer leur campement un peu plus haut, en venant à bout de leurs réticences par des promesses de terres gratuites et des pots-de-vin qui divisèrent la tribu comme jamais la faim et le froid n'auraient pu y parvenir. Je savais enfin que, juste en amont du barrage, tonnaient autrefois les chutes de Celilo, lieu sauvage et sacré où les hommes pêchaient le saumon depuis des millénaires. Ces chutes étaient désormais silencieuses, enterrées sous les eaux gonflées du barrage. Les Yakama attendaient encore des dédommagements pour compenser la perte de leur territoire ancestral. D'ailleurs, ils attendraient sans doute jusqu'à la fin des temps. Comme tout le monde, je ne savais trop qu'en penser. Le barman était amérindien, mais il arborait des cheveux blonds très courts, hérissés, comme une pop-star des années quatre-vingt, et beaucoup de maquillage. Je ne savais trop quoi penser de cela non plus.

Sur le tableau de bord, Zandt avait scotché une carte cornée, maculée de taches de graisse ; elle semblait avoir longuement séjourné dans les poches d'un gars aux mains crasseuses. Une petite croix était tracée au centre d'une grande zone vide, près d'une ligne bleue sinueuse appelée Dry Creek.

— D'où vient l'info ?

— Un message enregistré sur une ligne à caractère anonyme, pour citoyens délateurs. La transcription aurait dû finir à la poubelle – le type était vraiment bourré et assez incohérent – mais Nina l'a repérée.

— Par quel miracle ?

— Parce qu'elle n'est pas du genre à écarter a priori un témoignage qui paraît invraisemblable.

— Et comment tu es remonté jusqu'au type ?

— Les lignes anonymes ne sont pas aussi anonymes que le FBI le raconte. Nina a retrouvé l'origine de l'appel, un bar dans le Dakota du Sud. J'y suis allé et j'ai attendu que le gars se pointe de nouveau. Ça ne s'est pas fait tout de suite.

— Et puis ?

— Il s'appelle Joseph. Il est originaire de Harrah, un bled à quelques bornes de Yakima. Tu sais qu'on est dans une réserve ?

— C'est trop sinistre pour être autre chose. On a été si bienveillants avec les Indiens que c'est un mystère qu'ils ne nous aiment pas davantage.

— C'est pourtant là qu'ils habitaient, Ward. C'est pas notre faute si on se croirait sur la lune. Bref, la semaine dernière, Joseph a rendu visite à sa famille, et il est allé vadrouiller dans la nature. Plusieurs jours d'affilée. J'ajoute que d'après son allure, il doit boire comme un trou. Mais il s'est montré catégorique sur ce qu'il avait vu.

— Pourquoi il n'a pas alerté la police ?

— Je ne crois pas qu'il ait de bons rapports avec les flics du coin. C'est pour ça qu'il était dans le Dakota.

— Mais il a vu ta jolie barbiche toute neuve et il a décidé de te faire confiance.

Zandt hocha la tête :

— J'espérais que tu ne l'avais pas remarquée.

— Oh, que si !

— Nina aime bien.

— Elle aime aussi les sacs à main en cuir. Ça ne veut pas dire que tu dois t'en coller un sur la tête. Où est Joseph ?

— Parti. Il a deux cents dollars en poche, et je ne crois pas qu'il parlera à qui que ce soit. Il a assez eu la trouille comme ça. Il a cru voir un spectre.

Zandt secoua la tête, comme s'il trouvait cela parfaitement idiot.

Je détournai les yeux avant qu'il puisse surprendre l'expression de mon visage.

Une demi-heure après avoir traversé Toppenish, on se serait cru sur une autre planète. Peut-être venait-on ici autrefois pour une bonne raison – mais plus maintenant. Pas d'arbres, rien que des collines anguleuses, des canyons peu profonds, de maigres buissons et des herbes jaunâtres parmi des plaques de neige. Des rochers gris, ou d'un brun terne, qui auraient pu inspirer une aquarelle sans âme accrochée dans un couloir. Le ciel était d'un gris encore plus sombre, les nuages s'effilochaient sur les collines et les vallées comme une mousse blanchâtre.

Zandt gardait les yeux fixés sur la pendule. Au bout d'une quinzaine de kilomètres, il roula plus lentement, tout en observant le bas-côté. Dès qu'il repéra ce qu'il cherchait, il s'arrêta.

— C'est tout près.

Il braqua très sec et emprunta une piste cahoteuse que je n'avais même pas remarquée, dévalant la colline puis remontant le long d'un piton rocheux. De toute évidence, personne ne s'était aventuré ici depuis longtemps. Au bout de quelques centaines de mètres, la piste grimpa raide et je m'accrochai à mon siège des deux mains.

— C'est là ? demandai-je en regardant autour de moi.

— Non, il va falloir marcher.

— Je ne suis pas très porté sur la randonnée.

— Ça ne m'étonne pas.

Il extirpa de sa veste un appareil qui ressemblait à un agenda électronique dernier cri surplombé d'un gros renflement.

— GPS ?

Il opina de la tête :

— Je veux être capable de retrouver mon chemin.

Zandt enregistra la position du camion et désigna la colline du doigt. C'était le même paysage que nous avions parcouru tout l'après-midi, sauf qu'il n'y avait plus de route.

— Allons-y.

Nous suivîmes la piste jusqu'à ce qu'elle disparaisse, puis marchâmes à travers le néant jusqu'à une autre colline dont la pente escarpée menait à un canyon. Nous descendîmes au milieu d'un brouillard qui ne cessait de croître, avant de remonter de l'autre côté. Ensuite, le sol était plat sur une bonne distance, rocheux et nu, exception faite de plaques d'herbes et d'arbrisseaux d'un bleu-vert maladif. Pas d'arbres.

— Qu'est-ce que c'est ? s'exclama Zandt en donnant un coup de pied dans une plante.

— Je crois que c'est de la sauge, mais, à dire vrai, je ne connais rien à la flore des hautes plaines.

— Une foutue saloperie.

— C'est bien vrai.

Nous continuâmes d'avancer, tandis que la brume nous entourait. John consultait son gadget de temps en temps, mais notre marche ne semblait nous mener nulle part. Le temps était froid et sec. Je tentai d'imaginer que des gens avaient vécu là, en vain. Le lieu donnait l'impression de ne pas vouloir se laisser importuner.

Au bout d'un long moment, je regardai ma montre : 16 heures passées, le jour déclinait déjà. Un vent sournois s'était levé. Au milieu du brouillard, le soleil n'était plus qu'une pièce d'argent qui commençait à ternir.

— Je sais, lança John avant même que j'ouvre la bouche. Je n'ai que la croix sur la carte. On y est presque.

— Jamais vu un coin pareil de toute ma vie !

Nous continuâmes quand même. Le brouillard s'était épaissi jusqu'à former une chape qui, parfois, s'ouvrait brusquement en un tunnel que le soleil éclairait de l'intérieur, comme une vision aux reflets dorés. Nous cheminions le long d'une crête basse, au pied d'une autre colline qui grimpait ; à gauche, s'amorçait la bouche d'un canyon.

On n'avait pas l'air de toucher au but, mais je ne dis rien ; je n'avais nulle part où aller.

Pour finir, John s'arrêta.

— Quelle connerie ! lâcha-t-il, exaspéré.

Il paraissait à cran, sous l'emprise d'une fureur menaçante. Les cernes sous ses yeux trahissaient des nuits sans sommeil. J'espérais que son contact aurait la bonne idée de ne pas retourner trop vite dans ce bar du Dakota.

— Ton machin est rétro-éclairé ?

— Évidemment.

— Alors on aura le temps, dis-je en me remettant en marche.

— Ward, je ne crois pas que ça en vaille la peine. Même à vol d'oiseau, on est à trois quarts d'heure de la route, peut-être plus. On a fait le tour de toute la zone indiquée par la croix.

— Où a-t-il dessiné ce plan ?

— Dans le bar, répondit Zandt d'une voix qui perçait difficilement le brouillard.

— En d'autres termes, une semaine après, et à des centaines de kilomètres. Il était bourré ?

— Il a dit qu'il était sûr de l'endroit.

— Oui, comme il était sûr de tenir l'alcool. Tu croyais les témoins quand tu étais flic ?

— Bien sûr que non !

Il sortit son téléphone portable et le considéra d'un air mauvais :

— Pas de réseau. On n'est plus sur la carte, Ward.

— En effet. Mais...

Je m'interrompis : on aurait dit que le monde s'évaporait.

— Qu'est-ce que c'est, bon Dieu ?

Il arriva à ma hauteur et nous restâmes là, côte à côte, un instant. Puis il vit, lui aussi.

— Nom de Dieu !

Un homme se tenait devant nous, juste assez loin pour que sa silhouette paraisse un peu floue dans le brouillard. Son complet gris et ses chaussures noires juraient avec le décor. On entendait sa veste battre dans le vent. Il avait une allure décidée, comme s'il savait où il allait. Et pourtant, il ne bougeait pas.

J'avançai d'un pas puis m'arrêtai ; je pensai prendre mon arme, y renonçai, puis la sortis quand même.

Nous nous approchâmes de l'homme en nous écartant un peu l'un de l'autre.

On lui aurait donné soixante ans, avec ses cheveux gris récemment coupés, désormais collés sur sa tête. Il avait les mains et le visage d'une drôle de couleur, blanc strié de bleu et de rose vif, voire d'un brun rougeâtre. Une profonde entaille lui traversait le cou jusqu'à l'oreille gauche ; le couteau en avait arraché un bout, lui donnant un aspect bancal. Il n'avait plus de lèvre supérieure. L'odeur n'avait encore rien d'insupportable : il avait fait très froid.

Zandt contourna le corps et brandit son agenda électronique pour prendre des photos.

— Regarde !

Je le rejoignis, en restant à distance sans même m'en rendre compte, comme si l'homme allait poursuivre sa route. Une barre de métal – deux mètres de haut et cinq centimètres de diamètre – avait été plantée dans le sol. On l'y avait attaché, c'était cela qui lui donnait l'air de marcher. Dans quelque temps, son corps s'effondrerait, ses vêtements se désagrégeraient, la tige rouillerait.

— Bon Dieu ! répétai-je.

Zandt hocha la tête, apparemment à court de commentaires. Il fouilla la veste et les poches du pantalon, en vain.

Je reculai de quelques pas. En tenant compte du brouillard, on comprenait que la position du corps avait été choisie avec soin. La colline le dissimulait aux regards ; on ne pouvait le voir que de tout près, c'est-à-dire depuis un point d'observation où on n'avait aucune raison de se trouver.

Zandt fixa l'horizon.

— Il a dit qu'il y en avait deux.

— Réjouissante perspective.

— Mais il n'a pas précisé où.

— On dirait qu'il était censé aller quelque part, observai-je en penchant la tête vers le mort.

Nous suivîmes la direction qu'il semblait indiquer. Au bout d'une cinquantaine de mètres, nous distinguâmes l'entrée d'un canyon. Puis autre chose.

Assise au bord du ravin, elle devait avoir à peu près le même âge que l'homme, mais, vu l'état de sa peau, c'était difficile d'en jurer. Elle avait les coudes sur les genoux, le visage posé dans les mains. Une pose naturelle, donnée sans doute au corps avant qu'il ne raidisse. Seule fausse note, sa chevelure, dénouée et dressée en touffes grises. On aurait dit que des corbeaux s'étaient mis au travail puis avaient renoncé. Elle regardait devant elle, avec ses yeux vides et creux.

Zandt ne prit que deux photos, puis releva les coordonnées de la localisation du cadavre.

— Bon, dit-il. Allons-nous-en.

Je lui emboîtai le pas, m'arrêtai, sans savoir ce que je ressentais puis la regardai encore. Quelque chose dans sa posture m'intriguait.

— Ward, allons-nous-en ! Il fera noir bientôt !

Sans l'écouter, je m'approchai de la morte, m'accroupis d'aussi près que possible, levai les yeux dans la même direction qu'elle. Elle avait la tête penchée un peu en avant, comme pour contempler le canyon.

Il ne fut pas facile d'y accéder. Je commençai par descendre en marchant, mais je dus vite m'aider de mes mains. J'entendis Zandt jurer, puis me suivre ; heureusement, il eut la bonne idée de garder quelques mètres de distance : les rochers qu'il délogeait tombèrent loin de moi.

Une fois en bas, c'était comme en haut, mais davantage de rochers, un peu plus de végétation, quelques arbres courtauds. Le brouillard se dissipait vaguement tandis que le ciel se teintait d'un bleu plus sombre.

Puis j'aperçus un chemin devant moi, souvenir d'un petit ruisseau. Je le remontai sur une brève distance, et fus surpris de découvrir une vaste étendue. Zandt me rejoignit devant une forme trapue dissimulée derrière un affleurement rocheux.

À première vue, on ne savait pas trop ce que c'était. Puis on devinait une petite cabane, tout contre le flanc du canyon.

Nous nous avançâmes en restant à trois mètres l'un de l'autre. C'était une très vieille cabane en rondin, aux gros troncs gris avec des taches encore brunes. Des planches avaient été clouées de l'intérieur sur les fenêtres. La porte était fermée par un cadenas qui paraissait récent. Quelqu'un s'y était jadis acharné avec une hache ou une pelle. On apercevait, dans les traces des coups, des formes qui ressemblaient à des lettres.

Zandt sortit son arme et, de l'autre main, prit quelques clichés. Les fenêtres. Les murs. La porte.

Il rangea l'appareil dans sa poche et me dévisagea. Je hochai la tête.

Je m'avançai, donnai un coup de pied dans la porte. Zandt se tenait derrière moi, arme pointée.

J'entrai. Les fenêtres étaient obturées, il faisait sombre, mais l'ouverture de la porte laissait passer de la lumière. Je sentis mes cheveux se dresser sur ma tête.

Trois cadavres étaient assis sur un banc, effondrés contre le mur. L'un n'était plus qu'un squelette, les deux autres, tout noirs, émettaient une odeur infecte. Le premier n'avait plus de bras, le ventre du second avait éclaté voilà peu de temps. D'autres corps étaient regroupés, il y en avait au moins deux autres le long d'un mur. À les voir, aucun n'était mort récemment. Quelques-uns avaient des bouts de peau, de la chair qui pendouillait des os. D'un crâne dépassait le torse d'une poupée en plastique, dont les cheveux étaient gris de poussière.

À mesure que mes yeux s'habituaient à la pénombre, je vis d'autres corps desséchés, bien rangés en un petit tas le long du mur de gauche. Je le remuai du bout du pied et distinguai une couche d'os en dessous : elle était très épaisse.

Nous restâmes les bras ballants, à battre des paupières.

— Ce sont eux ? demanda Zandt.

— Les Hommes de Paille ? Peut-être. Mais certains cadavres sont là depuis très, très longtemps.

Zandt aurait voulu tout retourner, mais un regard me suffit pour comprendre qu'on ne trouverait rien. Ici, quand on tuait quelqu'un, on pouvait prendre tout son temps. Et, en plus, je ne voulais pas rester là. Plus le temps passait, plus on avait l'impression que la cabane respirait avec lenteur, exhalant un air rance. Je voulais sortir.

Je regagnai le seuil, en me retournant vers l'intérieur. Maintenant, j'étais moins surpris par la couleur des arbres. Ils avaient absorbé beaucoup, beaucoup de choses ignobles, qui les maintenaient humides, vivants. Ce qui s'était passé ici s'était déroulé sur une longue période. L'œuvre de plus d'une personne, peut-être de plus d'une génération. Était-ce simplement un repaire où abandonner les corps, ou bien leur présence silencieuse, leurs postures indiquaient-elles qu'on s'était livré à des activités plus nébuleuses ? Je songeai à cette immense région, avec ses espaces désertiques, et me demandai si c'était le seul endroit de ce genre.

Zandt sortit à son tour, puis s'arrêta net en contemplant quelque chose par-dessus mon épaule.

Je fis demi-tour. Il était à une dizaine de mètres, de l'autre côté du canyon, placé de telle façon qu'on l'apercevait en sortant de la cabane.

Le corps était beaucoup plus récent. On n'avait pas pris la peine de le disposer comme le couple que nous avions trouvé un peu plus haut ; on l'avait simplement jeté à terre. Il avait les bras étendus et une jambe repliée. Quelque chose de brun lui avait été cloué sur la poitrine, bien au milieu : on ne pouvait pas le manquer. La vacuité presque irréelle de la bouche béante m'apprit ce que c'était.

— C'est le type ? C'est Joseph ?

Zandt n'eut pas besoin de répondre.

Nous revînmes au camion après une longue marche et roulâmes en silence sur l'autoroute menant à Portland. À l'aéroport, nous prîmes des vols différents.

Nous ne nous revîmes qu'un mois plus tard, alors que tout avait changé.

PREMIÈRE PARTIE

HAVRES FROIDS

Je crois vraiment,
Sans les avoir trouvés,
Qu'il y a peut-être
Des mots qui sont des choses.

BYRON,
Le Pèlerinage
de Childe Harold

CHAPITRE 1

Il n'y a jamais de parking quand il le faudrait. Vous êtes là, fonçant à toute allure, la forêt des deux côtés, franchissant sans effort des côtes en pente douce et des descentes escarpées ; des rangées de bouleaux encadrent des perspectives fugitives d'une telle beauté enneigée que vous ne les voyez même pas, tout en pensant qu'au sortir du prochain virage vous tomberez forcément sur un endroit où vous pourrez vous garer, mais bizarrement il n'y en a jamais. On est milieu janvier, par un mardi après-midi nuageux, ce qui vous paraît étrange, car c'est un curieux moment : vous êtes seul sur la route, vous pourriez arrêter la voiture sur le bas-côté, mais ça ne vous semble pas correct. Ce n'est qu'un véhicule de location et il vous indiffère, mais c'est le dernier que vous conduirez, et vous ne voulez pas l'abandonner. Simple question d'ordre. Vous voulez que la voiture soit garée. Pour être en paix. Vous comprenez brusquement que c'est bien là le problème – un bon résumé de l'existence. Il n'y a jamais d'endroit où se reposer quand on en a besoin.

Merde.

Tom appuya sur la pédale de frein avec quelques secondes de retard, et beaucoup trop fort. La voiture dérapa sur une dizaine de mètres, fit un tête-à-queue puis s'arrêta en plein milieu de la chaussée, comme poussée par une main géante. Il resta immobile un moment, le cou brûlant. De l'air froid entrait par la vitre, avec le croassement d'un oiseau. Sinon, tout n'était que silence, Dieu merci. S'il y avait eu quelqu'un d'autre sur la route, ça se serait mal passé, ce qui n'aurait pas manqué de sel ; mais là encore, il ne le voulait pas. Il était déjà assez mal vu comme ça.

Il manœuvra le véhicule pour se retrouver dans la bonne direction, puis revint lentement en marche arrière vers le parking ; Sarah aurait su faire demi-tour directement, mais pas lui. Du moins n'en était-il pas sûr, aussi préféra-t-il s'en abstenir. Comme toujours. Cache tes faiblesses. Garde tes secrets. Ne cours jamais le risque de passer pour un imbécile, même si ça te donne l'air d'un idiot, et d'un lâche, de surcroît.

Il s'avança dans le petit parking, écrasant dix bons centimètres d'une neige que le vent avait balayée de la route. L'endroit correspondait au point de départ d'une piste de randonnée, fermée pendant la mauvaise saison. Ce n'est qu'une fois la voiture à l'arrêt que Tom se rendit compte que ses mains tremblaient violemment. Il s'empara de la bouteille sur le siège passager et but une longue gorgée, puis il regarda un instant dans le rétroviseur, mais n'y vit, comme il y s'attendait, qu'une peau livide, des cheveux bruns, des joues bouffies, un début de double menton. Le camouflage de l'âge mûr.

Ouvrant la portière, Tom laissa tomber la clé de contact dans le vide-poches. Pas la peine de se faire remarquer. Il s'extirpa du véhicule, glissa aussitôt sur une pierre et s'étala de tout son long.

Se redressant à genoux, il nota qu'une de ses paumes était entaillée ; son front et sa joue droite semblaient saigner un peu. Sa cheville lui faisait mal. De minuscules fragments de silex lui piquaient le visage ; ahuri, le souffle coupé, il reprit ses esprits.

Il sortit son sac à dos du coffre, qu'il referma avec un claquement sec, ce qui lui fit comprendre qu'après tout il éprouvait quelque chose pour la voiture. Il veilla à verrouiller les portières, puis enjamba la barrière de rondins et s'avança entre les arbres, dans la direction opposée à celle de la piste.

L'oiseau, ou un de ses congénères, faisait toujours autant de vacarme. Tom tenta de lui intimer silence, d'abord par des paroles puis par de simples cris. Le volatile se tut, mais ne tarda pas à recommencer. Tom comprit : ici, il n'était jamais qu'un animal bruyant, qui n'avait aucun droit de donner des ordres.

Il laissa donc l'oiseau tranquille et chercha surtout à ne pas tomber.

Avancer était difficile, la pente était raide. Tom comprit vite pourquoi il n'existait pas d'aires de repos : la forêt n'avait rien de paisible. Elle n'était pas là pour que chacun en profite : pas de chemins balisés, de lieux où s'attarder, de boutiques à hamburgers, aucun de ces médiateurs traditionnels entre le cru et le cuit. C'était parfait. Ses besoins était réduits, et assurés. Il n'y avait pratiquement que de l'alcool dans le sac à dos ; il rangea les bouteilles de façon qu'elles ne se heurtent pas. Il avait le ventre vide, la gnôle exceptée. Il doutait déjà que la vodka fût un mode de vie. Elle n'était pas faite pour les petites natures, en tout cas.

Il faut savoir accepter de se sentir très mal. Il y parvenait plus ou moins, il se montrait courageux.

Au bout de deux heures, il estima qu'il n'avait parcouru que cinq kilomètres à peine, bien qu'il ne fût plus entouré que d'épicéas et de cèdres. Le sol était désormais dépourvu de neige, mais aussi couvert de branches mortes et de buissons agressifs qui s'accrochaient à son jean et à sa veste. Les arbres étaient grands, paisibles, ils poussaient où ils voulaient. De temps à autre, il arrivait au bord d'un ruisseau. Au début, il les franchissait d'un bond mais, comme sa cheville le faisait souffrir, il se mit à effectuer des détours pour trouver des endroits plus faciles à traverser. Il marmonnait mais la plupart du temps il restait silencieux, pour épargner son souffle. Achevant la bouteille, il la laissa tomber et continua d'avancer. Quelques centaines de mètres plus loin, il regretta son geste et, faisant demi-tour, entreprit de la retrouver – en vain. Il devenait peu à peu complètement ivre, complètement perdu. Il marchait toujours d'un bon pas. L'examen des cartes lui avait montré que même les routes menant à des chantiers de bûcherons étaient rares dans la région, mais il savait par une expérience certes acquise dans les villes qu'il possédait un assez bon sens de l'orientation. Et aussi à quel point il était faible, à quel point une impulsion pouvait le conduire là où il ne voulait pas aller – avant de disparaître d'un seul coup et de l'abandonner, du sang sur les paumes. C'était bien pourquoi il lui était essentiel de se perdre. Sinon, il changerait d'avis, se dégonflerait, ferait traîner les choses et se planterait ; et il n'y a sans doute rien de plus dérisoire que de rater son propre suicide.

Tom Kozelek était venu dans le Nord-Ouest sans plan aucun, sinon le désir d'être ailleurs qu'à Los Angeles. Il avait choisi Seattle parce qu'il s'y était rendu récemment pour affaires et connaissait un bon hôtel. Il y passa une seule nuit, puis partit en voiture vers les Cascade Mountains : un endroit bizarre, avec des sommets, des vallées vertigineuses, des rochers déchiquetés aux multiples nuances de gris, et même un peu d'Histoire, du genre « et ils coupèrent encore beaucoup d'arbres ». Mais guère de routes, et les montagnes toujours à distance, au point qu'on pourrait croire qu'elles n'existent pas. Pendant deux jours, il erra vaguement d'une petite ville glacée à l'autre, passant ses soirées dans une chambre de motel, la télé éteinte. Il téléphona chez lui – ce qui avait été chez lui. On répondit à son appel, ce qui ne fit qu'aggraver les choses. La conversation avec sa femme et ses enfants fut brève, sans cris – c'était plus horrible encore.

Quelquefois, rien n'est pire que de se montrer raisonnable, parce que, si chacun se comporte en adulte alors que le monde est en miettes, où aller, ensuite ?

Pour finir, il arriva dans une ville nommée Sheffer et s'y installa. L'endroit se réduisait à une avenue principale et à cinq rues perpendiculaires qui prenaient fin très vite au pied des collines couvertes de sapins ; mais un ou deux endroits proposant un lit et un petit déjeuner, un café hippie où l'on trouvait de bons flocons d'avoine, et cinq exemplaires d'occasion de *Sur la route de Madison* laissaient penser que des gens venaient là de leur plein gré. Sans oublier un petit musée ferroviaire (fermé) et un bout de voie ferrée abandonnée le long de la grand-route, où s'entassaient les carcasses de vieux wagons à la rouille pittoresque. On était hors saison, la ville somnolait, les gens du coin sortaient du décor en ôtant des toiles d'araignée de leurs cheveux. Quatre jours avant d'entamer sa marche à travers les bois, Tom se retrouva au comptoir du Big Frank, le moins anodin des trois bars de l'endroit, à contempler à la télé un reportage sur un sport étranger dont il ne comprenait guère les règles, à la fois agité et apaisé à l'idée d'être ainsi perdu en territoire indien. Il avait quarante-trois ans, des cartes de crédit, une voiture à sa disposition. Il n'était pas limité par ce que l'on aurait pu connaître ou espérer de lui. S'il l'avait voulu, il aurait pu prétendre qu'il s'appelait Lance, qu'il était un ancien pilote de chasse devenu milliardaire grâce à Internet, ou un chorégraphe de jazz fusion nommé Bewildergob. Personne ne l'aurait contredit. Il pouvait raconter ce qu'il voulait – mais il comprit aussi que précisément il ne voulait rien du tout.

Plus rien ne changerait jamais. Il avait franchi la ligne.

Tom but jusqu'à en avoir la cervelle vide. L'idée lui vint aussi vite qu'une flèche tirée par un archer. Il se rendit compte qu'il existait un moyen de rendre les choses, sinon meilleures, du moins gérables. D'escamoter les problèmes. Il commanda une autre bière et alla s'asseoir dans un coin pour y réfléchir de plus près.

Il avait déjà pensé au suicide, comme tout le monde, mais jamais sérieusement ; un coup d'œil rapide, pour s'assurer que c'était une idée ridicule. Pas cette fois-ci. Il ne s'agissait pas d'un geste théâtral mais d'une décision rationnelle. Sa situation n'avait rien d'irrévocable. Son mariage était terminé mais il avait encore des amis. Il pourrait se trouver un nouveau boulot de concepteur de sites Internet pour d'autres. Dénicher un appartement. Laver son linge. Avoir un four à micro-ondes bien à lui. Dans un an, tout paraîtrait peut-être différent. Et alors ? Il serait toujours le même Tom, un homme aux maigres talents qui remettait tout au lendemain, lentement dilaté par l'âge comme par une pompe à vélo. Il serait toujours celui qui en était arrivé là. La vie était suffisamment moche comme ça – et si on découvrait le reste ? Les choix qu'il aurait souhaités n'appartenaient plus qu'au passé.

Alors, pourquoi ne pas en finir ? Tirer l'échelle, espérer que la réincarnation existait, tenter de faire un meilleur usage de sa prochaine existence...

Pourquoi pas ? Après tout, pourquoi pas ?

Il but jusqu'à la fermeture, puis tenta de bavarder avec les deux jeunes barmen tandis qu'ils le poussaient sans cérémonie vers la porte. L'un ne trahissait que l'ennui, l'autre un vague dégoût. Tom se rendit compte qu'il n'était sans doute guère plus jeune que leurs pères, sans doute des montagnards à la mâchoire carrée qui buvaient du bourbon une fois par mois. La porte fut vigoureusement refermée derrière lui. Comme il repartait en titubant vers le motel, il songea qu'il n'avait plus à se soucier de ce qu'ils pourraient penser de lui. Son nouveau projet le plaçait sur un plan supérieur. Il n'en avait rien à foutre. Il était si agacé qu'il fit demi-tour et revint vers le bar, comptant expliquer à ces deux crétins que s'il y avait peut-être de bons moments pour les gamins de vingt ans, c'était beaucoup moins gai pour les hommes d'âge mûr ; qu'un jour ils perdraient leurs abdos, oublieraient comment aimer, sans plus savoir qui ils étaient. Il avait l'impression que ce serait pour eux une leçon profitable. C'était d'ailleurs la seule qu'il pût leur offrir, de toute façon, et il tenait à ce qu'ils en bénéficient. Mais le temps qu'il arrive au bar, celui-ci était plongé dans l'obscurité. Il frappa à la porte un moment, en se disant qu'ils étaient peut-être encore à l'intérieur, mais surtout parce qu'il voulait cogner sur quelque chose. Puis il fut brusquement inondé de lumière. Se retournant, il aperçut une voiture de police garée dans la rue derrière lui. Un jeune type en uniforme s'appuyait sur une aile, les bras croisés.

— Je crois que c'est fermé, monsieur, dit-il.

Tom ouvrit la bouche, se rendit compte qu'il aurait trop à dire, et rien qui tienne debout. Il leva les bras, non pour se rendre, mais en une sorte de prière muette. Bizarrement, l'autre sembla comprendre, hocha la tête sans répondre, remonta dans la voiture et s'en fut. Tom rentra à son motel, avançant lentement au milieu de la grand-rue, sous le clignotement méditatif et résolu des feux de circulation, sans la moindre voiture en vue.

Le lendemain matin, il réfléchit longuement à tout ça. Ses options étaient limitées. Il n'y avait pas d'armurerie en ville, et il ne voulait pas rouler pour en chercher une. Même en partant du principe qu'on le laisserait acheter une arme à feu, elle lui aurait fichu la trouille. Sauter d'une falaise, à supposer qu'il en déniche une, était également exclu. L'idée était de toute évidence contre-révolutionnaire. Même si son esprit était bien décidé, son corps pouvait se montrer le plus fort, auquel cas il aurait un long trajet à parcourir jusqu'à la voiture, avec l'impression d'être le pire imbécile de la planète. Oui, je voulais me jeter de la falaise, c'est ça. Non, ça n'a pas marché. Désolé. Mais le paysage était superbe. Fais attention où tu marches. Tom ne voulait pas finir en cadavre disloqué qui serait retrouvé, photographié et renvoyé chez lui. Il ne voulait pas être brisé, mais effacé.

Le dimanche, il était chez Henry, le plus sympathique restaurant de la ville, devant un énorme hamburger, quand il entendit quelque chose qui mit tout en place. Un vieux briscard du coin se délectait à effrayer un couple de retraités de Winnebago en leur expliquant à quel point les bois étaient vastes et impénétrables. La répétition d'un chiffre attira l'attention de Tom. Soixante-treize.

Le couple se regarda et hocha la tête, très impressionné. Puis l'homme dit, de l'air du gars qui a trouvé la faille dans un raisonnement :

— Des grands ou des petits ? Quelle taille ils avaient, ces avions ?

— De toutes sortes, répondit le vieux schnock d'un ton un peu irrité. Des grands, des petits, des civils, des militaires... Les avions s'écrasent tout le temps, c'est un fait. Ce que je veux dire, c'est que de tous ceux tombés dans la région depuis la fin de la guerre, il y en a soixante-treize qu'on n'a jamais retrouvés.

C'est vrai ? songea Tom.

Il repoussa son hamburger, paya et partit acheter autant d'alcool qu'il pourrait en porter.

Tom fut surpris par la rapidité avec laquelle tombait la nuit. Il trébuchait plus qu'il ne marchait ; les muscles de ses cuisses et de ses mollets s'étaient transformés en plomb. Il n'avait fait qu'une douzaine de kilomètres, quinze au plus, mais il était épuisé. L'idée lui vint que s'il avait passé plus de temps dans un gymnase, il aurait été en meilleure forme pour mourir. Cela le fit rire, jusqu'à ce que sa bouche se remplisse de salive et qu'il soit contraint s'arrêter pour s'empêcher de vomir, plus ivre que jamais. Tout en se reposant un peu, les mains posées sur les genoux, à observer les taches flottantes qui passaient devant ses yeux, il réfléchit à la suite des événements. Il était déjà complètement perdu. Tout au long de l'après-midi, le terrain était devenu de plus en plus montagneux, raide, traître et glissant. Quand la nuit viendrait, il ferait vraiment noir, le genre d'obscurité qui engloutirait un gars de la ville. Il ôta son sac à dos et y chercha sa lampe de poche. En l'allumant, il comprit que la lumière n'était pas la seule à changer. Le brouillard se levait, il faisait aussi incroyablement froid. Pour le moment, la sueur qui lui coulait sur la peau se transformait en eau glacée, mais quand il serait gelé jusqu'aux os, ce serait difficile à supporter. Ce qui signifiait qu'il devait continuer à marcher.

Il fit pivoter sa cheville pour la réchauffer un peu, obliqua et avança péniblement. La forêt était tranquille : les oiseaux avaient croassé tout leur saoul avant de rentrer chez eux. S'agissant des autres animaux, Tom ne savait pas trop. Il avait déjà passé beaucoup de temps à ne pas penser aux ours. Tom ne croyait pas avoir l'air très menaçant pour tout grand mammifère qu'il pourrait rencontrer par hasard, et il n'emportait aucune nourriture qui puisse les attirer, mais peut-être que tout ça était bidon. Peut-être attendaient-ils les gens pour les attaquer, rien que par plaisir. En tout cas, il ne voulait pas y réfléchir. La torche comportait deux positions de réglage, lumineux et atténué ; il ne tarda pas à s'en tenir à la première. Le brouillard s'épaississait, lui renvoyant de plus en plus de lumière en plein visage, ce qui lui donnait le vertige. Et les ombres prenaient une allure inquiétante. En plein jour, les forêts sont des lieux amicaux, elles vous rappellent les promenades dominicales, les feuilles qui bruissent, la grosse main tiède de vos parents ou celle que vous tendiez à vos enfants. De nuit, elles tombent le masque, elles vous disent : « Va donc te trouver une grotte, homme-singe, ici ce n'est pas pour toi. »

Tom décida donc de rester insensible au brouillard, de s'assommer la cervelle à la vodka et de marcher toujours. Il se persuada que les craquements et les bruissements qu'il entendait étaient le fruit de son imagination. On pouvait continuer à avancer, en sûreté, dans un inconfort modéré mais sans plus : marcher jusqu'à ce qu'il fasse nuit noire et que le temps semble se réduire à rien, jusqu'à ce que la peur se replie sur elle-même, puis enfle peu à peu ; et il marcha de plus en plus vite, pour échapper à quelque chose qu'il portait en lui.

Sa chute le prit au dépourvu. Il était occupé à franchir une longue rangée de buissons, ce qui provoquait un nouvel accès de hoquets qui lui faisaient tressauter la tête, quand brusquement il n'eut plus rien sous le pied. Son corps fut projeté en avant, sans qu'il puisse garder l'équilibre.

Il dévala d'un seul coup une pente très raide, jambes écartées, battant des bras. Il alla donner tout droit contre un petit arbre, et en perdit sa torche et sa bouteille, non sans se heurter à tous les rochers rencontrés en route. Cela dura peu de temps et se termina par une réception à plat ventre qui lui coupa ce qui lui restait de souffle.

Tom eut un grognement de désespoir. Quand il en fut capable, il roula sur le dos. Sa douleur à la poitrine était si intense qu'il eut un sifflement involontaire. Cela lui faisait beaucoup plus mal que sa chute dans le parking, comme si on lui avait planté une lance dans le flanc droit et qu'on encourageait un enfant à en extirper la pointe. Ses testicules le torturaient, cela montait d'un petit creux brûlant dans son basventre.

Au bout d'un moment, il se redressa, se passa une main hésitante sur le flanc, sans vouloir regarder, au cas où. Rien. Il distingua la torche à trois mètres de lui, luisant faiblement dans la futaie, et rampa dans la boue pour la récupérer. Il voyait un peu double, mais comme c'était le cas depuis quelques heures, il ne s'en inquiéta pas outre mesure.

Retrouver la torche lui donna le sentiment d'avancer dans la bonne direction. Il semblait être tombé dans une vaste ravine rocheuse qui, au printemps, devait accueillir un ruisseau mais n'abritait plus qu'un mince filet d'eau, qu'il entendait devant lui, à quelques mètres. Sinon, tout était tranquille. Tout à fait tranquille, et très froid.

Tom estima qu'il était allé assez loin. Cette nuit même ferait l'affaire. Après tout, pas besoin d'attendre demain. L'heure de la sortie sonnait simplement un peu en avance, voilà tout.

Il recula, le dos contre un rocher. Puis il plaça le sac entre ses genoux et l'ouvrit. Une bouteille au moins s'était brisée ; le fond était trempé, l'odeur lui monta au visage. Allumant la torche, il constata que mieux valait ne pas y fourrer la main. Il préféra donc en retourner le contenu sur le sol. Cela prit un moment, mais il finit par trouver les boîtes de somnifères.

Tout en extirpant péniblement chaque comprimé de son emballage plastique, avant de les déposer en tas sur une feuille morte près de lui, il mena à bien une check-list intérieure.

Perdu, vérifié. Ivre, vérifié. Bon Dieu oui. Plutôt deux fois qu'une, souligné en rouge.

Il avait payé sa note d'hôtel, en mentionnant au passage qu'il retournait à Seattle. Vérifié.

Pour se promener dehors par un froid pareil, il fallait être complètement cinglé, et on était hors saison, en milieu de semaine. De plus, il s'était tenu à l'écart des pistes connues. Vérifié.

Pop, un comprimé. Pop, un autre. Il contempla la pile. Ça suffirait ? Autant être sûr. Il continua. Une overdose n'était pas signe de faiblesse si on s'y prenait comme lui. C'était viril.

Ah ouais.

La voiture serait peut-être retrouvée demain, et dans un jour ou deux quelqu'un enquêterait. Pas à pied, mais en hélico, un survol décousu, au mieux. Lors de sa dernière journée à Sheffer, Tom avait acheté des vêtements et un sac à dos aux couleurs automnales pour que d'en haut on ait encore moins de chances de l'apercevoir. S'il avait aussi banqué pour de bonnes chaussures de marche, sa cheville lui aurait fait moins mal, mais sur le moment cela semblait ne pas en valoir la peine. Comme quoi on doit toujours avoir l'équipement qu'il faut.

En tout cas, check-list générale. Vérifier les vérifications.

À mesure que les comprimés s'entassaient, il fut surpris de constater qu'il n'avait pas peur. Il avait pensé que la proximité de l'acte le ferait paniquer, qu'il chercherait à combattre la mort, mais il se rendit compte qu'il se sentait simplement très, très las. Quelque part sur le trajet entre la voiture et cette ravine, il avait perdu tout sentiment de sa propre existence. Ce n'était plus qu'un événement, ici et maintenant. Il faisait noir, il était tard. Tout était parfait. Tout allait bien.

Tom avait déjà froid, ses doigts étaient gauches et gourds. Il avala les comprimés, à raison de deux à la fois, qu'il fit passer avec de l'alcool. Il en perdit quelques-uns, mais il n'en manquait pas. Il en prit beaucoup, en marmonnant dans la pénombre. Adieu, Sarah, trouve-toi quelqu'un d'autre. Adieu, William, adieu, Lucy. Vous allez me haïr, je le sais, mais de toute façon vous en seriez vite venus à me détester.

À un certain moment, il parut accepter l'idée qu'il avait désormais dépassé la dose mortelle, après quoi tout devint beaucoup plus facile, à vrai dire. La forêt paraissait un peu moins froide. Tout devint flou et liquide tandis qu'il restait assis à tanguer dans le noir complet. Il avait froid et n'avait pas froid, il était mortellement las et parfaitement éveillé. La peur tournoyait dans les buissons mais restait hors d'atteinte, jusqu'à ce qu'il ne prenne plus la peine de se fourrer des choses dans la bouche. Vouloir suivre ses pensées, c'était comme marcher seul dans une rue déserte où tous les magasins fermaient l'un après l'autre.

Quand ses paupières commencèrent à battre, il tenta de garder les yeux ouverts, non par désespoir, mais un peu comme un enfant cherche à repousser le sommeil tout en sachant qu'il ne pourra le combattre. Lorsqu'ils se fermèrent enfin, il eut la tête un peu plus légère pendant un moment, puis tout sembla disparaître derrière un voile gris. Pendant un bref instant, il crut rêver, comme s'il s'inclinait lentement vers l'arrière, puis plus rien. Adieu.

Il ne s'attendait pas à se réveiller en pleine nuit, encore ivre, le corps parcouru de frissons. Il ne s'attendait pas à être toujours en vie, ravagé de douleurs. Et il ne s'attendait certainement pas à voir quelque chose au-dessus de lui, quelque chose d'énorme, à l'odeur de viande putréfiée.

CHAPITRE 2

Le restaurant était une grande salle divisée en deux sections : des tables au milieu, des alcôves sur trois côtés. À l'entrée de chacune étaient accrochées de petites lanternes qui ne marchaient plus. Les murs étaient ornés de grandes fresques dans le genre rétro : beaucoup de bleu, de rose pâle, de lignes noires écaillées. À l'entrée, de hautes fenêtres festonnées donnaient sur un parking envahi de feuilles mortes, et je regardai un moment le vent jouer avec elles. Je me trouvais à ma place habituelle, dans l'une des alcôves du fond. La banquette n'était pas trop près de la table, on ne sentait pas trop serré. Le menu était plein de calembours subtils, et de hamburgers, de burritos, de salades du chef et de chili – bref, mon genre de truc.

Tout bien considéré, c'était l'endroit idéal pour dîner, à un détail près. Le service était nul. J'attendais depuis longtemps, personne n'était venu me saluer, m'assurer que je faisais partie de la maison ou m'offrir de l'eau glacée que je ne boirais pas. Et en fait, ce n'était pas seulement le personnel qui déconnait. À mon arrivée, j'avais constaté que quelqu'un avait renversé presque toutes les chaises au centre de la salle, ce qui faisait désordre. Je les avais remises en place, mais après tout ce n'était pas mon boulot. Pas plus d'ailleurs que de remplacer les ampoules. Je songeai à passer à la cuisine, mais je savais que ça ne servirait à rien. C'était encore plus calme, là-bas, et plus sombre.

Je me penchai sur ma table en me demandant ce que je pouvais bien foutre là. Attendre trois jours un bol de chili, c'est long.

Je savais beaucoup de choses sur les villes de ce genre, parce que j'y avais passé le plus clair de mon temps ces derniers mois. J'avais erré en tous sens sur des kilomètres de campagnes paumées et de prairies, dans les États les plus mornes du pays. Au début, je descendais dans des motels puis, un après-midi, je voulus retirer de l'argent dans un distributeur et découvris que je ne possédais plus rien. C'est étonnant à quel point un petit rectangle aux couleurs vives peut influer sur votre bien-être, votre sentiment d'identité. Vous ne commencez à le comprendre que lorsque la machine recrache la carte de crédit et vous dit non – et ça ne signifie pas maintenant, pas plus tard, mais plus jamais ; cela vous rappelle tout d'un coup que la carte n'a jamais été un athanor magique produisant de l'or, mais un bout de plastique dont, légalement, vous n'êtes même pas le propriétaire. Je restai là, dans un parking du New Jersey, à la retourner entre mes doigts, jusqu'à ce qu'une femme dans un 4 × 4 chargé de trois enfants obèses me dise de dégager. Elle-même avait une carte, sûre qu'elle remplirait sa fonction. Je l'enviai pour ça – pas pour ses mômes, moches comme le péché.

Je revins à ma voiture et y montai avant de contempler un moment l'extérieur à travers le pare-brise. Il me restait dix-huit dollars, un peu de monnaie, un réservoir à moitié plein. Rien d'autre. Rien.

— Alors, Bobby, qu'est-ce qu'on va faire ?

Il ne répondit pas : il était mort. C'était mon meilleur ami, l'une des rares personnes qui aient compté pour moi. Il était mort dans un endroit appelé les Halls, alors que nous tentions de capturer un psychopathe qui se faisait appeler l'Homme Debout. Les Halls avaient été détruits par une explosion qui avait désintégré Bobby en même temps. Depuis, il était devenu un interlocuteur imprévisible. Parfois, il disait ce que j'avais besoin d'entendre : oui, Ward, c'est peut-être un bon endroit pour passer la nuit, ou : oui, j'avais bien besoin d'une autre bière, ou encore : oui, on a fait de notre mieux pour retrouver les gens qui avaient tué mes parents.

Puis il restait silencieux pendant de longs moments. Des semaines, parfois. Je ne savais pas où il allait pendant ce temps, quels changements survenus dans mon esprit faisaient que je ne pouvais plus l'entendre. Et je ne l'entendais que dans ma tête, je le savais.

Pour finir, je quittai le parking, et me trouvai, à trois villes de là, un boulot de plongeur de restaurant et d'éplucheur de patates. Le cuisinier équatorien me laissa dormir sur son plancher pendant deux jours à l'issue desquels j'avais assez d'argent liquide pour louer une chambre, pourvu bien sûr que j'accepte de la partager avec les punaises, le bruit et la poussière, et de ne pas manger. Travailler à la cuisine, c'est pas mal pour les gens qui sont dans ma situation, bien qu'on se lasse vite de la clientèle.

Les relations entre l'Équatorien et moi prirent fin une semaine plus tard, quand je tentai de prendre part à son petit trafic de coke auprès des autres membres du personnel et de certains résidents du lieu, jeunes ou moins jeunes, qui le soir venaient frapper à la porte de service. À l'aube, je quittai la ville à toute allure, saignant abondamment, avec la sensation d'être un parfait imbécile.

Le lendemain matin, en Virginie-Occidentale, je prenais un peu de repos à la terrasse d'un Burger King, saignant toujours, mais moins, quand j'entendis dans ma tête une voix qui répondait enfin à une question posée neuf jours plus tôt. J'allai me nettoyer dans les toilettes, m'offris un petit déjeuner mondialiste de matériaux d'allure alimentaire, et partis tout droit vers l'Arizona. Arrivé là, je repérai une résidence à Flagstaff, ce qui prit du temps parce que je n'y étais venu qu'une fois, passablement ivre, et en avais perdu l'adresse. Pendant vingtquatre heures, j'observai l'endroit avec soin avant de sortir mon petit rectangle de plastique désormais inutile et d'en faire usage pour entrer.

Et c'est ainsi que, cinq jours durant, je vécus dans la maison de Bobby Nygard.

Une fois que j'eus fait le tour des lieux, assuré que personne n'était venu cambrioler – si c'était le cas, ils avaient tout laissé en place, dont un matériel informatique valant des dizaines de milliers de dollars –, je commençai par me connecter au Net. Il y avait un certain temps que je ne m'y étais pas risqué, plus ou moins convaincu que toute tentative en ce sens serait remarquée quelque part et lancerait des gens à mes trousses. Bobby avait été un expert de l'effaçage des traces sur Internet ; en me servant de son système je serais en sécurité, au moins pour un petit moment.

J'examinai d'abord mon compte bancaire, constatai qu'on l'avait fermé ; son contenu avait disparu. Un autre compte dans une autre banque était toujours ouvert mais vide. C'est là qu'avait été transféré l'argent de l'héritage de mes parents. Quelqu'un l'avait nettoyé, en n'y laissant qu'un seul cent.

Je me déconnectai et restai là, la tête vide. Je n'étais pas franchement surpris, mais c'était quand même une très mauvaise nouvelle et j'avais envie de frapper quelqu'un. J'allai à la cuisine, pris une soucoupe pour me servir de cendrier et contemplai longuement la rue par la fenêtre. C'est alors que j'entendis Bobby. Il m'avait toujours cassé les pieds pour que je cesse de fumer et n'avait pas changé d'avis. Ce qui ne m'empêcha pas d'achever ma cigarette. C'était bon d'entendre la voix de quelqu'un, même si elle me gonflait, même si c'était la mienne.

Je restai dans la maison. Elle paraissait sûre et j'étais fatigué de bouger. Avec ses boîtes de conserve dans les placards, je n'avais pas à sortir. Je passai beaucoup de temps à lire les notes et les manuels de Bobby, puis fouillai les lieux, du sol au plafond, aussi respectueusement que je pus. Je découvris un ensemble de fausses identités dont je m'emparai, sachant que Bobby avait dû les acheter à quelqu'un en qui il avait confiance. Je trouvai également près de six mille dollars en liquide, cachés à la cave dans un boîtier d'ordinateur. Je m'assis pour les contempler un moment, pas fier de les avoir trouvés, et encore moins de ce que j'allais en faire. Bobby avait une mère ; je l'avais retrouvée un mois plus tôt pour lui apprendre sa mort. Ivre, elle m'avait jeté des choses à la tête, sans que j'aie su si c'était sa réaction à la nouvelle – ils n'étaient pas très intimes – ou sa pratique habituelle. Sans doute le magot aurait-il dû lui revenir, mais tant pis. C'était sans doute de l'argent sale, et je sentais que Bobby m'aurait approuvé. De toute façon, ce serait comme ça et pas autrement.

Quelques jours plus tard, je quittai les lieux, avec des vêtements à lui qui m'allaient plus ou moins, et chargé d'un petit sac contenant l'argent. J'avais aussi un de ses ordinateurs portables, ayant dû fourguer le mien quelque temps auparavant. Au milieu de la rue, je me retournai pour regarder la demeure, en me demandant combien de temps il faudrait pour que quelqu'un décide de la visiter. Des semaines, probablement. Beaucoup plus longtemps, même, tant que les factures seraient payées par prélèvement direct et que quelque chose ne sauterait pas en provoquant un incendie. Je me demandai aussi combien de chambres ou de maisons, dans tout le pays, étaient comme ça : une fois leurs occupants disparus, les machines continuaient à tourner toutes seules, sans personne pour s'en occuper.

Après cela, je m'arrêtai dans bien des lieux semblables. Parfois je puisais dans le magot de Bobby pour séjourner dans un endroit qui me rappelait qu'autrefois j'avais eu une vie, du genre chaîne d'hôtels de grande ville, où le matin il faut téléphoner à la réception pour se souvenir dans quel État on se trouve. Sinon, je prenais ce qui me tombait sous la main. Des motels aux fenêtres clouées de planches, à la sortie des villes, des bureaux de zone commerciale dont les vitres avaient viré au gris ; tout ce qui était oublié, négligé et portait un panneau *Entrée interdite*, seule chose qui puisse vaguement vous empêcher d'y pénétrer – avec la crainte de tomber sur quelqu'un qui pourrait recourir à la violence pour défendre son asile temporaire. Fort heureusement, je faisais moi-même partie du lot, si bien qu'une telle perspective ne m'inquiétait guère. Il y eut quelques confrontations, mais ceux qui ne possèdent rien se laissent facilement intimider, pourvu qu'on sache garder son calme et leur faire croire qu'on n'est pas comme eux.

John Zandt avait également survécu à l'explosion. Il m'appela un soir et nous nous retrouvâmes à Yakima. Nina rédigea un rapport interne au FBI sur ce que nous y avions vu et alerta le bureau local du Bureau mais cela parut ne déboucher sur rien. Nous étions seuls, en pleine cambrousse, ce qui nous permit de comprendre que la conspiration que nous avions découverte avait le bras plus long que nous ne pensions.

Ensuite, je m'essoufflai. Mes progrès, si on peut dire, se firent de plus en plus lents, jusqu'à ce que j'échoue à Relent, Idaho. J'avais un téléphone portable enregistré sous un faux nom, l'ordinateur d'un mort et une réserve d'argent sale qui commençait à fondre. Mes côtes me faisaient encore mal là où le couteau d'un dealer les avait frappées.

Mes parents auraient été fiers de moi.

Pour finir, je quittai le restaurant abandonné et entrai dans ce qui passait pour la rue principale de Relent. Les promesses du menu m'avaient donné faim et je n'avais en poche que quelques bâtonnets de bœuf teriyaki. Je dénichai un bar appelé le Cambridge, tenu par un couple d'âge mûr, très affable. Leur menu était moins appétissant que l'autre, toutefois, et je m'en tins donc à du scotch et à une quelconque bière locale qui donnait l'impression d'avoir été extraite des murs de vieux bâtiments mais se laissait boire une fois qu'on en avait consommé trois ou quatre. Je comptais m'en aller ensuite, mais il se mit à pleuvoir, un déluge qui s'en venait battre les vitres du bar ; on aurait cru que quelqu'un jetait des poignées de gravier. Je restai donc assis, effondré sur mon siège, à manger des olives sur un rythme lent mais soutenu, jusqu'à ce que je commence à redouter la crise de foie et que mes doigts aient pris une vague couleur verte.

À 21 heures, j'étais vraiment ivre. Une heure plus tard, ça ne s'était pas arrangé. Une jeune femme exaltée, aux cheveux frisés, chantait sur une petite scène des chansons dont le sens m'échappait. Je sentais que le monde avait été dur avec elle et sympathisais jusqu'à un certain point, mais sa voix me donnait mal à la tête. Il était temps d'aller voir ailleurs. Où ? Et il pleuvait toujours. De temps à autre, des gens entraient : à les voir, on aurait cru qu'ils venaient de sortir de l'océan tout habillés.

Au bout d'un moment, l'un d'eux retint mon attention : grand, mince, il alla s'asseoir seul à une table du fond. Je me rendis compte que je gardais l'œil sur lui dans le miroir derrière le comptoir. L'éclairage avait baissé au point d'en devenir ténébreux et je ne voyais pas bien le visage du gars, mais un picotement dans le cuir chevelu me disait qu'il regardait dans ma direction plus souvent que le hasard ne l'aurait permis. Je me levai pour aller aux gogues, ce qui était inutile, mais quand je passai à sa hauteur il avait tourné la tête, ostensiblement, pour regarder dehors.

Aux toilettes, je fis couler de l'eau que je me passai sur le visage. Il se pouvait simplement que ce type ait remarqué un étranger ; je pensais toutefois qu'il y avait plus que ça. Il existait une fenêtre très haute, mais rien pour y grimper, sinon un lavabo qui n'avait pas l'air de pouvoir supporter mon poids, et mes épaules avaient peu de chances de passer à travers.

Je décidai qu'il faudrait que j'affronte le gars. S'il devait se passer quelque chose, mieux vaudrait que ce soit dans un lieu public.

Quand je revins, la table était vide.

Maudissant ma paranoïa, je revins au bar et bus une gorgée de ma bière, désormais tiède. La chanteuse avait été rejointe par une amie dont la chevelure n'avait rien à lui envier. J'adressai un signe au barman et le patron m'apporta la note, qui n'était pas du genre coup de fusil. Je bavardai avec lui quelques instants et laissai un gros pourboire. Mon père m'a très bien élevé.

Dehors, il faisait encore plus froid que je ne l'aurais cru. Je faillis faire demi-tour, pour voir si, peut-être, ils pourraient m'adopter ou du moins me laisser dormir dans le bar, mais chaque fois qu'une porte se ferme derrière moi, je n'ai pas l'impression de pouvoir retourner en arrière. Pour me protéger de la pluie, je restai près des vitrines en avançant dans la rue déserte. J'aurais pu rentrer en voiture les yeux fermés, sans mettre en danger qui que ce soit, sinon moi-même.

Il me fallut une minute ou deux avant de comprendre que la raideur de mon dos voulait me dire quelque chose.

Je fis demi-tour. Il n'était pas facile d'y voir, mais j'aperçus quelqu'un dans l'encadrement d'une porte, à mi-chemin du trajet menant au bar. Je ne distinguais pas son visage et il ne bougeait pas, mais personne ne pouvait être dehors ce soir-là rien que pour le paysage.

— Je peux vous aider ?

Pas de réponse. Je glissai ma main dans ma veste. Bien entendu, j'avais laissé mon arme dans la voiture. Qui pouvait en avoir besoin à Relent, Idaho ?

— Qui vous envoie ?

Le gars sortit de l'ombre sur le trottoir. Il dit quelque chose que la pluie emporta.

J'étais las, ivre, j'avais la trouille. Tout me disait de dégager. Mais non. S'ils pouvaient me gauler ici, ils pourraient me gauler partout. Voilà à quoi se réduisait ma vie, désormais. Cela se produirait n'importe où, tôt ou tard. D'un seul coup, tout ce que j'ignorais, tout ce que je n'avais pas, fut là devant moi ; je me sentais vide, glacé intérieurement.

Je courus vers lui.

Il recula de quelques pas, mais pas assez vite. Je lui étais tombé dessus avant qu'il ait compris ce qui lui arrivait, et je me mis à le frapper. Je savais que j'aurais dû arrêter, que peut-être il connaissait des choses que je devrais savoir, mais je m'en foutais. J'y allais de la tête et des deux mains, si bien que nous tombâmes tous deux à terre. Je le repoussai pour me redresser, lui lancer un coup de pied, puis me penchai pour le saisir par les cheveux et le relever, prêt à lui cogner la tête par terre jusqu'à ce que tout soit terminé. J'étais vaguement conscient d'un bruit derrière moi, mais je ne fis pas le rapport jusqu'à ce qu'on me tire en arrière et que je comprenne quel idiot j'étais d'avoir cru qu'ils enverraient un seul homme et pas toute une bande, que ce dont j'aurais dû être surpris, c'est que l'un d'eux ne tire pas pour en finir.

Quelqu'un me saisit, en m'immobilisant chaque bras. Quelqu'un d'autre était agenouillé près du gars que j'avais frappé, en essayant de lui maintenir la tête au-dessus du sol trempé de pluie. Son visage était couvert de sang, mais je remarquai qu'il était beaucoup plus jeune que je l'avais cru, une vingtaine d'années, vingt-cinq au plus. Une femme le soutenait. Elle leva les yeux et je me rendis compte que c'était la patronne du Cambridge.

— Sale con ! dit-elle.

— Vous êtes un grand costaud, hein ? dit une voix derrière mon oreille droite, et je me tordis le cou pour constater que c'était le mari de la femme.

— Qu'est-ce qui se passe, merde ?

Deux ou trois personnes se trouvant au bar m'entouraient.

— Il m'observait au bar. Il était là à m'attendre.

La femme se dressa :

— Ricky est gay, dit-elle.

— Comment ? haletai-je, le visage brûlant.

Son époux me lâcha le bras :

— Vous vouliez lui donner une leçon ? Vous avez un problème avec les gens comme Ricky ?

Il s'éloigna d'un pas, comme si j'étais contagieux.

— Écoutez...

Ils n'en avaient aucune intention. Les chanteuses frisées avaient aidé le gars à se relever et le ramenaient vers le bar. La femme me jeta de nouveau un regard mauvais, parut vouloir dire quelque chose, puis se contenta de secouer la tête. Jamais je ne m'étais senti aussi petit. Elle partit avec les autres, la main posée sur le dos du jeune, et je compris beaucoup trop tard que Ricky était son fils.

Je me retrouvai seul avec son mari.

— Je ne savais pas, dis-je.

— Vous auriez pu lui demander.

— Vous n'avez aucune idée de ce qu'est ma vie.

— Non, en effet. Et je ne veux pas le savoir. Je ne sais pas non plus où vous logez. Mais vous devriez vous en aller. Vous n'êtes pas le bienvenu ici.

Il repartit vers le bar et, comme il en ouvrait la porte, se tourna pour ajouter :

— Je doute que vous soyez le bienvenu où que ce soit !

La porte se referma avec bruit, puis il n'y eut plus que la pluie.

Nietzsche disait que les hommes et les femmes de caractère connaissent des expériences typiques, des événements qu'ils semblent destinés à reproduire sans fin, dont ils peuvent dire : « Oui, c'est comme ça. » Du moins je crois que c'est lui, mais ça pourrait être Homer Simpson. En tout cas, l'un et l'autre pensaient sans doute à des choses plus positives que de se battre dans des endroits dont jamais personne n'a entendu parler ou d'imposer sa propre paranoïa à des gens qui ne le méritent pas. Chose incroyable, j'avais fait la même chose le soir des funérailles de mes parents, en sortant une arme dans le bar d'un hôtel, en flanquant la trouille à un groupe d'hommes d'affaires, et à moi aussi, par la même occasion.

Relent, Idaho, me montra enfin que ce n'était pas une façon de vivre. Comme me l'avait dit trois mois auparavant une fille qui savait par expérience de quoi l'Homme Debout était capable, il n'existait qu'une personne pour le boulot que je devais faire. Il fallait que je cesse de courir et me mette en chasse.

Le lendemain, à 16 heures, j'étais à San Francisco, et en fin de soirée j'avais enfin une piste.

CHAPITRE 3

L'aube trouva Tom blotti au pied d'un arbre, le regard fou, et gelé. Elle tenta de le renvoyer d'où il venait, mais il s'y refusa fermement. Il n'allait quand même pas se voir dénier le droit à un petit matin, même s'il ne s'était pas attendu à le voir.

Quand il était revenu à lui pendant la nuit, tout s'était passé très vite. Son esprit avait trouvé la pédale de fuite et appuyé dessus de tout son poids. Le tout accompagné du pénible sentiment qu'il s'était vraiment bousillé, mais alors l'odeur se fit de nouveau sentir : l'aire linguistique de son cerveau s'éveilla comme une sirène – UN OURS ! UN OURS ! UN OURS ! – et il fila comme une flèche.

Au début, il rampa sur les mains et les genoux, mais la crainte des griffes le remit sur pied. Il ricocha contre les parois de la ravine jusqu'à ce qu'il atteigne la forêt, puis remonta à toute allure la bordure boueuse et s'enfuit.

Ne pas regarder en arrière lui fut facile : il ne voulait pas voir. Des rapports lui parvenaient de lointains avant-postes – une tête en morceaux, une cheville qui hurlait, une torche disparue –, mais il n'en tint aucun compte et progressa dans l'obscurité en trébuchant. Toutes les douleurs n'étaient rien, comparées à ce que ce serait d'être dévoré par un OURS ! Il se précipita d'une manière qui court-circuitait tout ce que son espèce avait appris depuis l'avant-dernier âge glaciaire. Il courut comme un animal, poussé par une pure magie corporelle, il rebondit à travers des buissons, par-dessus des troncs, et finit par arriver dans un endroit où les arbres se raréfiaient.

Comme il avançait à tâtons, Tom remarqua qu'il avait neigé, bien après que l'information lui eut été transmise par les lourds crissements sous ses pieds. Cela, combiné aux gifles des branches, au geignement de ses poumons, formait une telle cacophonie qu'il lui fallut un moment pour se rendre compte que c'étaient les seuls bruits qu'il entendait. Rien d'autre, même en retenant son souffle pour mettre un terme à ses halètements. Sa poitrine le brûlait.

Il recula un peu pour se placer près d'un gros arbre. Essayer d'y grimper ne servirait à rien. L'ours serait beaucoup plus doué que lui en ce domaine, d'autant plus que le plantigrade ne serait pas sur le point de s'évanouir.

Il écouta. Tout restait tranquille.

Puis il crut avoir entendu quelque chose.

C'était en bas, au fond, parmi la noirceur d'encre et les ombres glacées. Un craquement de branchages.

Son corps se figea, accablé, mais il ne put bouger. Il avait couru par panique, et il ne lui restait plus que la terreur. Laquelle ne savait pas comment actionner ses membres.

Il resta là, immobile, mais n'entendit plus le bruit.

Pour finir, il tourna sur lui-même, en un cercle complet, aux aguets. Rien. Il ne voyait que de la neige et des ombres. Il n'entendait que des bruits d'eau qui goutte, un bruissement très doux quand une poignée de neige tombait d'une branche. Il ne savait que faire.

Il resta donc où il était.

Vers 6 heures du matin, il se sentit dans un état épouvantable. Une bosse sur sa tempe droite – sans doute due à sa seconde chute – lui donnait l'impression de tourbillonner. Ses côtes, sur le flanc droit, étaient atrocement douloureuses, qu'il remue ou non. La température glaciale portait l'effet au carré, jusqu'à l'inquantifiable. Il comprit que jamais il n'avait eu vraiment froid – et qu'il aurait préféré que ça continue. Pendant la nuit, il en était arrivé à un point où chaque centimètre de sa peau lui semblait être couvert d'insectes, et il avait passé les heures suivantes à tenter de bouger, en se déplaçant en silence d'une manière qui, espérait-il, demeurerait invisible. Il agitait les orteils, ou du moins s'y efforçait. Il lui était de plus en plus difficile d'en sentir la réaction. Tom gardait les mains enfoncées sous ses aisselles, il les en sortait parfois pour apporter une maigre chaleur à son visage et à ses oreilles en les frottant. Il somnola plusieurs fois, jamais longtemps. Il avait bien trop peur pour se rendre compte qu'à un moment il avait cessé de vouloir mourir.

Il se sentait également pris de nausées, de haut-le-cœur, visité par ces lambeaux de souvenirs que vous laisse un suicide raté, avec l'impression que des secteurs essentiels de son système étaient hors d'usage. Le foie ? Les reins ? Il ne se rappelait pas.

Dans un cas comme dans l'autre, ça n'était pas bon signe. Il avait compris très tôt pourquoi il était toujours en vie. C'était collé sur sa veste : une substance glacée, avec des débris de comprimés. Il avait vomi pendant son sommeil. Il était trop ivre. Son corps avait rejeté une partie de ce qui le faisait souffrir et qui était reparti avant d'avoir l'occasion d'agir. Qu'il soit resté assis bien droit l'avait aussi empêché d'étouffer à cette occasion.

À mesure que l'air alentour semblait se dilater pour permettre aux couleurs de revenir dans la platitude monochrome de la nuit, Tom commença à admettre qu'il allait survivre un jour de plus. Il ne savait pas ce qui se passerait ensuite. Il était mort de trouille, furieux contre lui-même, contre la vie et, plus que tout, monumentalement furieux contre le vieux con de Chez Henry. Si on voulait flanquer la frousse aux gens, quel meilleur moyen que de parler d'ours ?

Comme il sortait en titubant de derrière un arbre, Tom se rendit compte d'une chose. L'idée de s'en retourner gifler ce vieil enfoiré était la première à l'exciter depuis bien longtemps.

La neige n'était pas très épaisse, mais il lui fut facile de retourner sur ses pas en descendant la colline. En bas, il dut affronter des buissons enchevêtrés, couverts de givre. Il se retourna, en prenant garde à sa cheville enflée, et leva les yeux. Il se souvenait vaguement d'être venu de la droite. Il lui fallait donc tourner à gauche. Cela le conduirait à travers la partie la plus épaisse de la futaie – non, merci. Il fit un détour par un secteur plus élevé, marchant sur des rochers et escaladant des troncs d'un pas hésitant, jusqu'à ce qu'il puisse prendre la bonne direction.

Il n'avait pas d'idée précise de la distance qu'il avait pu parcourir en courant. Dans la belle lumière de « Un bon jour pour mourir + 1 », il ne savait même pas pourquoi il revenait sur ses pas. Cela tenait plus chaud que de rester debout, et s'il devait marcher, mieux valait avoir une destination, une vraie pour le moment, non pas l'endroit vague et sombre vers lequel il se dirigeait la veille en trébuchant. Il était encore là, et il restait sans doute suffisamment de choses dans son sac à dos pour le décider. Tom n'était plus trop sûr de ce que lui inspirait le projet, mais au moins il pourrait retrouver le sac.

Il marcha pendant vingt minutes. Le froid fondait ses innombrables tourments en une énorme douleur. Il passa une bonne part du temps à marmonner qu'il faisait vraiment froid, ce qui était inutile mais bizarrement réconfortant. Il s'arrêtait fréquemment, tournait la tête dans l'espoir de reconnaître son chemin et aussi pour s'assurer que l'environnement était dépourvu d'ours. Il allait renoncer quand il entendit quelque chose qui ressemblait à un bruit d'eau qui coule.

Il abandonna le chemin pour s'engager à travers la futaie, très prudemment. Une chute de plus, et il ne pourrait plus marcher du tout.

De l'autre côté des buissons, s'étendait une zone plus dégagée, puis une ravine. La sienne, espéra-t-il, bien qu'elle ne ressemblât pas à ce dont il se souvenait. Bien sûr, il n'y était entré que dans le noir, sans avoir le temps de l'observer avant de se retrouver au fond. Toutefois, ce qu'il avait pu apercevoir avec la torche montrait qu'elle était assez large : environ cinq mètres à l'endroit de sa chute. Ce qu'il voyait devant lui n'en mesurait guère que quatre, tout en étant beaucoup plus profond. Les bords en étaient très raides, bien trop pour qu'il songe à descendre.

Il devait avoir dépassé sa position de la veille.

Il jeta un coup d'œil sur sa droite, vers la direction qu'il devrait prendre. Des arbres et des buissons d'allure coriace grimpaient sur les parois. Il pouvait revenir au prix d'un détour, mais ce serait long. À gauche, ça paraissait plus dégagé, mais c'était la mauvaise direction. Et la pente était raide, qui plus est.

Bon Dieu, songea-t-il, lassé. Son estomac était plein de lames de rasoir, sa tête d'une avalanche de verre. Avait-il vraiment besoin du sac ? C'était peut-être l'odeur de l'alcool qui avait attiré l'ours. Peut-être était-il encore là, à attendre. Et complètement ivre. Tom resta sur place, ne sachant quelle décision prendre.

Trouve le sac, se dit-il. Qu'est-ce que tu pourrais faire d'autre ?

Il avança péniblement le long de la ravine. Elle commença à se rétrécir, mais pas assez pour qu'il puisse la franchir d'un bond. Vingt ans plus tôt, peut-être aurait-il essayé de sauter trois mètres. Aujourd'hui, non, d'autant plus que les deux bords étaient boueux et rocheux, qu'il n'y avait pas assez de place pour prendre de l'élan et que sa cheville était en compote. Il se heurta à un bosquet d'arbres qu'il contourna sur la gauche, avant de revenir vers la ravine.

Tom s'arrêta. Un arbre était étendu au travers, tombé depuis l'autre côté, et la chance avait voulu qu'il surplombe le vide.

Tom se dirigea vers lui en boitillant. Le tronc était assez grand, sans doute un peu moins d'un mètre de diamètre. Le bois paraissait en bon état. Pour voir, il tordit une branche, qui rebondit vivement, ce qui indiquait que l'arbre n'était pas tombé depuis très longtemps. Il ne serait donc pas pourri. Enfin, peut-être. Il n'aurait que quelques mètres à parcourir, et non des centaines.

Bon – mais quelques mètres pendant lesquels il n'y aurait rien en dessous de lui, sinon un espace vide, et, encore en dessous, des rochers acérés. Quelques mètres sur un tronc pas très large, peut-être glissant, sans doute couvert de neige. Quelques mètres difficiles, même s'il n'avait pas eu la cheville en si mauvais état.

La tête de Tom tournoya un instant, comme si un dépôt d'alcool oublié lui parvenait tardivement au cerveau. Quand le monde cessa de tourner, il s'avança jusqu'à l'arbre et posa son bon pied dessus. Le tronc ne bougea pas. Il était gros, solide, il pourrait supporter son poids.

Il avança d'un rien, non sans balayer un peu de neige. Intéressant, songea-t-il, voyant aussitôt la possibilité qui s'offrait : ne marche pas, glisse dessus. Comme ça, tu n'auras pas à mettre un pied devant l'autre, ce qui sera moins risqué, et enlever la neige rendra le pas suivant moins glissant. Il leva l'autre pied et le posa sur le tronc.

Il resta ainsi un instant, jugeant de son équilibre. Puis il s'avança, fit glisser son pied gauche d'une trentaine de centimètres, avança d'autant la jambe droite. Il se sentit en sécurité. Il se trouvait encore au-dessus du sol, certes, mais c'était un bon départ. Du pied gauche, de quelques centimètres. Puis le droit. Et de nouveau le gauche, juste au bord.

Plus il te faudra de pas, plus tu auras de chances de tomber.

— C'est toi, le patron ? s'exclama Tom à voix haute. Il avança un peu le pied gauche, puis le droit, pour qu'il soit à la même hauteur. Il était désormais en l'air. Il ne savait trop où regarder. Pas en bas, manifestement. Pas en haut. Donc droit devant lui. Par-dessus la ravine – non, non, pas par là. Non, merde.

À gauche. Vers là où tu vas.

Il tourna la tête. Bonne idée – l'autre côté n'était vraiment pas si loin que ça. Il fit de nouveau glisser son pied gauche. Puis le droit. Puis le gauche. Puis le droit. Il était désormais au milieu du tronc.

Il s'avança encore. Son pied heurta un nœud, il se cogna la jambe, crut d'abord que ça irait, mais il n'en était rien. Le reste de son corps était d'un seul coup indécis. Son torse s'affaissa lourdement en arrière. Il sentit, en dessous, la masse de la planète entière qui semblait vouloir qu'il la rejoigne.

À gauche. Regarde à gauche. L'espace d'un instant, il crut flotter, mais ne tomba pas. Il se reprit, s'immobilisa, contempla l'extrémité du tronc, à demi dissimulée dans les buissons couverts de neige, et il avança.

Il glissa une fois de plus, dépassa le milieu de l'arbre, glissa de nouveau, avec une étrange allégresse. Il avait l'impression d'être un personnage de jeu vidéo, le jour de Noël. Mais pour une fois...

Il glissa, tira, glissa, tira et ne tomba pas.

Il glissa de côté une dernière fois et se retrouva toujours sur le tronc, mais au-dessus du sol. Il s'accorda une pause, désormais en sécurité. Il regarda la ravine, avec l'impression d'être suspendu en l'air, puis marcha sur le sol.

L'espace d'un instant, la terre, elle aussi, parut immatérielle, comme si elle pouvait osciller, se renverser et disparaître. Il fit un nouveau pas, sur un sol enfin ferme. Il avait réussi.

Un coup d'œil à l'autre côté confirma ce qu'il soupçonnait : il était dur d'aller dans l'une ou l'autre direction. Tandis qu'ici, ce serait une vraie promenade – enfin, relativement.

Trois mètres, et non des centaines.

— Merci, dit-il dans le silence.

La voix ne répondit pas. Au-dessus de lui, le ciel virait au gris.

Tom marcha dix minutes, errant farouchement près du bord. Pour le monde, son petit monde, ici sous les arbres, tout allait bien. Chose incroyable, il paraissait faire plus froid, mais il pourrait le supporter. Il ne fut pas surpris d'apercevoir son sac à dos, plus bas, couvert de neige. La chance lui était revenue, voilà tout. Pour une fois, l'univers s'occupait de lui. Il s'accrocha à un petit arbre, se pencha en avant et contempla l'objet, rayonnant. La neige autour était toute chamboulée, sans doute par ses mains et ses pieds quand il avait voulu s'enfuir.

Et il n'y avait pas d'ours.

Il continua d'avancer, près du bord de la ravine, jusqu'à ce qu'il trouve un endroit où il pourrait descendre. Il remarqua des branches brisées et se dit que c'était sans doute là qu'il était tombé pendant la nuit. La seconde descente fut bien meilleure, avec simplement une petite glissade un peu précipitée à la fin. Il avait enfin atteint le fond, sur ses pieds. Avec le sentiment de refermer une sorte de cercle, il se dirigea en boitant vers le sac.

Celui-ci était ouvert, du verre luisait à l'intérieur. Une bouteille vide gisait à côté. Il y avait aussi quelques emballages dispersés et une poignée de comprimés, d'un bleu irréel. Le tout dans une sorte de petit nid, avec une paroi rocheuse derrière, le ruisseau un peu devant, des buissons des deux côtés. Tom contempla le tout et eut l'impression d'être un fantôme.

Au même moment, sa bouche se remplit de salive, et son estomac se tortilla en tous sens.

Il recula d'un pas en toute hâte, sans vouloir être trop près du sac, de peur qu'il ne le ramène à la nuit précédente. Puis il s'assit brusquement ; le choc se répercuta le long de son échine tandis que les buissons oscillaient devant ses yeux en clignotant.

Après quelques minutes à respirer profondément, la douleur s'apaisa un peu. C'était peut-être la gueule de bois, ou bien la vue des comprimés, qui suscitait un « plus jamais ça » de la part d'un cerveau bizarrement situé dans ses tripes. En fait, ce pouvait être, tout simplement, une faim de loup. Son corps s'était transformé en une sorte de tour de Babel : tout ce qui se trouvait en dessous de sa tête lui faisait l'effet d'avoir été remplacé par le système gastro-intestinal d'une autre espèce, fonctionnel mais incompatible avec le sien ; il disait des choses, et bruyamment, mais Tom ne comprenait pas quoi.

Oh, qu'il se sentait mal.

Il s'affaissa vers l'avant sans le vouloir. Il s'était mis à trembler de froid, très fort. Avec un frisson de peur véritable, il comprit qu'il se sentait brisé, amoché, tout au fond de luimême. Il leva les yeux vers le ciel et vit qu'il s'était encore assombri, d'un gris de plomb parsemé de taches. On aurait dit qu'il allait neiger de nouveau, et cette fois sérieusement.

Qu'allait-il décider ?

Même s'il restait suffisamment de comprimés, Tom ne pensait pas pouvoir les avaler. Ni faire quoi que ce soit, d'ailleurs. Sinon rester assis – mais comment, alors qu'il se sentait si mal ? Au moins la vodka lui réchaufferait l'intérieur. Cette perspective n'avait rien d'attrayant – à la lumière d'une relative sobriété, il était prêt à admettre qu'il la préférait avec du tonic et une tranche de citron, en quantité modérée et *dans un endroit chaud* – mais il n'avait rien d'autre. Meurs comme un homme, s'était-il dit, ou quelque chose comme ça. Il ne parvenait pas à se souvenir de ce qu'il avait pensé à Sheffer. Tout ça paraissait très loin.

Il s'avança à genoux, un bras sur l'estomac, comme si ça pouvait être utile, et tendit vers le sac une main qui tremblait violemment. Les frissons, les bons vieux frissons qui vous prennent quand on a passé la nuit dehors. Rien de pire. Pourvu que ce ne soit pas le signe que tout son système crépitait en lançant des étincelles, comme un câble électrique sectionné.

Il toucha le bord du sac et s'arrêta.

Tom retira sa main. Il y avait quelque chose de bizarre. Des taches de quelque chose sur le verre brisé, à l'entrée du sac. Elles avaient été de couleur vive puis avaient pris un aspect terne qu'il reconnut sans peine. Il y en avait beaucoup sur le dos de sa main.

Du sang ?

Il se rapprocha en grimaçant. Ça ressemblait bel et bien à du sang séché. Il leva la paume vers le haut : pas d'entailles supplémentaires. Il l'aurait senti, même par ce froid. Tom était également à peu près sûr que ça ne datait pas de la nuit précédente. Il lui aurait été inutile d'approcher la main du verre brisé.

Il prit le sac par le fond et le souleva. Des objets tombèrent en s'entrechoquant. Des morceaux de verre soudés ensemble par le froid. Une boîte entière de comprimés dont il ne s'était pas servi. Des fragments végétaux, sans doute accumulés au cours des errances de la veille. Ah ! une dernière bouteille, intacte.

Et quelques taches rouge-brun de plus sur un bout de verre.

Tom le ramassa avec soin. Du sang. Pas le sien, il en était certain. Il avait retourné le sac la nuit précédente, pour trouver ce dont il avait besoin. Il n'y avait pas plongé la main.

Mais l'ours, si.

L'animal n'avait pu sentir d'odeur de nourriture – il n'y en avait pas dans le sac, il n'y en avait jamais eu – mais celle de l'alcool devait être suffocante. Peut-être la connaissait-il déjà, à force d'avoir fouillé dans les poubelles à la sortie des petites villes. Et c'est sans doute pourquoi il n'avait pas pourchassé Tom : il était trop occupé à se servir à boire.

Tom jeta le débris de verre en toute hâte. La réalité de ce qui s'était passé pendant la nuit était jusque-là restée scellée derrière la gueule de bois, l'obscurité, quelques courtes bouffées de sommeil. Mais pas cette fois. C'était là, droit devant lui.

Il avait bien failli être attaqué par un ours.

Bon Dieu.

Il se remit sur pied. Ce n'était pas un endroit où s'attarder. Pas question d'être là quand un fauve énorme sentirait de nouveau l'odeur et déciderait de venir voir de plus près. Tom s'empara de la bouteille intacte et la fourra dans le sac. Puis, comme il s'apprêtait à partir, il remarqua quelque chose dans le buisson sur sa droite.

Il lui fallut un moment pour bien voir. Des poils très longs, d'un brun sombre. Quelques mèches épaisses, prises dans les branches.

Il tenta d'imaginer un ours : ces poils de quinze à vingt centimètres de long... Les ours pouvaient-ils être hirsutes à ce point ?

Tom éprouva brusquement une très forte envie de se trouver ailleurs, même si c'était très difficile.

Il quitta le nid qu'il occupait la nuit précédente, regarda autour de lui, en quête de sa torche. C'est alors qu'il vit les traces de pas dans la neige et comprit qu'il n'avait pas eu affaire à un ours.

CHAPITRE 4

À 8 heures juste, dans le nord d'Hollywood, l'agent de police Steve Ryan attendait, dans une voiture de patrouille, que Chris Peterson retraverse la rue avec du café. Il prenait son temps, parce qu'il s'empiffrait en douce au comptoir en croyant que Ryan ne s'en rendait pas compte ; mais au bout de deux ans, on a appris bien des choses sur celui avec qui on partage une voiture. Chris se livrait à ce petit manège chaque jour depuis six semaines, parce que sa femme était lancée dans un improbable régime magique, ce qui voulait dire qu'il n'y avait absolument rien de comestible dans la maison. Il était plus ou moins embringué dans l'affaire – peux pas manger ci, peux pas manger ça, peux pas manger grandchose, et pas tout au même moment. Il en était donc réduit à tricher en engloutissant une pâtisserie juste avant de prendre son service, et ça se voyait : il revenait en jetant un coup d'œil dans la rue avant d'essuyer des doigts collants sur son pantalon ; désormais, il était toujours prêt à aller chercher les cafés, alors qu'avant il fallait le virer de la voiture à grands coups de pied. Mais Ryan ne comptait pas faire d'histoires là-dessus. Il savait comment ça se passait. Il attendit, battant des paupières à cause de la lumière qui filtrait à travers le pare-brise, heureux d'avoir cinq minutes de plus pour se remettre les idées en place. Il se sentait fatigué, ses yeux étaient secs, ses épaules lui faisaient mal. Il avait palabré avec Monica jusqu'à 3 heures du matin. Le même sujet que d'habitude, discuté comme d'habitude, et comme d'habitude ne menant nulle part. Non qu'il n'ait pas voulu d'enfants, au contraire. C'était simplement qu'ils essayaient depuis plus de deux ans et que la chose commençait à devenir lassante. Devoir s'y livrer à des horaires précis – et uniquement à ces moments-là, le désir se réduisant à presque rien le reste du temps – perdait vite toute apparence de plaisir. Ça devenait un boulot, et il en avait déjà un. Certes, il n'avait pas beaucoup progressé dans la police, mais au moins il avait des espoirs.

Il s'était lié d'amitié avec certains inspecteurs. Sans s'imposer, juste en écoutant, en essayant de comprendre ce qu'ils faisaient, ce qu'ils pensaient. Que ça n'ait jamais marché pour son père ne signifiait pas qu'il en irait de même pour lui. Ça pouvait arriver. Il l'avait vu. Être au bon endroit au bon moment, deux mains menottées en guise de trophée, et on peut être intégré dans une équipe. D'un seul coup, vous n'êtes plus simplement un pantin dans une voiture, à guetter les fenêtres et à interrompre les querelles domestiques (Ryan en connaissait un rayon sur les épouses et il en avait beaucoup appris sur les maris), à poursuivre les fumeurs de crack dans les ruelles tandis que leurs copains s'esclaffent et vous conspuent en vous jetant des bouteilles. D'un seul coup, vous faites partie d'une unité, et là, vous avez des chances de pouvoir abandonner l'uniforme. Il suffisait d'avoir de la chance et de travailler dur, et dans ces deux domaines Ryan ne craignait pas grand-chose. Non, le plus épuisant, c'était quand le boulot ne servait à rien, quand la chance ne pointait jamais son nez, sans qu'on puisse expliquer ça à quelqu'un pour qui le monde devait être tel qu'il devait être, et non comme il était. Monica s'énervait quand ils en discutaient et il ne pouvait le lui reprocher. Lui aussi en était déprimé. Il voulait être père, il l'avait toujours voulu. Bon Dieu, il songeait même à ces foutues histoires d'insémination. Il l'avait dit la veille, ça avait aidé un peu, alors qu'ils savaient qu'ils ne pouvaient pas se le payer, si bien que tout le truc tournait en rond. Il avait expliqué qu'en faisant des économies, en se privant de vacances pendant deux ou trois ans, que s'il était promu... Elle avait dit non, il avait dit oui, peut-être. Elle s'était mise à pleurer... et ainsi de suite jusqu'à ce qu'il ne sache plus quoi dire, qu'il soit 3 heures du matin et qu'il lui faille vraiment aller au lit. Quand il était parti ce matin, elle n'avait pas semblé très causante. Sans doute étaitelle lessivée. Il lui téléphonerait un peu plus tard pour s'assurer qu'elle allait bien. Pourvu qu'il puisse décoller d'ici – qu'est-ce qui prenait si longtemps à Chris, merde ? Depuis qu'il y était, il aurait largement eu le temps d'engloutir un petit déjeuner complet. Se penchant, Ryan l'aperçut au comptoir, à se fourrer quelque chose dans la bouche. Il sourit. Et alors ? Qu'il mange ! Pour le moment, la radio restait silencieuse. Non que la ville soit à court de crimes au point qu'on les renvoie chez eux sans salaire : ça paraissait peu probable.

— Bonjour, dit une voix.

Tournant la tête, Ryan avisa un type sur le trottoir, tout près de la voiture, vêtu d'un pantalon de jogging vert et d'un gilet gris. Des cheveux courts, la peau bronzée, de petites lunettes rondes. Il avait l'air d'un de ces gars qu'on voit chanter au coin des rues – pas du tout d'un homme s'apprêtant à sortir une arme qu'il dissimulait dans son dos. Il tua Steve Ryan de deux balles dans la tête.

Le temps que Nina arrive, l'endroit avait été interdit d'accès et la foule s'était déjà agglutinée. Beaucoup de types en civil, beaucoup de flics, rassemblés en petits groupes, pleins d'une fureur impuissante, loin du banc où un gars en uniforme, aux cheveux roux, de grande taille, était assis, les yeux fixés sur le trottoir. Deux officiers l'entouraient : une femme qui avait posé la main sur son épaule, un homme qui lui parlait. Autant de gestes de bonne volonté, mais qui avaient peu de chances de remonter le moral de Peterson. Son coéquipier avait été abattu tandis qu'il était en face, à se remplir la panse.

Nina se gara et traversa la rue en hâte, constatant que Monroe était déjà là et qu'on le haranguait. Deux flics levèrent le bras en l'apercevant, mais elle avait déjà sorti sa carte :

— Nina Baynam, dit-elle. Fédés.

C'était un mot qu'elle employait volontiers, et parfois cela changeait tout. Mais pas ce matin, de toute évidence ;« FBI » n'était plus un terme qui imposait le respect, bien avant Waco et tous les plantages subséquents. Le langage corporel des deux flics ne transmettait qu'un message : qu'est-ce que vous foutez ici ?

Nina se posait la même question. Elle se dirigea vers Monroe, qui, abandonnant ses interlocuteurs, lui dit, d'un ton sec et précipité :

— Deux témoins. L'un a vu la scène depuis une chambre au premier étage (il désigna du doigt, de l'autre côté de la rue, un bâtiment d'allure délabrée avec des pancartes proposant des locations à la semaine, à des tarifs bizarrement bas),l'autre buvait un café. Ryan et Peterson arrivent à 7 h 30, Peterson traverse la rue tandis que son collègue reste dans la voiture. Ryan ferme parfois les yeux. Il ne voit pas un type aux cheveux courts, avec des lunettes, très mince, vêtu de vert et de brun, ou de brun et de gris, descendre de là et s'approcher du véhicule en gardant une main dans le dos.

Monroe ébaucha un nouveau signe de la main, cette fois en direction d'un parking légèrement surélevé menant à l'entrée du Knights, un motel à étage.

— Le type se dirige tout droit vers la voiture de patrouille, dit quelque chose et tire. Puis il disparaît.

— Comment ? demanda Nina. Le coéquipier du gars était à une dizaine de mètres !

Monroe eut un mouvement de menton en direction d'une petite ruelle un peu plus loin :

— Il a disparu à la vitesse du son ! On a retrouvé l'arme là-bas. Le temps que Peterson entende les coups de feu, voie ce qui était arrivé à Ryan et se mette à courir, il était trop tard. Le tireur s'était volatilisé.

Il s'avança vers le motel. Nina lui emboîta le pas.

— Personne ne sait rien sur Ryan, sinon que c'était un bon flic. Pas le plus doué, voué à l'uniforme pour le restant de ses jours, mais qui faisait bien son boulot. Rien n'indique qu'il en ait croqué ou qu'il ait trempé dans quoi que ce soit. On a pensé qu'on avait affaire à un simple tueur de flics jusqu'à ce que quelqu'un discute avec le manager du motel.

L'entrée du Knights était assez large pour qu'on y pénètre en voiture. Ce qui n'aurait d'ailleurs pas eu de sens, car l'intérieur n'abritait qu'une petite cour minable où se dressaient les restes d'une fontaine en ciment depuis longtemps défunte, et quelques plantes qui s'efforçaient de montrer que la vie peut triompher partout, sans avoir l'air d'y croire. Sur la droite, un abri en parpaings abritait un distributeur de Coca. Les flics s'affairaient de l'autre côté, et ils reculèrent de mauvais gré quand Monroe, suivi de Nina, entra dans le bureau à porte vitrée sur la droite, avec l'expression des gens qu'on empêche de faire un boulot leur revenant de droit. Quatre autres flics étaient déjà dans la pièce, avec un gros type en jean usé et tee-shirt blanc.

— Répétez-nous ce que vous leur avez dit, lui lança Monroe.

Charles était grand, dégarni, avec des cheveux coupés courts et des épaules d'ancien boxeur : en règle générale, quand il posait une question, les gens se hâtaient de répondre.

— Je ne sais rien ! geignit l'homme. Quand la fille du 12 est partie, elle m'a dit qu'il y avait du bruit dans la chambre à côté. C'était il y a deux jours. J'en ai parlé à vos collègues parce qu'ils m'ont déclaré que le tireur avait des cheveux courts et des lunettes, et je me suis dit qu'en fait il ressemblait un peu au gars du 11.

Nina hocha la tête tout en contemplant un magazine à demi dissimulé derrière le comptoir. Le manager s'en aperçut et parut trouver ça très excitant.

— J'adore ce genre de truc, lâcha-t-elle. Ça me donne envie de baiser la planète entière. Ça vous dirait, ici, tout de suite ? L'autre détourna les yeux.

— C'est bien ce que je pensais. En attendant, donnez-nous les clés du 10, du 11 et du 12.

Monroe les prit et adressa un signe à trois flics, qui les suivirent, Nina et lui, quand ils sortirent du bureau.

La chambre 11 était quatre portes plus loin, sur la droite. Deux des policiers reçurent les clés du 10 et du 12.

Ils sortirent leurs armes, puis ouvrirent les portes sans bruit et se glissèrent à l'intérieur.

Tous deux ressortirent quelques instants plus tard. L'un secoua la tête, l'autre dit :

— J'ai entendu quelque chose, peut-être quelqu'un qui parlait.

— Trois zones, intervint le premier. Séjour, salle de bains, chambre à l'arrière.

— Très bien, acquiesça Monroe.

L'espace d'un instant, Nina crut qu'il songeait à donner la dernière clé au troisième homme, puis se rendrait compte que ça ferait mauvais effet. C'est à cause de ce genre de trucs que les flics de base n'aimaient guère les gars du FBI. Elle sortit son arme, qu'elle tint des deux mains, loin du corps, en veillant bien à ce que personne ne la voie grimacer. Cela faisait trois mois, maintenant, et son bras droit lui posait toujours des problèmes. Deux médecins et trois kinés lui avaient affirmé que tout allait bien. Nina pensait que peut-être c'était la petite cicatrice ronde, en haut du torse, à droite, qui voulait s'exprimer, dire qu'elle connaissait tout des armes à feu et ne voulait plus en entendre parler. Dans ce cas, ce serait dur. Les agents du FBI sont tenus d'être armés en permanence. Elle-même dormait avec son arme sous son lit.

Monroe s'arrêta devant la porte, Nina juste derrière lui. Il enjoignit aux flics d'être prêts à les suivre, en leur laissant un peu de temps. Les autres hochèrent la tête. Ils semblaient plus décidés qu'elle – mais après tout, c'est ça, être un mec. Si jamais ils trahissaient la moindre faiblesse devant un collègue, plus personne ne voudrait d'eux.

Monroe glissa la clé dans la serrure, la tourna, attendit une seconde, puis poussa la porte, qui s'ouvrit sur une chambre plongée dans l'obscurité. Les rideaux étaient encore tirés. Il faisait chaud.

— Nous sommes le FBI, lança-t-il d'une voix ferme. Posez vos armes et sortez les mains en l'air. Il n'y aura pas d'autre avertissement.

Ils attendirent. Silence. La vieille devinette, les options opposées : ou bien il n'y avait personne, tout était cool et déjà terminé, ou bien se cachait là un type très méchant qui s'était mis dans l'idée de descendre un flic.

Nina était en position. Elle entra.

Épaisse pénombre. Air lourd. Vraiment, vraiment chaud, comme si quelqu'un avait coupé l'air conditionné vingt-quatre heures plus tôt. Pièce carrée, contenant un sofa délabré, deux fauteuils, un bureau, une vieille télé d'âge préhistorique. Pas d'effets personnels en vue. Une lumière clignotante venant de la porte entrouverte au coin, du côté de la cour.

Un bruit assez faible, sans doute une autre télé.

Mais qui la regarde ?

Nina se déplaça sur le côté pour laisser entrer Monroe, qui s'avança en silence, faisant signe de la main aux flics de rester où ils étaient. Une fois qu'il fut devant la porte donnant sur l'autre pièce, elle alla sans bruit jusqu'au placard et l'ouvrit sans cesser de tenir son arme près du corps.

Vide, hormis l'odeur de poussière. Même pas fermé. Elle tourna sur le pied droit pour faire face de nouveau à la pièce, eut un signe de tête à l'adresse de Monroe. À l'entrée, les flics demeuraient silencieux, prêts à intervenir. Charles s'avança, Nina le suivit, à un mètre cinquante derrière lui. Elle s'arrêta.

Tout se réduit à maintenant.

Monroe pousse doucement la porte de la main gauche. Le battant s'ouvre, révèle le mur de la chambre, une lumière grisbleu clignotante, le bruit se fait un peu plus fort. Il a un côté chuintant, aigu, par-dessus un bourdonnement grave. La télé, pas de doute. Parfois les gens allument le poste pour avoir un peu de compagnie. Après ils sortent et oublient d'éteindre. Ils se disent : et alors, c'est pas moi qui paie.

Monroe progresse d'un pas, ce qui le mène sur le seuil de la pièce. Il avance encore et se tourne rapidement, arme pointée vers une zone que Nina ne peut pas apercevoir.

Mais elle voit son dos s'affaisser.

Un autre long moment.

— Madame ? questionne-t-il.

L'estomac de Nina se crispe. Elle entend Monroe déglutir la bouche ouverte – un petit bruit sec. Il contemple quelque chose. Il est tendu, prêt à tirer. Il avance encore, semble se baisser et regarder en l'air. Puis il fait un pas de côté et disparaît. Petit silence, suivi d'un bruissement, puis de nouveau le silence.

— Nina, finit-il par dire, viens.

Elle savait que cela signifiait : elle seule, et elle fit donc signe aux autres de garder leur position. Elle baissa un peu l'autre bras, sans pour autant lâcher son arme.

Il faisait encore plus chaud dans la chambre à coucher. L'odeur était forte. La télé était fixée au mur par un support de métal, il en venait un faible gargouillis. Monroe était de l'autre côté du lit, qui était très grand.

Une femme y était assise, à regarder la télé. Proche de la trentaine, de longs cheveux bruns. Elle resta immobile quand Nina entra, parce qu'elle était morte. Assise bien droit, la tête un peu inclinée en avant, vêtue d'un pyjama usagé, bleu avec des motifs à fleurs. Son estomac commençait à se distendre. Son visage ressemblait à du plâtre peint, ses yeux étaient ouverts, comme sa bouche, dans laquelle on avait placé quelque chose.

— Bon Dieu ! s'écria Nina.

Elle se pencha. L'objet pouvait mesurer un demi-centimètre d'épaisseur, cinq centimètres de large, plusieurs de long – mais on ne le saurait que quand on l'aurait enlevé. Il était d'un métal brillant, avec une minuscule étiquette portant des chiffres et de tout petits caractères.

— Qu'est-ce que c'est, nom de Dieu ? demanda Monroe d'une voix rauque.

Il avait le front en sueur.

— Je n'en sais rien, répondit Nina.

Une demi-heure plus tard, elle sortit. La première vague de médecins légistes était arrivée, et avec les rideaux toujours tirés, la chaleur, on se serait cru dans un placard bondé. Nina veilla à passer les lieux au peigne fin, ce qui est toujours plus facile quand on est sûr de ne pas se faire tirer dessus, puis quitta la chambre. Monroe était resté sur place. Il faudrait l'arrivée des photographes pour le tirer de là.

Pas d'autres corps dans la pièce. Le bruissement entendu par Nina, c'était Monroe examinant la salle de bains. Là encore, pas d'effets personnels, aucun des vêtements que la femme devait porter en arrivant. On ne débarque pas en pyjama dans un motel, même au Knights. Et il aurait dû y avoir des objets de toilette, un sac à main. Des moyens d'identification. Les flics épluchaient déjà les listes des personnes disparues, mais quelque chose disait à Nina que ça ne donnerait rien avant un bout de temps.

Elle sortit dans une cour ensoleillée, pleine d'autres flics et de badauds qui pensaient pouvoir jeter un coup d'œil rapide sur le lieu du crime avant de retrouver leurs existences anonymes, mais qui allaient passer beaucoup d'heures à répondre à quelques questions. Ce soir, ils verraient, à la télévision, l'endroit où ils avaient dormi et dont les médias répéteraient le nom sans fin, jusqu'à ce qu'on finisse par s'en souvenir pendant des années, peut-être des décennies. Personne n'oublierait ce jour, et surtout pas la femme que Nina aperçut en sortant dans le parking. Deux de ses collègues s'efforçaient de la retenir. Elle se prénommait Monica : en arrivant, elle avait découvert que le corps de son mari avait déjà été emmené à la morgue et elle hurlait à l'adresse de Peterson, toujours assis sur son banc, parce qu'il n'y avait rien d'autre à faire.

Ce n'est que lorsque Nina se retrouva assez loin de la cohue qu'elle sortit son téléphone portable, en se plaçant à un endroit où on ne pourrait l'entendre. Elle composa le numéro de John Zandt. Au bout de douze sonneries, comme il ne répondait pas, elle fut aiguillée sur le répondeur.

— C'est moi, dit-elle à la machine. Je sais que tu ne veux plus entendre parler de ça, mais j'aurais besoin de ton aide.

Elle hésita, ne sachant trop que dire, puis ajouta :

— J'espère que tu vas bien.

Puis elle coupa la communication et resta là, indécise. L'espace d'un instant, elle eut l'impression de ressentir des picotements dans la nuque, comme si quelqu'un l'observait.

Elle fit volte-face, mais il n'y avait personne. Du moins, personne qu'elle puisse voir.

Peu après 14 heures, elle était assise, à touiller son café, tandis que son patron parlait au téléphone. Ils étaient à la terrasse d'un café miteux tout proche du Knights. Toutes les voitures de patrouille étaient parties, sauf une, mais d'où elle était Nina apercevait quatre véhicules banalisés qui prenaient part à l'enquête. Elle vit aussi qu'on emportait divers éléments de la chambre 11 pour les analyser en détail. On avait établi qu'elle avait été louée cinq jours plus tôt et payée d'avance, en liquide. Nina espérait qu'on interrogeait toujours le patron, de préférence dans un endroit brûlant et mal aéré, et que les gars prenaient leur temps.

Monroe mit un terme à la communication :

— C'est fait, dit-il avec une évidente satisfaction. Olbrich rassemble une force d'intervention regroupant toutes les agences, nous compris. Il va falloir que ça roule. Il y a beaucoup de flics furieux aux environs, et je ne voudrais pas me faire prendre à rôder autour d'une maison cette nuit.

— Descendre un flic en plein jour... même pour un cinglé, c'est quand même osé !

— Il va falloir aller vite, reprit Monroe. On a affaire à un type qui ne se contrôle plus. On devra s'y mettre en vitesse.

Nina roula des yeux. Qui ne se contrôlait plus... qui suppliait qu'on l'arrête. Et pourtant, qui restait introuvable. La seule véritable enquête de profilage à laquelle elle ait pris part – officiellement du moins –, c'était l'histoire du Garçon de Courses, en 1999-2000, déjà sous la direction de Charles Monroe. Puis Monroe était passé à d'autres tâches, plus loin, plus haut.

— La question, c'est : est-ce qu'il y en aura d'autres ?

— Ça se pourrait, oui. C'est bien ce que je suis en train de me demander. À moins que nous...

— Non, je veux dire : est-ce qu'il y en a déjà eu des comme ça ? Si c'est la fin, comme tu le penses, où est le début ? Qu'est-ce qui a amené ce gars ici ? D'où vient-il ?

— Il y a des gens qui y travaillent. La police de Los Angeles vérifie tout en ce moment.

— Et on ne sait toujours pas qui elle est.

— Pas de sac à main, pas d'affaires personnelles hormis son vieux pyjama, le connard derrière le bureau dit ne l'avoir jamais vue avant. On préparera une photo une fois qu'ils l'auront nettoyée un peu : des gens s'en iront dans les rues avec ça, dès la fin de l'après-midi. Tu sais ce qu'elle avait dans la bouche ?

Nina fit non de la tête. Elle avait déjà vu bien des cadavres, dont certains dans un tel état qu'elle devait édifier un mur dans son esprit pour que leur souvenir ne lui revienne pas à l'improviste. Mais mettre quelque chose dans la bouche de la victime... c'était autre chose. Les mutilations sexuelles, ça n'avait rien de surprenant. S'en prendre à certaines parties du corps – les yeux, la bouche, les mains – ressemblait davantage à une sorte de profanation sociale. Le sexuel était d'ordre privé, une agression personnelle ; publique, elle voulait dire : « Regarde, univers, ce que j'ai fait ! » C'était dirigé vers l'extérieur, une sorte d'opération magique destinée à changer le monde.

— Un disque dur, dit Monroe. Un petit, comme dans un portable. L'un des techniciens s'en est rendu compte avant même qu'on l'enlève.

— Pas d'empreintes ?

— Aucune. Mais quelqu'un au labo s'occupe de voir ce qu'il peut donner. Il y a un numéro de série, pour commencer. Il est venu de quelque part, on l'a acheté quelque part. Et, bien sûr, il est peut-être resté quelque chose dessus. On le saura ce soir.

Il surprit l'expression de Nina :

— Il l'a laissé là pour une raison bien précise. Remettons-nous au boulot.

Il se leva, composant déjà du pouce un nouveau numéro sur son portable. Clic, clic, clic. Nina n'aurait pas voulu être à la place du téléphone de Charles Monroe. Il fallait avoir de sacrés abdos.

Elle but le reste de son café, bien consciente qu'il la regardait d'un œil critique :

— Quoi donc, Charles ?

— Comment va ton bras ?

— Bien, répondit-elle d'un ton irrité.

En fait, il ne voulait pas parler de ça. Il entendait lui rappeler qu'ils devaient bosser, et les raisons pour lesquelles leurs relations professionnelles n'avaient cessé de se dégrader.

— Il est comme neuf, ajouta-t-elle.

Il fit mine de vouloir dire quelque chose, mais à ce moment son portable sonna et, faisant demi-tour, il s'éloigna. Quelqu'un était en train d'apprendre quel sacré flic était Monroe ; à quel point il contrôlait les choses, à quel point il les dominait.

Tout en le suivant, Nina, pour la vingtième fois, vérifia son propre portable. Il y avait enfin un message texte de Zandt.

Je suis en Floride.

— Bon Dieu de merde ! marmonna-t-elle, avant de remettre l'appareil dans son sac et d'avancer dans la chaleur.

CHAPITRE 5

Je m'installai à l'Armada, en plein centre de San Francisco, pas loin d'Union Square. L'endroit était agréablement coûteux, et un type vêtu en soldat espagnol se tenait devant, sur le trottoir. Les touristes qui passaient se prenaient mutuellement en photo avec lui, sans doute pour pouvoir dire, une fois rentrés, que c'étaient eux, là, avec un gars en costume, devant un hôtel où en fait ils n'étaient pas descendus. Le temps que je m'installe, il était trop tard pour le gros truc que j'avais prévu, aussi allai-je me promener.

Tout en marchant, je réfléchis à ce que je savais, qui se réduisait à ça : j'avais eu tort sur à peu près tout, s'agissant de ma vie. J'avais cru être le fils de Don et Beth Hopkins, qui menaient en Californie du Nord une vie bien rangée et raisonnablement fastidieuse. Ils tondaient la pelouse, lavaient la voiture, achetaient assez de biens matériels pour que les dieux du commerce continuent à leur sourire. Mon père s'était lancé dans l'immobilier et, après mon départ de la maison, avait continué à vendre des demeures de luxe – jusqu'à ce que ma mère et lui meurent dans un accident de voiture. Mais le lendemain de leurs funérailles, quand j'étais passé chez eux pour essayer de comprendre ce que j'étais censé faire, j'avais trouvé un message, caché de manière à ne pouvoir attirer l'attention que de quelqu'un qui connaissait très bien mon père.

Il disait que mes parents n'étaient pas morts.

C'était le genre de nouvelle que tout le monde voudrait apprendre. Cela suffit à me convaincre de passer le reste de l'après-midi à fouiller la demeure. Je trouvai la cassette vidéo que mon père avait dissimulée dans le magnétoscope de son cabinet de travail, ce qui me permit de découvrir à quel point je me trompais sur ma propre vie – ou à quel point on m'avait dupé.

J'avais cru être fils unique. Un passage de la cassette montrait que j'avais un frère du même âge que moi, délibérément abandonné, vers la fin des années 1960.

J'avais cru que la mort de mes parents était accidentelle. Il n'en était rien, et ils n'étaient pas mes parents. Ils avaient été assassinés par le groupe auquel avait appartenu mon vrai père, trente-cinq ans plus tôt : les Hommes de Paille. Ils prétendaient être les seuls membres de l'humanité non infectés par un virus faisant passer la conscience sociale avant l'individualisme glacé qu'ils croyaient inhérent à notre espèce. Qu'ils le pensent réellement, ou qu'il s'agisse d'un simple prétexte à des actes de violence et de destruction aveugles, cela demeurait incertain. Ce qui était clair, en revanche, c'est que le groupe ne manquait ni d'argent ni de relations. Il était également évident que leur tueur, qui se faisait appeler l'Homme Debout, mais qu'il était plus exact d'appeler Paul, le frère que j'avais perdu, était un individu infiniment dangereux. La veille de la mort de Bobby Nygard, nous avions regardé ensemble une vidéo d'origine gouvernementale, compilation des pires atrocités commises dans le monde au cours des vingt dernières années : fusillades, explosions, meurtres de masse. Nous avions aperçu l'Homme Debout à l'arrière-plan d'un certain nombre de ces événements. De plus, il avait servi de fournisseur aux occupants des Halls, un groupe d'hommes – et de femmes, avais-je cru comprendre – impliqués dans des meurtres en série méthodiques et répétés. Et pour couronner le tout, c'était mon jumeau.

Les premiers pas avaient été faciles. J'entamai mes recherches à cent cinquante kilomètres de Relent, assis dans un café avec un ordinateur portable. J'avais horreur de penser que quelqu'un puisse croire que j'écrivais un roman et ne cessais de jeter des regards mauvais aux gens qui me lançaient des sourires encourageants, mais j'avais besoin d'accéder au Net. Il fallait d'abord que je sache dans quelle ville mon jumeau avait été abandonné. Paul m'avait envoyé un message dans lequel il affirmait qu'il s'agissait de San Francisco, mais je n'étais pas disposé à le croire sans preuves. Je n'avais rien, sinon un bref passage à la fin de la cassette que m'avait laissée mon père et que j'avais transcrite sur un DVD.

Ce passage était en trois parties. La première montrait un voyage en train. Rien ne permettait de situer l'endroit, mais je connaissais assez mon père pour savoir qu'il n'avait pas intégré ce fragment rien que pour la couleur locale. L'allure délavée du film 8 mm transféré sur cassette, comme la coiffure et les vêtements de ma mère, m'avaient aidé à le dater, ce qui me fut d'autant plus facile que je pouvais m'y contempler, alors âgé de deux ans. Mon idée était donc que cette première partie avait pour fonction de signaler qu'un voyage avait eu lieu – assez loin de chez nous pour qu'il faille prendre le train, mais pas assez pour y aller en avion. Je n'avais plus qu'à choisir entre trente et quarante villes de Californie du Nord ou de l'Oregon.

La cassette montrait ensuite une large rue dans un quartier commerçant. La caméra suivait ma mère marchant sur le trottoir, tenant, comme la fin du film le montrait, les mains de deux petits garçons. Il n'y avait pas grand-chose d'autre à voir, hormis quelques exemples des modes de la fin des années 1960, vêtements ou voitures, et de fugitives devantures qui vous amenaient à vous demander comment diable on pouvait acheter quoi que ce soit à cette époque. Rien de remarquable, sauf...

Je figeai l'image. À droite de la rue, un grand magasin, en face d'une pelouse. Je pus en discerner le nom. Hannington's.

Dix minutes de recherches sur le Web m'apprirent qu'il n'y avait plus, aux États-Unis, de boutique de ce nom encore en activité, du moins aucune qui se soit fait connaître sur le Net.

Il me fallut donc renoncer momentanément aux méthodes scientifiques et en revenir à la fin du film.

Je me connectai à divers sites consacrés au « San Francisco d'hier » et passai un certain temps à éplucher des évocations du passé de la ville. Mes yeux commençaient à se brouiller quand je trouvai enfin une référence à ce qui, pour une petite fille aujourd'hui adulte, était un rituel du samedi matin : sa mère, depuis longtemps décédée, l'emmenait contempler des corsages hors de prix, dans une boutique appelée Harrington's. Elles ne pouvaient se les offrir et se contentaient donc de regarder. Je me rendis compte que je ne pouvais pas réfléchir à ça très longtemps.

J'en revins à l'image et vis qu'en effet j'avais pu mal lire le nom. L'angle de prise de vue n'était pas bon et le soleil avait surexposé le film, ce qu'on ne pouvait prévoir lors du tournage. De brèves vérifications m'apprirent qu'il n'y avait plus non plus de Harrington en activité, sur la côte ouest ou ailleurs. Il semblait très improbable que deux grands magasins aux noms presque identiques aient l'un et l'autre fermé leurs portes ; des recherches supplémentaires m'apprirent que l'établissement se trouvait autrefois sur Fenwick Street et qu'il avait connu un gros succès en son temps. Assez gros, sans doute, pour que mon père pense qu'il durerait toujours.

J'avais donc estimé que c'était bien San Francisco. De toute évidence, mon frère était capable de dire la vérité.

Fenwick Street se situait à dix minutes de marche de l'hôtel. Les rues étaient pleines de promeneurs et de gens qui faisaient leurs courses, jetant des ombres très longues sur des trottoirs immaculés. On avait élargi la rue, tout ou presque avait changé au niveau des rez-de-chaussée, mais je vis sans peine que j'étais au bon endroit.

Quand je parvins en face de l'énorme bâtiment qui avait autrefois abrité Harrington, j e m'arrêtai. Des gens m'esquivèrent, comme des feuilles mortes contournant un rocher dans un ruisseau. Le magasin d'autrefois avait été divisé en deux et abritait désormais un GAP et une vaste boutique de produits de beauté dont des femmes de tous âges émergeaient, l'air ravi, en tenant de très, très petits sacs dans chaque main. Les étages supérieurs semblaient accueillir des cabinets d'avocats.

Mes yeux furent attirés par le trottoir juste devant moi. Je ne me souvenais pas d'y avoir marché, et pourtant si. Je tenais la main de ma mère. Mon père nous avait filmés. Elle et lui étaient morts, mais l'endroit était toujours là, et moi avec. J'étais désormais plus âgé qu'eux à l'époque, mais j'avais alors le même âge que les mômes que je voyais passer à ma hauteur dans des poussettes : de petits êtres si différents de moi qu'il m'était difficile de croire que j'en avais été un.

C'est bizarre, le temps.

Le lendemain matin, j'étais au téléphone dès 9 h 05. À 10 h 30, je savais qu'on n'obtient pas de renseignements des Services sociaux au débotté. Au bout d'un moment, j'avais passé tant de temps à naviguer d'un interlocuteur à l'autre que je craignis qu'on finisse par me renvoyer à moi-même – ce qui, je le savais, me ferait péter les plombs. Je sortis donc pour m'y rendre à pied.

Moins de cinq minutes après, je regrettais déjà de ne pas m'en être tenu au téléphone. Rien de tel que la salle d'attente d'une agence publique pour vous rappeler quelle chance vous avez. Vous entrez dans un non-endroit, où le temps n'existe pas. Vous vous asseyez sur une chaise fatiguée, d'un bleu ou d'un vert crasseux qui ne seront jamais la couleur favorite de personne. Vous contemplez des pancartes qui n'ont rien à voir avec vous, des communiqués très vagues ignorant les règles de la ponctuation. Vous attendez jusqu'à en oublier ce que vous étiez venu faire, jusqu'à devenir une pierre oubliée là, mille ans plus tôt, par un glacier négligent. Vous êtes là. C'est tout ce que vous savez. Pour le reste, vous êtes dépouillé de tout sentiment d'individualité, de l'idée que vous pourriez être différent des gens présents dans la pièce, sinon par la nature de votre problème particulier, et vous devenez donc ce problème, dont vous vous fabriquez une identité, jusqu'à ce qu'elle enfle et suppure, et devienne tout ce que vous êtes. Dans notre espèce, nous tolérons la présence des autres, mais pas d'aussi près, dans de telles circonstances. Quand nous nous sentons si petits, nous devenons de simples rangées d'yeux secs et agités, en haïssant ceux qui nous entourent et en souhaitant sincèrement que le gars devant meure brusquement, histoire de gagner une place dans la file.

Ou peut-être que je me faisais des idées.

Je passai beaucoup de temps à attendre avant de pouvoir simplement exposer mes désirs à quelqu'un. Il fallut ensuite un certain temps pour surmonter le fait que je n'avais pas de véritable adresse et que le gars accepte celle de l'Armada. J'expliquai que j'avais un frère qui, pensais-je, avait été pris en charge par leurs services à San Francisco, dans la seconde moitié des années 1960, sans doute aux alentours de 1967, que je croyais qu'il se prénommait Paul, et que j'essayais de le retrouver ; je n'avais pas d'autre information, sinon que peut-être il portait un sweater avec son nom brodé dessus. L'autre nota tout ça, mais à son regard on voyait que la journée allait être longue. Enfin, il me tendit un numéro et je fus renvoyé dans le troupeau grouillant et toussotant des problèmes, des geignements et des psychoses.

Deux cents millénaires plus tard, mon tour arriva enfin. On m'invita à suivre un long couloir menant à une pièce tout au fond, où une Noire d'âge mûr était assise derrière un bureau envahi de papiers. Une pancarte annonçait qu'elle s'appelait Mme Muriel Dupree. Derrière elle, le mur était couvert d'affiches dont un mot sur trois était souligné, vous garantissant la confidentialité.

— Je ne peux pas vous aider, m'annonça-t-elle avant même que je m'asseye.

— Et pourquoi ? demandai-je en m'asseyant quand même.

— Parce que c'était il y a trop longtemps, expliqua-t-elle en montrant un papier devant elle. Selon vous, il s'agissait de votre frère, aux environs de 1967. C'était bien avant que je sois là, et que beaucoup de choses n'arrivent. Celle-là, par exemple.

Elle esquissa un signe de tête en direction d'un ordinateur d'âge antédiluvien.

— Tout cela n'a été informatisé qu'il y a une vingtaine d'années, puis il y a eu en 1982 un incendie qui a détruit les bandes et les dossiers au sous-sol, si bien que nous avons perdu la plupart des informations antérieures à cette date. Même si des documents écrits avaient survécu, il n'y aurait pas eu grand-chose, et vous auriez eu plus de chances de trouver Dieu que de les dénicher aujourd'hui. Je ne parle pas personnellement, bien sûr. Peut-être l'avez-vous déjà trouvé, ce qui est bien.

Elle vit que j'étais déçu et haussa les épaules :

— Les choses étaient différentes, alors. Aujourd'hui, on définit un plan d'adoption, on rédige des arrangements légalement contraignants, et tout le monde comprend qu'un enfant n'est pas une sorte de toile vierge, qu'il a besoin de disposer d'informations sur son propre passé, etc. Mais alors, c'était plutôt du genre : « Bon, tu as été recueilli ou adopté, bienvenue dans une vie nouvelle. Ne regarde pas en arrière, il n'y a rien de bien à y trouver. » Les gens changeaient les prénoms des gamins, leurs anniversaires, tout ça. Vous savez comment ça se passait autrefois ?

Je fis non de la tête. Je l'ignorais, et d'ailleurs j e m'en moquais éperdument, mais de toute évidence Mme Dupree me considérait comme une pause bienvenue entre deux personnes qui l'agonissaient d'injures.

— Il y a longtemps, ils prenaient les orphelins des villes pour les emmener en train sur les deux côtes. Ils s'arrêtaient à chaque gare, en pleine campagne, et les enfants étaient exposés sur le quai, dans l'espoir qu'un fermier ayant un peu de place – et manquant de bras, bien sûr, c'était donnant donnant – en prendrait un ou deux. Voilà l'enfant, nourrissez-le, et terminé. Tout ce qu'il y avait avant n'existe plus. Je ne veux pas dire que c'était encore comme ça dans les années 1960, attention, mais un peu quand même. Une fois sur deux, on ne disait pas aux mômes qu'ils avaient été adoptés. Ou les parents attendaient qu'ils soient assez grands, en gros arrivés à l'âge adulte, et ça les bousillait de découvrir que papa maman étaient à des centaines de kilomètres lors de leur naissance. Ce n'était pas un bon système, nous le savons aujourd'hui, mais à l'époque on croyait bien faire, et beaucoup de ces enfants ont eu des vies heureuses et productives. Vous allez bien ?

— Oui, répondis-je en levant les yeux de mes mains, que j'avais inspectées en me demandant si j'aurais jamais une vie de ce genre. Je ne m'attendais pas à me heurter si vite à un mur. Et... c'est très important.

— Je le sais. Je comprends.

Je secouai la tête en souhaitant être ailleurs :

— Non, j'en ai peur. Mais merci de m'avoir accordé votre temps.

Je me levai et me dirigeai vers la porte. J'avais la main sur la poignée quand elle demanda :

— Vous êtes malade ?

Je me tournai vers elle, perplexe. L'espace d'un instant, je crus qu'elle pensait à quelque chose de précis.

— Que voulez-vous dire ?

— Vous avez découvert que votre condition médicale est telle que quelqu'un d'autre doit l'apprendre, parce qu'il pourrait être dans le même cas ?

Je la regardai droit dans les yeux et songeai à mentir.

— Non. Moi, ça va. Mais il y a quelque chose qui ne va pas du tout chez lui.

Je la laissai assise derrière son bureau et, retraversant le long couloir, revins dans le monde extérieur, où je pourrais fumer, respirer de l'air frais, et où mon problème ne serait plus qu'une partie de ce que j'étais.

— Alors, Bobby, et maintenant ?

Silence. Il avait disparu, une fois de plus. Quelque part dans le monde des esprits, avec une bière et un grand sourire, flanquant la trouille aux autres fantômes.

On était en fin d'après-midi. J'allai boire une bière moi aussi, assis à une table en terrasse de l'Espresso, un café juste au coin de mon hôtel. Mes pieds étaient douloureux et pleins d'os. San Francisco est un endroit assez agréable mais, franchement, il y a beaucoup trop de collines.

Vu le foirage complet de la matinée, j'avais fait la seule chose à laquelle je puisse penser. Peut-être, me disais-je, peutêtre, Paul n'était-il même pas entré dans le système. Peut-être avait-il été ramassé dans la rue, recueilli par une quelconque femme de boutiquier au grand cœur. Je savais que c'était un fantasme, né du récit de Mme Dupree sur les trains d'orphelins du Midwest, mais je ne voyais pas d'autre issue, et il me fallait le retrouver. J'avais dérivé trop longtemps. C'était mon boulot, celui de personne d'autre.

En l'absence de toute référence visuelle utile, je tentai une autre méthode. Je savais que mes parents n'étaient pas du genre à jeter un enfant aux loups. Il était probable qu'ils l'avaient laissé dans un lieu qu'ils jugeaient sans danger, où il y avait suffisamment de piétons. Ils étaient à pied. On ne peut pas marcher au-delà d'une certaine limite quand on est avec des gamins de deux ans. Il était donc possible qu'il me faille chercher un quartier fréquenté, accessible à pied depuis Union Square, et d'où ils seraient partis. Au pire, ce serait un lieu du même genre, mais situé sur une ligne de tramway.

J'achetai une carte et je marchai. Et je ne trouvai rien, ce qui voulait dire que je n'avais plus nulle part où aller. J'avais essayé, quelques semaines plus tôt, de répondre à un e-mail que Paul m'avait envoyé. Le message m'était revenu moins d'une heure plus tard – adresse inconnue, jamais entendu parler, impossible à localiser. Les siens étaient des proclamations à mon intention, pas des tentatives pour communiquer. Là non plus, pas de piste.

Je finis ma bière et franchis les quelques mètres qui me séparaient de l'hôtel. Comme je traversais l'entrée, j'entendis quelqu'un appeler mon nom. Je me retournai lentement.

Un jeune type à l'accueil tenait un bout un papier :

— Un message pour vous.

Ça paraissait très improbable. Personne ne savait qui j'étais. Les rares personnes avec qui j'aurais été heureux d'entrer en contact auraient appelé sur mon portable. J'allai jusqu'au bureau, avec le sentiment d'avoir une cible épinglée dans le dos.

Je remerciai le gars, pris le bout de papier et m'éloignai. Quand je l'ouvris, je lus ceci :

Cette dame pourrait vous aider. Si elle le veut.

Il y avait le numéro de téléphone de cette inconnue, et le nom de la personne qui m'avait laissé le message : Mme Muriel Dupree.

Un coup de fil, une visite sur le Net, une douche rapide, puis je redescendis et hélai un taxi devant l'hôtel. Il me fallut un moment avant d'en trouver un qui accepte un tel trajet – de l'autre côté de la baie et encore un peu plus loin, et encore beaucoup plus loin. Celui que je finis par convaincre entendait bien qu'en échange je serve d'auditoire à une longue série de diatribes. Fort heureusement, il était lui-même trop empêtré dans sa propre dialectique pour que j'aie à lui donner la réplique. Je me contentai de grogner et de dire « C'est vrai » tout en regardant par la vitre à mesure que la ville, puis ses faubourgs, passaient à notre hauteur.

J'avais donné un coup de fil aux Services sociaux, en espérant parler à Mme Dupree. Espoir aussi vain qu'il y paraissait au premier abord. Je m'en serais mieux tiré en essayant de remonter dans le temps. Je n'avais donc aucune idée de celle que j'allais rencontrer, mais le Net m'avait appris que le numéro était celui d'une certaine Mme Campbell, et aussi où elle vivait. C'est une des choses que je sais faire. Oui, l'intention de Mme Dupree était manifestement que j'appelle avant de passer, que j'en obtienne la permission, que j'explique ce que je voulais, et plus généralement que je fasse ce qu'il fallait. Eh bien, je m'en abstins soigneusement. Je ne savais pas qui était cette personne, ni ce qu'elle pourrait me dire, mais mon expérience, certes limitée, m'avait appris qu'on arrive plus près de la vérité sans prévenir qu'on la cherche. Et je sais de quoi je parle. Bobby et moi nous étions rencontrés du temps où on travaillait pour la CIA.

Enfin, le gars cessa de parler et se mit à jeter un coup d'œil à une carte. Nous nous éloignâmes de plus en plus des grandes artères, pour arriver dans un quartier résidentiel, blanc, aux pâtés de maisons dispersés, plutôt décatis – pas un rêve d'agent immobilier. Nous y errâmes un petit moment avant que je prenne la carte et guide le chauffeur. Nous nous arrêtâmes au milieu d'une rue bordée de petites maisons de bois dont chacune était entourée d'un carré de terrain.

Je sortis et payai. Il n'y avait personne aux environs.

— Si vous vouliez faire la fête, c'est raté ! lança le chauffeur avant de repartir.

J'attendis qu'il ait disparu, puis remontai sur une cinquantaine de mètres le trajet que nous venions de suivre, car je lui avais délibérément donné une adresse inexacte. Deux autres rues, trois minutes de marche, et je fus devant la maison.

Je franchis le chemin d'accès, montai deux marches, me retrouvai sur le porche. Il avait été peint en blanc, fort bien, quelques années auparavant. Il faudrait bientôt recommencer. Je cherchai une sonnette et n'en trouvai aucune, aussi frappai-je à la porte. Je ne doutais nullement que cette femme serait chez elle.

Au bout de quelques instants, j'entendis un bruit, la porte s'ouvrit. Il y avait derrière un store et, dans la pénombre, une petite silhouette.

— Madame Campbell ?

Elle ne répondit rien, mais tendit lentement la main vers le store et l'entrouvrit. Ce qui me permit d'apercevoir une septuagénaire à la chevelure toujours soignée, mais au visage gris et bouffi, qui paraissait stupéfaite. Elle croisa mon regard, m'observa de la tête aux pieds, puis me contempla de nouveau.

— Mon Dieu ! finit-elle par dire. C'était donc vrai !

CHAPITRE 6

Quand le téléphone sonna, Nina était dans ce qu'elle appelait sa « chaise longue » ; ce mot suggère néanmoins un certain degré de confort et de relaxation qu'elle ne pouvait pas se permettre. Théoriquement, elle était en train de réfléchir : à la vérité, elle dormait. Au boulot, il était impossible de s'entendre penser, à cause du bruit : les gars entraient et sortaient, aboyaient dans des téléphones, se montraient vifs, professionnels. Un des grands trucs quand on est un mec, avait-elle remarqué, c'est qu'être bon, bien faire son boulot, ne suffisait pas. Il fallait que tout le monde sache que vous étiez un peu là. Elle préférait nettement réfléchir dans sa chaise longue. Il fallait qu'elle déménage, elle le savait. Surtout après que les choses eurent mal tourné avec John : la maison paraissait peu commode, lassée d'elle, inadéquate à tous points de vue. Elle était dans les collines de Malibu, ce qui était super, mais Nina ne pouvait la louer que parce qu'elle tombait en ruine. Il y avait en plein milieu du sol cimenté du séjour une fissure assez large pour y glisser trois doigts. Peu avant qu'elle emménage, la piscine avait succombé à un incendie de forêt. Une bonne secousse sismique, et la demeure finirait dans le Pacifique.

Elle avait passé le reste de la journée dans les rues, ou dans des bureaux, au téléphone, à trier des renseignements sans intérêt et à s'informer des résultats de l'autopsie. Personne n'avait rien apporté d'utile. Le pyjama avait été acheté chez Wal-Mart, ce qui n'est jamais bon signe quand on veut retracer l'histoire d'un objet quelconque. Le disque dur retrouvé dans la bouche de la victime était toujours en cours d'analyse ; inspecteurs et hommes de patrouille disposaient d'une photo qu'ils montraient aux gens. Il se pouvait que ça ne donne jamais rien. Une femme attirante, désormais morte... ça ne manque pas.

Nina revint à la maison pour trouver un message sur le répondeur. Elle appuya le bouton, pensant que c'était peut-être Zandt, avec une réponse plus encourageante. Mais c'était son amie

Meredith, une fille avec qui elle était allée en fac, qui disait que oui, il était temps qu'elles se retrouvent pour dîner et bavarder longuement. Nina ne se souvenait pas de quoi, mais supposait que c'était le moment. Cela faisait un an que son petit groupe de vieilles amies s'était reformé. Merry vivait dans la Silicon Valley et avait acquis, apparemment sans effort, un époux et trois enfants, comme s'il s'était agi d'une compétition sportive. Désormais, elle se préoccupait de choses que Nina jugeait incompréhensibles ou sans intérêt, et sa coiffure se faisait toujours plus implacable : bientôt il serait impossible de voir son visage et de se souvenir du bon vieux temps. Nina, d'un regard amusé, l'avait vue étrenner tout un assortiment de toilettes à l'occasion de vagues soirées chez des profs qui avaient toujours des appartements minuscules encombrés de livres. Sans doute était-elle tout aussi perplexe face à la présente incarnation de Nina, qui ressemblait bien à une femme, mais sans paraître savoir en quoi consistait son boulot. Elle savait qu'il fallait entretenir l'amitié, tout en se demandant souvent pourquoi elles en prenaient la peine. Peut-être Meredith se plaisait-elle à connaître une agent du FBI. Peut-être Nina aimait-elle croire qu'elle avait encore de vagues rapports avec le monde extérieur, hormis bien sûr les meurtriers, les flics en civil, les nuits blanches ; il y avait quelqu'un qui n'exigeait d'elle que des potins et un sourire.

Nina ne put se résoudre à répondre à l'appel et se remit à réfléchir. Elle finit par se demander quelle différence existait entre Merry, ou elle, et la fille retrouvée au Knights ce matin même. Combien de petits coups de pouce faut-il dans une existence pour finir dans un motel, dans une chambre empestant la fumée de cigarettes d'hommes venus étudier vos ultimes moments, tandis que vos oreilles mortes recueillent sans les entendre de longues discussions sur divers événements sportifs, et au moins une appréciation sur vos nénés ? John Zandt – membre de la Brigade des homicides de la ville avant que le Garçon de Courses n'enlève sa fille – avait fait remarquer à Nina, voilà bien longtemps, à quel point la vie d'une adolescente, à Hollywood, peut vite passer de A à B, puis de B à Z, puis à la morgue, avec une étiquette accrochée à l'orteil. Toutes se croient des stars, alors qu'elles ne sont que chair à canon dans la file, à attendre que les amants, les amis ou le destin ne tiennent pas leurs promesses.

Nina se servit un verre de vin. Un quart d'heure plus tard, elle dormait. Elle se réveilla en sursaut. Quand la sonnerie du téléphone lui parvint enfin, elle se leva du fauteuil en titubant :

on aurait dit qu'il sonnait depuis longtemps, incapable de l'arracher à un rêve dans lequel un vieillard la poursuivait dans une pièce obscure.

Elle se heurta à la porte vitrée en allant répondre, prête à passer un sérieux savon à Zandt. Mais ce n'était pas lui.

— Tu ferais mieux de revenir ici, dit Monroe. On a trouvé quelque chose.

Elle le rejoignit dans le bureau de Doug Olbrich. Celui-ci était lieutenant dans la Section spéciale n° 1, chargée des vols et des homicides ; il s'occupait du profilage et des rapprochements entre affaires de meurtres. Un type grand et maigre, avec des cheveux très courts.

— Salut, Doug.

— Nina ! Comment ça va ?

— Comme d'habitude. Ça fait un bon moment que je n'ai pas parlé à John, mais je suis sûr qu'il t'aura envoyé ses amitiés.

— Merci. Je les attendrai.

Devant Olbrich, elle aperçut plusieurs feuilles de papier, et quelque chose dans un sac en plastique transparent. Trois flics discutaient non loin de là. Un Noir efflanqué, en manches de chemise, que Nina reconnut vaguement, était perché sur le bureau de Doug.

— Nina, voici Vincent, annonça Olbrich.

Monroe lui tendit un café qu'elle accepta avec gratitude.

— Je me souviens, dit-elle. Un rat de laboratoire, c'est ça ?

Charles fronça les sourcils mais le technicien eut un grand sourire :

— Vince Walker, génie de la technologie !

— Ceux que je préfère, répondit-elle, soudain très lasse. Qu'est-ce que vous avez pour nous, Vince ?

— Ceci, répondit Olbrich en poussant vers elle le sac plastique. Et ce qu'il y avait dessus.

Nettoyé du sang, et ôté de la bouche de la victime, l'objet paraissait très quelconque. Cinq centimètres sur douze, cinq millimètres d'épaisseur. Sur le dessus, une plaque de métal avec deux stickers autrefois blancs, mais maintenant tachés çà et là de brun pâle. En dessous, les lignes vertes et sinueuses d'un circuit imprimé. Sur le côté, une étiquette proclamait : « Garantie annulée si le sceau est rompu ». Et si on le retrouve dans la bouche d'une morte, se dit Nina, elle est annulée aussi ?

— Le disque dur, dit-elle.

C'était la bonne réponse : les deux hommes étaient lancés, chacun prêt à tirer la couverture à lui.

— Oui, répondit Vince. Un Toshiba MK4309. Un peu plus de 4 gigas, ce qui est peu aujourd'hui. Le numéro de série confirme qu'il a été fabriqué voilà près de deux ans. Ça nous a aussi permis d'établir qu'il a été installé sur une machine assemblée au Japon et importée aux États-Unis en avril 2002.

— On s'en occupe, intervint Monroe. Ça pourrait nous dire qui était la femme... enfin, peut-être.

— Les gars sont dans les rues avec sa photo, ajouta Olbrich.

Nina l'avait rencontré plusieurs fois, du temps où Zandt faisait encore partie de la Brigade des homicides. Il l'avait impressionnée : c'était l'un des flics les moins frimeurs qu'elle eût jamais rencontrés.

— Nous savons que, le jour de sa mort elle n'avait guère mangé mais qu'elle avait beaucoup bu. Voilà deux heures, j'ai envoyé trois gars dans les bars et les clubs proches du Knights. On n'avait rien trouvé la première fois, mais...

— Rien sur le tueur dans la chambre ?

Olbrich haussa les épaules :

— Pas d'empreintes, pas de fibres de tissu, rien sur la victime... On croirait presque que le gars ne déplaçait pas d'air.

— Alors, le disque dur ?

— Deux choses, reprit le technicien, bien résolu à ne pas se laisser voler la vedette. Le plus gros est un fichier mp3 de sept mégas, un morceau de musique.

— L'Agnus Dei du *Requiem* de Fauré, précisa Monroe. Une œuvre très connue, apparemment. On s'occupe de savoir de quel enregistrement il s'agit, et bien sûr on s'efforcera d'éplucher les récents achats de CD, mais je n'ai pas beaucoup d'espoirs là-dessus. Après tout, ça pourrait bien avoir été téléchargé sur le Net.

— Et puis ?

— Tu m'avais demandé d'où le gars venait, reprit Monroe, en disant qu'il pourrait bien en sortir quelque chose. Il se pourrait que tu aies raison.

Il poussa les papiers vers elle :

— Lis donc ça.

Le sommeil est agréable, la mort bien meilleure. Ne jamais être né, voilà bien sûr le miracle.

Sa mère ne laissait pas sa propre mère fumer à la maison. Il y avait donc des jours où la vieille dame n'était pas de très bonne humeur, et d'autres où elle tenait à ce qu'on l'installe sur la véranda. On la laisserait là, même s'il faisait froid ou s'il pleuvait à verse. Sa fille ne l'aidait pas à rentrer, et donc, s'il lui désobéissait, gare à lui ! Grand-mère restait donc dehors jusqu'à ce que sa fille ait la bonté de la ramener à l'intérieur. Sans affection superflue.

Par une de ces journées, lors d'un après-midi si froid que des glaçons pendaient du toit, il demanda à sa grand-mère si cela valait la peine de rester dehors alors qu'il faisait bon à l'intérieur.

Elle regarda dans le vague un moment, à tel point qu'il se demanda si elle l'avait entendu.

— Tu connais la vieille blague, répondit-elle enfin. Pourquoi les poulets traversent-ils la route ?

Bien sûr, avait-il répondu. Pour être de l'autre côté.

— Les cigarettes, c'est pareil.

— Je ne te suis pas.

Elle réfléchit de nouveau :

— On se retrouve à vivre du mauvais côté de la route, si tu veux. Chaque nuit, il faut la traverser dans le noir, pour pouvoir rentrer. On ne peut pas savoir s'il y a des voitures qui arrivent, parce que le vent étouffe le bruit, mais ça va quand même, parce que ça n'est pas une route très fréquentée. Mais plus tu traverses, plus tu as de chances que tôt ou tard une voiture t'écrase. La voiture, c'est le cancer, il est grand, dur, et très rapide, et quand il t'a renversé, tu meurs.

— Mais alors, pourquoi continuer à traverser la route ?

— Pour aller de l'autre côté, répondit-elle avec un sourire pincé avant de hausser les épaules : c'est trop tard. L'important, c'est de veiller à ne pas se retrouver du mauvais côté de la route.

Elle toussa, puis alluma une autre cigarette, dont elle tira une longue bouffée avant d'en contempler la pointe rougeoyante :

— Ne te mets jamais à cette saloperie, tu m'entends ?

— Promis, assura-t-il.

Il fit de son mieux pour suivre ses conseils. Il se montra très prudent avec l'alcool, ne prit jamais aucune drogue, ne laissa pas la bouffe, la gym, la pornographie ou la collection de poupées de porcelaine lui prendre la main et lui dire qu'elle était son amie.

Et pourtant, une nuit, sept ans plus tard, il se retrouva avec du sang sur les mains et comprit qu'il avait enfin trouvé la route à traverser.

— Bon Dieu ! s'exclama Nina.

— Il a déjà tué, dit Monroe.

— Ou il veut nous le faire croire.
Charles eut un sourire crispé :
— Il est parfaitement capable de recommencer. Je pense que nous en sommes d'accord ?
— Oui. De qui est la citation ?
— Nous ne savons pas encore.
— Ça va, Nina ? intervint Olbrich.
Elle hocha la tête, contemplant toujours le texte :
— Ça me fout les boules, c'est tout. Un message bien en évidence et un requiem ! Nom de Dieu ! On a vraiment affaire à un cinglé !
— Il parle de lui à la troisième personne, remarqua Olbrich. Un peu bizarre, non ?
— Pas vraiment, rétorqua Nina. On a observé ça plus d'une fois lors des interrogatoires, notamment avec Ted Bundy. En théorie, ça les aide à décrire des crimes dont certaines parties de leur esprit veulent se dissocier. Dans le cas de Bundy, c'était aussi pour feindre d'avancer des hypothèses, du genre « j'imagine qu'un tueur dans cette situation ferait ceci ou cela », sans reconnaître réellement sa responsabilité. On peut tirer quelque chose du fichier lui-même, techniquement parlant ?
— J'ai bien peur que non, répondit Vince. Le disque dur est destiné à un PC, mais le fichier n'a aucune signature de système d'exploitation. Le texte aurait pu être écrit sur n'importe quelle bécane, du super-ordinateur à un Palm. Quelqu'un en bas travaille sur la signature du répertoire, mais nous ne sommes pas très optimistes. Le disque dur a été soigneusement nettoyé avant que ces fichiers n'y soient copiés. On a affaire à quelqu'un qui s'y connaît en informatique.
— Ce qui, en soi, est une information utile, nota Monroe.
— Tout à fait ! railla Nina. Ça veut dire qu'il a moins de cinquante ans et vit quelque part en Occident.
Charles pencha la tête et la regarda. Nina songea que rentrer chez elle sous peu serait une bonne idée.
— Une copie de ces fichiers est au profilage à Quantico, reprit-il. Ils ne devraient pas tarder à avoir des idées.
Il parlait un peu plus fort que d'habitude, d'un ton sérieux, studieux, professionnel, mais avec un peu d'excitation dans la voix. Il fallait s'y attendre : si on ne prend pas son pied à poursuivre les méchants, on n'a rien à faire dans la police. Mais depuis la toute première fois où Nina avait travaillé avec lui – pour capturer un tueur nommé Gary Johnson, qui dans le milieu des

années 1990 avait tué six femmes âgées –, Nina ne doutait nullement que Monroe n'eût d'autres projets en tête. Les crimes, comme leur élucidation, étaient des moyens en vue d'une certaine fin. Elle ne voyait pas trop laquelle – la politique ? Avoir le plus gros bureau du continent ? Mais elle savait que cela le motivait bien plus que de regarder les parents en face et de leur dire : « On a trouvé le gars et il va plonger pour l'éternité. » Peut-être y avait-il là quelque chose de pas trop bête. Lors des rares occasions où Nina avait pu dire quelque chose de ce genre, la morne expression des visages de ses interlocuteurs n'avait pas semblé beaucoup changer. Six femmes, mères et grands-mères, connaissent une mort prématurée, dans des conditions sordides ; le responsable est enfermé pour le restant de ses jours dans une cage en béton. Du point de vue châtiment, ça ne semblait pas vraiment réglo. Certes, personne ne veut aller en prison, surtout pas en Louisiane, après avoir tué, entre autres, deux vieilles Noires. On ne veut pas se lever chaque matin en se demandant si un fêlé quelconque aimant sa maman ne décidera pas d'égayer sa journée en vous cisaillant la tête avec le manche aiguisé d'une cuiller. Mais Nina ne croyait pas que les tueurs, dans leur majorité, ressentaient tout le poids de l'incarcération, parce que, tout simplement, ils ne voyaient pas les choses de la même manière. Dans un cas comme dans l'autre, ils devaient encore vivre. Ils mangeaient, dormaient, allaient aux chiottes. Ils suivaient des cours, erraient dans le labyrinthe des appels, ce qui faisait perdre du temps à tout le monde et coûtait de l'argent public qu'on aurait mieux employé à construire des écoles. Bien entendu, c'était leur droit. Cela leur épargnait d'être étendus, seuls, dans un trou creusé dans le sol, sans rien pour leur tenir compagnie, sinon le bruit de la terre qui se tasse. Ils ne dormaient pas, les bras allongés le long du corps, dans une boîte où ils pourraient commencer à pourrir.

Alors, oui, Monroe voyait les choses rationnellement. Mener le bon combat. Grimper l'échelle. Puis rentrer chez soi, retrouver son épouse, dîner en regardant le journal télévisé du soir. Qui sait, il pourrait peut-être même s'y admirer, occupé à sauver le monde. Ce serait sympa. À la vérité, le FBI n'était pas directement mandaté pour enquêter sur les meurtres en série. Charles s'y était lancé pour faire progresser sa carrière. Et alors ? Et Nina, quel prétexte avait-elle ?

— Rentre chez toi, lui conseilla Monroe. Dors un peu. J'aurai besoin de te voir fonctionner demain matin très tôt.

Elle leva les yeux, surprise par le ton et comprit qu'elle venait tout juste de somnoler pendant une trentaine de secondes. Vince la dévisageait avec curiosité, Charles sans grande affection. Seul Olbrich avait l'élégance de regarder ailleurs.

Monroe se leva et parla à Doug comme s'il suggérait qu'il n'avait plus besoin de Nina. Elle attendit qu'ils aillent se joindre aux autres flics au fond de la pièce, puis se tourna vers le cerveau de l'équipe et lui dit, de sa voix la plus douce, la plus affable, la plus séduisante :

— Vince, j'ai une faveur à vous demander.

Vingt minutes plus tard, elle quitta le bâtiment avec quelque chose dans son sac, et sortit dans la rue. La soirée était encore tiède ; Nina se demanda si elle tentait délibérément de bousiller sa carrière.

Il fallait qu'elle parle à quelqu'un, mais John ne répondait pas au téléphone et, à la vérité, il était encore plus cramé qu'elle. Il existait une autre option, une seule. Elle y réfléchit.

Oui, peut-être. Elle allait rentrer et voir comment elle se sentait.

Elle rentra en voiture paisiblement et, le temps d'arriver sur le chemin d'accès, elle avait pris sa décision. Dans la cuisine, elle composa le numéro. Les sonneries se succédèrent sans qu'on réponde.

Elle laissa un message, avec l'impression d'être une voix de plus sur un répondeur comme un autre.

CHAPITRE 7

La maison de Mme Campbell donnait, à l'arrière, sur une petite cour qui disait tout ce que l'avant s'efforçait de taire. J'attendis dans la cuisine, aussi patiemment que je pus, tandis qu'elle s'affairait à grand bruit. Je me souvenais que ma mère m'avait dit une fois que lorsqu'on refusait une boisson chaude offerte par une vieille dame, elle savait qu'on n'avait plus besoin d'elle. Toutefois, la vue depuis la fenêtre ne m'intéressait pas le moins du monde. Il me fallut toute ma patience pour ne pas saisir Mme Campbell par le col.

— Muriel a elle-même été adoptée, dit-elle en me conduisant enfin dans le séjour. Elle vous l'a dit ?

— Non, répondis-je en m'avançant en toute hâte pour m'emparer du plateau. Elle m'a simplement dit qu'elle ne pouvait pas m'aider et on en est restés là.

— Elle est comme ça, parfois. Je l'ai connue quand elle a commencé à travailler là-bas. Elle a eu de mauvais moments, au début. Son mari l'a quittée, a vidé la maison, il la rouait de coups. Mais, malgré tout, elle a fait son boulot et aidé bien du monde. Beaucoup de gens vont là-bas et oublient que le personnel est lui aussi composé d'êtres humains, qui ont leur vie à eux.

— Je comprends bien que c'est un boulot qui peut être difficile. C'est parfois dur de traiter avec les autres.

— Oh, que oui ! Bien sûr, certains de ceux qui travaillent là-bas sont des débiles, aussi.

Je ris. Elle eut un hochement de tête approbateur :

— Vous devriez sourire plus souvent, vous avez meilleure allure comme ça. Comme tout le monde, d'ailleurs, mais surtout vous. Quand vous ne souriez pas, on dirait que vous voulez faire du mal aux gens.

— Pas du tout.

— C'est vous qui le dites.

— Madame Campbell, j'ai le sentiment que...

— D'accord. J'y arrive. Vous cherchez un frère, c'est ça ? Muriel disait que, d'après vous, ça devait être en 1967. Ce qui est exact. En fait, si je me souviens bien, c'était en octobre, bien qu'à vrai dire ma mémoire ne soit plus ce qu'elle était. Ça va pour ce qui est des choses, mais moins pour les faits.

J'acquiesçai. J'avais l'impression d'étouffer.

— Un boutiquier chinois l'a trouvé dans la rue. Je ne sais pas depuis combien de temps il était là, mais il n'avait pas cessé de pleurer.

— Mes parents avaient leurs raisons, et l'histoire est compliquée, dis-je avec le besoin absurde de défendre une décision dans laquelle je n'avais eu aucune part et qui me restait incompréhensible.

— J'en suis sûre. Et ils ne l'ont pas laissé dans un endroit perdu, c'est déjà ça. Enfin, nous savions qu'il s'appelait Paul, parce que c'était le nom brodé sur son sweater. Le prénom lui est resté. Nous avons procédé aux vérifications habituelles, mais il n'y avait pas moyen de savoir d'où il venait, et il a donc été placé ici, en ville. Il y est resté quelques années. D'habitude, il n'est pas difficile de trouver une famille d'accueil pour un petit garçon très mignon, mais avec lui, ça ne marchait pas.

J'aurais voulu savoir ce qu'elle entendait par là, mais je ne tenais pas à l'interrompre.

— Je l'ai perdu de vue un moment. Il y a beaucoup d'enfants, il y en a toujours un nouveau pour qui il faut faire quelque chose. J'ai entendu reparler de lui au moment où il devenait un problème.

— De quel genre ?

— Il était dans une famille d'accueil pendant quelques mois, puis il revenait bien avant la date prévue. Au début, je n'ai pas fait attention, ça arrive. Mais ça se reproduisait. Paul était de retour, les gens n'arrivaient pas à... j'allais dire à faire face, mais ce n'était pas ça, enfin pas exactement. Il était revenu, voilà tout. Et il ne faut pas oublier qu'il s'agissait de familles qui s'occupaient de beaucoup d'enfants, qui savaient comment les mettre à l'aise. On le plaçait quelque part, et puis, quelques semaines après, je le retrouvais là, assis sur un rebord de fenêtre, à regarder dehors. Je lui demandais ce qui s'était passé et il me répondait la même chose que ceux qui l'avaient accueilli : ça n'avait pas marché.

Elle but une gorgée de café :

— Pour finir, nous avons décidé qu'il fallait chercher une vraie famille adoptive, trouver une solution à long terme. J'ai donc discuté avec Paul, je lui ai expliqué que c'était ce que nous allions tenter. Il a hoché la tête – et il n'avait que six ans, sept au plus ! – et quelque chose m'a dit qu'il n'était pas d'accord, mais qu'il se résignait à l'idée. Je lui ai demandé s'il ne voulait pas trouver une famille pour toujours, il m'a regardée bien en face et il m'a répondu : « J'en avais une, elle n'est plus là. Quand tout sera en place, je la retrouverai. »

J'eus comme un frisson glacé dans la nuque :

— Il se souvenait de nous ?

— Pas forcément, mais il savait qu'autrefois il y avait eu autre chose. Pas besoin d'être très futé pour comprendre qu'il était dans une position très artificielle et que c'était un gamin intelligent, ça se voyait. Les enfants ont souvent le sentiment qu'on les a abandonnés ou enlevés. C'est le syndrome du « je suis fils de roi, et quand je pleure la terre pleure avec moi ». J'ai cru que c'était ça.

J'avais regardé bien des fois, sur la cassette, le moment de l'abandon, sans jamais affronter l'idée de ce que cela avait pu être pour l'enfant que nous laissions derrière nous. Au cours des trois derniers mois, je ne m'en étais d'ailleurs guère soucié. Je m'efforçai de continuer.

— Est-ce que ça vous ennuie si je fume ? demandai-je.

— Allez-y, répondit-elle en souriant. Mon mari fumait, j'aime l'odeur. Mais vous savez que ça vous tuera ?

— Jamais de la vie ! C'est une rumeur répandue par les fanas de gym et les obsédés de la santé.

Elle cessa de sourire et hocha la tête :

— Oui, c'est ce qu'il disait aussi.

— Alors, que s'est-il passé quand vous avez cherché une famille d'accueil permanente ?

— J'ai fait ce boulot pendant longtemps, répondit-elle après un petit silence, j'y ai beaucoup réfléchi. Je crois que l'endroit où nous sommes nés s'infiltre en nous comme l'eau venue du sol, que nous avons des feuilles, comme les arbres, et que là où tombe la graine, c'est-à-dire nous, se trouve ce que nous sommes, et la couleur que prendront nos feuilles – même si un oiseau s'empare de nous et nous déplace de cent kilomètres. Mais une autre partie de moi-même pense : après tout, nous sommes tous les enfants de Dieu, non ? Alors, quelle importance qu'un enfant soit élevé par d'autres ? Qu'on leur donne un bon foyer, et ce sera comme si rien ne s'était passé. J'ai vu ça des centaines et des centaines de fois. Ce n'est pas toujours facile mais ça marche, et c'est l'une des choses qui me font penser que nous autres humains ne sommes pas si mauvais que ça.

Elle secoua la tête :

— Mais trouver quelqu'un pour adopter Paul n'était pas simple. Il a été placé dans trois familles successives. La première a duré un an, ils avaient déjà une fille à eux, plus âgée. À l'époque, j'avais des soucis, mon mari était malade... Un lundi matin, je suis arrivée et on m'a dit que Paul était là, que les gens étaient passés, qu'on l'avait retrouvé dehors, sur les marches. Il ne s'était pas enfui, sa famille d'accueil l'avait déposé là. Après ça, on lui a trouvé d'autres gens, au bout de quelques mois. Ça a duré deux années complètes, à l'issue desquelles il a eu neuf ans. Et puis un jour, voilà qu'on frappe à ma porte et que la mère est là. Elle m'a dit poliment qu'ils en avaient assez. Ce n'était pas du tout la faute de Paul, mais elle venait d'avoir une fille et ils pensaient ne plus pouvoir s'occuper de lui. J'étais furieuse, croyez-moi, j'ai failli lui arracher les yeux ! Ce n'est pas comme ça que ça se passe. Mais... on ne peut pas laisser un enfant chez des gens qui ne veulent plus de lui.

Elle prit sa tasse, constata que le café était froid, et la reposa.

— Est-ce que...

— Non, non, c'est bien, répondis-je. Continuez.

— J'ai revu Paul peu après. J'étais navrée pour lui. Je lui ai dit qu'on l'avait mal traité. Il a haussé les épaules et m'a répété une fois de plus : « J'ai déjà une famille. » Ça m'inquiétait qu'il pense ça et j'ai tenté de lui faire comprendre que non, qu'il devait nous aider à en trouver une autre. Il avait eu autrefois un vrai père et une vraie mère, mais désormais il devait être avec quelqu'un d'autre. « Pas eux, a-t-il répondu. Ils n'étaient pas vrais. Mais j'avais un frère, il était vrai, il était juste comme moi. »

Elle sourit :

— Je ne l'ai pas cru, bien sûr. J'ai pensé qu'il s'efforçait simplement d'imaginer... Il y avait quelque chose en lui qui était un peu... je ne saurais dire quoi. Mais quand vous êtes arrivé, j'ai compris qu'il avait raison. Il avait bel et bien un frère, semblable à lui.

Je hochai la tête, parce qu'il le fallait, mais je pensais qu'elle se trompait, et lui aussi. Je lui ressemblais physiquement, voilà tout. J'étais surpris qu'elle s'en soit rendu compte, si Paul était encore un enfant la dernière fois qu'elle l'avait vu.

— Enfin, nous avons trouvé une famille qui a tenu bon. C'était ici, en ville, et au bout d'un an ils ont quitté la Californie et il est parti avec eux. Cette fois, ça avait marché. C'est tout.

Je la regardai.

— Qu'est-ce qu'il y a ? demanda-t-elle.

Je continuai à la regarder. Elle baissa la tête pour contempler ses mains :

— Qu'est-ce qu'il a fait ? ajouta-t-elle d'une voix douce.

— Madame Campbell, dites-moi ce que vous ne m'avez pas dit. Il faut vraiment que je sache.

Elle leva les yeux et se mit à parler d'une voix rapide, le regard morne :

— Quelques années plus tard, j'ai rencontré par hasard le mari du couple qui l'avait laissé sur les marches. Je n'avais pas eu de nouvelles d'eux depuis lors, si on traite un enfant comme ça, on se retrouve au tribunal. C'est bien ce qui a failli se passer, mais la femme est tombée malade et ça c'est tassé. J'ai vu l'homme de l'autre côté de la rue et j'ai détourné les yeux, mais voilà qu'il a couru, il s'est mis devant moi et a commencé à parler. Il m'a dit que, quand Paul était avec eux, sa femme avait un chien. Selon lui, la plupart du temps, Paul était très gentil, comme s'il avait décidé que ce serait comme ça et qu'il fallait simplement s'habituer. Il s'entendait très bien avec leur fille, mais pas avec l'animal : il détestait l'entendre aboyer, et le regardait d'un drôle d'air. Le chien était très vieux, la femme l'avait depuis la fac, elle l'aimait plus que tout au monde – plus même que le mari, disait-il, mais il ne s'en formalisait pas, lui aussi aimait le chien. Qui ne faisait pas grand-chose, sinon dormir dans la cour et frapper le sol de la queue de temps en temps.

Mme Campbell s'interrompit et respira profondément :

— Et puis un jour, Paul est arrivé en courant et a dit que le chien venait d'avoir un accident. Ils sont sortis aussitôt. Il était étendu par terre, moitié dans la cour, moitié sur la petite route derrière la maison. Sa tête n'avait plus de forme, comme si elle avait été écrasée par une voiture. Paul était en larmes, tout ça, ils ont enterré la bête, et c'est seulement le soir, alors que les parents étaient au lit, que la femme a dit quelque chose, tout doucement, sans regarder son mari, comme si elle parlait au mur. Elle a dit que, depuis le temps qu'ils vivaient là, le chien ne s'était jamais approché de la route. Elle a ajouté qu'il serait bizarre que quelqu'un ait roulé trop vite pour s'arrêter. Et qu'il était étrange que seule la tête ait été touchée, que les deux yeux et la gueule soient si abîmés. Son mari a réfléchi à tout cela. Rien n'a été dit de plus cette nuit-là, et ils ont fini par s'endormir. C'était une semaine avant qu'ils ne ramènent Paul. L'homme admettait qu'ils savaient, sans avoir deviné comment, et qu'ils n'avaient pas de preuves. Cela aurait pu être un simple accident. Mais ça suffisait comme ça. Sa femme ne voulait plus de lui.

Mme Campbell leva un doigt pour m'empêcher de répondre quoi que ce soit :

— Écoutez-moi ! C'est juste ce que l'homme a déclaré. J'ai pensé que ça pouvait être un gros mensonge, pour excuser ce qu'ils avaient fait, et ça se voyait sans doute sur mon visage. L'homme a secoué la tête, et il a ajouté que je n'aurais qu'à regarder les yeux de sa femme, depuis toutes ces années : j'aurais su la vérité. Puis il s'est éloigné et je ne l'ai jamais revu.

— Bon Dieu !

— En effet. Je vais vous dire une dernière chose, et ensuite vous partirez. Six ou sept ans après ça, peu avant que je parte en retraite, il y a eu un incendie. Muriel m'a dit qu'elle vous l'avait appris. Beaucoup de papiers ont disparu.

— Oui, elle m'en a parlé.

— Il y a une chose qu'elle ignore. Ce matin-là, j'étais en retard au boulot, pour des histoires de tramway, j'avais dû terminer le trajet à pied. Le temps que j'arrive, la fumée montait déjà du bâtiment, les gens étaient sortis dans la rue, tout le monde courait dans tous les sens. Quatre personnes sont mortes, beaucoup d'autres ont été brûlées : le feu a commencé alors que l'édifice était plein. Et comme j'étais là, à essayer de comprendre ce qui se passait, j'ai ressenti quelque chose de bizarre dans la nuque, je me suis retournée et...

Elle déglutit péniblement :

— ... il était là. De l'autre côté de la rue, à regarder. Un peu plus grand, un jeune homme. Il vous ressemblait tout à fait, sauf qu'il était plus mince. Je l'ai vu une seconde, puis il a disparu. Ou peut-être que je ne l'ai pas vu du tout. Parfois je pense que j'ai vu son visage et que j e l'ai reconnu. La plupart du temps, je crois que c'était dans ma tête, et c'est pourquoi je n'en ai jamais parlé à personne, même pas à Muriel, qui était comme une fille pour moi – et qui l'est toujours, quand elle a le temps.

— C'était lui, dis-je doucement. C'était Paul.

Elle me prit par le bras :

— Il ne faut pas penser que c'est lié au fait qu'il était orphelin ou aux gens qui l'ont accueilli, qui ont tenté si fort de lui donner une vie. Ce n'est pas le cas. Des gens comme eux ont élevé Muriel et des milliers d'autres comme elle.

— Je sais. Mes parents n'étaient pas mes vrais parents non plus, et ils m'ont donné plus d'amour que j'en mériterai jamais.

Elle parut surprise, puis se reprit, se leva, et je compris que l'entretien était terminé.

À la porte, comme j'étais sur la véranda, elle posa la main sur mon bras :

— J'ai passé toute ma vie avec les jeunes, et dans l'ensemble ça m'a beaucoup plu. Mais avec le temps, mon idée du monde a changé sur un point, et pour de bon.

— Et lequel ?

— Je pense toujours que nous sommes tous humains. Mais je ne crois plus que nous soyons tous les enfants de Dieu. Plus du tout.

Je rentrai à l'hôtel parce que je ne savais trop quoi faire d'autre. En arrivant, je me sentis lessivé, et je finis assis au bar, à contempler la rue à travers des vitres teintées. Tout le monde a des expériences personnelles. C'est l'une des miennes.

J'étais à la fois ahuri et exaspéré. Tout se terminait en queue de poisson. Mme Campbell ne se souvenait pas du nom des gens qui avaient adopté Paul pour de bon. Les collègues qu'elle avait eus à cette époque étaient morts ou dispersés. La piste avait été coupée, notamment par l'incendie. J'étais persuadé que Paul était revenu mettre le feu au bâtiment, et que Mme Campbell le pensait aussi – tout comme je croyais qu'elle comprenait que le gamin retrouvé seul, dans la rue, n'avait toléré d'être ballotté d'une famille à l'autre que le temps d'être assez grand pour partir et faire son chemin dans le monde, pour devenir celui qui « remettrait les choses en place ».

Quand je sortis mon portefeuille pour régler ma bière, je me souvins que j'avais débranché mon portable. J'avais un appel. Je ne reconnus pas le numéro et je décidai donc de rappeler.

Elle répondit très vite :

— John ?

— Non, c'est Ward. Pour savoir qui appelle, Nina, il suffit de regarder l'affichage de ton téléphone.

— Exact. Je suis une idiote. Où es-tu ?

— À San Francisco.

— Oh ! Et pourquoi ?

— J'y ai laissé mon cœur et je suis venu le reprendre.

— Bonne idée. Dans quel état est-il ?

— Comme neuf, répondis-je. Qu'est-ce qui se passe ?

— Rien. Enfin, non, c'est la folie. Nous avons eu un double meurtre ce matin ; quelqu'un a laissé une morte anonyme dans un motel pourri, puis flingué un flic pour mieux se faire comprendre. Il avait placé un disque dur dans la bouche de la femme.
— Charmant.
— Pas vraiment. C'est du domaine de la police de L.A., bien sûr, mais Monroe est dans le coup, et par conséquent moi aussi. Je me demandais si tu accepterais de jeter un coup d'œil au disque dur. J'en ai obtenu une copie, tout à fait officieusement. Je sais que tu en es capable, en vrai pro.
— Sûr. Mais Bobby aurait mieux fait l'affaire. Et même une copie octet par octet n'est pas exactement la même chose qu'un original. Mais je regarderai ça.
— Ils ont déjà trouvé un texte et un fichier musical dessus. Ce gars-là a vraiment le sens du théâtre.
— Qu'est-ce que c'est comme musique ?
— Le *Requiem* de Fauré.
— Super.
— Je n'ai pas écouté.
— Tu devrais. C'est tout à fait revigorant, bien que destiné aux morts.
Elle resta silencieuse un moment. Je me gardai de l'interrompre.
— Ça va, Ward ?
— Comme ci, comme ça.
Je lui racontai, brièvement, ce que j'avais appris chez Mme Campbell.
— Ça m'a vraiment assommé. Et puis...
— Oui, dit-elle doucement. Je sais. Je... je fais souvent le même rêve. Je suis aux Halls, dans le lobby, après avoir été blessée. John et toi êtes dans les maisons, à chercher Sarah Becker. Bobby est parti je ne sais où. Je suis par terre, j'ai mal, et quelqu'un vient me faire la peau. Et cette fois je crois qu'il pourra.
— Merde ! C'est pas marrant.
— J'en ai encore rêvé il y a trois heures. Ça devient plus long à chaque fois. Je... parfois, je m'inquiète à l'idée qu'à un moment ça ne s'achèvera pas. Que le gars m'aura, que je ne me réveillerai pas.
— Les rêves durent aussi longtemps qu'on le leur permet. Les bons comme les mauvais.
— Voilà qui est profond, maître Ward.

— Ouais. Désolé. Je n'ai pas la moindre idée de ce que je voulais dire.

Elle rit, et ça paraissait convaincant.

— Bon, d'accord, repris-je, appelle quand tu as le disque dur. Je m'en occuperai. Je n'ai plus rien d'autre à faire ici.

— Il est sur ma table.

Je n'étais allé chez Nina qu'une fois, et brièvement, mais j'en gardais une image très nette. L'espace d'un instant, assis sur un tabouret avec un reste de bière, et les habituels bavardages autour de moi, je souhaitai être là-bas. Ou du moins ailleurs, quelque chose qui ressemble à un foyer.

— Et surtout que John n'y touche pas, recommandai-je. Je serai là demain soir. Tu peux le contacter ?

— Il est en balade. Je lui dirai que tu viens.

Je remontai dans ma chambre et fumai à m'en faire sauter la tête. Cela ne me remit pas de meilleure humeur mais suffit au moins à décramponner la guenon de la nicotine accrochée à mon dos. Je tirai le fauteuil jusqu'à la fenêtre, levai la vitre et, m'asseyant, regardai dehors un moment : de grands bâtiments sombres, des lumières, des bruits venus de l'extérieur et d'en dessous. J'avais l'impression d'être au bord d'un énorme continent, seul, sans tribu, sans foyer, sans terrain de chasse.

Je finis par m'endormir et je rêvai.

C'était dans une vieille ville, avec des rues pavées, des maisons branlantes, une place où s'entassaient un marché agricole et des étalages vendant des produits ménagers. J'étais jeune, adolescent, amoureux de la Gitane du marché, une fille aux longs cheveux, sûre de sa beauté, inaccessible mais en même temps si superbe qu'elle était aimée de tous. Je l'aperçus marchant au milieu des étalages, suivie d'autres filles plus ternes ; le plus beau visage du monde, entouré par une masse de cheveux bruns aux reflets auburn.

Plus tard, je revenais, devenu adulte, plus confiant mais plus sec, ayant perdu en magie ce que j'avais gagné en stature. Le marché s'était réduit à quelques étalages, dévoilant des rues là où autrefois il avait semblé constituer un royaume en soi qui n'avait nul besoin de cet environnement. Je n'entendais que des échos là où autrefois résonnaient des palabres et des rires.

Elle était à un étalage, vendant ceci ou cela : des bouts de tissu, des boutons dépareillés, des choses en plastique. Ses cheveux, coupés court, avaient viré au gris prématurément. Elle était toujours jeune de visage, mais s'était alourdie, semblait plus petite, se montrait plus quelconque.

Je passai devant elle et la vis fourrer quelque chose dans un sac en plastique, un achat de trois sous pour une vieille femme. Je compris qu'elle n'était plus qu'une boutiquière. La princesse que je revenais voir, pour lui montrer que j'étais désormais un homme, et par conséquent digne d'intérêt, digne de son regard, avait disparu et quelqu'un avait pris sa place. Si je ne l'avais pas vue, j'aurais pu continuer à croire qu'elle était toujours là, rayonnant de magie, de sexe et de sourires.

Mais c'était trop tard et je ne pouvais rien faire, sinon marcher un peu pour m'éloigner du marché, faire demi-tour et le regarder, en sachant que ma jeunesse, mon être même, tout ce qui m'avait poussé tout au long de ces années, n'était plus. Alors je me souvins qu'elle m'avait jeté un regard, sans me reconnaître ; si elle était une simple marchande, je n'étais rien et n'avais jamais rien été.

En me réveillant, je me tournai, l'esprit cotonneux, vers la pendule à côté du lit, et fus stupéfait de constater qu'il ne s'était écoulé qu'une demi-heure.

Mon portable sonna. Je reconnus le numéro.

— Alors, tu es de retour, dis-je.

Il y eut une pause :

— C'est Zandt.

— Je sais, répondis-je, un peu dans les vapes. Tu étais en balade.

Nouvelle pause :

— Ward, je suis en Floride.

Ce qui me parut absurde, mais je n'en dis rien.

— C'est bien pour toi. Alors ?

— Yakima.

Je me redressai :

— Quoi donc ?

— J'ai des renseignements. Enfin, peut-être. Ça ne veut pas dire grand-chose.

— J'ai dit à Nina que je passerais la voir demain à L.A. Pourquoi ne pas se retrouver là-bas ?

— Tu lui as parlé aujourd'hui ? Pourquoi ?

— Elle a appelé. Elle est sur une affaire de meurtre et veut que je jette un coup d'œil sur un disque dur.

— Où es-tu, en ce moment ?

— À San Francisco.

Encore une pause :

— Pourquoi ?

— Je tentais de retrouver l'Homme Debout. Sans grand succès.

— Reste où tu es, j'arrive.

— John, je te l'ai dit, je suis censé rencontrer Nina.

— Je ne veux pas aller à L.A.

Il y avait quelque chose d'un peu bizarre dans sa voix.

— Bon, d'accord, acquiesçai-je. Je reste un jour de plus et je te verrai ici.

— Je t'appellerai en arrivant.

Et là-dessus la communication fut coupée. J'étais à peu près sûr qu'il était ivre.

J'y réfléchis un moment, puis appelai Nina et lui dis que je ne pourrais la voir que le surlendemain. Sans lui expliquer pourquoi. Elle répondit qu'elle m'enverrait le disque dur.

— C'est bien, dis-je avant d'ajouter : John est revenu ?

— Oui, mais il est reparti.

— C'est dur de lui mettre la main dessus.

— Tu l'as dit !

On aurait cru qu'elle allait ajouter quelque chose mais elle se contenta de me dire au revoir.

Je retournai près de la fenêtre et contemplai encore un peu la cité. Comme toutes les villes, elle ignora mon regard.

CHAPITRE 8

Ce matin-là, Nina se dirigeait vers le quartier général de la police de L.A. quand un appel lui parvint : un enquêteur avait touché le gros lot avec la photo de la morte. Elle effectua un demi-tour brutal, salué par un concert de klaxon furieux, et partit vers un bar nommé le Jimmy's, pas loin de Hollywood Boulevard.

Il y avait déjà une voiture de patrouille, ainsi qu'une autre, banalisée, avec un gyrophare. Nina ajouta la sienne à la collection et entra en hâte. Le bar était sombre, il y régnait une odeur de bière, l'air paraissait croupi, comme s'il était passé dans les poumons de trop de gens incapables de tenir debout. Elle aperçut Olbrich parlant à un type aux cheveux longs, dont le sourire vitreux disait assez que, s'il avait su dans quelle merde il allait se fourrer, il n'aurait pas fumé cet énorme pétard avant de venir au boulot.

— Voici l'agent Baynam, annonça Olbrich comme elle approchait. Pourquoi ne pas lui répéter ce que vous venez de me dire ?

— Elle s'appelait Jessica, expliqua le barman. Ça, c'est sûr. Et je sais qu'elle vivait à West Hollywood. Je suis également à peu près certain qu'elle s'appelait Jones, j e crois qu'elle l'a dit une ou deux fois, et les gens l'appelaient JJ, mais tout le monde... euh...

— Ne donne pas toujours son vrai nom, je sais, répondit Nina. C'était une habituée ?

— Oui, souvent le soir, parfois l'après-midi.

— Elle tapinait ?

Il secoua la tête avec vigueur :

— Non ! Absolument pas ! Je crois qu'elle voulait être chanteuse ou quelque chose comme ça. Elle l'a dit une fois. Assez jolie. Je crois qu'elle est hôtesse. Enfin, était... et merde !

— Elle était là samedi soir ? demanda Olbrich.

— Ouais. Elle est venue vers 17 heures avec une copine. Je ne sais pas son nom, mais je l'ai déjà vue. Une Noire, avec de longs cheveux décrêpés. C'était une soirée à demi-tarif, et elles ont été beurrées très vite. Je suis à peu près sûr que la copine a fini à une table avec des types et qu'elle est partie en leur compagnie. JJ est restée à traînailler un moment, puis elle s'est assise avec un gars.

— Qui était-ce ? questionna Nina d'une voix égale, mais d'un seul coup elle avait l'impression d'étouffer.

Olbrich eut le bon sens de ne pas s'en mêler.

— J'étais en train d'expliquer ça à monsieur. Je ne le connaissais pas. Je l'ai remarqué parce que...

Il haussa les épaules.

Parce que tu étais branché sur Jessica, pensa Nina. *Je comprends.*

— Elle rencontrait souvent des types ?

— Assez souvent, répondit le barman en regardant ailleurs. Nina hocha la tête sans le quitter des yeux. *Et un soir, plusieurs peut-être, un baiser sur la joue lui a valu un verre de plus alors qu'elle était à court d'argent, c'est ça ? Et tu y penses toujours de temps à temps, alors que pour elle ce n'était rien et qu'elle n'y a plus songé dès la deuxième gorgée.*

— Celui-là avait quelque chose d'inhabituel ?

— Un type comme les autres. Des cheveux courts. Plutôt pas mal, c'est tout ce que je peux dire. Après ça, il y a eu du monde, et quand j'ai regardé il était tard, quelqu'un d'autre était assis à leurs places. Vous pourriez peut-être discuter avec les filles qui servaient ce soir-là, mais elles ne seront là qu'en soirée. Sauf Lorna, qui est de service pour le déjeuner.

On entendit un cri dehors, et un flic en uniforme passa la tête par la porte :

— Lieutenant ?

— Tu l'as ? s'enquit Olbrich en se tournant vers lui.

— Ouais.

— Nina, dit Doug, on a une adresse. Je viens avec toi.

— Elle est vraiment morte ? demanda le barman.

— Oui, répondit Nina, vraiment. Désolée.

Il acquiesça et s'éloigna.

Comme Nina arrivait à la porte, elle jeta un regard dans sa direction. Il nettoyait une table avec un chiffon, très lentement, dans un bar où il était bien obligé de travailler. Elle pensa : on ne sait jamais vraiment qui on laisse derrière soi.

C'était l'appartement n° 7, du 3140 Gardiner Street. Quand Nina arriva, Monroe était déjà dehors, avec deux flics.

— Il ne perd pas de temps, hein ? lança Olbrich.

— C'est le moins qu'on puisse dire.

Le bâtiment, d'un blanc sale, comptait deux étages. À chaque extrémité montait un escalier extérieur. Nina grimpa en compagnie de Monroe, et tous deux attendirent pendant que les autres flics cherchaient le responsable de l'immeuble.

— Tu vas mieux, ce matin ? demanda Charles.

— Très bien, répondit-elle. Le Profilage nous a donné quelque chose d'utile sur le texte ?

— Pas encore. Tu penses qu'il ne trouvera rien ?

— Ça n'a pas vraiment marché pour le sniper de Washington, non ?

— C'était différent.

— Oh, que non ! Ils ont décidé que c'était un Blanc, parce que les idées reçues – qui reposent sur une très vieille étude pas franchement scientifique – veulent que les serial killers, dans leur majorité, soient blancs, si bien qu'on n'a tenu aucun compte des coups de fil parlant d'un Noir. Dans le même temps, des gens disaient avoir vu une camionnette blanche, et voilà que tout le monde ne remarque plus que ça, bien que les autoroutes en soient pleines. C'est ne pas en voir qui serait surprenant ! Le numéro d'immatriculation de la voiture du tueur, qui était bleue, passe une demi-douzaine de fois dans le système informatique à cause de son comportement suspect, mais non, il n'est pas blanc et n'a pas de camionnette blanche, donc ça n'a pas d'intérêt. Les profileurs jurent que jamais les tueurs en série n'ont de complice – sauf que celui-là en avait un. On n'aurait pas dû les écouter, de toute façon, quiconque ayant un peu de cervelle savait depuis le début que ce n'était pas un serial killer, mais un meurtrier avec une mission politico-religieuse. Bref, tout ça n'a fait qu'embrouiller la question... et ça pourrait bien recommencer ici. Je ne suis pas sûre de croire encore à leurs salades.

— Alors, pourquoi me demander s'ils ont trouvé quelque chose ?

— Pour te tenir à l'écart de leurs sollicitations.

— Nina, quand vas-tu me dire ce qui s'est passé l'année dernière ?

— Je te l'ai déjà raconté, répliqua-t-elle avec un grand sourire.

Intérieurement, pourtant, elle se dit de prendre garde : on pouvait penser bien des choses de Monroe, mais il n'était pas idiot.

Olbrich fit son apparition à ce moment, avec un trousseau de clés :

— Zinman est en train de prendre une déposition, mais le gars n'a pas grand-chose à dire. Elle était très réservée, ce genre de truc... Et il est con comme un balai. On y va ?

Arme à la main, Nina et Monroe firent oui de la tête.

Olbrich frappa à la porte, attendit et, en l'absence de réponse, glissa la clé dans la serrure.

— Police ! Montrez-vous !

Silence. Il ouvrit un peu la porte, ce qui permit d'apercevoir une pièce assez grande, d'environ sept mètres sur sept. C'est du moins tout ce que distingua Nina, qui choisit de rester dehors tandis que les deux hommes entraient et répétaient leur mise en garde. Mais ils ne rencontrèrent pas âme qui vive.

Pénétrant à son tour dans l'appartement, elle aperçut, au milieu de la pièce, une table basse et un sofa rouge assez fatigué ; à l'autre bout, près d'une fenêtre, un ordinateur d'allure bon marché. Une petite lumière rouge était allumée en bas du moniteur, mais l'écran restait noir. À côté, un poste de télévision, qui devait être visible du sofa. Pour mieux voir encore, il aurait fallu le déplacer d'une cinquantaine de centimètres sur la gauche, mais cela aurait bloqué la porte donnant sur la chambre à coucher, où les deux hommes se trouvaient. Un câble noir partant de l'ordinateur s'y dirigeait. Nina alla jeter un coup d'œil dans la petite cuisine. Une grande fenêtre donnait sur la rue. Tout était bien rangé. Comme elle faisait demi-tour, elle avisa une vieille guitare derrière le sofa, couverte de poussière ; il y manquait une corde.

Dans le dernier coin du séjour se trouvait un petit bureau. Deux ou trois carnets... Nina en ouvrit un, avec précaution, et examina rapidement une page. Des gribouillis. Des trucs qui ressemblaient à des paroles de chansons. Un vers, « la pluie qui jamais ne lave », avait été barré.

— Viens voir ça ! lança Charles.

La chambre à coucher était exiguë. Un lit à deux personnes, une coiffeuse, rien de plus. Une minuscule salle d'eau à l'autre bout. Le lit était défait. Monroe et Olbrich regardaient un petit objet fixé sur un trépied et sur lequel le câble noir venait se raccorder.

— Une caméra, dit Olbrich.

— Une Webcam, rectifia Nina. Vous voyez où va le câble ?

Elle retourna dans le séjour, près de l'ordinateur, et déplaça doucement la souris, en prenant soin de ne pas la toucher des doigts.

L'écran du moniteur revint à la vie. En son centre apparut une fenêtre qui en occupait environ le tiers. On y voyait le lit, et Monroe juste à côté.

— Je ne toucherai à rien, dit-elle, mais vous allez trouver un modem derrière cette bécane. Jessica avait un site où les gens pouvaient la regarder.

— La regarder ? demanda Olbrich. Et d'où ?

— Du monde entier. Mauvaise nouvelle : nous avons maintenant des dizaines de millions de suspects.

Trois heures plus tard, elle était de retour au Jimmy's, dans une pièce à l'étage, avec le propriétaire des lieux, qui ne se prénommait pas Jimmy.

— C'est un nom de bar, contrairement au mien, avait-il expliqué.

Il s'appelait Jablowski.

Prévenu par Don, le barman, de la visite des policiers, il avait décidé de venir, contrairement à son habitude. Il était bizarrement bien vêtu pour quelqu'un possédant ce qui n'était jamais qu'une auge à bière pour les alcooliques de l'aprèsmidi, mais après tout beaucoup de trafiquants de drogue s'abstiennent aussi de consommer leurs produits. Don était rentré chez lui pour quelques heures, histoire de « décompresser ». Ils avaient son adresse. Nina ne pensait pas qu'ils auraient besoin d'aller le voir, mais elle n'était pas profileuse – et c'est bien pourquoi, à sa demande, un flic en civil avait été chargé de le suivre.

Deux autres évoluaient au milieu de la foule des clients. Une des serveuses, qui travaillait le soir où Jessica était passée pour la dernière fois, devait arriver bientôt ; on gardait également l'œil sur tous les hommes répondant à une description très générale. En d'autres termes, on n'allait nulle part, mais à toute allure. C'était le contraire dans l'appartement de la fille, qu'on retournait de fond en comble ; des enquêteurs de trois services différents examinaient tout ce qu'ils pouvaient trouver, lisaient, photographiaient, époussetaient.

Nina discuta avec une jeune Noire nommée Jean, venue là en quête de Jessica : toutes deux auraient dû se retrouver la veille au soir, mais son amie n'était pas venue. Et puis elle voulait boire un coup. Don lui avait montré les flics, d'un air insistant.

— Tapineuse de Webcam ? demanda Nina, répétant ce que Jean venait de lui dire.

— On appelle ça comme ça, répondit Jean en haussant les épaules. Ça ne signifie pas que vous baisez.

— Pour autant que vous sachiez, Jessica n'a jamais couché contre argent ?

— Foutre non ! Moi non plus, d'ailleurs, mettez-vous bien ça dans la tête.

— Les professionnelles ne sont pas admises ici, intervint Jablowski d'une voix douce. Je suis très strict là-dessus.

— Quand vous êtes là, c'est-à-dire pas souvent, monsieur. Pourriez-vous nous laisser seules un moment ?

L'homme s'en fut. Nina marqua une pause, puis reprit :

— Si je comprends bien, Jean, vous en êtes aussi ?

— Oui, je... c'est moi qui ai branché Jessica là-dessus. Mais je vous l'ai dit, ce n'est pas comme de...

Nina la regarda bien en face :

— Jean, c'est là un domaine dont j'ignore tout, mais j'ai besoin d'être informée, et vite. Jessica n'est plus là. Pourquoi ne pas me dire comment ça se passe ?

La fille se redressa, alluma une cigarette et se mit à parler.

Tapiner, c'était une chose, dit-elle. Tout le monde savait comment ça se passait. La Webcam, c'était différent. On ne rencontrait jamais personne, on ne prenait pas de risques. Et même, on ne faisait jamais rien, enfin pas vraiment. On enlevait ses fringues, on s'activait comme à l'ordinaire, mais nue. Regarder la télé. Nettoyer la cuisine. S'il y avait un petit ami aux environs, on laissait la caméra branchée, parfois tournée de l'autre côté. Comme on voulait. Le plus bizarre, c'est que pour certains des voyeurs, moins on en voyait, meilleur c'était. Un jour, Jean avait eu des montagnes de trucs à faire et ne s'était pas baladée en petite tenue ; elle avait tout simplement oublié la caméra – et le lendemain elle avait reçu une pleine pelletée d'e-mails torrides la félicitant de ses talents d'aguicheuse. Jean pensait que, question sexe, les mecs étaient complètement frappés. On croyait les avoir compris, et ils faisaient un truc qui montrait à quel point ils étaient tordus. Elle ressentait de temps à autre l'impression bizarre de baiser avec leurs têtes. S'asseoir puis montrer un bout de papier sur lequel on lisait : « J'ai préparé un machin végétarien immonde hier soir, et l'appartement pue encore comme un cul de vache. » Ou bien laisser échapper un pet d'ampleur continentale et rester là à sourire à l'objectif, sachant que, si grand que soit leur écran, il ne leur apprenait pas tout ce qu'il y avait à savoir de son monde.

— Vous dites que vous avez branché Jessica là-dessus, l'interrompit Nina. Comment ça s'est passé ?

— Il y a un an et demi, j'ai rencontré une fille dans une soirée. Elle s'y était mise, et m'a donné l'adresse e-mail du type qui s'occupe de ça, il se fait appeler Webdaddy. Vous lui envoyez une photo de vous, il vous répond et vous vous mettez à parler « paramètres » et « limites » – du genre vous vous mettrez à poil ou pas, ou vous avez un copain et vous ferez des trucs ensemble, on le verra ou pas, des machins comme ça. Si Webdaddy vous trouve sympa, il vous envoie un CD avec des programmes dessus. Vous vous offrez une connexion Internet par câble et vous allez vous acheter une Webcam pour cinquante dollars. Il s'occupe du reste, votre site, tout le bazar. Un chèque arrive à chaque fin de mois. C'est aussi simple que ça.

— Vous avez l'adresse postale de cette personne ?

— Non, rien qu'un e-mail. Pareil pour Jessica. Il est sur le Web, pourquoi ne pas aller le rencontrer pour de bon ?

— S'il y a un problème avec le système, il n'y a pas de chèque ?

— On lui envoie un mail et la réponse arrive aussitôt.

On mettait en place la Webcam – une simple caméra numérique basse résolution, très bon marché – qu'un câble USB raccordait à l'ordinateur, où un logiciel capturait ce qui était visible dans l'objectif, et le transmettait via Internet à un serveur. Peu après, l'image était remplacée par une nouvelle, et ainsi de suite. Pendant ce temps, dans l'univers lointain des hommes qui avaient le temps, l'utilisateur chargeait votre page, avec l'image bien au milieu. Voilà vingt ans, ç'aurait été de la science-fiction ; aujourd'hui, des inconnus, dans le Kansas, à Cardiff ou à Anvers, pouvaient se branler de temps en temps pendant que vous passiez l'aspirateur entièrement nue. Un monde tordu ? Sans aucun doute. Mais qui vous rapportait dans les deux cents dollars par semaine, et Jean n'avait pas à baiser avec des inconnus ni à tortiller du cul en faisant un strip-tease. Elle trouvait ça très bien et pensait que c'était un progrès pour les femmes.

— Jessica gagnait autant que vous ?

— Non. Ça faisait quelques mois seulement, elle n'avait pas beaucoup d'abonnés. Elle ne se donnait pas beaucoup de mal pour les distraire, si vous voyez ce que je veux dire. Les filles s'offrent en spectacle. Parfois elle ôtait son corsage – il le faut, ou on se fait jeter – mais ça ne lui plaisait pas. Et je ne crois pas qu'elle ait beaucoup aimé les trucs sexy. Elle disait qu'elle allait arrêter, se remettre à écrire des chansons. C'était vraiment un secret, personne n'était au courant, sauf moi.

— Les abonnés à votre site... quels contacts avez-vous avec eux ?

— Rien d'autre que des mails.

— Ils ne peuvent pas découvrir votre adresse ?

— Non, sauf si je la leur donne.

— Est-ce que Jessica a laissé entendre qu'elle l'avait fait ? Qu'elle avait des contacts particuliers avec un de ses abonnés ?

— Elle n'était pas vraiment dans tout ça, je vous l'ai dit. Elle avait besoin d'argent, mais c'était quelqu'un d'orgueilleux. Elle n'avait pas l'intention de faire des choses qui lui fassent honte. Sauf si elle était complètement ivre.

— Vous étiez un peu jetées, toutes les deux, samedi soir, non ?

Jean eut un sourire espiègle :

— Ça se pourrait.

— Et vous avez quitté Jessica pour aller à une soirée.

— J'ai rencontré des types. Quand je suis partie, elle était toujours là.

— Le barman dit qu'il l'a vue plus tard avec un type. Vous êtes au courant ?

— Je vous l'ai dit, j'étais partie.

— Elle n'avait personne que vous connaîtriez ?

— Pas ces temps-ci.

— Et récemment ?

— Elle a eu des petits copains. Mais c'étaient que des mecs.

Nina resta silencieuse un instant, puis regarda la femme assise en face d'elle. Après le choc initial en apprenant la nouvelle, Jean semblait s'être vite reprise. De toute évidence, Jessica était déjà passée par profits et pertes. Nina pensa de nouveau au trajet de A à Z, puis de Z à la morgue. Difficile de faire autrement, quand on avait affaire à une fille de vingttrois ans, qui voulait surtout se donner du bon temps et qui pensait que ça serait toujours comme ça, que croire en soi ferait office de sésame.

— Vous vous rendez compte que vous n'êtes pas invincible, quand même ? dit-elle à Jean.

Celle-ci lui rendit son regard, inclina la tête et eut un sourire glacé :

— Vous non plus, ma chère.

— On est dessus, dit Monroe. Dès que tu as appelé, on a mis un technicien sur l'ordinateur. Nous savons où est situé le serveur, et avons l'adresse e-mail du fameux Webdaddy.

Ils étaient sur le balcon de l'appartement de Jessica, qu'on fouillait toujours. Charles buvait de l'eau glacée, mais il avait l'air un peu fripé, ce qui ne lui ressemblait guère.

— Rien d'intéressant ici ?

— Non, hormis l'ordinateur. L'endroit était vraiment bien tenu. Pas beaucoup d'empreintes. La police de L.A. s'occupera de ce qu'on a trouvé, mais... Il y a des carnets avec des griffonnages et ce qui ressemble à de la très mauvaise poésie. Jusqu'à présent, pas de noms, pas de numéros de téléphone. Les médecins légistes sont dans la chambre à coucher, mais rien n'indique qu'elle y ait été tuée.

— Dans combien de temps irons-nous frapper à la porte de Webdaddy ?

— Dans pas longtemps. Jean et Jessica faisaient partie d'un groupe d'une quinzaine de filles : plusieurs ici, deux à San Diego, une à San Francisco, certaines dans des bleds perdus... À propos, le domaine s'appelle *daddysgirrls.net.*

— Charmant.

— S'il est ici, à L.A., on s'en occupera. Sinon, ce sera un local quelconque. Il est important d'aller vite.

— Que veux-tu que je fasse ?

— Le type chargé de surveiller le barman dit qu'il est rentré, a fumé des pétards trois heures durant en regardant le mur, puis qu'il est retourné au boulot, en servant héroïquement des bières. De toute façon, je crois qu'il n'est pas dans le coup. Tu pourrais m'épargner un coup de fil en allant casser les pieds aux gens de Quantico sur le profilage, mais sinon... tu as mangé ?

— Non.

— Vas-y. Pas loin d'ici. Si j'apprends quelque chose, tu le sauras.

Trois quarts d'heure plus tard, en plein milieu de sa salade, elle reçut un coup de fil. Pestant – c'était son premier repas depuis plus de vingt-quatre heures –, elle jeta de l'argent sur la table et sortit en courant.

Le temps qu'elle soit à mi-chemin, son portable sonna de nouveau. S'arrêtant sur le boulevard, elle écouta Monroe lui dire d'une voix atone :

— Ce n'est pas lui. Il s'appelle Robert Klennert, il a cinquante-huit ans et il est plus qu'obèse. Fondamentalement, c'est un tas de merde spécialisé dans les sites pornos. Il s'y connaît question technique, ce qui est intéressant vu le disque dur, mais j'ai du mal à l'imaginer en train de tuer une j eune femme, sans compter qu'il ne correspond pas aux descriptions des témoins.

— On en revient donc au scénario « un sur des millions ».

— Peut-être un peu mieux que ça. La police de L.A. a les archives de Klennert. Tous ses abonnés, et ceux qui ont simplement visité ses sites, sont logés. On emporte ses ordinateurs en ce moment même.

— Sous quel motif d'inculpation ?

— Aucun. Il coopère à fond. Bizarrement, il semble réellement avoir des sentiments paternels envers ses filles. Ou c'est du bluff de fort calibre, ou...

Monroe se tut un instant :

— Ou ce n'en est pas, ce qui est plus probable. Ce n'est pas lui. En attendant, on dirait bien que la musique sur le disque dur ne nous apprendra rien du tout. Je crains que ça ne parte en couille, Nina. À moins qu'il se passe quelque chose, ça pourrait tourner en eau de boudin.

Très juste, pensa Nina. Ou bien tu sens qu'il va falloir beaucoup bosser sur une piste virtuelle monstrueusement vaste, tu n'y comprends rien, et tu ne vois pas comment ça pourrait faire une bonne intrigue pour l'*Histoire de Charles Monroe* sur le câble.

Elle lui dit au revoir et s'en fut. De l'autre côté de la rue, une voiture s'arrêta dans un chemin d'accès et une famille en sortit. Le mari, l'épouse, une petite fille... Les adultes semblaient se disputer.

Nina baissa un peu la vitre pour écouter et entendit la fillette rire aux éclats. Ses parents firent de même presque aussitôt après.

Nina comprit qu'il s'agissait d'une fausse altercation, sans doute une imitation de gens à qui la famille venait de rendre visite. Elle songea un moment à sa propre enfance, qui en général avait été sans histoires, mais avait aussi compté suffisamment de vraies colères masculines pour qu'elle doute de pouvoir rire autant que la gamine.

Elle la regarda pendant que l'enfant suivait ses parents vers la maison, en se disant que si jamais elle était accueillie par un petit chiot tout mignon avec un ruban autour du cou, elle-même serait contrainte d'aller baffer la petite princesse.

Pas de chien. La fillette pourrait continuer à rire.

Nina mit le contact et partit vers l'océan.

CHAPITRE 9

La fille restait silencieuse. Avant, elle avait été très expansive – heureuse de vous rencontrer, eh, c'est pas mal ici, ooh, c'est bien, ici, ouais ! Maintenant, après, elle n'avait rien à ajouter. Peut-être pensait-elle que c'était ce qu'il voulait (et elle avait raison... pour le moment), peut-être jugeait-elle que c'était terminé, exception faite du petit cadeau (auquel cas elle avait tort). Se pouvait-il qu'elle ait eu une embolie et consacre toute son énergie à ne pas tomber dans les pommes ? Pete Ferillo n'en savait rien. Il s'en moquait d'ailleurs éperdument. C'est ce qu'il y avait de super là-dedans. Pete croyait qu'il fallait compartimenter. Ne pas être contraint de savoir. Ne pas se préoccuper.

Il tendit la main vers la table de nuit et prit un cigare dans son étui. Se le passa sous le nez. Sans raison – il savait quel en serait le parfum –, mais ça avait quelque chose de sensuel. Une bonne odeur.

Il en coupa le bout, puis se le fourra dans la bouche. L'alluma avec une allumette – récemment, quelqu'un qu'il respectait lui avait dit que c'était la meilleure méthode –, tira une bouffée et regarda monter l'épaisse fumée.

Il était nu, vautré dans un fauteuil, les jambes étendues. Chez lui, jamais il ne s'asseyait ainsi. Il aurait été trop conscient de sa panse, de ses cuisses ridées, du contraste brutal entre ses avant-bras perpétuellement bronzés et l'albâtre malsain du reste. Ici, cet après-midi, il n'avait pas à se soucier de ça, à penser que ça faisait de lui quelqu'un de vieillissant ou de peu ragoûtant. Il n'avait pas à écouter le déprimant message sur le passage du temps ou sur l'état probable de son système, à tenter de le mettre en œuvre pour faire reluire une épouse qui disait l'aimer mais dont les interminables séances sur le tapis roulant étaient autant de sarcasmes. Oui, Maria paraissait en meilleure forme que lui. Bien meilleure. Et alors ? Elle n'avait rien d'autre à faire qu'à courir les gymnases et les centres commerciaux. Si ç'avait été son boulot à lui, il aurait meilleure allure. Il l'aimait, bien sûr, et depuis vingt-cinq ans. On apprend à sourire quand on est furieux, à retenir sa main, et la plupart du temps tout le monde est content.

L'endroit appartenait à un très important client du Dining Room, quelqu'un avec qui Pete faisait des affaires depuis un bon bout de temps – ailleurs. Il venait dîner de temps à autre avec une dame qui n'était pas son épouse légitime. Pete restait discret, tout en se souvenant avec qui le gars était venu la dernière fois. Il y avait eu un accord amical, entre hommes, et désormais il avait ses propres clés. Un domestique venait chaque jour veiller à ce que la carrée reste nickel, et remplir le frigo d'eau minérale. L'appartement était simple mais bien meublé. Une chambre, un balcon, une salle de bains, un séjour. Ce dernier était une pièce assez vaste, dont une partie accueillait une petite table pour dîner, placée de manière qu'on ne puisse voir la porte quand on était dans la partie centrale, si bien que l'endroit paraissait plus grand. Très habile. Il était agréable d'être sur le balcon en peignoir, à savourer le plaisir d'une fin d'après-midi tandis qu'en bas les prolos de la ville avançaient au pas en klaxonnant. Plus tard, peut-être.

Pour le moment, le fauteuil travaillait pour lui. Il suivit des yeux la fille tandis qu'elle allait vers la petite cuisine – un vrai bijou. Il ne connaissait pas son nom de famille. Ni la couleur, la star de ciné, l'émission télé qu'elle préférait. Il ignorait tout de ses anciens amants. Elle était grande, bronzée, elle s'appelait Cherri – il aimait le côté bidon, cinéma, de ce prénom. Elle avait une chevelure de toutes les nuances du blond, qui lui tombait bien droit entre les épaules. Elle était mince (la minceur de la jeunesse, pas celle du genre cou de dinde décharné), elle avait des gros nénés, un joli visage et un très mignon tatouage – une rose noire – en bas du dos. En général, Pete n'aimait pas ça, du moins sur les femmes normales. Mais il aimait bien sur des filles comme celle-ci. Ça montrait qu'il s'agissait d'une nana consciente de son corps, qui en était maîtresse, qui s'en servait comme d'un moyen. Pete en connaissait, petites amies ou épouses d'amis, qui étaient tatouées depuis un ou deux ans, du temps où tout le monde s'y mettait. Maria en avait voulu un aussi – incroyable ! Un chat, ou Dieu sait quoi. Il avait dit non, et il avait raison. Ça vous donnait l'air d'être une strip-teaseuse, ce qui était idiot, sauf si on était strip-teaseuse. C'était comme le pole-dancing, la danse à la barre, bon sang de bonsoir : le gros truc chez les yuppies du coin, chaque femme devait apprendre, ou du moins suivre des cours avec une amazone de l'aérobic. Un machin d'une telle stupidité que Pete se sentait prêt à exploser. Les femmes mariées n'ont aucune raison de se mettre à ça. L'intérêt du pole-dancing, c'est que celles qui dansent ne sont pas votre épouse. Toutes celles qui se lancent là-dedans en croyant se montrer sexy ont toutes les chances (a) de se prendre trop au sérieux, ce qui n'est pas sexy du tout, et (b) de se croire bandante pour leur âge, ce qui est toujours fastidieux, même si c'est vrai, ou alors (c) elles ne sont pas franchement heureuses à la maison et voudraient faire l'amour avec quelqu'un d'autre. Avec n'importe qui, sans doute. L'exemple de Johnny, un ancien ami de Pete qui s'était occupé de ses comptes pendant onze ans, le montrait assez. Ça marchait à fond pour lui question boulot, tout ça. Puis son épouse était allée à un de ces cours, lui avait refait le truc à la maison. La première fois, ça avait marché, et puis après... bon d'accord, mais tu es toujours ma femme, et d'ailleurs tu as vraiment besoin de perdre quelques kilos. Quatre mois plus tard, elle se tapait un branleur boutonneux qui travaillait chez Barnes & Noble. D'une certaine façon, c'était la faute de Johnny, alors adieu mariage, bonjour pension alimentaire. Bientôt il passait ses après-midi à regarder des danseuses, des vraies, et à boire beaucoup trop. Pete avait confié ses affaires à un autre. Et tout le monde aussi.

Pete tira de nouveau sur son Don Tomas, en savourant la façon dont la fumée imprégnait la pièce. Ce n'était pas un cigare cubain, ni même particulièrement coûteux. Hondurien – il ne jetait jamais l'argent par les fenêtres –, mais qui avait bon goût. Ça faisait trois ans qu'il avait la permission de fumer à la maison. Ce n'était pas interdit avant, Maria n'était pas du genre à placer des snipers dans le séjour pour le descendre, mais il y avait toujours la Déception. La dissuasion silencieuse, l'arme de la destruction muette ; le regard disant que, malgré ses rêves, la vie avait tourné comme elle le redoutait.

Cherri coupa une tranche de citron qu'elle laissa tomber dans son verre. Gin tonic – Pete le sentait d'où il était. Il avait un très bon nez – il valait mieux, quand on travaillait dans la restauration. Maria buvait du chardonnay, elle avait toujours bu ça. La fille sentant qu'il la regardait, fit volte-face :

— Tu veux quelque chose ?

— Oh, que oui ! répondit Pete en riant de bon cœur. Mais laisse-moi une minute, je reprends mon souffle !

Elle eut un sourire professionnel :

— Non, je voulais dire : à boire.

— Oh ! Vodka. Pas de jus de fruits, beaucoup de glace. Et ensuite on s'y remet, fais-moi confiance !

— Je meurs d'impatience, dit-elle en se tournant pour lui préparer sa boisson.

Pete sourit. Il entendit un cliquetis venu du couloir – sans doute un besogneux quelconque revenant du boulot. Il tira une nouvelle bouffée, se rencogna dans son fauteuil. Il adorait l'idée d'être assis, dans toute sa hideuse nudité, tandis que dehors un consultant épuisé, à bout de souffle, rentrait péniblement chez lui, les bras chargés de dossiers. Et lui était là, les couilles au vent et un bon verre en route. *Je meurs d'impatience.* Sarcasme ? Sans aucun doute. Aucune importance. Elle attendait, ou pas. Elle trouvait son corps supportable ou non. Elle aimait faire ce qu'il demandait – rien de tordu, il n'avait pas besoin de ça, la routine de la part de quelqu'un de jeune et de nouveau, ça suffisait – ou pas. Rien de tout ça n'importait. Elle avait déjà quatre cents dollars à lui. Il avait toutes les chances de monter jusqu'à cinq cents. Maria pouvait claquer ça sans battre un cil et ne s'en privait pas. D'ailleurs, c'est tout ce qu'il fallait pour obtenir de quelqu'un comme Cherri qu'elle se lâche.

Tandis qu'elle versait de la vodka dans un verre et y ajoutait de la glace, Pete songea à la revoir. Bien qu'elle fût, très, très mignonne quand elle s'agenouilla pour ramasser un glaçon tombé à terre, il sut qu'il s'en abstiendrait. Une nouvelle à chaque fois, voilà l'important. S'il recommençait avec elle, se poserait la question de savoir si c'était mieux ou moins bien que la fois d'avant. Elle l'appellerait par son prénom, saurait ce qu'il voudrait boire, et la familiarité s'installerait. Il aurait le temps de remarquer des choses à son sujet, de se demander pourquoi elle n'avait pas eu le bon sens de mettre le glaçon dans le verre d'abord, ou pourquoi elle ignorait que le gin était meilleur avec du citron. Cet après-midi-là, quand ils avaient refait l'amour, il avait du se finir à la main parce qu'il bandait mou, c'était comme ça. Il le savait, mais pas la fille. La prochaine fois, elle serait au courant. Ne pas savoir, c'était ça l'important. On n'a pas à se préoccuper de ce qu'on ignore.

Elle était hors de vue, à faire un sacré boucan avec le bac à glaçons. Pourquoi diable, merde ? Le verre était là, plein à ras bord. Qu'on y ajoute une goutte, et il allait déborder sur... Un glaçon sur la pointe du néné... C'était une idée.

Il se pencha vers le cendrier afin d'y déposer le cigare, histoire de le réserver pour plus tard.

— Hé, dit-il d'un ton indulgent, ça va comme ça avec la glace. Mais amène-la quand même.

Il tourna la tête.

Il y avait un homme dans la pièce.

— Qu'est-ce que vous foutez ici ? demanda Pete.

Le sourire de l'autre disait assez qu'il ne répondrait pas. Ferillo comprit aussitôt que ce n'était pas un type qui, lui aussi, avait les clés du baisodrome. La fille fit son apparition, enfilant son corsage.

— J'ai fini, non ? demanda-t-elle à l'homme.

Il ne lui répondit pas non plus. Sans quitter Pete des yeux, il tendit la main, la saisit par les cheveux et, avant qu'elle ait le temps de glapir, lui écrasa le visage contre le mur. Elle eut un grognement et s'effondra comme une masse.

Pete comprit vite. Le cliquetis dans le couloir, le coup du bac à glaçons pour dissimuler qu'elle ouvrait la porte. Il ne savait pas qui était l'homme, ni ce qu'il voulait, mais il voyait désormais qu'il avait un couteau. Un grand, comme en ont les cuisiniers – sauf qu'il n'avait pas l'air très propre.

Tout d'un coup, la pièce parut froide, morne, pleine de fumée froide. L'homme enjamba la fille, regarda ailleurs un instant. Pete se rendit vaguement compte que c'était une occasion, qu'il fallait qu'il se lève, qu'il bouge, qu'il sorte de là. Mais il en paraissait incapable. L'homme était d'une taille un peu au-dessus de la moyenne, et mince. Pete était bien plus lourd que lui, et il avait une longue expérience du cassage de gueules ; mais il n'était pas convaincu que ça suffise à faire la différence. Il se sentit obèse, nu, et guère en position de bouleverser l'ordre du monde.

— Vous êtes Pete Ferillo, c'est bien ça ? demanda l'homme en prenant quelque chose sur le comptoir de la cuisine.

Pete vit que c'était un ouvre-bouteilles et, quand l'autre se tourna vers lui, il sentit toute idée disparaître.

— Écoutez, dit-il, je ne sais pas ce qui se passe. Mais j'ai de l'argent sur moi. Si c'est ça, pas de problème.

— C'est pas une question d'argent, répondit l'homme d'une voix douce, presque amicale – mais son regard ne l'était pas.

— Alors ? demanda Pete. Qu'est-ce que j'ai fait ?

— Il est pas question de vous.

— Mais qui êtes-vous, bon Dieu ?

— Mon nom est... l'Homme Debout.

Il regarda Pete en attendant sa réaction, tout en manipulant l'ouvre-bouteilles d'un air absent. Puis il hocha la tête, comme si, brusquement, il avait trouvé quoi en faire. Pete ne comprenait pas ce que ça pouvait être.

Au fil de l'heure qui suivit, il le sut.

DEUXIÈME PARTIE

LA ROUTE QUI FUME

C'est ce que je compte faire,
mais je ne sais pas pourquoi.

Gerard SCHAEFFER,
tueur en série

CHAPITRE 10

Quand le type fit son apparition, Phil Banner était appuyé contre la voiture de patrouille, juste devant chez Izzy, à manger un sandwich aux champignons et aux œufs qui ne lui avait rien coûté. Ce n'était pas sa faute – il proposait toujours de payer, et Izzy disait toujours non – mais il se sentait coupable à chaque fois. Ce qui ne l'empêchait pas de l'engloutir, ni de revenir tous les matins. Le sandwich était aussi épais que bon, et pas franchement prévu pour être mangé avec les doigts, si bien que le gars ensanglanté devait sans doute être visible depuis un moment avant que Banner, levant les yeux, ne l'aperçoive. Il l'observa pendant cinq bonnes secondes, sans cesser de mâcher, ni trop savoir ce qu'il voyait, puis se décida enfin à poser sa bouffe.

Le type marchait au beau milieu de la grand-rue. Elle était vide, parce qu'il était 8 h 30 et qu'il faisait très froid, mais apparemment il n'aurait pas changé de trajet même s'il y avait eu de la circulation. Il semblait ne pas savoir où il était, il avait un sac à dos qui paraissait à la fois neuf et complètement fichu. Il titubait comme s'il sortait tout droit d'un film d'épouvante, en traînant la jambe ; Phil, qui s'était avancé prudemment, vit qu'il était couvert de sang – certes séché, mais en abondance. Il avait sur le front une grosse bosse avec une vilaine entaille, et d'innombrables coupures sur le visage et sur les mains. De la boue séchée couvrait le reste.

— Monsieur ? lança Phil en faisant quelques pas.

L'homme continua de marcher, comme s'il n'avait rien entendu. Il respirait bruyamment – la vapeur se condensait autour de son visage – mais bien en rythme, comme si c'était important pour lui. Ou alors ça ne voulait rien dire du tout. Il finit par tourner la tête lentement, sans cesser d'avancer, mais en regardant Phil. Ses yeux étaient injectés de sang, il avait une barbe de deux ou trois jours, envahie par la glace. Cela faisait longtemps que Phil n'avait pas vu quelqu'un ayant autant souffert du froid.

Le gars s'arrêta enfin, cligna des yeux, ouvrit la bouche. La referma aussitôt, contempla la route un moment. Il parut si intéressé parce qu'il voyait que Phil regarda dans la même direction, mais on n'apercevait que l'extrémité de la ville.

— Ça va, monsieur ? demanda-t-il.

Il savait que c'était une question idiote. De toute évidence, non. Mais c'est ce qu'on dit dans ces cas-là.

L'expression de l'homme parut changer, et Phil comprit que ce devait être pour sourire.

— On est bien à Sheffer ? demanda-t-il.

Il parlait avec difficulté, comme si son visage était presque gelé.

— Oui, monsieur, en effet.

Le sourire s'élargit :

— Sans blague !

— Monsieur ?

Le gars secoua la tête, comme si brusquement il reprenait ses esprits, comme si traîner les pieds avait été une habitude. Phil se dit que l'homme lui paraissait vaguement familier.

— Ça, c'est du sens de l'orientation, reprit l'autre, dont les traits s'affaissèrent. On aura beau dire.

Phil vit qu'Izzy et un ou deux clients étaient sortis du restaurant, tandis qu'un autre petit groupe se formait de l'autre côté de la rue, sur le parking du marché. Il était temps de prendre la situation en main.

— Monsieur, est-ce que vous auriez eu un accident ? L'autre le regarda :

— Bigfoot, répondit-il en hochant la tête. Puis il se laissa lentement tomber en arrière.

Deux heures plus tard, Tom Kozelek était au commissariat, enveloppé dans trois couvertures et tenant à deux mains un bol de bouillon de poule. On l'avait installé dans la pièce réservée aux interrogatoires, lors des rares occasions où la police de Sheffer devait y procéder et qui le reste du temps servait à entreposer manteaux, bottes trempées et objets divers. Elle abritait une table, trois fauteuils et une pendule. Autrefois, elle accueillait la cuisine, désormais installée à l'étage, et comportait une paroi de verre qui l'aurait fait ressembler à un bureau si la vitre n'avait été couverte de stickers célébrant les quatre derniers défilés de Halloween dans la ville.

Phil se tenait de l'autre côté avec Melissa Hoffman, qui vivait à Ellensburg, à cinquante kilomètres de là, où elle travaillait au petit hôpital du comté. Le médecin de Sheffer, le Dr Dandridge, était fort aimé, mais aussi vieux que Dieu luimême, et beaucoup moins infaillible, si bien qu'on appelait Melissa de plus en plus souvent. Elle approchait la quarantaine, était fort jolie et ne semblait pas s'en rendre compte. Elle était aussi mariée à un type trapu qui tenait une petite boutique de bouquiniste et fumait des Marlboro Light à la chaîne. Ça voulait tout dire.

— Il va bien, déclara-t-elle. Sa cheville est un peu abîmée, il s'est cogné partout, et il a souffert du froid, sans pour autant avoir d'engelures. Il n'est pas très bavard, mais d'après ce qu'il raconte il s'est fait ses bosses voilà deux ou trois jours. S'il s'était cassé quelque chose, il ne serait pas là. Il a besoin de manger, de dormir, et c'est tout. Il a quand même eu de la chance.

Phil acquiesça, en souhaitant que le shérif ait été là, et pas à cent cinquante kilomètres chez sa sœur.

— Mais le reste...

— Il va bien physiquement, répondit Melissa en haussant les épaules. Mentalement, c'est une autre histoire.

Elle se tourna vers le bureau, où le sac à dos de l'homme était posé. Il dégelait et de l'eau glacée coulait dans les fissures du plancher. Prenant un crayon, elle s'en servit pour y fouiller, en tenant le sac de l'autre main.

— Ce truc empeste l'alcool, et vous disiez qu'il avait bu avant.

Il n'avait pas fallu longtemps à Phil pour comprendre pourquoi le visage de l'homme lui semblait familier :

— La semaine dernière, un soir, il voulait entrer dans le bar de Frank, qui avait fermé. Il m'a fallu lui dire d'arrêter.

Melissa contempla l'homme à travers la vitre. Il paraissait somnolent, incapable du moindre esclandre. Il clignait lentement des yeux, comme un vieux chien qui va s'endormir.

— Il avait l'air dangereux ?

— Non, plutôt triste. Le lendemain matin, j'ai croisé Joe et Zack par hasard, et ils m'ont dit qu'un type avait passé la soirée à boire, seul dans son coin. Ça avait bien l'air d'être le même.

— Quatre jours à picoler, sans doute sans rien manger, l'estomac plein de somnifères... Pour autant, il n'a pas l'air d'un fou.

— Personne n'en a jamais l'air, répondit Phil, qui ajouta d'un ton un peu hésitant : il dit qu'il a vu Bigfoot.

— Oui, répliqua-t-elle en riant, il y a des gens comme ça ! En fait, il a vu un ours, et vous le savez.

— Je crois bien.

Melissa le regarda fixement, et Phil rougit. Puis elle sourit :

— Vous en êtes certain ?

— Bien sûr, dit-il, un peu agacé.

Ce n'était pas le moment de discuter de ce que l'oncle de Phil, autrefois, avait déclaré avoir vu dans la forêt, en haut de la crête. Personne n'avait jamais pris ça au sérieux, sauf peut-être Phil lui-même, quand il était petit. L'oncle avait cessé de raconter l'histoire. Dans la région, plus d'une ville avait des légendes de ce genre, les caravanes en bordure des routes vendaient des muffins en forme de grosse créature poilue. Pas à Sheffer, où on considérait ça comme des pièges à touristes. Sheffer était une ville paisible, distinguée, qui avait autrefois servi de cadre à une série télévisée humoristique, qui possédait un musée ferroviaire, des restaurants sympathiques où ne venaient manger que des gens sympathiques. Tout le monde ici voulait que ça continue, et le shérif plus que tout autre.

Mais quand ce Tom avait raconté son histoire, il y avait bien des gens dans la rue, et tous n'étaient pas du coin. D'ici la fin de la journée, certains pourraient bien répercuter la nouvelle auprès de leurs amis et de leurs familles. Phil savait que le chef n'aimerait pas ça non plus, et il aurait bien voulu caser le gars quelque part avant qu'il ne prononce le nom fatidique. Un peu de publicité était une bonne chose : c'est toujours bien de pouvoir dire que des stars de la télé avaient passé deux ou trois nuits en ville, lors de la dernière décennie. Ce serait nettement moins bien de voir débarquer des journalistes dépeignant les gens du cru comme autant de péquenauds. Quand Banner l'avait appelé sur son portable, le shérif avait répondu qu'il serait de retour au plus tard en début d'après-midi. Phil en était ravi.

— Je vais voir s'il veut encore un peu de la soupe d'Izzy, dit-il.

Melissa approuva de la tête, et le vit s'asseoir pour parler doucement avec le gars. Elle était persuadée qu'il faudrait examiner Kozelek pour connaître les effets secondaires des somnifères qu'il avait pris, mais il avait fermement refusé d'aller à l'hôpital, et elle ne pouvait l'y contraindre. Il avait survécu à deux jours et deux nuits dans les bois, par un froid glacial, et marché longtemps en terrain très difficile. Tout compte fait, il paraissait en assez bonne forme pour un type qui avait voulu se suicider. Ce dont il aurait fallu discuter avec lui, d'ailleurs – mais là encore, pas moyen de l'y obliger. Elle pensait qu'une fois que sa cervelle aurait dégelé, toutes ces histoires d'êtres inconnus disparaîtraient. Ensuite, ils n'auraient plus qu'à le renvoyer à L.A. ou Dieu sait où, et la vie à Sheffer reprendrait son cours habituel.

Comme elle s'apprêtait à partir, elle remarqua quelque chose au fond du sac. S'arrêtant, elle regarda de plus près. Parmi les éclats de verre et les restes détrempés des boîtes de somnifères, il y avait des machins ressemblant à des fleurs séchées.

Elle en prit un et constata qu'il n'en était rien. Cela avait plutôt l'air de petites tiges enchevêtrées. Elles étaient sans doute tombées dans le sac quand le gars avait erré dans la forêt, heurtant des buissons et des arbres.

Ou alors il les avait achetées à quelqu'un au coin d'une rue.

Ce type déclarait donc avoir vu des choses, il avait tenté de s'introduire dans un bar, son sac était plein de débris végétaux. Alors ? En partie par réflexe professionnel, mais surtout par pure et simple curiosité, Melissa glissa les tiges dans son sac et sortit pour retourner à l'hôpital où, elle le savait, il y avait peu de chances de se passer quelque chose d'intéressant.

Vers midi, sa tête commença à lui faire vraiment mal. Comme avant, d'ailleurs, même pendant une bonne part de son errance dans les bois. Mais cette fois, il y avait une différence : c'était bien pire.

Tom était toujours assis dans son fauteuil près de la fenêtre. Du moins en était-il venu à penser qu'il lui appartenait, puisqu'il y avait passé toute la matinée. Depuis son Grand Voyage, les choses étaient devenues beaucoup plus simples pour lui. Il pensait en termes élémentaires. Ce fauteuil était le sien, et gare à celui qui tenterait de le lui prendre ! De toute façon, ça n'avait pas l'air d'intéresser grand monde. Le flic passait la tête par la porte de temps en temps, mais sinon, on l'avait laissé tranquille depuis le départ de la toubib.

Son mal de tête était de qualité professionnelle : il connaissait son boulot. Il lui emboîtait le crâne comme un couvre-lit glacé, lourd et insistant, et installait des avant-postes dans d'autres parties de son corps – en particulier ses tripes. Il avait dit à la femme qu'il ne voulait pas aller à l'hôpital, en partie pour tester sa réaction. Si elle avait aboyé : « Réfléchissez, espèce de crétin, vous êtes vraiment bousillé, on va devoir vous traîner par la peau des fesses dans un endroit plein de machines, et après vous mourrez ! », il y serait allé sans broncher. Mais elle n'avait rien dit, ce qui signifiait qu'il avait des chances d'être à peu près bien. C'était d'ailleurs son impression, pour l'essentiel, sauf le mal de tête et les tripes. Il avait lu quelque part qu'il y avait autour de l'estomac un coussin de tissus nerveux qui était le plus important de tout le corps – après le cerveau, bien entendu. Ce qu'il ressentait devait simplement être un signe que tous les deux étaient à l'unisson. Et si ça se produisait tout seul, Tom aurait peut-être mieux fait d'aller à l'hôpital.

Il espérait également que les antalgiques de la doctoresse allaient agir. À cause de sa tête, il commençait à loucher. Il s'accrochait encore à l'idée qu'il allait se lever, sortir et retrouver le vieil enfoiré qui avait parlé d'ours, mais pour le moment un tel plan ne paraissait pas réaliste.

Tom sourit brusquement.

Bien entendu, il n'était plus question d'ours. Plus maintenant. S'il voulait se sentir mieux, le plus tôt possible, c'est qu'il avait quelque chose d'intéressant à raconter. De très intéressant, même. Quelque chose qui l'avait maintenu en vie, qui avait sorti son corps de la forêt. Jusqu'à présent, il s'était tu, il avait pris son temps. Mais quand le moment viendrait...

Puis, tout aussi brusquement, son sourire s'éteignit. Il avait des informations nouvelles, oui. Mais rien qui puisse faire oublier la sombre lumière de ce qui s'était passé avant. Il était toujours compromis. Une fois que vous avez fait le truc, vous l'avez fait. Même si les gens n'en savent rien.

Il contempla avec lassitude, à travers la vitre, le bureau du shérif, en la personne de Phil, que Tom avait vaguement l'impression de reconnaître (mais c'était avant son Grand Voyage). Lequel vaquait à ses occupations. Il paraissait bien jeune et plutôt frêle : en ville, les flics donnaient toujours l'impression de passer leur vie dans les gymnases, pour que leurs biceps aient fière allure avec leurs chemises à manches courtes. Parfois Phil était dans la pièce, parfois il sortait. Rien de plus. Sans doute, hormis des accidents de voiture, des gens partant sans payer dans les bars, parfois une querelle domestique par les longues soirées d'hiver, ça ne devait pas être une vie très agitée – jusqu'à ce que quelqu'un sorte des bois avec une étrange histoire à raconter.

L'autre viendrait bientôt voir comment il était, peut-être entrerait-il dans la pièce. En attendant, Tom avala encore un peu de soupe. Froide, manquant de sel, mais sinon très bonne. Grâce à elle, il se sentait mieux.

Ce qu'il voyait vira peu à peu au blanc.

La voix venait de derrière lui.
— Monsieur ?
Tom secoua la tête, sachant qu'il ne pourrait pas couper à ça. Il y avait du rouge sur lui, des crissements sous ses pieds. Il savait déjà quelle serait la nouvelle, mais pas comment elle se logerait dans sa tête.
— Monsieur ?
Puis tout fut différent. L'esprit un peu nébuleux, il leva les yeux constata qu'il était toujours assis dans son fauteuil, dans un poste de police très loin de L.A. La pièce était vivement éclairée, il était enveloppé dans des couvertures ; un petit radiateur, posé par terre à moins d'un mètre, projetait vers lui un mince filet d'air tiède. Ça, c'est nouveau, pensa-t-il. M'en souviens pas.
Pas plus d'ailleurs que de l'homme de l'autre côté de la table. Tom battit des paupières :
— Quelle heure est-il ?
— 15 heures, répondit l'autre, plus âgé que le nommé Phil – et aussi plus grand, plus large, plus imposant à tous points de vue. Il s'assit dans l'un des fauteuils.
— Qui êtes-vous ?
— Je m'appelle Connelly. Je travaille ici.
— D'accord, répondit Tom d'un ton un peu agressif, avant de bâiller à s'en décrocher la mâchoire. J'ai chaud, maintenant.
— D'après mon adjoint, le médecin a dit de vous réchauffer, c'est donc ce que nous faisons. À moins bien sûr que vous ne préfériez passer la nuit à l'hôpital. J'ai l'impression qu'il y aurait des raisons à ça.
— Je vais bien, répliqua Tom.
L'homme se pencha par-dessus la table et le regarda :
— Vous êtes sûr ?
Maintenant qu'il était un peu mieux réveillé, Tom se rendit compte que ce Connelly ne semblait pas pressé d'être son ami. Il ne le traitait pas comme quelqu'un qui, miraculeusement, a réussi à échapper à une forêt enneigée.
— Tout à fait sûr, répondit-il en prenant cette voix à laquelle il recourait toujours dans les réunions, quand il fallait convaincre un client que ce qu'on lui offrait était exactement ce qu'il voulait, par-delà des différences purement superficielles.
Cela faisait très longtemps qu'il ne l'avait pas employée, depuis au moins deux semaines, et elle était un peu rouillée.
— Je vous remercie de la peine que vous prenez pour moi.
— Bien. Alors, pourquoi ne pas me raconter votre histoire ?

— Al, il l'a déjà fait, intervint Phil, qui entrait dans la pièce avec du café.

Connelly ignora son adjoint et, se rencognant dans son fauteuil, continua à fixer Tom.

— Je m'appelle Tom Kozelek. Je suis... en vacances. Il y a trois jours, je crois, je suis parti vers les montagnes, et je me suis garé près d'une piste de randonnée, je ne me souviens pas du nom...

— Howard Point, compléta Connelly. On en a ramené votre voiture hier après-midi. Votre réapparition a permis d'élucider un petit mystère.

— Oui. Je me suis donc garé, et je suis parti à pied pour une randonnée.

— Une randonnée. Qu'aviez-vous emporté avec vous comme provisions ?

— Je crois que vous le savez, marmonna Tom d'un ton froid. Je vois d'ici mon sac sur la table.

— En effet. Je ne crois pas que vous ayez eu l'occasion de regarder la télé pendant que vous étiez ici, mais on diffuse toutes les heures une mise en garde prévenant les gens de rester à l'écart des montagnes, à moins qu'ils ne disposent de l'équipement nécessaire. Vous regardez souvent la télévision, monsieur Kozelek ?

— J'étais en pleine confusion.

— Certes. Où étiez-vous donc, ces deux derniers jours ?

— Je revenais ici. Je me suis perdu. J'avais des cartes, mais je les ai oubliées dans la voiture. Quand je suis parti, j'étais un peu ivre, et d'habitude j'ai un bon sens de l'orientation, mais il neigeait, je suis tombé dans une ravine et, pour dire la vérité, je ne savais plus du tout où j'étais. J'ai tenté de regagner la route et je suis parti en sens inverse. Puis j'ai découvert quelque chose qui ressemblait à une piste, je l'ai suivie, mais elle ne semblait aller nulle part.

— Sans doute un vieux chemin de bûcheron, intervint Phil. Voire un bout de la vieille piste de montagne. On peut tout juste dire qu'il y a eu quelque chose là, parce que les arbres y sont plus clairsemés.

Connelly le regarda : Phil se tut. Puis le shérif se tourna vers Tom.

— Quel est le problème ? demanda celui-ci.

— Moi ? Aucun. Continuez.

À dessein, Tom but très longuement une gorgée de café. L'autre commençait vraiment à le gonfler. En définitive, ils étaient tous pareils. Si imbus de leur statut à part, toujours prêts à affirmer que, de toute leur vie, ils ne s'étaient jamais retrouvés dans une situation difficile.

— Alors, j'ai marché, reprit-il. Je ne sais pas où j'étais. La nuit dernière, j'ai enfin trouvé une route. Je suis resté au bord un moment, en pensant que quelqu'un passerait et me prendrait, mais il neigeait et personne n'est venu. Je me suis mis en marche et je suis arrivé ici ce matin.

— C'est une sacrée aventure, monsieur Kozelek, dit Connelly. Vous devez être heureux qu'elle soit terminée, et impatient de rentrer chez vous.

— Pas encore, répondit Tom.

Il se débarrassa de deux des couvertures ; non seulement il avait trop chaud, mais son côté « petit garçon perdu » ne devait pas pousser le shérif à le prendre au sérieux.

— J'ai des choses à faire ici d'abord.

— Et quoi donc ?

Tom le regarda bien en face :

— Je compte retourner dans la forêt.

Il respira profondément et dit quelque chose dont, il le savait, il se souviendrait pour le restant de ses jours :

— J'ai vu quelque chose, là-bas. Quelque chose de franchement stupéfiant.

— C'était peut-être Bigfoot ?

Tom le contempla, sidéré :

— Comment le savez-vous ?

— En arrivant ici, vous l'avez dit à mon adjoint, répondit Connelly en souriant. Ainsi qu'au Dr Hoffman, je crois. À la vérité, si j'ai bien compris, c'est même le premier mot que vous ayez prononcé, juste avant de vous effondrer.

Tom rougit. Il ne se souvenait pas de ça. Merde.

— Oui, répondit-il, je le savais. Mais je l'ai vu. J'ai vu Bigfoot. Il était juste au-dessus de moi. Je l'ai vu.

— Vous avez vu un ours, monsieur Kozelek.

— Non ! Je l'ai cru sur le moment, mais ce n'était pas ça. Il n'y ressemblait pas. Et quelle est l'odeur des ours ?

— Je n'en sais rien, je ne me suis jamais trouvé assez près pour le savoir. Ils sont un peu réservés.

— Celui-là avait une odeur atroce, vraiment atroce. Et non seulement ça, mais j'ai vu des traces de pas.

— Ah bon.

— Oui, bon Dieu ! Vous affirmez que c'était un ours, très bien, mais je les ai vues !

— Ce n'étaient pas les vôtres ? Quand vous vous êtes enfui devant l'ours ?

— Non. Je m'agitais dans tous les sens, ça n'aurait pas pu avoir cette forme-là. Et surtout, on voyait ses orteils, cinq orteils énormes et tout ronds. J'ai vu ça, bon sang !

— Oui, bien sûr, répondit Connelly, qui se tourna vers Phil : Fais donc entrer Mme Anders.

Perplexe, Tom vit le jeune homme aller chercher une femme assise dans l'autre pièce. Elle avait la soixantaine passée, des cheveux gris vaguement noués en queue-de-cheval. L'une de ses mains était plongée dans la poche d'un ciré jaune porté sur un épais pull-over. L'autre tenait un grand sac plastique. Elle avait l'air embarrassée, un peu gênée.

— Voici Patricia Anders, dit Connelly. Patricia vit à quelques kilomètres de Howard Point. Je ne sais pas si vous l'avez vu sur vos cartes, mais il y a au nord de la ville, un peu plus haut, un secteur en développement qui aurait dû être le prochain gros truc. Pour le moment, Mme Anders en est la seule occupante.

— Heureux de vous rencontrer, dit Tom, mais je ne vois pas le rapport.

— C'était moi, dans la forêt, dit-elle.

Tom la regarda fixement :

— Comment ça ?

— Je suis navrée, vraiment ! Je marche beaucoup. Je prends part à des programmes nationaux de suivi de la faune et de la flore, et j'en tiens un décompte tout au long de l'année. Je ne sais d'ailleurs pas trop si c'est utile, ça n'est pas très scientifique... J'en doute, mais...

Elle haussa les épaules :

— Enfin, je le fais quand même. L'autre jour, j'y étais, très tôt, et j'ai vu quelque chose dans la ravine, qui n'est pas trop loin de chez moi. C'est une bonne trotte quand même, mais j'aime marcher. Bref, je suis descendue et j'ai vu que c'était un sac à dos. Je ne savais pas trop si quelqu'un allait revenir le chercher, alors je l'ai laissé là.

Tom se tourna vers Connelly :

— Bon, et alors ?

— Les traces de pas que vous avez vues étaient celles de Mme Anders.

— Conneries ! Vous n'écoutez donc pas ce que je dis ? Les empreintes étaient énormes...

— Une heure de soleil, et les bords fondent, si bien que les traces ont l'air plus grandes qu'elles ne l'étaient.

L'espace d'un instant, Tom crut bien qu'il allait sauter pardessus la table et prendre l'autre à la gorge. Mais c'était une mauvaise idée, et pas seulement parce que Connelly était un représentant de la loi. Il préféra donc parler d'une voix égale. Après tout, il avait un argument décisif.

— Bien. Et le soleil fera aussi en sorte que les traces aient cinq orteils, c'est ça ? Il est bizarre, le soleil, chez vous.

Il y eut un silence, puis un léger bruissement : la femme sortait quelque chose de son sac.

Tom ne put d'abord comprendre ce qu'il voyait. Puis sa nuque se mit à le démanger.

— On les achète chez Cle Elum, dit-elle. Je sais bien que c'est un peu idiot, mais c'est amusant. C'est pour ça que mon mari me les avait achetées.

Tom contempla longuement les bottes, dont le dessus était couvert de fausse fourrure, et qui s'ornaient de pieds en plastique brun, avec cinq très gros orteils.

Phil emmena la femme. Tom eut l'impression – c'était peutêtre son imagination – que le jeune gars était un peu navré pour lui. Il l'espérait, en tout cas. Les gens compatissants promettaient d'être rares dans la région.

Connelly leva les yeux pour jeter un coup d'œil à la pendule, puis fouilla dans sa poche de chemise, en sortit un paquet de cigarettes et en alluma une.

— Une journée bien surprenante ! dit-il. Plus agitée que je ne l'aurais cru en me levant. Il ne se passe pas grand-chose ici, d'habitude, comme vous l'avez sans doute remarqué. Et vous avez dû deviner que je tiens à ce que ça continue.

— Je sais ce que j'ai vu, insista Tom en secouant la tête.

— Vous n'avez rien vu du tout, monsieur Kozelek, rétorqua le shérif, dont les yeux gris étaient glacés. Vous êtes entré dans la forêt avec une très mauvaise idée en tête, et je ne parle même pas de son caractère irresponsable, alors que d'autres ont pour tâche de vous retrouver, où que vous soyez. Vous vous êtes assommé à la gnôle et aux cachets, et soit vous avez vu un ours, soit vous avez eu des hallucinations.

Tom se contenta de faire non de la tête. Connelly écrasa sa cigarette :

— Comme vous voulez. Je ne vous dirai pas de partir ce soir même, parce que vous avez passé un mauvais moment et que, quoi que vous puissiez penser, je suis quelqu'un de raisonnable. Vous êtes en bien mauvais état et vous avez besoin de vous nourrir et de dormir un peu. Alors, pourquoi ne pas le faire, et songer demain matin à partir ailleurs ? Il y a beaucoup de jolies petites villes dans les environs. Snohomish, par exemple. Ou même Seattle, il y a un aéroport.

— Je ne vais nulle part.

Connelly se leva et s'étira avec de grands craquements d'os :

— Oh, que si ! Et bientôt. Vous voulez un conseil ?

— Certainement pas.

— Estimez-vous heureux de vous en être bien tiré. Vous auriez pu être attaqué par un ours et mourir dans les montagnes. Restons-en là. Car il y a autre chose.

Il jeta un coup d'œil à travers la vitre et vit que Phil enfilait son manteau, prêt, comme il en avait reçu l'ordre, à accompagner Kozelek quelque part en ville, pour qu'il y passe une nuit, une seule.

— En revenant ici, je me suis renseigné sur vous, poursuivit-il en baissant un peu la voix.

Il tournait le dos à Tom, qui comprit tout d'un coup que, si son Grand Voyage l'avait changé, le monde extérieur ne voyait pas la différence. Il y avait dans sa vie bien des moments qu'il n'aimait pas. Mais il n'avait pas songé à faire le ménage.

Connelly se tourna vers lui :

— Je sais ce que vous avez fait.

CHAPITRE 11

Un paquet envoyé par Nina m'attendait dès l'aube. Je dis aux gars d'apporter tout le café dont ils disposaient dans ma chambre, où je remontai. Je n'étais pas très optimiste et ne pensais pas pouvoir grand-chose – la police de L.A. comme le FBI avaient mis des gens sérieux sur l'affaire – mais ça pourrait m'occuper en attendant John Zandt.

Je déposai mon matériel sur la table et m'attelai à la tâche. Dans le paquet, je trouvai un petit sac de plastique translucide, conçu pour lutter contre l'électricité statique, ce qui est le meilleur moyen de bousiller tout appareillage électronique délicat. Hormis le laisser tomber par terre, bien sûr. Dedans, un petit disque dur, accompagné d'une note de Nina :

Sois très, très prudent avec. C'est l'original. Trouves-y quelque chose, puis rends-le-moi le plus tôt possible.

Avant de commencer, je l'appelai sur son portable. Elle paraissait crispée, un peu hagarde.

— Je suis heureuse de savoir qu'il t'est arrivé, dit-elle. Mais je ne crois pas que ça mènera quelque part. La police de L.A. vient tout juste d'en retracer l'histoire. Ils ont trouvé le gars qui avait acheté le portable d'origine, un nommé Nic Golson qui magouille vaguement dans le cinéma. Mais il avait un reçu prouvant qu'il l'avait vendu à une boutique d'occases de Burbank, en juillet de l'année dernière. Il avait cru un moment qu'on allait lui confier un gros boulot de scénariste, et comme ça ne s'est pas fait il a dû fourguer la bécane. Après quoi quelqu'un l'a achetée, en payant en liquide, puis l'a désossée, a gardé le disque dur et jeté le reste quelque part où nous ne le retrouverons jamais. On interroge les employés de la boutique en ce moment même, mais le tueur me fait l'effet d'être plus malin que ça.

— Et comment ça se fait que j'aie l'original ?

— J'ai eu recours à l'une de mes ruses féminines.

— Toi ? Des ruses ?
— Si tu savais, tu serais surpris... moi aussi, d'ailleurs !
Elle reconnut avoir sollicité un technicien du labo de la police de L.A. après que je lui eus fait remarquer qu'une copie n'était jamais qu'une copie. Le gars avait accepté, d'autant plus qu'ils avaient déjà essayé tout ce qu'ils pouvaient, en particulier une prise d'empreintes. Y toucher ne poserait donc aucun problème. Mais...
Je dis que j'en prendrais soin.
Puis je raccrochai et contemplai un objet qui avait passé un certain temps dans la bouche d'une morte. Je ne savais pas ce qui était le plus pénible : penser à ça, ou songer aux risques pris par Nina.
Le café arriva, j'en bus un peu en fumant une cigarette avec les résultats habituels : les défis posés par le monde me parurent un peu plus gérables. Je sortis un câble équipé d'une prise Firewire à un bout, et d'une Oxford à l'autre. Je raccordai la seconde sur le portable de Bobby, insérant prudemment sur la première les connecteurs du disque dur, qui apparut à l'écran.
Je l'ouvris, ce qui confirma ce que Nina m'avait dit. Deux fichiers seulement : un morceau de musique au format mp3, et un texte. Elle m'avait précisé que la citation du début était de Heinrich Heine. La version du *Requiem* de Fauré, fort cotée, datait du début des années 1960 – ce qui d'ailleurs ne signifiait rien. Toute interprétation de musique classique a quelque chose d'un peu intemporel, et la plus récente n'est pas forcément la meilleure. Tout ce que je pouvais tirer de celle-là, c'est qu'elle avait été numérisée à 192 K par seconde, en stéréo : c'était donc une conversion soignée. La plupart des gens ne pourraient pas faire la différence, de toute façon, ce qui suggérait que, peut-être, c'était destiné à un système audio de qualité, qui révélerait les défauts d'un sampling moins élevé ; ou, plus simplement, la musique avait de l'importance pour celui qui l'avait déposée là. Dans un cas comme dans l'autre, ça ne menait pas loin. J'écoutai le fichier plusieurs fois tout en poursuivant ma tâche et notai qu'on percevait un certain souffle, ainsi qu'un ou deux craquements. Il semblait improbable qu'un gars familier de l'informatique ait méprisé les CD ; peut-être possédait-il une version vinyle ayant pour lui une certaine valeur sentimentale. Là encore, ce n'était pas très probant.
Je lançai un programme de scanning et attendis. Beaucoup de gens semblent croire que les ordinateurs sont de simples machi-

nes, comme les aspirateurs ou les magnétoscopes. Ils ont tort. Dès le départ, nous avons eu avec eux des rapports particuliers. On a su tout de suite qu'ils avaient des droits. Quand votre machine à laver cesse de fonctionner ou que votre télé déconne, vous la faites réparer ou vous la flanquez à la décharge publique. Elles incarnent une technologie ancienne, transparente, sans aucune magie. Mais si votre ordinateur fait de même, à qui la faute ? Vous n'en êtes jamais sûr. Vous êtes impliqué, vous vous sentez vulnérable. C'est la différence entre un crayon et une voiture. Le premier est simple, prévisible, on ne connaît qu'un moyen de s'en servir et un seul type de panne : la mine est émoussée ou cassée. Avec la seconde, surtout s'il s'agit de ce tas de tôles rouillées par lequel nous commençons tous, c'est plus compliqué. Il faut savoir l'amadouer, surtout par les petits matins d'hiver. Et ce petit bruit qui ne cesse jamais, ces pannes aléatoires qu'on finit par attribuer aux phases de la lune. Elle n'est pas bousillée pour autant, mais il convient de prendre garde, elle a ses exigences. Peu à peu vous nouez avec elle une relation ritualisée, un lien né de l'imprévisible, de la nécessité de faire face. Après tout, c'est de la même manière qu'on finit par connaître les gens : non par ce qu'ils ont en commun avec les autres, mais en apprenant à se familiariser avec leurs particularités, leurs mauvais côtés et leurs faiblesses inattendues, les choses qui les différencient.

L'ordinateur se situe un peu entre les deux : il est comme une voiture, mais multipliée par mille. C'est une copie de vousmême, une représentation multicouches de ce que vous êtes, de ce qui vous intéresse, de vos péchés. Si vous passez la soirée à surfer sur le Web en quête de femmes nues, votre piste est là, dans l'historique du navigateur et le cache du disque dur – et votre adresse IP a été enregistrée par tous les sites que vous avez visités, afin qu'ils puissent vous inonder de spams jusqu'à la fin des temps.

Même si vous vous croyez malin et nettoyez tout, en vidant la corbeille et ainsi de suite, vous n'êtes pas pour autant tiré d'affaire. Quand vous « supprimez » un fichier, l'ordinateur se contente de faire disparaître toute référence à lui. Autant détruire une fiche dans le catalogue d'une bibliothèque : le livre correspondant est toujours sur un rayonnage, et on peut le trouver en cherchant bien. Les expériences passées de l'ordinateur, désormais coupées du monde extérieur, sont encore là, sur le disque dur, un peu comme des fantômes ou des souvenirs, mêlés au présent. Nous sommes comme ça aussi.

Il fallut une demi-heure au programme pour accomplir sa tâche. Sans rien trouver, ce qui confirmait simplement ce que le technicien enjôlé par Nina avait déjà établi : le disque dur avait été très soigneusement nettoyé avant que les deux fichiers n'y soient copiés.

Le café était froid. Je lançai un programme écrit par Bobby, qui analyserait la surface du disque dur, examinant ce qui y avait été écrit, en quête d'irrégularités ou de régularités inattendues. Hormis démonter l'appareil et le manipuler avec une pince à épiler, on ne pouvait aller plus loin. Le passé résiste toujours aux intrusions, même quand il est gravé sur silicium.

Une boîte de dialogue apparut à l'écran et me dit que l'examen prendrait un peu plus de cinq heures. Comme ce n'était pas très excitant à regarder, je vérifiai que le cordon d'alimentation était bien en place et sortis me promener.

À 15 heures, Zandt appela de l'aéroport. Je lui indiquai comment trouver l'Espresso et m'y rendis pour l'attendre. Trois quarts d'heure plus tard, son taxi arriva. John en sortit et jeta un regard mauvais au type en costume devant l'hôtel, avant de traverser la rue pour me rejoindre. Il ne perdait jamais son temps en salamalecs : il disait ce qu'il avait à dire, puis se taisait ou s'en allait. Zandt était ivre. Il faut avoir passé du temps en compagnie d'un buveur – comme moi autrefois, pendant un an – pour repérer les signes extérieurs. Les cernes sous ses yeux étaient plus sombres, il s'empara de son verre dès qu'il fut posé sur la table ; mais son regard était clair, sa voix calme et posée.

— Alors, qu'est-ce que tu as sur Yakima ?

— Comme je te l'ai dit, pas grand-chose. Je suis revenu à L.A. et j'ai expliqué à Nina ce que nous avions trouvé. Elle en a fait part à qui de droit et il ne s'est rien passé. Je m'y suis intéressé parce que...

Il haussa les épaules. Je comprenais. Il n'y avait rien d'autre ou presque. Il avait pris part à l'enquête sur les meurtres du Garçon de Courses, à la suite de quoi sa fille, Karen, avait été enlevée sans que personne la revoie jamais. Son mariage avait pris fin, il avait quitté la police. C'était à mes yeux un très bon flic : il avait établi que l'Homme Debout dirigeait un réseau chargé d'alimenter les psychopathes friqués des Halls, en kidnappant des victimes à la commande. Toutefois, même si Zandt avait voulu réintégrer la police, ce qui n'était pas le cas, il y avait peu de chances que celle de L.A. veuille le reprendre. Alors, que faire ? Devenir agent de sécurité ? Homme d'affaires – mais dans quel secteur ? Il était aussi peu employable que moi.

— On pourrait entrer au FBI, suggérai-je.

— Bien sûr ! Tu t'es fait jeter de la CIA, ça impressionne toujours. Enfin... tu te souviens des lettres gravées sur la porte de la cabane ?

— Pas vraiment. J'ai vu qu'il y en avait, mais ça me paraissait faire partie du foutoir.

Il sortit de sa poche un bout de papier brillant :

— C'est une des photos que j'ai prises, imprimée en fort contraste. Tu vois mieux, comme ça ?

J'examinai le cliché de près. En étudiant les lettres avec attention, on pouvait lire le mot, ou le nom, « Croatoan ». Il était là depuis longtemps, et en partie effacé par le passage du temps.

— Ce qui signifie ?

— J'avais d'abord cru que c'était le nom d'une ancienne compagnie minière ou autre, mais je n'ai rien trouvé. La seule référence que j'aie pu dénicher est plutôt bizarre.

Il me tendit une épaisse liasse de papiers. Un texte intitulé « Roanoke », composé en polices de caractères très diverses, divisé en paragraphes.

— J'espère qu'il y a un mode d'emploi !

— Tu as entendu parler de Roanoke ? Sur la côte est ?

— Oui, vaguement. Beaucoup de gens y ont disparu, voilà longtemps, un truc comme ça.

— Ils ont disparu deux fois, expliqua Zandt. Roanoke a été, pour les Anglais, la première tentative de création d'une colonie dans le Nouveau Monde. Élisabeth Ire avait accordé des terres pas loin à Walter Raleigh, l'explorateur. En 1584, il a monté une expédition, et ils ont exploré un endroit appelé l'île de Roanoke, au large des côtes de la Caroline du Nord actuelle. Ils ont jeté un coup d'œil, pris contact avec une tribu locale, les Croatoans, et sont repartis vers l'Angleterre. Deux ans plus tard, un deuxième groupe, composé d'une centaine d'hommes, est arrivé. Ça s'est beaucoup moins bien passé. Ils n'avaient pas assez de ravitaillement, ils ont eu des problèmes avec les Indiens, qu'ils traitaient plutôt mal, et pour finir tous, sauf quinze, sont montés à bord d'un bateau qui passait et sont rentrés. Mais Raleigh tenait à créer une colonie permanente, si bien que l'année suivante un troisième groupe est parti pour veiller à ce que la « Virginie » passe aux mains des Anglais. Il a désigné un nommé John White pour le diriger. Cent dix-sept personnes étaient du voyage.

Des hommes, des femmes, des enfants, l'idée étant que des groupes familiaux favoriseraient une implantation durable. On leur avait bien dit d'éviter l'île de Roanoke... et c'est là qu'ils se sont retrouvés. Ils ont découvert les fortifications édifiées par le groupe précédent, mais aucune trace des quinze hommes censés les garder. Disparus. Envolés. White a repris contact avec les Croatoans, qui ont dit qu'une « tribu ennemie » avait attaqué le fort et tué au moins plusieurs de ses défenseurs. Ça l'a agacé, et quand on a retrouvé mort un des nouveaux colons, il a décidé de s'en prendre à la tribu des Powhatans, les mauvais garçons du coin. Mais ses gars ont bâclé le boulot et ont en fait tué des Croatoans, sans doute en vertu du vieux principe : « C'est tous les mêmes, de toute façon. »

— Sympa.

— Bien entendu, les Croatoans, ce qui se comprend, n'ont plus voulu traiter avec eux ni leur fournir de nourriture. L'été arrivait, il était trop tard pour semer, et les colons n'avaient pas beaucoup de réserves, bref ça se présentait mal.

— Ils étaient vraiment pas malins.

— Pas malins ou courageux, ou les deux. En tout cas, White a décidé de retourner en Angleterre pour se procurer du ravitaillement. Il a été convenu que, si les colons devaient s'aventurer sur le continent, à l'intérieur des terres, ils laisseraient des marques indiquant où ils étaient allés et que, s'ils quittaient l'île en raison d'une attaque des Indiens, ils graveraient une croix dans un endroit facile à repérer. Le problème, c'est que, quand White est revenu en Angleterre, le pays était en guerre avec l'Espagne, et trois ans se sont écoulés avant qu'il puisse retourner en Amérique.

J'y réfléchis un instant. Être abandonné en terre étrangère, entouré de voisins qui vous détestent, à court de nourriture... Le chef s'en va et ne revient que beaucoup plus tard...

— Et quand il est revenu ?

— Plus personne. Tout le monde avait disparu. Aucune trace de cadavres. Toutes leurs affaires étaient là. Pas de croix en vue... mais le mot « Croatoan » était gravé sur un montant de porte.

— C'est franchement flippant. Qu'est-ce qui s'était passé ?

— Personne n'en sait rien. White aurait bien voulu l'apprendre, mais le capitaine et son équipage s'en foutaient, et il a dû rentrer en Angleterre. Peu avant 1600, il a tenté de monter une nouvelle expédition, mais Raleigh et ses commanditaires avaient perdu tout intérêt pour la colonisation. Depuis, des tas de gens ont cherché à comprendre, à commencer par un certain John Smith, qui, une vingtaine d'années plus tard, était un colon de Jamestown.

— Et alors ?

— Il a discuté avec les gens du coin, il a fait des hypothèses qui sont en gros celles qu'on avance encore aujourd'hui. Il semble que le mot « Croatoan » n'ait pas désigné simplement une tribu, mais aussi une zone géographique, assez vaste et mal définie. Le terme aurait donc pu être gravé pour indiquer une destination, comme cela avait été convenu avec White. Ou bien cela aurait pu vouloir dire que les Indiens, changeant d'avis, étaient venus en aide aux colons. Ou encore qu'ils les avaient attaqués, si bien qu'ils avaient été forcés de gagner le continent. Dans un cas comme dans l'autre, cela pourrait impliquer que certains colons, ou tous – plusieurs théories veulent que tous les hommes aient été tués –, se soient peu à peu intégrés à une tribu locale. Certains groupes indiens, en particulier les Lumbees, l'affirment depuis longtemps, et leurs arguments sont parfois solides. C'est une théorie qu'on prend au sérieux depuis le milieu du XIXe siècle. On raconte que, deux cents ans plus tôt, un pasteur avait rencontré des indigènes parlant anglais, on mentionne aussi un explorateur allemand dont je n'ai pu retrouver le nom, qui affirmait avoir rencontré une « puissante nation d'hommes barbus », c'està-dire de possibles descendants des colons.

Jusqu'à présent, les lettres gravées sur la porte de la cabane ne m'avaient pas laissé grande impression – ou du moins je le croyais. Mais les paroles de John me glacèrent, comme si brusquement je me retrouvais en compagnie des morts.

Il agita le bras pour attirer l'attention du serveur, qui tenta de lui expliquer qu'il était trop occupé, puis croisa son regard et partit nous chercher une autre bière.

— La question est de savoir pourquoi ce mot était gravé sur la porte de la cabane que nous avons découverte, conclut-il.

— Une citation ? Une référence au mystère de Roanoke ? Mais qu'est-ce que ça pourrait signifier ?

— Il veut nous dire quelque chose.

— Je ne crois pas que l'endroit ait le moindre rapport avec Paul, objectai-je. Rien ne permet de le relier à ça. Et d'ailleurs, pourquoi prendrait-il cette peine ?

— Quand Sarah Becker était sa prisonnière, il passait la moitié du temps à lui tenir de grands discours. Et puis il y a ce truc que tu as dégoté sur le Net il y a trois mois, selon lequel tout le monde, sauf les Hommes de Paille, est infecté par un virus social. Il est lancé dans une mission d'information.

Il se tut quand on nous apporta nos verres.

— Le plus important chez Paul, dis-je, c'est qu'il ne croit pas être fou.

— Comme tous les autres, Ward. Aucun d'eux ne se lève le matin en pensant : « Je vais faire quelque chose de mal. »

— Je comprends bien, répondis-je, un peu irrité par le ton qu'il prenait.

— C'est comme les camés qui se shootent. Ils ne tentent pas de se tuer ou de bousiller leurs vies ; il leur faut simplement un peu d'héro, comme toi tu as besoin d'une cigarette, comme d'autres doivent vérifier trois fois que la serrure est bien verrouillée avant de s'en aller. Tout le monde a ses formules magiques, ses rituels, tout ce qui fait tourner le monde.

— Et toi, qu'est-ce que c'est, en ce moment ? La bière ?

— Va te faire foutre.

— Comment ça va entre Nina et toi ?

— C'est pas tes oignons.

— Oh, que si ! m'écriai-je, furieux. Il y a trois personnes au monde qui connaissent l'existence des Hommes de Paille. Ça fait trois mois que je rôde dans tout le pays en essayant de ne pas me faire repérer. Dans l'Idaho, j'ai cassé la gueule à un pauvre gars parce que je croyais qu'il voulait ma peau. Je dois prendre beaucoup de risques, sans beaucoup de ressources.

— Et l'argent de tes parents ?

— Disparu. Nettoyé. Ils ont mis la main dessus.

— Merde ! Je suis navré de l'apprendre.

Il contempla la rue un moment, puis ajouta :

— Les choses ont mal tourné. Je me suis installé... Tu sais qu'elle et moi on a déjà été ensemble, du temps où j'étais marié. Je pensais que ça pourrait marcher, elle aussi. Mais... elle est vraiment un peu exaltée.

— C'est vrai. Et toi, tu n'es qu'un gros ours en peluche.

— Je l'ai toujours pensé.

— Qu'est-ce que tu fabriquais en Floride ?

Il se contenta de secouer la tête, agacé.

— Bon, bon. Qu'est-ce que tu as trouvé d'autre ?

— Rien.

— C'est tout ? Et tu traverses tout le pays pour me dire ça ? C'était ça, la grande nouvelle ?

— Je n'y ai pas consacré tout mon temps, Ward, et je ne suis pas en train de te faire un rapport. J'essaie de me créer une vie nouvelle. Il n'y a pas que les Hommes de Paille au monde, d'autres choses sont tout aussi importantes. L'Homme Debout est un tueur comme tant d'autres.

— Conneries ! Il a assassiné ta fille et mes parents ! Et le résultat de ton enquête, c'est je ne sais quelle ânerie survenue voilà quatre cents ans ?

— Pour faire ce qu'il faut, on doit parfois remonter très loin.

— Ce qui signifie ?

Il haussa les épaules : il n'avait rien à ajouter.

— Et qu'est-ce que tu comptes faire ?

— Me trouver un hôtel quelque part, sans doute.

— Celui-ci n'est pas mal, dis-je en regrettant aussitôt mes paroles.

Il sourit :

— C'est un peu cher pour les gens comme moi, Ward.

— Alors, accepte un prêt.

— Un prêt ? Je croyais que tu n'avais pas les moyens.

— John, pourquoi tiens-tu tellement à jouer au con ? Il se leva et jeta dix dollars sur la table :

— Parce qu'il faudrait plus que ça pour faire quelque chose contre eux.

Il s'éloigna, remontant la rue, sans regarder derrière lui. Je le suivis des yeux jusqu'à ce qu'il ait disparu, puis rentrai à l'hôtel préparer mon départ.

CHAPITRE 12

Il était un peu plus de 18 heures. Tom se tenait au balcon du petit motel à étage en forme de L, quand la voiture s'arrêta dans le parking. En gros, il se sentait mieux, mais plus mal à d'autres points de vue. Sortir du poste de police lui avait fait du bien, comme de changer de vêtements. L'adjoint du shérif avait attendu patiemment pendant qu'il achetait un jean, un gros pull et tout ce qui se trouvait en dessous. Ce qu'il avait possédé d'autre avant son Grand Voyage se trouvait dans le coffre de sa voiture de location.

Une bonne douche brûlante, un peu de repos dans le seul fauteuil de la chambre, et il se sentait à peu près capable de se mettre en quête de nourriture. Ses fringues d'avant étaient entassées dans le sac ayant contenu les neuves. Il y avait peu de chances qu'il puisse les porter de nouveau, mais il éprouvait envers elles une sorte de respect superstitieux. Une partie de lui-même – celle qui lui avait permis de conserver tous les portefeuilles qu'il avait possédés – était prête à accorder un certain pouvoir aux choses inanimées. Sans ces vêtements, qui sait ce qui se serait passé ?

Il n'aurait sans doute pas voulu l'admettre, mais il y avait autre chose. Des témoins. Ils se trouvaient là quand ça s'était produit, ils savaient ce qu'il avait vu ou ressenti. Tout en se débattant à travers la forêt, en cherchant désespérément la civilisation, Tom n'avait eu qu'une pensée : non seulement il voulait désormais rester en vie, mais en plus il avait une raison pour ça. Il savait quelque chose, il apportait une nouvelle.

Ça ne s'était pas franchement passé comme il l'espérait.

Il croyait toujours à ce qu'il avait vu, mais il était bien le seul. Le shérif avait fait connaître sa position, son adjoint l'avait imité. Le quart d'heure qu'il avait passé dans la petite boutique de fringues en face du marché lui avait montré que les nouvelles allaient vite. Il l'avait déjà deviné quand cette Patricia était venue avec ses révélations (elle s'en était ensuite excusée pendant cinq bonnes minutes, ce qui d'ailleurs n'avait fait qu'aggraver les choses). Les gens avaient vite appris l'histoire. Et le temps qu'il tende sa carte de crédit pour payer ses achats, il avait compris que tout le monde savait, ou croyait savoir, qu'il était cinglé.

Deux ou trois soirs auparavant, il était raide bourré chez Frank. Il avait tenté de se suicider en forêt, mais pas avec une arme. Des comprimés, paraît-il. Il est tombé dans les pommes, il a cru voir quelque chose. Puis il a passé deux jours complètement perdu. Bizarre, non ?

C'était drôle, ou triste. La fille à la caisse ne dit rien de tel, mais le sourire très gentil qu'elle lui lança s'en chargeait. Au motel, le type à la réception ne l'avait pas trop regardé en face, tout en ayant un petit sourire un peu narquois. Tom avait compris le message. Il était à un doigt d'être la risée générale, et à deux de quelque chose de pire. Si Connelly racontait ce qu'il avait découvert, les gens ne souriraient plus. Et le shérif était loin de tout savoir.

Tom avait passé beaucoup de temps, dans son fauteuil, à contempler le téléphone, en se demandant s'il devait appeler chez lui. Cela faisait trois, quatre jours... Il ne pouvait se souvenir s'il avait appelé la veille de son Grand Voyage, ce qui donnait une médiocre idée de son état d'esprit. Il croyait que non, aussi refusa-t-il de céder à la tentation de prononcer des paroles mémorables. Il avait l'impression de devoir contacter Sarah, pour lui faire savoir qu'il allait bien, puis il se rendit compte qu'elle n'avait aucune raison d'en douter. Le silence radio qu'il avait observé ne serait jamais qu'une preuve supplémentaire pour abonder dans le sens « Tom est un connard ». Il aurait pourtant voulu lui apprendre la nouvelle. Il fallait qu'il en parle à quelqu'un, et l'une de ses principales découvertes, pendant son Grand Voyage, était qu'il tenait encore beaucoup à Sarah. Pour commencer, il n'avait pas à lui expliquer pourquoi il errait dans les bois, il lui suffirait de lui expliquer ce qu'il avait découvert. Le problème était que, tandis qu'il tentait de s'accrocher à ce qu'il avait ressenti dans la forêt – courir un danger qui en valait la peine –, la nouvelle paraissait peu convaincante.

Et sans elle, il n'avait aucune raison d'appeler « chez lui », rien de neuf à dire. Je l'ai vu. J'ai été vraiment très près d'un animal mythique. Il était au-dessus de moi et j'ai senti son haleine atroce. Du moins... je pense que oui... tandis que j'étais ivre, à côté de mes pompes, à moitié endormi et tout près de la mort. Ensuite j'ai vu des traces de pas. Mais peut-être que non, et à dire vrai j'entendais parfois des voix. Voilà les nouvelles. PS : Je t'aime.

Ça lui vaudrait son respect séance tenante. Elle partirait d'un bond, laissant tomber le téléphone, rien que pour être de nouveau avec lui. Mon courageux explorateur. Mon... foutu petit crétin.

Non. Ce qu'elle savait était pénible, mais moins que ce que ce qu'elle pourrait découvrir un jour. Pour qu'ils aient la moindre chance de tenir contre ça, les faits devraient dorénavant être plus solides. Il faudrait qu'elle le croie sur parole, contre les autres. Pas question de l'appeler maintenant et d'avoir l'air d'un cinglé. Ni même de lui envoyer un texto. Quand il communiquerait avec elle, il faudrait que ça marque le début d'une remontée. Mais il avait beau réfléchir sur le balcon, il n'arrivait pas à voir comment.

La voiture décrivit un cercle dans le parking et s'arrêta en plein milieu ; la portière s'ouvrit presque aussitôt, un homme en sortit, d'une taille un peu au-dessus de la moyenne, avec des cheveux bruns bien coupés et une allure de citadin.

Levant les yeux, il lui adressa un petit signe de la main :

— Vous ne seriez pas Tom Kozelek, par hasard ?

Tom fronça les sourcils :

— Oui. Qui êtes-vous ?

L'homme eut un grand sourire :

— Incroyable ! Je viens de loin pour vous rencontrer, et vous voilà !

— D'accord, d'accord, mais qui êtes-vous exactement ?

L'autre sortit de son portefeuille une carte qu'il leva en l'air. Tom était trop loin pour la lire, bien sûr, mais le logo paraissait familier.

— Je voudrais vous entendre raconter votre histoire, dit l'homme. Je peux monter, ou me laisserez-vous vous offrir une bière ?

Il était plus de 19 heures et Al Connelly était toujours assis à son bureau. Sans raison particulière. Phil était rentré chez lui, Conrad, l'autre adjoint, tuait le temps dehors, devant la porte. Le shérif aurait pu retourner à la maison, mais il n'avait pas grand-chose à y faire. Il allait pourtant s'y décider quand on frappa. Levant les yeux, il aperçut Melissa Hoffman.

— Que puis-je pour vous, docteur ?

— Eh bien... ce n'est rien, mais je... j'ai trouvé un truc, et j'ai pensé que je pourrais vous en parler.

Jetant un coup d'œil à la cafetière, il constata qu'elle était à moitié pleine :

— Un café ?

Elle acquiesça et s'assit, un peu crispée. Comme tout le monde. Il servit le café, se rassit et resta silencieux.

— Bon, commença Melissa, j'ai fait quelque chose de mal. Ce matin, après m'être occupée du gars, j'ai remarqué quelque chose dans son sac.

— Quoi donc ?

— Ceci, répondit-elle en lui tendant quelque chose qu'il prit et retourna – on aurait dit de vieilles graines. Je n'aurais sans doute pas dû le prendre.

— Sans doute. Qu'est-ce que c'est ?

— Il y en avait plusieurs dans le sac, je me suis demandé ce que c'était. Et on avait affaire à un type qui tenait des propos invraisemblables.

— La question est réglée. Il s'est révélé qu'il y avait eu confusion.

— Oh ! s'exclama-t-elle, déçue. Alors, peut-être que ça ne vous apprendra rien. J'avais simplement pensé qu'il fallait que je vérifie.

— C'était une bonne idée. Donc...

— Il se trouve que j'ai une voisine qui s'y connaît en plantes et en herbes médicinales. Je lui ai donc apporté ça.

— Liz Jenkins, peut-être ?

— Oui, dit Melissa, un peu mal à l'aise.

— Elle sait beaucoup de choses là-dessus, c'est vrai. À propos, peut-être pourriez-vous lui suggérer de se montrer plus discrète sur l'usage qu'elle fait de certaines d'entre elles ? Et à son copain aussi.

— Je n'y manquerai pas. Je sais tout ça aussi, c'est en partie pourquoi je suis allée la voir.

— Ah bon ?

Elle rougit :

— J'ai pensé qu'elle saurait reconnaître ce genre de produit que les gens veulent fumer.

Connelly sourit :

— Tandis que vous n'y connaissez rien.

Melissa songea que le shérif était quelqu'un de meilleur, et de plus subtil, qu'on ne le pensait généralement.

— En effet. Dois-je continuer ?

— Je suis tout ouïe. Elle savait ce que c'était ?

— Il s'agit de deux plantes différentes.

Elle posa sur le bureau un bout de papier pour qu'il puisse lire, ou du moins déchiffrer, l'écriture assez baroque de Liz : « En regardant de près, on voit que l'une des tiges comporte encore de minuscules fleurs séchées. Je ne les ai pas aperçues tout de suite. Ça s'appelle Scutellaria Laterifoilia, ou scutellaire ou parfois bonnet de Quaker. »

Se penchant, Melissa sortit du tas une tige que Connelly n'aurait pas distinguée des autres.

— Et ça, c'est de la valériane, Valeriana Officinalis. La scutellaire pousse dans tous les États-Unis et au sud du Canada, elle n'a rien d'une plante rare. Ce qui est intéressant, c'est que, d'après Liz, il existait au XIXe siècle un groupe appelé les Éclectiques, qui s'en servaient comme d'un tranquillisant ou d'un sédatif pour traiter l'insomnie et les maladies des nerfs.

Connelly hocha la tête, pour l'inciter à poursuivre.

— Et la valériane est mentionnée par un herboriste d'avant la guerre de Sécession, un nommé Thompson. Selon lui, les premiers colons avaient découvert que plusieurs tribus indiennes en faisaient usage ; il y voyait – Liz m'a montré la citation – le « meilleur remède à la nervosité connu », c'està-dire un tranquillisant. Aujourd'hui encore, les homéopathes s'en servent pour traiter l'anxiété, les migraines, l'insomnie, et il paraît que les tests montrent qu'elle est supérieure au Valium.

— C'est étonnant ce qu'on peut trouver en errant dans les bois. Vous êtes donc en train de me dire que tout ça fait partie de la flore locale et que c'est tombé dans le sac du gars pendant qu'il errait dans la nuit.

— Non, Al, pas du tout. Et ce pour trois raisons. La première est que ce serait vraiment une coïncidence bizarre que deux plantes médicinales se retrouvent en même temps dans son sac, d'autant plus qu'elles auraient été parfaites pour son état mental du moment. La deuxième, c'est que si vous regardez de près, à la base, on dirait bien qu'elles ont été nouées ensemble.

— Je ne vois trop rien. C'est peut-être simplement qu'elles se sont mélangées dans le sac.

— D'accord. Mais il y a autre chose : la scutellaire est une plante vivace qui meurt en hiver.

Connelly garda le silence.

— Al, ce gars-là aurait pu traîner son sac d'ici à Vancouver sans que ces trucs-là se retrouvent dedans. Ce qui veut dire qu'on les y a mis volontairement.

Il la regarda, s'empara de la cafetière, mais elle fit non de la tête. Il prit son temps pour se servir, en souhaitant être reparti chez lui un peu plus tôt.

— Je ne vois pas où ça nous mène, commenta-t-il. Ce type est passé voir un herboriste récemment. Et alors ?

— Et alors, peut-être rien. Mais je ne l'imagine pas en train de les emporter alors qu'il s'apprêtait à se suicider. Ça vous paraît logique ?

— Je crains que non.

Connelly aurait pu dire que les plantes étaient depuis longtemps dans le sac à dos, mais celui-ci était d'un modèle qu'on vendait à Sheffer.

— Alors, où cela nous mène-t-il, docteur Hoffman ? Elle rit :

— Nulle part. J'ai simplement pensé qu'il fallait vous en parler. De toute façon, nous dînons chez les Wilson ce soir, et vous étiez sur le chemin. J'ai laissé Jeff chez Frank, et si je ne veux pas me brouiller avec nos hôtes, il va falloir que je le sorte de là avant qu'il ait trop bu.

Connelly la raccompagna, puis la regarda traverser la route en direction des néons de chez Frank en prenant bien soin de ne pas abîmer ses escarpins. C'était un bon médecin, et le shérif ne s'inquiétait guère si parfois Liz Jenkins et elle passaient une soirée un peu enfumée. Il avait fait de même en son temps. Sans doute cesserait-elle vite de penser à ces histoires de plantes qui ne menaient nulle part.

Pourtant, il revint dans son bureau, s'assit et réfléchit un moment.

Tom et le journaliste entamaient leur deuxième bière chez Frank quand le médecin entra chercher un type qui devait être son mari et discutait avec le barman, de l'autre côté de la pièce. Elle lui fit calmement mais fermement abandonner son verre sur le comptoir, puis l'emmena. Tom, qui les suivait des yeux tandis qu'ils traversaient le parking, la vit rire d'une chose que l'homme avait dite. Lui-même avait su faire rire les femmes, parfois. Ça lui manquait.

— Vous la connaissez ? demanda le journaliste.

— C'est le médecin du coin. La police lui a demandé de s'occuper de moi.

— Mignonne.

— Je le pense aussi. Mais elle est prise.

— Comme tout le monde de nos jours, Tom. Vous aussi, d'ailleurs, à en juger par votre alliance. Je pourrais en savoir plus ? Comment se fait-il que vous soyez venu ici seul ?

— Il y avait des problèmes à la maison. J'étais ici pour y voir un peu plus clair.

— D'accord. C'est suffisant pour le moment.

Tom se demanda combien de temps il faudrait à ce type pour vouloir en apprendre davantage, et comment lui-même pourrait ne pas lui répondre. Posant sa bière, il le regarda. Son costume et sa chemise bien repassée suffisaient à montrer que c'était un gars de la ville, qui peut-être n'était pas aussi malin qu'il le croyait. Il souriait en permanence, ce qui devrait être fort utile pour amener les gens à se confier. L'homme s'appelait Jim Henrickson et travaillait pour *Front Page,* dont Kozelek avait reconnu de loin le logo rouge et blanc sur sa carte. Mode, potins, célébrités – ainsi que La cachette d'Hitler en Antarctique, Des extra-terrestres ont volé ma carte de Sécu, Miss Idaho donne naissance à un enfant-poisson. Sans oublier Une rencontre avec Bigfoot.

À dire vrai, les manchettes du magazine n'étaient jamais aussi tape-à-l'œil, et ses rédacteurs se donnaient du mal pour avoir l'air de véritables journalistes. Même si certains de leurs articles erraient aux frontières du bizarre, ils étaient rédigés de manière très sobre, sur un ton impartial. *Front Page* était assez chicos. L'industrie du spectacle le prenait au sérieux, ses collaborateurs étaient invités à toutes les premières de mode et de cinéma. Si Henrickson avait travaillé pour une autre revue, Tom serait ailleurs à l'heure qu'il était, sans doute en train de manger. Mais il fallait bien annoncer la nouvelle quelque part, pour commencer, et depuis une demi-heure il en était venu à penser que peut-être il avait un public pour ça.

— Vous me croyez ? dit-il.

— En effet.

Tom se sentait épuisé, proche des larmes. L'autre s'en aperçut et lui tapa sur l'épaule :

— Tout ira bien, mon vieux.

— Pourquoi me croyez-vous ? Vous êtes bien le seul.

— Ma principale raison, c'est que vous n'avez pas l'air d'un menteur. Les âneries qu'on me raconte sont le plus souvent des bobards, pas des bourdes. De surcroît, ce n'est pas la première fois que je traîne par ici à cause d'une histoire de ce genre. Il y a neuf mois, à quatre-vingts kilomètres au nord-est de Sheffer, près de Mazama, trois chasseurs ont rapporté un incident similaire. Une apparition dans leur camp la nuit, une odeur très forte, ainsi que des bruits étranges, une sorte de gémissement paisible. Vous avez entendu ça ?

— Non. Je... je dormais à poings fermés et me suis éveillé.

— D'accord. Enfin, ça leur a fichu une trouille bleue. C'étaient trois vieux briscards qui connaissaient la forêt depuis leur enfance. Et voilà qu'ils arrivent en courant, morts de frousse.

— Je ne me souviens pas d'avoir entendu parler de ça.

— Vous nous lisez chaque semaine ?

— Non. D'ordinaire, c'est dans les salles d'attente. Désolé.

— Ne vous excusez pas, Tom. C'est important, les salles d'attente ! Grâce à nous, des hommes et des femmes oublient un peu leur angoisse avant d'affronter le dentiste. Enfin, vous n'avez pas entendu parler de ça parce que nous n'avons pas publié l'histoire. Les affirmations de trois barbus en chemises à carreaux pourraient suffire à nos concurrents, mais notre lectorat est un peu plus raffiné que ça. Notre point de vue est de couvrir aussi le bizarre, sans pourtant en rendre compte, sauf si nous pensons avoir affaire à quelque chose d'important.

— « La cachette d'Hitler en Antarctique » ?

L'autre rit et leva les bras au ciel :

— Que vous dire ? C'était une formation rocheuse d'allure étrange, et rien de plus. En ce qui me concerne, je n'aurais pas publié ça, mais je ne suis qu'un homme du rang. En tout cas, ça a fait un énorme tirage ! Hitler est le méchant par excellence, il nous manque depuis qu'il n'est plus là. Bref, je dis que si on doit trouver Bigfoot quelque part, c'est ici, dans le Nord-Ouest. Il y a des centaines de témoignages là-dessus, ça remonte à un nommé Elekah Walker, au début du XIXe siècle. Et beaucoup d'autres choses. Dans toute cette partie des États-Unis, on trouve des gravures rupestres d'allure assez simiesques, bien qu'il n'y ait pas eu de primates locaux – du moins c'est ce qu'on nous dit.

— Est-ce qu'il n'y a pas eu aussi des films ?

Henrickson secoua la tête :

— Le film Patterson ! Qui s'est révélé être un faux voilà peu. C'est le gros problème. Tout est faux, ou pourrait l'être, rien ne permet d'affirmer le contraire. Beaucoup de gens ne veulent pas que la vérité soit connue. Vous leur donnez le moindre indice et ils vous envoient promener. Vous voulez savoir la vérité ? Les théories de la conspiration sont bidon.

Tom hocha la tête :

— En effet. Lesquelles ?

— Toutes ! Elles ont été inventées par les autorités pour dissimuler ce qui se passe vraiment.

— Elle est bonne, celle-là ! dit Tom en éclatant de rire.

— Je ne plaisante pas. C'est la seule vraie théorie, parce que c'est moi qui l'ai trouvée. Plus une théorie paraît bizarre, plus elle a de chances d'être vraie. Elle n'a l'air tordue que dans le contexte des mensonges qu'on nous apprend à gober.

— Je ne vous suis pas.

— Les autorités contrôlent toutes les informations... donc elles doivent avoir aussi inventé les théories. Elles les diffusent parce qu'il serait encore plus grave que nous sachions la vérité. Vous connaissez celle qui veut que jamais nous ne soyons allés sur la Lune ?

— J'ai vu une émission télé là-dessus. Et il y a eu un film...

— En effet. Mais le fait est que cette idée est elle-même une théorie bidon, créée pour détourner l'attention de la vérité : la Lune n'existe pas.

— Pardon ?

— La Lune n'existe pas. Pas plus que les planètes et les étoiles. Tout le monde se demande si on y est allé ou pas, si bien qu'on passe à côté de la vérité vraie. Galilée se shootait ! Ça se réduit à ça : il y a cette boule rocheuse et rien d'autre. Vous connaissez l'histoire : l'État sait qu'il y a des extraterrestres et veut le cacher. Mais il n'y en a pas, parce qu'il n'y a pas d'univers. On a inventé ça parce qu'il devenait évident qu'il nous fallait un horizon nouveau, faute de quoi nous en viendrions à nous entre-tuer. Qui sera le premier sur la Lune, nous ou ces salauds de Rouges ? On y va, mais aussitôt ça nous ennuie, et plus personne ne s'y intéresse. N'est-ce pas un peu bizarre ? On y va grâce à une technologie qui a aujourd'hui quarante ans, et on laisse tomber maintenant qu'on est capable de faire tenir un ordinateur sur une tête d'épingle, alors qu'on avait réussi un alunissage presque en s'éclairant à la bougie ?

— Mais il y a les navettes spatiales.

— En effet. Et elles explosent de temps à autre. C'est pour ça qu'on n'est pas encore allés sur Mars, mesdames et messieurs : l'espace est dangereux ! Conneries, et c'est bien pourquoi on a besoin des petits hommes verts. On ne va pas dans l'espace, mais il vient à nous, donc il doit exister. Et il n'y a pas que ce genre de salades. Dites-moi : qui a tué Kennedy ?

— Je n'en sais rien. Mon impression est que c'est un vrai mystère.

— En effet ? Et pourquoi ? questionna Henrickson

— Je pense que vous allez me le dire.

— Pour dissimuler le fait que Kennedy n'est pas mort.

— Ah bon ?

— Bien sûr que non. C'est en fait une assez jolie histoire : il a été contraint de s'en aller par les gens que lui et sa famille avaient gonflés – la pègre, les Cubains, la CIA. C'était du genre : « Dégage ou on te bousille. » Il a donc conclu un marché aux termes duquel lui et la femme qu'il aimait (Marilyn, évidemment) pourraient disparaître. Leurs morts étaient bidon, aujourd'hui ils vivent ensemble en Écosse. Ils ont monté un élevage de moutons alpaga, l'un des meilleurs d'Europe, je crois. Rien de folichon, mais ils s'en sortent. C'est pourquoi les autres Kennedy continuent à avoir autant d'emmerdements. Certains d'entre eux sont au courant pour JFK. Ils sont censés la fermer, sinon toute l'histoire sera révélée et les gens penseront : « Merde, et si le reste aussi était faux ? » Dès que le moindre signe montre qu'un Kennedy pourrait l'ouvrir, pan ! Terminé. Discrédité, mort, ou les deux. Une rumeur veut que la princesse Diana ait eu vent de la chose, faut-il en dire davantage ?

— Vous ne croyez quand même pas à tout ça ?

— Non. Mais c'est la première chose qu'on apprend dans mon boulot : la vérité n'a pas d'importance ; ce qui compte, c'est ce que les gens croient.

Il y eut un petit bruit, et Tom vit qu'on venait de lui servir une nouvelle bière qu'il ne se souvenait pas d'avoir commandée. Sans doute un autre talent utile quand on faisait un boulot comme celui d'Henrickson.

— Jim, dit-il, ce n'est pas la peine de me saouler.

— Tom ! Bon Dieu ! Et vous me croyez paranoïaque ! Faites-moi confiance. Je suis partant pour quelques bibines, vous me tiendrez compagnie. Vous êtes dans le système, désormais, ce qui signifie que vous ne vous ferez pas avoir. J'espère que nous avons une histoire et que vous allez palper un max. Mais je veux votre parole, ici même, que vous ne parlerez de ça qu'à moi.

— Sûr, répondit Tom, sachant que personne d'autre ne l'écouterait.

— Excellent. Il ne nous reste donc qu'une chose à trouver.

— Une preuve quelconque.

— Pas de celles qu'on présente devant un tribunal, bien sûr. Sinon, on laisserait tomber *Front Page,* on irait voir CNN, le *New York Times* et la BBC. Mais il nous faut quelque chose. Vous me donnez une description qui ressemble assez à celle des chasseurs, mais après tout vous pourriez avoir trouvé ça ailleurs.

— Mais je ne savais pas que...

— Je vous crois, coupa Henrickson. Mais pas les autres. Vous avez aussi des traces de pas, mais elles auront disparu. Sans compter l'histoire de cette Mme Anders avec ses bottes à la noix.

— C'est tout ce que j'ai.

— Non. Pas d'après ce que vous dites. Il se pourrait que vous ayez quelque chose dont vous ne vous rendez pas compte. Nous y irons y voir de plus près demain.

Tom parut perplexe. Le journaliste cligna de l'œil :

— Faites-moi confiance.

Connelly quitta le poste de police. Une brève conversation avec Patricia Anders avait permis d'élucider la découverte de Melissa : c'est elle qui avait déposé les herbes dans le sac. Tout était donc bien en ordre. Il songea à se rendre chez Frank, mais la journée avait été longue ; une bonne bière chez lui, devant la télé, conviendrait tout aussi bien. Sa maison était grande et vide, mais tranquille, et le téléphone ne sonnerait pas.

Ça se présentait bien.

CHAPITRE 13

Dix minutes après sa conversation téléphonique avec le shérif, Patricia se tenait encore dans sa petite cuisine, ou plutôt dans le coin d'un mètre cinquante sur deux qui en tenait lieu dans le séjour. Une fenêtre donnait sur les arbres. Elle regardait dehors, sans d'ailleurs rien voir.

Bill et Patricia Anders avaient passé presque toute leur vie à Portland. Dans les années quatre-vingt, quand les enfants quittèrent la maison, ils tentèrent, non sans hésitation, de se souvenir comment occuper leur temps libre, comme s'ils étaient les gardiens d'un zoo abandonné dont les animaux auraient été relâchés dans la jungle. Ils commencèrent, de manière un peu futile, par passer des week-ends hors de la ville, mais ce n'est qu'avec Verona qu'ils découvrirent de nouveaux horizons.

Ce n'était guère qu'un point sur la route 101 qui longe le Pacifique. La ville comptait quelques rues, des maisons en bois, une épicerie, pas grand-chose d'autre ; on avait toutes les chances de la traverser sans songer à s'arrêter. Mais quand on descendait paresseusement vers le sud en gardant les yeux ouverts, on tombait, juste après le pont qui surplombait la crique, sur une pancarte gravée sur un vieux tronc et pointant vers les arbres, qui signalait les « Pavillons du Séquoia ». Patricia la vit, ils ralentirent pour jeter un coup d'œil, et ce hasard changea le reste de leur existence.

Les Pavillons appartenaient à une époque en voie de disparition : c'était un lieu de séj our à l'ancienne marquant la fin d'une matinée en voiture et le début d'un après-midi consacré à nager, à pousser des grand cris et à pagayer sur la mer, avant de rentrer avec du sable et des aiguilles de pin plein les chaussures. Maman était heureuse parce que l'endroit était agréable et disposait d'un lieu où laver le linge, papa soulagé de voir qu'ils n'avaient pas dépassé leur budget ; ce que les enfants savaient, même vaguement, tout en savourant la chaleur d'une réunion familiale unie, pour une fois, par des joies simples. Quatorze bungalows étaient dispersés sur un peu plus d'un hectare boisé, bordé par une côte rocheuse d'un côté, par la crique de l'autre. Lors de leur première visite, Bill tint à faire un croquis de la disposition des lieux (ils occupaient le bungalow n° 2), tant il était fasciné : séjour, kitchenette, chambre à coucher, salle d'eau, rangements occupaient chaque centimètre carré d'une cabane en rondin. Un fourneau à bois installé dans le séjour en faisait un endroit idéal pour les fraîches soirées de printemps, la chambre était douillette par les nuits d'hiver. En été et en automne, on vivait sous la véranda, écoutant les oiseaux et le lointain murmure des vagues, un livre sur les genoux, histoire d'avoir un prétexte pour ne rien faire, et surtout ne pas lire, tout en songeant à ce qu'on mangerait au dîner.

Le premier soir, ils traversèrent le pont pour se rendre dans la petite ville. Au bord de la baie, se dressaient un bar monté sur pilotis, avec des tables de billard et une musique bruyante qui leur était familière et, un peu plus haut sur la colline, un restaurant qui valait largement ceux de Portland. Ils burent du vin et de la bière de fabrication locale, et furent enchantés, ce qui ne leur était pas arrivé depuis longtemps. À leur âge, c'était plutôt difficile. Verona le leur permit. Bill et Patricia se retrouvèrent sur la plage, à se tenir la main, à sourire aux autres promeneurs et à regarder la mer. Ils prirent les mêmes apéritifs trois soirs de suite. Les Willard, le vieux couple qui gérait les pavillons, les appelèrent par leurs prénoms dès le lendemain de leur arrivée. Quand vint le moment du départ, Patricia dut être emmenée quasiment de force ; elle extorqua à son mari la promesse qu'ils reviendraient dès qu'ils pourraient.

Tout fut décidé à ce moment-là. C'est là qu'ils séjourneraient quand il leur faudrait s'éloigner du monde.

Dix ans passèrent, avec plus d'une vingtaine de visites. Les Willard prirent leur retraite en 1994, mais rien ne changea. Patricia et Bill revenaient, comme ces oiseaux de mer qu'on voit deux fois par an chevaucher les vagues. Ils faillirent amener leurs enfants, mais cela ne put se faire, ce qui n'avait rien d'étonnant. Une fois, comme ils parlaient de Josh et de Nicole, Bill avait dit qu'ils avaient avec eux une relation « cordiale » : tout le monde aimait tout le monde, certes, mais en gardant la tête froide, et personne ne débordait d'affection. Coups de fil réguliers, visites amicales, réunions lors des principales fêtes : on échangeait des cadeaux choisis avec soin, chacun se rendait utile à la cuisine. Leurs enfants travaillaient dur et, si leurs carrières passaient avant leurs visites, on n'y pouvait pas grand-chose. Cela n'empêchait pas Bill et Patricia de se rendre à Verona. Il était plaisant d'avoir l'endroit pour eux seuls, de ne pas s'inquiéter de savoir si d'autres le trouvaient aussi agréable qu'eux. Ils s'abstinrent de suggérer un nouveau voyage en famille.

Puis, un jour, lors d'un week-end à la fin du mois d'août, ils eurent l'occasion de discuter avec les nouveaux propriétaires. Contrairement aux Willard, Ralph et Becca semblaient les oublier après chaque séjour, et il fallait chaque fois renouer des relations amicales qui demeuraient un peu lointaines. Les Anders comprirent vite qu'il se préparait quelque chose. Ils posèrent la question, et Ralph confirma, apparemment sans grand regret, que c'était le dernier été des Pavillons.

Patricia en eut le cœur brisé. C'est à peine si elle entendit les explications : l'affaire ne rapportait pas assez, bien que la ville fût désormais plus courue. Cannon Beach, Florence et Yachats devenaient beaucoup trop chers ; les gens cherchaient ailleurs le long de la côte. Mais les jeunes n'aimaient guère les installations rustiques, ils voulaient des lecteurs de DVD et des jus de fruits bio. L'endroit était magnifiquement situé, un hôtel serait une mine d'or pour quiconque connaissant un peu le métier. En fait, un investisseur de San Francisco avait fait à Ralph et Becca une offre qu'ils ne pouvaient refuser.

Le soir, avant le dîner, les Anders se retrouvèrent à la terrasse du bar, lui buvant une bière, elle un Manhattan. Cela faisait longtemps que Patricia n'avait eu un tel cafard. Pourquoi la vie devait-elle être ainsi ? On aurait dit que chaque année le monde acceptait de plus en plus de choses qui n'avaient aucun sens pour elle, des innovations déroutantes que pourtant on saluait comme l'aube d'une ère nouvelle. Elle s'efforçait de comprendre l'intérêt des téléphones portables, de Windows ou d'Eminem, mais ce qui avait compté pour elle autrefois devait disparaître en même temps... Bill restait silencieux, avec ce regard qu'il prenait quand il ne voulait pas penser à quelque chose. Il se montra un peu renfermé durant le dîner et s'abstint de consulter la carte des vins, ce dont pourtant il avait pris l'habitude depuis qu'il avait plus ou moins renoncé à la bière. Il se posait la même question que sa femme : reviendraient-ils à Verona ?

Les Pavillons céderaient la place à un de ces hôtels comme on en trouve à foison dans les guides destinés aux couples d'un certain âge. Et eux ? Ils pourraient essayer le nouveau bâtiment une fois qu'il serait construit, mais ce serait se montrer infidèles au passé. Patricia ne pourrait se lever le matin et prendre le petit déjeuner sur un balcon surplombant un parking où autrefois se dressait leur bungalow.

Alors, que faire ? Trouver autre chose ? Elle n'y tenait pas : pas question de recommencer de zéro. Ils connaissaient chaque kilomètre de la route, s'arrêtaient aux mêmes endroits pour déjeuner. Ils allaient perdre tout cela, avec d'innombrables rituels, trop infimes pour avoir un nom, jusqu'au surnom qu'ils donnaient au vieux couple gay qu'ils saluaient sur la plage (les « deux gentilshommes de Vérone »). Bien entendu, Verona n'était pas le paradis sur terre – l'épicerie n'avait pas un très grand choix, ils achetaient tout à Cannon Beach –, mais il ne suffit pas de chercher un refuge pour en dénicher un.

Après dîner, comme ils marchaient le long de la route en se tenant la main, Bill la surprit en lui proposant un dernier verre. C'est ce qu'ils avaient fait au début : il fumait une cigarette, et ils regardaient passer les gens, puis peu à peu ils s'étaient rendu compte qu'ils se sentaient bien après avoir mangé, et ils se contentaient de rentrer tranquillement au bungalow.

Patricia sourit et accepta avec reconnaissance. Il ne parlait pas beaucoup (ce dont, au fil des années, elle s'était exaspérée plus d'une fois), mais il comprenait. Elle s'assit tandis qu'il allait chercher leurs consommations, et aperçut, comme d'habitude, des lumières dans certains des bungalows, de l'autre côté de la crique. Pour elle, c'étaient des étoiles permettant de naviguer à travers l'existence. Elle comprit que, la prochaine fois, elles se seraient éteintes, que c'était leur dernière visite. Quand Bill revint, elle était au bord des larmes.

— Je sais, dit-il en s'asseyant.

Il posa la main sur la sienne, regarda les lumières un moment, puis lui proposa de trinquer. Elle haussa les épaules : ce n'était pas le moment ; il n'y avait rien à fêter.

Il insista. Plus bizarre encore, il avait allumé une cigarette, alors qu'il ne fumait pratiquement plus. Patricia sentit que le regard lointain de son mari ne signifiait pas ce qu'elle avait cru. Elle leva son verre.

— J'ai une idée, dit Bill.

Près de la fenêtre, à regarder la forêt, Patricia se souvenait encore de cette soirée, avec une clarté dont sa vie avait été dépourvue depuis lors. La dernière grande décision, la dernière chose qui ait ressemblé à un pas en avant.

— Nous avons parlé d'acheter de la terre, dit Bill. Bon marché, avec des arbres.

C'était vrai – ou du moins il y avait pensé. Elle avait écouté, acquiescé, vaguement favorable, sans pour autant y croire. Ils avaient Verona, pourquoi chercher ailleurs ?

Sauf que... plus maintenant.

— Oui, répondit-elle.

— Nous pourrions nous y mettre.

— Mais nous n'avons pas assez...

— D'argent ? Si. Pour la terre.

— Mais pas pour construire une maison.

— En effet. Et si, demain matin, j'allais voir Ralph pour lui proposer d'acheter un de ses bungalows ?

Elle le regarda fixement.

— Le numéro 2, ajouta-t-il, et elle en eut les yeux humides. On s'entend avec Ralph, l'investisseur n'en voudra pas, ça le gênerait. Comme ça, il n'y aura pas à le détruire, et on le déplacera où on voudra.

— Tu peux vraiment ?

Ils en parlèrent pendant près d'une heure, jusqu'à en délirer, le regard fou. Et le lendemain matin, Bill tint sa promesse.

Ralph donna un coup de fil : une demi-heure plus tard, le marché était conclu. Pour autant, Bill restait songeur : en fin d'après-midi, ils étaient propriétaires non plus d'un seul bungalow, mais de trois. Il expliqua à Patricia qu'il en garderait un pour eux ; un autre servirait de bureau, le dernier étant destiné aux invités – peut-être leurs petits-enfants. Patricia y prit à peine garde. L'important était que le bungalow numéro 2 soit sauvé. Elle aurait souhaité qu'ils puissent rester à Verona, mais c'était impossible.

Une fois propriétaires, acheter un terrain – idée jusque-là assez vague – devint une nécessité. Ils passèrent plusieurs week-ends à chercher, et se décidèrent finalement pour un endroit situé juste au nord de Sheffer, sur le flanc est des montagnes. C'était à un après-midi de route de Portland, la ville avait un charme authentique, le terrain demeurait relativement bon marché. Des promoteurs avaient proposé des lotissements sur les routes menant à Sheffer, sans succès jusqu'à présent ; certaines pancartes

« À vendre » commençaient à s'écailler. Ils achetèrent une vingtaine d'hectares au sortir de la route d'accès, avec des tonnes d'arbres et même un petit lac glacé. Au-delà, c'était une réserve naturelle, personne n'y pourrait rien changer. Le bungalow numéro 2 aurait enfin un foyer permanent.

Ils firent déplacer les trois, en les suivant en voiture depuis la côte. L'un d'eux dut être presque entièrement reconstruit sur un côté, dépense à laquelle ils ne s'attendaient pas ; mais quand les bungalows furent en place, Patricia les contempla, le visage ruisselant de larmes. Elle ne se tourna pas vers Bill, sachant qu'il n'aimait pas qu'elle le voie pleurer.

Le bungalow numéro 2 fut installé près du lac, le « bureau » un peu plus loin, et le troisième de l'autre côté. Bill et Patricia surent que c'était là qu'ils vivraient désormais. Ils vendirent leur maison de Portland, se débarrassèrent de presque toutes leurs affaires et se mirent au travail. Il aménagea les deux bungalows supplémentaires, elle transforma un peu le décor avant les chutes de neige, puis s'assit près du feu avec des catalogues de plantes et de graines, en prévision du printemps. Ils passèrent Noël à Sheffer, pour se familiariser avec la ville. Les enfants téléphonèrent ce jour-là, ce qui était gentil de leur part.

Le 1er janvier 2001, Bill emmena Patricia hors du bungalow pour lui montrer le banc qu'il avait installé autour du plus gros arbre, à côté du lac ; il avait manipulé seul de lourds blocs de bois, en secret. Ils s'y assirent en frissonnant, avec une Thermos de vin chaud ; elle se serra dans ses bras et se dit que jamais elle n'avait été aussi heureuse.

En mars, il s'avéra que Bill avait un cancer du poumon. Quand il mourut, quatre mois plus tard, Patricia aurait pu le soulever d'une seule main.

CHAPITRE 14

Quelque part sur un écran, prisonnière du passé, une jeune femme pleure, assise sur son divan. Le divan défoncé est couvert d'une sorte de daim rouille sombre. Derrière, le mur blanc est décoré d'un miroir et d'un grand tableau représentant des tulipes, pas trop vilain. En bonne forme, la femme est bronzée, à l'exception de pâles triangles au niveau des seins ; presque nue, elle ne porte qu'une petite culotte blanche. De la main droite, elle tient une cigarette ; elle passe la gauche dans ses longs cheveux bruns. Elle a le visage mouillé et ratatiné ; les yeux sont ouverts, mais tournés vers l'intérieur. En face d'elle, sur la table basse, se trouvent un grand cendrier de verre, deux télécommandes et une tasse de café à moitié vide. Il est tôt, en ce dimanche matin, et elle semble en proie à une sacrée gueule de bois.

Elle fume sa cigarette, écrase le mégot. Les images sautent, car bien que tu sois abonné à ce site depuis trois mois, ton logiciel, un shareware joyeusement nommé CamFun à douze dollars quatre-vingt-quinze, est paramétré pour ne rafraîchir l'image que toutes les deux minutes. La plupart des gens se connectent à cette page avec un simple navigateur, comme Internet Explorer. Toi, tu utilises CamFun, qui te permet d'enregistrer les images plus facilement et de les sauvegarder sur le disque dur sous forme de fichier vidéo que tu peux regarder à loisir – d'ailleurs, en ce moment, tu repasses un film qui date de plusieurs semaines. Le site n'a pas été modifié depuis quelques jours, ce qui est étrange. Tu as aussi adopté ce logiciel, parce qu'il t'offre la possibilité de choisir la fréquence de rafraîchissement de l'image. Tu peux profiter de l'avantage réservé aux abonnés et la rafraîchir toutes les quinze secondes, ou toutes les minutes ou deux. Cela peut sembler regrettable quand on paie dix-neuf dollars quatre-vingt-dix-neuf par mois pour le haut débit, censé rendre l'expérience plus réelle. Mais pour toi, c'est l'effet inverse.

Le moment pendant lequel tu attends la nouvelle image charge la scène de désir. Deux minutes, c'est suffisant pour qu'une personne passe d'une extrémité du divan à l'autre, comme par magie, ou pour qu'elle ait allumé une nouvelle cigarette, déjà à moitié fumée, en une fraction de seconde. C'est assez pour qu'une femme disparaisse, pour qu'elle aille du salon à la cuisine. De la cuisine au salon, à nouveau. La voilà... clic, elle est partie. Où ? Hors de vue, hors de portée de radar, hors de la planète, et pourtant, sans doute toujours dans l'appartement. Clic... la revoilà. Deux minutes, c'est réel. En deux minutes, il peut se produire des choses.

La semi-nudité de la femme est presque immatérielle. Bien sûr, tu aimes voir son corps, de temps en temps. Mais tu n'as rien d'un pervers, car en fin de compte, c'est elle que tu observes, et non ses seins – ce qui est tout aussi bien, car elle ne les dévoile pas souvent. Cette femme qui a choisi d'organiser sa vie ainsi, de placer dans son appartement une fenêtre à travers laquelle des hommes, dans le monde entier, inondés par une lumière cathodique ou auréolés de la pâleur de leur écran plat, sur leur fauteuil ou leur lit, peuvent l'épier... cette femme qui joue de temps en temps de la guitare acoustique, jamais très longtemps, qui s'enfile sa demi-bouteille de Jack Daniels le soir ; qui a parfois des relations sexuelles indifférentes sur ce divan, rencontres auxquelles tu ne t'intéresses guère, voire pas du tout, bien que tu en aies enregistré certaines et que, lors de ces occasions, tu accélères la vitesse... Elle ne joue pas devant la caméra, et tu la soupçonnes à moitié d'oublier sa présence.

Cette femme qui, tu ne sais pourquoi, pleurait toute seule sur son divan un dimanche matin, voilà un mois... Ce film te fascine pour des raisons qui t'échappent un peu. Elle disparaît dans l'invisibilité pendant encore deux minutes puis revient sur le divan. Entre-temps, elle a allumé une autre cigarette et revêtu un peignoir en tissu-éponge bleu. Elle a repoussé en arrière ses cheveux, qui retombent derrière ses oreilles. Elle ne pleure plus, mais elle a les traits tirés et semble morose. Elle regarde sur le côté, par la fenêtre, crois-tu, bien que tu n'aies jamais vu ce mur de l'appartement. Deux minutes plus tard, elle a les pieds posés sur la table basse, elle considère ses genoux ; sa cigarette est presque terminée.

Combien de pensées ont traversé son esprit pendant ce temps ? Quelles pensées ? Tu ne pourrais le dire. Quelque part, entre elle et toi, cette information s'est perdue, extraite de la réalité par le processus de numérisation, le transfert, le stockage, et le nouveau transfert pour la projection en rouge, vert et bleu. Il paraît évident que cette perte s'est produite quelque part dans le circuit, mais ce n'est pas sûr : ce n'est peut-être qu'à la dernière seconde, lorsque l'information a tenté de franchir le gouffre qui sépare l'écran d'un autre esprit humain, que tout s'est perdu. Toutes les différences du monde ne sont rien comparées à cela : la différence entre toi, qui es toi, et moi qui suis moi. À côté, le hiatus qui sépare les dieux des hommes, les hommes des femmes, les morts des vivants semble presque trivial.

Tu regardes, tu spécules et tu réfléchis. Tout cela sans connaître les réponses, sans avoir à te soucier de la vérité de ce monde. C'est peut-être quelque chose de banal ou d'ennuyeux, un ongle cassé, un pare-chocs enfoncé ou la prise de conscience soudaine et vertigineuse qu'elle approche la trentaine et n'a toujours pas d'enfants. C'est peut-être autre chose, un événement plus sombre, plus aigu, hors de ta sphère de compréhension, une mauvaise expérience avec un client (tu supposes vaguement que c'est une pute) ; de mauvaises nouvelles d'un ami ; une autre mauvaise nouvelle comme ce monde en a toujours en réserve dans sa manche. Peu importe. C'est ça, la beauté de cette Webcam, de toutes les Webcams d'Internet – de notre monde tel qu'il est devenu. Tu peux observer, interpréter, ou te contenter de laisser défiler les images jusqu'à plus soif. Ensuite, tu peux fermer le fichier, le laisser reposer dans son dossier, t'en aller. C'est comme les infos, une vision furtive de l'Irak ou du Rwanda. Ce n'est pas ta vie, pas ton problème. Tu es à l'abri.

Du moins, c'est ce que tu t'imagines – sauf qu'une heure plus tard, deux agents du FBI se pointent chez toi pendant que tu dînes avec ta femme. Tu comprends, bien trop tard, que le regard est une voie à double sens, même sur Internet. Lors des derniers moments où ton mariage est encore sans tache, les joues cramoisies, tu écoutes la femme qui t'annonce que Jessica est morte et qu'au cours des trois derniers mois l'ordinateur de ton bureau a emmagasiné plus d'heures de connexion sur son site que tout autre.

En d'autres termes, tu es son plus grand fan, et le FBI veut te parler de ce qui lui est arrivé, tout comme les policiers qui attendent dehors ; on dirait que ton épouse a été sculptée dans du marbre blanc ; et pas moyen de cliquer sur « Quitter », pas d'issue possible.

Quarante minutes plus tard, Nina sortit du salon et laissa le fan de Jessica, un dénommé Greg McCain, avec Doug Olbrich.

Elle rejoignit Monroe, qui avait écouté la conversation du couloir. McCain se tenait raide sur le divan de cuir souplement tanné. Dans les trente-cinq ans, il avait une coupe de cheveux sophistiquée que Hugh Grant aurait appréciée. Il avait demandé la présence de son avocat. Peut-être aurait-on pu laisser McCain seul en attendant, mais Olbrich était assis en face de lui, en silence. Parfois, cela marchait.

Monroe se tourna vers elle.

— Qu'en penses-tu ?

— Je ne sais pas. Sa femme lui a donné un alibi à l'heure de la mort de Ryan. Elle dit qu'il est parti pour son travail vers 7 h 45 et elle est tellement furieuse qu'on a du mal à croire qu'elle l'ait couvert par pure loyauté.

— Découvrir que son mari regarde des nanas sur le Web, ce n'est pas la même chose que de le balancer pour le meurtre d'un policier. En tout cas, il n'est pas impossible de faire le trajet de chez eux au Knights en un quart d'heure.

— Non, mais cela serait difficile. J'ai pensé à autre chose.

— Quoi ?

— Il me semble que l'homme qui a tué Jessica est aussi celui qui a tué Ryan.

— Oui, bien sûr, mais je ne crois pas que...

— Charles, écoute-moi. Jessica était morte depuis environ quarante-huit heures quand on l'a retrouvée, c'est difficile à dire à cause de la chaleur. D'après ce que nous savons, un type a tué une femme, en privé, et un ou deux jours plus tard, il est sorti et a tué un policier, comme pour se faire remarquer.

— Trouve-moi une explication.

— Je ne peux pas. Pas encore. Je sais seulement que le seul lien entre les deux événements, c'est la proximité.

— Sacrée coïncidence quand même, tu ne trouves pas ?

— Non. Les deux événements sont peut-être liés malgré tout. Mais ce ne serait pas le même type. Ce qui signifie que l'assassin de Jessica pourrait être à l'autre bout du pays, à présent. Ou tranquillement chez lui, avec un alibi pour le mauvais jour.

Monroe détourna les yeux et parla d'un ton exceptionnellement doux.

— Pourquoi un autre type aurait-il tué un policier ?

— Je dis simplement que si nous gardons l'idée en tête, nous aurons une autre question à poser.

— Vas-y.

Gail McCain était à la cuisine. Le dos droit, elle regardait dans la cour par la fenêtre. Le couple n'avait pas d'enfant, et le dîner tranquille aurait dû être suivi d'une séance de télévision ou d'un peu de travail.

— Mon mari est en état d'arrestation ?

— Non, dit Nina. Pas encore.

— Alors, vous n'avez plus rien à faire ici ?

— Vous pouvez exiger qu'on s'en aille. Dans ce cas, la police de Los Angeles viendra vous arrêter pour qu'on vous interroge au commissariat. Tels que je connais ces types, ils allumeront tous les gyrophares, qui ne manqueront pas d'éclairer les fenêtres des voisins.

— S'ils avaient des raisons pour ça, ce serait déjà fait.

— Vous êtes avocate, madame McCain ?

— Non, je travaille à la télévision.

Quelque chose dans sa voix ou son visage échauffa les neurones de Nina d'un demi-degré. Elle se tourna vers la femme policier en faction devant la porte. Petite mais costaude, cette dernière surveillait le couloir, impassible. Ses cheveux étaient retenus en une petite queue-de-cheval qui paraissait assez rigide pour aplatir un nez.

— Vous entendez ? La dame travaille à la télévision. Cool, non ?

— Fait ce qu'elle veut, répondit la femme policier sans détourner le regard.

Nina haussa les épaules.

— L'agent Whalen est difficilement impressionnable. Moi, je trouve que la télé, c'est formidable.

— C'est un boulot comme un autre.

— Oui, mais c'est un travail important, non ? J'ai un ami, il s'appelle Ward, qui a une théorie : il dit que les producteurs sont les nouveaux prêtres et que leur travail consiste à servir de médiateurs entre l'homme de la rue et le royaume paradisiaque de l'autre côté de l'écran. Dites ce qu'il faut, faites ce qu'il faut, et vous entraînez le public dans la télé-réalité ou une série comme *Friends,* et vous vous retrouvez avec un Emmy, à la droite de Whoopie Goldberg. Vous vous êtes déjà sentie dans la peau d'un prêtre ?

— Je ne comprends pas de quoi vous parlez.

— Je ne vous le reproche pas. Je ne comprends pas non plus. C'est presque toujours comme ça avec Ward. Mais ce que je voulais dire, c'est qu'il vous serait beaucoup plus utile d'être avocate, en ce moment. Vous comprenez sans doute la situation ?

— Je crois.

— Vous savez que nous enquêtons sur le meurtre d'une femme appelée Jessica Jones, qu'on a retrouvée mardi matin. Jessica avait un site Web, dont votre mari était membre. Cela lui permettait de voir l'appartement de Jessica grâce à une Webcam où elle se promenait souvent nue.

La femme répondit entre ses dents.

— Oui, j'ai bien compris.

— Parfait. Diriez-vous que votre mari était techniquement compétent ?

— Qu'est-ce que vous entendez par là ?

— Les ordinateurs. Il y en a plusieurs dans son bureau. Il sait s'en servir ?

— Je crois. Il répare le mien quand il est en panne. Mais...

— Merci. Bon, dans l'ensemble, votre mari ne ressemble pas au suspect idéal. C'est pour ça que nous nous réjouissons que vous acceptiez de collaborer avec nous, et que nous sommes venus ici sans sirènes. Pour l'instant, je veux juste vous poser quelques questions, et c'en sera fini. D'accord ? Vous avez dit au lieutenant Olbrich que votre mari était parti travailler vers 7 h 45, c'est bien ça ?

— Non, rétorqua la femme d'un ton glacial. J'ai dit que Greg était parti à 7 h 45 pile.

— Comment pouvez-vous en être si sûre ?

— Greg s'en va tous les jours à 7 h 45. C'est son heure.

— Mais parfois, c'est sans doute un peu plus près de 8 heures, ou un peu plus tôt... Votre mari travaille aussi à la télévision, je crois ? Je suppose que quelquefois il doit arriver en avance ? Ce n'est pas comme s'il pointait.

— Non, mais...

— Donc il a des réunions de bonne heure, de temps en temps ?

— Oui, bien sûr.

— Et s'il s'en va en général à 7 h 45, il y a des moments où il est parti un quart d'heure après, ou même plus tard. Pourquoi affirmez-vous que, le jour en question, il est parti à l'heure habituelle ?

Gail paraissait exaspérée.

— Parce que je le sais. Écoutez, madame Baynam, vous êtes mariée ?

— Non.

Les neurones s'échauffèrent d'un demi-degré supplémentaire.

— Tout s'explique. Si vous l'étiez, vous sauriez de quoi je parle. Lorsqu'on est marié avec quelqu'un, on sait ce qui se passe dans son univers. Trop bien, peut-être. Vous avez votre vie, et la moitié de la vie de l'autre personne. Je sais quand Greg est très occupé, lorsqu'il est en difficulté au travail et qu'il commence à y avoir des réunions à tout bout de champ. Non, je n'ai pas son agenda dans la tête, je ne peux pas vous réciter son emploi du temps par cœur, mais je sais ce qui se passe.

— Alors, excusez-moi, mais vous étiez au courant pour la Webcam ? Vous saviez qu'il épiait de filles qui se trimballaient à poil, faisaient l'amour, en direct sur Internet ?

— Non, mais ce n'est pas...

— La même chose. Oui. Vous saviez tout de Greg, à part ça, et cela se comprend parfaitement. Les hommes sont cachottiers. On pouvait difficilement s'attendre que vous soyez au courant. Il y a peut-être un ou deux détails qu'il ignore à votre sujet aussi, non ? Ce n'est pas grave. C'est la vie de couple, d'après ce que je comprends. Enfin, je suppose. Je vois ça depuis le désert froid et sombre de la vieille fille.

— Je ne voulais pas...

— Je sais, je sais, ne vous en faites pas, Gail. Mais à part ces petits détails, vous êtes sûre d'avoir une connaissance précise de l'emploi du temps et de la vie de Greg ?

— Oui. Absolument.

— Parfait. Votre aide m'a été précieuse. (Nina entendit la sonnette, de l'autre côté de la maison.) On dirait que la grosse cavalerie a débarqué. Je crois que le lieutenant a terminé avec votre mari, alors nous allons vous quitter très bientôt.

Nina lui adressa un sourire chaleureux et s'éloigna.

Elle se retourna puis, sur le même ton qu'elle aurait demandé le nom du décorateur, elle posa encore une question.

— Que faisait votre mari, lundi soir ?

— Pardon ? répondit la femme en la regardant fixement.

— Que faisait votre mari lundi soir ? Puisque vous connaissez son emploi du temps ?

— Il...

Nina vit que Gail comprenait qu'elle avait hésité trop longtemps, que la question, sortie de nulle part, avait percé un fragile mur de défense.

— Il était sorti ce soir-là ?

— Oui.... Il... il était en réunion. Très tard.

— Tard ? Quelle heure ?

— Je ne me souviens pas. Il était tard.
— Une réunion à son travail ?
— Oui, je crois.

— On va s'en aller, dit Olbrich. (Lui, Monroe et Nina étaient seuls dans la cuisine.) Deux personnes ont confirmé qu'il était à son bureau ou dans les parages à l'heure habituelle. Il est sorti tard, la veille au soir, comme tu le savais, mais il n'était pas en réunion. Il prétend avoir accompagné un client dans un club de strip-tease. Ledit client est reparti pour l'Angleterre.

Nina bâilla.

— Nous n'avons rien pour le retenir, et il ne ressemble pas au type qu'on a vu avec Jessica chez Jimmy.

— Bon, d'accord, il matait Jessica, il fréquente des clubs de strip-tease. Pour le reste, c'est une impasse. Son avocat est arrivé, il est prêt au combat et il a raison sur un point. Ou on prend l'affaire au sérieux, ou on lui fiche la paix pour l'instant.

Monroe hocha la tête et avança dans le couloir.

Olbrich regarda Nina.

— Qu'est-ce qu'il y a ?

— Il n'aime pas partir les mains vides quand on arrive en force.

— C'était son idée. Je lui ai dit qu'il devrait la jouer plus cool.

— Monroe est plutôt du genre à foncer tête baissée.

Ils suivirent le patron et s'arrêtèrent devant la porte du salon. Nina s'attendait à une perfidie de l'un ou de l'autre, de l'avocat, surtout ; à la télévision ou au cinéma, tout le monde parlait aux flics et, dans la vie réelle, tout le monde se sentait obligé de faire de même, mais là rien ne venait.

Olbrich s'excusa sans s'excuser. Monroe leur demanda de ne pas quitter la ville pendant quelques jours. Nina allait partir sans un regard en arrière lorsqu'elle entendit une voix de femme prononcer son nom.

Vas-y, ma vieille, pensa-t-elle en se retournant, pousse-la un peu, on verra bien ce qui se passe.

L'un à côté de l'autre, les McCain lui faisaient face. Renvoyé au fond de la pièce, l'avocat n'avait pas l'air content.

— Ma femme me dit que je devrais vous donner ceci. Mon avocat n'est pas d'accord.

Le mari tendait un objet, un peu plus petit qu'un livre de poche, mais de la même épaisseur.

— Qu'est-ce que c'est ?

— Le disque dur de mon portable. Euh... Il y a des photos. Des vidéos, aussi. Tirées du site. Je ne sais pas si ça peut vous être utile, mais...

Sa femme termina pour lui.

— Nous n'en voulons plus chez nous.

Nina prit le disque.

— Cela nous aidera beaucoup.

Une fois qu'il en fut débarrassé, McCain laissa retomber ses épaules, soulagé. Nina comprit qu'au lieu d'être un désastre complet, la soirée pouvait tourner en sa faveur. Sa culpabilité de petit-bourgeois était étalée au grand jour, le destin l'avait libéré de son secret. Bien sûr, sa femme le lui ferait payer cher, elle serait blessée, et il devrait accepter le rôle de paillasson pendant un moment. Et cela allait alimenter les conversations !

Mais ce n'était plus un tabou, et pouvoir ouvrir les portes de ses placards secrets, cela valait la peine d'en payer le prix. Sa femme ne s'en irait pas : ils menaient la belle vie ensemble, et qui a envie de recommencer à draguer ? Dans quelques mois, la gêne de ce soir serait peut-être convertie en une vie sexuelle revigorée.

Certaines personnes flottent sur la vie.

— Je ne savais pas qu'elle était morte, reprit-il. Je suis désolé de l'apprendre.

— Les circonstances de la mort n'ont pas encore été divulguées, nous aimerions que cela ne change pas.

Il hocha la tête et détourna le regard. Gail fit un pas en arrière, comme si elle se détachait inconsciemment de la soirée, puis avança vers la porte pour raccompagner Nina : pour la voir quitter les lieux, en fait... une femme s'adressant à une autre femme, d'une manière telle que les hommes ne se rendent jamais compte de ce qui se trame.

En longeant l'allée qui menait aux voitures, Nina commit une minuscule transgression et glissa le disque dans sa poche avant que les hommes ne s'en aperçoivent. Demain, il irait rejoindre les autres pièces à conviction, intact.

Pas ce soir.

CHAPITRE 15

J'arrivai chez Nina en milieu de matinée. Le chauffeur qui me déposa observa la maison d'un air dubitatif.

— C'est là que vous habitez ?

— Une de mes amies.

— Charmant, grogna-t-il, et il s'éloigna.

Je longeai le chemin vertigineux qui menait à la porte d'entrée. Je n'étais allé chez Nina qu'une seule fois, brièvement, trois mois auparavant ; j'avais passé la nuit sur le canapé après avoir raccompagné Sarah Becker dans sa famille avec Zandt. Il ne semblait être rien arrivé de bon à la façade depuis. De vieux style californien, elle consistait en une rangée de pièces carrées, avec une avancée en forme de L pour la cuisine, comme dans un tout petit motel. Un truc extra à la fin des années cinquante, peut-être mais, même de loin, on voyait bien que ses jours étaient comptés.

Je frappai.

— C'est ouvert, me dit une voix lointaine.

En entrant, j'aperçus Nina qui téléphonait sur la terrasse. Elle m'adressa un signe distrait sans même me regarder.

Je posai mon sac et flânai un instant dans l'espace de vie principal. L'espace, en tout cas. Il ne semblait pas y avoir eu beaucoup de vie ces derniers temps. Ce n'était pas poussiéreux, ni désordonné, mais il n'y avait aucun objet personnel, à part les livres et les dossiers alignés sur les longues étagères. J'allai à la cuisine et ouvris le frigo : deux bouteilles de vin, un carton de jus d'orange, une brique de lait, et basta. Rien dans les placards non plus. Nina ne consommait que du carburant liquide.

Lorsque je retournai dans l'espace principal, tout semblait plus silencieux et plus figé. J'avais lu un jour qu'au milieu du premier millénaire, en Grande-Bretagne, lorsque les autochtones traversaient des contrées désertées, ils s'abritaient dans les ruines des villas romaines et des églises abandonnées. Ils désignaient ces lieux sous le nom de « ports froids », car s'ils pouvaient s'y protéger des éléments, ils n'y trouvaient ni vie ni chaleur humaine. La maison de Nina me donnait cette même impression, à moi, un homme qui avait passé des nuits et des nuits dans des motels et des usines désaffectées, aux fenêtres barricadées, avec des panneaux « démolition en cours » sur les murs.

— Ward ?

Nina avait raccroché et se tenait dans le couloir.

Ses cheveux étaient un peu plus longs, sa silhouette toujours mince semblait débarrassée de quelques kilos. Elle me rappela quelqu'un, mais je ne pus déterminer pourquoi.

— On devrait alerter les flics, quelqu'un a volé toutes tes provisions.

— Tu n'as pas bien cherché. Tout est rangé à sa place. Au supermarché.

— Tu as du café sur place, au moins, ou c'est Starbucks qui s'en occupe ?

Finalement, ce n'était pas cela qui manquait.

— Je l'ai passé à tous les logiciels possibles, dis-je en lui tendant le disque. Rien. Je possède d'autres outils d'analyse plus puissants, mais ils risqueraient de laisser des traces, alors je les passerai sur la copie, si tu l'as toujours. Bref, celui qui a effacé le disque a fait du bon boulot. Il est vierge de chez vierge. Je suis désolé. Parfois, il ne reste plus rien de rien.

— Ne t'inquiète pas. (Penchée sur la balustrade, elle regardait la mer brumeuse.) Je me doutais que ce serait difficile.

— Vous avez des pistes ?

J'avais placé ma chaise le plus loin possible sur la terrasse, afin de vaguement augmenter mes chances de survie lorsque tout s'écroulerait.

— Non. Les flics interrogent les habitués de son site. Il n'y en a pas beaucoup, et aucun n'a le profil. On a interrogé son fan numéro 1, mais je ne crois pas qu'il y ait grand-chose à attendre. Nous avons une vague description du type avec lequel elle a été vue la nuit de sa mort, nous savons qu'elle travaillait comme serveuse de temps à autre, les flics ont interrogé ses collègues de travail, et c'est tout.

— Vous savez qui c'était ?

— Elle vient de la Baie. La police de L.A. essaie toujours de retrouver sa famille à Monterey. On a une adresse sans doute valable, mais les parents doivent être en vacances. Ses rares relations à L.A. semblent ne rien savoir de son passé. Tu sais comment sont ces gens : c'était une mauvaise journée, alors pourquoi ne pas boire une bière et tout oublier ? Tu aurais dû voir son amie, Jean. Elles étaient très copines apparemment, leur prénom commençait par la même lettre, elles fréquentaient le même bar, les meilleures amies du monde. Et maintenant qu'elle est morte, pour Jean ce n'est plus qu'un mauvais rêve. À la suivante !

— Charmant.

— Qu'est-ce que tu crois ? Jessica était une fille qui habitait dans un appartement, qui se sentait parfois déprimée, buvait trop et qui est morte. Nous n'en saurons peut-être jamais plus.

Sa voix s'était étranglée sur les derniers mots pour ne devenir qu'un murmure.

— Nina ? Tu vas bien ?

Elle se tourna vers moi, ses yeux verts pétillaient.

— Oui, affirma-t-elle d'une voix plus ferme. Je ne connais pas la réponse à ta question, c'est tout. Qui était-elle ? Je te le demande. Elle avait un nom et une guitare. Elle était vivante, elle est morte. Le jour du Jugement dernier, c'est tout ce qu'il restera.

— Conception un peu déprimante, mais ce n'est pas ce que je voulais dire. C'était John au téléphone ?

Elle ouvrit la bouche mais se tut.

— Alors, où est-il maintenant ?

— Je ne sais pas. Il faut que je passe la journée à lui laisser des messages si je veux qu'il me rappelle et, ensuite, j'ai droit à cinq minutes de faux-fuyants avant le signal sonore. Ce n'est pas comme si je le traquais. C'est fini, et c'est OK pour moi. Je m'inquiète, c'est tout. Il se conduit bizarrement. Plus que d'habitude.

— Qu'est-ce qui vous est arrivé ?

— Tu lui as posé la question ? s'enquit-elle.

— Oui.

— Et il t'a répondu ?

— Un truc inintelligible.

— Tu n'as qu'à faire travailler ton imagination. Ça n'a pas marché, c'est tout, Ward. Comme il te l'a dit, on ne peut jamais revenir en arrière, et puis, il n'y avait pas tant de choses que cela à revisiter. Nous avions une chose en commun, deux, je crois : les moments passés ensemble avant le meurtre de Karen, et le fait que ni l'un ni l'autre nous ne prenions le départ pour l'escadron du grand amour...

— Et puis vous étiez plutôt terrifiants, tous les deux. Elle sourit vraiment pour la première fois depuis mon arrivée.

— Terrifiants ?

— D'une manière positive.

— Venant d'un type qui a des croûtes sur les doigts et un revolver dans la poche, je prends ça comme un compliment. Je planquai mes mains sous la table.

— Tu es très observatrice. Tu devrais t'engager dans la police.

— Tu veux me parler de cette bagarre ?

Je n'en avais pas envie. Je n'étais pas d'humeur à reconnaître devant Nina ce que j'avais fait ni à lui expliquer pourquoi.

— Un type n'arrêtait pas de me demander si je voulais des frites. Je l'ai rembarré. Tu sais comment ça se passe. Elle haussa les épaules.

— John est resté quelques semaines ici. Ça marchait assez bien. On sortait, on se promenait, on parlait de mon travail, bien sûr, il n'en avait pas. C'était en partie le problème. C'était peut-être même le seul problème. John était un excellent flic. Il fallait absolument qu'il trouve la solution. Mais il n'avait plus sa place dans la police de Los Angeles et il n'avait nulle part où aller. Assez vite, en rentrant du travail, je ne le trouvais plus ici. Il ne rentrait jamais avant minuit. Il ne voulait pas me dire où il était allé. En général, il avait bu, mais ce n'était pas ce qui m'inquiétait. Il a commencé à basculer. Il avait la tête ailleurs. Et tout d'un coup, il a disparu pendant cinq jours.

— Où était-il passé ?

— En Floride. Là où vit son ex.

Je savais que le ménage de Zandt s'était brisé après la disparition de sa fille. Pourtant, il était allé revoir son ex-femme lorsqu'on avait retrouvé le corps de Karen, dix-huit mois plus tard, et je me souvenais que, la veille au soir, il m'avait confié que les assassins n'étaient pas la seule chose importante dans sa vie.

— Il y est retourné voilà deux jours, enchaînai-je.

— Je sais, il m'a envoyé un texto, répondit-elle.

— Tu crois qu'il veut renouer avec elle ?

— Je ne sais pas. Je ne crois pas qu'il le sache non plus. Il n'a plus qu'une idée en tête, retrouver l'Homme Debout. Pour le reste, il ne sait pas où il en est.

— Bizarre, il m'a dit exactement le contraire.

— John est un menteur, observa-t-elle avec amertume, mais elle se ravisa. Parfois, il dit la vérité.

— Ses talents de détective commencent à se rouiller, j'en ai peur. Tout ce qu'il a à son actif depuis son passage à Yakima, c'est une non-information sur une vague colonie roanoke à la fin du XVIe siècle.

— Quoi ? Je lui relatai ce dont je me souvenais de la leçon d'histoire de John. À la fin, elle était livide et nous gardâmes le silence un instant.

Elle se leva enfin.

— Je dois aller bosser. Tu es pressé ?

— Je n'ai nulle part où aller.

— Tant mieux, je voulais te demander un autre service.

Nina partie, je refis du café. J'étais content d'être dans une maison, même aussi peu accueillante que la sienne. Chez soi, inutile de dépenser de l'argent ou de faire bonne figure en permanence. Mais je m'aperçus que, à flâner sans être observé ni importuné par d'autres êtres humains, je me sentais un peu bizarre. Je répondis donc à la requête de Nina.

Avant son départ, j'avais copié tous les fichiers du disque que Greg McCain lui avait confié. L'original était à présent entre les mains de la police, avec celui qu'on avait retrouvé dans le crâne de Jessica. Comment allait-elle expliquer le petit voyage clandestin, je l'ignorais, et je n'aimais pas la voir prendre autant de risques. Elle était la seule à rester liée au monde réel, et j'avais l'impression qu'elle dérivait, comme une prise qui glisse peu à peu en dehors des plots. D'expérience, je savais qu'ensuite la matière changeait subtilement de forme et qu'on risquait de ne plus s'adapter à l'ancien moule.

Les silhouettes recroquevillées à chaque coin de rue, sous chaque porche empestant la pisse, prouvent que la musique de la civilisation s'enraye souvent et qu'il ne reste jamais assez de chaises pour tout le monde.

Tout d'abord, je visionnai les films. Ce n'étaient pas de véritables vidéos, mais de longues séquences figées, qui changeaient à intervalles réguliers. Il y en avait six. Trois montraient Jessica ayant des relations sexuelles, avec à grand renfort d'alcool trois types différents ; deux fois sur le divan du minuscule salon, une fois dans le lit. Il y avait beaucoup de grain, peu de lumière et l'un des films était presque noir. Elle ne faisait aucun effort pour jouer devant la caméra, dont la position demeurait statique. On aurait cru regarder Ken et Barbie collés l'un à l'autre par une gosse qui n'avait aucune idée de la signification de ce geste. Sur un autre film, qui s'étalait sur quatre heures, on voyait Jessica regarder la télévision, faire un peu de ménage de printemps, grattouiller sa guitare et essayer de monter une étagère pas très compliquée. Pendant presque tout ce temps, elle portait un short orange, rien d'autre. Un autre film la montrait assise, après avoir beaucoup pleuré. Sur la dernière vidéo, aux intervalles beaucoup plus longs, cinq à dix minutes environ, Jessica dormait sur son divan, sous une couverture, à peine éclairée par l'écran d'une télévision hors champ. À la fin, elle s'assit et regarda une émission, devant une tasse de café. Nina m'avait dit que Jessica approchait la trentaine, mais, là, éveillée, elle en paraissait plus de quarante ans.

Ensuite, je travaillai sur les photos. McCain les avait rangées dans un unique dossier. J'utilisai un visualiseur graphique et cliquai au petit bonheur la chance. Les images montraient Jessica dans les mêmes situations que les vidéos, mais sans le sexe. Nue ou à moitié nue. Lisant un magazine. À table. Assise devant son ordinateur. Buvant du café ou du whisky. Dormant, fumant, regardant dans le vide. L'effet cumulatif était étrange, et je commençai à comprendre pourquoi Jessica avait séduit McCain. J'étais, moi aussi, un familier des Webcam, et j'avais passé des heures à observer les rues de La Nouvelle-Orléans, les rives du lac McDonald ou encore les vitrines des boutiques informatiques dans les rues de villes anonymes du Middle-West. Il m'avait fallu un moment pour comprendre ce que j'en retirais. On n'observe pas dans l'espoir de voir quelque chose de fascinant, bien au contraire. On regarde parce que le manque d'activité, l'absence totale de thème rendent la scène plus réelle. Si on observe quelque chose de particulier, la seule chose que l'on voit, c'est ce qui se passe. On voit le moment, l'événement, on oublie la lente marée dépourvue d'événements qui le submerge. Si l'on n'observe rien, on voit tout. On voit la chose telle qu'elle est.

Cette myriade de vues produisait le même effet. Aucune image n'était composée. Dans la plupart d'entre elles, Jessica se trouvait hors cadre ou dans une zone floue. En ne dévoilant rien de particulier, elle révélait tout. Notre vision de sa vie, devenait similaire à la sienne, une série interminable d'instants non prémédités, parfaitement neutres et, finalement, plutôt ennuyeux. La collection de McCain révélait la réalité de cette femme bien mieux que tout ce que j'aurais pu imaginer, en lui rendant gloire et en la capturant sous forme de pixels. Quinze mégaoctets de triomphe.

Ce fut seulement après avoir observé sa vie avant l'événement que je regardai les Polaroïd que Nina m'avait laissés. Ils montraient l'appartement de Jessica le jour où la police l'avait retrouvée. Deux images, plates, sans âme, mais pas neutres. Chaque millimètre carré disait quelque chose de direct : leur simple existence criait que la fille qui avait vécu ici était morte, c'était d'ailleurs pourquoi j'avais regardé les autres images en premier.

Je les examinai un instant, puis retournai aux premières images du disque dur ; je paramétrai le système pour les classer par ordre chronologique, et je regardai à nouveau.

Il me fallut un moment pour noter un détail.

— Tu vois ?

Nina hocha la tête.

— On ne peut pas avoir mieux ?

— Non, pas moyen. J'ai essayé d'agrandir... (J'ouvris une autre fenêtre que j'avais réduite.) Ce n'est pas un film, et l'agrandissement ne donne rien.

Nina se pencha pour observer l'image sur l'écran. Elle regardait une photo au grain grossier qui montrait Jessica allongée sur son lit. Un visage d'homme se penchait sur le sien.

Nous ne nous intéressions pas à l'homme. Les flics de L.A. allaient vite : ils possédaient déjà des tirages des trois hommes que l'on voyait sur le film de McCain et les avaient montrés aux copains de Jessica, en commençant par le bar de Jimmy. Le garçon avait dit qu'aucun ne ressemblait vraiment au type avec lequel on l'avait vue le dernier soir. Nina s'était chargée elle-même de ces investigations avant de rentrer en milieu d'après-midi. Nous observâmes la table de chevet de Jessica : une lampe, un radioréveil bon marché, une petite pile de livres dont les tranches criardes évoquaient des ouvrages de développement personnel, trois tasses de café et une petite photo encadrée.

Nina prit le Polaroïd de la chambre.

— Tu as raison, le cliché a disparu. Et je n'ai rien vu de tel dans tout l'appartement.

Dès que j'avais remarqué ce détail, je l'avais appelée pour lui décrire la photo, et elle était passée vérifier dans l'appartement de Jessica.

— Quand a-t-elle disparu ?

— Moins d'une semaine avant sa mort.

— À condition que la date soit correcte.

— Elle l'est. C'est confirmé par la date de création du fichier.

— Une semaine. Elle aurait eu le temps de la déplacer.

— Tu ne l'as pas retrouvée. Une photo assez importante pour qu'on la garde près de son lit, ça ne se jette pas comme ça.

— Si c'est un ex-petit ami ?

— Exact, mais regarde. (Je désignai une troisième image, avec le cliché en gros plan.) C'est un agrandissement encore plus important. J'ai utilisé un logiciel qui extrapole à partir de la couleur de chaque pixel, la compare à l'environnement et essaye de recréer une image cohérente. Ça ne ressemble à rien quand on l'applique à une photo de si mauvaise qualité, mais on discerne quand même un détail intéressant. On ne distingue aucun trait, mais on voit clairement deux personnes.

— Jessica et un ancien copain.

— Je ne crois pas. Quelle est la couleur au sommet de la tête ?

— Gris, répondit-elle.

— La couleur des personnes âgées, en d'autres termes. Des parents, sans doute.

— Tu crois ?

— Jessica ne rentrait peut-être pas souvent chez elle, mais j'aurais été surpris qu'il n'y ait aucune photo de famille dans l'appartement. Une jolie photo de papa et maman, ou bien, si elle avait un problème avec l'un ou l'autre, ou les deux, la photo d'une nièce. Une trace de la famille. Les filles sont comme ça.

— Ah bon ? Et tu en as déniché une ici ?

— Non, admis-je, mais je n'ai pas bien cherché. Et tu n'es pas une fille.

— Exact. Juste une femme terrifiante.

— Pas seulement. Mais ce que je veux dire, c'est qu'il manque quelque chose dans l'appartement de Jessica.

— Tu crois que l'assassin y était passé ? s'étonna-t-elle.

— Oui. Et j'en ai la preuve.

Je double-cliquai sur un autre fichier, une des images que McCain avait sauvegardées. On y voyait Jessica avachie sur le divan dans une pose peu élégante. Elle portait un pyjama bleu clair, avec des petites fleurs roses et blanches.

— Tu dis qu'on l'a trouvée... repris-je.

— C'est ça ! Les pyjamas ! Mon Dieu, tu as raison ! Il est venu.

— Je crois qu'il s'est approché d'elle, qu'il l'a traquée, si on adopte son raisonnement, et qu'il a passé un moment dans son appartement à construire son meurtre. Il a volé les pyjamas et un souvenir personnel. Il a dû comprendre qu'il s'agissait de sa famille et il voulait un objet qui la touchait de près, un objet important.

— Sans qu'elle l'ait remarqué ?

— Cite-moi un objet que tu regardes tous les jours, lui dis-je. Et puis, cette table, c'est un vrai foutoir. Et...

— Et les pyjamas ? Elle se serait aperçue de leur disparition, non ?
— Donc il était sans doute là le jour ou la nuit précédant le meurtre.
— Alors, pourquoi ne pas l'attendre tout simplement et la tuer chez elle ? questionna Nina.
— Parce que c'était sa maison à elle. Tu connais ces types. Ils veulent sculpter l'événement. Il faut que tout se passe selon leur plan. Il savait où elle habitait. Cela signifie qu'une fois au moins on a pu le voir près de l'appartement. Cela signifie qu'il a dû entrer. Mais comment ?
— La police a déjà passé le voisinage au peigne fin. Personne n'a rien vu.
— Mais comment a-t-il su où elle habitait ? insistai-je.
— Ward, tu as de bons yeux, mais tu n'es pas flic. Il l'a sûrement suivie à la sortie d'un bar. Je suis désolée, mais, même si tu as raison, ça ne nous mène à rien. Il a pris ses pyjamas et volé une photo. Peut-être. Tu parles d'une affaire ! On va ajouter ça aux charges, juste en dessous du meurtre !
Je me tournai vers elle, irrité, mais elle avait l'air fatiguée et je ravalai mes mots.
— Bizarre que ça n'ait pas marché avec John. Entre deux êtres aussi raisonnables et aussi conciliants...
Elle sourit.
— D'accord, je vais en parler.
— Merci, je me sens reconnu au-delà de toute espérance. Bon, eh bien, si on allait libérer un peu de nourriture du supermarché ?
— Laisse tomber. On va trouver un endroit où ils la font cuire en plus.

Nous finîmes à Santa Monica, dans un restaurant italien de la Promenade. Nous passâmes un moment à table avant de nous attarder plus longuement au bar. Nina était parfaite, un verre de vin à la main. Je lui racontai ce que j'avais fait ces derniers mois, puis, sous l'effet du vin, lui confiai que Bobby me manquait beaucoup, tout comme mes parents ; elle comprit et ne dit rien pour tenter de me consoler. Je me rendis compte que je savais peu de choses d'elle, je découvris qu'elle avait grandi dans le Colorado, fréquenté l'université à San Francisco, mais guère plus. Elle me parla d'une ancienne amie à elle qui l'avait appelée et qu'elle devait rencontrer, et nous nous accordâmes à dire que le passé était un autre continent dont les plaques tectoniques s'éloignaient d'année en année. Au fil de la soirée, le bar s'emplissait, Nina veillait sur mon siège pour empêcher les gens de le prendre durant mes petites expéditions pour aller fumer à l'extérieur. Avec Nina, un regard suffisait.

Avec l'ivresse, les gens me paraissaient de plus en plus bruyants, de plus en plus odieux. Les discussions parlaient de cinéma, bien sûr, de fric, de santé, de surpoids, de mode. Plus les sujets étaient superficiels, plus ils parlaient fort, interminable litanie adressée aux dieux du destin. J'étais de plus en plus exaspéré ; Nina m'écoutait fulminer en silence. La mode, ça m'énerve. Depuis toujours. Cet été, nous porterons tous de l'orange, vous savez ?

Nous avons perdu tout sens des proportions, pérorai-je, tout sens de ce qui est important, raisonnable ou sain : pendant ce temps, les pays qui n'ont ni le temps ni les moyens pour ces conneries sont de plus en plus furax de nous voir nous conduire comme si le monde était notre terrain de jeu. Mais qui s'en préoccupe ? Tiens, un nouveau film sur des ados débiles. Formidable, un nouveau régime vient de sortir. Jennifer Lopez vient d'acheter de nouveaux bij oux, t'as vu comme c'est beau ? Qui se soucie de ce qui arrive dans les trous à rats où on ne parle même pas anglais ? La vie est trop belle. Allez, ouvre donc un pinard sans caféine !

Je me retrouvai à bout de souffle et de carburant au même instant. Tout autour de nous, les jeunes me regardaient comme si je venais de déclarer que la structure en trois actes était nulle et non avenue.

— Allez vous faire foutre, suggérai-je à voix haute. Tout le monde détourna les yeux. Nina me regardait, elle aussi, sourcil levé.

— Le Prozac ne marche pas très bien...

— Le monde est pourri, murmurai-je, embarrassé. Tout le monde est pourri. Bienvenue à Armageddon.

— Oh, je sais encore ce que c'est d'avoir quinze ans ! Ne t'énerve pas. Ça passera. Allez, viens, dit-elle en se levant. Je suis saoule, tu es complètement bourré. Il est temps de rentrer.

Je vis la note sur la table et compris que, au cours du dernier quart d'heure, elle avait tout payé.

Je glissai de mon tabouret et la suivis, un peu penaud.

Au moment où le taxi nous ramena chez Nina, le vin s'était diffusé dans mes veines et je commençais à me sentir fatigué.

La plus grande partie de la journée s'était déroulée dans le silence. J'insistai pour payer la course, puis trébuchai en sortant de la voiture.

Une fois à l'intérieur, je me précipitai vers la machine à café. Pour cela, je dus passer devant le répondeur.

— T'as un message.

Nina toucha un bouton et examina le numéro qui s'affichait.

— C'est Monroe.

Le message était court. Une voix d'homme demandait à Nina de rappeler, à n'importe quelle heure. Elle leva les yeux au ciel mais s'exécuta aussitôt.

— Bureau de Charles Monroe.

La voix était puissante et claire.

— Nina Baynam, dit-elle en se frottant les yeux. On m'a laissé un message.

La personne ne répondit pas, mais moins de trois secondes plus tard, le patron de Nina prit l'appel.

— Nina, où étais-tu passée ?

— J'étais sortie. Pourquoi tu ne m'as pas appelée sur mon portable ?

— Je l'ai fait. Trois fois.

— Oh, pardon, il y avait beaucoup de bruit, dit-elle en me regardant. Qu'est-ce qui se passe ?

— Je viens de recevoir un appel de Portland.

— Un autre meurtre ?

— Oui et non. Pas d'autre disque dur. Pas d'autre fille.

— Alors, quoi ?

Monroe parla lentement, avec précaution.

— Une prostituée du nom de Denise Terrel est entrée dans un commissariat avant-hier soir. Elle prétendait être allée à un rendez-vous, l'après-midi, où il s'était passé quelque chose. Ensuite, en pleine nuit, elle s'est retrouvée assise contre une poubelle. On s'est aperçu qu'elle avait de sérieuses contusions et on l'a emmenée à l'hôpital. Le lendemain matin, ses souvenirs sont revenus, et elle a raconté que son agence avait pris rendez-vous pour elle avec un de ses réguliers, mais qu'elle avait passé un accord avec un autre type qui savait que l'agence était en contact avec ce client. Il l'avait appelée directement et lui avait proposé de l'argent pour qu'elle lui dise où et quand devait avoir lieu le rendez-vous. Le gars lui aurait promis un beau paquet pour pincer son client dans un endroit privé, où il ne serait pas sur ses gardes. La fille, Cherri pour les intimes, a accepté.

— Charles, où veux-tu en venir ?

— Hier matin, les flics de Portland sont allés à l'adresse qu'elle avait indiquée. Ils ont trouvé un cadavre. Celui de Peter Ferillo. Il possédait un restaurant et avait des liens avec la mafia de Los Angeles. Il était nu, plutôt amoché. On lui a tiré une balle dans la tête et on l'a laissé affalé sur une chaise. La police a passé la pièce au peigne fin, du sol au plafond, mais on n'a rien trouvé. Plus tard, un agent a découvert un objet dans une plate-bande, à une trentaine de mètres. Un ouvrebouteilles, avec des traces de sang. Le sang de Ferillo, on l'a fait analyser. Et une belle empreinte. On l'a identifiée.

Les effets du vin semblaient totalement dissipés. Nina et moi nous regardions.

— Nina, dit Monroe, ce sont les empreintes de John Zandt.

CHAPITRE 16

Tout en conduisant, il avait conscience du réseau qui l'entourait. Le réseau de rues, le réseau d'individus, de lieux, d'objets. De l'autre réseau aussi, celui du monde moderne. Ce lieu parallèle, avec ses chemins privés d'adresses électroniques, ses sites de commerce en ligne. On y trouvait tout, on touchait la réalité du doigt, comme un dieu. Sur le Net, tout est information, et tout est sur le Net ; aujourd'hui, donc, le monde s'est transformé en information. Tout témoigne de ce phénomène, de cette banque de mots et d'images : tout est quelque chose dont il parle, dont il a parlé. C'est une histoire d'achats et de regards, une histoire d'habitudes et de désirs, de contacts avec les autres, de voyeurisme, d'ambitions, de dépendances. Cela raconte notre histoire, et parfois cette histoire a besoin d'un peu de travail. Parfois, il faut en extraire des éléments. Trouver Jessica par ce biais, c'était un nouveau commencement. Bien sûr, il y a pléthore de Jessica ; mais il y a aussi une Jessica unique. Une fois qu'on l'a repérée, il suffit d'ouvrir une fenêtre sur sa vie, pour confirmer son existence ; mais on peut tout aussi bien la refermer, quitter le programme, défaire sa naissance. On peut éteindre la machine, la réinitialiser, et le passé a disparu, tout est propre. La touche « suppr », ce n'est pas fait pour les chiens ! Parfois, il faut que tout soit comme neuf.

Une de ses séquences de Webcam favorites vient de Pittsburgh, ville où il n'est jamais allé. Elle consiste en trois images, couvrant la tranche horaire de 5 h 43 à 6 h 14 du matin, fin mai 2003. Toutes sont prises par la même caméra, malgré la variation de l'angle de vision et du réglage du zoom entre les poses, qui donnent des cadrages différents.

Sur la première, le ciel couleur d'aurore, bleu et rouge, sillonné de nuages fantastiques, occupait la moitié supérieure du cadre. En dessous, l'Allegheny se courbait vers la gauche après les ponts des 6e, 7e et 9e Rues, dont les lumières se reflétaient dans le sombre miroir d'eau. Partout, le long des rues, de chaque côté de la rive, tout autour de la fontaine et du bassin, au bout du parc de Point State et de Gateway Center, on percevait d'autres lumières. Des petits points blancs, colorés d'or ou de rose par l'obscurité évanescente et les imperfections de la caméra. L'angle de la deuxième vue était plus serré et, pendant ce quart d'heure, la caméra avait zoomé et pivoté dans une direction différente. Impossible de dire comment cette petite section s'insérait dans l'ensemble de la ville. Le cadre était essentiellement occupé par des arbres entre lesquels on apercevait le virage d'une nationale qui traversait la ville et quelques oiseaux matinaux qui se mettaient en route pour le labeur. Avec la dernière image, on se retrouvait au confluent des fleuves, vu en grand angle. La position n'était pas tout à fait la même que sur la première image. On était un peu plus au sud, on voyait la Monongahela qui se jetait dans l'Allegheny, alors que Fort Pitt Bridge demeurait dans l'ombre. On ne discernait plus aucun point lumineux : ou la ville éteignait tous les lampadaires sans exception à 6 heures pile, ou le ciel plus clair avait poussé la Webcam à sous-exposer les zones terrestres.

Il avait passé beaucoup de temps à étudier ces images, pour comprendre ce que le réseau racontait sur les gens qu'il épiait. Vous pouviez vivre dans une ville, être un de ses habitants sans comprendre son cadre général, sans en faire partie. Comme des souris qui habitent dans une maison : c'est leur adresse, mais cela ne signifie pas qu'elles y ont des droits. De même, vous pouvez passer toute une journée dans un restaurant en n'étant qu'un type qui occupe temporairement un espace qui appartient à un autre, un espace qu'il loue en payant son café et ses hamburgers. Même si vous avez votre jolie petite maison en banlieue, vous payez partout votre dîme ; vous grignotez l'emprunt contracté pour acquérir votre propriété, vous taillez dans les intérêts pour l'appareil dentaire de votre fils et le trousseau du futur mariage de votre fille, vous payez l'assurance qui couvrira les frais de chimiothérapie de vos parents mais ne leur sauvera pas la vie.

Et tout cela jusqu'à ce que votre vie commence à rebrousser chemin ; d'une maison, vous passez à une seule pièce chez vos enfants, à condition qu'ils vous acceptent ; vous finissez dans un étrange bâtiment, entouré de vieux cinglés que vous n'avez jamais vus auparavant et que vous n'auriez peut-être pas aimés de toute façon ; les jeunes ne comprennent pas que les similitudes physiques de la vieillesse n'impliquent pas que tout le monde soit identique à l'intérieur. Ils ne comprennent pas non plus le rythme de la vie, d'ailleurs. Tout ce temps passé à acheter une maison, tous ces emprunts et ces aspirations se sont envolés, effacés du disque dur de votre vie... Toutes les choses que vous avez acquises sont données, vendues, jetées !

À présent, d'autres gens empruntent votre route pour aller et venir, vivent dans votre maison, repeignent les murs et démolissent vos étagères, et la planète tourne toujours.

Un jour, après un voyage particulièrement douloureux aux toilettes, une fois réinstallée sur sa chaise, minuscule, l'air épuisée et honteuse, la grand-mère l'avait regardé et lui avait dit :

— Quel dommage qu'Il ait mis le pire à la fin !

Il n'avait pas compris tout de suite, mais seulement sept mois plus tard, blotti derrière l'un des fauteuils du salon, deux heures après l'enterrement de sa grand-mère. Il était resté là un moment, à penser à la vieille dame, lorsque sa mère était entrée au salon, un disque à la main. Elle s'était approchée du tourne-disque, l'avait allumé, s'était assise et avait écouté.

Il avait une peur bleue. Il ne savait pas quoi faire. Il surprenait sa mère dans un moment d'intimité et elle ne serait pas contente de découvrir qu'elle n'était pas seule. Il en fut d'autant plus persuadé quand il entendit quelque chose qui aurait pu être des pleurs. Il n'avait jamais entendu sa mère pleurer auparavant. Il ne l'entendit jamais plus.

Il resta donc assis et écouta.

Sa mère ne fit pas un mouvement pendant toute la durée du disque. Puis elle se leva, ôta le disque de la platine, et le jeta méchamment dans un coin, où il éclata en mille morceaux.

Elle sortit, brusquement. La porte d'entrée claqua.

Il quitta sa cachette derrière le fauteuil. Son corps lui disait de ficher le camp, de monter dans sa chambre, d'aller dehors, de faire quelque chose, mais son esprit estimait que sa mère était déjà presque arrivée dans un bar, et il tenait à savoir quelle était cette musique. L'esprit gagna.

Il s'approcha du tourne-disque et regarda la pochette. C'était le *Requiem* de Fauré, qu'il reconnut pour l'avoir vu dans la chambre de sa grand-mère, un des rares objets qu'elle avait apportés chez eux. La pochette était vieille et tout abîmée, on avait l'impression que le disque avait été maintes fois sorti et rangé. C'était peut-être ce qui l'avait poussé à ramasser l'un des plus gros morceaux du disque brisé et à l'emporter dans sa chambre : il avait compris que viendrait un jour où, une fois de plus, il serait contrôlé par les autres.

Il avait douze ans. Quatre ans plus tard, le *Requiem* de Fauré fut le premier disque qu'il acheta. À la fin de l'adolescence, c'était déjà une musique qu'il n'écoutait qu'en privé. Fauré était un compositeur très connu, avait-il appris. C'était comme de mettre les *Quatre Saisons* de Vivaldi ou la *Cinquième* de Beethoven. Vous finissez par passer pour un ignare, même si vous l'aimez, car vous êtes entouré de gens qui accordent plus de valeur à l'idée qu'à l'expérience. Des gens qui pensent qu'il vaut mieux admirer quelque chose que l'aimer.

Des gens qui n'ont pas le courage de comprendre que, s'ils se conduisaient avec assez de volonté, ils pourraient renverser des montagnes.

Peu de temps avant de laisser ces gens, tous les gens, loin derrière lui, il trouva la « route qui fume ». Écouter sa mère, entendre ces sons étranges qui venaient du profond de son corps, cela avait été réel. Ça s'était produit, c'était un événement en temps réel, un changement dans la couleur du monde, tandis que la mort de sa grand-mère marquait sa trace indélébile sur la réalité. C'était une vision furtive d'un lac lointain, d'une fille endormie, d'une ruelle sordide, dépouillée et vulnérable dans sa vérité toute simple.

La mort est réelle. La mort transforme. Tout le reste n'est que du remplissage, un simple message de notre sponsor.

La mort de la vieille dame avait un sens, d'autant plus qu'il savait que la chute qui l'avait finalement emportée n'était pas un simple accident. On l'aidait à descendre l'escalier, après tout, et il ne l'avait entendue dire : « Non ! », d'un ton suppliant, qu'une seule fois, juste avant la chute.

Ensuite, tout fut silencieux pour elle ; elle ne pleura plus dans la nuit, ne se souilla plus, sa respiration rocailleuse se tut. On la mit en terre, elle dormit aisément, elle devait savoir, d'une manière ou une autre, que sa fille avait pleuré après son départ.

Le pire n'avait pas besoin d'être réservé pour la fin, ça au moins, c'était évident. La fin ne devait pas être silencieuse et vaine. Tant qu'il y avait quelqu'un pour s'en occuper, cela ne devait pas finir si mal.

Alors, pourquoi attendre ?

Arrivé en ville, il se gara. Il marcha jusqu'à sa destination. Il était perpétuellement en mouvement, car le mouvement, c'est ce qu'il y a de mieux. Aujourd'hui encore, cet aspect du processus restait étrange et incontrôlable, et un être inférieur aurait bêtement cru que c'était parce que la compulsion ne provenait pas de son esprit conscient. Mais pas lui. Il savait que tout avait une signification, que parfois c'est pour cela que nous existons.

Il marchait. Il attendait la nuit. Il attendait pour que quelqu'un de très particulier n'ait plus à attendre. Il agissait pour son propre bénéfice, bien sûr, tout autant que pour des buts pédagogiques plus vastes, mais cela n'empêchait pas que cela soit juste pour elle aussi. Tout serait comme neuf, tout serait tranquille.

Il n'y avait aucun moyen de perdre.

CHAPITRE 17

L'ascenseur s'ouvrit. Burt se trouvait à l'intérieur. Il sourit et recula pour laisser entrer Katelyn, puis comprit qu'il devait sortir avec son gros chariot, et donc passer le premier, quelles que soient les règles de son code de galanterie. Il hésita, recula et avança de quelques centimètres puis roula les yeux et haussa les épaules. Cette scène, à peu de chose près, se renouvelait presque tous les soirs.

Il s'écarta dans un cliquetis métallique en s'excusant et se retourna pour lui tenir la porte.

— Vous allez chercher les menus, m'dame ?

— Exact, Burt. Comment s'est passée la soirée ?

— Ça se termine.

Burt était le seul employé noir du Fairview de Seattle, en dehors du concierge de jour, le très apprécié Big Ron. Katelyn aimait bien Burt. Il était deux fois plus âgé que les autres employés et travaillait deux fois plus dur, même après 3 heures du matin. Quand on voyait Burt, il était toujours en train de faire quelque chose. La seule idée de le voir au repos était absurde.

Rassuré de la savoir en sécurité dans l'ascenseur, il alla réparer un objet, attacher un truc ou gratter quelque chose. Katelyn le regarda tandis que les portes se refermaient. C'était un travailleur de la nuit, lui aussi, et intérieurement, elle savait qu'il éprouvait le même sentiment qu'elle, la même impression d'occuper une place à part. Elle ne lui avait jamais posé la question, parce que cela ne se faisait pas. Ou était-ce trop simple ? Croyait-elle qu'une telle observation aurait dépassé les limites de leur relation professionnelle, et si oui, pourquoi ? Cela trahissait-il un trait déplaisant de sa personnalité ? La hiérarchie était-elle plus importante qu'elle ne l'imaginait ? Était-elle condescendante sans s'en apercevoir, ne le prenaitelle pas au sérieux parce qu'il était vieux ou...

Mon Dieu, qu'il était tard !

Ce n'était pas un travail pour le directeur de nuit. Certains hôtels laissaient ce soin au garçon d'étage, une dernière course avant de partir, ou, s'il y avait un service 24/24, le cuisinier de nuit mettait le répondeur aux heures creuses, vers 4 heures, et allait ramasser lui-même les menus. À un endroit, on lui avait demandé d'accrocher les menus aux portes à 6 heures, au lieu de 2, c'était le premier travail de la journée pour l'équipe qui monterait les petits déjeuners à l'étage. Cela ne lui paraissait pas normal. On pourrait penser que le petit déjeuner est le premier événement de la journée, mais c'est faux. Pas pour les clients. C'est le dernier. Ils reviennent de leur soirée, dans une ville inconnue, grincheux, ou plus tard, faisant poliment la gueule. Katelyn aimait se les imaginer affalés sur le lit, pris de hoquets, attrapant un stylo publicitaire, les sourcils froncés par la concentration, pour cocher et annoter leur commande. En vacances ou en voyage d'affaires, le petit déjeuner prend une signification existentielle.

Katelyn tenta de l'expliquer à l'un des types de la réception, qui la regarda comme si elle parlait chinois. Quelques-uns la traitaient comme une nulle. Les directeurs de nuit étaient rarement des femmes. Sans doute à cause de la fonction, parce qu'il fallait faire face à d'étranges situations... expliquer aux intrus qu'il n'y avait pas de service de taxis pour la banlieue ; dissuader des hommes d'affaires aux yeux exorbités de ramener des femmes qui étaient de toute évidence des prostituées ; trouver quelqu'un pour nettoyer le vomi dans l'ascenseur du milieu (les gens vomissaient toujours dans celui du milieu, personne ne savait pourquoi, pas même Burt). La plupart des directeurs de nuit n'avaient pas le vent en poupe. Ils arrivaient à 21 heures, ou à l'heure où l'hôtel estimait que la véritable activité était terminée, buvaient du café dans leur petit bureau. Avec un peu de chance, ils continuaient ainsi jusqu'au lever du soleil, s'interrompaient une minute ou deux pour vérifier le nettoyage et les stocks dont s'occupaient des employés deux fois moins payés qu'eux. S'il y avait le feu, ils houspillaient les gens qui les entouraient jusqu'à ce que le problème ait disparu, soit oublié ou réglé, puis retournaient à leurs magazines. À l'aube, ils se dissipaient comme la rosée, retournaient dans leur petite maison pour y dormir le jour, tels des vampires angéliques.

Katelyn était différente. Tandis que l'ascenseur montait dans la nuit, son reflet sur les miroirs des parois la rassurait : une femme, jeune, belle et séduisante. Bon, pas si jeune. Oublions.

Elle avait une belle peau, néanmoins, et ses cheveux nécessitaient peu de soins. Elle faisait très femme d'affaires dans son tailleur anthracite. Elle n'était pas obligée d'être là. Elle n'aurait pas dû, peut-être. On pouvait être un directeur d'hôtel sans expérience, mais elle avait travaillé sur assez de listes de procédures pour savoir que rien ne valait la formation sur le tas. La journée, un hôtel ressemble à une grosse machine, actionnée par des principes internes. Bien sûr, dès que vous franchissez le bureau de la réception, et que vous avez ouvert quelques portes marquées « privé », vous vous rendez compte qu'il n'en est rien. Un hôtel est une collusion frontale entre un milliard de listes de « procédures » à accomplir à des vitesses variables ; c'est un ordinateur de chair et de pierre, qui fait appel à dix-sept logiciels différents (des nouveaux, des potentiels, des vieux bourrés de bogues, avec des fuites mémoire dans tous les coins), sous la menace constante d'un crash imminent. Il existe une certaine inertie, néanmoins, un écosystème qui tient le choc, une équipe de relais qui dispute une course sans fin.

La nuit, tout est différent. Le logiciel est en sommeil, on prend conscience des problèmes de matériel : le bureau, les chaises, les ampoules. Les ascenseurs qui se mettent à voyager à leur propre guise, sans raison apparente, qui ronflent dans le silence du petit matin. Et surtout, le bâtiment lui-même, ses longs corridors, ses piliers massifs, imprégnés du bruit blanc de la panne. Les hôtels en voient de toutes les couleurs. Ce qui se passe dans un hôtel filerait une dépression nerveuse à une maison normale en une seule journée. Au petit matin, le bâtiment a un peu de temps pour lui, pour remâcher ses pensées lentes et pesantes.

C'est peut-être pour cela que la plupart des directeurs ne sont pas des femmes. Katelyn savait qu'elle aurait dû être chez elle, endormie ou en train d'écouter la respiration d'un autre être vivant. Un chat, cela ne comptait pas, même si elle l'aimait beaucoup. Il aurait fallu que ce soit un enfant ou un homme, au moins. On pouvait écouter tant qu'on voulait dans son appartement, on n'entendait rien. Elle devait cesser de se faire des illusions.

C'est pourquoi elle était ici.

Les portes s'ouvrirent au sixième étage et, en bon directeur de nuit, elle sortit. Six étages, ce n'est pas beaucoup, et c'était tout ce que le Fairview avait à proposer.

L'air était doux et calme. Elle avança dans le couloir moquetté, suivit les trois côtés du petit carré. À droite, en face, à gauche.

Il n'y avait pas beaucoup de menus. À cette époque de l'année, les week-ends étaient calmes. Un couple de touristes résidait à la 5 – comme elle les avait vus rentrer pas mal éméchés après minuit, Katelyn était curieuse de savoir ce qu'ils avaient commandé – mais la plupart étaient des hommes d'affaires. Ceux-là se lèveraient tôt, avaleraient leur café Starbucks gratuit et le croissant servis en salle entre 7 heures et 8 h 30. Sur tout l'étage, elle ne trouva que douze commandes, la plupart pour la version maison du petit déjeuner, avec deux œufs. Pas grand-chose d'intéressant, bien que la commande de flocons d'avoine nature la fît sourire. Le client était un grand costaud. Ce n'était sans doute pas ce dont il avait envie. Il voulait bien se tenir. Sa femme aurait été fière de lui. Il aurait dû demander le déjeuner complet, celui dont il rêvait. Enfin, c'était un bel effort de sa part.

Au bout du couloir, elle se retourna pour vérifier qu'elle n'avait rien oublié et ouvrit la porte de l'escalier. Le riche tapis s'arrêtait de l'autre côté de cette porte, mesure d'économie qu'elle approuvait entièrement.

Elle arrivait au demi-palier lorsqu'elle entendit un bruit audessus. Elle leva les yeux, prête à sourire, pensant que Burt descendait pour accomplir une tâche quelconque.

Personne. Bizarre. Le son ne pouvait pas venir d'en bas, car elle voyait la porte du cinquième. Elle se pencha sur la balustrade. Aucun mouvement en bas non plus.

Peu importait. Il y avait du bruit dans les hôtels. Peut-être un membre de l'équipe de nettoyage qui arrivait. Elle regarda sa montre : 3 h 15, non, ce n'était pas possible.

Elle ouvrit la porte, s'attendant plus ou moins à rencontrer Burt et songeant à lui dire quelque chose. Une parole amicale, pour lui montrer qu'elle n'avait aucun comportement antivieux ou raciste.

Le couloir était vide.

Burt ne saurait jamais ce qu'il avait raté.

Au cinquième, l'activité était également ralentie. Quelques toasts et quelques cafés, mais... ah... œufs, saucisses, bacon, supplément saucisses, flocons d'avoine, fruit, café et thé pour combien, quatre ? Et un petit déjeuner continental avec toasts. Et un muffin, sans doute. Parfois, c'était plus de toasts. Ou du bacon. Un jus d'orange. Le tout pour 7 h 30.

Katelyn sourit : sans doute les touristes éméchés. Elle sortit un stylo de sa poche et pratiqua quelques modifications, réduisant leur commande à quelque chose qui ne les terroriserait pas le lendemain matin. Elle reporta l'heure à 7 h 45, ils la remercieraient.

Elle poursuivit. D'autres toasts, d'autres œufs. Katelyn essaya de se souvenir de la dernière fois où elle était partie en vacances. Cela faisait un moment, c'était sûr, cela remontait à avant la mort de ses parents, c'est-à-dire cinq ans. C'était drôle, les souvenirs. Des instantanés. Un endroit où on aimait prendre le café en lisant un roman de gare. Un bijou de pacotille, tant convoité, acheté et abandonné dans un tiroir. Des amours de vacances. Des garçons, devenus des hommes, comme elle était sans doute devenue une femme. Toutes les plus de quarante ans qui pensaient à elle comme à une fille se leurraient, malgré tout ce que prétendaient les magazines qui survivaient grâce aux publicités pour les crèmes antirides.

Une porte s'ouvrit. Elle se retourna.

— Burt ?

Pas de réponse. Elle parlait à voix basse, personne n'avait envie d'être réveillé à une heure pareille, mais il aurait entendu, et répondu.

Un client, peut-être. Elle avait ajouté le menu des touristes sur la pile et rebroussait chemin. En passant devant la porte, elle remarqua qu'elle était ouverte. Pas beaucoup.

Elle ne l'avait pas laissée ainsi. Il fallait la fermer. Les consignes anti-incendie étaient très strictes, une pancarte les rappelait. Burt les observait toujours. Drôle de moment pour recourir à l'escalier, de toute façon.

Elle ouvrit un peu plus et dit :

— Bonjour.

Le son résonna dans la cage d'escalier mais ne sembla atteindre personne, ni en haut ni en bas. Rien qu'un nouveau son dans sa tête. Sauf que...

Elle se retourna brusquement.

Le couloir était vide. Naturellement. Mais il lui semblait que tel n'était pas le cas un instant plus tôt.

C'était un peu effrayant. Burt ne ferait jamais de farces pareilles. Un client tardif non plus.

Il n'existait qu'une autre issue. Katelyn passa devant les ascenseurs. Les voyants lumineux indiquaient qu'ils étaient tous au sous-sol. Ce qui laissait...

Elle regarda dans l'autre couloir.

Vide. Deux portes vers l'extérieur. Silencieux, comme il se devait.

Puis elle entendit un clic. Très faible, venant de l'extrémité.

Bon, probablement un client très tardif. Qui était passé par l'escalier pour des raisons mystérieuses. Claustrophobe. Oui. Pas de quoi en faire un plat.

Sauf que... quelque chose ne tournait pas rond.

Le client avait dû passer derrière elle. Bizarre qu'il ne l'ait pas saluée, même s'il était ivre mort et un peu honteux en face du personnel.

À moins qu'il n'ait eu aucune raison d'être là.

Cela arrivait tout le temps. Les portes de l'hôtel s'ouvraient toute la journée et la moitié de la nuit. On entre, on avance vers le bureau de la réception, personne ne vous demande rien. En arrivant au bon moment dans l'après-midi, on pouvait tranquillement visiter quelques chambres.

Katelyn avait deux solutions. Descendre, prendre la radio qu'elle aurait dû porter sur elle – merde ! – et prévenir Burt ou essayer de secouer le type de la sécurité qui passait son temps à se branler au sous-sol. Burt, de préférence, qui ne lui demanderait pas pourquoi elle était directrice si elle avait besoin qu'on lui tienne la main dans le noir. Qui ne se poserait même pas la question. Mais d'autres s'en donneraient à cœur joie s'ils avaient vent de l'affaire !

Ce qui la conduisait à la seconde solution.

Elle s'éloigna des ascenseurs et s'engagea dans le couloir. Très calme, maîtresse de la situation, elle ramassa quelques menus en chemin. Déjeuners continentaux.

Derrière elle, elle perçut le bruit d'un ascenseur en mouvement.

Elle s'arrêta, se retourna, espérant qu'il s'arrêterait à l'étage et qu'un employé en sortirait. Elle en profiterait pour l'appeler, sous un prétexte ou un autre.

Les portes ne s'ouvrirent pas. Elle hocha la tête, irritée.

C'était son hôtel. Personne ne lui collerait la trouille.

Un autre menu. Quelques portes vides. Un autre menu.

Katelyn s'arrêta au milieu d'un couloir, se retourna.

Étrange. La chambre 511 n'avait pas de menu, mais l'affichette : « Vous pouvez faire la chambre.

C'était stupide. Qui mettait ça avant de se coucher ?

Elle poussa la porte qui s'entrouvrit. Il faisait sombre à l'intérieur. Bizarre. Le battant aurait dû être fermé, bien sûr. La fermeture automatique était une règle de sécurité fondamentale dans un hôtel moderne.

Elle frappa doucement. Pas de réponse.

Katelyn ne savait pas si la chambre devait être occupée. En plus de la radio, elle aurait dû avoir une liste. Elle n'en voyait jamais l'intérêt. Ou les gens voulaient déjeuner, ou ils ne voulaient pas. À quoi cela lui aurait-il servi ? À réveiller les clients, pour leur demander s'ils n'avaient pas oublié ?

Elle passa la main à l'intérieur et appuya sur l'interrupteur. Rien ne se produisit. Ah. Soudain, cela devenait plus compréhensible. Il y avait un problème dans la chambre 511 : l'interrupteur... Cela arrivait. L'affichette était sans doute là pour qu'on n'oublie pas de réparer.

Mais pourquoi ne lui avait-on rien dit ? C'était le genre d'information qui devait lui être transmise. Si on ne la prenait pas au sérieux, comment pouvait-elle accomplir son travail ?

Katelyn pinça les lèvres. Elle ne supportait pas qu'on la sous-estime.

Elle entra. L'atmosphère était confinée. Autour d'elle, l'air semblait affluer et refluer, telle une marée soulevée par le souffle de tous ceux qui dormaient autour d'elle. La rue et les lumières d'ambiance auraient dû donner assez de clarté pour qu'elle distingue les formes, mais les rideaux étaient tirés. Elle voyait que le lit était vide et encore fait, mais guère plus.

Elle s'approcha du bureau et tenta d'allumer.

Cela ne fonctionnait pas non plus. Bon, plus d'électricité. Elle ne comprenait pas comment cela pouvait se produire dans une seule chambre, mais...

Soudain, la pièce parut encore plus sombre, et elle entendit un petit clic. Le rectangle de lumière jaune de la porte avait disparu.

Katelyn entendit quelque chose qui pouvait être un bruit de pas sur la moquette. Elle recula un peu, se cogna dans le bureau. Elle avala sa salive.

— Il y a quelqu'un ?

Il ne répondit pas, mais il était là. Il sortit de l'ombre dense, le visage très doux dans la lueur sombre.

Katelyn essaya de reculer, mais sans savoir où se réfugier. Il avança d'un pas, et elle aperçut un éclat lumineux près de sa main.

Elle se reprit pour crier, mais à ce moment précis le visage passa sous un rayon de lumière filtrée, un nuage qui sortait de l'ombre d'un autre, plus sombre. Quelque chose dans ses traits la retint, et elle le regarda.

— Non, vous ne me connaissez pas. Personne ne me connaît.
Il approcha d'elle, à travers le temps, à une vitesse que personne n'aurait pu arrêter.

Personne n'eut ses œufs, ses toasts ni ses flocons d'avoine nature à l'heure le lendemain. Il y eut beaucoup de plaintes, surtout des deux derniers étages, où les menus avaient disparu de manière inexplicable. Ce ne fut qu'en début d'après-midi qu'un client entra dans la chambre 511 et les trouva répandus sur le sol d'une chambre vide, sans électricité.

L'hôtel s'arrangea pour que la disparition passe le plus inaperçue possible. La police interrogea d'abord Burt, mais il était aussi traumatisé que les autres et plus bouleversé que la plupart. Il aimait beaucoup Mlle Katelyn. La veille, il avait failli lui dire quelque chose lorsqu'ils s'étaient croisés près de l'ascenseur, il avait essayé de lui dire « b'soir » d'une manière plus personnelle, au cas où elle l'aurait trouvé trop lointain, parce que c'était la patronne et qu'elle était blanche. À présent, elle avait disparu. La plupart des gens pensaient qu'elle avait fait une fugue et qu'elle reviendrait dans quelques jours, un peu penaude. Une directrice de nuit, cela signifiait « pas de mec », et ce genre de bonne femme n'était jamais loin de la ferme Lunatique ou de la plage Prozac.

Burt, lui, savait que Mlle Katelyn n'était pas comme ça, et quand les portes de l'ascenseur s'ouvrirent le lendemain soir et qu'il ne la vit pas, il sut qu'elle était partie pour toujours, et pas pour le meilleur.

CHAPITRE 18

Quand Nina se réveilla, peu avant 5 heures, elle savait qu'il était inutile d'essayer de se rendormir. Pendant deux heures après le coup de téléphone de Monroe, elle avait cherché à comprendre ce que tout cela signifiait. Pour l'instant, elle ne voyait qu'une chose. D'une certaine manière, quelque part, Zandt avait piétiné les plates-bandes d'une personne proche des Hommes de Paille. Comme ils n'avaient pu s'en prendre à lui directement, ils lui avaient tendu un piège. Elle avait essayé toute la nuit d'entrer en contact avec lui. Son téléphone était coupé.

Vite dessaoulé, Ward avait fini par émettre une suggestion à prendre au sérieux. Elle devait informer Monroe de certains détails, en privé. Pas au téléphone. Face à face. Pour le convaincre que des hommes et des femmes opéraient derrière le visage de ce que la plupart de gens prenaient pour l'Amérique, qu'ils tuaient et avaient à présent son ex-amant dans le collimateur, il fallait qu'ils soient dans la même pièce. Elle aurait dû tout lui expliquer trois mois plus tôt, mais, en proie à une crise de paranoïa et avec plusieurs morts sur les bras, ni Ward ni elle n'avaient trouvé que c'était une bonne idée.

À présent, ils comprenaient leur erreur.

Elle but cinq tasses de café tout en réfléchissant à ce qu'elle allait dire. Que pouvait-elle révéler de ce qui s'était passé aux Halls sans qu'elle et Ward finissent en prison ? Elle attendit 7 heures, lorsque Monroe serait réveillé et levé. Si elle pouvait le coincer avant qu'il parte travailler, ils pourraient peut-être se rencontrer. Elle s'approchait du téléphone quand il sonna.

Monroe. Il était déjà au bureau. Il lui ordonna de le rejoindre immédiatement, et il n'avait pas l'air de quelqu'un à qui on pouvait confier un quelconque secret.

Le visage de marbre, il attendait devant l'ascenseur du cinquième étage.

— Charles, dit-elle, rapidement, il faut que je te parle.

Il hocha la tête sèchement et se tourna vers le couloir. Un peu plus loin, il ouvrit une porte et se retourna pour l'attendre. Elle pressa le pas pour le rattraper et entra.

La salle 521 était le genre d'espace anonyme qu'on trouve dans toutes les grandes entreprises américaines. Dans le monde des affaires, cela signifiait : « Vous voyez, on peut s'offrir ce qu'il y a de mieux. Nous n'avons peur de rien ! » Quel message cela devait transmettre devant les forces de la loi, Nina n'en avait aucune idée. Une grande table de bois poli trônait au milieu de la pièce, entourée des chaises les plus chères et les moins utilisées de tout le bâtiment. Un des murs de verre donnait sur le parking ; les autres, en dur jusqu'à hauteur de la taille, étaient totalement nus. Il y avait néanmoins une photo d'une personne recevant une décoration, pas très récente. Rien d'autre.

Un homme en costume anthracite était assis sur une chaise disposée de telle manière qu'elle sortait de l'angle gauche de la table. D'une taille supérieure à la moyenne, il avait ce genre de peau qui laisse croire, sur les hommes d'un certain âge, qu'on a pratiqué des injections. Les cheveux bien coupés, il avait des yeux ternes, d'un bleu très clair, et des cils longs. Il ne portait pas de cravate, mais sa chemise proclamait que cet accessoire ne lui était pas indispensable. Il avait une bonne cinquantaine. Bien qu'il eût été reconstitué en harmonie parfaite avec les canons de l'esthétique conventionnels, Nina trouvait que c'était l'homme le moins inoubliable qu'elle eût jamais rencontré. Rien ne précisait que ce n'était pas un agent, mais cela n'en était pas un. En tout cas, ce n'était pas le chef de la police de Portland, qu'elle avait rencontré.

— Bonjour, dit-elle, en lui tendant la main.

Il ne la lui serra pas. Il ne se présenta pas, il ne sourit pas. Nina garda la main levée pendant cinq secondes avant de la laisser retomber. Elle tint son terrain encore quelques instants, lui donnant une chance de cesser de se conduire comme un crétin. Il ne la saisit pas. Elle soutint son regard aussi longtemps qu'elle l'estima nécessaire, puis baissa les yeux.

Elle aussi pouvait jouer à ce petit jeu.

— Comme vous voulez.

— Assieds-toi et reste tranquille, aboya Monroe. Tu es là pour écouter. Tu parleras quand on t'interrogera. Sinon, tu la boucles. Pigé ?

Quelque chose devait avoir mal tourné. Monroe avait des défauts. Il se trouvait plus intelligent que les autres, il croyait que les criminels – ou les autres agents – obéissaient aux mêmes règles de gestion des ressources humaines que les représentants de commerce. Sur le reste, il était très professionnel, mais son ton exprimait une hargne personnelle.

Il la regardait toujours.

— Alors, pigé ?

— Bien sûr, dit-elle en écartant les mains. Qu'est-ce...

— L'affaire Sarah Becker, dit-il, et Nina eut un pincement au cœur.

Même si c'était lié à ce qu'elle avait à lui raconter, ce n'était guère une bonne entrée en matière. Pas en présence d'un témoin, et surtout pas du type assis dans le coin. D'ailleurs, pourquoi ne pas s'asseoir d'un côté ou de l'autre ? Il avait tout tenté pour se faire remarquer, et pourtant Monroe ne l'avait pas présenté. Il semblait ignorer sa présence. Un peu comme s'il y avait eu un fantôme au bout de la table, visible aux seuls yeux de Nina.

— Bon, dit Monroe en ouvrant son dossier. La famille Becker affirme que leur fille s'est retrouvée sur le palier, comme ça, après avoir été portée disparue pendant une semaine. Elle prétend avoir été relâchée tout près du lieu où on l'avait enlevée, à Santa Monica, et être rentrée par ses propres moyens. Un voisin soutient le contraire. Il dit que la fille a été ramenée par un homme et une femme, alors qu'une voiture conduite par un autre type attendait de l'autre côté de la rue. Le voisin est assez âgé, et je n'y aurais pas prêté attention si une adolescente de l'âge de Sarah et répondant au même signalement n'avait été admise aux urgences d'un hôpital de Salt Lake City la veille au soir. Elle a été soignée en même temps qu'une femme qui souffrait d'une blessure par balle sur le côté droit de la poitrine. Les deux patientes avaient disparu le lendemain matin. Et tout cela au moment où tu as eu une blessure identique, lors d'un accident de chasse au Montana.

Nina avait mal à la tête et son cœur était lourd comme la pierre. Elle haussa les épaules, sachant qu'elle ne pourrait plus rien dire à Monroe. Ni maintenant, ni jamais.

— La position de l'hôpital m'intéresse parce que, entre cet endroit et une ville appelée Dyersburg, dans le Montana, la ville dans laquelle tu avais filé la veille, il y avait une résidence, du nom de Halls, qui n'est plus qu'un trou dans la terre, phénomène que tout le monde, des flics du coin à la NSA, aimeraient bien expliquer. Les flics surtout, car ils ont sur les bras un officier disparu, le cadavre d'un agent immobilier et quelques victimes non identifiées.

Nina se tut. Monroe l'observait. L'homme aussi. Finalement, il commençait à l'énerver.

Elle se tourna vers lui et demanda :

— Vous, qui êtes-vous, exactement ?

L'homme la regarda un peu comme un planning de vacances d'une société pour laquelle il ne travaillait pas.

— Nina, tu me prends pour un idiot, c'est ça ? lâcha Monroe.

— Non, Charles, bien sûr que non. C'est bizarre. Je n'en sais pas plus que toi sur le retour de Sarah Becker. J'étais allée voir John dans le Montana, comme je te l'ai dit à l'époque et plusieurs fois depuis.

— Oui, approuva-t-il, d'une amabilité neutre qui la déconcerta.

L'intonation de sa voix lui permit de comprendre qu'il y avait quelque chose derrière tout cela et qu'elle n'allait pas tarder à le découvrir.

Ce ne fut pas Monroe qui prit la parole, mais l'homme mystérieux. Il avait une voix sèche, sans accent, bien qu'un peu nasale.

— Vous parlez de John Zandt, n'est-ce pas ?

— Oui, acquiesça Nina.

Monroe la donnait en pâture à ce type. Il ne semblait pas troublé.

— Un ancien policier de la Crime de Los Angeles, compromis dans une affaire de meurtre à Portland. Dont la fille, enlevée en mai 2002, n'a jamais été retrouvée. Il a quitté la police et disparu, avant de refaire surface il y a trois mois et de devenir votre amant.

— Situation qui a totalement changé. Et je me demande en quoi cela vous regarde.

L'instant de pause qu'elle avait marqué avant de poser sa question était censé lui donner du poids. Sans succès, car elle semblait être devenue inaudible. Aucun des hommes ne lui répondit.

Elle regarda Monroe en luttant pour garder son calme.

— Qu'est-ce que c'est que cette histoire ? Un coup de règle sur les doigts ? J'ai parlé à John de l'affaire du Garçon de Courses, je sais, je n'aurais pas dû. Tu le sais déjà. Il me semblait qu'il méritait d'être informé parce que sa fille avait disparu et parce qu'il nous avait aidés à coincer un homme qui tuait des enfants noirs à un moment où on n'aboutissait nulle part, avec les médias sur le dos. Tu m'as dit que mon attitude était contraire au protocole du FBI, tu m'as expliqué ta conception de la discrétion, et depuis, plus personne ne me traite de la même manière. J'ai déconné. J'ai compris le message. Je croyais qu'on en avait fini avec ça. On peut peut-être passer à autre chose.

Monroe regardait par la fenêtre.

— Nous ne sommes pas ici pour passer à autre chose, mademoiselle Baynam, reprit l'homme du coin. Nous sommes ici pour revenir sur l'affaire.

— De quoi voulez-vous parler ?

— Nina...

— Ça suffit, Charles ! J'en ai assez ! Je ne sais pas qui est ce type ni pourquoi il s'imagine qu'il a le droit de me parler sur ce ton.

Monroe posa un attaché-case sur la table dont il sortit un portable. Il l'ouvrit et orienta l'écran vers Nina. Ni lui ni l'homme ne tentèrent de changer de position et Nina comprit qu'ils avaient déjà vu ce qu'ils allaient lui montrer. L'écran s'alluma automatiquement et laissa apparaître une fenêtre noire au centre. Monroe tapota quelques touches, la fenêtre s'anima de couleurs changeantes. Il fallut un moment avant qu'elle réalise que c'était une image prise par une caméra vidéo, de l'autre côté d'une rue.

La rue resta vide pendant une seconde, révélant l'arrière d'une rangée de maisons. Puis le cadre se resserra sur une habitation. Une maison de bois à deux étages, aux peintures couleur de sable, avec des bordures blanches, pas très récentes. On la voyait de trois quarts, vue qui montrait les fenêtres arrière et une fenêtre latérale aux doubles rideaux tirés, ainsi que la porte arrière.

La scène resta statique quelque temps. Des voitures passèrent, une de la droite à la gauche, deux autres en sens inverse. Il n'y avait pas de son, mais Nina ne savait pas si la bande était absente du fichier ou si le haut-parleur était coupé.

Le plan de la caméra se resserra encore. Il lui fallut une seconde pour voir ce qui avait attiré l'œil du cameraman. La porte : entrouverte de quelques centimètres, révélant l'obscurité de l'intérieur. Elle se referma pendant une seconde, puis s'ouvrit pour laisser passer un homme. Un peu plus grand que la moyenne ; les épaules larges. Il referma derrière lui et longea l'arrière de la maison. Il se déplaçait de telle manière qu'un passant peu attentif n'aurait pas discerné son visage ni même remarqué sa présence.

Le vidéaste était plus observateur. Nina se mordit les lèvres. C'était John Zandt.

Il avança vers la route et la caméra le suivit jusqu'à une voiture que Nina reconnut, une voiture qui n'était plus à lui mais qui avait passé quelques après-midi garée devant chez elle quelques années auparavant. Il ouvrit la portière du conducteur, et, juste avant qu'il s'installe, la caméra eut une magnifique vue de son visage, au-dessus de la carrosserie. Pâle, il avait les yeux cernés. Il ressemblait à de nombreux hommes qu'elle avait vus en photographie, les mains menottées devant eux. Il n'avait plus grand-chose à voir avec celui qu'elle avait brièvement cru aimer.

La caméra reprit un angle large sur la moitié de la rue, puis le film s'arrêta brusquement.

Le visage d'une neutralité étudiée, Nina se cala sur sa chaise.

— D'où ça vient ?

— On nous l'a envoyé par courrier électronique. C'est arrivé aujourd'hui, au petit matin.

— Étrange coïncidence. Juste après qu'on eut retrouvé le corps à Portland.

Les deux hommes l'observaient attentivement. Allez vous faire voir, pensa-t-elle. Si c'est ce que vous voulez, il faudra vous débrouiller tout seuls.

— Et alors ?

— Eh bien, dit l'homme, votre petit ami s'est rendu dans la maison d'un type qu'on a interrogé à propos de l'affaire du Garçon de Courses, enquête à laquelle vous avez pris une part non négligeable. Stephen DeLong a été interrogé, il a fourni un alibi solide et on l'a éliminé de la liste des suspects.

— Les indices circonstanciels nous permettent de dater cette scène à l'époque de l'affaire, précisa Monroe.

— C'est ce qu'on veut nous faire croire ! s'exclama Nina. Comme ce grand travelling à la fin, pour que n'importe quel idiot sache bien d'où ça a été filmé.

Monroe cligna des yeux. L'homme ne sourcilla pas.

— Une semaine plus tard, des voisins ont signalé une odeur nauséabonde en provenance de la maison que nous venons de voir. On a retrouvé DeLong dans sa chambre, tué d'une seule balle. Il n'y avait aucun autre signe de violence physique. La maison présentait toute la panoplie du parfait petit dealer, ce qui avait conduit les policiers à penser que c'était le résultat d'un trafic qui avait mal tourné. DeLong a été rayé de la carte et oublié. Tout le monde s'en fichait et personne n'a relié sa mort à l'enquête en cours.

— On aurait dû ?

— À l'époque, il n'y avait aucune raison d'y penser. Mais comme vous le voyez, nous en avons une excellente maintenant. Nous avons besoin de votre aide sur un point, mademoiselle Baynam. Nous aimerions parler à John Zandt. Où est-il ? insista l'homme en se penchant en avant.

Quinze minutes plus tard, Nina sortit du bâtiment. Elle marchait le dos droit, avec des pas d'une égalité parfaite. Elle ne leva pas les yeux vers la salle 521, bien qu'elle soupçonnât fort Monroe de la guetter par la fenêtre. Si elle le voyait, elle risquait de rebrousser chemin, de remonter l'escalier à toute vitesse pour lui casser la figure. Elle était costaude. Elle y arriverait peut-être. Ça la soulagerait, mais ça risquait de jeter sa carrière par la fenêtre.

Elle rejoignit sa voiture et sortit du parking. Elle prit tout son temps et conduisit lentement, sans but. Moins de dix minutes plus tard, elle était toujours furieuse, et un peu effrayée de constater qu'on la suivait.

Nina s'arrêta à la première cabine téléphonique. Elle s'avança, se sentant dans la peau d'une actrice, et passa deux appels. À la première personne, elle demanda un service, attendit qu'on lui explique pourquoi ce n'était pas possible, puis fournit un motif succinct mais efficace pour prouver le contraire.

Tandis qu'elle attendait qu'on décroche, elle observa la rue et vit la voiture banalisée se garer à une vingtaine de mètres. Ou le type était un débutant, ou on lui avait donné l'ordre de ne pas se cacher. L'un comme l'autre l'énervait.

Au bout d'une dizaine de sonneries, on décrocha.

— Bon, ça se gâte sérieusement, dit-elle à un répondeur. Reste à l'écart et fais gaffe à tes arrières.

Elle raccrocha et retourna à sa voiture. En passant devant la berline grise, elle adressa un bras d'honneur au conducteur. Il la regarda d'un air impavide mais ne la suivit pas. En rentrant chez elle, elle se sentait déçue d'avoir les yeux humides : des larmes de fureur tout autant que de tristesse. La fureur, c'était parfait. La colère, cela menait quelque part.

— Tu vas le regretter, Charles, marmonna-t-elle.

Cela la soulagea un peu, mais pas pour longtemps. En tant qu'agent suspendu, avec un ex-petit ami soupçonné de deux meurtres et un patron qui ne croyait plus un mot de ce qu'elle disait, elle se demandait comment elle pourrait inspirer des regrets à qui que ce soit.

— On file d'ici, dit Ward.

Il fourrait un ordinateur dans le sac avec lequel il était arrivé. Il avait attendu sagement que Nina hurle pour la deuxième et troisième fois dans le répondeur de Zandt avant de lui arracher le téléphone.

— Peu importe de savoir qui est ce type en costume. On sait pourquoi il est là. Il s'occupe de l'enquête sur John et il est assez puissant pour avoir ses entrées au FBI et s'arranger pour que le patron te traite de cette façon. Tu es sûre que ce n'est pas une huile du Bureau ?

— Ça n'y ressemblait pas.

— Peu importe. Il s'occupe de sécurité quelque part, et, soit il appartient aux Hommes de Paille, soit il fait ce qu'ils lui demandent. Ça signifie que nous ne sommes plus en sécurité, ni dans cette maison ni dans cette ville.

— Mais où veux-tu aller ?

— Ailleurs. Tu parles russe ?

— Ward, il faut retrouver John. Il court un plus grand danger que nous. Ils essaient de le coincer pour un truc qu'il n'a pas commis.

— Peut-être ben que oui, peut-être ben que non.

— Comment ça ? s'étonna-t-elle.

— La seule chose qu'on sait, c'est ce qu'il nous a dit. Il t'a raconté qu'il était en Floride, et à moi aussi. Il avait de bonnes raisons. Ni l'un ni l'autre n'allons essayer d'obtenir un mandat pour retracer son appel.

— Mais pourquoi aurait-il éliminé ce Ferillo ?

— Tu crois que c'est impossible ? Il a bien tué l'homme qu'il a pris pour l'assassin de sa fille. Et à l'époque, il était flic.

— Il devait avoir une bonne raison, répliqua Nina.

— Peut-être. Nous ne le saurons pas tant qu'il ne nous répondra pas. Entre-temps, tu as un moyen de retracer les appels sur son portable ? Si nous pouvons retrouver la ville d'origine, ça confirmera son alibi.

— Je m'en suis occupée. J'ai passé un coup de fil avant de venir.

— Parfait. Mais en attendant, fais ta valise.

— Ward, je ne veux pas quitter ma...

Il cessa de préparer ses bagages, lui posa les mains sur les épaules. Il la regarda dans les yeux. Elle se rendit compte qu'ils n'avaient jamais été si proches l'un de l'autre. Elle comprit aussi que c'était un homme qui avait passé trois mois sur la route, non pour le plaisir, mais parce qu'il savait que ce moment ne manquerait pas d'arriver.

— Si. Nous savions qu'ils finiraient par s'intéresser à nous pour de bon. C'est fait. Ça commence.

Une heure plus tard, ils passèrent devant le 99, Bakersfield, en direction du nord. Ward conduisait vite et sans un mot. Le portable de Nina sonna, et elle se cassa un ongle en le sortant de son sac. Elle jura et regarda l'écran.

— C'est John ? demanda Ward.

— Non. Je ne reconnais pas le numéro. C'est peut-être le coup de fil que j'attends. Ou...

— Si c'est Monroe, ne lui dis rien et coupe tout de suite.

Elle décrocha. Elle écouta la voix de Doug Olbrich, qui avait fait ce qu'elle lui avait demandé. Elle lui posa trois questions qu'elle avait déjà formulées mentalement. Lorsqu'elle obtint les réponses, elle coupa la connexion et mit sa tête dans ses mains.

Ward lui accorda exactement vingt secondes.

— Alors ?

Elle ne bougea pas.

— C'était un type de la police de L.A. que je connais. Il dirige la brigade qui s'occupe de l'assassin au disque dur. Je lui ai demandé de vérifier des trucs. Il a un type très fort pour ça. (Soudain, sans avertissement, elle cogna le tableau de bord de toutes ses forces.) J'ai merdé, Ward.

— Pourquoi ?

— Olbrich a réussi à obtenir le compte du mobile de John. Il a retracé certains appels. John a passé quelques appels sur ton numéro de portable avant-hier.

— La belle affaire ! On devait se retrouver à San Francisco. Mais il m'a dit qu'il était en Floride.

Elle hocha la tête, silencieuse. Elle regardait ses mains, sur ses genoux. La cuticule de son ongle cassé saignait.

— Nina ?

— Il a menti. Il n'a pas mis les pieds en Floride depuis six semaines. Il était à Portland le jour de la mort de Ferillo.

TROISIÈME PARTIE

UN JOUR DE PLUIE

Le sens de la vie, c'est qu'elle prend fin.

Franz KAFKA

CHAPITRE 19

On la retrouva dans les buissons. Ça arrive. On retrouve aussi des gens dans les bois, dans des chambres surchauffées et désordonnées, dans des ruelles, des parkings, des sorties de cinéma, on en retrouve dans les piscines ou les voitures. On retrouve des morts presque partout, mais les buissons, c'est souvent le pire. Leur état et leur situation laissent peu de place à l'idée réconfortante qu'ils sont peut-être simplement endormis, ivres, évanouis, inconscients pour une raison ou une autre mais capables de revenir dans le monde des vivants. Dans les buissons, les morts sont vraiment morts.

Ces buissons se situaient au fond du parking de Cutting Loose, un salon de coiffure sur la grand-rue de Snoqualmie. Le corps fut découvert, comme c'est souvent le cas, par un homme qui promenait son chien tôt le matin. Il avait monté la garde assez longtemps pour passer un coup de fil sur son portable et attendu près de l'endroit, mais pas trop, pour éviter d'attirer les curieux, avant d'indiquer le buisson aux deux policiers envoyés par le shérif ; à présent, il était prostré, dos à la palissade un peu plus loin, la tête entre les genoux. Un peu troublé par l'odeur de vomi, mais fidèle et loyal, son chien était à ses côtés. Lorsqu'ils seraient rentrés, l'animal resterait confiné dans sa niche pendant une longue journée où son maître serait absent. Il n'était donc pas pressé de rentrer. Si le prix de la liberté était un peu de pluie près d'une flaque de vomi, tant pis. Il lécha la main de son maître pour lui manifester sa sympathie. La main le caressa, mollement.

L'un des policiers avertissait le central par radio. Le deuxième se tenait à quelques mètres du corps, les mains sur les hanches. Il n'avait pas encore vu beaucoup de cadavres, et il se sentait soulagé de savoir que son collègue allait bientôt revenir et reprendre la situation en main, que ce ne serait pas à lui que reviendrait la responsabilité, pendant les prochains jours, les prochaines semaines ou toute l'éternité, de déterminer par quel processus on avait transformé un être vivant en cette créature livide.

Il ne voulait pas trop creuser l'esprit de l'homme – en admettant que ce soit un homme, puisque c'était presque toujours le cas – qui estimait normal, voire commode, de jeter une personne à quelques mètres de la route, comme un vulgaire déchet. Pire encore, parce que les ordures, on les met dans des sacs. Ce corps avait été abandonné comme s'il valait moins que cela, comme s'il ne méritait pas l'enterrement temporaire, en surface, qu'on offrait aux boîtes de conserve et de céréales vides.

Il entendit son collègue couper la communication et décida qu'il en avait assez vu. En se détournant, néanmoins, il remarqua un objet brillant au sommet du crâne. Contre toute logique, mais pour agir en bon policier, il s'approcha d'un pas et se pencha pour regarder de plus près.

Il ne faudrait pas longtemps pour déterminer les causes du décès. La femme était vêtue d'un tailleur élégant ou plutôt de ce qu'il en restait. Elle avait un objet enfoncé dans la tête, couvert de brun, du sang séché sans doute, et d'un autre matériau plus sombre, si noir qu'il était difficile d'en deviner les formes. C'était au milieu, juste au-dessus du front.

— Fais gaffe, dit son collègue. Si tu abîmes les lieux, on t'arrache le trou du cul et on t'en fait une bague !

— Je sais, je sais.

Il se pencha encore un peu plus. Il n'irait pas plus loin.

L'odeur était étrange, la vue épouvantable. C'était affreux.

Dans le carnage du front, un objet semblait déplacé.

Il retint son souffle. De si près, on ne pouvait éviter de voir les fourmis et autres insectes qui accomplissaient leur tâche, en hâte, comme s'ils savaient qu'on allait bientôt les priver de leur proie. On distinguait un objet enfoncé dans le front. La partie qui dépassait avait la largeur d'une carte à jouer, bien que plus épaisse... un demi-centimètre, un peu plus, peutêtre.

— Qu'est-ce que c'est que ce truc ?

Peu après 9 heures, on sut que l'objet était un disque dur, de petite taille, comme on en trouve dans les portables. Il ne fallut pas longtemps pour que l'information soit transmise au FBI à Everett, puis à Los Angeles. Ensuite, ce fut un vrai boxon.

Charles Monroe essaya tous les numéros dont il disposait, mais Nina Baynam ne répondait pas. Il insista néanmoins, à intervalles réguliers. Quelque chose avait mal tourné dans sa vie, il ne comprenait pas comment, mais la situation empirait de minute en minute.

CHAPITRE 20

Henrickson coupa le moteur et se tourna vers Tom avec un sourire. C'était bien, estima Tom, le quinzième de la matinée, et il n'était que 10 heures du matin.

— Vous êtes prêt ?

Tom attrapa son sac à dos.

— Je crois.

Il s'était écoulé quarante-huit heures depuis son retour à Sheffer. La veille, après une nuit d'insomnie, il s'était levé trop patraque pour une longue marche dans les bois. L'adrénaline qui l'avait ramené à Sheffer était tarie, le laissant épuisé, perclus de douleurs et nauséeux. Mais il était temps de réfléchir pour de bon.

Henrickson ne l'avait pas bousculé et l'avait laissé se reposer. C'est ce que Tom avait fait au début, installé dans le fauteuil de sa chambre, emballé dans autant de couvertures que possible ; il avait tenté de remettre ses idées au clair. Au début de l'après-midi, il avait fait une longue promenade en voiture et n'était rentré qu'à la tombée de la nuit. Il s'était enfin senti assez bien pour aller boire un verre avec le journaliste. Ce matin, il allait mieux, même si ce n'était pas encore la grande forme. Il se sentait plus calme, peut-être, plus distancié.

S'engager dans l'aire de repos devant la piste de Howard's Point provoqua une réaction plus vive qu'il ne l'aurait cru. À l'idée de retourner vers son nid dans le ravin, il avait l'impression d'être un esprit qui rentrait chez lui et, en sortant de la Lexus d'Henrickson, il croyait être dans la peau de son propre grand-père. Le journaliste s'était garé de l'autre côté du parking, à l'opposé de l'endroit où Tom était allé se reposer – et était tombé pour la première fois – mais il en fut d'autant plus troublé. Lorsque l'écho de la portière qui se refermait se propagea dans les arbres, le paysage lui parut empreint d'une fragilité vacillante, comme si on l'avait peint à la hâte, pardessus une autre scène. La charge émotionnelle s'était transformée. Bien sûr, la dernière fois, il était ivre mort alors qu'il n'avait plus aujourd'hui qu'une légère gueule de bois, et la neige était plus épaisse.

— Jim, ce sera très difficile de retrouver l'endroit précis.

— Bien entendu.

Le reporter s'était débarrassé de son costume, et portait un vieux jean et une veste éreintée. Ses bottes témoignaient d'une longue expérience de randonneur. Il était bronzé, affûté, bien mieux préparé que Tom.

— Il faisait presque noir. Ce n'est pas la fin du monde si vous ne retrouvez pas l'endroit exact. Mais... ce serait bien quand même.

— Vous ne voulez pas me dire ce que vous cherchez ? Sourire numéro 16.

— Vous n'aimez pas les surprises ? rétorqua Henrickson.

— Pas vraiment.

— Faites-moi confiance. Ce sera formidable : « Kozelek nous mène à l'endroit qui a changé le cours de l'histoire, de la biologie et de je ne sais quoi, tels que nous les connaissions. Ses écrits audacieux apportent une preuve irréfutable. Ils partagent une accolade virile. » Ça, c'est un truc de potes. L'accolade, ce n'est pas obligatoire.

Tom hocha la tête en regrettant une fois de plus d'avoir évoqué cette idée de livre. Henrickson avait prétendu ne pas vouloir le saouler, et Tom l'avait cru, mais à la fin de la seconde journée, il lui avait tout raconté. Presque tout.

— Je n'ai pas envie de me perdre.

— Nous ne nous perdrons pas. Je fais de la randonnée. J'ai une boussole et je sais m'en servir. Et si vous n'aviez pas le sens de l'orientation, vous ne seriez plus là pour vous plaindre.

— Sans doute.

Tom fit doucement pivoter sa cheville. Ses nouvelles chaussures amélioraient la situation. Il enfila son sac à dos. Cette fois, il contenait une bouteille d'eau, une Thermos de café sucré, quelques biscuits. Il restait sans doute encore un peu de verre brisé au fond, mais peu importait. Tom avait justement pris ce sac parce qu'il l'avait la dernière fois. Le verre venait de là aussi. Il avait l'intention de le jeter quelque part dans la forêt, de laisser derrière lui tout ce qu'il représentait.

Il s'approcha de la piste, hésita un moment, puis enjamba le tronc épais qui délimitait le parking.

Henrickson attendit qu'il avance de quelques mètres et se retourna. Pendant une fraction de seconde, il sentit quelque chose derrière sa nuque, un peu comme si on l'épiait. Il scruta les environs mais ne vit personne. Étrange. En général, il ne se trompait pas sur ce genre de chose.

Quand il se retourna, il constata que Kozelek s'était arrêté. À présent que le périple avait commencé, l'enthousiasme du bonhomme augmentait, comme prévu.

— C'est par là !

Henrickson enjamba le tronc et suivit Tom dans la forêt.

Malgré un banc de nuages à l'ouest, le soleil fort et lumineux projetait des ombres sur la neige vierge. Les deux hommes marchaient lentement le long de la pente, sans dire grand-chose. La route était déjà loin derrière eux, et on n'entendait plus que le son de leur souffle et de leurs pas.

— Vous semblez bien sûr de vous, Tom. Vous vous souvenez d'être passé par ici ?

— Ce n'est pas que je m'en souvienne. Je... reconnais les formes générales. Ça paraît idiot, et je n'ai pas l'habitude de la nature, mais... (Il s'arrêta pour indiquer la silhouette des arbres et des collines.) Vous voudriez passer par un autre chemin ?

— Je comprends ce que vous dites. Certaines personnes n'ont aucun sens de l'orientation, un peu comme des jouets mécaniques. On les remonte et elles vont tout droit, jusqu'à ce qu'elles tombent sur un obstacle. D'autres sentent les choses. Cela se travaille avec le temps. Quelle heure est-il, à votre avis ? Réfléchissez une seconde. En fait, non, répondez d'instinct.

Tom réfléchit. Il devait s'être écoulé une demi-heure depuis qu'ils s'étaient mis en route.

— Environ 10 h 30.

— On serait plus près de 11 heures ; 11 heures moins cinq, je dirais. (Il tendit son poignet et regarda sa montre. Un sourire, et il montra le cadran à Tom.) Pas loin. Moins quatre.

Tom s'arrêta. Ils arrivaient au bord d'une crête, et il hésitait. Henrickson recula de quelques pas et regarda en arrière. Tom comprit qu'on lui donnait une chance de réfléchir, de ressentir, et fut envahi par un absurde élan de gratitude. William et Lucy étaient assez grands pour voir ses défauts plus que ses qualités. Sarah le connaissait trop bien. La malédiction de l'homme mûr. Dès qu'on a cette idée en tête, on tient à prouver le contraire : et c'est là qu'on commet des erreurs, là que tout se gâte.

— Par là, dit-il en tournant à droite.

Les vingt minutes suivantes furent assez rudes, et il fallut un moment avant qu'ils récupèrent assez de souffle pour parler. Le terrain entamait une descente de l'autre côté, mais une ascension encore plus ardue les attendait. Le paysage ne lui rappelait plus rien, mais il lui semblait que c'était le bon chemin.

Tom jeta un coup d'œil vers le journaliste, qui marchait à côté de lui d'un pas ferme.

— Vous devez chercher Bigfoot depuis un sacré bout de temps ?

— C'est sûr, répondit Henrickson.

— Comment ça se fait que personne n'y croie ?

— Oh, tout le monde y croit ! Mais c'est une vérité difficile à admettre et personne n'a envie de passer pour un illuminé.

— Alors, qu'est-ce que c'est ?

— À votre avis ?

— Un grand singe, je suppose, dit Tom. Une créature qui vivait ici avant les hommes et qui s'est réfugiée dans la forêt. Ce n'est pas l'espace qui manque par ici, pas vrai ?

— À moitié vrai. Moi, je suis persuadé qu'il s'agit de l'un des derniers spécimens de l'homme de Neandertal.

Tom s'arrêta net.

— Quoi ?

Henrickson avançait toujours.

— Ce n'est pas une théorie nouvelle. Le seul problème, c'est de faire coïncider les détails. Vous savez comment sont les archéologues... Bla-bla-bla, pas de preuves. Bla-bla-bla, les fossiles montrent que... Bla-bla, mon professeur dit que c'est impossible. De la manière dont je vois les choses, vous avez un homme de Neandertal, l'une des espèces les mieux adaptées que le monde ait jamais connues. Ces types fabriquaient des lances voilà plus de quatre cent mille ans ! Ils se sont dispersés sur la moitié du monde, y compris les deux côtés de l'Europe, à une époque où il ne faisait pas bon y vivre. En pleine période glaciaire, avec des pléthores d'animaux aux dents acérées, ils ont survécu pendant des centaines de milliers d'années. Ils pratiquaient des rites funéraires. Ils avaient des dentistes. Ils commercialisaient des ornements et des bijoux, et disposaient d'un réseau pour les distribuer dans toute l'Europe. Ensuite, l'homme de Cro-Magnon est arrivé : c'est nous. Pendant un moment, les deux espèces ont coexisté. Et après, l'homme de Neandertal aurait disparu d'un seul coup, en laissant tout juste assez d'os pour remplir un sac à main ?

— Et alors ? Qu'est-ce qui se serait passé, d'après vous ?

— Il n'a jamais disparu. Ils n'ont jamais été très nombreux. Ils sont devenus experts dans l'art du camouflage.

— Du camouflage ? Mais où ?

— Dans deux types d'endroits. Tout d'abord dans les forêts et les montagnes, au nord de l'Europe de l'Est, en Finlande et dans l'Himalaya, mais ici aussi, dans notre bonne Amérique. Les paléontologues prétendent que l'homme de Neandertal n'a pas pu se rendre en Amérique du Nord. La théorie, c'est que l'homme y est arrivé par voie terrestre et que cela ne s'est produit que quinze mille ans avant Jésus-Christ. C'est sousestimer le Neandertal. Il n'y a aucune raison pour qu'il n'ait pas eu de petits bateaux. Il aurait atteint la côte en partant de Russie, se serait arrangé pour traverser les eaux glacées près des territoires du Nord, puis descendre le long de la côte jusqu'à une région habitable. Puis, une fois arrivés en force, les Neandertal se seraient aventurés dans les forêts. C'est l'endroit rêvé. Des milliers de kilomètres carrés de terres vierges que personne ne vient troubler, pas même aujourd'hui. On en trouve des traces dans toute la culture amérindienne. Chez les Chinooks, des légendes évoquent le « peuple fantôme » qui vit dans des lieux à lui et avec lequel ils ont des relations. Et il y a le « peuple animal » des Okanogans : il habite dans les montagnes, et on dit que ce sont d'anciens animaux qui avaient une culture avant que le « vrai peuple », c'est-à-dire l'homme, se soit formé.

— Et l'autre endroit ? L'autre cachette ?

— Sous notre nez. Quel est le type de légende le plus répandu en Europe ?

— Je ne sais pas.

Tom ne savait pas non plus s'il suivait toujours le bon chemin. Ils avaient dépassé le fond du vallon et recommençaient à monter. Le terrain de plus en plus accidenté lui était familier, mais sans plus, et, en toute direction, la pente était très escarpée. Pour l'instant, il se contentait d'avancer, et Henrickson de parler, avec le débit régulier de celui qui a longuement réfléchi à la question. Et avec la confiance de celui qui n'est pas aussi intelligent qu'il le croit.

— Les fées. Les ogres. Les lutins. Les trolls. Pour moi, ce sont tous des survivants de l'homme de Neandertal. Des créatures qui vivaient avant nous et observaient d'étranges coutumes. Assez communes au début, puis de plus en plus rares, jusqu'à ce que personne ne les croise plus jamais. Elles restent dans les mémoires. Le langage fonctionne d'une étrange manière. Vous avez bien entendu des trucs du genre : « À cette époque, vivaient des géants ?« Il me semble que » géant « n'implique pas qu'ils étaient grands de corps. Cela signifie que les premiers explorateurs ont rencontré une population puissante et civilisée, comme le peuple animal des Okanogans : une espèce géante, sur le plan culturel.

— Mais ils se sont éteints.

— Pas complètement. Qu'est-ce qu'on entend d'autre, dans le monde entier ? Des histoires de fantômes. Des présences intangibles. Et quoi encore ? Des extraterrestres. Des hommes verts. Qui, comme par hasard, font atterrir leurs vaisseaux dans des forêts, ce qui est un endroit plutôt étrange, vous ne croyez pas ? Des hommes verts, des fées, des fantômes, cela explique toutes les bizarreries que l'on rencontre de temps en temps. Cela explique l'existence de toute une espèce que l'on prétend éteinte mais qui s'est juste réfugiée dans l'ombre, qui rôde autour de nous sans jamais croiser notre chemin.

— Je ne vois pas le rapport avec l'homme de Neandertal, objecta Tom.

— Non, et pour deux bonnes raisons. D'abord, les légendes exagèrent toujours avec le temps. En des centaines de milliers d'années, elles prennent leur propre poids, elles obéissent à leurs propres règles. Les fées ressemblent à ci ou ça, les lutins sont habillés en vert, les fantômes trimballent toujours une triste histoire derrière eux. Et puis le Neandertal a ses propres méthodes pour troubler nos esprits.

— Lesquelles ?

— On dit que la bouche et la gorge de l'espèce n'étaient pas vraiment aptes à la parole articulée. Et pourtant, ils ont fabriqué des tas de trucs qui prouvent qu'ils communiquaient, d'une manière que le langage corporel et quelques grognements n'auraient jamais permise. Ma théorie, c'est qu'ils recouraient au moins partiellement à la télépathie. Ils continuent, et nous aussi, d'ailleurs, de temps en temps. La télépathie, c'est simplement de l'empathie poussée à l'extrême. Lorsqu'ils sont confrontés à un danger, à nous par exemple, ils envoient des images dans notre tête. Nous voyons des formes dans notre esprit. Elles nous renvoient à notre propre imagination.

— Ça ne tient pas debout, dit Tom, distrait. Je suis désolé, mais je n'en crois pas un mot.

— Pensez à notre petite expédition. Si j'ai raison et qu'on cherche un Neandertal, pourquoi tous ceux qui ont vu Bigfoot disent-ils qu'il mesure deux mètres cinquante ? Ils nous font croire qu'ils sont grands, parce que cela nous fait peur. Et pourquoi tant de gens, comme vous, Tom, parlent-ils d'une odeur nauséabonde ? Pourquoi sentiraient-ils mauvais ? Il n'y a aucune raison. Ils nous font simplement croire qu'ils puent. C'est un mécanisme de protection, l'un des plus simples qui soient. Ils se cachent en crachant un écran de fumée. C'est pour cela qu'ils sont si difficiles à trouver. Près des zones civilisées, on croit apercevoir un fantôme. Dans la forêt, on ne s'attend pas à voir des esprits, sauf sur une route déserte, c'est pour ça qu'on voit Bigfoot. On voit quelque chose de plus proche de leur véritable forme, parce que, au fond, nous nous doutons qu'ils sont toujours là.

Tom s'arrêta et regarda le journaliste. Il ne souriait pas, pour une fois. Il restait sérieux comme un pape. Même si Tom se réjouissait d'avoir quelqu'un à côté de lui, il aurait préféré de beaucoup que l'homme soit persuadé d'avoir affaire à un étrange primate en liberté plutôt que de divaguer sur les lutins et la prise de contrôle de l'esprit.

Pour l'instant, c'était secondaire. Tom avait une mauvaise nouvelle à annoncer.

— Je suis complètement perdu.

Une heure plus tard, la situation ne s'était guère améliorée. Henrickson s'était montré patient, avait souvent pris ses distances pour que Tom se repère, l'avait encouragé à aller de l'avant, en lui disant qu'il le rattraperait dès qu'il aurait retrouvé sa piste. Mais Tom ne retrouvait rien. Plus il avançait, moins il savait où il allait. Il finit par s'arrêter.

Henrickson l'appela.

— Alors, on chauffe, mon vieux ?

— Non, je ne sais pas où on est.

— Ce n'est pas grave, dit Henrickson lorsqu'il le rattrapa.

Il fouilla dans sa poche et en sortit une carte de randonnée. Il la déplia, consulta la boussole attachée à sa veste par un cordon puis dessina un petit cercle.

— On est là, reprit-il.

Tom regarda.

« Là », c'était une zone blanche avec quelques lignes topographiques serrées... La dernière demi-heure avait été une suite de montées et de descentes périlleuses.

— Au milieu de nulle part, commenta Tom.

— Pas tout à fait. Il y a un courant par ici, insista le reporter en indiquant une ligne sinueuse. Vous pensez que nous sommes assez près pour que ce soit votre ravin ?
— Je n'en sais rien. On pourrait toujours aller voir.
— Alors, allons-y.
Vingt minutes plus tard, ils perçurent un bruit de ruissellement. Près d'une formation rocheuse, ils découvrirent un ruisseau de montagne, d'un mètre cinquante de large, qui coulait entre des petits bancs de mousse irréguliers.
Henrickson regarda en amont.
— Ça devrait être plus raide de ce côté.
— Peut-être. Je ne sais pas.
Henrickson, qui paraissait en forme et plein d'allant au début de l'expédition, n'avait plus souri depuis un moment.
— Je sais à quoi vous pensez, mon vieux, mais ça ne pose pas de problème. Je vous l'ai dit, j'ai vraiment envie de trouver cette créature. Autant être ici, non ? Alors, suivons ce chemin. On cherche quelque chose qui y ressemble et il n'y a rien d'autre sur la carte. Mais d'abord, j'aurais besoin d'un peu de carburant.
Tom commença à se débarrasser de son sac à dos mais Henrickson leva la main.
— Inutile, je m'en occupe.
Il défit les lanières, et Tom sentit qu'une main fouillait le haut du sac.
— Faites attention, il y a du verre à l'intérieur.
— Ah bon. Comment ça se fait ?
— Il reste quelques bouteilles cassées de la dernière fois. Je ne l'ai pas bien vidé. J'aurais dû le retourner, mais...
Le journaliste ne l'écoutait plus. Il avait retiré ses mains du sac.
— Ça va ?
Pas de réponse. Tom se retourna et s'aperçut que Henrickson tenait quelque chose qui n'était pas la Thermos de café.
— Qu'est-ce que c'est ?
— À vous de le dire, c'était dans votre sac.
Tom regarda de plus près et vit un petit bouquet de plantes fanées.
— Aucune idée.
— Sans doute rien. Cela a dû tomber dans votre sac. Il leva les yeux vers Tom, avec cette fois un sourire d'une oreille à l'autre.
— Allez, on y va !

Tandis qu'ils avançaient en buvant le café chaud, Tom remarqua que le pas du journaliste avait redoublé d'énergie.

Quarante minutes plus tard, ils avaient grimpé plusieurs centaines de mètres. Ils suivaient le courant sur un parcours accidenté. Les rives ne semblaient pas monter plus haut. Cette fois, ce fut le journaliste qui s'arrêta.

— Cela ne me dit rien qui vaille. (De nouveau, il sortit sa carte.) Nous devrions être par ici, dit-il en montrant d'autres lignes, ce qui est bien plus à l'est que je ne l'aurais voulu.

— Qu'est-ce que c'est, cette ligne noire ?

— Une route. Vous n'avez pas pu la rater lorsque vous avez essayé de rebrousser chemin, mais regardez les lignes topographiques. Ça descend jusqu'ici, c'est sans doute là que vous avez atterri. Dans ce cas, il ne vous aurait pas fallu deux jours pour rentrer chez vous. Alors... Vous allez bien ?

Tom avait la bouche légèrement ouverte. Il la referma lentement et parla à contrecœur.

— Oui, c'est juste que...

— Je vous sens un peu noué. C'est mauvais pour la digestion.

— La femme, Patricia. Celle qui avait les bottes.

— Oui, eh bien ?

— Elle était là. Elle a vu mon sac et, d'après ce qu'elle raconte, c'est elle qui aurait laissé les empreintes. L'adjoint du shérif a dit qu'elle habitait quelque part par ici. Ce qui signifie...

— Qu'elle sait où se trouve l'endroit et qu'elle pourra peut-être le retrouver ? C'est ça ?

Tom acquiesça.

— Vous n'y avez pas pensé plus tôt, ou vous n'aviez pas envie que quelqu'un d'autre intervienne dans l'histoire ?

— Non, franchement, je n'y ai pas pensé. J'étais malade lorsqu'elle est venue au commissariat.

— Allons-y ! décréta Henrickson, les mains sur les hanches, en regardant un instant de l'autre côté. Bon, j'aurais pu y penser tout seul. Et puis, c'est vrai, ça aurait été chouette d'y aller tout seuls. Mais on n'y arrivera pas, c'est ça ?

— Je suis désolé.

— Pas grave. Mais maintenant, il faudrait retourner à la voiture et aller chercher de l'aide. Si cette femme peut nous accompagner, ça nous fera gagner beaucoup de temps. Et le temps, c'est primordial.

Henrickson sortit sa carte encore une fois et consulta sa boussole.

— On va couper par là. Le sens de l'orientation, c'est bien beau, mais on va faire court, non ?

Il rebroussa chemin et Tom le suivit.

Il leur fallut un peu plus d'une heure pour regagner la piste, grâce à un chemin plus direct en descente. Lorsqu'ils franchirent à nouveau le tronc d'arbre, Tom comprit que la situation avait changé. C'est lui qui suivait. Cela ne devait pas se passer ainsi. Si nécessaire, il interviendrait pour inverser la situation.

Henrickson reprit la route et parcourut quelques kilomètres en direction de Sheffer. Il s'arrêta à une station-service et posa quelques questions en faisant le plein. En s'asseyant dans la voiture, il fit un clin d'œil.

— Je crois avoir trouvé notre bonheur. À quelques kilomètres, dans l'autre sens. Un nouveau lotissement, appelé Cascade Falls. Ça n'a jamais marché. Mais il y a au moins un habitant. Le type pense que la femme s'appelle Anders.

— C'est elle !

— Formidable, les affaires reprennent !

Il leur fallut presque une demi-heure pour se diriger vers le nord puis bifurquer vers les montagnes. La route se rétrécit soudain. Créée par le promoteur, elle se contentait de mener aux terres qu'il avait essayé de vendre. De chaque côté, elle était bordée d'une rangée d'arbres touffus.

— C'est sûrement la route la moins fréquentée du monde, opina Henrickson, tout joyeux.

Tom se demandait quelles raisons pouvaient bien inciter quelqu'un à s'installer ici. De temps à autre, on voyait une pancarte, clouée à un arbre. On pouvait acheter une parcelle de terrain et venir vivre ici. Mais pour quoi faire ?

Henrickson s'arrêta et coupa le moteur. Juste devant, à gauche, le nom « Anders » était peint sur une plaque de bois.

Ils sortirent, ouvrirent le portail et suivirent un sentier sinueux à travers les arbres. Deux cents mètres plus loin, ils aperçurent un bungalow. Tom se demandait s'ils étaient au bon endroit. La bâtisse paraissait minuscule, glaciale, malgré la lumière allumée au-dessus de la porte.

— Ça ne ressemble guère à une maison.

On aurait plutôt dit une cabane, une grange de bois, avec une entrée de garage sur le côté. L'entrée de la maison proprement dite, en contrebas, faisait face à la piste : une simple porte marquée « 2 », à hauteur de la taille, avec quatre panneaux vitrés sur la moitié supérieure, mais un rideau qui obstruait la vue.

Henrickson frappa.

— Un peu compact, pour le moins.

Comme il n'obtenait pas de réponse, il frappa de nouveau. Tom alla explorer un peu plus loin. À une vingtaine de mètres, un autre cabanon caché dans une végétation envahissante était plongé dans le noir. En approchant, il distingua le faible reflet d'une mare gelée, qui appartenait sans doute à la propriété. Plus loin, on discernait la ligne des arbres, mais...

Il avança encore et crut entrevoir une troisième cabane, de l'autre côté. Il songea à appeler Henrickson mais y renonça, sans savoir pourquoi. Il fit demi-tour.

Henrickson frappait pour la quatrième fois.

— Personne. Elle est sans doute à Sheffer, pour profiter de la ville et des lumières. Pas de chance. Mais... (Il regarda sa montre) Le temps passe. D'après vous, elle a dit que l'endroit où elle vous avait trouvé était à bonne distance de sa maison ? Nous pourrons peut-être tenter l'aller-retour dans la journée, quand même.

Il s'éloigna de la porte et s'approcha d'une des deux petites fenêtres sur le côté, également fermées par un rideau, mais plus fin. Tom regarda à l'intérieur mais ne put deviner grandchose.

— C'est fichu pour aujourd'hui, décida Henrickson. On va retourner en ville, voir si on peut trouver son numéro de téléphone, afin de pouvoir faire les choses correctement demain. Pour l'instant, j'ai une faim d'ours... Ne le prenez pas mal.

Ils regardèrent une fois de plus par la fenêtre et se dirigèrent vers le portail. Une fois qu'ils furent remontés en voiture et que le bruit du moteur s'estompa derrière les arbres, le rideau s'écarta un peu.

CHAPITRE 21

Une fois certaine que les hommes étaient partis, Patricia déverrouilla la porte et sortit. Elle attendit un instant, en écoutant attentivement, mais n'entendit que les bruits habituels : le silence. Elle ne tenait jamais compte du vent d'automne, des oiseaux ou des insectes agités en été. Ce n'étaient pas des bruits.

Les traces dans la neige indiquaient que les hommes avaient remonté l'allée pour faire le tour du bungalow. Elle comprit aussi que l'un d'eux avait sans doute...

Patricia Anders suivit les pas traînants qui menaient vers la petite crête et le lac. Les traces s'arrêtaient au bout de quelques mètres. À moins d'être très peu observateur, l'homme avait dû repérer l'autre construction, un peu plus loin. Elle ne l'avait pas entendu appeler ni en parler à son compagnon. Cela ne signifiait peut-être rien. Il pouvait avoir froid ou faim. Peu importait. Il n'y avait rien dans cette cabane, à part des outils, de l'humidité, et un souvenir d'une scène d'amour avec Bill, par une nuit d'hiver, alors qu'ils étaient censés réparer le toit.

Elle alla vers la mare qui délimitait la frontière entre la nature et sa propriété, s'assit sur le banc qui encerclait le grand arbre, à quelques mètres de la rive et contempla la surface glacée.

— Ils arrivent, dit-elle d'un ton calme. Qu'est-ce que je fais ?

Il ne répondit pas. Il ne répondait jamais. Il ne savait même pas de quoi elle parlait. Mais elle lui posait la question, au cas où. Les hommes aimaient bien se sentir impliqués.

Dans les mois qui avaient suivi la mort de Bill, Patricia avait erré dans un monde étrange où tout semblait avoir été brisé et remonté de travers. Elle apprit qu'un réfrigérateur paraissait froid si on ne le remplissait qu'avec ce dont on avait besoin. Elle se souvint que les bouts de papier ne se couvraient pas tout seuls de gribouillis, que les enveloppes, les factures et les reçus n'étaient pas spontanément décorés de dessins d'arbres, de chats ou de bateaux. Ils paraissaient bizarres, nus. Le plus dur, c'était de s'apercevoir que certaines informations ne trouvaient plus leur place. Elle pouvait passer le temps avec le facteur, bavarder en faisant la queue au marché, mais impossible de se retourner vers quelqu'un et de chantonner l'air d'une publicité débile. Le genre de chose qui fait penser : « La pauvre vieille perd la boule, une bien triste histoire, il faudrait faire quelque chose pour elle. » Il s'était produit un événement, puis tout s'était évaporé, comme une goutte d'eau qui tombe sur l'asphalte. À part elle, personne ne regardait un vieux magnétoscope qui fonctionnait mal.

La journée passe, et on se demande quelle est sa récompense. Il devient vite évident que le seul trophée à gagner, c'est le droit d'avoir à supporter le lendemain. On survit, heure par heure, mais à la fin, on n'a plus grand espoir. On commence à comprendre les règles. Ce qui est sûr, c'est que le prix accordé aujourd'hui est identique à celui d'hier. Apparemment calme, mais avec un cri intérieur qui monte, tel le bruit d'une vieille machine à vapeur oubliée au sous-sol, on traverse ce lendemain aussi, et toute une série d'autres lendemains, encore. Assez de lendemains pour se rendre compte qu'on s'est fait avoir, que ce ne sont pas des lendemains mais simplement un éternel aujourd'hui. La rébellion ne mène nulle part. Si on arrête de fumer et si on se rend compte que l'espoir de ne pas fumer demain n'est pas une récompense suffisante pour s'en abstenir aujourd'hui, on peut s'acheter un paquet, l'ouvrir, se sentir à la fois heureux, déçu, méfiant et coupable. Avec la mort, aucun moyen d'éprouver un tel échec triomphant. On ne peut pas dire : « Tu m'emmerdes, rends-moi mon mari. » On le comprend vaguement. On ne met pas le monde entier à l'épreuve car formuler cette demande et se la voir refuser, ça rend cinglé. À la dure, on apprend qu'il n'y a aucune issue, qu'on ne peut pas renoncer à renoncer, ni aller chercher la trousse d'urgence de l'être aimé ; qu'il est impossible d'aller le dénicher là où il se cache, en haut du placard de la cuisine ou derrière l'escalier de la salle de bains ; qu'on ne peut pas l'épousseter, passer ses doigts dans ses cheveux et l'embrasser sur les lèvres pour le réveiller et ramener le monde à la normale, comme si tout cet épisode n'avait été qu'un cauchemar, une farce stupide.

Après avoir passé sa vie à défendre le point de vue des libéraux dans tous les débats, Patricia était devenue la proie des pensées les plus politiquement incorrectes. Elle regardait les gens qui encombraient les rues du marché, les vieux grincheux insupportables. Six mois plus tôt, elle se serait demandé ce qui les avait rendus si malheureux et aurait tenté l'impossible pour voler à leur secours. À présent, elle trouvait injuste qu'ils soient encore en vie. Lorsque, à la télévision, on lançait un appel à la solidarité pour les enfants hospitalisés, elle se demandait pourquoi les gens s'attendrissaient tant sur les enfants qui n'avaient encore rien apporté au monde, alors que son Bill ne faisait même plus partie de la vie des autres. De la sienne, par exemple.

Et lorsqu'un garçon avait essayé de lui épingler un badge contre le sida, un jour à Snohomish, elle l'avait enguirlandé. Le jeune homme aux yeux innocents s'était tourné vers sa camarade, une adolescente d'une beauté impressionnante, qui débordait de compassion, et avait lancé une remarque.

Patricia le fixa.

— Elle te baise bien ?

Le garçon avait rougi. En arrivant à sa voiture, Patricia était écarlate, furieuse contre elle-même, mais une petite voix intérieure insistait. Ce n'était pas juste ! Il y a des gens qui se tuent aux sports d'hiver, après avoir choisi de descendre sur des planches glissantes, et c'est une tragédie ! Si quelqu'un a un cancer du poumon, bien fait pour lui, ce salopard de fumeur qui m'intoxique ! Même sa veuve est responsable, elle aurait pu l'en empêcher, non ? Elle ne l'aimait pas ou quoi ? Nous cherchons des responsables, ça laisse Dieu en dehors du coup, et sans Lui, nous ne savons pas où nous tourner.

Patricia en connaissait un rayon en matière de responsabilité. C'était elle qui avait tué son mari, après tout.

Cinq ans avant qu'ils s'installent au Nord, Bill avait pris conscience du poids des ans et arrêté de fumer. C'était difficile, horrible même, mais il s'y tenait. Cinq mois plus tard, ils se retrouvèrent à la terrasse d'un bar à Verona, où ils passèrent une journée formidable. Un bon repas les attendait, là, seuls devant le coucher de soleil, tout semblait presque parfait. Ils bavardaient, souriaient, contemplaient la petite île. Tout allait bien, mais ce n'était pas ça, et Patricia regardait Bill en sachant qu'il acceptait à regret que certains de ses moments les plus agréables soient de brefs supplices : le supplice du plaisir manqué, ce qui en Occident remplace la souffrance.

Alors elle avait plongé dans son sac et en avait sorti quelque chose qu'elle avait acheté à Cannon Beach le matin même. Ce n'était pas la fin de l'abstinence, mais, face à l'expression de son visage, elle avait posé le paquet sur la table.

Il l'avait regardée avec un sourire complice, comme si on lui avait montré la photographie d'un ami décédé depuis des mois.

Il s'était penché et l'avait embrassée.

Il ne fuma pas ce soir-là. Il fuma le lendemain soir.

Le paquet dura un mois. Aucun moyen de savoir si la cigarette qui avait déclenché le cancer provenait de celui-là ou de ceux qui suivirent. Mais elle supposait qu'elle venait de ce paquet-là. On trace des lignes, on établit des relations, on expose son cœur aux reproches, on croit qu'il pleut parce qu'on est méchant. Les années suivantes, il ne fut qu'un fumeur occasionnel, et il paraissait bien plus heureux. Les cigarettes n'y étaient peut-être pour rien. C'est ça, le problème avec la mort : on ne sait jamais. On ne sait jamais ce qu'on aurait dû faire avant qu'il soit trop tard... et, même là, on ne sait pas. C'est un grand jeu de la vérité et de ses conséquences, sauf qu'il n'y a pas de vérité, simplement des conséquences. La vérité est une fiction que l'on complète après coup pour rendre les circonstances moins épouvantables, plus explicables, pour en rejeter la faute sur un autre, même si c'est forcément la sienne.

Elle ne songeait pas souvent à ce paquet de cigarettes, mais lorsque cela lui arrivait, il ne fallait pas longtemps pour qu'une autre pensée lui vienne à l'esprit : elle se demandait si elle avait acheté le paquet pour Bill, ou si elle avait refusé de gâcher ses propres vacances à Verona en sachant qu'il n'était pas heureux. La seule chose qui rendait ces moments supportables, c'était de savoir que Bill ne lui en aurait pas voulu, même dans le second cas. Il l'aimait à ce point-là. Et il était mort.

Pendant un moment, quelques mois plus tard, Patricia estima que les choses devenaient un peu plus faciles. Puis, par une longue nuit de décembre 2001, alors qu'approchait le premier Noël sans lui, quelque chose se produisit. Elle possédait un CD de ses airs favoris qu'il avait choisis pour qu'on les passe le jour de ses funérailles à Portland. Des chansons qu'elle avait aimées en sa compagnie, des œuvres classiques qu'elle n'avait jamais entendues mais qui lui tenaient à cœur dans sa vie à lui, sa vie d'avant leur rencontre. Elle n'avait plus écouté ce CD depuis. Cette nuit-là, elle le passa pour la seconde fois, l'écouta jusqu'au bout. Elle trouva une grande bouteille de scotch que Bill avait laissée derrière lui et la but en entier. Elle n'avait jamais rien fait de tel auparavant.

À minuit, chancelante au beau milieu des arbres, les cheveux balayés par les rafales d'un vent glacial, pieds nus, insensible au froid ou presque, elle avait parlé, crié, hurlé, pleuré, la gorge irritée, la bouche sèche. Elle avait envie d'arracher les yeux de tous les êtres vivants, de trouver quelqu'un, n'importe qui, et de lui écraser la cervelle avec une pierre. Elle était prisonnière d'une spirale de l'horreur et, cette nuit-là, elle comprit l'essence du monde. L'essence, la vérité, la voici : l'enfer, c'est la vie, et la vie, c'est tout ce qu'il y a.

Se tuer, cela aurait été renoncer. La mort est plus méchante et plus forte que tout. La mort, c'est la clé, ce n'était pas à remettre en question, mais ce n'était pas une raison pour se ranger de son côté.

Elle prit donc une décision tout en hurlant au bord du lac d'un froid de glace, toujours chancelante après avoir englouti la bouteille de son mari. Elle n'allait plus, comme on disait dans le langage populaire, se faire baiser. Elle n'aurait plus de comptes à rendre. À personne, ni Dieu, ni maître, ni idéal, ni vérité, ni promesse. Rien ne valait la peine, on ne pouvait avoir confiance en rien. Avant, il y avait Bill. Maintenant, il n'y avait plus rien.

Mais, deux semaines plus tard, elle avait découvert quelque chose dans la forêt, si bien qu'elle changea d'avis.

Le ciel était sombre, le lac ressemblait à une plaque de marbre noir. Il faisait froid. Il était temps de rentrer. Patricia resta assise un peu plus longtemps, malgré tout, car elle aimait cette vue et craignait que les choses soient sur le point de changer. Elle craignait que les hommes ne reviennent et qu'elle soit obligée de défendre la seule chose qui lui tenait à cœur.

Eh bien, qu'il en soit ainsi !

CHAPITRE 22

Nous nous étions réfugiés au Morisa, un vestige de la grandeur d'antan, près du centre de Fresno. L'hôtel semblait avoir été conçu pour résister à une pluie de bombes. C'est ce qui nous avait séduits. Nous étions arrivés en ville la veille, tard dans la nuit, et avions préféré ne pas aller plus loin. Sans plan, ni endroit où aller, nous risquions de prendre une des nombreuses directions à éviter. Nous nous rendîmes à la réception séparément, prîmes des chambres à des étages différents et montâmes nous coucher. Le lendemain matin, nous allâmes au centre-ville à pied. Nous marchâmes des heures, sans parvenir à nous décider. Il y a quelque chose d'absurde dans les magasins lorsque vous n'éprouvez aucun intérêt pour le shopping. Bizarrement, je crus reconnaître des initiales sur une porte, mais, en y regardant de plus près, je notai que la deuxième lettre était un « B » et non un « R ». Je n'en étais pas certain. Je traversais une crise de paranoïa.

En fin de matinée, nous nous étions retrouvés dans ma chambre. La pièce n'était pas grande et n'avait pas été décorée récemment. Je m'assis sur le fauteuil, et Nina sur le lit. Nous bûmes le café que nous avions commandé.

Nina regrettait d'avoir quitté Los Angeles. Elle voulait revenir. Il n'en était pas question. Je savais ce que cela signifiait d'être en cavale. Elle avait un travail, même si on venait de lui demander d'y renoncer. Se retrouver dans cette situation à cause d'une relation, d'une ancienne relation, de surcroît, était propre à énerver n'importe quelle femme. Et Nina n'était pas n'importe quelle femme. Elle était si furieuse contre Zandt, après ce mensonge, qu'elle ne voulait plus allumer son portable. Je tentai de l'appeler à plusieurs reprises, mais n'obtins qu'une voix métallique me disant que le téléphone était coupé. Il pouvait être n'importe où dans le pays, à faire Dieu sait quoi, ou avoir de sérieux ennuis. Il était peut-être mort, aussi.

Nous ne pensions pas pour autant que Zandt était incapable de tuer Ferillo. Nous savions tous les deux que, dans la phase initiale de la recherche de sa fille, il avait personnellement coincé et tué l'homme qu'il croyait coupable. Mais il y eut un autre enlèvement après cet incident. Nous avions à présent un nom pour le ravisseur, Stephen DeLong, et nous savions qu'il n'était qu'un des kidnappeurs qui officiaient pour le compte des Hommes de Paille, mon frère jouant un rôle majeur dans ce groupe. L'arrivée soudaine du fichier vidéo qui accusait John du meurtre de DeLong, longtemps gardé en réserve, prouvait qu'ils le cherchaient pour lui causer des ennuis dignes de ce nom. Restait à savoir si la mort de Ferillo représentait un échantillon des ennuis à venir, ou en était la cause.

Nina passa deux appels de la chambre, qui nous permirent d'apprendre que Ferillo possédait un restaurant sur Stark Street, à Portland, appelé le Dining Room. Quatre ans auparavant, il s'était fait arrêter à L.A. dans le cadre d'une enquête sur un racket et avait failli se retrouver à l'ombre. Il s'en était sorti et avait monté une usine à bouffe fréquentée par les grands et les vertueux du nord-est de l'Oregon. Du petit truand au riche restaurateur, il y avait un pas de géant, mais rien n'expliquait pourquoi Zandt avait débarqué dans sa vie... ni pourquoi quelqu'un aurait voulu nous le faire croire.

Après ces appels, nous gardâmes le silence. Le café refroidissait, mais nous continuâmes à boire, jusqu'à ce que je commence à avoir l'estomac noué. Par la fenêtre ouverte, je regardais les bâtiments délabrés, battus par une pluie persistante. Je trouvais idiot de ne rien essayer, mais je n'avais pas d'autre idée. Nous n'avions aucun moyen de retrouver John, aucun moyen de faire avancer l'enquête sur Ferillo.

Soudain, une faible lueur s'alluma dans mon esprit. Elle vacilla, s'éteignit pendant une seconde, puis revint avec plus d'intensité.

— Téléphone à Monroe.

— Pas question.

— Considère les choses de son point de vue. Ce n'est pas un imbécile. Il sait qu'il t'est arrivé quelque chose de grave à la fin de l'année dernière. Tu as été blessée, et Sarah Becker a retrouvé ses parents. Tu refuses d'en parler, et maintenant tu es liée à un type qui fait n'importe quoi.

— C'est ce qu'on voudrait faire croire.

— Comme tu veux, répondis-je. Même si Monroe n'était pas poussé par quelqu'un, tu te retrouverais sur la corde raide.

— Il y a quelque chose dans ta voix que je ne comprends pas.
— Raconte-moi encore ce qui s'est passé lorsque tu es allée au Knights. Le jour où on a retrouvé le corps de Jessica.
— Ward...
— Raconte-moi.
— Charles m'a appelée. Sur mon portable. Quelqu'un venait de tirer sur un policier dans sa voiture de patrouille et avait disparu.
— Et ensuite ?
— Rien. Il m'a demandé de venir le rejoindre.
— Pour le meurtre d'un policier.
Elle hésita.
— Oui.
— Ce qui ne devrait pas intéresser le FBI. À moins que...
Elle garda le silence pendant près de vingt secondes et réfléchit.
— Oh, mon Dieu.
— Oui, peut-être.
Elle cligna des yeux.
— Et pourquoi veux-tu lui parler, nom d'un chien ?
— Parce qu'il n'y a personne d'autre. Lorsque tu lui poseras la question, on verra bien ce qu'il répondra.
Nina avait pris sa décision avant que je m'explique. Elle se leva et sortit son portable de son sac. Quelques secondes plus tard, il bipa.
— Des messages.
Elle écouta, puis écarta le téléphone et se leva, une expression bizarre sur le visage.
— John ?
— Monroe. Quatre fois. Pas de message. À part : « Appelle-moi. »
— Alors, appelle-le. Pas au bureau. Sur son portable.
— Oui, mais il saura où nous sommes.
— Il saura où nous étions ! Allez, vas-y.
Elle composa le numéro et écouta la tonalité, les yeux fixés sur moi.
— Charles, c'est Nina.
À deux mètres, j'entendais le torrent de paroles. Nina écouta.
— Qu'est-ce que tu... Charles, je te rappelle.
Elle coupa la communication et resta sans voix.
— Nina ?
— Ils ont trouvé une autre femme avec un disque dur.

À 17 h 15, il faisait déjà sombre. Nous étions dans la voiture, à une cinquantaine de mètres du Daley Bread. Je l'avais repéré la veille : un gigantesque restaurant anonyme que j'avais choisi parce qu'il donnait sur une grande rue, à quatre carrefours de la 99, qui menait vers le nord ou le sud. Facile à trouver, facile à fuir. Nous étions arrivés en avance pour voir si quelqu'un essayait de placer des hommes, si les autorités locales avaient été prévenues. Si on pouvait faire confiance à Monroe, en d'autres termes.

En une demi-heure, nous ne vîmes personne, à part une poignée de passants hirsutes, une couverture déchirée autour des épaules, intercalés entre des petits groupes de jeunes bien nourris. Les deux communautés semblaient indépendantes et il était difficile d'imaginer comment elles pouvaient cohabiter, telles deux espèces différentes, par pur hasard. Nous observions chaque groupe qui s'approchait avant de s'éloigner. Certains regardaient dans la voiture et se demandaient sans doute pourquoi un couple attendait ici, par une soirée glaciale. Nous les regardions également. Nous étions aussi méfiants que possible. Lorsqu'il n'y avait personne, nous observions la rue, dans les deux sens.

À 18 h 15, quinze minutes avant le rendez-vous, j'ouvris la porte et sortis.

— Fais attention

— Ça ira. Il ne sait pas à quoi je ressemble.

— Non, mais d'autres te connaissent.

J'avançai d'un pas modéré, essayant de trouver une place entre les clochards et la jeunesse branchée. J'attendis un moment de l'autre côté de la route, du côté du Diner, n'aperçus rien de particulier.

En traversant la rue, je me rendis compte qu'un type un tant soit peu intelligent aurait gardé secret le lieu du rendez-vous jusqu'à ce que Monroe soit arrivé en ville, afin de limiter les possibilités de contacter les autorités locales. Plus que jamais, j'aurais voulu que Bobby soit près de moi. Ou ma mère. Privé des deux, je savais que je resterais toujours exposé au danger.

Je posai une question silencieuse, sans bouger les lèvres.

— C'est une idée stupide ?

Pas de réponse.

À l'intérieur du restaurant, il faisait un peu plus chaud ; l'air était confiné. Une fille à l'air las, en uniforme, s'approcha de moi, un menu à la main.

— Je m'appelle Britnee, dit-elle, inutilement, car elle portait un badge de la taille d'un plateau. Vous dînez seul ?

Je répondis que oui et dirigeai mon regard vers un box plutôt qu'au milieu de la salle. Comme il n'y avait que deux couples, elle n'avait aucune raison de refuser que je m'installe où je voulais.

Je commandai un chili sans même consulter le menu. Lorsqu'elle alla réveiller le cuisinier, je me mis dans la position que nous avions décidée, avec Nina. Je m'assis du côté droit du box, le dos au muret qui le séparait de son jumeau. La table n'était pas visible de l'autre côté, mais je devais pouvoir tout entendre.

Je sortis un magazine gratuit que j'avais pris dans le hall de l'hôtel, baissai les yeux et commençai à lire.

Cinq minutes plus tard, la porte du restaurant s'ouvrit. D'un coup d'œil rapide, je vis Nina entrer. Britnee essaya de l'envoyer à l'une des tables proches de la fenêtre, sans doute pour l'agréable vue sur la route froide et trempée, mais Nina refusa. Je la perdis de vue lorsque la serveuse la conduisit de l'autre côté, mais une minute plus tard j'entendis quelqu'un s'installer derrière le muret.

Nous gardâmes le silence pendant un moment. Une autre serveuse s'approcha de Nina et lui demanda si elle voulait boire quelque chose. J'entendis également la réponse de Nina. Pour le son, tout fonctionnait.

J'observai les publicités pour les magasins locaux dont je me fichais éperdument et les restaurants de famille qui ressemblaient à ceux qu'on trouve dans tout le pays. Nina faisait la même chose. De temps en temps, j'observais la rue. Il ne se passait rien.

Finalement, Nina annonça doucement :

— Le voilà.

Je jetai un coup d'œil vers la porte et avisai un homme athlétique, proche de la cinquantaine, portant un costume et un long manteau de cuir. Il entra d'un pas vif et dépassa Britnee avant même qu'elle puisse lui proposer de s'installer sur la jolie terrasse. Il avait repéré Nina de l'extérieur.

— Bonjour, Charles.

J'entendis quelqu'un s'asseoir.

— Pourquoi tu n'as pas voulu qu'on se voie à ton hôtel ?

— Comment sais-tu que je suis à l'hôtel ?

— Où veux-tu que tu sois ?

Il y eut une longue pose.

— Charles ? Tu vas bien ?

— Non. Et toi non plus. On a vérifié la vidéo. C'est bien John, et elle est authentique. L'empreinte du doigt sur l'ouvrebouteilles de Portland aussi, et maintenant on a un témoin oculaire qui a vu un homme sortir du bâtiment en transportant une femme. Il avait raconté qu'elle était saoule et qu'il la raccompagnait chez elle. La photo ressemble tellement à Zandt qu'on a du mal à y croire, et la fille confirme la ressemblance. J'ai également parlé à Olbrich, et je sais qu'il a cherché un renseignement pour toi. John était à Portland ce soir-là.

— Merci, Doug.

— Il est policier, ce n'est pas ton indic personnel. Zandt a tué Ferillo, autant l'accepter. Il a également frappé la fille assez fort, elle a eu un traumatisme crânien. Je ne sais pas ce qui lui passe par la tête, mais c'est une mauvaise idée de vouloir le protéger.

— Le poursuivre ne servira à rien non plus. Tu es mouillé.

— Comment ça ?

À cet instant, deux choses se produisirent. Tout d'abord, la serveuse arriva avec mon chili, et prit le plus de temps possible pour le poser sur la table en faisant le plus de vacarme possible. Elle voulait aussi me poser des tas de questions. Où je logeais, si je me plaisais ici, si je ne voulais vraiment pas d'oignons, si elle pouvait m'en rapporter. Je répondis par monosyllabes, le plus vite possible.

Ensuite, Nina se dégonfla.

Je n'avais pas besoin de la voir pour savoir qu'elle fixait la table, incapable de passer à l'étape suivante. Alors, je me décidai. Lourde erreur. Je me levai, laissai mon assiette et me dirigeai de l'autre côté du muret.

Je tirai une chaise vers l'autre côté du box où Nina et Monroe n'avaient pas encore touché à leur soda.

— Je peux vous aider ? demanda Monroe.

— J'espère. Je suis un ami de Nina. Je vais vous poser la question qu'elle n'ose pas vous formuler.

— Nina, tu le connais ?

— Oui.

— Vous êtes Charles Monroe. Je m'appelle Ward Hopkins. Je suis l'une des deux personnes qui peuvent confirmer ce que Nina va vous raconter. Probablement la seule que vous écouterez, car vous ne semblez pas avoir envie de prendre Zandt au sérieux.

— Je n'ai pas l'intention de vous écouter non plus. Nina...

— Oh ! si, vous allez nous écouter. Dès que vous nous aurez expliqué comment vous saviez que vous alliez découvrir un cadavre au Knights.

Il ne s'y attendait pas. Il tenta de me toiser, mais c'est bizarre, depuis la mort de mes parents, je ne me laisse pas facilement impressionner. Cela n'avait jamais été le cas, mais à présent, c'est pire. Comme si une partie de moi, tout au fond, se fichait de tout.

Nina l'observait attentivement.

— Tu vas lui répondre ?

Il se taisait, et à son expression je compris que Nina croyait soudain à l'hypothèse que j'avais émise.

— Espèce d'enfoiré ! lâcha-t-elle.

— Nina... Je ne sais pas ce que ce type t'a raconté, mais..

— Ah oui ? s'écria-t-elle. C'est noir ou blanc. Si un flic se fait tuer, l'enquête, c'est le boulot de la police de L.A. Ça ne regarde pas le FBI à moins que les flics le demandent, ce qu'ils ne font jamais. Les Feds, on n'a pas envie de les voir. Les flics ont une section spéciale pour ce genre de meurtres, des brigades entières qui laisseraient tout tomber pour retrouver celui qui a tué l'un des leurs. Alors, qu'est-ce que tu fichais là-bas ? Et si vite ? Comment se fait-il que tu te sois trouvé sur les lieux avant qu'on ait examiné la chambre ? Avant même qu'on sache qu'il y avait quelque chose à voir ?

— C'est ridicule ! J'en ai assez...

— Charles, regarde-moi et ferme-la.

Je ne reconnaissais plus la voix de Nina. C'était un son entre le sifflement et le grognement, le cri d'un chat sauvage, longtemps retenu en cage, fatigué de s'être fait avoir.

Monroe la regardait. Moi aussi.

— Charles, où sont mes mains ?

— Sous la table.

— Et d'après toi, qu'est-ce que je tiens ?

— Oh, mon Dieu, Nina...

— C'est ça. Et je n'hésiterai pas à tirer si tu ne racontes pas une histoire plausible.

— Les gens savent où je suis.

— Sûrement pas. Tu n'aurais jamais compromis ta précieuse réputation en racontant que tu traversais tout l'État pour venir me parler, avec cette sale histoire de John qui se trimballe partout. À moins que tu ne te sois fait accompagner, bien sûr, ce qui apparemment n'est pas le cas.

— Bien sûr que non, rétorqua Monroe, si furieux qu'il était difficile de ne pas le croire. Nom d'une pipe, on a travaillé ensemble pendant des années. Nous sommes liés.

— C'est ce que je croyais. Jusqu'à ce que je sois suspendue hier. Grâce à toi.

— Je n'avais pas le choix, tu le sais. Tu es trop compromise avec Zandt.

— Compromise ? C'est moi que tu accuses d'être compromise ? Commence donc par répondre à nos questions. J'ai toujours les mains là où je t'ai dit, et je ne plaisante pas.

Monroe garda le silence, les yeux fixés sur le set de table. Je savais qu'il ne pourrait pas conserver sa concentration longtemps.

— Ça se passe mal, finit-il par avouer d'une voix calme. Et pas seulement pour toi. Mais c'est ta faute. C'est à cause de ta croisade personnelle. Pourquoi avoir refusé de me raconter ce qui s'est passé l'an dernier ?

— Pour te protéger. Tu ne pouvais rien pour moi, et on ne savait pas à qui faire confiance. Si on pouvait faire confiance à qui que ce soit.

— Excuse-moi, mais c'est de la paranoïa.

— Pas du tout !

Nous avions parlé à l'unisson.

Monroe me regarda vraiment pour la première fois.

— Qui vous en voulait ? À qui aviez-vous affaire ?

Je hochai la tête.

— Les Hommes de Paille, dit-elle. Nous ne savons pas combien ils sont ni qui ils sont. Ils possédaient un grand territoire dans le Montana, là où tout a explosé.

— C'est vous ?

— Non, c'est eux, expliquai-je. C'était piégé. C'était un champ de preuves. Des corps. Des tas de corps. Des gens tués pour le plaisir. Ils disposaient d'une chaîne de fournisseurs, grâce à des intermédiaires comme Stephen DeLong. Le type que vous appelez le Garçon de Courses est un de leurs fournisseurs aussi, le plus prolifique, un tueur en série qui fait partie de l'organisation. C'est également mon frère. Il se fait appeler l'Homme Debout. C'est un rouage clé dans une autre de leurs activités. Vous vous souvenez de l'explosion de l'école d'Evanston, l'année dernière ?

— Oui, on a coincé deux gosses.

— Ils sont innocents. C'était lui. Et des meurtres et des fusillades en Floride, en Angleterre, en Europe, qui remontent à une vingtaine d'années. Plus, peut-être. Le groupe existait déjà dans les années soixante. Ils tuent, et ils s'arrangent pour en faire accuser d'autres à leur place.

Monroe paraissait sidéré.

— Nina, tu y crois ?

— Croire, ce n'est pas le mot. C'est la vérité. Il s'agit d'un groupe qui prospère dans ce pays et qui sévit depuis longtemps. Ils sont très puissants et ils tuent. C'est eux qu'on a foutus en pétard. Et maintenant, pour la dernière fois, parle-moi de Jessica.

Il n'hésita qu'un instant. Il avait pris sa décision.

— J'ai reçu un appel.

Même si elle savait qu'il céderait, elle avait failli tirer. Je crois que Monroe en était persuadé, lui aussi.

Ensuite, il y eut un long silence.

Monroe finit par ouvrir la bouche. Sa voix s'érailla. Il but une gorgée de soda, si bien qu'elle se transforma en gargouillis.

— J'avais reçu l'appel la veille au soir. Sur mon portable, ma ligne perso. Il n'y a pas beaucoup de gens qui connaissent ce numéro. En fait, j'ai cru que c'était toi. J'étais au théâtre avec Nancy. C'était l'entracte. Au foyer, il y avait beaucoup de bruit. C'était une voix d'homme, mais je n'entendais pas bien et, le temps que j'aille au calme, il avait raccroché. Je n'avais aucune raison de... Le lendemain matin, au boulot, deuxième appel. Un homme aussi, qui m'a demandé pourquoi ça ne m'intéressait pas. Je lui ai répondu que je ne savais pas de quoi il parlait. Il m'a annoncé qu'un flic venait de se faire trucider et que ce serait bien d'aller au Knights. Ce serait...

— Bon pour ta promo... compléta Nina.

— Oui, c'est exactement ce qu'il a dit.

— Le même numéro que le type qui t'avait appelé la veille ?

— Oui, ça aurait même pu être un type du service.

— Qui serait resté anonyme ?

— Si c'était bon pour ma promo, ce serait bon pour tout le service.

— À d'autres... Je ne te crois pas, Charles, et je m'en fiche. Tu y es allé parce que tu avais un tuyau qui pourrait faire avancer ta carrière, et tu m'as mouillée dans une sale histoire. Tu n'as dit à personne que tu avais un tuyau. Tu t'es arrangé pour qu'Olbrich te trouve des hommes, et tu as magouillé quelques jours pour qu'on croie que ça ne menait nulle part. Quand on était chez

McCain et que je t'ai demandé si le tueur de flic avait aussi assassiné Jessica, tu savais déjà qu'il n'y avait peut-être pas de rapport.

— Cela ne voulait pas dire qu'il n'y en avait pas.

— Écoute ! s'écria-t-elle. Tu as même essayé de m'en dissuader ! Et le lendemain, alors que John était soudain devenu le suspect numéro 1 dans l'affaire Ferillo, tu as reçu un autre message électronique. Intraçable, je suppose.

— Peu importe la manière dont il est arrivé. Il est authentique. Et inutile de monter sur tes grands chevaux, nom d'un chien ! Tu savais. Tu savais que Zandt avait tué DeLong, tu as dissimulé les preuves.

— Je ne le savais pas à l'époque. Il ne me l'a dit qu'à la fin de l'année dernière.

— Peu importe. À l'instant où tu l'as appris, tu devenais un élément de l'affaire, alors ne viens pas...

— Qui était le type avec vous lorsque vous avez montré la vidéo à Nina ? questionnai-je.

— Je ne sais pas, dit-il, amer. Quand il est arrivé le matin, il était déjà au courant de tout. Il avait une accréditation de la NSA, mais hier, lorsque j'ai essayé de le retrouver, on m'a dit qu'il n'existait pas. J'ai insisté, j'ai pas mal râlé...

— Et maintenant, tu es dans une mauvaise passe, toi aussi, conclut Nina.

— Indirectement, corrigea-t-il, la respiration lourde. On rouvre le dossier Gary Johnson.

— Quoi ?

— Un avocat en Louisiane prétend tout à coup qu'il a la preuve qu'on a trafiqué le rapport du légiste. Plus précisément, que tu l'as trafiqué et que j'ai fermé les yeux. Quelqu'un veut te discréditer. Et comme agent responsable, ça ne va pas manquer de m'éclabousser. Tu es contente ?

— C'est toi qui t'es compromis, Charles. Tu n'as rien à me reprocher.

— Ce n'est pas la peine de jouer les parangons de vertu. Tu as dissimulé un meurtre, tu as menti sur les événements de l'an dernier, et si tu crois que je ne me suis pas aperçu que tu avais piqué le disque dur de Jessica pendant vingt-quatre heures... Ça suffirait largement à ruiner ta carrière, et ça, c'est bien ta faute.

— Bon, il y a eu un autre meurtre au disque dur, dis-je. Vous aviez reçu un avertissement pour celui-là ?

— Non, et puis vous, qui êtes-vous ?

— Les parents de Ward ont été tués par les Hommes de Paille, dit Nina. Il nous a aidés à sauver Sarah Becker, et c'est la seule personne au monde à qui je fais confiance en ce moment. Je pense que ça suffit. Parle-moi du dernier meurtre.

— Nina...

— Tu as été mêlé à l'affaire avec Jessica. Si c'est un autre meurtre du même type, nous avons une petite chance de résoudre l'affaire, ce qui est la seule issue possible si tu veux conserver un espoir de retrouver une vie normale.

— Et toi aussi.

— La mienne est déjà en vrac, répliqua Nina. Ça me rend furieuse. Je veux retrouver ceux qui m'ont fait ça. Ward et moi, on a un compte à régler avec eux.

— Elle s'appelait Katelyn Wallace, expliqua Monroe. Elle travaillait dans l'équipe de nuit au Fairview de Seattle. Quelqu'un l'a enlevée alors que l'hôtel était plein et que le concierge était présent, lui aussi. On l'a retrouvée dans un buisson, à Snoqualmie, une petite ville à soixante kilomètres à l'est. On a la moitié du numéro minéralogique d'une voiture qui a été repérée tard dans la nuit, mais c'est une voiture de location dans une zone de villégiature. Le corps de Katelyn était plus mal en point que celui de Jessica. On croit, c'est l'avis d'un profiler mais les photos le confirment, que l'assassin a perdu son sang-froid. Il ne s'est pas amusé à l'habiller et, cette fois, le disque dur n'était pas dans la bouche. Il a percé un trou dans le crâne, pour l'enfoncer. On a découvert le même morceau de musique que sur celui de Jessica.

— Il y avait un mot ?

— Non. Trois photos de paysage, de mauvaise qualité. Des images de Webcam. De Pittsburgh, croyez-moi ou pas. Alors le FBI est en état d'alerte, mais vous savez ce que cela signifie...

— Qu'est-ce qu'on sait sur la femme ? demandai-je.

— Originaire de San Francisco. Quarante-deux ans, elle était à Seattle depuis douze ans. Célibataire, mais beaucoup d'amis et un chat, et personne n'a la moindre idée sur l'identité du coupable. Elle semble avoir été choisie au hasard.

— Je ne crois pas, répondis-je. Pourquoi parcourir la moitié du pays afin de prendre une victime au hasard et signer le meurtre en utilisant le même mode opératoire ? Il y a forcément un lien. Nina vous a parlé de la photographie manquante dans l'appartement de Jessica ?

— Oui. Nous avons retrouvé les trois hommes que l'on voit sur les vidéos. Deux sont des clients de chez Jimmy, elle avait rencontré le troisième dans une fête, à Venice Beach. Aucun ne fait un bon suspect, bien que l'un d'eux ait confirmé avoir vu une photo de ses parents près de son lit ; ça l'avait fait rigoler. Mais le pervers de papy du Net, Robert Klennert, se souvient plus ou moins que quelqu'un a essayé de dénicher Jessica en envoyant un e-mail au Webmaster du portail, voilà deux mois environ. Ça arrive tout le temps, apparemment. Toutes les filles en sont victimes. Il se contente de les renvoyer sans répondre. Il ne se rappelait pas en avoir eu pour Jessica en particulier avant de fouiller ses fichiers. Ça ne veut peut-être rien dire.

— Ou bien l'assassin cherchait un moyen de s'introduire, objectai-je. C'est un peu ténu. On a volé quelque chose chez Katelyn Wallace ?

— On n'a pas la chance d'avoir des images. Katelyn n'était pas une prostituée d'Internet. C'était une femme rangée, qui travaillait dur.

— Elles meurent aussi. Mais on suppose que l'assassin a pris la photo pour avoir un souvenir. Un objet personnel, pour s'approprier la vie de la femme qu'il avait l'intention de tuer. Et si c'était plus que ça ?

Nina me regarda.

— À quoi tu penses ?

— Ils veulent faire tomber le tueur, dis-je en parlant lentement pour ne pas interrompre le flot de mes pensées. C'est pour cela qu'ils ont renseigné Charles. C'est forcé. Mais pourquoi ? Qui les Hommes de Paille veulent-ils faire tomber ?

Je levai les yeux. C'est à cet instant que je le vis.

Si j'avais obéi aux consignes que je m'étais fixées et m'étais contenté d'observer Nina pendant qu'elle parlait, je l'aurais aperçu plus tôt. Là, je n'eus qu'une vague impression d'un homme mince aux cheveux courts qui portait des lunettes, à l'extérieur du restaurant. Il regardait à l'intérieur, droit vers nous.

— Merde !

Ce fut tout ce que je pus dire avant d'entendre un grand bruit de verre, deux clics et le vacarme de la balle qui s'enfonçait dans le mur derrière nous.

Je me jetai hors du box et attrapai mon arme. J'étais rapide, mais Nina était en avance sur moi car elle avait déjà son revolver à la main.

Nous tirions tous les deux avant que Monroe eût une vague idée de ce qui se passait. De ma main libre, je saisis une chaise et la jetai maladroitement à la fenêtre, essayant de gagner assez de temps pour qu'ils puissent s'extirper du box.

La chaise retomba dans le vide, l'homme continua à tirer par le trou dans la baie vitrée. Des tirs précis, successifs.

Je m'accroupis pour disparaître de son champ de vision, agrippai Nina par le bras et la tirai derrière une table. Des cris retentissaient autour de nous. Britnee était allongée par terre, des morceaux de verre sur le visage.

Ombre furtive, l'homme courait le long de la vitre, mais il ne s'enfuyait pas. Il se dirigeait vers la porte.

— Mon Dieu ! s'exclama Nina.

En me tournant, je vis Monroe affalé sur la table. Nina commença à reculer vers lui, mais je l'attrapai de nouveau par le bras pour la plaquer au sol.

— Laisse...

La porte s'ouvrit, les cris redoublèrent d'intensité.

— Ward, il est blessé.

— Je sais.

L'homme entra dans notre aile. Je m'attendais plus ou moins à voir mon frère, mais ce n'était pas lui. Il était plus jeune, en assez bonne forme, malgré son torse épais. Il portait une tenue de combat et un manteau sombre. Il se tenait à l'extrémité de l'allée, sans la moindre crainte, et visait Nina.

Je tirai. Je lui plantai une balle dans la poitrine.

Il tomba en arrière, sur une table.

Il n'y avait pas de sang. Je compris qu'il portait un gilet. Je reculai, essayant de nous cacher derrière quelque chose avant qu'il tire à nouveau. Je tirai encore, visant plus haut, mais ratai mon coup. C'est difficile de toucher un homme à la tête. Rien que viser n'est pas facile. Il faut une sacrée envie de tuer. Mais ça, je l'avais.

Un bruit de tir vint d'un autre angle, et je pensai : merde, en voila un autre. Mais c'était Monroe. Son imperméable était couvert de sang, il était coincé dans son box, mais il s'était un peu redressé et vidait son chargeur.

Je saisis l'occasion, entraînai Nina et l'attirai de l'autre côté du muret. Ma serveuse était là, le souffle coupé, essayant de crier, mais n'émettant que le bruit d'une souris frappée par un coup de marteau.

Au fond, se trouvaient des portes saloon.

— Ward, il faut s'occuper de Charles...

— Trop tard.

Je l'emmenai de l'autre côté des portes, dans la cuisine. Au début, elle se débattit, mais elle finit par me suivre, devant les deux cuistots terrifiés. Nous sortîmes par la porte arrière. Je glissai sur la première marche d'un petit escalier mais me rattrapai à la balustrade.

Nous courûmes le long du restaurant. Les tirs avaient cessé. Je distinguai l'homme près du box où Monroe se tenait à présent, la tête sur la table.

Il se tourna et nous vit. Aussitôt, il se mit à courir. À courir vite.

— Va chercher la voiture.

Nina courait, elle aussi.

Je pointai mon arme vers la rue, en reculant aussi vite que possible. Il avait tiré avant même que je comprenne qu'il était déjà dehors.

Je tirai moi aussi, dans le ventre ; il fit un pas en arrière. Je sprintai jusqu'à la voiture à l'instant même où les phares s'allumaient et où le moteur se mettait en route.

J'eus l'impression qu'on me donnait un coup à l'épaule. Cela me déséquilibra, quelqu'un me retourna et je tombai sur le trottoir. Je me relevai, sans comprendre pas ce qui se passait, mais tirai encore.

La voiture avança vers moi, la portière s'ouvrit, et je me jetai à l'intérieur. J'avais toujours les jambes à l'extérieur lorsque Nina passa la marche arrière et fila. Une fois la portière fermée, elle fit demi-tour et s'engagea sur la route.

— Où on va ?

La manière dont elle écarquilla les yeux me confirma ce dont je me doutais déjà.

Je posai la main sur mon épaule. C'était mouillé et chaud.

— N'importe où, dis-je alors que la douleur me transperçait d'un coup de poignard.

CHAPITRE 23

Ils sortirent de chez Henry's Diner sous une bruine légère mais persistante. Tom tremblait de froid. Il avait réussi à manger la moitié de son repas, le nez dans son assiette, en face d'Henrickson, à une table d'angle. Tom avait remarqué qu'on l'observait. « Tiens, le type de Bigfoot », lisait-il dans les regards, et cela n'avait guère creusé son appétit. Henrickson s'était montré peu loquace durant le dîner, et il s'était écoulé un bon bout de temps depuis le dernier sourire. Le journaliste était peut-être fatigué, lui aussi, mais cela ne se voyait pas. Ses mouvements restaient précis, il mastiquait rapidement, méthodiquement, et il avait fait un sort à son steak de poulet. Il l'avait demandé saignant, une première pour Tom, et pour la serveuse aussi, à en juger la manière dont elle l'avait dévisagé. Quand il ne mangeait pas, Henrickson regardait par la fenêtre, comme s'il souhaitait que l'obscurité se lève.

— Bon, dit-il alors que Tom essayait de s'enfoncer dans son manteau pour se protéger du froid, je pense que je vous ramène à l'hôtel.

Tom avait pensé qu'ils iraient faire un tour au bar. Ce n'était pas qu'il eût envie de boire. Il était épuisé par la journée de marche, et le restaurant surchauffé l'avait engourdi. Le lit était tentant. Mais s'il se retrouvait seul dans sa chambre, il songerait à appeler Sarah, et il n'avait toujours aucune preuve.

— Je vous offre une bière ? dit-il, un peu gauche.

— Pourquoi pas ? répondit Henrickson.

Au ton de la réponse, Tom se demanda s'il n'acceptait pas pour des raisons très personnelles, sans aucun rapport avec l'envie d'une bière. Mais lorsqu'ils s'assirent au comptoir de Big Frank, bar vide et d'une froideur de pierre, Henrickson trinqua avec lui.

— Excusez-moi si je vous parais absent, dit-il. Je ne peux pas m'empêcher de penser au temps qui passe. C'est primordial pour moi.

— Je sais. On trouvera demain, je vous le promets.

— Excellent. Mais voyons d'abord ce qui se passe ici. Tom se retourna vers un grand homme qui s'approchait d'eux, lentement mais d'un pas décidé.

— Merde, fit-il. Le shérif !

Tom observait Connelly et le journaliste, qui se toisaient mutuellement. Le policier se tourna vers Tom.

— Monsieur Kozelek, dit-il, je vois que vous n'avez pas encore renoncé à l'hospitalité de Sheffer.

— Qui était-ce ? demanda Tom. La serveuse ? Un des vieux assis dans le coin ?

— Je ne comprends pas bien de quoi vous parlez, riposta Connelly.

— Je crois qu'il insinue que quelqu'un vous a informé de sa présence en ville, expliqua Henrickson, et je suis enclin à croire qu'il a raison.

— Je passais simplement par ici et je vous ai vus entrer tous les deux.

Henrickson but une gorgée de sa bière et scruta le policier par-dessus son verre.

— Vous avez un problème avec nous, shérif ?

— Je ne sais même pas qui vous êtes.

— Je suis écrivain.

— Et que pouvez-vous bien trouver à écrire ici ?

— Un article sensationnel. Les jolies villes de vacances du Nord-Ouest.

— Et M. Kozelek vous aide, c'est ça ?

— On peut le concevoir ainsi.

— Je n'ai jamais eu beaucoup de temps à consacrer à la lecture, lâcha Connelly. La plupart du temps, c'est des conneries.

Tom n'aimait pas la manière dont les deux hommes se regardaient. Il essaya de penser à quelque chose qui pourrait détendre l'atmosphère. Il entendit la porte s'ouvrir. Deux personnes entrèrent et s'ébrouèrent pour ôter la pluie de leurs cheveux.

— Bonsoir, dit une femme.

Tom la reconnut : c'était le médecin qui l'avait examiné. Elle approcha et rejoignit le groupe :

— Ne vous inquiétez pas. Vous étiez assez déboussolé la dernière fois qu'on s'est rencontrés. Comment vous sentezvous ?

— Mieux, dit Tom.

Son mari était avec elle. Il adressa un signe à Connelly et se dirigea de l'autre côté du bar, vers la table de billard, avec l'air d'un homme fort peu intéressé par les échanges de politesses.

— Parfait, approuva Melissa en observant Tom à la façon des médecins : avec un splendide détachement, comme si l'opinion du patient sur son propre état n'avait aucune conséquence sur le diagnostic. Pas de nausées ? Pas de maux de tête ?

— Non, mentit-il. Je vais bien, merci.

— Excellent. Oh... à votre place, j'irais mollo sur les herbes médicinales. On ne connaît jamais assez l'effet de certaines d'entre elles.

Connelly sembla se raidir.

— Le mystère est éclairci, dit le policier. Elles n'appartenaient pas à M. Kozelek.

— Des herbes médicinales ? répéta Henrickson.

Melissa esquissa un sourire, comme si elle hésitait à s'aventurer sur ce terrain.

— J'en ai retrouvé un bouquet dans le sac de M. Kozelek.

— Melissa, vous voulez me rendre un petit service ? demanda Connelly. Je serai très heureux de vous rejoindre dans un instant tous les deux, mais je dois discuter un peu avec ces garçons avant.

— Bien sûr, acquiesça-t-elle en reculant poliment.

Normalement, elle aurait dû être vexée ; or ce que Tom avait décelé dans son regard, ce n'était pas un avis très professionnel, mais l'effet apaisant d'un bon gros joint.

— Vous voulez une bière ?

— Avec plaisir.

Les trois hommes la regardèrent s'éloigner.

— Donc, si ces plantes n'appartenaient pas à Tom, reprit Henrickson, comment se sont-elles retrouvées dans son sac ?

— Je croyais que vous ne saviez pas de quoi il s'agissait.

— Désolé de vous avoir donné cette impression. Je crois que vous parlez de la valériane et de la scutellaire dans le sac de Tom.

— Quoi ? dit Tom en se tournant vers le shérif.

— Aucune idée.

— Ça m'étonnerait. (Henrickson fouilla dans sa poche et en sortit un petit sachet de plastique qu'il posa sur le comptoir.) C'est ce qu'a découvert le médecin ?

Connelly détourna le regard.

— Les plantes se ressemblent toutes pour moi.

— Pas pour moi. Ce sont deux herbes médicinales qui étaient utilisées par un certain groupe.

— Les Indiens.

— Bien avant. Alors, dites-moi, shérif, à en juger à la réaction de Tom, je ne crois pas que ce soit lui qui les a placées dans son sac. Mais vous pourrez peut-être m'apprendre comment elles y sont arrivées.

— Patricia Anders les y a mises.

Henrickson sourit.

— Ah oui. La femme aux bottes, je présume.

— Lorsqu'elle a trouvé les affaires de M. Kozelek dans la forêt, elle a tout de suite compris qu'elles appartenaient à quelqu'un qui n'avait pas tous ses esprits. Mme Anders s'intéresse aux médecines douces. Elle a laissé les herbes dans son sac avec l'espoir qu'il les reconnaîtrait et pourrait se soigner.

Cette fois, Henrickson éclata de rire.

— Vous plaisantez, j'espère.

— C'est ce qu'elle m'a dit.

— Mettons les choses au clair. Elle se baladait dans le coin avec des bottes toutes neuves – bien pratiques – et elle tombe sur le petit campement de Tom. Elle voit vite qu'il va mal et elle dépose des herbes médicinales dans son sac, au cas où il devinerait à quoi elles servent et déciderait de se soigner ? Des herbes qu'elle avait sur elle au cours d'une petite promenade dans les bois ? Des herbes que l'on consommerait aujourd'hui en décoction ou en infusion ?

— Les gens font de drôles de choses.

— Je n'en doute pas. Eh bien, merci. Ces plantes m'intriguaient depuis le début. Je suis ravi que vous m'ayez fourni une explication aussi limpide. (Henrickson se leva et sourit à Tom.) Bien, mon vieux. Dommage que nous n'ayons pas rencontré ce monsieur plus tôt. Il a réponse à tout. Je suis fatigué par notre longue marche. Il est temps de retrouver mon plumard.

Connelly ne bougea pas.

— J'aimerais autant que vous choisissiez une autre ville du Nord-Ouest, lança-t-il.

— Et moi, j'aimerais autant que vous arrêtiez de vous moquer de mon ami. Il sait très bien ce qu'il a vu, et vous aussi. Il a vu Bigfoot.

— Bigfoot n'existe pas. M. Kozelek a rencontré un ours.

— Croyez-le si ça vous amuse. Mais, à moins que vous trouviez une raison officielle pour l'interroger, il serait temps de le laisser tranquille.

Henrickson se dirigea vers la porte sans se retourner. Confus, ne sachant si la situation s'était améliorée ou le contraire, Tom le suivit.

Dès qu'ils furent dehors, le journaliste marcha rapidement en direction du motel sous ce qui devenait un rideau de pluie.

— Jim, haleta Tom en essayant de le suivre, qu'est-ce que c'est que cette histoire ?

— J'ai compris que j'avais touché un point sensible, mais je n'espérais pas qu'on m'offrirait la solution sur un plateau. Vous savez ce que sont les herbes médicinales, non ?

— Oui, on s'en sert pour soigner les maladies, à la place des médicaments. Comme pour... je ne sais pas, l'aromathérapie.

— Non, c'est différent, répondit Henrickson en franchissant la petite barrière pour entrer dans le parking. Les gens ont recours aux plantes depuis très longtemps. La médecine, ce n'est qu'une forme de nourriture spécialisée, guère plus. Dans les années soixante-dix, on a trouvé une tombe de Neandertal au nord de l'Irak. Le corps avait été enterré avec huit plantes différentes, toujours utilisées par les herboristes modernes. Les hommes de Neandertal connaissaient déjà ces plantes il y a soixante mille ans, plus sans doute. C'est pour ça qu'on les a retrouvées dans votre sac.

— Je ne comprends pas. Pourquoi ?

— Parce que la créature que vous avez vue est revenue. Elle a posé les plantes à un endroit où vous pourriez les trouver.

— C'est un homme de Neandertal qui m'a prescrit ces herbes ?

Henrickson sortit ses clés et appuya sur un bouton. Les phares de sa voiture s'allumèrent.

— Montez, je vais vous expliquer.

Tom s'installa sur le siège du passager. Henrickson fit demi-tour, s'engagea à vive allure sur la route principale et passa devant chez Frank, en direction de l'est.

Tom crut apercevoir Connelly qui les observait derrière la vitre.

— Jim, où on va ?

— Discuter avec quelqu'un. Quelqu'un qui en sait plus que ce qu'on a bien voulu nous dire.

Le journaliste ne prononça plus un mot pendant la demiheure de trajet. Tom savait où ils allaient, avant même que la voiture tourne sur la route solitaire qui menait au lotissement dont personne ne voulait. Henrickson se gara sur la route déserte balayée par le vent, à cinq mètres du portail de la propriété Anders. Il laissa tourner le moteur mais éteignit les phares. L'obscurité tomba comme une pierre.

— Attendez-moi ici.

Tom le regarda descendre de la voiture. À peine avait-il approché de la pancarte qu'il était déjà difficile à distinguer. Il revint dix minutes plus tard.

— Il y a quelqu'un, cette fois. (Il avait le visage glacé, et des cristaux de glace se formaient dans ses cheveux.) Qui ne se cache pas assez bien pour avoir pensé à éteindre les lumières.

En voiture, il franchit le portail et avança lentement dans l'allée bordée d'arbres.

— Vous n'avez pas allumé les phares.

— Je sais.

Après le dernier virage, le lac glacial dans le clair de lune diffus devint visible. Presque surnaturel, il paraissait hautain, comme si rien n'avait jamais changé pour lui, comme s'il était ainsi depuis la nuit des temps. Tom distinguait la forme sombre du bungalow, blotti dans les arbres, avec deux petits rectangles de lumière jaune pâle.

Henrickson arrêta la voiture, coupa le moteur. Il observa la maison.

— Bon, on y va. Refermez la portière doucement.

— Écoutez, Jim, on peut pas faire ça maintenant. On aurait dû appeler avant. On ne peut pas se pointer comme ça. Deux types à sa porte, on va lui flanquer une peur bleue.

Henrickson se tourna vers lui et esquissa un mouvement avec sa bouche. Ce n'était ni une grimace, ni un sourire. Cela ressemblait à ce qu'il avait arboré depuis le début, et Tom se demanda avec une inquiétude sourde si c'étaient des sourires, finalement.

— Descendez.

Si Henrickson ne se trompait pas, cette femme avait menti afin de le faire passer pour un idiot. Une fois au moins, peutêtre deux. Bien sûr, Connelly la croyait, elle, plutôt que lui, d'autant plus qu'il ne supportait pas les allusions à Bigfoot. Avec ses mensonges, cette femme avait anéanti son histoire. Elle l'avait privé du seul élément susceptible de remettre un peu d'ordre dans sa vie.

S'il fallait lui réserver une mauvaise surprise pour sauver la situation, cela en valait la peine.

Henrickson ouvrit le coffre de la voiture, en tira un grand sac et le passa sur son épaule en un mouvement lisse. Puis il se pencha à nouveau et plongea les deux mains dans le coffre. Quand il se redressa, Tom resta bouche bée.

— Qu'est-ce que c'est que ce truc ?

Question idiote. On voyait bien que c'était un fusil. Il était aussi évident que l'objet plus court, plus émoussé qu'il portait à la main était une arme de poing à gros calibre. Ni l'un ni l'autre ne ressemblaient à des articles de chasse mais plutôt aux armes qu'on voit aux informations, avec des nuages de fumée à arrière-plan.

— La forêt, c'est parfois dangereux, marmonna Henrickson en fermant le coffre.

— Ça l'est sûrement maintenant. On ne pourrait pas laisser ces trucs dans la bagnole ?

Le journaliste se dirigeait vers le bungalow. Se demandant ce qui se passait, Tom le rattrapa. Lorsqu'il arriva à son niveau, Henrickson frappait déjà à la porte. Ils attendirent. Henrickson leva de nouveau le bras mais interrompit son geste, l'oreille tendue.

On entendit le son de deux verrous, puis la porte s'ouvrit.

Patricia Anders resta à l'intérieur. Derrière elle, la pièce semblait confortable. Elle paraissait un peu plus âgée que dans le souvenir de Tom et encore plus petite. Mais elle ne semblait pas effrayée. Ni surprise.

— Bonsoir, monsieur Kozelek, dit-elle. Qui est votre ami ?

— Vous savez très bien qui je suis, répondit Henrickson.

— Non, mais je sais pourquoi vous êtes ici.

— Ça va faciliter les choses.

Elle haussa les épaules.

— Pour moi, oui. Je n'ai rien à vous dire.

— Oh, que si ! s'écria Henrickson.

Sa voix paraissait lointaine. Il passa devant la vieille dame,

entra, scruta les murs. Il arracha le téléphone de sa prise, trouva le portable de Patricia, le jeta par terre et le piétina.

— Jim, s'exclama Tom, horrifié, ce n'est pas le bon moyen !

— Le bon moyen pour quoi ? demanda-t-elle. (Elle tentait de paraître imperturbable, mais sa voix était rauque et son visage tiré.) Qu'est-ce qu'il est venu faire ici ?

— Il est journaliste, expliqua Tom. Il veut écrire un article sur ce que j'ai vu.

Patricia le regarda.

— Mon pauvre, vous êtes complètement stupide.

— Qu'est-ce que vous voulez dire ? aboya-t-il, excédé de voir que tout le monde comprenait, à part lui.

— Il n'est pas là pour écrire. C'est un chasseur. Il est là pour tuer.

— Pour tuer quoi ?
— Des ours, je suppose. Il n'y a que ça ici.
Tom regarda Henrickson et dut admettre que son compagnon ne ressemblait plus guère à un journaliste. C'était à la fois dû aux armes et à la manière dont il ouvrait et fouillait les placards.
— Jim, dites-moi que ce n'est pas vrai ! supplia-t-il.
— Mme Anders frime, mais à part cela je suis entièrement d'accord avec elle, répondit Henrickson sans se retourner. Sur mes intentions et sur votre intelligence. Ah. (Il sortit un gros paquet de cordes et le jeta à Tom.) Attachez-lui les mains dans le dos.
— Vous plaisantez ! Je ne peux pas faire un truc pareil !
Le canon du fusil de Henrickson balaya un petit arc qui se termina face au visage de Tom. Il n'avait rien vu venir.
Il recula, heurta un appareil ménager, glissa sur le tapis et tomba. Il avait vaguement conscience que Henrickson l'enjambait. Celui-ci alla fermer la porte et attrapa la vieille femme par les cheveux. Tom secoua la tête pour s'éclaircir les idées, avec l'impression qu'on lui avait enfoncé un tournevis dans chaque narine.
— Vous feriez bien d'obéir tout de suite, dit Patricia, mais ne comptez pas sur moi pour vous aider.
Pour toute réponse, Henrickson lui envoya un coup qui la fit valdinguer sur le divan. Ensuite, il se tint au-dessus de Tom, la corde à la main.
— On va trouver ce maudit truc, lança-t-il calmement.
Tom le regarda. Il saignait du nez mais il avait compris pourquoi la voix de Henrickson paraissait différente. Il avait perdu son accent, ses intonations populaires, son parler péquenot. À présent, on aurait dit un étranger. Tom avait la sensation de n'avoir jamais rencontré cet homme, mais tous ceux qui avaient entendu cette voix devaient s'en souvenir. Sa voix disait qu'il vous connaissait. Qu'il connaissait tout de vous, qu'il connaissait tout de tout le monde.
— Tu vas m'aider, parce que, sinon, j e t'oblige à la tuer et j'ai comme l'impression que ça ne va pas te plaire. Tom secoua la tête.
— Bien sûr que si. Après tout, ça ne sera pas la première fois. D'accord, les circonstances sont différentes.
— La ferme ! s'exclama Tom.
La vieille femme le scrutait à présent.

— Tom est sur la liste, reprit Henrickson. Il était associé dans une boîte de design à Los Angeles. Tout allait pour le mieux : belle bagnole, belle petite famille, belle petite baise avec l'une des jolies filles de la boîte derrière son Mac à écran géant. Un soir, ils travaillent tard, prennent un verre, il la raccompagne et, juste au coin de la rue, Tom grille un feu... Peut pas se permettre d'arriver encore en retard ! Une Porsche s'enfonce dans la porte du passager. La fille crève. Le petit garçon qu'elle a dans le ventre aussi. Tom est juste en dessous de la limite, et, par chance, le conducteur de la Porsche est bourré comme pas permis. Alors, Tom s'en tire bien.

— C'est ce que vous croyez ? hurla Tom. (Il se redressa sur ses pieds, s'essuya le nez d'un revers de manche, sans tenir compte de la douleur.) Vous croyez que je m'en suis bien tiré ?

— Ils sont morts, tu es vivant. Fais le calcul.

Tom commença à bouger, mais Henrickson s'en aperçut avant même qu'il commence. D'un mouvement rapide, il planta son fusil sur le front de Patricia.

— Tu la tues et ce sera terminé. Je te libère. Tu ne t'es pas suicidé la dernière fois. Je te laisserai tranquille pendant un an ou deux, et ensuite je te retrouverai pour mettre fin à tous tes malheurs. Peut-être. Ou alors, je déniche ce truc, on le photographie, et ensuite il s'échappera, du moins c'est ce que tout le monde croira. Tout sera parfait. Tu obtiendras les honneurs qu'on ne peut pas trouver dans la culotte d'une jolie fille. Sarah te reprendra peut-être.

— Comment savez-vous tout ça ?

— Parce qu'il n'est pas humain, répondit la vieille femme d'un ton paisible.

Henrickson eut un petit rire.

— Tom, alors, ces mains, c'est pour aujourd'hui ou pour demain ?

Tom regarda Patricia. Son visage était rouge, mais elle avait le regard clair et elle l'observait.

— N'en faites rien. Pas pour moi. Pour eux.

Mais il détourna les yeux, et, lorsque le paquet de cordes heurta de nouveau sa poitrine, il l'attrapa.

CHAPITRE 24

— Ward, reste tranquille, je t'en prie.
— Ça fait mal !
— Sois cool.
— Ah, ça suffit ! C'est bon pour les jeunes. Je suis assez vieux pour reconnaître que ça fait un mal de chien.

J'étais installé sur le siège du passager avec les pieds à l'extérieur. Accroupie en dehors de la voiture, Nina nettoyait mon épaule avec une compresse imbibée de désinfectant. Nous ne savions pas où nous étions, à part sur le parking d'une station-service près d'une ville dont nous ne connaissions pas le nom.

— C'est propre. Je crois.

Je regardai mon épaule et vis une vilaine déchirure le long du deltoïde. Cela saignait toujours, mais un peu moins que pendant les cinquante premiers kilomètres après Fresno. Cela faisait très mal quand même, malgré les analgésiques puissants que nous avions trouvés sur le marché où nous avions acheté les compresses. Cela me faisait aussi mal que si j'avais huit ans et qu'un grand costaud ne cessait de me donner des coups de poing dans l'épaule, si fort et si vite que les impacts se confondaient en une douleur continue.

Nina me regardait, avec l'air de quelqu'un qui a bien fait son travail, en espérant que je n'allais pas me plaindre trop longtemps. Je compris que ma blessure à l'épaule n'était rien par comparaison à celle qu'elle avait reçue aux Halls. J'avais aussi de la chance que la balle n'ait pas atterri à vingt centimètres plus à droite.

— Merci, dis-je. Ça va mieux.
— Menteur !

Elle se leva et, par-dessus le toit, regarda la station-service où un barbu se tenait derrière la vitrine.

— On nous observe.

— C'est juste un curieux, rétorquai-je. Qui se demande si on va lui acheter de l'essence. Tout le monde ne nous en veut pas.
— Théorie séduisante, commenta-t-elle. Tu as des preuves ?
— Pas vraiment.
— Qu'est-ce qu'on va faire ?
— Il faut appeler quelqu'un, suggérai-je. Faut les prévenir, pour Monroe.
— Ils sont déjà au courant, objecta-t-elle. Il avait certainement ses papiers sur lui.
— Je ne parlais pas de ça. Il faut leur expliquer ce qui s'est passé. Et ce que ça signifie.
— Nous l'ignorons, répliqua Nina.
— Mais non.
— Je n'ai pas vu l'homme qui est sorti du Knights et qui a tué le flic. Je n'ai que les déclarations des témoins.
— Je sais, lui répondis-je. Mais apparemment, il ressemblait beaucoup au type qui a essayé de nous liquider. Il portait les mêmes vêtements.
— C'est une description très vague. L'employé d'ici n'est sûrement pas très différent.
— Je ne parlais pas que du physique, insistai-je. C'est le genre de type qui entre dans un restaurant et qui continue à tirer en présence de témoins, avec trois personnes qui répliquent. Ne coupe pas les cheveux en quatre. Je crois pas qu'il y en ait beaucoup comme ça dans le coin.
— Qui est-ce, alors ? Tu as quelque chose en tête, et j'aimerais bien savoir ce que c'est.
— Il faut continuer à rouler, dis-je. Pas seulement parce qu'il faut s'éloigner le plus possible du lieu du désastre. Mais parce qu'il y a une femme qu'on doit rencontrer ce soir, et on n'est pas arrivés.
— Où ?
— Au nord. Passe-moi mon sac, j'ai l'adresse.

Mme Campbell n'était pas chez elle.

Cette fois, j'avais prévenu, bien avant d'approcher de San Francisco. Je n'avais pas obtenu de réponse. C'est bizarre comme on s'habitue à l'idée que les maisons ont une mémoire, sont liées en permanence à l'extérieur et transmettent les messages. Cette maison-là n'était pas à notre service. Nous nous y rendîmes donc directement. Nina refusait toujours de contacter le FBI à Los Angeles. Ils étaient au courant pour Monroe ou ne tarderaient pas à l'être. Elle n'avait guère envie de leur faire confiance. J'estimais qu'elle avait tort, qu'il était raisonnable de donner notre position et d'affirmer notre innocence le plus vite possible. Un type bizarre hantait les couloirs de la justice mais cela ne signifiait pas pour autant que toute l'institution était pourrie. Je ne parvins pas à la convaincre. Finalement, nous cessâmes d'en parler. Plus je passais de temps avec Nina, plus je me rendais compte qu'elle possédait des défenses internes, une véritable enceinte de château, avec douve, herse et sans doute réserves de poix bouillante.

La douleur demeurait supportable tant que je me bourrais d'analgésiques. Le problème, c'est que cela commençait à tirer. À l'approche de la banlieue de San Francisco, j'avais l'impression d'avoir été recousu par un type qui ne voulait pas savoir ce qui se passait en dessous du bandage. J'étais donc limité à la lecture de la carte, ce qui constituait sans doute une bonne division du travail. Nina conduisait bien mais n'avait guère le sens de l'orientation. Les inconvénients de l'espace à trois dimensions semblaient l'agacer. Je n'aurais pas aimé la voir au volant d'un tout-terrain. Elle aurait sans doute foncé droit dans tout ce qui lui barrait la route.

— Pourquoi maintenant ? questionna-t-elle enfin. Pourquoi attendre trois mois avant de nous tomber dessus ? Bon, d'accord, tu étais en cavale. Mais ils pouvaient s'en prendre à John et à moi tout de suite.

— Ils ont dû rassembler leurs troupes après l'explosion des Halls.

— Ils n'étaient pas tous sur place. S'ils sont aussi puissants qu'on l'imagine, il y en avait d'autres. Tu crois vraiment que le type que j'ai vu avec Monroe était un des leurs ?

— Oui, et ça me fiche la trouille, admis-je.

— À moi aussi. Mais il est d'autant plus difficile de croire qu'ils ne pouvaient pas nous éliminer avant.

— Ils ont bien essayé ce soir.

— Oui, mais pourquoi pas avant ? réitéra-t-elle.

— Tu travailles pour le FBI. Si on te retrouve dans une décharge, on va commencer à se poser des questions. Je vois bien Monroe se lancer dans une croisade.

— Dans l'intérêt du service, bien sûr. Mais je serais morte quand même.

— Ces gens réfléchissent à long terme, répondis-je. La cabane qu'on a découverte près de Yakima montre bien qu'ils n'ont pas commencé hier. Ils nous auraient laissés nous acharner tant que nous n'étions pas trop dangereux et se seraient débarrassés de nous à la première occasion. Mais tout s'est accéléré lorsque John a liquidé Ferillo. Il a dû tomber sur un sacré machin, et leur foncer dessus. Ils le surveillaient très certainement après la disparition de sa fille. Ils l'ont suivi quand il est sorti de chez DeLong. Peut-être que DeLong était bon pour la retraite, de toute façon, mais à présent John a fait un truc qui les oblige à agir. John, c'est la clé de tout.

— S'il n'appelle pas bientôt, c'est moi qui le tuerai de mes propres mains.

— Chouette, je te donnerai un coup de main.

Il était près de 21 heures lorsque nous approchâmes de la maison. J'appelai de nouveau. Toujours pas de réponse. Ou elle ne répondait pas volontairement ou elle n'était pas chez elle. La première solution n'avait aucun sens. La seconde m'inquiétait.

Nina se gara devant la maison où une seule lampe était allumée, au-dessus de la porte.

— Personne.

— Pas sûr.

Je grimpai les marches et sonnai. Aucune lumière ne s'alluma. Personne ne vint ouvrir.

— Ça ne me dit rien qui vaille. Les personnes âgées ne sortent pas beaucoup.

— On devrait peut-être aller voir les voisins.

Je baissai les yeux sur ma tenue, puis sur la sienne. Son chemisier présentait une tache de sang. La manche de ma veste ne tenait plus qu'à un fil et paraissait souillée à la lumière du lampadaire.

— Excellente idée !

— T'as raison. Alors, qu'est-ce qu'on décide ?

Je sortis une carte de crédit devenue inutile mais que je n'avais pas eu le courage de jeter.

— Oh, super !

Elle se retourna et observa les fenêtres voisines pendant que je glissais la carte dans l'encoignure de la porte.

Cinq minutes plus tard, nous avions la certitude que Mme Campbell ne se trouvait pas chez elle. J'avais eu peur de l'apercevoir avec une hache plantée dans la tête. Toutes les pièces étaient vides, propres, bien rangées.

— Elle est sortie, conclut Nina. Elle a peut-être une vie sociale plus riche que la tienne.

Nous l'attendîmes jusqu'à 21 h 30. Je me levai et fis les cent pas, j'allai dans le couloir, où je vis une petite table de téléphone. Un petit carnet, avec une couverture à motif floral.

Un carnet d'adresses.

Je consultai la lettre « D ». Pas de nom connu. Puis, comprenant que j'aurais sans doute fait de même, je regardai à« M ».

Voilà.

Je composai le numéro. Il était tard. Mme Campbell m'avait dit que Muriel avait des enfants, mais je ne connaissais pas leur âge. J'allai sûrement entendre parler l'un d'eux quand on me répondrait.

— Résidence Dupree.

— Muriel ?

— Qui est à l'appareil ?

— Je m'appelle Ward Hopkins. Nous nous sommes rencontrés...

— Je me souviens de vous. Comment avez-vous obtenu mon numéro ?

— Je suis chez Mme Campbell. Je l'ai trouvé dans son carnet.

— Qu'est-ce que vous faites chez elle ?

— J'avais besoin de lui parler de toute urgence. Je suis venu la voir. Elle n'était pas là. Je me suis inquiété et je suis entré.

— Qu'est-ce qui vous inquiète ? Vous savez quelque chose que je devrais savoir ?

— Muriel, savez-vous où elle est ?

Une pause.

— Attendez un instant.

Le son devint plus assourdi. Des bruits de voix, mais je ne comprenais pas les mots.

— Elle accepte de vous parler, dit Muriel en me laissant bien sentir qu'elle n'était pas d'accord. Vous feriez mieux de venir.

La traversée de la ville dura vingt minutes. Muriel Dupree n'était guère accueillante en ouvrant la porte, mais elle finit par s'écarter. Elle regarda Nina d'un air soupçonneux.

— Qui est-ce ?

— Une amie.

— Elle sait qu'elle a du sang sur son chemisier ?

— Oui, répondit Nina. La journée a été longue. Ward est taché, lui aussi.

— Pour un homme, c'est presque normal.

Soignée, la maison de Mme Dupree était l'une des mieux décorées que j'avais vues depuis longtemps. Le décor fonctionnel et simple d'une personne qui menait une vie rangée. Elle nous conduisit vers une grande cuisine, avec un petit coin salon. Mme Campbell était installée près du chauffage, plus frêle que dans mon souvenir.

— Si ce n'est pas trop indiscret, pourquoi êtes-vous ici ?

— Quelque chose devrait m'en empêcher ?

En jetant un coup d'œil vers Muriel, je compris que Mme Campbell comptait beaucoup pour elle, et que son apparence bourrue cachait quelque chose. De l'inquiétude, sans doute. De la peur, même.

Je m'assis à l'extrémité du divan.

— Madame Campbell, je dois vous poser une question...

— Je sais. Eh bien, allez-y !

— Pourquoi êtes-vous ici ?

— Il se passait de drôles de choses, répondit Muriel. Joan entendait des bruits sous ses fenêtres, la nuit. Dans son quartier, ça n'a rien d'anormal, mais ensuite, un homme est venu chez elle et lui a posé des tas de questions.

— Quand ?

— Le lendemain de votre visite.

— À quoi ressemblait-il ?

— Votre taille. Les épaules un peu plus larges.

— John. Du moins, j'espère. C'est un policier. Il a dû retrouver une vieille liste d'employés.

— Il savait que je travaillais là-bas, c'est sûr, dit Mme Campbell. Mais je n'avais pas la réponse à ses questions. Il est parti. Il a été poli. Pourtant, il n'avait pas l'air de quelqu'un qui traite tout le monde comme ça.

— Qu'est-ce qu'il a demandé sur vous ?

— Ce que vous avez l'intention de me demander. Mais maintenant, je connais les réponses.

— La dernière fois, vous m'aviez parlé d'une famille qui avait accueilli Paul. Dont le chien était mort dans de drôles de circonstances.

— Je m'en souviens.

— Ils s'appelaient Jones ?

Nina se tourna brusquement vers moi.

— Non, répondit Mme Campbell. C'étaient les Wallace. Les Jones, c'est l'autre famille. Celle qui l'a renvoyé lorsqu'ils ont eu une petite fille.

J'étais pris de vertige.

— Comment se fait-il que vous vous en souveniez maintenant ?

— Joan m'a demandé de faire des recherches, répondit doucement Muriel. Après votre départ, elle m'a appelée. Je pensais qu'elle serait furieuse que je vous aie mis en contact avec elle, mais ce n'était pas le cas.

— J'ai suggéré à Muriel de mener une petite enquête pour moi, continua la vieille dame. De retrouver deux de mes anciennes collègues. J'en ai localisé une en Floride, qui se faisait griller la peau comme un crocodile. Et une autre dans le Maine. Elle avait voulu se rapprocher de sa famille, quand les enfants sont morts les premiers. C'est la vie. Avec trois mémoires, on a reconstitué les événements. (Elle se mordit les lèvres.) Alors, que s'est-il passé ?

— Paul les a tuées toutes les deux. On a retrouvé Jessica Jones dans un motel à Los Angeles il y a quatre jours. Et Katelyn Wallace hier matin.

— Où ?

— Au nord. À l'est de Seattle. Il a laissé des disques durs effacés dans leur corps. Comme s'il voulait défaire le passé, tout nettoyer, une sorte de purification.

— Oh, mon Dieu !

Ses mains tremblaient. Muriel approcha et les prit dans les siennes.

— Jessica et Katelyn étaient des enfants dans ses familles d'accueil ? s'enquit Nina. Il les a tuées rien que pour cela ?

— C'étaient des familles qui avaient essayé de l'adopter pour de bon, de lui offrir un véritable foyer, expliquai-je. Mais avec lui, c'était impossible. Il en rejette la faute sur quelqu'un. Il efface son propre disque dur. Il... Madame Campbell, vous savez où vivent les parents de Katelyn Wallace aujourd'hui ?

— Ils sont morts, répondit Muriel. De mort naturelle, voilà cinq ans. Enfin, dans la nature. Sur un voilier qui a coulé dans la baie. Personne n'y a rien vu d'étrange à l'époque.

— Et les Jones ?

— Je l'ignore.

— La police de Los Angeles les a cherchés à Monterey, dit Nina. Ils n'ont trouvé personne. Les voisins leur ont dit qu'on ne les avait pas vus depuis six semaines. On a supposé qu'ils étaient partis en vacances.

— J'espère que c'est ça, répondis-je.

Je songeai cependant à deux personnes, du même âge, dont on avait trouvé les corps dans une plaine déserte, à sept cents kilomètres au Nord de l'endroit où nous nous trouvions à présent. Que John avait photographiées et qu'il aurait pu retrouver... s'il avait poursuivi une enquête qu'il avait gardée secrète. En l'absence de certitudes, je préférai me taire. John aurait pu aller en Floride, contacter la collègue de la vieille dame et les retrouver grâce à cette piste.

Nina me regardait.

— Comment savais-tu tout ça ?

— Je ne le savais pas, répondis-je d'un ton distrait. Je me demandais simplement pourquoi l'assassin avait piqué une photo des parents de Jessica. Quand on prend un souvenir, un talisman, on préfère quelque chose de plus personnel. Un morceau de corps, un vêtement. Et il s'est contenté d'une photo qui n'appartenait même pas à la victime. D'après Monroe, on avait déjà essayé de la localiser voilà plusieurs mois. Cela ressemble plus à une traque qu'au mode opératoire d'un tueur en série. Et si l'assassin de Jessica n'était pas le même que celui du flic ? Quelles seraient les motivations du tueur de flic ? Simplement de mettre en valeur l'assassin de Jessica. Avec un cadavre de femme dans un motel poussiéreux, les flics n'auraient pas perdu beaucoup de temps sur l'affaire, même avec le disque dur dans la bouche. Mais ça, plus un flic assassiné au grand jour, et toute la Crime débarque, le FBI s'en mêle... d'autant plus que le FBI a déjà été prévenu par un appel anonyme...

— Qu'est-ce qui nous dit que c'est l'Homme Debout qui a tué Jessica ?

— Qu'est-ce qu'il te faut de plus, Nina ? Mme Campbell vient de confirmer le seul lien qui unissait les deux femmes. C'est Paul. Dès que le type a essayé de nous descendre à Fresno et que j'ai pensé que c'était le même qu'à Los Angeles, je ne voyais plus d'autre solution.

— Il y a des milliers d'autres solutions, Ward. Bon, d'accord, le tueur de flic travaille pour les Hommes de Paille. Peut-être. Ou il essaie d'attirer l'attention sur un autre meurtrier. Mais de là à affirmer que c'est ton frère...

Je ne comprenais pas où elle voulait en venir.

— S'ils s'arrangent pour que quelqu'un le coince à leur place, dis-je, c'est qu'il est tellement dézingué, tellement dangereux qu'ils ont besoin de l'aide de la police pour s'en débarrasser.

— Mais pourquoi s'en débarrasser ? Il est des leurs. Il leur fournit des victimes, il les aide à faire sauter des immeubles, il organise des tueries. Pourquoi...

— Parce qu'il fait d'autres trucs... assassiner mes parents, enlever la fille de Zandt, ce qui fait que plusieurs personnes le traquent une arme à la main. Il a fait tuer leur avocat. Il a réduit en fumée leur nid douillet du Montana. Et qui sait ce qu'il manigance ? Si Paul s'intéresse à toi ou si tu l'énerves un peu, je suis sûr que tu ne tarderas pas à entendre parler de lui.

Soudain, je m'aperçus que les deux femmes nous regardaient et que nous criions. Je repris plus calmement :

— Nina, je ne vois pas le problème. Tu as bien entendu que...

— Ward, ça pourrait être John...

— Comment ça ? m'exclamai-je, suffoquant soudain.

— Tu sais à qui veulent s'en prendre les Hommes de Paille ? John. Qui est impliqué dans la vidéo qu'ils ont envoyée ? John. Qui a tué l'un des leurs ? Alors, comment affirmer que ce n'est pas John qui a tué ces femmes ?

— Et pourquoi ?

— Parce qu'elles appartenaient à la vie de l'Homme Debout. Tu sais ce que ton frère lui a fait. Il lui a volé Karen. Il l'a tuée, et il a pris tout son temps. Il l'a kidnappée, et il n'a avoué qu'elle était morte que lorsqu'il a disposé ses os sur une piste destinée à piéger John pour le tuer aussi. Il a détruit la vie de John. Tu crois que John va mettre des limites à sa vengeance ?

J'ouvris la bouche. La refermai aussitôt.

Nina se leva. Elle était furieuse ; je n'avais jamais vu personne dans une telle colère.

— Va au diable, Ward. Je t'attends dans la voiture. Elle sortit, claqua la porte derrière elle. Je me tournai vers les deux femmes, qui m'observaient, tels deux chats curieux.

— Merci. Il faut que j'y aille.

J'entendis le cri d'un enfant qui appelait à l'étage.

J'étais à la porte avant que Mme Campbell reprenne la parole.

— Vous ne m'avez pas posé la question à laquelle je m'attendais.

— C'est-à-dire ?

— Je ne sais pas comment on attrape les gens, mais je m'imaginais que vous aimeriez savoir où il était allé en dernier.

— Quand ? dis-je sans savoir de quoi elle parlait, m'attendant plus ou moins à entendre Nina partir sans moi.

— À l'époque. La famille qui l'a recueilli. C'est mon amie qui s'en occupait. Elle m'a dit qu'ils étaient allés à Washington, car la mère de la femme se faisait vieille et avait besoin qu'on s'occupe d'elle. La dernière fois que Dianne a eu de leurs nouvelles, c'était un an après leur déménagement. Le mari était parti avec une jeunette qu'il avait rencontrée dans un bar.

— Elle se souvient de leur nom ?

— Oui. Parce qu'il ressemblait au nom du guitariste qui était mort quelques années plus tôt. Mais ça ne s'écrivait pas pareil.

— Qui ?

— Ils s'appelaient Henrickson. Ils habitaient à Snowcalm, quelque chose comme ça, près des Cascades.

Nina conduisit jusqu'à l'aéroport dans un silence sombre et venimeux. Je tentai de lui parler, mais c'était un fantôme, occupé ailleurs ou prisonnier du passé. Je pensais à John Zandt et à ce dont il était capable. Je me souvins d'une phrase qu'il avait prononcée lorsque nous nous étions rencontrés devant l'hôtel à San Francisco, une phrase à laquelle je n'avais pas trouvé beaucoup de sens, sur le coup.

« Parfois, on doit rebrousser chemin. »

Je comprenais mieux maintenant.

Nina se gara et nous descendîmes de voiture. Elle avança droit vers l'escalier et je la suivis en me débattant avec mon sac.

— Nina !

Ma voix résonna contre les murs de béton et me revint droit aux oreilles, plate et forte.

Nina se retourna et me gifla en pleine figure. Surpris, je trébuchai. Elle s'approcha, me frappa, me frappa encore, en criant des mots incompréhensibles.

Je tentai de lever le bras gauche pour la repousser, mais la douleur me rendait maladroit et entravait mes mouvements. Elle le remarqua, voulut me viser droit sur l'épaule, mais se retint au dernier moment.

Elle me regarda, de ses yeux si verts, si brillants que j'avais l'impression de les découvrir pour la première fois.

— Ne recommence jamais ! Ne me cache plus jamais rien !

— Nina, je ne sais même pas de quoi tu...

— Je m'en fiche ! Ne recommence jamais. Ne me traite pas comme si j'étais une... gamine à qui on raconte ce qu'on veut. C'est ce que John a fait, et si jamais je le revois, je lui casse la gueule.

— D'accord, mais ne te venge pas...

— Sur ta petite personne ? En deux jours, j'ai été suspendue, mon ex s'est mis à tuer des gens, personne ne sait combien, et mon patron s'est fait descendre sous mon nez ! J'ai encore son sang sur ma chemise, comme on me l'a gentiment fait remarquer. Alors, si jamais tu oses...

Elle cessa de crier, cligna des yeux, et je compris que, si ses yeux paraissaient si brillants, c'est qu'ils étaient humides. Je pris un risque et posai la main sur son épaule. Elle se dégagea méchamment, mais ses yeux redevinrent secs.

— Nina, je suis désolé.

Je comprenais que John l'avait profondément blessée. Elle compensait sur moi. Étant donné la fureur de Nina, il avait de la chance d'être ailleurs !

Elle recula d'un pas, les mains sur les hanches. Elle détourna le regard et poussa un profond soupir.

— Je t'ai fait mal à l'épaule.

— C'est le dernier de mes soucis. J'ai l'impression de m'être écrasé le nez contre un mur. Tu n'y vas pas de main morte avec les gifles.

Elle me regarda, la tête penchée.

— Exact. Maintenant que tu le sais, ne m'oblige pas à recommencer.

— J'essaierai.

— Ça ne suffit pas. Essayer, c'est à la portée de tout le monde, il faudra faire mieux.

— D'accord, répondis-je, très sérieux. Fais-moi confiance. Je ne recommencerai pas.

— Bien, dit-elle, et m'adressa un sourire plus fugitif qu'un battement d'ailes qui me fit néanmoins hérisser les poils de la nuque. Parce que, n'oublie pas, je suis armée.

Elle s'écarta brusquement et se dirigea vers l'escalier.

— Franchement, tu n'es pas comme les autres filles.

— Oh, que si, dit-elle sans que je sache si elle plaisantait ou non. C'est juste que vous, les hommes, vous n'en avez aucune idée.

De justesse, nous réussîmes à prendre le dernier vol pour Seattle. Il était minuit lorsque nous finîmes par louer une voiture. Une carte et deux hamburgers achetés à Tacoma, et nous étions prêts pour le départ.

Je conduisis, en essayant d'oublier mon bras et en laissant Nina faire ce que nous avions décidé dans l'avion. Elle refusait de contacter le FBI, car l'homme qu'elle avait rencontré avec

Monroe risquait d'être encore là, mais elle voulait bien appeler quelqu'un.

Elle téléphona à Doug Olbrich et lui parla pendant cinq minutes. J'étais assez occupé par l'échangeur de l'autoroute Seattle-Tacoma pour ne pas comprendre tout ce qui se disait, bien que la conversation me parût positive dans l'ensemble.

Elle raccrocha, regarda dans le vide et, comme la veille, mais un peu moins furieuse, elle tapa sur le tableau de bord.

— Qu'est-ce que ça donne ?

— Ça pourrait être pire. Monroe n'est pas mort.

— Tu plaisantes ?

— Non. Ce saligaud est toujours vivant. Surprenant. Il a plus de cran que je ne le croyais. Il a cinq trous dans le corps, il a passé six heures sur la table d'op, il va très mal. On lui donne vingt pour cent de chances, au mieux. Mais il n'est pas encore mort.

Je me sentais horriblement coupable d'avoir abandonné Monroe, de l'avoir laissé pour mort.

— Tu as eu raison de m'entraîner, conclut-elle. Sinon, je ne serais sans doute plus là.

— Tu as d'autres mauvaises nouvelles ? Je le sens.

— Doug est passé chez moi. Quelqu'un était venu avant lui.

On a tout saccagé et pris tous mes dossiers. Tu avais raison, il était temps de filer.

— Navré.

— Peu importe. L'affaire Gary Johnson commence à peser lourd. Apparemment, l'avocat de Louisiane laisserait un sacré paquet de fric derrière lui, et ça soulèverait pas mal de vent.

— Ah oui, et je me demande bien d'où il souffle.

— Effectivement. Monroe est dans de sales draps, même s'il survit. Tu sais comment ça se passe. Une fois qu'on a soulevé le rocher, il faut bien trouver quelque chose dessous pour se justifier. Je sais que je n'ai rien laissé filer dans l'affaire Johnson, mais Monroe est peut-être tombé sur un nouveau truc. Il tenait à cette affaire. C'est comme ça qu'il a été promu.

Elle se tut et resta silencieuse. Je la laissai en paix, une cigarette à la main, jusqu'à ce que je sois engagé sur la 18, avec la 90 en vue.

— Tu ne lui as pas dit ce que nous savions, repris-je.

— Ce que nous croyons savoir.

— Peu importe. Tu ne lui as rien dit.

— Je sais. Je ne lui ai rien dit parce que je crois qu'à part nous personne n'ira là où il faut.

— Et c'est où ?

— Il y a un endroit pour les types qui enfoncent des trucs dans le crâne des femmes, et je ne parle pas de la prison.

— Tu n'y songes pas sérieusement.

— Pour l'instant, si. Même si c'est John. Et je n'ai rien dit à Doug parce qu'il a mentionné quelque chose en passant... Euh, tu as de l'essence ?

— Je crois.

— La voiture que Monroe a mentionnée, celle qu'on a repérée à Snoqualmie la veille de la mort de Katelyn...

— Oui ?

— Il y a trois heures, un shérif a procédé à des vérifications. Cela n'a rien donné, parce que c'était une voiture louée sous un faux nom, mais Doug a noté les coordonnées et dit que quelqu'un pourrait peut-être aller voir. C'est à environ quatrevingts kilomètres dans les montagnes, après Snoqualmie. Je crois qu'on devrait commencer par là.

— Alors, où on va, exactement ?

Elle regarda la carte, posa un doigt sur un point situé au beau milieu des montagnes.

— Là. Sheffer.

Vers 1 heure du matin, Nina s'endormit, les bras croisés devant elle. Tout en roulant vers l'est, j'écoutais sa respiration. Le paysage était trop sombre mais mon horloge interne ou je ne sais quel organe m'indiquait que l'altitude grimpait. De temps à autre, nous croisions une voiture, un voyageur embarqué vers une autre aventure.

Nous montions toujours et je ralentis à soixante-dix kilomètres heure, puis soixante, car la route devenait de plus en plus sinueuse. Je commençai à avoir froid, des fantômes brumeux étaient suspendus entre les arbres, illuminés par les phares à iode et par la lune qui jouait avec les nuages. Je m'arrêtai un moment pour me repérer. Nina bougea un peu mais ne se réveilla pas et je repartis aussi doucement que possible.

Au-dessus de la crête des montagnes, j'empruntai une sortie vers une route secondaire qui annonçait Sheffer à quinze kilomètres. Après avoir eu l'impression que les montagnes et la forêt n'étaient qu'un décor, je me sentis bientôt dans la peau d'un intrus.

À Sheffer, tout était fermé à 3 h 15 du matin. Je longeai lentement la rue principale, avec le sentiment d'être un envahisseur qui avait choisi le bon moment pour préparer son attaque. Je longeai un marché, un bar, quelques restaurants. Tout au bout, je repérai la pancarte d'un motel.

J'entrai dans le parking et contournai un grand rond-point pour me garer. Aucune lumière à la réception. Hors saison, dans un trou pareil... Pas de sonnette de nuit. Quelques heures de froid dans ma voiture m'attendaient.

Je coupai le moteur et ouvris la portière, me glissai à l'extérieur aussi vite que possible pour ne pas laisser pénétrer le froid, dans l'intention de fumer une dernière cigarette avant d'essayer de m'endormir.

Je m'aperçus alors que quatre voitures étaient garées de l'autre côté du parking. Bien sûr, quoi de plus normal, dans un motel ? Mais je m'intéressais à une en particulier.

Je ne connaissais pas le numéro de plaque que nous cherchions. Nina ne me l'avait pas donné et je l'aurais probablement oublié, de toute façon. Et puis, avions-nous une chance de la voir juste devant un motel ?

Je m'approchai de la première voiture et regardai à l'intérieur. Le siège arrière était couvert d'accessoires de vacances : parkas chaudes, cartes de randonnée et toute une collection d'objets de couleurs vives.

La deuxième se trouvait à une dizaine de mètres. Il faisait un froid glacial et j'avais terminé ma cigarette. Je songeai à rebrousser chemin. Finalement, je m'approchai. Elle ne ressemblait pas à une voiture de location. C'était une énorme bagnole, toute rouillée, couverte de boue. Je regardai quand même à l'intérieur.

Au dernier moment, j'entendis des pas tranquilles et me retournai.

Ensuite, j'eus des étoiles dans les yeux qui s'assombrirent aussitôt.

— Oui, quand j'ai su qu'à Los Angeles il avait aidé à organiser le transport des jeunes filles jusqu'aux assassins. Il croyait peut-être qu'on voulait les former comme putes... c'est ce qu'il m'a raconté. Mais tu sais quoi ? Ça suffisait.

Je voyais dans son visage qu'il était incapable de reconsidérer la mort de Ferillo.

— John, détache-moi, nom d'un chien !

— Pas question ! Tu te mettrais en travers de mon chemin. Tu n'es pas prêt.

— Va te faire foutre !

Soudain, il brandit un doigt vers mon visage.

— Qu'est-ce que tu as fait la dernière fois ? Tu as tué l'homme qui a démembré ma fille quand il était dans ta ligne de mire ?

Je ne pouvais que me taire. Il savait que je n'avais pas tiré.

— Il est là, c'est ça ?

— Oui, répondit John. Il cherche quelque chose. Il est persuadé qu'il va tout arranger.

— Il a merdé, c'est ça ? questionnai-je. Ce n'est plus le psychopathe au service des méchants ? Ils l'ont exilé, et maintenant ils veulent sa mort ?

— Pas si bête, il faut le reconnaître.

— Dis-moi, John, j'ai le droit de savoir. Alors, détache-moi ou offre-moi quelque chose à boire, on se gèle ici.

Il alla à la salle de bains. Quelques cliquetis dans le noir, et il réapparut avec un petit verre contenant dix centimètres de liquide ambré. J'ouvris la bouche et il me fit boire. Je toussai mais la chaleur se diffusa dans ma poitrine.

— Il n'est pas dans ce motel ?

— Il y était, avec un autre type. Je suis arrivé en milieu de soirée, et il était déjà parti. Mais il rôde toujours dans le coin.

— Comment tu le sais ?

— Parce qu'il est cinglé. Il pense qu'il a trouvé une baguette magique qui va transformer le monde à son image.

— Quoi ?

— Tu ne me croiras jamais.

— Tu sais que les deux femmes venaient des familles adoptives où il a vécu quand il était gosse ?

— Oui, j'ai retrouvé les gens qui l'avaient placé. J'ai parlé avec la vieille femme de San Francisco. Et j'ai additionné deux et deux.

— Pourquoi Ferillo ?

— C'était une façade pour les Hommes de Paille, une des nombreuses qu'ils ont dans tout le pays. Ils se sont arrangés pour lui éviter un procès il y a quatre ans. Je ne crois pas qu'il ait vraiment compris quels étaient leurs buts, mais il blanchissait de l'argent pour leur compte dans son restaurant. L'appartement dans lequel il est mort appartenait à un certain George Dravecky. Un promoteur bourré de fric. Il ne possédait aucune maison aux Halls, mais c'est lui qui a initié le projet. Il a couvert la mise de fonds de départ. C'est un des leurs.

— Comment tu l'as découvert ?

— Je suis un bon.

— Tu n'es plus flic et tu as refusé de faire appel à Nina. Alors, quelles sont tes sources ?

— Je travaillais souvent avec la police de L.A. Avec un type qui piquait un petit sachet par-ci par-là pour son usage perso. Pas grand-chose, mais, maintenant qu'il a vieilli, il s'est rangé et il ne voudrait pas que ça se sache. Il fait ce que je lui demande.

— Ce ne serait pas Doug Olbrich, par hasard ?

John sourit brièvement. Ce n'était pas agréable à voir.

— Pas si stupide que ça.

— Non, tout juste tendance à faire confiance à des gens qui ne le méritent pas, surtout quand je les prends pour des amis. Olbrich est au courant pour le reste ?

— Non, ce n'est qu'un flic.

— Tu as retrouvé Dravecky ?

— Oui. Il a confirmé des trucs que j'avais déjà devinés. Tu n'as aucune idée de ce à quoi on se frotte.

— Je crois que si.

— Sûrement pas ! J'ai prononcé le nom de Roanoke, pour voir si ça te disait quelque chose. J'ai bien regardé ton visage, pour voir si tu en étais arrivé là tout seul, et je n'y ai rien vu. Comment ça se fait, Ward ? Qu'est-ce que tu as fichu pendant tout ce temps ?

— J'ai essayé de rester en vie.

— Tu t'es planqué ? Et pourquoi ? Une fois que tu es au courant pour ces dingues, il n'y a pas moyen de faire marche arrière. Autant se branler devant la télé. Il n'y a plus de famille heureuse, plus de promenades sur la plage, plus de vie normale. Plus rien à faire, nulle part où aller.

— John, qu'est-ce que tu crois savoir ?

— Les Indiens n'ont pas tué les colons à Roanoke. Ce sont les Hommes de Paille les coupables.

— Quoi ?

— Les Croatoans les connaissaient déjà. Ils ont dit à l'autre groupe de colons que c'était une « autre tribu » qui avait fait le coup, qui avait tué les quinze personnes restées sur place après la seconde expédition. Que cette autre tribu, c'étaient les Hommes de Paille. Leurs ancêtres, en tout cas, qui essayaient d'éliminer tous les Européens avant qu'ils ne prennent possession d'un pays qui leur appartenait depuis bien longtemps. Ils ont exterminé l'expédition suivante, mais ils ont gardé certaines femmes et certains enfants... devine pourquoi. Ils ont fait la même chose avec les Espagnols et les autres. C'est pour ça que le mot Croatoan était gravé sur la cabane. C'était une manière de les accuser. Et maintenant c'est un moyen de dire : « Nous sommes passés par ici » ou « C'est chez nous ».

— Les Hommes de Paille, déjà présents en 1500 ? ironisai-je. Formidable.

— Ils étaient là bien avant. Ils sont arrivés les premiers. Ils ont volé l'Amérique aux autochtones bien avant que le reste du monde sache qu'elle existait. Tu as entendu parler de Oak Island ?

Il avait envie de raconter son histoire et je voulais l'entendre, mais surtout, je souhaitais qu'il continue à parler, au cas où sa voix porterait jusqu'à la chambre voisine, pour que Nina comprenne que j'étais toujours en vie, si elle était éveillée.

— C'est un minuscule rocher, au large de la Nouvelle-Écosse. En 1975, un certain McGinnis a découvert une vieille galerie couverte de dalles taillées dans une roche qu'on ne trouve pas sur place. Depuis, des centaines de gens ont essayé de découvrir ce qu'il y avait au fond. Lorsqu'on l'a creusée, tout a été inondé, car elle donnait dans une autre galerie qui ouvrait sur la mer. On cherche toujours. On a découvert six plates-formes de chêne, à soixante mètres les unes des autres, et on n'a toujours pas touché le fond. Personne ne sait qui l'a creusée, et on a émis toutes sortes d'hypothèses, des pirates aux Vikings ou aux chevaliers du Temple.

— Alors, d'où ça vient ?

— Des Hommes de Paille. Le puits remonte à un millier d'années. C'est là qu'ils ont planqué leur argent lorsqu'ils ont compris que le continent ne leur appartiendrait plus très longtemps.

— Mais qui sont-ils ? m'énervai-je.

— Tout le monde et personne. Ils sont venus de partout, à des époques différentes. Des Phéniciens, des Romains, des Irlandais, des Égyptiens, des Portugais, des Norvégiens. Les Romains ont conquis la moitié du monde, ils ont déplacé des dizaines de milliers d'hommes d'un continent à l'autre. Tu crois qu'ils auraient été incapables de faire un saut à quelques centaines de kilomètres vers l'Atlantique nord ? Ils sont arrivés par poignées, des asociaux qui n'acceptaient pas les nouvelles règles, surtout lorsque le christianisme a mis à mal les vieilles croyances. On retrouve des signes de leur présence dans tout le pays, des indices dont on n'a pas tenu compte. Des objets d'artisanat occidental dans les mauvaises strates, d'anciennes pièces chinoises dans le Nord-Ouest, des légendes sur des gens qui parlaient anglais ou gallois, un tombeau égyptien dans le Grand Canyon, de vieilles inscriptions celtiques dans les roches de Nouvelle-Angleterre, des mégalithes dans le New Hampshire, des légendes sur des Indiens aux cheveux roux en Oregon. Le Nouveau Monde a toujours attiré ceux qui n'aimaient plus le vieux continent, qui estimaient qu'il se laissait corrompre par le virus de la civilisation... Peu à peu, ces groupes sont entrés en contact les uns avec les autres et ont collaboré. De temps en temps, il y avait des fuites... La légende de saint Brendan ou la carte de Piri Reis, qui montrait des zones géographiques dont on prétend qu'elles étaient inconnues à l'époque... Mais tout a toujours été démenti. Les Hommes de Paille voulaient le pays pour eux tout seuls, pour en faire leur royaume, leur refuge, d'autant plus que c'était un bon moyen de s'enrichir.

— Et comment, précisément ?

— Le cuivre. À partir de l'an 3000 avant Jésus-Christ, un demi-million de tonnes de cuivre a été extrait de la péninsule du Michigan. Cinq mille mines, réparties sur plus de deux cents kilomètres, exploitées sur un millénaire.

— Jamais entendu parler.

— Bizarre, hein ? Bien qu'ils aient laissé des millions d'outils et des milliers de trous. Où sont passées ces cinq cent mille tonnes de cuivre ? Elles ont été exportées dans le monde entier, et c'est comme ça que les Hommes de Paille se sont enrichis. Ils avaient le pouvoir de garder leur secret. Lorsque quelqu'un venait les perturber, ils s'en débarrassaient. Ils ont exterminé les Anasazis lorsque leur civilisation leur a semblé trop avancée. Ils ont rayé Roanoke de la carte. Ils ont failli en faire autant avec Jamestown. Ils ont éliminé autant de pionniers que possible. Dans son journal, Patrick Breen, un membre de l'expédition Donner, a fait une drôle de remarque pour le vendredi 18 : « Je n'ai vu aucun étranger aujourd'hui. » Quels étrangers ? Dans tout le reste du journal, il n'est jamais question de ces « étrangers ». Qu'est-ce qu'ils fichaient là bas, dans un endroit si isolé que les membres de l'expédition tombaient comme des mouches et... chose intéressante, commençaient à s'entre-dévorer ? Qui était-ce ?

— Des Hommes de Paille, sans doute, d'après toi.

— Oui. Ils étaient là avant nous. Ils ont toujours été là. Les gens le savaient, les croisaient de temps en temps, mais cela ne coïncidait pas avec nos mythes de la genèse du pays et, peu à peu, on a cessé d'en parler.

— Ils ont renoncé ?

— Bien sûr que non. Mais on ne peut pas se battre contre un afflux de millions de personnes saines, et les Hommes de Paille n'ont jamais été très nombreux. Ils se sont réfugiés dans l'ombre, ils ont mené leurs affaires tranquillement. Ils gagnent leur vie, font ce qu'ils aiment, et, de temps en temps, ils commettent des atrocités, pour garder la main et honorer leurs dieux. C'est leur mode de vie.

— Le meurtre n'est pas un système de croyance.

— Si, Ward. C'est de cela qu'il s'agit. Nous en sommes tous là. Aujourd'hui, on ne tue plus que par haine, par cupidité ou pour infliger un châtiment, mais pendant cent mille ans notre espèce croyait en un type de meurtres liés à la vie et à l'espoir.

— Quels meurtres ?

— Le sacrifice. Nous avons sacrifié des animaux, nous avons sacrifié des hommes. Le sacrifice, c'est un meurtre à visées magiques, et les tueurs en série reproduisent une version erronée de cet instinct. Ils transforment les jeunes adolescentes et les jeunes gens perdus en symboles des dieux : parfaits, intouchables, cruels, et tout leur mode opératoire n'est qu'une version pervertie d'anciens rituels.

— Je ne comprends pas.

— La démarche est la même. Ils accomplissent des préparatifs, choisissent leur victime, l'emmènent dans un endroit secret, puis se lavent, mangent, tentent de communiquer avec elles, en bref les honorent avant le sacrifice. Ils peuvent avoir des relations sexuelles, en partie pour tenter de s'accoupler avec les dieux, mais aussi parce que les dérèglements sexuels sont la seule chose assez forte pour renvoyer l'homme moderne à ses instincts les plus élémentaires. Ensuite, ils les sacrifient, ou les tuent, pour user d'un autre terme. Parfois, ils consomment une partie du corps, pour absorber son pouvoir. Ils conservent souvent un morceau de la victime ou ses vêtements, un peu comme une peau d'ours ou une dent de loup, ils les placent dans un endroit particulier, afin que le défunt reste en vie. Cela ne te rappelle rien ?

— Si, je dois l'avouer.

— Ensuite, ils enterrent le reste, le rendent à la terre ou le distribuent... Le démembrement est un trait caractéristique du sacrifice. Ils restent en sommeil pendant un moment, et le cycle recommence... jusqu'à un nouveau sacrifice.

— Les tueurs en série ne sont pas des prêtres.

— Non, c'est une bande de cinglés, et donc il arrive un moment où le cycle s'emballe. La plupart des assassins savent qu'ils ont tort, tout au fond d'eux. Ils comprennent qu'ils sont victimes d'un dysfonctionnement névrotique qu'ils essaient en vain de comprendre et de maîtriser. Ils s'emballent à la fin, parce qu'ils renoncent à renoncer. Mais pour les Hommes de Paille, cette conduite est acceptable. C'est là toute la différence. Ils pensent que ce qu'ils font est non seulement normal, mais essentiel. Ils estiment que si on tue la bonne personne au bon moment, tout ira bien. C'est l'acte magique par excellence. Ils en sont restés à la croyance originelle qui dit « tuer, c'est bien ».

Il se tut. Belliqueux, il avait le menton en avant, et tout son corps vibrait du refus de voir le monde d'une autre manière. Je le regardai, ne sachant que répondre. Je ne savais pas comment lui expliquer qu'une connaissance partielle est dangereuse, ni que ce qu'on lisait sur Internet était parfois loin de la vérité, ni que vouloir que toutes les informations entrent dans un cadre prédéterminé était une sorte de manie obsessionnelle. Je ne savais pas comment lui expliquer que s'il croyait tout ce qu'il venait de me raconter, c'était qu'il avait perdu la boule. On n'a pas envie de dire ces choses-là à un homme armé qui vous a attaché à une chaise.

— C'est Dravecky qui t'a raconté tout ça ?

— En partie. Il m'a également confirmé que certains au sein de la « tribu », c'est le mot qu'il a utilisé, estimaient que l'Homme Debout était devenu un handicap, et il m'a expliqué ce qu'il était venu faire ici. Un sacrifice qui n'a pas été accompli depuis longtemps, très longtemps.

— Paul croit que les Hommes de Paille le reprendront s'il se calme ?

— Je crois qu'il s'en moque. Il vole de ses propres ailes. Il est du genre à penser que même les Hommes de Paille y vont trop mollo.

— Où est Dravecky ?

— Dans la rivière.

— Formidable. Encore toi, je suppose. Dis-moi, John, c'est toi qui as tué les femmes ou non ?

— Non.

Le mot avait été prononcé immédiatement et simplement. Je ne savais toujours pas quoi penser.

— Alors, qu'est-ce que Paul fiche ici ?

— Tu ne crois pas un mot de ce que je t'ai raconté ? Ça m'est égal.

Il se leva et sortit un objet de sa poche. Un morceau de tissu épais, de soixante centimètres de long.

— Ne fais pas ça...

En un mouvement rapide, il m'avait bâillonné. Il serra fort. Ensuite, il contourna la chaise et s'assit en face de moi, me regardant droit dans les yeux. Je n'avais même pas remarqué que, pendant qu'il parlait, les rideaux avaient commencé à s'éclaircir. L'aube se levait. Dans la lumière grise, je distinguai le bleu intense de ses iris, le cercle noir de la pupille, mais rien de plus.

— Ne te mets pas en travers de mon chemin, Ward. Pour moi, sa mort est bien plus importante que ta vie.

Il vérifia les nœuds, se redressa et se mit à rire.

— Tu veux connaître le clou de l'histoire ? Voilà quarante ans, ils estimaient que la politique démocrate conduisait le pays à la ruine. Cela exigeait le sacrifice suprême, le sacrifice du roi. Le 22 novembre 1963.

Je le regardai. Il m'adressa un clin d'œil.

— Ils ont assassiné Kennedy.

Il se dirigea vers la porte, sortit dans le couloir bleu nuit et disparut.

CHAPITRE 26

Pendant toute la nuit, l'homme armé resta assis, le dos raide, face à la porte. L'autre homme, Kozelek, essaya de lui parler à deux reprises, mais n'aboutit nulle part et renonça. Il était effondré sur une chaise, les yeux dans le vide. Il se leva, alla à la cuisine, où il trouva une bouteille de vin. Il la liquida en vingt minutes et s'endormit. Il ne fit pas de beaux rêves. Il prononça un prénom de femme, deux fois.

Patricia resta allongée sur le divan. Les mains liées derrière le dos, elle ne pouvait guère faire autre chose. Pendant un moment, elle avait gardé les yeux ouverts, mais s'était rendu compte que cela ne changeait rien, si bien qu'elle les referma. Elle ne dormit pas, néanmoins. Elle ne somnola même pas.

Ils se mirent en route dès les premières lueurs. L'homme au fusil, Henrickson, la fit marcher la première. Kozelek trébuchait derrière elle. Il avait la gueule de bois et souffrait à la cheville, mais surtout il semblait avoir renoncé à tout.

Patricia les conduisit vers la rive nord du lac. En passant devant le deuxième bungalow, elle leva les yeux et aperçut son reflet dans la fenêtre poussiéreuse. Elle sourit, au cas où un peu de Bill y planerait encore et où elle n'aurait pas l'occasion de revenir.

— J'espère que vous ne me faites pas tourner en bourrique, dit Henrickson.

Tom s'arrêta, heureux d'avoir un prétexte pour se reposer. Après deux heures de grimpette, il se sentait à bout. Le ciel, un voile bleuté entre les arbres, s'était couvert de sombres nuages qui s'amoncelaient, telles des mottes d'argile. Il avait mal à la tête, mais ne pouvait s'empêcher de faire la comparaison avec la première fois où il était retourné à l'endroit vers lequel ils se dirigeaient. Bien sûr, il n'avait pas envisagé les choses ainsi. Il voulait simplement s'endormir, et l'alcool le lui avait permis.

À présent, aussi, il aurait voulu dormir. Ou s'en aller. Il avait cessé de croire bêtement qu'il allait s'en sortir comme une fleur, que sa découverte allait lui sauver la vie.

Henrickson se tenait face à la vieille femme.

— Vous avez raconté aux flics que c'était à une heure de marche de chez vous. À moins que vous ne soyez propriétaire de tout l'État, ça commence à faire beaucoup.

— J'ai menti, dit-elle simplement.

— C'est encore loin ?

— Assez.

— Vous essayez peut-être de nous perdre, grommela Henrickson. Vous trouvez peut-être que c'est une bonne idée. Mais je peux marcher deux fois plus que vous deux mis bout à bout, et je serai encore debout bien après vous. Oh, bien sûr, ça m'empêchera de trouver l'endroit aujourd'hui, mais maintenant que je sais où c'est, j'insisterai. Je finirai par trouver, et tout ce que vous aurez gagné, c'est d'être morts tous les deux et de m'avoir fait perdre un peu de temps.

— Qu'est-ce que ça peut foutre ? s'écria Tom. Si vous voulez le tuer, si vous voulez tuer cette créature, qu'est-ce que ça peut foutre que ce soit aujourd'hui ou la semaine prochaine ?

— Qu'est-ce que vous cherchez ? demanda Patricia en le dévisageant d'un air curieux.

— Vous le savez très bien.

Elle haussa les épaules.

— Il n'y a que des ours. Des animaux qui vivent là depuis la nuit des temps et méritent qu'on les laisse en paix.

Tom regarda Henrickson.

Sans rien dire, celui-ci leur fit signe d'avancer.

Ils se remirent en marche. Finalement, Tom réussit à marcher au même rythme que la vieille dame. Il commença à lui parler et elle semblait écouter. Il lui raconta de sa balade dans les bois, les raisons qui l'avaient amené ici, et il finit par lui confier ce que personne ne savait. Cela sortait lentement, mais il lui était impossible de s'arrêter. Il avoua s'être retourné pour voir la fille sur le siège du passager, salement esquintée ; il lui avait fallu se battre pour ne pas mettre fin à ses jours. Il évoqua les petites indélicatesses avec les comptes de la société pour laquelle il travaillait, qui remonteraient à la surface un jour ou l'autre.

Patricia l'écouta sans dire grand-chose. Il se sentait un peu mieux, mais pas beaucoup, et il comprit que la seule chose qui aurait pu changer, c'était de tout avouer à Sarah. La société, il l'avait volée, sa femme, il l'avait trompée. Le mensonge était pire. Ce soir, quoi qu'il arrivât, il téléphonerait. Elle l'avait aimé, elle l'aimait peut-être encore. Au moins, elle lui dirait que faire, et ce serait peut-être l'absolution qu'il pouvait espérer.

Finalement, tandis que les entrailles de Tom lui signalaient qu'il était midi, ils arrivèrent à destination.

Tom ne savait plus où ils étaient. Pendant un moment, il avait cru que Henrickson avait peut-être raison, que la vieille femme essayait de les perdre. Mais elle n'hésitait jamais un seul instant sur la direction à prendre. Ils avaient progressé lentement mais régulièrement. Elle avait tourné à droite et à gauche, leur avait fait contourner certains obstacles et franchir d'autres. Pour une personne de son âge, elle paraissait dans une forme exceptionnelle. Elle flancha une fois ou deux cependant, glissa et fut incapable de retenir sa chute. Enfin, elle avait commencé à ralentir.

Lorsqu'elle s'arrêta, elle était essoufflée. Elle ébaucha un signe de tête.

— C'est en bas.

Henrickson passa devant elle pour aller au bord du ravin et appela Tom.

— C'est là ? interrogea-t-il.

Tom s'approcha de lui et regarda le lit du courant. Au début, il ne remarqua rien de différent. Puis il repéra le petit coin dans lequel il s'était assis dans le noir, avant d'y retourner le lendemain matin, moins d'une semaine plus tôt.

— Oui, c'est là. C'est là que cela s'est passé.

— Bon, dit Henrickson en s'éloignant du bord et en revenant vers Patricia. Merci.

Ils retournèrent vers le ravin. Cinq minutes plus tard, Henrickson bifurqua vers la gauche, à travers le bosquet. Il s'arrêta quelques instants plus tard.

Tom resta bouche bée. L'homme les avait conduits au tronc tombé en travers du ravin.

— Madame Anders, pourriez-vous dire à Tom ce que nous avons là ?

— Un arbre tombé.

Henrickson hocha la tête, accomplit les derniers mètres qui le séparaient du bord et posa le pied sur le tronc. Il examina l'extrémité et examina l'autre côté, comme si c'était un large pont.

— Les deux extrémités ont été aménagées, constata-t-il en s'accroupissant. Et les branches ont été élaguées. On a modifié l'angle de la chute. Ça m'étonne que vous n'ayez rien remarqué, Tom.

— Je n'étais pas en forme.

Il disait vrai, bien sûr, mais honnêtement il ne comprenait pas comment cela avait pu lui échapper. C'était évident.

— On peut traverser la rivière à gué pour l'instant, expliqua Henrickson, mais au printemps, le chemin se rallonge drôlement, d'un côté comme de l'autre. C'est un pont. Quelqu'un l'a fabriqué. Un de nos amis de la forêt l'a placé ici. L'a renforcé. On est ici, on veut aller de l'autre côté. Alors on construit. La voilà, la preuve, Tom. Je vous avais dit que ça valait le déplacement.

— Comment savez-vous que ce n'est pas un type quelconque ? L'œuvre d'un bûcheron ?

— Il n'y a jamais eu de bûcherons par ici, et je vois mal un homme utiliser des outils de pierre. Un arbre tombé, hein ? reprit-il en regardant Patricia.

— Pour moi, oui. Vous voyez ce qui est dans votre tête, pas devant vos yeux. Vous n'êtes pas le seul, d'ailleurs. Il refit le chemin en sens inverse et sourit.

— Si vous voulez. Mais allons un peu plus loin, on verra bien ce qu'on trouve.

Ils marchèrent pendant dix minutes, le long de la crête du ravin.

Les flancs étaient de plus en plus escarpés, de plus en plus profonds, le courant s'élargissait et grondait, alimenté par les cascades d'hiver, parcimonieuses mais régulières.

Enfin, ils arrivèrent au bord de la crête. Tom suffoqua.

Sous leurs pieds, le sol plongeait. À gauche, la rivière sortait soudain au grand air pour retomber dans un large puits rocheux à soixante mètres en contrebas. Immense tapis caillouteux de conifères enneigés, la forêt s'étendait à l'infini, vers le Canada et au-delà. Tout en haut, on discernait la faible trace d'un réacteur dans la dernière bande de ciel bleu. C'était la seule trace de civilisation. Ailleurs, tout était resté comme si l'homme n'avait jamais existé. Tom observa les nuages qui comblaient l'espace clair puis tourna la tête vers la forêt.

— C'est magnifique.

— Imaginez un peu quand il n'y avait que ça, dit Henrickson d'un ton calme, juste à côté de lui. Quand il n'y avait personne.

Tom se contenta de hocher la tête, paralysé devant ce monde comme il l'était devant les mots. Il secouait la tête lentement, les yeux remplis de larmes, il ne savait pourquoi.

— Je dois vous remercier, Tom, enchaîna Henrickson, retrouvant soudain son accent péquenot. Vous vous êtes donné du mal, mon vieux, et ça n'a pas été facile pour vous. Vous savez le plus drôle ? Ça me faisait du bien d'avoir quelqu'un à qui parler.

Tom secouait toujours la tête, mais il acquiesça vite. Il leva les yeux, aperçut la forme floue de la vieille dame, les mains toujours liées derrière le dos. Elle lui sourit, tristement, et détourna le regard.

Ensuite, Henrickson posa la main sur l'épaule de Tom et le poussa.

Tom se sentit basculer, puis il prit soudain conscience de ne pas savoir ce qu'il y avait en dessous de lui, comme s'il se retrouvait sur le pont improvisé et que la petite voix intérieure n'était plus là pour le guider. Ensuite, ce fut l'apesanteur de la chute, rapide et brève, avant qu'il commence à heurter des objets. Cette fois, la collision ne s'effectua pas contre des buissons. Dans une succession d'impacts à se briser les os, il n'était plus qu'une poupée de chiffon tourbillonnante. Encore une chute, et il s'écrasa comme un verre qu'on laisse tomber.

Tom se retrouva coincé entre deux gros rochers cachés sous une végétation mousseuse. Il essaya de crier, mais n'émit qu'un gargouillement humide. Son corps était distordu, aplati, ses vêtements déchirés et tachés de sang, et sa jambe gauche avait subi un sort terrible. De l'eau froide coulait sur son pied et sa main gauches, mais il ne sentait rien. Bien qu'il eût eu le crâne et la mâchoire brisés, il voyait toujours et son bras droit fonctionnait encore... un peu.

Pendant les vingt minutes suivantes, il ne parvint qu'à faire une seule chose : sortir son portable de sa poche. Laborieusement, il se déplaça dans le menu et, d'un pouce tremblant et raidi, il réussit à composer :

Vu Bigfoot. Je t'aim...

Ensuite, il mourut. Il n'avait pas envoyé le message.

On ne retrouva jamais son corps.

À soixante mètres plus près des cieux, Patricia regardait l'homme, abasourdie.

— Vous étiez obligé de faire ça ?

— Oui, et je n'espère pas que vous compreniez.

— Vous allez me pousser dans le ravin, moi aussi ?

— Un seul, ça suffit, et vous avez encore un travail à accomplir.

— Cet endroit, c'est le seul que je connaisse, et je ne peux pas aller plus loin. Si vous voulez un ours, il va falloir aller le chercher vous-même.

— Si on en arrive là, je saurai bien vous forcer à me dire où ils habitent. Mais pour l'instant, on descend près de la rivière, là où Tom a vu le sien, et on attend.

— Vous croyez qu'ils se promènent dans le coin ?

— Non, mais ils comptent beaucoup pour vous, ce qui me laisse penser que vous comptez pour eux. Quand ils vous sauront ici, ils viendront peut-être vous rendre une petite visite.

— Comme si j'étais une maman ours ? Vous pouvez courir ! Mes propres enfants ne sont pas venus me voir depuis dix-huit mois.

— Patricia, vous commencez à m'énerver avec vos mensonges.

— Ils sauront que je ne suis pas seule.

— Bien sûr. Surtout si je m'occupe un peu de vous. D'après ce que j'ai vu, vous essaierez de ne pas crier, mais ils comprendront quand même que ça ne va pas fort. Et ils accourront.

Elle fixait le sol, terrifiée.

— Je savais que quelqu'un viendrait, finit-elle par dire, mais je pensais que ce serait un chasseur. Un crétin qui veut faire fortune et passer à la télé. Mais vous n'êtes pas là pour ça.

— Non, effectivement.

— Alors, qui êtes-vous ?

— Je m'appelle Paul. Parfois, on m'appelle l'Homme Debout.

CHAPITRE 27

Je m'étais endormi. Incroyable à première vue, mais, comme le coupable qui s'assoupit dans sa cellule, toutes les tensions soudain libérées par l'incarcération à laquelle on ne peut plus échapper, bien attaché sur une chaise, on en est réduit à la passivité.

Une fois réveillé, je ne pus me rendormir. C'était pire. Cela me laissait le temps de penser et de chercher à m'échapper. J'essayai de basculer la chaise, en tentant de la soulever à la force du dos. Un mouvement brusque faillit me faire basculer en avant – la porte ouverte à une mâchoire brisée et à un nez cassé ! Je renonçai. Je ne m'appelle pas Jackie Chan.

Ne rien tenter était encore pire. Je voyais le rideau s'éclaircir, j'entendais le bruit des graviers sous les pneus, des éclats de rire lointains, des claquements, des toux. La douleur au bas du dos augmentait, j'avais l'épaule en feu. J'observais le réveil près du lit, en attendant que chaque chiffre accroisse d'une unité... parfois, je le croyais arrêté tant c'était long, mais, lorsqu'il passait au chiffre supérieur, rien n'avait changé.

L'attente fut interminable jusqu'à 12 h 51, lorsque Nina défonça la porte, avec deux hommes que je n'avais jamais vus.

— C'est sûr qu'il vous ressemble, déclara le plus costaud. (On m'avait dit que c'était le shérif Connelly. L'autre, un jeune homme audacieux, s'appelait Phil.) Mais on voit bien que vous n'êtes pas lui.

— Il s'appelle Paul.

— On a entendu M. Kozelek l'appeler Jim.

— Il utilise peut-être le nom de Henrickson.

— Oui, ça doit être ça, admit le shérif.

Phil écarquillait les yeux.

— C'est un tueur en série ?

— Oui, et pas des moindres.

Nous étions au poste de police. On nous avait offert du café. Les doigts toujours engourdis, j'avais du mal à tenir ma tasse. Nina n'allait guère mieux. La femme de ménage l'avait trouvée attachée et avait prévenu la police avant de penser à la détacher. Livide, elle paraissait exténuée, décharnée. J'avais envie de retrouver John Zandt et de lui écrabouiller la tête, et pas seulement pour la veille au soir.

En une demi-heure, nous fîmes un résumé succinct de ce qui était arrivé et de ce que nous savions. Dans cette version, c'était l'Homme Debout qui nous avait attaché, et non John. Nina avait bien fait comprendre qu'elle était agent fédéral et dissuadé le shérif d'appeler le FBI. Pour l'instant. Une femme médecin au joli sourire nous avait examinés et m'avait bandé l'épaule avant de partir. J'avais les yeux secs, irrités, et la lumière m'éblouissait.

Phil hocha la tête.

— Bordel de merde !

— Alors, qu'est-ce qu'il fiche à Sheffer ? demanda Connelly. Où est-il parti ?

— Je ne sais pas... mais il m'a raconté des drôles de trucs hier soir. Des histoires de sacrifice. Une sorte d'étrange rite de purification. Il a déjà éliminé tous ceux qui appartenaient à son passé, alors je ne vois pas qui pourrait être le prochain sur la liste. Ou bien des gens avec lesquels il a travaillé.

Connelly regardait par-dessus mon épaule, une drôle d'expression sur le visage.

— M. Kozelek a passé quelque temps dans la forêt, dit-il en se tournant vers moi. Il en est revenu dans un triste état, en prétendant avoir vu quelque chose.

— Quoi donc ? s'enquit Nina.

— Bigfoot.

Surpris, j'éclatai de rire.

— Pour de vrai ?

— Oui, répondit Connelly avec un sourire pincé. C'était un ours, bien sûr. Mais votre frangin a passé beaucoup de temps avec Kozelek. Et je ne vois pas pourquoi, à moins que les histoires de Kozelek l'aient intéressé. Vous voyez une raison à ça ?

Je ne comprenais pas. Je hochai la tête.

Le shérif se mordit les lèvres.

— Phil, tu pourrais passer un coup de fil de ma part à Mme Anders ?

— Pourquoi ?

— Appelle-la, c'est tout. 384931.

Le jeune policier prit un appareil et composa le numéro. Il laissa sonner un instant.

— Pas de réponse.

— Essaie son portable.

Il lui donna le numéro. Phil appela, attendit, hocha la tête.

— Tu l'as vue en ville ce matin ?

— Non.

— Moi non plus. Et j'ai mentionné son nom hier soir. Je crois qu'on devrait aller voir. Phil, va chercher des parkas et des gants pour nos invités. Je vais voir si je peux leur trouver des bottes. Et puis va dans l'armoire et prends quelques armes.

— Lesquelles ?

Connelly me regarda, j'acquiesçai.

— Des gros calibres.

Rapidement, nous nous rendîmes au parking, pour nous apercevoir qu'il avait commencé à pleuvoir. Les policiers ne semblèrent pas le remarquer. Dans le Nord-Ouest, la pluie, c'est la routine. Connelly se dirigea vers un véhicule et son assistant vers un autre.

— N'essaie pas d'arriver avant moi, lui dit-il. Tu restes derrière, compris ?

Nina et moi montâmes à l'arrière. Connelly s'installa et ferma la portière. Il démarra puis se tourna vers nous :

— C'est bizarre, j'ai vu Henrickson et Kozelek quitter la ville hier soir, vers 20 h 30. C'est là que j'ai relevé le numéro. J'ai vérifié plus tard, je n'ai pas vu la voiture. Mais vous êtes arrivés au petit matin et il était dans le coin pour vous ligoter.

Nous nous taisions.

— C'est ce que je craignais, soupira le shérif. L'autre type. Il va nous poser des problèmes ?

— Je ne sais pas, répondis-je.

— Il est avec vous ou avec eux ?

— Il n'est avec personne.

— Et le reste, c'est la vérité ?

— Pour l'essentiel, murmura Nina.

Connelly passa une vitesse.

— Super. Je suis content que vous soyez venus chez nous.

Il sortit rapidement du parking et s'engagea sur l'asphalte mouillé de la nationale, attendit que son adjoint le rattrape et accéléra. Plus tard, par radio, on lui annonça que, deux minutes après le départ des voitures de patrouille, une serveuse avait aperçu une voiture qui sortait de la cour arrière de Big Frank et nous suivait.

Je passai quinze minutes à essayer de retrouver un peu de sensation dans mes doigts. Nina fit de même. J'avais envie de lui raconter tout ce que John m'avait dit, mais ce n'était pas le moment. Connelly conduisait vite sur la route presque déserte. La pluie s'arrêta, mais ce n'était pas de bon augure. Il faisait de plus en plus froid.

Juste après un petit café, nous bifurquâmes vers une route étroite qui ne semblait pas porter de nom. Nous y étions engagés depuis moins de trente secondes lorsque la voix de l'assistant crachota dans la radio.

— Chef, vous avez raté la route. Cascade Falls est juste derrière...

— Continue à regarder la route et suis-moi, on passe par un autre chemin.

Le trajet fut beaucoup plus long que je ne croyais. D'après ce que j'avais compris, la femme habitait dans un lotissement non loin de la nationale. Et cette route ne semblait mener nulle part. Vingt minutes plus tard, elle se réduisit à un simple chemin enneigé, si bien que le shérif ralentit. Les arbres poussaient jusqu'au bord. Il continuait à rouler. Par la lunette arrière, je voyais l'assistant qui suivait nos traces. Il tenait ses distances de sécurité mais restait malgré tout assez proche pour que je lise la perplexité sur son visage.

Puis, sans raison, Connelly freina. Il regardait vers la droite.

— Shérif, vous êtes sûr de savoir où on va ?

— Oui. En fait, on y est.

Il coupa le moteur et descendit. Lorsque je me retrouvai dehors, avec Nina, l'endroit nous sembla encore plus isolé. De chaque côté, arbres et buissons obstruaient la vue et tout était couvert de neige immaculée. La route se terminait une cinquantaine de mètres plus loin.

Phil s'était garé juste derrière nous.

— Chef, où sommes-nous ?

— Au bout du vieux chemin de chantier. Tu vois ? ajouta-t-il en indiquant un point par-dessus mon épaule.

Avec beaucoup d'efforts, on discernait la silhouette d'un vieux bâtiment en ruine, caché dans les arbres, à une dizaine de mètres.

— D'accord, mais pourquoi ? m'étonnai-je.

Connelly passa son fusil sur son épaule et se mit en marche.

— J'ai parlé avec Mme Anders avant-hier soir. Elle m'a avoué ne pas avoir dit la vérité en indiquant l'endroit où elle avait retrouvé les affaires de M. Kozelek. Elle jugeait qu'il n'avait pas toute sa tête et ne souhaitait pas le voir revenir. Elle m'a appris où se situait le véritable endroit. Si Henrickson l'a kidnappée, ce que je crois, il lui a sans doute demandé de l'y conduire.

— C'est près ?

— Non.

Je remarquai une zone où les arbres paraissaient moins denses, plus jeunes : sans doute un vieux sentier de bûcheron, qui n'était plus entretenu. Plus exploité.

Pour Nina et moi, la marche s'avéra difficile dès le début. Nous ne faisions que monter. Une heure plus tard, plus rien ne laissait penser que nous suivions une piste quelconque. Je n'avais pas remarqué à quel moment le chemin s'était arrêté. Nous étions cernés par des grands arbres, très épais, et le terrain était escarpé. Je n'ai rien du randonneur, comme je l'avais dit à Zandt, et je peinais. Avec la neige, il était difficile de savoir où on posait le pied. Parfois, on tombait sur de la roche, parfois sur quelque chose de solide, et, sans avertissement, on se retrouvait sur les genoux. Il faisait de plus en plus sombre, en grande partie à cause du plafond de sombres nuages. Il ne pleuvait toujours pas. Il faisait froid dès le départ, mais bientôt ces premiers moments m'apparurent comme le comble du confort. Si Kozelek avait passé deux jours dans ce froid, il avait eu de la chance d'y survivre. J'étais sidéré par l'acharnement des pionniers qui avaient creusé des routes dans ce paysage.

— Ça va ?

— Plus ou moins !

Nina et moi marchions quelques mètres en arrière.

— Et vous ?

— Je crois. Il fait un froid... !

Sans compter la fatigue, la faim.

— On est encore loin ?

— Non, marmonna le shérif sans se retourner. À mi-chemin à peu près.

— Merde ! dit Nina, doucement. J'ai horreur de la nature. Ça craint !

Nous continuâmes à marcher. Tranquillement, je confiai à Nina le reste des propos de John. Elle aussi en conclut qu'il avait perdu la tête.

Nous cessâmes bientôt de parler, tant nous étions essoufflés. Phil semblait peiner, lui aussi, mais Connelly conservait un pas régulier. Nous faisions beaucoup de bruit avec nos quatre paires de bottes dans la neige, nos respirations haletantes. La fatigue, le manque de sommeil, la couleur blanche presque en permanence devant mes yeux commencèrent à avoir un effet hypnotique. Je cessai de réfléchir, ne pensant qu'au pas suivant, au prochain rocher, à la pente, aux creux, à l'odeur des aiguilles de pin et de l'écorce dans l'air d'une pureté extravagante. Mon visage perdait de son élasticité, paraissait engourdi au toucher. Quand je clignais des yeux, je voyais un flash lumineux. Je trébuchais de temps à autre. Nina aussi.

— Stop.

Connelly s'exprima d'une voix basse, calme, déterminée.

Tiré de ma rêverie, je levai la tête et m'immobilisai.

— Quoi ? Nous y sommes ?

Il se tourna vers nous mais ne répondit pas. Il scruta la forêt, du côté où nous étions venus puis sur la gauche. Après notre escalade, le silence était bruyant, mes oreilles sifflaient.

— Vous avez entendu quelque chose ? demanda Nina.

Connelly garda le silence pendant vingt secondes.

— Non, rien. J'ai cru voir quelque chose. Je me suis retourné pour m'assurer que vous suiviez et j'ai cru apercevoir une ombre, à une vingtaine de mètres, de ce côté.

— Il y a pas mal d'ombres, rétorquai-je, il commence à faire sombre.

— Peut-être. Un autre type pourrait s'intéresser à Henrickson. Il risque d'être dans les parages, lui aussi.

— Ah bon ? Et qui donc ? questionna Phil, soupçonneux.

— Un ancien flic, expliqua Nina. L'Homme Debout lui a salement gâché la vie. (Elle fit quelques mètres dans la direction que Connelly avait indiquée et scruta les arbres.) Il veut sa peau tout autant que nous.

— Il est dangereux ?

— Hum, dis-je. Mais pas pour nous, j'espère.

Soudain, Nina se mit à crier, nous faisant sursauter.

— John ? John, c'est toi ?

Quatre paires d'yeux grands ouverts observaient le paysage. Rien ne semblait bouger.

— Si tu es là, John, viens avec nous. Nous voulons le pincer, nous aussi. Fais les choses correctement. Viens avec nous. Rien ne bougea. Nina hocha la tête.

— Ce ne sont que des ombres. Oh, non, voilà qu'il se met à neiger.

Elle avait raison. Des petits flocons blancs descendaient en spirale.

— J'aurais préféré que vous vous absteniez de crier. Les sons voyagent loin par ici. Je n'ai pas envie que ce type sache qu'on arrive.

— Je sais comment voyagent les sons, répondit Nina. Il y a longtemps qu'il sait qu'on est là. Pas vrai, Ward ?

— Si. Et on doit vous prévenir : ce n'est pas ça qui changera grand-chose. Il ne s'enfuira pas, il ne se cachera pas. Il agira.

Le flic passa la main sur son épaule et prit son fusil, le tint en position de tir et me dévisagea. Bien que Connelly eût dix ou quinze ans de moins que lui, il avait quelque chose de mon père dans le regard : une approbation tranquille, l'impression de ne pas vraiment comprendre ce que signifiait battre en retraite.

— Bon, dit-il, on est prévenus.

Le vent se levait, la neige tourbillonnait autour de son visage.

CHAPITRE 28

Patricia n'avait jamais eu aussi froid de toute sa vie. L'homme lui avait laissé le temps d'enfiler une parka avant de partir et, pendant presque toute la journée, elle l'avait regretté. Quand on bouge, un manteau ne sert à rien, ce sont les parties qu'il ne couvre pas, le visage et les mains, surtout lorsqu'elles sont attachées dans le dos, qui souffrent le plus. Cela ne sert qu'à vous faire transpirer. Mais depuis deux heures qu'ils s'étaient arrêtés, elle était contente de l'avoir. Elle serait probablement morte, sinon. Son nez avait un peu coulé, cela avait formé de minuscules stalactites. Elle avait demandé qu'il lui attache les mains devant, pour avoir moins froid, mais il avait refusé. Elle savait pourquoi. Ses bras et ses épaules commençaient à lui faire mal. Très mal. C'était la première étape de ce qu'il ne manquerait pas de faire s'il n'obtenait pas ce qu'il voulait.

La neige tomba peu après quatre heures. La lumière commençait à décliner et, si certains flocons étincelaient encore, d'autres n'étaient plus que des ombres flottantes. Certains habitants de la région considéraient la neige comme une croix à porter. Pas elle. Même au bout de trois ans, elle lui semblait toujours aussi magique.

L'homme l'avait installée contre la paroi du ravin, ce qui était un bien. Le vent ne venait que d'une seule direction. Il était allé s'asseoir sur la rive basse, de l'autre côté du courant, son fusil sur les genoux, dans un silence total. S'il avait froid, il n'en montrait rien.

Il neigeait depuis vingt minutes déjà, et de plus en plus fort, lorsqu'elle le vit lever les yeux. Il écouta attentivement.

— Vous avez entendu quelque chose ? interrogea-t-elle.

— Au loin.

— Je ne vois pas de quoi vous parlez. Tom a aperçu un ours, c'est tout. Je vous ai conduit ici parce que vous êtes un méchant homme et que je préférais vous emmener dans un endroit où vous gèleriez à mort, avant qu'on vous retrouve.

— Peut-être. Je vous vois bien faire un truc pareil. Je vous aime bien. Vous me rappelez quelqu'un.

— Votre mère ?

— Non, pas elle.

— Elle est toujours vivante ?

Il ne répondit pas, et soudain elle sut que la mère de cet homme était morte, qu'elle n'était pas enterrée dans un lieu conventionnel et qu'il savait où se trouvait la dépouille.

— Vous étiez enfant unique ?

La tête de Henrickson bascula vers elle.

— Je fais bouger mes lèvres pour que mon visage ne se prenne pas en glace, reprit-elle.

Patricia disait vrai. Elle avait parfois constaté qu'on pouvait émouvoir un enfant en lui parlant, en lui parlant sans relâche. Cet homme n'était pas un enfant. C'était un psychopathe. Il fonctionnait peut-être de la même manière.

— Ils vont peut-être nous entendre et venir écouter ce qu'on raconte. Alors, vous étiez enfant unique ?

— Je le suis devenu, dit-il sans émotion. J'ai eu trois mères. Elles sont toutes mortes, c'est ce qui m'a donné de la force. Je suis né dans une forêt, mon père a tué ma mère, et ensuite des gens sont venus et l'ont tué, lui aussi. Ils nous ont gardés, mon frère et moi pendant un moment, et après, ils l'ont gardé lui, et ils se sont débarrassés de moi. Les gens ont essayé de me caser dans des endroits, mais je ne voulais pas, jusqu'à ce que je sois venu vivre près d'ici, avec ma dernière mère.

— Elle vous traitait mal ?

— Patricia, je suis tellement loin de la psychologie de comptoir que vous ne le croiriez pas !

— Alors, je vous rappelle qui ?

— La femme qui a été ma grand-mère pendant un moment.

Patricia supposa qu'il s'agissait d'une sorte de compliment.

— Pourquoi tout cela ?

— Tuer, c'est ce que font les animaux. Les carnivores tuent pour se nourrir. Les chiens sauvages tuent les petits des autres chiens sauvages. Les mouches pondent leurs œufs dans la chair des mammifères morts. Elles s'en fichent pas mal. Les esclavagistes arabes de Zanzibar jettent les malades, hommes ou femmes, dans les eaux de la baie, pour ne pas payer de taxes sur des marchandises invendables. Les paysans russes ont vendu des morceaux de corps humains en Sibérie, pendant les hivers terribles des années 1920. Nous sommes des animaux qui inventent des machines volantes pour les lancer contre des immeubles où vivent nos congénères. C'est nous qui commettrons un génocide en bonne et due forme, un jour. Les hommes sont des animaux qui tuent et qui détruisent.

— J'aimerais que vous ayez l'air de trouver ça mal.

— Ce n'est ni mal ni bien. C'est la simple vérité. Notre espèce a envahi l'Europe, où des créatures vivaient depuis des centaines de milliers d'années, et en quelques millénaires elle se l'est appropriée. Comment c'est arrivé, d'après vous ?

— On était mieux adaptés.

— Sur un seul point. Nous avions la volonté de tuer des créatures proches de l'être humain. On a exterminé l'homme de Neandertal, et ensuite on s'est attaqués les uns aux autres. On ne respecte pas les animaux comme les hyènes et les vautours, les charognards. On glorifie les lions, les tigres, les requins, les animaux qui ont du sang frais qui leur dégouline des mâchoires. Le fait que nous ayons des pouces, la parole et de grandes illusions spirituelles n'y change rien. Il n'y a ni mal ni bien. Il n'y a que le comportement qui compte, et le nôtre, c'est de tuer.

— Alors, tuez quelqu'un. Vous l'avez sûrement déjà fait. Il ne répondit pas, ce qui était peut-être pire. Frigorifiée comme elle l'était, Patricia sentit qu'elle avait la chair de poule. Elle était avec quelqu'un qui n'interprétait pas le monde de la même manière que les autres.

— Alors, tuez un autre être humain, nous sommes des millions. Pourquoi ne pas en tuer d'autres ?

— Parce qu'il est temps de faire ça.

— Ce sont vos voix qui vous le disent ?

— Personne n'a fait cela depuis des lustres. On a tué d'autres choses. Des symboles du pouvoir, des femmes, des bébés. Ce sont des succédanés de l'homme sauvage, du véritable sacrifice.

— Nom de Dieu, parce que vous croyez que ça peut marcher ?

— Ça marche.

— Vous tuez un être vivant, et par on ne sait quel miracle, la musique des sphères se réajuste ? Vous y croyez vraiment ?

— C'est la vérité, et si vous étiez née quelques siècles plus tôt, vous le sauriez. Mais on a oublié. Tout le monde... même ceux qui n'auraient pas dû. Maintenant, on ne croit plus qu'aux câbles... On croit que c'est important de choisir un opérateur téléphonique plutôt qu'un autre. On essaye de ne pas marcher sur les failles.

— Vous êtes malade.

— Je ne crois pas. (Ses yeux brillaient dans le crépuscule tombant.) Et votre opinion ne m'intéresse pas.

— Alors, ne me dites plus rien. Je ne veux plus rien entendre.

— Parfait. Mais il faut quand même que vous sachiez quelque chose. La grand-mère dont je vous ai parlé... Elle déglutit.

— Je l'ai tuée. On l'aidait à descendre l'escalier et j e l'ai poussée, avant de me rendre compte que j'en avais l'intention. J'avais douze ans. Je sais que c'est ce qu'elle voulait. Elle est morte très vite, c'était propre. Si vos amis ne se pointent pas très vite, vous allez mourir aussi. Mais ce sera très, très lent. À dix mille kilomètres de là, les gens se retourneront dans leur tombe.

Sans même s'en rendre compte, Patricia s'était recroquevillée et écartée de quelques centimètres. Elle était toujours beaucoup trop près. Parfois, ces dernières années, elle s'était crue prête à mourir. Elle ne voulait pas jouer son jeu, mais, sans Bill, il n'y avait plus grand-chose pour la retenir ici, et il était peut-être temps de danser avec la mort. Blottie dans la neige avec quelqu'un qui semblait à la fois plus et moins qu'humain, elle comprenait son erreur.

La neige tombait dru, il faisait presque nuit, elle était piégée dans la forêt, les mains liées, avec un détraqué.

Elle décida de se taire.

Soudain, il leva les yeux vers la crête du ravin, derrière elle, puis il regarda derrière lui. La bouche légèrement entrouverte, il avait la tête penchée. Il sauta par-dessus le courant et se hissa en haut du ravin, comme s'il n'avait eu besoin que d'un seul pas.

— Ils arrivent.

Cela ne semblait pas l'enchanter. Patricia ignorait de qui il parlait. Il resta sur place un moment, comme pour flairer le vent, puis disparut, telle la lune qui se cache derrière un nuage.

Patricia songea à courir, mais ses jambes étaient trop engourdies et il n'y avait nulle part où aller. Elle se recroquevilla, ferma les yeux et pensa à Verona.

CHAPITRE 29

Cette fois, nous entendîmes tous.

Un bruit sec, pas très proche. Assez distinct pour couvrir les rafales de vent et le son rauque de ma respiration qui résonnait dans ma tête. Connelly se tourna aussitôt.

— Baissez-vous !

Nina posa les mains sur mon dos et me poussa. Nous glissâmes sur le côté, courbés en deux, essayâmes de courir, mais finîmes dans un terrain chaotique avec de la neige jusqu'aux chevilles. Nous nous cachâmes derrière un arbre, tout près d'une excroissance rocheuse, l'arme à la main.

Connelly et son adjoint reculèrent vers nous, fusils en position de tir. La voix de Phil était grave et chancelante, mais ses pas restaient mesurés et stables.

— Vous l'avez vu ?

Le shérif fit un signe négatif mais balaya un arc de trente degrés avec son arme.

Ils passèrent de notre côté du rocher. Lorsqu'ils furent en place, je me retournai... Il n'est pas toujours facile de deviner la direction d'un son dans les bois, et j'avais vu trop de films. Je ne distinguais pas grand-chose. Le flanc de la montagne, des arbres, des rochers, des buissons, de la neige. Le contraste transformait tout en une sorte de dessin d'Escher qui offre plusieurs interprétations, qui se transforme sous vos yeux puis s'amalgame en un tout incompréhensible. Je ne perçus aucun mouvement.

De nouveau, je regardai devant moi. Rien ne bougeait de ce côté-là non plus, à part les flocons de neige. Nous tournions la tête, tout doucement, rien que les yeux et les oreilles. Les secondes s'écoulaient.

Soudain, la tension se relâcha dans mes jambes. Ma main droite, dégantée était glacée, inutile. Je passai mon arme dans la gauche et frottai ma main droite sous le creux de mon bras, en serrant les dents, car le mouvement me fit mal à l'épaule. Je me sentis mieux lorsque je repris l'arme dans la main droite, même si j'avais l'impression que la peau allait geler sur le métal.

— Ce n'est pas John. J'en suis sûr.

— Non, nous sommes tout prêts. C'est l'Homme Debout.

— Qu'est-ce qu'on fait ? murmura Phil.

— On continue à avancer, dit Connelly.

Il révéla un petit instrument qu'il avait dissimulé dans la paume de sa main. Je me demandais comment il se repérait dans le noir. Il poussa un bouton, et un minuscule écran s'éclaira un bref instant.

— Il ne doit pas être à plus de deux ou trois cents mètres.

— Il a dû nous entendre.

— Nous sommes quatre et il est seul, rétorqua Nina. Il ne va pas venir vers nous. Il va attendre que nous nous séparions ou que nous fassions une bêtise quelconque. Et il nous prendra un par un.

Connelly acquiesça.

— Alors, on y va ?

— Restez groupés. Vous croyez qu'il est droit devant nous ?

— Presque.

— Bon, alors, on passe par là, on arrive par la gauche et on le prend de biais. On va où, exactement ?

— C'est une sorte de ravin. On arrive par le haut. Ça descend doucement par le nord de ce côté et beaucoup plus vite de l'autre. Les rives sont assez basses sur la droite, plus hautes vers l'est.

Nina me fixa :

— Qu'est-ce que tu en penses : on le contourne par la droite et on essaye de remonter ?

— Ça devrait marcher.

— Allons-y.

Nous avançâmes plus lentement, en essayant de respirer en silence. En carré, à deux mètres l'un de l'autre, nous veillions à conserver notre formation.

Connelly nous fit prendre à gauche. Le terrain accidenté montait en pente escarpée, et je devais m'aider des mains pour me stabiliser dans les rochers. Exténué, j'avais la tête vide, lessivée. Tout mon corps me faisait mal. Une fois arrivé au sommet, je pivotai au niveau de la taille et tendis la main vers le bas ; Nina la saisit et se hissa.

Le terrain descendait des deux côtés, comme si nous marchions sur la colonne vertébrale d'un animal géant. Nous progressâmes parmi les arbres, accroupis, respirant lentement.

Soudain, dans un froid glacial qui transperçait les oreilles, une bourrasque de vent tonitruante monta de la forêt en contrebas et secoua les branches tout autour de nous.

— Mon Dieu ! s'exclama Nina.

Le vacarme persistait, un vaste séisme qui encerclait un cri glacial. On aurait dit une force à laquelle on ne pouvait s'opposer, et peut-être que l'un de nous se redressa un peu trop, plusieurs d'entre nous, peut-être. Cela suffit.

On entendit un claquement sec et un grognement ; le shérif se retourna brusquement et retomba à plat dos.

J'avais vaguement l'impression que Nina et Phil se resserraient autour de moi, pour venir se protéger derrière les arbres. Je me jetai en avant sur le sol et me glissai au niveau du shérif.

Connelly avait les traits tirés.

— Je vais bien, dit-il.

J'ouvris le haut de sa parka et discernai une tache sombre au niveau du diaphragme, à gauche. Je mis sa main sur la blessure et appuyai très fort. Sa respiration restait régulière. C'était un dur.

En baissant les yeux, je vis Nina accroupie, à quelques mètres, pointant son pistolet dans la direction du coup de feu. L'assistant était assis, le dos à un arbre. Les rafales de vent ne faisaient plus qu'un bruit de vagues.

— Phil, venez ici ! criai-je.

Il se leva. Il y eut deux autres claquements secs.

— Baissez-vous !

Il se laissa tomber sur le ventre et rampa vers moi, à la manière des commandos.

— Chef ! Merde, dit-il en voyant le sang.

— Restez avec lui. Tu l'as vu ? demandai-je à Nina.

— Trop sombre. Il peut très bien nous avoir traqués depuis une demi-heure, pour avoir sa chance.

— À la manière dont Connelly a été touché, le tir devait venir de cette direction, repris-je en indiquant le bas, à droite. Je vais monter le long de la crête, redescendre en boucle et essayer de le surprendre. Si tu vois quelque chose bouger, tire !

— Soit prudent.

— J'essaierai.

J'adressai un signe à Phil pour lui indiquer mes intentions. Il acquiesça et pointa son fusil dans la même direction que Nina.

J'escaladai rapidement les rochers. En haut, j'entendis un autre claquement sec venant du bas, suivi de deux coups de feu tirés par Nina. Elle jura et rechargea son arme.

Je rampai sur une dizaine de mètres, allongé sur le ventre. Froid et nu, le flanc de la montagne descendait à pic. Aucun détail, aucun trait caractéristique auquel se raccrocher. Les formes des troncs, des branches et des roches se répétaient à l'infini et, dès que l'on tournait les yeux, on perdait ses repères. Il fallait absolument garder le contact visuel, tourner la tête lentement.

Je le vis.

L'image était si indistincte que cela aurait pu être une ombre, une illusion d'optique, créée par l'obscurité et la neige. Mais je le revis une fois encore, et j'aurais juré qu'il avait bougé.

Il se trouvait à une trentaine de mètres, exactement là où nous le pensions.

Je rampai encore de quelques mètres le long de la crête, jusqu'à ce que je sois dissimulé par un buisson. Je m'appuyai sur un pied et un genou. S'il ne m'avait pas vu me mettre en position, j'avais mes chances. Je pouvais surgir du bosquet, descendre à toute vitesse vers la droite, me diriger vers de gros arbres touffus devant moi. Et vider mon chargeur en chemin. Et si rien ne me faisait tomber, je pourrais recharger à l'abri des arbres pour l'étape suivante.

Un contre un : aucune raison pour que cela ne tourne pas en ma faveur. Peut-être. Je glissai la main dans ma poche droite afin de m'assurer que les chargeurs étaient en place. Pas de problème. Mon cœur tambourinait, car je savais que c'était un moment où il faut y aller, où la réflexion est moins importante que la rapidité.

Je me déplaçai vers la droite, lentement, un mètre, un mètre cinquante... J'hésitai à me lancer dans la course. Je jetai un dernier coup d'œil sur le côté.

Il y avait quelqu'un.

Une jeune femme. À trois mètres, sur une partie légèrement surélevée. Pieds nus, elle portait un pyjama à fleurs. Elle se tenait entre deux arbres, à demi dans l'ombre, avec les flocons de neige qui tourbillonnaient autour d'elle. Certains se posèrent sur ses épaules et sur ses longs cheveux. Je devinais tout juste ses yeux, la ligne du menton. Jessica Jones...

— Attention, dit-elle, ils sont nombreux.

Et elle avait disparu.

J'étais en déséquilibre, prêt à aller de l'avant, mais je tombai en arrière contre le rocher. Je me figeai un instant, en clignant des yeux, scrutant l'endroit où je l'avais vue. Je regardai à gauche, à droite. Rien.

Je crapahutai jusqu'à l'endroit où elle était apparue. Il n'y avait personne, mais la neige était piétinée. Je crus apercevoir une empreinte, deux peut-être, mais beaucoup trop grosses.

Tout d'un coup, je fus incapable d'agir comme prévu. Je fis demi-tour, me glissai de l'autre côté de la crête et allai rejoindre Nina.

— Qu'est-ce qui t'est arrivé ?.

— Je crois qu'il n'est pas seul, répondis-je, évitant son regard.

— Comment le sais-tu ? Qui est avec lui ?

— Je ne sais pas.

— Qu'est-ce que tu as vu ? Ward, que s'est-il passé ? Je ne répondis pas. J'en étais incapable. Je ne savais que dire. Je m'approchai de Phil, qui était resté près du shérif.

— Comment ça va ?

— Bien, répondit Connelly, qui mentait visiblement. Je n'ai pas besoin de baby-sitter. Coincez-moi ce crétin.

— Ils sont au moins deux, objectai-je. Alors, Phil, on va avoir besoin de vous en bas.

Phil jeta un coup d'œil vers son patron, qui lui adressa un bref signe de tête.

— Arrange-toi pour ne pas te faire descendre, murmura Connelly. La journée a été assez dure comme ça sans que j'aie besoin d'aller prévenir ta mère.

Phil s'approcha de moi, à quatre pattes.

— Il m'a semblé sentir une drôle d'odeur. Vous n'avez rien remarqué ?

— Non, c'est bizarre.

Il haussa les épaules.

Lorsque nous revînmes au niveau de Nina, elle me lança un regard dur.

— Qu'est-ce qu'il y a ? Tu as l'air bizarre.

— Rien. Mais...

Et voilà, un autre tir venu d'en haut et un autre de la gauche.

— Il y a quelqu'un avec lui ? demanda Phil. Qui ?

— Je ne...

Pendant une seconde, l'idée ridicule de John et de Paul s'associant m'effleura l'esprit. Mais non. Alors, qui... ?

Je cessai vite de réfléchir car, telle une ombre fulgurante, un homme remontait la colline vers nous, en tirant pendant sa course.

Nina et moi fîmes feu au même instant, mais nous le manquâmes tous les deux. Phil se roula en boule et se cogna durement contre un arbre. Il le contourna pour tirer mais hésita une seconde de trop. Je me levai et tirai deux fois.

L'homme pivota comme une toupie et tomba à terre. Je tirai encore deux fois, entendit un gémissement. Nina sortit de sa cachette, l'arme à la main, marchant en crabe.

— Nina, reste ici ! Phil, vencz avec moi.

Je fis signe à l'adjoint d'avancer le long de la crête. Je courus derrière lui, courbé en deux, et nous nous séparâmes au niveau de Connelly. Une nouvelle série de claquements éclata en provenance de la position du premier homme.

— Merde, je croyais que vous l'aviez eu !

— Alors, c'est qu'ils sont trois.

Nous restâmes immobiles un moment. La forêt semblait de plus en plus sombre, de plus en plus épaisse. Je tremblais, je me sentais vieux. Mon cou me grattait, je tournai vivement la tête à gauche et crus distinguer des silhouettes à travers les arbres, à une vingtaine de mètres, mais c'était impossible, car de nouveau elles ne portaient que des pyjamas, ce qui aurait été pure folie dans un tel endroit, par un tel froid. J'étais exténué, en plein délire, je créais des ombres dans les arbres, j'inventais des images absurdes. Je devais faire attention. Je baissai la tête et pris plusieurs inspirations profondes.

Je levai de nouveau les yeux et, de nouveau, un claquement retentit, juste en face de nous ; quelque chose vint siffler entre nos têtes pour aller rebondir sur le rocher derrière nous. Phil et moi répliquâmes.

Ensuite, j'entendis Nina tirer derrière nous.

— Merde ! m'exclamai-je, pris de panique.

— Phil, restez en position ici. Liquidez-moi ce type, si possible. Je reviens.

— Je ne le lâche pas, dit Phil.

Il s'allongea et rampa un peu plus loin. Il donnait l'impression d'avoir regardé pas mal de films de guerre. Tant mieux.

Je me redressai un peu plus que je n'aurais dû et courus en trébuchant vers l'endroit où devait se tenir Nina. Je ne la voyais pas mais je l'entendais tirer dans les arbres, sur la gauche. Je passai devant le corps du premier homme et aperçus son visage : froid, émacié, dur. Je ne le connaissais pas.

D'autres coups de feu éclatèrent devant moi, plus difficiles à distinguer, car le vent les amalgamait en un son unique. Je me précipitai vers l'origine des tirs sans savoir s'il y avait un ou deux tireurs.

Je plongeai près d'un rocher et faillis me briser la cheville, mais réussis à me retenir de justesse. Je m'enfonçai dans une épaisse couche de neige et j'y pataugeai comme dans du sirop gelé.

Finalement, je rencontrai un sol plus rocailleux. Les tirs avaient cessé, mais je ne voyais toujours personne. – Nina ?

Pas de réponse. Je fis un tour complet et, trois mètres plus loin, alors que je prenais de la vitesse, je me rendis compte que mes poumons étaient vides, que j'étais allongé sur le dos, avec de la neige dans les oreilles et un rocher enfoncé dans le dos.

Quelqu'un surgit de l'ombre d'un arbre. Un pied se posa lourdement sur ma poitrine. Je luttai pour respirer, la douleur de mon dos se propageait dans tout le corps. Je hurlai malgré moi. Le pied appuya un peu plus fort et un visage apparut au-dessus du mien.

Des cheveux courts, des lunettes rondes.

Le type de Fresno. Il plaça le canon froid de son fusil au milieu de mon épaule, pesa de tout son poids.

— Salut, connard, dit-il.

Nina était à une cinquantaine de mètres. Elle avait entendu courir quelqu'un qui ne semblait ralenti ni par les rochers ni par la neige, ni par le terrain accidenté. Cela devait être Paul.

Elle avait donc descendu la pente pour le suivre, tirant sans discrimination, et aperçu quelque chose qui bougeait en contrebas. Quelques instants plus tard, elle s'arrêta, essoufflée, mais ne distingua plus rien.

Elle perçut un cri derrière elle.

— Ward, dit-elle en remontant la pente.

Elle glissa, se coupa la joue sur le rocher. Elle continua.

L'homme enfonça le canon sur ma tempe.

— Alors, c'est toi, le frangin. T'as eu de la chance au restau. Tu t'en tireras pas aussi bien cette fois. T'es pas aussi doué que lui. Un simple amateur.

Je toussai. Je ne pouvais rien faire d'autre.

— Il va mourir ce soir, ajouta le type en faisant tourner le canon, grâce à ton copain.

— Quel copain ?

— Ce type, Zandt. Comment tu t'imagines qu'on vous a retrouvés ? Il a passé un accord.

— Il n'a pas tué Dravecky ?

— Le patron est toujours vivant et en pleine forme. Ton pote pense qu'il va s'en tirer comme une fleur. Il se trompe méchamment.

Il appuya encore plus fort sur ma poitrine. Ses yeux pétillaient derrière les petits cercles de verre. Visiblement, il prenait grand plaisir à me voir suffoquer.

— Alors, adios, tas de merde. Le temps passe...

Je voyais son doigt se resserrer lentement sur la détente, je sentais le sol s'aplatir sous moi, comme s'il se transformait en pierre tombale.

Je fermai les yeux. Je ne voulais pas garder le visage de cet homme pour dernière image.

J'entendis un coup de fusil, tout proche. Puis deux autres, peu après.

J'ouvris les yeux au moment où l'homme tombait en arrière. Je tournai la tête. Nina apparut dans mon champ visuel. Elle s'agenouilla près de moi.

— Ça va ?

Du sang coulait sur une de ses joues.

Je me hissai sur mes coudes. Oui, ça allait, dans le sens où je pouvais encore marcher et où tous mes membres me faisaient mal, ce qui signifiait que je n'avais rien de cassé et que je pouvais me lever.

— Qu'est-ce que tu as au visage ?

— Rien. Qu'est-ce qu'il disait ? Il parlait de John ? J'ai cru entendre son nom ?

— Non, ils en ont après Paul.

Elle m'attrapa le bras et m'aida à me relever. Je vacillai, à peine capable de me tenir debout. Je recouvrai mon équilibre, respirai profondément, les mains sur mes genoux.

Lorsque je me redressai, Nina se pencha vers l'autre type. J'entendis trois coups de feu.

— Nina !

— Un instant...

Le type essayait de se redresser. Du sang coulait de sa cuisse et de la base de son cou. Nina avait pensé qu'il risquait de porter un gilet. Il se déplaçait lentement mais semblait tenir le coup.

Nina lui assena un coup de pied.

— Prends ça pour Monroe, dit-elle d'une voix dure. C'est un connard, mais c'est mon connard.

— C'est un pourri, haleta l'homme avec un vague filet de voix.

— Qui ne l'est pas ? Et puisque vous l'aviez déjà mis au parfum, pourquoi avoir tué un flic ?

— Une assurance. Monroe n'avait pas bougé la première fois.

— Il s'appelait Steve Ryan.

— De toute façon, dit-il en grimaçant, je fais mon boulot, c'est tout.

— Bien, dit Nina.

Elle se retourna et lui tira trois fois dans la tête.

— Tiens, dit-elle en se penchant vers lui, ça, c'est de la part de sa femme.

CHAPITRE 30

Patricia se tenait en chien de fusil depuis près de dix minutes lorsqu'elle entendit courir au-dessus de sa tête. Elle ne savait que faire. Ne pas bouger lui semblait la meilleure idée. Au fond de soi, nous sommes tous persuadés que si nous restons bien tranquilles et que nous ne regardons pas, le monstre ne nous voit pas.

Mais elle voulait savoir.

Elle leva la tête et vit la Faucheuse sauter dans le lit du courant. Indécise au milieu de l'eau, elle semblait se demander ce qu'elle était venue faire ici. Patricia la voyait peser les différentes possibilités.

Puis elle remonta le courant en trottinant et disparut derrière deux grands arbres. Elle n'était pas partie bien loin...

Je fouillai les poches de l'homme et pris toutes ses cartouches. Puis je me rendis compte que je ne voulais pas utiliser l'arme de ce monstre et je les laissai tomber sur sa poitrine.

— Il s'est passé quelque chose.

— Je sais, dit Nina, j'ai entendu les coups de feu.

Nous remontâmes par le même chemin. Il faisait froid, le vent gémissait dans les arbres et je me sentais très loin de chez moi. Je boitais à présent, et l'affreuse douleur du côté droit de mon dos me disait assez que j'avais quelques côtes brisées. J'étais allé plus loin que je croyais. Il me fallut cinq minutes avant que Nina se redresse. Quelqu'un se tenait derrière elle sur la crête.

— Ne tirez pas ! s'exclama Phil. Vous allez bien ? Que s'est-il passé ?

— On en a eu un. Et vous ?

Il hocha la tête et regagna en hâte l'endroit où se trouvait Connelly. Nous le suivîmes.

— J'ai essayé de le suivre, j'ai perdu sa trace. Ensuite, il a tiré, il a failli me décapiter. J'ai riposté et j'ai essayé de trouver un rocher de l'autre côté, mais j'étais devant un ravin profond et j'ai cru que c'était la fin. Je n'avais nulle part où aller... J'aurais peut-être dû tirer plus tôt. Je n'avais jamais essayé de tuer un homme avant. Alors, je me suis redressé à moitié, pour chercher un autre chemin, et c'est là que j'ai vu l'autre type.

— Quel autre type ?

— Je ne sais pas. Il est sorti de nulle part. Je ne l'ai aperçu qu'une fraction de seconde. Il a fait ça... (Phil imita quelqu'un qui appuie un fusil sur son épaule) et tiré avant d'être en position. Un seul coup. J'ai plongé, comme si je tombais. Je n'ai plus rien entendu pendant quelques minutes. Alors, j'ai relevé la tête. Le type au fusil avait disparu. Et il y a un corps, à trois mètres de moi...

— Ce n'est pas vous qui l'avez eu ?

— Non, je viens de vous le dire... Mais quelqu'un s'en est chargé. Je suis allé voir. Une seule balle, en plein milieu du front, comme si on y avait dessiné une cible. Qui ça pouvait bien être ? Qu'est-ce qui se passe ici ?

— Ça doit être John, avançai-je.

— Non, répondit Nina. John est un type de la ville. Je ne le vois pas crapahuter par ici et descendre un de ces types d'une seule balle. Je crois qu'il n'a jamais tiré avec un fusil.

— Qui, alors ?

— L'Homme Debout. C'est forcé. C'est lui que les autres types sont venus tuer, pas nous.

— Je n'y crois pas. Il les aurait laissés nous tuer avant.

— Tu es son frère, Ward, répliqua Nina.

Je ne voyais pas quelle différence cela pouvait bien faire.

Lorsque nous arrivâmes au niveau de Connelly, nous le trouvâmes debout. Appuyé contre un arbre, mais debout.

— Mon Dieu, shérif, asseyez-vous !

— Je vais bien.

— Monsieur, avec tout le respect que je vous dois, vous n'en avez pas l'air. Vous saignez comme un goret...

Le policier baissa les yeux, vit les taches sombres qui commençaient à maculer son pantalon.

— C'est vrai. On ferait mieux de se dépêcher.

Il sortit son GPS de sa poche. Sa main tremblait, pas trop. Une petite lumière sur l'écran, et il nous fit un signe de tête, vers la droite.

— Autant y aller maintenant.

Nous avançâmes à travers les arbres, passâmes devant le corps de l'autre tireur allongé sur le dos. Phil avait raison, celui qui l'avait descendu connaissait son affaire.

Le terrain s'aplanissait un peu et s'incurvait vers la crête, des deux côtés, comme si nous entrions dans un demi-tunnel bordé d'arbres et d'ombres : un vieux lit de rivière, sans doute, ou, mieux encore, un ancien chenal glaciaire. Le vent se renforça de nouveau et nous poussa encore plus vite. Avec un peu de chance, il couvrirait le bruit de nos pas.

Connelly trébucha, bascula en avant et tomba. Je me penchai vers lui.

— Non, allez-y.

J'ôtai mon manteau et le couvris.

Les buissons étaient touffus, immenses boules de coton glaciales et hérissées de pics. On entendit une sorte de cri, en haut, sur la gauche. Je crus que c'était le vent. Les branches basses des arbres oscillaient sans cesse, comme secouées par un dingue.

Nina tendit le bras :

— Par là...

Je regardai. À une cinquantaine de mètres, les troncs laissaient place à un vide. Le bord du gouffre. Forcément.

Phil murmura :

— On descend tout droit ?

— Non, dit Nina. Vous contournez par la droite, assez large. Je fonce tout droit. Ward, tu passes par la gauche. Dès que ça bouge, on tire et on crie.

Nous acquiesçâmes. Phil partit tout de suite, se faufilant dans les buissons aussi silencieusement que possible.

Nina me lança un signe d'avertissement puis se mit en route. Je fis demi-tour et me dirigeai le long de la pente.

Tout allait bien, me dis-je avant d'entendre un coup de feu.

J'étais entre les mains des dieux. J'espérais qu'ils me prêteraient attention et se montreraient cléments.

Nina ralentit. Les cinq minutes de progression l'avaient conduite à une trentaine de mètres. À gauche, elle discernait une ombre vague qui se dirigeait vers le côté du haut vallon. Phil. Il disparut de son champ de vision quelques instants plus tard, derrière les arbres ou sur un terrain en contrebas. Elle ne distinguait pas Ward à sa droite. Le terrain était très escarpé et accidenté de ce côté. Il lui faudrait décrire un grand détour. Elle espérait que personne ne s'était perdu en route. Que personne n'allait mourir. Pas ici, il faisait trop froid.

Il faisait sacrément noir aussi. Les arbres ne lui laissaient plus qu'une voie possible, mais les buissons rendaient l'accès difficile.

Elle se baissa derrière un tronc penché, s'appuya sur des arbres. Malgré le vent, elle percevait le bruit de l'eau, un petit gargouillis d'éclaboussures solitaire.

Soudain... bang, le bruit d'un coup de feu.

D'où venait-il ? Pas de la gauche, sinon...

Puis elle perçut un cri, étouffé, indistinct. Il venait de la droite, elle en était sûre. Phil, il était touché.

Oubliant toute prudence, elle avança. Plus de temps à perdre. Elle espérait que Ward avait entendu, lui aussi. Il viendrait vite, elle en était sûre.

Elle tenait son arme devant elle, la tête penchée, à l'abri de la végétation, essayant d'éviter les branches et leurs griffures glaciales.

Quatre pas de plus, et elle s'écroula.

J'étais allé trop loin. Beaucoup trop loin. J'avais bien estimé la distance au début, mais chaque fois que j'essayais de descendre vers le ravin, quelque chose obstruait le chemin. Des arbres, debout ou couchés. Des jeunes plants, trop hauts pour être enjambés. Des roches glissantes qui s'ouvraient soudain en hiatus béants que je ne pouvais franchir. J'étais sans cesse contraint de m'éloigner, sur une crête de plus en plus étroite qui ne menait pas à l'endroit où je devais me rendre.

À la fin, je renonçai, essoufflé, et coupai la pente encore plus loin jusqu'à ce que je trouve un contrefort qui me laissait le champ libre pour un moment. Je ne voyais toujours pas comment descendre, et le temps passait. J'étais trop lent. J'aurais voulu être plus léger. J'aurais voulu que Nina fasse appel aux Feds, à l'armée, aux Éclaireuses. Nous n'avions pour nous couvrir que deux flics, dont l'un tremblait au pied d'un arbre à cinquante mètres de nous.

Finalement, il me sembla trouver un petit passage qui menait de manière erratique vers une étendue de roches nues, vers un espace au sommet que je pensai pouvoir escalader. Ensuite, un coup de feu.

Un cri peut-être aussi, quelques secondes plus tard, mais je n'en étais pas certain.

Je glissai mon arme dans ma poche et me raccrochai aux rochers. J'allais passer par-dessus, quoi qu'il advienne. Je me hissai et me laissai glisser de l'autre côté, où s'étendait un terrain dégagé. Enfin.

Je tombai sur le sol et courus, courus...

Elle tomba aussitôt, essaya de se rattraper, perdit son arbre. La chute fut brève et bruyante, mais elle lui parut longue. Nina se heurta, le ventre en avant, contre quelque chose qui la retourna si violemment qu'elle en eut des vertiges. Elle atterrit sur le sol.

Elle s'assit immédiatement, roula sur le côté et se projeta en avant lorsqu'elle fut certaine de savoir où elle était. À quatre pattes, elle regarda à droite et à gauche, devant et derrière, à la recherche de son arme.

Nina se trouvait dans un endroit sombre et rocailleux, beaucoup plus près de l'eau. Mais où était son calibre ?

Elle espérait qu'il n'était pas resté coincé en haut, dans une crevasse quelconque. Elle en avait besoin, de toute urgence.

À tâtons, elle rampa. Elle avait toujours des vertiges après sa chute. Sous ses mains, le sol froid était couvert de galets. Mouillés. Pointus. Ses yeux se perdaient dans le noir. Il lui était difficile de déterminer ce qui se trouvait devant elle : l'obscurité ou un mur de pierre.

Elle entendit une sorte de grognement. Pas tout près. Un grognement, c'était mauvais signe. À moins que ce soit lui. À moins que Phil ne l'ait eu. Ou le vent. Si ce n'était ni le vent ni l'Homme Debout, cela ne présageait rien de bon.

Et si c'était Ward ? Est-ce que je suis près du ravin ?

Où est mon arme ?

Elle discerna une silhouette devant elle, pâle, mais ce n'était pas de la neige. Elle se concentra. Une vieille femme, emmitouflée dans un grand manteau, de l'autre côté d'un cours d'eau, appuyée contre un haut rocher.

Les yeux écarquillés, elle regardait Nina sans faire le moindre bruit. Elle avait la tête et les épaules couvertes de neige, telle une statue dans un cimetière abandonné, loin des sentiers battus.

La forme et la position de la femme donnèrent enfin à Nina un repère visuel, une clé pour comprendre l'espace environnant. Elle se trouvait au fond d'un ravin, forcément, un vallon aux flancs escarpés et au fond plat, de cinq mètres de large, qui rétrécissait rapidement de chaque côté.

Nina cligna des yeux, pour stabiliser ses vertiges, puis chercha de nouveau son arme en s'efforçant de prendre son temps.

Ouf !

Nina rampa vers le courant, ramassa son pistolet dans l'eau. Elle le secoua, changea de chargeur. Courbée en deux, elle traversa et s'approcha de la femme. Elle parla à voix basse, respirant calmement.

— Vous êtes Patricia Anders ?

La femme la scruta. Elle avait des glaçons dans les cils. Encore un peu, elle était transformée en sucette glacée. Elle sembla bouger la tête, imperceptiblement.

— Madame ?

— Oui, répondit-elle à voix haute.

— Chut... Vous êtes seule ? Il est toujours là ?

— Oui, quelque part.

— Qui ? L'autre ? Tom ou Henrickson, je crois ?

— C'est lui, mais ce n'est pas son vrai nom.

— En fait, si.

Nina s'accroupit près d'elle et suivit le regard de la vieille femme. Elle ne voyait rien que le cours d'eau, qui se nivelait un peu sur le côté gauche.

Un autre gémissement de douleur.

— Ne bougez pas, dit-elle.

Elle comprit soudain pourquoi la position de la vieille dame lui paraissait si étrange. Elle avait les mains liées derrière le dos. Les doigts gourds, Nina défit les nœuds. La corde avait gelé, et il lui fallut un temps fou pour desserrer les liens. Elle finit par y parvenir. Lentement, très lentement, la femme ramena ses bras devant, comme si elle avait peur qu'ils se brisent en mille morceaux.

— Ne bougez pas, répéta Nina.

Courbée en deux, elle contourna le buisson et longea le bord du ravin. Elle ne lâcherait plus son arme, plus jamais, mais elle glissait sur les rochers humides, avec une seule main pour se retenir. Elle s'agrippait aux branches, essayait de se tracter, mais sa progression restait lente. Les ruisselets d'eau transformaient ses mains en glace. Elle mit de longues minutes à parcourir les cinquante mètres vers l'amont, chaque pas était une aventure en soi.

Pourvu que Ward ne tarde plus.

Devant elle, le mur de roches ne s'élevait qu'à deux mètres, deux mètres cinquante, pas plus. Une forme était étendue au pied de la paroi.

C'était Phil, toujours vivant, mais il se tenait la cuisse à deux mains, le corps roulé sur lui-même. Les yeux exorbités par la douleur, il essayait de ne pas faire de bruit, mais un nouveau gémissement lui échappa lorsqu'il la vit.

— Il m'a tiré dessus, lâcha-t-il dans un sifflement éraillé. Henrickson. Il a pris mon fusil.

D'un signe de tête, il indiqua le chemin par lequel Nina était venue.

Mais elle regarda plus loin derrière lui, scrutant les parois du ravin. Peu importait le chemin par lequel la silhouette avait disparu. Il pouvait être revenu par en haut depuis longtemps.

Elle envisagea de longer le courant, d'essayer d'escalader l'une des parois pour remonter, dans l'espoir que l'Homme Debout redescendrait. Ce serait lui qui serait pris au piège, et non elle.

Mais elle s'en savait incapable, une arme à la main, de plus, son dos offrirait une cible idéale à qui savait tuer.

— Gardez le point de compression, conseilla-t-elle.

Elle s'éloigna des parois, cette fois, et marcha au beau milieu du courant, avec de l'eau jusqu'aux genoux, glaciale. Des eaux froides et bruyantes, des eaux tumultueuses, dans le vent hurlant qui charriait des rideaux de neige.

Elle ne pouvait pas se retourner, car le courant était trop imprévisible sous ses pieds. Elle essayait de marcher, au centre, en regardant devant elle, tout en tâchant de ne pas perdre Mme Anders de vue, afin de savoir à quelle distance de son point de départ elle se trouvait. Elle songea à crier, au cas où Ward pourrait l'entendre, mais l'Homme Debout risquait d'être encore plus près et, dans un éclair de lucidité douloureux, elle comprit que son plan « tire et crie » était stupide.

Puis, du coin de l'œil, elle avisa une silhouette sur la gauche. En une fraction de seconde, elle aperçut un fusil sur l'épaule et sut que ce n'était pas Ward. Avec une rapidité qui dépassait sa conscience, elle fit volte-face, leva la main et tira trois fois.

Deux des coups éclatèrent comme des battements de main. Le dernier renvoya un bruit sec de gifle dans les branchages. La forme glissa, tomba le long de la paroi du ravin.

Nina courut dans l'eau, oubliant le froid, oubliant tous les détails, ne songeant qu'à l'homme allongé devant elle. Elle garda son arme pointée vers lui, jusqu'à ce qu'elle ne soit plus à trois mètres.

Une fois, ce n'était jamais assez. Elle devait tirer encore. Son doigt se resserrait sur la détente lorsqu'il se redressa et montra son visage.

— Mon Dieu ! John !

Elle entendit quelqu'un atterrir légèrement à côté d'elle. On lui arracha son arme des mains, un bras se glissa autour de son cou et elle sentit la froideur du métal contre sa tempe.

— Bonsoir, agent Baynam. Excellent travail.

CHAPITRE 31

Je faillis aller au bout du monde.

Si, au dernier moment, je ne m'étais pas retenu de la main gauche, ça n'aurait pas manqué. Je serais allé jusqu'au rocher, j'aurais fait un pas de trop et j'aurais plongé dans la nuit éternelle. Là, suspendu dans le vide, j'eus une vision fugitive de la chute vertigineuse, je sentis la branche se plier et j'entendis le rugissement de l'eau qui bouillonnait, très loin, très loin, tout en bas.

Je me rattrapai et me tournai, dos à la chute, désespéré, terrifié. Mes poumons en feu semblaient remplis de verre pilé. Mes poumons me haïssaient.

Je me penchai et vis que oui, j'étais arrivé au ravin, mais pas au bon endroit. Il mesurait douze mètres de large ici, les parois étaient si abruptes qu'elles paraissaient avoir été tranchées d'un seul coup, avec une hache géante.

Pourtant, c'était le bon ravin.

Je devais rebrousser chemin.

Je restai à quelques mètres de la crête et passai par les buissons. Les arbres étaient un peu plus petits, mais cela ne m'était pas d'un grand secours : la végétation avait eu tout le loisir de prospérer. Avant longtemps, je m'éloignai à nouveau du ravin, forcé de reprendre le chemin par lequel j'étais venu.

Je luttai toujours pour avancer, au pas de course le plus souvent possible. À travers les arbres, je crus apercevoir quelque chose, une vision fugitive, sur le bord. J'avançai, sachant que le gouffre serait encore trop large.

Une fois sur place, je compris.

Il y avait un tronc en travers du ravin, tombé là, qui formait un pont. Difficile de ne pas y voir une invitation. L'autre côté semblait beaucoup plus hospitalier.

J'allai jusqu'à l'extrémité du tronc, lui donnai un coup de pied. C'était solide. De l'autre côté, la rive semblait offrir un passage dégagé vers l'endroit où je devais aller.

À condition que je puisse franchir le gouffre de trois mètres au-dessus du courant glacial et rocailleux, sur un tronc recouvert de dix centimètres de neige.

Laisse tomber. Tu ne serviras à rien le crâne brisé.

Trois coups de fusil retentirent, puis un cri, une voix qui ressemblait à celle de Nina mais n'avait rien d'un cri de triomphe.

Je sautai sur le tronc, pris une profonde inspiration.

Je ne savais que faire, à part traverser au pas de course.

Patricia regardait ce qui se passait devant elle. Henrickson était revenu près du courant, comme sur un film qu'on passe à l'envers. Dans un mouvement lisse, il désarma la jeune femme et lui braqua un pistolet sur la tempe.

Du pied, il poussa le fusil de l'autre homme dans le courant, puis entraîna la femme au milieu de l'eau.

L'homme à terre semblait souffrir mais s'efforçait de le cacher. Les hommes étaient comme ça.

— John, comment tu es arrivé ici ? questionna Nina.

— Dravecky, répondit John non sans une certaine satisfaction... Même les plus grands psychopathes du monde veulent se débarrasser de toi, ajouta-t-il à l'adresse de Henrickson. Tu es le paria des parias. Tu n'as plus nulle part où aller.

— Il y a toujours un endroit, riposta Henrickson. Trouver Dravecky et le zigouiller, pour commencer. Et ensuite, son pote de la NSA à Los Angeles. Tu l'as rencontré, Nina ?

— Oui.

— Je m'en doutais. Ne t'inquiète pas, il est beaucoup moins important qu'on ne le croit.

L'homme à terre bougea brusquement, et soudain il eut une arme à la main. Mais au même instant, Henrickson avait reculé de deux mètres et placé la femme juste devant lui, en bouclier.

— Qu'est-ce que tu vas faire ? La tuer pour m'avoir ?

La jeune femme essaya de s'écarter pour laisser un angle de visée qui ne soit pas abrité par son propre corps, mais l'homme qui la tenait était rapide et efficace.

— Qu'est-ce qui est plus important pour toi ? Me planter une balle dans le corps pour venger Karen, et tuer ta copine flic par la même occasion ? Je devrais peut-être t'épargner la décision et la tuer tout de suite.

L'homme à terre s'était redressé. La main qui tenait l'arme ne semblait pas très stable.

— Tu la tues, et je te tue.

Patricia estima que ses chances étaient quasi nulles, même si le type se découvrait un peu et elle savait que Henrickson en était persuadé, lui aussi, mais que cela ne l'empêcherait pas de tenter le coup.

Soudain, elle se dit : il ne m'a pas regardée.

Henrickson n'avait pas j eté le moindre coup d'œil en sa direction depuis son retour. Cela ne signifiait pas pour autant qu'il l'avait oubliée.

Pouvait-elle réussir ? Pouvait-elle se jeter en avant, se jeter sur lui ? Le déséquilibrer suffisamment pour que le type au fusil puisse tirer ?

Elle n'en était pas certaine, mais cela valait la peine d'essayer.

Lentement, elle déplia ses bras, avec l'impression qu'on enfonçait un câble brûlant dans ses os. Elle tenta de bouger les pieds, il ne se passa pas grand-chose, mais peu importait. Elle n'avait pas besoin d'arriver jusqu'à lui. Il suffisait qu'elle crée la surprise.

Elle se propulsa en avant.

Son corps ne bougea pas d'un pouce. Elle essaya à nouveau. Elle était incapable de bouger, comme si quelque chose la retenait. Elle était gelée sur place, ses jambes étaient si bloquées que...

Non. Quelque chose la retenait.

Elle tourna les yeux. Quelqu'un lui posait la main sur les épaules. Lentement, elle tourna la tête.

Tom Kozelek était accroupi à côté d'elle. Il semblait étrange. Il avait posé ses grandes mains sur ses épaules et la retenait.

— Vous êtes sauvée, lui murmura-t-il à l'oreille. Quelqu'un arrive.

Il la lâcha, se fondit dans le noir et disparut. Il lui sembla entendre un bruit de pas dans l'eau derrière elle.

Pourtant, elle ne pouvait toujours pas bouger. Elle avait les jambes gelées, après tout.

Je fis les trois quarts du chemin, et mon pied glissa comme si j'avais marché sur la neige avec des chaussures de glace. Je jetai les deux mains en avant et fis ma prière.

Je m'écroulai de l'autre côté, des branchages de buisson dans chaque main. Je me hissai, en m'agrippant avec mes pieds et mes mains, comme un chien avec ses pattes. Je poursuivis à travers les rochers, les racines, la neige... et quelque chose pour poser les pieds.

Je courais. Mes poumons ne me faisaient plus mal, ni mon dos, ni mon épaule. Mes pieds se posaient à chaque pas comme si je courais sur une pelouse bien tondue ; les buissons fondaient tels des rêves brumeux et les arbres révélaient un chemin qui avait toujours été là, comme si les montagnes s'étaient galbées pour l'accueillir. Je ne voyais pas grandchose à travers le rideau blanc, mais je savais où je devais être... si j'arrivais à temps.

Je dus monter en zigzag, sur une cinquantaine de mètres seulement. Je contournai par la droite et me dirigeai vers le bord du ravin. Je courais vite, me baissant le plus possible, sans me soucier du bruit. Il était trop tard pour s'inquiéter de cela, comme du reste d'ailleurs.

En haut, je me glissai derrière un arbre et m'accroupis. Je mis un nouveau chargeur, pris une inspiration et me redressai.

— Ward ! dit une voix en bas. Je t'attendais.

J'avançai d'un demi-pas, puis reculai d'autant pour me rapprocher de l'arbre. Je baissai les yeux vers le fond du ravin, pistolet en avant. Au début, je crus que c'était Paul, mais je vis John et compris que ce n'était pas lui qui m'avait appelé.

Nina se trouvait à une trentaine de mètres, sur une ligne diagonale par rapport au ruisseau. Elle se tenait au beau milieu du courant, dans une posture bizarre. Puis je vis un bras d'homme autour de son cou, et un fusil contre sa tempe. C'était Paul.

— Lâche-la.

— Pas avant de la descendre.

— Je te tuerai.

— Je ne crois pas. John ne peut pas et toi non plus.

Il avait raison. Il s'était placé le dos à l'autre flanc du ravin. Comme John et moi étions du même côté, nous ne pouvions l'atteindre sans toucher Nina d'abord.

— Tire, Ward, cria-t-elle.

Je reculai pour me couvrir. Paul tira. Je crus qu'il avait tué Nina et mon cœur cessa de battre, mais je compris qu'il m'avait visé, car la balle vint siffler près de mon visage. L'arme avait aussitôt retrouvé sa place sur la tempe de Nina.

— Oui, vas-y, c'est ton tour...

— Ward ! Tue-le, bon sang ! cria John.

— Je n'ai pas le bon angle.

Je tentai d'avancer le long de la rive, mais Paul pouvait me voir. Il modifia légèrement sa position, tout en restant bien protégé.

— Qu'est-ce que tu vas faire ? criai-je. Retourner à Seattle ? C'est pas la porte à côté, je te préviens.

Il se contenta de rire.

Ce n'était qu'un jeu. Il savait que j'allais venir. Il m'attendait. Il voulait que l'un de nous en arrive là, qu'il soit poussé à la faute irréparable.

Sinon, il le ferait, lui, sans sourciller ; il ne resterait plus que lui et moi, et un homme allongé par terre, qui semblait blessé. Je ne nourrissais pas des sentiments très positifs envers John, mais je risquai de commettre une erreur qui lui coûterait la vie.

John tira.

Il rata sa cible. L'Homme Debout recula d'un pas, entraînant Nina avec lui.

Je scrutai le ravin et compris que, s'il la tuait maintenant et remontait le ruisseau en courant, il pouvait s'échapper avant que je le rattrape. Le temps pressait.

Il allait tuer Nina et s'enfuir.

Elle me regardait toujours. Je sentais qu'elle me disait qu'il était temps d'agir pour le mieux.

Je reculai d'un pas, laissant retomber mes bras un instant. J'avais froid aux mains. Ma tête avait trop froid, elle était à la fois vide et totalement emplie de la simple décision que j'avais à prendre.

Je ne voyais que le visage de Nina.

Soudain, je perçus un mouvement dans la périphérie de mon champ de vision, tout en haut, sur l'autre crête du ravin. Pas sur le bord, juste un peu plus loin. Quelqu'un s'approchait, marchant très doucement.

Je me redressai :

— Va te faire foutre. Ça te ferait trop plaisir !

— Comme tu voudras. (Il me regarda droit dans les yeux, pressa le canon un peu plus fort sur la tempe de Nina.) Je vais le faire à ta place.

De l'autre côté, la silhouette s'approcha encore un peu, presque jusqu'au bord du ravin. Je regardai toujours Paul, m'interdisant de cligner des yeux.

— Ward, tire, sinon, je m'en charge.

— John, surtout, ne bouge pas !

J'attendis un instant. Puis je me déportai vivement sur la gauche.

— Maintenant ! criai-je.

Paul se retourna et recula pour que Nina reste entre lui et moi.

Connelly tira. Il visa avec précision, planta une seule balle au sommet de l'épaule, et disparut de l'autre côté du ravin.

Paul se retourna, le fusil devant lui et, pendant un précieux instant, il se trouva dans ma ligne de mire, sans aucun obstacle. Je tirai à trois reprises. L'épaule, le bras, la jambe.

Il se retourna maladroitement et tenta de retenir Nina, mais elle s'esquiva, lui donna un coup de pied et réussit à lui échapper. Il essaya de se retourner encore mais ne réussit qu'à avancer de quelques pas avant de tomber.

Je descendis la paroi. Je tirai encore en chemin, touchai le corps cette fois, et Paul fut projeté contre le mur. Son arme lui échappa.

Je me mis rapidement entre lui et John. Je ne savais pas si cela changerait quelque chose, mais John ne tira pas.

Je traversai le ruisseau, marchai dans l'eau glaciale jusqu'à l'autre rive. Je m'arrêtai à deux mètres.

Je levai le bras. Pointai mon arme vers lui.

Paul gisait au pied de la paroi. Il saignait abondamment. On avait du mal à croire que c'était lui.

Il me ressemblait tant.

Cannon Beach

Quatre jours plus tard, nous suivîmes le conseil de Patricia. Arrivés à Portland, nous prîmes la direction de l'ouest, sur la route 6. Il plut tout le long du chemin jusqu'à Washington et il pleuvait encore lorsque nous arrivâmes sur la côte par la forêt de Tillamook. C'est une forêt magnifique. Avec beaucoup d'arbres. Ce n'avait pas toujours été le cas. Elle avait été exploitée pendant des décennies et, en 1933, un gigantesque incendie avait creusé une tranchée en plein milieu. À la fin, plus de cent vingt mille hectares de bois avaient disparu en fumée et l'on disait que des cendres brûlantes tombaient sur les bateaux à cinq cents milles de la côte. Mais le feu finit par être maîtrisé et l'on planta d'autres arbres. Par un étrange coup du sort, la forêt brûla encore en 1939, 1945 et 1951, telle une malédiction qui frappait tous les six ans. Alors la population vint planter d'autres graines : les clubs de jardinage, les scouts, les écologistes, tous passaient leurs week-ends à réparer les dégâts. À présent, elle ressemble à une forêt ordinaire. Si on ne sait rien, on pense qu'elle a toujours été ainsi. Nous sommes comme ça. Parfois.

Il ne nous vint pas à l'idée d'aller faire une promenade. Nous n'y aurions pas pensé, même s'il n'avait pas plu. Nous avions vu assez d'arbres pour un moment.

Nina ne m'avait pas laissé tirer.

J'allais le faire. Vraiment. Je ne voyais pas quel sens cela pourrait avoir autrement. C'était l'homme qui avait tué mes parents et détruit ma vie. Il avait tué la fille de l'homme qui gisait de l'autre côté du ravin. Il avait tué des gens dont je ne saurais jamais le nom, des gens dont on ne saurait jamais plus rien.

Elle arriva derrière moi. Elle ne dit rien, n'essaya pas de me prendre la main, assez proche pour que je sente la chaleur de son souffle sur ma nuque. J'observais l'homme à mes pieds qui essayait de bouger, avec ses mains qui glissaient faiblement contre la roche, telles deux pâles créatures, proches de l'agonie.

Je ne sais pas comment fonctionnent les fous, mais ils ont certainement une sacrée force de volonté. C'est sans doute parce qu'il leur manque le contrôle et l'équilibre dont nous jouissons, mais je me leurre peut-être : leur esprit ne se laisse pas impressionner par les angoisses qui nous entravent. Pourtant, la force de volonté ne lui suffisait plus. Il ne pouvait pas bouger, il n'avait pas d'arme.

Je pouvais quand même le tuer. Je le savais. Personne ne me le reprocherait. Connelly observait du haut de la crête. Il avait le visage blafard et j'entendais sa respiration d'ici, mais le canon de son fusil restait stable. Il avait l'air capable de tirer si je ne le faisais pas. Je savais ce que John voulait.

Enfin, je laissai retomber mon bras.

— Abruti de première ! grommela John.

Nina s'approcha de lui, s'accroupit et lui murmura quelque chose à l'oreille. Elle continua à parler pendant un moment et lui prit son arme.

Elle revint près de l'Homme Debout, pendant que j'aidais Connelly à descendre au fond du ravin. Il était horriblement touché, mais moins que je ne le craignais.

En boitant, il m'accompagna jusqu'à l'endroit où Nina disait que se trouvait Phil. L'adjoint essaya de nous venir en aide mais je dus à moitié le porter pour qu'il puisse repartir avec nous. J'appuyai Phil contre la paroi, en face de Paul. Le shérif s'écroula à côté de lui, son arme toujours braquée vers Paul.

Je ne savais pas ce qui allait se passer ensuite. Il neigeait toujours. Un peu moins fort, mais la neige ne semblait pas vouloir s'arrêter. Ni Phil ni Connelly ne pourraient rentrer seuls, et la radio du shérif ne donnait rien. John avait l'air en meilleure forme. À en juger d'après sa manche, Nina n'avait guère fait que lui égratigner le bras. Il refusait de me parler, néanmoins. Il ne me regardait même pas dans les yeux.

Nina alla chercher la vieille dame. Je ne m'étais même pas aperçu de sa présence. Elle semblait avoir aussi froid que si on l'avait découverte dans le permafrost, à côté d'un mammouth laineux. Elles discutèrent un moment, et Nina s'approcha de Connelly et lui demanda son GPS.

— Vous n'en avez pas besoin, dit Patricia, je connais le chemin.

Nina le glissa tout de même dans sa poche. Elle s'approcha de moi, enleva sa parka et me la donna.

Puis, avec la vieille femme, elle commença à remonter le courant.

— Je viens avec vous, déclara John en se redressant sur ses pieds.

— Ça va aller, merci, répliqua Nina.

— Peut-être, mais il y a des ours dans le coin. J'en ai vu un tout à l'heure. Enfin, j'ai vu quelque chose.

Je m'installai sur une grande roche plate, à quelques mètres de Paul, et les regardai partir.

Dans la nuit, il se produisit deux choses que je ne compris pas.

La première fut un incident mineur. Phil et Connelly discutaient à voix basse et je tendis l'oreille.

— Vous avez toujours su, c'est ça ? demanda Phil.

— J'étais avec ton oncle, ce soir-là, dit Connelly.

Il avait l'air de vouloir en dire plus, mais il regarda dans ma direction et s'aperçut que j'écoutais.

Il lança un clin d'œil à Phil et hocha la tête.

Une demi-heure plus tard, ils s'endormirent. Je ne savais pas si c'était une bonne idée, mais ils étaient allongés l'un contre l'autre et se tenaient aussi chaud que possible. J'étais si fatigué que je ne pouvais pas jurer de ne pas m'endormir, moi aussi. Tous deux respiraient fortement. Il faudrait que je surveille bien ça.

J'avais l'impression d'avoir une tête en pierre, en équilibre sur une autre pierre, d'avoir couru pendant trois mois et d'être arrivé au bout de la piste pour constater qu'il n'y avait pas de ruban à franchir. Paul semblait inconscient, mais il tremblait très fort. Il me vint à l'esprit que j'avais toujours une arme à la main. Et que Nina ne savait sans doute pas combien de trous il avait dans le corps. Un de plus, cela ne se verrait pas. Peut-être avais-je le moyen de trouver cette fichue ligne d'arrivée. Tuer Paul serait peut-être ce qui marquerait la fin pour moi.

Je me levai, m'approchai de lui.

Un coup.

Les autres se réveilleraient, mais je pourrais dire qu'il avait bougé.

Je savais pourquoi Nina m'avait retenu. Elle ne voulait pas que je commette un meurtre de sang-froid. Elle estimait aussi que les familles des victimes de l'Homme Debout... les familles des jeunes filles qui avaient disparu à Los Angeles deux ans auparavant et de toutes les autres avaient droit à mieux qu'une exécution sommaire dans un bois, à l'abri des regards. C'était en grande partie ce qui la motivait dans son travail au fil des ans, ce qui la poussait à mettre les méchants hors d'état de nuire. D'accord, nous avions gardé le secret sur les événements des Halls, mais nous n'avions pas d'otages entre les mains.

Cependant, ce ne fut pas pour ces raisons que je baissai mon arme. D'ailleurs, pour être honnête, je ne sais pas pourquoi.

Je me levai, enlevai la parka de Nina et en couvris Paul, en prenant garde de bien le border. Son visage était livide, ses lèvres bleues.

Je m'aperçus que je pleurais.

Je m'assis près de sa tête. Je la mis sur mes genoux, pour lui tenir chaud, et passai mon bras autour des épaules.

Je ne sais pas pourquoi. Je ne comprends pas pourquoi. Je savais combien de gens il avait tués. Je sais qu'il aurait tué Nina, John et moi. Mais c'est comme ça.

Connelly se réveilla mais ne dit rien. Maladroitement assis contre la paroi, je m'endormis moi aussi, d'un sommeil interrompu par les crampes. Ensuite, je fus réveillé par un grondement au-dessus de ma tête.

J'ouvris les yeux : Connelly et Phil se soutenaient l'un l'autre, les yeux levés vers le brancard qui descendait doucement de l'hélicoptère.

Je fus le dernier à être hélitreuillé, le dernier à quitter cet endroit glacial. J'avais une horrible migraine et j'étais si fatigué que ma vision se brouillait. Je montai en spirale dans le vacarme et les tourbillons de neige.

Mal m'en prit, je baissai les yeux une fois et, pendant un instant, dans un nuage de lumière, je distinguai un petit groupe de silhouettes en contrebas, dans le ravin, qui me regardaient m'élever dans le ciel. Je clignai des yeux, essayai de discerner les détails, mais c'était un peu comme s'il n'y avait plus rien à voir.

Ensuite un tourbillon de neige dissimula le sol, et des mains m'attirèrent dans la machine métallique.

Une fois arrivés à l'océan, nous prîmes à droite le long de la côte. La côte n'appartient à personne en Oregon, elle est restée sauvage et on y assiste parfois à d'étranges phénomènes. Autrefois, on trouvait souvent des pains de cire dans le sable et à l'intérieur des terres ; on en découvrit plusieurs tonnes en tout. Certains semblaient porter des symboles qui, je crois, auraient pu être du chinois ancien. Cette partie du récit de Zandt au moins était confirmée, mais je ne croyais pas le reste pour autant. Des symboles, ce ne sont que des symboles. Ils ne décrivent pas forcément la réalité.

Nous ne savions pas où était John. La nuit fatidique, il avait accompagné Patricia et Nina pendant une grande partie du trajet, sans dire mot. Je suppose qu'il avait surveillé leurs arrières. Une pénitence, en quelque sorte. Mais lorsqu'elles s'étaient approchées de la civilisation, il avait disparu. Nina l'avait appelé dans la forêt, mais il n'avait pas répondu.

C'était toujours le même problème avec lui, ce type ne répondait jamais !

Je ne parlai pas à Nina de ce que m'avait raconté le type aux lunettes rondes, ni de ce que John avait fait. Il m'avait sans doute dit la vérité, mais cela ne changeait rien. D'ailleurs, Dravecky risquait de recevoir une nouvelle visite bientôt. John n'aurait jamais dû tuer Ferillo. Il avait franchi une vilaine frontière et ne reviendrait plus jamais de notre côté.

Le reste du trajet ne prit que quarante minutes. Pendant tout ce temps, les pieds sur le tableau de bord, Nina contemplait la mer. Son téléphone sonna juste après Nehalem.

— Doug, dit-elle après la communication.

— Alors ?

— Il n'est pas mort.

— Qui ?

— L'héroïque Charles Monroe accomplit des progrès gigantesques, apparemment. Je l'avais mal jugé.

— Non, c'est faux. Son heure n'était pas arrivée, c'est tout.

De toute façon, c'était une bonne nouvelle.

Nina avait été folle de rage en apprenant que Doug traitait avec John derrière son dos, mais ce n'était rien en comparaison des avantages que cela nous procurait. Nous étions déjà blanchis des incidents qui s'étaient produits dans la forêt au nord de Sheffer. Connelly avait tout arrangé avant que d'autres s'en mêlent. Aux yeux de la loi, nous étions partis juste après que le médecin eut soigné mon épaule. Le shérif et son assistant s'étaient rendus seuls dans la forêt. L'un des types de l'hélicoptère était le neveu de Connelly, il tiendrait sa langue. Connelly avait nos armes, qu'il avait confiées à la balistique afin d'élucider la mort des tireurs et les blessures de Paul. Une arme retrouvée dans la voiture des Hommes de Paille relierait sans doute le type aux lunettes à Charles Monroe. Patricia Anders confirmerait le récit de Connelly. Sacrée bonne femme. J'avais comme l'impression qu'elle manigançait quelque chose avec le shérif, quelque chose dont ils assumaient seuls la responsabilité. Je me suis toujours demandé comment il connaissait l'endroit exact. Peu importe. Tout le monde a ses petits secrets.

— Où est Paul ?

— Sous surveillance policière, dans un hôpital de Los Angeles. Les médecins se demandent encore par quel miracle il a survécu.

— Dieu veille sur les enfants, les ivrognes et les cinglés. Nina sourit.

— Je crois que ce qui sauve Monroe, c'est de savoir qu'on a le Garçon de Courses, bien au chaud, en mille morceaux, avec des gardiens armés qui le surveillent en permanence. Charles a fini par résoudre l'affaire et ses problèmes vont être réglés.

— Ce qui signifie que tout ira bien pour toi aussi ?

— On verra...

Sa voix était posée. Je regardai la route puis me tournai vers elle.

— Qu'est-ce qui ne va pas ?

— Rien, en fait. Doug m'a parlé d'une certaine Jean que j'avais interrogée la semaine dernière. Avant-hier soir, elle est allée à une fête dans une grande maison de Mulholland Drive. Elle est à l'hôpital, avec des brûlures de cigarettes et une mâchoire brisée.

À travers le pare-brise, Nina regardait la route.

— Pourquoi sommes-nous comme ça ?

Je ne détenais pas la réponse.

Nous arrivâmes à Cannon Beach juste avant 17 heures. Nous traversâmes lentement la ville, qui n'était guère plus qu'une rangée de maisons de bois, avec une rue principale, un marché, quelques boutiques d'artisanat. Il faisait sombre et il pleuvait toujours, en cette demi-saison tranquille, mais au nord, nous trouvâmes l'hôtel appelé Dune qui arborait une pancarte « chambres libres ». À en juger d'après le parking vide, nous aurions même l'hôtel à nous seuls.

Nous prîmes deux chambres.

La mienne, très vaste, avec une cheminée, se situait au troisième étage. Une baie vitrée donnant sur la mer couvrait tout le mur d'en face. Je ne voyais que du noir, mais je m'installai à la table près de la fenêtre pour boire une bière. Sur un coup de tête, je sortis mon ordinateur portable – le portable de Bobby – et branchai le câble du modem dans la prise du téléphone. J'ouvris mon navigateur et tapai une adresse.

Quelques secondes plus tard, je l'avais à l'écran. Le site de Jessica. Il était toujours actif. Le Webmaster n'avait pas pris la peine de l'effacer. Il ne s'en soucierait peut-être jamais : quelques mégas de plus sur un serveur, ça ne gênait personne. Cela rejoindrait tout le reste, les souvenirs éphémères, les mots et les images du Net. Était-ce l'immortalité ? Non.

La page d'accueil montrait le beau sourire de Jessica. Le lien sur la Webcam était rompu, il restait une autre page, consacrée à ses hobbies, la composition de chanson, ce qui expliquait la guitare et quelques photographies. Une seule la montrait à moitié nue, et je passai rapidement. Les autres étaient parlantes. Des photos d'une jeune femme qui menait sa vie, regardait la télévision, lisait des magazines. La véritable jeune femme qu'elle avait toujours été présente, un peu plus que le corps froid dans l'un des tiroirs de la morgue de Los Angeles. J'avais toujours du mal à me défaire de l'idée que je l'avais vue dans la forêt, mais je savais que ce n'était qu'un vilain tour que m'avait joué mon esprit.

J'usai de mes talents de pirate pour entrer dans le serveur lui-même. J'en recopiai le contenu sur mon propre ordinateur. Pour mettre les données à l'abri, au cas où le type déciderait de tout effacer. Lorsque j'eus terminé, je remarquai la présence d'un fichier texte. Je l'ouvris. Il était court. Quelques entrées de journal, qu'elle avait décidé de ne pas relier au site. Les Feds l'avaient sûrement entre les mains. La dernière entrée datait de trois jours avant sa mort. Elle parlait d'un certain Don, qui semblait bien l'aimer et se demandait si elle ne devait pas l'appeler, un jour.

Je fermai l'ordinateur de Bobby et pensai à lui pendant un instant, dans son petit coin secret, tout au fond de ma tête. C'est là qu'ils vont tous ; dans le cimetière de nos esprits. Là, derrière les yeux, on peut toujours les voir. Ce n'est pas forcément un endroit solitaire. On peut aller leur rendre visite de temps en temps.

Le lendemain matin, je me levai tard. Il avait cessé de pleuvoir, mais le vent s'était levé en force. Derrière ma fenêtre, entre des falaises découpées, s'étendait une longue plage, avec du sable gris, de l'eau grise, un ciel gris.

Un peu plus tard, Nina frappa à ma porte.

— Tu es partant pour une petite promenade ?

— Pourquoi ? Parce qu'il fait un temps magnifique ?

Nous errâmes dans les rues désertes, bûmes un ou deux cafés, passâmes quelques heures sur la plage, seuls au monde, parfois ensemble, parfois chacun de son côté. Nous contemplions les vagues qui battaient les rochers, encouragions les oiseaux intrépides qui voguaient dans le chaos orageux. En milieu d'après-midi, le vent devint si violent qu'on pouvait écarter les bras, se pencher en avant et compter sur lui pour vous redresser. C'est ce que nous fîmes, au milieu des tourbillons de sable.

Il se remit à pleuvoir, nous trouvâmes une sorte d'abri au pied d'un rocher, où nous nous réfugiâmes, à une petite distance l'un de l'autre pour contempler la mer. Je compris alors pourquoi nous sommes si sensibles au son des vagues, à la pluie, au vent dans les arbres. Parce qu'ils ne veulent rien dire. Cela n'a rien à voir avec nous. Les éléments échappent à notre contrôle. Ils nous rappellent un temps où, tout au début de notre vie, nous ne comprenions pas les bruits et nous nous contentions de les accepter. Ils nous soulagent un peu de nos tentatives perpétuelles pour changer le monde par des actes magiques ou des pensées sans fin. Des sons absurdes que nous aimons parce qu'ils nous protègent de l'angoisse de l'action, des supputations, des tentatives d'interprétation.

Pendant une heure, nous restâmes sans rien faire ; deux personnes au bout du monde, auquel elles tournent le dos. À la nuit tombante, nous retournâmes à l'hôtel. Je pris une douche, me changeai puis allai frapper à la porte de Nina.

— Oui ?

— Tu veux aller boire un verre quelque part ?

Elle leva les sourcils.

— C'est quoi ? Un rendez-vous galant ?

— Non, pas du tout.

Quelques rues plus loin, nous trouvâmes un bar appelé Red's Tavern où on servait des bières corsées fabriquées au grenier. Un peu plus tard, il commença à se remplir d'habitués, et un orchestre improvisé se forma au fond de la salle. Quelques guitares, une lap-steel, un violon et un washboard ; les gens jouaient un moment, partaient et revenaient, à leur gré. Sous l'éclairage doux et chaud, je m'aperçus pour la première fois que la femme à côté de moi avait des reflets auburn dans les cheveux. Nous écoutions la musique, battions des mains avec les autres, tout en observant les serveuses qui riaient en remplissant les pintes d'une bière aussi claire que de l'eau de source. Je commandai un plat de chili. Cela n'allait pas si mal.

L'orchestre jouait toujours, mais plus doucement lorsque nous partîmes. Nous rentrâmes à l'hôtel après avoir acheté une bouteille de vin en chemin, allumâmes le feu dans ma chambre et entrouvrîmes la fenêtre pour entendre à la fois le bruit des vagues et le crépitement du feu. Assis par terre, le dos contre le lit, nous discutâmes jusque très tard dans la nuit sans nous en apercevoir.

Nous alimentions le feu sans cesse car nous ne voulions pas qu'il s'éteigne, mais à la fin il faisait noir dans la chambre et assez chaud pour que nous n'ayons plus besoin de mots.

Elle fit le premier pas.

Elle est comme ça.

LE SANG DES ANGES

Traduit de l'anglais
par Jean-Pascal Bernard

À Ralph Vicinanza

Avec quelle rapidité le temps effacera tout cela !
Et combien de choses n'a-t-il pas déjà effacées !

Marc AURÈLE,
Pensées pour moi-même

Key West

Ils sont venus le chercher pendant son travail. Ils sont venus par un bel après-midi, très chaud, quand les affaires allaient bon train et que Jim se donnait encore une heure, maximum, avant de fermer boutique pour commencer la soirée tôt. Le quai était inondé de touristes, de toutes formes et de toutes tailles, voguant tel un banc de poissons bariolés sans destination précise. En mangeant. En mastiquant. En dévorant tout ce que les cafés et les marchands ambulants de Key West avaient à offrir, des hamburgers et burritos jusqu'aux crèmes glacées, sans oublier les churros nappés de sucre et brûlants d'huile de friture. Ceux qui avaient encore une main libre n'omettaient pas de boire, sirotant cafés frappés, thés glacés, ou des sodas si volumineux qu'on aurait pu y plonger des bébés nageurs. Il était à peine 15 heures. La nourriture engloutie ne pouvait tenir lieu de déjeuner tardif ni de dîner précoce. Jim se doutait que tous ces gens s'étaient restaurés à la mi-journée et qu'ils se poseraient à nouveau sur le coup de 19 heures, pour passer aux linguinis, à l'espadon grillé et aux burgers plus sophistiqués, descendus à grandes lampées de chardonnay glacé. Dans l'intervalle, elles broutaient, ces sauterelles géantes et affables, ces heureuses vaches dans un immense pâturage bon marché. Leur dévotion laissait rêveur. Avec un peu d'imagination, on pouvait voir dans ces corps de simples véhicules pour systèmes digestifs en goguette, surmontés de dents faucheuses. À se demander ce qu'il adviendrait si le stock de bouffe venait à s'épuiser ; à se demander si, passée la surprise, les têtes hébergeant ces gueules voraces n'allaient pas se mettre à scruter les gens alentour, à y voir leur nouveau gibier.

C'est une question que Jim se posait sans arrêt, malgré lui. Il était appuyé au garde-fou, du côté nord de Mallory Square, une large promenade en terrasse qui reliait à la mer les hôtels et restaurants de cette partie de l'île. Les paquebots y mouillaient la nuit, ces monstres multiniveaux qui tuaient l'hôtellerie et déversaient leurs troupeaux de ruminants à cartes bancaires. Dans le dos de Jim se levait, depuis les eaux peu profondes, une brise faible mais bienvenue. Il tenait un appareil photo dans sa main droite. Le sac à son épaule renfermait des pellicules Polaroïd et une boîte cartonnée contenant les pochettes de présentation qu'il avait fait réaliser.

Jim Westlake photographiait les touristes depuis plusieurs années maintenant. Il était censé détenir une licence officielle, mais il ne se l'était jamais procurée, et visiblement cela ne posait aucun problème. Il ne dérangeait pas les gens à table, ne remontait pas la promenade en criant ou en sautant au visage des badauds, toutes dents dehors. Il n'avait jamais eu l'âme commerçante et n'inspirait aucune méfiance. Il avait soixante et un ans ; ses joues s'affaissaient peut-être un peu, mais il demeurait grand et large d'épaules. Il était vêtu d'un pantalon bleu ciel et d'une chemisette blanche, qu'il détestait, mais qui correspondait au rôle. Ses cheveux étaient en majorité gris, plaqués en arrière, et il portait des lunettes noires pour contrer l'éblouissement du quai. Il passait ses journées à tourner dans Mallory ou à arpenter Duval Street et, lorsqu'il repérait le type de touristes idéal, il proposait simplement de les prendre en photo. Beaucoup secouaient la tête, contrariés à l'idée d'une dépense imprévue, d'autres allaient leur chemin comme s'il n'existait pas. Quelques-uns s'arrêtaient, réfléchissaient un instant, se disaient : allez quoi, c'est les vacances ! À ce stade, ils avaient déjà visité la maison d'Hemingway, foulé le point le plus au sud et admiré les récifs à bord d'un bateau à coque vitrée. Bref, ils passaient une chouette journée, et c'est à cela que servent les photos : à prouver aux autres comme à soi-même qu'on a pris du bon temps, quelques coups de soleil, et oublié le turbin quotidien.

Le plus souvent, ils avaient leur propre appareil autour du cou, parfois même un numérique, et Jim n'ignorait pas que sa profession était condamnée. Grâce à lui on pouvait toutefois se faire prendre ensemble sans solliciter d'autres promeneurs, et puis Jim n'avait pas son pareil pour arracher un sourire aux enfants. Le cliché était ensuite inséré dans une carte au grammage épais et à la police classieuse, dont le rabat indiquait : « On tombe la veste à Key West. » Le parfait cadeau à envoyer à sa maman : personnalisé, sans exiger le moindre effort. Les touristes pouvaient constater que le photographe se donnait du mal, lui : quand la première pose lui déplaisait, il recommençait sans facturer de supplément. Quinze dollars, ce n'était pas donné, mais

Jim s'était aperçu qu'il obtenait plus de clients à ce tarif-là qu'à cinq dollars. Cinq billets, c'était juste un type qui prenait une photo. Quinze, c'était un véritable souvenir.

Cette activité ne l'enrichirait pas, mais Jim ne voulait pas être riche. Les gros chiffres n'avaient pas d'intérêt à ses yeux. Il s'en sortait pas mal, et par définition le « pas mal » est suffisant. C'est de cette façon qu'il envisageait de finir ses jours, en se débrouillant. Mais quand il vit les deux types remonter les planches sans la moindre barquette de frites à la main, son petit doigt lui dit qu'ils étaient là pour lui.

Le premier paraissait la quarantaine, à un ou deux ans près, le second une petite vingtaine d'années. Sveltes, en forme. Le plus jeune portait un tee-shirt noir et un treillis kaki, son visage affichait la certitude d'être une force avec laquelle il fallait compter. Le plus vieux arborait un costume anthracite sur une chemise blanche. Il semblait à l'aise malgré la chaleur et avait l'air de se foutre qu'on le prenne au sérieux ou non.

Le jeune s'approcha le premier. Jim sourit, brandit le Polaroïd.

— Une petite photo ?

Les habitants de Key West croisaient beaucoup de couples homos.

Le jeune se contenta de le toiser. Il mesurait dix ou douze centimètres de moins que Jim. Il y avait de l'évaluation dans l'air, mais la finalité de cet examen restait obscure.

Le gamin finit par lâcher :

— C'est vous, James Kyle ?

Jim secoua la tête.

— Tu t'es trompé de bonhomme, fiston. Moi, c'est Westlake. Désolé.

Le jeune ne cilla pas.

Jim la joua penaud :

— Vous êtes là pour la licence ? Je m'étais dit que ça n'avait plus trop d'importance, de nos jours.

D'un signe de tête, il montra le petit groupe de portraitistes agglutinés en bas de Duval, ceux qui promettaient aux clampins le physique de Brad Pitt ou de Dolly Parton.

— Des tas de gens s'installent et tentent leur chance. Mais si ça pose vraiment problème, dans ce cas, je serai ravi de...

Il se tut, laissant un blanc. Le blanc s'étira, sans que personne vienne le combler. Le jeune restait planté devant lui, avec son air indéchiffrable. Une vieille cicatrice d'environ deux centimètres soulignait sa pommette droite. Jim se demanda d'où elle provenait, tout en ayant une pensée reconnaissante pour son auteur.

— Écoute, fiston... Tu cherches quoi, au juste ?

Le jeune homme se détourna.

— J'en reviens pas que ce soit lui.

Jim comprit que ces mots s'adressaient à l'autre, qui venait de surgir à son côté. Pourquoi ne l'avait-il pas senti se rapprocher ? Comme Jim se trouvait vieux, d'un coup.

— C'est bien lui, répondit le type. Tu te souviens de moi, James ?

Jim pivota en prenant son temps. Oui, bien sûr qu'il se souvenait. Ça faisait un bail, et l'homme avait pris de l'âge, mais juste à l'extérieur. Tout était dans les yeux, des yeux qui semblaient normaux de prime abord, mais qui se révélaient dénués de chaleur, d'émotion, de vie. Jim lui aussi avait été un animal froid en son temps, aucun doute là-dessus, mais ce type l'était depuis toujours : comme s'il était sorti du ventre maternel la tête pleine de mauvaises pensées. Si Jim ignorait son nom, il savait qui c'était. Un homme que Jim avait espéré – et cru, à la longue – ne plus jamais revoir.

C'était le Garçon précoce.

— Oui, je me souviens. Qu'est-ce que tu veux ?

— J'ai du boulot pour toi.

— J'ai déjà ce qu'il faut.

— On a passé un accord, James.

— Ça remonte à loin. J'ai fait ce que tu voulais et tu ne m'as plus rien demandé. Je pensais que c'était fini.

— Faux. Tu savais que je reviendrais.

— On n'a qu'à dire que j'oublie tout. Que je m'en vais sans faire d'histoires.

— Dans ce cas, tu vas passer la nuit en taule, si t'es pas mort avant.

Jim regarda au loin, par-delà le quai. Des oiseaux marins tournoyaient dans le ciel. L'un des bateaux à coque vitrée klaxonnait dans le port, à une cinquantaine de mètres, les hublots embrasés de soleil. Des gens attendaient de pouvoir monter, d'autres de pouvoir descendre. Beaucoup léchaient une glace. Toujours la même rengaine, mais maintenant c'était différent. Malgré la sueur dans sa nuque, Jim était frigorifié.

Lorsqu'il leva le regard, le jeune le fixait toujours de ses yeux bleu clair, et il se produisait quelque chose à la commissure de ses lèvres. Ses muscles frémissaient, comme s'il réprimait un sourire.

— Qu'est-ce que vous attendez de moi ? demanda Jim d'une voix morne.

Le plus âgé plongea la main dans sa poche et en tira une enveloppe matelassée. Jim s'en empara.

— Tu as trois jours pour prendre ton poste. Tu recevras des instructions détaillées.

Jim rangea l'enveloppe dans son sac.

— Pourquoi moi ?

— Parce que tu nous es redevable, et parce que je sais que je peux compter sur toi. T'inquiète pas, ça va te plaire.

— Je ne fais plus ce genre de truc.

L'homme le regarda comme le dernier des imbéciles. Il avait raccroché ? Plus maintenant. Il était prêt à rempiler. Le gamin le dévisageait toujours. Avec ce petit rictus sur le point de lâcher.

Prenant ses ennemis de vitesse, Jim cala sa main gauche contre les lombaires du gamin et, d'un coup trop ramassé pour attirer l'attention des passants – quand bien même il eût placé son dos en paravent –, il lui planta son poing droit dans le ventre, tel un piston expulsé par une charge explosive. L'effet revenait à écraser le rein entre deux briques. L'expression du gamin valait le détour.

— Ravale ça, murmura Jim à l'oreille du gosse. Si t'es aussi fort que tu le crois, ça devrait pas être trop dur.

Il lui tapota l'épaule avant de reculer d'un pas, sans se presser, ravi de voir le gamin blêmir, les tendons de son cou se raidir comme des cordes d'amarrage. Ses lèvres ne seraient plus seules à frémir. Le garçon finit par émettre un son, celui d'un mourant cherchant son dernier souffle.

Jim se tourna vers l'autre type, qui affichait un air impavide, pour ne pas dire amusé.

— Je vais le faire, déclara Jim. Mais après c'est terminé.

Et sur ces mots, il s'en alla.

La maison de Jim se trouvait plus au nord, à quarante minutes à pied, dans les broussailles d'une petite île que personne ne remarquait en quittant Key West. Un petit pavillon sans piscine, qui n'offrait même pas de vue exceptionnelle, si ce n'est depuis tel point précis du porche où, en position assise, on devinait l'océan à travers les arbres – à certaines périodes de l'année.

Son fauteuil placé dans le bon axe, Jim se désaltérait d'un verre de thé glacé garni de feuilles de menthe. Sa voisine Carol fit une arrivée mouvementée avec ses filles ; elle le salua d'un geste, derrière les dix mètres qui séparaient leurs pelouses respectives, surprise de n'obtenir aucune réponse. Tout en restant réservé, M. Westlake était d'une politesse sans faille, jamais avare de sourires pour les deux petites chéries. Il les avait même photographiées, et Carol conservait l'adorable cliché sur sa table de chevet. Peut-être avait-il appris une mauvaise nouvelle, songea-t-elle ; il était âgé, après tout, et cette enveloppe dans ces mains... Elle se promit de venir lui faire coucou plus tard, sous un prétexte ou un autre, puis Amy et Britney commencèrent à se chamailler, et elle mit toute son énergie à les ramener à l'intérieur, là où les attendait un comique audiovisuel (la télévision, cet objet béni, l'infatigable providence des mères) qui lui permettrait de se concentrer sur sa tâche prioritaire par cette chaleur : retarder le moment du premier verre de vin.

La moustiquaire claqua derrière la petite famille, arrachant Jim à sa rêverie. Il considéra l'enveloppe entre ses mains, sans éprouver le besoin d'en réexaminer le contenu. Elle renfermait un téléphone portable, et un bout de papier où figuraient un nom de ville et deux lignes de consignes.

Malgré lui, une partie de son esprit entreprenait déjà le voyage. Il maudissait cet homme pour avoir su que ça se passerait comme ça, que Jim n'oserait pas jeter l'enveloppe et prendre la tangente. Si les autres nous connaissent mieux qu'on ne se connaît soi-même, que devient-on ? Il chercha l'océan dans le trou du feuillage, mais la vue n'était d'aucun secours. S'était-il vraiment mis à croire que c'était fini, qu'une décennie avait en quelque sorte dissous le passé, l'avait réduit à néant ? Si oui, Jim était un idiot.

Il avala le reste de thé et retourna dans la maison, où il lava le verre et le mit à sécher sur le bord de l'évier. Il possédait en tout et pour tout un verre, un jeu de couverts, un bol à soupe et une assiette. Il ne lui avait jamais fallu davantage. Le mobilier et la déco suffisaient juste à ne pas intriguer les rares visiteurs. Ce dépouillement datait de son arrivée dans les Keys, à l'époque où il risquait de devoir fuir à tout moment. Puis, au fil des huit années suivantes, à mesure que son âme s'apaisait et qu'il trouvait un équilibre, il s'était aperçu que ce mode de vie lui plaisait.

Pourquoi avoir quelque chose en double quand un seul exemplaire convient ?

Pourquoi posséder quelque chose quand ce n'est pas nécessaire ?

Il gravit l'escalier jusqu'à sa chambre, où une petite valise se tenait prête, contenant ses vêtements et son appareil. Restait un espace libre d'environ dix-huit centimètres sur vingt-cinq. Jim se rendit devant le placard vide, s'accroupit. Ses genoux lui dirent tout le mal qu'ils pensaient de ce mouvement, mais ils lui demeuraient fidèles. Comme le petit con en treillis l'avait découvert, le corps de Jim défiait le temps.

Il souleva le tapis du placard, puis la planche. On faisait moins bateau comme cachette, mais, s'il avait redouté quoi que ce soit, il n'aurait rien planqué à cet endroit. Le plancher abritait une boîte à chaussures. Il l'exhuma, avant de replacer lattes et tapis.

Il glissa la boîte dans sa valise, verrouilla le bagage et quitta la chambre sans se retourner. Il en avait envie, mais résista. James Kyle n'avait pas à voir cet endroit. Lequel appartenait à Jim Westlake.

En bas, il s'assura que les fenêtres étaient fermées, et la porte du fond verrouillée. Il descendit l'allée jusqu'à sa voiture blanche immaculée et cala sa valise à l'arrière.

Il resta prostré derrière le volant, les yeux rivés sur son petit pavillon. Et s'il demandait à Jim de l'attendre sagement à l'intérieur ? Il s'acquitterait de sa mission en tant que James, puis reprendrait le cours de sa vie. Ce maudit après-midi avait peut-être une seule et unique cause : un certain bien appartenant à James, un objet qu'il aurait dû bazarder de longue date. Une petite casserole cabossée. Pas grand-chose à première vue, mais...

Celui qui ne veut rien *être* ne doit rien *posséder*. Il le savait. Depuis longtemps. Et pourtant... il l'avait gardée. Incapable de s'en débarrasser.

Voilà pourquoi Jim était un si bon photographe.

Il connaissait le sens du mot souvenir.

Il se résolut à démarrer. S'engageant sur la route, il avisa la voisine à la fenêtre de sa cuisine. Il la salua du geste habituel.

Carol sourit et lui rendit la politesse, heureuse que le vieux bougre eût retrouvé son allant, loin d'imaginer que ce n'était pas Jim Westlake qu'elle regardait ainsi mais un inconnu du nom de James Kyle ; un homme en partance pour le passé, en route vers l'enfer.

PREMIÈRE PARTIE

LES RIENS

Il existe une allégeance entre les morts et les êtres à naître, dont nous autres vivants ne sommes que la ligature.

Richard POGUE HARRISON

CHAPITRE 1

Je m'appelle Ward Hopkins, et une partie de cette histoire m'est arrivée. J'y étais, et les événements ont fendu l'air devant mon visage telle une volée d'oiseaux fuyant en ordre dispersé. Me serais-je conduit autrement, les choses auraient pu prendre un autre tour. Meilleur. Pire. Je n'en sais rien. Je décide de croire au libre arbitre, mais aussi à l'idée que nos mouvements ont un rayon d'action restreint : que des arcs prédéterminés nous guident dans le chaos de la vie, que des forces invisibles se dressent sur notre route. Tous, nous nous sauvons, nous nous cachons, nous traversons des nuits d'insomnie, nerveux et troublés, amoindris par les ombres qui hantent nos existences.

Le temps est un lac qui se creuse d'année en année, goutte après goutte. La tension superficielle, la présence électrique de nos actes saccadés nous permettent de courir en surface, à la manière de puces d'eau, insouciants des profondeurs qui nous portent. Nous sommes en sécurité, à flot dans le présent, puis nos battements cessent et nous sombrons lentement dans le passé. Là seulement, nous mesurons à quel point comptaient tous ces hier, comment ils portent au soleil chaque instant du présent ; et combien de personnes nous laissons derrière nous, gisant dans le temps tels des insectes desséchés. Toutefois, de loin en loin, une chose s'agitera dans les profondeurs du lac, une chose enfouie dans le passé et néanmoins toujours vivante.

Il arrive cependant que les créatures du dessous fusent de leur repaire pour crever la surface. Elles nous assaillent dans la nuit comme le sifflet strident d'un train fou, filant dans le noir des collines vers une destination qu'on ne verra jamais, même si le monde entier l'entend se crasher en gare.

Une partie de cette histoire m'est arrivée, mais une partie seulement.

Je vais vous dire ce que nous savons.

Le premier mail nous parvint dans les derniers jours de l'été. Nina et moi avions passé l'après-midi à Sheffer, la solution shopping la plus proche de la maison où nous vivions. Minuscule ville de la chaîne des Cascades, dans l'État de Washington, Sheffer possède une grand-rue bordée de façades en bois, croisée par cinq artères qui se dissolvent dans les sapins sans cérémonies ni regrets. On y trouve un marché, une brasserie qui sert une bonne carte, des livres et des CD d'occasion, des babioles. Il y a une pharmacie, un droguiste-caviste et un salon où les dames se font couper les cheveux comme au temps de Jimmy Carter. La ville propose en outre deux bed-and-breakfast haut de gamme et trois bars, ainsi qu'un motel judicieusement placé au bout de la grand-rue, au cas où l'on aurait perdu la notion du temps ou trop picolé pour reprendre le volant. En ce qui nous concerne, nous avions testé les deux. Il existe aussi un petit musée du Chemin de fer, un commissariat dirigé par un brave homme, et c'est à peu près tout. C'est un coin correct, et on y croise des gens sympas, guère plus qu'un gros nid-de-poule dans la chaussée.

Notre résidence provisoire était encore plus petite, un bungalow de rondins issu d'un village à l'ancienne sur la côte de l'Oregon. À la fin des années 1990, un couple de retraités de Portland acheta trois de ces baraques, les chargea sur des camions et les installa sur un terrain de seize hectares en bout de forêt, à une demi-heure au nord-est de Sheffer. Le mari était mort peu après, mais Patricia se maintenait. Parfois elle me rappelait ma mère, et si Beth Hopkins avait toujours été de ce monde, bien malin qui aurait su prédire l'issue de leur bras de fer. Patricia nous avait proposé l'un de ses bungalows après que nous l'eûmes tirée d'un mauvais pas. Après réflexion, nous avions pris quelques dispositions et accepté son offre.

Adossée à une réserve naturelle, la propriété de Patricia possédait un vaste étang. Quand le regard s'y perdait par un après-midi d'automne, on imaginait sans regret que l'humanité n'avait jamais existé. Notre bungalow se trouvait en périphérie, à un petit kilomètre de la route. Il se composait d'un séjour avec cheminée et coin cuisine, d'une salle de bains et d'une chambre. C'était largement assez grand. Ma vie s'était condensée au point que l'ensemble de mes biens tenait dans le coffre d'une voiture moyenne. Ça aussi, nous en avions une. Elle appartenait à la femme qui me fut d'abord présentée comme l'agent spécial Baynam.

Nina. Elle se trouvait présentement sous le porche, à l'avant du bungalow. L'air était froid sans être glacial, empreint d'une modération promettant que l'hiver allait attendre. Nina était censée admirer le coucher du soleil, mais je savais qu'elle n'en faisait rien. Il ne suffit pas de braquer les yeux dans la bonne direction. Le soleil lui savait sans doute gré de ce répit ; essayer de franchir l'horizon avec grâce sous le regard de Nina, c'est une grosse pression, même pour un corps céleste.

J'étais dans le coin cuisine, à confectionner une salade, et j'en faisais tout un plat au sens propre comme au figuré. Nina était mutique depuis le matin, mutique à la façon d'un rocher gisant à mi-flanc de coteau. Quand je lui demandais si ça allait, j'obtenais des oui peu convaincants mais non négociables. J'ignore pourquoi les femmes font ça, mais je sais qu'on ne peut rien en tirer tant qu'elles ne sont pas décidées à parler. Nous allions au-devant d'une conversation – une semaine que ça couvait – et je n'étais pas tellement pressé. En conséquence de quoi la salade prenait des proportions baroques. L'esthétique culinaire était loin derrière nous, et le but semblait désormais de débarrasser le comptoir en regroupant toutes les crudités dans un unique saladier. J'avais poussé l'effort jusqu'à faire bouillir des haricots verts, qui refroidissaient dans un bol d'eau glacée au milieu de l'évier.

Pour tuer le temps, je me suis aventuré au living et j'ai ouvert l'ordinateur portable de Nina. J'en avais un à moi, mais ce n'était pas vraiment le mien et il était caché dans les combles du bungalow. Son contenu avait été copié, crypté et stocké sur un serveur lointain. Mais la bécane contenait les toutes premières versions des fichiers, et mon esprit leur conférait une sorte de primauté. Étrange, cette façon dont le cerveau humain accorde statut et antériorité même à des données numériques, à des électrons qui peuvent être partout à la fois, autrement dit nulle part. Peut-être a-t-on besoin de croire que toute chose a un commencement. Comment pourraient-elles cesser, sinon ?

J'interrogeais ma messagerie une fois toutes les deux semaines, et encore. Je n'avais plus beaucoup de liens avec le monde extérieur. Le seul type qui m'écrivait régulièrement était mort. C'est son portable que je gardais planqué sous le toit. Les seuls mails que je recevais dorénavant me proposaient tour à tour de durcir ou de rallonger mon pénis, de souscrire des prêts étudiant ou de regarder quelque starlette se faire mousser dans des films amateurs à l'éclairage étonnamment professionnel. L'anonymat de ces invitations les rendait plus vides encore que le silence. Peut-être était-ce là mon attente : un regain de calme, un rab de bruit blanc – et à travers eux, la promesse que ce truc appelé la vie allait encore durer un temps.

Alors, quand j'ai vu que j'avais un message, et qu'il me semblait personnellement adressé, je me suis comme qui dirait figé.

L'objet indiquait juste : « WARD HOPKINS ? »

Je ne reconnaissais pas l'émetteur. Inscrit chez Hotmail, le repaire préféré des spammeurs – à défaut d'être le seul. C'était arrivé à l'une de mes vieilles adresses, qui n'avait pas servi depuis deux ou trois ans. Je doutais qu'un tel message puisse de près ou de loin concerner ma vie actuelle.

Je l'ouvris. Une seule et unique phrase :

Il faut que je vous parle.

Il ne faut rien du tout, pensai-je. Bye-bye.

Je fis glisser le message vers la corbeille, avant d'hésiter. Ne devais-je pas au moins enregistrer l'adresse d'expédition ? Effacement ne vaut pas négation, j'en savais quelque chose.

Un craquement dehors ; en me retournant, je vis Nina s'approcher de la porte. Elle portait un jean noir et un gros blouson marron que je lui avais acheté à Yakima deux mois plus tôt. Elle avait une allure certaine, mais ronchonne.

Je rabattis l'écran et regagnai la cuisine.

— J'ai maté ce foutu coucher de soleil aussi longtemps que je pouvais, Ward. Elle arrive, cette bouffe ?

— Y en a pour une seconde.

Je sortis les haricots de l'eau, les secouai vite fait et les répartis artistiquement sur les autres éléments du saladier. Nina observait en silence ; le résultat semblait lui inspirer une profonde perplexité.

— Voilà, lançai-je. La salade du siècle.

— Au moins. Et si on ajoutait quelques pommes de pin ? Un couple d'écureuils, pour un tableau vivant. Ou carrément un arbre entier. Je te ramène ce que tu veux. Tes désirs sont des ordres, maestro.

— Assez, assez. Je te jure, je ne fais pas ça pour les remerciements. La joie sur ton visage vaut toutes les récompenses.

Elle sourit, un peu.

— T'es grave.

— Peut-être. Allez, goûte quand même. T'as pas le choix, de toute façon. Y a rien d'autre. J'ai mis tout ce qu'il restait.

Elle se fendit d'un sourire plus franc. Versa une cuillerée de salade dans une assiette, puis une deuxième en gage de bonne volonté. M'embrassa sur la joue, avant de retrouver le porche avec son assiette et la bouteille de vin.
Je l'ai suivie.

Nous avons entamé nos assiettes.
L'air demeurait doux mais se chargeait de lames fraîches venues des sommets montagneux. Ce n'était pas une soirée à salade. Au bout d'une dizaine de minutes, Nina reposa sa fourchette selon un angle qui disait : « Ça suffira comme ça. » Elle n'avait presque rien avalé.
— Désolée, dit-elle en voyant mon regard. Tu t'es donné du mal.
— Beaucoup trop. Ça craint. C'est la Salade de la Honte. Je savais bien qu'il fallait juste prendre une grande boîte de poulet frit chez Izzy.
— Peut-être bien.
— C'est très clair. Tu devrais m'écouter pour ce genre de trucs. Je suis le grand maître de la malbouffe. C'est un don. N'importe quel jour de l'année, je sais prédire les meilleures cochonneries à grailler – non seulement pour moi, mais pour toute la tribu. En d'autres temps, j'aurais été le chaman du snack. J'aurais consulté des osselets, déchiffré des présages dans le ciel, avant de déclarer : « OK, les gars, tout à l'heure vous aurez des envies de tacos, alors tâchez de chopper un mammouth quand vous irez en ville. » Et j'aurais eu tout bon.
Nina m'observa.
— T'es encore en train de parler ?
— Ce devait être le vent.
Le lac revêtait son habit crépusculaire, noir et moiré. Nina le considéra un moment, puis enfin elle parla :
— Qu'est-ce qu'on va faire ?
Et voilà. Ce que j'appréhendais n'était pas une conversation, mais juste ça : une question. *La* question.
J'allumai une cigarette.
— Qu'est-ce que t'aimerais faire, toi ?
— Le problème n'est pas là et tu le sais bien. Le problème, c'est que... ça ne peut pas durer. Ce n'est pas une vie.
— Vraiment ?
— Pas dans ce sens-là. Tu m'as comprise. Les circonstances, le fait de ne pas avoir le choix...

Je lui pris la main. L'été nous avait souri, envers et contre tout. Nous avions à peine quitté la région de Sheffer. Fait connaissance avec quelques figures du cru, tout en gardant profil bas. L'un avec l'autre, aussi : Nina et moi étions ensemble depuis une petite semaine lorsque nous avions accepté le bungalow de Patricia – même si nos existences étaient déjà liées depuis six mois. Elles l'étaient, en gros, depuis que mes parents avaient trouvé la mort – assassinés, apprit-on par la suite. De mon passé, Nina savait l'essentiel. Moi aussi, je savais certaines choses à son sujet. Plus que quiconque, à mon avis, y compris un certain John Zandt, qui fut notre ami avant de tourner casaque.

Chacun de nous trois détenait des secrets qu'il eût mieux valu ne pas connaître. Voilà pourquoi nous vivions ainsi.

Nous faisions de longues, très longues promenades dans les bois, cuisions des aliments sains sur mon barbecue de pierres plates. Nina s'était marrée devant mon œuvre et son style « campagnard à deux balles ». Le système fonctionnait, pourtant. Entre la marche à pied et les travaux manuels auprès de Patricia et des autres, mon corps rajeunissait de plusieurs années. Ma blessure à l'épaule, contractée cinq mois plus tôt, ne me gênait plus. Si j'avais perdu quelques kilos, Nina en avait pris un ou deux et elle les portait bien. On s'était également offert quelques virées en voiture, vers l'est ou le sud, plus ou moins au hasard. Comme deux fugitifs. Nous avions rallié Los Angeles à deux reprises, dans des circonstances tendues, la seconde fois pour récupérer quelques affaires dans la maison de Nina, une construction précaire nichée sur le versant le moins prisé des collines de Malibu. Nina l'aurait bien mise en vente – à supposer qu'il y eût des candidats – mais nous ne pouvions nous attarder dans le secteur, alors nous l'avions bouclée et laissée en l'état.

Ce qui, d'une certaine façon, motivait la question de Nina. Au fond, tout était comme avant, en dépit de mes efforts pour affirmer le contraire. Nous avions barricadé nos fenêtres mais le monde continuait de tourner. Nina avait un boulot, pour commencer. Un boulot sérieux. Le congé qu'elle avait négocié grinçait à force de s'étirer : elle allait devoir démissionner, ou reprendre. Ma situation personnelle était plus abstraite. J'avais travaillé pour la CIA quelques années auparavant, comme spécialiste en surveillance média. Puis on m'avait laissé partir. En fait, ainsi qu'aimait à le rappeler mon ami Bobby, j'avais repris mes billes – juste avant de devoir passer au détecteur de mensonges.

Nous avons tourné autour du pot pendant un moment, mais nous sentions la parenthèse se refermer, ce qui nous déplaisait. Comme si un géant nous avait accueillis dans la chaleur de sa paume géante. À présent, nous sentions la main redescendre pour nous poser à terre.

— On prendra une décision, dis-je. Mais ce soir, la vie pourrait être plus moche. Je suis assis avec toi, et je n'ai pas l'impression de manquer de grand-chose.

— Normal, avec une merveille comme moi : belle, intelligente, facile à vivre... La compagne idéale à tous points de vue.

Je levai un sourcil.

— Je n'irais pas jusque-là.

— Ah bon ? Trouve-moi un seul défaut.

— Eh bien, t'as flingué ton dernier petit ami en date.

— C'était un accident.

— Ouais, c'est ce qu'on dit.

— Parce que c'est vrai. (Un clin d'œil.) Mais la prochaine fois, ce sera délibéré.

J'ai rigolé, et la question s'est provisoirement retranchée sous sa pierre. Nous sommes restés sur le porche, à bavarder en regardant le lac se fondre, puis le froid nous a repoussés à l'intérieur. Un peu plus tard, dans notre lit, j'ai écouté les arbres et la respiration régulière de Nina, jusqu'à ce que les sons se mélangent et que je sombre à mon tour.

Cinq mois, cela peut sembler court, mais cette relation était l'une des plus longues que j'aie jamais connues. Elle gardait comme un goût de miracle. Je n'ai rien dit à propos du mail, ni du suivant – même objet, même message –, qui me parvint le surlendemain. J'ai effacé le premier et enfoui le second au fin fond d'un dossier. La simplicité ne fut jamais le trait dominant de ma vie, pas plus que le sentiment d'avoir un chez-soi. Je tâchais de retenir l'un et l'autre par le seul moyen à ma disposition : en me planquant.

Pendant quelques jours, cela suffit. Ce fut uniquement lorsqu'une voiture monta vers le bungalow de Patricia que je me rendis à l'évidence : rien – ni sourire, ni pieux mensonge, ni salade hypertrophiée – n'empêcherait le monde de venir nous chercher.

CHAPITRE 2

Nous avons reçu l'appel peu après 11 heures du matin. Je me trouvais au lac, à jouer avec de vieux morceaux de poutres récupérés un après-midi où j'avais aidé à retaper une grange. Je caressais l'idée de fabriquer une table rustique, pour nous éviter de descendre celle du porche quand nous voulions manger au bord de l'eau. Même si, en mon for intérieur, je doutais qu'il nous reste beaucoup d'occasions de dîner dehors, ou qu'un objet de ma fabrication puisse supporter de lourdes charges, par exemple un verre de vin. J'ai entendu au loin les trois petits bips du cellulaire de Nina, sans y prêter plus d'attention. Nous n'avions pas le téléphone au bungalow. Si le voisinage souhaitait nous joindre, il composait le numéro de Nina. Je poursuivis mon remue-ménage, cherchant à assembler mes bouts de bois, jusqu'à ce que j'aperçoive Nina sous le porche, livide.

— C'est Patricia ! cria-t-elle. On a de la visite !

Je lâchai mes outils et remontai d'un pas vif. L'esprit alerte, mais la poitrine froide.

— Combien ?

— Un seul véhicule. Pour l'instant.

Pendant que Nina filait à l'intérieur, je glissai le bras sous le porche pour sortir une chose longue et lourde, enveloppée d'un épais plastique. Je chargeai le fusil en vitesse, guettant le moindre bruit de l'autre côté du lac.

Je rentrai à mon tour. Nina avait posé nos deux flingues sur la table.

— Toujours rien ?

— Non, répondis-je en empochant mon pistolet.

J'embrassai Nina et m'éclipsai. Me figeant une seconde, je regardai derrière l'étang. Une voiture s'était arrêtée devant le bungalow de Patricia. Elle ne pouvait aller plus loin. Toute personne désireuse de nous rendre visite devait parcourir les deux cents derniers mètres à pied, et il n'y avait que deux itinéraires possibles : suivre le sentier qui contournait le lac ou couper à travers bois pour atteindre l'arrière du bungalow. Les deux parcours étaient placés sous ma responsabilité.

J'attrapai le fusil et fis le tour du bungalow pour gagner une étroite corniche camouflée par les arbres. Restant baissé, je longeai le lac, d'environ un quart de tour. J'avais repéré un point d'observation sur le chemin situé trente mètres plus bas, sans qu'on puisse me voir dans l'autre sens. Je m'accroupis derrière le tronc le plus large, pris mes appuis. Positionnai le fusil, le calai contre mon épaule. J'avais passé des heures à m'entraîner au fond de la propriété. À cette distance, j'étais sûr d'atteindre ma cible, et archisûr d'en avoir envie.

Une voiture. Quatre hommes, probablement. À supposer qu'il n'y eût pas de second véhicule posté en retrait. Ni d'autres hommes lancés à nos trousses dans la forêt. Auquel cas j'entendrais les tirs de Nina avant qu'elle n'entende les miens.

Au bout de six minutes, je perçus, en bas à gauche, des bruits dans l'air immaculé, un crépitement de semelles sur les feuilles sèches du sentier. J'attendis, enjoignant calme et silence à mon cœur, évitant d'imaginer les conséquences d'une fusillade. Où nous irions, ce que nous ferions. Ou de me demander si la question se poserait, si nous n'allions pas nous dissoudre sur place, dans ce sol vierge, rouges la première heure, bruns les quelques jours suivants, puis irrémédiablement mêlés aux autres boues et poussières. Cette dernière pensée fut la plus facile à chasser. Je n'ai aucune intention de mourir. Ni maintenant, ni jamais. Je n'en vois pas la nécessité. Et, tôt ou tard, le monde devra bien se ranger à mon point de vue.

Encore deux minutes, puis un mouvement furtif au pied de la pente. Juste assez pour voir qu'il y avait deux hommes, trois au maximum. Soit c'était bon signe, soit ils avaient des renforts quelque part. Me renversant en arrière, je scrutai la corniche sur ma gauche. Personne. Alors je repris ma position, leur donnant vingt secondes pour atteindre ma ligne de mire suivante. Ils en mirent plutôt quarante, ce qui me laissa songeur. Ils n'étaient guère pressés. Comment interpréter cette approche mesurée ? On ne traite pas de la même façon un ennemi prudent. Et puis je ne savais toujours pas combien ils étaient, car, au moment où ils traversèrent ma deuxième fenêtre d'observation, un rayon de soleil ricocha sur l'étang, qui me noya la rétine de blanc et de jaune.

Ils disparurent à nouveau. Il me restait une lucarne, la dernière. Deux ou trois types, cela revenait à peu près au même.

Un tir surprise par-derrière pouvait m'offrir deux secondes de rab, sachant que l'écho de la forêt dissimulerait ma position.

Je posai un genou à terre et visai. Des vêtements sombres dansaient derrière le feuillage, cinq mètres avant le point d'impact. Trente secondes s'écoulèrent.

Puis j'entendis quelque chose.

Je me demandai quoi, avant de reconnaître le portable de Nina.

Je me pétrifiai. Grave erreur de la part de Patricia. Très grave erreur. Elle n'avait pas à venir aux nouvelles, on s'était bien mis d'accord. Si elle voyait quelqu'un, elle appelait Nina, et, en cas de coup de feu, elle appelait un ami à Sheffer. Pour le reste, elle se terrait chez elle et faisait mine d'être sortie, ou morte, ou les deux.

Les visiteurs eux aussi s'étaient figés. Ils avaient dû entendre la sonnerie. Hélas, je manquais de visibilité pour être sûr de ma première balle.

Quelques murmures. Les hommes décidèrent de poursuivre, comprenant peut-être que ce coup de téléphone ne faisait qu'aggraver notre cas.

Ils se remirent en marche, et apparurent enfin au grand jour. Deux hommes, de grande taille. En tenue sombre.

Distance : quarante mètres. Oublions la tête, visons la masse corporelle. Pas droit à l'erreur. Je pris une brève inspiration, bloquai mon souffle, projetai mon regard dans l'axe du canon. Pressai doucement la détente...

Un mouvement furtif : quelqu'un marchait d'un pas vif, de notre bungalow vers les deux hommes, dans un éclair marron – un marron riche, celui de...

Je redressai la tête, juste à temps pour entendre un cri de femme en contrebas. Le blouson de Nina – donc sa voix ?

Sans même réfléchir, je me retrouvai sur mes deux jambes, dévalant la pente avec maîtrise mais vitesse, le fusil en joue et l'index prêt à tirer, dépassé par les événements.

Je courus me planter vingt mètres derrière les intrus, atteignant le sentier au moment précis où, cent mètres plus loin, Nina déboulait du tournant, de sorte que les hommes se retrouvèrent pris en sandwich. Ils s'étaient arrêtés. L'un tenait un téléphone. L'autre se retourna vers moi et leva lentement les mains.

— Salut, Ward, dit-il. Pas de bêtise, hein ?

— Bordel de Dieu, Charles ! cria Nina au type avec le téléphone. T'es malade ou quoi ?

Ils étaient assis sous le porche. Nina et moi restions debout, un peu à l'écart. Ni elle ni moi n'étions assez sûrs de nos réactions pour oser nous rapprocher, même une fois le café servi, Patricia démobilisée et la sueur refroidie sur mon front.

— On aurait pu vous tuer, lâcha Nina pour la énième fois.

Les mains sur les hanches, elle ne décolérait pas.

— Je peux encore m'en charger, grommelai-je.

Ils avaient pris les chaises sur lesquelles nous dînions d'habitude. Le deuxième type, celui qui m'avait accueilli, ne décrochait pas un mot sauf pour refuser du sucre. Grand et sec, le cheveu court. C'était le lieutenant Doug Olbrich, de la Section spéciale n° 1 de la police de Los Angeles – la division Vols et Homicides, chargée du profilage des affaires de meurtre.

Le premier type, qu'on voyait boiter même de loin, était Charles Monroe, directeur régional du FBI de Los Angeles et patron de Nina. Nous nous étions rencontrés très brièvement, juste avant qu'il ne se fasse cribler de balles par un ennemi commun, cinq mois plus tôt, dans un petit restaurant de Fresno. Il avait de la chance d'être en vie, bien qu'il eût sans doute passé de nombreuses semaines à penser le contraire – et, vu les précautions qu'il devait prendre pour simplement s'asseoir, cette période n'était pas tout à fait révolue. L'homme qui l'avait blessé – en même temps que moi, à l'épaule – était mort. Nina l'avait abattu en forêt, à trois kilomètres d'ici, le jour où nous avions rencontré Patricia. Depuis, Nina n'avait revu son patron qu'à deux reprises, lors de nos deux passages éclairs à L.A., lorsqu'on l'avait convoquée pour un débriefing, puis pour des auditions en vue du procès du tueur que nous avions capturé ce jour-là.

— Tu ne pouvais pas t'annoncer ? fulmina-t-elle. T'annoncer la veille, j'entends, pas quand tu as déjà longé la moitié du lac ! Te tuer est le dernier emmerdement dont on ait besoin... aussi séduisant soit-il !

Monroe posa son café sur la table.

— Vous seriez restés pour nous accueillir ?

— Bien sûr que oui.

Il n'en croyait pas un mot. Pendant ce temps, Olbrich contemplait les arbres, ravi de ne pas être mêlé à cette discussion. Mais je lisais une grosse tension sur son visage.

— Aurais-tu décroché ?

— Mais bon sang, Charles...

Nina se frotta le visage, puis saisit sa tasse posée sur la balustrade. J'intervins :

— Qu'attendez-vous au juste, Monroe ?

— De vous, rien du tout. Vous n'êtes pas membre du FBI, cette agence qui m'emploie et qui emploie Nina, si par bonheur elle s'en souvient. Vous n'êtes même pas flic, et ne l'avez jamais été. Je crois savoir que vous avez jadis travaillé pour une autre organisation, dans ce qu'on appelle communément le « renseignement ». Mais d'après la rumeur cela remonte à loin et vous n'êtes pas tellement regretté. En ce qui me concerne, vous pourriez disparaître dans cette forêt que ça ne me ferait ni chaud ni froid.

— Ward t'a sauvé la vie, objecta Nina.

— Vraiment ? La dernière chose que j'ai vue, c'est qu'il t'évacuait par l'arrière du restaurant alors que je pissais le sang dans un box. Le tireur vous a suivis. J'ai survécu par défaut.

— C'est une version très subjective, commentai-je, même si au fond j'étais de son avis.

C'est vrai, je m'étais davantage soucié de la vie de Nina, et de la mienne, que de celle de Monroe – surtout après ses blessures que je pensais mortelles. Pire, je m'étais rendu compte que je pouvais l'assumer.

— Allons, messieurs dames, lança Olbrich, tout ceci ne mène nulle part.

— Mais Nina et moi ne *comptons* aller nulle part, rétorquai-je. Elle a passé un accord qui interdisait à ce connard de révéler notre localisation, y compris à vous. Il vient d'enfreindre cette règle, et m'est avis que c'est juste un début. Vous n'êtes pas venu apporter le courrier de Nina, Monroe, alors vous voulez quoi, bordel ?

— Nina, répondit-il. Il est temps de rentrer, maintenant.

La question de Nina trouvait subitement sa réponse. Mais elle était écrite depuis le début. Je secouai la tête et m'éloignai de quelques pas.

— Je ne suis pas certaine de revenir, protesta-t-elle. Je me plais ici.

— C'est une démission ? Sérieusement ? Dans ce cas, prends un stylo et une feuille. Il me faut une trace écrite.

Nina me regarda. Je haussai les épaules : c'était son job, son choix, sa vie.

— Allons, Nina, fit Monroe, entre agacement et fausse douceur. Tu connais la musique. Je t'ai obtenu un congé car les circonstances étaient exceptionnelles.

— Et que j'ai environ deux millions d'heures de récup à prendre.

— Ça fait des semaines que tu t'es sauvée. Comme tu le sais.

— OK, soupira-t-elle. Peut-être bien que je vais raccrocher. Peut-être que j'ai fait le tour de ce boulot. De toute manière, pour ce que ça change, qu'on soit là ou pas...

— Tu ne peux pas dire ça, Nina. Ça fait dix ans que tu envoies des tueurs en taule.

— Enfermer quelqu'un alors qu'il a tué deux, quatre, six personnes, tu parles d'une victoire ! Moi, j'appelle ça éponger le lait renversé. Super, le sol redevient propre, momentanément. N'empêche que t'as toujours pas de lait. Les familles des victimes continuent de se réveiller avec un goût de mort dans la bouche. Le drame a quand même eu lieu, et on arrive après la bataille.

— À moins que tu ne trouves le moyen de remonter le temps, c'est le principe même de la police, devisa Olbrich.

Nina rougit. Elle n'avait pas développé un argument, mais un sentiment personnel. Seulement le flic ne l'avait pas compris, et du coup elle se sentait bête.

Monroe s'en rendit compte, qui ajouta aussitôt :

— Et puis il y a autre chose. J'ai besoin de toi.

— Arrête ton char, dit-elle. T'as plein d'autres femelles sous la main. Même des génies qui comprennent la notion de causalité, je parie.

— Écoute, Nina, j'ai fait beaucoup de chemin pour te voir et j'ai perdu assez de temps comme ça.

— Eh bien, reprends l'avion pour L.A. et fais bon voyage. La prochaine fois, tu nous préviendras et t'apporteras quelque chose : des fleurs, des muffins... On ne t'a pas appris les bonnes manières, visiblement.

— Je ne rentre pas en Californie. Le FBI a été appelé en renfort sur une histoire d'homicide en Virginie. C'est assez obscur, et on pense à un tueur en série. C'est l'avis des flics locaux, en tout cas. J'aimerais que tu m'accompagnes.

— Je ne...

— Il s'agit d'une femme.

Nina se tenait au bord du lac. Dix minutes déjà. À chaque nouvelle seconde, ses chances de départ augmentaient, je l'avais bien compris. Sa posture suggérait que tout ce qui l'entourait – les arbres, l'étang, les montagnes, moi-même sûrement –, tout cela cessait d'exister. J'étais resté au bungalow avec les autres. Aucun des deux types n'essaya d'engager la conversation. Olbrich surveillait l'heure de près.

— Expliquez-moi un peu, lançai-je à Monroe. Vous trouvez un cadavre dans l'est du pays, et c'est peut-être du ressort des fédés, ou peut-être bien que non. Ce qui m'échappe, c'est pourquoi vous avez ramené votre pote Olbrich. Je me fiche de savoir comment vous définissez la juridiction de Los Angeles, mais quand même, la Virginie, c'est pas la porte à côté.

Les deux hommes échangèrent un coup d'œil.

— Dis-lui, fit Monroe avant de se relever avec précaution et de descendre du porche. Ça le regarde, pour une fois. Et on n'a plus le temps d'attendre.

Il partit rejoindre Nina.

— Le fameux Henrikson, commença Olbrich quand son collègue fut assez loin. C'est votre frère, pas vrai ?

Il faisait référence au tueur que nous avions arrêté dans les bois, un homme qui serait bientôt jugé pour le meurtre d'une habitante de Los Angeles, Jessica Jones, et d'une autre de Seattle, Katelyn Wallace, dont on avait retrouvé le corps à soixante kilomètres d'ici. Le dossier était en béton armé. Viendrait ensuite un second procès, concernant la mort, cinq ans plus tôt, de cinq adolescentes de Los Angeles – les meurtres du Garçon de Courses, selon les titres de l'époque. Là, la partie s'annonçait plus délicate.

— Nous sommes jumeaux, confirmai-je. Mais on ne s'est jamais connus. Son vrai prénom est Paul, mais il se fait appeler l'Homme Debout – cette connerie de Garçon de Courses était une idée de Monroe, je vous le rappelle. Paul ne travaille pas seul, du reste. Mais vous savez tout ça. C'est dans le rapport de Nina.

— Ça n'y figure pas, avoua Olbrich en détournant la tête. On a estimé que vos allégations brouillaient le dossier.

— Je n'ai dit que la vérité. Nina le sait très bien. Paul travaillait pour une conspiration de tueurs, il leur fournissait des victimes sur commande. Il ne faisait pas que ça...

— C'est Monroe qui s'occupe de l'enquête, pas Nina.

— Monroe s'occupe surtout de sa carrière. Mais si vous en veniez au fait ?

— À sa sortie de l'hôpital, votre frère a été transféré à Pelican Bay. La prison supermax près de la frontière de l'Oregon.

— Je croyais qu'elle était réservée aux gangs de fous furieux, genre Confrérie aryenne, Jeunes Caïds, Guérilla noire...

— En temps normal. Mais Monroe tenait à placer notre homme en quartier de haute sécurité jusqu'à son procès – seul, en consigne maximum, dans une cellule sans fenêtre, avec des matons qui considèrent les pertes humaines comme de la paperasse. Vu ce qu'il avait fait à ces femmes, ça paraissait justifié. Monroe a donc obtenu gain de cause. Emporté le morceau. Les centres de Corcoran et de Tehachapi ne voulaient pas du prisonnier, alors on l'a envoyé dans le nord, au Pelican. En l'espace de trois mois, il a survécu à trois tentatives de meurtre, dont l'une venait d'un membre du personnel, qui se trouve toujours à l'hosto. Puis là-dessus...

Olbrich expira lourdement, et cela suffit à lui faire perdre la vedette. Surtout lorsque je vis Monroe soliloquer d'un air grave au bord de l'eau, et Nina relever brusquement la tête en me cherchant du regard. Elle reprit aussitôt le chemin du bungalow.

— Ne me dites pas ça... murmurai-je.

J'entendais le sang jaillir dans mes oreilles, le sentais hésiter dans mes veines.

— Il y a deux jours, il a quitté la prison pour être transféré à Los Angeles. On ignore ce qui s'est passé en route, mais, au bout de cent cinquante kilomètres, il s'est bel et bien passé quelque chose. On a retrouvé le fourgon blindé quasi entier, et les corps de deux gardes dans un rayon d'un kilomètre. Les deux autres ont disparu. L'hypothèse, c'est qu'ils sont morts eux aussi.

— Dites-vous plutôt qu'ils ont aidé Paul à s'évader.

— Monroe était persuadé que vous répondriez ça. Il m'a prévenu de vos penchants conspirationnistes. Un tueur sous chaque lit, etc.

— Vous savez parfaitement de quoi je parle. Vous avez aidé John Zandt. Il vous a transmis des infos. Il était sur la trace de Paul.

— J'ai aidé John parce que je le connaissais lorsqu'il bossait à la crime, et que c'était un flic brillant. Il ne l'est plus. Je vous signale qu'il a deux avis de recherche aux fesses.

— Ouais, il a tué un mec qui conduisait des fillettes à la mort, et un autre qui participait à leur enlèvement.

— Prudence, Ward. Si jamais Zandt refait surface, vous pourriez être obligé de répéter cette déclaration devant un tribunal.

— Je ne cautionne pas ce qu'a fait John. Mais quand j'aurai le temps et que je me sentirai en sécurité, j'irai danser sur les tombes de ces deux ordures.

— En sécurité vis-à-vis de qui ?

— À votre avis ? Des gens qui sont derrière tout ça. Je ne délire pas, Olbrich. Pourquoi croyez-vous qu'on s'est retranchés ici, avec Nina, sous des noms d'emprunt ? Parce qu'on serait dans un trip retour à la nature ? Ou par timidité ? Sans déconner !

— Je pensais que c'était lié à votre frère. Je sais que vous avez participé à sa capture.

— Non, rétorquai-je froidement. Ce n'est pas à cause de lui. Nous imaginions que le système pénal californien maîtrisait la situation.

— Ward ? lança Nina.

Monroe remontait lui aussi, avec quelques foulées de retard, les mains dans le dos. Il affichait l'air du type qui a bouclé sa mission, obtenu le résultat voulu, et pense déjà à la suite.

— T'es au courant ? demanda-t-elle.

— Eh oui. Félicitations, Charles. Vous venez juste de perdre l'un des types les plus dangereux au monde.

— On le trouvera, jura-t-il.

— Aucune chance. Si jamais quelqu'un doit trouver quelqu'un, c'est lui qui vous trouvera. Et là, je ne donne pas cher de votre peau.

— Je doute que ce soit moi qui l'intéresse.

— En effet. Mais dans ce cas, n'auriez-vous pas dû nous prévenir un peu plus tôt ? Il y a un jour et demi, par exemple ? Ou bien vous préfériez poster des agents à Sheffer et nous utiliser comme appâts ?

— Vous perdez la tête.

Je refusais de le croire, et cela signifiait que Sheffer n'était plus un lieu sûr. Paul ne viendrait pas nécessairement me chercher. Mais d'autres pouvaient s'en charger pour lui.

— Si je comprends bien, Olbrich veut me demander si je sais où Paul se cache ?

— C'est le cas ?

— Non. Mais en l'état actuel des choses, même si je le savais, je ne vous dirais rien.

Je fumai sous le porche pendant que Nina bouclait ses bagages. Monroe et le flic restaient plantés un peu plus loin, bouillant d'impatience. J'ai passé un moment à scruter la nuque de Monroe. Un tir propre et net ne suffisait plus. Je voulais le saisir à la gorge et le noyer dans le lac. En public, avec des billets bon marché et des sandwiches gratuits.

— J'ai fini, annonça Nina.

Je me retournai ; elle se tenait dans l'entrée, un sac à la main. Elle avait ressorti le vieux tailleur de fédé. Ce n'était plus la même. Elle avait un air sérieux, professionnel, un air... en fait, elle avait un air assez cool.

Je me levai.

— Agent Baynam, fidèle au poste.

— Ça m'emmerde autant que toi, dit-elle en se rapprochant. Tu me crois, au moins ?

— Oui, je te crois, répondis-je à mi-voix. Car je pense qu'une certaine personne a ignoré les coups de fil de Monroe depuis un petit moment. Je me trompe ?

— Pas sûr.

— Tu aurais dû me dire qu'il s'était manifesté.

— C'est vrai. Mea culpa. Et toi, qui te réclame ?

— Comment ça ?

— Le soir où tu as fait cette salade ridicule. Tu as refermé l'ordi sans quitter la messagerie.

Que pouvais-je ajouter ?

— Grillé.

— Alors, qui essaie de te joindre ?

— C'est parvenu à une vieille adresse. Ce pourrait être n'importe qui.

— J'en doute. T'es un chic type, quand on creuse bien, mais t'as pas non plus un succès fou. M'étonnerait qu'on te propose un bowling. Dehors, il y a un monde hostile qui nous veut du mal, Ward. Tu dois découvrir qui t'a écrit.

— C'est toi le boss.

— Exact. Ne le perds jamais de vue. (Elle se pencha pour m'embrasser.) À plus.

Elle marcha jusqu'au lac et suivit les deux hommes. Je la sentis disparaître bien avant que mes yeux ne la perdent.

J'ai consacré l'après-midi à fermer le bungalow. J'ai fait le ménage, coupé le chauffe-eau, tiré les volets. J'ai passé une bonne partie de ce temps à me demander où aller, sans grand résultat. Le mieux qui me soit venu à l'esprit était de rouler vers l'est. Je suis monté dans les combles pour récupérer la bécane de Bobby. Je l'ai mise à recharger pendant que j'acheminais quelques affaires vers la voiture de Nina, garée après le lac derrière chez Patricia. J'ai expliqué à notre amie que Nina était partie et que je n'allais pas tarder moi-même. Je lui ai fait promettre d'être prudente, de se méfier des inconnus, et d'alerter le shérif au moindre signe suspect. Elle m'a offert un café qui n'a fait qu'alourdir mon sentiment de solitude.

De retour là-haut, j'interrogeai ma messagerie. Il n'y avait rien, ce qui, d'une certaine façon, me compliquait la tâche. Nina était bien gentille de réclamer l'identité de mon correspondant, mais le mail contenant l'adresse d'expédition se trouvait sur *son* ordinateur. Pour ma part, je restai sceptique quant à la nécessité de ce contact. Elle voyait peut-être la main de Paul, qui cherchait à me débusquer. Auquel cas elle raisonnait de travers : Paul était encore détenu à la date du mail.

J'allais éteindre la machine lorsque je vis qu'elle téléchargeait quelque chose. Une fenêtre d'avancement annonçait du courrier sur l'un des comptes de Bobby. J'avais maintenu ses adresses électroniques, par respect ou par superstition, me refusant à clore ce dernier vestige de sa vie.

Trois messages dans la boîte aux lettres. Tous avaient pour titre « APPELLE-MOI » et le dernier datait de trois semaines. J'aurais spontanément cru à des spams si leurs adresses Hotmail ne m'avaient paru familières.

J'ouvris le plus récent :

T'es là, Bobby ? Il y a de drôles de bruits qui courent et j'ai besoin de tes lumières. Vite. Toujours pas signé. L'émetteur devait penser que Bobby et moi reconnaîtrions son adresse. Je réfléchis quelques instants, puis recopiai celle-ci dans un nouveau mail, sur mon propre compte. Avec pour texte :

Ici Ward Hopkins. Bobby est mort. Qui êtes-vous et que voulez-vous ?

Je cliquai sur « envoyer » avant de me poser trop de questions.

La sagesse commandait de partir au plus vite, mais je trouvais cela difficile. Je jetai un dernier coup d'œil dans la maison, comme mon père me l'avait enseigné dans les locations de vacances, en vérifiant les pièces l'une après l'autre, en refermant chaque porte derrière soi. Je cherchai mon manteau partout, avant de conclure qu'il devait être accroché à la porte d'entrée, encore ouverte. Je ne me souvenais pas de l'avoir suspendu, mais la journée avait été si mouvementée...

En rabattant la porte, je découvris non pas un manteau, mais deux. Le blouson marron de Nina pendait juste à côté du mien.

Elle l'avait abandonné.

L'espace d'un instant, je retombai en adolescence, puis tâchai d'être philosophe. Trop gros pour son sac. Ne cadrait pas avec le look propret de fédé. T'as trente-neuf ans, Ward. Ressaisis-toi.

Puis de l'une des poches je vis dépasser un coin de feuille pliée. Un petit mot, écrit de la main de Nina.

Quelle gourde. Je vais attraper la mort.

La voilà, ma destination.

Je verrouillai le bungalow et pris la route de Yakima, qui possédait un aéroport. Je prévoyais d'y passer la nuit, avant de m'envoler pour la Virginie.

CHAPITRE 3

Après coup, Lee John Hudek déclarerait avoir tout de suite flairé l'embrouille. Il ne savait dire à quel moment précis, mais l'impression était nette. Il s'en souvenait d'autant mieux qu'il ne croyait pas à ces conneries de prémonitions. Prémonitions, mon cul. C'était bon pour les charlatans de Venice Beach, pour ces soiffardes avec leurs stands de tarots pourris, ou pour Pete Voss à la rigueur – Pete le Dormeur semblait convaincu de capter des ondes vaudou du fait que sa mère avait un huitième de sang Blackfoot ou bouddhiste ou Dieu sait quelle merde. Il y croyait surtout lorsqu'il était défoncé, c'est-à-dire les trois quarts du temps, mais Hudek restait persuadé que tout ça c'était du flan – chez Pete en particulier comme dans l'univers en général. Hudek avait compris dès son plus jeune âge que le monde suivait des lignes fermes et droites, et que ces foutaises étaient faites pour les losers. Dormeur était une sorte de bourrin fini, c'était clair, mais il était sympa et loyal, c'était un revendeur lent mais régulier, et puis c'était un des anciens de l'équipe, alors peu importait.

Voici ce dont Hudek se souvenait :

Installés dans la caisse avec Pete à sa droite et Brad Metzger à l'arrière, garés après la sortie de Tujunga Canyon Drive, ils avaient ouvert le toit pour profiter du soleil pendant qu'ils patientaient. Et patientaient. La voiture appartenait à Hudek, réglée cash. Dans sa situation, la plupart auraient opté pour une vieille relique au moteur gonflé – comme lui, un temps – mais il considérait aujourd'hui qu'une Merco de businessman envoyait de meilleurs signaux. Et les flics vous donnaient moins de fil à retordre. Pas tout à fait vingt ans, et Lee Hudek possédait aussi sa propre maison. Elle n'avait rien d'une baraque de mac, juste une discrète villa de trois chambres à Summer Hills, mais avec piscine et double garage, et il fallait voir la taille du téléviseur. Pete et Brad habitaient à Simi Valley, au sud de Santa Barbara, chez leurs parents respectifs. Toutes leurs connaissances créchaient encore chez papa et maman, ayant accompli peu de chose depuis la fin du lycée, l'année précédente, sinon appris que le boulot ne tombe plus du ciel en Californie et qu'avoir été une demi-pointure dans l'équipe de foot valait que dalle à l'extérieur.

La voiture avait le capot pointé sur la route ; la paroi sèche et broussailleuse du sous-canyon se dressait côté passager. On crevait de chaud. Dormeur jouait avec sa console de poche, ping, ping, bip. Brad déblatérait sur la poule qu'il s'envoyait, se demandait s'il devait lui payer un nouveau nez. Ayant lui-même tringlé la poule en question, Hudek estimait que Brad se lasserait d'elle avant la fin de la période post-opératoire, d'autant que, à court terme, les perspectives de pipes seraient sensiblement réduites. Mais il laisserait Brad le découvrir par lui-même. Si le keum comptait vraiment gâcher tout ce fric sur une gonzesse, c'est qu'il était trop payé.

Trop payé par Hudek, s'entend.

— Je l'ai tringlée, mec, lâcha-t-il spontanément. Y a un an de ça.

Brad se tut. Puis embraya sur une autre fille.

Quand Hudek consulta sa montre, il était 13 h 17. Les autres avaient plus d'une demi-heure de retard. Du jamais vu. C'était casse-couilles de traverser la Valley, mais on procédait toujours ainsi. Comme Hernandez s'était toujours montré réglo, Lee acceptait ses conditions sans broncher. Ce retard n'avait rien de réglo, par contre ça aggravait sensiblement le côté casse-couilles de la chose. Surtout maintenant. Maintenant que Lee avait le Plan, la fiabilité devenait le maître mot.

Il revérifia sa montre, pour rien. Toujours 13 h 17. L'espace d'un instant, il lui sembla que la grande aiguille ralentissait, comme si cette dix-septième minute enflait, menaçait d'exploser.

Ce fut peut-être là, le moment où il tiqua, où lui vint le curieux pressentiment que cette journée prenait la route d'un avenir absent des panneaux.

Il fut 13 h 18.

Un cliquetis de briquet dans son dos.

— Fume pas dans la caisse, Brad.

— Le toit est baissé, putain.

Hudek se retourna sur son siège et dévisagea son pote.

Brad secoua la tête, ouvrit sa portière, descendit. Il s'éloigna de quelques pas, jusqu'au mur du canyon. S'assit sur une pierre et grilla sa cigarette. Ce dont il avait vraiment envie – *besoin*, ouais –, c'était un joint. Mais si Lee soufflait le chaud et le froid à propos de la clope, il était intraitable sur la fumette au boulot.

Vivement que cette mission se termine, se dit Brad. Ces opérations le faisaient toujours flipper. Il n'y avait jamais eu de pépin, mais les transactions de drogue restent des transactions de drogue – un truc éprouvant, même sans gueule de bois quatre étoiles comme celle qu'il se tapait en ce moment. Brad avait passé la soirée de la veille autour de la piscine avec les mômes Reynolds et leur bande de copains, à Santa Barbara, pendant qu'à Hollywood les parents Reynolds s'enfilaient cocktails, petits-fours et blagues sur le nouveau proc entre confrères et consœurs du barreau. Les gosses avaient leur propre menu de fête : vin, herbe et XBox. Brad avait fini dans les vapes, au lit avec une meuf. Pas celle qui voulait changer de nez, Karen – retenue par un dîner de famille méga-important – , mais une quelconque lycéenne de quinze ans, dont il ne sut pas le prénom mais qui à l'évidence raffolait de la cocaïne et qui, il l'espérait de toutes ses forces, aurait tout oublié le lendemain. À 7 heures du matin, M. Reynolds était entré dans la chambre pour ouvrir la penderie et choisir une cravate.

En ressortant il avait dit :

— Bonjour, Bradley.

— 'jour, m'sieur Reynolds, avait marmonné Brad en retour.

Aussi simple que ça. M. Reynolds avait paru préoccupé. Par une grosse affaire à plaider, peut-être. Et non par le fait que son fils et sa fille passaient la moitié de leur temps défoncés au Valium, Valium fourni par Hudek aux bons soins de Brad.

Assis sur sa pierre, Brad se demandait surtout si Hudek avait réellement baisé Karen Luchs. Possible. On ne savait jamais, avec lui. Ils avaient beau se connaître depuis toujours, avoir traîné dans les mêmes salles et dribblé les mêmes balles, Brad ne cernait pas tout à fait son ami. Certaines questions restaient sans réponse. Où il puisait son énergie, par exemple. D'où il tenait son savoir-faire. Ou pourquoi il était devenu le type pour qui on bosse, et non l'inverse. Peut-être avait-il tringlé Karen. Peut-être pas. La plupart d'entre eux avaient sauté la plupart d'entre elles à un moment ou à un autre. Au final, ça importait peu, sans doute.

Brad releva les yeux en entendant quelques mesures polyphoniques du générique des *Simpson*. Lee Hudek porta le Motorola à son oreille. La conversation dura à peu près une minute. Hudek n'éleva pas la voix, mais il le faisait rarement, donc ça ne prouvait rien. Pour finir, il referma l'appareil et convoqua Brad avec son index.

— C'est retardé, expliqua Hudek quand Brad regagna la voiture.

— Comment ça, retardé ?

Hudek se contenta de hausser les épaules.

— C'est quand même craignos, maugréa Brad, dont la migraine empirait au soleil. Y a plein de gens qui attendent.

— Je sais bien, répondit Hudek. On peut pas faire ça tout de suite, c'est tout. On reporte à ce soir. Rien de grave.

Mais si, c'était grave. Et d'une, ça ne leur était encore jamais arrivé, et de deux, on était samedi et des tas de clients attendaient leur shit et leur coke. Or qui dit livraison tardive dit distribution tardive, et donc bringue tardive. Hudek avait l'air assez furax comme ça, les yeux plantés dans le pare-brise, alors au lieu de protester Brad sortit le flingue de sa place stratégique – sous le siège passager, prêt à l'emploi –, fit le tour de la voiture et le rangea avec soin dans le coffre, enveloppé dans un drap de plage, en vue du retour à la maison. Brad respirait toujours mieux une fois l'arme enfermée.

Il s'assoupit lors de la traversée de la Valley par la route 118. Hudek maintint un petit 90, sans décrocher un mot. Dormeur joua avec sa console d'un bout à l'autre du trajet. Bip, bip, ping.

Le soleil écrasait le paysage de tout son poids.

Après avoir déposé les autres au centre commercial de Belle Isle, Lee Hudek roula sans but dans la Valley. Il échoua à la terrasse du Frisbee, devant un cheeseburger et des frites au piment, à regarder passer les bagnoles. Les badauds voyaient quant à eux un jeune homme blond, athlétique et légèrement bronzé, vêtu d'un jean baggy soigné et d'un tee-shirt de marque. Un unique tatouage barbelé autour du biceps gauche (le folklore gang n'était pas son truc) et un physique qui, sans être à tomber, pouvait passer pour satisfaisant d'un bout à l'autre du continent. Très faune locale, en somme : un jeune spécimen mâle de l'espèce peuplant cette plaine.

Son téléphone sonna à quelques reprises, et chaque fois il décrocha, répondit méthodiquement aux interrogations des gens étonnés de n'avoir pas reçu leurs friandises. Il connaissait toutes ces personnes de visu, et aucune ne prit la mouche. Tout baignait. Lee honorerait la commande. Comme toujours. Jusqu'à présent, en tout cas. Le seul à ne pas se manifester était le mec le plus indispensable, celui qui devait fixer le nouveau rendez-vous.

La situation devenait, pour reprendre une expression paternelle, « très inhabituelle ».

Comme beaucoup de ses amis, Hudek s'entendait bien avec ses vieux. Mais, alors que la plupart toléraient leurs parents parce qu'ils ne faisaient pas d'histoires, se contentaient de flotter à l'arrière-plan de leur existence, de les conduire à droite et à gauche, de leur filer du fric pour les fringues, le psy ou la désintox, et de ne s'abstraire que sporadiquement de leurs carrières pour se demander à voix haute si leur fils ou leur fille avait, à terme, un quelconque projet en matière de boulot, Hudek vénérait son père. Quand ce dernier était dans les parages. Hudek senior travaillait dans l'immobilier et voyageait beaucoup. Il avait un carnet d'adresses volumineux et gagnait un fric fou, sans jamais l'étaler. Sa femme et lui possédaient des tas de trucs hors de prix qui ne sortaient jamais de la baraque et qu'ils planquaient lorsqu'ils recevaient. Il avait de l'humour, aussi. Pour Lee, la sonnerie de portable *Simpson* tenait du clin d'œil, en souvenir de ce jour lointain – il était encore gamin – où il avait vu son père se tordre de rire devant cette réplique d'Homer à ses gosses : « N'oubliez pas : pour les autres, on est une famille normale. » La devise de Ryan Hudek était du même acabit, bien qu'insolite en ce lieu et à cette époque : si tu sais ce que tu vaux, le monde entier n'a pas besoin d'être au courant.

Lee en avait pris bonne note, comme de la leçon véhiculée par le cinéma et démontrée toutes les semaines dans les reality-shows policiers : ce qui est voyant attire l'attention, et l'attention attire les ennuis. Pour l'heure, Lee incarnait le jeune qui ne sait pas quoi faire de son samedi après-midi. Et ça lui convenait parfaitement.

Au bout d'un moment, il se souvint que c'était l'anniversaire de sa mère. Il acheta une carte et un flacon de parfum, avant de se pointer chez ses vieux. Son père était sorti – une partie de golf, sans doute –, alors il passa par-derrière.

Lisa Hudek était à demi allongée sur un transat au bord de la piscine, oisive sous ses lunettes noires. Elle accepta le cadeau non emballé ainsi qu'un baiser, et sourit plus ou moins dans la direction de son fils.

Il lui tint compagnie pendant une heure, à regarder trembloter l'eau du bassin, pendant que sa mère vidait consciencieusement une grande bouteille givrée. Puis son portable sonna enfin ; c'était Hernandez.

Le nouveau rendez-vous était fixé à 19 h 30. Très tard, en effet, mais c'était comme ça. À lui de gérer. Le lieu serait précisé ultérieurement.

Au fait, pouvait-il se pointer seul, cette fois ?

— Ouais, d'accord, marmonna Hudek avant de rentrer dans la maison pour passer quelques coups de fil.

À 19 h 15, il s'arrêta en douceur, à mi-hauteur du pâté de maisons au croisement de Roscoe Boulevard et de Sennoa Avenue. Cinquante mètres après le carrefour se trouvait une station-service. En face, un bâtiment bas que Hudek avait croisé des centaines de fois sans même y jeter un œil. Lorsqu'il était enfant, ces murs avaient abrité une cafétéria, et il était sûr d'y avoir mangé en famille à deux ou trois reprises. Puis le restau avait mis la clé sous la porte, était devenu un magasin de moquettes, puis un centre de pièces détachées, puis avait abrité divers commerces avant d'atteindre l'invisibilité pure. L'un de ces endroits tantôt murés, tantôt ouverts, qui peu à peu échappent à nos radars mentaux.

— OK, dit-il en coupant son moteur.

— C'est très loin d'être OK, objecta Brad.

Ce coup-ci, Brad était assis à l'avant. Pete le Dormeur occupait l'autre voiture avec Steve Verkilen, le quatrième membre du groupe. Eux étaient garés à deux pas de la station-service. Brad n'apercevait que leur calandre, ce qui n'était pas très rassurant. Le pistolet reposait sous le siège de Brad, mais ça ne le consolait pas davantage. Son petit doigt lui disait qu'on risquait de l'utiliser, et ça craignait. Le flingue n'était pas censé servir. Il suffisait de le posséder, et que les autres le sachent. Une fois, une seule, ils avaient tiré sur la maison d'un gus, la nuit, en guise d'avertissement, après s'être assurés que la baraque était vide. Mais s'en servir pour de vrai, c'était un tout autre univers, où le pistolet perdait soudain sa vertu protectrice. Brad pouvait presque sentir l'arme refroidir le siège, s'étirer, s'assouplir, et siroter un café en demandant : « Sans déconner, gamin, tu me prenais pour qui ? »

La présence d'un second flingue sous le siège de Hudek ne faisait qu'aggraver la situation. Il côtoyait le sac de fric.

Brad alluma une cigarette, dans l'espoir qu'elle l'apaiserait.

— La façade est bouclée avec des chaînes et des planches. T'es supposé entrer comment ?

— La porte de derrière.

— Tu comptes sérieusement y aller ?

— Pas moi, répondit Hudek. *Nous* y allons.

— Attends, Lee. S'ils t'ont demandé de venir seul, comment ils vont réagir si on s'amène tous les deux ?

— C'est ce qu'on est venus découvrir.

Face au regard de Hudek, Brad ne put que se taire : il avait des dettes, et Hudek était sa seule source de revenus. Si Brad n'était pas content, il lui restait une solution : ouvrir la portière et se tailler, auquel cas le futur devenait une zone obscure, sans drogue ni fric ni statut social.

— J'aurais pu t'emmener comme d'hab, souligna Hudek. Sans te répéter ce qu'ils m'avaient demandé. Mais je t'ai mis au parfum.

— OK, fit Brad. OK.

Hudek adressa un coup de klaxon à Pete, qui accusa réception. On vit alors Dormeur et Steve s'extirper de leur caisse et traverser le carrefour pour disparaître dans l'autre rue. Lorsque la voie fut libre, Hudek et Brad quittèrent la Mercedes, sortirent les flingues et les coincèrent en vitesse à l'arrière de leurs jeans. Ils enfilèrent leurs blousons, Hudek prit le sac de fric dans sa main gauche et fit un signe de tête impérieux. En avant. Deux petits gars s'en allant en balade. On pouvait presque entendre démarrer la BO.

Mais ouais, se dit Brad, légèrement écœuré. *On est pas des blaireaux. Tout baigne, putain, qu'est-ce qui pourrait nous arriver ?*

Il traversa la chaussée sur les talons de Hudek, la tête droite, le pas ferme et régulier. Ils coupèrent par le petit parking pour atteindre la droite du bâtiment. Un étroit passage menait à l'arrière vers un second parking, plus grand et désert, en dehors d'une vieille bâche recroquevillée sur le bord. Tout au fond se dressait un petit mur de béton, que Hudek avait repéré deux heures plus tôt en passant en voiture. De l'autre côté, cachés derrière une benne rouillée, seraient postés Pete et Steve.

À mi-chemin du parking, Hudek s'arrêta. Il fallait admettre que le lieu était bien choisi : en plein jour, mais invisible depuis la route. On distinguait une porte dans le dos trapu du magasin, mais Hudek ne s'y rua pas, préférant scruter le parking. Écouter. Il prit son temps, parfait prédateur surveillant son territoire.

Brad admirait son calme. L'admirait à mort. En espérant vraiment, vraiment, que cela suffirait.

— On t'avait dit de venir seul.

Le cœur de Brad lui défonça le thorax. La voix venait de derrière.

— Retournez-vous. Lentement.

Hudek et Brad s'exécutèrent, pour découvrir trois hommes plantés à quatre mètres de distance. Minces, bruns, les yeux foncés. Ils avaient dû guetter leur arrivée sur le grand parking avant de les suivre dans l'allée. Seigneur.

Le seul visage connu était celui d'Hernandez, leur contact habituel. Une trentaine d'années, une gueule de vieille hache. Il secoua la tête à l'intention de Hudek.

— On t'avait dit de venir seul.

— J'ai compris, rétorqua Hudek. Et si tu me disais de te sucer la bite, je le ferais pas non plus.

Hernandez pinça les lèvres et remua faiblement la tête, comme s'il appréciait cette réponse, avec son allure de méchant de jeu vidéo.

Il se tourna vers son voisin de droite, hocha la tête de plus belle. Le voisin hocha la tête à son tour. Puis ils hochèrent la tête ensemble tout en considérant Hudek.

La scène dura un temps de trop aux yeux de Brad, et cette histoire commença à lui foutre les boules.

— Tu voulais que je vous retrouve à l'intérieur, rappela Hudek. Ça rime à quoi de nous chopper ici ? Ça rime à quoi de nous prendre par surprise ?

— C'est pour s'assurer que tu as suivi les instructions, répondit Hernandez.

— Tu croyais sérieusement que j'allais le faire ?

— On espérait.

— La vie est pleine de déceptions, mec. À toi de gérer. On peut passer à la suite, maintenant ? J'ai la moitié de West Valley qui attend son trip.

— On devrait peut-être trouver un autre gosse de riche pour écouler notre came. Un gentil garçon obéissant.

— Tu vois, y a un truc que t'as pas l'air de piger, répliqua Hudek. Je te jure, t'y es pas du tout, là. T'as l'air de croire qu'on va rester plantés là à gober vos salades. Mais ça va pas se passer comme ça.

— Ah non ?

Hernandez avait retrouvé le sourire. Or Brad n'aimait pas ce rictus. Il s'aperçut que l'un des types tenait un flingue, avec naturel, le long de sa cuisse. L'arme était sûrement là depuis le début, mais sous l'effet du stress il n'avait rien remarqué... *Gobe leurs salades, Lee,* pensa-t-il, fébrile. *Je t'en supplie, on gobe leurs salades et on s'arrache.*

— Pete ! lança Hudek un ton au-dessus. Si tu venais nous rejoindre ?

Un long silence. Pete le Dormeur n'émergea pas du cube métallique.

Scrutant Hudek d'un air piteux, Brad vit pour la première fois une nuance de confusion dans le regard de son pote.

Hudek rouvrit la bouche :

— Steve ?

Le même silence, que seul vint rompre un coup de klaxon là-bas sur le carrefour, c'est-à-dire au bout du monde.

Hernandez inclina la tête.

— Pete ? Steve ? De qui peut-il s'agir ? D'autres pique-assiettes ? D'autres rats de la Valley ?

Hudek se taisait, les yeux rivés sur la benne.

Hernandez interpella de nouveau son voisin de droite :

— Tu connais un Pete ou un Steve, toi ? T'as croisé un Pete ou un Steve ?

— Oui, répondit le gars. Je crois que oui.

Il se dirigea vers la grande bâche avachie au pied du mur latéral. En souleva un coin.

Brad écarquilla les yeux. Couchés sous la toile goudronnée, se trouvaient Pete et Steve. Bâillonnés avec du ruban adhésif argenté. Et sans doute ligotés, puisqu'ils étaient immobiles. Du moins était-ce une explication possible à leur immobilité.

Le type laissa retomber la bâche.

— Vous auriez pas dû faire ça, menaça Hudek d'une voix basse et monocorde. (Il glissa la main droite sous son blouson, puis dans son dos.) Vous auriez pas dû.

— Fallait venir seul.

— Détache-les.

— Va chier.

Hudek sortit sa main. Elle tenait le pistolet automatique.

— Maintenant tu les détaches, connard !

Ils remuèrent la question pendant un petit moment, mais Brad les entendait à peine. Des paroles tournaient à vide dans sa tête, recouvrant tout le reste. *J'ai vingt ans*, bégayaient-elles. *C'est trop tôt. J'ai pas eu mon compte. Je croyais être grand, mais je me trompais. Ça m'intéresse plus, la baise, la drogue et la bière. Ça m'intéresse plus d'être ici. Je veux rentrer à la maison, regarder de vieux épisodes de X Files en bouffant de la glace. Je veux retrouver mes dix ans.*

— Gros sac à merde, dit Hudek, sans doute en guise de conclusion.

Brad sortit de sa rêverie. Les calibres ennemis n'étaient plus baissés vers le sol, mais braqués sur Lee et lui. Ça dégénérait salement, et Brad savait ce qu'on attendait de lui, alors il dégaina aussi. La fin était proche. Cinq mecs armés. Le calcul était vite fait.

Hudek leva son flingue, le pointa sur le torse d'Hernandez. Le type resta parfaitement impassible, et Hudek fut enfin fixé : cet homme avait déjà tué, et pas qu'une fois. Hudek possédait une clairvoyance dont Brad ne pouvait pas même rêver, mais en cet instant leurs pensées se rejoignaient. Nous y voilà. Le grand final.

— OK, dit Lee, si ça doit finir comme ça...

Soudain, un grand « clic » dans leur dos. Suivi d'un cognement sourd, comme du bois contre un parpaing.

Hernandez décala son regard. Hudek et Brad se retournèrent.

La porte arrière du bâtiment était grande ouverte, révélant un homme en costume.

— Dépêchez-vous d'entrer ! On vous attend.

Brad était aussi surpris que Hudek, mais incapable de prononcer le moindre mot. À Hudek, donc, de reprendre ses esprits, et il eut toutes les peines du monde. Il ne vit même pas qu'Hernandez en profitait pour frapper le crâne de Brad, pas plus qu'il n'entendit son ami tomber en vrac sur le bitume.

Bouche bée, il fixa l'homme au costard pendant cinq bonnes secondes, avant d'articuler :

— Monsieur Reynolds ?

CHAPITRE 4

L'intérieur du bâtiment était sombre, étouffant. Les ampoules nues semblaient suspendues au hasard, révélant çà et là des débris d'activités passées : rouleaux de moquette moisis, pièces mécaniques non identifiables, et des équipements que Lee Hudek associa vaguement aux cuisines des restaurants. Ça sentait la poussière et le chaud. Talonnant M. Reynolds, Hudek traversa une grande salle, enfila un couloir, franchit la porte d'un espace encore plus vaste et plus obscur qui occupait toute la surface restante. Là, M. Reynolds s'écarta, et Lee se retrouva en tête. À l'évidence, il était censé poursuivre son chemin. Une lumière brillait tout au fond. Sans doute indiquait-elle sa destination. L'espace d'un instant, il envisagea de ne pas obtempérer, de rebrousser chemin et de foncer vers la sortie. Mais ça ne rimait pas à grand-chose. Brad était resté sur le parking, Pete et Steve également, après avoir subi Dieu sait quoi de la part d'Hernandez et consorts. Ce que l'avenir réservait à Hudek n'était pas tellement plus clair.

Autant le découvrir.

Il s'avança dans le noir. Ses pas auraient dû résonner dans l'immensité, et pourtant non. Peut-être l'obscurité renfermait-elle d'autres rebuts, qui étouffaient le bruit. Peut-être était-ce dû à cette putain de fournaise. L'air semblait vieux, très vieux ; épais à en bouffer les sons. Et les gens, qui sait.

La lumière émanait d'une unique lampe, posée sur un coin de sol dégagé. L'objet provenait d'un motel bas de gamme ou d'un film : un pied droit, en bois, coiffé d'un grand abat-jour blanc que l'âge et la poussière avaient rendu crème dégueu. À côté trônait un fauteuil : massif, râpé, d'une couleur indéfinissable. Dedans se trouvait un homme.

— Salut, Lee. Tu te souviens de moi ?

Hudek stoppa à quelques mètres du type. Non par déférence, mais parce qu'il préférait, pour des raisons difficiles à expliquer, garder une certaine distance.

Le type était en costume et chemise sombres. Il paraissait proche de la quarantaine, bien bâti quoique maigre. Les cheveux courts, le teint pâle. Son visage était si mal éclairé qu'on peinait à distinguer ses traits, mais Hudek savait une chose : c'était la première fois qu'il le voyait. On aurait dit un gros chien au caractère incertain, assis bien droit dans son fauteuil, aux aguets. Prêt pour la promenade. Ou pour le dîner.

— Non, répondit Hudek.

— Bien.

L'homme l'observa en silence. D'un air impersonnel, comme face à une toile de peintre du dimanche à laquelle il eût fallu trouver un emplacement – dans une pièce secondaire ou dans le couloir où l'on remisait les vieux anoraks et les raquettes fendues.

— Alors, comment ça va, Lee ?

Hudek haussa les épaules.

— Ça peut aller.

— Bien. Très bien. Installe-toi.

Hudek resta perplexe. Puis il vit que l'homme montrait quelque chose de la main gauche. Se retournant, il aperçut une chaise en bois à quelques centimètres de ses mollets. Il lui suffisait de plier les genoux pour s'asseoir. Ce qu'il fit.

Il tenait toujours le sac de fric pour la transaction qui, de toute évidence, n'aurait jamais lieu. Il le posa par terre. Son cœur lui faisait l'effet d'un marteau ; les coups n'étaient pas extrêmes mais suffisaient à lui tordre les côtes.

— On t'avait demandé de venir seul, glissa l'homme comme s'il venait de se rappeler un détail mineur. Tu n'en as rien fait. Pourquoi ?

Hudek chercha la meilleure réponse.

— Ça ne me disait rien qui vaille.

— Je te comprends. Tes fournisseurs te demandent d'apporter le fric, sans escorte, et ils te font lanterner. Forcément, ça rend nerveux. Alors tu te dis : « Et puis merde, je vais ramener des potes, je vais pas me laisser dicter ma conduite. C'est moi le client. Lee John Hudek. »

— Exactement, fit Hudek, ravi de reprendre pied.

Quelle que fût l'identité de ce type, il saisissait parfaitement la situation.

— Si tu refais ça, poursuivit l'homme, si tu désobéis aux consignes, qu'elles soient simples ou compliquées, la police ne retrouvera jamais ta tête. Je te tuerai, ainsi que tous ceux qui te sont chers, et là tes ennuis ne feront que commencer. Pigé ?

Hudek cligna des yeux.

— Est-ce que c'est pigé ?

— Oui, putain. Bien sûr. Cinq sur cinq. Je vous assure. Je suis désolé.

— Excellent. (L'homme hocha la tête, plein de douceur retrouvée.) Vois-tu, c'est fondamental, Lee, car je dois sentir que tu es digne de confiance. *Nous* devons le sentir. OK ?

— OK, OK, répondit Hudek en opinant frénétiquement, désormais persuadé de mourir. Mais... quand vous dites « nous », vous pensez à qui, au juste ? Je veux dire, je croyais qu'Hernandez était...

Sans un mot, l'homme ôta ses mains des accoudoirs et leva les paumes vers le ciel.

De l'obscurité surgirent quatre hommes. Deux avaient la cinquantaine, les deux autres étaient nettement plus jeunes. Chaque paire se composait d'un type tiré à quatre épingles, et d'un autre, vêtu de manière à se fondre dans la foule.

— Nous sommes les gens qui te vendent la drogue, déclara l'homme. Nous la distribuons par le biais d'Hernandez, entre autres. Bienvenue à l'étage supérieur, Lee.

Il lui fallut une bonne minute pour retrouver un pouls normal. Les types avaient regagné l'ombre, le laissant seul avec l'homme assis.

— Dis-moi un truc, Lee. Tu as des idées ?

Hudek bloqua. Ça voulait dire quoi, ça ?

— Quel genre d'idées ?

— Eh bien, tu es doué. Nous sommes contents. Un gros turnover, une réelle discrétion. C'est tout ce que tu vises ? Tu estimes avoir atteint tes limites ?

Hudek hésita de nouveau. Il pensait avoir compris la question, mais craignait de se fourvoyer.

— Ben, ouais. Enfin, j'ai pensé à un truc.

— Explique-moi ça.

— La Floride, lâcha Hudek.

— Mais encore ?

— J'ai un plan. Tous les ans, des millions de jeunes déferlent en Floride pour les vacances de printemps, vous voyez ? Ça grossit d'année en année, un vrai parc d'attractions, avec une fiesta sponsorisée par-ci, un concert MTV par-là, et tout le business qui s'engouffre dans ces conneries. Cette année, j'ai passé une semaine à Panama City, pour me faire un avis. Et franchement, ça déconne à la base. T'as quatre cent, cinq cent mille étudiants qui débarquent dans cette ville chaque saison, de mars à avril. Ils veulent de la bière, ils veulent des tee-shirts mouillés, ils veulent s'envoyer en l'air. Ils veulent de la drogue. Même s'ils le savent pas encore, c'est ça qu'ils veulent. Alors on en trouve, bien sûr, et sans difficulté, mais c'est pas organisé. On pourrait mieux faire. Beaucoup mieux.

— Je vois, répondit l'homme. Et tu te dis : « Pourquoi seules les grosses boîtes comme IBM mettraient le grappin sur ces jeunes lions ? Si les entreprises américaines partent à la conquête de la Floride, pourquoi Lee John Hudek n'en ferait-il pas autant ? Pourquoi bouder cette clientèle ? »

— Exactement. Avec une équipe soudée et travailleuse, un mec malin écoulerait des quantités carrément démentes. Je veux être ce mec-là.

Voilà, c'était dit.

Il avait dévoilé le Plan. C'était la première fois qu'il l'énonçait à voix haute, et il en ressortait plus confiant que jamais.

— C'est une idée, approuva l'homme. Mais elle n'a rien d'original, et elle se heurte à trois obstacles que je vais t'exposer, si tu veux bien.

Hudek se figea, ses espoirs subitement douchés.

— D'abord et avant tout, les flics. Il y a déjà des flopées de dealers sur le terrain, comme tu t'en doutes. Ils jouent petit, ou ils arrosent les autorités locales. Ces bringues de printemps représentent une manne pour ces villes, Lee, elles amassent jusqu'à deux cent cinquante millions de dollars par saison, selon le nombre de connards en mal de bière qu'elles attirent dans leurs boîtes, venus s'éclater à mort avant de décrocher leurs jobs sinistres et de s'enterrer dans la normalité. Les municipalités savent que la drogue circule. Ça fait partie du jeu. C'est encadré, toléré. Mais si la ville est inondée de came, et que c'est exposé en une des journaux, c'est fini. Le boulot des flics, c'est de maintenir le juste milieu – et de grappiller quelques miettes pour l'entretien du yacht et le plan de retraite. Tu connais beaucoup de flics en Floride, Lee ? Tu as ce genre de relations ? Tu as une expérience du bakchich pour grosses légumes ?

Hudek secoua la tête.

— Je me disais bien. Le deuxième problème, c'est l'offre. À supposer que tu te mettes les flics dans la poche et que tu places ton personnel – et il t'en faudra beaucoup, et de gros moyens en termes de communication, de transport et de stockage, même si, là aussi, les flics peuvent t'aider –, la question du produit se posera toujours. Tu finances comment ? Je ne sais pas quel pourcentage de ton apport contient ce sac, mais ça ne suffira pas.

Hudek savait que c'était le point faible du projet.

— J'avais pensé... j'avais pensé à un accord du genre paiement différé.

— Quelqu'un te prête la drogue, tu vois si tu peux l'écouler, et sinon tu la retournes avec la facture et dix pour cent d'intérêts ? Et on ferme les yeux sur la came manquante, les pilules et la coke que tes revendeurs auront prises pour leur consommation perso ou marchandées contre une partie de baise sur la plage ? C'est vraiment à ça que tu pensais ?

Hudek haussa les épaules, les joues en feu.

L'homme du fauteuil garda son sérieux, mais un rire fusa dans l'ombre. Des rires.

— Pour finir, conclut l'homme, où comptais-tu trouver ces généreux mécènes ? Laboure le Sud en quête de contacts, et les gangs de Miami t'auront dépecé avant que t'aies pu ouvrir la bouche. Les Crips[1] sont des putains d'enfants de chœur à côté de ces racailles cubaines.

— D'accord, c'était une mauvaise idée, capitula Hudek en regardant ses mains.

— Non, Lee. C'est une *bonne* idée. Qui commence déjà à prendre forme. Mais elle n'est pas aboutie, tu as raison sur ce point. Il faudrait des stocks fiables et conséquents, et une filière de blanchiment. Il faudrait profiter de relations préexistantes avec les forces de l'ordre et avoir les moyens d'en créer de nouvelles. Il faudrait quelqu'un qui puisse t'aider à éviter les erreurs stupides et/ou de finir noyé dans les marais. Il te faudrait, en somme, de bons commanditaires.

L'homme le fixait intensément.

— Des commanditaires qui croient en toi, qui sachent qu'on peut te faire confiance. Que tu fais toujours ce qu'on te demande.

Hudek n'osait plus parler.

— Des commanditaires susceptibles de vouloir te mettre à l'épreuve, toi et ton groupe, de réclamer une ou deux faveurs, au préalable, en gage de bonne foi. Tu saisis ?

1. Gangs ultra-violents spécialisés dans le trafic de crack. (N.d.T.)

— Je crois. Qu'est-ce que vous voulez ?

L'homme sourit.

— On n'est pas obligés de voir ça tout de suite. Bientôt, mais pas dans la minute.

Il dévisagea Hudek un instant.

— Ravi de te revoir, mon ami. En sortant, tu recevras ce que tu es venu chercher. Cette fois, tu pourras remporter ton propre sac : c'est notre geste de bonne foi. Répartis le cash avec tes copains. Fais des heureux. On te contacte bientôt pour l'autre truc.

Hudek se leva.

— C'est avec vous que je vais bosser ? En direct, je veux dire ?

L'homme secoua la tête, et Hudek se surprit à penser « ouf ».

— Je repars dès ce soir. D'autres affaires m'appellent. Tu bosseras avec Hernandez. Joue le jeu. Regarde et enregistre. C'est un bon. Tu apprendras quelques ficelles. Utiles, pour de futurs postes.

Il lui adressa un clin d'œil. Hudek risqua un sourire.

L'homme pencha légèrement la tête. Hudek comprit le message et tourna les talons.

M. Reynolds attendait Lee dans le couloir. Il le guida à travers le bâtiment, puis dans la grande salle aux ampoules suspendues. Hudek éprouvait un mélange de fébrilité, de stupeur et d'euphorie, tout en peinant à admettre qu'il se faisait reconduire par le père de Stacy et Josh Reynolds. Il avait occulté cette bizarrerie-là pendant sa confrontation avec l'homme du fauteuil. Comme quoi ce type savait accaparer votre attention.

Juste avant la sortie, M. Reynolds s'arrêta.

— Ne mentionne ma présence devant personne. Je me contente de donner des avis juridiques, rien de plus. Tout le monde a le droit d'être conseillé.

— Pas de problème, monsieur Reynolds.

— Veille à ce que Bradley comprenne la même chose.

— Sans faute. Il fait ce que je lui dis.

— J'en suis certain. Je suis sûr qu'ils t'obéissent tous. Les armes devront rester ici, je le crains. (Il plongea la main dans la pénombre et produisit un petit sac. Rouge vif, frappé du logo Nike.) Mais ceci est pour toi.

Hudek baissa la glissière du sac de quelques centimètres. Il contenait la substance habituelle, mais en plus grande quantité, une vraie moisson.

— Merci.

— De rien, Lee. Mais si jamais j'apprends que tu as vendu cette merde à mes gosses, ne serait-ce qu'un gramme, tu regretteras d'être venu au monde.

Sur ces mots, il se retourna et disparut.

Hudek parcourut les deux derniers mètres et ouvrit la porte. Posa le pied sur le parking.

Il faisait encore jour, ce qui le fascina. D'après sa montre, seules trente-cinq minutes s'étaient écoulées depuis qu'ils avaient garé la voiture. Incroyable.

Au centre du parking poireautaient trois gars. Les siens.

Il les rejoignit. Brad semblait un peu vaseux. Pete et Steve avaient des marques rouges autour de la bouche. Tous trois paraissaient choqués, muets derrière leur cigarette. Pour une fois, Hudek aurait aimé en griller une avec eux.

— Ils nous ont dit d'attendre ici, expliqua Pete. Hernandez et les deux autres connards. Ils... Je sais pas. Ils ont attendu avec nous un certain temps, puis ils se sont cassés, comme ça. C'est l'hallu totale, mec.

Brad cligna des yeux, comme s'il revenait seulement à lui. Il comprit ce que Hudek tenait dans chaque main et fronça les sourcils devant le sac rouge.

— T'as la drogue ?

— Ouais.

— Et t'as toujours... la thune ?

— Exact. C'est un bonus, en quelque sorte. Tout baigne.

Brad secoua la tête. Son esprit nécessitait un redémarrage.

— Mais alors, c'était quoi, ce putain de traquenard ?

— Tout baigne, répéta Hudek. Je sais rien de plus.

L'équipe regagna la route. Lee n'était pas sûr d'avoir bien assimilé l'incident. Sa seule certitude, c'était d'en ressortir grandi et de traiter désormais avec des professionnels d'une tout autre envergure. S'il y avait une chose à ne pas faire, c'était de sous-estimer ces gens. Surtout le gars du fauteuil. Ses menaces pouvaient passer pour des phrases creuses, de celles qui servent à impressionner l'adversaire. Surenchère, cinoche. Mais Hudek n'y croyait pas. Pour lui, le type pensait ce qu'il disait, et peut-être même qu'il avait adouci le propos.

En tout cas, ce visage ne lui évoquait absolument rien.

— Je savais qu'y se passerait un truc chelou aujourd'hui, dit Dormeur en se frottant les poignets.

— Ta gueule, Pete, répliqua Brad.

— Je vous jure, les gars. J'ai fait un rêve super bizarre juste avant de me réveiller. J'étais au centre commercial et j'entrais au McDo pour déjeuner. Sauf qu'il faisait nuit, ça se voyait au fait que les grandes vitres derrière les tables étaient toutes noires, mais je sentais que c'était l'heure du déjeuner. Et alors là, au lieu de prendre un hamburger ou quoi, j'ai commandé une salade, ce qui est quand même délirant, non ?

— Ça défie l'imagination, dit Brad. Ça mérite les gros titres. Alerte CNN, mec.

— Non, attends. J'ai commandé une salade, d'accord, mais c'est pas ce qu'on m'a servi ! J'ai eu droit à un énorme sachet de Fritos. Et Dieu sait si je suis fana de chips, mais sur le moment j'avais pas envie de ça. Je voulais ma putain de salade ! Alors je lance : « Vas-y, raboule ta daube verte, mec. C'est quoi, ton problème ? » Là, le serveur me fait une espèce de sourire, et c'était pas le genre de robot habituel, il était beaucoup plus vieux, avec des cheveux gris, et il était gros et il avait un côté zarb et flippant. Il a repris les chips et m'a filé un sac en papier avec le haut roulé. Je me suis barré, et l'instant d'après j'étais sur le parking et j'ouvrais le sac, et c'était toujours pas ma putain de salade !

— C'était quoi ? demanda Brad malgré lui.

Hudek n'avait pas suivi un traître mot de la conversation. Le regard plongé au fond de la Valley, il savourait ce moment, conscient qu'ici et maintenant sa vie changeait de braquet. L'épisode du bâtiment paraissait incroyable – digne d'un rêve ou d'une série télé – mais les sacs qu'il portait dans chaque main prouvaient sa réalité.

— Une tarte aux pommes ! répondit Pete. J'allais pas refaire tout le chemin pour l'échanger, alors je l'ai ouverte et j'ai mordu dedans. Mais la garniture était toute rouge. Et elle était carrément glacée !

Brad le considéra.

— C'est ça, ton histoire ?

Tous les portables se mirent à sonner en même temps.

— OK, dit Lee. Les clients s'impatientent. Au boulot.

À l'intérieur du bâtiment, les autres hommes avaient quitté la pénombre. Quelqu'un actionna un interrupteur et la salle s'emplit de lumière.

— Il fera l'affaire, déclara l'homme du fauteuil. (Il se releva en douceur, s'étirant les épaules et le dos.) Mais titillez-le quand même. Et on pourra y aller.

CHAPITRE 5

Jim était installé derrière la vitrine de chez Marsha, en Caroline du Sud. Il avait suivi la route 95 jusqu'à Savannah, avant de bifurquer vers la 321 au bout de vingt-cinq kilomètres. Il avançait à un rythme d'escargot depuis son départ de Key West, après avoir déjà perdu une journée par rapport à ce qu'exigeait ce voyage. Tout le temps de sa remontée vers le nord, il s'était cantonné à la file de droite, comme n'importe quel papy dans un tacot, le genre qu'on double à fond sur toutes les autoroutes du pays. La raison première était simple : cela faisait des lustres qu'il n'avait pas entrepris un tel périple, et depuis huit ans la voiture servait uniquement à des trajets courts. Mais il se réhabitua vite à sentir la route filer sous ses pneus, et dès lors sa prudence n'excusa plus sa vitesse. Pas plus que son caractère de cochon, même si cela jouait également.

De gros nuages s'amoncelaient dans le ciel. À peine 17 heures, et il faisait déjà sombre. D'après les rumeurs du comptoir, que Jim pouvait entendre sans effort, on attendait de sérieuses averses, imminentes.

La serveuse s'approcha et remplit sa tasse sans rien lui demander. Il lui sourit, elle fit de même, puis elle partit en se dandinant accomplir de nouvelles BA. Jim suivit son reflet dans le chrome du distributeur de serviettes. Ses cheveux étaient d'un drôle de blond criard, et elle eût fait un réfrigérateur de grande capacité. Elle n'en possédait pas moins un réel sex-appeal, quelque chose d'authentique dans sa banalité. Étonnant, comme phénomène. Des mains habiles et de la gentillesse séduisaient plus qu'on ne l'imaginait. Jim s'aperçut alors, sans grande surprise, qu'il avait passé une décennie entière sans faire l'amour. Une pensée plus réconfortante qu'autre chose.

Il délaya une cuillerée de sucre dans son jus et scruta au-dehors. L'heure avait sonné. Jim avait traîné les pieds, décrété la route du nord trop raide pour son moteur. Mais maintenant qu'il se trouvait à une heure de sa première destination, il était temps de voir les choses en face. Il allait là où il allait, à moins de renoncer tout de suite. Il restait à peine quelques kilomètres, et cet intervalle était le moment clé : s'il avait l'intention de ne rien faire, il n'y avait pas une minute à perdre.

Ses tripes lui infligeaient une sensation connue. Une tension creuse, si sourde qu'il aurait pu s'agir d'une simple faim. Jim jeta un œil sur le menu calé contre le rebord et le refusa une fois de plus. Il fallait bien se nourrir, pourtant. Deux tables plus loin, un client avait pris un sandwich au corned-beef, et l'odeur ne semblait pas déplaisante : le pain légèrement grillé, la choucroute chaude et riche, la sauce savoureuse et épaisse, comme Jim les aimait autrefois. Il avait depuis toujours des goûts culinaires très précis. La sensation disparaîtrait peut-être s'il avalait un morceau.

Était-ce son souhait ?

Il n'en savait rien. Sincèrement rien.

Alors il termina lentement son café puis s'en alla, laissant un pourboire d'un dollar sur une addition d'un dollar cinquante, en espérant qu'il irait à la bonne serveuse.

De retour à sa voiture, il avisa le sac sur le siège passager et resta perplexe un instant. Ah, oui. Il se souvenait à peine des emplettes qu'il avait faites dans un centre commercial au sud de Jacksonville. Mais il avait bel et bien acheté le contenu de ce sac, pour sûr, en même temps qu'un objet plus lourd, juste avant de quitter les Keys, et cela signifiait sans doute qu'il avait pris sa décision.

On s'en moquait, de toute façon. D'autres avaient déjà tranché pour lui. Après tout la terre tourne, le cœur bat et le sang circule sans qu'on nous demande notre avis.

Au bout de cent trente kilomètres, Jim prit la direction de Benboro. Peu après il se trompa d'embranchement et rebroussa chemin. Ce secteur ne lui était pas familier, et ici on ne faisait guère d'efforts pour aiguiller les étrangers. Le genre de coin qu'on se contente de traverser. Hormis quelque futaie de loin en loin, la plaine était nue des deux côtés de la route.

Passé Benboro, c'était plus simple : une seule route quittait la ville. Au bout d'un kilomètre se dressait un grand panneau indicateur. Jim se rappelait qu'il était autrefois vert, mais entre-temps on l'avait repeint en rouge – voilà plusieurs années, à en juger par son état. Visiblement, la personne chargée de cette tâche considérait les lettres comme de vagues motifs géométriques et non comme des symboles chargés de sens.

BENBORO PARK, indiquaient-elles.

Il suivit la flèche et enfila la route d'accès, sûr de la trouver inchangée : des coups de fil impulsifs, tous les deux ou trois ans, lui avaient confirmé que le numéro du village de mobile-homes restait valide. À chaque fois, Jim raccrochait trop tôt pour savoir si le parc était encore géré par la même bonne femme ou s'il avait changé de nature. Mais pourquoi aurait-on mis à la rue tant de familles, de couples retraités et de célibataires hirsutes ? La ville de Benboro possédait le dynamisme d'une vieille chaussette : personne n'y implanterait de lotissements ou de zones d'activités avant longtemps.

Qu'en avait-il à faire, de toute manière ?

N'empêche qu'il avait téléphoné, tous les deux ou trois ans.

La voie suivait un tracé sinueux qui, croqué à la va-vite dans une cabane de promoteur, avait sûrement paru léché, mais qui n'était qu'interminable. À mi-chemin s'étendaient une soixantaine de caravanes, dans autant de styles et d'états différents. Chose assez rare, les rues qui les reliaient ne décrivaient pas un quadrillage standard. Le type de la cabane aimait les fioritures. Les résidents seraient d'autant plus difficiles à localiser, ce qui offrait sans doute des avantages et des inconvénients. Mais Jim savait où il allait. Aux confins du parc, là où un bouquet d'arbres marquait le début d'une friche courant jusqu'au bord d'une triste rivière, s'alignaient quatre baraquements en bois. Aussi vastes que déglingués. Deux servaient à stocker de la ferraille et du matériel pour l'entretien du parc. Les autres étaient subdivisés en hangars de location.

Jim se rangea devant l'entrée du parc. Une sorte de portique – deux poteaux métalliques soutenant une planche à hauteur de caravane – confirmait que l'on était à Benboro Park et non à Bel Air, au paradis ou dans le meilleur des mondes. Derrière, la route partait en fourche. Entre les deux branches reposait un mobile-home du même rouge que le panneau de la route principale. C'était l'emplacement n° 1, où habitait la vieille gérante. Hannah, qu'elle s'appelait. Si elle était encore de ce monde.

Jim sortit de la voiture. Les nuages s'étaient alourdis, anthracite, figés, menaçants, mais la pluie se retenait. Jim espérait qu'elle finirait par tomber, ne serait-ce que pour les gaillards perchés au comptoir de Marsha – ils avaient tellement l'air d'y tenir. Quitte à gâcher le plaisir de cette fillette qui jouait seule devant une caravane, dans la ruelle de droite. Elle fredonnait pour elle-même, et c'était assez joli.

Marchant jusqu'au n° 1, il révisa la petite histoire servie à l'époque. Il sortait d'un long et pénible divorce, oh oui, et voilà tout ce qu'il avait réussi à sauver : des misères, mais il y était attaché. Il voulait les mettre à l'abri, loin des avocats et des huissiers. Il partait pour Miami. Un ami pensait lui trouver du boulot dans un hôtel. Sinon, il filerait peut-être en Arizona, ou au Texas, pour pousser sa chance vers l'ouest.

Il frappa à la porte, tendant l'oreille aux bruits d'un téléviseur. On finit par lui ouvrir.

— Oui, m'sieur ?

La femme dont il se souvenait. Après tant d'années à mariner dans un bocal de fumée, sa peau avait pris la non-couleur d'un ex-torchon blanc. Tirés en queue-de-cheval, ses cheveux secs et gris-bruns clamaient qu'elle se savait à vomir et n'en avait rien à cirer.

— Bonjour, dit Jim dans un grand sourire. Hannah, c'est exact ? Je ne sais pas si vous me remettez.

— Je peux pas dire ça, non. Z'êtes pas du parc.

— En effet. Je vous ai loué un hangar il y a un petit moment de ça. J'aurais besoin d'y accéder.

— D'accord. Quel numéro ?

— Dix-sept, répondit Jim.

Elle s'éloigna vers un bureau où régnait une pagaille cataclysmique. Croisant les doigts, Jim attendit à l'extérieur du mobile-home, les yeux sur la route. La fillette avait disparu.

Hannah revint au bout de deux minutes.

— Un petit moment, comme vous dites ! Douze ans. Vous n'avez payé que pour cinq.

— J'ai été retenu, expliqua-t-il.

Elle hocha la tête.

— C'est vous, le gars qui partait en Australie ?

— Miami. C'est bien moi.

— Z'avez pas aimé ?

— Comme ci comme ça. Un peu trop chaud.

— Chaud ? M'en parlez pas. On a eu un de ces étés, et il a toujours pas plu. Vous me devez de l'argent.

Il lui remit la liasse qu'il avait préparée. Elle compta les billets.

— Je n'ai pas tenu compte de l'inflation, glissa-t-il.

Elle rit.

— Y a pas d'inflation, par ici. On n'a pas les moyens !

Jim sourit.

— J'aimerais prolonger d'un an, si possible.

— Aucun problème, et je vois que l'argent est là. (Elle lui confia une petite clé rouillée.) Bonne nuit. Vous la laisserez sur le perron.

Puis la porte se referma, et Jim en eut terminé.

Remontant en voiture, il s'enfonça dans le parc en direction des hangars, bluffé par tant de facilité. Voilà douze ans, il s'était pointé très tard, à cran, avec des fables qui tenaient à peine debout. Et malgré cela, Hannah l'avait reconduit à Benboro pour qu'il attrape le car de Miami. Il avait réglé cinq années d'avance, puis disparu plus du double. On se serait attendu à ce qu'elle... enfin bref. Il avait juste fait le bon choix, en flairant que ces hangars connaissaient un faible roulement. À moins que la vieille n'eût vendu ses affaires depuis belle lurette et ne fût présentement tordue de rire sur son pognon, derrière sa porte verrouillée.

Il se gara devant la troisième remise et marcha jusqu'à la cinquième porte. Il tourna la clé et pénétra à l'intérieur.

L'espace 17 consistait en un simple box de trois mètres sur six. Jim constata que son contenu n'avait pas bougé.

Il ôta la bâche et la laissa par terre. Puis resta planté devant la bête. Il s'était interdit tout sentimentalisme, mais c'était plus fort que lui.

Un vieux Combi Volkswagen, de couleur blanche, dont l'aspect ni trop neuf ni trop amoché n'attirait pas les regards. Un immense pare-brise, pour une visibilité maximale, et des flancs aveugles. Une petite lunette arrière, aux trois quarts du coffre, voilée d'un épais rideau blanc. Invisibles de l'extérieur, une kitchenette et une minuscule couchette séparée, dans la largeur de l'habitacle, qui obligeait à dormir sur le côté avec les jambes légèrement relevées. Tout ce dont un voyageur avait besoin. Ce voyageur-ci, du moins.

Jim regagna sa voiture pour extraire les deux sacs. Il ouvrit sa petite valise, plongea la main dans la boîte à chaussures et sortit les vieilles clés de voiture. Ça faisait tout drôle de les reprendre en main, avec ce vieux porte-clés en plastique râpé, un objet publicitaire signalant une fête scolaire vieille de dix-huit ans. Il resta immobile, se remémorant ce fameux après-midi et l'achat de cette breloque. Dans une autre vie.

Revenu à l'espace 17, il déverrouilla la camionnette et jeta sur la banquette le sac le plus léger. Puis il emporta le plus lourd à l'arrière du véhicule. Il purgea le fond du réservoir et versa de l'essence neuve. Il retira la batterie, la remplaça par celle du sac, puis rangea l'ancienne dans son coffre de voiture. Avant de rejoindre le van.

L'instant de vérité. L'humidité avait pu gripper le système électrique. L'huile avait stagné. Longuement.

Il grimpa derrière le volant, sentit le siège se creuser comme un vieil ami. Il inséra la clé et la tourna sans cérémonie.

Un clic, puis rien.

Il réessaya. Le Combi toussa, péta, et ressuscita joyeusement. Jim secoua la tête, tout attendri. Il n'était pas le premier à s'ébaubir du génie des ingénieurs Volkswagen.

— Content de te revoir, vieux cheval.

Dix minutes plus tard, il reposait la clé sur le perron de l'emplacement n° 1 avant de retrouver le van et son doux teuf-teuf. Il laissa un couple de quinquagénaires traverser la rue. Ni l'un ni l'autre ne lui prêta attention. Une camionnette blanchâtre, et alors ? Sans compter que Jim avait plus de soixante ans et qu'on prête rarement de grands projets aux hommes de cet âge. Sa voiture avait pris la place du Combi dans l'espace 17, sous la bâche. Elle renfermait la tenue précédente de Jim. Il portait maintenant un jean noir et une chemise en chambrai achetés au centre commercial. Ce n'était pas le genre d'habits qu'aurait arborés Jim Westlake, mais plutôt le style d'un certain James Kyle, enseignant, propriétaire, un type comme vous et moi.

La fillette avait repris ses jeux solitaires dans la rue. Jim fronça les sourcils. Il aurait fallu qu'un adulte la surveille, qu'un adulte s'installe sur les marches, avec une bière si nécessaire, mais qu'on garde un œil sur elle. Les résidents de Benboro Park se connaissaient entre eux, mais ça ne suffisait pas. Le mal frappe facilement les enfants. Trop facilement. Si facilement que c'en est déprimant. Dans un monde normal, on se serait organisé de façon à protéger les innocents, afin que chaque individu vive sa vie et se dise en bout de course : « En fin de compte, ce n'était pas si mal. » Mais était-ce jamais le cas ? Chacun passait son temps à regarder du mauvais côté. Au lieu de penser plans de carrière et pelouses impeccables, de se focaliser sur les chaussures à la mode, les régimes minceur ou la célébrité, au lieu d'être obsédés par l'image qu'ils renvoyaient ou qu'ils se faisaient d'eux-mêmes, les gens devraient veiller les uns sur les autres, sur leurs enfants, leurs parents, leurs conjoints et leurs animaux de compagnie. Se consacrer à protéger ces trésors que sont les êtres aimés, car il fallait toujours un deuil ou une cassure pour mesurer tout ce qu'ils avaient de merveilleux, d'unique. Mais les gens n'y pensaient jamais à l'avance, parce qu'ils étaient stupides. Parce que la vie était un puits de distractions. Parce que c'était comme ça.

C'était l'une des raisons qui avaient motivé ses actes, jadis. Pour donner au monde le sens des priorités. C'est du moins ce que Jim s'était dit, parfois, mais il se disait beaucoup de choses à l'époque, et la plupart étaient fausses. Sur ce point, il ne valait pas mieux que les autres. Au fond, pensait-il, chacun possède deux personnalités qui se racontent sans arrêt des mensonges. Des mensonges plus ou moins gros et plus ou moins mortels.

Au bout de quelques kilomètres, le moteur était bien chaud et semblait apprécier le retour sur la route. Jim rattrapa la 321, qu'il suivit vers le nord dans la lumière du crépuscule, toujours talonné par les nuages orageux.

CHAPITRE 6

Dans la forêt de Raynor, Nina tentait de se concentrer malgré le babillage des collègues. Elle avait passé la matinée au commissariat de Thornton, à prendre connaissance du dossier et à regarder les innombrables clichés en noir et blanc d'un homme retrouvé mort. Tout cela s'était révélé en grande partie superflu. Après qu'Olbrich les eut quittés pour regagner Los Angeles, Monroe et Nina avaient parlé de l'affaire tout au long du voyage. Elle était au parfum. Les faits étaient limités, et plus on s'y attardait, plus ils devenaient flous, tel du pain sous la pluie qui gonfle et perd toute consistance.

Monroe se trouvait six mètres plus loin, parmi un troupeau de flics, à répéter les mêmes infos. Elle essayait d'occulter sa voix lorsqu'une autre surgit, bien plus proche.

— Vous voyez ces buissons ? C'est pour ça que personne n'a rien vu plus tôt.

C'était Joe Reidel, un jeune trapu de la crime. Il faisait partie de l'armada dépêchée à Thornton par le shérif du comté de Cathridge le matin précédent. La présence de la police judiciaire ne semblait pas contrarier les agents locaux. Celle du FBI non plus, bien qu'il eût été sollicité par Joe Reidel. On comprenait vite que la municipalité ne goûtait guère ces apparitions inopinées de cadavres et se réjouirait qu'on y mette fin. Reidel était le seul à ne pas avoir livré sa version des faits à Nina. Si elle le laissait parler, peut-être que les autres se décideraient à fermer leur gueule et qu'elle pourrait enfin méditer ses propres pensées.

— Je vois, dit-elle.

La forêt de Raynor enveloppait le nord de la ville. Le ruisseau qui la traversait rendait la terre marécageuse, pleine de flaques survolées par des nuées de moucherons. On avait retrouvé le cadavre à demi plongé dans l'une d'elles, à quelques mètres d'une grosse bosse dans le sol. Un bouquet d'arbustes l'isolait du chemin qui longeait le ruisseau. Nina examina cette configuration.

— Quand même, reprit-elle, on ne peut pas dire qu'on ait tout fait pour cacher le corps.

— En effet. Et c'est une promenade très prisée. (Reidel indiqua la côte où le bois se réduisait. On apercevait au sommet un petit parking.) On n'ose pas l'appeler le Chemin des Amoureux, mais c'est le haut lieu du batifolage.

— Elle n'a rien de très charmant.

— C'est une petite ville, ici. On fait ce qu'on peut avec ce qu'on a.

— Toujours aucune trace de ses fringues ? Pas de sang ?

— Non. (L'inspecteur indiqua les alentours.) C'est délicat, à cause des feuilles, des brindilles et de toutes les saloperies de la forêt, mais je suis certain que rien n'a bougé avant l'arrivée des technos. Le couple qui a trouvé le corps est resté en retrait.

— Les analyses des entailles sur le torse et sur le ventre indiquent-elles que la victime est morte habillée ? demanda Nina.

— Très probablement. Plusieurs présentent des traces de fibres. Même si deux autres, à l'aine, sont immaculées. Ce qui nous porterait à croire...

— Qu'un acte sexuel était en cours au moment de l'agression. Noté.

Nina plongea la main dans l'enveloppe coincée sous son bras et produisit deux photos de la victime in situ. La première était une vue d'ensemble, sûrement très proche de ce qu'avait découvert le couple de témoins. Un corps, dans les bois, à la tombée du jour. Grand, livide, étendu en silence. Un spectacle si inattendu qu'on ne remarquait pas le sexe sur-le-champ, bien que le cadavre fût entièrement nu. Couché sur le dos, les jambes raides. Des marques de couteau sur le ventre et la poitrine. Une poitrine d'homme.

Le second cliché était plus rapproché. La tête baignait dans l'eau ; seuls émergeaient le menton et le nez. Les yeux étaient noyés mais ouverts, tout comme la bouche. La netteté de la photo montrait que les régions glabres avaient une texture moite, comme une pièce de viande crue restée trop longtemps sur le comptoir. Le bras gauche était couché sur le côté, et le droit légèrement relevé, fendant l'eau telle une branche morte. Il n'y avait rien au bout.

— Aucune nouvelle de la main ?

Reidel secoua la tête. Il n'avait pas l'air d'un type sur le point de rentrer chez lui.

— Et puis c'est propre, ajouta Nina. Dix-sept coups de poignard, plus l'amputation, sans que le corps soit barbouillé de sang. Sachant qu'il n'a pas plu depuis.

— Ce qui suggère que la victime a été exécutée ailleurs, avant d'être dénudée, mutilée et amenée ici. Simple hypothèse.

— La mort remonte à l'avant-dernière nuit ?

— Tout à fait. Comme l'abandon du corps, d'après les pathologistes, qui se basent sur la concentration de bestioles et de micro-organismes sur le cadavre. L'amputation, quant à elle, est post mortem, ce qui explique mieux l'absence de sang. Attendre que le type soit mort pour lui couper la main, c'est s'épargner beaucoup de peine.

— Le travail n'en reste pas moins soigné, commenta Nina.

Elle se replongea dans ses notes. Main sectionnée par deux ou trois coups d'un objet lourd et tranchant, et non par un mouvement de scie. Examen sanguin complet à venir.

— Et la zone de stationnement la plus proche, c'est ce parking en haut de la côte ? Ça fait une sacrée distance pour porter un corps de cent kilos. Surtout pour une femme.

— Le cadavre présente des coupures et diverses contusions dans le dos. L'idée, c'est qu'on l'a traîné.

— Même. Je n'aimerais pas devoir le tirer sur cette longueur, et pourtant je suis sportive. Je l'étais, en tout cas.

— Ce meurtre pue le sexe à plein nez, et la victime n'était pas homo. C'était un père de famille de quarante-six ans.

— Les hommes mariés peuvent être homosexuels. À moins que ce ne soit plus rare dans le Sud ?

Il sourit.

— Je suis né à Washington, madame. On voit toutes sortes de choses là-bas, pour peu qu'on sache où regarder. Dans le cas présent, la victime est un homme réputé pour ses mains baladeuses – sur le derrière des femmes. Et qui fréquentait les bars du coin. Personne ne se souvient de l'avoir vu repartir avec une nana, mais il passait son temps à les draguer. Y compris celles qui tenaient le bar, comme cette serveuse qui n'hésite pas à le qualifier de « chasseur de chattes ». Il nous reste à interroger l'ensemble du personnel. Rien ne permet d'affirmer qu'il n'était pas gay, certes, mais pour ce qui est des preuves directes, à vous de jouer. Tant que vous y serez, vous aurez sûrement à cœur d'expliquer ces traces de rouge à lèvres sur le cou de la victime.

— Je chercherai aussi des preuves que ce meurtre est l'œuvre d'un tueur en série. Car, pour l'instant, je ne vois rien de plus qu'un crime isolé.

— C'est vous l'experte. (Reidel éteignit son mégot et le rangea dans son paquet de cigarettes.) Bon, je vous laisse cogiter en paix, agent Baynam. S'il vous faut quoi que ce soit, vous savez où me trouver.

Il descendit retrouver ses camarades. Au bout de quelques instants, un rire s'éleva du groupe.

Nina se détourna. Considéra longuement la pente.

Nina et Monroe partirent seuls interroger Julia Gulicks et Mark Kroeger. Ces deux habitants de Thornton travaillaient ensemble à Owensville, la ville la plus proche. C'est à la faveur de leur cinquième rendez-vous qu'ils s'étaient promenés dans la forêt de Raynor. Ils n'avaient pas encore couché ensemble. Ils avançaient pas à pas, voulant croire au grand amour. C'étaient encore deux gosses, même si, objectivement, à vingt-cinq et vingt-neuf ans, ce n'étaient plus des gamins.

On les avait réunis dans la salle de conférence de l'entreprise. Ils étaient mal à l'aise, mais Nina n'y voyait rien d'étonnant. Au bout de trois semaines de rancards clandestins après le boulot, dans un café du coin de la rue, leur idylle naissante devait déchaîner les langues devant la fontaine à eau. Nina crut déceler chez Julia Gulicks une soif de discrétion qui n'était pas sans rappeler la sienne. Passé l'âge de vingt ans, les rapports sentimentaux n'ont plus rien d'un jeu, encore moins d'un spectacle.

Monroe conduisait l'entretien :

— À quelle heure avez-vous quitté le bar ?

— Vers 21 heures, répondit Mark Kroeger, qui avait une voix posée et quelques cheveux gris sur les tempes. Un peu plus tôt que d'ordinaire, parce que... Disons que ces derniers temps on avait pris l'habitude de se rendre à l'Italian Kitchen. C'est à deux rues d'ici.

— Et pourquoi n'y êtes-vous pas allés jeudi dernier ?

Kroeger sembla rougir. Il coula un regard vers Julia Gulicks, puis vers le sol.

— Eh bien... fit la jeune femme. (Ses cheveux étaient d'un roux flamboyant, son teint pâle et tavelé.) On a déjà évoqué tout ça, n'est-ce pas ?

— Je sais, répondit Monroe. Mais s'il vous plaît...

Réprimant un sourire, Nina cessa d'écouter. Tout était consigné dans ses notes, arraché par Joe Reidel lors des auditions de la veille. Kroeger avait livré l'essentiel. La nuit de jeudi soir devait être le grand soir. Il le savait. Et il pensait que Julia le savait aussi. Ce postulat tacite avait empreint la soirée d'une certaine solennité et bridé la conversation. Ils s'étaient retrouvés après le boulot pour filer au bar d'Union Street, comme d'habitude. C'est là que s'étaient déroulées leurs trois premières moitiés de rancard, les trois autres moitiés se concluant chez l'Italien, où le personnel était sympa et savait vous traiter comme un couple. Du deuxième au quatrième rancard inclus, ils s'étaient embrassés avec une ferveur grandissante, de sorte que le cinquième rancard monta sur le ring avec la ferme intention d'en découdre. Aucun des deux protagonistes ne savait s'il y aurait un restau et aucun n'osait le demander. Nina aurait parié que, ce soir-là, on aurait trouvé à Thornton deux appartements exceptionnellement bien rangés. Julia Gulicks aurait même changé les draps. Kroeger n'aurait pas poussé le bouchon aussi loin, mais il aurait au moins fait son lit. Chaque frigo aurait recelé une unique bouteille de bon vin – pas plus, puisque tous deux se disaient petits buveurs. On aurait rajusté le sofa, remis de l'ordre dans la bibliothèque, avec les ouvrages intellos bien en évidence... Ni l'un ni l'autre n'eut pourtant le courage de lancer : « Tiens, et si on allait chez moi ? » Pas plus qu'ils n'osèrent suggérer un Italien, puisqu'on ressortait toujours de là avec le ventre lourd, ce qui n'est pas idéal pour, enfin bref...

Alors ils s'étaient rapprochés sur la banquette, embrassés un peu, un peu seulement car ce n'était pas l'endroit le mieux indiqué et qu'il y avait vingt-cinq minutes de voiture jusqu'à Thornton – or personne ne voulait conclure trop vite. Petit à petit, la conversation avait glissé des anecdotes de la journée vers le temps qu'il faisait, et vers la clémence du ciel ces jours-ci, et à ce propos – mais oui, en voilà une idée : si on allait faire un tour, histoire de profiter des derniers jours de l'été ? Kroeger avait réglé l'addition, et ils avaient marché dans la rue un petit moment, mais à vrai dire il n'y a pas grand-chose à voir à Owensville, et cette lacune les ramena vite à la voiture de Kroeger. Direction Thornton, même si, de l'avis général, il était encore un peu tôt. Puis voilà que, juste après la sortie, moins d'un kilomètre avant la ville, Julia avait eu une idée géniale : « Et si on se garait pour aller se promener ? »

Monroe hocha la tête.

— C'est un itinéraire que vous connaissez bien ?

— Non, répondit Julia Gulicks. Enfin, si. Je connais le bois, comme tout le monde par ici. On y croise plein de familles la journée. Mais je n'y étais jamais allée... la nuit.

— Moi non plus, fit Kroeger.

Qu'ils sont choux, songea Nina.

— Mais les habitants du coin en font couramment usage ?

Gulicks et Kroeger opinèrent de conserve.

— Donc vous vous êtes mis en marche, puis vous avez stoppé, et c'est là que M. Kroeger a repéré quelque chose derrière les buissons.

Kroeger expliqua qu'ils avaient parcouru une centaine de mètres après le parking, avant de longer le ruisseau quelque temps. Levant les yeux alors qu'il tenait Mlle Gulicks dans ses bras, il avait aperçu une forme blanche, étendue avec un bras déployé. Le couple s'était approché et, comprenant la nature de cette découverte, avait appelé la police avec un téléphone portable.

Quatre heures plus tard, chacun des tourtereaux regagnait son appartement, seul. Le grand soir n'avait pas eu lieu, et Nina savait que ça risquait d'attendre. L'image d'un cadavre s'accroche longtemps à la rétine. Rien que d'en reparler, Kroeger semblait malade.

— On a identifié la victime, annonça Nina. Un homme du cru nommé Larry Widmar. L'un de vous le connaît ?

Ils secouèrent la tête, et ce fut terminé.

Pour rejoindre Thornton, Monroe prit le volant. Nina avait souvent roulé avec lui, et elle constata qu'il avait réduit sa vitesse de croisière d'au moins dix pour cent. Ses blessures par balle avaient sans doute émoussé son goût du risque, comme si son corps lui lançait des signaux d'alerte. Il paraissait plus vieux, du reste, et Nina comprenait ce phénomène. Elle aussi s'était fait tirer dessus, voilà presque un an, peu après sa rencontre avec Ward. Cela s'était produit dans un lieu baptisé les Halls, dans les montagnes près de Yellowstone. L'un des types compromis dans le meurtre des parents de Ward l'avait touchée à la poitrine, juste sous la clavicule. Pendant les mois suivants, elle avait accusé le coup, comme si les vents froids lui traversaient le corps. Et aujourd'hui... elle ne savait pas trop. Curieuse sensation que de retrouver le monde extérieur, de reprendre son travail. Un côté virtuel, irréel. Et puis elle avait mal au crâne.

Les flics du coin allaient passer la soirée à quadriller les bars, en quête d'un témoin ayant vu Widmar mercredi soir ou à tout autre moment. D'autres monteraient la garde en forêt de Raynor pour terroriser tout couple convaincu que ce soir serait *son* soir. Monroe et Nina filaient accomplir la dernière tâche de la journée : entendre l'autre personne supposée impliquée.

— Reidel t'a raconté quoi à mon sujet ? demanda soudain Nina.

— Quand ça ? répliqua Monroe.

— Tu le sais très bien. Quand il est revenu au ruisseau après m'avoir embêtée alors que j'essayais de réfléchir.

— Que tu semblais persuadée que le tueur n'était pas une femme. Ce qu'il trouvait curieux, vu que tu semblais capable de castrer un homme sans le moindre état d'âme.

— Et ça t'a fait rire ?

— Pas moi, Nina.

— Tu sais comme moi que c'est débile, cette théorie de femme.

— Les femmes aussi sont capables de tuer, Nina.

— Pas de cette façon.

— Je peux t'en citer plusieurs qui ont été condamnées.

— Condamnées ne veut pas dire coupables.

— Eh si, figure-toi. C'est comme ça que ça marche.

La conversation s'éteignit rapidement, ce qui permit à Nina d'admirer la ville à mesure qu'ils s'y enfonçaient. Non que le panorama fût si savoureux. Elle n'aurait su l'expliquer, mais Nina n'aimait pas Thornton. Certes, la bourgade avait tout pour plaire, dans le sud-est de l'État, à trente minutes de Smith Lake et à une heure du parc national de Blue Ridge. La route d'Owensville vous menait en douceur vers un petit quartier commerçant. On pouvait avaler un hamburger chez Renée, faire une vidange, stocker des trucs ou les expédier, acheter une tondeuse ou prendre une chambre dans un hôtel de chaîne. On pouvait aussi passer directement de l'autre côté, et ne rien voir de la vieille ville. Mais si l'on prenait à gauche, au niveau du Ponderosa, la route serpentait dans les collines jusqu'à une grande église ancienne et un lycée. On découvrait ici un ensemble d'édifices plus ou moins gothiques, trônant fièrement derrière une grande pelouse boisée. De l'autre côté de la rue, une construction plus modeste abritait le jardin d'enfants de Sleepyheadz. Après quoi la route redescendait vers des rues bordées d'arbres, très respectables, qui proposaient un café Starbucks, des restaurants, ainsi que des boutiques où l'on réalisait de jolis encadrements. Les immeubles ne faisaient pas plus d'un étage, toutes les façades étaient en bois, les pavés couverts de feuilles étaient en brique rouge. Les gens déambulaient avec le journal local sous le bras. De jeunes et sveltes mamans promenaient leurs bambins et s'arrêtaient pour échanger quelques mots. Un fourgon UPS ronronnait çà et là, distribuant ses petits cadeaux. Puis le secteur perdait son prestige sur quelques pâtés de maisons, jusqu'à ce que les rues débouchent sur de grandes zones pavillonnaires cramponnées à leurs rêves. Peu après, c'était à nouveau la campagne, un tiers de la ville cerné par la forêt de Raynor sur plusieurs kilomètres vers le nord.

Le genre d'endroit qui chaque année loupe de justesse le classement des Petites Villes les plus charmantes d'Amérique. Il n'empêche qu'au bout de vingt-quatre heures Nina s'y déplaisait déjà.

Il fallait tenir compte du fait qu'elle avait connu Thornton en tant que lieu du crime, ce qui teintait la bourgade d'une ambiance particulière : aussi coquette fût-elle, elle avait opposé deux individus dans un contexte de meurtre. Comment en arrive-t-on là ? La ville était-elle impliquée dans les actes commis sur son sol ? On s'attend que les gens s'entre-tuent dans les mégapoles : en notre for intérieur, nous savons qu'elles sont trop vastes et qu'elles agglutinent des inconnus sans jamais préciser le pacte moral qui les lie. Mais les petites villes... N'étaient-elles pas censées générer de l'entraide, incarner le type même de communauté qui empêche de tels drames ?

Cela dit, Nina était dans le métier depuis trop longtemps pour se bercer d'illusions sur les vertus sociales de la ruralité. Si les centres urbains fournissent le gros de la criminalité, les petites villes en offrent souvent les expressions les plus baroques.

Non, il y avait quelque chose de particulier à Thornton. Quelque chose de pas net.

De taille moyenne, la maison de Widmar datait du dix-huitième siècle. On y accédait par une terrasse de brique couverte. Gayle Widmar approchait de la cinquantaine. Une femme très chic, dont le brushing semblait hors de prix, même après trente-six heures de deuil. Ses enfants se trouvaient chez sa sœur, à quarante minutes de là, où Gayle les rejoindrait sitôt l'entretien bouclé. Son sac de voyage l'attendait dans le hall à cet effet. La maison trempait dans le silence pantois d'un domicile où tout vient de basculer.

Mme Widmar s'installa dans un fauteuil à haut dossier, au centre du grand séjour, pendant que Nina la faisait récapituler. Le mari de Gayle, Lawrence de son prénom – pas une seule fois elle ne l'avait appelé Larry –, possédait une « chaîne » de pressings (c'est-à-dire deux boutiques, comme Nina le savait déjà) et la moitié d'une pizzeria en vogue dans le centre historique. Il siégeait au conseil de l'école. Ils auraient pu habiter un quartier plus chic, mais ils aimaient celui-ci, où ils avaient tous deux grandi. Ils étaient mariés depuis vingt ans.

Mme Widmar se montra sèche et tendue, comme souvent lorsqu'on perd brutalement son conjoint. On a besoin de soutien mais on ne fait plus confiance au monde. On se sent vaguement mis en cause. On se dit que les voisins pensent qu'on l'a bien cherché : que si l'on s'était montré plus vertueux, ce drame n'aurait jamais frappé l'autre. Chaque fois qu'il y a meurtre, cinq pour cent des critiques visent les proches de la victime, et nous lui en voulons, inconsciemment, de nous mettre dans cette position, dont il faudra se dépêtrer tout seul.

Pendant que Monroe faisait confirmer à Mme Widmar les derniers mouvements connus de son mari – parti mercredi soir dîner avec son associé de la pizzeria, jamais rentré, déclaré disparu par son épouse à 7 heures le lendemain matin –, Nina examina les photos sur la cheminée. Lawrence Widmar était à ce point normal qu'on l'aurait presque remarqué. Sourire artificiel, cheveux bouffants et grisonnants, un aplomb très pilier-de-la-communauté. On l'imaginait bien dans une banque. On l'imaginait bien produire un beau discours lors d'une réunion de parents d'élèves, et soutenir avec flamme l'équipe de base-ball de l'école. On l'imaginait bien dans un bar, également, rapprochant son tabouret pour demander ce que la petite dame prendra. C'était peut-être injuste, mais la mort soulève de si grandes interrogations que presque toutes les réponses semblent convenir.

— L'associé de votre mari dit que chacun est reparti de son côté aux alentours de 22 heures, poursuivit Nina. Vous ne savez pas où Lawrence aurait pu aller ensuite ?

— Non. Il lui arrive... lui *arrivait* de faire de longues promenades le soir. Il avait pris cette habitude voilà trois ou quatre ans. C'était sa manière à lui de garder la ligne. Il n'aimait pas les salles de sport.

— À votre connaissance, avait-il des lieux de prédilection ?

— Non. Il tournait dans la ville.

— La forêt de Raynor, par exemple ?

Gayle la dévisagea froidement.

— La dernière fois qu'il s'est rendu là-bas de nuit, ça remonte à plus de vingt ans. Avec moi.

— Les flics du coin semblent convaincus que votre mari a été assassiné par une femme, madame Widmar. Quelle réaction cela vous inspire-t-il ?

— La même qu'à vous.

— Mais encore ?

— Foutaises.

— Autant que vous sachiez, il n'a jamais eu de liaison ? Je suis navrée de vous demander ça, mais...

— Je sais. Vous êtes obligée. Et ma réponse – la même qu'à tous ceux qui me posent directement ou indirectement la question depuis deux jours – est non.

— Et personne ne lui voulait du mal ?

Mme Widmar secoua vigoureusement la tête et parut sur le point de craquer. Puis elle se moucha avec bruit et cligna des yeux en regardant ses genoux.

— J'aimais mon mari. Je l'aime toujours. C'était un homme intègre, et un bon père. Ça n'a rien d'original, mais c'est la vérité. Il va beaucoup manquer aux enfants. Énormément. Mais... c'était juste un homme. Un homme ordinaire. C'est incompréhensible. Pourquoi voudrait-on tuer un homme comme lui ?

— Les « pourquoi » ne trouvent pas toujours de réponse.

— Et comment se fait-il que le FBI soit sur le coup ?

— Certains aspects du meurtre ont attiré notre attention, répondit Monroe.

Un sourire crispé fendit le visage de Gayle.

— Dans ce cas, restez vigilants. Et trouvez la personne qui nous a fait ça.

— On fera tout notre possible, promit-il.

Mais la veuve ne s'adressait pas à lui.

Nina et Monroe redescendirent la route à pied. Le soleil déclinait, projetant une lumière oblique et dorée.

— T'en penses quoi ?

— On sait que ce n'est pas elle, répondit Nina, et je ne vois pas quel rôle elle aurait pu jouer. L'associé n'avait aucune raison de le buter ?

— C'étaient de vieux potes. Le gus n'avait rien à y gagner, et beaucoup à perdre. Il paraît encore plus bouleversé que l'épouse.

— Crois-moi, elle est chamboulée.

Ils atteignirent la voiture. Nina attendit que Monroe débloque les portières, mais il fixait une maison sur le trottoir d'en face. Plus petite que celle des Widmar, et en bien plus mauvais état. Elle était à vendre, mais on ne pouvait pas dire que l'agence y mettait beaucoup de cœur. Monroe finit par se retourner.

— Widmar avait une certaine réputation... quand on grattait un peu. Une des employées de sa boutique, et une serveuse du restaurant. Des gestes déplacés, à l'occasion...

— Je sais. Reidel m'a informée. Et à l'évidence certaines de ses promenades le conduisaient dans les bars. Mais le fait que sa femme n'en sache rien ne prouve pas qu'il ait mené une double vie, ni que ce soit un beau salaud. Les hommes d'un certain âge aiment conter fleurette aux serveuses. Le harcèlement reste du harcèlement, mais des mains peuvent s'égarer sans que leur propriétaire soit un valet de Hitler. Tout le monde n'a pas ta droiture, Charles, ni ton self-control.

— Je n'ai jamais dit que Widmar était un sale type. Et je n'apprécie pas tes sarcasmes.

— Quels sarcasmes ? Ta droiture est légendaire !

— Pourquoi tu me casses les couilles, Nina ?

— Pour me divertir, je pense.

— À d'autres. Tu ne fais jamais rien sans raison. C'est le moment ou jamais de m'expliquer, parce que je ne te le redemanderai pas.

— OK. Ward m'a dit que les infos qu'on t'a données ont été virées du rapport officiel Jones-Wallace. Et que Paul a repris son existence de psychopathe errant.

— Putain d'Olbrich... C'est un bon flic mais qu'est-ce qu'il est bavard !

— Il estimait peut-être que Ward avait le droit de savoir qu'on avait égaré son cinglé de frère. Dont nous n'avons, je présume, toujours aucune nouvelle. Malgré toute la confiance que tu affichais hier.

Monroe garda le silence.

— Alors, vrai ou faux, Charles ?

— Rien ne prouve qu'il y ait eu des complicités dans les meurtres Jones et Wallace, qui sont l'unique objet du procès. Les allégations de Hopkins sur une prétendue conspiration de tueurs en série auraient seulement brouillé les pistes.

— Tu sais comment ils s'appellent, Charles. Ils s'appellent les Hommes de Paille.

— Je sais très bien ce que tu m'as dit. Mais j'ignore si c'est vrai. Et ne compte pas sur moi pour démêler cette histoire.

— Donc tout ceci n'a rien à voir avec le fait qu'en évoquant cela devant la cour, on risque d'apprendre que tu as été tuyauté au sujet de l'emplacement du corps de Jessica Jones ? Ça ne laisse pas envisager des complicités, *ça* ? Mais il ne faut surtout pas brouiller les pistes, pas vrai ?

Nina remarqua un homme dans le jardin de la maison délabrée ; il l'observait avec intérêt tout en arrosant son gazon. Nina baissa brusquement la voix.

— Rentrons à l'hôtel, Charles. Allons dans un endroit neutre. J'ai assez vu cette ville pour aujourd'hui.

CHAPITRE 7

La voiture s'arrêta devant le Holiday Inn peu avant 19 heures. J'étais planté sur le parking, en partie car je pouvais y fumer sans m'attirer de regards noirs, en partie parce que je n'avais pas trop envie de tomber nez à nez avec Monroe. Il descendit avec Nina, et ils furent rejoints par un individu qui les attendait dehors.

Je regardai le trio franchir les portes et disparaître dans le hall. C'était plutôt bizarre. Le troisième larron faisait à peu près ma taille, à peine plus baraqué. J'avais presque l'impression d'avoir quitté mon corps, ou ma vie, et de les suivre en spectateur. Cette sensation n'a rien d'agréable. Elle s'intensifia tout le temps que je passai à surveiller les trois silhouettes de la chambre 107, au rez-de-chaussée. Plus jeune, j'aurais pu trouver cool l'idée de rôder sur un parking avec un flingue dans la poche.

Au bout de quarante minutes, l'homme ressortit de l'hôtel et partit en voiture. Peu après, on ne distinguait plus qu'une silhouette dans la chambre de Nina. Immobile derrière le rideau.

Je pénétrai dans l'hôtel, passai au large du comptoir et remontai le couloir. Je frappai à la porte, et il fallut une bonne minute pour qu'on m'ouvre.

Nina avait ôté ses pompes et semblait avoir perdu cinq centimètres alors qu'elle faisait peu ou prou la même taille. Elle avait l'air naze et méfiante.

— Comment as-tu trouvé mon hôtel ?

— J'ai appelé les flics en me faisant passer pour un sous-fifre du FBI chargé de te remettre un colis important.

— Seigneur ! Et la chambre ?

— J'ai demandé à la réception. La sécurité laisse à désirer dans cette ville. Je dois te prévenir que si Al-Qaida décide d'attaquer le Crédit municipal de Thornton, ils ont toutes les chances d'y arriver.

Même pas un sourire.

— J'ai eu tort de venir ? J'ai juste... cru qu'on m'avait laissé un petit mot.
— Excuse-moi, dit-elle en s'écartant.
Je m'avançai dans la pièce. Les chambres d'hôtel des autres sont étranges. À moins d'y être entré avec eux, d'avoir assisté à la dispersion initiale des valises, des vestes et de la petite monnaie, d'avoir glissé plein d'espoir la tête dans la salle de bains, et tiré son rideau pour décréter que la vue n'est pas si démente, on a toujours l'impression de violer un nid. Une serviette humide, c'est privé.
C'était peut-être seulement ça.
— Tu vas bien, Nina ?
— Ça va, lâcha-t-elle. C'est ma première sortie dans le monde réel depuis une éternité. Je ne mesurais pas combien je m'étais faite à notre petite vie.
Une cafetière sur le secrétaire. Je me servis une tasse et m'installai dans un objet dont, quelque part, un designer avait sans doute théorisé la qualité de siège.
— Il n'y a rien d'autre ?
Elle s'assit jambes croisées au bout du lit.
— Possible.
Le café n'était pas terrible, mais je m'accrochai. Nina contemplait le miroir au-dessus du secrétaire.
— Dis-moi, Nina. Dis-moi ce qui t'amène ici.
— Je fais mon métier.
— Mais non. Enfin si, mais ce n'est pas la raison première. Monroe ne s'est pas déplacé pour rien, il savait que tu le suivrais. Pourquoi ?
Elle sourit.
— J'oublie tout le temps que tu n'es pas idiot.
— Moi aussi. C'est une erreur tentante.
Elle me regarda, roula les yeux, et l'espace d'un instant elle eut l'air bien. Puis elle se rembrunit. Elle se laissa lentement retomber, jusqu'à reposer sur le lit, le corps droit, les yeux au plafond.
Je bus en silence pendant quelques minutes, jusqu'à ce qu'elle se décide à parler :
— Quand j'étais petite, il y avait une femme...

Fille unique, Nina avait grandi à Janesville, dans le Wisconsin. Ses parents s'entendaient bien, entre eux comme avec elle. Elle était brillante et sportive, mais pour une raison ou pour une autre, cela ne lui avait pas valu beaucoup d'amis. Après l'école, au lieu de monter dans le bus de ramassage avec ses camarades, elle marchait jusqu'au boulot de son père pour l'attendre sur un banc public. Puis elle grimpait dans sa voiture et il racontait sa journée, ou bien, en quelques rares occasions, il se terrait dans un silence ruminant. À l'âge de treize ans, elle se trouva enfin une bande et devint un peu plus sociable, mais pendant de nombreuses années tous ses après-midi se terminèrent ainsi. La sortie des classes, la marche à pied, puis l'attente sur le banc, où elle entamait ses devoirs ou se contentait d'observer le monde. Ça lui plaisait, et elle n'acceptait un siège dans le hall de l'entreprise qu'en cas de grand froid ou de pluie. Ce n'était pas un quartier formidable, mais son père distinguait le banc depuis la fenêtre de son bureau et le vigile de l'entrée gardait un œil sur elle. Les choses seraient peut-être différentes aujourd'hui, mais à l'époque cet arrangement ne posait aucun problème.

En face de l'entreprise se trouvait un bar, au rez-de-chaussée du dernier immeuble victorien d'une rue envahie de barres de béton. Nina ne se lassait pas des allées et venues des clients. Des hommes d'affaires en costume qui se rendaient là comme à une réunion mais qui s'asseyaient seuls derrière la vitre et ne s'attardaient pas. Et aussi les vieux, qu'on voyait soit entrer, soit sortir, mais jamais les deux car l'intervalle était trop long – des années, peut-être. Les vieux portaient de gros manteaux, marchaient d'un pas lent et mesuré, et leur barbe de trois jours leur grisonnait le menton. On voyait aussi des types moins âgés, en tenue de tous les jours. On se demandait ce qu'ils pouvaient fabriquer quand ils n'étaient pas au bar. Dormir, peut-être, ou vaquer à cette mystérieuse activité qui leur rapportait juste de quoi se payer une nouvelle bière.

Sans oublier les femmes. Moins nombreuses, mais quelques-unes quand même. À la longue, l'une d'elles devint une figure familière.

Nina la remarqua pour la première fois vers l'âge de onze ans, puis la revit régulièrement pendant deux ans. Elle l'avait d'abord repérée parce qu'elle semblait un peu plus jeune et plus jolie que les autres clientes du bar, qui dans leur grande majorité étaient franchement des cageots. Les cheveux bruns et fournis, celle-ci portait des jeans moulants et un pull sans manches. À son deuxième ou troisième passage, elle aperçut Nina et lui adressa un clin d'œil depuis le trottoir d'en face. Par moments, des clients de l'autre sexe épiaient la fillette assise sur son banc ; ils la dévisageaient au point que Nina rougissait. Elle n'aimait pas ça. Mais le clin d'œil de cette femme ne l'incommodait nullement. Elle se sentait un peu plus grande.

Jamais elles n'échangèrent un mot. Pas une fois la femme ne traversa la rue, n'agita la main. Mais à vingt ou trente reprises au cours des années suivantes, son regard captura celui de Nina. Sur toute cette période, Nina nota que la femme changeait. Ne l'ayant jamais approchée de près, elle l'avait peut-être rajeunie du fait de ses vêtements. Toujours est-il qu'elle vieillissait à grande vitesse. À chaque apparition, il semblait s'être produit quelque chose d'irréparable. Elle prit quinze kilos. Ses cheveux virèrent au blond, puis au roux, puis encore au blond, avant de retrouver une sorte de brun désolant. Son teint rosissait par endroits, pâlissait à d'autres. Seule sa démarche était immuable, et cette façon d'atteindre le bar comme un lieu inconnu dont on lui aurait dit beaucoup de bien et où elle avait bon espoir de s'amuser. Elle garda cet air jusqu'au bout, même lorsqu'elle prit l'habitude d'arriver en titubant. Mais à ce stade Nina n'appréciait plus de la voir. C'était comme de regarder une vie s'écouler en accéléré, comme si chaque pas de cette femme valait un millier des nôtres. Quoi qu'il en soit, la femme reconnaissait toujours la fillette, de manière sporadique, et alors elle lui décochait un clin d'œil. Un clin d'œil lent, qui disait simplement : « Salut, je te vois, tu me vois, y a pas de lézard. » Les toutes dernières fois, Nina se demanda si la femme la reconnaissait réellement ou si elle agissait par pure routine. Mais le rituel perdurait.

Pendant ce temps, on découvrait des corps.

Trois en deux ans. Puis un quatrième et un cinquième. Des cadavres d'hommes dans des voitures en stationnement, tués par balle. Des hommes qui partaient le soir en quête de plaisirs faciles et pas chers, et qui se retrouvaient le lendemain délestés de leur fric et de leur vie.

Trois semaines avant le treizième anniversaire de Nina, on arrêta une femme. Quand Nina regarda le bulletin d'informations, assise avec ses parents un jeudi soir, elle resta bouche bée.

C'était elle.

La femme aux clins d'œil.

L'affaire fit grand bruit dans le Wisconsin, comme un avant-goût d'Aileen Wuornos, célèbre tueuse qui s'illustrerait en Floride quelques années plus tard. On surnomma la meurtrière de Janesville la Veuve noire, bien qu'elle ne fût ni veuve, ni noire, ni arachnide. On apprit qu'elle avait subi des sévices sexuels durant son enfance, de la part d'au moins deux membres de sa famille. On raconta que, depuis quelques années, des hommes la violaient en bande jusqu'à ce qu'elle perde conscience, et même au-delà. On évoqua à mots couverts ce qu'elle était capable de faire pour obtenir ne fût-ce qu'un verre ou la promesse d'un verre. Rien de tout cela n'était présenté comme des circonstances atténuantes, mais comme des preuves à charge.

La femme clamait son innocence et Nina la croyait. Elle l'avait vue remonter la rue par des après-midi d'été, avec son pas mal assuré. Une personne qui marchait de cette façon était incapable de tels crimes. Quelque part, quelqu'un mentait.

Puis il apparut qu'elle pouvait être coupable, en définitive – c'était du moins ce que comptait plaider son avocat. Oui, sa cliente avait brandi le pistolet. Mais toujours pour se défendre, toujours. Elle s'était retrouvée en mauvaise posture face à des hommes et n'avait eu d'autre issue.

Cela non plus, Nina n'y croyait pas. Entre-temps, l'adolescente s'était prise de passion pour l'affaire, épluchant les quotidiens et les magazines en quête de nouveaux éléments, surveillant le journal télévisé. Dès qu'on en parlait autour d'elle, au collège ou dans la rue, elle ralentissait pour tendre l'oreille. Elle devint une éponge, absorbant tout, jusqu'à ce que cela fasse partie d'elle-même.

Un mois plus tard, nouvel épisode du feuilleton. La femme revenait sur ses aveux. Elle était blanche comme neige, n'avait jamais frayé avec aucun de ces types. Son avocat avait également tenté de la violer, jurait-elle. Ainsi que le juge. Le monde entier cherchait à la baiser, au sens propre comme au figuré. Tous les hommes et toutes les femmes.

Mais pas moi, pensait Nina devant ce triste spectacle. Pas moi.

— Et puis... lâcha Nina avant de se taire.

Je restai muet, comme depuis le début. Quand Nina rouvrit la bouche, sa voix était nouée.

— Et puis il y a eu ces cinq secondes d'images, où on l'a vue quitter le tribunal sous escorte et s'engouffrer dans une voiture. Il pleuvait, et la pluie plaquait ses cheveux sur son crâne. Elle avait perdu du poids, mais aux mauvais endroits. Comme si elle avait maigri du cerveau. Avant de plonger la tête dans la bagnole de police, elle a regardé droit devant elle, vers la caméra. Et on pouvait le lire dans ses yeux.

— Lire quoi ?

— Elle l'avait fait. À cet instant j'ai su qu'elle les avait bel et bien tués. Il suffisait de voir son regard pour comprendre qu'elle était coupable, qu'elle avait appuyé sur la détente. Mais je savais en même temps qu'elle n'était *pas* coupable. Je savais qu'elle avait commis ces crimes sans vraiment les commettre. Et je me demandais comment c'était possible. Et ce qui faisait qu'un jour sa tête s'était retrouvée à court de clins d'œil.

Je méditai ces paroles quelques instants.

— Qu'est-elle devenue ?

— Elle a plongé pour les cinq meurtres et s'est suicidée huit mois plus tard. Elle s'est procuré une cuiller, a cassé la tête, et s'est enfoncé le manche dans la gorge après l'extinction des feux. On a dit qu'elle avait dû mettre trois heures à mourir.

Nina se tut cinq bonnes minutes. Puis je me rendis compte qu'elle respirait différemment, et qu'elle dormait.

Je la contemplai un moment, avant d'ouvrir mon ordinateur et de le brancher sur la prise téléphonique. Je n'avais pas encore eu l'occasion de vérifier mon courrier, de voir si mon correspondant mystère avait quelque chose à ajouter.

La connexion de l'hôtel était lente. Pendant l'attente, je me surpris à penser à mon père. Ces mails tombés du ciel me rappelaient son dernier message à lui, un billet d'une seule phrase dissimulé à l'intérieur d'un fauteuil, dans la maison qu'il occupait avec ma mère dans le Montana. Quand on meurt, ce sont les petits détails qui prouvent qu'on était vivant. Les boîtes de conserve qu'on était le seul à aimer. La carte de vœux jamais envoyée, sous sa Cellophane poussiéreuse et jaune, avec son étiquette délavée et son prix dérisoire. Mes parents avaient laissé derrière eux des tas de petits détails. Grâce à eux, j'ai découvert que, bébés, mon frère Paul et moi avions été officieusement adoptés suite à un affrontement avec mon père biologique, un homme qui avait agressé ma mère. Mes parents avaient abandonné Paul dans les rues de San Francisco deux ans plus tard, convaincus qu'il valait mieux nous maintenir séparés et que cet abandon était la seule solution. Trente ans plus tard, l'organisation dont avait fait partie mon géniteur était toujours active et mon père adoptif – qui n'était jamais qu'agent immobilier – remonta sa trace jusqu'à un complexe de luxe niché dans les montagnes au-dessus de Yellowstone. Cela lui coûta la vie, et celle de ma mère. Le groupe demeurait restreint et très bien caché, mais il possédait de l'argent et de l'influence. Je le connaissais désormais sous le nom d'Hommes de Paille. D'après mon ancien ami l'inspecteur John Zandt, les Hommes de Paille évoluent en Amérique depuis au moins trois mille ans et ont bâti leur fortune sur les mines de cuivre préhistoriques de la région des Grands Lacs. C'est là, toujours selon Zandt, que s'est établi un vaste agrégat d'hommes et de femmes venus des quatre coins du globe au fil des ères successives, tous unis dans une même haine de la civilisation. Partant, Zandt les voyait derrière toute chose – de la disparition des premiers colons de Roanoke aux tribus barbues et violentes décrites dans les vieilles légendes indiennes –, guerriers luttant contre l'invasion d'une terre qu'ils estimaient la leur. Cela dit, je concevais de sérieux doutes sur la santé mentale de Zandt, de même que lui s'interrogeait sur mon compte. Par deux fois, j'avais eu l'occasion d'abattre le meurtrier de sa fille, et par deux fois je ne l'avais pas saisie. Ce meurtrier était mon frère Paul.

Mais j'imagine qu'on trouve toujours que sa vie est plus compliquée que celle des autres.

La machine bipa enfin. J'avais un message. Que j'ouvris.

Ward,

Ici Carl Unger. Bobby est mort ?
C'est quoi ce BORDEL ?

Je suis tombé sur un truc bizarre et ça pourrait concerner un truc que Bobby m'a soumis il y a environ un an.
C'est à cette occasion qu'il a mentionné votre nom.
C'est très important.

Appelez-moi vite. 07589576543.

C.

Je lus le message à deux reprises, puis me renversai sur mon dossier pour réfléchir. Je ne me souvenais pas d'un Carl Unger, mais cela ne prouvait rien. J'avais quitté la CIA depuis quelques années maintenant, et la maison n'est pas du genre à envoyer des lettres d'information pour annoncer le mariage de Machin ou les exploits de l'équipe de foot. Sans compter que je ne retiens pas les noms. Je les considère sans doute comme indépendants des individus, comme un blouson fétiche qu'ils porteraient souvent, rien de plus. Peut-être avais-je connu cet Unger par ouï-dire : le mail sous-entendait qu'il était plus proche de Bobby. Moi aussi, je peinais à croire que Bobby Nygard était mort, alors même que j'avais assisté au drame.

Je vérifiai le carnet d'adresses de Bobby. Aucun Unger. Cela dit, je n'y figurais pas non plus. Bobby était un professionnel de la surveillance, spécialisé en informatique. Allez savoir par quelles manips il camouflait sa vie. Je ne trouvai aucun mail d'Unger antérieur à ces deux messages récents, ni d'autre adresse utilisant ce patronyme – avec un nom de domaine plus officiel comme « gov », par exemple. Mais ça ne prouvait toujours rien. Il allait de soi que Bobby archivait et purgeait ses mails de manière régulière. Le plus ancien en date n'avait pas un mois lorsque nos routes s'étaient recroisées. Il avait trouvé la mort moins d'une semaine après.

Dans ses tout derniers jours, Bobby avait sollicité plusieurs contacts, afin d'explorer certaines pistes que nous avions ouvertes. Unger pouvait être l'un d'eux. Mais étais-je prêt à m'y risquer ? Et pour gagner quoi ? Si on lisait le mail en boucle et dans un esprit optimiste, on pouvait se prendre à rêver que cet Unger disposait d'infos utiles. Mais le contraire était tout aussi envisageable. Or l'optimisme n'est pas mon fort.

— Dis-moi, Bobby, ce mec existe-t-il vraiment ?

Aucune réponse. J'avais déjà essayé de l'interroger, et pas une fois il ne s'était manifesté. Par esprit de contradiction, sans doute. De son vivant, ce mec était une mine de conseils.

Je rouvris le mail et recopiai le numéro dans mon téléphone. Je n'allais pas appeler Unger. Pas encore. J'allais d'abord recueillir l'avis de Nina. Si son existence et la mienne faisaient cause commune, ce que j'espérais, elle avait son mot à dire.

Je fus réveillé par une série de bruits sourds. Quelqu'un tambourinait à la porte, méchamment. L'espace d'une seconde aussi confuse qu'amère, je crus que c'était Paul, qu'il était venu nous débusquer en Virginie. Mais il n'aurait jamais pris la peine de frapper, fût-ce comme une brute.

Je me redressai, le dos en compote. Il faisait noir dans la chambre, hormis le halo du parking à travers les rideaux. Je tournai les yeux vers le lit : Nina n'y était plus. Le réveil indiquait 23 heures et des poussières. J'aurais dit davantage. Genre après-demain.

Je me levai péniblement et vis que la salle de bains était fermée. Un filet de lumière sous la porte. J'allai m'y coller.

— Nina ?

— J'en ai pour une minute. C'est qui à la porte ?

— Un service de chambre qui se la pète.

— Tu commandes à une heure pareille ?

— C'était une blague.

Boum boum boum de plus belle. Je visai dans le judas, avant de me souvenir de tous ces films où on écope d'une balle dans l'œil. Par chance, je reconnus le visage de l'autre côté, même déformé.

— Monroe, annonçai-je à Nina tout en essayant de me réveiller. Et il a pas l'air de vouloir partir.

— Seigneur ! Laisse-le entrer.

J'ouvris brusquement, alors que son poing s'apprêtait à remettre ça.

— On a entendu, Monroe.

Il déboula dans la pièce.

— Qu'est-ce que vous foutez là ? grommela-t-il sans avoir l'air très surpris.

— J'équipe gratuitement les suites de luxe. Regardez sous votre lit, il y en a un comme moi dans chaque chambre.

— Entravez cette enquête, et je vous fais arrêter.

— C'est noté.

Nina sortit de la salle de bains. J'avais cru qu'elle se douchait, mais elle était encore habillée. Elle semblait bien plus fraîche que moi. Comme toujours.

— Quoi de neuf ? demanda-t-elle. On a fait mouche dans un des bars ?

Monroe attrapa la veste de Nina sur le lit, la lui tendit.

— C'est un peu plus lourd que ça. On a un nouveau mort sur les bras.

CHAPITRE 8

Il était un peu plus de 20 heures, et la fête commençait à prendre. Y avait à boire, y avait de l'ambiance, et on approchait de la fin du début, qui est souvent la meilleure partie. Les vioques allaient rentrer, mais d'après leurs antécédents, les Luchs étaient des hôtes relax. Ils viendraient traîner dans le jardin, joueraient les enthousiastes pendant un quart d'heure, puis se replieraient dans leurs quartiers en tirant les rideaux. Tant que rien ne prenait feu, ils étaient cools. La piscine était remplie de mecs hilares et de meufs gloussantes, et un gus faisait danser la faune en mixant les titres de son iPod sur la sono qui jouxtait le baril de bière. Ça peut être casse-couilles, un dingue de zique qui vous matraque les tympans avec ses goûts persos, mais ce keum avait une bonne play-list nu-indie/country alternative.

Brad était allongé dans un transat, sous le vieil arbre centenaire au centre de la pelouse, et tout baignait pour lui. Il y avait eu un moment de panique quand la poule de l'autre soir l'avait surpris avec Karen et semblait à deux doigts de péter les plombs, mais depuis elle avait visiblement dégotté de la poudre et elle était trop jouasse pour se vexer. Karen se trouvait maintenant au bord de la piscine, à discuter avec des amies que Brad ne connaissait pas. Elle avait parlé à sa mère, et si elle tenait vraiment à se faire opérer du nez, alors ses vieux lui paieraient l'intervention, ne serait-ce que pour avoir leur mot à dire sur le choix du chirurgien : la plupart des copines de Mme Luchs étaient bioniques à un endroit ou à un autre du corps, et cette femme avait une approche intelligente du sujet – sans être jamais passée sur le billard elle-même. Quand Karen lui expliqua l'arrangement, Brad éprouva un soupçon de regret, comme si on lui retirait quelque chose. Il lui fallut deux secondes pour comprendre quoi : une responsabilité et une grosse dépense. Bah ! il pourrait s'en dispenser. Par ailleurs, Lee ayant redistribué à l'équipe une grosse part du super bonus de came, les finances de Brad s'étaient nettement redressées. D'accord, elles avaient refondu depuis, mais il arborait un nouveau bermuda Zpatuula, un polo Penguin et un caleçon CK, sans parler de l'étagère entière de CD et de jeux vidéo qui avaient fleuri dans sa chambre. Il s'était payé un iMac pour... pour diverses choses, c'était sûrement très pratique, et puis comment résister à une boutique d'informatique qui vous ouvre les bras ? Sinon, quand Karen s'était retournée pour lui faire coucou, le pendentif qu'il lui avait offert avait chatoyé dans les derniers rayons du soleil. Il avait hâte de le voir osciller, ce petit « K » doré. D'avant en arrière.

Il se redressa et but sa bière tout en écoutant la musique, satisfait de ce bref moment de solitude. Parfois, il se disait que c'était ça, l'intérêt d'avoir plein d'amis : on pouvait prendre plaisir à se retrouver seul.

Même lieu, une ou deux heures plus tard, Brad assis de l'autre côté de la piscine, avec Karen sur les genoux. Elle s'était fait couper les cheveux en prévision de la fête : mi-longs, soyeux. On discutait de tout et de rien, et on écoutait un gus raconter que son oncle allait lui filer du pognon pour monter un site Web qui apprendrait comment devenir cool aux gamins pas cools. Brad trouva le principe drôlement cool et se dit qu'il devrait lui aussi creuser une idée de ce genre. Il s'y attela quelques minutes, tout en caressant le cou de Karen, mais ça ne donnait rien. Plus tard, peut-être. Il avait levé le pied sur la bière, car il venait d'avaler une des nouvelles pilules ; elles n'avaient pas de nom mais elles étaient rouges et frappées d'un « À » – peut-être qu'on les appelait les A. Ça détendait vachement, mais on restait vif, du moins quand on regardait des formes aux angles nets, sauf si l'on remuait la tête trop vite, auquel cas elles étincelaient un peu. Du coup, les guirlandes électriques suspendues aux arbres rendaient super bien. Beaucoup de gens carburaient aux À ce soir. Elles remportaient un succès bœuf. Bonne nouvelle pour Lee, et bonne nouvelle pour Brad. Bonne nouvelle pour tout le monde.

Au bout de deux minutes, le keum de la start-up tenait le crachoir un peu trop longtemps, et Brad sentait son esprit divaguer. Soit le DJ était tombé dans les vapes, soit il avait trouvé un pote pour le relayer. On entendait un truc assez mid-tempo, mais potable. Ça lui disait vaguement quelque chose, mais quoi ? Les Luchs avaient montré leur nez, en coup de vent, et la fête culminait en mode milieu de soirée, quand on a l'impression qu'elle durera toujours. La meilleure phase, sans aucun doute, encore meilleure que la fin du début. Brad envisageait de se lever d'ici peu, d'aller chercher quelque chose à bouffer, quand il avisa une présence inattendue de l'autre côté de la piscine.

Lee. Brad ne savait même pas qu'il était là. Lee n'était pas un fêtard. Plutôt du genre sérieux. La preuve : il était suspendu à son téléphone. Tout autour, les gens s'éclataient, bien décidés à se lâcher à fond, mais Lee restait planté là, l'air grave, comme en pleine transaction boursière. Achète à la baisse ! Vends à la hausse ! Tu parles d'un déconneur.

Lee rangea son portable. Il parut fixer un point lointain, avant de balayer l'assemblée du regard. Il devait chercher quelqu'un. Puis son visage s'immobilisa, ce qui sous l'effet de l'A prenait un aspect robotique. Lee aperçut Brad, hocha la tête.

Brad fit de même, avec le sourire. Son vieux pote Lee.

Alors Lee secoua la tête, pour préciser que le mouvement précédent n'était pas un simple salut, et pour mieux se faire comprendre il courba son index.

La personne recherchée n'était autre que Brad.

Il embrassa Karen dans le cou.

— Je reviens tout de suite, bébé. Faut que je touche deux mots à Lee.

Elle sauta à terre ; il se leva et s'éloigna dans la foule. Sa jambe s'était endormie sous le poids de Karen, et il contourna la piscine en boitant légèrement.

— Salut, dit-il en retrouvant Lee. Sympa comme fête, non ?

— T'es très chargé ?

Brad cligna des yeux. Bonjour la conversation.

— À peine, répondit-il. Deux bières.

— Tant mieux. J'ai besoin que tu m'accompagnes.

— Aucun problème. Tu vas chercher des hamburgers ?

— Non. Pete et Steve sont là ?

— Pas Steve. Il est... je sais pas où. Mais Dormeur doit être dans les parages. J'ai dû le croiser tout à l'heure.

— Je vais voir sur les côtés, tu regardes à l'intérieur. Essaie de les trouver, et on se rejoint à la voiture. Profites-en pour te passer un coup d'eau sur le visage.

— Qu'est-ce qui se passe ?

— On a un truc à faire.

— J'ai compris, mais quoi ?

— Fais ce que je te dis, Brad.

Lee s'éclipsa au milieu des danseurs. Brad secoua la tête, comme pour y remettre de l'ordre ; tout compte fait, il se sentait plutôt bien. Il se dirigea vers la piscine, cherchant Pete d'un œil et Steve d'un demi-œil seulement, car il restait persuadé que Steve n'était pas venu du tout. Il scruta aussi le coin où il avait laissé Karen, pour la prévenir qu'il devait s'absenter, mais elle avait disparu.

La maison demeurait assez vide, plus calme que d'ici quelques heures, lorsque la fièvre serait retombée et que les convives seraient d'une humeur plus horizontale. Aucune trace de Pete ni de Steve. Brad trouva la porte d'entrée, en ne se perdant qu'une fois. Il oublia de se rafraîchir le visage, mais il se sentait frais.

L'allée était bondée de voitures, dont la nouvelle BMW bleu vif de Karen. Deux grappes de personnes bavardaient. Brad localisa la bagnole de Lee. Il patienta quelques minutes avant de sortir une cigarette et de s'apercevoir qu'il avait laissé son briquet au bord de la piscine. Merde. Fait chier.

— C'est ça que tu cherches ?

Il se retourna. Karen lui tendait une flamme.

Il sourit, alluma sa clope.

— Mon ange de miséricorde... ou de feu. Mais à coup sûr un ange.

— T'es trop mignon. Alors, t'en es où ? Tu cherches un job de voiturier ?

— J'attends Lee.

Ces mots à peine prononcés, des bruits de pas sur le gravier. Ceux de Hudek.

— Il arrive, lança-t-il à Brad. Salut, Karen. Chouette teuf.

— Merci ! (Elle s'étira.) On fait ce qu'on peut. Alors, où vont nos deux lascars ?

— Juste faire un tour, dit Lee. Chopper un peu de bouffe.

— Mais on a déjà tout ce qu'il faut. Y a assez de guacamole pour planquer un bébé !

— Je sais. J'en ai déjà pris. Mais j'ai des envies très spécifiques. Tiens, je crois qu'on te fait signe, là-bas.

Karen se retourna vers la Porsche rouge qui ronronnait devant la porte. Deux individus restaient plantés, avec l'air de ceux qui ne veulent pas repartir sans dire au revoir mais qui doivent impérativement filer, genre *tout de suite*. Karen plissa les yeux pour les identifier.

— Ah oui, Sara et Randy. Ils ont un impératif. Bon, eh bien, le devoir m'appelle. Soyez prudents sur la route.

— Comme toujours, répondit Lee. Tu le sais bien.

Elle répondit d'un bref sourire un peu crispé, puis se pencha vers Brad et l'embrassa sur la joue.

— À plus, dit-elle avant de courir vers la maison, les bras prêts à étreindre ses amis en partance.

Encore deux ou trois minutes, et Pete le Dormeur apparut. Il avait l'air bien raide, mais c'était souvent pire, malgré ce grand sac de Doritos qu'il grignotait goulûment.

— Qu'est-ce qui se passe, les gars ?

— Un petit boulot rapide. T'en es ?

— J'en suis, mec. J'ai l'éthique du travail chevillée au corps. Comme tu le sais.

— T'es un type bien, répondit Hudek.

Il pressa la télécommande et la voiture émit un petit couinement. Les portières se débloquèrent.

— Viens t'asseoir devant, Brad.

Lee alluma l'autoradio et poussa le volume. Franchit la grille, remonta la route qui serpentait à travers le ranch, dépassa les autres grilles privées jusqu'à celle, encore plus haute, qui protégeait toute la communauté de Faircroft Ranch. Les vigiles dans leur guérite appuyèrent sur le bouton sans même regarder. Le but des résidences protégées est d'empêcher les gens d'entrer, non de sortir, et puis il n'y avait pas l'ombre d'un Black dans la voiture de Lee.

Hudek traversa Santa Barbara par la grande avenue, en conduisant de manière responsable. Au bout de dix minutes il prit la 192 en direction du nord. Brad regardait défiler les lumières – phares, lampadaires, signaux lumineux. Le poste diffusa un morceau dont il avait oublié le titre. Il était obnubilé par la question suivante : que cachaient ces « Soyez prudents/Tu le sais bien » entre Karen et Lee, à supposer qu'il y eût quelque chose à cacher ? Et par celle-ci : qu'allait-on faire dans le nord ?

— Eh, monte le son, dit Pete. C'est de la bombe !

Hudek augmenta les enceintes arrière et réduisit celles de devant.

— C'est quoi, le plan ? finit par demander Brad. Tu nous emmènes où, Lee ?

— Hernandez a appelé, précisa Hudek. Il a conclu un truc de dernière minute. Mais son staff n'est pas disponible. Il a besoin de deux types pour l'escorter.

— C'est-à-dire nous ? Tu te fous de notre gueule ? Putain, Lee, la dernière fois que j'ai croisé ce connard, il m'a assommé avec son flingue !

Lee hocha la tête, les yeux dans le rétro.

— C'est vrai. Mais cette requête est bon signe, Brad. C'est capital.

— Et depuis quand on est à son service ? Faites ci, faites ça...

— On peut pas se défiler. Alors on va faire ce truc, puis on ira s'acheter de gros hamburgers, on regagnera la teuf et on décompressera à fond. C'est d'accord ?

Pete chantonnait sur sa banquette, étranger à la discussion. Brad secoua la tête, mais ce n'était pas un refus. Il sortit une cigarette et l'alluma.

— Eh, mec...

— Va chier, Lee. Le toit est ouvert et on roule. J'ai envie d'une putain de clope et j'ai pas besoin de ton avis.

Hudek sourit.

— OK, mec. Pète pas les plombs. Je voulais juste que tu me dises oui ou non.

— Eh bien ouais, t'as gagné, putain. Mais franchement, je le sens pas.

Hudek lui adressa un clin d'œil, puis se rangea brusquement sur le bas-côté. Brad paniqua, avant d'apercevoir Hernandez posté au coin de la rue, un sac à l'épaule.

— De quoi ? s'exclama Pete, la bouche pleine de chips. Qu'est-ce qu'il fout là, lui ?

Hernandez remonta jusqu'à la voiture, baissa les yeux sur la banquette arrière.

— Où est l'autre mec ?

Hudek prit son temps avant de répondre. Ça faisait du bien d'être en position de force. S'il le voulait, il pouvait repartir sur-le-champ. Le laisser planté comme un con.

Il coupa le poste et répliqua :

— Le préavis était un peu court, tu crois pas ?

— Quel préavis ? lança Pete. Sans déconner, qu'est-ce qu'on va fricoter avec cette tête de pine ?

— On va lui filer un coup de main, expliqua Lee. Le conduire à un rendez-vous important. Pigé ?

— Ouais, ouais, concéda Pete d'un air sceptique. On va dire que oui.

— Parfait. (Hudek fixa Hernandez et sourit.) T'attends quoi pour monter ?

Dirigé par Hernandez, Lee prit le chemin des collines. Terrains de golf, ranchs sécurisés. Il ne connaissait pas bien le secteur, qui ne méritait même pas le nom de secteur, sinon d'un point de vue immobilier. Ils s'enfoncèrent dans la vallée de Santa Ynez, puis encore un peu plus loin.

— On va rouler comme ça jusqu'au Nevada ? s'enquit Pete.

Hormis ces mots, lui et Brad restaient muets à l'arrière.

Hernandez montra enfin une petite route sur la gauche. Aucun panneau n'indiquait sa destination. Lee la suivit sur trois ou quatre kilomètres, dans une sorte de forêt de broussailles. Ils atteignirent le sommet de la colline et amorcèrent une lente descente. La route devint rudimentaire, et déboucha bientôt sur un grand parking de gravier et de poussière, au milieu d'arbres qui se fondaient dans un crépuscule d'encre. L'endroit semblait désert.

— C'est là ?

Hernandez confirma.

— Ils ont rendez-vous à quelle heure ? Et qui sont ces gens ?

— Ils sont exactement comme vous. Y a rien à craindre.

— Alors pourquoi t'as besoin qu'on te tienne la main ?

Apparut soudain une paire de phares à l'autre bout du parking. Brad sentit son cœur dédoubler un coup. Lee, quant à lui, jubilait de se trouver dans la voiture de la came. D'être celui qui *reçoit* les sacs de fric. C'était un pas dans la bonne direction, pour sûr.

— C'est eux ?

— Ouais, répondit Hernandez avant de lancer à Pete : Toi, tu restes ici.

— Pas question.

Lee observa Hernandez.

— Quel est le problème ?

Ça faisait drôle de parler à ce type d'égal à égal. Ça faisait surtout du bien.

— Toi et l'autre gamin, je vous ai vus nous tenir tête dans le parking. Mais ce loser, je me souviens juste de sa gueule couverte de Scotch.

— Vous étiez trois contre deux, rétorqua Pete. (Il avait l'air furax, ce qui était rare.) Mais je te prends quand tu veux. Tout de suite, tiens.

Ce que tu peux être con, Pete, se disait Brad. Au loin, trois silhouettes se dessinaient dans le contre-jour des phares. *Si tu veux mon avis, rester dans la bagnole est une très bonne idée.*

— Sans façon, répondit Hernandez à Pete en lui servant l'un de ses vilains sourires. Un gros gaillard comme toi, je n'ai pas la moindre chance...

Pete la boucla.

— Passe derrière le volant, lui conseilla Lee. On sait jamais.

Lee fut flatté de voir Hernandez approuver. Ce dernier ouvrit son sac et sortit un flingue, qu'il tendit à Hudek.

Brad secoua la tête :

— Pourquoi on aurait besoin de ce truc, si c'est soi-disant... ?

— Ferme-la, Brad.

Lee coinça le flingue à l'arrière de son pantalon. Il vit Hernandez se palper les reins, comme pour vérifier la position de sa propre arme. C'est bon, on pouvait y aller.

Le trio quitta la voiture. Pete se glissa sur le siège conducteur.

— Faites gaffe, murmura-t-il.

Hernandez ouvrit la marche, avec Lee sur sa droite, légèrement en retrait, et Brad sur sa gauche.

L'un des types d'en face lança :

— Salut, Emilio. C'est qui, ceux-là ?

— Des amis, répondit Hernandez. Y a pas de souci.

Brad examina chacun des visages. L'un des types devait avoir son âge et les mêmes origines sociales. Les deux autres faisaient plus vieux, tel celui au crâne rasé. Ces mecs puaient l'embrouille. Pourquoi restaient-ils en retrait, d'abord ? Selon l'usage, on s'avançait, ils s'avançaient, on se rencontrait à mi-chemin et on échangeait les sacs. Éventuellement on s'en grillait une ensemble, ou on partageait un rail pour se la jouer conviviale, puis chacun repartait de son côté.

Pourquoi n'avançaient-ils pas ?

Lee se posait la même question, mais ces inconnus voulaient peut-être les employer. Une petite démonstration mesquine, pour se donner de l'importance. Même s'il n'y croyait guère, au fond. Hernandez s'arrêta. Lee et Brad aussi.

— Alors, les gars ? s'écria Hernandez. Vous avez les pieds coincés, ou quoi ?

Pas de réponse, mais une évidence qui frappa Lee comme une massue.

Aucun des trois hommes ne tenait de sac.

— Hernandez...

Puis ils ouvrirent le feu.

Ni sommation, ni déclaration. Juste des bras tendus et des balles qui fusent. Clac clac clac.

Lee trébucha en arrière, cherchant son flingue dans son dos. Hernandez fut bien plus rapide. Déjà il canardait la voiture adverse, en se déportant rapidement sur la droite, vers les arbres.

Brad restait figé, et Lee se souvint qu'il n'était pas armé. Il vit Brad se demander comment sortir de ce putain de merdier, puis piquer un sprint vers la gauche du parking.

Lee tira sur son flingue. Qui resta coincé dans le pantalon. Il tira de nouveau, le libéra et l'actionna.

Deux des types s'étaient abrités à l'arrière de leur caisse. Le troisième s'affairait derrière le coffre. Il visa Hernandez mais le manqua.

Lee fit feu. Un coup. Deux coups. Sans succès.

Soudain le mec pivota pour cueillir Brad, dont la fuite ne menaçait personne, qui n'était pas armé et qui se trouvait là uniquement parce que Lee le lui avait demandé.

Lee vit Brad trembler, perdre l'équilibre, se manger un arbre et s'étaler de tout son long.

Puis la voiture le frôla à toute bombe, l'aspergeant de graviers. Un dernier coup de feu, et Lee sentit l'air se fendre juste au-dessus de sa tête.

Quarante-cinq secondes, grand maximum. Et c'était terminé.

Lee resta pétrifié un instant, avec l'impression que le monde avait basculé, viré au négatif.

— Oh merde, Brad...

Il courut au fond du parking, où Brad gisait face contre sol, pour constater, éberlué, que son ami bougeait encore et n'était pas couvert de sang. Brad roula sur le dos, fixa Lee dans les yeux. Son visage exprimait beaucoup de choses, sans rapport avec la douleur.

Lee l'empoigna.

— Putain, mec, j'ai cru qu'ils t'avaient eu. J'ai cru que t'y étais passé.

Brad se redressa.

— Moi aussi. Mais j'ai juste trébuché. Un gros caillou. J'ai trébuché.

— Ça t'a sûrement sauvé la vie, mec. La vache.

— Ouais. Un caillou porte-bonheur.

Ils se regardèrent, les yeux écarquillés, et rigolèrent. Mais d'un rire nerveux, qui n'en était pas vraiment un.

— Il est touché ? lança Hernandez de loin.

— Non ! répondit Lee, speedé par son adrénaline comme par une dose de coke pure. C'était quoi, ce putain de guet-apens ?

— J'en sais rien, marmonna Hernandez. Mais on se casse. Tout de suite. (Il s'éloigna d'un pas vif vers la voiture de Lee.) On a des coups de fil à passer.

Lee tendit la main à Brad, qui se laissa hisser sur ses jambes. Bloqué dans un monde plus jeune de soixante secondes, le cerveau de Brad n'avait pas intégré les derniers événements. À l'évidence, du grabuge s'était produit entre-temps, mais il avait dû louper l'essentiel. Et n'en revenait pas d'être encore là.

— Allez, viens, dit Lee. On se casse.

Il savait que cet épisode revêtait une importance majeure. Hernandez voudrait trucider ces types. Et Lee Hudek serait à ses côtés. Depuis ce soir, il n'était plus un simple gamin qui se pointe avec le blé.

Il prit Brad par l'épaule pour l'inciter à presser le pas.

Brad boitait, mais il avait hâte de mettre les voiles. Il aurait aimé que l'A se fasse oublier un instant, juste le temps de remettre de l'ordre dans ses idées. Il vit qu'Hernandez avait stoppé quelques mètres avant la bagnole.

— Qu'est-ce qui t'arrive ? demanda-t-il avant de regarder la voiture.

Il y avait quelque chose sur le siège conducteur.

Une chose affreuse. Une chose sortie de la forêt et de la nuit pour prendre possession de leur caisse. Elle était monstrueuse, mais immobile.

C'était Pete le Dormeur.

— Oh mon Dieu, expira Brad.

Il se tourna vers Lee, mais Lee fixait Pete. Brad se força à l'imiter.

— Oh, non...

En dessous des épaules, tout était normal. C'était encore Pete. Assis sur le siège, le sachet de Doritos à la main. Mais l'une des balles destinées à Hernandez, à Lee ou à Brad les avait croisés tel un oiseau migrateur ignorant des centaines de kilomètres de mer et avait trouvé le point de chute que lui destinait le canon. Elle avait arraché environ un quart de la tête de Pete. Elle était entrée dans la partie supérieure droite du visage, pour lui défoncer la pommette et la cervelle, avant de ressortir par l'arrière via un trou grossier. L'œil restant était encore ouvert. Les déchets de son nez ressortaient n'importe comment.

Brad crut d'abord que l'œil gauche était simplement luisant, humide, avant de comprendre qu'il essayait de bouger.

Puis la mâchoire de Pete s'affaissa. Une tache sombre apparut sur son entrejambe. Et Pete fut parti pour de bon.

CHAPITRE 9

— On n'a pas le choix, insistait Hernandez. Écoutez-moi.

Brad était assis par terre, les bras noués autour des genoux. Il en était à sa troisième cigarette d'affilée, ce qui signifiait que la conversation durait depuis au moins dix minutes. Brad écouta, mais il ne voulait jouer aucun rôle actif. Lorsqu'il essayait de réfléchir, il avait l'impression de marcher pieds nus sur des braises, des braises s'étirant à l'infini, dans toutes les directions. Sauf que c'étaient des braises glacées.

Lee rétorqua :

— J'ai connu ce type à l'école primaire ! Il y a forcément une autre solution.

La position d'Hernandez était simple. Le cadavre devait disparaître. On ne pouvait le laisser ici. Dès qu'ils le découvriraient, les flics remueraient ciel et terre. Un gosse de riches avec la tête explosée n'était pas le genre d'affaire qu'on classe sans suite. Non, il fallait effacer l'incident. Brad était scié par le détachement dont Lee faisait preuve. Certes, sa voix n'était pas d'une constance absolue, il se frottait la lèvre avec l'index et ses coups d'œil sur le siège avant n'étaient pas plus fréquents que ceux de ses complices. Mais il avait eu l'idée de déplacer le cadavre pour faire croire à une fusillade en voiture, à une agression ayant mal tourné. Il s'accrochait à son idée, mais Hernandez ne voulait rien savoir.

— Écoute-moi, Lee. (Hernandez parlait avec calme, et Hudek songea que c'était la première fois que le type l'appelait par son prénom, plutôt que par « gamin » ou « eh, toi ».) On n'a plus le temps. Quelqu'un aura forcément entendu les coups de feu, cambrousse ou pas. Faut s'en débarrasser au plus vite.

Lee acquiesça :

— OK. Allons-y.

Ils ouvrirent la portière et sortirent le corps de Pete, d'abord en le portant, mais là des restes de tête tombèrent en franges et une substance visqueuse se mit à couler sur la main de Brad, qui lâcha prise en même temps que Lee, de sorte que la carcasse s'écrasa sur le gravier. Lee ôta son tee-shirt pour débarbouiller le siège, pendant que Brad et Hernandez traînaient le cadavre vers le coffre, chacun par un pied. Lentement, pour éviter que la tête ne rebondisse. Ils ouvrirent la malle, Hernandez sortit le drap de plage, enveloppa la tête et le cou, puis ils soulevèrent le corps et le plièrent de diverses façons pour le faire entrer, ce qui ne fut pas simple. Enfin ils refermèrent le coffre.

Le corps hors de vue, on respirait un peu mieux. Brad s'accorda une pause pendant qu'Hernandez éparpillait le gravier tout autour de la voiture et ramassait les morceaux sanguinolents pour les fourrer dans le sac de Doritos. Très pro.

Là-dessus ils remontèrent en voiture. Ils rebroussèrent chemin, avant de virer à gauche pour s'élever dans les collines. Le parc national était fermé, et c'eût été une mauvaise idée de toute façon, mais ils trouvèrent une route de service qu'ils suivirent assez longuement. Puis ils se garèrent, sortirent Pete ou ce qu'il en restait, et le portèrent le long d'un sentier pédestre, sur huit cents mètres environ. De son vivant, Pete était assez gros. Mort, il était très lourd. Lourd, difficile à manœuvrer, et encore chaud, avec des mains démesurées et pleines de doigts. Puis on quitta le sentier pour s'enfoncer au hasard. Le dos de Brad le mettait au supplice, comme s'il avait eu un clou planté dans le bas de la colonne.

Quand Lee estima le coin suffisamment reculé, ils s'arrêtèrent et posèrent le cadavre au pied d'un arbre. Lee n'ayant, bien entendu, pas de pelle dans son coffre, ils creusèrent à la main, en s'aidant du cric de la bagnole. Ce fut long et laborieux, même sur un sol relativement tendre, même en s'y mettant à plusieurs. Ils poussèrent Pete dans le trou, mais il était trop grand, alors ils le firent rouler sur le côté et agrandirent la tombe. À force de ténacité, ils parvinrent à le caser. S'ensuivit une discussion sur le sort de la serviette. Lee jugeait qu'elle était aussi bien ici. Hernandez aurait préféré que Lee la remporte et la brûle. Brad souhaitait qu'on la laisse en place, pour ne pas avoir à regarder Pete une dernière fois, même s'il s'en voulait de penser ça. Lee obtint gain de cause. C'était une serviette bas de gamme. On trouvait les mêmes partout. Brad fut soulagé.

Pouvait-on laisser des empreintes sur un cadavre ? Dans le doute, Lee essuya Pete avec le tee-shirt ensanglanté, qu'il jeta dans la fosse. Puis il se ravisa, retourna le corps et se servit du tee-shirt pour lui attacher les mains, histoire de brouiller les pistes au cas où. Ils recouvrirent le corps de terre, puis chacun partit de son côté chercher les bouts de bois les plus gros possible. Ils les disposèrent sur la tombe de manière anarchique, et Lee s'éloigna de dix mètres pour voir si cela faisait illusion. Malgré l'obscurité, il lui sembla que oui. Après avoir déblayé l'excédent de terre, Lee contempla le travail en secouant lentement la tête.

Ils regagnèrent la voiture sans un mot.

Lee attendit le retour à la civilisation pour allumer ses phares. Il rejoignit la ville sans commettre d'excès de vitesse et se rangea pour déposer Hernandez à l'endroit désiré.

Le gars mit le pied sur le trottoir.

— Je vais tout de suite appeler certaines personnes. L'homme que tu as rencontré. Je te ferai savoir quelle est la riposte. Cette nuit.

Lee acquiesça, le regard fixé sur la route. Hernandez claqua la portière et repartit d'un pas soutenu.

Lee pressa le champignon, fonça jusqu'à la 101, traversa en trombe Ventura et Oxford. Puis freina fortement et prit à droite jusqu'à la plage.

Ils se garèrent, toujours silencieux. Ils ne trouvaient rien à se dire, sinon de brusques exclamations qui n'étaient d'aucun secours. Lee emporta le sachet de Doritos ; pendant qu'ils grimpaient les dunes, il en répandit le contenu avec parcimonie, laissant le vent en emporter le plus possible. Le sac était vide lorsqu'ils atteignirent le rivage. Lee s'avança dans l'eau et rinça l'intérieur avant de réduire le plastique en une pluie de lambeaux qu'il lâcha dans la mer, happés par le vent tels des fragments de clair de lune.

Il rejoignit Brad, qui se balançait d'un pied sur l'autre. Ils se regardèrent dans le blanc des yeux.

— Pete le Dormeur, dit Lee.

— Merde, Lee. Merde.

Ils repartirent à travers les dunes. Lee enfila un vieux sweat-shirt qui traînait dans son coffre et mit le cap sur le centre-ville. Ils s'arrêtèrent au premier Starbucks venu pour commander des *lattes* à la vanille, qu'ils burent sur le chemin du grand Frisbees de Jolacha Avenue. Là, ils achetèrent trois grands menus hamburgers, effrayés par les lumières agressives et par l'étrange bruit de la caisse enregistreuse, comme par la façon dont les autres clients papotaient, riaient ou commandaient de la sauce barbecue comme s'il ne s'était rien passé. Ils ressortirent les jambes molles.

Traversée du centre, et retour à la grille principale de Faircroft.

— Prends l'air cool, murmura Lee.

Brad sourit dans le vide.

Un vigile apparut à la portière. Lee lui avait déjà parlé en arrivant à la fête. L'échange fut bref et amical. Lee lui proposa un sandwich, et le type faillit dire oui avant de se souvenir qu'il était au régime, que les cadeaux étaient interdits ou bien qu'il venait de s'enfiler une pizza, et il se contenta de leur ouvrir la barrière.

Lee s'enfonça dans la copropriété, vit défiler les grilles.

— Arrête-toi ! enjoignit Brad au bout de quelques centaines de mètres.

Lee s'exécuta. Brad descendit pour vomir au bord de la route. Ça puait la bière et le sang fermenté, l'eau croupie et la tourbe.

Il remonta en voiture et ils parcoururent le dernier tronçon.

Les voitures étaient encore nombreuses dans l'allée des Luchs. Lee se gara et coupa le moteur. Serra les poings plusieurs fois de suite.

— OK, dit-il. Maintenant il faut réintégrer cette teuf. Et s'amuser.

— Tu déconnes ?

— J'en ai l'air ? Nous devons être *présents*, si tu vois ce que je veux dire.

Brad voyait. Ils sortirent avec leurs burgers et firent le tour de la maison. Il y avait encore une trentaine d'invités, la fête continuait et l'espèce de DJ était reparu, mais sa musique avait changé. Un truc monotone et arythmique. Les sacs Frisbees attirèrent quelques têtes.

— Personne n'aurait vu Pete ? demande Lee d'un ton désinvolte. Il y a un burger à son nom – et un deuxième, je parie.

Les gens rigolèrent et répondirent que non, ça faisait un bout de temps qu'ils ne l'avaient pas vu. Quelqu'un crut même savoir qu'il s'était rendu à une autre fête.

C'était bien résumé, songea Brad avec dégoût. Une tout autre fête. Heureusement qu'il avait déjà vomi...

Soudain une main creusait la sienne, et Karen était là.

— On croyait vous avoir perdus, dit-elle.

— Mais non, sourit-il avant de lui tendre des frites.

Une heure et demie plus tard, Hudek s'arrêta devant la maison de ses parents. Il y chercha de la lumière, tout en méditant son plan d'action.

Il revenait du lavage automatique. En fait, il en avait visité deux – celui où il avait ses habitudes, puis un deuxième, vingt rues plus loin, où personne ne le reconnaîtrait. Il n'avait nettoyé que la carrosserie, bien sûr, et n'était passé au premier centre qu'après s'être assuré qu'aucun employé connu n'était de service : il ne serait qu'un jeune-à-belle-bagnole parmi tant d'autres.

L'extérieur était impeccable. Contrairement au reste. Le siège conducteur et le haut de la portière semblaient nets, mais juste en apparence. Lee pourrait revendre la voiture d'ici quelque temps, mais pas tout de suite ; ça faisait une sacrée différence de savoir qu'un de ses meilleurs amis était mort à l'endroit même où on posait ses fesses.

Il aurait pu remettre ça au lendemain, mais il préférait en finir ce soir, pour se dire au réveil que l'incident était clos. Bien entendu, ce ne serait jamais aussi simple. L'absence de Pete allait vite intriguer. Mais au moins Lee pourrait-il vaquer à ses occupations en faisant celui qui ne sait rien.

Pour cela, il devait accomplir plusieurs tâches.

Vérifier que les rainures des pneus ne retenaient ni cailloux ni poussières associables à ce putain de parking ou à la route menant à la tombe de Pete.

Nettoyer le coffre de fond en comble, pour chasser toute trace de gravier et de sang. On avait bien fait d'envelopper la tête de Pete, mais pendant le trajet le drap de plage s'était gorgé d'hémoglobine. Au besoin, détruire et remplacer le revêtement.

Laver le siège avant, et tout l'intérieur. Avec soin. Les reportages de *Police scientifique* qu'il avait vus par hasard s'avéraient – qui l'eût cru ? – fort utiles. Le garage de Lee se prêterait bien à ces travaux, avec son éclairage vif. Mais il comptait d'abord faire un saut chez ses vieux, en espérant y trouver quelqu'un. Deux raisons à cela. En premier lieu, son père possédait plein de trucs utiles : détergents, solvants, outils... La berline de Hudek senior brillait comme un sou neuf, et il y avait assez de stocks pour que Lee se serve incognito. En second lieu, il semblait judicieux de montrer son nez. *Ouais, je l'ai vu ce soir-là. Non, je l'ai trouvé en forme, très détendu, pourquoi ? Ah non, c'est impossible. Tout à fait impossible.*

Pas de lumière dans la chambre parentale, mais des signes de vie dans le bureau de papa. Lee se gara, verrouilla la voiture et pénétra en silence dans la maison. Il passa la tête dans le bureau ; l'ordinateur était en marche, mais il n'aperçut personne. Puis il trouva les portes-fenêtres de la cuisine ouvertes.

Et une silhouette solitaire dans un fauteuil près de la piscine.

Il posa le pied sur la terrasse.

— Papa ?

La silhouette se retourna, et Lee reconnut son père, un cigare à la main.

— Salut, fiston, dit-il à mi-voix. Quel bon vent t'amène ?

— Je passais dans le coin. Je me suis dit que j'allais faire coucou, si vous étiez encore debout.

— C'est gentil, ça. Je te sers quelque chose ?

— Volontiers. Je prendrai une bière. Sans alcool, de préférence.

— Bien sûr. Tu n'avais pas une fête, ce soir ?

— Exact. Chez les Luchs.

— Jolie baraque.

— Mouais. Un peu tape-à-l'œil, quand même.

Ryan Hudek sourit et s'éloigna vers la cuisine, pressant au passage l'épaule de son fils.

C'est à cet instant que Lee fut le plus près de craquer, mais il serra les dents. Une fois franchi ce cap, il s'interdirait à jamais de faire machine arrière. Il tint compagnie à son père une bonne demi-heure, avant de prendre congé. Son paternel le salua sans quitter son fauteuil. Avant de ressortir, Lee fit un crochet par le garage et prit ce dont il avait besoin, en arrangeant les étagères de façon à cacher son larcin.

Puis il rentra chez lui, rangea la voiture et se prépara un café corsé. Le flingue d'Hernandez était planqué sous une pile de bricoles, dans un tiroir du garage. Épuisé ou pas, il fallait rester dispo, et ce n'était pas le moment de se droguer. Lee avait surtout besoin de caféine. Regardant le jus noir goutter, il voulut penser à Pete, puis préféra s'abstenir. Pete était mort. Point barre. Et personne ne l'avait obligé à faire quoi que ce soit.

Tout en lessivant la voiture, Lee se surprit à regarder son téléphone toutes les deux minutes. Trois heures s'étaient écoulées, et Hernandez n'appelait pas.

CHAPITRE 10

Je ne saurais comment le décrire. Ce n'était ni une forêt, ni un champ, à la pointe nord-ouest de la forêt de Raynor. Une sorte de grand néant, par hectares entiers, vaguement parsemé d'arbres et de buissons, sillonné de ridules tel un visage parcheminé. Il y avait en principe une petite ville à un kilomètre de là, mais on ne l'aurait pas deviné. Cette zone ne servait à personne, sinon aux insectes, aux oiseaux et aux bestioles à fourrure qui s'en nourrissaient. J'avais croisé une friche comparable dans le New Jersey, baptisée la Lande. Je ne sais comment les indigènes appelaient celle-ci, mais je ne voyais pas de meilleur nom, surtout sous une froide lune grise.

J'avais suivi Monroe et Nina sur le parking, avant de monter à l'arrière de leur voiture. Cela avait soulevé une vive polémique, mais Monroe était trop pressé pour s'offrir le luxe d'une bagarre. Il traversa la ville en trombe, continua pendant une dizaine de minutes, jusqu'à repérer sur le bas-côté les gyrophares des flics locaux. Il se plaça derrière le véhicule, puis le suivit au long d'une suite de routes qui ne semblaient mener nulle part, jusqu'à celle qui décrivait une longue ligne droite. Le véhicule de tête finit par se ranger sur les gravillons. Deux autres voitures étaient garées un peu plus loin, ainsi qu'un fourgon, style police scientifique.

Deux flics s'extirpèrent de l'habitacle, l'un en uniforme et l'autre en costume. Ce dernier ressemblait au type que j'avais repéré à l'hôtel auprès de Nina et de Monroe.

— C'est qui, lui ? demanda-t-il comme nous venions vers eux.

— Un collègue, répondit Nina. Ça pose problème ?

— Non, madame. Vous pouvez aussi appeler votre mère, pour qu'elle nous apporte des sandwiches.

— Ça suffit, Reidel, ordonna Monroe. Conduisez-nous là-bas.

Le dénommé Reidel me toisa. Je le toisai. Puis il se retourna et nous précéda sur un sentier.

Après deux cents mètres, nous distinguâmes au loin des faisceaux de lampes torches. Le sol se mouillait, poisseuse course d'obstacles de fougères et de boue qui nous suçaient les chevilles.

— Qui l'a trouvé, cette fois ? s'enquit Nina. Personne ne serait assez désespéré pour vouloir baiser ici.

— Un vieux gars du coin, qui se promenait pour ramasser du bois. Il sculpte des serpents dans des branches mortes, paraît-il, pour les revendre au marché artisanal. Il a découvert le corps en début de soirée, en fait, mais le type a des antécédents psychiatriques, et il s'est mis à flipper, au point de ne plus savoir ce qu'il avait réellement vu. Il lui a fallu plusieurs heures pour se décider à appeler. Depuis il se sent mieux.

— Qui voudrait d'un serpent en bois ? demandai-je.

— Bonne question, monsieur.

Le ruban d'interdiction commençait un peu plus loin. Il délimitait un ruisseau peu profond, d'environ un mètre de large, même si les remous brouillaient la séparation entre l'eau et le sol. Ensuite le ruisseau se divisait en deux, pour former une île imprécise, large d'une quarantaine de mètres. Il y avait quelque chose tout au bout, mais le premier point d'observation se trouvait plus près de nous. Une poignée d'hommes encerclait une forme à terre. On avait monté sur pied une grosse lampe, qui frappait de lumière blanche les têtes et les épaules.

Reidel se glissa sous le ruban pour nous conduire aux deux grosses planches qui enjambaient le cours d'eau.

— Une personne à la fois, conseilla-t-il.

Le pont de fortune ploya sensiblement sous notre poids ; il semblait condamné à périr avant l'aube. Suivaient deux marches mouillées, et le terrain redevenait stable.

Les experts reculèrent à notre approche. On pressentait un spectacle terrible. L'air stagnant exhalait une irrésistible note funeste.

— Vous tombez à point, dit l'un des techniciens. On s'apprêtait à l'emmener...

Il braqua sa torche sur la chose, créant un jeu d'ombres.

Monroe fut le premier à voir.

— Seigneur, souffla-t-il avec calme.

Un homme, couché sur le flanc droit, un bras coincé sous lui. Comme s'il avait fait une chute. Il portait un jean et une chemise foncée à carreaux, fort tachés l'un et l'autre. Sa tête était tournée d'une façon étrange, comme s'il avait cherché la lune à travers les arbres. Il avait les yeux ouverts ; quelque chose comme trente-cinq ans, même si les cadavres ne disent jamais vraiment leur âge. Parfois la mort semble ôter quelques années de soucis et lisser le caractère. Un petit animal l'avait en outre allégé d'un morceau de joue.

La lumière frontale donnait à la scène une allure de photographie, dénuée d'émotion, jusqu'à ce que l'odorat nous rappelle à la réalité. Mais Monroe réagissait surtout à ce que révélait la manche de chemise gauche. Elle était roulée au-dessus du coude, et le bras avait perdu l'essentiel de sa chair. Non par décomposition, même si l'odeur âcre et humide montrait que cela avait joué, mais par décharnement, jusqu'à l'os. L'œuvre d'une main étrangère. Puis on vit que le reste du corps semblait rongé de la même façon. Le visage était celui d'un homme de corpulence moyenne, voire bien en chair. Mais la manière dont les vêtements flottaient sur la carcasse, associée aux taches sombres et sèches, suggérait que l'ensemble du corps avait subi le sort du bras. On pouvait incriminer les animaux, mais pourquoi auraient-ils fureté sous les habits et laissé le gros du visage ?

Je me retournai, pour constater avec bonheur que je n'étais pas le premier. Nina avait les yeux rivés sur l'extrémité de l'île.

— C'est quoi, là-bas ?

— Venez voir.

Reidel nous guida sur le sol inégal. La forme que nous distinguions restait mystérieuse, comme détachée dans le clair de lune. Mais il suffisait d'approcher pour comprendre.

Le premier élément que j'avais remarqué en arrivant ici était un tee-shirt blanc. Propre, de taille moyenne, suspendu à trois branches d'arbuste comme à une corde à linge. L'arbuste côtoyait l'un des cinq arbres de l'île.

Nina, Monroe et Reidel contemplèrent la mise en scène quelques instants, avant de grogner plus ou moins à l'unisson.

L'image avait effectivement ce pouvoir-là, encore plus marquante, peut-être, que celle du corps. Je me demandais si le sculpteur de couteaux avait vu ce tee-shirt et si ce n'était pas la cause première de sa terreur.

Un bruit sourd du côté du cadavre, suivi d'un juron dans la bouche d'un expert. Il releva la tête pour nous appeler.

— Eh, les gars ! Faut que vous voyiez ça !

En professionnels consciencieux, nous fîmes demi-tour. On venait de retourner le corps avec maintes précautions. Le bras droit s'avérait dans le même état que le gauche : manche retroussée, chair retranchée. Mais avec une différence notable.

Il manquait la main.

— Bon, murmura Reidel après un silence. On peut parler de tueur en série, maintenant ?

— Au moins quatre ou cinq jours, affirma le légiste tandis qu'on chargeait le corps sur une civière. (Avec son espèce de crème Vicks sous les narines, le toubib avait l'air du type le plus enrhumé du monde.) Peut-être une semaine, ajouta-t-il, même si l'ablation des tissus et des organes rend le verdict plus délicat. J'en saurai plus quand il aura intégré le labo, mais ne vous attendez pas à une précision d'horloger. Ce n'est pas le meilleur environnement pour être mort.

Il disait vrai. Même le si bleu nuit du ciel commençait à s'adoucir, cet endroit n'avait rien d'accueillant. Cela faisait quatre heures que nous avions quitté l'hôtel ; j'avais les yeux rouges, et plus beaucoup de clopes. J'étais resté en retrait, muet, de peur qu'un grand se souvienne de ma présence et m'envoie au lit.

Je regardai la dépouille traverser le pont précaire. L'ambiance de l'île changea du tout au tout, comme si on avait coupé une inaudible bande-son. Un corps sans vie fascine toujours. Il pare les lieux d'insolite. C'est sans doute devant un cadavre que l'on a admis pour la première fois la possibilité de choses puissantes, importantes et néanmoins invisibles. On pouvait observer maints aspects de la réalité – rivières, animaux, arbres, soleil – mais le phénomène le plus radical (la différence entre un être animé et un corps sans vie) était impalpable. Le rite et ses condensés – les dieux – ne sont qu'une manière d'enrober ce vide, cette chose insaisissable. Je me demandais qui pouvait être cet homme, et dans l'existence de qui il laisserait le plus grand trou. Pas dans la mienne, me dis-je avec soulagement, avant de me sentir nul et seul. Où que j'aille sur cette planète, mes parents ne m'attendraient nulle part. Plus jamais Bobby ne me paierait une bière. Un an après les faits, je peinais toujours à les assimiler. Je les oubliais pendant des heures, parfois même des jours, puis ils me revenaient à l'esprit, toujours aussi inexplicables. Je ne parvenais jamais à me dire : « Ah oui, ils sont morts. Bien sûr. OK. C'est enregistré. » Et je n'y parviendrais peut-être jamais.

J'allumai l'une de mes dernières cigarettes et tendis l'oreille aux discussions alentour.

Monroe continuait à prendre des notes :

— Vous êtes sûr que le décharnement ne s'est pas produit après l'abandon du corps ?

— Sûr et certain. Il a perdu quelques bouts après coup, au visage par exemple – ces marques sont à l'évidence celles d'un rongeur –, mais l'essentiel a été débité avant. Avec un couteau, un couperet, un instrument de ce genre. L'humérus est entaillé. Et le tibia est rayé. On devrait en voir beaucoup plus lorsqu'on l'aura déshabillé. Un objet lourd et tranchant. Le même qui aura servi à couper la main, j'imagine.

— Je suis surprise que les bêtes aient fait si peu de grabuge pendant tout ce temps, observa Nina.

— Quel temps ?

— De quatre jours à une semaine, comme vous venez de le dire.

— Non, sourit-il en levant l'index. Je n'ai pas dit ça. Cela fait quatre ou sept jours qu'il est *mort*. Mais pas sur cette friche désolée. Je pense qu'il est resté ici vingt-quatre heures, grand maximum.

Nina le dévisagea en silence. Je connaissais ce regard. Il signifiait : « Ne m'oblige pas à te soutirer les infos, ou tu vas le regretter. »

Le légiste portait une alliance : lui aussi connaissait ce regard.

— La chair du dos est relativement intacte, expliqua-t-il. Et les surfaces dorsales présentent des signes de stase sanguine, le sang s'accumule dans les parties les plus basses après la mort. Elle est répartie sur tout le dos, et sur les restes de fesses et de mollets, alors qu'on a découvert le corps...

— Couché sur le côté.

— Exactement. C'est donc ailleurs qu'il a passé les premières heures de sa mort, étendu sur le dos.

— La question est de savoir où, commenta Reidel. Mais là, j'imagine que vous n'avez aucun indice à nous livrer ?

— On verra ce que dénichent les experts. Mais ne vous faites pas trop d'illusions. Avec un peu de chance, on trouvera des éléments provenant de son environnement précédent – poussières, débris, terre. Mais s'il était nu là-bas, ces traces auront disparu avec les chairs ou été balayées lors du rhabillage.

Monroe opina :

— En somme, le décharnement pouvait servir à dissimuler la provenance du corps.

— Possible.

— Ou pas, nuançai-je.

Je fixai de nouveau le bout de l'île, frappé par une évidence.

— Qu'y a-t-il, Hopkins ?

— Que signifie ce tee-shirt, d'après vous ?
— Aucune idée pour le moment. Ça n'a peut-être rien à voir.
— Bien sûr que si. Un cadavre et un vêtement tout près, quand il n'y a rien d'autre sur un kilomètre à la ronde ? Pour moi, les deux sont forcément liés.
— Le type qui a déposé le corps aurait jeté ce truc en partant, suggéra Reidel. Ce qui nous indiquerait dans quel sens il est parti.
— Ça non plus, je n'y crois pas. Pourquoi s'en débarrasser ? Le tissu est propre. Pas de sang, rien du tout. On croirait qu'il sort de l'emballage.
Nina m'observait.
— Tu penses à quoi ?
— Viens que je te montre.
Je me mis en marche. Un temps d'hésitation, et ils me suivirent.
Ayant pris toutes ses photos, un technicien s'apprêtait à décrocher le tee-shirt.
— Une seconde ! lançai-je. (Je fis signe à Nina de se positionner derrière le vêtement.) Tu ne remarques rien ?
Elle examina le tissu, en vain. Je l'aidai un peu :
— Ce n'est pas le tee-shirt en tant que tel. Place tes épaules parallèles aux siennes, regarde droit devant toi et dis-moi ce que tu vois.
Elle pivota légèrement. Leva les yeux.
— La lampe.
— Là où se trouve le corps, en d'autres termes. Le tee-shirt tient une pose. Il représente quelqu'un. C'est un témoin.
Le trio promena les yeux entre les deux points.
— Bon, peut-être, acquiesça Nina.
Monroe ne semblait qu'à moitié convaincu :
— En quoi cela nous dit-il si le décharnement sert à chasser les traces ?
— Je ne sais pas, avouai-je. Mais si on a pris la peine d'accrocher ce tee-shirt, c'est dans un but précis. Pour créer un tableau. Ou pour le *re*créer. On ne traîne pas un corps jusqu'ici sans raison particulière. On l'a détenu dans un endroit qui permettait de le découper en toute discrétion. Puis on a choisi de l'amener ici, où il serait forcément retrouvé. Donc son emplacement est déterminant. Et si on l'avait décharné pour réduire son poids ?
Trois individus fronçaient leurs sourcils. On aurait dit une rangée de points d'interrogation.

— Vous m'avez l'air bien schlass, les gars. Plus le corps est léger, plus il sera facile à transporter. Le tueur avait une idée en tête, et il avait choisi ce lieu. Mais il se savait trop faible pour porter la carcasse sur une telle distance. Alors il a retiré le plus de masse possible, sans rompre l'intégrité globale du corps.

— Donc une personne qui, se sachant incapable d'amener le corps ici, aurait fait en sorte de rendre la chose faisable, paraphrasa Reidel avant de marquer une pause. Une femme, par exemple.

— Eh bien, ouais. Sauf que...

Monroe me fixait d'un drôle d'air. Nina regardait ses pompes. Reidel arborait un sourire bizarre.

— Qu'est-ce que j'ai dit ?

La voiture nous déposa sur le parking de l'hôtel. Monroe et Reidel filaient directement à la morgue pour assister au début de l'autopsie. L'aube commençait à poindre.

— Je suis désolé, Nina. Je ne savais pas que je m'immisçais dans un débat sensible.

— C'était bien vu. Je crois même que t'as épaté Charles.

— Ça ne prouve pas que ce soit une femme, en tout cas. Si je devais parcourir une telle distance avec ce truc sur le dos, moi aussi je chercherais à l'alléger. Ce type n'était pas petit. Et même les nains morts font leur poids.

— La voix de l'expérience ?

— Non, je crois que ça vient de Confucius, ou de cette bande-là. Viens, allons voir s'il y a moyen de boire un café.

— Il n'est même pas 5 heures, Ward.

— T'as ton flingue sur toi ?

— Bien sûr.

— Moi aussi. Ils nous serviront. Ils n'ont pas le choix.

Nous sommes rentrés dans l'hôtel, et j'ai fait du gringue à une employée jusqu'à ce qu'elle accepte de me trouver du café pour que je lui fiche la paix.

Le restaurant décrivait un L. Nous avons emporté nos tasses au fond de la grande longueur. Il eût sans doute été plus logique de regagner la chambre, mais la fréquentation des morts donne envie de grands espaces. On s'est assis pour siroter notre jus dans un lointain bruit d'aspirateurs.

Puis le bipeur de Nina a sonné. Elle l'a consulté.

— On a les résultats sanguins du premier gus. Des traces d'une substance de type Rohypnol. Ce qui signifie préméditation.

— Mais pas forcément meurtre. Est-ce qu'une femme ferait ça ? Recourir à une drogue de violeur ?

— Pas si elle veut se faire sauter.

J'informai Nina du mail d'un certain Carl Unger, qui disait avoir connu Bobby et voulait me contacter d'urgence. Nous tournâmes et retournâmes la question, puis Nina réfléchit un instant.

— Appelle-le.

J'acquiesçai, et il y eut un silence.

— J'ai vérifié, reprit-elle soudain. Ce comté n'a pas connu le moindre meurtre au cours des dix dernières années. Pas un seul. Et voilà deux cadavres la même semaine, aux abords de la même ville.

— Un citoyen qui ne rigole plus.

— Ouais.

Nous sommes restés là, avec nos cafés tiédis, tandis que dehors le monde peu à peu s'éclaircissait.

CHAPITRE 11

Pendant ce temps, un homme poursuivait sa route, ayant vaincu ses hésitations.

La veille, en fin d'après-midi, Jim Westlake avait atteint Petersburg, conformément aux instructions reçues à Key West. Il s'était garé sur le parking d'un supermarché Publix, à l'orée de la ville. Il descendit de voiture pour se dégourdir les jambes, en attendant les 17 heures tapantes. Puis il maintint enfoncée la touche 1 du téléphone portable, programmée sur un numéro dont les chiffres restaient masqués. S'il avait été jeune et féru de technologie, il aurait peut-être pu retracer ce numéro. Mais, l'utilité d'une telle recherche n'étant pas prouvée, il avait laissé tomber.

On décrocha au bout de trois sonneries.

— C'est moi, dit Jim. Je suis arrivé.

— Bien, bien, répondit une voix.

Ce devait être le plus âgé des deux visiteurs de Key West. Le Garçon précoce. Toujours lui.

— Et maintenant ? Tu me donnes une adresse ou quoi ?

— Tu vas devoir la trouver toi-même.

Jim émit un rire acerbe.

— T'as déjà foutu les pieds à Petersburg ? C'est pas New York, mais c'est pas non plus le trou du cul du monde. Si vous êtes pressés, il va m'en falloir plus que ça.

— Justement, poursuivit la voix, c'est là que je t'ai un peu menti. Il te reste encore un peu de route.

— Je suis censé aller où ?

La voix indiqua une autre ville. Jim resta muet comme une tombe.

— T'es toujours là, Jim ?

— Ouais, et je t'emmerde.

— Je m'y attendais plus ou moins. C'est pour ça que tu es assis là où tu es, et pas là où tu devrais être.

Jim regarda sa main gauche pendre le long de sa cuisse. Elle tremblait un peu. Il serra le poing.

— Je n'irai pas là-bas, déclara-t-il.

— Eh si, répondit la voix. Tu es obligé, et pas seulement parce que je l'exige. Si je t'avais expliqué la situation dans ton nid des Keys, à tous les coups tu aurais refusé. Ce qui aurait été idiot, car si tu te défiles t'es foutu. Tu as laissé traîner des saletés dans ta vie antérieure. Elles viennent de resurgir. Il va falloir nettoyer, et tant que tu seras là-bas, tu vas aussi bosser pour nous. Un travail important, qui me tient personnellement à cœur, et que tu ne vas pas foirer.

Jim posa les yeux au fond du parking. Le supermarché étincelait au soleil. Les familles s'affairaient, poussant des chariots vides ou pleins. Cherchaient de la bouffe, rapportaient de la bouffe. Chasseurs de rayonnages, glaneurs de barquettes allégées et d'emballages sous vide.

— OK. Crache. Je veux la vérité, maintenant.

— Reste où tu es. De nouvelles infos suivront. Puis tu passeras à l'action. Cette affaire compte beaucoup à mes yeux, à *nos* yeux, mais c'est ta vie qui est en jeu, pas la mienne.

La ligne coupa.

Jim resta près de la camionnette quelques minutes, puis le téléphone lança une série de sons bizarres. Après un bref répit, il remit ça. L'écran priait Jim d'appuyer sur une certaine touche. Ce qu'il fit.

Une photo envahit l'écran. Un visage de femme. Sur demande, il enfonça une autre touche et découvrit un second mini-portrait. Encore une femme. Sous chacune figurait une consigne. Deux fois la même.

TUE-LES.

Perdu dans le temps. Perdu dans les chiffres. Perdu dans le souvenir des choses perdues.

Tiré de sa rêverie par des braillements, Jim se rendit compte qu'il était demeuré adossé au Combi pendant une éternité. Planté sur le parking, un peu sonné, les yeux baissés sur un téléphone portable.

— Il m'a donné un coup de pied ! Maman, il m'a donné un coup de pied !

Jim tourna la tête. De l'autre côté de l'allée se trouvaient deux garçonnets, debout à l'arrière d'un 44. Environ sept ans et déjà costauds. L'arrière du véhicule béait comme une mâchoire géante, rempli de sacs et de produits bariolés. De quoi nourrir un petit pays d'Afrique pendant un mois. Les gosses étaient livrés à eux-mêmes, le temps que leur mère ramène le chariot cinquante mètres plus loin.

Stupide. Vraiment stupide.

Écarlate, le gamin cogné hurlait. Son frère l'observait d'un air concentré, comme s'il analysait l'expérience. Un coup de pied, il chiale. Très intéressant.

Remontant dans son fourgon, Jim songea que, pour celui qui entend hurler, l'agresseur est moins pénible que l'agressé, celui qui crie en boucle : « Il m'a donné un coup de pied. » Certes, le gosse qui a frappé est un connard, et certes c'est lui qui a commencé, et les coups de pied c'est très vilain. Pourtant, c'est le cul du chialeur qu'on a soi-même envie de botter. On voudrait le punir pour cette larmoyante complaisance de victime, pour cette volonté de convertir la faiblesse en pouvoir. Ce que veut le gosse frappé, c'est qu'un adulte vienne gifler le frappeur. Il veut commettre un acte violent *sans l'avoir fait* – tirer avantage du geste sans devoir l'assumer. Voilà le genre de gosse qui, plus tard, portera plainte contre son voisin pour s'être cassé la gueule dans son allée, après avoir passé l'après-midi à lui bouffer ses hot-dogs et à lui vider ses bières.

Jim manœuvra le Combi avant de baisser sa vitre. Quarante mètres séparaient encore les gosses de leur maman, qui amorçait son retour.

— Eh, gamin, lança Jim.

Les deux gosses se retournèrent. Jim s'adressa au pleurnicheur :

— Soit tu lui rends ses coups, soit tu fermes ta gueule.

Le gosse cligna des yeux, entre deux sanglots automatiques.

— T'as compris ? T'arrêtes de chialer, maintenant.

Les sanglots cessèrent. Le gosse était blême.

— Parfait. Ne recommence jamais. Ou je reviens te chercher.

Jim remonta sa vitre et quitta le parking, puis la ville, pour reprendre l'autoroute.

Il roula toute la nuit. Plein ouest, sans sortir de Virginie, enchaînant les routes secondaires après l'agglomération de Petersburg. Il arrivait. Il n'avait plus le choix.

Il rentrait à la maison.

DEUXIÈME PARTIE

LA MULTITUDE

Les lieux laissent toujours une empreinte, pour le meilleur et pour le pire.

David M. SMITH

CHAPITRE 12

Oz Turner remplit ses poumons avant d'expirer lentement, en claquant des lèvres tel un cheval déprimé ; 10 h 45, un lundi matin de septembre froid et pluvieux. Septembre est un mois pourri. À peine commencée, la semaine traînait déjà un goût de vieux, et l'heure présente précédait de quinze minutes le sempiternel « Quand est-ce qu'on s'amuse ? ». Ajoutez la précipitation, légère mais réelle, et vous obteniez une matinée bonne à jeter.

Depuis son bureau collé à la fenêtre, Oz considéra d'un œil torve la grand-rue de Lincoln. Deux riverains marchaient tête baissée. Quand ils eurent poussé la porte de Jayne's Stores, les trottoirs furent à nouveau déserts.

Salut l'ennui, vieille branche.

Il se leva pour se rendre dans la kitchenette, derrière deux bureaux vides. Le journal abritait rarement plus d'un employé à la fois. Pour réunir l'effectif total, s'élevant à trois personnes, il eût fallu le déclenchement d'une guerre, un débarquement d'extraterrestres ou la foire du comté. Pendant que l'eau bouillait, Oz se versa quatre cuillerées de café soluble dans une tasse, et trois autres de sucre. Sa dernière ex-femme le tarabustait sans cesse sur sa consommation de caféine, et ne plus l'avoir sur le dos était un plaisir chaque jour renouvelé. Une vraie plaie, cette femme. Bonne sous la couette, pour sûr, mais vu qu'elle en faisait désormais profiter un autre (cela durait depuis trois années, dont une antérieure à leur divorce), c'était un avantage à double tranchant. Oz avait compris de longue date que la gent féminine se divisait en deux catégories : les femmes timides et coincées qui restaient fidèles et que l'on quittait, et les femmes chaudes et aguicheuses qui partaient les premières, en général vite et loin. Il avait essayé – *épousé* – les deux, sans succès. Il voulait encore croire à l'existence d'une troisième voie, mais il craignait d'être trop vieux pour la découvrir un jour.

Cela faisait cinq minutes qu'il observait son reflet dans la bouilloire – du moins supposait-il qu'il s'agissait de son reflet, bien que les cheveux et la barbe lui parussent très gris, très broussailleux – quand il s'aperçut que l'eau était prête. De retour à son bureau, il fixa son écran avec dépit, en calant son café sur son ventre. Depuis le matin, il avait ajouté cinq cents mots à son article, mais ceux-ci ne valaient rien. Ainsi la tour Newport de Rhode Island révélait une forte influence nordique. Et après ? Ça n'avait rien d'un scoop. Oz avait choisi pour angle d'attaque « L'indécrottable aveuglement de l'establishment et des pouvoirs publics ». Sa philosophie de la vie, en somme. Mais Oz avait aussi une conscience professionnelle, et on ne peut ressasser les mêmes rengaines *ad vitam aeternam*. Quand bien même le lectorat eût ignoré sa rubrique (ce qui semblait fort possible), ou l'eût à peine survolée en croyant à une blague (et il savait de source sûre que cela arrivait), il fallait bosser proprement.

Que restait-il, sinon ?

Il sélectionna son texte, l'effaça. La demi-heure suivante s'écoula sans grandes fulgurances.

Un bruit dehors. Il releva les yeux.

Une voiture était apparue dans la rue du Jayne's. On la remarquait parce qu'elle s'était garée en marche arrière – téméraire infraction qui eût ruiné la plupart des indigènes –, et parce qu'il s'agissait d'un véhicule de location, gros et noir, immatriculé dans le Colorado. Un type se tenait devant le capot. De taille moyenne, large d'épaules, le cheveu noir et ras. Ses joues semblaient râpeuses et légèrement creusées. Il fixait les fenêtres du *Lincoln Ledger*.

Puis il balaya la rue du regard. À gauche, à droite. D'un air très pro.

Oz se pencha pour mieux l'observer, amusé de repérer chez ce type des faux airs de Terminator. Non qu'il fût aussi baraqué et gonflé que le Schwarzie de l'époque, ni vêtu de cuir noir des pieds à la tête. Un vrai clone du Terminator dans le comté de Webster, Massachusetts ? Vous pensez, on en parlerait déjà aux infos (infos diffusées depuis un petit studio situé deux villes plus loin, d'un fauteuil qu'Oz occupait en personne une fois par semaine, sur la morne tranche de minuit à 1 heure, pour « Oz-Zone », une émission qui rassemblait une poignée d'insomniaques bileux et dont les droits de retransmission cherchaient encore preneur). Non, ce gars-là était mince et compact, et la ressemblance tenait à sa démarche ; à une apparente indifférence à la pluie, à l'impression que l'averse, comme sans doute beaucoup d'autres choses, ne lui faisait ni chaud ni froid. Appelez-moi Woodward ou Bernstein si ça vous chante, mais ce type n'est pas là pour une pause café, pensa Oz en se retournant pour empoigner son Caméscope. Il lui fallut trente secondes pour localiser l'appareil sous des piles d'ouvrages de référence, et, lorsqu'il revint à la vitre, il n'y avait plus personne en bas.

— Et merde, grogna-t-il en scrutant la rue.

Tout compte fait, le type avait dû entrer dans la supérette. Un voyageur en manque de cigarettes ou de soda. Lincoln pouvait dormir sur ses deux oreilles : tout risque d'animation se trouvait écarté.

Puis on frappa à la porte.

Violemment.

Oz pivota sur son siège, la nuque parcourue d'un frisson. C'était impossible. Anormal, en tout cas.

Les coups reprirent. Trois lourdes charges.

Oz se leva en douceur. Recula de deux pas, pour mieux examiner le battant. La moitié supérieure était en verre dépoli.

On distinguait une silhouette.

Oz attendit : le visiteur allait-il insister ? Puis il songea qu'il faisait plus clair dans le bureau que dans la cage d'escalier. En d'autres termes, si Oz pouvait voir l'invité surprise, sûr que ce dernier voyait Oz. Et pourquoi se remettre à cogner quand on sait que l'autre a entendu ?

Oz redressa les épaules – l'effet était minime, mais il se sentait un peu plus mâle – et alla ouvrir.

C'était bien le type de la rue.

Des gouttes d'eau parsemaient son long manteau. Il se tenait les jambes écartées, enraciné dans le sol, immobile, prêt à bondir dans n'importe quelle direction. Ses yeux verts et perçants dévisagèrent Oz.

— Oswald K. Turner ?

— Comment êtes-vous monté ?

— J'ai pris la porte, puis l'escalier.

— Vous avez juste ouvert, comme ça ?

— Voilà.

— Elle était verrouillée, très cher.

— J'ai pas fait gaffe.

— Pas fait... Bon, admettons.

Oz n'était pas coutumier de ce genre de sujet. Plus il considérait ce type, plus il se demandait s'il était flic. Ou soldat. Voire un authentique Man in Black. Un truc balaise, en tout cas.

— Je peux entrer ?

Que de politesse pour un gus qui avait l'air d'aller où bon lui semblait...

— Ça dépend, répondit-il avait méfiance. Que désirez-vous ?

L'homme plongea la main dans son manteau et sortit un bout de papier. Le brandit sous le nez d'Oz. C'était une page intérieure du *Ledger*, vieille de trois semaines. Signé Oswald K. Turner, l'article s'intitulait : « ET EUX, QUAND NOUS PRENDRONT-ILS AU SÉRIEUX ? »

— Je suis venu pour parler de ceci, expliqua l'homme. Je veux que vous m'y conduisiez.

— Là, tout de suite ? Mais il pleut !

— J'avais remarqué.

— Ma voiture est chez le garagiste.

— Pas la mienne.

— Il y a une sacrée trotte, une fois qu'on est garé. Ça grimpe. Et c'est un terrain privé. Et le proprio ne m'apprécie pas des masses.

L'homme ne dit rien, se contenta d'incliner la tête d'environ cinq degrés, sans ciller. Oz comprit le message. Ils allaient voir ce machin, et ils y allaient maintenant, qu'il grêle, qu'il neige, ou qu'il pleuve des météorites.

— Vous faites partie de... quelque chose ?

— Moi ? (L'homme sourit, plus ou moins. Il semblait manquer d'entraînement.) Non, je ne suis qu'un lecteur assidu.

— Et vous avez un nom ?

L'homme le considéra quelques instants, comme s'il le jaugeait.

— Je m'appelle John Zandt.

L'homme conduisit sans un mot, sinon pour demander l'itinéraire. Oz n'était pas plus bavard, qui préférait exercer son sens de l'observation. La banquette arrière était jonchée de cartes et de livres. La voiture affichait 10 400 kilomètres au compteur fixe, et 9 920 au compteur d'étape. Le loueur avait peut-être omis de le remettre à zéro, mais Oz en doutait. Hormis une légère odeur de tabac froid, le véhicule semblait sortir de l'usine. Oz se dit que M. Zandt sillonnait les routes depuis un bon bout du temps, ou bien qu'il roulait très vite, ou encore les deux à la fois.

Oz l'orienta sur la 112, puis à droite sur la 51 C – appelée route du Vieil Étang pour une raison qui n'allait pas de soi : elle longeait bel et bien un étang sur la gauche, mais le secteur était truffé de plans d'eau, et on se demandait bien ce que celui-ci avait de plus que les autres. Le type y jeta un œil, et Oz supposa qu'il se faisait la même réflexion. Comment aurait-il su que cette mare rappelait à Zandt un petit lac du Vermont visité un an plus tôt ? Et qu'il tâchait de se remémorer sa vie de l'époque, sans y parvenir ?

Au bout de vingt-cinq kilomètres, Oz indiqua de nouveau la droite, et ils s'engagèrent sur une route à voie unique bordant la forêt de Robertson, près de quatre cents hectares d'arbres et de contreforts qui attiraient quelques randonneurs avertis.

Au bout se trouvait une barrière.

— Va falloir se garer ici, dit Oz. Faire le reste à pied, comme je vous le disais. Trois bornes.

L'homme coupa le moteur. Descendit. Attendit qu'Oz fasse de même. Verrouilla la voiture. Attendit qu'Oz montre le chemin.

Les premières centaines de mètres rendirent Oz nerveux, car il s'agissait bel et bien de l'allée de Frank Pritchard. La ferme Pritchard était retranchée dans le parc, au milieu des arbres. Pour atteindre le lieu qui intéressait Zandt, le mieux était d'emprunter ce raccourci, avant de rejoindre le domaine public, puis de fouler à nouveau le terrain privé. Le seul problème était Frank, qui se foutait pas mal du patrimoine régional. La dernière fois qu'il avait surpris Oz dans les parages, il avait menacé, avec beaucoup de conviction et en agitant un fusil, de détruire la chose.

Ils quittèrent la route sans incident et s'enfoncèrent dans les bois, où Oz perdit vite l'impression de mener la marche. Il lui fallait donner les indications, mais son accompagnateur avançait à un rythme de cinquante pour cent supérieur au sien, avalant les côtes et enjambant les ruisseaux aussi facilement que du plat. Oz se retrouva vite en nage.

— Vous savez, ce truc est là depuis des lustres, ahana-t-il dans le dos du type. Ce ne sont pas vingt minutes de plus ou de moins qui changeront quelque chose.

L'homme s'arrêta, le regarda froidement. Mais ensuite la cadence se réduisit.

Oz suivit un cap nord-est à travers les collines.

— Ah zut, fit-il un peu plus loin. (Il stoppa et tendit l'index.) Cet enfoiré a réparé la clôture.

Il s'appuya sur ses talons pour reprendre son souffle, plutôt soulagé. Leur destination se situait derrière un grillage tout neuf de deux mètres cinquante. Bref, ils étaient bloqués.

La fête est finie, mon coco. Ravi de t'avoir connu.

Le type avança jusqu'au treillage et l'examina, avant de puiser dans sa poche un petit outil. Il lui fallut à peu près quarante-cinq secondes pour découper une fente d'un mètre cinquante de haut.

— Ah, bravo ! pesta Oz, mais le type se faufilait déjà de l'autre côté.

Oz le suivit.

Une fois revenu sur les terres de Pritchard, on n'était plus bien loin de la destination. On parcourait environ trois cents mètres au pied d'une crête, puis on bifurquait à gauche pour gravir une butte. L'autre versant était assez raide, et les deux hommes descendirent avec prudence. Six mètres de plat, puis un second talus, fort similaire, cerné d'arbres, et recouvert de végétation dense. Oz conduisit le type sur la droite, puis s'immobilisa.

— Vous y êtes, dit-il avec une pointe de fierté. Le fameux grenier à légumes.

Il s'apprêtait à en dire plus, à dérouler son laïus, mais l'homme lui coupa le sifflet en levant sa paume.

Zandt se rapprocha de l'objet de la visite. Celui-ci n'évoquait rien à l'œil profane, et l'on pouvait facilement passer devant sans rien remarquer. D'ailleurs, pendant de nombreuses années, l'existence de la chose avait échappé à tout le monde.

Au milieu des racines et des cailloux, une petite niche rectangulaire affleurait dans le flanc de la butte. Sa base était le sol ; ses parois légèrement arquées étaient constituées de pierres comparables à celles dispersées tout autour, mais empilées en deux murs de trente centimètres d'épaisseur et d'un mètre de haut, espacés de soixante centimètres. Une grosse pierre plate faisait office de linteau. Au-dessus, la terre.

L'homme s'accroupit pour examiner l'ouverture, tout en promenant son doigt sur la jointure des pierres. Puis il se releva, considéra le talus un instant, sortit un appareil photo et prit deux clichés, avant de se décaler de quelques mètres pour étudier l'agencement.

Puis, sans lâcher l'ouvrage des yeux, il retourna auprès d'Oz.

— C'est bon, dit-il. Je vous écoute.

En 1869, l'occupant d'une ferme située près de Lincoln – alors un obscur hameau – ramassait du bois en forêt lorsqu'il fit une étrange découverte. Une construction de pierre, semblable à un soupirail. Revenant le lendemain pour en débroussailler l'accès, lui et son fils retirèrent de gros cailloux enfoncés dans la gorge d'un tunnel. Le garçon se glissa à l'intérieur, muni d'une bougie, et constata que le tunnel courait sous la butte sur une longueur d'environ deux mètres cinquante, avant de déboucher sur une chambre circulaire et voûtée de quelque trois mètres de diamètre. Ce soir-là, le fils, qui avait un bon coup de crayon, exécuta un dessin fidèle de la chambre souterraine. Le lendemain, son père explora la galerie à son tour, non sans difficulté du fait de sa carrure, et resta quarante minutes à l'intérieur. Sitôt ressorti, il annonça à son fils qu'ils allaient reboucher l'entrée, pour empêcher qu'un enfant ou du gibier se retrouvent piégés à l'intérieur – bien que ces deux engeances fussent plutôt rares dans le secteur. Il tint en outre à garder le secret sur cette découverte. Le fils relata l'événement dans son journal intime, qui fut récupéré, avec le dessin, parmi les archives du petit musée de Lincoln après plus d'un siècle.

Une curiosité locale, rien de plus. Sûrement un grenier à légumes abandonné, qui en raison de sa structure en pierre avait survécu aux derniers vestiges de la cabane avoisinante.

Mais voilà. Le 1er février 1876, le *Journal* de Boston faisait état d'une autre galerie souterraine à deux pas de Dedham, au sud de la grande ville. Au cours des cinq décennies suivantes, des centaines de caves analogues furent mises au jour dans toute la Nouvelle-Angleterre – dans le Massachusetts, le Vermont, le New Hampshire et le Connecticut. Souvent creusées dans des talus, et de deux sortes : de simples passages couverts, aménagés dans une crevasse rocheuse naturelle, ou des structures plus complexes, en forme de ruche, telle celle de Lincoln, avec des murs de pierre sèche et des plafonds en encorbellement. Dans un cas comme dans l'autre, elles étaient en général recouvertes de terre, de manière à se fondre dans les talus.

Bien vite, certains notèrent leur ressemblance avec certaines galeries de pierre bâties en Europe très, très longtemps avant la découverte théorique de l'Amérique. Mais les archéologues autorisés restaient formels : il s'agissait ici de greniers à légumes, conçus par des lignées de colons éteintes ou parties depuis lors. Certaines de ces galeries (dont la cave de Dedham) retombèrent dans l'oubli, et le phénomène avec.

On en resta là jusqu'aux années soixante, quand, face à la recrudescence d'hypothèses alternatives défendues par des amateurs passionnés, les experts durent remonter au créneau. Ils mirent dans la balance des montagnes de données concernant l'absence de vestiges précolombiens dans la région, ajoutées au fait que la répartition de ces structures épousait sensiblement la carte des peuplements de colons recensés en Nouvelle-Angleterre, et que les traditions orales et historiques confirmaient l'existence de caves à légumes en pierre. CQFD, et maintenant bas les pattes ! Mais les amateurs soulignèrent que certaines galeries semblaient répondre à des critères d'ordre astronomique : ainsi une cave du Gungywamp, dans le Connecticut, présentait la même orientation qu'une autre du site préhistorique de New Grange, en Irlande. Pure coïncidence, répondirent les archéologues. La datation au carbone 14 d'un morceau de charbon extrait d'une cave de Windham Country, dans le Vermont, l'avait cependant situé aux alentours de 1405, insistèrent les amateurs. Mais la technique du carbone 14 est notoirement capricieuse, se gaussèrent les experts. Sauf lorsqu'elle étaye vos propres affirmations, répliqua-t-on. Et puis, même si la plupart étaient bel et bien des greniers à légumes, les autres s'avéraient ridiculement larges pour un tel usage, assénèrent les amateurs, qui commençaient à s'échauffer. On s'en fiche, tranchèrent les archéologues en se bouchant les oreilles. Du balai !

— Mais quelque chose me dit que vous savez tout ça, marmonna Oz en guise de conclusion. La moitié figure déjà dans l'article du *Ledger*.

L'homme le toisait, et pour la première fois de sa vie Oz eut envie de se taire. On a tous entendu parler du regard du vétéran du Vietnam, qui vous foudroie à mille mètres. Eh bien, celui de ce type semblait dix fois plus puissant.

— Qu'y a-t-il là-dedans ? s'enquit-il.

— Vous, alors ! ricana Oz. Vous croyez que j'y suis entré, peut-être ?

— Quoi ? Vous avez écrit au moins trois papiers sur cet endroit sans y avoir jamais mis les pieds ?

— Voyez par vous-même. L'une des grandes questions est justement de savoir pourquoi les ouvertures sont souvent très étroites. Vous m'avez regardé ? Vous avez un sérieux problème de perception spatiale si vous croyez que je passe là-dedans ! Même sans tous ces satanés cailloux qui obstruent le passage...

Un long silence, puis d'un air morne Oz regarda le type enlever son manteau.

Cela prit une grosse demi-heure. La pluie n'était plus que crachin, et Zandt s'enfouissait jusqu'au torse pour attraper de nouvelles pierres. Oz faisait ce qu'on lui disait : réceptionner celles-ci et les empiler en un joli tas sur le côté. Puis le souffle de l'homme se réverbéra de manière un peu différente, et il émergea lentement, en tirant à bout de bras une pierre beaucoup plus large – et plate – que les précédentes.

— La voie est libre, dit-il en fixant Oz.

— J'aurais bien envie de vous suivre mais je suis trop vieux, trop gras, et bien trop claustrophobe. J'ai visité l'un de ces trucs dans le Vermont il y a dix-huit mois. L'entrée était carrément plus large, et j'ai quand même eu l'impression d'y être enterré vivant.

Le type esquissa un sourire, toutefois Oz n'y vit aucune moquerie. C'était plutôt comme si, par pur hasard, sa remarque avait mis dans le mille.

Là-dessus, l'homme disparut ; on distinguait encore ses pieds, puis plus rien.

Oz s'assit sur le tas de pierres et patienta.

Au bout de dix minutes, il se pencha pour scruter le tunnel. Deux mètres après l'entrée, il avisa le faisceau furtif d'une lampe torche puissante. Le type avait prévu son coup, il fallait au moins lui reconnaître ça.

Oz reprit son siège de fortune. La pluie récidivait et il rêvait d'un café, mais à part ça il éprouvait une certaine excitation. Quelques heures plus tôt, il maudissait tous les lundis matin, priant pour qu'il se passe quelque chose. Eh bien, il était servi. Il se passait vraiment un truc, là. C'était la quinzième, la vingtième fois qu'il venait jusqu'ici. Qu'il s'asseyait à peu près au même endroit, s'interrogeant sur la nature de cette construction. Mais aujourd'hui un type était à l'intérieur pour découvrir la réponse.

Un nouveau quart d'heure s'écoula. Émergèrent d'abord les mains, chacune chargée d'un sachet transparent. Ils renfermaient quelque chose, mais Oz ne sut dire quoi. Le type remonta sur ses jambes, frotta ses vêtements boueux et ramassa son manteau. Les sachets disparurent dans ses poches.

— Alors, qu'avez-vous trouvé ? Qu'avez-vous vu ? Ça ressemble au dessin ?

— Le dessin était assez fidèle.

— Allez, dites-moi ce que vous...

Un craquement, très bruyant. Et des cris. Oz se retourna pour voir Frank Pritchard remonter la butte, en furie, un fusil dans les mains.

— Merde, on est mal ! C'est la dernière personne avec qui j'aie envie de me fritter.

Le type observa la progression du fermier, tout en essuyant la terre de ses mains. Les braillements de Pritchard devinrent plus intelligibles. Il était un peu bourré, comme d'habitude, et refusait que des fumiers d'enfants de salauds débarquent sur sa terre quand ça leur chantait. Il avait sa bon Dieu de carabine et cette fois il allait s'en servir, et aucun jury n'y trouverait à redire.

Il s'arrêta et braqua son canon.

— Et merde ! paniqua Oz. Écoutez, Frank...

Pritchard agitait son fusil comme un dément. S'il pressait la détente, il allait arracher une tête.

— On est occupés, lança le dénommé Zandt avec flegme.

Le vieux ferma son clapet. Après quelques instants, Oz s'aperçut que le type au long manteau était armé lui aussi. Sa main droite tenait un flingue surgi de nulle part, pointé droit sur la tête de Pritchard. L'arme ne tremblait pas d'un millimètre.

— Je reconnais que nous violons une propriété privée, dit Zandt d'une voix posée. Ce qui, en soi, justifie que vous vous sentiez lésé. Mais maintenant vous allez partir et nous laisser tranquilles. Et si vous revoyez M. Turner dans les parages, vous passerez votre chemin en regardant ailleurs. Cet arrangement vous convient-il ?

— C'est ma propriété, articula Pritchard avec une étonnante clarté.

— Nous le savons. Nous n'abîmerons rien.

Le vieux fit mine de capituler et tourna brusquement les talons. Au bout de deux mètres, il se retourna.

— Ce truc est un nid d'emmerdes, fit-il. C'est moi qui vous le dis.

— Oui, répondit l'autre. En effet.

Pritchard jura dans le vide et repartit d'un pas fumasse.

Pendant qu'Oz le regardait s'éloigner, Zandt se pencha sur le tas de pierres pour reboucher le tunnel.

Dix minutes passèrent, sans qu'on revienne les embêter. La grande pierre plate fut posée en dernier. En la voyant sceller l'entrée, on comprenait que c'était là sa fonction initiale.

L'homme se redressa.

— Ne donnez jamais de détails supplémentaires sur l'emplacement de ce truc, compris ? Que ce soit dans votre canard, sur votre site Web, ou dans cette épouvantable émission de radio.

— Mais pourquoi ? Maintenant que vous êtes entré... (Les yeux du type le dissuadèrent de continuer.) D'accord.

Zandt sortit l'appareil photo de sa poche et le lui remit.

— Il renferme quinze clichés. Le reste de la carte contient le même genre de vues. Dont l'intérieur de la cave de Dedham.

— Quoi ? Vous rigolez. On l'a perdue...

— Je l'ai retrouvée. Mettez-les en ligne, pour que je les retrouve en cas de besoin.

— Mais... pourquoi faites-vous tout ça ?

— Il faut que les gens sachent. Mais vous êtes venu seul, et vous y êtes entré seul. Pigé ?

— Je vous oublierai sitôt que j'aurai retrouvé le monde réel, bonhomme. Et ce sera avec plaisir, croyez-moi.

L'homme sourit, et cette fois cela parut presque sincère. Puis il s'en alla.

Quand Oz rejoignit la barrière, la voiture noire avait disparu et la pluie noyait déjà ses traces. Par chance, une berline surgit cinq minutes après qu'il eut atteint la route du Vieil Étang.

Manque de pot, c'était celle de son ex-femme.

CHAPITRE 13

Ce matin-là, les choses étaient allées très vite. Nous avions été réveillés à 8 heures et des poussières par un coup de fil de Reidel : un flic du coin tenait une piste concernant Larry Widmar, qu'on avait vu discuter avec une femme dans un bar de la route d'Owensville le soir de sa disparition. Nina prit quelques notes et promit de retrouver Reidel au bar d'ici quarante minutes.

Pendant qu'elle se douchait, j'ai composé le numéro fourni par Unger dans son mail, non sans avoir vérifié dans un recoin d'Internet que cette ligne était bien attribuée à un certain C. Unger, adresse non communiquée. C'était bon signe, y compris le fait qu'Unger utilise sa supposée appartenance au monde du renseignement pour soustraire des infos à un fichier qui n'était pas censé exister. Je restais réticent à l'idée de l'appeler avec mon portable, mais avais-je vraiment le choix ? Le téléphone de la chambre, celui de la réception ou même une cabine téléphonique auraient trahi ma position encore plus facilement qu'un cellulaire. Toute autre approche relevait de la manipulation : emprunter l'appareil d'un quidam, avec ou sans son accord, cela revenait à transférer le danger sur un innocent, ce que même ma morale élastique tendait à réprouver.

Alors je me suis décidé à l'appeler, et pour finir il n'a même pas décroché. Je suis tombé sur une boîte vocale, où l'on m'a proposé de laisser un message à Carl Unger. La voix ne m'était pas familière, mais elles le sont rarement. J'ai décliné mon nom et l'ai invité à me rappeler.

Puis nous sommes sortis, avons grimpé dans ma voiture de location et roulé jusqu'au bar.

Reidel était déjà sur place, en compagnie d'une femme au milieu d'un parking froid, vide, léché par la brume. Le café se trouvait en bord de route, quelque huit cents mètres après la ville, une longue structure plate dont les maigres fioritures étaient vaguement censées évoquer un bateau. Curieuse idée, quand l'océan le plus proche se trouvait à une journée entière de voiture. Il s'appelait le Mayflower.

Reidel nous présenta la jeune Hazel, avant d'expliquer qu'ils étaient à la porte car le patron avait du retard.

— Très bien, Hazel, fit Nina en montrant sa plaque. Auriez-vous l'obligeance de me répéter ce que vous avez dit à l'inspecteur ?

La trentaine, Hazel avait une maigreur de fumeuse, et de jolis traits en voie de calcification. Sa voix laissait penser que sa gorge pouvait servir de râpe et que sa propriétaire était plutôt du soir. Son langage corporel était celui d'une femme dont la bonne humeur n'excède jamais les deux minutes.

— Le type sur la photo qu'on m'a montrée... Il passe de temps en temps, pas souvent. Mais il était là mercredi, en fin de soirée, et je me souviens qu'on était mercredi parce que je devais être de repos sauf que Gretchen nous a encore plantés à la dernière minute mais bon, vu que ce gros tas se fait sauter par Lloyd, personne ne la fait chier.

— Lloyd est le patron, précisa Reidel. On ne l'a pas encore entendu. Il n'était pas là hier soir, ni mercredi.

— Ben, tiens, commenta Hazel. Trop occupé à tremper son biscuit. Il est marié, vous savez. Trois gosses. Mignons comme tout, d'ailleurs.

— Mercredi, donc, reprit Nina.

— J'ai pas parlé à ce type, je le connais pas vraiment. Tout ce que j'ai dit au flic, c'est qu'il était là et que je l'ai vu causer avec une poule aux cheveux courts qui vient des fois s'envoyer des vodkas. Moi, j'ai toujours cru qu'elle était, comment dire, une femme à femmes, mais qu'est-ce que j'en sais, hein ?

— Pourriez-vous nous la décrire ?

— Même taille que vous, vingt kilos de plus, une mine de papier mâché. On n'a pas envie de l'embrasser.

— Pensez-vous que d'autres collègues sauraient nous dire son nom ?

— Peut-être Donna. Mais elle ne reviendra que jeudi.

Nous nous tournâmes vers le véhicule qui entrait dans le parking. Un pick-up rouge, qui s'arrêta devant l'entrée du bâtiment.

— Ou lui, ajouta Hazel en resserrant ses bras sur sa poitrine. S'il a pensé à prendre sa cervelle.

Le type qui descendit du camion approchait de la cinquantaine. Mince, gagné par les cheveux blancs. Il visait le renard argenté, mais faisait plutôt fouine grisonnante et bronzée.

— Qu'est-ce qui se passe ?

— Police, fit Reidel. C'est au sujet d'un de vos clients. Un certain Lawrence Widmar.

Lloyd parut aussitôt sur ses gardes.

— Exact. Le type qui s'est fait tuer.

Hazel ouvrit de grands yeux.

— Il est mort ?

— Oui, répondit Nina. C'est l'homme qu'on a retrouvé en forêt de Raynor. Le policier ne vous a rien dit ?

— Non. Il m'a juste demandé si je le reconnaissais... Je savais pas. (Elle sortit ses cigarettes, chercha son briquet. Je lui tendis le mien.) Merci. La vache...

— Je ne le qualifierais pas d'habitué, ajouta le patron comme pour prendre ses distances. On le voyait peut-être une ou deux fois par mois. Ce n'était pas un gros buveur. Il prenait une table, et souvent il ouvrait un bouquin. Il lui arrivait aussi d'aborder quelqu'un.

— Des femmes ?

— Ouais.

— Hazel dit qu'il était en grande conversation mercredi soir.

Lloyd considéra Hazel d'un œil vide.

— Tu sais bien, dit-elle. Grosses joues. Cheveux courts. Je me disais que c'était peut-être une gouine.

— *C'est* une gouine, Hazel. Je te l'ai déjà expliqué. Diane Lawton, qu'elle s'appelle. Il parlait à cette nana-là ?

— Sûre et certaine.

— Alors il perdait son temps, jugea le patron d'un air déconcerté. Y a de ces débiles, des fois.

Hazel lui jeta un regard propre à décaper les murs.

— Vous avez le bottin dans votre bar, monsieur ? demanda Nina.

Le patron la conduisit à l'intérieur avec Reidel. Je restai dehors avec la serveuse.

Elle accepta une nouvelle cigarette, visiblement peu pressée d'entrer.

— Vous croyez que c'est un habitant de Thornton qui a fait le coup ? interrogea-t-elle.

— Ou des environs. Cette Lawton en serait capable, d'après vous ?

Elle fronça les sourcils, surprise par la question.

— Eh bien, non, pas du tout. Mais bon, les gens qui tuent d'autres gens, c'est une autre planète pour moi. Ils ont un truc particulier, non ? Vous savez les reconnaître ?

— Pas vraiment.
— Sérieux ? Vous connaissez des tueurs ?
— Quelques-uns.
— Ils sont comment ?
— Comme vous et moi. Sauf qu'ils tuent des gens.
— Mais alors... pourquoi ?
— Bonne question. C'est comme ça.
— J'avais un petit copain dans le temps, confia-t-elle après une pause. On sentait qu'il allait blesser quelqu'un un jour ou l'autre. Il dégageait un truc étrange ; par exemple, il avait une drôle de façon de regarder ses mains. Mais il n'a jamais rien fait, autant que je sache. On a cassé, et un soir, il est venu s'asseoir dans mon jardin. Il n'a pas crié ni rien. Il est juste resté assis. J'ai cru qu'il allait y passer la nuit, mais il a fini par repartir.
— Et ensuite ?
— Rien. Je le croisais de temps en temps. Deux ans plus tard, il s'est tué dans un accident de voiture. Je sais pas pourquoi je vous raconte ça.
Nina et Reidel ressortirent du bar. Nina vint remercier Hazel pour sa collaboration.
— Et j'apprécierais que vous ne parliez de ceci à personne. À ce stade, nous cherchons simplement à rayer Mlle Lawton de la liste des suspects.
— Pas de problème, répondit Hazel en piétinant son mégot. Merci pour la clope, me lança-t-elle avant de partir vers le bâtiment.
— Tu te fais des amies ? demanda Nina.
— Tu me connais. Je suis irrésistible.

Diane Lawton vivait dans une petite maison à deux kilomètres du Mayflower, dans Thornton intra muros. Il était déjà 9 h 20 lorsque nous arrivâmes là-bas, mais la grosse voiture garée dans l'allée suggérait une présence. Ainsi que la musique provenant de la fenêtre de la cuisine. Légère, baroque, égayée par un hautbois prolixe. Du Bach, probablement, qui devait suffire à s'attirer les regards du voisinage. Des jouets d'enfant jonchaient les pelouses de part et d'autre de la maison.
Nina frappa trois coups à la porte latérale, que vint aussitôt ouvrir une femme conforme à la description.
— Mademoiselle Lawton ?
Elle nous considéra en hochant la tête. Elle avait environ trente ans, le visage rond et l'une de ces bouches qui semblent

en permanence esquisser un mouvement. Il fallait y regarder de près pour repérer les cernes sous ses yeux et comprendre que les coins de ses lèvres n'allaient pas forcément piquer vers le haut.

— Oui, dit-elle. Et vous êtes... ?

Nina montra son badge.

— On peut entrer quelques instants ?

— Je partais au boulot.

— Ça ne devrait prendre que cinq minutes.

— « Devrait » ?

Nina la fixa droit dans les yeux.

— Comme l'indique ce badge, j'appartiens au FBI, madame. Nous tolérons l'impertinence à la télévision pour des questions de mise en scène. Mais beaucoup moins dans la vraie vie.

Lawton se recula pour nous accueillir. Sa cuisine était chaleureuse, pleine de pots à fleurs en macramé et de jattes en terre cuite réalisés soit par des passionnés, soit dans le cadre d'une ergothérapie entre quatre murs élevés.

Elle saisit une télécommande et coupa la musique.

— De quoi s'agit-il ?

Nina montra une photo.

— Reconnaissez-vous cet homme ?

Un sourire pincé.

— Il s'appelle Larry. M'étonnerait pas qu'il soit marié.

— Il l'était. Mais il est mort.

Un temps d'arrêt.

— Vraiment ? Eh ben...

— Vous n'avez pas entendu parler du corps retrouvé en forêt de Raynor ?

— Non. Je n'étais pas là ce week-end. Que s'est-il passé ?

— C'est ce que nous aimerions découvrir, répondit Reidel. Certaines personnes du Mayflower disent que vous lui avez parlé mercredi soir.

— Les gens du coin s'intéressent beaucoup aux affaires des autres, n'est-ce pas ? Oui, bien sûr, nous avons discuté un petit moment. Il est venu s'asseoir à côté de moi.

— Vous êtes repartis ensemble ?

— Non, pas du tout.

— Serait-ce, comment dirais-je, par incompatibilité ?

Lawton soupira.

— Je travaille dans un foyer pour femmes de Dryford, inspecteur, car il arrive que les femmes aient besoin d'un refuge. Je ne suis pas d'ici, et je porte les cheveux courts car je déteste les avoir dans les yeux – et parce que j'ai la faiblesse de croire que ça me va bien. J'ai sûrement tort, mais, non, je ne suis pas lesbienne. Entendu ?

— Cela ne nous regarde pas, répondit Nina en foudroyant Reidel du regard, mais personnellement je vous comprendrais. Vu la pauvreté du choix...

— Je n'ai pas quitté le bar avec lui parce que je ne recherche pas les histoires d'une nuit. Sans ça, je serais repartie au bras de suffisamment de braves gars pour que les gens soient fixés sur mes préférences. Et aussi parce que c'était un connard.

— C'est-à-dire ?

— Il est venu me voir et m'a payé un verre, sans que je lui demande rien, puis il a passé son temps à en reluquer une autre par-dessus mon épaule. Pour finir, il s'est levé, il m'a remerciée – même là, il a pas été fichu de me regarder – et il est parti illico brancher l'autre nana. Ça vous ira comme histoire triste ? Une femme se fait accoster mais le type s'aperçoit que c'est un boudin, alors il reprend ses billes pour les placer ailleurs ?

— Cette histoire n'est pas aussi triste que ça, répondis-je. Les types mariés qui draguent dans les bars ne cherchent pas l'élévation intellectuelle. Il a repéré une proie plus facile, voilà tout. De toute façon, ce mec est un cadavre démembré à l'heure qu'il est, alors considérez que vous avez gagné.

Elle cligna des yeux.

— À quoi ressemblait cette autre femme ? questionna Nina.

— Elle avait les cheveux longs, marmonna Lawton comme si cela expliquait toute la laideur du monde. Elle était mince, et *mignonne*, et elle avait de longs cheveux roux comme de la braise. Une vraie bombe, je parie.

Reidel leva les yeux de son carnet :

— Répétez-moi ça ?

— Mais elle était bourrée, poursuivit-elle sans cacher son dédain. Alors vous avez peut-être raison. Je suppose qu'elle était plus facile que moi. Un jeu d'enfant...

— Madame, coupa sèchement Reidel, vous avez parlé de cheveux roux ?

— Oui, c'est ce que j'ai dit.

— Et vous ne savez pas comment elle s'appelle ?

— Non. Mais je l'avais déjà croisée au bar. À quelques occasions. Elle s'installe tout au fond et picole toute la soirée.

— Son âge ?

— J'en sais rien. Vingt-quatre, vingt-cinq ans ?

— Vous l'avez vue ressortir, mercredi soir ?

— Évidemment, fit-elle comme face à une bande de demeurés. Elle est partie avec le tombeur de ces dames. Je pensais que vous étiez là pour ça.

Un instant de flottement, et Reidel prit la porte.

Nina conclut en quatrième vitesse avant de lui courir après. Je suivis le mouvement. Reidel était déjà au téléphone, près de sa voiture.

— Qu'est-ce qui se passe ? demandai-je.

Nina avait une mine défaite.

— Rends-moi service, s'il te plaît. Retourne au bar pour voir si notre serveuse se souvient d'une femme correspondant à ce signalement, une femme présente mercredi ou tout autre soir. Appelle-moi dès que t'as la réponse.

— Reidel a une piste ?

— Non. J'espère que non, en tout cas. Mais si j'éteins pas ce truc dare-dare, la rumeur va se répandre et briser la vie d'une pauvre fille.

Reidel referma son téléphone.

— J'ai eu Monroe, dit-il. On le retrouve là-bas.

Sur ces mots, ils mirent les voiles, dans le véhicule de Reidel.

Je regagnai le bar, où il faisait froid et sombre. Les murs étaient festonnés de filets, sûrement dans l'idée que le *Mayflower* avait joué un rôle clé dans l'histoire de la pêche américaine. La salle était aussi longue qu'étroite, et je comprenais qu'on la choisisse pour prendre un verre sans ameuter toute la ville.

Hazel nettoyait le bar ou tout du moins redéployait la crasse.

— Re-bonjour, dit-elle. Si vous cherchez Lloyd, il téléphone dans l'arrière-boutique. Question de sûreté nationale, sans doute.

— En fait, j'ai une question pour vous.

— Je vous écoute.

— Si je vous dis : « mince, jolie, rousse », vous pensez à quelqu'un de particulier ?

— Bien sûr, opina-t-elle. On en a une comme ça.

— Causante, la fille ?

— Pas vraiment, non.

— Elle est venue, mercredi dernier ?

Un temps de réflexion.

— Peut-être. Ouais, c'est possible. Mais elle aurait squatté vers le fond, et moi j'étais coincée ici. Alors je suis pas sûre à cent pour cent.

— J'imagine que vous ne savez pas son nom ?

— Désolée. Parfois on capte des bribes de conversation, et parfois les gens vont serrer des pognes pour faire salut-je-m'appelle-machin, comme si ça changeait quelque chose au fait qu'ils boivent tout seuls, mais les gens d'humeur conviviale préfèrent les bars du centre.

— Tête-de-nœud saurait ça, lui ?

Elle sourit.

— Possible. Il aime la jouer « je suis votre hôte » quand il est d'humeur, surtout si elles sont mignonnes. Elle aurait toutes ses chances. Allez lui demander.

J'enfilai le couloir jusqu'à la porte entrouverte d'un petit bureau. Le patron était vautré dans un fauteuil, les pieds sur la table. Il parlait la bouche collée au combiné, avec un sourire pas très sexy. En me voyant, il se renfrogna et couvrit le micro.

— Quoi encore ?

— Une de vos clientes. Rousse, séduisante. Son nom ?

Il me soumit deux possibilités, émit des doutes sur leurs orientations sexuelles respectives, puis fit valoir qu'il était occupé, que l'établissement n'avait pas encore ouvert et que je serais gentil de partir.

— Ce fut un plaisir, répondis-je. À propos, les femmes ont le droit de vous dire non, même si elles ne sont pas lesbiennes. Et quand elles le sont, elles méritent votre respect.

Il me regarda comme si j'avais parlé swahili avec un sale accent.

— Peu importe, dis-je. Bonne continuation.

De retour sur le parking, je contactai Nina.

— Alors ?

— Une habituée, le genre qui aime se faire oublier. Mais Hazel ne peut affirmer l'avoir vue mercredi.

— T'as obtenu un nom ? demanda-t-elle d'une voix tendue.

— Juste un prénom, mais le patron n'est pas catégorique. M'est avis qu'il s'est fait rembarrer une ou deux fois.

— Je t'écoute, Ward.

— Il hésite entre Julie et Julia. (Pas de réaction.) Nina ? Ne perds pas de vue que ce type est un crétin.

— Je te rappelle, dit-elle avant de couper.

Silence radio pendant quatre heures. Je me suis garé dans le quartier historique, j'ai tué le temps en m'offrant un petit-déjeuner et une série de cafés. La gazette locale publiait un non-article sur Lawrence Widmar, insistant sur les détails macabres de sa découverte et enjolivant ses triomphes de businessman. Trop tôt pour que l'on parle du corps trouvé la veille au soir. Trop tôt, mais il y avait peut-être autre chose. Parmi les personnes interrogées ce matin-là, deux ignoraient tout du type retrouvé nu dans les bois deux jours auparavant. Diane Lawton était peu impliquée dans la vie de la communauté, soit. Mais les barmaids, elles, savent plus ou moins ce qui se passe en ville. Comme Nina l'avait remarqué aux petites heures du matin, ce bled avait déjà deux cadavres dans le frigo, et pourtant, à regarder vaquer ses piétons, rien ne laissait penser qu'il était frappé en plein cœur.

J'essayai de joindre Nina, sans succès. Je n'avais pas le numéro de Monroe ni celui de Reidel, et de toute manière je doutais qu'ils fussent disposés à me parler. Comme je faisais défiler, désœuvré, le maigre répertoire de mon téléphone, je tombai sur John Zandt. Ce nom me laissa songeur. Nina et moi avions tous deux tenté de le contacter au cours des cinq derniers mois, mais il ne décrochait jamais. On lui laissait des messages, en vain. Je n'étais toujours pas enchanté à l'idée de communiquer avec M. Unger. Serait-il judicieux d'en parler à Zandt, pour vérifier s'il avait déjà croisé ce type quelque part ? La dernière fois qu'on s'était vus, j'avais pu constater que Zandt avait mené de vastes recherches sur les Hommes de Paille – même si nombre de ses conclusions paraissaient délirantes. Au pire, je ne serais pas le seul à connaître le nom d'Unger s'il devait m'arriver quoi que ce soit.

Je lui envoyai donc un bref SMS. Deux minutes plus tard, le téléphone m'annonça d'un *ping* que le message était bien arrivé. Mais je n'obtins aucune réponse.

— Va te faire foutre, maugréai-je, offusquant une serveuse.

Alors je quittai la boutique pour acheter un autre jus ailleurs.

Assis en face d'un square sur un banc, avec mon gobelet, j'étais sur le point de regagner l'hôtel lorsque mon téléphone sonna. Ce n'était pas Nina, et ce n'était pas Zandt.

— Allô ? fis-je en me redressant.

— Vous êtes bien Ward Hopkins ?

— C'est ça.

— Carl Unger à l'appareil. Écoutez, tout d'abord je suis vraiment peiné d'apprendre la mort de Bobby. C'était un mec bien. C'est arrivé comment ?

— Il a été assassiné.

— D'accord, dit-il d'un ton neutre. Vous m'expliquerez tout ça de vive voix. Vous êtes où ?

— À vous de me le dire.

— Je n'ai pas fait pister votre portable, Ward. Je vous demande ça parce que j'ai à vous parler et que je refuse de le faire au téléphone.

— Pourquoi ?

— Drôle de question pour un type qui a été payé à surveiller les communications ! Ne compliquons pas les choses, Ward. J'essaie juste de trouver un lieu de rendez-vous.

— Pourquoi pas Greensboro ? En Caroline du Nord.

— Pas de problème. Ce soir, ça vous va ? Je pense pouvoir être là vers 19 heures.

— Appelez-moi quand vous atterrissez. Au fait, vos cheveux sont de quelle couleur, aujourd'hui ? Je crois me souvenir que vous grisonniez un peu, la dernière fois.

— Monsieur Hopkins, dit-il d'une voix patiente, nous ne nous sommes jamais rencontrés. J'ai rejoint la boîte un an après votre départ. Je connais votre nom uniquement grâce à Bobby, vu ? Si j'en trouve un, j'arborerai un œillet rose à la boutonnière. Sinon, je me contenterai d'agiter la main.

Je raccrochai en me sentant nul. Si le type venait pour me tuer, il avait toutes les chances de réussir avec mes ruses à deux dollars. J'avais choisi Greensboro parce que c'était facile d'accès en avion, et surtout parce que c'était dans un autre État. J'avais souhaité entendre un brin de surprise dans sa voix, la preuve qu'il savait ma véritable position, que mon téléphone était bel et bien pisté. Mais je n'avais rien perçu de tel. Pire, il y avait au bas mot deux heures de route de Thornton à Greensboro, ce qui me gonflait royalement. Me repassant la conversation, je notai qu'il connaissait mes états de service au sein de l'Agence. Et songeai qu'il lui serait aussi facile de me flinguer à Greensboro qu'ailleurs.

J'enfonçai la touche de rappel.

— Salut, fit-il. Quoi de neuf ?

— Changement de programme. Retrouvons-nous plutôt dans un bled du nom d'Owensville. À deux heures de Greensboro, au nord-est, en Virginie. C'est possible ?

— Pas de problème. Mais vers 20 h 30 ou 21 heures, alors.

Je restai assis un moment, les yeux rivés sur le trottoir opposé, espérant que je ne commettais pas une grave erreur.

Le téléphone sonna de nouveau. C'était Nina. Elle semblait furax.

CHAPITRE 14

— Elle sera là dans moins d'une heure, indiqua Nina tout en me guidant à travers les modestes locaux du shérif de Thornton. (Elle s'arrêta devant une porte fermée.) Les flics cherchent des figurants pour un tapissage, mais va trouver des gens avec cette couleur de cheveux !

— Je pense avoir croisé une candidate en chemin. (Derrière la vitre armée, une femme d'une vingtaine d'années baissait la tête.) La nana du hall a trente balais de plus et pèse dans les cent dix kilos. Personne ne va la confondre avec celle-ci. Ni croire que Larry Widmar lui aurait fait du gringue.

— Je sais bien.

— Et surtout pas Diane Lawton.

— Sans déconner. C'est pourquoi Julia Gulicks sera très bientôt en état d'arrestation.

Un peu plus loin dans le couloir se trouvait une porte sans inscription. Nina l'ouvrit, et nous pénétrâmes dans la pénombre d'une petite pièce cachée derrière une glace sans tain. Reidel était adossé au mur du fond. Nina et lui se consultèrent à mi-voix.

J'examinai le visage de celle qui allait devenir le principal suspect d'un double homicide. Assise derrière une banale table en métal et contreplaqué, elle semblait fixer les rayures du plateau comme pour déchiffrer leur message. C'était la table des mauvaises nouvelles. Une table qui signifiait votre présence dans un endroit où vous n'aviez aucune envie d'être. Même la plus infâme gargote l'aurait couverte d'une nappe à carreaux.

La demoiselle finit tout de même par relever la tête. On était d'abord frappé par ses cheveux. Longs et ondulés, d'une couleur inhabituelle : un roux naturel injecté d'une chaleur sombre, moins proche du feu que du sang désoxygéné. Sa peau avait la pâleur de pêche qui accompagne cette teinte, et ses traits étaient francs. Elle était relativement mince, vêtue d'un tailleur sombre et chic. L'air calme et réservé. Au pire, on l'eût imaginée commettre une fraude industrielle mineure et compliquée, un montage incompréhensible. Mais jamais ce que j'avais vu la veille.

En même temps, comme je l'avais signalé à Hazel, on ne pouvait jamais savoir.

— Pas d'avocat ? demandai-je.

Reidel haussa les épaules.

— Elle dit que c'est inutile.

— Elle n'a pas le look d'une personne qui traîne dans les bars.

— C'est vrai, admit Nina. Et Mark Kroeger affirme qu'il l'a rarement vue boire. Deux petits verres par soir, maximum, et elle ne finit jamais le second.

— Un vieux truc d'alcoolo, commenta Reidel. On se retient en public, et dès qu'on regagne ses pénates, on extirpe les bouteilles de la corbeille à linge sale.

— Qui est Kroeger ?

— Le type qui l'accompagnait quand elle a découvert le corps en forêt de Raynor.

— Il est ici ?

Reidel secoua la tête.

— Si je comprends bien, vous êtes quasi persuadé que cette femme a fait le coup, lançai-je. Et sans aucune complicité ? Sans personne pour l'aider à porter deux cadavres sur une grande distance ?

— Ce genre de crimes est le fait d'individus isolés, répondit Reidel. Mais merci pour votre contribution.

— Cette femme a les cheveux les plus roux que j'aie jamais vus, insistai-je. Comme si elle avait le crâne en sang. Vous ne croyez pas qu'elle aurait tenté d'atténuer un signe aussi distinctif ?

— Pas si le meurtre de Widmar était un acte impulsif, non.

— Mais ça ne l'était pas. L'analyse sanguine a révélé des traces de drogue, de celles qu'utilisent les violeurs. Ça cadre mal avec un coup de tête. Et si le corps décharné lui est également imputé, vous pouvez tout de suite écarter la thèse du meurtre de convenance. Sachant qu'il précédait celui de Widmar, pas vrai ?

— Rappelez-moi un truc, dit Reidel avec froideur. Qui vous a autorisé à venir ?

— Moi, fit Nina.

— Je serais curieux de savoir ce qu'en pense Monroe.

Nina se retourna. Même dans le clair-obscur, elle paraissait lasse et frustrée. Ça clochait quelque part.

Vingt minutes plus tard apparut un Monroe écarlate, prêt à crever l'écran. Nina le persuada d'attendre que Diane Lawton ait jeté un œil sur la suspecte et de prendre toutes les mesures pour contenir les médias.

Ils firent patienter Lawton encore quarante minutes, pour cause de dissensions sur le choix des figurantes. Une seule était vraiment rousse, l'obèse que j'avais croisée. Deux étaient plus ou moins auburn, les autres carrément brunes – et aucune n'avait la crinière fournie de Julia Gulicks. Nina ne se laissa pas faire, et obtint qu'on demande à ces dames de tirer leurs cheveux en arrière.

— Me Baynam défend sa cliente avec brio, lâcha Reidel avec une colère manifeste. Je croyais qu'on était là pour confondre des coupables. À moins que j'aie raté un épisode ?

Monroe pénétra dans la salle d'interrogatoire pour recommander à Julia Gulicks de prendre un avocat. Elle refusa de plus belle. Il lui expliqua qu'elle avait le droit d'être assistée, qu'en aucun cas la présence d'un pénaliste ne passerait pour un aveu de culpabilité et qu'elle risquait vite de perdre pied faute de conseils avisés. Mais elle secoua la tête, et je découvris sa voix dans l'Interphone :

— Je n'ai rien fait de mal. Je ne suis même pas en état d'arrestation, n'est-ce pas ?

Monroe nous rejoignit derrière la vitre.

— On commence, dit-il. Elle a eu sa chance.

Les cinq autres femmes furent amenées et placées dans la salle. Julia Gulicks se retrouva deuxième en partant de la gauche. On baissa le store devant la glace.

Diane Lawton fut conduite dans la pièce d'observation. Monroe lui rappela son rôle, en précisant qu'elle ne devait pas se sentir obligée de désigner quelqu'un – mais que, le cas échéant, elle devait indiquer en toute franchise son degré de certitude. Elle acquiesça. Elle avait pigé.

Reidel tendit le bras et leva le store.

On avait fait au mieux. La lumière était tamisée pour atténuer l'éclat des cheveux qui, plaqués en arrière, ramenaient l'attention sur les traits du visage. J'avais beau savoir quelle place occupait Julia Gulicks, il me fallut un temps pour la remettre.

Le regard de Lawton balaya lentement la rangée, de gauche à droite, s'arrêtant deux secondes sur chaque visage. Puis repartit en sens inverse.

Monroe l'observait de près.

— Êtes-vous capable d'identifier quelqu'un ?
— Oui, dit-elle.
— Et vous êtes sûre de vous ?
— À cent pour cent. Ce n'est pas comme si je l'avais vue pour la première fois mercredi.
— La question n'est pas de savoir si vous avez croisé cette personne au Mayflower, précisa Nina, mais de nous dire avec qui se trouvait Lawrence Widmar mercredi soir, et avec qui vous pensez l'avoir vu repartir.
— J'ai bien compris, répondit Lawton. C'est elle. La numéro deux.
Monroe hocha la tête. Reidel sourit jusqu'aux oreilles. Nina regarda ses pompes.

On conduisit Lawton dans une autre pièce pour prendre sa déposition. La police d'Owensville avait été chargée de localiser Mark Kroeger et de l'emmener au poste, en vue d'un nouvel interrogatoire concernant sa dernière sortie avec Julia Gulicks. Monroe était allé se procurer un mandat de perquisition pour l'appartement et la voiture de la suspecte, et organiser la fouille. Deux flics locaux étaient dépêchés au Mayflower pour glaner de nouveaux témoignages sur la soirée de mercredi.
Tous avaient reçu des consignes de discrétion très fermes. Jusqu'ici, le secret tenait. Mais il céderait tôt ou tard, et dès que la presse s'en serait emparée, on pourrait dire adieu aux pistes sérieuses. De nos jours, ce sont d'abord les rédacs-chef et les téléspectateurs avachis qui disent le droit. Caché derrière ma vitre, je suivis en direct l'inculpation de Julia Gulicks pour présomption de meurtre sur Lawrence Widmar. On lui proposa encore d'appeler un avocat, offre qu'elle déclina. À sa demande, on lui apporta un verre d'eau.
Reidel s'assit à un bout de la table. Nina prit l'autre. Quoique pâle, Julia Gulicks se dominait. Sa blancheur pouvait être due à l'éclairage. Et la maîtrise semblait être son état naturel. On pouvait même parler de froideur.
Reidel se chargea de l'interrogatoire :
— Vous avez bien compris les raisons de votre arrestation ?
— Un témoin affirme que je me trouvais dans un bar avec un homme mercredi soir.
— Nous pensons que vous avez rencontré M. Widmar au café Mayflower en fin de soirée. Que vous avez bu et discuté avec lui

un petit moment. Notre témoin dit que vous êtes repartis ensemble aux alentours de 23 heures.

— Et après ?

— Je vous demande pardon ?

— Que s'est-il passé ensuite ?

— C'est à vous de nous le dire.

— Je n'en ai *aucune idée*. Je n'ai jamais mis les pieds dans ce café. Je n'ai jamais rencontré cet homme de son vivant. Votre témoin a tout faux.

— Vous n'êtes jamais allée au Mayflower ?

— Je n'ai pas l'habitude de boire seule dans les bars.

— Oui ou non, mademoiselle Gulicks ?

— Non.

— Vous tenez vraiment à être aussi catégorique ?

— Je ne comprends pas ce que vous entendez par là.

— Ce qu'il entend par là, intervint Nina, c'est que nous tenons votre prénom de la bouche du patron. Ce n'est pas lui qui vous a identifiée, mais un témoin supplémentaire dont nous recueillerons la déposition dans la soirée. Si vous soutenez n'avoir jamais poussé la porte du Mayflower, alors que deux personnes affirment le contraire, vous êtes en mauvaise posture. Admettre que vous fréquentez ce bar ne revient pas à avouer quoi que ce soit d'autre. Et vous vous défendrez d'autant mieux que vous éviterez les petites contre-vérités, que ce soit maintenant ou plus tard.

— D'accord, fit Julia Gulicks en hochant lentement la tête. Merci pour la traduction, effrayante madame. Il n'empêche que je n'ai jamais mis les pieds là-bas.

— Vous devriez écouter ce que vous dit l'agent Baynam, reprit Reidel. C'est un peu votre conseillère officieuse, ici. Pour avoir vu le cadavre, je serais personnellement moins enclin à vous faciliter la tâche.

— Je l'ai vu comme vous, rétorqua Julia Gulicks. Et j'attends toujours qu'on m'explique pour quelle raison je l'aurais montré à un tiers. Puis appelé la police et attendu qu'elle arrive. Qu'on me dise pourquoi j'aurais fait une chose pareille !

— Pour vous couvrir, répliqua Reidel. Vous rencontrez Lawrence Widmar au bar. À un moment donné, vous glissez un produit dans son verre pour altérer son discernement. Vous repartez ensemble, montez dans votre voiture. Vous vous arrêtez quelque part, vraisemblablement sur le parking de la forêt de Raynor. Vous tuez la victime et traînez son corps en bas de la pente jusqu'à l'endroit où elle a été retrouvée. Puis, cette nuit-là ou le lendemain, vous avez l'idée d'y conduire votre petit ami du bureau, sous prétexte d'une balade romantique, ce qui permet de tomber sur le cadavre en tant que promeneuse lambda. C'est plutôt malin, mais en fait c'est assez bêta.

— Bêta, en effet, répondit la jeune femme sans ciller. Je ne sais pas ce que vous avez fumé, mais je n'ai jamais rien entendu d'aussi stupide. Et vous allez me poursuivre sur la seule foi de mes cheveux roux ?

— Pas tout de suite, dit Reidel en se levant. Nous avons trois jours pour ça. Et s'il nous faut une prolongation, nous l'obtiendrons.

Julia Gulicks se tourna vers Nina.

— Mais... vous n'avez aucune preuve !

— Nous en trouverons, promit Reidel avant de la toiser un instant. Puis nous passerons au second cadavre.

Elle balbutia, à peine :

— Quoi ?

— L'autre macchabée. Vous ne pensiez pas qu'on l'avait découvert, lui aussi ?

— Je ne sais pas de quoi vous parlez.

— Votre disque est rayé, Julia. Mais ne vous en faites pas, j'aurai largement l'occasion de m'expliquer. Tôt ou tard, vous saisirez où je veux en venir, et vous verrez qu'on ne plaisante pas.

Soudain, Julia Gulicks eut une tête d'enfant perdue, comme si elle comprenait que cet après-midi allait s'éterniser ; qu'au moins une personne dans la pièce la prenait pour une meurtrière et n'en démordrait pas.

Elle se renversa sur son dossier et croisa les bras.

— J'aimerais m'entretenir avec un avocat, maintenant.

Elle resta seule, avec la porte verrouillée et un flic en faction. Reidel s'en alla quérir le procureur, laissant Gulicks mariner dans un lieu confiné au-dessus d'une table déprimante.

J'emmenai Nina dans la rue, histoire de respirer et de m'en griller une. Quand on ne peut même plus fumer dans un comico, c'est que l'air du temps a gagné. Nous touchions le trottoir quand le portable de Nina sonna. Elle décrocha, écouta.

— Ils ont le mandat, annonça-t-elle en rangeant l'appareil. L'appartement de Julia va être mis sens dessus dessous.

— Avec un peu de chance, on ne découvrira rien.

Nina secoua la tête, mais sans préciser sa pensée. Il serait bientôt 18 heures.

— Il va bientôt falloir que je file, dis-je. J'aimerais arriver en avance, pour trouver un coin sûr avant que cet Unger ne débarque en ville.

— Je ne peux pas t'accompagner.

— Je sais. Je n'y comptais pas.

— J'aurais aimé couvrir tes arrières.

— Mais tu le feras. J'entendrai ta voix.

— Dis-moi que tu seras prudent.

— Je serai prudent. Les gens me surnommeront Ward « Prudent » Hopkins. Je deviendrai synonyme de précaution. Des bêtes nocturnes et timides avec de grands yeux me montreront du doigt, se moqueront de moi et me traiteront de tous les noms pour stigmatiser ma prudence excessive.

— Sérieusement, Ward.

— Ouais, je sais.

— Bipe-moi juste avant.

— Promis. (Je lui pris la main.) Te bile pas pour moi. Retourne là-dedans et règle cette histoire.

— Tu crois qu'elle a fait le coup ?

— J'en sais rien, Nina. Je n'ai aucune compétence en la matière. Je ne suis qu'un civil.

— Tu te souviens de cette femme dont je t'ai parlé ?

— Eh bien ?

— J'ai dû m'endormir avant de finir mon histoire. Un an et demi après son suicide à la cuiller, ils ont arrêté un automobiliste en état d'ivresse dans un bled situé à quatre-vingts bornes de Janesville. Là-dessus, on a découvert dans son coffre un stock de drogue ainsi que deux flingues enveloppés dans un manteau taché de sang. Alors ils l'ont embarqué, et pour finir ils l'ont épinglé pour un meurtre commis dans le Dakota du Nord.

— Et alors ?

— Puis ils lui ont également attribué l'un des meurtres de Janesville, qui s'est révélé être une pauvre affaire de cambriolage foireux. Elle n'y était pour rien, en fin de compte.

— Elle n'avait pas commis ce crime-là... mais les autres ?

— Elle était innocente d'une des charges qui l'ont fait plonger. C'est suffisant. On aurait dû...

— D'accord, Nina, mais c'était trop tard, et par sa faute. Dis-moi un truc, mon cœur : pourquoi t'es venue ici ?

— Tu le sais très bien.

— Je t'assure que non. Tu es entrée au FBI à cause de ce fameux feuilleton, si j'ai bien compris, et ça te travaille encore. Monroe le sait, et pourtant il choisit cette affaire-ci pour te ramener au bercail. Pourquoi ?

— Pour éviter qu'une autre femme ne subisse le même sort.

— Peut-être. Ou parce qu'il veut une poulette à bord, pour se couvrir au cas où l'affaire prendrait un tour politique.

Je détournai les yeux vers la rue. À quelques pas de là, deux types, à l'évidence des journalistes, attendaient dans une voiture. Trente mètres plus loin, j'avisai un van blanc sans vitres, qui pouvait facilement cacher des caméras, ou une armée de reporters prêts à bondir.

— Regarde, Nina. Il y a déjà des fuites. D'ici peu de temps, tout le pays sera au courant.

— Raison de plus.

— Si tu veux. Tout ce que je te demande, c'est de ne pas te laisser bouffer.

— Bien sûr. Allons-y mollo. Pensons d'abord à nous. Et tant pis si un innocent tombe.

— Ce n'est pas ce que je voulais dire, et tu le sais parfaitement. Mais le monde ne changera pas.

— Merci, Yoda. C'est très réconfortant. Toi aussi, tu devrais peut-être oublier ce qui est arrivé à tes vieux. Tu peux rien y changer, après tout.

— Attends, c'était ma famille !

— On est toujours la famille de quelqu'un, Ward. Chacun de nous, partout, à tout moment. Ça n'est pas une affaire de sang. Impossible. Ou alors nous sommes tous des Hommes de Paille. Soit les gens veillent sur les inconnus et font au moins semblant de s'intéresser à eux, soit tout l'univers s'écroule.

— T'as raison, capitulai-je. Tu feras comme tu l'entends. Mon avis est notoirement stérile, de toute façon. Je te dirai ce que ça donne, OK ? Je t'appelle dès que c'est terminé, et je rentre à l'hôtel aussi vite que possible.

Elle n'ajouta pas un mot. Je m'éloignai, avec l'impression de prendre la mauvaise direction.

Juste avant de tourner au carrefour, je regardai en arrière. Nina restait plantée sur le trottoir du commissariat. Au bout d'un moment, elle agita discrètement la main, et je vis scintiller son poignet. Je lui rendis son geste.

Elle articula deux mots. J'articulai : « Promis. »

CHAPITRE 15

Le cœur de Brad se fit lent, poussif, sitôt que Mme Luchs sortit de la maison. D'ordinaire, la mère de Karen se déplaçait à petites foulées. C'était l'un de ses traits distinctifs, comme le fait qu'elle eût – cas d'espèce parmi les parents des copains – quelques kilos en trop. Non qu'elle fût grosse, bien sûr, et ce léger embonpoint lui conférait une présence que n'avaient pas les malingres esclaves du tapis de jogging. Comme une façon de proclamer qu'elle était assez équilibrée pour ne pas avoir à se traîner tous les jours à la salle de gym – ou, dans son cas, à descendre les quelques marches menant à son sous-sol.

Mais là elle marchait d'un pas résolu, un tantinet trop hâtif. Peut-être l'avait-il également décelé sur son visage, ce signe subliminal indiquant que l'imprévu surgissait dans la vie policée des Luchs.

Ou alors, c'est qu'il s'y attendait.

Il était un peu plus de 16 heures et Brad était assis près de la piscine. Karen nageait consciencieusement d'un bord à l'autre. Après quarante longueurs intensives, elle commençait à fatiguer. Ex-pilier de l'équipe de natation du lycée, elle s'entraînait encore avec beaucoup de sérieux, ce que Brad n'était pas sûr de comprendre.

— Karen ? appela sa mère. Des policiers sont là.

L'acide envahit l'estomac de Brad. Karen regagna le bord et sortit de l'eau d'un mouvement fluide. Le soleil s'incrusta dans la gerbe de gouttelettes, comme dans les oscillations du pendentif en K.

Elle attrapa une serviette.

— Des policiers ? Pourquoi ?

— Je n'en sais rien, trésor. Mais ils aimeraient te parler.

Deux types en costume de ville s'étaient avancés sur la pelouse, quelques mètres derrière madame.

— Bon, dit Karen. Allons voir ça.

Les policiers ressemblaient à deux clones, si ce n'est que l'un avait une moustache et l'autre un teint blafard, comme s'il évitait le soleil. Le premier brandit un badge.

— Inspecteur Cascoli, déclina-t-il. Vous êtes Karen Luchs ?

— Oui.

Le flic regarda Brad :

— Et vous êtes... ?

— Brad Metzger, dit-il d'une voix normale.

— Tiens donc, fit l'inspecteur. (Il sortit un carnet, le feuilleta, tapota une page.) Voilà : Bradley M. Vous figurez également sur la liste.

Mme Luchs croisa les bras sur sa poitrine.

— De quoi s'agit-il ?

Le flic l'ignora, mais poliment, comme il se doit dans une demeure de cette taille.

— Vous êtes tous deux amis avec un certain Peter Voss ?

— Ouais, fit Karen. Pete, bien sûr. Pourquoi ? Il a un problème ?

— Nous l'ignorons, mademoiselle Luchs, car nous ne savons pas où Peter se trouve. Il n'est pas rentré chez lui samedi soir. Il lui arrive de passer la nuit chez des copains, donc rien d'inquiétant jusque-là. Sauf qu'il n'est pas reparu hier et n'a pas appelé. Ses parents nous ont alertés hier soir.

— Mon Dieu ! fit Mme Luchs en agitant sa main devant sa gorge. C'est vrai ?

— Quand l'avez-vous vu pour la dernière fois ?

— Samedi soir, répondit Karen d'une voix blanche. On faisait la fête ici, comme vous le savez sûrement. Pete est arrivé assez tôt, puis il est parti ailleurs.

— Vous savez où ?

Karen secoua la tête et regarda Brad.

— Aucune idée, dit ce dernier. Je l'ai vu en début de soirée, mais après ça, introuvable. On est partis acheter des hamburgers, puis on a voulu lui en proposer, mais il avait déjà dû se barrer.

— La mère de Peter dit qu'il s'est rendu à votre fête dans la voiture d'un certain Andy.

— C'est ça, confirma Karen. Ils sont tous arrivés ensemble. Mais Andy est resté jusqu'à la fin. Comme toujours.

— Ensuite il est reparti chez lui ?

— Il ne boit pas beaucoup, s'empressa-t-elle d'ajouter. D'ailleurs, je crois qu'il a laissé le volant à Monica. Ouais, je m'en souviens. C'est elle qui conduisait.

Le flic la considéra.
— Très bien.
Pendant qu'il prenait quelques notes, son collègue prononça ses premiers mots :
— Monsieur Metzger, des personnes que nous avons interrogées nous ont dit que vous et Pete étiez de grands amis. C'est exact ?
— Ouais, ouais. Je veux dire, Pete est pote avec tout le monde. Mais ouais, je crois qu'on peut dire ça. On se fréquente.
— Vous ne savez pas où il s'est rendu après la fête ?
— Non.
— Et vous n'avez aucune nouvelle depuis ?
— Non. Mais bon, je me disais qu'il avait dû aller à une autre soirée, qu'il avait picolé et qu'il avait passé la journée d'hier à récupérer au plumard. J'allais l'appeler ce soir, sûrement, ou demain.
— Ce garçon prend des cuites régulières ?
— Peter Voss est un jeune homme charmant, protesta Mme Luchs. Il est toujours d'une parfaite courtoisie.
— Ce que nous tentons d'établir, expliqua Cascoli, c'est le risque qu'il se soit attiré des ennuis. Qu'il soit monté dans la voiture d'un type qu'il connaissait mal, qu'il ait atterri au mauvais endroit, se soit mis dans de sales draps...
— C'est possible, estima Brad. (Tous les regards fondirent sur lui.) Enfin, je dis juste que c'est un mec chaleureux. Pete cause avec tout le monde. C'est son côté chouette. Mais bon... il aurait pu s'adresser aux mauvaises personnes, à un moment ou à un autre. Ça ne m'étonnerait pas plus que ça.
— Cela s'est déjà produit, à votre connaissance ?
— Non, déclara Brad. (Ses mains devenaient moites contre ses cuisses. Il croisa les bras.) Non, je dis juste que c'est possible. Mais il doit simplement traîner quelque part, pas vrai ?
— Espérons, répondit le flic. Car sa mère semble à deux doigts de perdre les pédales.
Il referma son calepin, le rangea dans sa poche, puis sortit son portefeuille et remit à chacun une carte.
— S'il se manifeste, appelez-moi. Dites-lui qu'on ne cherche pas à savoir s'il a des ennuis. On veut juste rassurer sa maman.
Brad et Karen hochèrent la tête de concert. Mme Luchs raccompagna les policiers à l'intérieur. Derrière cette femme, on aurait dit de vulgaires livreurs.

— C'est drôlement inquiétant, lâcha Karen. Seigneur, j'espère que Pete n'a rien.

— T'en fais pas pour lui. Tu connais Dormeur.

— Je devrais peut-être faire le tour des copains. Demander si personne ne l'a vu. (Elle empoigna son téléphone posé sur la table, prête à parcourir son répertoire.) T'en penses quoi ?

— Les flics s'en chargent déjà.

— Mais on pourrait inciter les gens à le chercher. À vérifier certains endroits, comme les maisons des amis ou ses boutiques préférées. Des lieux auxquels les flics ne penseront pas forcément.

Brad opina.

— Ouais, pourquoi pas ? Bonne idée.

Karen s'assit en tailleur sur le gazon et sélectionna un premier numéro, persuadée que Pete avait seulement besoin d'être débusqué – et qu'elle était la fille de la situation.

Brad patienta pendant vingt minutes atroces. Au sixième appel de Karen, il prit congé en prétextant une commission urgente pour son père – ça lui était sorti de la tête – et promit de la rappeler dans la soirée.

Il n'avait pas atteint la maison que Karen était de nouveau en ligne.

— On est cuits.

— Mais non, on n'est pas cuits.

— Si, Lee, on est cuits. Complètement grillés.

— Et pourquoi on serait grillés, Brad ? Dis-moi un peu. Explique-moi pourquoi et comment on serait grillés.

Ils se tenaient dans la cuisine de Lee Hudek, qui était d'une éternelle et inquiétante propreté, une vraie maison témoin. Brad ne comprenait pas comment Lee faisait son compte, même s'il ne cuisinait jamais. Le quotidien finit toujours par causer du désordre. Par laisser le chaos s'installer.

— On est cuits, répéta Brad à mi-voix.

La voiture de Lee trônait dans l'allée. Elle aussi ressemblait à une pub.

— Le keum est porté disparu, résuma Lee. Les flics allaient nécessairement interroger ses amis. Ensuite, Karen donne une fête, où il a été vu en dernier. Donc ils s'adressent aussi à elle. Tout ça était prévisible. Les flics en viendront vite à la conclusion qu'il s'est volatilisé et qu'il réapparaîtra un jour, mais en attendant ils ont une procédure à suivre.

— Sauf qu'il ne va *pas* revenir. T'as oublié ou quoi ? *Il ne reviendra pas*.

— Je sais, merci. Mais tant qu'il est absent, il est absent. Rien de plus. Pete est un shitman de première. Ils sont déjà au courant. Et je connais d'avance leurs conclusions : le keum se sera éclipsé. Aura décidé de devenir snowboarder professionnel, et se sera barré dans le Colorado. Ou bien il se sera endormi sous une poule rasta dont personne n'aura entendu parler, et refera surface quand il aura besoin de cash. Les flics sont obligés de gesticuler, mais ils vont vite lâcher prise. Ils sont pauvres, ils vivent dans des petites baraques merdiques et ils détestent les gens comme nous.

— Mais pas elle, rétorqua Brad. La mère de Pete ne lâchera jamais l'affaire.

Soudain il la vit dans sa tête, une image assez forte pour éclipser le monde réel. Maria Voss était une femme menue – Pete tenait sa taille et sa carrure de son père –, avec de longs cheveux noirs et de grands yeux marron. La vision dura une seconde, et l'instant d'après ce regard était plein de larmes, plein comme peut l'être l'océan. Brad n'avait jamais vu un tel spectacle en vrai, mais il savait à quoi cela ressemblait. Le visage de Maria se décomposa, et il entendit presque le cri s'arracher de sa bouche.

— C'est super craignos, Lee.

— Mais rien n'a changé. Écoute-moi, Brad. Rien n'a changé depuis l'instant où la balle lui a troué la tête.

Brad tressaillit.

— Putain, mec. T'est vachement... glauque.

— Écoute-moi ! Il est temps que tu regardes les choses en face. C'était écrit dès le départ. Et puisqu'on ne peut pas remonter le temps, il va falloir vivre dans le monde d'aujourd'hui. On n'y est pour rien.

— À peine !

— On l'a pas tué.

— On l'a emmené là-bas. On aurait dû, on aurait dû...

— Quoi ? On aurait dû *quoi* ? Qu'est-ce qu'on pouvait faire ?

— On n'aurait pas dû s'en débarrasser comme ça.

Lee Hudek secoua la tête, en homme sûr de son fait, ou agacé par la contradiction.

— Hernandez a tranché. Et il n'y avait rien d'autre à faire. Pete était déjà mort. Ça ne servait à rien de tomber avec lui, et c'est tout ce qu'on aurait obtenu.

— Tiens, Hernandez ! Notre vieux pote Hernandez. Il a appelé, au fait ? On a des nouvelles de tes soi-disant amis ?

— Non. Mais ça va venir.

— Tu rêves, Lee. On est devenus un problème pour ces mecs-là. Une menace. On est cuits.

Hudek lui empoigna les épaules. Brad soutint son regard, n'y vit que force et détermination. Peu à peu, il reprit une respiration normale.

— Rentre chez toi, dit Lee. Fais une sieste. Branle-toi un coup. Joue à la Xbox. Fais ce que t'as à foutre, mais détends-toi.

Brad retrouva ses pénates. Il tenta la Xbox, pour s'apercevoir que la quasi-totalité de ses jeux consistait à dégommer des gens. Il se rabattit sur un jeu de course automobile, mais il fallait suivre les méandres d'un circuit et la tête lui tournait déjà pas mal. Alors, pour finir, il s'allongea sur son lit. De là, il distinguait ses panneaux d'affichage, recouverts de photos retraçant les cinq dernières années, les plus anciennes alignées avec soin, les plus récentes punaisées les unes sur les autres, parfois sur quatre épaisseurs. Fêtes, événements scolaires, portraits de groupe en virée. Des nuits sympas, des jours heureux. Pete y figurait, bien sûr. On le voyait avec Brad à l'arrière de l'ancienne caisse de Lee. On le voyait dans le jardin de Brad. On le voyait avec l'équipe du lycée, après un grand match, accroché aux épaules des potes, beuglant victorieusement.

Dix-huit mois plus tôt. Dix-huit mois seulement ?

Son portable sonna. C'était Steve Verkilen, le mec qui s'était fait bâillonner en même temps que Pete sur le parking. Il était hors d'haleine.

— Putain, mec, t'es au courant ? Je viens d'avoir la mère de Pete au téléphone. Elle a l'air grave flippée.

— Je suis au courant, répondit Brad avec calme. Les flics sont passés chez Karen. Tu saurais pas où il est ?

— Aucune idée, mec. Aucune. Ça fait plusieurs jours que je l'ai pas vu. On était censés se retrouver chez Karen, mais j'étais trop décalqué pour bouger.

— Il est venu, lui, répondit Brad. Puis il s'est barré, et plus personne n'a eu de nouvelles.

— C'est zarb.

— Tu m'étonnes.

— Je comptais appeler Lee, voir s'il était au courant.

— Inutile, je reviens de chez lui. Tu connais Pete, Steve. Il est forcément quelque part. Si ça se trouve, il a juste perdu son putain de téléphone.

— Ouais. (Une pause.) On maintient l'approvisionnement cette semaine ? Si Pete est pas revenu ?

— Je sais pas.

— Ben, tu me diras, OK ? J'ai besoin de thune.

Steve laissa Brad sur cette interrogation : quid de la livraison hebdo ? Hernandez était aux abonnés absents. Steve n'était pas le seul à avoir besoin de ce fric. Mais pouvait-on continuer ? Faire tourner la boutique comme si de rien n'était ?

Il cogita quelques minutes, allongé sur le dos, puis se retourna pour ne plus voir le mur tapissé de photos. Le visage enfoui dans le matelas, les yeux fermés, il respira l'odeur familière des draps. Comme disait Lee, tant qu'on ne découvrait rien, il n'y avait pas de crime. Brad se frotta le front contre le tissu, et peu à peu le moral revint. Il se remit sur le dos, fixa le plafond pendant un moment, puis se redressa et posa les pieds par terre. En se levant, il se sentit à la fois naze et reposé ; il avait même un petit creux. Il décida de descendre, pour regarder s'il y avait des Fritos dans les placards. C'était généralement le cas. Ces choses-là repoussaient toutes seules. Comme il longeait la mezzanine et prenait l'escalier, la maison lui parut déserte. Il avait pourtant croisé sa mère en rentrant, ainsi que sa sœur, qui écoutait Smash Mouth à fond. Il pénétra dans la cuisine, fut surpris de la trouver si bien rangée. La famille Metzger était plutôt adepte du bordel organisé. Mais cet après-midi-là, on se serait cru chez Lee : pas la moindre miette, un frigo propre comme un sou neuf... Même la table était vide, alors qu'elle servait de point de chute à tous les objets n'ayant pas de place attitrée. Le placard des chips avait été vidé. De fond en comble, jusqu'à la dernière poussière. C'était donc ça : elles étaient parties au supermarché. Un grand nettoyage de printemps, puis on refaisait le plein. Ouais, ça collait. Il ouvrit le placard suivant. Vide également. Il les essaya tous, pour le même résultat. À croire que le printemps donnait des ailes. Bien qu'on fût en septembre. Un nettoyage d'automne, alors. Il se retourna en entendant un bruit. Un bruit indéfinissable, comme une sorte de mâchonnement. Cela semblait provenir du jardin. Brad se rendit à la fenêtre et constata qu'il faisait nuit. Il avait dû dormir sans s'en rendre compte. Pourtant... N'avait-il pas vu le soleil cinq minutes plus tôt en cherchant la voiture de sa mère dans l'allée ?

Brad regagna en vitesse l'avant de la maison. Là aussi, c'était nickel. Nickel chrome. Pas un magazine, pas un journal, pas une télécommande, et dans la rue il faisait jour. Ça clochait quelque part, mais il n'aurait su dire où. Alors il se retourna pour traiter l'autre problème, l'espèce de gargouillement qui perdurait vers le fond. Le bruit n'augmentait pas, mais ne semblait pas près de cesser. Brad fila dans la cuisine, ouvrit les portes-fenêtres et sortit dans le jardin. Il faisait très sombre et très froid, mais sans vent. Il y avait des arbres, une rangée de grands arbres dont la cime pointait derrière la maison. Quelques-unes semblaient même sortir du toit. Brad crut également percevoir le bruit d'un ruisseau, non loin de là. L'odeur était inhabituelle. Un mélange de cannelle, de sucre et d'un troisième truc mystérieux. Il s'avança au milieu des arbres, mais le paysage se dérobait devant lui. Il y avait, en tout cas, un sérieux problème de taupes. Où que l'on regarde, des galeries couraient sous la surface du sol forestier, comme un réseau de veines enflées. Elles bougeaient. C'étaient elles qui produisaient ce bruit de mastication, et à mesure qu'elles grouillaient le sol devenait transparent. On découvrait alors des gens, étendus, les yeux fermés, et la plupart avaient perdu quelque chose. L'odeur gagnait en puissance, et Brad reconnut de la pomme. Oui, ça sentait la tarte aux pommes, la fine tarte aux pommes de chez McDonald's, précisément, dont l'emballage stipulait que c'était super chaud. Ces corps étendus n'avaient rien dans les mains ni dans la bouche. La tarte était bien quelque part, pourtant. On pouvait la sentir. Tout le monde pouvait la sentir ! Brad se glaça. Il suffisait de venir ici pour connaître le fin mot de l'histoire.

Puis un bruit de phalange contre une vitre, et il se retourna pour voir sa mère et sa sœur dans la cuisine. Sa mère déchargeait les courses et sa sœur frappait au carreau, essayant de lui dire qu'elles avaient racheté des Fritos, qu'il n'avait pas besoin de chercher des tartes, que c'était même déconseillé. Il voulait lui répondre que tout allait bien ; tant que les tartes restaient cachées, il était en sécurité – tout le monde était en sécurité. Mais à mesure qu'il essayait de regagner la cuisine, celle-ci semblait rapetisser. Sa sœur se mit à frapper sur la vitre de plus en plus fort, mais le son se rapprochait de celui d'une cloche, selon un rythme familier et chargé de sens. D'un coup, l'odeur de pomme devint surpuissante, écœurante, et alors...

— Merde, grogna-t-il en se redressant.

Endormi sur le flanc, il venait de se cogner la tête contre le mur. Il tapota la couette, comprenant que l'espèce de tintement de cloche provenait de son téléphone. L'appareil avait glissé par terre.

L'écran indiquait : « K PORTABLE ».

— Salut, marmonna-t-il.

Il tenta de se réveiller en ouvrant grand la bouche et les yeux, la moitié du cerveau retenue ailleurs.

Karen se taisait. On aurait dit qu'elle pleurait.

— Qu'est-ce qui va pas, bébé ? Y a un problème ?

Elle renifla, fort. Déglutit.

— Oh, mon Dieu, gémit-elle. Ils ont retrouvé Pete.

CHAPITRE 16

Lee occupait l'un des canapés du salon parental. En face de lui se trouvaient les deux inspecteurs, rendus anonymes par le contre-jour des portes-fenêtres – ce qui était somme toute appréciable. Il avait déjà déduit que le moustachu était le chef, et c'est tout ce qu'il fallait savoir.

— Tu préfères que je reste ?

Ryan Hudek se tenait sur le seuil de la pièce, en pantalon de coton et polo Lacoste bleu ciel. Il s'était montré imperturbable lors de l'arrivée inopinée des policiers. Apprenant qu'ils cherchaient son fils, il demanda pourquoi, et les laissa entrer dès qu'ils se furent expliqués – non sans avoir examiné leurs badges. Lee se réjouissait de la présence de son père. Curieusement, il se sentait très jeune cet après-midi-là.

— Non, ça va aller, répondit-il.

Hudek senior donna un petit coup de menton.

— Je serai dans les parages. Si jamais tu changes d'avis...

Le moustachu regarda ses mains, le temps que les pas de Ryan Hudek déclinent dans le couloir. On entendit coulisser la porte vitrée du jardin, puis elle se referma dans un bruit mat. L'inspecteur leva les yeux pour dévisager Lee.

— Bon. Désolé de venir vous débusquer chez vos parents, mais vous n'étiez pas chez vous.

— Exact, répondit Lee d'une voix posée. J'étais ici.

— Je vois. C'est très compréhensible. Si un de mes copains était retrouvé mort, un bon copain, qui plus est, je chercherais le soutien de mes amis et de ma famille.

Lee resta muet. Son instinct le dissuadait de répondre à des questions qui n'en étaient pas.

L'inspecteur marqua une pause, avant de reprendre :

— Je suis désolé de vous apprendre cette triste nouvelle au sujet de Peter Voss.

— Mais vous ne m'apprenez rien. Ça fait une heure et demie que je suis au courant.

— Quelqu'un vous a appelé ? Qui donc ?

— Deux ou trois personnes. Le bruit a vite circulé. J'en reviens toujours pas.

— Vous ne vous rappelez pas qui vous a informé en premier ?

Lee feignit de réfléchir.

— J'étais assez remué, vous comprenez.

— Bien sûr. Et quand avez-vous vu Peter pour la dernière fois ?

— Pete, corrigea Lee. Personne ne l'appelait Peter. Je l'ai vu, que je me souvienne... vendredi, en fin de matinée.

L'inspecteur fronça les sourcils.

— Vous ne l'avez pas vu à la soirée des Luchs ?

— Pas que je me rappelle. Je suis arrivé assez tard. Il avait déjà dû repartir.

— Vous seriez donc surpris si l'on vous disait qu'un témoin vous a vu parler à Peter lors de cette fête.

— Pas plus que ça, mais ça ne m'évoque rien. Vous savez, Pete et moi, on discutait tout le temps ensemble. Si je devais me souvenir de chaque occasion... Pourquoi, quelqu'un dit ça ?

— Non.

Lee haussa les épaules.

— Ah bon. Drôle de question.

— Vous l'avez donc vu vendredi. Dans quelles circonstances ?

— C'était au centre commercial, peu avant midi. On s'est partagé une grande frite, puis il est reparti. Il devait retrouver un type, je sais pas trop qui. Visiblement, il n'avait pas envie d'être en retard.

Le premier point était vrai : Lee avait bel et bien vu Pete au centre commercial de Belle Isle le vendredi précédent – de loin, sans lui adresser la parole. Mais vu que Voss était seul ce jour-là, personne n'allait contredire cette version.

— Avait-il l'air un peu distrait ? Avez-vous remarqué quoi que ce soit d'inhabituel chez lui ? De l'inquiétude ?

— Pas vraiment. Je veux dire, Pete était toujours un peu à l'ouest. On a parlé de la fête chez Karen, celle du lendemain soir, et on prévoyait de s'y retrouver tôt. Il avait une deuxième fête juste après, mais il ne m'a pas donné de détails.

Intervint l'autre inspecteur :

— Pensez-vous qu'il puisse y avoir un lien entre cette deuxième fête et la personne qu'il devait rencontrer vendredi ?

Lee cogita.
— J'imagine que oui. Mais il ne m'a rien dit sur cette fête.
— Et après ça, vous n'avez plus eu de nouvelles ?
— Aucune.
— La maman de Peter dit que son fils semblait détenir une somme de liquide inhabituelle la semaine dernière. Il s'est acheté de nouveaux vêtements, a offert des cadeaux onéreux à ses parents. Vous pourriez nous éclairer là-dessus ?
Lee secoua la tête. Le flic reprit :
— Vous avez une bien belle maison, à Summer Hills.
— Merci.
— Elle doit valoir vachement cher.
Lee haussa les épaules. *Si t'as une question, tu la poses franchement.*
— Et que savait exactement votre interlocuteur ? Celui qui vous a appris pour Peter.
— Personne ne savait grand-chose. Juste que Pete était mort.
Moustache fit mine de consulter ses notes :
— On a retrouvé le corps dans les montagnes de Santa Ynez, en lisière de la forêt nationale de Los Padres. Il était caché à quelques centaines de mètres d'un sentier de randonnée. On lui a détruit les trois quarts de la tête, avant d'essayer de l'enterrer, mais c'était du travail de cochon. Le trou n'était pas assez profond.
— C'est toujours pareil, considéra le collègue d'un air méditatif. De vraies feignasses ces gens-là.
— Bref, les coyotes l'ont flairé. Le temps qu'ils l'exhument et l'entament, il s'était encore dégradé. Heureusement, des gamins à vélo l'ont découvert assez vite. Mais quand même, il était dans un sale état.
— M'en parle pas, fit l'autre. Tout déchiqueté. Chaleur, mort et vermine. Une combinaison redoutable.
— Vous permettez ? s'insurgea Lee. C'est de mon ami que vous parlez !
Moustache leva les yeux.
— Oh, je suis désolé...
— Non, vous ne l'êtes pas. Je vous demande juste un peu de respect. Pour moi, et pour lui.
Le flic le fixa. Lee soutint son regard.
— Je vous prie d'excuser notre froideur, dit le collègue après une longue, très longue pause. Déformation professionnelle.

— Je vous déconseille d'employer ce ton-là avec ses parents, dit Lee, ou vous allez passer un sale quart d'heure.

Un doute le saisit : Moustache était-il vraiment le chef ? Ou bien était-ce le pâlichon ?

— Nous les avons déjà rencontrés, répondit le flic. Mais... dont acte.

— Très bien, dit Moustache d'une voix douce, nous vous recontacterons si nous avons d'autres questions. Une dernière petite chose avant qu'on parte : auriez-vous la gentillesse d'interroger votre portable, pour voir de qui émanait ce premier appel ?

— Bien sûr.

Lee sortit son téléphone, enfonça les touches avec un détachement parfait, comme si la réponse lui était égale. Il trouva et hocha la tête.

— Voilà. C'était Brad. Bien sûr.

— Bradley Metzger ?

— C'est ça. Karen l'a prévenu, et il m'a aussitôt appelé.

— Parce que tous les trois vous étiez amis. De bons amis.

— Absolument.

Les flics se levèrent de conserve et Lee les reconduisit. Comme ils passaient la porte d'entrée, Moustache se retourna :

— Ah, oui, vous saurez peut-être nous confirmer un détail. Peter gardait-il son portable sur lui, en général ?

— Il ne s'en séparait jamais. Pourquoi ?

— Pas moyen de mettre la main dessus. Ni sur le cadavre, ni près de lui, pas plus que chez lui. Sa mère a tenté de le joindre, et ça sonne dans le vide, aussi j'imagine que la batterie n'est pas encore morte. Vous n'auriez pas retrouvé l'appareil chez vous ou ailleurs ? Il ne l'aurait pas perdu récemment ?

— Non. Il l'avait sur lui vendredi dernier. Au centre commercial.

— Que voulez-vous, ces choses-là ont vite fait de nous échapper. Elles peuvent glisser n'importe où. On continue d'ouvrir l'œil.

Lee regarda les deux poulets rejoindre leur voiture. Il attendit qu'ils aient redescendu l'allée, puis tint la pose quelques instants.

Tu as menti aux flics. Le monde ne sera plus jamais comme avant.

— Ils n'étaient pas si coriaces que ça, hein ?

Lee sursauta, avant de découvrir sa mère plantée dans son dos, en peignoir argenté. Il ne savait même pas qu'elle était à la maison.

Elle n'avait pas ses lunettes noires et son regard était brumeux, mais elle semblait le voir.

— Tu t'en es très bien sorti, dit-elle. La tirade sur le respect était un joli coup.

Elle passa le doigt sur la joue de son fils, très brièvement, avant de se couler hors de la pièce, vers Dieu sait où.

— C'est moi.

— Salut, bébé. Comment ça va ?

Karen soupira.

— On fait aller, quoi.

— C'est clair. T'es où, là ?

— Près de la piscine, histoire de changer d'air. Quand je suis à l'intérieur, maman vient sans arrêt me demander comment je vais, ce qui est très gentil de sa part, mais bon, tu vois...

— Ouais.

— Elle compte rendre visite à la mère de Pete.

— Je savais pas qu'elles se connaissaient bien.

— Ben, pas tant que ça, en fait, mais bon, Pete est mort, quoi.

Brad hocha la tête, même si Karen ne pouvait l'entendre. Pete était mort, en effet. Devenue publique, la nouvelle était un peu plus soutenable. Ce n'était plus une sombre affaire qu'il fallait occulter dans le vain espoir que la vérité du cauchemar s'estomperait, mais une chose irrévocablement précise et nette. Dorénavant, ce que les gens savaient correspondait à la réalité : les faits étaient là. Restait juste à empêcher ces faits de vous écraser comme des mouches.

Après le départ des flics, Lee avait appelé Brad pour lui répéter ses déclarations. Brad avait transmis la question de Steve au sujet de la prochaine livraison, et Lee de répondre qu'il ne savait rien. Hernandez demeurait injoignable. À priori, ce n'était pas une riche idée, mais on verrait. Brad fut rassuré par ces paroles, convaincu en son for intérieur que dealer était, jusqu'à nouvel ordre, la dernière connerie à faire.

— T'es toujours là ? demanda Karen.

— Ouais, ouais. Je réfléchissais à tout ça.

— Ouais.

Elle se tut un moment, et Brad se dit qu'elle voulait peut-être raccrocher. Mais elle reprit la parole :

— Je peux te demander un truc ?

— Bien sûr.
— Tu n'es au courant de rien, n'est-ce pas ?
Brad ouvrit la bouche, mais il n'en sortit rien. Il toussa un coup et réessaya :
— Comment ça ?
— Je veux dire... Pete, Lee et toi vous étiez tellement proches, tu vois ? Je me demandais juste si Pete n'avait pas une sorte de secret, que vous deviez garder pour vous.
— Non, fit Brad avec soulagement. C'est vrai, on était proches. Comme des frères de sang. Mais, autant que je sache, il n'y avait rien de louche dans la vie de Pete.
— D'accord.
Pour une raison ou pour une autre, le ton de ce « d'accord » raviva les craintes de Brad. Comment ça, d'accord ? D'accord, merci pour l'info ? D'accord, je te crois ? D'accord, je ne te crois pas du tout, mais on en reparlera plus tard ?
— C'est juste que...
— C'est juste quoi, bébé ?
— Tu sais, lors de la fête, quand tu es parti chercher des hamburgers avec Lee, et que je vous ai vus devant la maison avant que vous ne partiez...
— Eh bien, quoi ?
— Eh bien, j'y repensais tout à l'heure, et quand Lee t'a rejoint en passant par le côté de la maison, il n'aurait pas dit une phrase comme « Il arrive », « Il est en route », « Il se ramène », quelque chose dans ce genre ?
Brad redoubla de prudence.
— Je me rappelle pas.
— Je suis sûre d'avoir entendu ça. Car quand je suis partie saluer Sara et Randy, vous êtes restés plantés là, Lee et toi, comme si vous attendiez quelqu'un.
— Mais non, répondit Brad. (Après une demi-seconde, sa cervelle trouva la parade :) Si tu te souviens bien, je fumais une clope, pas vrai ? Or Lee n'aime pas qu'on fume dans sa caisse. Je voulais juste la terminer avant qu'on décolle.
— Ah, d'accord.
Là-dessus ils bavardèrent de choses diverses, puis juste avant de prendre congé Karen ajouta :
— Brad ?
— Ouais ?
— Tu crois qu'on trouvera les coupables ?

— J'en sais rien.
— Moi, si, dit-elle à mi-voix. Je pense qu'on les coincera.

Après le coup de fil à Brad, Lee avait repris sa voiture et roulé au hasard avant de regagner son domicile. Il prépara du café, qu'il but à sa table de cuisine immaculée. Il résista à l'envie de rentrer la Mercedes dans le garage pour chercher le téléphone de Pete. Si l'appareil était tombé dans l'habitacle, Lee l'aurait trouvé en nettoyant les sièges la nuit de la fusillade. Mais tout de même – à supposer que le flic n'ait pas inventé un bobard pour le piéger –, cette histoire était bizarre. Pete ne pouvait vivre sans son portable ; il se le serait fait greffer si ç'avait été possible. Il l'avait forcément sur lui ce soir-là. Mais où était-il passé, bordel ?

Peu importait. L'appareil avait probablement glissé de sa poche, entre le parking et les fourrés où on l'avait enterré. En traînant le corps dans le noir, personne n'aurait vu ni entendu chuter l'objet. Mais tant pis. Même s'il réapparaissait aujourd'hui, il ne risquait en rien de compromettre Brad et Lee.

N'empêche.

Lee restait prostré, la tête baissée. On aurait pu tout éviter. Rien de plus facile, à vrai dire.

Il avait pris l'appel d'Hernandez. Il aurait pu s'abstenir.

Il avait dit oui. Il aurait pu dire non.

Une nuance capitale.

On aurait pu tout éviter.

Il ressortit, rangea la voiture et la fouilla de fond en comble.

Pas de téléphone.

Il se lavait les mains à l'évier lorsque la sonnette retentit. Ce devait être Brad, pris d'un nouveau coup de flippe. Mais en ouvrant il découvrit un homme d'un tout autre âge.

— Monsieur Reynolds ? fit-il, étonné de n'apercevoir aucune voiture dans la rue. Que faites-vous ici ?

L'avocat pénétra dans la maison sans attendre d'y être invité.

— Il faut que tu me racontes ce que tu leur as dit.

— À qui ?

— Aux policiers, pardi.

— Pourquoi me demandez-vous ça ?

— Parce que je suis désormais ton conseiller juridique. En cas de besoin, ce que je n'espère pas.

— C'est mon père qui vous a embauché ?

— Pas du tout.
— Mais alors qui...
— Je t'écoute, Lee. Chaque question, chaque réponse. Dans les moindres détails.

Alors Lee conta son entretien avec les deux inspecteurs. Reynolds resta très concentré. À la fin du récit, il affichait une mine grave.

— Tu as conscience d'être désormais le complice d'un meurtre ? Metzger et toi, vous encourez de nombreuses années de prison, rien que pour ce crime. Aux yeux de la justice, vous n'êtes plus des enfants, aussi dingue que cela paraisse. Et alors s'ils établissent le lien avec la drogue et si le bureau fédéral des Narcotiques s'en mêle, autant balancer la clé de la cellule.

— Mais comment feraient-ils le rapprochement ?

— Sinon, pourquoi se rendre sur un parking désert, éloigné de tout, au beau milieu de la nuit ? Et pourquoi cacher le cadavre ?

— C'était un accident, ou un truc du genre. On a pris peur.

— Sans compter que tu as menti à un officier de police. Bradley aussi. Et puis es-tu certain qu'il n'y aura personne pour déclarer que vous deux – ainsi que Peter Voss, bien sûr – avez passé les six derniers mois à dealer dans toute la vallée ?

Lee songea à toutes les personnes qu'il avait fournies. À toutes les baraques qu'il avait visitées, à toutes les fêtes où on l'avait invité, à toutes les poignées de main, les bières gratuites et les « Eh, mec, content de te revoir » – de pures foutaises lorsqu'on se retrouve face à des flics dans le salon de ses vieux et qu'on pige que lâcher le nom de Lee évitera de sérieux ennuis.

— On est très mal, murmura-t-il.

— C'est assez bien résumé. Tu as besoin de régler cette affaire, et vite. Tu comprends ça, au moins ? Tu as conscience de la gravité de ce merdier ?

Lee était vanné. Nauséeux. Il verrait très vite le tableau, sans aucun doute, mais pour l'heure il avait juste envie de s'allonger face contre terre et de ne jamais se relever.

— J'en ai conscience.

— Suis-moi, dit Reynolds avant d'ouvrir la porte.

Une voiture stationnait dans la rue. Noire, jusqu'aux vitres fumées. Le moteur tournait, mais il était très, très silencieux.

Lee Hudek verrouilla la maison et suivit l'avocat dans l'allée. Ce dernier lui ouvrit la portière du véhicule. En se voûtant, Lee distingua deux banquettes en vis-à-vis. L'intérieur était aussi sombre que la carrosserie, et l'odeur était celle de l'ombre.

Un homme était assis au milieu de la banquette orientée vers l'avant. C'était le gars du hangar désaffecté. Celui à qui il avait exposé le Plan. Le type que, pour être franc, Lee aurait préféré ne plus jamais revoir.

— Salut, Lee John. Viens donc t'asseoir. Il faut qu'on parle.

Hudek hésita, mais il n'avait pas d'autre endroit où aller. Il monta, s'installa en face du type. M. Reynolds vint s'asseoir à côté de Lee. La portière se referma dans un doux et luxueux *clic*, et la voiture se mit à flotter sur la route, comme aspirée par une légère brise.

— Comment tu te sens, Lee ? demanda l'homme.

— Ça va.

— Tant mieux. Un peu secoué, peut-être ?

— Sans blague. Ça fait trois jours que j'essaie de joindre Hernandez. Pourquoi il rappelle pas, putain ?

— Parce qu'il est mort, Lee.

Hudek cligna des yeux, toute colère envolée.

— Il est mort la même nuit que ton ami. Il est parti chercher les mecs qui vont ont canardés. Il n'en est pas revenu. Heureusement, nous avons mieux caché son corps que vous n'avez caché celui de M. Voss. Ce que nous n'avons pas réussi, par contre, c'est à localiser le portable de M. Voss, auquel la police semble s'intéresser.

— Seigneur... expira Lee. (Inutile de demander d'où l'homme tenait l'info. Il la tenait, point barre.) On est dans la merde.

— C'est sérieux, oui, mais ça peut s'arranger. Il faut regarder vers l'avenir, toujours. La disparition d'Hernandez libère un poste et te place en position idéale pour ton programme sur les vacances de printemps. Le « Plan », c'est ça ? Vive le Plan. Si toutefois nous surmontons ces petits soucis locaux. Mais nous y arriverons sûrement, grâce ton indéfectible coopération.

— Ce serait bien, répondit Lee. Ce serait... très bien.

— Tout est compris dans le service.

L'homme se carra dans son siège et fixa longuement Lee, qui devint très mal à l'aise.

— Quoi ? lâcha-t-il.

L'homme portait une chemise à col ouvert, et Lee releva cette cicatrice sur le haut de son torse – une blessure par balle, à l'évidence. Et cette marque de la taille d'une pièce de monnaie, cinq centimètres plus bas, n'en était-ce pas une deuxième ?

— Tu ne te souviens toujours pas de moi ?

— Non. Sincèrement, non.

— Eh bien, on va devenir de vrais associés, maintenant. Cela mérite des présentations en bonne et due forme.

Il tendit sa paume à Hudek. Un bracelet de cuivre pendait à son poignet.

— Enchanté, Lee. Moi, c'est Paul.

CHAPITRE 17

Jim s'était cru capable de remonter la rue sans sourciller. Cela faisait si longtemps, et il s'était employé à croire que ce passé n'était pas le sien. Mais peu après avoir doublé l'ancienne adresse, il s'aperçut qu'il roulait de moins en moins vite, comme si sa batterie se vidait. Par deux fois, il fit une pause sur le bas-côté. Avant de repartir, lentement, au hasard. À travers les bois. Puis retour en ville. Resserrant la spirale.

Il échoua sur le parking de chez Renée. Maintenant il avait faim, plus de doute possible. Il n'avait toujours rien avalé depuis Key West. Il sortit de la camionnette, et, sitôt le pied posé, la sensation de ventre vide s'estompa. L'arrière du bâtiment recrachait une odeur d'huile de friture. L'estomac de Jim réclamait quelque chose, mais pas ce qu'on vendait là-dedans.

Il resta planté quelques instants, les mains dans son blouson, à savourer l'air frais, en espérant que le vent fendrait les nuages qui s'amoncelaient sous son crâne, plus orageux d'heure en heure. La situation ne cessait de se compliquer depuis qu'il avait récupéré le van et s'était nettement dégradée après Petersburg. Le point critique se rapprochait.

Tiens... Quel était cet objet dans sa poche ?

Il le sortit, fronça les sourcils. Un paquet de cigarettes.

Cette découverte était si déroutante qu'il craignit d'avoir pris le blouson de quelqu'un d'autre. Mais il ne voyait pas où l'erreur aurait pu se produire, et un bref examen des autres poches lui confirma qu'il en était bien l'heureux propriétaire. Mais d'où sortaient ces clopes, alors ?

Jim n'avait pas fumé depuis une éternité. Il n'avait jamais fumé, en fait. James avait fumé, lui, durant toute son adolescence et ses années à l'étranger. Tout le monde fumait dans l'armée. À son retour, aussi. Rien de tel que la salle des profs pour s'en griller une. Et, enfin, pendant cette autre période. Mais jamais depuis. Il avait arrêté en s'installant à Key West, rompant avec cette manie comme on se coupe un doigt : on s'adapte, on vit sa vie à un rythme légèrement différent – à la manière calme et posée de Jim Westlake. Et pourtant, au cours des dernières quarante-huit heures, il s'était procuré un paquet de Marlboro.

Il rangea les cigarettes dans sa poche. Il n'en avait pas envie. Il n'était pas fumeur. Mais il savait qu'elles se trouvaient là.

D'un coup, il se sentit las et furieux contre lui-même. Comme si ces dissociations imbéciles le lavaient de ses péchés. Ce n'était pas moi, mais l'alcool, monsieur le juge. Mon geste a été dicté par mes hormones, monsieur le juge. Il n'y a que des poules mouillées dans ma caboche, et les poules mouillées ne portent pas de poignard.

Il s'éloigna du van et flâna dans le quartier. Il n'était pas prêt à remonter tout de suite.

Il avait relevé d'autres altérations que le paquet de cigarettes. Ce matin-là, il avait découvert une petite poêle à frire dans le sac calé au pied de la banquette. Encore une emplette dont il ne se souvenait pas. Or cette poêle ne provenait pas de la camionnette, car il l'avait vidée de fond en comble. Et elle paraissait neuve, contrairement à celle de la boîte à chaussures. Il n'y avait rien d'étrange ou de magique à cela : il avait dû l'acheter, voilà tout, de la même façon qu'on se surprend parfois à vider un paquet de chips juste après l'avoir rangé dans le placard, sans s'être aperçu qu'on avait regagné la cuisine entre-temps. Il faut croire que les mains jouissent d'une certaine autonomie. S'il existait une vraie dissociation, c'était peut-être celle-là. L'âme et le corps, unis par un ennemi commun : l'esprit du mal.

Il se débarrassa de la nouvelle poêle. Une seule suffisait.

Ce quartier était sans intérêt, et parvenu aux trois quarts du pâté de maisons, Jim se sentait prêt à reprendre la route. Il repassait derrière chez Renée lorsqu'il entendit un bruit. Il stoppa, se retourna.

Il n'y avait rien à voir, sinon l'arrière ombrageux et lézardé d'un fast-food, un bloc de béton aux portes cadenassées à moitié cachées par de grosses bennes métalliques, dans des odeurs de matières pourries dispersées au jet d'eau. Un grillage de deux mètres cinquante séparait cet établissement du suivant, un magasin de pneus. Jim longea une allée étroite, qui desservait sans doute quelque unité de stockage. L'endroit où poussaient les steaks hachés, par exemple.

Le bruit provenait de ce coin-là.

Jim l'entendit à nouveau. Il pouvait s'agir d'un petit animal heurtant le bas du grillage. Blessé, peut-être.

Jim aimait les animaux. Il fallait vérifier.

Il traversa le trottoir, puis quelques mètres de béton défoncé avant de fouler l'allée. Elle était longue d'une dizaine de mètres et menait à un mur.

Oui, voilà : une petite chose, tout au bout. Qui cognait contre la clôture.

Il fit deux pas de plus. La chose s'agitait, comme prise au piège, quand il aurait sans doute suffi de contourner l'obstacle.

Elle avait une drôle de forme, du reste. Comme dressée sur ses pattes postérieures. Environ un mètre de haut.

Jim fit un dernier pas, se pencha pour mieux voir.

Un visage blême se tourna vers lui.

C'était une enfant. Une fillette revêtue d'un minuscule imperméable foncé, le crâne découvert, les cheveux balayés en tous sens au gré de ses mouvements de tête. Elle avait les deux mains agrippées au grillage, qu'elle secouait de toutes ses forces. Son visage était flou, livide, souillé de terre.

La clôture claquait au vent, mais la gamine se taisait.

Jim regagna le trottoir d'un pas chancelant, puis s'arrêta pour inspirer profondément. Le grillage continuait à tinter dans le vent. Rien de plus. Il n'y avait rien de plus qu'un grillage.

Il enfila la dernière portion de trottoir, jusqu'à repérer le Combi. Alors il sortit les cigarettes, ainsi qu'un briquet jetable rangé dans l'autre poche, et se remit à fumer, comme ça.

Passées les premières taffes, c'était comme une seconde nature, et cela masquait un peu l'odeur d'huile expulsée par les conduits du bâtiment, une huile ayant servi à chauffer les morts pour nourrir les mourants. Ça le rendait malade. Tout le rendait malade. Il se sentait vieux, mauvais, et néanmoins plein d'énergie, les mains envahies d'une force dont il ne savait que faire.

CHAPITRE 18

Je me suis décidé pour un endroit baptisé Lucy's, sur Union Street à Owensville. Ce café avait autant de caractère qu'un hall d'aéroport, mais il acceptait les fumeurs, et c'était ça ou un Denny's. J'ai choisi une table près de la vitre, et en observant le carrefour j'ai soudain pris conscience que c'était ici que Julia Gulicks et Mark Kroeger s'étaient donné leurs premiers rendez-vous, dont celui du fameux soir où ils avaient découvert le cadavre de Widmar dans la forêt. J'ai songé à demander au barman s'il se souvenait d'eux, en m'appuyant sur la couleur de la chevelure de Julia Gulicks, mais c'était le coup à passer pour un flic, ce qui n'est pas idéal quand on veut consommer incognito dans un bar. Les clients n'aiment pas. C'est comme d'avoir sa mère postée dans un coin. Sa mère, avec un flingue. Personne n'a besoin de ça.

Unger appela dès qu'il atterrit, et je lui indiquai le chemin. Je l'attendis pendant deux heures, à me tourner les pouces en essayant de ne pas trop boire. Après cette nuit dans la lande, j'étais épuisé, déphasé, en proie au mal de tête sec et lancinant qui vient quand on se lève trop tôt en sachant qu'on se couchera tard. Je pris de l'aspirine achetée à la pharmacie d'en face et fis quelques sauts aux toilettes afin de m'asperger le visage, pour un résultat minime et de courte durée. Je m'affalai dans mon siège, espérant que ce n'était pas là la fin que me réservait le destin : stagner dans un bar, crevé, à me demander si un type n'allait pas venir me tuer. Je réfléchis à l'affaire qui occupait Nina, sans parvenir à la moindre conclusion. J'espérais que Julia Gulicks était coupable, pour qu'on en finisse au plus vite. Cela paraissait un peu facile, un cadavre, une rouquine et hop ! en prison, mais il arrive que ça se passe comme ça. Quatre-vingt-cinq pour cent des crimes élucidés le sont au cours des premières quarante-huit heures. Au-delà, les souvenirs s'effacent, les gens se trompent, et le monde passe à d'autres méfaits.

Vers 21 h 15, un taxi s'arrêta de l'autre côté du carrefour. Je vis un homme régler le chauffeur tout en examinant le bar. Il était petit et n'avait rien d'un athlète, mais c'était forcément lui. La preuve : sitôt descendu de voiture, il se dirigea vers la porte du Lucy's.

Une fois à l'intérieur, c'est à peine s'il scruta les lieux. J'étais la seule personne isolée à me désintéresser tant du match à la télé que de la clientèle du sexe opposé. C'eût été un jeu d'enfant. Au lieu de sortir son flingue pour me descendre, il se planta devant ma table, les bras ballants.

— Ward Hopkins ?

On s'est serré la main. La sienne était chaude et moite. Il s'est assis en face de moi, se glissant sur la banquette avec une saccade disgracieuse. Ses cheveux noirs et rares lui collaient au crâne. Son costard était fripé. Il devait avoir mon âge, peut-être un ou deux ans de plus. Ses traits étaient remarquablement fins pour un type accablé de vingt-cinq kilos de trop. Il attira du premier coup l'attention d'une serveuse et commanda deux bières. Il me fixa de plus belle et sourit.

— Je sais, dit-il. Avec un nom pareil, tu t'attendais à voir Dolph Lundgren. T'inquiète pas. Je suis habitué à cette lueur de déception.

— T'es un bel homme, répondis-je. Ne laisse personne te dire le contraire.

Il rit.

— Grand-père a laissé tomber le « stein » à son arrivée aux USA dans les années trente, expliqua-t-il. Mon père parlait sans cesse de se refaire appeler Ungerstein, mais ce n'était jamais le bon moment. Moi, je suis né Unger, alors peu importe. Je sais que la dernière partie est là.

Nos boissons arrivèrent. Unger vida d'un trait la moitié de sa bière.

— Ah ! je revis, expira-t-il. Dieu ce que je déteste les avions... Bon, avant toute chose, qu'est-il arrivé à Bobby, bordel de merde ? Je le croyais invulnérable.

Je ne savais par où commencer, ni si j'étais prêt à en parler.

— C'est pas facile, Carl.

— Je sais. Tu ne me connais ni d'Ève ni d'Adam, et je me pointe en parlant de ton meilleur ami comme si on était amants. J'imagine que tu as vérifié mon numéro et eu confirmation de mon identité, mais sans obtenir davantage. C'est fait exprès, bien sûr. Alors oui, je porte un flingue, parce que moi non plus, je ne sais rien de toi. Tu as sûrement pris les mêmes précautions. Mais il faut bien que l'un de nous deux parle le premier, pas vrai ? Si tu préfères que ce soit moi, y a pas de problème.

— J'apprécierais.

— Très bien. Alors voici ce dont je dispose. À Langley, comme tu le sais, un groupe de gens est chargé de ratisser les réseaux Internet pour guetter les signes d'activité terroriste. Il y a quelques mois, une nana de l'équipe qui surfait un peu en dehors de son temps de travail a décidé de s'intéresser aux spams. Tu te souviens de cette période où on a vu apparaître des mots bizarres dans le titre ou le corps des messages intempestifs ?

— Pour déjouer les logiciels de filtrage.

— Exact. C'est du moins ce que tout le monde croyait. Tu bourres ton spam de mots pris au hasard, de façon à déjouer les filtres à spam. Étant donné que les spams contiennent plus souvent les mots « sexe », « Viagra » et « prêt » que « bison », « fraise » ou « dindonneau ». Mais venons-en aux faits. Cette collaboratrice, Ramona pour les intimes, décide de compiler tous les exemples qu'elle trouve – soit littéralement des dizaines de milliers – et de lancer une analyse statistique. Elle n'en attend rien, c'est juste pour tuer le temps.

— Je vois le genre. Il fut une période où je faisais le même genre de boulot.

— Les premiers résultats ne révèlent pas grand-chose : la distribution lexicale était assez aléatoire. Alors elle s'est proposée, par simple curiosité, de décomposer les mots en unités sémantiques. « Garde-boue » devient « garde » et « boue », « international » donne « inter » et « national », etc. Et là, les choses commencent à se préciser. Parmi les meilleurs scores, on trouve « Hommes » et « Paille ».

Ma réaction n'aurait pas échappé à un passant posté sur le trottoir d'en face, même par une nuit d'encre. J'en perdis carrément ma clope.

— Voilà où je veux en venir. Ce ne sont pas les seuls mots – il y a environ cent cinquante réponses concluantes –, mais ils figurent dans le haut du panier.

— Mais alors... (Je me tus, peinant à saisir d'emblée les implications de cette découverte.) Seigneur...

— Comme tu dis. En vérité, personne n'aurait tiqué sur ces mots si je ne m'étais pas souvenu que Bobby m'avait commandé une recherche sur l'expression « Hommes de Paille ». Alors on est remontés à la source de tous les expéditeurs ; on a extrait leurs codes, sans résultat. Puis j'ai poussé un peu ma réflexion sur le spam. Pour une bonne part, ils sont très basiques. On a les Nigérians avec leur « J'ai un milliard de dollars à donner, et je t'ai choisi *toi* pour m'aider », l'arnaque type qui vise les débiles mentaux. On a les marchands de Viagra en ligne et les organismes de crédit – qui te vendent des appels surfacturés sans même se soucier si tu habites dans le bon pays, vu que ça leur coûte pas un rond. Celui qui m'a fait cogiter est un message qu'on ne voit plus aujourd'hui : « Britney Spears nue ». Ça m'a toujours interloqué, car je ne comprenais pas qu'on puisse marcher. Y aurait-il ne serait-ce qu'un clampin pour croire qu'on allait vraiment trouver des photos érotiques d'une vierge de renommée internationale – à l'époque, en tout cas – pour seulement cinq dollars quatre-vingt-dix-neuf ? Et sinon, quelle était la finalité de ces messages ? Donc, Ramona et moi avons sélectionné une série de spams – des normaux, sans mots aléatoires – et on les a étudiés de près. C'est-à-dire qu'on les a tous fourrés dans une bécane, pour voir ce qui en sortait.

— À savoir ?

— Rien dans un premier temps. Un catalogue de ce que les gens sans scrupule tâchent de fourguer à leurs contemporains sans espoir. Des trucs pour te durcir la bite. Des photos de nanas aux nichons impossibles. Des diplômes pour analphabètes. Du sexe, encore du sexe et toujours du sexe. Puis Ramona a eu une intuition : elle a réservé le système cryptographique central et y a balancé toutes les données. Rien de concluant pendant plusieurs jours, et là je commence à me dire que ces occurrences d'« hommes de paille » sont une simple coïncidence. Et puis, voilà trois semaines, on a eu une prise. (Il siffla une grande gorgée de bière, ne laissant qu'un petit centimètre de fond.) La chose avait tout d'un spam pour médocs, mais... Tu sais ce que sont les codes-livres, n'est-ce pas ?

— Chaque mot ou lettre renvoie au mot ou à la lettre qui tient la même position dans un livre donné. Par exemple, le premier mot de la première ligne peut correspondre au premier mot du premier chapitre, le troisième mot de la cinquième ligne au troisième mot du cinquième chapitre, et ainsi de suite.

— Exactement, avec des milliers de variations possibles. Complètement grillé une fois que l'ennemi a repéré le bouquin en question, mais facile d'utilisation et dur à briser sans tuyau. Donc, quand toutes nos moulinettes de décryptage standard ont fait chou blanc avec nos spams, on les a bennés dans un logiciel qui recherche des constructions grammaticales basées sur les quelques centaines de milliers de bouquins stockés en mémoire. J'ai consulté les résultats un soir, et une phrase ressortait. Ça disait : DEMAIN CE N'EST PAS LES HOMMES DE PAILLE, MAIS RÉJOUISSEZ-VOUS.

— D'accord, il y a « Hommes de Paille », mais après ? Ça ne veut rien dire, a priori.

— Effectivement, sauf quand on sait que ce spam-là ne fut émis qu'une seule fois, un seul jour, vers des millions d'adresses à travers le monde. En fin d'après-midi, le 10 septembre 2001.

Je le dévisageai.

— Eh oui, dit-il. Ces gens étaient au courant de ce qui couvait. Ils savaient que les Twin Towers allaient tomber, et ils n'ont rien fait pour empêcher ce drame. Ils ont précisé qu'ils n'étaient pas les auteurs des attentats, mais qu'ils les approuvaient.

— Bordel de Dieu...

— On croit toujours qu'une pub est une pub, mais une fois sur un million, c'est autre chose. Quand tu sais que le contenu de ton message intéresse de près les services de sécurité, tu as intérêt à *éviter* tout aspect confidentiel. Et donc, au lieu d'envoyer un courrier à une adresse précise, tu balances ce qui ressemble à un non-message à une foultitude de personnes. Le ou les vrais destinataires ont juste besoin a) de figurer sur la liste de diffusion et b) de connaître le code. Tous les autres jettent le spam à la poubelle. Mais toi, tu reçois l'info. Et même si par chance on parvient à briser le code et à repérer la présence d'un message, il est difficile de prouver qu'il y a bien communication, du fait que ça arrose des tas de gens d'un coup. Mais le pire, c'est que, même si on trouve un truc du genre « L'assassinat du chef de l'État est fixé à mercredi 16 heures », on n'est pas plus avancé ! Car ça ne conduit à personne. Comment veux-tu vérifier les millions d'adresses qui ont reçu le mail, sachant que la moitié seront de banals comptes Hotmail tout à fait innocents ? Impossible de découvrir qui est vraiment visé, qui est la véritable cible.

— Si je comprends bien, ils peuvent envoyer des messages au grand jour, à qui ils veulent, tout en préservant l'anonymat des destinataires.

— T'as tout pigé. C'est un putain de cauchemar. Au mieux, on soupçonne que certains des mails les plus récents proviennent de Californie du Sud – de L.A. ou ailleurs dans la Valley. C'est là que je commence à me biler et que j'essaie de joindre Bobby. On a manifestement affaire à un système trop rapide pour qu'on le coince –, surtout quand on découvre la situation au dernier moment et qu'on tente désespérément de rester dans la course.

— Quelle situation ? Quelle course ?

Unger ordonna d'un geste deux bières supplémentaires.

— C'est ce qu'on ne sait toujours pas. Et c'est pourquoi je prends l'avion pour rencontrer un type comme toi au pied levé. Les codes changent sans arrêt. On a eu quelques succès les premiers temps, mais il est impossible de vérifier tous les bouquins du monde, et l'analyse voyelle-phonème ne mène pas bien loin. Ça fait deux semaines qu'on n'obtient que du charabia, ce qui laisse penser qu'ils se savent observés. Ou qu'ils ont un ou plusieurs sbires à l'intérieur de l'Agence, et là je préfère ne même pas y penser.

— Tu devrais, répondis-je. Une amie du FBI a été suspendue alors qu'elle commençait à gratter dans la bonne direction. Et voilà une semaine, un détenu a pu s'échapper d'un fourgon blindé en Californie. Ces gens-là ont de puissantes relations.

— Mais qui sont-ils, alors ? Que sais-tu sur leur compte ?

— Mes parents sont morts l'an dernier, dans le Montana. Ça ressemblait à un accident de la route. Mais quand je suis remonté pour l'enterrement, j'ai découvert un truc qui m'a poussé à y regarder de plus près. J'ai trouvé une vidéo réalisée par mon père, qui mentionnait un groupe appelé les Hommes de Paille. J'ai aussitôt contacté Bobby, uniquement pour savoir comment convertir la cassette en DVD. Sa participation devait s'arrêter là.

— Il ne savait jamais s'arrêter.

— Il a entrepris quelques recherches et constaté qu'il n'y avait aucune trace de ma naissance dans ma ville natale. Pour faire court, on a fini par apprendre que j'avais été recueilli après que mon père a buté un gars qui avait agressé ma mère. Il n'avait pas voulu le tuer, à mon avis, mais toujours est-il que mon père biologique faisait partie d'une bande d'énergumènes terrés dans les bois. Et on était ses gosses.

— « On » ?

— J'avais aussi un frère.

— Et tu n'en savais rien ? Vous vous êtes rencontrés depuis ?

— Si on veut. Il travaillait pour les Hommes de Paille, et c'est lui le prisonnier qui s'est fait la belle la semaine dernière. C'est un tueur en série. Il pratique aussi le rapt, pour le compte de camarades qui tuent par plaisir. Il soutient une théorie selon laquelle l'humanité a choppé un virus voilà plusieurs dizaines de milliers d'années, un virus qui nous a rendus plus sociables et a permis l'avènement de la société moderne en étouffant l'hostilité naturelle de l'homme vis-à-vis de ses congénères. Les groupes humains se sont sédentarisés, ont inventé l'agriculture et développé le monde moderne. Ce que les Hommes de Paille déplorent. Ils veulent retrouver la planète telle qu'elle était avant.

Unger était suspendu à mes lèvres. Je poursuivis :

— Il y a pire. On a découvert que les Hommes de Paille étaient à l'origine de la tuerie de l'école d'Evanston, l'an dernier, dans le Maine. Comme de nombreuses autres atrocités commises ces dernières années. Si l'attentat d'Oklahoma City n'avait pas été élucidé, j'y aurais bien vu leur signature. Ils ne connaissent aucune limite. Absolument aucune.

Unger resta immobile un instant, puis tendit le bras pour me taxer une cigarette. J'ignore s'il avait conscience de son geste tant il semblait absorbé. Il alluma la clope, puis me considéra.

— Je vois, fit-il dans un murmure. Maintenant, j'ai un deuxième élément à t'expliquer. Avant que les codes retournent se planquer, une phrase émergeait constamment. On l'a dénichée dans deux spams de premier plan, et, une certaine nuit, elle est apparue simultanément sur des systèmes téléphoniques d'entreprises dans une trentaine de grandes villes américaines.

Il sortit de sa poche un bout de papier où étaient imprimés les quatre mots suivants : LE JOUR DES ANGES.

— Ça t'évoque quelque chose ?

— Non, répondis-je, la nuque parcourue d'un frisson glacé. Mais ça n'a rien de rassurant.

Nous continuâmes d'échanger le peu que nous savions, puis Unger me montra mon verre et je secouai la tête.

— Je dois rentrer, expliquai-je. Il vaut mieux que je roule lentement.

— J'espérais un débriefing complet.

— Pas ce soir. Je dois parler à quelqu'un.

— Elle est au courant de tout ceci ?

— Comment sais-tu que c'est une femme ?

Unger leva les mains d'un air innocent.

— Au son de ta voix, sans doute.

— Oui, *elle* est au courant.

— Je pourrais lui parler, moi aussi ?

— J'en sais rien. Il faudrait que je lui demande.

— Entendu. (Il sortit un stylo et griffonna une adresse sur le bout de papier posé sur la table.) Voici où je crèche ce soir.

Le Days Inn, à environ cinq rues d'ici, vers... vers l'est. Chambre 211. Tu m'y trouveras jusqu'aux alentours de 9 h 30 demain matin. T'as mon numéro, alors n'hésite pas. Si tu peux venir me voir, je resterai aussi longtemps que nécessaire. Personne d'autre dont il faudrait me parler ?

— Non.

— Ce serait vraiment super si je pouvais m'entretenir avec vous deux, dit-il, le visage enfin débarrassé de son espèce de rictus machinal. J'ai un mauvais pressentiment pour notre sûreté nationale. Je pense qu'il se trame un sale truc, et l'Agence n'a pas besoin de ça en ce moment. La dernière fois, on s'est fait trucider.

— C'est sûr que l'Irak ne fut pas vraiment votre heure de gloire.

Il prit un air piqué.

— Oh, on a plutôt assuré. Les bavures n'étaient qu'une boulette médiatique, le fait de petits merdeux qui tournaient en rond sous leurs tentes. D'accord, ça fait tache, mais ces dérapages ont toujours existé. Ce qui change, c'est qu'aujourd'hui on a des appareils numériques qui permettent de montrer ses exploits au pays. L'armée bossait très bien dans d'autres zones. L'Agence aussi. Seulement la presse ne sait rien de tout ça. Elle n'a pas à le savoir. C'est un *secret*. Mais le fond du problème, c'est que le 11 septembre n'aurait jamais dû se produire et que les informations ayant motivé la deuxième guerre du Golfe ont été glanées par des esprits un peu trop imaginatifs. Alors on a désigné un coupable – une fois la poussière retombée et notre objectif atteint. Peu importe que le contre-espionnage militaire ait été réduit à peau de balle au début des années 1990. En tout et pour tout, c'est tout juste si on disposait de cent arabophones dans nos services. On n'était pas prêts pour le nouveau désordre mondial. Personne ne l'était. Ce ne sont plus les têtes nucléaires et les régiments de soldats qu'il faut redouter, mais des armées qui tiennent dans une voiture. Le terrorisme, ce n'est pas James Bond ou Tom Clancy. Même Al-Qaida fait vieillot aujourd'hui. Il suffit d'un piéton avec une bombe. Il emprunte les mêmes trottoirs que nous. Il entretient les mêmes pensées. Mais il possède une bombe. Le seul espoir, c'est de mettre un agent derrière chaque type, pour le travailler au corps, se glisser dans sa tête. Pour dire si c'est un péquenot ou un fanatique. Deviner où il va frapper. Or c'est précisément le personnel doué pour ce taf hors norme qu'on a dégraissé, à la CIA : on a lourdé des gens comme Bobby, même s'il faut reconnaître qu'aucune langue étrangère n'aurait sauvé sa putain de vie. Désolé, j'ai mal choisi mes mots. Mais c'est tellement plus facile de cogner sur la CIA que sur un enturbanné qu'on n'est même pas foutu de coincer...

— Que la CIA n'est pas foutue de coincer, rectifiai-je. (Il me tardait de partir, soudain.) Et si tu essaies de me convaincre que l'Agence mérite le Nobel de la Paix, tu frappes à la mauvaise porte. N'oublie pas que j'y ai bossé. Ça grouille de mecs complètement handicapés sur le plan moral, et on a fait notre lot de conneries pendant des années. Pourquoi tu crois qu'on nous déteste tant ?

— Ça me dépasse. Je te jure que nos intentions sont très louables.

— Peut mieux faire. Et puis je vais te confier un truc : les vrais méchants sont déjà dans la place. Si ça se trouve, ils y étaient même avant nous.

— Comment ça ?

— Je vais retrouver mon amie, dis-je en me levant. À demain, peut-être.

— Je l'espère. T'inquiète pas, je ne bouge pas de ce siège tant que tu ne t'es pas évaporé. Mais si ce que tu dis est vrai, un de ces jours, il faudra bien que tu fasses confiance à quelqu'un, sans quoi ta vie ne sera qu'un long voyage vers les ténèbres.

— Confiance ? Oui, je crois me souvenir de ce mot.

Je lui ai serré la main et on s'est séparés. En traversant la rue, j'ai jeté un œil derrière moi : Unger était toujours sur sa banquette. J'avais pris soin de garer la voiture à une certaine distance, au bout d'un dédale de rues impossible à déjouer, à moins de m'être relié par une corde, et j'ai quitté la ville par une route qui menait partout sauf là où je me rendais. Tout en roulant vers Thornton, je m'efforçai non sans mal de me faire un avis sur Unger. Une partie de moi voulait lui faire confiance, sentir qu'il y avait, au sein d'une agence officielle, un allié possible. Mais l'autre partie de moi doutait. Avait-il réellement deviné à mes seules intonations que je parlais d'une femme ? Et lorsqu'il demandait si nous étions plusieurs sur le coup, dans quelle mesure s'agissait-il d'une question innocente ? S'il était de mèche avec les Hommes de Paille, n'était-ce pas logique de chercher à recenser nos forces ?

Le problème, avec la paranoïa, c'est qu'on ne sait plus où s'arrêter. Dès que l'on se met à douter des principes élémentaires qui régissent les rapports humains, tout devient envisageable. Si les images de tortures pratiquées en Irak par les soldats américains ont tant choqué l'opinion mondiale, ce n'est pas seulement en raison de la nature des événements. Les exactions au Vietnam sont largement documentées. Nous savons ce qui se passait dans les camps de prisonniers de la Seconde Guerre mondiale. On se rappelle les têtes plantées sur des lances au Moyen Âge, les chevaliers brûlés vifs à Azincourt, l'inventivité des Romains et des Carthaginois lors des guerres puniques. Non, ce qui scandalise, dans ces clichés, c'est le fait même qu'on les ait pris, le désir affiché d'immortaliser ces actes et la certitude que certains de nos congénères s'en délecteront. Est-on tellement loin du tueur qui conserve la photo ou une mèche de cheveux de ses victimes ? Parce qu'il se vit comme un marginal, le tueur en série est capable de sévir sur le sol national – quand la plupart des tueurs ont besoin de l'anonymat et de l'éloignement d'un pays étranger, du vague mandat conféré par une déclaration de guerre. À part ça, rien ne les distingue vraiment.

Les services américains n'ont pas échoué à prévenir le 11 septembre par incompétence. On se croit toujours plus doué que le reste du monde. Ce ne sont pas les autres qui ont remporté le match, c'est nous qui avons perdu le ballon. Bel aveuglement. En réalité, les méchants gagnent parce qu'ils sont aussi doués que nous. La force de la volonté et la pureté de la haine suffisent à combler un fossé technologique immense. Une nation qui reste persuadée de sa toute-puissance malgré le coup qu'on lui a porté est restée coincée sur ses soirées étudiantes, les yeux écarquillés devant les tee-shirts mouillés de notre culture, seul choc visible à l'horizon.

Il se mit à pleuvoir.

Ce fut un trajet déprimant.

J'ai rejoint Thornton peu après minuit. La ville dormait sous un timide clair de lune, lisse et insaisissable comme le rêve d'un étranger. Roulant au ralenti devant l'immeuble du shérif, j'envisageai d'appeler Nina, avant de me dire qu'elle serait soit occupée, soit déjà repartie. La voiture des présumés journalistes était toujours là, mais vide. Ils devaient être à l'intérieur du bâtiment, et l'affaire Julia Gulicks paraîtrait le lendemain matin. La fourgonnette blanche avait disparu.

Le véhicule de Reidel était stationné devant l'hôtel, celui de Monroe deux places plus loin. Je suis entré dans le bâtiment, en espérant qu'ils tenaient leurs mystérieux conciliabules ailleurs que dans la chambre de Nina. Mais le bar et le restaurant étaient vides. Personne dans le hall, pas plus qu'à la réception.

Je m'engageai dans le couloir en me demandant si je pouvais dire aux deux autres flics de foutre le camp. Nina ne m'avait jamais paru aussi fatiguée que cet après-midi. Elle avait besoin de repos. Moi aussi.

Je frappai à la porte, à titre de sommation, puis ouvris avec ma clé magnétique. Je refermai derrière moi.

— Nina ?

Pas de réponse. La conférence se déroulait donc ailleurs. Savais-je seulement quelle chambre occupait Monroe ? J'enfilai le minuscule couloir longeant la salle de bains, dans l'idée de m'allonger sur le lit et de laisser Nina me rejoindre quand elle aurait fini.

Mais je me figeai sur le seuil de la pièce, comme bloqué par une vitre.

Je ne vis d'abord que du sang.

CHAPITRE 19

J'avais conscience de produire des sons. J'en avais conscience, car ils me brûlaient la gorge. J'ignorais seulement leur puissance.

On aurait dit qu'un animal ou un humain était passé à la tronçonneuse. Le sang avait giclé sur les murs, sur la télé, sur les fauteuils, le couvre-lit, la grande glace murale. La chambre puait le cuivre et la mort, et ce rouge mat omniprésent tenait presque du bruit. Je restai paralysé une seconde, incapable d'organiser la scène en image cohérente.

Le grabuge était tel qu'il me fallut une longue minute pour comprendre que je ne voyais qu'un seul corps.

Celui de Reidel.

Il gisait sous la fenêtre côté parking, tordu sur lui-même comme s'il avait heurté le mur avec une force à même de briser le plâtre. À demi scalpé, le visage trempé couleur rouge-brun, les yeux ouverts. On aurait dit que l'agresseur avait essayé de découper ses fringues avec une hachette, avant de perdre patience et de tailler dans le vif. Un trait de sang barrait le papier peint derrière sa tête. Mais les plaies les plus profondes se concentraient sur la gorge, les bras et la droite du torse. Une flaque sombre s'étirait sur soixante centimètres de part et d'autre du cadavre.

Un cadavre.

Un seul.

Je m'en retournai vers la salle de bains, j'ouvris d'un grand coup de pied, mon flingue au poing, prêt à le vider. La propreté clinquante me parut irréelle après ce que je venais de voir, et il n'y avait personne, ni mort ni vivant.

Je retournai dans la chambre et poussai le lit, sans doute dans l'idée que Nina pouvait se cacher en dessous – elle n'aurait pas reconnu la voix qui braillait son nom – ou qu'on y avait glissé, morceau par morceau, les restes de son corps. J'avais dégagé la moitié du sommier quand j'entendis bouger dans mon dos.

Je virevoltai pour distinguer dans l'entrée une employée en uniforme.

Elle se mit à hurler comme un Jumbo au décollage.

— Trouvez-moi l'agent Monroe ! criai-je. De toute urgence !

Un mouvement de recul, et elle trébucha. Elle tremblait de tout son corps. La cervelle aveuglée par le bruit et l'hémoglobine, il me fallut une seconde pour comprendre que c'était moi qu'elle fuyait. Je tenais un flingue, j'étais couvert de sang, et il y avait une chose horrible avachie au fond de la pièce. Moi aussi, j'aurais pris peur.

Je rangeai l'arme dans ma poche et parvins à lui saisir le bras, avant de l'immobiliser en lui agrippant l'épaule de mon autre main.

— Je n'y suis pour rien, dis-je en essayant de maîtriser ma voix et de ne pas lui briser les os. (Ses yeux naviguaient dans leurs orbites, voyant tout sauf moi. Je rapprochai mon visage et répétai un ton plus haut :) Je n'y suis pour rien. Maintenant prévenez les flics et *trouvez l'agent Monroe*.

Je la repoussai vers la porte. Elle prit ses jambes à son cou.

Procéder avec méthode. Je savais qu'il ne fallait pas déranger le lieu du crime, mais je l'avais déjà bien remué, et je devais finir l'inspection. Si Nina se trouvait là, c'était à moi de la découvrir.

Je m'allongeai par terre et regardai sous le lit afin d'être définitivement fixé. Je me relevai ensuite pour fouiller la petite armoire. Rien, sinon les quelques effets de Nina. Je laissai les portes ouvertes afin d'empêcher son corps de s'y matérialiser, ruisselant et disloqué. Je regardai derrière le gros téléviseur, écartai les tentures. Ce faisant, je dus enjamber Reidel à plusieurs reprises, et je sentais bien qu'il y avait une anomalie dans le tableau, mais je ne savais dire laquelle et ce n'était pas prioritaire.

Je regagnai la salle de bains et la scrutai de nouveau, sans but, ouvrant le rideau de douche avec le coude pour ne pas importer le sang de l'autre cauchemar.

Nina n'était pas là. Elle ne se trouvait nulle part dans cette chambre. Mais je n'en tirai aucune conclusion. Il fallait que je la retrouve, point.

Je me ruai vers la réception. Au bout de dix mètres, je tombai sur Monroe, qui débouchait dans le couloir en bras de chemise, l'air vieux et perdu.

— Qu'est-ce qui se passe ?

— Reidel est mort.

Sa mâchoire se décrocha, et l'instant d'après je déboulais sur le parking refroidi par la nuit.

Je le traversai à toute vitesse et de long en large pour regarder dans chaque voiture, mais toutes étaient vides et aucune ne quittait l'endroit, alors j'ai fini par ralentir et m'arrêter. Tout était immobile, hormis les nuages dans le ciel.

Il n'y avait personne à pourchasser, et je ne pouvais rien faire. Quoi qu'il fût advenu, j'arrivais trop tard. Au loin s'élevaient les sirènes.

Elles aussi arrivaient trop tard.

Au bout d'une heure, je fumais assis sur le trottoir. La main qui tenait la cigarette était incrustée de sang. Tout comme mon jean. Je fixais le goudron cloqué entre mes pieds, pour que mon esprit se raccroche à quelque chose. J'avais passé la majeure partie de ces soixante minutes dans la chambre de Nina, jusqu'à saturation. La fureur des flics locaux se mêlait à la mienne, et ma tête était en proie à une douleur indicible et glaciale.

L'heure écoulée se résumait à une ellipse. Elle n'avait rien donné d'utile. Le temps s'était contenté de filer, emportant avec lui toute forme d'initiative. Je voyais deux flics arpenter le parking, penchés en avant, cherchant des traces de sang. J'avais déjà essayé.

J'entendis coulisser les portes automatiques de l'hôtel. Apparut Monroe, seul. Derrière lui, le hall grouillait d'employés et de clients, que les flics essayaient de renvoyer dans leurs chambres, dans leurs bureaux, n'importe où pourvu qu'ils libèrent la voie. La moitié des résidents paraissaient effrayés. Les autres semblaient les heureux figurants d'un reality-show gore. J'avais envie de les frapper. Méchamment.

— Toujours rien ?

Monroe secoua la tête.

— On retourne tout l'hôtel. Sous-sol, toit, tous les lieux de stockage. Mais elle n'est pas là.

Mes yeux retombèrent sur le goudron.

— Tous les flics de cette ville se trouvent ici ou patrouillent les rues, reprit-il. On a rappelé tous les agents en repos. Les shérifs d'Owensville, d'Andley et de Smithfield sont sur le pied de guerre. J'ai avisé nos deux bureaux les plus proches. Les renforts sont en route.

— C'est trop tard.

— Vous ne pouvez pas dire ça. Un agent fédéral vient d'être enlevé. Nous avons une longue expérience de ce type d'incident. Nous la ramènerons, quel qu'en soit le prix.

— Et par où allez-vous commencer les recherches ?

— On a déjà installé des barrages sur les trois grandes routes qui desservent Thornton. Dès l'arrivée de mes collègues, nous ferons boucler toute la ville. On fouillera chaque maison si nécessaire. Nous la retrouverons, même s'il faut démonter ce patelin brique par brique.

Monroe ne devait pas entendre ses accents d'héroïsme désespéré.

— À quelle heure êtes-vous rentrés à l'hôtel, cette nuit ? Combien de temps avant que je ne découvre le grabuge ?

— Environ une heure, estima-t-il. Peut-être un peu plus.

— Ce qui en fait déjà deux. Il a eu le temps de changer d'État.

— « Il » ? Vous pensez à qui ?

— À votre avis ? On vient d'attaquer les deux têtes pensantes de l'enquête. Reidel a été dépecé avec le même genre d'arme que les victimes des bois, une sorte de gros couperet. Vous n'avez pas vu la balafre dans le mur ? (Dans l'intervalle j'avais en partie compris ce qui clochait dans le corps de Reidel.) Sans compter que sa main gisait à un mètre du bras. Ce qui désigne clairement votre tueur. Qui d'autre, sinon ?

— Julia Gulicks est toujours en cellule. Sous la surveillance d'un gardien.

— Bien sûr qu'elle y est, Monroe, puisque *ce n'est pas elle la coupable* ! Nina avait raison. Votre thèse est une coquille vide, basée sur le fait que Gulicks a trouvé Widmar, et étayée par les allégations d'une femme jalouse dans un bar. Même si elle était coupable et qu'elle se soit téléportée hors du commissariat, on ne me fera pas croire qu'elle ait pu réussir un coup pareil. Reidel avait une carrure de boxeur et Nina savait se défendre. Vous voyez Gulicks les démolir l'un après l'autre ? Sincèrement ? Vous la voyez, oui ou non ?

— Non, admit-il.

— Donc c'est un homme, et un homme qui n'en est pas à son premier meurtre. Julia Gulicks est innocente, le vrai tueur vient maintenant s'en prendre à nous, et nous ne savons pas qui c'est, ni ce dont il est capable.

Monroe passa la main dans ses cheveux ras. Certes, il tenait beaucoup à Nina, mais il devait aussi se demander de quoi il avait l'air avec un agent kidnappé sous ses yeux, sans parler d'un flic massacré. Il fut obligé d'émettre une hypothèse détestable :

— Ce n'est pas forcément le tueur. Et si c'était lui ?

— Qui ça ?

— Votre frère.

Je le dévisageai.

— Pourquoi s'en prendrait-il aux flics locaux ? Et il était encore à Pelican Bay quand l'autre gus est mort.

— Je sais. On peut toutefois imaginer que Julia Gulicks soit bien la tueuse de Thornton, mais que votre frère soit venu vous cueillir, Nina et vous. Nina lui a tiré dessus voici quelques mois. Il sera peut-être venu se venger.

Je n'y avais même pas pensé.

— Et la main coupée serait une simple coïncidence ? Ça m'étonnerait. Mais si vous avez des doutes, je vous conseille de prier très fort. Car Paul est capable de tout.

— J'essaie juste de chercher des pistes.

— Mais on n'a rien vu ni entendu dans l'hôtel ? Ni à la réception, ni dans les couloirs ? Quelqu'un s'introduit dans la chambre, la transforme en abattoir et emmène Nina de force, sans que personne remarque quoi que ce soit ?

— On continue d'interroger les clients, mais ça ne donne rien. Le service de chambre cesse à 22 h 30. La réceptionniste qui assurait ce tour de garde est la fille que vous avez terrorisée. Elle a passé la soirée dans le bureau de service – pratique courante : on prépare le boulot du lendemain, et on rejoint le comptoir seulement si ça sonne. Entrer incognito était un jeu d'enfant. Quant à ressortir...

— Nina était inconsciente. C'est obligé.

Monroe détourna les yeux.

— Inconsciente, ou...

— Non. Pas ça. Elle était inconsciente.

Je me levai. J'avais la frousse, et besoin de m'activer.

— Je me tire.

— Où allez-vous ? s'enquit Monroe.

— Je vais faire ce que j'aurais dû faire depuis une heure. Chercher. Elle n'est pas ici, donc elle est forcément ailleurs.

Je trouvai un bout de papier dans ma poche et griffonnai mon numéro de portable.

— Appelez-moi. Dans la seconde. Quelles que soient les nouvelles.

— D'accord. (Il me fixa d'un œil sévère.) Faites-en autant. Et ne vous avisez pas de régler cette affaire en solo.

— Vous croyez que je vais vous promettre ça ?

— Non. Mais si vous trouvez le type, je tiens à être là.

— Vous ne m'empêcherez pas de le tuer.

— Qui parle de vous empêcher ?

J'ai sillonné la ville au hasard, à toute allure, les vitres grandes ouvertes, à l'affût des sirènes. J'ai songé à vérifier les sites où l'on avait découvert les deux cadavres, mais le parking de la forêt de Raynor était désert, et je n'avais pas été assez vigilant pour me remémorer le trajet vers le second lieu du crime. J'appelai donc Monroe, qui promit d'envoyer quelqu'un sur place.

J'intensifiai l'effort de quadrillage. De loin en loin je croisais des gyrophares, mais personne ne fit attention à moi, malgré ma conduite de cinglé et – m'eût-on seulement accordé un regard – mes fringues souillées de sang.

Au deuxième passage devant le poste de police, j'eus l'instinct de piler. Je sautai de la voiture et gagnai l'entrée avant même de savoir ce que j'avais en tête. Un seul flic assurait l'accueil, d'un air tendu. Par chance, c'était l'un de ceux qui m'avaient vu circuler dans l'après-midi.

— Vous allez bien, monsieur ? Vous êtes blessé ?

— Non, répondis-je. J'arrive du lieu où Reidel s'est fait buter. Où est Julia Gulicks ?

— En cellule. Vous ne pouvez pas y aller.

— Si, je peux. Appelez donc l'agent spécial Monroe, de la part de Ward Hopkins.

Sans attendre le feu vert, je filai dans la partie arrière du bâtiment et parcourus le couloir de la salle d'interrogatoire. Tout au fond, un panneau indiquait le quartier des gardes à vue, où s'alignaient trois portes sinistres. Devant celle du milieu se tenait un planton à cran.

— Ouvrez la porte, ordonnai-je.

— Pas question, monsieur.

Je me plaçai de manière à voir dans l'œilleton. La pièce était minuscule, trois mètres sur trois ; seuls les rais de lumière du couloir venaient percer l'obscurité. Une couche étroite occupait tout un côté, et l'autre était muni d'un lavabo et d'une cuvette. Entre les deux, une chaise adossée au mur.

Julia Gulicks y était assise, dos droit et tête baissée. Voyant que je la regardais, elle releva le menton. Elle était pâle, les yeux grands ouverts.

Elle me fixa. Et l'expression de son visage se modifia.

Je doute que ce fût un sourire. Mais sa bouche avait effectué un mouvement curieux.

— Monsieur ? (Le flic de l'entrée remontait le couloir d'un pas vif.) Vous devez partir. Sur-le-champ.

— Je veux lui parler, protestai-je.

Elle me dévisageait toujours.

— Ce ne sera pas possible, monsieur Hopkins. Je viens d'avoir l'agent Monroe. Il ne souhaite pas qu'on vous arrête mais nous demande de vous sortir illico. (Il me toucha le bras.) Je n'aimerais pas devoir...

Je me dégageai d'une secousse. Le planton me toisait.

— C'est bon, je m'en vais.

Le flic me reconduisit jusqu'à la rue. Il semblait hors de lui, et je savais que je ne lui referais pas le coup deux fois. Non que la première m'eût apporté grand-chose, du reste.

Sauf que...

Le regard que m'avait lancé cette femme était lourd de sens. Personne ne sourit ainsi sans raison précise. Julia Gulicks savait sans doute qu'il était arrivé quelque chose en ville : le commissariat avait dû s'enflammer à l'annonce du meurtre de Reidel, et les rumeurs avaient pu traverser sa porte. Réagissait-elle à cela ? À la mort du flic qui l'avait harcelée dans l'après-midi ?

Je restai planté sur le trottoir, conscient que le flic me surveillait depuis son bureau. J'étais déjà vanné lors du rencard avec Unger ; maintenant j'arrivais à peine à réfléchir.

Je me laissai choir sur les marches.

Hypothèse, ou espoir : Nina n'était pas morte. Car si le tueur devait les supprimer tous deux, le plus commode eût été de s'en acquitter dans la chambre.

D'où la question : pourquoi tuer Reidel mais épargner Nina ?

Soit le meurtrier était venu dans l'intention d'éliminer les deux, avant de se raviser et d'enlever Nina. Soit son but initial était de kidnapper Nina, et Reidel avait juste eu le malheur de se trouver sur son passage. Ou bien le tueur projetait dès le début un meurtre et un enlèvement. Dans tous les cas de figure, il avait réussi son coup.

Supposons que Nina ait été la cible. Cela signifiait quoi ? Et désignait qui ?

Tous les regards se tournaient vers le tueur fou de Thornton. Depuis sa cachette, il avait observé le travail de Nina et conclu qu'elle menaçait sa sécurité ou qu'il aimerait en faire sa prochaine proie. Cette supposition me terrifiait, mais, Dieu merci, elle était bancale, puisqu'elle bouleversait le mode opératoire et le profil de la victime : du piège mortel tendu à deux hommes mûrs, on passait à l'enlèvement musclé d'une femme agent fédéral.

Donc : partons du principe que c'était l'œuvre d'un autre.

Monroe m'avait ébranlé en évoquant Paul, mais là encore ça ne tenait pas la route. Certes, mon frère était parfaitement capable d'un tel acte, tant au plan physique que moral. Sauf que ça ne lui ressemblait pas. D'une part, il m'aurait sûrement attendu dans la chambre. D'autre part, je n'avais aucune envie d'y croire : s'il s'agissait bien de Paul, je savais que je ne reverrais jamais Nina, quoi que je fasse.

Qui, alors ? D'autres séides des Hommes de Paille ? Beaucoup étaient capables de tuer un flic de sang-froid : six mois plus tôt, j'en avais vu un vider son chargeur sur Charles Monroe dans un troquet bondé, et j'avais découvert leurs jardins remplis de cadavres, dans un parc immobilier de luxe. Ces gens-là ne s'embarrassaient pas de convenances.

Puis j'eus comme un déclic.

Quitter Thornton fut un jeu d'enfant : j'ignore quelles étaient les trois issues théoriquement barrées par la police, mais celle que j'empruntai ne présentait aucun obstacle. Pied au plancher, je fus vite à Owensville. Dans la grand-rue, la vue des bagnoles de flics m'incita à ralentir, l'alerte de Monroe avait visiblement porté. Je trouvai le Days Inn et m'y introduisis avec une lenteur innocente.

À 3 heures du matin, il y avait encore un employé à la réception. Somnolent, mais présent. Si l'hôtel de Thornton avait fonctionné comme celui-ci, Nina aurait peut-être été dans sa chambre à l'heure qu'il était, et moi à ses côtés.

— Je suis dans la 211, annonçai-je d'un air naze et pompette. Je crois que j'ai paumé ma clé. Vous en auriez une autre ?

— Désolé, monsieur, mais je n'étais pas là lorsque vous avez pris la chambre. Je vais donc vous demander une pièce d'identité.

Je fis mine de fouiller ma poche.

— Attendez... C'est bon, je l'ai retrouvée.

— Bonne nuit, monsieur.

Je me ruai dans l'escalier puis dans le couloir jusqu'à la 211. Je frappai à la porte et mis la main dans mon blouson.

Aucune réponse. Je recommençai, plus fort. Approchai mon oreille. Aucun bruit à l'intérieur. Mon sang se glaça.

— Carl ? C'est Ward.

Rien. Je reculai d'un pas et flanquai un grand coup de pied dans la porte. Elle ne tressaillit pas. Je récidivai, sans plus de résultat. Je ne sais pas bidouiller ce genre de serrure et n'étais pas assez stupide pour l'attaquer au flingue.

Je courus jusqu'à la réception.

— Je crois que ma clé déconne. Il m'en faudrait une autre.

— Et moi, une pièce d'identité.

Je sortis mon arme.

— Cela suffira-t-il ?

Je vis le type tendre le bras vers sa gauche.

— Lâchez ce téléphone. J'enquête sur un homicide à Thornton.

— Vous êtes flic ?

— FBI.

— Il fallait le dire.

Se voyant déjà défrayer la chronique au titre du Citoyen le plus coopératif d'Owensville, le type devint une crème. Il fit crépiter son clavier d'ordinateur.

— La 211, c'est ça ?

— C'est ça. Carl Unger.

Il fronça les sourcils.

— Ah, non.

— Je dois à tout prix voir cette chambre.

Il attrapa une carte vierge, la programma, mais au lieu de me la remettre il quitta son guichet, décidé à m'accompagner. N'ayant ni le temps ni l'envie de cogner, je le suivis. Il monta l'escalier en hâte et fondit sur la 211. Il glissa la carte dans le lecteur et ouvrit la porte.

Je pénétrai dans la pièce. Elle était vide. Le lit intact. La salle de bains immaculée, les toilettes stérilisées.

Jurant, je me tournai vers le réceptionniste :

— Expliquez-moi.

— M. Unger a réservé une chambre en fin d'après-midi. Puis il a rappelé peu avant 22 heures pour annuler. Le délai était dépassé, alors il a quand même dû payer, mais il a réglé la note sans voir l'hôtel.

Je me laissai choir sur un coin du lit, incapable de prononcer le moindre mot tant je me sentais con. Unger avait réservé une chambre pour simuler un point de chute, ou pour nous attirer Nina et moi le lendemain matin, au cas où la mission de cette nuit échouerait.

Mais elle avait réussi. Et donc il avait disparu. Ainsi que Nina. Ne restait plus que moi.

De retour à Thornton, le parking de l'Holiday Inn grouillait de voitures de police, talonnées par deux camions de télé. Un cordon de badauds matinaux battait le pavé. Je roulai sans m'arrêter. Je ne trouverais rien ici.

Direction : un site visité un peu plus tôt, le parking situé en amont de la forêt de Raynor. Je restai assis sur mon siège, les vitres baissées, à écouter bruisser les bois. Le prénom de Nina tournait en boucle dans mon crâne, et la panique me comprimait le cœur.

— Je fais quoi, Bobby ? Comment je la retrouve ?

La question quitta mes lèvres à mon insu, et ne fut suivie que de silence.

Sans réfléchir, j'empoignai mon portable et sélectionnai le numéro de Nina. Comme prévu, je tombai sur la messagerie : Monroe avait déjà tenté de localiser son téléphone, sans succès. L'appareil était éteint. Les fous de complots vous diraient qu'on pouvait quand même le pister. Hélas, ils avaient tort.

Alors j'ai tiré ma dernière cartouche, en composant un autre numéro. En essayant encore, et encore, et encore, toutes les cinq minutes, jusqu'à ce que me réponde, sur le coup de 7 h 03, une voix que je n'avais pas entendue depuis cinq mois.

— Qu'est-ce que tu veux, Ward ? J'ai reçu ton SMS et je ne connais pas ce gars-là.

— Il faut que je te parle, John.

— On n'a rien à se dire.

— Ils ont pris Nina.

Un long silence.

— Qui ça, ils ?

— J'en sais rien.

— T'es où ?

— À Thornton, en Virginie. Grouille-toi, John.

CHAPITRE 20

— Les affaires continuent, je te dis. On fait ce qu'on a toujours fait.

— Là, tu rêves, mec. J'ai pas reçu un seul coup de fil en vingt-quatre heures. Rien que cette semaine, deux fêtes ont été annulées. Les parents font dans leur froc, quelque chose de bien, et avec les flics qui posent leurs questions dans tous les coins... L'ambiance n'est plus à la teuf, Lee. Les gens restent chez eux pour mater la télé.

Ils étaient assis dans la voiture de Lee, devant le Starbucks du soir où Pete était mort, à engloutir comme d'habitude des quadruples *latte* parfumés à la vanille. Mais à 8 h 40 ce matin, le café avait un goût écœurant. Cette virée était une idée de Lee, au motif que, selon lui, il fallait fréquenter les lieux en rapport avec la nuit du drame, afin d'aplanir les souvenirs des gens : que la dernière impression laissée au personnel ou aux clients – à supposer qu'ils en aient quelque chose à foutre – soit celle de deux jeunes relax, en phase avec leur monde, et non de deux flippés qui viennent d'enterrer un pote. Pour Brad, anticiper l'attitude des flics était un jeu réservé à des gens expérimentés. L'assurance de Lee le tracassait. Depuis son tête-à-tête avec le fameux Paul, il semblait un peu déconnecté de la réalité.

— Ils pouvaient pas se douter que ça allait se corser ?

— Bien sûr que si, répondit Lee. Mais c'est une question de turnover. De fric qui entre et de fric qui sort. Ils ont un méga stock de pilules et elles ne durent pas éternellement.

Brad jugea l'argument pipeau, mais répondit néanmoins :

— Bon, je peux éventuellement brancher Matt et Dena. Ils connaissaient pas tellement Pete. Peut-être qu'ils continuent à teufer.

— Non, pas les Reynolds, lâcha Hudek.

— Pourquoi ?

— C'est comme ça.

— Eh bien, je sais pas quoi te dire, Lee. Ramène la came à tes amis et explique-leur que depuis qu'un abruti a explosé la *tête* de Pete, le marché fait légèrement la gueule.

Lee se tourna vers Brad.

— Ça va, frangin ?

— Non, ça va pas, Lee. Et Pete me manque. Grave.

— Je sais. Moi, c'est pareil.

Brad n'était pas sûr de le croire. Dans l'univers de Lee, Pete semblait se réduire à un simple contretemps.

— Sa mère discutait au téléphone avec la mienne hier soir, pour lui demander si elle savait quoi que ce soit.

— Mais elle appelle tout le monde.

— Garde tes putains de laïus, Lee. Pour l'instant, je m'en branle de toi, de Steve ou du clampin de Bagdad. Je te parle de *moi*. Hier soir, c'est *ma* mère qu'elle a appelée. Et là-dessus maman vient dans ma piaule et s'assoit sur la chaise, et bref, tu vois le topo. On est dans la merde.

— Tout va bien se passer.

— Non, Lee. J'en suis pas sûr du tout. (Il hésita.) Et puis j'ai aussi eu un coup de fil de Karen.

— Quand on saute une meuf, on lui cause un peu. À toi de gérer, mec.

— C'est pas le problème.

— Je sais. Je te faisais marcher.

— Parce que tu trouves ça drôle ?

— Bon, où tu veux en venir, Brad ? Je suis peut-être futé mais pas extralucide.

— Elle me pose des tas de questions.

— Du genre ?

— Dès qu'ils ont retrouvé le corps de Pete, elle m'a appelé, on a parlé de comment ça craignait et tout, puis tout à coup elle m'a demandé si je savais quoi que soit sur sa mort.

— Et t'as répondu quoi ?

— J'ai dit non, bien sûr. Mais aussi... elle a entendu un truc, Lee. Le soir où je t'attendais devant sa baraque, tu te rappelles ? Quand tu t'es pointé en disant « Il arrive bientôt », « D'ici une minute » ou un truc dans ce goût-là. Ça m'a pas marqué plus que ça, mais elle, si. Elle sait qu'on attendait quelqu'un, et elle pense que c'était Pete.

— Putain. Et pourquoi t'as pas dit qu'on attendait Jed, ou Matt, ou Greg ?

— Parce que j'ai pas eu la présence d'esprit, d'accord ? Je lui ai juste dit qu'on était partis acheter des hamburgers tous les deux. Et puis d'abord, pourquoi mêler d'autres mecs à cette histoire, surtout si on sait qu'ils lui jureront le contraire ?

— Ouais, d'accord. Et hier soir, qu'est-ce qu'elle disait ?

— Rien de plus. Sauf que... Elle a dit qu'elle espérait que tous ceux qui avaient des infos iraient trouver les flics. Elle a dit que les flics avaient l'air sensés, et qu'il valait mieux, au final, dire tout ce qu'on savait, même si c'était pas très reluisant.

— Elle pensait à toi en disant ça ?

— Je crois. J'ai répondu qu'elle avait raison et on en est restés là. Tu peux pas savoir comment ça me rend nerveux, mec. Et je me sens... coupable.

— Mais tu l'es pas, et je le suis pas. Karen fait simplement ce qu'elle croit juste. Nos amis vont régler cette histoire, Karen verra qu'on sait que dalle, et voilà.

— Ils nous aideront, nos « amis », même si on leur dit qu'on arrête de dealer ?

— C'est pas une question de came. Ils ont autre chose en tête. Je connais pas les détails. Mais il se trame un truc. Ces types-là ont des relations, mon pote. Une sorte de mafia, mais ni italienne ni colombienne, rien de ce genre. Des Blancs. Ils ont un gros projet en vue, et on va y participer. Tout ça n'aura plus aucune importance.

— Pete aura toujours de l'importance, Lee. Pete le Dormeur comptera *toujours* pour nous.

— Ouais, bien sûr.

Brad comprit soudain que Lee avait quasiment supprimé Pete de sa mémoire.

Le bipeur de Hudek sonna. Il baissa les yeux pour lire le message et démarra.

— C'est eux. On y va.

Se retrouver dans l'espace restauration de Belle Isle avait quelque chose d'irréel. Brad connaissait bien l'endroit. Il y avait ingurgité des monceaux de taquitas et de pâtés impériaux, acheté d'innombrables sodas et autres granités aux fruits rouges à siroter tout en flânant avec la bande, avec Pete, ou seul avec Karen. En remontant un peu plus loin dans le temps, on l'y voyait également en compagnie de papa, maman et sœurette, fuyant les mortelles séances de shopping après avoir demandé la permission de patienter devant un milk-shake au chocolat, ce qui la plupart du temps lui était accordé.

À cette heure matinale, l'activité débutait à peine et la salle était déserte, hormis quelques ménagères faisant salon et un monstre briseur de sièges, penché sur des reliefs de hamburgers assez nombreux pour nourrir une petite famille entière, des frites dépassant de sa bouche telles des pattes d'insecte.

Il y avait aussi un mec, attablé comme un ermite au centre de l'espace. Brad fut surpris, jusqu'à ce que son regard embrasse une nouvelle fois les lieux. Disséminés dans le restaurant, se trouvaient trois autres types, qui n'avaient pas l'air d'être venus pour les calendriers animaliers en promo. Aucun ne mangeait ni ne buvait. Leurs regards convergeaient sur Brad et Lee. Avec décontraction. Enfin, plus ou moins. Le plus vieux avait une quarantaine d'années, le plus jeune une petite vingtaine. C'était ce dernier qui les fixait avec le plus d'insistance. Il y avait quelque chose chez ce gamin que Brad n'aimait pas du tout.

Lee avait le sac de drogue. Brad n'aimait pas ça non plus, mais telle était la consigne affichée sur le bipeur. Parvenu à la table centrale, Lee s'apprêtait à remettre le sac quand le type secoua la tête, très brièvement.

— Attends un peu, conseilla-t-il. Asseyez-vous.

Lee et Brad s'installèrent en face à face. Brad s'était imaginé rencontrer un type au physique d'acteur ou de vedette, le genre de gus qu'on remarque au milieu d'une foule. Mais celui-ci semblait pouvoir se fondre dans tous les paysages.

— Brad, c'est ça ?

— Ouais.

— Ravi de te connaître, Brad. Je m'appelle Paul.

— Très bien, répondit Brad tout en pensant : *T'as pas l'air net comme mec*.

Le type ramena son attention vers Hudek :

— Alors comme ça, vous avez du mal à écouler la came ?

— On fait de notre mieux, mais... La mort de Pete a plombé l'ambiance.

— J'imagine que vous ne fournissiez pas seulement les amis proches ?

— Non, bien sûr que non. Mais on couvrait un secteur bien délimité, et une certaine classe sociale. Mais on va persévérer, si tu veux.

— T'inquiète pas. On va se rabattre sur West Hollywood.

Hudek était déçu. Cette décision ne semblait guère confirmer sa promotion.

— Je suis désolé, dit-il. Mais les circonstances sont vraiment pourries.

— Ce n'est pas bien méchant, répondit l'homme. Et ne t'en fais pas, ça ne menace en rien le Plan. (Il se tourna vers Brad :) Tu veux me rendre un service, Brad ? Va me chercher un expresso.

Brad faillit lâcher : Va chier, je suis pas ta bonniche. Mais avec ce type, hélas, on l'était tous plus ou moins. Lorsqu'il ouvrit la bouche pour répondre, une crampe d'estomac lui arracha une grimace.

L'homme l'observait.

— Tu te sens pas très bien ?

— Je me traîne un putain de mal de bide, expliqua Brad. J'arrive pas à le faire passer.

— Le stress, sans doute.

— On peut le dire.

— Et tu te soignes avec quoi ?

— Oh, comme d'hab. (Toujours prompte à dorloter son fils, la mère de Brad avait sorti le nec plus ultra de sa pharmacie. Mais l'intérêt que lui manifestait ce type était plus suspect.) Des tonnes de médocs...

— Tu devrais essayer les herbes. L'échinacée, par exemple.

Brad hocha la tête, son irritation retomba, et il comprit qu'il allait bel et bien aller chercher ce café – pire, que ça lui était égal. Voilà ce qui rendait le bonhomme si particulier. Ce n'était pas dans son physique. Il avait ce don auquel Lee aspirait – et qu'il possédait en partie – mais à la puissance mille. On faisait ce qu'il disait, sans discuter.

Brad se rendit au premier comptoir libre. Tandis que la serveuse actionnait la machine, il se retourna vers Lee, qui était en grande discussion avec le type. Sûrement au sujet de ce « Plan » à la noix – la descente en Floride, un truc comme ça. Mais il n'en avait rien à battre. Brad n'aspirait qu'à en finir avec ce rendez-vous, après quoi il irait peut-être chez Karen. Au moins, avec elle, il se sentait en terrain connu. Deux des mecs repérés plus tôt s'étaient volatilisés. Seul le jeune était resté. Il avait l'air imbu de lui-même, et prêt à vous agresser à la moindre occasion. Brad se demanda en quoi Lee et lui pouvaient bien aider ces gens-là, sans trouver de réponse crédible – à croire qu'ils cherchaient surtout de la chair à canon. De bons petits soldats pour exécuter les deals à risque, comme celui de l'autre soir. En fait, le milieu de la drogue ressemblait en tout point à celui du cinéma, que le père de Brad – avocat dans le show-biz – lui avait appris à démystifier. Les masses semblent persuadées que la gloire est à portée de main, qu'il suffit de saisir l'aubaine, de montrer qu'on en veut – et toutes ces salades rabâchées dans les talk-shows. Mais ces deux industries sont d'énormes prédatrices, qui nous considèrent comme du menu fretin. Avec un peu de chance, Lee finirait par le comprendre. Brad avait beau le mettre en garde, Lee n'entendait que les phrases nées de son propre esprit.

Brad reprit le chemin de la table, en veillant à ne pas renverser le gobelet de café.

— C'est toi qui décides, disait Lee à Paul.

— Tu vois le grand magasin de sport, au niveau 2, près de l'escalator ? Je me rappelle plus son nom.

Lee s'en souvenait. L'équipe y avait acheté beaucoup de matos au fil des ans.

— Serious About Sport, répondit-il. Je connais bien.

Paul prit le café des mains de Brad.

— C'est ça, le Serious. Il y a des rangées de sacs sur le mur latéral. Accroche-le là. Vers le fond, là où personne ne le remarquera. Un de nos gars passera le cueillir d'ici peu.

Brad fronça les sourcils :

— C'est pas un peu risqué ?

— C'est un matin de semaine, et un sac de la saison dernière. Les princes de la Valley comme vous n'achèteraient jamais ça, pas vrai ?

Il fit un clin d'œil et porta le gobelet à sa bouche. Il le vida d'un trait gracieux. Brad cacha sa stupeur. Il avait vu la machine cracher un jus bouillant, à la limite de fondre le plastique.

Puis le type se figea, le récipient au bord des lèvres. Il fixait quelque chose. Lee et Brad se retournèrent.

Au mur du Branigan's Irish Bar était fixé un grand écran plat. Le pub n'ouvrait que dans plusieurs heures, mais le téléviseur fonctionnait afin que personne ne perde une miette de réclame. Réglé sur CNN, sans le son, il montrait un essaim de flics regroupé sur... un parking d'hôtel – oui, on apercevait une pancarte Holiday Inn. Un correspondant local parlait dans son micro, la mine grave. Il semblait y avoir un mort – concept que Brad trouvait moins abstrait qu'autrefois.

Lee regarda Paul.

— On se voit plus tard, hein ?

L'homme opina, obnubilé par l'écran. Ils étaient congédiés. Lee crevait d'envie de lui serrer la main, de sceller par un geste leur association, mais l'attention de Paul s'était fixée ailleurs.

— À plus, fit Lee sans obtenir de réponse.

Brad suivit son ami à travers la salle, jusqu'aux escaliers mécaniques. Arrivés au deuxième niveau, ils entrèrent chez Serious About Sport. Brad se présenta au comptoir pour poser une question longue et ardue sur les arcanes du snowboard, et les vendeurs furent ravis d'étaler leurs compétences de fumistes.

Ils lui montraient une troisième planche lorsque Lee reparut, sans le sac.

— On doit y aller, décréta-t-il.

Brad regarda les vendeurs en haussant les épaules, et ils quittèrent le magasin.

— C'est quoi, ces méthodes de barbouzes ? s'énerva Brad. Qu'est-ce qui nous empêchait de lui remettre le sac au restau ?

— Il a ses raisons.

Qu'il n'a même pas eu à t'expliquer, supposa Brad alors qu'ils retrouvaient la chaleur du parking, puisque maintenant tu lui obéis au doigt et à l'œil. Je me demande bien ce que tu comptes gagner. Et s'il est prévu que je touche ma part. Et si oui, à combien elle se monte.

— Et maintenant ?

— Profil bas, expliqua Lee. Les flics vont recevoir un tuyau dans l'après-midi. À propos d'un deal entre Pete Voss et deux mômes du nord de l'État rencontrés dans une teuf vendredi soir, la veille de la fête chez Karen. Une transaction de mille dollars, qu'ils auront raflés en le tuant samedi soir. On a déjà piégé les mômes.

— Et pourquoi les flics croiraient-ils à cette histoire ?

— Parce qu'elle n'est pas entièrement fausse, et parce que ce tuyau leur livrera l'emplacement du corps d'Hernandez, qui était le complice de Pete, et qui a été lui aussi tué à l'occasion de ce deal. C'est du clés en main.

Brad n'était guère emballé.

— Si je comprends bien, la mère de Pete va apprendre que son fils a été buté en achetant de la came ?

— Ouais. Mais c'est pas faux. Rappelle-toi.

— On peut pas inventer un accident ?

— Non. C'est trop tard. En si c'était un accident, seuls ses amis auraient pu l'enterrer, c'est-à-dire nous.

— Je ne suis pas d'accord. C'est...

— Écoute, Brad, c'est la carte qui va t'éviter la taule. On te demande pas de choisir la couleur.

— Et ça va nous coûter quoi ?

— Rien.

Brad secoua la tête.

Sitôt que Lee l'eut déposé chez lui, Brad grimpa dans sa propre voiture. Baissa la capote et fila dans le matin.

La voiture de Karen n'était pas dans l'allée des Luchs. Il avait tenté de la joindre en chemin, mais soit ça sonnait occupé, soit il tombait sur la messagerie. Il songea à faire demi-tour, mais c'eût été malpoli, et puis la BM était peut-être chez le garagiste.

Mme Luchs vint lui ouvrir. Elle semblait un peu tendue.

— Bonjour, Bradley.

— Salut, madame Luchs. Votre fille est dans les parages ?

— Elle est partie à la pharmacie. Mais elle t'a laissé un mot.

Il la remercia et regagna sa voiture pour prendre connaissance du message. C'était une feuille blanche pliée en quatre, dans une enveloppe fermée. Où figuraient ces mots :

Brad,

Maman t'a sûrement dit que j'allais faire une course mais il n'en est rien. Je lui ai raconté ça parce que je voulais me sauver. J'ai beaucoup réfléchi, et il faut qu'on parle. Je sais ce que tu m'as dit, mais je sais aussi ce que j'ai entendu. Mme Voss n'arrête pas d'appeler ma mère pour lui répéter que Pete a été vu pour la dernière fois chez nous, comme si ma mère y était pour quelque chose, et ça commence à l'énerver. Et puis, si quelqu'un sait quoi que ce soit, ce n'est pas bien de le cacher à la police. Or je pense que Lee et toi savez où Pete est allé cette nuit-là. Pete mérite qu'on retrouve ses assassins. J'ai écouté ma conscience, et je crois sincèrement devoir parler à la police. Discutons-en avant.

Je me rends dans un endroit paisible pour méditer. Je t'en prie, viens m'y retrouver.

Je t'aime,

Karen

Sa lecture terminée, Brad resta tétanisé pendant dix secondes, à regarder le texte se diluer sous ses yeux. Dans une demi-transe, il saisit son téléphone et réessaya de la joindre. Ce n'était plus occupé, mais après quelques sonneries, il tomba sur la boîte vocale.

Il jeta le portable sur le siège et redescendit l'allée à quatre-vingts kilomètres/heure.

Il la chercha. Et la chercha encore. Cramponné à son volant. Priant des dieux auxquels il ne croyait pas.

Luttant contre les larmes.

Au bout d'une heure et demie, son esprit s'embourbait dans une sorte de purgatoire diffus. Le soleil et la gamberge lui comprimaient le crâne. Il avait ratissé tous les lieux qui lui venaient à l'esprit. Tous. Chaque centre commercial, sauf celui de Belle Isle car il savait que Karen n'y était pas. Il avait gravi la colline de Santa Monica, où elle aimait se promener avec ses copines, il avait parcouru de long en large la promenade de la 3e Rue, et même poussé jusqu'à la jetée puisqu'ils y avaient dîné un soir. Il avait téléphoné chez lui pour s'assurer qu'elle ne l'y attendait pas, à l'intérieur ou assise sur le perron. Il avait appelé tous les amis de Karen dont il possédait le numéro, et vu qu'il fournissait la plupart d'entre eux, ça faisait pas mal de monde. Mais personne ne savait où elle était passée. Deux ou trois lui demandèrent s'il avait des pilules en stock, mais c'était foutu, maintenant.

Et Karen ne décrochait toujours pas.

Que faire ? Il reprenait le chemin de la Valley via Universal City, faute de mieux. En se concentrant, en se concentrant vraiment très fort, il finit par se convaincre qu'ils s'étaient croisés un peu plus tôt, qu'elle avait atteint le parking de Belle Isle au moment même où Lee et lui le quittaient, et qu'elle était restée là-bas à l'attendre. Ils y avaient passé tellement de temps, après tout. C'était peut-être leur lieu fétiche. Peut-être bien. Comment savoir ? Ce sont toujours les filles qui décident de ce genre de choses.

Le trafic était chargé, mais il ne baissa pas les bras. Et si elle n'était pas à Belle Isle... il donnait sa langue au chat.

Avaient-ils jamais, l'un ou l'autre, écrit « je t'aime » auparavant ? Couché l'aveu sur le papier ? Il en doutait. Après avoir retourné dans tous les sens le contenu de la lettre, il était convaincu que Karen tenait vraiment à lui parler avant de trouver les flics. Il la connaissait et lui faisait confiance. Quelle importance, finalement, qu'elle soit sortie avec Lee ? Ça ne l'avait jamais gêné, en fait, même si on n'aime jamais apprendre que d'autres vous ont précédé. L'important, c'était qu'ils s'aimaient.

Ils auraient une grande conversation, il lui montrerait qu'il n'y avait rien de méchant, puis en fin de journée les flics recevraient le scénario bidon, et tout rentrerait dans l'ordre. Brad s'en fichait, à ce stade, qu'on salisse l'image de Pete. C'était un dealer, oui ou non ? Il l'avait payé, certes très cher, mais à quoi bon poursuivre les survivants ? Brad allait en ressortir plus vieux, plus triste et plus sage, mais il s'en accommoderait. Le destin lui disait qu'il était grand temps de renoncer au commerce de substances illicites pour trouver un vrai job. Et il pourrait compter sur le soutien de Karen. Tout irait bien. C'était jouable.

À condition qu'il la trouve avant qu'elle alerte les keufs.

Des bouchons se formaient déjà. Pour ne pas disjoncter, Brad examina toutes les ramifications du problème, pour la trente-cinquième fois environ. À l'évidence, Karen n'avait rien dit à sa mère, sans quoi l'accueil de Mme Luchs aurait été fort différent. Karen avait beaucoup d'amis, mais sur un sujet aussi grave, elle aurait sûrement tenu sa langue : après tout, n'était-ce pas Brad, aujourd'hui, son meilleur ami ? Elle avait peut-être coupé son téléphone pour réfléchir en paix, non ?

Non. L'idée lui semblait absurde depuis déjà trois tours. Si Karen avait souhaité lui parler, elle aurait laissé son mobile branché.

Plus de batterie, alors ? Peu probable.

Peut-être qu'elle se trouve dans un secteur sans réseau.

Mais oui, pourquoi n'y avait-il pas pensé plus tôt ? Elle avait pu gagner les montagnes de Santa Monica, et les parcs nationaux de Malibu ou de Topanga. Elle aimait ces coins-là. Depuis toujours. Que disait la lettre, déjà ?

Un endroit paisible...

Putain. Et lui qui empruntait la direction opposée !

Se dressant au-dessus du pare-brise, Brad vit que la circulation reprenait un peu plus loin. Dilemme : rejoindre la Valley jusqu'au centre commercial – un choix qui prenait valeur de talisman, notamment parce qu'il permettrait de vérifier si le sac avait bien été cueilli au magasin de sport, car Brad n'aimait pas savoir l'objet suspendu au grand jour, même s'il ne l'avait jamais touché et que le sac ne pouvait pas le trahir – ou suivre sa nouvelle intuition ?

De toute manière, il n'y avait pas de sortie avant le prochain pâté de maisons. On avait tout le temps d'aviser.

Le trafic se déporta peu à peu sur la gauche, et Brad constata qu'on avait fermé la file de droite. Cool. Une fois passé le goulot, on pourrait remettre les gaz.

Montagnes ? Belle Isle ? Montagnes ? Belle Isle ? L'option centre commercial semblait la plus judicieuse : il pourrait toujours rebrousser chemin s'il faisait chou blanc.

Il distinguait à présent la ronde paresseuse d'un gyrophare. Une bagnole de flics. Son cœur avait tellement l'habitude de s'emballer devant cette image qu'il mit un moment à comprendre qu'il y avait deux voitures arrêtées sur la file de droite. Ainsi qu'une ambulance.

Brad parvint à hauteur de l'accident, le dépassa lentement tout en écarquillant les yeux.

Une voiture s'était emplafonnée contre un poteau. Heurtée par l'arrière. L'avant était déchiqueté, le pare-brise pulvérisé, les débris barbouillés de rouge foncé.

Il s'agissait d'une BMW flambant neuve. Bleu électrique.

Derrière lui, les avertisseurs hurlaient, mais le pied de Brad avait glissé de l'accélérateur. Il n'allait nulle part.

Pas après avoir reconnu la BMW de Karen.

CHAPITRE 21

Elle se trouve dans une pièce où l'air est opaque et doux. À l'étage d'une maison de bois, pleine de poussière. La porte ouverte révèle un couloir. Un rai de lumière vaporeux s'échappe par une fenêtre sale voilée d'un store. On y a calé une planche étroite, brisée à son extrémité comme par un coup violent.

On a peint en blanc, comme les murs, la porte de la chambre. Avec le temps, et dans cet éclairage précaire, ces murs revêtent la grisaille d'une tempête couvant en mer : encore lointaine et inexorable. Malgré sa taille, la pièce ne semble guère exister par elle-même. Comme si elle n'était qu'une division arbitraire d'un espace plus vaste ; un espace s'étirant au-delà des limites du lieu, sans toutefois atteindre les arbres que l'on devine alentour. Il s'en échappe parfois des bruits – ululements, grognements, bruissements –, mais si faibles et si distants qu'ils n'ont plus rien de réel et ne prouvent en aucune façon l'existence d'un monde extérieur.

Le sol est tapissé d'une vieille moquette poussiéreuse, jonchée d'éléments hétéroclites. Tessons de verre, bris de plâtre, la moitié d'un miroir et quelques feuilles d'arbres dont la présence intrigue, puisque les deux grandes fenêtres de la façade sont fermées et leurs vitres intactes. Un abat-jour repose au pied de la cloison. Sévèrement passés, ses tons violets des années 1970 devaient à l'origine se marier avec les fines rayures de la moquette, tout aussi délavées.

Une chaise de bois trône en solitaire. Disposée en travers, tournant le dos à la porte. Des années d'usage ont écaillé la peinture vert foncé et ce siège n'aurait rien d'exceptionnel s'il n'était pas le seul objet entier de la pièce, en dehors du cadre de lit repoussé dans un coin. La chaise n'occupe pas le centre exact ; il semble qu'on ait voulu laisser davantage de place à l'avant qu'à l'arrière. Est-on venu s'y asseoir pour regarder quelque spectacle, ou bien fut-on forcé de regarder ?

Puis elle s'aperçoit qu'il y a bel et bien quelqu'un sur ce siège.

Une femme. Le visage tourné. Tordue dans une position d'aspect inconfortable : dos droit, genoux pliés. Nina contourne la chaise à pas de loup, pour identifier l'inconnue.

Elle n'est que légèrement surprise de se reconnaître elle-même.

Elle est un peu plus mince qu'elle ne le pensait, mais surtout livide et fatiguée. Ses yeux sont ouverts et elle semble fixer un coin de moquette. Au bout d'un moment, son œil droit semble frémir. Le mouvement gagne en force et en détermination, jusqu'à devenir un battement.

Voilà. Elle cherche à cligner de l'œil, à dire : « Bonjour, je sais que tu es là et ça ne me dérange pas. » Ce qui, Nina le sait, ne signifie pas qu'elle va bien.

Puis elle est revenue en elle-même, dans son corps, sur la chaise.

L'atmosphère devient plus sombre, plus dense. Elle ne peut pas bouger et ses pieds sont glacés. Son poids lui paraît insoutenable, les crampes la taraudent. Elle ne regarde pas le coin de la moquette, en fait, mais le bas du mur. Il y a quelque chose qui remue. Elle pense à un oiseau, avant de reconnaître une main. Elle étire et replie ses doigts comme pour attraper un objet – ou pour comprendre sa position, attachée au mur à une vingtaine de centimètres de hauteur. Sa main s'ouvre, se referme.

Puis elle s'immobilise.

Et pivote légèrement, telle une bête tendant l'oreille.

Nina aussi a entendu le bruit. Une porte qui s'ouvre et se referme. Épaisse, apparemment. La porte d'entrée.

On entre.

Les pas sont lourds. Ils résonnent dans la maison, et là encore c'est un son curieux. Chacun des pas achève de la réveiller, en même temps qu'il lui ôte la vue – les images précédentes, la pièce elle-même ne sont plus que des fantômes sur une rétine troublée. Elle voit une dernière fois bouger la main du mur, puis c'est le noir.

Dans son effort pour séparer le sensé de l'absurde, Nina ne peut établir que ceci :

Ce qu'elle observait n'était qu'un souvenir, l'instantané d'un endroit oublié. Soit elle est quasi aveugle, soit elle a les yeux bandés. Elle est attachée. Pieds nus.

Et elle se sent très mal en point.

Elle essaie de respirer normalement. Il est déjà tout près. Cela indique que la pièce n'est pas si grande que ça ou bien qu'on l'a transportée ailleurs. S'est-elle encore évanouie ?

Il se tait. Il ne tient pas à lui parler. Chez certains criminels, la voix d'une victime est un pénible rappel à la réalité.

— Je suis réveillée, dit-elle.

Du moins essaie-t-elle de le dire. Le résultat est une bouillie de syllabes. Elle réessaie deux fois, jusqu'à former des mots intelligibles.

Il ne répond rien.

Elle suppose que c'est bien un homme. Il n'émane pas d'un homme, même propre, la même odeur que d'une femme, mais Nina n'est pas certaine de son jugement tant l'odeur de poussière et d'essence est forte. Elle sent les mains du type se promener sur elle, mais il ne fait que vérifier ses liens. Lorsqu'il resserre le nœud derrière sa nuque, elle est soulagée : on lui a bandé les yeux – c'est toujours mieux que d'être aveugle. Derrière le tissu, le monde n'apparaît pas entièrement noir, mais il ne vaut pas mieux que des ombres à minuit, que la suie la plus sombre sur un fond de jais. L'homme s'attarde sur l'intérieur du coude gauche. Il la maintient fermement, sans doute entre le pouce et l'index. Il applique une pression énorme. Ce doit être un type costaud, puissant. Elle n'a aucun souvenir de son visage. Il a surgi dans la chambre d'hôtel comme une lame de fond sur une cabine de plage. À peine avait-elle perçu cette présence étrangère qu'elle a perdu connaissance. Elle ignore ce qui s'est passé ensuite, mais elle est lucide : à en juger par sa présence ici, ce n'est guère positif. Sa seule certitude, pour l'heure, c'est d'être encore en vie.

— Je m'appelle Nina Baynam, lance-t-elle.

Il semble amorcer un recul. Elle sent qu'il la regarde. La position de Nina est inconfortable : allongée sur le dos, les mains nouées derrière la nuque. Ses cuisses semblent pointer vers le sol, avant que ne fléchissent les genoux.

— J'ai froid aux pieds.

Pas de réponse, alors elle décide de se taire. Elle pense avoir reçu un coup à la tête, après quoi on l'aurait droguée. Elle n'a jamais testé le Rohypnol et ne connaît pas bien ses effets, mais le produit qu'on lui a administré doit être autrement puissant. Des bulles d'étrangeté continuent d'affleurer. Elle ne parvient pas à chasser l'impression d'être dans la chambre du haut, celle de la chaise, bien que tout indique le contraire. Peut-être était-ce avant. Ou peut-être jamais.

Plutôt que de parler dans le vide, autant se faire oublier, afin de dissuader son geôlier de la droguer à nouveau.

L'homme semble s'éloigner de quelques pas, afin de s'asseoir – elle entend une matière gémir sous son poids. S'ensuit un moment de silence, qui souligne les bruits du dehors. Ces mêmes bruits qu'elle ne savait interpréter en reprenant connaissance et qu'elle croyait issus de la forêt. Faute d'être fixée sur leur vraie nature, elle n'essaie pas de les atteindre. L'heure n'est pas aux théories hasardeuses, ni aux constructions illusoires. Elle n'a pas le droit à l'erreur.

Soudain jaillit un autre bruit, dont le volume la choque.

La sonnerie d'un téléphone portable, d'un prosaïsme détonnant. Nina connaît cette mélodie, un air préprogrammé, comme tous ceux qui sont entrés dans le paysage sonore de la vie moderne.

Le bruit stridule dans le vide, puis se tait.

Peu après, Nina entend le siège se distendre lorsque l'homme se relève. Elle se raidit. Se demande finalement s'il ne faudrait pas lui parler, tenter de le raisonner. Pas de l'implorer. Pas encore.

Il s'approche. Elle l'entend respirer.

— Ouvre la bouche.

Une voix basse, calme. Assez grave. Sans âge. Celle d'un homme, assurément.

Elle n'a aucune envie d'ouvrir la bouche, et serre les lèvres. Une tentative désespérée pour conjurer son impuissance, bien sûr, mais elle s'en moque. Il ne lui reste que ça.

— Ouvre.

Il lui fait peur. Mais ce qu'elle redoute par-dessus tout, c'est sa réaction si elle persiste à lui tenir tête. Il arrivera de toute façon à lui ouvrir la bouche – le premier marteau venu fera très bien l'affaire.

Alors elle s'exécute.

On y introduit un corps sec, légèrement spongieux. On le tend autour de sa tête, on le noue dans sa nuque. Il s'agit d'un bâillon. Elle déglutit un coup, sans salive.

Dix minutes passent et il s'en va. Elle l'entend ouvrir et refermer la porte. Là encore, le son est particulier.

Elle s'en veut moins que tout à l'heure. Elle aurait peut-être dû simuler l'inconscience, mais ça n'aurait pas empêché l'homme de prendre ses précautions. Le vrai mystère, c'est qu'il ne l'ait pas fait plus tôt. Soit il pensait que la drogue agirait plus longtemps – ce qui serait bon signe –, soit il a du métier et sait qu'une personne droguée risque de s'étouffer avec la bouche encombrée. Ce qui serait mauvais signe.

Dans un cas comme dans l'autre, rien ne prouve qu'elle ait provoqué le bâillon. Excellent. Un point pour elle. Que sait-elle d'autre, à part ça ?

Elle sait que Reidel a toutes les chances d'être blessé, voire mort. C'est mauvais. Elle sait que son ravisseur a pu entrer dans l'hôtel et ressortir avec elle sans que personne l'arrête. Ça aussi, c'est mauvais.

Elle imagine à présent une chronologie virtuelle. Disons qu'elle est restée évanouie pendant deux heures, même si cela lui a paru plus long. À mesure qu'elle recouvre ses facultés, se succèdent des bribes d'un temps passé. De choses qui ne furent pas des rêves mais des souvenirs vécus, dont celui, récurrent, de cette nuit où un homme s'en est pris à elle, un an plus tôt, dans les montagnes du Montana, et a failli la tuer. Elle chasse l'événement de sa tête, mais n'en sait pas moins qu'elle est ici depuis au moins deux heures. Et personne n'est venu la sauver. C'est mauvais. D'autant plus mauvais que l'agresseur a eu ainsi tout loisir de fuir le lieu du crime.

Ça suffit comme ça. C'est mauvais sur toute la ligne. Sauf que...

Ward n'était pas dans la chambre au moment de l'enlèvement, ce qui laisse à penser qu'il est sain et sauf. Un bon point.

À moins...

Une violente douleur lui tord l'estomac. Si le rendez-vous de Ward n'était qu'un leurre, conçu pour les séparer, la colonne des mauvais points peut enfler jusqu'au ciel. Et si ce n'est pas l'œuvre du tueur de Thornton, comme elle l'a cru jusqu'ici, il peut s'agir d'un coup des Hommes de Paille.

Auquel cas Ward est peut-être...

Mort dans une chambre, étalé dans son sang. Mort dans une ruelle, à moitié couvert de détritus. Avachi dans une voiture, la cervelle répandue sur la vitre, le teint livide et cireux.

Non.

Non. Elle contracte chaque muscle de son corps, concentrant toute son énergie dans une convulsion totale. Ses entraves ne bougent pas d'un pouce, mais la douleur est telle que son esprit quitte le puits où il menaçait de sombrer.

Elle se résout à ne plus bouger, le temps de se vider la tête. Ruminer le malheur ne l'a jamais corrigé. L'occulter non plus.

Il faut juste essayer de penser à autre chose.

Elle s'y emploie de son mieux, avant de faire une découverte. Une présence dans sa tête. Une chose venue se nicher en elle, une émotion qu'elle a combattue presque toute sa vie. Pour l'instant la chose n'est qu'une novice sournoise, qui sait qu'elle a du pain sur la planche, des sillons à creuser. Mais elle est là.

Elle essaie de respirer profondément, calmement. Ça l'aide, mais à peine. Elle doit accepter cette chose comme un fait établi.

Elle a peur.

Elle a très, très peur.

CHAPITRE 22

Je fumais clope sur clope, garé sur le parking du Mayflower. En regagnant le centre-ville j'avais revu l'Holiday Inn, qui fourmillait déjà d'hommes et de femmes en coupe-vent du FBI. Il y avait pléthore de bagnoles de flics, et les journalistes seraient bientôt assez nombreux pour couvrir les JO. La crise était reprise en main par une corporation que je ne comprenais pas et dont je me méfiais. Je n'aurais pu approcher Monroe, même si je l'avais voulu. Au cas où, j'avais toujours son numéro. Mais pour l'heure j'étais seul avec moi-même, avachi dans une caisse sur le parking d'un bar pourri.

J'avais tenté de grappiller un peu de sommeil aux premières heures du jour. Ça ressemblait à une trahison, mais je n'avais nulle part où exercer ma conscience, et mon esprit se grippait comme un avion à court d'essence. Les quarante minutes de repos ne lui furent pas d'un grand secours. Il ressassa les mêmes pistes, comme si la répétition était un facteur de réconfort. Mais quel réconfort espérer lorsque les questions sont : « Ai-je facilité les choses en rencardant Unger en dehors de Thornton, quand j'aurais pu rester plus près de Nina ? », « Quel genre de type envoie-t-on accomplir ce genre de boulot ? » et « Que risque-t-il de faire ensuite ? » Le pire, c'était de savoir que j'aurais dû retourner vers Nina : après notre prise de bec en face du commissariat, j'aurais dû revenir sur mes pas pour l'embrasser, lui dire au revoir dignement. C'était l'affaire de dix mètres. Plus maintenant.

Bien entendu, j'avais tenté de joindre Unger. Et dès que j'en aurais l'occasion, j'enverrais un mail à ce connard. Mais hormis le frisson de le menacer à distance, ce serait aussi efficace que d'aboyer en direction du ciel.

— Mon Dieu... Vous allez bien ?

Je faillis me déchirer le cou en me tournant vers la vitre embuée. Une silhouette était plantée là. Il me fallut ouvrir la portière pour l'identifier.

Hazel me fixait d'un air dubitatif.

— J'ai connu des jours meilleurs, admis-je.

— Il y a un rapport avec ce qui est arrivé à l'Holiday Inn ?

Inutile de répondre : mon visage parlait pour moi.

— J'ai les clés, ce matin. Venez donc à l'intérieur.

Je me suis extirpé de la voiture pour la suivre.

En entrant, j'ai surpris mon reflet dans la glace murale et compris l'étonnement de Hazel. Je faisais peine à voir. Alors j'ai filé aux toilettes me laver le visage et les mains, à l'eau froide pour me fouetter les neurones. Quant au sang séché sur mes fringues, il n'y avait rien à faire. J'évitais de regarder le miroir, craignant de ne pas me reconnaître en monstre.

À mon retour, un café m'attendait.

— J'ai mis plein de sucre, dit Hazel. Je vous conseille de le boire comme ça, même si vous le préférez autrement.

C'était parfait. Je sentis affluer l'arôme et la chaleur jusque dans ma poitrine. Je repris du poil de la bête.

— C'était votre copine ?

— *C'est* ma copine.

Un regard sceptique.

— Aucune idée de l'endroit où elle peut se trouver ?

— Non.

— C'est un coup du type qui a tué les deux gus d'ici ?

— Je pense pas.

Elle me jaugea quelques instants.

— Vous n'êtes pas flic, n'est-ce pas ? Ni du FBI.

— Non.

Elle fronça les sourcils tout en fixant un point par-dessus mon épaule.

— Lloyd n'arrive pas avant midi, s'étonna-t-elle. Et puis ce n'est pas son camion.

Je me retournai. Une voiture noire s'était engagée sur le parking. Elle décrivit un lent arc de cercle pour s'arrêter à l'autre bout, près de ma propre bagnole.

Je sortis mon flingue. Vérifiai le chargeur.

— Restez ici, dis-je. Et n'approchez pas des fenêtres.

Je regagnai le parking, la main droite baissée contre ma cuisse, un peu en retrait. Je pris une trajectoire légèrement courbée, pour m'approcher du véhicule par l'arrière, histoire de compliquer la tâche d'un éventuel tireur.

Je m'arrêtai à cinq mètres.

Le moteur se tut. La portière s'ouvrit et un homme descendit. Les cheveux courts, les joues creuses, le regard affûté.

C'était John Zandt.

— Salut, Ward. T'as une mine atroce.

— S'il n'y avait que ça. (Je m'avançai vers lui, la main en avant, sans savoir s'il accepterait de la prendre.) Ça fait du bien de te revoir, John.

Imperceptible hochement de tête. Il me serra la pogne.

— Je pensais pas dire ça un jour, mais moi aussi je suis content de te revoir.

Nous avons pris une table vers le fond du bar. Hazel nous a servi du café et proposé de nous préparer un casse-croûte, mais l'appétit me semblait perdu à jamais.

— Explique, fit John.

Je lui racontai tout ce qui me venait à l'esprit, de la visite inopinée de Monroe à Sheffer jusqu'à l'enquête de Nina. Je lui parlai d'Unger, de ses efforts apparents pour nous contacter, Bobby et moi, et de notre rencontre dans un troquet d'Owensville. Je lui décrivis l'expression de Julia Gulicks derrière l'œilleton de sa cellule, l'état dans lequel j'avais trouvé Reidel, les deux cadavres de la forêt de Thornton.

Il écouta, les yeux baissés, les mains jointes sur la table. Quand j'eus fini, il resta silencieux, puis me considéra :

— As-tu songé qu'Unger n'était peut-être pas mêlé à cette histoire ?

— Pas encore. Mais je ne demande qu'à le croire.

— Disons que je ne vois pas pourquoi il t'aurait confié tout ça s'il bossait pour les Hommes de Paille.

— Pour gagner ma confiance.

— Mais pourquoi te livrer des infos dont tu ne disposais pas ? La technique du spam semble réaliste. D'ailleurs, je pense qu'elle sert également de filet de pêche, en partant du principe qu'un zozo assez stupide pour y répondre – attiré par l'appât du gain – sera assez con pour retrouver un type dans une impasse sombre sans en informer personne. Bref, j'ai plutôt l'impression qu'Unger te filait des biscuits pour te délier la langue. S'il comptait te descendre, il t'aurait abattu là, en plein bar, sans se soucier des conséquences. Et puis, il ne t'a pas fait mauvaise impression ?

— Il a l'air réglo. Un défenseur acharné de la CIA, du reste, même si là encore il pouvait jouer la comédie. Cerner les personnalités n'est pas une science exacte.

— Et quand t'es-tu gouré pour la dernière fois ?

Je réfléchis.

— M'en souviens pas.

— Tu vois. Pour ce que j'en sais, t'as plutôt du flair.

— Peut-être, même si je m'interroge encore sur toi. Comment tu vois les choses ?

— Le seul élément qui puisse incriminer Unger, c'est le rapt de Nina. Mais ça ne prouve rien : le kidnappeur vous observait peut-être depuis un moment, et il aura profité de ton absence. Si ça se trouve, ce n'est qu'une coïncidence. Vu ses méthodes, ta présence ne l'aurait pas nécessairement freiné. Nous avons affaire à un type entraîné et dingue. Si ça se trouve, il se serait fait une joie de te massacrer comme l'autre.

— Ce qui m'amène à l'élément que je n'ai pas encore mentionné : Paul s'est évadé.

Quelle réaction attendais-je ? Rage, perplexité, départ en trombe ? John se contenta de secouer la tête.

— Je pensais que tu serais aussi furax que moi, confiai-je. Mais tu dois te dire que ça te laisse une nouvelle chance de le buter, pas vrai ?

— Dans le fond, t'es plutôt un mec brillant.

— Et ça ne t'inquiète pas qu'un type comme Paul puisse se faire la belle au nez et à la barbe des fédés ?

— Ça ne me surprend pas du tout. Ça fait un an que je me renseigne sur ces gens. Je sais des choses que tu ne voudrais même pas croire.

— Épargne-moi tes délires, John. Je suis pas d'humeur. Si t'as découvert que les Hommes de Paille sont des extraterrestres en provenance de la ceinture d'Orion, garde ça pour toi. Ces tarés me font bien assez flipper sans que tu leur colles des tentacules.

— Ils sont tout ce qu'il y a de plus humain, Ward. C'est ça, le pire. T'as essayé d'appeler Unger depuis hier soir ?

— Évidemment.

— Il fait le mort ?

— Messagerie.

— T'as envoyé un mail ?

Je secouai la tête.

— Fais-le, Ward.

— Pourquoi ?

— T'as rien à perdre si c'est un des leurs. Et beaucoup à gagner si ce n'est pas le cas.

— Il n'avait pas l'air de savoir grand-chose, John.

— Il connaît leur nom, ce qui le place d'ores et déjà en quatrième position des gens les plus informés de notre cercle, voire de tous les États-Unis. Ça le rend utile – surtout si Monroe n'est pas disposé à leur filer le train. Unger est déjà sur l'affaire, et je parie qu'il en sait davantage. En outre, il a accès à l'appareil d'État. Et si c'est effectivement un méchant, alors on aura l'air encore plus bêtes en l'appelant.

— D'accord. On ira au Starbucks pour utiliser la wi-fi.

— Y a un Starbucks dans ce trou ?

— Y en a partout, John. T'étais où, pendant tout ce temps ? Ils construisent de nouvelles villes rien que pour y installer des cafés. Allez viens, assez traîné.

Avant de sortir, je suis allé remercier Hazel. J'ai voulu la payer, mais elle a refusé. Elle m'a souhaité bonne chance, et promis de prier pour mon amie. Je me suis dit que si son Lloyd de patron avait eu la moindre jugeote, c'est Hazel qu'il aurait draguée.

Nous avons pris la voiture de Zandt, après que j'eus contacté Monroe et compris à ses réponses qu'il n'avait pas progressé d'un poil. L'agent spécial me parut bouleversé, à la fois fumasse et vanné, et je sentais bien qu'il remuait ciel et terre.

J'ai guidé John jusqu'à la vieille ville, et nous nous sommes arrêtés juste en face du café. C'était plein d'individus aux allures de journalistes. Un vieux costaud aux cheveux gris était assis près de la fenêtre, qui regardait la rue avec des yeux las. Par chance, le signal wi-fi était assez puissant pour que nous accédions au Web depuis la voiture. J'envoyai un mail à Unger, où j'exposais la situation et lui demandais de rappeler dare-dare. Je n'excluais pas qu'il eût joué un rôle dans l'enlèvement de Nina, mais, comme le disait John, on n'avait rien à perdre. Quand j'eus terminé, je vis Zandt se renfrogner devant son propre écran.

— Des ennuis ?

— Je sais pas, répondit-il. Tu pourrais vérifier un truc pour moi ?

Il me dicta une adresse Internet. S'ouvrit une page mal foutue, avec un nombre hallucinant de polices, de tailles et de couleurs. Elle était censée contenir de nombreuses images, mais les cases demeuraient vides.

— Pas terrible, ton truc. C'est quoi exactement ? Et c'est qui, Oz Turner ?

— L'absence de photos, là, c'est dû à un signal trop faible ?

Je fis quelques manips avant de secouer la tête.

— Non. Les espaces réservés sont inscrits dans l'HTML de la page, mais les images semblent absentes du serveur.

John referma sa machine et sortit son téléphone. Il obtint quelqu'un, demanda Oz Turner. On dut lui répondre que le type n'était pas arrivé. Il laissa un message du style « z'avez intérêt à transmettre », enjoignant à Turner de l'appeler de toute urgence. Puis il replia son mobile, le visage sombre.

— Qu'est-ce qui se passe ? m'étonnai-je.

— Rien, j'espère. On ferait mieux d'y aller.

— Aller où ? J'ai cherché, John. Crois-moi, notre homme n'est pas en train de casser la graine sur le trottoir.

— Je sais. On est coincés. On ne sait pas où il la retient – à supposer que ce soit juste un bonhomme, et non tout un groupe. On ne sait même pas par où commencer. Mais je veux vérifier un autre truc.

J'attendis qu'il m'explique, mais il la boucla. Il regardait quelque chose à travers le pare-brise. Cinquante mètres plus loin, des gosses du lycée de Thornton stagnaient sur la grande pelouse fermée par les grilles, tuant le temps entre deux cours, laissant filer les heures comme seuls les jeunes savent le faire.

— John ?

Il ne m'entendait pas, et je compris qu'il cherchait Karen, sa fille, que l'Homme Debout avait enlevée le 15 mai 2000, sa fille qu'il n'avait jamais revue entière. Elle aurait été dans sa vingtième année si on ne l'avait pas assassinée. À première vue, les gamins de ce lycée n'avaient pas plus de dix-sept ans, autrement dit Karen n'aurait pu apparaître à leurs côtés ce matin – à moins qu'on ne l'ait retenue longtemps, très longtemps à l'intérieur, pour peaufiner un devoir, entretenir un enseignant sur les costumes de la pièce de fin d'année, et prendre sous son aile, deux années durant, une camarade de classe un peu moins douée qu'elle. Pour être en tout point la jeune fille parfaite qu'elle était libre d'être, puisqu'elle n'était plus en vie.

— Ça va, John ?

— Super.

Il détourna les yeux de la route, le temps de dépasser le lycée.

Zandt m'indiqua où il voulait aller et je le dirigeai de mon mieux. En chemin il joignit le bureau du shérif de Thornton en se faisant passer pour un agent du FBI. Ayant officié à la crime de Los Angeles, il connaissait bien mieux que moi le vocabulaire et le protocole. Je n'en étais pas moins bluffé par son aisance.

— Tu uses souvent de ce stratagème, pas vrai ?

Il ne répondit rien, concentré sur son cap nord-est. Nous atteignîmes enfin la grande ligne droite qui s'enfonçait dans les bois humides, et le paysage redevint familier.

— Qu'est-ce qu'on est venus faire ?

— Un truc que tu m'as dit à propos de la seconde victime, expliqua John. Que révèle l'autopsie ?

— Je ne sais rien de plus que ce que j'ai vu cette nuit-là. Ma présence était seulement tolérée, et Nina n'a pas eu le temps de me briefer avant que n'éclate la tempête Julia Gulicks. Si ça se trouve, les analyses ne sont même pas terminées. Les légistes ont sûrement tout lâché pour examiner les restes de Reidel.

Avisant une bande de gravier sur le bas-côté, je fis stopper Zandt une centaine de mètres plus loin.

Je descendis, contemplai la forêt bourbeuse.

— C'est bien ça.

Zandt fit le tour de la voiture, ouvrit le coffre et attrapa quelque chose.

Je pivotai face à la route.

— Rien ne t'a frappé durant le trajet, John ?

— Pas vraiment. Sinon que je n'étais pas mécontent de quitter ce bled.

— On n'a croisé aucun barrage.

Je rappelai Monroe. Il n'avait ni le temps ni l'envie de papoter, mais j'exigeai qu'il m'écoute. S'il avait ordonné le bouclage de la ville, c'était raté. Il finit par me raccrocher au nez.

Zandt referma le coffre. Un long sac de toile pendait à son épaule.

— Y a quoi, là-dedans ?

— Des outils. C'est par où ?

Nous nous sommes mis en marche. Le chemin devenait vite marécageux, suite aux pluies de la veille. Me dirigeant au petit bonheur la chance, je commençais à douter lorsque j'aperçus au loin le ruban d'interdiction. Nous gagnâmes l'endroit où les planches formaient un pont vers l'îlot. L'une d'elles s'était brisée. Nous traversâmes sans incident.

John s'arrêta pour scruter l'île.

— Le tee-shirt se trouvait par là, dis-je en tendant l'index.

Je le conduisis au point exact, et le fis pivoter pour qu'il voie ce que j'avais montré aux autres, deux nuits plus tôt.

— Ça paraît assez abstrait aujourd'hui, mais sur le moment c'était parlant.

— Je te crois. Le tee-shirt était suspendu à ces branches ?

— Exactement. Orienté comme ceci.

John observa l'arrière de l'îlot. Rien de plus que des arbres gravissant une colline, même s'ils semblaient ensuite s'éclaircir.

— Il y a quoi là-bas ?

— Une petite ville, je crois.

— Les flics ont examiné toutes les issues partant de cette île ?

— J'imagine. Mais s'ils ne l'ont pas fait le soir même, tu peux être sûr qu'ils ne le feront jamais. Les effectifs sont mobilisés ailleurs. Sérieusement, John, qu'est-ce qu'on fabrique ici ? Je m'en fous de savoir comment le macchabée est arrivé sur l'île. Je veux juste trouver Nina, et j'ai l'impression d'avoir des araignées sous la peau.

— Je sais ce que tu ressens. Mais tu avais raison. La position du tee-shirt n'avait rien de fortuit. Le tueur nous disait quelque chose. Question : pourquoi ici ?

Il rejoignit l'endroit où l'on avait découvert le corps. La zone était largement défrichée, et les techniciens avaient creusé le sol par endroits, dans le vain espoir de déterminer la provenance du cadavre.

Je regardai Zandt décrocher le sac de son épaule et tirer la longue glissière.

— Alors, ta théorie ?

Il plongea la main dans le sac et produisit une pelle.

— Attends, John, si on bousille le lieu du crime, on ira en taule !

Il se mit à creuser.

Le sol était mou comme du beurre, et en l'espace de quinze minutes il fut complètement saccagé.

— Génial, dis-je. T'as remué la boue pour trouver encore plus de boue. Je me casse. C'est une perte de temps et...

— Et Nina est en danger, je sais. (Il continuait à creuser, une vraie machine.) C'est pas pour tes beaux yeux que j'ai fait tout ce chemin.

— Si ton idée est tellement géniale, pourquoi les flics n'y ont-ils pas pensé avant ?

— Parce qu'ils n'ont aucune raison pour ça.

— Mais nous, si ? Le sol avait un aspect normal, John. Personne n'avait remué la terre dernièrement.

Il se redressa, comprenant peut-être que j'étais à deux doigts de me tailler.

— C'est ainsi que fonctionnent les enquêtes, Ward. On fait ce qu'on peut, en espérant que ça nous mènera au but. Si tu veux partir, pars. Tu peux aussi rester ici à te rendre malade pour un problème sur lequel tu n'as aucune prise, ou tu peux attraper une pelle et me filer un coup de main.

Je n'en croyais pas mes oreilles.

— Tu veux dire que t'as une deuxième pelle ?

Il se remit au travail.

— Bien sûr.

— Mais pourquoi s'encombrer de deux pelles ?

— Je possède tout en double, Ward. J'ai deux pelles, deux appareils photo, deux armes à feu dans presque toutes les catégories. J'ai deux cartes routières, deux ordinateurs portables et toute une panoplie de faux papiers.

— Je te demandais la raison, pas l'inventaire.

— La raison ? Quand on est seul sur la route, on ne peut pas se permettre de manquer du nécessaire. Alors on amasse tout en double, pour compenser le fait qu'on est soi-même unique.

La voûte de ses épaules témoignait de voyages interminables dans une voiture vide, de soirées entières assis dans la cour de quelque motel minable, à l'extérieur d'une chambre silencieuse, et de sombres heures perdues à méditer. Je ne connaissais pas bien John, mais le changement était patent, comme s'il avait fait le tour de son âme et jeté tout ce qui ne le rapprochait pas de son objectif. Il devenait une sorte de patrouille à une tête, un mercenaire solitaire luttant pour sa propre cause.

— Rien ne t'obligeait à vivre en ermite.

— Et toi, comment as-tu passé les six derniers mois ? répliqua-t-il.

— Planqué. Dans un bungalow qu'on nous a prêté, près de l'endroit où on t'a vu pour la dernière fois.

— Je me doutais que ce serait un plan de ce genre. Et vous aviez vraiment besoin d'un voisin comme moi, coupable de trois homicides ?

— Je croyais que c'était deux.

— Tu te souviens de Dravecky, le promoteur immobilier ?

— Le type auquel tu t'es vendu pour retrouver la trace de Paul ?

— Je ne me suis pas vendu. C'est ce que je lui ai fait croire. Je suis revenu à lui plus tard.

— Et tu l'as tué.

— Un homme très malfaisant.

— Je ne suis pas sûr que tu sois bien placé pour en juger.

— Et moi je pense l'être. (Il se tut, leva les yeux.) Trois aussi, c'est un mensonge. Il y en a eu quatre autres depuis.

— Bon sang, John. Tu devrais carrément postuler chez les Hommes de Paille. T'es mûr, là. Sept meurtres, rien que ça ?

— Ces types *étaient* des Hommes de Paille. Quand j'ai refroidi Dravecky, j'ai raflé ses fichiers informatiques. Ces quatre gars étaient tous membres de l'organisation, et c'étaient des monstres. Oui, de vrais monstres – sanguinaires, cinglés, mais trop riches et trop puissants pour être inquiétés. Alors je m'en suis chargé. Et je n'hésiterai pas à recommencer.

— J'imagine qu'ils donneraient cher pour avoir ta peau.

— C'est bien ce que je dis : vous n'aviez pas besoin de moi dans les parages.

— On a quand même cherché à te joindre.

— Ouais, ouais. La femme qui a partagé ma vie m'a téléphoné. L'homme qui n'a pas abattu le meurtrier de ma fille m'a téléphoné. Mais certains jours, on n'est pas d'humeur à répondre à ce genre de coups de fil.

Il pivota pour reprendre sa besogne. Je respirai profondément, deux ou trois fois, et m'emparai de la deuxième pelle.

Les vingt minutes suivantes ne firent qu'ajouter au grabuge sans rien révéler d'utile. La boue était lourde et collante, de plus en plus dure à soulever.

Je levai la tête, depuis ma portion de carnage paysager, pour constater que John avait fini sa fouille.

— Y a rien ici, trancha-t-il.

Me revint alors cette excursion vers une haute plaine désolée au sud de Yakima, où John et moi avions dégotté une cabane ayant longtemps servi à entasser les morts. C'est John qui avait entrepris ce périple, sur la foi d'un tuyau que je ne voulais même pas connaître. Mais c'est grâce à moi que nous avions tenu jusqu'en haut. On pourrait appeler ça l'obstination du paresseux : poursuivre la tâche en cours dispense d'échafauder d'autres projets. Présentement, j'avais chaud malgré le froid, et le rythme des coups de pelle aidait ma tête à faire le vide.

Alors je me suis décalé de deux mètres, et j'ai entamé un nouveau trou.

Après une petite pause, John m'a donné la réplique.

— Eh, Ward, viens voir.

Cinq mètres nous séparaient, et nous creusions depuis plus d'une heure. Je le rejoignis sans attendre.

Il se trouvait à trois mètres de son point de départ, au bord d'un trou d'environ deux pieds de profondeur. Au fond stagnaient déjà trois centimètres d'eau. Mais on distinguait quelque chose d'autre.

Je me penchai pour regarder de plus près. Puis fixai Zandt.

— C'est quoi, ce truc ?

Nous avons repris les pelles, en accélérant la cadence. L'eau s'infiltrait par les côtés presque aussi vite qu'on la purgeait, mais au bout d'une poignée de minutes l'objet s'avéra de belle taille. John alla chercher dans son sac deux déplantoirs, et nous passâmes les dix minutes suivantes à genoux, à dégager une découverte dont la nature ne faisait plus de doute.

— C'est bien ce que je pense ?

— Je crois bien, répondit Zandt. Une cage thoracique.

— Nom de Dieu... Humaine ?

— On dirait.

D'après la position et la courbure des sept côtes exhumées, j'imaginai l'orientation de l'éventuel squelette, pour enfoncer mon déplantoir cinquante centimètres à gauche.

— Qu'est-ce que tu fais ?

Pour toute réponse, je continuai de creuser jusqu'à trouver l'os du bras. En poussant vers la gauche, je révélai le radius et le cubitus, brisés comme deux baguettes.

Puis plus rien.

— OK, soupirai-je. Pas de main.

Zandt comprit l'idée. Nous nous relevâmes.

— Bon sang, John. Qu'est-ce qui se passe, ici ?

— J'en sais rien. Mais t'as remarqué ? Pas d'odeur. Aucune. Et tu vois la couleur des os, leur texture ?

— Des taches brunes. Un aspect poreux. Ce qui signifie que ça croupit là depuis un moment, non ?

— Au moins dix ou douze ans. Quel âge a la suspecte des deux meurtres ? La rouquine ?

— Vingt-cinq ans.

Alors nous nous sommes tus, et nous avons fait le calcul.

CHAPITRE 23

Lee était assis dans sa cuisine. Un modèle de propreté. La raison pour laquelle son intérieur était toujours nickel – et il savait que Brad trouvait cela plutôt louche – était fort simple : il consacrait beaucoup de temps au ménage. Dès son installation, il avait compris qu'il en serait ainsi. Depuis tout petit, il avait toujours connu une ou deux bonnes à la maison, et la pire corvée qu'il eût vu sa mère accomplir était de rincer un verre à Martini, dans des cas de force majeure. Mais Lee ne voulait pas d'une bonne. Il avait vingt ans. C'eût été ridicule, et puis les produits qu'il planquait chez lui n'étaient pas à mettre sous le nez d'une fouineuse de Mexicaine, des fois qu'elle décide de monnayer l'info en échange de la bienveillance des services d'immigration.

Alors il nettoya lui-même. Se découvrit des facilités, et finit par se prendre au jeu. Ça faisait un peu tapette, sûrement, mais au moins on savait que c'était propre. Désormais, chaque fois qu'il devait méditer un problème, il lessivait. C'était son petit secret. Mais bon, qui n'en avait pas ?

La maison baignait dans un silence parfait. Comme il l'aimait. La plupart de ses amis – ainsi que leurs vieux, sans parler des petites sœurs – étaient incapables de vivre sans bruit de fond. Il fallait qu'ils écoutent la télé ou la radio. À défaut, des conversations. N'importe quoi, mais quelque chose. Lee n'était pas comme ça, de même qu'il n'avait pas besoin de se défoncer pour faire la fête. Il sniffait un rail de coke à l'occasion, pour montrer sa bonne volonté. Sinon il restait clean. Cette vie exigeait un esprit vif. Des idées claires. De la maîtrise.

Vivement qu'on lui dise que l'épisode Pete était clos. Il serait alors le roi du pétrole, et cette stupide soirée ne serait plus qu'une péripétie l'ayant rapproché des grosses huiles et – par un heureux hasard – débarrassé d'Hernandez. Il fallait fêter ça ! Il envisagea d'appeler Brad, mais son ami était drôlement ronchon ce matin.

Le problème quand on devient proche des gens, des gonzesses en particulier, c'est qu'on les met en position de nous démonter la tête. Puis c'est tout notre univers qui est ravagé.

Lee décida de coucher sur papier quelques notes en vue de sa prochaine rencontre avec Paul. Des idées pour booster le Plan, maintenant que les choses rentraient dans l'ordre. Comme il se levait pour prendre son bloc dans la moitié droite du deuxième tiroir sous les couverts – chaque chose avait sa place –, il vit une bagnole s'approcher à toute blinde.

Il reconnut la caisse.

Elle s'arrêta en dérapage. La portière s'ouvrit ; Brad descendit, hébété. C'est tout juste s'il marchait droit. Il fondit sur la porte d'entrée en criant vaguement.

Hudek alla ouvrir avant que Brad n'ait pu frapper. Il avait une mine rouge, humide, les cheveux en bataille.

— Espèce de... cria-t-il. Espèce de...

Puis il lui tomba dessus. Pétage de plomb intégral. Comme une attaque de tigre sous amphets. On l'oubliait facilement, mais Brad mesurait cinq centimètres de plus que Lee, et le dépassait en muscle de quelque cinq ou dix pour cent.

Lee tituba en arrière et tomba sur le dos. Hystérique ou pas, Brad lui cognait le visage, le cou, la poitrine en hurlant. Lee rendait les coups comme il pouvait, tout en essayant de repousser Brad avec les genoux et de s'esquiver en roulant, de s'échapper afin de reprendre le combat dans une position plus équitable – et si possible avant que Brad ne lui attrape la tête pour la fracasser contre le carrelage, ce à quoi il semblait fort résolu.

Il parvint à aligner une phrase, expirée morceau par morceau :

— Qu'est-ce qui t'arrive, putain de merde ?

Brad n'expliqua rien.

Pour finir, Lee décocha une droite assez puissante pour repousser Brad, ne serait-ce qu'une seconde, puis il lui latta le flanc, deux fois de suite, et balança un revers de bras. Le crâne de Brad heurta le mur, et Lee en profita pour remonter sur ses jambes.

Il espérait obtenir un temps mort, mais Brad bondit de plus belle.

Quel que soit le problème, c'était sérieux. Lee se retourna pour détaler dans la maison. Atteignant le séjour, il vit que les portes-fenêtres étaient restées ouvertes. Il n'avait jamais tenté d'escalader la clôture du jardin, et cela ferait baliser les voisins, qui le prenaient pour un jeune homme bien sous tous rapports. Mais bon, à eux de gérer, après tout...

Surgissant de nulle part, Brad lui sauta dans les côtes et le plaqua au milieu du salon. Les coups se remirent à pleuvoir, de moins en moins précis mais toujours aussi forts.

Lee se libéra de nouveau ; sûr de perdre au sprint dans le jardin, il changea d'objectif et se retrancha vers la cuisine, tout en continuant de gueuler sur Brad, d'exiger qu'il lui dise à quoi rimait tout ce bordel.

C'est là qu'il comprit, au regard et aux râles de Brad, qu'il n'y avait rien à négocier. Ce n'était pas une saute d'humeur : Brad cherchait vraiment à lui faire la peau.

Il obliqua vers le couloir, gagna la porte du fond et se rua dans le double garage. Il entendait Brad se carapater derrière lui, mais s'il maintenait l'effort jusqu'au meuble de rangement...

Il l'atteignit à temps. Ouvrit le tiroir, sortit le flingue.

Et virevolta pour le braquer sur la tronche de Brad.

Il crut d'abord que Brad n'allait pas s'arrêter, qu'il allait foncer droit sur l'arme. Mais il hésita.

Lee soufflait comme un bœuf.

— C'est quoi, c't'embrouille, Brad ?

Brad faisait pitié à voir. Il pleurait, dégoulinait de morve, mais il ne s'en rendait même pas compte, ou bien il s'en moquait.

— Tu l'as tuée, grinça-t-il.

— Mais de quoi tu parles ?

— Me pipeaute pas la gueule. Tu l'as tuée. Tu l'as fait tuer.

Lee maintenait l'arme fermement pointée sur le visage de Brad. Deux mètres à peine les séparaient. Si Brad bougeait, il faudrait tirer très vite.

— Écoute, Brad. Je comprends rien à ce que tu me racontes. Qui est mort ?

— Tu le sais très bien. Kar...

Le visage de Brad se chiffonna et ses paroles furent momentanément brouillées, coulées dans une longue plainte qui déboucha néanmoins sur un prénom reconnaissable.

Lee s'étonna :

— Karen ? Karen est morte ?

Alors Brad hurla :

— Bien sûr qu'elle est morte, connard ! Tu croyais peut-être qu'elle allait s'en sortir ? Ou bien ils t'ont pas dit comment ils allaient s'y prendre ?

— Écoute, Brad, il faut que tu te calmes et que tu m'expliques tout ça, parce que, là, je pige que dalle.

Brad plaqua ses mains sur ses yeux, renifla à fond. Il saignait d'une narine.

— Tu leur as dit.

— Dit quoi ?

— Que Karen avait des doutes. Qu'elle pensait qu'on savait des choses sur la mort de Pete.

Hudek ouvrit la bouche, pour la refermer aussi vite. Grillé.

— Je le savais, fit Brad. Tu l'as balancée. Tu t'es dit qu'elle allait nous foutre dedans, alors tu l'as donnée.

Lee se lécha les lèvres pour mesurer ses mots :

— Je vais être honnête avec toi, mec. Ouais, j'en ai parlé à Paul. Je lui ai exposé la situation, pendant que t'allais lui chercher son café. Je pensais qu'il fallait le mettre au courant, c'est tout. Mais je lui ai dit qu'il n'y avait pas de souci, qu'elle ne ferait rien qui puisse te nuire, et il a répondu que ça allait, que tout serait réglé d'ici ce soir et qu'on n'avait rien à craindre.

— Rien à craindre mon cul ! À quoi ça servait de fourguer aux flics une version bidon ? Si Karen leur racontait qu'on avait menti, ils nous seraient tombés dessus de toute façon. Ils nous seraient tombés dessus, et notre version se serait dégonflée dans la seconde. Tu le savais, et Paul le savait, alors tu l'as fait tuer pour l'empêcher de parler.

— C'est faux, Brad.

— Tu sais comment ils s'y sont pris ? Devant tout le monde. En pleine rue. Un mec en 4 × 4 la tamponne au beau milieu de la journée et passe son chemin. Ça lui a écrabouillé le visage. Et arraché le bras, Lee. Son putain de bras a été arraché !

— Écoute, Brad. Je te jure, j'ai rien à voir là-dedans. J'ai parlé à Paul, mais c'était juste pour l'informer. Rien ne prouve que ce soit eux, d'ailleurs. Et si c'était un accident ? T'y as pensé, au moins ?

— Me prends pas pour une buse.

— Je ne voulais pas qu'ils lui fassent du mal.

— Je te crois pas. Et je puis je m'en branle. T'as parlé. Tu l'as fait buter. Volontairement ou non, c'est toi qui l'as tuée.

— C'est pas ma faute !

— Rien n'est jamais ta faute, hein ? C'est quoi, ton problème, Lee ? Tu supportais pas qu'elle soit avec moi et pas avec toi, c'est ça ?

Lee ricana.

— Quoi ? J'en avais rien à secouer, mec.

— Menteur. Putain de menteur. Depuis que je sors avec elle, t'arrêtes pas de faire chier. De raconter que tu l'as tringlée. De faire tes petites allusions. OK, tu craignais qu'elle parle et t'as voulu t'en débarrasser pour ça, mais c'était aussi une vengeance personnelle, pas vrai ? Reconnais-le, bordel !

Lee se lâcha :

— Mais je te la laissais avec plaisir, abruti ! Cette espèce de glaçon ambulant. Même pas foutue de sucer.

Brad se terra dans le silence.

Un silence menaçant. Les larmes aussi avaient cessé.

— T'es mort, déclara-t-il d'un ton sec.

Hudek vit le corps de Brad se raidir. Les tendons de son cou saillaient comme des câbles. Si Brad déconnait maintenant, Lee serait obligé de l'abattre.

— M'oblige pas à faire ça, dit Lee tout en gardant le flingue dressé et en regrettant de ne pas avoir compté les coups tirés sur le parking. Ne m'oblige pas, putain.

— Je l'aimais, dit Brad avec un calme à donner la chair de poule. C'est un truc que tu pourras jamais comprendre parce que t'es complètement baisé de la tête. J'aimais Karen. Et ils l'ont tuée par ta faute.

— Écoute, Brad, il faut qu'on...

Mais Brad se jeta sur Lee.

Lee pressa la détente comme on le fait à la télé. Deux coups rapides, boum boum.

L'espace confiné rendit le bruit assourdissant. Pointé droit sur le visage de Brad, le flingue dévia tout au plus d'un degré.

Brad percuta Lee tel un train, le projetant contre le mur. Les deux têtes s'y fracassèrent, avant de cogner le sol, et Lee voulut aussitôt se débattre, paniqué par ce corps affalé sur le sien.

Puis il vit qu'il n'y avait pas de sang.

Et que Brad bougeait encore ; ce n'étaient pas des convulsions, mais des mouvements francs.

Ils se dégagèrent l'un et l'autre, pour se retrouver assis à deux mètres de distance, sur le béton froid.

Leurs regards se croisèrent. La droite du visage de Brad montrait des traces de projection, semblables à de la suie. Le flingue à la main, Lee faisait des yeux ronds.

Puis Brad se mit à rire.

C'était un son rentré, affreux, le bruit d'une âme qui s'envole en battant des ailes.

— Des balles à blanc, dit-il. C'étaient des putains de balles à blanc !

Lee regarda bêtement son arme.

CHAPITRE 24

— Nous ne savons pas encore qui est la seconde victime, dit Monroe. Aucune disparition n'a été signalée. On a fait circuler une photo, mais sans résultat. Le cadavre n'avait pas de papiers, ni de tatouages ou de signes distinctifs, et ses empreintes ne donnent rien. Ses vêtements ne sont d'aucune aide et le labo n'a rien trouvé quant à la précédente localisation du corps. C'est soit un voyageur de passage, soit un habitant d'une ville voisine, soit un type tombé du ciel.

— On a montré la photo au Mayflower ?

— Hier après-midi. Mais les gens ne sont pas doués pour reconnaître les morts. En tout cas, il est certain qu'il n'aura pas marqué les esprits de la même façon que Widmar.

Monroe et moi étions assis dans le hall de l'Holiday Inn. La foule du matin s'était dispersée. Les techniciens s'étaient retirés, les flics locaux étaient repartis, et s'il restait deux ou trois reporters sur le parking, la plupart considéraient l'incident de cette nuit comme de l'histoire ancienne : ils étaient partis chercher un endroit plus sympa pour poser leurs sacs et se dragouiller entre pairs dans l'attente de développements croustillants. Des fédéraux s'étaient cloîtrés dans la salle de conférence de l'hôtel, reconvertie en QG. Le seul corps de métier à occuper le terrain était le personnel, qui paraissait drôlement enjoué en cette fin d'après-midi, comme si les accueils bruyants et les sourires éclatants pouvaient conjurer le drame qui s'était produit dans ces murs. Il n'empêche que la clientèle avait déserté en masse.

Monroe avait bien voulu m'accorder dix minutes. Il paraissait encore plus schlass que moi. Il ne s'était pas changé depuis le matin, quand j'avais envoyé la réceptionniste le tirer du lit. J'avais l'impression que cela remontait à un siècle – et encore, je n'avais pas eu à me taper toutes ses emmerdes à lui. Il tenait une grande tasse de café, qu'il vidait par petits traits réguliers, et seul l'aspect robotique du geste laissait deviner qu'il n'en sentait plus le goût.

— On va la trouver, Ward. Il faut me croire.
— Vous ne savez même pas qui la détient.
— Avez-vous le moindre indice ?
— On procède maison par maison. Si Nina est en ville ou dans les environs, elle est forcément dans un bâtiment, un appartement. On la trouvera. Il faut bien que le ravisseur la garde quelque part.
— Ce n'est pas Julia Gulicks, en tout cas.
— Vous n'auriez pas dû aller la voir hier soir.
— Je devais m'assurer qu'elle était là. Et puis j'ai vu un truc.
— Quoi donc ?
— Elle devait savoir ce qui venait d'arriver. Les flics étaient tellement retournés qu'elle aurait pu le sentir à travers les murs. Elle a levé les yeux vers l'œilleton et m'a lancé un regard que j'ai été incapable d'interpréter.
— Elle est coupable de deux meurtres et elle a consacré les deux dernières semaines à découper des types. Qui sait ce qui peut se passer dans sa caboche ?
— La fouille de son appart n'a rien donné ?
— C'est en cours.
— Ce qui veut dire non. Ni cave où conserver un corps, ni morceaux de chair humaine, ni pic à glace planté dans l'escalier.
— Elle a manifestement opéré ailleurs.
— A-t-on déjà vu une femme commettre ce genre de crime ? Y a-t-il ne serait-ce qu'un seul précédent ?
— Je sais ce que pensait Nina de cette affaire. Je ne vous demande pas de reprendre le flambeau.
— Pourquoi l'avez-vous amenée ici ?
Monroe hésita, une fraction de seconde.
Je me penchai en avant.
— Je pige pas, Charles. Elle m'a expliqué, et je pige toujours pas. Vous serez peut-être plus convaincant. Vous connaissez cette fameuse affaire de Janesville qui a marqué Nina quand elle était gamine, donc vous savez que c'est un sujet qui la perturbe. Alors pourquoi choisir exprès cette enquête pour la ramener dans le grand monde ?
Monroe ouvrit la bouche, mais j'étais trop furax pour le laisser parler.
— Un monde où trois jours plus tard elle sera kidnappée à son hôtel par un buteur de flic ? Expliquez-moi, s'il vous plaît, et faites simple.

Monroe secoua la tête et, l'espace d'un instant, il me fit de la peine. Il n'avait aucun moyen de revenir en arrière. Il pouvait juste attendre pendant qu'on fouillait les maisons.

À travers la baie vitrée, il scruta le parking où s'invitait le crépuscule. La pluie avait repris. J'espérais que Nina l'entendrait, où qu'elle se trouve, et qu'elle saurait au bruit des gouttes que le temps s'écoulait toujours – et que, s'il s'en écoulait assez, j'aurais une chance de la retrouver.

— C'est la meilleure enquêtrice que j'aie jamais connue, dit-il. Je lui dois tout. Je m'occupais de la procédure et du protocole pendant qu'elle résolvait les crimes. Vous vous souvenez quand l'affaire Aileen Wuornos est partie en vrille, avec les pétitions, les documentaires et tout le bastringue ? Ce coup-ci, je voulais qu'on soit irréprochables, au cas où les médias monteraient le truc en épingle. Nina allait se montrer d'emblée sceptique face à une femme suspectée de meurtre, ce qui nous obligeait à trouver des preuves en béton.

— Et maintenant Reidel est mort, Nina a disparu et tout ça ne rime à rien.

— Non, vous ne pouvez pas...

— Il se passe des choses incompréhensibles dans ce bled. Je ne sais pas si elles ont un rapport avec Nina, mais pour l'instant c'est tout ce que j'ai.

Plongeant la main dans le sac posé au pied de ma chaise, je sortis une série de clichés en couleur réalisés en début d'après-midi avec un numérique.

Je lui mis le premier sous les yeux.

— Vous reconnaissez ?

Il fronça les sourcils.

— On dirait le deuxième lieu du crime.

— Exact. (Je lui glissai la photo suivante.) La même, vue de l'autre bout de l'île, à l'endroit du tee-shirt.

— Je ne vois rien qui...

Troisième image.

— Voici à quoi ça ressemble maintenant.

Il fixa le cliché.

— Nom de Dieu. Mais qu'est-ce qui s'est passé ?

— J'ai sorti la pelle.

Ses yeux étaient exorbités.

— Vous me faites marcher. C'est pris ailleurs. Vous n'avez pas pu faire ça.

— Je vous jure que si. Et c'est encore pire en vrai.

— Je peux pas croire que vous ayez fait ça. Putain !

— Il fallait bien que quelqu'un se dévoue. Ce tee-shirt avait forcément une fonction. Mais vous ne regardiez que la surface des choses. (J'abattis une nouvelle photo.) Et là, vous voyez quoi ?

— De la boue, Ward, dit-il d'une voix morfondue. Et un lieu du crime massacré.

— Encore un effort.

À contrecœur, il se pencha sur le document pour l'examiner.

— Il y a quelque chose dans ce trou.

J'alignai les quatre clichés restants.

Il parcourut la série, cligna des yeux, revint en arrière.

— Nom de Dieu... (Ses yeux refirent l'aller-retour.) Pas de main.

Il se leva, comme prêt à filer dans trois directions différentes.

— Vous auriez dû me prévenir... Il fallait dépêcher une équipe autorisée. Vous avez tout saccagé !

— Si ça peut nous dire où se trouve Nina, je n'ai pas le temps d'attendre qu'on y mette les formes. J'ai autre chose à vous montrer. Dehors.

— Mais je dois...

— Monroe, vous n'avez plus la main. Alors maintenant vous me suivez, ou je pars en ville trouver le journaliste le plus con pour lui raconter tout ce que je sais.

Sur ces mots, je m'éloignai vers l'entrée principale. Monroe m'avait rejoint lorsque je posai le pied sur le parking. Je gagnai en hâte l'allée où j'étais garé.

— Montez.

Quand nous fûmes tous deux à bord, je démarrai.

— Qu'est-ce que vous faites ?

Il entendit s'ouvrir la portière arrière et quelqu'un monter. Monroe se tordit sur son siège, et considéra Zandt comme si c'était le diable en personne, avec des hot-dogs à la main.

— Salut, Charles. Ça fait une paye.

J'actionnai le verrouillage centralisé, écrasai une plate-bande et m'arrachai du parking.

La pluie avait redoublé et la grand-route devenait une patinoire. Cela eut toutefois l'avantage de distraire Monroe, cramponné des deux mains au tableau de bord. Zandt en profita pour lui faire les poches, et lui confisquer son arme et son téléphone.

Monroe boucla sa ceinture.

— Je ne sais pas ce que vous avez en tête, dit-il, mais c'est une très mauvaise idée.

— Je ne crois pas, rétorquai-je. Ça ne va pas vous plaire, mais c'est votre problème, et vous avez dix minutes pour vous faire à ce qui va suivre.

— Rendez-moi mon flingue.

— Pas question, lâcha John. Prends ça, plutôt.

Monroe vit atterrir un objet sur ses genoux.

— C'est quoi, ce truc ?

— À toi de me le dire.

Monroe saisit la chose du bout des doigts. Cela ressemblait à un petit bâton recourbé, mais pas besoin d'être médecin légiste pour savoir que c'était autre chose.

— Je suis allé à Richmond cet après-midi, expliqua Zandt. Des paléontologues sympas ont bien voulu m'aider en jetant un œil sur ceci, comme sur d'autres ossements qu'on a trouvés.

— Ça sort d'où ?

— Quoi, Ward ne t'a pas montré les photos ?

— Bon Dieu ! pesta Monroe. Alors non seulement vous saccagez les lieux, mais en plus vous profanez le corps. T'as déraillé à ce point ?

Zandt ne releva pas.

— L'échancrure sciatique du pelvis indique un corps de femme. Ce genre de truc est assez coton à dater. Mon ami aurait préféré le voir in situ et juge idiot de l'avoir exhumé, ce en quoi il te donne raison. Sauf que, à son avis, il a bien fallu une quinzaine d'années, à deux ans près, pour que cet os prenne cet aspect.

— Répète un peu ?

— T'as très bien entendu, Charles. Cet os est sorti de terre tout près d'un tee-shirt laissé sur les lieux d'un meurtre que tu attribues à la suspecte bouclée à Thornton. Seulement voilà : ce corps a été enterré à une époque où ton unique suspecte avait entre huit et douze ans.

Monroe considéra l'objet dans sa main.

— À supposer que je puisse vous croire, dit-il.

— Et pourquoi on mentirait ? objectai-je tout en prenant la courbe qui menait au centre de cette prétendue ville. La seule chose qui m'intéresse chez votre tueur, c'est de savoir en quoi il peut nous mener à Nina. Pour elle, c'était clair depuis le début : une femme n'avait pas pu tuer ces deux hommes. Et ceci nous montre qu'elle était peut-être dans le vrai.

— Et qu'est-ce que je fabrique dans cette voiture ? Vous ne pouviez pas me parler de tout ça à l'hôtel ?

— Justement, c'est là que vous allez devoir prendre sur vous, dis-je. On s'apprête à enfreindre la loi, et vous allez nous aider. John souhaite s'entretenir avec Julia Gulicks.

— Hors de question, répliqua-t-il. C'est totalement exclu.

Ce refus ne me surprenait pas, mais sa véhémence, si. J'avais espéré que l'ossement lui ferait sauter le pas, comprendre qu'il n'y avait pas trente-six voies possibles.

— On sera chez le shérif d'ici trois minutes, ajoutai-je. Vous allez vite devoir changer d'avis si on ne veut pas perdre de temps. D'après moi, même un type comme vous admettra que John a été un bon inspecteur de la crime. D'après ce que j'en sais, il vous a dépatouillé d'au moins une affaire par le passé. Alors qu'est-ce qui vous gêne ?

— Laissez tomber, décréta Monroe. Il n'est plus de la maison. Il fait l'objet de plusieurs avis de recherche pour meurtre, c'est un homme dangereux, et jamais de mon vivant je ne le laisserai s'approcher...

— Écoutez, commençai-je avant de percevoir une sorte de gazouillis mélodieux.

Je mis du temps à identifier ce bruit, comme s'il ne m'était pas familier.

Je me rangeai brusquement sur le bas-côté, manquant de peu de bloquer les roues. Je sortis mon portable de ma poche. Lus le numéro affiché.

Ce n'était pas Nina.

— C'est Unger, annonçai-je avec le sentiment de marcher au-devant d'un précipice.

— Prends-le dehors, conseilla John.

J'ouvris la portière et sortis sous la pluie.

— Ward ? entendis-je aussitôt. C'est Carl.

Je restai muet. Qu'étais-je censé faire ?

— Ward ? T'es là ?

— Je suis là.

— Ton amie, l'agent Baynam. Pas de nouvelles ?

— Comment sais-tu qu'il s'agissait d'elle ?

— Dès l'instant où t'as quitté le bar j'ai passé quelques coups de fil.

Comme j'aurais aimé l'avoir devant moi...

— Tu l'as mise en danger, accusai-je. Tu l'as peut-être tuée.

— C'était un risque nécessaire. Je protège mes arrières, c'est tout, et tant pis pour ta vie privée. Tu m'as caché des trucs hier soir. Alors j'ai fait ce que je devais.

— Et t'as gagné quoi ?

— Rien. Je sais toujours pas ce qui se trame. Je me trouve à Langley, actuellement. On peut se revoir ? Je serai là en moins de deux heures. Il faut qu'on parle. Je peux peut-être t'aider.

Je lui raccrochai au nez.

En remontant dans la voiture, je m'attendais à plonger dans un concours de hurlements. Mais ce fut un silence de mort qui m'accueillit. Monroe rangeait son arme et son portable dans sa poche de veste.

— Alors ? me demanda John.

— Unger a lâché nos noms sur les ondes hier soir, mais c'était après l'enlèvement, car le sang de Reidel commençait à sécher quand je suis arrivé à l'hôtel. La question est de savoir si Unger nous ment, et je ne peux pas en juger au téléphone. Il veut me revoir.

— Et ?

— Possible qu'il vienne. (Je frappai rageusement le volant avec mes poings. Dans ma tête résonnait encore l'espoir que l'appel fût de Nina.) Putain. Je ne sais plus par quel bout prendre cette histoire. Impossible de savoir qui fait quoi...

— Conduis-nous au comico, dit Zandt.

J'observai Monroe.

— Vous le laissez parler à Julia ?

L'agent du FBI garda le silence. Se contenta de scruter la nuit, le teint cireux dans la lumière des lampadaires. J'ai démarré et avalé le dernier kilomètre.

Le flic de l'accueil était celui que j'avais blousé la veille. Il se levait déjà de son siège quand il vit que cette fois Monroe m'accompagnait.

— Monsieur, lança-t-il, c'est lui le type qui...

— Je sais, coupa Monroe. La salle d'interrogatoire est prise ?

— Pas pour l'instant.

— Amenez-y Julia Gulicks.

Nous franchîmes les portes de sécurité et suivîmes Monroe dans le couloir.

— Tu veux la précéder dans la salle ? demanda-t-il.

On pouvait penser ce qu'on voulait, mais quand la règle du jeu changeait, Monroe savait s'adapter.

— Non, répondit John. Je veux d'abord la voir.

Monroe nous conduisit dans la pièce d'observation. Nous patientâmes dix minutes, le temps que John parcoure la transcription des précédents interrogatoires. Puis une porte s'ouvrit. Deux flics amenèrent Julia Gulicks et la firent asseoir. L'un repartit, l'autre se posta devant le battant.

Comme la veille, Julia était installée dans la longueur de la grande table, pile au centre. Sa pâleur tirait sur le diaphane. Ses mains tremblèrent un peu lorsqu'elle les posa l'une sur l'autre, puis se figèrent. Elle leva les yeux, fixa la glace sans tain. Inclina la tête sur le côté.

Je regardai John l'observer. Il se tenait debout, le coude calé sur son autre main, l'index dressé contre son nez. Je ne sais combien de temps je l'ai épié, mais il n'a pas cillé.

— Toujours pas d'avocat ?

— Eh non, indiqua Monroe. Elle en a pris un, mais elle refuse de lui parler.

— Pas de visites familiales ? Pas d'amis ?

— Personne. Le pseudo-fiancé est passé hier en fin d'après-midi pour répondre à quelques questions. Il n'a pas souhaité lui parler. D'après lui, Julia n'évoquait jamais sa famille. Je te préviens, John, c'est pas le moment de déconner.

Zandt quitta la pièce. Une minute plus tard, le flic de la salle pivota pour ouvrir la porte. S'avançant, John le pria de patienter dehors, et attendit d'être seul avec la suspecte pour prendre la chaise en bout de table.

Après quelques instants, elle tourna la tête vers lui.

— Vous êtes plutôt bel homme.

— Merci. Ça fait un bail qu'on m'a pas fait de compliment.

— Oh, le pauvre.

— Bah, je survis. Et vous ?

— Quoi, et moi ?

— Vous allez tenir le coup ? Vous n'avez pas de famille ou de proches qui pourraient vous soutenir ?

— Non. Je suis fille unique. Mes parents sont morts.

— Triste histoire...

— Ça dépend comment on la raconte.

— Vous avez grandi à Boulder, n'est-ce pas ?

— Exact.

— Et vous avez emménagé à Thornton voilà six ans ?

— Un coin charmant, vous ne trouvez pas ?

— J'ai pas eu le temps de visiter, mais j'avoue que ça ne m'a pas frappé.

— Mais tous ces arbres ! Ces petites maisons pleines de cachet !

— Je ne confonds pas pittoresque et beau.

Elle sourit.

— Alors vous éviterez peut-être de confondre ravissante et idiote.

— Vous avez conscience d'être dans de sales draps, n'est-ce pas ? Vous n'essaieriez pas de plaider la démence, par hasard ?

— J'aurais des arguments, croyez-moi.

— Ce genre de tirade est sûrement censé servir votre ligne de défense. Mais moi, ça me renforce dans l'idée que vous vous fichez de moi.

— Je vais tâcher de faire mieux.

— J'ai un truc à vous montrer.

John puisa un objet dans sa poche et le posa au centre de la table.

Monroe inspira en sursaut.

— Seigneur, gémit-il. Ce type est incroyable.

Pour ma part, j'approuvais l'approche de John. L'attitude de Julia Gulicks s'était métamorphosée depuis son premier passage dans cette pièce. Ses yeux étaient creusés de cernes, à croire qu'elle n'avait pas dormi depuis son interpellation. Creusés, et teintés de doute.

Elle regarda l'objet devant elle.

— J'ai déjà mangé.

— Vous savez ce que c'est ?

— C'est un os.

— Une côte. Une côte de femme. D'où provient-elle, à votre avis ?

— De votre collection personnelle ?

— Elle provient de la forêt située au nord-est de la ville. On y a découvert un corps il y a deux nuits de ça. Certaines personnes ici voudraient vous imputer ce meurtre, ainsi que celui de Lawrence Widmar. Mais je vois d'après mes notes que l'inspecteur Reidel a déjà évoqué la question.

Gulicks se taisait, sans détacher les yeux de l'os.

— Mais cette côte n'appartient pas à cette victime-là, poursuivit John. Car il s'agissait d'un homme. Or cette femme fut enterrée par cinquante centimètres de profondeur.

— Alors ça vient d'où ? Venez-en au fait.

— Vous vous souvenez de l'agent Baynam ?
— Tout à fait. Une fille sympa.
— C'est vrai.
— Vous l'avez sautée ?

Je ne sais pas ce que l'on éprouvait à se trouver dans cette salle, mais de mon côté de la glace, l'ambiance était glaciale.

Elle fit un grand sourire :

— Je parie que oui. Entre nous, elle n'avait pas l'air commode, mais c'est peut-être votre genre.

— Non, répondit John. Je préfère les femmes comme vous. Rien de tel qu'une nana qui passe ses journées à ronger son frein et ses soirées tapie dans le noir. Qui se trouve présentement devant moi, les mains jointes pour cacher le fait qu'elle a besoin d'un verre. Ce genre de nana, vous n'avez pas idée... C'est le rêve de tout mec.

Le visage de Gulicks s'éteignit. Ni colère, ni détresse. Neutre.

— Seuls les faibles renoncent, dit-elle en baissant brusquement la voix d'un quart d'octave. C'est bien connu.

— Si je vous parle de l'agent Baynam, reprit John en accordant sa voix à la sienne, c'est parce qu'elle vient d'être appelée sur une autre affaire. Ce qui n'arrange pas les vôtres. L'agent Baynam était convaincue de votre innocence. Contrairement aux autres. À commencer par l'inspecteur Reidel.

— Quelle importance, puisqu'il est mort ?
— Comment le savez-vous ?

— Toute la ville en parle. Je n'ai rien à faire, à part écouter aux portes. Je les observe attentivement. Vous voulez savoir quel adjoint manque à son devoir ? Qui fait sauter les PV contre quelques billets ? Vous voulez connaître leur point de vue sur mes nibards ? Parce qu'ils en ont un, figurez-vous. Ils s'intéressent aux nichons d'une meurtrière qui ne s'est pas douchée depuis quarante-huit heures. Mais bon, c'est toujours des nichons, hein ? Et ils sont comment, ces NICHONS ? Les hommes sont spéciaux, n'est-ce pas ? Pas assez gros, en fait, tel est l'avis général.

— Les femmes aussi peuvent faire le mal.

— C'est faux, réagit-elle en montrant ses premiers signes de ferveur et de souffrance. Les représailles ne sont que des ricochets. Ça ne compte pas.

— Et qui est l'objet de votre vengeance ?
— Je parlais sur un plan général. Rien de personnel.
— Vraiment ? On aurait cru.

Elle avait retrouvé son calme.

— Il n'y a rien de personnel chez moi, jeune homme. Je n'ai pas de talents propres. Demandez à mon patron. Ça revient de temps à autre, dans les évaluations. Bonne petite fourmi, mais on n'aimerait pas l'avoir pour voisine de table, et patati et patata.

— Mark Kroeger pensait manifestement le contraire. Il est sorti avec vous, n'est-ce pas ? À quelques reprises...

— En effet.

— Il s'est promené dans les bois avec vous. C'était votre idée, mais il a suivi.

— C'est exact.

— Vous protégez quelqu'un, Julia ?

— Qu'est-ce qui vous fait croire ça ?

— Voyez-vous, je ne suis pas persuadé que vous ayez tué ces gens. Je me dis que c'est peut-être quelqu'un d'autre, et qu'il y a un rapport avec ce troisième cadavre. Alors je me demande si, pour des raisons qui vous appartiennent, vous ne cherchez pas à le couvrir.

— Eh bien, vous vous trompez.

— Sur quel point ?

— Sur tout.

— Vraiment ? Et merde. (Même dans l'Interphone son agacement semblait sincère.) Ça me dérange pas d'être un peu con-con. Mais complètement à côté de la plaque ? Ça craint.

— Je sais un truc sur vous, dit-elle en souriant à la table. Vous devriez être capable de lire la même chose sur mon visage.

— Et de quoi s'agit-il ?

Un moment de silence, et elle le dévisagea de nouveau.

— Vous avez regardé.

— Pardon ?

— Ça suinte, puis ça s'arrête. On sait qu'ils sont partis lorsque ça sèche.

— Mais de quoi parlez-vous ?

Elle leva lentement la main pour pointer son doigt sur John.

— Un tueur, dit-elle. Vous êtes un tueur.

Je croisai le regard de Monroe.

Dans la salle, John restait impassible.

— Qu'est-ce qui vous fait penser que j'ai tué des gens ?

— Alors vous niez ?

— Je vous pose une question.

— Ça saute aux yeux. Ces taches-là ne partent jamais.

John s'accouda sur la table, l'air amical et ouvert.

— Je vois bien que vous essayez de dire quelque chose, Julia. J'aimerais savoir quoi. Mais j'ai peur de me méprendre, aussi je vous demanderai un petit effort de simplicité. Vous voulez bien ?

— D'accord, dit-elle avant d'énoncer très clairement : je les ai tués.

John cligna des yeux.

— Répétez ça ?

— J'ai tué Larry dans la forêt. Je l'ai emmené faire un tour, et j'ai sorti mes nibards assez longtemps pour qu'il se dise : « C'est dans la poche. » Il n'était pas très clairvoyant car je l'avais drogué avec un produit commandé sur Internet. Pas du Viagra, soit dit en passant...

— Si vous l'avez tué en forêt de Raynor, comment se fait-il que le cadavre n'ait pas été ensanglanté ?

— Je l'ai tué au bord du ruisseau. Il y est tombé comme une masse. Je pensais qu'il serait plus beau lavé. Pas vous ?

— Pourquoi l'avoir déshabillé ?

— Pour l'humilier. C'était un connard. Je l'ai vu jeter une autre fille pour tenter sa chance avec moi. Quant au deuxième gus, j'ignore tout de lui, désolée. Je l'ai déniché à la station de bus d'Owensville, il y a deux semaines. C'était un coup d'essai, et putain ce qu'il était lourd, quand j'y repense ! Larry était un peu plus maniable.

John s'efforçait de paraître moins surpris que Monroe et moi, et c'était plutôt réussi.

— Nous avons caché à la presse un détail concernant la deuxième victime, dit-il. Vous aimeriez peut-être me le confirmer ?

— Ce ne sera pas facile, vu que je n'ai pas ouvert un journal depuis un moment, mais ne serait-ce pas une histoire de mutilations massives ? Une vraie boucherie, je sais, mais sans ça je n'aurais jamais pu le traîner là-bas. Il existe sans doute une meilleure méthode, mais *Le Manuel des Castors juniors* ne disait rien à ce sujet.

— Vous avez conscience que nous parlons devant témoins ?

— Le contraire m'aurait étonnée. (Elle regarda la glace et agita la main.) Salut, les gars.

— Vous avouez donc le meurtre de ces deux hommes ?

— Je préférerais le mot « confirmer ». Car je n'ai rien à me reprocher, si ce n'est mon amateurisme. Deux types, c'est un score minable. Quand on pense à toutes ces femmes qui se font assassiner, tout le temps, partout. Ça me fout la haine.

— Alors pourquoi m'en parler ?

— Parce que vous êtes tellement mignon. Et puis vous m'avez vraiment eue avec votre numéro du bon flic et du mauvais flic réunis en un seul.

— Vraiment ?

— Ah ouais... Et puis vous sentez aussi bon que moi.

— Ça m'aiderait beaucoup si vous vous en teniez à des déclarations simples, Julia.

— Non, répondit-elle. Non, non, non.

— Vous brûlez d'en parler. C'est la seule chose qui vous intéresse dans l'immédiat. Ne serait-ce que pour m'expliquer le coup de la main sectionnée.

— Allez vous faire foutre, rétorqua-t-elle d'une voix blessée. J'en sais rien. À vous de me le dire. À vous de trouver.

John baissa la tête au niveau de la sienne.

— Et supposons que je m'en fiche ?

Elle lui bondit au visage.

Une vingtaine de secondes s'écoulèrent avant que les deux flics du couloir surgissent dans la pièce, laps de temps dont Julia fit le meilleur usage. Réfléchie, concentrée, silencieuse. Tout en poings, coudes, ongles.

John subit l'assaut sans quitter sa chaise, sans même lever les mains. Il ne pouvait se permettre de lui infliger la moindre marque – pas ici, et pas lui. Fût-ce par autodéfense.

Lorsque les agents la maîtrisèrent, Julia parut enfin craquer. Elle émit un son que je peinerais à décrire, entre cri perçant et rire hystérique, et trois costauds ne furent pas de trop pour l'évacuer de la salle.

Monroe s'attarda dans les locaux pour prolonger de trois jours l'incarcération de Julia Gulicks, désormais inculpée de double homicide, cependant que John et moi retournions à l'air libre.

Je lui ai offert une cigarette et on a fumé en silence.

— Et ça nous apprend quoi, ces conneries ? ai-je demandé.

John se frotta le front, étalant par mégarde le sang d'une longue balafre. Il ôta sa main et l'examina.

— Aucune idée, Ward.

CHAPITRE 25

Hudek était aux commandes. Brad avait trop mal au crâne pour conduire, et maintenant que la crise était dissipée, Lee avait repris ses galons de chef, donc le volant. Les rues paraissaient vides, comme dégagées exprès pour leur montrer la voie. Chaque visage arborait de luisantes contusions, là où s'étaient écrasés les poings de l'autre. Sans rien en montrer, Hudek pensait avoir au moins deux côtes fêlées et n'entendait plus très bien d'une oreille. Brad avait éliminé la poudre et le sang de son visage, mais rien n'y faisait : on l'aurait cru passé sous une moissonneuse-batteuse.

Depuis que Lee avait téléphoné pour connaître leur destination, ils se taisaient. De quoi auraient-ils discuté ? Sur tous les tons, Hudek s'était défendu d'avoir voulu nuire à Karen. Il s'était aussi excusé pour ses autres vacheries. Brad n'avait rien offert en retour. Son esprit était un terrain vague. Au bout d'un moment, son portable sonna ; il l'éteignit en reconnaissant le numéro de ses vieux. Il les affronterait bien assez tôt, comme il venait d'affronter les Luchs : en feignant de croire à la thèse d'une mort accidentelle, d'un tragique carambolage avec délit de fuite, d'un destin foudroyé et tout le tralala. Mais en serait-il capable ? Pourrait-il trahir la mémoire de Karen, lui tourner le dos et vivre sous le poids des mensonges ? Seul le temps le dirait.

Pour l'heure, il y avait plus urgent.

Certains individus allaient devoir comprendre que ces deux gosses de la Valley n'étaient pas du genre à tendre l'autre joue. Que « riche et privilégié » ne signifiait pas « stupide et faible », et que jeunesse ne rimait pas avec « impuissance » et « docilité ». On allait apprendre que la détermination qui avait contribué à tous les succès sportifs de la promo de 2003 opérait également dans le monde extérieur ; que cette détermination représenterait une arme redoutable pour des conflits d'adultes.

Voilà ce que Lee et Brad allaient leur montrer. Ils partaient à leur rendez-vous avec le fameux flingue, et cette fois-ci les balles étaient réelles. Brad les avait chargées lui-même.

Parvenus au bâtiment à l'angle de Roscoe et de Sennoa, ils allèrent se garer sur le parking, non loin de l'endroit où Pete et Steve avaient séjourné sous une bâche, ligotés et bâillonnés.

— On aurait dû piger tout de suite, marmonna Brad, une nouvelle clope au bec. Tu connais beaucoup de gens qui pratiquent de telles méthodes ? (Ils descendirent de voiture.) Ça me rassure pas de revenir ici.

— C'est le seul endroit où il acceptait de nous recevoir. (Lee respira un grand coup.) On va vraiment faire ça ?

Brad se souvint d'avoir posé la même question lors de leur précédente visite, un soir qui ne paraissait pas si lointain. Puis lui revint l'image d'un petit K oscillant au cou d'une jeune fille, et il redouta le moment où, tôt ou tard, on lui restituerait le bijou.

— Ouais, Lee, on va le faire.

— Mais s'ils sont tout un groupe, on se la joue cool, on attend le bon moment et on fait ceux qui ne savent rien. Promets-le-moi.

— Magne-toi, Lee. Moi non plus, j'ai pas envie de crever. Je veux juste entendre ce que ce connard trouvera à dire.

Ils gagnèrent l'entrée latérale. Lee tira sur la poignée. La porte n'était pas verrouillée.

Brad le suivit à l'intérieur. Sans un bruit, ils traversèrent une grande salle, puis un couloir desservant un espace encore plus vaste.

Fait nouveau, les lieux étaient éclairés par une série de néons, ce qui conférait à l'entrepôt désaffecté des allures de QG high-tech. Contre le mur s'alignait une rangée de bureaux, équipés d'ordinateurs et de moniteurs plats, d'une machine semblable à un serveur Internet, de deux téléviseurs, et d'un nombre suspect de téléphones portables. Tous les appareils étaient éteints et des câbles couraient dans tous les sens.

Il y avait du déménagement dans l'air.

Devant l'une des tables se tenait l'homme qu'ils venaient voir. Il ne les regarda pas d'emblée, trop occupé à ranger des CD informatiques dans un gros sac de voyage.

Lee et Brad s'approchèrent un peu.

Paul leva enfin les yeux.

— Salut, les garçons. Quel est le problème ?

Brad avait prévu de se jeter sur lui, comme sur Lee précédemment. Mais il comprit très vite que ça ne se passerait pas comme ça. L'autorité naturelle de ce type inhibait toute idée d'attaque.

— T'es tout seul ? demanda Lee.

— Rien que vous et moi, confirma l'homme. On a du boulot. Alors je vous demanderai de faire court.

Brad ne pouvait plus attendre :

— Pourquoi t'as fait ça ?

— Fait quoi, Brad ?

— Tuer Karen.

— Tuer qui ? Il va falloir que tu synthétises.

— Te fous pas de notre gueule ! lança Lee avant de dérouler leur plan : Écoute, on n'est pas venus faire des histoires. Visiblement, vous préparez un coup. C'est pas grave, pour la nana. Mais dis-nous de quoi il retourne. On a juste besoin de comprendre.

L'homme finit de remplir son sac, le ferma et le posa par terre. Puis croisa les bras et observa ses visiteurs.

— Explique-moi ce qui te tracasse, Lee.

— Ce matin, je t'ai dit que la copine de Brad avait tenu certains propos suspects et risquait de nous mettre en difficulté. T'as répondu que c'était pas un souci. Et voilà que deux heures plus tard, alors qu'elle se rend chez les flics, elle se fait emboutir et tuer en pleine rue !

— Je suis désolé de l'apprendre, dit Paul en se tournant vers Brad. Tu dois être bouleversé.

— Je t'emmerde, répondit Brad. (Il se mit à trembler, et décida qu'après tout il allait quand même se le faire.) Je t'em...

— Du calme, réagit Lee en lui attrapant le bras. Alors là-dessus Brad et moi on se frite méchamment parce qu'il croit que je l'ai balancée exprès, ce qui est faux. Ça dégénère, et... (Lee sortit le pistolet de son blouson.) Tu reconnais ceci ?

L'homme haussa les épaules.

— J'en vois tellement...

— Il s'agit du flingue qu'Hernandez m'a filé la nuit où Pete s'est fait buter et que j'ai gardé planqué chez moi. Je l'ai ressorti cet après-midi pour tirer deux coups, et devine ce qui s'est passé.

— J'ai hâte de le savoir, Lee.

— Le flingue était chargé à blanc !

— Tiens donc. Je me demande bien pourquoi Hernandez aurait fait une chose pareille.

— Tu parles ! Ce type ne faisait rien sans ton accord.

— Alors quelle est ta théorie ? Je tiens à ce que tu comprennes par toi-même, Lee. Mais je pense que c'est déjà fait.

— C'était un coup monté, depuis l'instant où Pete est mort, répondit Brad d'une voix ferme. Tu nous as envoyés sur un deal foireux, mais t'avais pas assez confiance pour nous filer de vrais flingues. Pete s'est fait tuer, et Hernandez en a profité pour nous pousser au crime en exigeant qu'on enterre le corps. Après ça, on était à ta merci.

Paul hocha la tête.

— Excellent.

Hudek dévisageait l'homme d'un drôle d'air, et Brad de comprendre – avec un sentiment proche de la joie – que pour la première fois de sa vie Lee était hors de lui.

— C'est vrai ?

Paul sourit vaguement.

— Oui, Lee, c'est vrai.

Suivit un long silence. Puis Lee dressa lentement son arme.

— Toi, mon gars, t'es fini.

— Tu vas m'abattre ?

— Exactement.

— Alors que des tas de gens savent qui tu es, où tu habites, et n'auront aucun scrupule à éliminer chaque personne à qui tu auras adressé la parole ?

— Je m'en fous. Tu nous as baisés, mec. Pete est mort par ta faute, et t'as fait liquider Karen. Tu nous mènes en bateau depuis le début. Alors je vais te buter.

— Excellent, dit Paul. Je savais que t'en serais capable.

Lee était écarlate.

— Ne me prends pas de haut, connard.

— C'est loin d'être le cas, Lee. Je suis sincèrement ravi.

Lee coula un regard vers Brad, et Brad devina que Paul était finalement parvenu à ouvrir une trappe dans l'esprit de Lee, à enfoncer une porte qui aurait pu rester fermée. Ses yeux brillaient de haine, et Brad connaissait assez son pote pour savoir que Paul passerait les dernières secondes de sa vie à comprendre ceci : lorsque Lee prenait une décision, il allait jusqu'au bout.

Lee plaça sa main gauche sous la crosse, pour stabiliser l'arme.

— Adieu, connard.

Brad se préparait pour la suite. Mais une voix féminine s'éleva :

— Non, mon cœur.

Lee pivota avec toute la lenteur d'un homme luttant contre le vent, comme s'il avait conscience qu'en cet instant sa vie basculerait à jamais.

Sa mère, debout au fond de la salle, à quelques mètres de l'entrée. Flanquée de son mari.

Mme Hudek avait l'air calme, sereine. Le visage de monsieur était plus énigmatique.

— Maman ? fit Lee d'une voix juvénile. Papa ? Mais qu'est-ce que vous faites là ?

— Baisse ce pistolet, fiston. Tu n'en as pas besoin dans l'immédiat.

Cette recommandation ne cadrait pas avec sa mère. Lee maintint sa position.

Brad était pétrifié. Il ne rêvait pas, et pourtant... Si M. et Mme Hudek se matérialisaient ici, pourquoi Karen ne serait-elle pas en vie ? Brad aurait décroché un bon job à la télé, Karen serait son épouse, il n'aurait pas vu Lee depuis des années et il allait bientôt se réveiller avec une légère gueule de bois, qu'il chasserait d'un petit footing pour bien commencer la journée.

Les bras de Lee amorcèrent leur descente.

— Qu'est-ce qui se passe, ici ?

Lisa Hudek s'approcha de lui.

— Nous savions que tu serais déboussolé, et nous voulions t'empêcher de commettre une bêtise.

— Vous... vous savez que ces types ont fait tuer Karen ?

— Oui, nous sommes au courant.

— Mais alors... qu'est-ce que vous faites ici ? Comment vous savez... ? Qu'est-ce que... ?

Mme Hudek lui ôta doucement l'arme des mains pour la remettre à Paul. Avant d'expliquer à son fils :

— Nous sommes venus te dire au revoir. Tu vas faire un long voyage. Nous tenions à te souhaiter bonne chance.

— Pardon ?

— C'est exact, dit Ryan Hudek d'une voix résignée. L'aboutissement d'un long processus.

Brad éprouva le besoin d'intervenir :

— Je sais pas ce que vous mijotez, tous, mais ce type (il désigna Paul) n'ira nulle part.

Lisa Hudek le regarda comme un vulgaire nettoyeur de piscine, un employé impossible à virer eu égard à son ancienneté, mais d'une incompétence crasse.

— Tu peux rester en dehors de ça, Brad ?

— Non, je peux pas, madame Hudek. Quel que soit le « ça » en question.

— Je n'ai pas beaucoup de temps, dit Paul. Et vous devez vite rentrer chez vous, pour vous tenir prêts. Le spectacle va commencer.

— Prêts pour quoi ? questionna Lee.

Paul secoua la tête.

— Pas toi. Tes vieux. Ils regagnent leur domicile après une séance de shopping sur Rodeo Drive. Et se retrouvent pris dans les bouchons.

— C'est quoi, ce shopping ? insista Lee. C'est quoi, ces bouchons ?

— C'est la raison pour laquelle ils n'auront pas vu les images à la télé, ni entendu la nouvelle avant de retrouver leur rue et d'apercevoir les flics devant chez eux.

— Quelles images ?

Paul tendit le bras pour allumer un téléviseur. Fit défiler les programmes jusqu'à trouver la chaîne d'infos 24/24.

Lee et Brad fixèrent l'écran. On voyait l'extérieur d'un lieu si familier qu'ils mirent du temps à le reconnaître. Ils s'y étaient rendus le matin même. Un vaste complexe de boutiques, centré sur un réseau de galeries.

— Ce serait pas Belle Isle ? demanda Lee. On dirait presque...

Les journalistes avaient monté une petite boucle, prête à être coupée dès qu'il y aurait du nouveau. Le récit était succinct : voilà une heure et demie, une bombe de petite taille avait explosé au deuxième niveau du centre commercial de Belle Isle. Les pompiers étaient facilement venus à bout de l'incendie. Les dégâts se limitaient au magasin Serious Sport et aux boutiques attenantes ; on ne déplorait que quelques blessés, des passants atteints par des éclats de verre. Les vendeurs et trois clients étaient actuellement soignés pour des coupures sans gravité et pour avoir inhalé des fumées toxiques.

Le présentateur apparut pour détailler les premières conclusions de l'enquête. L'engin avait vraisemblablement été caché dans un sac de sport en toile et placé dans la matinée. La vidéosurveillance du magasin désignait un suspect, dont l'identité était en cours de vérification.

Une image fixe. Grise, floue, prise de face et de haut. Un jeune homme suspendait un sac à un portant, le dissimulait avec soin, sans savoir qu'il se trouvait dans la mire de deux caméras.

Il s'agissait de Lee John Hudek. Qui devenait ipso facto un nouvel homme, prenait une tout autre stature.

— Mais le sac était bourré de drogue ! s'étonna-t-il à mi-voix.

Paul éteignit la télé.

— Bien sûr. Mais celui d'après ne l'était pas. Celui que nous avons placé nous-mêmes, avec un peu plus de précautions, la veille au soir. Ils vont débarquer chez toi très vite, Lee, s'ils n'y sont pas déjà. Ils vont voir la voiture de Brad et se dire que lui aussi était dans le coup. Ils décideront de fouiller la maison des Metzger, et c'est alors qu'ils retrouveront le portable de Pete Voss, dans les buissons du jardin.

— Mais les flics viennent de recevoir...

— En fait, on a procédé autrement.

Brad le dévisageait. Chaque fibre de son corps hurlait de froid.

— Mais à quoi vous jouez ? Pourquoi... Je comprends pas.

— C'est un coup dur, Bradley, répondit Ryan Hudek avec sympathie. Nous en sommes bien conscients. Mais on n'échappe pas à certains sacrifices.

— Mais... vous êtes ses parents ! Et vous avez laissé faire ? Vous resterez les parents d'un jeune terroriste...

Mme Hudek sourit :

— Il y aura toujours moyen d'accuser la société. Et les jeux vidéo, les féculents, la famille Bush, Ben Laden, sans oublier ce pauvre vieux Charlton Heston. Mais pas nous. Jamais nous. Seuls ceux qui nous connaissent sauront notre mérite.

— C'est qui, ce « nous » ? s'impatienta Brad.

Mais personne ne daigna répondre.

— Vous savez quoi ? lança Lee, silencieux depuis un moment. Brad n'est pas le seul à halluciner. C'est quoi, ce bordel ?

— La seule chose qui compte est la suivante, répondit Paul. Ton ancienne vie est révolue. Ce n'est pas une grosse perte. Tu n'allais nulle part, de toute façon. Ton « Plan » concernant les vacances de printemps est la chose la plus fumeuse et la plus naïve que j'aie jamais entendue. Aujourd'hui, on peut t'emmener dans un coin qui vaut le coup.

Lee regarda ses pieds.

— On s'était déjà rencontrés, pas vrai ? Ça remonte à quand ?

— Il y a bien longtemps. Mais tu devais comprendre par toi-même que tu n'as pas d'autre endroit où aller. Et, ce faisant, tu as prouvé que tu étais des nôtres.

— Comment ça ?

— Toi et moi, on part en voyage. Je t'expliquerai en route.

— Et moi alors ? réagit Brad. C'est quoi, mon rôle dans l'histoire ?

Paul le regarda.

— T'en as pas.

Et de lever le flingue d'un geste souple pour lui trouer le front.

Brad partit en arrière et s'effondra lentement. L'espace d'un instant, il parut vouloir se relever, avant de s'écrouler sur le côté. Ses jambes suivirent le mouvement, au ralenti, entraînant son tronc le long d'un arc paresseux, dans une écume de sang.

Dans son esprit jaillissait un incendie de couleurs, de lumières et de souvenirs ; le temps et l'espace s'amalgamaient, compactaient l'éternité en un point pivot. Il se sentait lucide, au chaud, comme si tout allait bien se passer. Puis il perçut l'écho d'une voix familière. Il tenta d'ouvrir la bouche pour lui dire quelque chose, sûrement son prénom, ou juste « hé ».

Paul tira une seconde balle et le corps se figea.

Un petit silence, le temps que l'écho meure à son tour.

— Putain, expira Lee.

Ses parents demeuraient impassibles, eux qui avaient servi de l'orangeade à Brad avant même qu'il ne sache tenir un verre. Lee considéra le corps de son ami. Ce devait être un choc, sans doute, mais curieusement il se sentait peu touché. Il y avait toujours eu de la faiblesse chez Brad. Une faiblesse d'ordre mental. Au lycée, il était plus costaud que Lee, plus beau, sans conteste, et même plus futé, s'il avait seulement ouvert les yeux. Mais il se la coulait douce. Brad n'avait jamais songé à voir loin, à voir grand. Aurait-il eu les tripes de dégainer son flingue face à des agresseurs ? Peu probable. Quoique, il avait quand même essayé de tuer Lee dans l'après-midi. Alors bien fait pour lui.

— Ça vaut mieux comme ça, conclut Paul. Sa vie serait vite devenue un enfer.

— Peu importe, opina Lee avant de se passer la main dans les cheveux.

Il se sentait prêt. Il venait de connaître des jours mouvementés, à chercher son chemin dans le brouillard. Mais soudain tout s'éclaircissait. Il n'y avait qu'une seule issue.

— Alors, c'est quoi la suite ?

— Ça fait longtemps qu'on attend ce moment, confia sa mère. Le Jour des Anges. Et tu vas y prendre part.

Comme à la maison après le départ des flics, elle leva la main pour lui toucher la joue. Son regard était limpide.

— Tu seras notre grande fierté.

TROISIÈME PARTIE

L'UNIQUE

L'erreur la plus fréquente et la plus grave de la modernité consiste à croire que l'Antiquité est morte.

Clarke ASHTON SMITH

CHAPITRE 26

Taraudé par la faim, Jim reprit le chemin de chez Renée. Avant d'entrer, il jeta un œil dans l'allée de derrière : aucune gamine en vue, bien sûr – elle n'y était pas la première fois. Il commanda le « petit-déjeuner du gourmand », s'installa près de la fenêtre et considéra la pitance qu'on lui servait. Ça ressemblait trop à la photo du menu et ça sentait mauvais. Il avala cinq ou six bouchées, puis se rua aux toilettes pour vomir.

S'essuyant la bouche avec sa manche, il revint s'asseoir dans la salle. Inutile de reprendre sa fourchette, rien que l'odeur l'écœurait. Autour de lui, personne ne semblait incommodé. Deux camionneurs et une poignée de lève-tôt engloutissaient leur assiette comme si c'était la dernière. La faim empêchait Jim de réfléchir correctement. Ça, et le fait que James ne brillait pas par sa finesse. La finesse, c'était plutôt le truc de Jim. James, lui, avait dix-sept ans d'âge mental et n'en faisait qu'à sa tête. Maintenant, quand Jim regardait un plat, il ne savait même plus si c'était de la nourriture. Une seule chose était sûre : il ne pouvait pas avaler cette merde. Il lui fallait quelque chose, mais ce n'était pas ça.

Jim savait à quoi pensait James. Deux jours qu'il y pensait. Pas forcément en grande quantité. Juste assez pour apaiser son estomac. Mais quitte à s'aventurer aussi loin...

Non. Pas question.

Il but son café, paya et s'en alla.

On avait repris la route.

Mais Nina s'était préparée. La première fois, le véhicule s'était mis en mouvement alors qu'elle somnolait dans sa position tordue. Somnoler n'est pas le meilleur terme : c'était plutôt un état semi-conscient, une peur sombre et cotonneuse, mais préférable à celle de la veille. Elle avait vaguement entendu la porte s'ouvrir, avec ce curieux timbre métallique, puis un moteur avait démarré.

Sans même lui laisser le temps d'analyser cette donnée, le véhicule s'était ébranlé. Elle fut projetée au sol, dans un espace exigu, et se cogna la tête. Elle cria car sa jambe était coincée et sa rotule menaçait de lâcher, mais le bâillon retenait les sons dans sa bouche. Réduite à prendre son mal en patience, elle parvint petit à petit à décaler son poids, de sorte que la douleur se fit un peu moins cinglante. La voiture s'arrêta enfin. L'homme avait sans doute remarqué l'incident, car il vint rejeter Nina sur le lit, sur la table ou sur Dieu sait quoi, sans douceur aucune. Il la traitait comme un vulgaire objet. Le genou de Nina la fit souffrir un certain temps, mais le pire, c'étaient ces nausées permanentes, à force d'être ballottée les yeux bandés – ou bien étaient-elles dues aux fumées d'échappement ? Nina réussit néanmoins à ne pas vomir. Elle resta allongée, sans rien faire. Trop malade pour reprendre son jeu, qui consistait à se remémorer tous les ingrédients de la salade ridicule concoctée par Ward l'autre soir à Sheffer, et à imaginer qu'elle en mangeait davantage, qu'elle finissait son assiette et que Ward était aux anges. Elle préférait se voir assise avec lui au bord du lac, en bas du bungalow qui fut un temps le leur. Puis elle comprit que cela n'arriverait plus jamais et le chagrin mit fin à sa rêverie.

L'homme remonta dans la camionnette. Oui, c'était sûrement une camionnette, avec ce moteur qui broutait comme un vieux Combi Volkswagen. Anticipant le mouvement, elle laissa l'accélération la repousser en arrière, puis négocia le lent reflux. Avec brio. Elle apprenait toujours très vite.

Le fourgon roula un petit moment. Une vingtaine de minutes, peut-être. Puis le type s'arrêta, coupa le contact. Ouvrit sa portière, la referma. Faisait-il une nouvelle pause ?

Non. La porte latérale coulissa de plus belle. Nina grappilla un peu d'air, au parfum très pur, et entendit des chants d'oiseaux. Il l'avait conduite hors de la ville. Pourquoi ? Était-ce le moment fatidique ? Son heure avait-elle sonné ?

La camionnette tangua sous le poids de l'homme, et il referma la porte. Il se tenait tout près. Elle se raidit de tout son corps. Par quoi commencerait-il ? Par où ?

— N'aie pas peur.

Très bien, pensa Nina. Je n'ai pas peur. Comme quand on dit : « Je ne suis pas ivre. »

Elle émit tous les bruits qu'elle pouvait, espérant qu'ils seraient explicites.

— Non, je ne te laisserai ni voir ni parler, car tôt ou tard tu prononceras la phrase de trop. J'ai déjà commis des erreurs. Ça s'est mal passé à l'hôtel, et je sais que le Garçon précoce va me le faire payer. Mais... Écoute, reste tranquille, d'accord ? Ne bouge pas, ne dis rien, et tout se passera bien.

Nina en doutait. En doutait fort. Ce n'était pas à elle qu'il s'adressait, mais à lui-même. L'intérêt de se rassurer sur ses propres intentions, c'est qu'on fait preuve d'indulgence une fois qu'elles se révèlent mensongères.

Malgré tout, elle se tut, et l'écouta parler lorsqu'il rouvrit la bouche après un bref instant. Il commença de manière lente, comme quelqu'un qui n'a pas eu d'auditoire depuis des lustres – en dehors de lui-même.

Un jour, un prof au lycée avait dit une chose marquante. Comme il tentait d'illustrer une notion quelconque (James ne se souvenait jamais de laquelle, juste que c'était un après-midi chaud et que personne n'écoutait vraiment), l'enseignant s'était mis à digresser sur le fait qu'une différence donnée pouvait représenter divers types de différences. Tout absconses qu'elles étaient, ces paroles s'infiltrèrent dans la conscience de James, qui entendit alors ceci :

— Exemple. Quelle est la différence entre deux et trois ?

Un silence, que seule rompait une mouche s'excitant contre la vitre.

Un élève, sans doute une fille, finit par répondre :

— Un.

— Exact, dit le prof. Les deux valeurs diffèrent d'une unité, et une fois qu'on a dit ça, il n'y a pas grand-chose à ajouter. Maintenant, si je demande : quelle est la différence entre un et deux ?

Un élève, sûrement la même élève et après le même laps de temps, donna une nouvelle fois le résultat « un ». L'enseignant opina de plus belle, mais avec ce rictus espiègle annonçant qu'il avait une idée derrière la tête, un truc à même, croyait-il, de faire passer M. Chips pour un mec super cool, quand cela ne faisait qu'aviver l'envie de le voir s'écrouler dans la seconde, foudroyé par un infarctus.

— C'est toujours exact. Mais pensons-y un instant. Passer de deux à trois, cela signifie juste qu'on possède davantage. Trois dollars en poche, c'est mieux que deux. Trois dissertations en retard au lieu de deux – là, c'est pire. (Personne ne rit. Une nana se fendit d'un sourire, à la rigueur. Les filles sont gentilles. Elles le font croire, en tout cas.) La différence est unitaire, et donc c'est un peu mieux, ou un peu moins bien, selon ce que l'on compte. Mais pas de quoi fouetter un chat, n'est-ce pas ?

Aucune réponse. Le prof regarda par la fenêtre d'un air las, comme s'il comptait les années le séparant de la retraite et qu'il les jugeait trop nombreuses. Mais il persévéra :

— L'écart entre un et deux, c'est beaucoup plus sérieux. C'est la différence entre un seul et beaucoup, entre l'unique et le commun. Si un gars vous dit qu'il existe deux dieux, et qu'un autre – ou une fille, bien sûr – rétorque qu'il y en a trois, ou cinq, chacun gardera son calme. En vérité, les polythéistes sont tous dans le même camp. Maintenant, si vous mettez un monothéiste face à un polythéiste, tous aux abris ! Un seul et unique dieu contre une bande d'idoles païennes ? C'est ce qu'on appelle un désaccord de fond. Et ça promet de barder. Vous couchez avec un type ou vous couchez avec deux. Ce sont des différences de taille, n'est-ce pas ? Vous voyez ce que je veux dire ?

Non. Pas au point de l'exprimer, en tout cas. La mouche bourdonnait toujours. Elle s'arrêta un instant, puis recommença, comme le font les mouches.

— Ce qui nous amène à un point plus crucial. La différence entre zéro et un. Là encore – c'est bon, Karla, je me charge du calcul, c'est pour ça qu'on me paie si cher –, la différence est d'une unité. Vous avez zéro, vous ajoutez une simple unité, et ça fait un. Vu ?

À présent, James le regardait. Ce que racontait le bonhomme commençait à l'atteindre, et il était à deux doigts d'écouter. Une expérience inédite. Fascinante.

— Mais en vérité, asséna le prof en levant l'index, *ça ne fait pas un du tout*. On dit ça pour des raisons de commodité mathématique, mais c'est complètement faux, pour la simple raison que nous avons quitté le monde des sommes pour ce que les philosophes appellent l'« ontologie ». On ne parle plus de nombres, ni de quantité, mais de *qualité* – on affirme quelque chose sur la nature du monde.

— Comme ? intervint une voix. Comme quoi ?

— C'est très variable. Exemple : la différence entre avoir un enfant et avoir des jumeaux n'est pas énorme...

— Vous croyez ça ? s'indigna une fille.

— Pas dans le sens où je l'entends, répliqua-t-il. Un ou deux gosses, c'est une question de degré, et certes la nuance est considérable en termes de dépenses, d'organisation, de poussettes et de machins, mais : enceinte ou pas enceinte, c'est ça qui bouleverse l'existence. La différence entre être femme et être mère. Quand le zéro devient un, c'est là que l'univers bascule et que la vie change. Tu comprends ?

— Ouais, ouais, marmonna la fille, blasée ou rendormie.

Le prof enfonça le clou :

— Vous voyez où l'on se situe, jeunes gens ? Au cœur des choses. Il y a un dieu, il n'y a pas de dieu. L'existence face à la non-existence. La vie ou la mort.

— Le vrai ou le faux, murmura un garçon.

— Mais oui, James, pavoisa le prof, et James de comprendre que cette réponse venait de lui. Merci beaucoup. Et moi qui te croyais dans le coma... Zéro, un. Marche, arrêt. Vrai, faux. Si telle chose n'a jamais eu lieu, le monde présente tel visage. Mais si elle s'est produite, il en revêt un autre. Le pas qui va de zéro à un est un moment de création et provoque un changement irréversible.

James regardait le prof. Il comprenait.

Puis la sonnette retentit et la classe se dispersa.

L'homme se tut, comme pour ruminer l'épisode. À sa façon de le raconter, on devinait qu'il ne l'avait pas évoqué depuis longtemps.

— Tout le monde se souvient de sa première fois, reprit-il au bout d'un moment. Karla, en ce qui me concerne. Ne va pas croire que je n'aime pas les femmes. Je les aime. Mais certaines seulement. Je m'entends bien avec celles que j'apprécie. J'ai eu une femme, j'ai eu... J'ai eu une femme. Seulement, ce n'est pas tous les jours qu'une fille me fait de l'effet. Elle doit être très spéciale. Ça m'embêtait, avant, quand les autres gars reluquaient la serveuse ou je ne sais quelle nana, en disant qu'ils la trouvaient chaude. Je voyais qu'elle avait un visage correct, de beaux seins ou un beau cul, tout ce que tu voudras, mais ça s'arrêtait là. Comme quand on t'offre un sandwich et que tu te dis : « Sûr, ce pain est frais et moelleux, croustillant sur les bords, la garniture est belle, épaisse, avec une pincée de poivre pour relever le tout. C'est un superbe sandwich qu'on me présente là. Sauf que... j'en veux pas. Non que je déteste les sandwiches. Au contraire. C'est juste que... j'en ai pas envie. » Là, c'est pareil. Jusqu'à ce que t'en voies une qui t'attire vraiment. Alors il faut que tu la possèdes. Une avec des ailes. Et ça finit toujours mal. (Il respirait lourdement.) J'ai faim, confia-t-il. J'essaie, vraiment très fort. Tu n'aurais même pas mal, mais ce serait une connerie de relancer la machine.

Nina ne comprenait rien à cette logorrhée. La déshydratation rend la concentration difficile, a fortiori quand l'ennemi risque à chaque seconde d'illustrer son propos en vous plantant une lame dans la chair, sous un ongle, ou dans l'œil. Nina n'avait pas l'intention de frayer avec cet univers-là. Elle en avait passé, des heures, à entendre des psychopathes, et derrière une vitre blindée ce pouvait être fascinant – même si la plupart vous entraînaient inexorablement sur les mêmes voies, tels des wagons branlants aiguillés vers une station sombre et sanglante. Bloqués dans l'enfance, persuadés que la leur fut prégnante comme nulle autre. Le tic mécanique des blessures et des offenses non digérées. On tourne autour d'un événement capital tel un chien enragé attaché à un poteau, incapable de comprendre qu'il ne s'agit, pour les autres, que d'un point lambda du passé, alors même qu'au fond de soi il bat comme un cœur psychique.

L'homme avait cessé de parler. Au bout de quelques minutes, Nina reconnut l'odeur du tabac sec.

— Je me contenterai de cette connerie-ci, reprit-il, mais sa voix était celle d'un homme en train de perdre une bataille.

Un craquement d'allumette, suivi d'effluves de cigarette. Nina n'en fut guère gênée – ça lui rappelait Ward.

Puis il se tint un peu plus près, et elle se crispa. Il parut hésiter, avant de lui glisser les mains dans la nuque. D'un mouvement rapide, il lui ôta le bandeau.

Il lui fallut de longues secondes pour s'habituer à la lumière, aussi sombre que fût le fourgon. Derrière le pare-brise se dressaient des arbres. Entre eux et Nina, un homme. Plus tout jeune. Large d'épaules. Des yeux tristes. Mais leur compassion s'adressait à lui-même, pas à elle.

Battant des paupières, Nina scruta l'intérieur de la camionnette. Dépouillé, sans vitres sur les côtés. Quelques griffures rouillées dans l'intérieur de la porte, brèves et parallèles. Comme si on avait gratté avec des ongles, en cherchant désespérément à fuir.

Et qui sait, peut-être que plusieurs s'y étaient succédé.

Il reprit la parole, avec lenteur, et malgré le bandeau défait, il évita soigneusement de la regarder dans les yeux.

On se souvenait de la dernière, expliqua-t-il d'une voix heurtée, mais par-dessus tout on se souvenait de la première. Quand le zéro devient un. C'était comme de descendre sa première bière, ou s'allonger à côté d'une fille après l'avoir fait pour la première fois – perdu, excité, un peu déçu : maintenant qu'elle paraît un peu plus femme, on se sent encore plus jeune et plus petit qu'en début de soirée. Toutes ces nuits étaient écrites, ces batailles décisives dans la conquête des collines inconnues de la maturité. On ne sait pas très bien où l'on va. Mais on y va, comme tous les autres. Ça commence par l'alcool, quand on s'aperçoit que les grands boivent des trucs qui nous sont défendus. La gorgée occasionnelle, sifflée en douce, révèle un goût bizarre, mais c'est plus ou moins fait exprès, se dit-on, et il y a un côté adulte et grisant à boire quelque chose qu'on n'adore pas. Le comble de la sophistication, l'antithèse du puéril ! Puis soudain tel camarade de classe saute le pas phylogénétique, vous en met plusieurs mois dans la vue. Des rumeurs envieuses circulent sur une certaine fête samedi dernier : un ado plus vieux aurait fait circuler un pack de bière, et le camarade en question en aurait englouti la moitié, sans vomir, avant d'embrasser une fille...

La partie bécots est fausse. Les jeunes gens profèrent toujours un mensonge de trop ; les hommes aussi, bien sûr. Mais le reste est vrai, et l'identité du héros ne vous étonnera point.

C'est le Garçon précoce.

Chaque classe a le sien. Celui qui arrive toujours le premier, qui laissera toujours ses poubelles au sommet de votre montagne ; celui qui prend la file express vers le monde adulte.

Une fois qu'il a ouvert le bal, le tabou saute pour tous, et vient un soir où l'on se retrouve entre copains à la porte d'un bar ; le premier parvient à entrer sans qu'on lui demande ses papiers, et l'instant d'après chacun tient un grand verre glacé, et ça n'a rien à voir avec la bouteille tiède qu'on vous a accordée l'été dernier lors d'un barbecue. On avale une gorgée métallique, mousseuse, qui d'après son goût pourrait avoir fui d'une machine. Mais quoi, c'est de la bière, et on sait – tout en vidant laborieusement ce verre qui, dans quelques années à peine, disparaîtra en quelques lampées machinales –, on sait qu'on a franchi une étape.

On détient sa première carte magique. On connaît les charmes de l'alcool.

Du jour au lendemain, on devient un gars qui boit et qui chopine à tour de bras – à se demander parfois si on ne va pas virer pochard. La saveur est toujours aussi savonneuse et amère, en réalité, mais on ne dit rien car les autres ne font aucune remarque à ce sujet, et puis ce serait le coup à passer pour une chochotte, alors qu'on vient de prouver le contraire.

À ce stade, le Garçon précoce dont on recherche tous l'estime (en même temps qu'on lui voue une crainte mêlée de haine) aura conquis sans sourciller de nouveaux horizons. Il sentira régulièrement la cigarette ou aura exploré le corsage d'une jolie fille. Et puis, enfin, il sera passé à l'acte. Le grand jour, le grand final de l'adolescence, l'événement qui transforme les garçons en hommes, qui vous hisse dans le carré VIP des ados : démarqué par l'expérience, magnifié par l'action, instantanément promu en taille et en « cool attitude », d'une manière que vous ne connaîtrez plus jamais par la suite.

Mais on ne le comprend pas, ça, pas encore, et l'heure est à engranger des points. Alors un après-midi on essaiera la cigarette, et on trouvera ça infect, ou non, sans se dire que cette petite nuance vous coûtera des dizaines de milliers de dollars, d'innombrables pauses café debout sous la pluie entre collègues parias, et, pour finir, la vie. Ensuite, à la faveur de telle ou telle circonstance, on sentira dans le creux de sa main toute la chaleur et la douceur d'un sein, en se voyant accorder, contre toute attente, le droit de caresser ce sommet mythique et chauve. Ce cap franchi, on se trouve un peu désœuvré – aucun progrès ne s'impose spontanément – mais c'est toujours ça de pris. Puis vient enfin le jour où l'on baise, moment intimidant et bref qui nous éjecte de l'autre côté du rideau, dans un pays où il ne reste plus grand-chose à accomplir, toutes les étapes importantes ayant été franchies avec succès, à l'exception de deux.

Tôt ou tard, on s'inventera de nouveaux objectifs, histoire de tuer le temps. Des plus sommaires et ostentatoires – grosse voiture, grande maison, belle situation – aux plus minutieux et intimes. Tout est dans la main. Quand on regarde la main d'un individu, qu'on l'examine avec soin, on discerne ce qu'il est, ce qu'il a été, ce qu'il a fait. Les mains sont l'action. Quand on prend la main de quelqu'un, on le possède entièrement.

Si le premier mégot peut valoir la peine capitale, il en va de même pour l'autre chose. C'était sympa, cette première fois, mais elle avait comme un goût d'inachevé. Il doit y avoir davantage, un petit plus qui accorderait la réalité à l'attente, le monde extérieur avec l'intérieur de sa tête.

La plupart des hommes trouveront des moyens avouables de poursuivre cette quête, et leur existence suivra son cours éternel, toujours dans le sillage du Garçon précoce. Car il sera encore le premier à franchir l'avant-dernière étape, celle de mettre une fille enceinte – avant de délaisser les trucs de garçons précoces pour renouer avec les joies adultes comme gagner sa vie, monter des étagères ou s'asseoir seul sur la terrasse pour boire une bière qui a désormais une saveur de flotte, le regard perdu dans le jardin, cherchant une raison de se lever le lendemain.

C'est le premier, aussi, à redevenir poussière, à franchir l'ultime horizon avant la mort, mais ça, l'école ne vous l'avait pas dit. À l'époque, la place de premier était toujours enviable. C'est plus tard que l'on comprend l'intérêt de flâner en queue de peloton. D'un autre côté, ce garçon-là ne meurt jamais vraiment, même lorsqu'il finit broyé dans un amas de tôle, sur une route de campagne, avec plusieurs grammes d'alcool dans le sang. Il est immortel, une graine noire plantée au plus profond, dans l'inaltérable image de soi. C'est lui qui nous a fait comprendre qu'on n'avait rien d'exceptionnel. Il n'est plus là, mais on le croisera de nouveau. Il aura pris de l'âge, et perdu ses joues, mais un jour on comprendra que nombre de nos actes étaient en fait les siens. Il aura toujours cette foulée d'avance, nous connaîtra mieux qu'on ne se connaît soi-même. Il tirera les ficelles, nous attirera dans des ruelles sombres, inventera à notre intention de nouvelles étapes à franchir, aussi étranges qu'effroyables.

Et quand, ayant exécuté ses ordres, on se retrouvera pantelant dans le noir, face au reflet d'un monde devenu sans issue, c'est son visage à lui que l'on verra dans le miroir.

Son téléphone l'interrompit au beau milieu d'une phrase.

Il laissa sonner. Il avait déjà réagi de cette façon, plus tôt dans la journée... à moins que ce ne fût la nuit ? En entendant l'appareil, Nina s'était dit que son propre portable se trouvait peut-être dans ce fourgon. Mais éteint. Or elle était ligotée. Bon, le téléphone était une impasse, mais elle gardait l'idée dans un coin de sa tête.

Quelques minutes de silence, puis une nouvelle sonnerie. Cette fois l'homme décrocha. Il écouta quelques instants, pour conclure d'un simple « D'accord ».

Il reprit une cigarette. Nina perçut tout de suite que l'atmosphère avait changé.

— Ça y est, il est en route. (Sa voix avait recouvré sa dureté.) Quand on parle du loup. Le Garçon précoce en personne. Et donc... je vais le faire, en fin de compte.

Nina voulut répondre quelque chose. N'importe quoi. Mais son bâillon ne permit que des gargouillis. L'homme lui remit rapidement le bandeau, et tout redevint d'encre.

Il se releva pour faire deux pas, ce qui ébranla la camionnette. Nina entendit coulisser ce qui devait être un tiroir, juste au-dessus de sa tête. Une série de sons quelconques, et il revint auprès d'elle. Puis un déclic, légèrement grésillant.

Comme celui d'un Polaroïd.

Il reposa l'objet et se rapprocha encore. Des deux mains, il lui prit le bras droit. Il respirait vite, et ce n'était pas rassurant.

Il lui noua quelque chose autour du biceps, très serré. Elle voulut taper du pied pour se dégager, avant qu'une chose terriblement pointue ne lui perfore l'intérieur du coude. Elle se pétrifia.

Et toujours ce souffle, superficiel et rapide.

L'aiguille resta en elle un certain temps, peut-être cinq, dix minutes.

L'homme s'immobilisa quelques instants, penché sur elle, comme s'il jaugeait cette chance ultime de renoncer. Il finit par s'éloigner.

Et maintenant ? Qu'allait-il faire, maintenant ?

Il sortit du matériel d'un placard. Des couteaux ? Un *cling*, un bruit de vissage, un bref chuintement suivi d'une odeur de gaz. Puis un craquement d'allumette. Pas de cigarette.

Nina fit tout son possible pour penser à autre chose. Se réfugier près du lac. Admirer sa surface moirée sous un ciel de nuages, croire qu'il suffirait de se réveiller et de tourner la tête pour trouver Ward assis à côté d'elle, un demi-sourire aux lèvres, amusé par les cris qu'elle aurait poussés en dormant.

Mais c'était trop loin.

Il lui fallait rester ici, dans ce fourgon avec son geôlier. Impossible de ne pas comprendre ce qu'il lui faisait. Passe encore le bruit du sang coulant dans un récipient métallique, bien qu'elle frémît en mesurant l'importance du volume prélevé. Cela demeurait supportable.

Mais l'odeur de cuisson...

CHAPITRE 27

J'avais très mal dormi. Je m'échinais à trouver du repos, puisque je ne connaissais pas de meilleur moyen pour faire venir le lendemain. Je pris une chambre à l'Holiday Inn. Couché sur un grand lit plat, je fixai le plafond en lui enjoignant de se dissoudre, de me laisser léviter vers un lieu où l'absence cesserait de me broyer le crâne. Il s'y refusa, mais y étais-je moi-même disposé ? Une période d'inconscience m'aurait donné l'impression de m'éloigner de Nina, de laisser le temps délaver le réel, comme le vent balaye les feuilles d'automne. À 2 heures et des brouettes, je me suis levé, j'ai sorti mon téléphone de mon blouson, et j'ai réessayé son numéro. Toujours la boîte vocale.

Au petit matin, j'ai dû sombrer dans un état proche du sommeil, car j'ai visité des lieux fort éloignés de Thornton. Je me suis planté sur le balcon précaire de la maison de Nina, à Malibu, pour attendre son retour. En vain. Regagnant l'intérieur, je me suis introduit dans la maison de mon enfance, plusieurs centaines de kilomètres au nord de Los Angeles, dans la ville de Hunter's Rock. Le pavillon était glacial, vide, les coins et les murs gagnés de moisissures. Dans l'une des pièces j'ai aperçu, posé sur la table de chevet, un téléphone obstinément muet. Je suis resté là un certain temps, me disant qu'en dernière instance je recevrais un coup de fil de ma mère. Puis j'ai marché dans les bois, qui ressemblaient moins à la forêt chétive de Thornton qu'à celles, immenses, qui entourent Sheffer. Il avait neigé, et des ombres se mouvaient derrière les hauts troncs silencieux. Ces ombres possédaient une âme et elles connaissaient mon prénom, mais aucune ne souhaitait parler. Elles se contentèrent, sans hostilité particulière, de me regarder m'enfoncer parmi les arbres.

Ensuite je me suis retrouvé sur le canapé d'un appartement de Seattle où j'avais habité dix ans plus tôt. Cinq étages, avec vue sur Elliot Bay, l'un des meilleurs logements que j'aie jamais occupés. J'y ai vécu avec une femme, ma relation la plus longue à ce jour. Les trois quarts du temps, ma compagne se montrait dynamique, affairée, ennemie jurée de l'indolence et du pessimisme. Je lui avais donné un surnom : Hope. L'espoir. À cause de sa ressemblance avec le personnage de la série *Nos meilleures années*, mais aussi parce qu'elle en avait à revendre, de l'espoir. L'espoir – ou la certitude – que le monde était un lieu sûr et raisonnable, qu'il était au service des valeureux et des justes, et qu'au final il leur serait toujours favorable.

Mais de temps à autre, quelque chose se brisait en elle. Au café, son regard devenait subitement flou, ou bien elle étudiait ses mains ; à la maison elle fixait l'écran de télé sans voir la moindre image. Ses mouvements devenaient brusques, farouches, ses pupilles s'écarquillaient. Quand je le remarquais, je lui demandais si elle avait des soucis. Elle répondait par la négative, alors je reprenais ma bière, riais aux blagues de Chandler ou vidais mon paquet de chips – bref, les trucs importants.

Elle prit l'habitude de m'interroger, pour un oui ou pour un non :

— Tu crois que tout ira bien ?

— À quel propos, chérie ?

— D'une manière générale, répondait-elle à mi-voix. Tu crois que tout ira bien ?

Alors je lui disais que oui, je la serrais dans mes bras, nous revenions en douceur à des considérations plus terre à terre, et souvent le lendemain je l'entendais chanter sous la douche. Elle chantait comme un crapaud, mais j'avais plaisir à l'écouter.

Notre couple a tenu dix mois, puis s'est délité. À la fin, tout n'allait pas bien. Ni pour nous deux, ni en général.

Rien ne va jamais bien sur cette terre. Mais je suis content de ne pas l'avoir su à l'époque, et de ne jamais le lui avoir dit.

Je me suis réveillé à 5 heures, en pensant avoir perçu du bruit dans la salle de bains. J'ai dégringolé du lit pour m'y traîner, mais il n'y avait personne. Ces fredonnements avaient dû provenir d'une autre chambre, ou d'une autre époque. Devinant qu'il serait vain de me recoucher, je pris une longue douche. Les informations circulent très vite dans un hôtel et, lorsque j'arrivai au réfectoire, bien avant l'heure d'ouverture, les employés interrompirent leurs préparatifs pour me servir du café. Mes remerciements furent sans doute un peu chiches, mais je n'étais pas prêt à recevoir la sollicitude que l'on réserve aux endeuillés et, face aux marques de gentillesse, je perds toujours ma langue. J'ai préféré boire mon café dehors.

Telle la mer en hiver, le parking froid, gris et plat, était peu rempli. Je le scrutai désespérément, comme pour y faire réapparaître Nina. J'étais sûr d'une chose : si je ne la retrouvais pas, rien n'irait plus jamais bien.

Je savais aussi que le temps s'épuisait.

— T'as réussi à la faire parler. T'y arriveras peut-être une deuxième fois.

John secoua la tête.

— M'étonnerait.

Je n'avais pas bougé du parking, où John m'avait rejoint dix minutes plus tôt. Après plusieurs cafés et autant de clopes, je me sentais un peu plus vivant, sans être sûr de gagner au change. Le matin était froid, piquant, et la température avait chuté de cinq degrés par rapport à la veille. Si nous avions attendu ce jour pour creuser dans la forêt, il aurait fallu briser le sol à coups de marteau-piqueur.

— Je ne vois pas en quoi ça nous aiderait, ajouta John. Cette femme n'a pas enlevé Nina.

— Une fois, un type m'a dit qu'on menait une enquête en poussant dans toutes les directions possibles, dans l'espoir que l'une d'elles mènerait à bon port. Mais attends, c'était pas toi, ce type ? Pas plus tard qu'hier ?

— Tu sais que t'es le roi des casse-couilles, des fois ?

— Vu que les gens essaient régulièrement de me descendre, oui, je m'en doute. Écoute, John...

— Monroe ne me laissera pas recommencer.

— Pourquoi ? T'as obtenu des aveux. C'est-à-dire davantage que lui ou Reidel. Ou Nina, du reste.

— Je n'ai rien obtenu de cette fille, Ward. Pour des raisons que j'ignore, elle a choisi de me dire des choses qui peuvent être vraies comme elles peuvent être fausses. Et elle l'a bouclée. T'as bien vu la suite. Même si je lui tenais la jambe toute une semaine, elle ne me confierait plus rien.

Je détournai la tête.

— Mais qu'est-ce qu'on peut faire d'autre ?

— Y a fort à parier que Nina n'est même plus à Thornton. Pourquoi le ravisseur la garderait-il en ville, d'abord ?

— Parce que c'est ici qu'il habite.

— Oui, si on considère que c'est un cinglé du cru. Sauf que Julia vient d'avouer les meurtres. Par contre, si c'est un coup des Hommes de Paille, Nina doit être loin à l'heure qu'il est.

Tu comprends ça, j'espère ? (Il me considéra d'un air grave et choisit d'aborder une autre question :) Tu dois aussi te préparer à l'idée qu'elle puisse...

— Ferme ta gueule, répliquai-je, ne retenant plus ma colère. Chaque fois que je ferme les yeux, je la crois morte. Mais tant que je n'en ai pas le sentiment, ni la certitude, elle ne l'est pas.

— J'essaie juste de...

— T'es pas le seul à avoir perdu des proches, d'accord ? Descends de ta putain de montagne, un peu.

— Ça veut dire quoi, ça ?

— Cette façon que t'as de toiser la terre entière, sous prétexte que t'as perdu ta fille. Je comprends que ce soit horrible, John, sincèrement, et je regrette que mon frère ait fait ça. Mais les Hommes de Paille ont aussi tué mes vieux, tu te rappelles ? Ils m'ont amputé d'une partie de moi-même, et je ne laisserai personne recommencer. Nina est en vie tant que je ne suis pas obligé d'admettre le contraire, vu ? Alors tu l'élimines gentiment dans ta tête si ça te chante, mais tu gardes ça pour toi.

John pinça les lèvres, détourna le visage et disparut dans l'hôtel. Le temps que je me calme avec une nouvelle clope, il était déjà réapparu, une feuille de papier à la main.

— C'est le bordel, là-dedans. Ils doivent régler à la fois une affaire de tueur en série, un meurtre de flic et un enlèvement, sachant qu'ils sont largués sur les trois fronts. Sans compter la mini-bombe qui a sauté hier dans un centre commercial de LA. Bonjour l'ambiance.

— Une bombe ?

— Rien de méchant, visiblement. Un gamin sans histoires qui préférait jouer au petit chimiste plutôt qu'au sniper. Quoi qu'il en soit, Monroe est parti aux aurores, et personne n'a voulu me dire où.

— Il suffit de l'appeler sur son portable. Je pouvais le faire sans que t'embêtes les fédés.

— J'étais pas rentré pour lui. J'ai une autre idée.

— Ah ouais ? fis-je d'un air sceptique. Laquelle ?

L'immeuble se trouvait à quelques encablures du Starbucks, de l'autre côté du vieux centre, mais c'était déjà moins coquet. Le programme de reconversion de ces anciens bureaux avait dû prendre fin avec les années 1990. Nous sommes arrivés peu après 8 heures, et nous avons attendu dix minutes. Aucun flic en vue.

— Ils n'auraient pas dû poster un agent ?

— Ils sont sûrement occupés, expliqua John.

Nous descendions de voiture lorsqu'il reçut un appel. Il décrocha, écouta en silence.

— Non, répondit-il. Ce n'est pas moi. (Nouvelle pause.) Le plus loin sera le mieux. Ne dites rien aux flics du coin. Ne les appelez même pas. Et ne retournez pas chez vous ni au bureau avant de m'avoir consulté.

Il raccrocha, se mordit la lèvre.

— C'était qui ?

— Tu te souviens du site Web où il manquait les photos ? C'est Oz Turner qui le gère. Les images sont absentes car son serveur a été effacé. Hier, un Latino d'allure louche est venu le demander au boulot. Heureusement, Oz était en pleine crise de parano et s'était fait porter pâle. Il a passé la nuit chez un ami. En rentrant chez lui ce matin, il a découvert que sa porte arrière a disparu, ainsi que tous ses dossiers et son ordinateur.

— Mais c'est qui, ce type ? Et c'est quoi, cette histoire de photos ?

— Ça concerne les Hommes de Paille. Leurs origines.

— Lointaines, ces origines ?

— Assez pour que tu cries au délire.

— Alors ça ne m'intéresse pas.

On accédait à l'appartement de Julia Gulicks par un escalier séparé, sur le côté du bâtiment. La porte était barrée par un ruban jaune et placardée d'écriteaux ENTRÉE INTERDITE.

— Monroe affirme qu'ils s'en sont déjà occupés, rappelai-je.

— Les flics ont peut-être fouillé les lieux après l'arrestation, mais ils ne cherchaient que des preuves directes. Depuis hier matin, les priorités ont changé.

Je surveillai la rue pendant que Zandt lacérait le ruban et maniait un trousseau de tiges métalliques. La serrure céda en deux minutes. Nous nous glissâmes à l'intérieur et je refermai derrière nous.

Le logement n'était pas grand. La porte donnait directement sur la pièce principale, bordée d'une kitchenette. L'un des murs était recouvert de livres, et la grande table de bois servait aussi bien aux repas qu'au boulot. Les deux portes du fond desservaient la salle de bains et la chambre à coucher. Cette dernière était minuscule : le lit double occupait tout l'espace, et l'on pouvait à peine ouvrir le placard. Chaque pièce dénotait de gros efforts d'agencement.

Il nous fallut moins d'une minute pour scruter les lieux.

— Ça t'inspire quoi ? demanda Zandt en me rejoignant dans le séjour.

— Ça m'inspire qu'on ne va rien apprendre ici. Les flics ont bien fait leur boulot. Et puis, d'après le légiste, la victime anonyme a croupi quelque part avant de se faire mutiler. Mais je ne vois pas comment elle aurait tenu ici.

— Sur le plancher du salon. Sur la table ou sur le lit, couchée à côté de Gulicks. On ne sait jamais comment vivent les cinglés.

— À supposer qu'elle ait fait le coup. Ce dont Nina doutait.

— Mais Nina avait tort, Ward.

— J'ai hâte de t'entendre le lui expliquer en face, répliquai-je avant de m'approcher du mur rempli de bouquins.

Je frappai trois coups, qui sonnèrent creux.

— Les cloisons sont minces comme tout. Je n'aimerais pas devoir tuer quelqu'un dans un lieu si mal isolé. Non que je sois expert, tu me diras.

— Elle a drogué Widmar, répondit John. L'autre aussi, sûrement.

— Mais elle n'a pas tué Widmar entre ces murs, n'est-ce pas ? Ce qui, d'une certaine façon, me donne raison. Et puis, tant qu'un type n'est pas refroidi, il y a toujours le risque qu'il se mette à crier. Là-dessus les voisins appellent les flics et le projet tombe à l'eau. Autrement dit, dans l'hypothèse où elle a bien liquidé ces types, c'est un autre lieu qui nous intéresse en premier chef.

— Et tu connais l'adresse ?

— Non.

— Alors on fouille ici.

Zandt prit la chambre, tandis que j'inspectais rapidement la cuisine et le séjour. Les tiroirs renfermaient une poignée de couverts de facture honnête, les placards quelques aliments non périmés. La surface et le bas des meubles cachaient de la poussière et trois bracelets élastiques. Je déplaçai les meubles, les palpai, sans rien noter d'intéressant. Puis je parcourus la bibliothèque, ouvrant et secouant chaque ouvrage, retournant chaque vase ou bibelot. Quatre trombones ; sept coupons d'abonnement à des revues, utilisés comme marque-pages ; l'épingle d'une broche. Les livres étaient assez standards : romans en format poche, grimoires non lus sur la réussite au travail et/ou les régimes minceur et/ou la vie en général, avec en couverture les mêmes tronches dégarnies et véreuses ; un guide d'initiation à Windows et deux manuels pointus sur un logiciel de comptabilité d'entreprise. Dans un tiroir du bas, je débusquai une petite boîte de bois remplie de photos de gens tout à fait ordinaires, posant avec ou sans Julia. Je l'emportai sur le canapé, en attendant que Zandt termine la salle de bains.

Il ressortit bredouille.

— Rien de probant, dit-il.

— Il faudrait déterminer ce que les flics ont déjà pris.

— Je n'ai rien vu dans le genre carnet d'adresses ou journal intime.

— Voici ce qui s'en rapproche le plus.

Nous étudiâmes les clichés pendant quelques minutes. Pour se persuader que l'on est soi-même et non un autre, il suffit de regarder les photos qui vous sont chères : on les croirait fausses tant elles sont banales. Des femmes aux cheveux bouclés et aux yeux rouges, des types à la peau grasse levant leur bière, des vieux au regard mi-figue mi-raisin... Tous se fendaient du sourire réglementaire, comme pour se fortifier face à la douleur. Je fus vite lassé, emporté dans une vilaine rêverie. Existait-il la moindre photo de Nina et de moi ? N'avions-nous pas laissé une marque aussi triviale de notre histoire ?

— Regarde un peu, lança-t-il.

J'avisai la photo brandie par John. Un bout de forêt, par temps couvert.

— Eh bien, quoi ?

— Tu ne trouves pas que ça ressemble au site du deuxième cadavre ?

— Mouais, pour les arbres. Mais je ne pense pas qu'il s'agisse du même endroit. La végétation est différente.

— Pas sûr. Regarde les tons. Soit cette photo est restée longtemps exposée au soleil, soit c'est l'une des plus anciennes de la boîte.

Il avait raison. J'examinai le cliché de plus près, avant de le retourner.

— Tu crois que ces chiffres au dos peuvent nous indiquer une année ?

— Ça vaut le coup d'essayer. (Il se leva.) C'est bon, on a fait le tour. Je passerai quelques coups de fil en route.

— On va où ?

— L'interroger une nouvelle fois.

— Je croyais que c'était peine perdue.

— Plus maintenant. Si cette photo nous montre effectivement le lieu du crime, on peut prouver la préméditation. Or, si son attitude brindezingue et la façon dont elle s'est jetée sur moi visaient à plaider la démence, cet élément risque de faire très mal. De quoi l'achever.

— M'étonnerait que ce soit notre cartouche miracle. Et je doute qu'il s'agisse du même endroit.

— Peu importe. L'essentiel, c'est que ce soit possible. (Il passa la porte.) Quand on sait bien viser, même les cartouches à blanc font l'affaire.

— Elle n'est plus ici, indiqua le flic de l'accueil.

C'était l'un de ceux qui avaient arraché Zandt aux griffes de Gulicks. Il semblait sur la défensive.

— Où est-elle ?

— À l'hosto.

Je le dévisageai.

— Pourquoi ?

— Écoutez, les gars, je ne devrais pas...

— C'est votre faute ? coupa Zandt.

Il parlait comme s'il avait son casier dans ce comico et qu'il faisait l'appel du matin.

— Non, monsieur, je vous jure, débita le type, soudain disposé à nous dire tout ce qu'il savait. Écoutez, j'ai... J'assurais sa garde. J'ai jeté un œil à 4 heures du matin et elle dormait. Tout allait bien.

— Et ensuite ?

Le type lâcha dans un souffle :

— Peu après, elle a tenté de se supprimer.

— Comment ça ?

— Elle s'est mise debout sur le lit, les mains dans le dos, et s'est laissée tomber en avant. La tête la première. Trois fois de suite, semble-t-il. Nous n'avons rien entendu, à aucun moment.

— Génial, pesta Zandt. Et un point de plus pour la thèse de la démence !

— Vous plaisantez, objecta le flic. Vous seriez capable d'un tel geste, vous ? Deux fois, trois fois de suite ? Pas moi. Elle a voulu se tuer, je vous dis. Cette femme avait décidé d'en finir.

— Quel est son état ?

— Comme pour n'importe qui à sa place : pas brillant.

Le flic nous laissa pour répondre au téléphone. Nous nous écartâmes du guichet. Je me frottai le visage, comme pour empêcher ma tête de se décrocher.

— On peut dire adieu à notre plan d'enfer. Et maintenant, on fait quoi ?

Pour la première fois, Zandt semblait dépassé.

— J'imagine qu'on part à l'hosto. Je parie que c'est là que Monroe a filé aux aurores. Ce qui signifie sans doute que la donzelle n'en a plus pour longtemps. Si ça se trouve, elle ne peut même plus parler.

— Il le faut. Sinon... On perd pied, John. Cette histoire nous file entre les doigts.

Mon portable sonna. J'aurais voulu que ce soit Monroe, mais le nom sur l'écran anéantit cet espoir.

— C'est Unger, indiquai-je avant de répondre. Qu'est-ce que tu veux, Carl ?

— T'es où ?

— À Thornton.

— Je me trouve tout près. J'aimerais te parler.

Je n'en étais plus à me demander si Unger était un Homme de Paille ou assimilé. Ça me faisait une belle jambe. Soit John et moi l'aurions, soit c'était lui qui nous aurait.

— Très bien, répondis-je. J'ai d'abord une course à faire, je serai de retour dans une heure. T'as qu'à me retrouver au Mayflower, un bar sur la route d'Owensville.

— J'y serai.

— Juste une chose, Carl. Je te conseille de la jouer réglo, ou on te butera sur-le-champ. Crois-moi.

Refermant l'appareil, je vis que John et le flic me dévisageaient.

Je haussai les épaules.

— Autant qu'il sache à quoi s'en tenir.

CHAPITRE 28

Ils avaient atterri à Huntsville, en Alabama, dans les premières heures du matin. Trois véhicules les attendaient sur le parking de l'aéroport, le moteur en marche. Un type ouvrit la portière pour les accueillir. Une petite vingtaine d'années, l'air sûr de lui. Lee l'avait déjà vu quelque part. Soit lors de sa première rencontre avec Paul, soit la veille à la cafétéria de Belle Isle, parmi les mecs disséminés dans la salle. Il n'aurait su dire.

Lee s'assit à l'arrière, à côté de Paul. Le jeune prit la banquette d'en face, flanqué d'un autre homme, plus vieux, élégant, au regard bleu perçant. Ce dernier salua Paul d'un signe de tête avant de lancer :

— Salut, Lee. Bienvenue à bord.

Le cortège quitta le parking et s'enfonça dans la campagne en un convoi lâche. Pour ne pas attirer l'attention, les conducteurs laissèrent d'autres véhicules s'intercaler entre eux. C'était la moindre des choses, jugea Lee, qui, à leur place, aurait varié la couleur et le modèle des bagnoles, ou essayé d'emprunter trois itinéraires différents. Mais il avait la vague impression que ces types s'en foutaient.

Ils atteignirent le Tennessee par la 231, puis bifurquèrent à l'est vers des collines escarpées. Lee contemplait le paysage. Le voyage depuis LA l'avait lessivé, et il devait finir de digérer les événements de la veille. Il se trouvait assis à côté du meurtrier de Brad – et de Karen, indirectement. C'était un peu bizarre.

Les deux autres passagers consacrèrent la première partie du trajet à passer de brefs coups de fil plus ou moins sibyllins. Paul émit un seul appel, pour annoncer son arrivée à quelqu'un. Lee ne pouvait comprendre de quoi il retournait, mais il n'en éprouvait pas le besoin. Il allait où ils allaient. Il faisait ce qu'ils faisaient. Et tout s'éclaircirait, pensait-il.

Au bout de deux heures, ils sortirent de l'autoroute pour prendre une série de voies secondaires menant à une petite maison forestière délabrée, au jardin terreux infesté de hautes herbes et cerné de vieilles épaves. Les trois véhicules se rangèrent côte à côte, et tout le monde descendit.

Les deux autres voitures libérèrent chacune cinq individus. Vêtus de costumes et de manteaux sombres, effectuant des mouvements de soldats. Ils suivirent Paul et son escorte à l'intérieur de la maison, qui semblait avoir été bâtie par un type aussi débrouillard que cinglé, à partir de matériaux de récupération. Lee alla se soulager dans les buissons, puis revint s'adosser à la voiture.

Le groupe ressortit au bout de quinze minutes, lesté de matériel. Des caisses de bois, portées à deux telles des petites civières, ou des sacs à bandoulière visiblement très lourds. Rien n'était étiqueté et rien ne semblait avoir été acheminé par des voies légales. On chargea le tout dans les coffres, puis on remonta à bord et les trois moteurs redémarrèrent à l'unisson.

Les voitures rebroussèrent chemin vers l'autoroute, où elles se dispersèrent sur un petit kilomètre de circulation, comme précédemment. Elles maintinrent le cap à l'est, avant de pousser vers le nord, pour le Kentucky. Peu après la frontière de l'État, on répéta le même scénario en suivant le véhicule de tête sur des routes de campagne. La maison suivante était bien plus cossue, un édifice aux colonnes hautes de deux étages, avec une petite mare entourée d'une pelouse immaculée. Les voitures s'arrêtèrent et l'équipée descendit. Ce coup-ci, Lee attendit vingt-cinq minutes, à regarder deux jeunes cavalières parcourir la prairie au petit galop. La queue des chevaux voletait dans le soleil du matin. Les adolescentes n'eurent pas un regard pour Lee. C'était comme si les voitures n'étaient même pas là.

Revinrent les autres, avec des sacs moins nombreux que tout à l'heure, et portés avec délicatesse. Il y eut une dernière et brève halte, dans une station-service désaffectée où l'on ramassa deux sacs anguleux cachés à l'arrière des locaux. Au moment de repartir, le jeune et le type aux yeux bleus changèrent de voiture, laissant Lee et Paul en tête à tête. Le soleil arrivait au zénith.

Peu après, ils entrèrent en Virginie.

Paul scrutait le paysage d'un air méditatif. Il n'avait pas bougé depuis une demi-heure, pas un muscle, et ça devenait flippant. Mais Lee avait appris à se taire devant un interlocuteur muet

– les Hudek cultivaient les bonnes manières d'antan – et ne souhaitait pas énerver son compagnon de route. Il se souvenait aussi d'une certaine fête chez les Metzger, où il avait discuté avec un client du père de Brad, un scénariste dénommé Nic Golson – patronyme qu'il lâchait à tout bout de champ, comme pour aider les gens à le mémoriser, preuve que c'était efficace. Ce type avait exposé à Lee la théorie suivante : dans les réunions importantes, au lieu de jacasser d'emblée afin de passer pour un mec chaleureux et conciliant, on obtenait de meilleurs résultats en jouant les bourrus. N'importe qui pouvait se montrer cordial, alors qu'une attitude fermée forçait les gens à venir sur votre terrain, vous aidait à gagner des places dans la hiérarchie, vous rendait incontournable. C'est du moins ce que le gus avait prétendu, avant de repartir les poches pleines de petits-fours.

Lee avait testé cette méthode au cours des six derniers mois, et c'était assez concluant. Néanmoins, après une heure supplémentaire de route et de silence, il finit par craquer :

— Alors comme ça, les flics me recherchent, à L.A. ?

Paul tourna lentement la tête.

— J'imagine.

— Tu sais, tu aurais aussi bien pu me proposer de vous suivre. De vous aider sur ce deal.

— Ce deal ?

— Ouais. Cette mission, quoi. Y avait pas besoin de faire tout un barouf. Me griller auprès des keufs, etc.

— Et si tu avais refusé ?

— Tu connais mes vieux, apparemment. Comment ça se fait, d'ailleurs ? Comment vous vous êtes connus ?

— De vieux amis.

— Ils auraient pu me convaincre, non ? Ils savaient réellement que t'allais m'infliger tout ça ?

Paul éluda la question :

— Les hommes de ton âge ne se laissent pas facilement influencer, c'est bien connu. Il faut un cas de force majeure.

— En tout cas, vous avez amassé une sacrée cargaison de came. J'ai jamais vu autant de sacs et de caisses, et les mecs qui les portent ont pas l'air de rigoler. (L'homme souriait un brin.) La question, c'est qu'est-ce que ça peut bien foutre, que je sois là ou pas ?

— Tu joueras un rôle clé, t'inquiètes pas.

— Tu pourrais peut-être me dire de quoi il retourne ?

— Eh bien, à ton avis ?

— Vous avez un énorme deal en vue. Un gros festival ou quelque chose. Vous emmagasinez un gros stock de matos, et...

— Il ne s'agit pas de drogue, Lee. Essaie de penser à plus long terme.

— Je croyais que vous faisiez uniquement dans la drogue.

— Disons que ça nous a bien aidés au cours des dernières décennies. Ça rapporte du fric et c'est beaucoup moins fatigant que la mine.

— La mine ? (Lee tâcha de reprendre le fil de ses pensées.) Alors c'est pas une histoire de dope ?

— C'est une histoire de passé et de nouveau départ. Une forme plus personnelle de rupture. Ramener le combat sur son vrai terrain.

— Tu vois, t'arrêtes pas de balancer des trucs, et je comprends que dalle !

— Pas grave. On s'en remettra.

Lee cogita quelques instants, puis attrapa la poignée et ouvrit la portière à fond.

Il n'avait aucune intention de sauter – la voiture faisait un bon 110 – mais il obtint l'effet voulu. Le véhicule zigzagua quelques secondes, le temps que le chauffeur corrige le changement aérodynamique.

Imperturbable, Paul se pencha et referma la portière.

— Ne fais pas ça, conseilla-t-il.

— Alors me traite pas comme un gosse.

— Justement, Lee, tu ferais bien de me considérer comme un père – qui te cramera la cervelle si tu le baises. (Un grand sourire.) Est-ce que c'est clair ?

Son ton affable n'empêcha pas Lee de frissonner.

— Pigé. Mais j'ai déjà un père, merci.

Paul hocha la tête, toute contrariété envolée.

— C'est vrai, répondit-il. Tiens, Lee, je vais te raconter un truc, qui explique en partie le choix de notre destination. Quand j'étais tout petit, j'habitais en Californie du Nord. Mon père était un homme important. Pas riche comme ton père, mais important. Une nuit, une bande d'hommes et de femmes est venue dans les bois où il vivait et ils l'ont assassiné. Ils m'ont enlevé et m'ont gardé quelque temps, avant de se dire qu'ils ne m'aimaient plus tellement et de m'abandonner au coin d'une rue.

— Putain, ça craint, mec.
Paul opina.
— Merci, Lee. Oui, ça craint. Par chance j'ai survécu, notamment grâce à certaines personnes que j'ai rencontrées. Nous avions des intérêts communs, mais c'est devenu bien plus que ça. Ils sont comme une grande famille. Ton papa et ta maman en font partie. Toi aussi. On s'était déjà rencontrés, Lee, c'est vrai, même si tu ne t'en souviens pas. Peu après ta naissance, et quand tu avais trois ans.
Il sortit quelque chose de son blouson, un objet qu'il avait dû garder sur lui en attendant le moment propice. C'était une vieille photo, où l'on voyait les parents de Lee dans un jardin en compagnie d'un jeune homme. Il s'agissait de Paul, âgé d'une petite vingtaine d'années. Un bambin se tenait fièrement debout, agrippé à la main de Paul.
— C'est moi, ça ?
— Tu faisais déjà partie de la bande.
— Quelle bande ?
— On veille les uns sur les autres. Parfois on se dispute, comme dans toutes les familles, mais fondamentalement on est tous du même côté. On est forts en ce moment, mais des ennemis rôdent sans cesse. Un homme, en particulier, qui a tué plusieurs d'entre nous.
— Faut le bousiller, dans ce cas.
Paul sourit.
— C'est peut-être ce qu'on fera. T'es un gosse bien, Lee. Tes parents t'ont bien élevé. Tu t'entends bien avec ton paternel ?
— Ouais. Il est cool. Il a pas mal d'humour.
Paul éclata de rire, comme malgré lui. Lee sourit, surpris mais content que l'ambiance se réchauffe un peu.
— C'est rien de le dire, répondit l'homme après une pause. Tu ne t'es jamais interrogé sur ton nom ?
— Hudek ? Non, pourquoi ?
— Et tes prénoms ?
Lee fronça les sourcils.
— Qu'est-ce qu'ils ont de spécial ?
— Dis-les, pour voir.
— Lee John.
— C'est ça. Alors, dis-moi comment tu t'appelles.
— Ben, Lee John.
Paul le dévisagea, dans l'attente d'une réaction, puis sa voix redevint glaciale :

— Bon Dieu, ces gosses. Vous êtes vraiment des putains d'ignares[1].

Enfin les voitures ralentirent.

Cela faisait une demi-heure qu'ils suivaient une route monotone à travers des kilomètres et des kilomètres de forêt rabougrie. Paul reçut un appel, qui semblait provenir de la voiture de tête. Il approuva ce qu'on lui disait, et peu après les trois voitures se rangèrent à l'approche d'un tournant. Paul ouvrit sa portière et descendit, talonné par Lee.

Les véhicules étaient ouverts. Lee se préparait une nouvelle fois à attendre sans comprendre, lorsqu'un détail le frappa.

Les passagers des bagnoles avaient changé.

Il les regarda, bouche bée. Leurs habits, leur attitude... Il n'en reconnaissait pas un seul. Il y avait même deux femmes. Ce n'était pas avec ces gens qu'il avait voyagé.

À moins que...

Il reconnut soudain le jeune qui leur avait tenu la portière devant l'aéroport de Huntsville. Il portait à présent un sweat-shirt à capuche, des baskets délacées et un jean baggy, ainsi qu'un petit sac à dos rouge. Il passa devant Lee pour sortir du matériel de leur voiture. Le type même de l'ado qu'on croise dans tous les centres commerciaux du continent, avec son skateboard et son roulement d'épaules trop-cool-pour-se-tenir-droit.

Lee examina les autres et reconnut quelques traits par-ci par-là. Le type propret – celui qui les avait accompagnés au début – avait revêtu un uniforme de flic, si convaincant que Lee eut un instant de panique. Deux types arboraient des costards bon marché, un autre un bleu de travail crasseux. Ce dernier ouvrit le coffre de la deuxième voiture pour s'enduire les mains d'huile. La première femme qu'avisa Lee était vêtue en Banana Republic de la tête aux pieds, dans un gros pull moelleux. Avec ses cheveux tirés en arrière, Lee n'avait même pas même vu qu'il s'agissait d'une poule. Elle ôta une barrette et secoua sa crinière. Sa transformation achevée, on l'aurait imaginée déambulant dans la grand-rue d'un charmant village, avec une poussette haut de gamme, partant retrouver ses copines pour prendre un petit crème – voire, dans un violent accès de goinfrerie, un muffin

1. Allusion voilée à « Mon nom est Légion » dans la Bible (Marc 5, 9), jouant sur l'homonymie entre « Lee John » et l'anglais Legion. (N.d.T.)

allégé histoire de culpabiliser jusqu'au soir. Elle sourit brièvement, à personne en particulier, comme pour s'entraîner. La seconde femme avait passé une sorte de blouse vert clair, façon épicière.

Tous portaient quelque chose. Une sacoche. Un cabas. Une boîte à outils. Un carton rectangulaire frappé du sigle UPS. Un sac de courrier.

Deux minutes plus tard, les malles et portières étaient refermées. Deux véhicules reculèrent lentement sur la route pour opérer un gracieux demi-tour. Puis ils mirent les gaz et disparurent au loin, abandonnant la troupe sur la chaussée. En l'absence des voitures, la présence de ces gens semblait encore plus saugrenue. Comme si des extraterrestres avaient décidé de relâcher arbitrairement quelques otages, en les renvoyant d'un coup de laser sur le bas-côté de nulle part.

Lee regarda Paul évoluer dans le groupe. Il eut un mot pour chacun. On hochait la tête, on se serrait la main... La mère au foyer décocha un sourire plus sincère que les précédents.

Sans signal visible, tous se mirent en marche vers la forêt. D'abord ensemble, puis par deux ou par trois, dans des directions divergentes, afin d'atteindre leur mystérieuse destination en ordre dispersé.

Revenant sur ses pas, Paul fit signe à Lee de remonter en voiture.

Ils roulèrent pendant vingt minutes avant d'atteindre une ville. Pas grande. Un bled comme on en trouve partout, où l'on s'arrête pour faire le plein et boire un café avant de repartir, tout en bénissant le ciel d'habiter un endroit où la parade de Halloween n'est pas l'événement le plus remarquable de l'année. La voiture se faufila dans des rues longées de pavillons en bois, puis entra dans le supposé centre-ville. Des immeubles anciens ; un commissariat, une école maternelle, un Starbucks...

Lee observait en silence. Cet endroit n'avait rien de palpitant. D'après son système de valeurs, en tout cas.

— C'est le trou du cul du monde, ici.

— Tu exagères, Lee.

— Alors tu dois voir un truc que je ne vois pas.

— Moi, je vois des logis d'abeilles travailleuses, de bons téléspectateurs et d'excellents consommateurs. Des industries de commerce et de services qui tournent tranquillement. Et de longues années d'histoire, avec quelques zones d'ombre. Cette ville est plus intéressante qu'il n'y paraît. Plus vieille, aussi.

— N'empêche. Ça reste un trou.

— Des gens y vivent, Lee. De vraies gens. Bien de chez nous. De purs Américains, du moins à les entendre. Les petites villes sont la colonne vertébrale de cette nation.

— C'était peut-être vrai il y a cent ans. Aujourd'hui, c'est dans les grands centres que ça se passe, mec.

— Beaucoup le pensent, visiblement.

À l'approche d'un Holiday Inn, Paul demanda au chauffeur de ralentir, ce qui piqua la curiosité de Lee. Il crut alors reconnaître l'hôtel aperçu à l'écran du centre commercial de Belle Isle – celui qui avait capté l'attention de Paul. Deux voitures de flics stationnaient près de l'entrée avec quelques berlines anonymes. Puis il vit un type en coupe-vent FBI.

— Euh, s'inquiéta Lee, on tient vraiment à traîner ici ?

— Peut-être plus tard, dit Paul, et la voiture reprit de la vitesse.

Ils tournèrent à gauche pour ressortir de la ville. Au bout de quelques minutes, ils s'engagèrent sur une petite route décrivant une longue courbe, à la lisière des bois. Elle débouchait enfin sur un patelin nommé Dryford, où les maisons étaient de taille honnête, mais dont l'heure de gloire remontait bien à cinquante ans.

Un nouvel embranchement les aiguilla vers le fond d'un cul-de-sac où s'élevait une grille. Juste avant se trouvait un portail. Le chauffeur sortit pour l'ouvrir. Derrière s'étendait un jardin profond, envahi d'herbes folles. Tout au bout, au milieu du chiendent et d'un bouquet d'arbres livrés à eux-mêmes, se dressait un pavillon. Un peu plus gros que ceux de la « ville », et assez délabré. Le chauffeur remonta et les conduisit dans la propriété. On voyait qu'elle était restée inhabitée pendant des années. Des planches se détachaient de la façade ; le toit perdait ses bardeaux et les carreaux semblaient renfermer une vieille brume. Un antique Combi Volkswagen stationnait dans l'allée, qui lui aussi semblait stagner depuis des années, bien que ce fût peu probable.

— Oh, le joli cheval blanc, railla Paul.

Ils quittèrent le véhicule. La porte coulissante de la camionnette s'ouvrit à leur approche. L'intérieur était sombre. Un homme en sortit, referma la portière presque à fond. Il était grand et plutôt vieux, à en juger par son visage et ses épaules légèrement voûtées. Mais costaud.

— Salut, James, dit Paul d'une voix froide. Je te faisais confiance.

— C'était un accident, répondit l'homme. Ça fait un bail et je manque d'entraînement. La situation m'a échappé.

— T'as peut-être fait ton temps.

— Peut-être bien. T'aurais dû confier ça à un de tes bébés zombies. (Il posa les yeux sur Lee, qui se surprit à frémir. Quand il vous matait, ce type n'avait plus rien d'un vioque.) Ce n'est pas le même, dis-moi ?

— En effet, répondit Paul. Voici notre nouvelle vedette. (Il leva le menton et renifla.) T'as cuisiné, James ?

— C'est ta faute si j'ai craqué.

— Erreur, James. Tu te souviens de notre petite discussion, la nuit où je t'ai sauvé la mise ? Tu es ce que tu es. Tu le sais. Un « un » ne redevient jamais un « zéro ».

Le type âgé considéra Paul un instant, comme s'il se pliait aux conseils d'un être qu'il connaissait depuis toujours.

— Non, finit-il par admettre. J'imagine que non.

Paul marcha jusqu'au fourgon et rouvrit la porte latérale. Intrigué, Lee le suivit. Le vieux aurait sans doute préféré qu'ils restent dehors, mais on ne lui demandait pas son avis. À son tour, Lee remarqua l'odeur. Un peu terreuse, légèrement suave. Il n'aurait su dire ce que le vieux avait mitonné, mais malgré sa faim, il savait d'ores et déjà qu'il refuserait d'y goûter.

Puis il s'aperçut d'une présence.

Tout au fond, ligotée dans une drôle de position, reposait une grande fille maigre. Bâillonnée, les yeux bandés.

Lee était mal à l'aise. Ça lui rappelait certains documentaires du câble déconseillés aux âmes sensibles.

Paul grimpa dans le fourgon et s'accroupit devant la femme. Il lui retira son bandeau. Lee se tenait assez près pour la voir écarquiller les yeux. Alors Paul lança :

— Salut, Nina. Tu te souviens de moi ?

CHAPITRE 29

L'hôpital avait dû ouvrir ses portes en fanfare dix ans plus tôt, pour être aussi vite oublié. C'était l'œuvre d'un architecte rêvant de postérité et muni d'une règle bien droite, qui se mit sans doute des baffes en s'apercevant qu'il avait oublié les fenêtres, ainsi que le fait que des gens risquaient d'y séjourner. Le hall sentait les produits chimiques et grinçait au rythme des talons de caoutchouc sur le lino, les murs étaient bardés d'affichettes glauques ou barbantes. Je n'aime pas les hôpitaux. Ils ne m'ont jamais réussi. On y va pour se rendre malade ou pour apprendre qu'on l'est.

Et pour mourir, bien sûr.

Dès que j'aperçus Julia Gulicks, je sentis que cette troisième option était la sienne. Nous avons descendu un long couloir gris, dans l'un des derniers étages, jusqu'à une chambre individuelle située en fond de bâtiment. Un policier était assis devant la porte, mais le hublot carré au-dessus de sa tête laissait entrevoir un corps allongé sur un lit, immobile telle une maquette de montagne faite de chutes de bois et d'un drap. La majeure partie du visage était bandée. Le seul signe distinctif était la chevelure rousse dispersée sur l'oreiller, qui perdait elle aussi sa vigueur et son éclat, signe que la vitalité se faisait la belle.

Avant que le flic ait eu le temps de nous rembarrer, des pas claquèrent dans notre direction.

— Qu'est-ce que vous fichez là ?

C'était Monroe, muni d'un gobelet de polystyrène fumant. Je revis soudain Nina me confier combien son boss était accro. Mais ce souvenir ne m'était d'aucune aide.

John brandit la photo.

— On est venus pour ça.

— C'est quoi ?

— On pense qu'il s'agit du coin de forêt où a été retrouvé le second corps.

— Et ça sort d'où ?

— De son appartement. En venant ici, j'ai demandé à quelqu'un de déchiffrer le code inscrit au dos. Ce cliché a été tiré il y a cinq ans. Il y a préméditation, Monroe. Et ton affaire est bouclée.

— Ce n'est rien de plus qu'un paysage.

— J'aimerais...

— Elle ne peut rien voir, Zandt. Tu ne l'as pas regardée ? Ses yeux sont recouverts de gaze, et quand bien même, ça ne changerait rien. Elle a le crâne en miettes, et de sévères lésions aux deux lobes frontaux. Ils l'ont mise sous perf et la maintiennent en vie, mais pour l'instant c'est tout ce que la médecine peut faire.

— Tant pis si elle ne voit pas, insista John. Il suffit qu'elle entende. Je veux juste lui parler.

— Mais elle n'écoutera pas. La prochaine chose qu'elle va faire, c'est mourir.

Puis il réfléchit et s'écarta.

— Si quelqu'un pouvait parler à sa place, dit-il, je ne te laisserais jamais y aller.

— Je crois qu'elle n'attend que ça, répondit John.

Monroe s'arrêta au bout du lit. Je restai en retrait, près du mur. Julia Gulicks était reliée à une batterie d'appareils. Leur but était sans doute de rassurer, mais si vous voulez mon avis ça ne marche pas. Être connecté à une machine n'est jamais bon signe. Même les comateux le savent.

Quand John se percha au bord du lit, il me sembla que la jeune femme sentit le matelas tanguer. Elle tourna imperceptiblement la tête et ses lèvres se décollèrent avec un petit bruit sonore.

— Julia ?

Elle referma la bouche et ramena sa tête de l'autre côté.

— C'est à moi que vous avez parlé hier soir. Au poste de police. Vous reconnaissez ma voix ?

Aucune réaction.

— Julia ? Pourquoi ce geste désespéré ?

Je ne m'attendais pas à cette question-là. Monroe non plus, à voir son visage. Elle avait fait ça parce qu'elle était folle. Ou parce qu'elle était coupable et redoutait le procès. Mais peut-être pas...

Elle se retourna vers John. Puis sa langue se mit à s'agiter dans sa bouche, un peu trop vite. John saisit un verre d'eau sur la table de chevet et lui versa quelques gouttes entre les lèvres. Sa langue s'arrêta peu à peu.

— Canot, dit-elle d'une voix limpide. Pingouin dans le hall de la banque. C'est un mélange merdique, faut rajouter du beurre.

Je regardai mes pompes. Ce n'étaient pas tant ses paroles que les mouvements de sa langue. Trop autonome. Comme un rat voulant quitter le navire.

Il lui posa d'autres questions. Si elle se rappelait les photos qu'elle conservait dans une boîte, à son domicile. Si elle se souvenait d'avoir photographié la forêt près de Thornton et, le cas échéant, à quel endroit exactement. Si elle avait jadis gardé ce cliché en évidence, et pourquoi.

Il n'obtint aucune réponse. Rien ne prouvait qu'elle n'était pas rendormie, ou, fût-elle éveillée, que son esprit ne s'était pas perdu dans quelque abîme. Je n'y connaissais rien en lésions cérébrales, et j'ignorais si elles étaient irréversibles, mais je pressentais que Julia était partie et qu'elle ne reviendrait pas. À la place de Mme Lawrence Widmar, j'aurais sans doute enragé : on a toujours besoin d'un coupable sur qui décharger sa haine. Mais Gulicks se trouvait désormais au-delà du bien et du mal.

— Il est ici, lâcha-t-elle subitement.

Je relevai la tête. Pour absurdes qu'ils fussent, c'étaient ses premiers mots depuis dix minutes.

— Que dites-vous ? demanda John et se penchant plus près.

La seconde fois fut plus précise :

— Il est ici, pas vrai ?

— Qui est ici ? intervint Monroe.

— Où suis-je ?

John leva la main pour que Monroe cesse de la troubler.

— Vous êtes à l'hôpital, Julia.

— C'est ma naissance ?

— Non.

— Vous avez parlé d'une photo ?

— Nous avons trouvé un cliché de forêt. Vous vous souvenez de l'avoir pris ?

— Il y a plein de forêts. Depuis toujours. Depuis que je suis toute petite, petite.

— Pourquoi avez-vous pris cette photo, Julia ?

— Quelle photo ?

— La photo de la forêt.

— Garder les souvenirs. Hein ? Hein ? Hein ? Magnifique.

— Je ne comprends pas ce que vous dites.

— Vous êtes déjà allé à Disneyland ?

— Oui, répondit John. Il y a longtemps.

— Personne ne voulait m'emmener quand j'étais petite. J'y suis allée par mes propres moyens dès que j'ai eu l'âge. Mais c'est pas pareil.

— Je comprends.

— Personne ne devrait avoir à visiter Disneyland tout seul.

— Julia...

— Dites donc, vous ne m'auriez pas chié dessus ? Y a une de ces odeurs ! (Sa langue se remit à pendre. Son menton tremblait.) Oh, c'est mauvais...

— Pourquoi, Julia ? Pourquoi les avez-vous tués ?

— Je me suis plantée, d'accord ? Ça aide pas, vous savez. C'est une putain, une putain de maladie. (Puis sa voix baissa, de presque une octave.) Donne-moi l'enfer en douze étapes, sale nègre.

Elle éclata de rire, en cambrant le dos. Puis toussa et vomit.

Monroe pressa le bouton d'appel et John la fit rouler sur le côté. Bientôt la pièce grouillait de toubibs et nous étions refoulés dans le couloir.

Nous avons patienté en silence, pendant qu'à l'intérieur on faisait le nécessaire. Au bout d'une demi-heure, les médecins repartirent au compte-gouttes, laissant une infirmière dans la pièce. La dernière blouse blanche à ressortir fut une femme qui nous fusilla du regard tout en claquant la porte derrière elle.

— J'avais dit que c'était trop tôt pour lui parler.

— On la soupçonne de deux meurtres, expliqua Monroe.

— C'est votre problème. Elle est mourante. Les blessures de son cerveau compromettent le système nerveux autonome. Il s'agit de la partie qui contrôle normalement les fonctions indépendantes de notre volonté, comme la respiration. Je me moque de savoir ce qu'elle a pu faire ou pas : vous n'avez pas le droit de précipiter sa mort. Embêtez-la encore et j'appelle la police.

— Euh, madame, fit le type en uniforme assis sur sa chaise, c'est moi, la police.

— Vos supérieurs, alors, répliqua-t-elle avant de tourner les talons.

Monroe considéra Zandt.

— Alors ? C'était vraiment utile ?

— Elle a bien pris cette photo. Elle a tué ces hommes à cause d'un événement ancien. Un événement qui l'aura poussée à photographier un certain bout de forêt, pour garder une trace. Elle a vu quelque chose là-bas.

— Elle a vu le meurtrier de la victime enterrée, avançai-je. Et Julia est la seule personne qui puisse éventuellement nous révéler son identité.

— Le squelette était bien celui d'une femme, annonça Monroe. Environ quarante-cinq ans. On a exhumé les derniers ossements cette nuit. Et il s'avère que vous aviez raison concernant la datation des faits. Mais qui ça peut être, bon sang ? Et qui l'a tuée ?

— J'en sais rien, répondis-je. Mais à mon avis, il y a une chance sur deux pour qu'il s'agisse du ravisseur de Nina.

— Et si ce n'est pas lui ?

— Des fantômes, dis-je amèrement. Des gens auxquels vous ne croyez pas et que personne ne peut coincer.

Nous avons quitté l'hôpital. Monroe devait estimer que nous n'apprendrions rien de plus, et je partageais son point de vue. On n'avait plus rien à attendre de Julia. Plus maintenant.

John se dirigeait vers ma voiture. Je restai en arrière pour toucher un dernier mot à Monroe. J'étais frappé par sa mine : non seulement crevée, mais défaite.

— Il faut trouver quelque chose, Charles.

— Je fais tout mon possible.

— J'espère. Le temps file. Si vous ne sauvez pas Nina, votre vie n'en sera plus une.

Sur ces mots, je regagnai la bagnole, tout en composant le numéro de Nina. Toujours rien.

Tôt ou tard, allumé ou non, son téléphone n'aurait plus de jus.

CHAPITRE 30

La cave, c'était pire que la camionnette. C'est dire si la situation était désespérée... Au moins, avant, il y avait du mouvement. Et donc l'espoir qu'on la relâche, qu'on l'abandonne en pleine nature ou qu'on la passe par-dessus bord. Aucune de ces options n'était mirifique, mais elles avaient le mérite de déboucher sur un après – pourvu qu'on soit forte, et chanceuse. Un poisson suspendu à l'hameçon pensait peut-être pareil, jusqu'à ce qu'on l'assomme d'un coup sur la tête.

Couché sur le sol d'une cave, on ne sent guère pointer la force et la chance. Il n'y a que l'humidité, le froid, et la redoutable sensation d'être sous terre.

Être sous terre n'est pas bon. Normalement, c'est là qu'on passe sa mort.

Conduite en bas d'un escalier de bois, Nina avait entrevu un espace d'environ trois mètres carrés, avant que la porte ne se referme et ne ramène le noir. Elle avait alors fermé les yeux, pour ne pas se laisser abuser par les ombres. Elle s'imprégna de l'atmosphère, se figura la disposition des lieux, réfléchit aux moyens d'atteindre les marches – dans l'hypothèse abstraite où elle ne serait pas couchée sur le dos, pieds et poings liés. Elle voulut s'accrocher à ces cogitations, mais elle n'avait pas dormi depuis longtemps et ses oreilles bourdonnaient. Son corps était à bout. Perdre un demi-litre ou un litre de sang n'aidait pas, même si cela n'expliquait pas ses vomissements cinq minutes plus tôt... Les fumées d'échappement devaient être un sacré poison.

Il fallait rester en alerte. Son ravisseur avait merdé à un moment ou à un autre. Paul avait chargé ce « James » d'une autre mission, à la place, ou en plus, de l'enlèvement. Mais laquelle ? S'il s'était agi de la tuer, Paul aurait obtempéré sur-le-champ. Il n'était pas du genre à épargner les gens pour le plaisir. Au contraire.

La question n'avait rien d'agréable, mais il fallait bien se la poser : pourquoi Nina devait-elle rester en vie ? Et que pouvait-elle faire pour ne pas...

Un détail lui revint en mémoire. Dans la camionnette, elle avait songé que son téléphone devait se trouver non loin. Or elle se souvenait de l'avoir éteint juste avant de retrouver Reidel, ce qui signifiait qu'il restait sans doute un peu de batterie – avec un peu de veine.

Dans le fourgon, Nina était trop vaseuse et trop entravée pour espérer trouver le portable. Mais depuis, la donne avait changé. Il suffisait de faire vite. Même pas besoin de passer un coup de fil : si les autres cherchaient sa trace radio, ou si Ward tentait de la joindre...

Nina s'étira le cou, tourna la tête sur le sol terreux. On l'avait enfermée en même temps qu'un gros sac contenant tout le barda du Combi. Celui-ci devait être calé contre le mur. Lorsque l'agresseur avait fait irruption dans la chambre d'hôtel, Nina portait sa veste, avec le téléphone au fond de la poche intérieure. Autrement dit, l'appareil devait se trouver dans ce sac.

Après une longue phase d'impuissance à ruminer son inconfort, Nina s'était enfin trouvé un boulot, un vrai.

Mettre la main sur ce téléphone. Et ne pas mourir, bien entendu.

Le mieux était sûrement la position assise, qui permettrait de se tracter avec les talons. Ce n'était pas la méthode la plus rapide, mais elle offrait un semblant de contrôle.

Il fallut cinq minutes à Nina pour se redresser sur son séant, au prix de violentes contorsions. Puis elle pointa ses pieds en direction du sac et se mit à gigoter.

Ça fonctionnait, plus ou moins. C'était lent. Le sol était inégal, jonché d'objets. On ne les repérait qu'en s'y heurtant, après quoi il fallait les contourner.

Mais elle persévéra, centimètre après centimètre, jusqu'à ce que ses semelles touchent le mur. Là, elle pivota légèrement sur la droite, s'étira sur le côté. Perçut un bruissement de toile.

Le sac.

Et maintenant ? Elle ne pouvait utiliser ses mains ni ses pieds. La veste se trouvait peut-être au fond du sac, et elle était épaisse – Ward la lui avait achetée en prévision des rigoureux hivers du Nord-Ouest pacifique. Il n'y avait donc aucun moyen de sortir le téléphone. Mais, en exerçant une pression sur le sac...

Une série de pas au-dessus de sa tête. Sans attendre, elle retomba sur son flanc, tendit les jambes et roula. Roula vite, sans se soucier des obstacles sur son chemin. Elle devait regagner le point de départ, pour ne pas montrer que le sac l'intéressait.

Elle y parvint juste à temps, au prix de bleus et de griffures. En chemin, son poignet accrocha quelque chose, et elle dut tirer de toutes ses forces pour le libérer. Elle avait mal, et le souffle court. Mais elle reprit sa place. Se coucha de tout son long. En refrénant les mouvements de sa cage thoracique.

La porte de la cave s'ouvrit. Levant la tête, Nina distingua deux silhouettes en haut des marches : l'Homme Debout et le type âgé. Le jeune passa furtivement derrière eux.

Ce gamin pourrait-il l'aider ? Elle avait vu à son visage qu'il n'était pas dans son élément. Saurait-elle l'amadouer en jouant les grandes sœurs ? Tu rêves, Nina. Ses parents, alors : quel âge pouvaient-ils avoir ? Et si l'on tentait le registre de la mère sans défense, de la gentille maman gâteau ?

Mais le garçon ne suivit pas les deux adultes dans l'escalier ; il resta là-haut et disparut dans une autre pièce.

Au lieu de fondre sur Nina, Paul s'accroupit au pied d'un mur. Il examina le sol quelques instants, en passant l'index dans la terre.

— Notre rouleur de mécaniques ne devrait pas nous embêter longtemps, dit-il au vieux. Encore une fois, je ne te remercie pas. (Il se releva, regarda Nina.) Notre cher agent peut rester encore un peu. Mais pas indéfiniment. Je te préviendrai en temps utile, et si je te dis de la tuer, tu le feras sur-le-champ. Fini de jouer. Compris ?

L'autre hocha la tête.

— Il était censé nous buter tous les deux, murmura Nina. C'est bien ça ? Il a tué Reidel en le prenant pour Ward, puis il a préféré me garder pour lui.

Paul se rapprocha, la toisant à la verticale.

— Tu es perspicace, agent Baynam. Perspicace et super maligne. Mais tu te trompes. James devait te liquider, en effet. Mais la deuxième cible n'était pas Ward.

— Qui, alors ?

— Voilà un bon exercice pour notre étudiante.

— Vous êtes fou à lier.

— Nullement. Et puis ce n'est pas moi qui suis ligoté dans un sous-sol, à schlinguer la sueur et la peur. À ce stade, neuf propriétaires de chat sur dix préféreraient ma place à la tienne.

Les hommes remontèrent ; peu après, la porte d'entrée claqua, et une voiture s'en alla. Nina attendit avant de repartir vers le téléphone, pour s'assurer qu'ils étaient tous absents. Dans l'intervalle, elle tâcha de réfléchir calmement, de voir si les mots de Paul lui apprenaient quoi que ce soit. Jadis elle avait lu, dans un texte de référence datant des années 1930, que le psychopathe était une « machine à reflets », capable d'imiter l'esprit humain de manière si parfaite qu'il devenait impossible de distinguer la part de simulation. Il y avait de ça dans le regard des types qui se bagarraient dans les bars. Si l'on passait en début de soirée – en admettant que l'on fût assez imprudent pour les dévisager –, on décelait une certaine dureté dans leurs yeux, un vide derrière leur masque fragile de gaieté. Les taciturnes sont en général des misanthropes, des dépressifs ou de gros buveurs. Les plus dangereux dégagent un charme aussi froid qu'irrésistible, tels des chiffres flous sur un listing informatique : tendus vers la résolution d'une équation complexe, mais sans jamais atteindre de résultat. Résultat que serait une personnalité fixée, avec laquelle on peut discuter. Voilà ce qui manque à ces hommes-là. Ils sont des poches de violence attendant le premier prétexte pour se libérer, des tornades démoniaques sous enveloppe humaine.

Les vrais cinglés sont encore différents. Ils sont habités par quelque chose, mais on ne sait trop quoi. La « machine à reflets » du Dr Cleckley appelait une question : quel était le moteur de l'imitation ? Quelle était cette « machine », et que faisait-elle lorsqu'elle ne singeait pas l'humanité ? Quelles étaient ses réactions normales ? D'où venait-elle ? *Que voulait-elle ?*

Était-ce chaque fois un système à part, ou pouvait-il s'agir d'une seule et unique entité, d'une même substance diabolique, d'un même esprit détraqué derrière tous ces regards ? La formation et les convictions de Nina privilégiaient plutôt la thèse d'humains abîmés, manifestant chacun ses propres psychoses et pathologies. Mais quand un type comme Paul posait les yeux sur vous... vous ne pouviez que vous interroger.

Alors même qu'elle estimait avoir assez patienté, la porte se rouvrit, laissant passer un rai de lumière. Les espoirs de Nina s'effondrèrent. Les enfoirés n'étaient pas tous partis.

Le visiteur descendit d'un pas lourd. Ce ne pouvait être que James, celui qui avait pris son sang. Arrivé en bas, il s'assit sur une marche, alluma une cigarette et fuma en silence.

— C'est quoi, la suite, James ? demanda Nina d'une voix lasse. On attend que votre chef vous dise de me tuer ?

— Ce n'est pas mon chef. Et je ne suis pas James.

— Il vous appelle comme ça, pourtant.

— Je suis Jim Westlake. Je bosse dans la photo.

— Ah oui ? Mon père était photographe. (Elle mentait, bien entendu.) Quel genre de photos ?

Après une hésitation, l'homme alla ouvrir le sac. Il sortit d'abord la veste marron, la posa par terre. Le cœur de Nina fit un bond.

La voilà donc. Excellente nouvelle. Sauf qu'on risquait de shooter dedans, et ce faisant de dévoiler le téléphone. Si seulement elle pouvait l'atteindre... ou se la faire apporter...

— J'ai froid, dit-elle. Le sol est glacial.

Il n'eut pas l'air d'entendre. Lorsqu'il se redressa, il tenait une petite boîte, qui semblait conçue pour une paire de chaussures d'enfant. Reprenant sa place au pied de l'escalier, il ouvrit la boîte et parcourut son contenu pendant quelques minutes, comme s'il avait oublié la présence de Nina. Puis, sans la regarder, il lui montra quelques Polaroïd.

Nina ne voyait pas grand-chose, sinon qu'il s'agissait de femmes ou de filles, d'âges divers, dans un endroit lumineux.

— Je n'ai rien fait, dit-il. À aucune d'elles. Pendant des années. J'ai vécu à côté de... Regarde.

Il feuilleta rapidement la pile avant d'isoler un cliché.

— Regarde.

Il lui fourra la photo sous le nez, si près que Nina put à peine la regarder. Deux petites filles, âgées de quatre ou cinq ans, tout sourire.

— Mignonnes.

— Ce sont mes voisines.

— Vraiment ? Elles habitent juste à côté ?

— Non. Je ne vis plus ici depuis longtemps.

— Alors comment y avez-vous accès ?

— La maison m'appartient toujours mais... J'y ai vécu avec ma femme.

— Vous êtes marié ?

— Plus maintenant.

Nina allait poser une nouvelle question, mais elle se ravisa. Ce n'était pas elle qui menait la danse. Elle n'était pas en train d'interroger un prévenu. Alors le silence reprit ses droits.

L'homme finit par se confier :

— On s'est rencontrés quand j'ai quitté l'armée. On a passé quelques années à bouger un peu partout. Puis on a trouvé cet endroit et on s'y sentait bien. J'ai passé le certificat d'enseignant. J'avais toujours eu le goût des chiffres. Je suis devenu prof de maths. Seulement... (Il marqua une longue pause.) À l'étranger, je n'avais jamais eu de problèmes. Dans l'armée j'aurais pu... mais non. Et ensuite... après quelque temps ici ça n'allait plus. Je n'arrivais plus à mettre de l'ordre dans ma tête. Le compte n'était jamais bon.

Nina ne put se retenir :

— Pardon ? Vous prétendez que c'est la faute à Thornton ? C'est la ville qui vous a forcé la main ? Sincèrement, cette ligne de défense ne passera jamais.

— Je m'en fiche. Pendant longtemps, j'étais comme tout le monde. Je savais que je pouvais déraper si je me laissais aller. Mais je ne voulais pas. J'ai... j'ai fait de mon mieux. Et puis un jour... (Il plongea la tête dans ses mains.) Une élève. À l'école. Elle me rappelait Karla. Il a suffi de ça. Elle ressemblait à Karla, et boum !

— Qui est Karla ? Votre épouse ?

— Ma femme s'appelait Laurie. T'as pas écouté ?

— Pardon. Qui est Karla, donc ?

— Une fille que j'ai connue dans le passé. Au lycée. Ce fut ma première.

— La première avec qui vous ayez couché ?

— Ouais.

— Mais ce n'est pas à ça que vous pensiez...

— Non.

— Vous l'avez tuée.

— Oui.

Alors il lui parla de cette fille, de cette Karla. Il se souvenait de chacun de ses traits. De sa démarche. Et des préparatifs, fomentés en secret cependant qu'au grand jour il menait des activités tout autres, normales. Il se souvenait également s'être reposé après coup au bord de la rivière, sur une berge où il avait beaucoup joué étant gamin. C'était une nuit sombre et froide, sous une pluie éparse. Il était assis sur la terre dure et criblée de gouttes, à côté de la main sectionnée de Karla, et il n'y avait pas une once de lumière, sinon quelques points vacillants dans les maisons de l'autre rive. En leur tournant le dos, pour regarder au loin parmi les bruits du vent, on pouvait croire que le monde entier s'était dissous, qu'on avait retrouvé l'époque où les choses précieuses d'aujourd'hui restaient à engendrer, où les hommes et les garçons étaient encore libres d'être eux-mêmes. Les détails de son geste s'émoussaient déjà, et le risque de se faire pincer était le cadet de ses soucis (un état d'esprit voué lui aussi à perdurer). Lorsqu'il se retourna pour observer une dernière fois les lueurs des habitations, il sut que personne ne le voyait et que les occupants ne l'auraient pas davantage remarqué s'il avait frappé à leurs portes. Leurs existences lui étaient désormais fermées. Il n'avait nulle part où aller. Alors il resta là, sous la pluie, à écouter chanter la nature sauvage, jusqu'à ce que le froid le ramène chez lui, à deux kilomètres de marche, où il avala un reste de poulet avant de se coucher.

Suivirent de longues années sans la moindre rechute. On aurait pu en rester là : un incident réel mais isolé, une parenthèse. Mais Karla était la première, en aucun cas la dernière.

Il fit de son mieux. Il intégra l'armée. À sa sortie, il sillonna le pays et finit par s'établir dans une ville sympa, avec un boulot stable, une épouse et un gosse.

Mais c'était trop tard. Depuis le début.

— Une fille stupide, de toute façon. Celle qui ressemblait à Karla. Une parfaite idiote. Les gens ont été ravis de penser qu'elle avait pris la tangente, sans doute pour la Californie, comme toutes ces feignasses de traînées.

— Ça remonte à quand ?

— Quinze ans. Ce n'est pas arrivé souvent. Jamais des femmes de Thornton ou de Dryford, du reste. Une serveuse d'Owensville, une fois. Elles reposent dans les bois, maintenant. Personne ne les trouvera jamais. Mais une fois que j'avais commencé... Je pensais que je maîtrisais la situation. C'était mal, bien sûr, mais je me contrôlais.

— Votre femme n'était pas au courant ? Elle n'a jamais deviné ?

Une petite pause.

— Elle est partie.

— Vraiment ?

Il détourna les yeux.

— Non, admit-il.

Soudain il fronça les sourcils, leva le menton.

— T'as entendu ?

— Non, répondit Nina. Ça ressemblait à quoi ?

Mais il secoua la tête.

— Parfois, quand il y avait de l'orage, la nuit, elle se mettait dans tous ses états.

— Votre femme ?

— Elle pensait que le ciel criait. Que la nuit était furieuse, qu'elle voulait blesser quelqu'un. Je lui répondais que non, que ce bruit ressemblait davantage à des jeux d'enfant. Que la foudre était le bruit que faisait le ciel quand il jouait, tout là-haut.

Il se tut quelques instants, et Nina s'aperçut qu'il pleurait.

— Jim ? fit-elle d'une voix douce. J'ai vraiment très froid. Je ne me sens pas bien. Pourriez-vous... Pourrais-je avoir mon blouson ? Vous n'avez qu'à l'étendre sur mon ventre. Pour me qu'il me tienne chaud. Je suis sûre que Paul serait d'accord.

Mais il restait sourd, et Nina rêva de lui briser le crâne avec une barre à mine. Ces malades étaient tous les mêmes : ils n'entendaient rien d'autre que leurs jacassements internes. Ils parlaient sans fin, mais pas un mot ne passait en sens inverse. Il n'entrait jamais rien.

— J'ai perdu... C'en est arrivé à un point... bredouilla-t-il. Tu me comprends, n'est-ce pas ?

— Non, répliqua-t-elle froidement. Et je ne comprendrai jamais.

Paul revint un peu plus tard. En entendant le claquement de portière, les pas dans l'allée, puis l'ouverture de la porte par un sbire, Nina craignit que ce ne soit la fin.

Mais une fois dans la cave, Paul se pencha pour l'aider à se relever. D'abord sur les genoux, puis sur les pieds.

— C'est quoi, la suite du programme ? demanda-t-elle, exaspérée par l'insolente douceur de Paul.

— On t'emmène. (Il lui remit le bâillon.) Je n'ai pas le temps de savourer tes pensées profondes. Pardon si tu en ressens une certaine impuissance.

— Elle a froid, lâcha soudain Jim/James.

Le vieux ne regardait personne, ni rien de visible. Il semblait absorbé par sa nouvelle cigarette, au point que Nina se demanda s'il parlait bien d'elle, et à Paul.

En tout cas, ce dernier se pencha pour ramasser le blouson, celui que Ward lui avait offert voilà une éternité, et il lui recouvrit les épaules.

— C'est mieux comme ça ?

Nina hocha la tête : un mouvement rapide, chaleureux et reconnaissant, le plus courtois de toute sa vie.

Elle sentait un poids dans la partie gauche de la veste. Seuls quelques centimètres la séparaient de son téléphone. Il ne restait plus qu'à trouver le moyen de presser la touche ON à travers le tissu, puis celle du numéro préenregistré. Une tâche ardue, mais pas impossible. Ward n'aurait même pas à entendre sa voix, juste à lire le numéro inscrit à l'écran...

— Ah oui, dit soudain Paul. J'allais oublier.

Il plongea la main dans la veste et sortit le portable. Ce faisant, il examina le visage de Nina, qui malgré ses efforts ne put cacher son désarroi.

— Tu ne me croyais pas aussi con, tout de même ?

Comme il la poussait dans l'escalier vers une lumière incertaine, Nina reconnut, la mort dans l'âme, que depuis le début Paul avait une longueur d'avance sur elle, sur elle et sur les autres.

Et qu'il risquait fort de la conserver jusqu'au bout.

CHAPITRE 31

Unger nous attendait devant le Mayflower. Il était habillé plus chic que la dernière fois, costume griffé anthracite et sobre cravate foncée. Ce n'était plus le même homme. J'emmenai John à sa rencontre.

— On est synchro, se félicita Unger.

Même sa voix était plus guindée.

— T'es venu comment ?

— En tacot. J'aime pas conduire.

— Sérieux ?

— C'est dangereux.

— Plus que l'avion ?

— Les pilotes ont de l'entraînement. Pas moi. (Se tournant vers John :) Carl Unger, enchanté.

— John Zandt.

— Je sais qui vous êtes.

John fronça les sourcils.

— Qu'entendez-vous par là ?

— Plusieurs choses. (Il regarda le café.) On va discuter à l'intérieur ? Ça pèle, ici.

Le Mayflower venait d'ouvrir et une fausse blonde tatouée avec un piercing à la langue remplaçait Hazel. Elle faisait sa blasée. À fond. Ce devait être la fameuse Gretchen dont Lloyd s'était amouraché. Vous parlez d'une flèche : le simple fait de servir le café semblait être le bout du monde.

Nous prîmes l'un des box du fond. Unger posa les mains sur la table.

— Aucune nouvelle de ton amie ?

— Aucune.

— Tu m'en vois sincèrement navré. Mais pour être honnête, je suis venu te parler d'autre chose.

Il nous fit glisser une feuille de papier, que je lus en même temps que John.

C'EST NOTRE JOUR. NOUS SOMMES EN AMÉRIQUE, MAIS NOUS NE SOMMES PAS AMÉRICAINS. NOUS SOMMES EN EUROPE. NOUS NE SOMMES PAS EUROPÉENS. NOUS NE SOMMES NI ASIATIQUES, NI ARABES, NI AFRICAINS ; NOUS NE SOMMES NI CHRÉTIENS, NI MUSULMANS, NI JUIFS. NOUS SOMMES LES VRAIS HUMAINS. IL EST TEMPS D'ÉDUQUER LES MASSES.

— Encore un spam codé ?

— Expédié à 6 heures ce matin. Codé à partir du Coran, rien que ça.

— Très messianique. Moi aussi, je vais être franc, Carl : ma priorité, c'est Nina.

— Hier après-midi, une bombe a explosé dans une boutique de sport d'un centre commercial californien.

— On est au courant. Et alors ?

— Les bandes vidéo ont fourni un suspect, mais il a disparu. Un gosse de riches de la Valley. Lee John Hudek.

John tilta :

— Hudek ?

— J'imagine que ce nom vous dit quelque chose ?

— Oui, mais c'était Ryan Hudek. Il figurait dans les dossiers que j'ai dénichés dans la maison d'un promoteur, un certain Dravecky.

— Que vous avez tué.

Il était rare de voir Zandt pris au dépourvu.

— Comment le savez-vous ?

— Une intuition, disons. J'ai remarqué plusieurs morts suspectes ces derniers mois, des hommes dont j'ai croisé le nom dans des contextes bien particuliers. J'en ai déduit que quelqu'un traquait de supposés Hommes de Paille de premier plan. J'ai raison ?

John ne répondit rien.

À mon tour de prendre un regard méfiant :

— J'ai comme l'impression que t'en sais bien plus que ce que t'as lâché la dernière fois.

— Mon boulot consiste à glaner des infos, pas à les répandre. Au jour d'aujourd'hui, nous ne sommes que treize à connaître toute l'histoire. Nous ne faisons pas confiance à n'importe qui. Et moi encore moins que les autres. Mon père et mon grand-père y ont laissé leur peau.

— C'est qui, « nous » ? Si la CIA sait...

— Oublie l'Agence, coupa Unger. Elle est trop infiltrée, comme le FBI.

— Alors qui ? insistai-je. Écoute, Carl, on ira plus vite si tu veux bien répondre.

— J'y viendrai, promit-il avant de lancer à Zandt : Ça fait un moment que j'essaie de vous rencontrer. Mais on ne vous coince pas comme ça.

— À qui le dites-vous ! grommelai-je.

John n'accueillit pas Unger à bras ouverts :

— Pourquoi voulez-vous me parler ?

— Aidez-moi d'abord à vous faire confiance. Je sais seulement que vous êtes un ami de Ward, que vous étiez un bon inspecteur de la crime et que vous êtes devenu un homme dangereux pour ceux qui ne sont pas de votre côté. J'ai lu le rapport d'autopsie de Dravecky.

John cogita, avant de sortir de sa poche deux sachets en plastique, du genre de ceux qu'on utilise pour ramasser des pièces à conviction. Chacun contenait une petite forme indéfinissable. J'en attrapai une. On aurait dit une motte de terre, veinée de bleu et de vert.

— En Nouvelle-Angleterre se trouvent certaines structures de pierre que l'on décrit généralement comme des greniers à légumes. Le contenu de ces sachets provient de l'un d'eux, situé dans le Massachusetts.

— Y aurait-il un rapport avec Oz Turner ? demandai-je.

John acquiesça.

— J'ai obtenu qu'il me conduise à celle qui fut découverte en premier, dans le comté de Webster. Elle est tombée dans l'oubli depuis des lustres, peu de temps après qu'on l'a retrouvée. Du coup, c'est de loin la mieux préservée de toutes. Certaines des photos ayant disparu de son site Internet ont été prises ce jour-là, par votre serviteur. Quand je suis passé voir mon contact à Richmond au sujet de l'os qu'on a dégotté en forêt, j'en ai profité pour lui montrer le contenu de ces sacs.

Je saisis la deuxième pièce.

— Là aussi, on dirait un bout d'os.

— C'en est un. Un fragment de crâne. (John regarda Unger.) Saurez-vous me dire ce que contient l'autre sachet ?

— Un petit morceau de cuivre, répondit Unger. Extrait de la région des Grands Lacs, environ deux mille ans avant notre ère.

— Pitié, soupirai-je. Tu vas pas t'y mettre aussi...

— Exact, dit John en ignorant mon commentaire. L'os est également très vieux. Et humain. Je n'ai pas eu le temps de le faire dater au carbone 14.

J'avais le rôle de l'idiot de la classe, ce que je déteste.
— Et que faisaient ces machins dans une cave à légumes ?
— Il n'y a jamais eu de légumes, m'expliqua Unger. Ni là, ni sur les autres sites. Il s'agit de chambres funéraires antiques.
— Comment ça, « antiques » ?
— Certaines datent de seulement cinq ou six siècles. Mais d'autres ont quelques milliers d'années, parfois plus. Il se peut qu'elles aient servi de sépultures aux grands chefs des Hommes de Paille.
— Quoi ?
— Je doute que ce soit aussi simple, tempéra Zandt. Je crois qu'on y enterrait aussi leurs victimes. Comme une collection, une exposition. Une œuvre. Le tueur repose parmi les vestiges d'une vie de meurtres. (Il se tourna vers moi.) Tu te souviens de cette cabane qu'on a visitée au sud de Yakima ? C'est assez comparable. Par ailleurs, j'ai appris dans une revue qu'on avait trouvé sur un site archéologique allemand une série de corps vieux de dix mille ans, agencés de manière très précise. Il n'est pas impossible que certaines des sépultures les plus anciennes, et en apparence les plus anodines, relèvent du même principe : il ne s'agirait pas de chefs enterrés parmi leurs serviteurs afin qu'ils disposent de valets dans l'au-delà – opinion communément admise – mais de tueurs en série au milieu de leurs victimes.
— Et dans quel but ? m'étonnai-je. Pour marquer leur territoire ? On enterre nos morts ici, donc c'est chez nous ?
— Sans doute, mais pas seulement, répondit Unger. Si l'on étudie la répartition des chambres en Nouvelle-Angleterre, on a l'impression qu'elles étaient disposées de façon à circonscrire certaines zones.
— Tu veux dire qu'ils pensaient créer une sorte de champ de force autour de leurs habitations ? Comme un mur de sang ?
— C'est un peu l'idée, oui.
— Tu te fous de moi ?
— À ma connaissance, les tueurs en série actuels pratiquent ce genre de rituels. C'est la même chose, mais érigée en mode de vie.
— Bon, à vous de me donner des gages, enchaîna John. Je pensais être le seul à connaître ces éléments sur les Hommes de Paille. Alors dites-moi ce que vous pensez savoir.
— D'accord, opina Unger. Mais il faudra me croire sur parole, parce que je n'aurai pas le temps de citer mes sources, encore moins de vous donner des preuves.

Je plongeai la tête dans mes mains et laissai parler le bonhomme.

— Version courte, annonça Unger en allumant une de mes cigarettes. Comme vous le savez, la dernière période glaciaire s'est terminée vers 11 000 avant Jésus-Christ. Fini de cailler, la planète redevient enfin un lieu confortable, et le genre humain peut s'offrir le luxe de prospérer. Au cours des cinq millénaires suivants, nous avons commencé à réarranger le monde. On a mis au point l'agriculture. Il y a eu le commerce, la communication, et on a définitivement creusé l'écart avec les Néandertaliens. Voilà pourquoi on trouve des vestiges d'anciennes cultures dans des coins inattendus du globe : il ne s'agit pas d'anomalies, mais d'indicateurs d'une civilisation mondialisée. Toutes les civilisations ont bâti des légendes sur cette période heureuse : le jardin d'Éden, Dilmun, l'Airyana Vaejo, le Dreamtime. Mais il y a toujours eu des hommes et des femmes différents des autres, regroupés en une force qui fut l'ancêtre des Hommes de Paille. Vers l'an 6000 avant Jésus-Christ, ceux-ci décidèrent d'attaquer les centres névralgiques de la culture. Leur dessein était hallucinant ; il disait en substance : « Nous détestons ce nouvel ordre mondial et nous allons le bousiller. » Ça leur a pris cinq siècles pour assaillir la planète entière, continent après continent, telles des fourmis tueuses.

— Ils ont conquis le monde entier ?

— Non. Ce n'était pas une invasion. Ils ne cherchaient pas à dominer, seulement à tuer ceux qui n'étaient pas comme eux. Ils détruisaient, massacraient, et passaient leur chemin. Le truc avec les Hommes de Paille, c'est qu'ils ne renoncent jamais – et il faut garder à l'esprit que, même individuellement, ces gens peuvent être des fous furieux, mais comme vous n'avez pas idée. Vous voulez savoir à quoi ressemblait la dernière bataille des Hommes de Paille ? Imaginez Jérôme Bosch multiplié par l'Holocauste. Ils ont incendié nos villes et nos campagnes de manière si monstrueuse que tous les mythes de l'enfer sont apparus au cours de la même ère. L'anéantissement de plusieurs millénaires de civilisation fut un tel traumatisme que son souvenir a fini par se confondre avec d'autres grands tournants historiques – comme la grande inondation qui marqua la fin de la dernière période glaciaire, et qui restait jusque-là comme la pire catastrophe de l'histoire humaine. Voilà, par exemple, pourquoi la légende de l'Atlantide se referme sur un déluge – sans oublier l'arche de Noé – et pourquoi l'histoire de l'Atlantide fut à tort située vers 11 600 avant Jésus-Christ, quand les eaux montèrent partout sur la Terre. Ajoutez à cela les cataclysmes volcaniques et autres chutes d'astéroïdes recensés au fil des millénaires (voir Sodome et Gomorrhe), et on obtient le mythe d'un désastre généralisé à l'échelle de toute une ère. Les défaites infligées par les Hommes de Paille au reste du monde étaient si effroyables et si radicales qu'on a voulu rétrospectivement y voir la main des dieux – des dieux courroucés par la mauvaise vie des hommes. Le monde a passé les *huit derniers millénaires* à se remettre de cette apocalypse. Il y a trois ou quatre mille ans, on a repris l'avantage sur eux, et ils se sont fait oublier. Dans le *Timée* – la première trace attestée du mythe de l'Atlantide – , Platon parle d'un « cuivre des montagnes » qui revêtait les murs de la cité et composait les colonnes où était inscrite la loi. Le philosophe en fait quasiment un élément central. Et pourquoi, d'après vous ?

— Les mines de cuivre préhistoriques des Grands Lacs, répondit John.

— Exactement. Cela renvoie, de manière erronée, au fait que les Hommes de Paille étaient partis fonder leurs mines sur le sol américain, cette terre qu'ils revendiquaient. Fait correct, mais époque fausse, car souvenons-nous que Platon a écrit vers – 400, soit deux mille ans après les débuts de l'activité minière ici, et cinq mille après la guerre. Déjà on commençait à confondre les deux. Enfin, Ward, il suffit d'ouvrir les yeux ! Comment appelle-t-on l'étendue d'eau qui a maintenu l'Europe à bonne distance du repaire des Hommes de Paille ? L'Atlantique.

— C'est une coïncidence. Ou une...

— Il n'y a pas de coïncidences. La majorité des mythes de la planète, jusqu'aux histoires de vampires, de loups-garous et de démons, visent à rendre compte de cette lutte sans fin – pour nous rappeler qu'il existe des gens qui agissent contre nous, la nuit, et qui nous traquent avec la volonté de tuer. Le Christ ne chassait pas les démons au sens propre du terme. Il s'agit ni plus ni moins d'un code glissé dans la Bible, pour certifier que l'Église vit le jour afin de tenir tête aux Hommes de Paille et d'ériger un rempart contre les *vrais* démons, ces gens qui prennent pour ennemie l'humanité tout entière.

Il y eut une pause, pendant laquelle j'essayai d'intégrer tout ce déballage.

— J'en déduis que tu ne bosses pas pour la CIA, finalement.

— Si, je t'assure. C'est un moyen comme un autre d'épier la pègre mondiale. Mais oui, c'est une couverture. Je bosse pour les francs-maçons. Je collabore officieusement avec le Bohemian Club, et je m'efforce de temps à autre de maintenir une bonne entente avec les Jésuites. L'objectif de tous ces gens, y compris ceux du groupe de Bilderberg – bien qu'ils soient effectivement devenus une racaille mondialiste, avec une fâcheuse tendance à gouverner le monde en sous-main –, leur objectif commun, c'est la stabilité. Or le plus dingue, c'est qu'ils n'ont même pas conscience de servir ce but. Chacun est persuadé d'être là pour le fric et le pouvoir. Mais le résultat global, comme le point de départ de tout l'édifice, c'est de dresser des digues contre les Hommes de Paille.

— Si je comprends bien, même les conspirateurs ignorent le véritable objet de la conspiration ? Prodigieux...

— Ils n'ont pas besoin de savoir. Il suffit que ça marche. Tu ne peux pas saper un truc dont tu n'as jamais eu vent. Comme disait Héraclite – oui, c'était l'un des nôtres –, un lien caché est toujours plus fort qu'un lien visible.

Je me tournai vers Zandt :

— Ce qui est génial avec Carl, c'est qu'il te ferait passer pour un mec parfaitement équilibré.

Mais Zandt fixait la table, avec l'air de celui dont l'opinion se voyait enfin validée, fût-ce au-delà de ses espérances.

— Rien de tout ceci ne me paraît inconcevable.

— Mais tu croirais n'importe quoi, John !

Unger m'agrippa l'épaule.

— Attends, Ward, tu penses aussi que c'est une coïncidence si l'Empire romain a embrassé le christianisme alors qu'il touchait le fond ? Bien sûr que non. C'était un transfert de pouvoir délibéré, négocié par l'empereur Constantin et sa mère. L'Église catholique romaine fut créée pour prolonger les structures de l'Empire romain, comme celui-ci avait pris la relève des Grecs et des Égyptiens. Elle fut constituée par les dominants de l'époque, de même que l'Union européenne est l'œuvre des maîtres cachés de la nôtre. Comme toutes les institutions, l'Église a décliné et fini par mal tourner, mais à ses débuts, Paul – celui de la Bible, bien sûr, pas ton frère – a vu dans les enseignements d'un rebelle juif le parfait outil pour bâtir une organisation internationale. Et il avait raison. Un pouvoir centralisé, démultiplié par des unités de plus en plus petites qui placent un prêtre dans chaque village. L'Église catholique a façonné l'Europe quinze siècles durant, l'a maintenue en alerte face au danger des Hommes de Paille. La maçonnerie est la fille de l'ordre du Temple, l'une des dissidences qui s'engouffrèrent dans la brèche lorsque l'Église marqua le pas. L'Église avait encore les moyens de niquer les Templiers, au nom de l'hérésie, alors ils se sont discrètement reconvertis en francs-maçons, et huit cents ans plus tard ils sont toujours actifs. Mais parmi les maçons d'aujourd'hui, combien savent que leur confrérie est davantage qu'un club de copinage ? Seuls trois membres du 33e degré comprennent son but initial. L'éternelle guéguerre entre la franc-maçonnerie et l'Église tient à ce qu'elles furent toutes deux conçues – à des époques différentes et par des gens très dissemblables – pour combattre la même menace. L'Église s'est offert une longue croisière et s'en est mis plein les poches – pas question de céder les rênes à quiconque –, et c'est pourquoi l'Occident se farcit leur bras de fer depuis cinq siècles. Il n'y a pas assez de place pour toutes les élites occultes quand elles se disputent les mêmes cœurs, les mêmes âmes et le même fric – a fortiori lorsqu'elles ont oublié leur raison d'être initiale. Oui, les maçons en font partie, tout comme Bilderberg, la Commission trilatérale, la BRI, La Défense, el-Rachid, Yom Pek et d'autres dont, avec un peu de chance, tu n'auras jamais entendu parler. Dans chacun de ces groupes, un seul membre connaît la vérité, en plus de nos trois maçons. Nous sommes ce qui se dresse entre ce monde et les Hommes de Paille. (Unger avait soudain l'air épuisé.) Le problème, c'est qu'ils sont carrément plus forts que nous, car tôt ou tard nous succombons à la cupidité et nous oublions notre mission. Le temps obscurcit tout. Bien sûr que le Christ a étudié en Chine, qu'il a fui en France après la crucifixion, et qu'il s'est retrouvé au Cachemire, où il est mort. Mais tout le monde s'en cogne. La question n'est pas là. C'était juste un brave type, comme Mahomet, Bouddha ou James Stewart. Ceci concerne la multitude. Nous devons contenir les Hommes de Paille, et c'est tout ce qui compte. Le voilà, le sens de l'existence. Jusqu'ici, on s'en était pas mal sortis, mais la technologie nous a donné le coup de grâce. Avec Internet, on ne peut plus empêcher les Hommes de Paille de se rassembler. Ils sont plus motivés. Plus haineux. C'est comme ça.

Il se renversa sur son dossier pour faire place au silence.

— Mais alors, demandai-je, et les ovnis dans tout ça ?

— Ils n'existent pas, Ward. Arrête de faire le con.

Il ne sera pas dit que je suis fermé aux idées neuves. J'ai passé les cinq minutes suivantes à ruminer les propos de Carl, des fois qu'ils m'auraient aidé à retrouver Nina. Mais ce n'était pas le cas, alors j'ai laissé tomber. Quand je suis sorti de ma rêverie, la discussion battait son plein entre John et Unger.

— Et qu'y avait-il au juste dans ces dossiers ? demandait Carl.

— Des contrats, essentiellement. Ryan Hudek n'avait pas le profil d'un gros poisson. Disons que Dravecky faisait son possible pour lui apporter des clients. Une sorte de faveur.

— Ou de paiement, suggéra Unger. Je pense que la bombe d'hier à LA était une simple mise en jambes.

— En vue de quoi ?

— Je sais pas, mais...

Un voyant s'alluma dans mon crâne. Je levai les yeux.

— Voilà pourquoi ils ont délivré Paul !

Les deux visages se tournèrent vers moi.

— Réfléchissez une seconde. Il y a six mois, Paul a pété les plombs, au point de devenir un boulet pour les Hommes de Paille. Dravecky a même envoyé des gars pour le liquider dans les bois, rappelez-vous. Et aujourd'hui, ils s'embêteraient à attaquer un fourgon blindé pour le ramener au bercail ? Comment expliquer ce revirement ?

— Il sait tout sur eux, dit John. Il fallait le déloger avant qu'il essaie de monnayer ses infos contre un allégement de sa peine dans les affaires Jones et Wallace. Il aurait pu anéantir les Hommes de Paille. Alors ils se sont débarrassés de lui.

— Tu parles ! Il ne les aurait jamais trahis, et ils le savaient. Toi aussi, tu le sais. Tu ne serais pas là devant moi si tu le croyais mort. S'ils avaient voulu le tuer, il leur suffisait d'un coup de fil pour le faire buter à Pelican Bay. Non, ils ont besoin de lui. C'est la seule explication valable.

John se tut quelques instants, avant de tressauter comme s'il avait reçu un coup.

— Quoi ? fit Unger.

— Comment a-t-on pu passer à côté de ça ? Ce « Jour des Anges », c'est Los Angeles ! Le Jour des Anges, dans la Cité des Anges.

Unger le dévisageait.

— Ils vont attaquer L.A. ?

— Paul connaît la ville, précisai-je. Il y a des chances pour qu'il ait commis à L.A. ses tout premiers meurtres. Il y a collecté des victimes pour les Hommes de Paille, en tout cas, et tué Jessica Jones.

— Dravecky, Hudek et au moins quatre de leurs potes sont ou étaient basés dans cette ville, ajouta John. Leurs taupes au FBI ont déjà obtenu la suspension de Nina il y a quelques mois.

Je me levai, le regard rivé au parking, le sang glacé par la panique.

— Et voilà quatre jours, Nina, moi et le seul agent du FBI à qui nous ayons parlé de ces histoires – Charles Monroe, qui bosse lui aussi à LA –, nous sommes envoyés à l'autre bout du pays pour enquêter sur un meurtre qui semble être le fait d'une tueuse en série. Le surlendemain, Nina disparaît. J'appelle John, qui débarque à son tour. Tous réunis, et tous au mauvais endroit !

John et Carl étaient déjà debout.

— C'est jouable, dit Carl. Je vais trouver un aérodrome et réserver un avion pour dans une heure. Vous avez l'adresse de Ryan Hudek, John ?

— Ouais. Mais la police de Los Angeles y sera passée avant nous.

— Eh, les gars...

— Évidemment, répondit Unger à John, mais ils auront juste interrogé les parents pour savoir où se trouve leur fils, puis ils auront mis les voiles. Qui va imaginer qu'ils puissent être de mèche avec leur môme ?

— Vous pouvez les faire retenir chez eux ?

— Je peux essayer.

— Je n'irai pas à LA, déclarai-je enfin.

Zandt s'agaça :

— Mais enfin, Ward, tu viens toi-même de dire...

— Je sais ce que j'ai dit.

L'oreille collée à son portable, Carl m'observait du regard fixe et froid du zélote. Une porte s'était ouverte dans sa tête comme dans celle de John, et ils avaient franchi le pas.

— Y a pas à hésiter, Ward. Trop de vies sont en jeu.

— Je préfère en sauver une autre. LA devra se passer de moi.

John eut l'air navré.

Je l'abandonnai pour remonter les rangées de tables où de pauvres hères venaient cacher leur premier verre derrière un sandwich. Ressortant d'un pas rageur, je manquai de culbuter un type dans un costard à cinquante dollars. Il n'osa pas soutenir mon regard.

Comme je repartais vers ma voiture, j'entendis appeler mon nom. Je scrutai le parking ; Hazel venait de se garer de l'autre côté. Je savais si peu où j'allais que la moindre hésitation pouvait me bloquer à jamais. Alors je lui répondis d'un simple signe. Qu'elle imita. Puis je sautai derrière le volant.

Au bout d'un kilomètre, mon portable sonna. Je le sortis de ma poche, pensant tomber sur John ou Carl – dans l'espoir qu'ils avaient choisi de laisser LA se démerder. Mais c'était Monroe.

— Où êtes-vous ? dit-il d'une voix pleine de vigueur.

— Je roule dans Thornton. Du nouveau ?

— Venez me prendre au commissariat. Je crois qu'on a quelque chose.

CHAPITRE 32

J'ai laissé la caisse devant le comico pour courir à l'intérieur. Monroe était au fond du bâtiment, en train de causer à un policier que je n'avais jamais vu. Dès qu'il m'aperçut, l'agent spécial s'interrompit pour me rejoindre. L'autre flic me toisa de sa supériorité de flic. Même à dix mètres, sa tête ne me revenait pas.

Monroe me prit le bras et me ramena dans la rue.

— C'est votre voiture ?

— Ouais. Vous discutiez avec qui, à l'instant ?

— Avec le shérif d'un bled situé après Owensville.

— Qu'est-ce qu'il fait ici ?

— Honnêtement, j'en sais rien. Mais il a l'air beaucoup plus au parfum que les locaux.

— C'est pas dur...

— N'empêche qu'on vient enfin d'obtenir du bon boulot. Lors de son interrogatoire face à Reidel, Julia Gulicks a déclaré qu'elle s'était installée à Thornton voilà six ans. Ce qui cadre avec la date de signature de son bail et son arrivée dans l'entreprise d'Owensville. Mais elle disait également avoir grandi à Boulder.

— Et... ?

— Elle mentait. On a passé un jour et demi là-dessus, sans trouver aucune trace d'elle là-bas. Naissance, lycée, boulots, que dalle. Et puis, il y a une demi-heure, le seul flic de Thornton qui planche encore sur le dossier tombe sur une certaine Jane Gillan, née et scolarisée à Dryford.

Monroe y voyait manifestement une belle prise. Mais ça ne me parlait pas.

— Dryford, hein ? Toujours dans le Colorado ?

— Non. Ça se trouve à dix bornes d'ici, de l'autre côté de la forêt où on a découvert les deuxième et troisième corps.

J'ouvris la mâchoire, la refermai d'un clac.

— C'est bien ce que je me suis dit, fit Monroe. Alors en avant.

J'attendis qu'on soit en route pour demander :

— Et on a quoi sur cette Gillan ?

— Pas grand-chose, mais ça colle. Les derniers éléments remontent à sa treizième année. Son père est mort en 1992 dans des circonstances assez louches, et c'est ce qui a fait tilter notre flic. Papa est tombé dans l'escalier, un après-midi qu'il était soûl. Nuque brisée. Sa jeune fille était le seul témoin présent au moment du drame. À son retour, la mère dit avoir trouvé un corps inanimé dans le couloir, et Jane qui faisait ses devoirs dans la cuisine. Walter Gillan était un alcoolique violent et raciste, un fumier qui battait régulièrement sa femme, et c'est pourquoi personne n'a vraiment cherché à approfondir la thèse du crime. La veuve et l'enfant ont aussitôt quitté Dryford, et on n'en a plus jamais entendu parler. Mais en 1998, une certaine Julia Gulicks emménage à Thornton.

— Elle n'est quand même pas douée...

— Comment ça ?

— Cette fille a la chevelure la plus rousse du monde, mais elle ne pense pas à la couvrir pour rencontrer sa deuxième victime au Mayflower. Elle croit malin d'être celle qui découvre le cadavre de Widmar, alors que ça ne fait qu'attirer les regards sur elle. Et quand elle change de nom, c'est pour garder les mêmes initiales ! Sans déconner...

— Ça arrive souvent avec les fugitifs, ou les gens normaux qui veulent prendre un nouveau départ. C'est plus facile pour signer des chèques, ou pour réagir à son nouveau nom.

— Personnellement, je risquerais un changement radical.

— Mais les meurtriers ne sont pas tous des pros ; il y a même une majorité d'imbéciles. Très peu réfléchissent à long terme.

— C'est à se demander combien de tueurs en série se font pincer au tout début, avant d'avoir trouvé leur rythme de croisière...

— Je n'ai pas envie de penser à ça pour l'instant. Pas plus que je n'ai envie de mourir avant d'arriver là-bas, alors pourriez-vous s'il vous plaît *ralentir* ?

Je levai le pied, mais seulement pour traverser le grand carrefour en face de chez Renée. Sitôt arrivé à la route forestière menant à Dryford, je remis les gaz sans me soucier des sueurs froides de Monroe.

— J'ai l'adresse de la maison d'enfance de Gillan, ajouta-t-il en évitant de regarder le paysage. Vous me direz : ça n'explique pas ce qui s'est passé après l'arrestation de Gulicks.

— Elle a toujours agi dans l'ombre de quelqu'un. Vous vous souvenez quand John lui a demandé pourquoi elle coupait les mains ? C'est là qu'elle a pété un câble.

Monroe me sembla un peu largué.

— Elle n'en sait rien, expliquai-je. Elle ne sait pas à quoi rime cette histoire de mains. Il y a longtemps, elle a dû voir un truc affreux. Elle aura voulu en parler, mais n'aura récolté que des coups ou des cris, et depuis ça l'obsède. Aujourd'hui elle exprime la psychose d'un autre, mais elle est déboussolée car ce n'est pas vraiment la sienne.

— Elle a quand même tué deux personnes.

— Ah oui ? Comme cette femme que Nina croisait à Janesville, celle qui a tué ses violeurs ? Le meurtre est-il toujours le pire acte qui soit ? Quel que soit le contexte ?

— Jusqu'à preuve du contraire, ni Widmar ni le tartempion n'ont fait de mal à Gulicks.

— Eux, non. Mais quelqu'un d'autre, si.

Dryford était une petite ville décatie qui n'en revenait pas d'être encore debout. La route forestière coupait sa grand-rue à angle droit. Encore deux virages et nous atteignîmes Jefferson Avenue, une longue voie étroite bordée de petites maisons. Environ cent mètres avant la fin, Monroe me fit stopper en face du numéro 23.

Nous examinâmes par la vitre la façade jaune du pavillon. Ce n'était pas grand, mais bien entretenu.

— Qu'est-ce qu'on cherche, au juste ?

— N'importe quoi, Ward. Tout ce qu'on pourra trouver.

Nous prîmes l'allée ; Monroe sonna. La porte s'ouvrit sur une femme entre deux âges, le genre à mitonner des brownies d'enfer. Monroe lui demanda depuis combien de temps elle vivait ici, et elle répondit huit ans. Nous apprîmes que ni elle ni son mari n'étaient apparentés aux Gillan, dont ils n'avaient du reste jamais entendu parler. Monroe eut le tact de ne pas préciser qu'un gus de ce nom s'était tué dans l'escalier que l'on apercevait derrière elle. Cette femme n'avait pas reçu de visite inhabituelle ces derniers jours – en dehors de la nôtre – ni rien remarqué d'étrange dans le quartier. Bref, elle ne voyait pas de quoi nous voulions parler, et dès lors rien n'eût justifié que l'on inspecte les lieux – qu'aurions-nous pu dénicher ?

Nous l'avons donc remerciée, avant de redescendre l'allée. Songeur, j'observai la rue par où nous étions arrivés.

— Je me demande combien de temps il faudrait pour rejoindre la forêt à pied. Si j'étais une fillette libre de flâner à sa guise.

— Pas beaucoup, supposa Monroe. Mais dans ce sens-là, on part vers l'est.

— On pourrait faire un crochet. Mais oui, vous avez raison. (Je me retournai pour considérer la dernière portion de route.) Allons jeter un œil là-bas.

Nous remontâmes la rue jusqu'au bout. Derrière une haute grille, une friche s'étirait en pente vers les bois, traversée d'un vague sentier.

— On peut aussi passer par là.

— Ça fait une sacrée trotte, jugea Monroe.

— Vous seriez étonné de voir jusqu'où peut s'éloigner un gosse qui ne se sent pas en sécurité chez lui. Or, en suivant ce chemin, on arrive tout près du cadavre et du tee-shirt.

— Bon, il se peut qu'on en sache un peu plus sur Mlle Gulicks, conclut Monroe en se frottant le visage. Et maintenant ?

— Et maintenant rien. Un mur de brique. Encore une fois.

Repartant vers la voiture, je m'arrêtai au bout de dix mètres, face au dernier pavillon de la rue.

Monroe stoppa quelques pas derrière moi.

— À quoi pensez-vous, Ward ?

Souvent, dans les petites villes, plus on s'éloigne du centre et moins les maisons sont soignées. Dans Dryford, qui tenait presque du village, le phénomène était accentué : en deux minutes de marche, on passait d'un numéro 23 impeccable à un 25 apocalyptique. Les arbres se pressaient contre le flanc. Le jardin de devant était submergé de hautes herbes et de broussailles. À première vue, c'était inhabité depuis un bail.

Cependant une camionnette blanche stationnait devant. Un mini-camping-car, sans vitres sur les côtés.

— J'ai vu le même van quelque part.

— On en voit partout, répondit Monroe. Mon beau-frère en possède un. Un Combi, ça s'appelle.

— C'est celui-là que j'ai vu. Dans un endroit précis. Avant-hier, je crois.

Monroe me rejoignit en deux foulées.

— Où ça ?

— Quelque part à Thornton. Attendez... Dans la rue du commissariat, la nuit où Nina a disparu. J'ai cru qu'il s'agissait des premiers journalistes ameutés par l'arrestation de Gulicks.

— Vous êtes sûr de vous ?

— Quasi.

— Il appartient peut-être à un riverain. On peut vérifier la plaque.

— Faites ce que vous voulez, dis-je en m'approchant de la maison. Moi, je vais jeter un œil.

J'ouvris le portail, guettant une réaction dans la baraque. Elle semblait morte, avec ses fenêtres sombres et opaques. L'herbe était aplatie sur deux lignes parallèles.

Brossés par la végétation, nous traversâmes la moitié du jardin avant de nous retourner. Un oiseau pépiait dans les arbres, au-dessus du trottoir.

— Le van est sorti plus d'une fois au cours des derniers jours, nota Monroe.

— Mais si son propriétaire habite ici et se déplace quotidiennement, comment expliquez-vous que l'herbe soit si haute ?

Monroe soupesa la question et plongea la main dans sa poche. Il sortit son flingue, vérifia le chargeur. Je fis comme lui.

— Soyons prudents, dit-il.

— Vous ne voulez pas de renforts ?

— Pour quel motif ? Herbe écrasée ?

Il sourit, et je compris qu'il aurait volontiers demandé de l'aide mais craignait de passer pour un froussard ; qu'il voulait prouver, aux autres comme à lui-même, que ses blessures de début d'année ne l'avaient pas achevé ; et que lui aussi tenait beaucoup à Nina.

— Après vous, chef. Moi, j'ai appris ça dans les séries télé.

Il fondit sur la porte. J'approchai en décrivant un arc, afin d'examiner le côté de la maison. Rien à signaler, sinon les mêmes planches décollées et les mêmes buissons sauvages. Ou bien le proprio abhorrait le jardinage, ou bien il revenait d'un très long congé.

Je rejoignis Monroe sur le perron. À travers leur voile de poussière, les fenêtres sur la droite révélaient des toiles d'araignées. Je me couchai contre le mur, prêt à bondir en cas de besoin. Monroe frappa.

Pas de réponse. Il recommença. Rien.

Alors il posa doucement la main sur la poignée. La tourna. Un clic et la porte céda. La retenant du bout des doigts, il me regarda.

Je savais ce qu'il pensait. Soit les occupants refusaient de répondre, soit la baraque était vide.

Je lui adressai un signe de tête : *on entre.*

Comme s'il avait besoin de mes conseils.

Je m'écartai du mur pendant qu'il poussait la porte. Ce geste n'étant suivi d'aucun effet fâcheux, nous pénétrâmes à l'intérieur. Je refermai délicatement.

J'écoutai la poussière.

Devant nous courait un couloir sombre ; des portes ouvertes se succédaient de part et d'autre. Monroe m'indiqua une pièce d'un mouvement de tête avant de prendre celle d'en face, son flingue pointé vers le bas. J'ignore ce qu'il découvrit ; pour ma part je vis des meubles et des cadres ensevelis sous la poussière. Une bibliothèque agonisait contre le mur du fond, garnie de romans et de vieux manuels de maths.

De retour dans le couloir, nous poursuivîmes l'exploration, en évitant de faire craquer le plancher. La salle à manger et le petit bureau étaient dans un état comparable. Puis le couloir débouchait sur une cuisine. Elle aussi semblait avoir connu des années d'ordre et de propreté, avant d'être laissée en plan. Aucune trace de vandalisme. Soit les gosses du coin étaient des anges, soit ils n'aimaient pas cette adresse.

Je suivis Monroe au pied de l'escalier. Je fis le guet pendant qu'il montait à pas de loup, le dos près du mur, regard et flingue pointés en l'air. J'attendis qu'il fût en haut pour le rejoindre.

Il y avait trois chambres, que nous nous répartîmes. Chacune possédait un lit, dont l'un privé de matelas. Au milieu de cette pièce-là trônait une chaise en bois, et un abat-jour gisait par terre. Le sol était parsemé de feuilles d'arbres et d'éclats de verre.

Ayant fini le premier, je vérifiai la salle de bains, tout aussi muette et défraîchie. Je ressortis, et à peine commençais-je à me détendre que je crus entendre un bruit.

Il provenait de l'une de mes chambres. Celle de la chaise.

Comme un reniflement, ou une déchirure.

Je raidis mes bras, relevai le flingue de quelques degrés. Retournai sans un bruit vers la pièce, restai un instant sur le seuil.

Monroe émergea de la chambre principale. Je levai l'index. Il se figea, et me donna son feu vert d'un coup de menton.

Je m'avançai d'un pas, le canon braqué vers le fond de la pièce. Virevoltai pour couvrir le mur de la porte. Pivotai de plus belle, refis un pas. Posai un genou à terre pour regarder sous le lit simple, celui qui avait perdu son matelas. Il n'y avait personne dans cette chambre.

— Vous n'avez rien entendu ? demandai-je à Monroe, posté à la porte.

— Non. C'était quoi ?

— Rien, sans doute.

Je soufflai un grand coup. Toute la maison parut expirer avec moi.

Nous redescendîmes. Jetâmes un dernier coup d'œil avant d'échouer dans la cuisine.

— Je crois qu'on a fait chou blanc, Ward.

— J'ai quand même reconnu la camionnette. Le fait qu'il n'y ait personne ne change rien.

— Mais ça nous avance à quoi ? Nous...

Il s'interrompit, fixant un point par-dessus mon épaule. Je me retournai pour aviser une porte discrète dans le mur du fond.

— Placard ?

— M'étonnerait, dit-il en se remettant à chuchoter. Elle donne sous l'escalier.

Je l'ouvris sans attendre. L'obscurité, une bouffée d'air frais... Je palpai le mur à la recherche d'un interrupteur. En trouvai un et l'actionnai.

Un escalier de bois menait à une cave.

Nous nous y engouffrâmes. Le sous-sol était un grand espace rectangulaire. Et vide. J'allais rebrousser chemin lorsque Monroe me retint :

— Attendez. Visez un peu le sol.

La lumière crue de l'unique ampoule ne laissait aucune place au doute, dès lors qu'on savait ce qu'on cherchait : la terre n'était pas égale, mais griffée de volutes larges et irrégulières. Un peu comme si un corps y avait récemment roulé, un corps entravé. Je m'agenouillai pour effleurer la terre avec ma paume. La scruter, l'écouter.

Monroe m'imita un peu plus loin.

— Très bien, déclara-t-il. On tient peut-être quelque chose.

Je levai les yeux, à contrecœur.

— Quoi donc ?

— Une trace de sang, très légère. On a pris soin de la nettoyer. Mais il y a aussi des taches d'acides gras volatils. Comme si un corps avait stagné ici.

— Mort ou vivant ?

— Juste après sa mort. On vient peut-être de trouver l'endroit où Julia Gulicks a conservé sa première victime avant de la décharner.

— C'est donc le corps de Widmar qui aurait produit ces traces ?

Je ne voulais pas qu'il réponde oui, et il le savait.

— Je l'ignore, Ward. Mais c'est probable. Rien ne permet de penser que Nina ait séjourné ici.

— Sauf que le Combi est revenu *après* l'arrestation de Gulicks.

Il réfléchit.

— Exact. Allons l'examiner.

Il remonta les marches en trottant. Je restai en arrière pour observer une dernière fois les marques dans la terre. Étaient-elles l'œuvre de Nina ? N'est-on pas censé déceler la présence de l'être aimé, même après qu'il s'en est allé ? Nos cinq sens ne servent-ils pas à cela ? J'essayai, mais en vain. Je captais des ondes, sans savoir s'il s'agissait de son écho. Repartant vers l'escalier, j'eus soudain l'idée de frapper l'ampoule du plafond, afin de modifier l'éclairage. Elle oscilla d'avant en arrière, projetant son feu dans différents recoins. Les taches relevées par Monroe prirent un nouveau relief, et je dus admettre que l'hypothèse d'un corps d'homme était aussi valable que la mienne.

Je vis aussi scintiller quelque chose près du mur du fond. En quelques enjambées, je traversai les ombres épaisses et mouvantes, et m'accroupis pour tâter le sol. Mes doigts effleurèrent un objet anguleux.

Un bracelet. Du toc, avec des anneaux en plaqué argent et de fausses turquoises. La chaîne était brisée, comme après s'être accrochée quelque part.

Mais le bijou conservait un certain éclat, ce qui signifiait qu'il n'était pas là depuis des lustres. Il ressemblait pas mal à celui que Nina avait acheté pour six dollars lors d'une de nos virées à l'est de Sheffer.

Je n'étais pas formel. Loin de là. Mais tout à coup les formes tracées dans la terre me semblaient familières.

Je courus au rez-de-chaussée. Monroe jugerait cela insuffisant, mais pour la première fois en deux jours, j'avais l'impression de m'être rapproché de Nina.

Sitôt sur le perron, je me figeai. Il y avait quelque chose dans le jardin. Une longue forme, à demi cachée dans les broussailles.

Mon flingue avait retrouvé mon poing. Je me décalai sur le côté du porche, essayant d'identifier la chose sous un meilleur angle, tout en gardant un œil sur la camionnette.

— Charles ? Où êtes-vous ?

Pas de réponse, juste le bruissement des feuillages. Je traversai le porche de bord en bord. Rien non plus de ce côté-ci.

Je posai le pied dans l'herbe pour atteindre la forme avachie, la tenant en joue jusqu'à ce que je reconnaisse la couleur du costard.

Monroe était couché sur le ventre. Inerte. Les brindilles alentour étaient mouchetées de rouge, comme de minuscules fleurs sauvages.

Je le retournai. Il y avait du sang partout. Des entailles profondes et droites au front, au visage et au cou. De la chair à vif et des bouts d'os. Une dent brillait, nette et polie, à travers la joue trouée.

Sa mâchoire s'affaissa lentement sur le côté, relâchant un gros caillot de substance interne, et son dernier soupir forma un mot.

Pardon.

Je prononçai son nom, mais son regard était déjà vide, et divergent.

Je restai agenouillé, sans savoir quoi faire, ni qui l'avait tué. Je cherchai un pouls dans son cou, mais il était parti battre ailleurs. Le dénommé Charles Monroe n'était plus de ce monde. Seule restait une carcasse qui lui ressemblait fort, une carcasse morte et déchue, à deux mille bornes de chez elle.

Les herbes s'agitèrent. Bruyamment.

Je levai les yeux.

Un homme accourait depuis la gauche du pavillon. Massif, armé d'un immense couteau.

Je lançai mes bras en l'air, puis à droite, et tirai avant même de comprendre ce qui fonçait sur moi.

La balle l'atteignit à l'épaule. Je me projetai en arrière, parvins de justesse à me relever.

L'homme continuait d'avancer, et sa lame menaçait de m'atteindre. Reculant toujours, je lui tirai dans la cuisse. Il pivota, tomba, glissa.

Avant qu'il ne puisse se relever, je lui écrasai la main pour lui faire lâcher son arme, que je balançai loin dans les broussailles.

Reculant d'un pas, je pointai mon canon sur sa tête. Il avait les cheveux gris. Les mains et la figure aspergées de sang, celui de Monroe et le sien.

— Dis-moi qui tu es. Et où elle se trouve.

Il me dévisagea, l'air perdu.

— C'est toi, fit-il. C'est toujours toi.

— On ne se connaît pas.

— Ça a toujours été toi.

— Je ne comprends pas ce que tu me racontes, et je ne suis pas...

— Finissons-en, d'accord ? Par pitié, finissons-en.

— T'inquiète pas, je vais te buter. (Posant un pied sur son torse, j'appuyai lourdement le flingue sur son front.) Mais tu vas d'abord parler.

Il s'appelait Jim Westlake. Il s'appelait James Kyle. Il avait habité cette maison, et tué dix-huit femmes en vingt ans, la dernière étant son épouse. Il l'avait inhumée dans les bois, à quinze minutes à pied d'ici – c'était elle que John et moi avions découverte. On avait atteint un certain point. Pouvais-je comprendre ça ? On avait atteint le point critique. Il n'avait pas conscience, à l'époque, que des gens suivaient ses agissements depuis longtemps, des gens qui n'étaient pas de la police et qui comprenaient pourquoi des femmes d'Owensville, de Rackham et des environs disparaissaient de loin en loin. Ils savaient même où il les enterrait, et ils l'approuvaient. Un type qu'il surnommait le Garçon précoce s'était présenté une nuit, la pire de toutes, la dernière de cette ancienne vie, la nuit où il se trouvait seul dans sa maison, sans épouse mais avec une enfant qui courait se cacher après avoir compris que son père ne tournait pas rond, qui se réfugiait dans sa chambre et plongeait sous son lit, comme devant une tempête venue l'enlever. Sa femme était morte depuis seulement deux jours, et il se savait perdu ; la situation avait dégénéré jusqu'au point critique. Alors qu'il enterrait Laurie, il avait relevé les yeux et cru déceler dans le clair de lune une flaque de sang noir, suspendue à un mètre du sol, de l'autre côté de l'îlot. Du sang, comme celui qu'il leur avait pris à toutes, le sang de ces anges dont il s'était nourri. On ne peut pas manger une main, c'est plein de petits os, mais on peut manger du sang et se délecter du goût. Il pourchassa l'apparition, mais elle s'enfuit dans les bois, alors il termina sa besogne.

Que faire ensuite ? C'est insoluble. La première fois amène la deuxième, qui amène la multitude. Dans une autre ville, il s'en serait peut-être sorti, et Karla serait restée l'unique, la Karla du passé, de l'époque bénie des premières fois. Mais lorsque la terre vous demande d'alimenter sa force, vous n'avez d'autre choix que d'obtempérer, et le soir où vos proches découvrent le pot-aux-roses, il n'y a qu'un dénouement possible, et c'est ainsi que vous

vous retrouvez piégé dans la gangue brûlante d'une vie où votre propre fille panique à votre vue...

— Je ne comprends pas, frissonna-t-il, le visage blême et perlé de sueur glacée, tordu par je ne sais quelle indicible réalité. Je lui ai parlé des orages. Je l'ai rassurée. Je l'ai couverte d'amour. Mais j'ai quand même fait ce que j'ai fait. Je n'arrive pas à comprendre.

— Que s'est-il passé ?

— Quand ça ?

— Après avoir tué la personne dont tu parles.

— Tu t'es pointé.

— Je ne suis pas... Bon, d'accord, qu'est-ce que j'ai fait ?

— On l'a enterrée près d'ici et tu m'as aidé à tout nettoyer, à déménager, et tu t'es servi de tes relations pour étouffer l'affaire. Tu m'as fait supprimer des gens, puis tu ne m'as plus rien demandé et je n'ai pas eu de nouvelles pendant des années et des années. Je n'ai plus tué personne. Je suis devenu photographe. J'étais comme tout le monde. Je pensais que tu avais disparu pour toujours. Chaque année qui passait, je retranchais mentalement un ange à mon bilan. En espérant un jour revenir à zéro. Mais tu ne t'en vas jamais. Tu es toujours là. Toujours.

— C'est pas moi. C'est quelqu'un d'autre.

— Si, c'est toi. T'as l'air différent, mais c'est toi. C'est toujours toi.

— Pourquoi t'es revenu ?

— Parce que tu m'as fait revenir. J'étais censé m'occuper de la fille qui m'a vu à l'époque et qui s'est mise à reproduire mes meurtres. Mais elle venait d'être arrêtée. Et puis tu m'avais chargé d'une autre mission.

— C'est toi qui as tué le flic à l'Holiday Inn.

— Oui.

— Et t'as enlevé l'agent du FBI. Tu l'as amenée ici.

— Pour quelques heures, puis on a pris la route. On aurait pu rouler jusqu'à la fin des temps. Personne ne nous aurait jamais retrouvés. Je n'allais pas... T'es arrivé, et il a fallu que je revienne ici. Tu m'as forcé.

— Qui ?

— Toi.

— Mais c'est pas moi, d'accord ? (Un détail me fit tiquer.) Attends un peu. Tu dis qu'il est arrivé ici ? À Dryford ? Mais quand ça ?

— Ce matin.

— Paul est venu ici ce matin ? Par ses propres moyens ?

L'homme me considéra, et parut enfin comprendre que je n'étais pas celui dont il parlait.

— En bagnole, avec un jeune. Ils ne connaissent pas son vrai visage. Ils ne comprennent pas qu'il n'est même pas réel.

— Il est où, maintenant ?

— Je sais pas. Il est parti.

— Quand ?

— Il y a deux heures.

— Il a Nina ?

Soudain il tenta de se relever, en me prenant par surprise. Je le repoussai avec ma semelle.

— Réponds ou je te jure que...

— Oui.

— Elle est vivante ?

— Elle l'était quand ils sont partis.

J'augmentai la pression du flingue, sur le crâne d'un type à peine plus jeune que mon père au moment de sa mort.

— Où sont-ils ? Où sont-ils allés ?

— Je sais pas.

— Ils ont quoi, comme bagnole ?

— Grosse. Noire.

Je resserrai mon doigt sur la détente, mais je vis dans ces yeux qu'il n'y avait plus rien à tuer.

Alors je l'ai laissé ramper dans les herbes et je suis retourné auprès de Monroe. Je m'apprêtais à lui fermer les paupières quand je me suis dit qu'il y avait pire, comme spectacle éternel, que le mouvement paisible des branches dans un beau ciel bleu.

Puis j'ai couru jusqu'à la voiture.

CHAPITRE 33

Lee en avait sa claque. Lee se sentait paumé. Lee commençait à se dire que c'était vraiment n'importe quoi.

On avait installé la femme à l'arrière de la bagnole, et Paul s'était assis avec elle, dans le secret des vitres teintées, reléguant Lee auprès du conducteur. C'était un type courtaud, avec de gros sourcils et un nez crochu, le genre que Hudek faisait tout pour éviter. Le sourire pincé du gars montra que c'était réciproque.

Ils reprirent la route forestière jusqu'à la ville principale et sillonnèrent les rues. Lorsque son téléphone sonnait, Paul fixait l'écran sans répondre, et quand il voulait quelque chose, la voiture s'arrêtait pour que Lee fasse les commissions. Le chauffeur se plantait alors devant le capot, sûrement pour s'assurer que le jeune n'allait pas fuir, ce qui mettait Lee hors de lui. Il ne comptait aller nulle part. Il voulait juste qu'on lui dise de quoi il retournait, et quel était son rôle dans l'histoire. Ils se rangèrent devant le Starbucks, où Lee fit la queue pour commander les cafés. Ils se rendirent à l'épicerie pour qu'il achète des clopes. On lui remit un appareil numérique afin de photographier l'intérieur de l'église et la façade du commissariat, ce qu'il fit sans rechigner. Il acheta des piles chez Radio Shack, dans une galerie marchande à l'orée de la ville, puis un peu plus tard des beignets d'oignon chez Renée, pour le conducteur. Chaque fois les commerçants l'accueillaient avec chaleur. Ils souriaient, hochaient la tête, lui souhaitaient une bonne journée – un vrai spot de pub pour la vie en province. Mais son jugement ne varia pas : c'était un bled pourri.

Ils finirent par regagner le centre-ville, et la voiture s'arrêta le long d'une grille à barreaux munie d'un grand portail en fer forgé. Elle entourait la grande pelouse d'un lycée. Des gosses de différents âges s'y attardaient, prolongeant de quelques minutes la pause déjeuner. De l'autre côté de la rue se dressait un bâtiment du même style, mais plus petit. Un panneau super chaleureux indiquait un jardin d'enfants.

Paul pria Lee de revenir s'asseoir à l'arrière. La femme occupait toujours la même place, toujours ligotée. Lee fut surpris de voir si peu d'angoisse dans ses yeux.

— Tu vas me demander de la tuer ?

Paul leva les sourcils.

— Qu'est-ce qui te fait croire ça ?

— Je sais pas. Une sorte de baptême du feu.

— Ce n'est pas ce que je prévoyais. Dans un combat au corps à corps, je ne te donnerais qu'une chance sur deux. Au mieux.

— Je t'écoute, alors.

Paul attrapa sous son siège un petit sac. Le genre d'accessoire pour balader un iPod ou un lecteur CD. Il le remit à Lee. S'y trouvaient quatre flacons de pilules.

— C'est quoi, ça ? (Lee se tourna vers le lycée.) Tu veux que j'aille vendre de la drogue ?

— Non. Je veux que tu l'offres.

Lee allait dispenser Paul de ses sarcasmes, quand il comprit que c'était sérieux.

— Pourquoi ?

— Pour amorcer la pompe.

— C'est ça, le gros boulot que tu me réservais ?

— Une simple mise en jambes. Je reviens dans une heure, et là, nous aborderons la grande leçon du jour.

Lee descendit et regarda la voiture s'éloigner.

Il allait faire ce qu'on lui demandait – ce qu'on lui *ordonnait*, plutôt – mais sa patience était à bout. Comment se contenter de si peu après avoir traversé toutes ces épreuves, après avoir été arraché à son existence ? Lee voulait réussir. Jusqu'à cette dernière semaine, il avait pris la bonne direction, s'était lancé sur une voie sûre et rectiligne, tracée par lui seul. Il avait une équipe. Des amis. Il avait une vie. Il avait un plan.

C'était lui qui conduisait.

Abandonné sur le trottoir avec ses doutes, Lee mesura combien Brad lui manquait. Pete le Dormeur également, ainsi que deux ou trois autres potes, mais surtout Brad – même s'il pouvait encore sentir ses poings contre ses côtes. Ils étaient restés amis pendant des lustres, alors que Lee n'avait jamais eu beaucoup de copains. Brad et sa façon de dire les choses, à la fois lucide et directe. Lee aurait tant aimé que Brad soit là pour lui glisser : « Laisse béton, mec, ça craint » ou bien : « D'accord, c'est zarb, mais on y va. » Ou entendre Pete jacter sur les tableaux cachés dans tel jeu vidéo, ou même apercevoir Karen au bout de la rue. Il regrettait d'avoir fait si peu d'efforts après l'unique fois où ils avaient couché ensemble. Mais il s'était senti tout chose, exposé à des émotions qu'il ne maîtrisait guère. Réaction naturelle : ne pas la rappeler. Il fallut qu'elle se tourne vers Brad pour que Lee se dise qu'il aurait aimé la revoir. Mais comme disait Hudek senior, on ne rejoue pas les matches perdus.

L'espace d'un instant, il eut envie d'appeler ses vieux. De leur dire : *Je comprends rien à tout ce cirque. Je suis censé faire quoi ?*

Mais ils se seraient sans doute contentés de reposer le combiné. Lee n'était plus leur gosse, mais celui de Paul. Il appartenait désormais – sinon depuis toujours – à cette mystérieuse famille élargie. Non, la vraie question était : *Vous discutiez de quoi sur la photo que j'ai vue ce matin ? Quel était l'enjeu ? Quel a été le marché ? Qu'est-ce que vous avez gagné ?*

J'étais quoi, pour vous ?

Mais à quoi bon ? Saisit-on jamais ce qui se passe dans la vie HBO de ses vieux ? Eux-mêmes peuvent-ils jamais comprendre nos grands moments MTV ? Ou bien suit-on des chemins parallèles mais séparés dans le temps, pour se faire signe de manière épisodique, d'une distance longue comme un stade de foot embrumé ?

Peu importe, conclut-il. Au final, on marche toujours seul.

Lee ouvrit le sac et sortit un flacon. Ôta le capuchon et le glissa dans sa poche. Il marcha jusqu'à la grille et promena son regard sur les quelques gamins qui traînaient dans l'enceinte. Ils étaient sapés comme tous les gosses du pays, et Lee ne risquait guère de détonner – ce que Paul avait nécessairement calculé. Dans le pire des cas, un prof lui demanderait ce qu'il fabriquait là, et il se ferait passer pour le grand frère d'un élève, venu transmettre un message parental. Il l'avait déjà fait, à ses débuts, quand il avait dix-huit ans. Paul le savait-il, d'ailleurs ? Paul était-il au courant de tout ?

Reconnaissant dans son dos le bruit râpeux de petites roues dures, Lee pivota pour voir s'approcher un jeune à skateboard. Il comptait l'aborder quand il s'aperçut que le skateur était plus vieux qu'il n'y paraissait, et qu'il portait un sac à dos rouge déjà vu quelque part. Le keum lui fit un clin d'œil, avant de disparaître dans la pente.

Alors Lee se retourna et franchit la grille.

Paul la débâillonna peu après.

— Je vais te demander un petit service, dit-il. Tu ne seras pas d'accord, ce que je respecte, mais si tu n'obéis pas je te tuerai sur-le-champ. Pas de deuxième chance.

Lee parti, Paul s'était métamorphosé. Nina comprit qu'il changeait de rôle avec chaque interlocuteur. Même là, il devait jouer la comédie, sans forcément s'en rendre compte. La fameuse machine, en plein travail d'imitation.

— Qu'est-ce qui vous amène ici ? demanda-t-elle. Dans une ville pleine de gens innocents ? Vous êtes venus faire quoi ?

— La récupérer. Tu ne sens rien ? Tu ne vois rien ?

— Non, juste une ville ordinaire. Vous avez dû rêver. Un effet secondaire de la folie, sans doute.

Il sourit froidement.

— Si, je sais bien que tu l'as senti. Il y a beaucoup de villes semblables. Des lieux où nous vivions il y a deux mille ans et plus. Avant que nous ne bougions. Nous aimons bien vagabonder. Il nous arrive de revenir, mais on ne s'établit jamais nulle part. Il y avait plein de place avant que tous les autres ne débarquent. Mais ils ont afflué en masse, et ils ont trouvé des amas de pierres fort utiles, ainsi que des pistes, et bon sang, se sont-ils dit, comme c'est gentil de la part des Indiens d'avoir laissé ces choses pour qu'on puisse y bâtir nos petites fermes, et nos petites villes à la con. Sans se douter qu'ils pillaient les biens d'un *autre* groupe. Que *nous* les avions mis là dans un but précis. Que tout cela était *à nous*.

— Il faut vraiment que je vous fasse rencontrer John Zandt. Lui aussi a des théories de maboule sur votre bande. Évidemment, il y a des chances pour qu'il préfère d'abord vous tuer.

— T'inquiète pas, je compte bien le revoir. Je me suis donné du mal pour organiser la rencontre. Même si je crains qu'elle ne soit fort brève.

Son téléphone sonna ; il se tut quelques instants pour lire le message affiché.

— Il n'y en a plus pour longtemps, promit-il.

Il sortit un flingue de sa veste, le chargea lourdement, défit le cran de sûreté et le garda en main.

— Quel que soit votre plan, déclara Nina, je ne ferai rien.

— Oh que si, répondit-il sans s'énerver. Sans quoi je m'arrangerai pour expédier ton cœur à Ward, avec un mot disant que tu ne l'aimais pas assez pour rester en vie. Que tu préférais avant tout jouer les héroïnes. Que tu as couché avec lui dans le seul but de te rapprocher de moi. Sur une idée de Charles Monroe.

Nina regarda par la fenêtre, se concentrant sur la ville qui défilait doucement.

— Tout cela est faux.

— Possible. Mais il ne le saura jamais.

CHAPITRE 34

Ils ont d'abord refusé de me croire. Unger, l'homme à la tête remplie de chronologies millénaires, m'accusa sans détour d'essayer de les retenir en ville quand leur présence était requise ailleurs. Je fus obligé de hurler pour l'empêcher de me raccrocher au nez. Finalement, ils consentirent à m'attendre vingt minutes, pas une de plus. J'ai regagné Thornton aussi vite que je pouvais.

Poussant la porte du café, je les repérai, tendus à souhait, deux carcasses improbables sur un sofa bordeaux près de la fenêtre, cernés de gens normaux. Unger s'excitait au téléphone. Il s'était tourné vers la fenêtre par souci de discrétion, mais ses mains bougeaient dans tous les sens et ses joues étaient en feu. L'image même d'un homme que personne ne prend au sérieux. Rien de très surprenant.

— Alors ? lança John. Qu'est-ce qui s'est passé, bordel ?

Je m'assis tout près et débitai :

— On a trouvé l'endroit où Julia Gulicks a grandi. Ainsi que la baraque d'un type qui, si j'ai bien compris, aurait tué sa femme sous les yeux de Gulicks quand elle avait onze ans. On s'est introduits dans la maison et on a découvert un cellier, où Nina était enfermée ce matin. Le reste du temps, elle voyageait à bord d'une camionnette. Le FBI n'avait aucune chance de la localiser par le porte-à-porte : le kidnappeur la déplaçait sans cesse.

— Pas si vite, dit John en levant la main.

Je repris mon souffle, conscient qu'il fallait à tout prix le persuader de rester.

— J'ai tiré sur le ravisseur de Nina. Il jure que Paul est ici, à Thornton, en ce moment même. Il se promène avec un jeune, et il détient Nina. Elle était encore vivante il y a deux heures.

— Et pourquoi croirais-tu ce type ?

— Parce qu'il voulait que je le tue, et j'ai promis de le faire s'il me racontait la vérité.

— Tu l'as buté ?

— Non. Qu'il crève tout seul.

Unger raccrocha.

— Ramona n'est pas venue bosser aujourd'hui. Vous savez, la fille avec qui j'ai remué cette histoire de mails. Et ça ne répond pas chez elle. Je n'arrive à joindre personne à Langley. Ils sont tous absents, et j'ai été coupé deux fois.

— Et pour L.A. ? lui demanda John. Ils s'organisent en vue de notre arrivée ?

— Nous allons devoir gérer ça sur place. Ils ne veulent rien entendre au téléphone.

— Ce qui veut dire non, explicitai-je. Ils ne vous croient pas et ils ont raison. Il ne va rien se passer à L.A. Nous devons...

— On va pas remettre ça sur le tapis, interrompit Carl. J'ai des ordres venus d'ailleurs. Je vais aux chiottes et après je me casse.

Sur ces mots, il s'éclipsa au fond du bâtiment.

Je me retournai vers John.

— Pour l'amour du ciel...

— Désolé, Ward, mais on ne me fera pas croire que Paul est ici.

— Mais puisque je te dis que je viens de parler à un type...

— Qui est un cinglé de meurtrier. Ces tarés ont toujours du mal avec la réalité. Sans compter qu'ils mentent. Au fait, j'ai reçu un coup de fil d'Oz Turner. Il a vérifié une nouvelle fois son serveur, qui est à présent infesté d'images pédophiles. Il a tout effacé, débranché la machine, et s'est dépêché de quitter l'État. Ces types passent à l'offensive. Ils préparent un gros coup. Je me demande bien pourquoi Paul se baladerait ici aujourd'hui.

— Parce que nous y sommes, nous ! Il ne cherchait pas à nous éloigner, mais à nous réunir ! Tu veux connaître la dernière parole de Monroe ?

— Attends, là. Sa *dernière parole* ?

— Oui, le tueur l'a eu, expliquai-je d'un air détaché. Il lui a massacré la tronche avec un énorme couteau. Je pensais que tu l'aurais compris, vu qu'il n'est pas à mes côtés. Essaie de suivre, John. Charles Monroe *est mort*.

— Nom de Dieu ! Écoute, ralentis un peu, Ward. Il faut que...

— On n'a pas le temps. (Autour de nous je voyais les gens commander une rincette, lire la page des sports, mener une vie rationnelle.) Le dernier mot de Monroe fut : « Pardon ». Tu ne saurais pas pourquoi il a dit ça ?

— C'est que...

Je levai les mains au ciel.

— Alors tu le sais ! Bien sûr que tu le sais, mais tu n'as pas jugé bon de m'en informer. Le soir où tu voulais parler à Julia Gulicks, et que j'ai reçu le coup de fil de Carl, tu m'as dit de prendre l'appel hors de la voiture. Parce que tu soupçonnais Monroe d'avoir reçu un ordre d'on ne sait où, l'ordre de ramener Nina ici.

— C'est vrai.

— Alors tu l'as questionné, et il n'a rien avoué, mais soudain il voulait bien te laisser interroger la première suspecte d'un meurtre !

— Je lui ai juste fait comprendre que si jamais tu étais amené à croire qu'il avait fait venir Nina sur ordre, quelle que soit l'origine de cet ordre, tu serais tenté de le buter dans la seconde. Mais je n'étais sûr de rien. Ce n'était qu'une simple intuition. Je ne t'ai rien dit parce que...

— Parce que tu ne fais confiance à personne, que tu me prends pour un crétin, et tu n'as peut-être pas tort. Mais ce coup-ci c'est *moi* qui ai raison. Dis donc, il en met, du temps, le Carl. On a des recherches à mener, nous.

— Quelles recherches ? Même si t'avais raison, comment veux-tu qu'on les débusque ?

— Une grosse voiture noire. Dans un bled plein de pick-up et de petites citadines, ça doit se voir. On ne peut pas rester inertes.

Zandt fronça les sourcils.

— C'est vrai qu'il est long, notre ami.

Nous attendîmes trente secondes de plus, avant de remonter les cinq mètres de couloir menant aux toilettes. Nous entrâmes chez les hommes. Trois lavabos, trois urinoirs, trois cabines. Aucune trace de Carl.

— Bizarre, fit John.

— Il nous a lâchés. Il savait que tu canerais en apprenant que Paul pouvait être en ville, alors il s'est barré à L.A. tout seul.

— Impossible. Ce mec a besoin de nous. On le croit.

— Non, *tu* le crois. De toute façon, n'a-t-il pas l'appui de toutes les élites occultes du monde ? Des francs-macs jusqu'aux membres fondateurs du G8 ? Pourquoi il n'appelle pas SOS-Gentils ?

— Mais non, Ward. Ils n'ont aucune idée de... T'as rien écouté ou quoi ? (Il poussa la porte de la première cabine, vide.) Il a besoin de nous. (Idem pour la deuxième.) On a combattu ces gens en face, toi et moi. Mais les autres n'y voient qu'un mythe.

De toute façon, je ne sais pas comment Carl serait ressorti. Il était obligé de repasser devant nous.

Il ouvrit la troisième cabine. Carl Unger y était assis.

Les jambes tendues, les bras ballants. Adossé au réservoir de la chasse d'eau, la tête renversée en arrière. Un beau trou au milieu du front, laissé par un petit calibre, de quoi bousiller la cervelle sans rejaillir par l'arrière.

Nous eûmes un geste de recul, avant de nous rapprocher d'un pas. Le bonhomme était mort, ça ne faisait pas un pli.

— Mais... comment ? bredouillai-je dans un accès de terreur.

La mort me suivait partout depuis ce matin.

Nous avons vite refermé la cabine et regagné le couloir, sur nos gardes. John inspecta les toilettes des femmes pendant que je faisais le guet. Il ne vit personne.

Nous vérifiâmes qu'il n'y avait aucune sortie possible : le couloir finissait sur un mur. La seule et unique issue était l'entrée du café.

Dans la salle, deux types grisonnants se prenaient le bec. Les jeunes mamans bavardaient par deux ou par trois, admirant leurs emplettes respectives de chez Baby Gap. Quelques ménagères isolées bullaient devant un magazine et des scones à la canneberge, carrées dans leur cocon pour regarder passer les heures. Un quinquagénaire griffonnait sur son bloc-notes. Un couple de touristes stressés étudiait une carte routière. Une vieille femme lisait la gazette locale. Personne ne faisait attention à nous. Tous semblaient vaquer joyeusement dans un rêve.

— On se tire d'ici, murmurai-je.

— Avec plaisir.

Nous avons traversé l'espace, collés l'un à l'autre, prêts à bondir. Soudain une femme en pull duveteux éclata de rire, et je fus à deux doigts de dégainer. Mais elle ne faisait que jouer avec le bambin d'une autre. En arrière-plan, les percolateurs sifflaient et crachaient, et les employés criaient à propos de doubles express, de lait au soja, ou du café du commerce équitable.

En regagnant le trottoir, nous nous sommes retournés pour voir si quelqu'un nous épiait.

Mais personne ne regardait dans notre direction. À croire que nous n'étions pas ressortis, ni même entrés. Alors nous avons remonté la côte d'un pas rapide, les mains dans les poches, agrippés à nos flingues.

John ne pouvait s'empêcher de regarder derrière lui.

— Qu'est-ce qui s'est passé, là-dedans ?
— Tu n'avais remarqué aucun comportement bizarre ?
— Non, mais je ne cherchais rien de *bizarre*. On était chez Starbucks, bon sang !
— C'est forcément Paul.
— Et il serait passé devant nous, en tirant Nina par les cheveux ? Je l'aurais vu, ça.
— Il est accompagné d'un jeune.
— Hudek, sûrement. J'imagine que...
— Mais personne n'est entré ni sorti, John. J'étais assis face à la porte. J'aurais remarqué.

J'ouvris la portière de Zandt avant de m'installer de l'autre côté. Nous restâmes transis sur nos sièges, sonnés.
— Unger est mort.
— Et merde pour la cavalerie.
— On ferait mieux de quitter ce quartier, dis-je en démarrant. D'un instant à l'autre, quelqu'un va aller aux gogues et trouver un macchabée avec qui t'es resté assis pendant une demi-heure. On devrait alerter les fédés.
— Laisse tomber. Monroe est mort, et sans lui on ne nous connaît plus.
— Pas faux. Et je doute que les flics soient d'une grande aide. Au fait, quand je suis passé prendre Monroe au comico ce matin, il y avait un petit nouveau, un flic d'une autre ville. J'ai tout de suite senti qu'il m'avait dans le pif.
— Faut dire que t'es le roi des paranos.

Je l'ai fusillé du regard.

Après un rapide demi-tour, j'ai ralenti l'allure en repassant devant le Starbucks. On aurait dit un joyeux aquarium. Pour tous ces clients, rien n'était arrivé. Ils évoluaient dans une carte postale, le regard tourné vers l'extérieur.

Zandt s'était vite remis ; le flingue sur les genoux, il semblait brûler de s'en servir.
— On devrait y retourner, Ward.
— Pour quoi faire ? Celui qui a buté Carl nous zigouillera avant qu'on ait le moindre indice sur son identité. La moitié de la salle sera prise dans la fusillade, et ça ne sauvera pas Nina.
— Tu proposes quoi ?
— On reste en mouvement.

Je repris de la vitesse à l'approche de la grande pente. Des deux côtés de la route allaient et venaient les piétons. Les arbres cédaient quelques feuilles à la douce brise d'automne. Un employé d'UPS avait arrêté son fourgon pour livrer un long colis plat dans une boutique de cadeaux. La ville entière était un panneau animé, une image infranchissable, un monde qui nous tenait à l'écart.

— Où tu vas comme ça ? questionna Zandt.

— Aucune idée. Il faut découvrir où Paul a emmené Nina. Et d'ici là, éviter de se faire buter. On arrive à court de gentils.

— C'est ça qu'on est devenus ? On est les gentils ?

— On est ce qui s'en rapproche le plus.

Dix minutes plus tard, nous avions traversé la ville, et je montais un côté avant de bifurquer sur une voie unique. Au bout se trouvait le parking dominant la forêt de Raynor. Par chance, il était désert.

Je me garai tout au fond. Quittai la bagnole et fis les cent pas. Il me restait du sang de Monroe sur les doigts. Je les frottai nerveusement.

John me rejoignit au bout de deux minutes.

— On devrait quand même appeler les flics, considéra-t-il. Leur signaler le cadavre, et le décès de Monroe. Les prévenir qu'il se passe un truc.

— Et depuis quand tu fais confiance aux flics, toi ? En une seule semaine, cette ville totalise un policier mort et un agent du FBI enlevé, plus deux citoyens refroidis dans les bois. Si avec ça ils ne sont pas en état d'alerte, ils ne le seront jamais !

— Y a un truc pas net, dans ce bled. Comment dire...

Il s'engagea sur le sentier, comme attiré par quelque chose.

— Qu'est-ce que tu regardes ?

— J'aperçois une sorte de monticule, là-bas. Au milieu des arbres.

— C'est là qu'on a trouvé Lawrence Widmar. Dans ce secteur, en tout cas. Pourquoi ?

— Ça ressemble aux trucs que j'ai cherchés dans toute la Nouvelle-Angleterre.

— Pitié, John, c'est pas le moment de...

Il leva brusquement sa paume.

— C'est quoi, ce bruit ?

Je plongeai la main dans ma poche. Sortis mon téléphone.

« NINA », indiquait l'écran.

Mes doigts étaient du caoutchouc, et je dus m'y reprendre à trois fois pour enfoncer la touche. Je portai lentement l'appareil à mon oreille. Le sang affluait dans ma tête, sans savoir où jaillir.

— Nina ? fis-je d'une petite voix. C'est toi ?
— Salut, Ward.
C'était un homme.
— Qui est à l'appareil ?
— À ton avis ?
Ce ne pouvait être que lui.
— Paul ?
John me dévisagea. Je lui fis signe de rester tranquille.
— Gagné, répondit mon interlocuteur. J'avais envie de prendre des nouvelles. Tu n'écris pas, tu n'appelles jamais...
— Où est Nina ? Où es-tu ?
— À ton avis ?
— Il y a une théorie qui circule selon laquelle il se trame un incident majeur à L.A.
— Waouh, vous êtes malins, les gars.
— Mais ce n'est pas la mienne. Je pense que vous êtes bien plus près.
— Alors t'es encore plus futé que je ne pensais. Quelqu'un aimerait te dire bonjour.
Je me cramponnai au téléphone.
— Salut, chéri, dit Nina.
— Salut. (Un poing invisible m'étranglait.) Tu vas bien ?
— Ça va, répondit-elle faiblement.
— T'es où ?
— J'ai un flingue sur la tempe, Ward.
— Alors ne dis rien. Qu'est-ce qu'il veut ? Qu'est-ce que je dois faire ?
— Il veut John.
— Reste en vie. Reste en vie pour moi.
— Je ferai de mon mieux. Je t'assure. On apprend à tout âge. On apprend à tout âge. Je t'...
Elle n'était plus là.
— Voilà pour la carotte, dit Paul en reprenant l'appareil. L'ex-inspecteur Zandt est-il avec toi ?
— Non. Il est parti pour L.A.
— Vraiment ? Quel dommage ! Il s'est planté de direction. Il va se sentir tout bête. J'avais tellement envie de lui parler.
— C'est le risque avec les jeux idiots. Et si tu n'avais pas mis la main sur le portable de Nina ? Tu m'aurais contacté comment ?
— J'avais déjà ton numéro, Ward. Attends, écoute.
Une pause de deux secondes, puis une nouvelle série de notes, non loin de moi. C'était le téléphone de John.

J'entendis Paul s'esclaffer.

— Ben ça alors ! Figure-toi que je viens d'entendre sonner de ton côté. Ce qui, à mon avis, prouve que vous êtes bel et bien ensemble, Zandt et toi. Il va de soi que je possède aussi son numéro. Vous n'avez pas l'air de comprendre à qui vous avez affaire.

— Personne ne comprend. Pas même les francs-maçons, visiblement.

— Ah, M. Unger. Un drôle de type, bercé d'illusions. Mais bon, il est mort, alors peu importe. Tu ne m'as pas vu ? Je suis toujours là, en toile de fond. Et je le serai toujours. Nous sommes légion, frangin. Passe-moi John.

Je tendis le portable à Zandt, qui sans un mot, sans un cri, sans même une menace, trouva la force d'écouter l'assassin de sa fille.

Puis il me rendit le téléphone.

— Si tu préviens la police, m'avisa Paul, je le saurai immédiatement. Essaie de me baiser, et je tue Nina – et ce ne sera même pas rapide. Tu sais que je tiendrai parole.

— Oui, je sais. Mais toi aussi, il faut que tu saches un truc. J'ai tué ton maboule de pote, il y a une heure de ça.

— C'est-à-dire ? demanda-t-il, effleuré d'une hésitation.

— Du côté de Dryford. Jim, James, que sais-je. Le type pissait le sang dans son jardin quand je suis reparti. Plutôt bavard, du reste. Ça risque de faire des vagues, même jusqu'ici.

— Ça ne fera pas trois lignes dans le journal, rétorqua Paul avant de couper.

Tourné vers la forêt, John alluma une cigarette, avec le visage posé d'un homme dont la décision est prise. Le mien rayonnait d'avoir entendu Nina. Tout restait à faire, mais au moins elle était en vie. Pour le moment.

— Qu'est-ce qu'il t'a dit ?

John répondit sans bouger :

— Que si je me livrais à lui, il annulerait l'autre chose. Et relâcherait Nina. Toi, de toute façon, il te veut vivant.

— C'est quoi, cette « autre chose » ?

— Il n'a pas précisé. Peut-être qu'il bluffe.

— J'en doute. Ce n'est pas toi, la cible de l'opération. Ça ne justifierait pas tout ce remue-ménage.

— Il veut que je me rende à l'Holiday Inn. Il passera me prendre.

— Et tu comptes y aller, n'est-ce pas ?

— A-t-on vraiment le choix ?

— Tu te dis surtout que tu vas enfin le chopper.

— L'hôtel est à moitié rempli de fédés. Quand aurai-je une meilleure occasion ?

— Et tu crois qu'il ne le sait pas ?

— Si on part tout de suite, on pourra leur dire ce qui est arrivé à Charles. Et ils auront le temps de se préparer.

— Mais ils ne nous croiront jamais. Ils vont sûrement t'arrêter, et... (Je me tus pour laisser germer un songe.) Mais oui, c'est ce qu'il cherche ! Il veut que tu te pointes là-bas, en sachant très bien que je viendrai te couvrir. Ce qui signifie qu'il sera ailleurs. Il nous veut, c'est sûr, mais ce n'est pas sa motivation première. (Je plongeai mon visage dans mes mains.) Nina a dit un truc...

— Quoi ?

— Elle a dit... un truc étrange. Tout à la fin. Elle a dit : « On apprend à tout âge. » Deux fois de suite.

— Et ?

— Tu trouves que ça ressemble à Nina, toi ? Elle parle par énigmes, d'habitude ?

— Non, mais...

— Et puis c'était hors contexte. Apprendre à tout âge, rester en vie. Je ne vois pas le rapport. Pourquoi a-t-elle...

— Elle n'a rien ajouté ?

Je me répétai mentalement notre échange. Elle allait bien... Tenue en joue... Rester en vie...

— Non, rien de plus. Elle disposait d'environ trente ou quarante mots. Elle en a gaspillé dix là-dessus.

Zandt restait suspendu à mes lèvres. La réponse finit par venir :

— Le message dont Carl nous a parlé ce matin, John. T'en as retenu quoi ?

— Ils expliquaient qu'ils n'étaient pas américains. Ni européens, ni quoi que ce soit. Et puis il y avait un couplet sur...

— *Il est temps d'éduquer les foules*, quelque chose dans ce goût-là.

— Ils ont un jeune dans leur équipe. Un gamin qui pourrait se fondre dans le décor pour poser une bombe, comme il l'a fait hier à LA. Et une grosse, ce coup-ci. Un engin qui va vraiment faire du bruit. Voilà pourquoi Paul nous veut à l'autre bout de la ville ! (John s'élança vers la voiture.) Ils vont s'en prendre au lycée.

CHAPITRE 35

Je me suis garé cinquante mètres avant le lycée, en face de la grande église. Nous sommes restés plantés près de la bagnole, battus par le vent froid remontant la colline. D'ici, le panorama embrassait la moitié de la ville, mais je ne vis personne – je ne parle même pas d'individus suspects : il n'y avait vraiment pas un chat, comme s'ils se cachaient tous au détour d'une rue ou derrière les murs. Je consultai ma montre : 3 heures moins le quart, un horaire de boulot par excellence, une heure qui vient toujours avant ou après l'événement. Une parenthèse où chacun bosse, fait des courses, étudie ou se morfond. Si vous deviez poser votre vaisseau, 3 heures pétantes serait le créneau idéal pour envahir la planète Terre.

Nous vérifiâmes nos flingues. Quinze minutes plus tôt, j'avais conduit John à l'Holiday Inn. Nous avions bougé son véhicule pour le rendre visible de la grand-route – au cas où Paul aurait posté des observateurs. Puis nous sommes discrètement revenus vers ma voiture, avec un sac noir contenant les munitions et les armes de poing de John. Le déplacement de sa bagnole ne servirait sans doute à rien, mais sur le moment l'idée nous parut bonne.

J'éjectai mon chargeur entamé pour refaire le plein de cartouches, et nous remplîmes nos manteaux noirs d'accessoires dont il ne sort jamais rien de bon.

John me regarda.

— Prêt ?

J'acquiesçai.

— Mais au premier indice laissant penser que Nina se trouve ailleurs, je me tire illico.

Nous prîmes à droite, dans une allée qui longeait l'église et débouchait sur un axe parallèle à la grand-rue. Nous remontâmes celui-ci pour nous arrêter cinquante mètres derrière l'école. Toujours rien, ni action ni gens.

Nous avons traversé la rue, et poursuivi la marche tout en examinant l'arrière du lycée, un groupe de bâtiments de la taille d'un pâté de maisons. La structure principale était séparée de la route par un grand rectangle d'asphalte. Contrairement aux pelouses à l'avant, celui-ci ne semblait pas destiné aux récréations. La grille s'ouvrait de chaque côté pour laisser entrer les camions de livraison, comme celui que déchargeait présentement un homme en bleu de travail. Mais sa cargaison semblait des plus normales : des pizzas surgelées, descendues en chariot vers une aire souterraine. Quand le type réapparut, il engagea la discussion avec le récipiendaire de la livraison, et ils s'accotèrent au mur pour s'en griller une. Ils se connaissaient depuis longtemps.

Nous avons atteint l'extrémité du périmètre, puis tourné à gauche et pris l'allée latérale jusqu'à retrouver la grand-rue. Les vieux bâtiments pointaient en haut de la colline, protégés par leur esplanade jonchée de feuilles mortes.

Nous vîmes également un gars posté devant la grille. Un jeune, qui n'était pas là quelques instants plus tôt. Il avait un skateboard sous le bras, et son âge se situait entre dix-sept et vingt ans, voire un peu plus. Hélas, ni John ni moi ne connaissions la bobine de Lee John Hudek.

— Et qu'est-ce qui nous étonne ? murmurai-je.

— La cloche n'a pas sonné. Qu'est-ce qu'il fait là ?

— Un rendez-vous chez le dentiste. Il attend que sa mère vienne le chercher. Il a insulté un prof qui l'a envoyé au piquet.

— Il est quand même bronzé, non ?

— T'as raison, John. Allons le plaquer sur le trottoir. Désolé, fils, mais t'es pas assez blême. On va devoir t'abattre.

Malgré mes réserves, je remontai le côté pour me planter deux mètres devant lui. Je pris tout mon temps, et il évita de lever les yeux à mon approche, mais n'importe quel ado aurait fait de même.

— T'as du feu ?

— Fumer tue, répondit-il.

— T'as besoin qu'on te dépose quelque part ?

Il me regarda enfin. Sa peau était nette, son regard bleu glacé. Il m'examina d'un air mauvais.

— J'imagine que vous plaisantez.

— L'école est pas finie, lançai-je.

— Et qu'est-ce que ça peut vous foutre ?

— Ce que ça peut nous foutre, dit John en surgissant à mon côté, c'est qu'on se demande ce que tu fabriques.

— Vous êtes genre la police du trottoir ?

— Ouais. Genre.

— Branleurs.

Puis il remarqua la façon dont John le fixait et tourna les talons, nonchalamment, pour se retrancher derrière la grille.

— Ça s'est bien passé, marmonnai-je.

John s'abstint de répondre. Resta immobile, à regarder le môme arpenter la pelouse. À aucun moment le gamin ne se retourna pour nous défier du regard ou nous montrer son majeur. Je me demandais s'il fallait s'en étonner, quand il fit une chose surprenante.

Il s'était dirigé vers une annexe à droite du bâtiment central. Mais lorsque deux élèves en sortirent, il dévia de sa trajectoire pour ne pas les croiser.

Zandt me regarda.

— Ouais, John, j'ai vu.

Nous franchîmes la grille. Notre bonhomme avait accéléré le pas et filait vers une arcade menant à la cour centrale.

— Eh ! Lee ! cria John.

Le gosse ne se retourna pas. Mais un autre réagit.

L'un des mômes qui venaient de sortir dressa la tête, interrompant ses activités avec son camarade. Mon cœur s'emballa.

— John...

— J'ai vu.

Il s'élança en direction de l'autre jeune. Celui-ci rangea quelque chose dans son blouson et replongea dans le bâtiment, avec souplesse et assurance. Ce petit gars avait du métier. Il n'était pas lycéen. Jusqu'à récemment, peut-être, mais plus aujourd'hui. Ce ne pouvait être que Lee Hudek.

John le suivit à l'intérieur. Le troisième larron s'éloignait sur le côté en forçant son air innocent. Je m'empressai de l'intercepter :

— Ce gosse à qui tu viens de parler, tu le connais ?

Le garçon secoua la tête. À l'évidence il se piquait d'être cool, mais il n'avait pas plus de quatorze ans et paraissait inquiet.

— Pas vraiment.

— Alors vous parliez de quoi ?

— De rien.

— Mon œil. Je l'ai vu glisser un truc dans sa poche. C'était quoi ?

— Je vois pas de quoi vous voulez parler.

— Bien. Je suis donc certain que ce n'était pas de la drogue.

Son visage m'avoua tout ce que je devais savoir.

— Ne touche pas à cette merde, l'avisai-je avant de me ruer dans le bâtiment.

L'entrée débouchait sur un grand hall austère, au départ de deux longs couloirs. Au fond s'élevait un escalier de pierre, d'où filtrait une lumière grisâtre. Je crus effectuer un bond dans le passé. Quels que soient les matériaux utilisés, une école sentira toujours l'école. On ne s'en rend pas compte lorsqu'on y passe ses journées, mais quand on y retourne en adulte, on a l'impression de visiter une écurie pour extraterrestres crados. De chaque côté se succédaient des portes en verre dépoli ; tout au fond s'étirait un mur de casiers cabossés, et le fond sonore résonnait de laïus peu ouverts à la discussion.

Je m'arrêtai, tendis l'oreille. Un bruit de pas à l'étage. Je m'y ruai. John errait au bout du couloir, près d'une rangée de fenêtres.

— Il est passé où ?

— J'en sais rien, répondit-il en revenant vers moi. Je croyais qu'il était monté ici.

— Il n'avait pas de sac, notai-je. Soit on s'est plantés et c'est juste un mioche qui deale de l'herbe, soit il a planqué je ne sais quel engin avant de ressortir. On s'y prend mal. Il est temps d'avertir un responsable, puis on décampe pour chercher Nina.

— D'accord. Allons trouver le bureau du prov...

Il s'interrompit. Je vis alors par la fenêtre une voiture de police devant l'école. Un flic baraqué traversait la pelouse à grands pas.

— On n'avait pas besoin de ça, maugréai-je.

Je dévalai les escaliers, talonné par John. Nous voyant ressortir du bâtiment, le policier obliqua vers nous.

— Intéressant, murmurai-je. (Nous avancions toujours.) Je suis presque sûr que c'est le flic qui discutait avec Monroe ce matin.

— Il m'a tout l'air d'un flic normal.

— Moi aussi, avec un uniforme.

— Non, toi tu serais plutôt du genre à intéresser les Affaires internes.

L'agent stoppa à quelques mètres de nous. C'était bien celui que j'avais aperçu au comico. Distingué, large d'épaules, le regard froid.

— Qui êtes-vous et que faites-vous là ? demanda-t-il.

— C'est quoi, votre problème ? répliqua John sans se rendre compte qu'il parlait comme le skateur.

— Il ne vous aura pas échappé que je suis officier de police ?

— Qu'y a-t-il, monsieur l'agent ? tempérai-je.

— Un enseignant vient de m'appeler. Un élève dit que deux types louches l'ont accosté devant la grille. C'était vous ?

— Nous pensons qu'un grave danger menace cette école.

— En effet, grogna le flic. C'est vous, ce danger. Et il ne s'agit pas d'une question, mais d'une affirmation. Allez, suivez-moi.

Nous n'avons pas bougé un cil.

— Où étiez-vous lorsque vous avez reçu l'appel ? demanda John d'un air anodin. Au poste ?

Le flic le dévisagea.

— Vous êtes drôlement rapide, ajoutai-je. Ça fait quoi, cinq, six minutes qu'on a parlé à ce gosse ? Il a fallu qu'il entre dans le lycée, qu'il trouve un prof, le persuade que c'était bien la peine de déranger les flics. Puis le prof contacte le comico, le message est transmis à qui de droit, et vous voilà.

— Service express, commenta John.

— Ou alors le gamin n'a rien rapporté du tout, suggérai-je. Peut-être qu'on nous observe. Mais dans ce cas, pourquoi monsieur l'agent nous mentirait-il ?

— Vous allez appeler des renforts ? demanda John. C'est ce que je ferais, à votre place. Deux types louches sur la voie publique ? J'exigerais des hommes sur-le-champ. À supposer que vous soyez ici pour les raisons que vous dites.

Le regard du type se refroidit un peu plus.

— Le prénom « Paul » ne vous évoque rien ? lançai-je. On s'est laissé dire qu'un certain Paul traînait dans le secteur. C'est peut-être lui que vous recherchez.

— Jamais entendu parler, répondit le flic en portant la main à son flingue. Et je vous ai assez écoutés.

— Vous avez raison, dis-je en levant les mains. (C'était la seule méthode possible, et nous perdions un temps dont Nina ne disposait pas.) Ce serait dommage de partir sur un malentendu. C'est une affaire sérieuse. Nous devrions aller au commissariat. Nous expliquer tranquillement.

— À la bonne heure.

Nous le suivîmes à travers la pelouse, au-delà des grilles et jusqu'à sa voiture.

Comme il commençait à ouvrir la portière arrière, John me fit un petit signe de main. Je scrutai la rue en vitesse. C'était le bon moment.

J'attrapai la portière et l'envoyai de toutes mes forces dans le ventre du flic. Anticipant le coup, il s'écarta partiellement, mais pour mieux s'écraser le visage contre le poing de John. Ses yeux tournoyèrent, mais il se ressaisit à temps pour que je lui en colle une deuxième. John l'enfourna sur la banquette.

J'ouvris la portière conducteur et m'installai derrière le volant.

— Pas de clés.

Je détournai les yeux, le temps qu'une série de coups mats fassent taire des cris étouffés. Puis le silence revint.

— Attrape, dit John en me tendant les clés.

Nous roulâmes doucement jusqu'au coin de la cité scolaire, puis je virai à gauche.

— Son badge a l'air vrai, dit John. Si c'est un faux, c'est très bien imité.

— Faux ou pas, c'est un connard.

Je me rangeai à mi-hauteur de la rue pour aider John à extraire l'homme inconscient de l'habitacle. J'ouvris le coffre et nous y rangeâmes sa carcasse. C'était un balèze, et il rentrait de justesse.

— Tu veux le bâillonner ?

— Laissons-lui le bénéfice du doute.

Je claquai le coffre.

— C'est un pourri, tu le sais comme moi.

— Dans ce cas, il se trame bien quelque chose, ici.

— Allons faire boucler ce bahut.

CHAPITRE 36

Nous négligeâmes le premier bâtiment pour nous engager dans la cour intérieure, où un escalier nous conduisit en territoire plus prometteur. C'était l'immeuble le plus grand et le plus ancien, ce qui laissait supposer qu'il abritait l'administration. Au rez-de-chaussée, des salles de cours bordaient deux longs couloirs adjacents et vides. Dans mon souvenir, on trouvait toujours quelqu'un pour errer dans ces halls, qui séchant un cours, qui prétextant un bobo, qui zonant purement et simplement. Mais le lycée de Thornton savait tenir ses élèves. Je devinai enfin une silhouette tout au bout du couloir ; elle ignora mes cris et disparut vite fait. Le premier étage était augmenté d'une grande salle déserte aux allures de laboratoire. Au dernier nous découvrîmes des murs lambrissés, et une porte dont la taille ne laissait aucune place au doute. Légèrement entrouverte, elle indiquait : « A. L. Singer, Proviseur ».

Nous pénétrâmes dans un vestibule où une espèce de matrone s'affairait derrière une machine à traitement de texte préhistorique. Elle nous darda un œil réprobateur qui me donna l'impression d'avoir douze ans.

— Qui êtes-vous ?

— Nous devons parler au proviseur, déclarai-je. Tout de suite.

— Elle est au téléphone.

Je suivis John jusqu'à la porte suivante, qu'il ouvrit. Une pièce plus grande, au centre de laquelle trônait un immense bureau. Derrière se trouvait une autre femme, qui bien entendu n'était pas au téléphone. Les cloisons étaient recouvertes d'ouvrages sérieux, et de photos en noir et blanc de valeureux prédécesseurs ou de poignées de main officielles. Dans son dos, la fenêtre offrait une vue sur la grande pelouse.

— Il faut évacuer le lycée, lança John.

La femme nous observa. Elle était grande, la coiffure bouffante, et je l'aurais bien vue botter des culs dans une multinationale.

— De quoi parlez-vous ? Qui êtes-vous ?

— Des gens s'apprêtent à attaquer le lycée.

La femme se leva lentement.

— Des gens ? Comment ça ? D'où sortez-vous une idée pareille ?

— On n'a pas le temps d'entrer dans les détails, répondis-je. On vous demande juste de croire que c'est très probable.

La femme tendit la main vers le téléphone au coin de son bureau. Mais celle de John l'y précéda. Pinçant les lèvres, le proviseur lança par-dessus mon épaule :

— Jane ! Veuillez appeler la police, et faites monter Ben sur-le-champ. Dites-lui que nous avons des intrus.

Me retournant, j'avisai la matrone dans le cadre de la porte.

— S'il vous plaît, ne faites pas ça.

Distraite par autre chose, la vieille disparut de ma vue. Je revins au proviseur.

— Écoutez, madame Singer, je sais que c'est bizarre, mais il faut nous croire.

— Ben voyons. Si l'établissement était menacé, la police m'aurait contactée.

— On vient d'en parler à un agent. Mais il n'est pas sûr qu'on puisse leur faire confiance.

— Leur faire conf... Dieu du ciel ! Faites monter Ben, Jane ! Tout de suite !

Pas de réponse dans le bureau voisin. Puis on entendit la matrone élever la voix contre quelqu'un, et la porte s'ouvrit en grand.

Entra un gosse. Le skateur que nous avions abordé devant la grille – qui n'avait rien d'un gosse et n'était pas plus lycéen que moi. Il prit la parole d'un ton offensé, sûr de son bon droit :

— Madame, j'étais devant la grille quand deux types sont venus me harceler. Ces deux types-là.

La femme l'observa. Je sentais bien que son quotidien ne ressemblait guère à ça.

— Qui es-tu ? Je ne me souviens pas de toi.

— Jason Scott, madame. Ces deux hommes voulaient m'embarquer dans leur voiture.

La femme ramena son regard sur John.

— Lâchez ce téléphone. C'est du ressort de la police.

John maintenait sa main sur le combiné. Alors elle ouvrit un tiroir et sortit un cellulaire.

— Eh bien, allez-y, pestai-je. Appelez les flics si ça vous chante, mais faites évacuer les lieux, bordel ! C'est quoi, votre problème ? On vous dit que le lycée est en danger. Vous voulez vraiment tenter le diable ?

— À votre place, je ne les écouterais pas, dit le gamin avant de pouffer. L'un d'eux m'a montré sa bite...

Je tentai de la raisonner :

— Vous n'avez jamais vu ce jeune auparavant. Réfléchissez. Il n'est pas scolarisé ici. Vous le reconnaissez, oui ou non ?

— Je ne peux pas mémoriser chaque...

— Eh bien, vérifiez son nom ! Voyez s'il existe un Jason Scott dans vos fichiers. Mais faites vite.

Elle se tourna vers l'ordinateur perché sur le bord du bureau, sans conviction. Je rassemblai toute ma patience pour ne pas lui sauter à la gorge.

— Évacuez les lieux, ou on s'en chargera nous-mêmes.

— Madame le proviseur ? intervint la vieille d'à côté. Il y a un problème avec le téléphone. (On l'entendait s'acharner sur son poste.) Pas moyen d'obtenir une ligne.

Le proviseur me fixa. Soudain je n'étais plus la pire curiosité de sa vie.

— Je vous jure, lui dis-je droit dans les yeux. Faites-le, et au plus vite.

— Très bien, Jane, ordonna-t-elle. Déclenchez l'alarme anti-incendie.

Le skateur esquissa un sourire.

— Ce que vous pouvez être lourdingues, les mecs.

Lâchant sa planche, il sortit de sa poche de sweat un flingue muni d'un silencieux. Il tira dans la tête du proviseur et s'enfuit par la porte.

Le temps que John et moi sortions nos armes, une deuxième balle claqua. Nous déboulâmes dans le vestibule, pour trouver la matrone avachie contre le mur, les jambes remuantes mais le visage dans la moquette. Il lui manquait l'arrière du crâne.

— Trouve l'alarme, dit John avant de détaler.

Je retournai le bureau sans débusquer la moindre commande, avant de songer qu'il devait y avoir des boîtiers d'alerte dans les couloirs. Je remontai le hall au pas de charge, en frappant à chaque porte. Lorsque j'atteignis l'escalier, des têtes médusées émergeaient de la plupart des salles.

— Il y a un incendie, expliquai-je de la façon la plus sereine et convaincante possible. Faites sortir tout le monde.

— Où ça ? demanda quelqu'un. Quel incendie ?

Je pointai mon flingue sur lui.

— Ferme ta gueule et obéis.

Je me précipitai au premier, où je vis John disparaître à l'intersection suivante. M'élançant à sa suite, j'avisai vite un boîtier rouge. Je brisai la vitre avec le coude et cognai le gros bouton.

Sans résultat.

Je recommençai, avec la paume. Puis le pressai plus doucement, mais toujours à fond.

— Seigneur, soupirai-je en reprenant ma course.

Un nouveau boîtier. Je brisai la vitre, pressai le bouton, encore et encore.

Rien. Reparut Zandt, pistolet au poing.

— Ils ont désactivé le système d'alarme, John. Oublie le gosse. On va devoir avertir les gens nous-mêmes.

S'éloignant au galop, il poussa la première porte venue et aboya ses ordres. Je partis l'imiter dans l'autre allée.

J'ouvris une salle remplie de regards hypnotiques.

— Évacuation, annonçai-je avec calme. On pense qu'il y a le feu. Quittez le bâtiment sans attendre.

L'école avait soigné ses entraînements : la classe se leva de manière prompte et disciplinée.

— Prévenez tous ceux que vous croiserez, pensai-je à ajouter. Ne courez pas. Sortez et restez vigilants.

Ça va bien se passer, me dis-je en les voyant partir en file. *D'ici dix minutes, les lieux seront déserts et je m'occuperai du plus important.*

Je constatai bientôt que les autres pièces se vidaient. Puis, par la vitre, je vis John traverser la pelouse en direction du premier bâtiment. Peut-être voulait-il simplement poursuivre l'évacuation, mais, à en juger par son regard, il valait mieux que je le suive.

Fendant le brouhaha des gamins excités, je dévalai les marches aussi vite que possible. Au rez-de-chaussée, la foule grouillait et se déplaçait bien plus vite, dans une certaine confusion.

J'entendis crier et compris qu'on avait découvert les corps du proviseur et de sa secrétaire. Un quinquagénaire en veste de tweed descendait l'escalier tout en jappant sur un collègue. Le genre de type qui, même face à un risque d'embrasement, exigeait l'aval de la hiérarchie.

Je remontai le flot, l'empoignai par les revers et plaçai mon visage à quelques centimètres du sien.

— Ferme-la, sifflai-je. Boucle-la, maintenant.

Il ouvrit de grands yeux.

— Quoi ? Qu'est-il arrivé à...

D'une voix basse et rapide :

— Mme Singer est morte. Mais personne n'a besoin de le savoir pour l'instant. Ne va pas semer la panique. Contente-toi d'aider à l'évacuation.

L'homme m'examinait toujours.

— Vous ne travaillez pas ici. Qui êtes-vous ?

— Un type quelconque, répondis-je en le poussant vers la foule. Et maintenant, bouge !

CHAPITRE 37

Lee se sentait désœuvré. Il avait rempli sa mission : démarcher quelques jeunes, jouer les mecs sympas et leur filer quelques friandises à l'essai. Il se demandait à quoi cela rimait, puisqu'il doutait d'être encore là pour prendre les commandes, mais bon, il avait fait ce qu'on lui disait – pour mieux refuser la prochaine fois. Il se dirigeait vers la sortie, prêt à retrouver Paul devant la grille et à réclamer un rôle plus gratifiant, lorsqu'il entendit son prénom. Levant les yeux, il avisa deux types ultra-louches, tout en noir, qui traversaient la pelouse d'un air pressé.

Ils poursuivaient d'abord un autre môme – impossible de dire, à cette distance, si c'était l'un de ceux qu'il venait de fournir –, et Lee se serait mis des baffes pour avoir réagi comme un amateur. Même Pete le Dormeur aurait mieux assuré. Mais le mal était fait, et il fallait rattraper le coup. Lee regagna les locaux avec le plus grand naturel, puis s'éclipsa dans le dédale de couloirs. Il entendit l'un des types le prendre en chasse, mais ses quelques notions des lieux lui permirent de consolider son avance.

Accroupi dans un recoin du rez-de-chaussée, le temps de choisir la direction suivante, il crut avoir une hallucination en voyant voir Paul s'engouffrer dans le hall. Il eut beau cligner des yeux, le type ressemblait toujours à Paul, bien qu'il s'agît à coup sûr d'un des deux types en noir. Le sosie se statufia quelques secondes, comme un animal aux aguets, puis sprinta jusqu'à l'étage rejoindre son complice.

Les deux gus repartis, Lee se glissa là-haut, résolu à rester planqué jusqu'à ce qu'ils aient mis les voiles. Visiblement, ni l'un ni l'autre ne gagnaient à être connus.

Lee allait ressortir lorsqu'il prit conscience que le monde ne sonnait plus pareil. L'instant d'avant, tout était silencieux – comme son ancien bahut ne l'avait jamais été –, mais voilà que bourdonnaient des voix. Au loin. Dehors ?

S'approchant d'une fenêtre, il vit un groupe de mômes émerger de l'arrière d'un bâtiment pour se rassembler dans la cour. D'après sa montre, ce n'était pas la fin des cours. Et pourquoi sortir par-derrière ? Ce devait être une sorte d'exercice, même s'il n'avait perçu aucun signal sonore.

Ça devenait carrément zarb. Il valait mieux changer d'air.

Il approchait de la cage d'escalier lorsqu'il vit monter quelqu'un. Trop tard pour rejoindre sa cachette. Cela dit, les deux hommes ne le reconnaîtraient pas forcément. Il expliquerait qu'il revenait des toilettes, qu'il avait aperçu ses camarades dans la cour et qu'il s'en allait à son tour, comme l'exigeait la procédure.

Il faillit bousculer l'arrivant. C'était le skateur de tout à l'heure, mais sans sa planche. Il semblait très à cran.

— Qu'est-ce qui se passe ? demanda Lee.

— On a un blème, répondit le keum en faisant son possible pour rester cool. Le plan traverse une zone de turbulences. Faut que je consulte Paul.

Il sortit de sa poche un mini talkie-walkie, tourna le dos à son complice, et se mit à parler vite et bas. Il s'éloigna dans le couloir pour écouter la réponse, puis regarda Lee.

Et c'est là que Lee pigea.

Qu'il pigea pourquoi ce visage lui semblait familier. Oui, c'était bien l'un des types de la cantine de Belle Isle, et sans doute l'un de ceux qui étaient sortis de l'ombre lors de la première rencontre avec Paul. Mais il l'avait aussi vu ailleurs.

Sur un parking isolé, la nuit, dans la lueur des phares d'une voiture. L'un des trois gus qui ne s'étaient pas avancés comme prévu, mais étaient restés en retrait avant de se déchaîner.

Lee continua de le dévisager. Tâcha d'imaginer cette face sous un autre jour. En vain. L'image s'était fixée.

Le jeune homme rangea sa radio et sourit. Il était parfait dans le rôle du skateur glandeur, à sautiller d'un pied sur l'autre. Il déclara :

— Il semble que ce soit ça, le plan. Alors on va dire que ça roule.

— Et ce plan a débuté quand, au juste ?

— Comment ça, frangin ?

— Tu vois, jusque-là je pensais qu'il avait commencé quand un de mes potes s'est fait buter et qu'on a déconné en cachant son corps comme on nous le demandait. Hernandez nous avait poussés à la faute, et tout est parti de là.

— Ouais, c'est ça. Écoute, on n'a pas le temps de...

Lee s'avança vers lui.

— Mais ça a commencé bien plus tôt, pas vrai ? D'accord, mes parents y sont pour quelque chose. Ça, je l'ai compris. J'ai du mal à piger, mais c'est enregistré. Non, ce dont je te parle, c'est de cette fameuse nuit, quand je me suis rendu avec des amis sur un parking de Santa Ynez et qu'un mec que je connais depuis l'enfance a eu la tête explosée. C'était pas je ne sais quel gang à la con du nord de l'État. C'était toi.

Le skateur se tenait immobile, le dos droit. Il n'avait plus rien d'un môme.

— Où tu veux en venir ?

— Je veux en venir que c'est toi ou un de tes connards qui a buté mon pote !

— Bien joué, petit futé. Ouais. C'était moi. Ma balle.

— Et depuis ce soir-là, vous m'avez fait danser. À me pousser dans une direction, puis dans une autre. À bousiller ma vie.

— Quelle vie ? ricana l'autre, et Lee comprit que ce type l'avait toujours méprisé. T'as été acheté dès le premier jour. Depuis le jour de ta...

Lee se jeta sur lui.

Lee Hudek savait se battre. Il possédait les qualités requises : de la force, de l'assurance, l'envie de faire mal. Mais son ennemi avait bien davantage – il y prenait du plaisir. Il réagit instantanément, comme au déclenchement d'un interrupteur.

Ils s'affrontèrent en silence, concentrés, émettant tout au plus quelques grognements. Les deux premiers coups de Lee furent cadrés, mais ensuite l'autre enchaîna des mouvements que seule une pratique intensive pouvait rendre si naturels. Lee s'y connaissait assez pour ne pas tenter le style hollywoodien. La baston, ce n'est pas se donner de grands airs. La baston, c'est niquer l'autre. On lui accroche les cheveux, les fringues, les yeux, et on essaie de l'envoyer au tapis le plus vite possible, puis on le salit avec les pieds, les poings, ou le premier objet contondant qui vous tombe sous la main. Mais Lee n'avait aucune prise sur ce type. Il vous glissait entre les doigts, et à ce rythme c'était lui qui allait mater Lee, à lui marteler la tronche et le bide avec les poings, les coudes et le tranchant des mains.

L'une d'elles l'atteignit en pleine gorge, lui coupant le souffle. Lee tomba sur un genou, et eut à peine le temps d'aspirer une bouffée rauque qu'un pied lui fauchait le visage.

Il tomba sur le dos, à demi groggy. Vit le jeune se planter au-dessus de lui, un flingue à la main.

— Je suis pas censé t'abattre, dit-il sans même haleter. Mais j'exclus rien. Je pourrai toujours extraire la balle.

— Je t'emmerde.

— C'est ça, répondit-il avant de lui piétiner le visage.

Puis il remit ça, deux fois de suite.

Lee inconscient, le type rangea son flingue dans son sweat, se pencha et glissa les mains sous les aisselles de Hudek. Il le traîna rapidement jusqu'au fond du couloir. Il connaissait la disposition des lieux sur le bout des doigts, pour avoir potassé le plan pendant des heures la semaine précédente. Il savait donc où se rendre : au cagibi qui faisait l'angle. Ce n'était pas prévu au programme, mais il n'avait pas le choix. On ne lui en tiendrait pas rigueur. On tolérait certaines pertes.

Il ouvrit le placard et y jeta le corps de Hudek. Referma la porte et baissa le loquet extérieur. Personne ne viendrait le chercher ici.

Il termina juste à temps, une dizaine de secondes avant que l'un des emmerdeurs du bureau de la dirlo ne déboule en criant au feu. Mauvais signe. L'Homme Debout n'avait pas prévu ce genre d'aléas, et ces deux gars refusaient de décamper. Ils semblaient décidés à en découdre, notamment celui qui ressemblait à Paul, et que la consigne interdisait de tuer.

Il s'isola pour recharger son flingue. Les instructions étaient toujours sujettes à interprétation. Quelle que fût la suite, il n'était pas question d'accepter une mort minable.

Ce n'était pas pour les gens comme lui.

CHAPITRE 38

Quelques mètres me séparaient de John lorsqu'il surgit dans le premier bâtiment. Il prit à droite et moi à gauche, pour ouvrir les classes et déloger les gens. On nous obéissait. C'était chouette. John boucla sa partie très vite, et je persistais à me demander s'il n'était pas venu courser quelqu'un. Je me contentai de frapper à la dernière porte, sans l'ouvrir, persuadé que le message passerait grâce au bruit des gosses en partance, puis je me faufilai dans la foule pour rejoindre Zandt à l'étage. Comme j'arrivais dans l'escalier, un flot d'élèves se déversa sur les marches, et je dus baisser la tête pour charger à contre-courant.

— John ?

— Il est ici ! cria-t-il en filant comme une fusée.

Il parlait sans doute du type croisé dans le bureau du proviseur. Je courus derrière John, pour achopper sur une lourde porte.

Je découvris de l'autre côté un réseau de couloirs et de pièces symétrique au précédent. De chaque côté de l'allée se succédaient trois longues vitres opaques. De nouveaux labos.

John était plaqué contre le mur, le pistolet devant la poitrine. Il agita sa main libre devant sa cuisse, m'assignant le mur opposé.

— Il est dans le labo de gauche, chuchota-t-il.

— Eh bien, qu'il y reste !

John me regarda froidement.

— Je m'en fous, de ce mec, expliquai-je. L'important, c'est d'évacuer le lycée. La police ou le FBI peuvent prendre la relève. J'ai fait ma part, et maintenant je vais...

— Nulle part, Ward. Tu ne sais pas où aller. Mais ce type peut nous conduire à Paul.

J'hésitai, avant d'entendre un son métallique à l'intérieur du labo. Nous retînmes notre souffle, les canons braqués sur la dernière porte. Il me parut plus judicieux de prendre deux cibles différentes, et je visai, sans tirer, la deuxième des trois vitres opaques.

— D'accord, soufflai-je. On le choppe. Mais fais gaffe à toi.

Zandt faillit sourire.

— J'ai toujours apprécié tes conseils.

— Très bien, je retire. Fais-toi buter si ça te chante.

Si je pulvérisais le carreau, nous pourrions voir le type, ou du moins aurait-il plus de mal à se cacher. Mais cela nous rendrait aussi plus vulnérables. L'autre solution était de remonter le couloir, baissés à ras des vitres, pour l'arroser par la porte.

Nous avons choisi les deux. John s'élança pendant que j'abattais l'un après l'autre les carreaux du labo, en commençant par le plus éloigné, espérant ainsi repousser le mec vers moi. Comme le deuxième carreau s'effondrait, je vis l'ennemi filer dans la bonne direction. Sitôt brisée la dernière fenêtre, j'accourus auprès de John.

J'étais à mi-chemin quand John lança la fusillade, planté dans l'encadrement de la porte. Puis il fut projeté contre le mur, et je pris son relais en vidant mon flingue dans la pièce. Mes deux premières balles partirent au hasard, après quoi je vis le jeune chercher refuge derrière un banc. J'ajustai ma position et tirai sans relâche.

Quand je me suis arrêté, l'autre ne répondait plus.

John leva le bras ; il avait une simple éraflure au poignet. J'ai rechargé mon arme, puis nous nous sommes avancés dans la salle.

Nous avons lentement remonté les travées extérieures, pour atteindre en même temps le dernier banc. Le flingue insonorisé gisait sur le carrelage ; je l'écartai d'un coup de pied.

Le skateur était assis par terre, adossé au mur du fond, les bras raides le long des flancs. Dans une flaque de sang qui continuait de s'étendre.

— Où est-il ? demandai-je.

Le type secoua la tête, intransigeant.

— Vous arrivez trop tard. C'est déjà commencé. Le Jour des Anges.

— Pas pour toi, répondit John. Ta fin est proche. Tu vas nous dire où est Paul. Peu importe ce qu'on devra faire pour te soutirer l'info. Tu vas parler, même si je dois t'exploser les membres un à un.

Il secoua la tête de plus belle. John lui pointa le flingue sur la jambe.

— Je plaisante pas, dit-il.

Le jeune ferma les yeux, comme pour rassembler ses forces. Je le tenais en joue.

Il leva péniblement le bras.

— Ne fais pas ça, conseillai-je. Laisse tes mains où elles sont.

Mais il accéléra le mouvement, plongea la main droite dans son sweat-shirt et produisit un couteau.

— Tu n'arriveras à rien avec ce truc, dit John.

— Profitez de votre monde pendant qu'il est temps, déclara-t-il avant de se planter la lame dans le cou.

John voulut arracher l'arme, mais le type y avait mis ses dernières forces. Le poignard enfoncé à fond, il le décala d'un coup sec.

Ce fut un beau désastre, et il mourut très vite.

John fouilla le cadavre. Il trouva un portefeuille contenant un peu d'argent mais pas de papiers, un demi-paquet de cigarettes – il s'en alluma une –, ainsi qu'un talkie-walkie compact. L'objet avait dû recevoir une de mes balles car il était fendu, tordu, et je n'en tirai aucun son. Il avait sans doute servi à contacter le faux flic, qui croupissait actuellement dans le coffre de sa voiture.

Rien qui pût nous mener à Nina.

En rage, je m'éloignai de la boucherie pour regarder dehors. La fenêtre donnait sur l'arrière de la cité scolaire, qui, Dieu merci, continuait à se remplir d'élèves.

On mettait les lycéens en rang, derrière une ligne tracée au sol, dans l'ombre de l'établissement. Ils étaient quelques centaines. La partie avant de l'aire était dégagée.

Et une voiture venait d'entrer par le portail.

Je commençai par louer le ciel : les flics arrivaient enfin et nous pouvions ficher le camp ! Mais cette pensée stupide ne dura qu'une nanoseconde. Les flics du coin ne roulaient pas en berline noire aux vitres teintées. Contrairement à d'autres.

— Eh, John, viens voir. Dépêche.

La voiture traversa l'aire au pas. Personne n'y prêtait attention ; on était occupé à peaufiner les rangs et à savourer la poilade que constituaient toujours ces exercices.

— Nom de Dieu, fit John.

La voiture s'engagea sur la rampe par laquelle on avait acheminé les pizzas. Et disparut en quelques secondes, comme si elle n'avait jamais existé.

— Le sous-sol, dis-je. Ou la zone de chargement. Pourquoi emmène-t-il Nina dans...

— Ils n'avaient pas encore posé leur bombe, lâcha piteusement John. Elle vient d'arriver sous nos yeux.

Une voiture bourrée d'explosifs ferait des ravages considérables. Au mieux, elle projetterait sur les mômes les débris incandescents du lycée. Au pire, selon la disposition et la superficie du sous-sol, elle exploserait sous leurs pieds.

Nous n'avions rien réussi du tout. Paul conservait au moins une longueur d'avance.

J'agrippai la poignée de la fenêtre, mais la peinture l'avait scellée de longue date. Je martelai la vitre avec mes poings, mais personne ne regardait.

— Laisse, dit John. On sait où aller, maintenant.

Nous nous sommes rués dans le couloir pour regagner l'autre aile. Pendant que nous courions, je crus entendre des cris, mais je ne savais d'où ils venaient, et n'avais plus le temps de m'en soucier.

Nous avons dévalé les marches jusqu'au rez-de-chaussée, puis cherché un accès vers le sous-sol. Il n'y en avait pas. Alors nous avons coupé par la pelouse jusqu'au bâtiment principal, et enfilé le couloir de gauche qui menait vers un grand espace, visiblement la cafétéria.

Ce qui signifiait cuisines. Nous avons cavalé jusqu'aux comptoirs. Derrière se trouvaient des fours cabossés et d'énormes réfrigérateurs. Tout au fond, la porte des escaliers. Nous avons sorti nos flingues, et pris le chemin du sous-sol.

CHAPITRE 39

Lorsque Paul dit au chauffeur d'entrer par l'arrière du lycée, Nina devina que toute parole serait vaine. Le choix de ce sous-sol ne pouvait signifier qu'une chose. Elle savait que rien ne lui ferait entendre raison, non parce que Paul était trop fou pour saisir ses arguments, loin de là, mais parce qu'il ne serait pas d'accord. Comme l'avait dit Wittgenstein, si le lion pouvait parler, nous ne le comprendrions pas. C'était la rationalité personnelle de Paul qui l'avait conduit dans ce lieu, une suite d'étapes simples et logiques à l'intérieur d'un esprit aux connexions différentes, centré sur d'autres valeurs. L'événement allait bien se produire. Il émanait de la détermination de Paul une énergie quasi sexuelle, et Nina prit conscience qu'elle entamait sans doute les dernières minutes de sa vie. C'était bien d'avoir pu parler à Ward, même brièvement. Dommage qu'il n'ait pas capté son allusion au lycée, mais elle n'avait eu d'autre choix que de faire dans l'évasif : elle avait bel et bien improvisé avec un pistolet sur la tempe.

Elle ne pouvait plus changer grand-chose maintenant, ni dans un sens ni dans l'autre. Elle ne pesait plus sur rien. Il fallait se contenter d'être, pour le peu de temps qu'il restait.

Parfois, renoncer aux causes perdues est la plus courageuse des décisions.

La voiture s'enfonça dans un espace large comme quatre véhicules, bordé d'escaliers de service et de box de six mètres de profondeur. Les murs étaient partiellement recouverts de petits carreaux brillants, sous une lumière glauque et tremblante. Le conducteur se gara dans l'un des derniers box, près d'un mur qui s'ouvrait sur une autre portion de sous-sol. Le moteur se tut.

Paul se tourna vers Nina.

— L'avantage, dit-il, c'est que ce sera très rapide. Boum, pulvérisée. L'inconvénient, c'est que tu vas devoir attendre. Je pourrais te tuer maintenant, bien sûr. C'est vrai que je n'ai pas particulièrement envie de te faciliter la tâche, et en même temps je me dis que tu aimerais profiter de ces instants, aussi imparfaits soient-ils. J'ai raison ?

Nina le regarda sans ciller.

— Merci, Paul. Oui, j'aimerais bien.

— Marché conclu. Je ne peux pas t'indiquer de délai précis, car je déclencherai l'explosion moi-même. Mais ça devrait te maintenir en alerte. Fais de chaque songe une pensée agréable, parce que bon, ce pourrait être la dernière.

Le chauffeur sortit. Le claquement de sa portière avait quelque chose de définitif. On entrait dans la phase des dernières fois.

Nina continuait de fixer Paul.

— Vous vous rendez compte que, malgré tous vos discours, le fait de me laisser en vie est fondamentalement un geste de compassion ?

— Et après ?

— Vous en êtes donc capable. Vous savez compatir. Faites-le de manière régulière, et vous pourrez devenir un vrai garçon. Comme Pinocchio.

— Je suis plus vrai que tu ne le comprendras jamais.

— Non, c'est faux. Vous êtes comme moi, et comme n'importe qui. Il n'y a aucune différence génétique entre vous et Ward. Ce qui vous pousse à agir ainsi, ce n'est pas votre supposée appartenance à une race supérieure, mais ce que vous avez vécu. Vous pourriez être comme tout le monde.

— Tiens donc, répondit-il avec intérêt. Et qui seraient ces gens normaux ? Tu aurais dû passer plus de temps devant le journal télévisé, au lieu de traquer mes semblables. Votre espèce pille et tue depuis qu'elle est bipède. Elle ment depuis qu'elle sait parler. Nous ne sommes pas les seuls à faire la guerre, à violer et à tuer. La seule différence, c'est que vous prétendez que c'est mal.

— Mais, Paul, sincèrement, vous ne formez pas une espèce différente. Vous le savez, non ? En votre for intérieur, vous devez bien comprendre que vous êtes humain.

— Non. Nous sommes les fils de Dieu. Nous sommes liés à vous depuis beaucoup trop longtemps, et aujourd'hui nous coupons le cordon. Tu ne seras plus là pour le voir, mais crois-moi, ça va être terrible.

Là-dessus, il sortit. Il échangea quelques mots avec le chauffeur, puis passa derrière la voiture pour traficoter dans le coffre. Nina entendit un bip – l'amorçage d'un dispositif. Le conducteur s'éloigna sans attendre.

Paul reparut, glissa la tête à l'intérieur pour regarder Nina.

— Tu penses peut-être parvenir à ouvrir une de ces portières, bien que tu sois ligotée et que je m'apprête à les verrouiller. Ça me paraît illusoire, mais l'espoir fait vivre. Aussi dois-je t'expliquer que chaque portière est branchée au bazar du coffre. Ouvres-en une, et tout pète. Alors sois gentille, laisse-moi le temps de quitter le bâtiment avant d'essayer.

Nina le dévisagea, dans l'espoir de trouver, enfin, quelle âme habitait ce regard.

— Paul ? dit-elle. Te rappelles-tu seulement qui tu es ?

Il claqua la portière. Suivit le clic du verrouillage centralisé.

Puis un bruit de pas, déclinant.

CHAPITRE 40

Le sous-sol était un entrepôt labyrinthique conçu à une époque où l'on ne pouvait se faire livrer sept jours sur sept. L'escalier des cuisines menait à un bloc rempli d'étagères métalliques. Il y avait une pile d'emballages de pizzas, ainsi qu'une pièce pleine de vieilles chaises en bois. Aucun employé en vue.

Nous traversâmes la zone comme des flèches, cherchant le chemin de l'aire de chargement. Nous finîmes par trouver un couloir qui s'enfonçait vers la lumière. Deux mètres avant la fin, John me fit signe de me plaquer au mur.

Il passa le seuil et se rabattit sur la droite. Aucun tir. Je le suivis.

L'espace semblait aussi étendu que la cité scolaire. Des néons grésillaient au plafond bas ; les petits carreaux et les plaques de moisissures donnaient à l'ensemble une couleur verdâtre, qui s'estompait sur la gauche, là où la rampe remontait vers la cour. On entendait pépier au loin quelque deux cents gamins. Mais pas la moindre sirène. Que fabriquaient les flics ?

— Tu devrais les prévenir, dit Zandt.

— Quand j'aurai trouvé Nina. Mais tu peux t'en charger, toi.

— Si Paul est ici, la bombe ne va pas sauter dans la seconde.

— On ne sait pas ce qu'il est prêt à risquer. Et s'il se planque ailleurs, on risque d'exploser d'un instant à l'autre, et on aura la réponse trop tard.

Il haussa les épaules avant de se tailler vers la droite. Et lui, qu'était-il prêt à risquer ? J'aurais préféré la compagnie de Bobby, qui fut un homme plus intègre que John et plus gentil que moi. Bobby prenait toujours les bonnes décisions, faisait toujours passer les autres avant lui, quoi qu'il lui en coûte, et maîtrisait les situations violentes de manière prodigieuse. Moi, je n'étais qu'un amateur qui n'avait plus le droit à l'erreur.

Il fallait alerter les gens.

Et j'allais devoir m'y coller.

Maugréant, je m'élançai vers la rampe. J'avais accompli le quart du trajet lorsque je distinguai dans le fond une silhouette râblée.

Elle me tira dessus. Trois coups détachés, avec silencieux.

Pris par la vitesse, je ne pus que plonger dans le renfoncement le plus proche. Je me heurtai contre le mur et glissai à terre. Je tâchai de rouler, de façon à pouvoir riposter, mais le tireur se contentait de canarder les abords du box.

— Ward ?

La voix de John se répercutait sur toutes les parois, m'empêchant d'évaluer sa distance.

— Ça va ! répondis-je. Tu peux l'avoir ?

La réponse prit la forme d'une volée de plomb qui dura bien dix secondes. Les coups sonores de John le disputaient aux clics sourds du type, dans le sifflement des ricochets.

La dernière balle de John résonnait encore qu'il cria :

— Maintenant, Ward !

Sans même réfléchir, je jaillis du renfoncement pour me replier sur la gauche, le flingue tendu à l'opposé, shootant sans relâche. John me couvrait de son feu. Parvenu au centre, je le repérai enfin, accroupi à l'entrée d'un box, à une trentaine de mètres. Mes dernières foulées furent rythmées par une nouvelle salve ennemie.

Puis je me trouvai dans la niche, scié d'être encore de ce monde. J'étais entouré de vieux bureaux. Mes oreilles bourdonnaient.

— Seigneur...

— C'est qui, ce con ? demanda John.

— Aucune idée, ahanai-je. Mais on n'atteindra jamais cette issue tant qu'il ne sera pas mort.

— Ça va pas être de la tarte, garantit John. Il t'a manqué de peu.

— Ravi de l'apprendre.

Il avança son flingue pour rouvrir le feu. On lui répondit dans la seconde.

Au fond de notre box se dressait une porte. J'allai l'ouvrir. Un corridor partait vers la droite.

— Ça ne nous rapprochera pas de lui, mais ça peut nous conduire dans la direction opposée. Ce qui me convient parfaitement.

Je m'y engouffrai le premier. John suivit, quittant le box à reculons, au cas où le gars aurait décidé de nous courser.

Je refermai dernière nous. Nous remontâmes le couloir tout en rechargeant fébrilement. Tous les dix mètres se succédaient de nouvelles portes, qui donnaient sur des box identiques à celui que nous venions de quitter, à moitié remplis de matériel inutilisé. Je les essayai toutes, sans leur trouver le moindre intérêt. Si ce n'est que le couloir finissait en cul-de-sac.

— Merde, pesta John.

— J'imagine qu'on n'a pas le choix, tout compte fait.

— Il faudrait déjà savoir où le type est passé. On est coincés. S'il déboule par la première porte, on est faits comme des rats.

J'ouvris la dernière porte. Une fois dans le box, j'aperçus celui d'en face. Il ne contenait ni cartons ni chaises, sans être vide pour autant.

Il abritait une grosse voiture noire.

— On a trouvé, déclarai-je.

Les phares étaient coupés. Entre les vitres fumées et les néons rasants, impossible de voir à l'intérieur. Je posai un pied en dehors du box, mais dus aussitôt le retirer afin d'éviter une balle. Le type au flingue avait gardé son poste. Il était sans doute là pour protéger le véhicule, d'où le fait qu'il ne nous eût pas suivis dans le corridor. Impossible d'atteindre la voiture sans qu'il nous voie. Cela dit, il se trouvait assez loin ; en courant vite et en croisant les doigts, nous pouvions y arriver. Peut-être.

John était déjà prêt à bondir.

Puis je compris que ce n'était pas si simple. Sur la droite du box convoité se trouvait un passage en voûte, débouchant sur une nouvelle section de sous-sol. Il faisait trop sombre pour distinguer quoi que ce soit, et...

— Attends un peu, murmurai-je. Paul est forcément là, lui aussi.

— Qu'est-ce que t'en sais ?

— S'il était ressorti, l'autre gus l'aurait suivi. Donc, si on admet que Paul était dans la voiture lorsqu'elle s'est garée ici, ou bien il y est toujours, ou bien il se trouve là-bas, quelque part sur la droite.

John comprit le raisonnement. Si Paul était resté au sous-sol, la voiture pouvait couper deux lignes de feu distinctes. Soit le complice nous épinglait dans notre course, soit nous nous jetions dans la gueule de Paul, planqué après la voûte.

— Ça passe ou ça casse.

Que faire ? Nous nous trouvions à six mètres d'une bombe, à laquelle Nina était peut-être ligotée, et nous ne pouvions quitter le bâtiment pour prévenir qui que ce soit. Il n'y avait qu'un trajet possible : tout droit. La seule inconnue était de savoir où nous mèneraient nos pas, et combien il nous en restait. Alors j'émis une pensée, une requête, dans l'espoir qu'un être plus malin nous trouverait une solution.

Là, ce serait super, Bobby. Ce serait le moment idéal.

Les secondes défilaient. John avança une jambe. Une nouvelle balle nous frôla.

— Il va falloir risquer le tout pour le tout, Ward...

Et pour finir mon vieux pote me répondit.

— Son téléphone ! murmurai-je. Paul t'a contacté. Tu as donc son numéro en mémoire. Tu peux le rappeler.

— Il aura coupé la sonnerie.

— Peut-être que non.

— Pour capter quelque chose ici...

— On n'a plus que ça, John.

Il sortit son portable et consulta le journal d'appels. Le dernier numéro en date était inconnu du répertoire, et l'heure indiquée correspondait aux faits.

— Je l'ai, annonça John.

— Alors attends. (Je pris une grande inspiration, resserrai mes doigts sur le flingue.) C'est du tac au tac. T'appelles. Si ça sonne, et que le bruit semble indiquer que Paul n'est pas posté dans l'encoignure, on décolle aussitôt. Je cours jusqu'à la caisse. Tu fonces là-bas et tu trouves Paul.

Il jaugea le plan, convint que c'était le bon.

— OK.

— Reste en vie, chuchotai-je.

Ses yeux m'envoyèrent le reflet d'êtres disparus de longue date. Son sourire était triste, mais réel.

— De quelle vie tu parles ?

— C'est mieux que rien, non ?

— Sors Nina de ce traquenard, si tu y arrives. Et t'avise pas de me rejoindre.

— Je savais que tu dirais ça.

Il souffla un grand coup.

— Donne-moi le signal.

J'attendis un instant, puis :

— Top.

John pressa la touche de rappel. Une série de bips imperceptibles, le temps que l'appareil compose le numéro choisi. Deux ou trois secondes de silence. Puis on entendit sonner. Dans l'espace de droite, mais assez loin. Au moins dix mètres après l'entrée, à vue de nez.

— On fonce.

Je traversai l'allée à toute blinde. John fit de même, en se déportant vers la droite pour se jeter sous l'arche. Trois ou quatre balles fendirent l'air autour de nous. Aucune ne nous happa.

Sitôt dans le renfoncement, je m'arrêtai dans une glissade, et virevoltai au cas où le sniper aurait eu l'ordre de venir en découdre sur place.

— John ? lançai-je.

Mais il était parti. Parti dans l'autre section, sur une route qu'il s'était juré de suivre jusqu'au bout.

Je me laissai trente secondes, puis osai quitter le recoin pour la voiture. Je fondis sur la portière arrière droite, et j'avais la main sur la poignée lorsque j'entendis un bruit mat, étouffé.

J'eus un geste de recul. Le bruit se répéta, et je compris qu'on frappait au carreau depuis l'intérieur.

Je lâchai la poignée, collai mon nez à la vitre, les mains en œillères pour contrer la lumière.

Je reconnus les traits de Nina.

Je ne voyais pas ses mains, et elle semblait en porte-à-faux, le corps tendu vers moi. Elle avait cogné son front contre la fenêtre pour m'alerter.

Rapprochant son visage au mieux, elle articula une phrase d'un air catastrophé. Je compris à la troisième tentative.

Elle me disait que les portières étaient piégées.

CHAPITRE 41

De la poussière, d'abord, puis une odeur d'humidité. Le noir, et une méchante douleur au crâne.

Lee se redressa lentement, douloureusement, sans savoir où il se trouvait. Il voulut se relever, mais l'espace était réduit et ses jambes flageolaient. Il tomba à la renverse, déséquilibrant un tas d'objets bruyants sur les étagères de bois.

Il réessaya, en s'aidant des deux mains. Se hissa sur ses jambes en luttant contre le tournis. Des lumières blanches jaillissaient sous ses yeux. Une sensation étrange, claustrophobe. L'endroit paraissait minuscule.

Il tendit la main sur la droite. Risqua deux pas dans cette direction, avant d'atteindre de nouvelles étagères. Elles soutenaient des masses molles, comme du linge ou des serviettes. Son pied percuta un corps métallique. Puis un autre objet, qui ne pouvait être qu'une serpillière.

Il était dans un placard.

Voulant repartir dans l'autre sens, il se cogna de plein fouet dans la porte. Il avait une vague impression de déjà-vu, comme s'il s'était réveillé un peu plus tôt pour crier et tenter de sortir. Sa tête menaçait de se décrocher, et putain ce qu'il avait mal.

Il palpa la porte, cherchant un moyen de l'ouvrir. Localisa la poignée, mais elle était inopérante. Ce devait être verrouillé de l'extérieur.

Il tambourina un moment, mais les coups se propageaient dans ses bras et jusque dans sa tête, pour former une mer d'encre où les lames se brisaient sans bruit.

Aucune réponse de l'extérieur.

Alors il se détacha de la porte, résolu à marquer une pause, à respirer profondément pour reprendre ses esprits. Il ne parvint qu'à perdre l'équilibre et à heurter le sol, aux trois quarts sur le cul, un quart sur le flanc.

On était mieux par terre, en vérité. Lee n'irait nulle part tant qu'on ne l'aurait pas délivré. Alors autant rester assis. Quelqu'un finirait bien par partir à sa recherche. Paul, par exemple. N'apercevant pas Lee devant le lycée, il viendrait jeter un œil. Ou bien son père. Mais pas sa mère. Aucune chance. Elle avait toujours paru inaccessible, comme si elle avait toujours eu l'esprit ailleurs. Mais son père, pourquoi pas ? Son père l'avait toujours aimé. Ou alors Brad. Ouais, Brad. On pouvait compter sur lui. C'était l'ami de toute une vie. En dernier ressort, Brad trouverait ce placard, débloquerait la porte, sortirait Lee de ce traquenard et l'emmènerait ailleurs. Prendre un hamburger, ou voir la mer.

Il suffisait d'attendre sagement. Il ne faisait pas froid ; les murs vous enveloppaient. C'était douillet.

On faisait pire.

Je m'écartai de la portière. J'avais demandé à Nina de répéter une dernière fois, pour être sûr d'avoir compris. Paul avait piégé la voiture, et toute tentative d'ouverture déclencherait la bombe.

Et Nina était aux premières loges.

Je plaquai ma paume contre la vitre. Nina n'imita pas le geste, et j'en déduisis qu'elle était ligotée. Mais elle y appuya son front, le plus près possible de mes doigts, et sans cette foutue vitre je lui aurais touché les cheveux.

Je ne connaissais rien aux bombes. Connexion, déconnexion... N'importe quel môme là-haut était aussi calé que moi pour désarmer cette bagnole – sinon plus, pour peu qu'il ait suivi ses cours de physique. Ça pouvait péter à tout moment. À chaque seconde qui passait, le monde pouvait virer au néant.

Mais Nina était à l'intérieur.

Le seul moyen d'ouvrir le coffre était de désagréger la serrure. En m'éloignant suffisamment pour éviter le rebond de la balle, mais cela m'aurait ramené dans la mire du sniper. Sans compter que tirer dans une malle remplie d'explosifs était hautement risqué.

Je me reculai pour considérer le véhicule, dans l'espoir qu'il me soufflerait une idée fiable, et vite.

Quatre portes à éviter. Un coffre inaccessible, bourré d'engins que je ne savais pas neutraliser. Et il nous restait très peu de temps. Si Paul avait quitté la voiture, c'est qu'elle était prête à sauter.

En fait, les données du problème se résumaient à ceci :

Les déclencheurs des portières fonctionnaient-ils par contact électrique ou par détection de mouvement ? Ne reconnaissaientils que les ruptures de courant, ou réagissaient-ils aux chocs dans toute partie du véhicule ?

Il n'y avait aucun moyen de le savoir. Les coups de tête de Nina n'avaient rien provoqué. Mais elle n'y avait pas mis beaucoup de force. L'action qui m'était demandée serait autrement violente.

Je n'avais plus qu'à prier pour que ce ne soit pas électrique. L'hypothèse était juste, ou fausse. Notre avenir serait long, ou infiniment court. Je n'avais d'autre choix que de jouer à pile ou face.

Je revins à la vitre. Y plaquai mon visage, distinguai l'œil de Nina dans le noir de l'habitacle. Je lui dis que je l'aimais, puis lui fis signe de se coucher sur la banquette, le plus loin possible du pare-brise, et le plus près du sol, mais sans s'éloigner de la portière. Elle ne posa pas de question, mais soutint mon regard avant de s'enfoncer dans le noir.

Je reculai. Déglutis. Me plantai devant le capot. Levai mon arme à deux mains et visai la droite du pare-brise.

Pressai la détente.

John Zandt arpenta l'obscurité d'un pas prudent, dans un sous-sol semblable à tous les lieux qu'il fréquentait depuis trois ans. Cette aile-ci était également subdivisée en box, mais plus petits, destinés aux objets de rebut. C'était une caverne silencieuse, moite, froide et négligée, plus proche d'une forêt que d'une construction humaine.

Il s'arrêta au centre. Un peu plus loin filtrait une lueur laiteuse à travers un carreau crasseux. Il y avait sûrement une issue, un escalier de secours vers le monde extérieur. Auquel cas Paul avait déjà pu se sauver. Mais John en doutait. Il flairait sa présence. Il avait toujours su que ce moment arriverait, du jour où l'Homme Debout lui avait pris sa fille. Toujours su, à moins que d'autres n'éliminent Paul avant lui, que ça se terminerait ainsi. Qu'ils se retrouveraient face à face, et qu'il n'en resterait qu'un. Au mieux.

Voilà pourquoi il n'éprouvait ni peur ni frustration : il n'avait plus rien à perdre. Il arrive qu'une mort soit la seule réponse valable à une question jamais formulée. Il faut juste espérer que c'est l'autre qui tombera.

Il devait procéder par tâtonnements, et l'heure était venue. Il entendit Ward tirer dans l'autre section, produisant de curieux bruits d'impacts. Le temps filait.

Alors il s'avança jusqu'au premier box et ouvrit le feu.

Le pare-brise était coriace : il fallut plusieurs balles pour perforer ses strates blindées. Mais aucune ne déclencha la bombe.

Des coups fusaient dans le tréfonds du sous-sol, comme si John progressait méthodiquement, pied à pied.

Je grimpai sur le toit, ôtai mon blouson, enveloppai mes mains dans l'étoffe pour arracher les bris de verre. Lorsque le trou fut assez grand, je me glissai à l'intérieur, préférant m'érafler de tous les côtés que de retomber lourdement. Nina était là, dans le fond, et sitôt auprès d'elle je l'embrassai avec fougue, au cas où ce serait là notre dernier instant, puis je déliai ses membres avant de l'amener vers l'ouverture. Elle était faible et livide, les muscles gourds d'être restés si longtemps entravés. Chaque seconde menaçait de rendre caduques nos interrogations sur le dispositif des portières, dès lors que quelqu'un, quelque part, appuierait sur un bouton. Mais peu importait. Je hissai Nina jusqu'au toit, m'entaillant de plus belle. J'essayai de la protéger, mais les éclats l'écorchèrent elle aussi. Puis nous glissâmes à terre, et la voiture n'explosa toujours pas.

J'aidai Nina à se relever.

— Et John ?

— Je sais pas, répondis-je. La bombe a un minuteur ?

— Paul la déclenche à distance.

— Alors il faut partir, Nina.

— D'accord.

J'extirpai des restes de mon blouson un deuxième flingue, que je chargeai et lui remis.

— Tu peux courir ?

— Je peux essayer.

— Alors on y va.

Je lui entourai les épaules, de façon à m'intercaler entre elle et le tireur, le temps de traverser l'allée centrale. Il nous canarda, mais je ripostai, et nous atteignîmes l'autre côté sains et saufs.

Je guidai Nina vers le fond du box, ouvris la porte et l'emmenai au fond du corridor – en espérant que le tireur n'aurait pas le temps de nous y accueillir. J'envoyai ma semelle dans la porte du premier box exploré avec Zandt, puis j'appuyai Nina contre le mur, dans l'angle mort du tireur, le temps qu'elle reprenne son souffle.

— Bon, chuchotai-je. Maintenant il faut se jeter à l'eau.

Elle sourit.

— Comme toujours.

Et de nous élancer en pilonnant l'ennemi.

Zandt avait fait une bonne moitié du chemin lorsqu'il perçut un bruit dans le dernier box de gauche. Comme le frottement d'un bureau ou d'une chaise sur le béton, heurtés par une personne se repositionnant dans le noir. Paul n'ignorait pas ce que John avait en tête : il savait qu'il approchait, qu'il vérifiait chaque case en y vidant un demi-chargeur, qu'il ne ferait pas machine arrière. Et donc il se préparait à le recevoir.

John se demandait si Paul avait la trouille. Si oui, il risquait à tout moment de sprinter vers l'échelle de secours. Auquel cas John était sûr de l'avoir. Mais s'il n'avait pas peur et décidait de l'attendre, alors le dernier mot reviendrait aux esprits désinvoltes qui tenaient d'une main distraite le gouvernail du monde. Ces esprits qui avaient laissé mourir Karen, mais également permis à John de retrouver l'Homme Debout. Des esprits qui soit s'en foutaient, soit mettaient un point d'honneur à ne pas jouer les favoris.

Zandt rechargea son arme. Ferma les yeux une seconde, pour se remémorer certains visages, certains lieux, certaines époques. Il entendait Ward actionner son propre flingue, une longue salve qui semblait plus lointaine que les précédentes. John voulut y voir un signe encourageant.

Il marcha jusqu'au dernier box.

— Salut, John, fit une voix dans le noir. J'ai ta fille à côté de moi.

Puis seules les armes parlèrent.

Nina eut le tireur, je crois. L'un de nous le toucha, en tout cas : je le vis tournoyer et tomber avant même de m'apercevoir qu'aucune balle n'arrivait plus.

Les genoux de Nina tendaient à la trahir, aussi resserrai-je mon étreinte pour accélérer l'allure jusqu'à la rampe d'accès. Malgré la pente, nous nous sentions aspirés par la lumière, comme par les rapides d'un torrent.

Sitôt dans la cour, je me mis à hurler en agitant mon bras libre. Au début, personne ne comprit ce que je disais, puis le message passa et les gens rompirent les rangs pour se ruer vers les portails. On aurait dit un mur de verre explosant à l'horizontale. Lentement, puis très, très vite.

Dès que je les vis détaler, j'évacuai Nina par la sortie la plus proche. Ses muscles étaient déjà plus agiles, et elle scrutait la foule en se demandant ce qu'elle devait faire, comme s'il en allait de sa responsabilité. Je dus la pousser pour l'empêcher de ralentir. Nous franchîmes la grille en courant, pour nous arrêter deux rues plus loin.

Nous nous retournâmes vers le lycée. Les élèves et les profs continuaient de se déverser sur la chaussée, mais il n'y avait plus personne dans l'enceinte. Les bâtiments me faisaient l'effet de ballons de baudruche gonflés à leur extrême limite.

— Qui sait ? fit Nina. John a peut-être...

Puis tout a sauté.

On aurait dit que la planète entière tremblait, comme une balle frappée contre un mur.

Je crus percevoir deux déflagrations, espacées d'une fraction de seconde. La cour arrière de l'école fut pulvérisée vers le ciel, en même temps qu'étaient soufflées toutes les vitres. À peine le verre commençait-il à voler que le toit et les étages explosèrent et s'effondrèrent en même temps, la destruction empruntant toutes les directions à la fois.

Ceux qui couraient derrière nous s'arrêtèrent pour regarder, avant de fuir la pluie de brique, de bois, de verre, de terre et de feu. M'abritant contre un mur avec Nina, je m'aperçus qu'elle criait :

— Téléphone ! Passe-moi ton téléphone !

Je lui filai mon portable, cependant que la terre semblait prise d'une nouvelle secousse. La tour du lycée se renversa telle une masse, dans un mouvement lent et inexorable, alimentant le panache de fumée noire qui s'élevait des ruines, densifié par quelque sombre dessein, comme pour former un visage maléfique. Les gens continuaient de nous dépasser, livides quand ils ne saignaient pas encore, dans un tonnerre constellé de hurlements.

— Je ne capte rien, pesta Nina avant de s'exciter sur le clavier.

— T'appelles qui ?

Pour toute réponse, elle m'empoigna le bras et me ramena vers le lycée. Nous nous élançâmes contre les flots d'élèves paniqués. Tout autour avaient jailli des flammes de six mètres de haut, et d'autres surgissaient de manière spontanée, comme soulevées par une main invisible. À trente mètres de distance, elles nous brûlaient déjà.

— Où est ta voiture ?

J'indiquai la gauche, et nous longeâmes l'arrière de l'établissement, en zigzaguant entre les débris en feu. La poussière et les cendres chaudes retombaient en neige. Nous tournâmes à l'angle de la cité scolaire, pour remonter vers l'entrée principale.

Les arbres s'étaient embrasés. Une autre portion de bâtiment s'effondra. Je me demandais pourquoi le sol continuait de trembler, à moins que ce ne fût dans ma tête.

Nous dépassâmes la vieille église en bout de rue, pour gagner le sommet de la colline et la route du lycée avant de replonger dans Thornton. Nos pas faiblirent, et nous finîmes par nous arrêter.

Nous pivotâmes, lentement, pour regarder la ville, et là je compris pourquoi l'explosion semblait se prolonger. C'était bien le cas.

Toute la ville était en feu.

Des colonnes de fumée se dressaient de toutes parts, de tous côtés. Je courus jusqu'au virage suivant : le commissariat était désintégré. Le centre historique était la proie des flammes. À l'opposé, je vis gonfler un nuage noir dans le secteur de l'Holiday Inn.

Nina essayait toujours de joindre quelqu'un, quelque part, d'alerter des autorités qui n'existaient même plus. Je tournais sur moi-même, incapable de m'arrêter, quand je vis un immense brasier s'élever de la forêt de Raynor.

Deux secondes plus tard, nous fûmes projetés au milieu de la route : l'église venait d'exploser en une joyeuse tempête de pierres.

Sheffer

À présent, il fait froid. Nous avons repris possession du bungalow de Patricia, au cœur de la forêt. Cela fera bientôt trois semaines. L'hiver a eu le temps de descendre des montagnes, qui chaque nuit profite de notre sommeil pour se rapprocher un peu plus. Les dîners sous le porche ne sont plus de mise, mais il nous arrive encore de passer la soirée près du lac, pour regarder les nuages sombres se réfléchir sur l'eau glacée. Nous restons assis l'un contre l'autre, mais dans l'ensemble nous parlons peu.

Nina se trouve au bord de l'eau. Je l'observe depuis la cuisine, mais je ne prépare pas le repas.

Ce soir, nous dînons en ville.

Je suis de près les reportages sur les événements de Thornton. On ne peut y échapper, et c'est souvent l'unique sujet traité par les médias. Les estimations sur le nombre de morts varient dans de telles proportions qu'elles en deviennent absurdes. On a largement dépassé le millier, mais tout porte à croire que le bilan final sera bien plus élevé. On sait que des engins explosifs ont été placés au commissariat, à l'église, au lycée, chez Starbucks, chez Renée, au jardin d'enfants, à la caserne de pompiers, dans deux restaurants du centre historique, à la grande épicerie, à Radio Shack dans la galerie marchande de la route d'Owensville, à la bibliothèque municipale et au temple maçonnique. Entre autres. Le Mayflower fut également visé, où périrent Hazel, Lloyd et Gretchen, ainsi que tous les clients présents dans le bar. Diane Lawton fut de ceux-là, venue prendre un café après le travail pour se prouver qu'elle pouvait y mettre les pieds sans passer la soirée à boire. Elle aurait peut-être gagné son pari. La réceptionniste que j'avais effrayée à l'Holiday Inn est morte, en même temps que les sept agents du FBI qui s'y étaient établis, et qu'un jeune gars venu vérifier le minibar de Nina le lendemain de mon arrivée. Il avait paru tétanisé devant cette tâche somme toute rudimentaire, mais je l'avais trouvé affable et cordial. Deux qualités qu'il ne possède plus que dans ma mémoire. Le nombre et la nature des autres morts dépassent l'entendement. Je soupçonne Thornton d'avoir des antécédents, et certaines théories de Nina, qu'elle-même tient de Paul, me confortent dans cette idée. Une longue tradition de meurtres. Qui remonterait à la nuit des temps.

Nous avons fait ce que nous pouvions, cet après-midi-là, et jusque dans la soirée. Nous avons couru au centre-ville pour aider les gens à quitter les immeubles, les éloigner des feux qui se déclaraient de partout. Rien n'était épargné. Aucun système, aucune manière d'être. Ce fut un enfer sans fin. La ville se délita, masse éclatée d'individus brûlés et ensanglantés, cherchant à fuir dans toutes les directions. Même ceux tentés par l'héroïsme perdaient leur sang-froid, lâchaient les blessés pour prendre leurs jambes à leur cou. Il ne restait plus de policiers. Ni de pompiers. Les flammes se propageaient rapidement, sans personne pour les éteindre. Il fut vite impossible de dire d'où elles étaient parties, où avaient eu lieu les explosions initiales. Tout brûlait sans distinction.

Je me raccroche au fait que nous avons évacué le lycée. Les chutes de gravats firent une vingtaine de blessés, et l'on déplora une crise cardiaque, mais sans notre intervention c'eût été pire, bien pire. De l'ordre de plusieurs centaines. Cette pensée me réconforte, certains jours. Nous y sommes retournés à deux reprises pour chercher John. En vain. Le feu nous tenait à distance.

Les premiers secours organisés furent le fait d'habitants des villes voisines, avant l'arrivée de l'armée. Celle-ci débarqua en nombre, avec force mitraille. Nul ne savait à qui se fier, l'ennemi pouvait être n'importe qui. Pendant ce temps, nous avons tenté de mettre des gens à l'abri. D'aucuns s'en sortirent. Beaucoup n'eurent pas cette chance.

Nous avons remué ciel et terre, jusqu'à tomber de fatigue, puis les renforts ont pris le relais. Alors nous sommes revenus ici, après un voyage long de plusieurs jours à travers un pays dont de nombreuses routes étaient coupées, les avions cloués au sol, où les téléviseurs diffusaient tous la même image, où chaque citoyen de chaque ville se demandait qui seraient les suivants.

À ce jour, l'événement reste sans suite. Mais pour combien de temps ? Nul ne le sait.

Personne ne s'explique comment un groupe terroriste a pu infiltrer une communauté et semer autant de bombes sans soulever le moindre soupçon. Les théories conspirationnistes vont bon train. Les médias parlent régulièrement de deux hommes, parfois d'un homme et d'une femme, qu'on aurait aperçus au lycée juste avant et juste après la déflagration. Certains disent qu'ils ont lancé l'alerte, mais la plupart les ont vus brandir des armes à feu en criant des slogans de type islamiste. On a retrouvé les restes d'un inconnu dans le coffre d'une voiture de patrouille carbonisée, garée près de l'église d'où partit l'incendie. On suppose, à son uniforme, qu'il s'agissait d'un policier, mais vu l'ampleur des pertes parmi les autorités locales, cela reste à confirmer. Il y a beaucoup, beaucoup de cadavres, et il faudra du temps pour identifier tout ce monde.

Mais l'un d'eux, bien sûr, est déjà connu.

Dans un bâtiment du lycée, parmi les vestiges d'un placard du premier étage, les pompiers mirent au jour les maigres restes d'un jeune homme. Sa denture prouva qu'il s'agissait de l'auteur de ce que l'on considère désormais comme une attaque de diversion, à savoir l'attentat du centre commercial de Los Angeles, commis la veille du carnage.

Il s'appelait Lee John Hudek.

Un nom qui restera gravé dans les mémoires. Son visage était connu avant même qu'on ne trouve sa dépouille : des rescapés jurent l'avoir vu rôder en ville le jour du drame. Des dizaines de témoignages dignes de foi attestent de sa présence à l'épicerie, dans la galerie marchande, chez Starbucks, aux abords de l'église, et dans une série d'endroits ravagés par les premières explosions – toujours accompagné, semble-t-il, d'un homme de petite taille et de type arabe. Il fut également aperçu par de nombreux lycéens, dont certains se virent offrir de la drogue. L'opinion s'accorde à penser que ce trafic servait de couverture pour poser des bombes. On le soupçonne également du meurtre d'un de ses amis, un certain Bradley Metzger, dont le corps fut retrouvé dans la Valley. Il y aurait un lien avec la mort d'un autre garçon, ainsi que d'une adolescente, mais rien n'est établi. L'entrepreneur local Emilio Hernandez aurait surpris des conversations laissant croire qu'il s'agissait de complices, éliminés par Hudek juste avant l'assaut final.

Une chose est sûre ; l'opinion peut désormais mettre un visage sur l'horreur de Thornton. Cela évitera de voir au-delà, de se demander comment cette chose fut possible. Alors qu'il se trame peut-être autre chose. Comme toujours, sans doute.

Ryan et Lisa Hudek ont fait plusieurs apparitions télévisées, lors de conférences de presse soigneusement encadrées par leurs avocats. S'ils inspirent beaucoup de haine, de nombreux Américains les considèrent comme des victimes, de bons citoyens ayant perdu leur fils unique au profit de forces obscures que personne ne comprend. Les Hudek sont les premiers à se demander ce qui a poussé leur fils dans les bras d'une organisation terroriste. Mme Hudek a confié, en larmes, que Lee John avait un comportement bizarre depuis quelques mois, et a avoué dernièrement l'avoir entendu fustiger des « actes de guerre » commis au nom de la politique étrangère américaine.

Julia Gulicks est toujours en vie, quoique à peine. Elle ne se réveillera jamais, mais on ne peut la débrancher puisqu'elle a reconnu deux meurtres : il faut bien quelqu'un ou quelque chose à juger. Lui arrive-t-il, dans son sommeil, de revoir cet après-midi où son père dégringola l'escalier du pavillon de Dryford ? Et dans ces rêves interminables, reçoit-elle la visite d'une autre fillette, un peu plus jeune, celle qui n'aura vraisemblablement jamais quitté leur rue ?

Car on a aussi arrêté un dénommé James Kyle, dans le jardin d'une maison où il vécut un temps avant de quitter Dryford du jour au lendemain, voilà douze ans. On pensait, à l'époque, qu'il avait déménagé avec son épouse et leur jeune fille, mais à présent rien n'est moins sûr. Le jour des attentats de Thornton, un voisin signala un homme au comportement étrange dans l'ancien jardin des Kyle. Il l'avait aperçu depuis la route, trop loin pour repérer dans les hautes herbes le corps d'un agent du FBI. Dommage, car la police aurait peut-être atteint le lieu du crime plus tôt – à supposer qu'elle n'ait pas été appelée à Thornton pour une urgence autrement importante. Il pleuvait des flics morts cet après-midi-là. Lorsqu'une patrouille arriva enfin, on découvrit un sexagénaire ensanglanté, assis au bord d'un trou superficiel qu'il venait de creuser à la main dans un coin du terrain. Il était bouleversé, et brandissait des ossements semblables à ceux d'une enfant de huit ans.

L'affaire Gulicks s'est compliquée du fait que Kyle revendique désormais le meurtre des deux hommes découverts dans les bois la semaine précédente. S'il s'avère incapable d'étayer ces aveux, il clame que tout est sa faute. Et il affirme que d'autres corps croupissent dans le secteur. En grand nombre.

Pour l'heure, les gens s'intéressent peu à cet aspect de l'enquête. Thornton ressemble assez aux enfers sans qu'on aille déterrer de pauvres diables morts il y a dix ans. Jusqu'à nouvel ordre, Julia restera donc à hôpital, en attendant son heure, ni coupable ni innocente, ni morte ni vivante. J'évite de prononcer son nom devant Nina.

Charles Monroe fut inhumé, parmi tant d'autres. Nous avons envoyé des fleurs.

J'ignore si Paul est toujours vivant. Je sais seulement qu'en quittant le sous-sol du lycée avec Nina, j'entendais John à ses trousses. Je ne vois pas comment Paul aurait eu le temps de fuir avant de déclencher la mise à feu – non seulement celle de la voiture piégée, mais de tous les engins disséminés dans la ville.

S'il est mort, gageons qu'il sera mort heureux. Les Hommes de Paille ont eu leur grand jour et versé le sang de nombreux anges. Le massacre de Thornton monopolise les conversations. Le fait qu'on ait visé exprès une petite ville, refuge des gens ordinaires, n'est pas l'élément le moins traumatisant. Ce n'est pas un symbole qui fut attaqué, ni l'un de ces endroits qu'on voit seulement à la télé. Les ténèbres sont venues faucher le petit peuple chez lui. En outre, la présence massive du FBI au moment des attentats discrédite encore la force publique, qui paraît incapable de nous protéger malgré des déclarations chaque jour plus tonitruantes. Là aussi, c'était peut-être voulu, tout comme le nombre élevé de journalistes déjà sur place, qui a permis de montrer l'horreur dès les premières secondes, dans toute sa vérité. Faut-il y voir un simple coup de chance, ou bien les pièces d'un puzzle savamment agencé ? Avec Paul, on ne sait jamais. Comme James Kyle l'a dit à Nina, c'était un garçon précoce.

Tous les jours, j'ai tenté de joindre John sur son portable, et je sais que Nina en fait autant. Mais ça sonne dans le vide. John n'était pas du genre à rappeler, mais on pouvait au moins lui laisser un message.

J'imagine qu'on finira par abandonner. Il faudra bien s'arrêter.

L'agent Nina Baynam figure sur la liste des morts de Thornton, étant donné qu'elle était portée disparue depuis plusieurs jours au moment du drame. Elle ne sait pas encore si elle va démentir.

En attendant, nous aidons Patricia et les autres voisins dans leurs travaux manuels. J'empêche Nina de porter de lourdes charges. Sauf quand elle insiste. Nous dînons au calme dans le centre de Sheffer, et parfois nous finissons un peu éméchés autour du billard de chez Bill. On s'assied tous les deux sur le banc près du lac, ou dans les fauteuils de notre bungalow.

Et nous passons du temps ensemble.

Je n'ai pas entendu les médias évoquer le meurtre d'un agent de la CIA dans les toilettes d'un café. Le Starbucks a été réduit en cendres : on ne parvient même plus à le situer sur les photos. Les restes de Carl Unger sont sûrement mêlés à ceux des autres clients, retournés à la poussière dans le courant souterrain de l'histoire.

J'ai passé du temps à méditer son récit, à envisager s'il était plausible. Le problème, c'est que je ne crois pas plus au paradis qu'à l'enfer. Ceux-ci visent à expliquer ce que nous sommes, or l'identité humaine n'est pas une question, mais un fait. Dans le noyau sombre de notre être gravite un rêve de mort, que vénèrent isolément des tueurs solitaires. De loin en loin, un fou furieux essaiera de commettre un génocide au nom de notre espèce et fera trembler le monde pour cinquante ans ; pendant ce temps, les flingueurs solitaires poursuivront leur besogne avec succès. Régulièrement, nous en attrapons un ou une, avant de l'exécuter ou de l'enfermer. Mais il y en aura d'autres. La mort ne s'effacera pas car elle est dans nos cœurs. Ce n'est pas par stupidité ni par manque de clairvoyance que nous déclarons la guerre, que nous exterminons nos voisins ou détruisons d'autres espèces – du moins ce ne sont pas là les seules raisons. Nous fûmes les premiers animaux à comprendre la mort, et nous avons besoin de montrer que nous ne sommes pas impuissants face à elle. Nos trente mille ans de meurtres ne constituent peut-être qu'un acte de défi, un cri de vengeance : nous savons que la mort approche, alors parfois nous croisons le fer. Ou alors peut-être Paul et ses amis sont-ils dans le vrai, peut-être n'y a-t-il rien de mal à cela. Tuer serait notre nature.

Mais notre nature ne se résume pas à si peu, et la mort n'est jamais simple. Qu'aurais-je fait si j'avais dû choisir entre Nina et le lycée ?

Ne me posez pas cette question.

Pour l'heure, je suis perplexe. Contrairement à ce que Carl semblait penser, je ne crois pas que l'histoire se répète. L'histoire n'obéit à aucune sorte de cycle, mais elle produit indéfiniment les mêmes choses. Ce qui se passe échappe parfois à notre vigilance, voilà tout. Et parfois nous sommes bien obligés de voir. Chaque attentat frappe des innocents. Chaque riposte aussi.

Nous nous replions au centre d'un cercle de rancune, sans trop savoir où donner de la tête. Des tueurs de toutes les religions, de toutes les obédiences et de toutes les époques se plantent sur les lisières de notre monde et le mitraillent. Ils ne voient que ceux qui leur sont opposés, les ennemis choisis, les adversaires diabolisés. Nous autres, pris entre deux feux, restons invisibles.

Comme ils doivent nous trouver irréels, nous, les gens normaux, comme ils doivent railler chez nous l'absence d'ardeur purificatrice. Comme ils doivent nous trouver pleutres avec nos rêves étriqués : aller au bout des courtes vies allouées par nos divers dieux, faire notre temps sans être fusillés, affamés ou pulvérisés au nom d'intérêts ou d'idéaux qui nous dépassent.

Il n'empêche qu'ils nous tuent. C'est leur manière de vivre, leur manière d'être. Il nous incombe de leur résister, en permanence et pour toujours. À charge pour nous de trouver le moyen de faire retentir un « non » plus vibrant.

Voici trois jours, une de nos questions trouva sa réponse. Dans les deux semaines suivant notre retour à Sheffer, Nina souffrit régulièrement de crises de nausées, qui la laissèrent faible et migraineuse. Nous avons voulu y voir le contrecoup temporaire de sa séquestration dans un vieux véhicule imprégné de gaz d'échappement, mais devant la persistance des symptômes, j'ai persuadé Nina d'aller chez le médecin.

Où nous apprîmes qu'elle était enceinte.

Je vais sortir, maintenant, pour rejoindre Nina et l'emmener dîner. Pas une salade. Quelque chose de consistant, de nourrissant. Elle peut toujours faire la difficile, mais moi, je suis capable de me montrer impitoyable. Irrésistible. Nous évoquerons peut-être la suite, ce futur que nous allons devoir inventer.

Nous y allons pas à pas, sans précipitation. Suis-je prêt à être père ? Je verrai bien quand j'y serai. Drôle de monde pour élever un enfant.

Mais je suppose que c'est ainsi depuis la nuit des temps.

Retrouvez le dernier titre de Michael Marshall aux Éditions J'ai lu :

Les Intrus

Après avoir quitté la police de Los Angeles, Jack Whalen s'est retiré dans une petite ville paisible sur la côte nord du Pacifique. Mais quand Gary Fisher, un ancien copain de lycée, lui demande son aide pour élucider un double meurtre, Jack a un dossier plus urgent à résoudre : la disparition de sa femme Amy. Et si ce bon vieux Gary n'était pas celui qu'il prétend être ? Qui est cette fillette errant dans les rues de Seattle ? Quel sombre secret abrite ce vieil immeuble du centre-ville ?

N°8760

Composition
PCA

Achevé d'imprimer en Italie
par GRAFICA VENETA
le 2 septembre 2008.

Dépôt légal septembre 2008.
EAN 9782290011911

ÉDITIONS J'AI LU
87, quai Panhard-et-Levassor, 75013 Paris

Diffusion France et étranger : Flammarion